AF883605

Late Paleocene–Early Eocene Climatic and Biotic Events in the Marine and Terrestrial Records

Late Paleocene–Early Eocene Climatic and Biotic Events in the Marine and Terrestrial Records

Marie-Pierre Aubry, Spencer Lucas,
and William A. Berggren,
Editors

Columbia University Press
New York

Columbia University Press
Publishers Since 1893
New York Chichester, West Sussex
Copyright © 1998 Columbia University Press
All rights reserved

Library of Congress Cataloging-in-Publication Data

Late Paleocene–early Eocene climatic and biotic events in the marine and terrestrial records / editors, Marie-Pierre Aubry, Spencer G. Lucas, and William A. Berggren.
p. cm.
Includes bibliographical references and index.
ISBN 978-0-231-10238-4
1. Paleontology—Paleocene. 2. Paleontology—Eocene. 3. Paleoclimatology—Paleocene. 4. Paleoclimatology—Eocene. I. Aubry, Marie-Pierre. II. Lucas, Spencer G. III. Berggren, William A.
QE736.8.L37 1998
551.7'83—dc21 98–20969

Cover Illustration: Logo of UNESCO-IGCP Project 308 (Paleocene/Eocene Boundary Events in Space and Time)

The Paleocene/Eocene boundary interval was characterized by a number of events recorded in marine and terrestrial stratigraphies. Major extinction/turnover events in the deep sea are symbolized by the last appearance datum (LAD) of the benthic foraminifera *Stensioina beccariiformis* (upper right) concurrent with a major 3–4‰ excursion in $\delta^{13}C$ in organic carbon. The first appearance datum (FAD) of the calcareous nannoplankton *Tribrachiatus digitalis* (upper left) symbolizes the Ypresian transgression and the current definition of the base of the Eocene. A major turnover in terrestrial mammalian faunas is represented by the FAD of *Hyracotherium* (lower center) holding the first planispiral planktonic foraminifera *Pseudohasterigerina wilcoxensis* in its jaws, symbolic of the approximate contemporaneity of these events recorded in disjunct stratigraphic settings. The temporal framework of these events is provided by correlation to a common paleomagnetic stratigraphy.

Casebound editions of Columbia University Press books are printed on permanent and durable acid-free paper.
Printed in the United States of America

CONTENTS

CONTRIBUTORS

Marie-Pierre Aubry
Institut des Sciences de l'Evolution
Université Montpellier II
Place Eugène Bataillon
34095 Montpellier Cedex 05
tel.: 33-467-14-47-16
fax: 33-467-14-36-10
e-mail: aubry@isem.univ-montp2.fr

Richard A. Beck
Department of Geology
Miami University
Oxford, OH 45056-1618
tel.: 513-529-7578
fax: 513-529-1542
e-mail: beckra@muohio.edu

William A. Berggren
Department of Geology and Geophysics
Woods Hole Oceanographic Institution
Woods Hole, Ma 02543
tel.: 508-289-2593
fax: 508-457-2187
e-mail: wberggren@whoi.edu

Anne Boersma
Microclimates
14 Dogwood Lane
Pomona, NY 10970
tel.: 914-942-1394
fax: 914-352-2280

Henk Brinkhuis
Laboratory of Paleobotany and Palynology
University of Utrecht
Heidelberglaan 2
Utrecht 3584 CS
The Netherlands
tel.: (31)-30-25-32-799
fax: (31)-30-25-35-096
e-mail: HENKB@boev.biol.ruu.nl

Jonathan Bujak
The Lexis Group
Albion House
9 Albion Avenue
Blackpool, Lancashire FY3 8NA
England
tel./fax: (44)-12-53-39-31-61
e-mail: kac89@dial.pipex.com

Douglas W. Burbank
Department of Geoscience
Pennsylvania State University
University Park, PA 19802
tel.: 814-863-6712
e-mail: burbank@geosc.psu.edu

Richard Corfield
Department of Geology
Oxford University
Parks Road
Oxford OX1 3PR
England
tel.: (44)-865-27-20-41
fax: (44)-865-27-020-72
e-mail: Richard.Corfield@earth.ox.ac.uk

David Dockery
State of Mississippi
Department of Environmental Quality
Office of Geology
P.O. Box 20307
Jackson, MS 39287-1307
tel.: 601-354-6328
fax.: 601-961-5521

John Flynn
Field Museum of Natural History
Department of Geology
Roosevelt Road at Lake Shore Drive
Chicago, IL 60605-2496
tel.: 312-922-9410, x291
fax: 312-427-7269
e-mail: flynn@fmnh.org

Gregg Gunnell
Museum of Paleontology
The University of Michigan
Ann Arbor, MI 48109-1079
tel.: 313-764-0489
fax: 313-763-690
e-mail: ggunnell@umich.edu

Pamela Hallock
Department of Marine Sciences
University of South Florida
St. Petersburg, FL 20132
tel.: 813-893-9567
fax.: 813-974-2654
e-mail: pmuller@seas.marine.usf.edu

Joseph H. Hartman
Energy and Environmental Research Center
University of North Dakota
15 North 23rd Street
P.O. Box 9018
Grand Forks, ND 58202-9018
tel.: 701-777-8000
fax: 701-777-5181
e-mail: jhartman@plains.nodak.edu

Jeremy Hooker
Department of Palaeontology
The Natural History Museum
Cromwell Road
London SW7 5BD
tel.: (44)-071-938-9408
fax: (44)-071-938-9277
e-mail: jjh@nhm.ac.uk

Howard Hutchison
Museum of Paleontology
University of California
Berkeley, CA 94720
tel.: 510-642-3733
fax: 510-642-1822
e-mail: howardh@ucmp1.berkeley.edu

Asrar Khan
Hydrocarbon Development Institute of Pakistan
Sector 18, H/9
P.O. Box 1236
Islamabad, Pakistan
tel.: (92)-51-853964
fax: (92)-51-217385

Robert W. O'B. Knox
British Geological Survey
Kingsley Durham Centre
Keyworth
Nottingham NG12 5GG
tel.: (44)-11-59-36-31-00
fax: (44)-11-59-36-32-00
e-mail: rwok@wpo.nerc.ac.uk

Spencer Lucas
New Mexico Museum of Natural History
1801 Mountain Road NW
Albuquerque, NM 87104-1375
tel.: 505-841-5837
fax: 505-841-5866
e-mail: lucas@darwin.nmmnh-abq.mus.nm.us

Catharine Nigrini
161 Rorris
Canrore
Alberta
Canada T1W 2R7

Isabella Premoli Silva
University of Milan
Via Mangiagalli
Milan, Italy
tel.: (39)-2-23698248
fax.: (39)-2-70638261
e-mail: micro@imiucca.sci.unimi.it

David Rea
The University of Michigan
Department of Geological Sciences
Ann Arbor, MI 48109-1063
tel.: 313-936-0521
fax: 313-763-4690
e-mail: davidrea@umich.edu

Barry Roth
Museum of Paleontology
University of California
Berkeley, CA 94720
tel.: 415-387-8358
fax.: 415-863-4305
e-mail: barry@ucumpl.berkeley.edu

Annika Sanfilippo
A-015, University of California
San Diego
La Jolla, CA 92093
tel.: 619-534-2049
fax: 619-534-0784
e-mail: annika@ucsd.edu

William J. Sercombe
Amoco Production Company
501 Westlake Park Boulevard
Houston, TX 77253
tel.: 713-366-3542
fax: 713-366-7648
e-mail: wjsercombe@amoco.com

Ashish Sinha
University of Southern California
Department of Geological Sciences
University Park
Los Angeles, CA 90098-0740
tel.: 213-740-5120
fax: 213-740-8801
e-mail: sinha@coda.usc.edu

Lisa C. Sloan
Institute of Marine Sciences
Applied Sciences Building
University of California
Santa Cruz, Ca 95064
tel.: 408-459-3693
fax: 408-459-4882
e-mail: lcsloan@earthsci.ucsc.edu

Lisa Tauxe
Scripps Institution of Oceanography
La Jolla, CA 92093-0220
tel.: 619-534-6084
fax: 619-534-0784
Web page: http://sorcerer.ucsd.edu

Ellen Thomas
Earth and Environmental Sciences
Wesleyan University
265 Church Street
Middletown CT 06459-0139
tel.: 860-685-2238
fax: 860-685-3651
e-mail: ethomas@wesleyan.edu
and
Department of Geology and Geophysics
Kline Geology Laboratory
P. O. Box 6666
New Haven, Connecticut 06511-8130
tel.: 203-432-3114
fax: 203-432-3134

Scott Wing
National Museum of Natural History Smithsonian Institution
Washington, DC 20560
tel.: 202-357-2649
fax: 202-286-2832
e-mail: wing.scott@simnh.si.edu

PREFACE

Although it has been known for over a century that a major turnover occurred in terrestrial mammalian fauna near the Paleocene/Eocene Epoch boundary, it is only relatively recently that apparently simultaneous, equally important evolutionary turnovers among marine invertebrates and protists have been recognized. Subsequent to the discovery of a sharp late Paleocene extinction event in deep sea benthic foraminifera—the most notable evolutionary event among marine taxa—sharp variations in the oxygen and carbon isotopic composition of upper Paleocene marine carbonates were identified. Late Paleocene–early Eocene variations in the isotopic composition of organic carbon of similar amplitude as those described from marine carbonates have now been described from teeth of late Paleocene land mammals and from upper Paleocene carbonate concretions in paleosols. In addition, marked sedimentologic features such as a decrease in grain size of atmospheric dust and an increase in kaolinitic content of marine deposits, in particular at high latitudes and on epicontinental shelves, have been documented.

These findings have led to the upper Paleocene–lower Eocene stratigraphic record becoming the focus of considerable attention in Earth science circles during the past few years, and to the recognition that the latest Paleocene–earliest Eocene epochs constituted the most critical turning point in the Cenozoic history of our interstellar home. The new Global Change initiatives sponsored by the Earth and Ocean Science Divisions at NSF recognize the early Paleogene, and the Paleocene–Eocene transition specifically, as a crucial period if we are to better understand the dynamics of the earth environment and the impact climatic and oceanographic changes have on earth's biota. The suddenness and amplitude of the events that occurred then somehow dwarf the significance of the events associated with the Eocene/Oligocene Epoch boundary, and this volume is designed to present an up-to-date overview and synthesis of our knowledge of global biotic, climatic and oceanographic changes across the Paleocene/Eocene boundary.

This book describes and further documents the now well-established turnovers that occurred in major marine and terrestrial groups, among bathyal and abyssal deep-sea benthic foraminifera and terrestrial mammals. It also documents lesser known, if not unexpected, turnovers, as in the calcareous nannoplankton and terrestrial mollusks. Although not all groups are considered and geographic coverage is uneven, two significant conclusions can be drawn from the 12 chapters devoted to the description of these changes. First, the greatest changes (evolutionary radiations, extinctions, and immigrations) did not occur synchronously in different groups, indicating that the response of organisms to environmental changes is extremely complex. Most relevant is the fact that the paleontologic record of the Bighorn Basin shows that the pronounced increase in diversity in the nonmarine mollusks in the mid-Clarkforkian Age preceded by a few hundred thousand years the sharp turnover in the terrestrial mammals at the Clarkforkian/Wasatchian Age boundary. Second, whereas a multitude of mechanisms are proposed to explain these changes, a consensus appears that increased global warming and changes in fertility/productivity patterns best explain them. However, major questions remain to be answered; in particular, was there a long episode of consistent warmth near the Paleocene/Eocene Epoch boundary or were there several successive warm episodes separated by cooler intervals?

The late Paleocene–early Eocene biotic turnovers occurred as the earth was undergoing major physical and geochemical changes. These are explored in five chapters that deal with the possible climatic consequence of the India–Asia collision, the profuse volcanism in the North Atlantic associated with the isolation of the northern latitudes (North Sea area) from the North

Atlantic Ocean, the increase in ocean productivity through the Paleocene Epoch and the probable change in the source(s) of deep water formation near the Paleocene/ Eocene Epoch boundary, ocean warming at the high latitudes, and weakening of atmospheric circulation. Neither Himalayan tectonics nor North Atlantic volcanism alone can explain the latest Paleocene–earliest Eocene warming, but each may have contributed to it. However, modeling fails to create the earth conditions at the time of the short abrupt warming (the Late Paleocene Thermal Maximum [LPTM]). The production of greenhouse gasses has been invoked to explain the warming, and massive production of methane through the dissociation of oceanic methane hydrate is hypothesized. Whereas CH_4 release from oceanic hydrates as a result of ocean bottom warming may well explain the perturbations in the carbon cycle at the time of the LPTM, the cause for warming remains to established.

Because reconstructions of earth history must rely on chronologic successions identified in the stratigraphic record, three chapters of the book deal with means to establish a sound interpretation of stratigraphic sections and with the geochronology of the Paleocene–Eocene boundary interval. Extracting data from the stratigraphic record may not be as simple as often thought and the correct identification of the temporal sequence of events requires a careful examination, and judicious evaluation, of stratigraphic sections and their correlations.

We hope that this book, with its multidisciplinary approach, will provide a sense of the complexity of the earth system and the relationships between biotas and their environment. Although the changes that occurred near the Paleocene/Eocene Epoch boundary are more comprehensively described, the fundamental mechanisms involved remain elusive.

This book is also the byproduct of the activities of IGCP Project 308. In an attempt to delineate clearly the location/distribution in space and time of these changes, and to decipher the cause–effect relationships involved that resulted in global warming, IGCP Project 308 was initiated in 1990 and funded for a period of 5 years through UNESCO. Simultaneously, the Paleogene Subcommission of the International Commission on Stratigraphy (ICS) designated a Working Group on the "Paleocene/Eocene boundary." ICS has been engaged for two decades in the search for stratigraphic sections that may serve as standard bearers for boundaries between series, the so-called Global Stratotype Points and Sections (GSSPs) and their "Golden Spikes." In this context, IGCP Project 308 has also been active in seeking to provide the appropriate criteria by which the Paleocene/Eocene Epoch boundary may be characterized and the Paleocene/Eocene Series boundary recognized and correlated.

As is the case with many such epoch/series boundaries, there is considerable controversy over the exact definition and location of the Paleocene/Eocene Epoch/Series boundary and its estimated age. One of the two primary reasons for this controversy is that the chronostratigraphic subdivision of the stratigraphic record is solely based on the marine record and that marine stratigraphers have paid insufficient attention to the continental record. Following the recommendation of ICS (Hedberg 1976) and the agreement of a large part of the international stratigraphic community, the base of the Eocene Series is currently defined by the base of the Ypresian Stage, i.e., by the base of the shallow marine Ieper Clay Formation in Belgium that constitutes the stratotype of the stage. Meanwhile, stratigraphers studying the continental record saw the profound turnover in terrestrial mammals as the most appropriate criterion to characterize the Paleocene/Eocene Series boundary. Yet this mammalian turnover occurs at a level demonstrably older than the base of the Ieper Clay.

The second primary reason relates to the problem of correlation. The northwest European stratigraphic record, which has constituted the basis for our chronostratigraphic framework (a situation that is only progressively changing with the new ICS regulations [Cowie et al. 1986]), is notoriously poor in the two microfossil groups, calcareous nannofossils and planktonic foraminifera, which provide the best means for global correlations. In addition, it is riddled with unconformities. Even though a firm magnetostratigraphic framework has been established for the lower Eocene succession of Belgium, England, and France (partim), the base of the Ypresian Stage (located in Chron C24r) has remained difficult to characterize in sections outside of NW Europe. Thus criteria have been proposed to *approximate* the base of the Ypresian Stage (and by inference that of the Eocene Series), and the NP9/NP10 (calcareous nannofossil) zonal boundary has become the criterion most commonly used to approximate the Paleocene/Eocene Series boundary. However, this criterion, proposed for the sole purpose of (approximate) correlation, soon became used by many workers as the *defining* criterion for the Paleocene/Eocene Series boundary, thus adding to the confusion already existing and due to the dual definition of the series boundary for the marine and continental records. The NP9/NP10 (chrono)zonal boundary is demonstrably older than the base of the Ypresian Stage (= the Paleocene/Eocene Series boundary in the marine record) and younger than the mammalian turnover (= Paleocene/Eocene Series boundary in the continental record). To avoid the confusion that would result from a multiple definition of the Paleocene/Eocene Series boundary, we have requested that authors in this book respect the current chronostratigraphic definition of the base of the Eocene Series (i.e., = the base of the Ieper Clay), thus ensuring homogenous concepts for Paleocene and Eocene Series.

The definition of the Paleocene/Eocene Epoch/Series boundary is likely to change. The new regulations of ICS (Cowie et al. 1986) propose that series boundaries be defined independently of marine stages with the consequence that the base of the lower stage of a series will then be subordinated to the base of the series. A consensus appears to be developing that the major $\delta^{13}C$ isotope spike that has been found in marine and terrestrial carbonates and in vertebrate tooth enamel at stratigraphically correlative levels in lower Chronozone C24r would provide a unique signature/criterion for linking marine and terrestrial stratigraphies in the form of a Paleocene/Eocene boundary GSSP. (However, until a new criterion is accepted by ICS to define the Paleocene/Eocene Series boundary, the base of the Eocene Series remains subordinate to the base of the Ypresian Stage).

In order to avoid the situation in which the redefined Paleocene/Eocene Series boundary violates the current relationships between the Thanetian and Ypresian Stages and their stratotypes and the Paleocene and Eocene Series, we have determined a stratigraphic interval comprised between the top of the Thanetian Stage s. st. (i.e., corresponding to the top of the Thanet Beds in England) and the base of the Ieper Clay, in which to locate the Paleocene/Eocene Series boundary. This interval is often referred to as the "Paleocene/Eocene boundary interval." Loosely constrained at its base, it spans an approximately 2-my interval (from latest Chron C25r to mid Chron C24r). (Previous ICS rules [1976; under which we are working until a Paleocene/Eocene GSSP is established] determined that the base of a stage [as defined by the base of its stratotype] defines the top of the preceding stage. All strata comprised between the top of the Thanetian stratotype and the base of the Ieper Clay are thus currently assigned to the Thanetian Stage. However, this is a broad definition of the Thanetian Stage, and placement of the Paleocene/Eocene Series boundary in this interval will not violate the extent of the Thanetian stratotype or that of the Ypresian stratotype).

The ultimate achievement of IGCP Project 308 and the Working Group on the "Paleocene/Eocene boundary" is to select the most appropriate criterion(a) to characterize the Paleocene/Eocene Epoch/Series boundary, and to choose the most suitable section to serve as boundary stratotype. However, such selection/choice cannot be appropriately made without a clear knowledge of the pre- and post-boundary interval geological history. We hope this book will serve that purpose.

REFERENCES

Cowie, J. W., Ziegler, W., Boucot, A. J., Basset, M. G., and Remane, J., 1986. Guidelines and statutes of the International Commission on Stratigraphy (ICS). *Courier Forschungsinstitut Senckenbergiana,* v. 83, p. 1–14.

Hedberg, H. D. (ed.), 1976. International stratigraphic guide. A guide to stratigraphic classification, terminology, and procedure. New York: Wiley, 200 pp.

ACKNOWLEDGMENTS

The editors wish to acknowledge the many participants of IGCP Project 308 (Paleocene/Eocene Events in Space and Time) for their enthusiastic involvement and contributions to this project during its 5-year (1989–1994) tenure. We thank Isabella Premoli Silva for having encouraged our submission of the original project proposal to UNESCO along the lines of the highly successful IGCP Project 174 (Terminal Eocene Events), which eventually led to the establishment of a GSSP for the Eocene/Oligocene Epoch/Series boundary. The continued financial support of UNESCO (and the personal support of Endre Dudich and Vladimir Babushka, successive Secretaries for IGCP Projects), is gratefully acknowledged. Continued financial support by French and U.S. IGCP Programs made possible the IGCP 308 Working Group activities in France (under the leadership of M.-P. Aubry) and the United States (W. A. Berggren), respectively. Financial contributions by national IGCP 308 committees made possible the participation and contribution of scientists from many countries.

We thank the contributors (and the reviewers) to this volume; their cooperative spirit has rendered our task much easier as we passed through various iterations of manuscript editing (for stylistic, nomenclatural, and terminological consistency), indexing, and, finally, the preparation of the overview of their contributions. It has resulted in the timely appearance of a volume that clearly reflects our present understanding of the Paleocene–Eocene transition.

Finally, we thank Ed Lugenbeel, Executive Editor at CUP, for his constant support and encouragement; and Leslie Kriesel, Assistant Manuscript Editor, whose efficient editorial counsel guided this volume to completion.

Marie-Pierre Aubry
Spencer Lucas
William A. Berggren

Late Paleocene–Early Eocene Climatic and Biotic Events in the Marine and Terrestrial Records

1

Late Paleocene–Early Eocene Climatic and Biotic Evolution: An Overview

WILLIAM A. BERGGREN
SPENCER LUCAS
MARIE-PIERRE AUBRY

ABSTRACT
The late Paleocene to early Eocene Epochs witnessed significant changes in global climate and ocean circulation as reflected in evolutionary turnovers and extinctions recorded in marine and terrestrial stratigraphic records. These events occurred during a ~2-my interval of global warming in contrast to the ~10-my interval of climatic cooling and attendant biotic change that characterized the late Eocene–early Oligocene Epochs.

A revised geochronologic scale provides an integrated framework for ranking various bio- and chemostratigraphic events in a relative temporal sequence and estimating the timing and rates of processes associated with the Paleocene–Eocene transition. Yet, weaknesses and inconsistencies inherent in some paleomagnetic and radioisotopic data sets preclude unequivocal temporal correlation of some parts/events of the marine and terrestrial stratigraphic record, particularly over the critical interval of the ~2.5-my-long Chron C24r. In particular, the temporal relationship between the "Paleocene–Eocene" transition in terrestrial vertebrate faunas in North America (Clarkforkian/Wasatchian North American Land Mammal Ages [NALMA]) and Europe ("Sparnacian/Neustrian European Land Mammal Ages [ELMA]) and the Paleocene/Eocene Epoch/Series boundary in the marine stratigraphic record as denoted by the base of the Ypresian Stage (base of the Ieper Clay, Belgium) remains controversial, although the consensus among authors treating this subject in this volume is that both terrestrial events predate the Paleocene/Eocene Epoch marine boundary.

This chapter presents an overview/summary of the 21 chapters in this volume to provide an integrated perspective on the Historical Geology of this geologically fascinating interval in earth history.

The late Paleocene–early Eocene (Thanetian–Ypresian Ages) interval was a critical episode in earth history because major changes in ocean circulation and global climate, reflected in significant turnovers in marine and terrestrial biotas, occurred during approximately 2 my of this interval. This deliberate paraphrase (with only temporal terms and duration changed) of the Introduction (Berggren and Prothero 1992) to a recent monographic treatment of late Eocene–early Oligocene climatic and biotic evolution (Prothero and Berggren 1992) reflects the role played by these factors in the original delineation of many of the divisions of the chronostratigraphic record. Indeed, the abstract to that same Introduction stated that "The late middle Eocene–early Oligocene interval was the most significant episode of climatic change and extinction since the end of the Cretaceous "*with the exception of the Paleocene/Eocene boundary event*" (italics ours; Berggren and Prothero 1992:1).

During the Eocene–Oligocene transition, climatic and biotic evolution was played out against a background of globally lowering atmospheric temperatures, oceanic cooling associated with stronger thermal stratification and the initiation of the oceanic psychrosphere and the initiation of Antarctic glaciation. In contrast, the Paleocene–Eocene transition was characterized by marked global atmospheric warming, particularly at high latitudes. This was punctuated by a brief (~0.01 my), extreme warming associated with a large (~3–4‰) negative excursion in $\delta^{13}C$ in planktonic and benthic foraminifera and a concomitant, albeit temporary, reduction in global productivity and relocation of deep water sources from high to low latitudes, and decreased vertical thermal gradients. Higher levels of greenhouse gases and increased transport to high latitudes have been invoked to explain the conditions that characterized this anomalous interval.

The chapters in this volume treat a broad spectrum of topics dealing with late Paleocene–early Eocene climatic and biotic evolution as recorded in marine and terrestrial stratigraphies. Ocean history, climate modeling and the possible effect of regional tectonism on global climate and the carbon budget are also discussed. The various scenarios are set within a geochronologic framework based on the recent Integrated Magnetobiochronologic Time Scale (IMBS) for the Cenozoic of Berggren et al. (1995). Nearly all the chapters in this volume are recalibrated to this chronology. Chapters on marine micropaleontology and marine geology include empirical observations on the geologic record, (inter)regional marine (and terrestrial) correlations, faunal stratigraphic and geographic distribution, the role of trophic resources on faunal distribution patterns, and climatic modeling. Those that deal with marine invertebrates, terrestrial vertebrates and plants focus on (inter)regional stratigraphic correlations, biogeographic distribution patterns based on an essentially global stratigraphic data base, and calibration to the IMBS. Together, these chapters provide a clear look at the intimate link between marine and terrestrial ecosystems during the ~2-my-long late Paleocene–early Eocene interval examined in this volume.

In this chapter we review each of the contributions to this volume against the background of late Paleocene–early Eocene epochal chronology. The chapters are loosely grouped into the following categories (although distinct, and in some instances, considerable overlap occurs in the treatment of different subjects by several authors): Chronostratigraphy and Chronology; the Record of Global Climatic Change; Marine Biotic Response; Terrestrial Biotic Response. If we have piqued the interest of the reader in the world of 54–56 my ago, provided an insight into the dynamic coupling between Earth atmosphere–ocean–biotic evolution and contributed at the same time to the opening of a window into the Historical Geology of this fascinating interval in earth history we shall have achieved in some measure the goals we set ourselves in preparing this volume.

Chronostratigraphy and Chronology

Imprecise and overlapping original definitions of the terms Paleocene and Eocene and subsequent failure to determine an appropriate and internationally acceptable boundary (the Global Stratotype Section and Point, or GSSP) between them has resulted in a century of inconclusive (and often rancorous) debate. As a result of a 5-year investigation by working group members of IGCP Project 308 (Paleocene/Eocene Boundary Events in Space and Time) the boundary interval has been narrowed down to a stratigraphic interval between the top of the Thanetian Stage (London Basin) and the Ypresian Stage (Belgian Basin). Berggren and Aubry (chapter 2) review the classic NW European upper Paleocene–lower Eocene stratigraphic interval (and associated chronostratigraphic units and terminology) in its historical context, delineate the main bio-chemo- and magnetostratigraphic events that occur within the ~2-my span of this stratigraphic interval, and present an estimated chronology of some 18 events between (and including) Chron $C25n_{(o)}$ and $C24n.3n_{(o)}$.

Several boundary definitions currently vie for attention/ approval within this stratigraphic interval, ranging from the base of the Sparnacian Stage of NW Europe (favored by European mammalian paleontologists) that lies at a level low in Magnetozone C24r to the base of the Ypresian Stage (which lies at a level high in Magnetozone C24r). Between these levels/events are 1) a major $\delta^{13}C$ excursion/spike (which has been identified in both marine and terrestrial stratigraphies and has the virtue of providing a unique means of marine–terrestrial correlation; and appears to lie at the boundary between the Clarkforkian and Wasatchian North American Land Mammal Ages, long considered the Paleocene/Eocene Series/Epoch boundary by North American vertebrate paleontologists), 2) a correlative major extinction/ turnover in deep water (and to a lesser extent shallow water) benthic foraminifera; and 3) sequentially occurring calcareous nannoplankton (NP9/NP10) and planktonic foraminiferal (P5/P6) zonal boundaries denoted by stratigraphic ranges of nominative taxa. The authors note, indeed, *insist,* that until such time that a GSSP is appropriately established, the correct boundary definition of the Paleocene/Eocene Series boundary is "the base of the Ypresian Stage" (i.e., the base of the Ieper Clay in Belgium).

The chronology of the various microfaunal evolutionary events spanning the ~2 my of the late Paleocene–early Eocene interval is estimated by calibration to the recently developed Geomagnetic Polarity Time

Scale (GPTS) of Cande and Kent (1995) and the Integrated Magnetobiochronologic Time Scale (IMBS) of Berggren et al. (1995). A modified "short" chronology is developed for the mid-part of Chron C24r by the authors based on the realization that in order to remain consistent with the GPTS of Cande and Kent (1992; = CK92) the calibration of 55 Ma at the NP9/NP10 zonal boundary must be retained and the $^{40}Ar/^{39}Ar$ age determination of 54.5 Ma on the −17 Ash used in estimating the chronology of events within the ash series at Hole 550. This results in: a) an order of magnitude shorter estimate (50,000 yr vs. 467,000 yr) for the duration of the ash series present in DSDP Hole 550 compared with that estimated by Berggren and Aubry, 1996; = BA96); b) a revised estimate for the P5/P6 zonal boundary of 54.48 Ma (vs. 54.7 Ma in BA96); c) the realization that the ash series present in Hole 550 is probably incomplete, particularly the lower (negative) series (2a = 2.2 of Knox, chapter 5) inasmuch as the base of the ash series in Hole 550 is within (lower) Zone NP10, whereas the ash series present in the Sele Formation in subsurface cores of the North Sea Basin are correlative with the Woolwich–Reading Beds of the Lambeth Group (= Zone NP9) in outcrop in SE England.

Aubry (chapter 3) demonstrates the importance of sound temporal interpretation of the stratigraphic record for reconstructing earth history. Based on theoretical cartoons, she first shows how undeciphered unconformities in stratigraphic sections may produce apparent patterns of diachrony or synchrony, and may result in the correlation of actual events with pseudoevents. She shows that, although more subject to scrutiny than any other events, biostratigraphic datums are not alone in being affected by hidden unconformities: Isotopic events are as vulnerable, and, as she showed in an earlier paper (Aubry 1995), magnetic patterns may be affected as well. She insists thus that stratigraphic sections must be submitted to rigorous temporal analysis before claims of diachrony or synchrony can be made and before apparently correlative events can be compared. She illustrates her point with two examples.

In the first example, she explains how an undeciphered unconformity in the NP10 zonal record of DSDP Hole 550 has created problems for the current GPTS and IMBS. This has resulted in the assignment of an age of 55 Ma to two distinct stratigraphic levels, an actual level that corresponds to a (then) undeciphered unconformity (Swisher and Knox 1991) and the NP9/NP10 biochronozonal boundary used by Cande and Kent (1992, 1995) as a calibration tie point in their GPTS, with the consequence that if the age of 55 Ma is assigned to the level that was actually dated (through extrapolation of sedimentation rates) there is a stratigraphic interval around the NP9/NP10 chronozonal boundary that cannot be discriminated using the current GPTS. A provisional solution is proposed that consists in assigning the age of 55 Ma to the NP9/NP10 biochronozonal boundary while estimating a new age for the upper surface of the unconformity in Hole 550.

In the second example, she discusses the upper Paleocene–lower Eocene stratigraphic record in DSDP Hole 577. She shows that disturbance in the upper part of the interval assigned to Zone NP9 and an undeciphered unconformity at the NP9/NP10 zonal boundary result in misinterpretation of the chemostratigraphic and benthic foraminiferal records (Pak and Miller 1992), grain size record (Miller et al. 1987), and evolutionary patterns among planktonic foraminifera (Lu and Keller 1995a, 1995b). She recognizes that poor recovery and drilling disturbance associated with a core break (at the NP9/NP10 zonal boundary) may be responsible for the records that are observed rather than an unconformity, but she notes that either would produce the same effects either associated with an artificial stratigraphic gap (as an artifact of drilling) or with a true stratigraphic gap (an actual unconformity). She demonstrates that the isotopic records from the southern Ocean and the Atlantic and Pacific Oceans do not cover the same temporal intervals, and that there are currently no data on the isotopic composition of the southern Ocean during the earliest Eocene.

After remarking on the importance of composite reference sections, she concludes by stating that "we must think of stratigraphic sections not as proxies for time, but as windows on specific intervals of time, and realize that stratigraphy is to time what fossils are to species," and by remarking that, based on the construction of composite sections, we have reached a point in history when we can predict a temporal succession of events and search for them in the stratigraphic record. With regards to the Paleocene/Eocene Epoch/Series boundary, she notes that we are now in a position to determine the best criterion(ia) to characterize the Paleocene/Eocene Epoch boundary to which the series boundary will be subordinate.

Flynn and Tauxe (chapter 4) review the published paleomagnetic data base from marine and terrestrial stratigraphies that span the Paleocene–Eocene transition and regional correlations of the associated terrestrial vertebrate faunas from various areas of the world. They point to the shortcomings of magnetostratigraphic data sets that do not contain criteria of reliability (i.e., fold tests, conglomerate tests, thermal demagnetization, i.al.) and the resulting danger of translation into the familiar black/white polarity patterns upon which the unsuspecting pounce so eagerly. The authors use two simple means of assessing the reliability of magnetostratigraphic data sets: a) histograms of the parameter used to determine polarity (usually the virtual geomagnetic polarity pole (VGP) latitudes or inclinations) should exhibit characteristically two "humps" corresponding to the two polarities. If the histogram plots do not allow discrimination of the polarities, the magnetostratigraphic patterns may not be recording reliable polarity information; and b) the jackknife parameter ("J")—a quantitative assessment of the dependence of a given polarity stratigraphy on sampling density. Sections with high "J" values (0–0.5) are considered relatively robust; low "J" values (high negative values, below –0.5) are interpreted as representing numerous polarity events based on only one or two sampling horizons, with the resulting probability that the magnetostratigraphy is incomplete or incorrect.

In applying their criteria of reliability the authors indicate that DSDP/ODP Sites 577, 690, 752 and the Contessa Road outcrop section (northern Apennines, NW Italy) are essentially reliable, whereas Sites 528, 549 and 550 are less so. Flynn and Tauxe (chapter 4) suggest that nannoplankton datum events between low northern latitude Site 577 and high southern latitude Site 690 may be time transgressive/diachronous (an interpretation strongly rejected by Aubry, chapter 3). However, the authors do not rule out the possibility that the "anomalous" normal polarity pattern in Chron C24r in Hole 690B may be due to overprinting and, together with unconformities in that Hole and Hole 550, may account for the seemingly anomalous location of the $\delta^{13}C$ spike within different parts of Zone NP9 (cf. Aubry, chapter 3).

Flynn and Tauxe then review correlation of mammalian faunas of North and South America, Europe, Africa and Asia. (see also Lucas, chapter 21). North America sections (Ellesmere Land, the northern Big Horn Basin and the San Juan Basin) are considered to have the most complete magnetostratigraphic records spanning the P–E transition. They conclude by stressing the potential value (if it can be verified elsewhere) of a short normal "event" (cryptochron), detected in at least two terrestrial sections in the Big Horn Basin and at least one marine section (the authors do not discuss to which marine sequence they are referring; from our knowledge only NE Atlantic Site 552 contains this event; in fact three normal events have been recorded in mid–Chron C24r at that site), within mid–Chron C24r at a level close to the Paleocene/Eocene boundary in regional and global correlation and for age determination of events within the relatively long Chron C24r (2.5 my).

Knox (chapter 5) discusses the role of late Paleocene regional North Atlantic tectonism and, in particular, the influence of the uplift of the Iceland plume during the latest Paleocene Epoch, in modifying the contemporary geography of NW Europe and the concomitant effects on marine and terrestrial environments and associated faunas and floras based on a consideration of outcrop and well data. A relatively rapid (~0.5-my) regional sea level fall of ~100 m is suggested to have led to much of the North Atlantic Ocean having been subaerially exposed around the time of Chron C25n resulting in deposition of the well known marginal marine and paludal sediments of the London Basin (Lambeth Group) and Paris Basin (Sables de Bracheux and Argiles plastiques). Dysaerobic bottom water conditions in the deeper, and partially isolated, parts of the North Sea and Shetland Basins resulted in a notable reduction in abundance and diversity of agglutinated benthic foraminiferal faunas. Eventual isolation from oceanic waters owing to volcanic outpouring, sediment aggradation and local uplift led to anoxic bottom conditions resulting in virtual (if temporary) elimination of these faunas and provided a route for intercontinental migration of terrestrial mammal faunas between North America and Eurasia, probably via the Greenland–Scotland land bridge (if not a more extensive uplift associated with the eastward tilting of eastern Scotland). Palynomorph and clay mineral associations indicate climatic warming at this time, which supports a causal relationship between the regional sea level fall in NW Europe and the environmentally related changes in marine and terrestrial faunas and floras.

Knox (chapter 5) resists the temptation to draw a direct relationship between the penecontemporaneous global changes in oxygen and carbon isotopes (CIE) and extinction of benthic foraminifera (BFE) in deep and marginal marine environments and those observed in NW Europe–North Sea areas via changing patterns in global ocean circulation resulting from separation of Atlantic and Arctic water masses. He rightly points out that this separation continued well after the global isotopic excursions and benthic extinctions. Nor is it likely according to Knox (chapter 5) that the gas/dust emissions associated with North Atlantic volcanism had any significant effect on global warming or ocean chemistry at this time (as some authors have suggested) inasmuch as they occurred over a considerably longer interval of time. In fact the isotopic excursions and extinctions took place during an interval of only minor pyroclastic activity in the North Atlantic.

The Record of Global Climate Change

The late Paleocene–early Eocene interval was an anomalous episode in earth history. Following the dramatic and relatively short-term perturbation in climate, biota and productivity caused by the impact event(s) at the Cretaceous/Paleogene Period boundary, the Paleocene Epoch was characterized by gradual restoration of equable climates, recovery, evolutionary radiation and increased diversity of marine and terrestrial biotas, and by increased oceanic productivity. By the late Paleocene Epoch the earth experienced globally uniform warm temperatures; equatorial sea surface temperatures (SST) were in excess of 20° C (Pacific) to 26° C (Caribbean; comparable to, but somewhat cooler than, the Holocene–present day), whereas high southern ocean (Antarctic) SSTs were on the order of 10–12° C. The latitudinal temperature gradient (10–15° C) was minimal and equal to or less than half of today's value (~28–30° C). Relatively uniform bottom water temperatures (7–10° C) were produced primarily in the western subtropical Atlantic (paleo-NADW) during the early Paleocene Epoch; by the late Paleocene Epoch, the influence of the western subtropical Atlantic was replaced by a high latitude Southern Ocean (Antarctic region) source. Marine productivity reached its Cenozoic maximum during the late Paleocene Epoch (early Chron C25r) as reflected by flourishing calcareous plankton and benthic foraminiferal faunas.

During the latest Paleocene Epoch the earth experienced a relatively brief (in geologic terms) global warming, referred to as the Late Paleocene Thermal Maximum (LPTM; Zachos et al. 1994). Marine SSTs at high latitudes increased by ~4–6° C (from ~10–12° C to ~14–16° C), whereas equatorial temperatures remained relatively unchanged; bottom waters warmed to ~10–14° C as the normal thermohaline driven circulation with a high latitude source(s) of bottom water formation was suddenly, if only temporarily, shifted to a halothermal driven circulation (with the formation of warm saline bottom waters (WSBW) at low latitude site(s), probably linked with Tethys); marine and terrestrial productivity collapsed; the formation of uniformly warm, low oxygen, corrosive bottom waters resulted in widespread $CaCO_3$ dissolution and a dramatic extinction event (BFE) among deep water, and to a lesser extent, neritic benthic foraminifera. Whereas the high latitude SSTs and bottom water temperatures remained relatively constant during the early Eocene Epoch, the thermohaline driven bottom water circulation (with a high latitude Southern Ocean source(s)) resumed relatively rapidly following the LPTM. The early Eocene global warming is known as the Cenozoic Global Climate Optimum (CGCO) and this is reflected in widespread expansion of equatorial marine and terrestrial biota into high northern (Ellesmere Land, Spitsbergen, Norwegian–Greenland Sea) and southern (Kerguelen Plateau, Antarctic continent) regions.

This, then, is the setting explored in the chapters of this volume. An earlier study (Prothero and Berggren 1992) examined the climatic and biotic changes that accompanied the transition from the Eocene "doubthouse" world (following the late Mesozoic–Paleocene "hothouse") to the Oligocene–Neogene "icehouse" world associated with the ?late Eocene–early Oligocene initiation/development of Antarctic glaciation. The late Paleocene–early Eocene climatic and biotic changes took place against a background of global warming in an ice-free world, the antithesis of the climatic framework of the Eocene–Oligocene transition. While the latter interval was characterized by sequential (~10 my), but not dramatic, global extinctions superimposed on a gradually cooling (particularly at high latitudes) earth and increased eutrophication associated with increased biosiliceous productivity at high southern latitudes, the Paleocene–Eocene transition was characterized by a major, abrupt (~0.01 my) extinction/turnover among deep water benthic foraminifera, major evolutionary turnover among terrestrial mammals faunas, and accelerated evolution (~1–2 my) among calcareous plankton (nannoplankton and foraminifera) and tropical larger benthic foraminifera (nummulitids, discocyclinids, alveolinids) during an episode of enhanced global warming. In the following section we review the nature of, and evidence for, global climate change(s) associated with the late Paleocene–early Eocene transition and their putative effect on concomitant biotic changes.

Himalayan Tectonism

The role of major tectonism—in this instance the late Paleocene India–Asia collision—in inducing changes in the geochemical cycling of carbon and oxygen in marine and terrestrial environments and attendant climatic and biotic changes is explored by Beck et al. (chapter 6) in the following scenario: 1) gradual suppression of Trans-Himalayan arc volcanism as a result of the collision may have led to passive global warming during the mid-Paleocene Epoch. Concomitantly, stagnation and anoxia in remnant foreland basins (such as the Indus Basin) marginal to the gradually closing Neo-Tethys, may have led to enhanced preservation of marine C_{org} in sediments, lowered marine $\delta^{13}C_{carb}$, and increased global cooling by means of lowered pCO_2; 2) during the late Paleocene–early Eocene interval $\delta^{13}C_{carb}$ decreased (reflecting decreased rate of net organic burial), which suggests to the authors that the collision may have warmed the earth by exhuming organic carbon from the colliding continental margins, eliminating extant C_{org} sinks and, concomitantly, enhancing the greenhouse effect; 3) during the early Eocene Epoch $\delta^{13}C_{carb}$ increased, suggesting an increased rate of net organic carbon burial due, possibly, to C_{org} burial in Himalayan foreland basin(s), among others. Beck et al. (chapter 6) recognize that some of these mechanisms exhibit temporal overlap and, in fact, may have operated in opposite directions relative to the isotopic record and climate; however, they stress the magnitude of associated chemical and heat fluxes relative to the earth's long term biogeochemistry and climate in evaluating the feasibility of implicating the India–Asia collision in the significant climatic and biotic changes that characterized the late Paleocene–early Eocene Epochs.

An (editorial) note of caution might be appropriate here. While Beck et al. (chapter 6) provide an impressive list of (primarily, but not exclusively) geotectonic–geochemical data in support of a (late) Paleocene collision of India with Asia, faunal evidence in the form of two anurans (a pelobatid frog and a discoglossid frog) discovered at Taklii and Naskal, India, respectively, and a palaeoryctid mammal (the eutherian *Deccanolestes*) of Asian affinities, discovered at Naskal in beds interstratified between Deccan Trap flows dated at 66 ± 1 Ma and placed in Chron C29r, suggest an intra-Maastrichtian (latest Cretaceous Period) "landfall" between India and Asia (Rage and Jaeger 1995). Non marine ostracodes and charophytes in Indian Maastrichtian deposits are cited as further support of a pre-Paleocene initial contact and development of a terrestrial connection between the two land masses inasmuch as salt water serves as an effective barrier to dispersal of nonmarine amphibians, charophytes and ostracodes. These data do not invalidate the scenario proposed by Beck et al. (chapter 6) regarding the influence/relationship of tectonism on late Paleocene global climates and biotic events; they merely serve to indicate that different disciplines have their own criteria of determining the recognition/initiation of the same event in the geologic record. Indeed, it must be acknowledged that the Asia–India contact may have been of Late Cretaceous age allowing terrestrial, fresh water organisms to move freely between the two plates. However, this would definitely predate the suturing of the two plates, and the suppression of arc volcanism, which probably postdates initial contact. The issue remains equivocal.

Reduced Atmospheric Circulation

A significant reduction in late Paleocene–early Eocene atmospheric circulation (and concomitant increased global aridity) is reflected in reduced flux and grain size of eolian dust in North and Central Pacific and Indian Ocean pelagic sediments (Rea, chapter 7). While high latitude temperatures increased by ~6° C, tropical temperatures (somewhat lower than present day values) remained stable or may have even cooled somewhat. A significant lowering (~7° C) of the sea-surface temperature gradient from ~17° C (equator) to ~10° C accompanied these changes in atmospheric circulation.

An enigmatic sequence of events spanning ~4 my in the late Paleocene Epoch and involving a warming of (primarily) high latitude waters leading to partitioning of carbon between ocean and sedimentary reservoirs, leading to a marked reduction in global wind intensity (EDE) leading to a deep sea benthic foraminiferal extinction (BFE) defies adequate explanation by any single scenario as explained by Rea (chapter 7). Rather, Rea supports the recently developed model(s) that implicates methane exsolution from sub-seafloor hydrates resulting from deep sea warming of ~5° C; a further corollary of this scenario is that it adequately accounts for the shift to the light carbon isotope values in the oceanic reservoir associated with the multiple climatically related events during the latest Paleocene global warming (see also discussion by Sloan and Thomas, chapter 9, and Thomas, chapter 12). Rea concludes by noting that exsolution of methane from subseafloor hydrates helps explain the extreme shift (3–4‰) in $\delta^{13}C$ (as well as the rapidity of the major, short-term excursion associated with the BFE we might add) without having to resort to explanations that require extreme continental weathering (an unlikely development given increased global aridity during the latest Paleocene–early Eocene Epoch) and the retention of exceptionally warm polar regions via polar cloud mechanisms proposed in some scenarios.

Oceanic Circulation

Corfield and Norris (chapter 8) review Paleocene–early Eocene oceanography in terms of stable isotope history, with particular emphasis given to the LPTM. They observe that the ~10 my-long Paleocene Epoch was bracketed by two significant, transient negative $\delta^{13}C$ excursions of comparable magnitude (~3–4‰), indicative of global reduction in productivity, one at the Cretaceous/Paleogene Period boundary due to exogenic causes (bolide impact(s)), the other coincident with the LPTM and due to endogenic causes (feedback effects from changing ocean/atmospheric dynamics when threshold limits may have been exceeded). The Paleocene Epoch itself was characterized by the most isotopically positive $\delta^{13}C_{carb}$ values of the Cenozoic Era. This has been interpreted as reflecting a return, at least in part, to normal, if not enhanced, global ocean productivity following the essentially total extinction of oceanic plankton at the Cretaceous/Paleogene Period boundary.

Recent studies by the authors on the Paleocene record of subtropical DSDP Site 384 (NW Atlantic) have provided much needed isotopic data allowing comparison with equatorial and high southern latitude regions. Site 384 contains the most positive $\delta^{13}C$ benthic values when compared with contemporaneous, more negative, values at high southern (ODP Site 689, Maud Rise) and equatorial (DSDP Site 577, Shatsky Rise, NW Pacific) localities. This strongly supports the suggestion that the temperate North Atlantic (rather than the high southern latitude region(s) adjacent to Antarctica) was the source of deep water formation during the latest Cretaceous Period and early Paleocene Epoch. During the mid-late Paleocene Epoch the Southern Ocean became a source for deep water formation as well at the expense of other (low latitude) locations. By later Paleocene time, the circum-Antarctic region had become the main (if not sole) locus of bottom water formation of essentially globally uniform similar temperature. The LPTM was associated with a geologically brief (transient) change from (high latitude generated) thermohaline to (low latitude generated) halothermal mode of deep water formation and concomitant, transient climate changes registered at low and high latitudes in the $\delta^{18}O$ record. ^{13}C values returned rapidly to pre-excursion values at high southern latitudes and a marked latitudinal gradient was once again developed, signifying reestablishment of a predominantly Antarctic source for bottom water formation that continued throughout the early Eocene Epoch. The results/scenario of Corfield and Norris (chapter 8) are generally supportive of the modeling experiments of Sloan and Thomas (chapter 9) and provide hope that a unified explanation for the enigmatic LPTM (and its associated phenomena) may soon be forthcoming.

Climate Modeling

In order to obtain a better understanding of the "how" and "why" of the Late Paleocene Thermal Maximum (LPTM; Zachos and others 1993), Sloan and Thomas (chapter 9) conducted climatic model experiments using an NCAR GENESIS (Global Environmental and Ecological Simulation of Interactive Systems) climate model. Integrating data from stable isotopes on surface (morozovellids) and intermediate (subbotinids) dwelling planktonic foraminifera, using global land/sea paleogeometries, reconstructed land vegetation (including the specification of deciduous forests at high latitudes and substitution of (sub)tropical ground cover for all present day desert regions as suggested by available paleobotanical information), and assuming an ice-free (non glacial) world, atmospheric pCO_2 values comparable to today (340 ppm) and present day orbital geometries, the LPTM results were compared with those obtained from modeling a "background climate."

While results are of considerable interest in understanding the events associated with the LPTM, they leave unanswered several important questions regarding the nature and relative rapidity of the LPTM event itself. Major conclusions are that: 1) although high latitude surface waters and global deep waters warmed by ~4–6° C and these temperatures persisted for several hundred thousand years, global mean surface temperatures were > 2° C warmer than the "background" case during the LPTM, while at the same time low latitude surface waters remained at ~24–26° C; 2) continental warming took place primarily above 40° latitude with winter warming magnitude of land masses generally exceeding summer warming; 3) more even latitudinal distribution of precipitation and greater precipitation over land masses were observed for the LPTM model; 4) LPTM wind velocities were notably decreased (consistent with analytic data reviewed by Rea, chapter 7); 5) model results do not unequivocally support subtropical source for WSBW during the LPTM but rather imply that the "background" model may have been more appropriate, whereas the surface water salinity conditions predicted by the subsequent LPTM SSTs would have been unable to sustain deep water formation at mid latitudes. The authors note that the conditions required to establish the LPTM remain enigmatic and unresolved. In the absence of high latitude ice, it is difficult to conceive of adequate mechanism(s) of increasing high latitude temperatures (particularly in the absence of evidence of contemporary low latitude warming). Poleward oceanic transport under "background climate" conditions (particularly from the Pacific) may have led to a small amount of high latitude warming, concomitant precipitation and associated lowering of surface water salinities, returning the system full circle to increased formation of deep waters at high latitudes during the early Eocene Epoch (see supporting $\delta^{13}C$ data for this scenario in Corfield and Norris, chapter 8). Indeed, the LPTM model results exhibit a greater subtropical negative moisture flux than the "background climate," supporting, at least indirectly, the indication that LPTM (sub)tropical formation of bottom waters was not sustained very long in geologic terms. The LPTM event was a geologically rare, if not unique, event, which, as the authors emphasize in conclusion could only have occurred as the result of a complex chain of interrelated events. Stay tuned for further modeling experiments!

Marine Biotic Response

In this section we review the response of marine invertebrate biota to climate changes associated with the late Paleocene–early Eocene transition.

Protists

Calcareous nannofossils

Aubry (chapter 10) discusses the modalities and rates of diversification of the coccolithophorids during the early Paleogene. She shows that 50% of the Cenozoic genera evolved during the first 15 my (Paleocene to early Eocene Epochs) of the Cenozoic Era, and that 75 % of the Cenozoic families were established by the beginning of the Eocene Epoch. Paleocene diversification occurred in three waves: in the earliest Paleocene Epoch (65–63.8 Ma), in the earliest late Paleocene Epoch (60.6 to 59.4 Ma) and in the latest Paleocene Epoch (55.5 to 53.4 Ma) in an interval that she refers to as the Paleocene/Eocene Boundary Interval (P/E BI) and that spans the interval between the late Paleocene thermal maximum (LPTM, Zachos et al. 1993) and the beginning of the Ypresian transgression as recorded in Belgium and England (Aubry 1996). According to her, Paleocene diversification resulted in the transformation of the early Paleocene nutrient-rich low-productivity "Strangelove Ocean" (Hsü and McKenzie 1985) into a predominantly late Paleocene oligotrophic ocean. She suggests that Paleocene diversification in the coccolithophorids was accelerated through co-evolution with the photosymbiont-bearing planktonic foraminifera *Morozovella–Acarinina–Igorina* group. Co-evolution in the two groups resulted in rapid consumption of the nutrients, increased transparency of oceanic waters and thus vertical thickening of the photic zone, which itself led to further diversification and evolution of K-mode specialists. Thus as oligotrophy was developing, the Paleocene ocean was strongly productive as inferred from the strong and rapid increase in $\delta^{13}C$ at all depths that occurred from early Chron C26r to mid Chron C25n and reached a late Paleocene maximum (Shackleton et al. 1985; Corfield and Cartlidge 1992; Zachos et al. 1993).

Selectivity in the extinction of taxa occurred at the Cretaceous/Paleogene Period and around the Eocene/Oligocene Epoch boundary. In both cases, selectivity is easily understood. Selectivity also occurred in the P/E BI turnover, but is enigmatic. The taxa that became extinct are those that were dominant and radiated during the Paleocene Epoch and evolved with increasing oligotrophy. There seems to be a strong relation between selectivity and structure of the coccoliths, which suggests that the undetermined event that triggered the turnover, and is expressed by sharp carbon and oxygen isotopic anomalies, affected biomineralization.

Planktonic foraminifera

Late Paleocene–early Eocene evolution of planktonic foraminiferal communities is treated by Boersma et al. (chapter 11) in terms of the Trophic Resources Continuum (TRC) model. The TRC is a spectrum of chemical conditions in the global ocean spanning the extremes of maximum nutrient deficient areas (subtropical gyres) to highly productive areas (regions of upwelling and continental margin drainage). The model is predicated on the TRC controlling the diversity of oceanic euphotic habitats and serves to predict/reconstruct global diversity cycles in marine organisms. The authors review the TRC in modern oceans and examine the different responses by biologic communities to well-mixed and reduced-mixing (sluggish) circulation.

Muricate, non-spinose morozovellids and microperforate biserial chiloguembelinids are interpreted as having been oligotrophic and eutrophic indicators, respectively. The non-spinose, bluntly muricate acarininids exhibit wider geographic distribution than their isotopically similar, shallow water dwelling "cousins," the morozovellids, and are suggested to have occupied a mesotrophic habitat and to have possessed symbionts. (Sub)tropical-adapted, muricate planktonic foraminifera exhibited an increase in abundance and expanded into high southern (Kerguelen Plateau) as well as northern (England) latitudes near the Paleocene/Eocene Epoch boundary, whereas the microperforate chiloguembelinids were essentially restricted to cool temperate, higher latitudes. Late Paleocene (Biochrons P4–P5) communities were of lower diversity than their earliest Eocene (Biochrons P6–P7) counterparts. The low-latitude oligotrophic morozovellids did not extend into high-latitude areas, whereas the eutrophic chiloguembelinids were geographically more widespread. By the early Eocene Epoch, diversification of the oligotrophic muricates led to their dominance in low-latitude communities and expansion into high-latitudes. Southern Ocean early Eocene faunas bore a resemblance to late Paleocene counterparts in temperate eastern gyres.

The authors conclude that late Paleocene faunas suggest an intermediate TRC, whereas the abundance and diversification of the muricate oligotrophs indicate the widespread development of warm, low nutrient surface waters during the early Eocene Epoch. High latitude concentration of the eutrophic chiloguembelinids indicates areal restriction of higher nutrient concentrations in intermediate waters. The Paleocene Epoch is suggested to conform to a moderately well mixed, warm ocean: marked surface–deep water thermal gradient, efficient nutrient cycling throughout the water column and relatively high fertility. The early Eocene data, by contrast, suggest a poorly mixed ocean (and concomitant sluggish atmospheric circulation): low vertical as well as horizontal (geographic) thermal gradients, reduced fertility, and concentration of nutrients at the base of the thermocline. The interpretations of the TRC model are seen to be consistent with stable isotope data and global atmospheric circulation patterns described earlier and reviewed in this volume (Corfield and Norris, chapter 8; Rea, chapter 7, among others).

Benthic foraminifera

Coincident with the LPTM, deep-sea (bathyal and abyssal) benthic foraminifera underwent the most dramatic (up to 50%), essentially global extinction/taxonomic turnover in more than 70 my (since the Cenomanian/Turonian Age boundary; Thomas, chapter 12). A lesser but decidedly recognizable turnover event has been recognized in neritic environments as well (particularly southern Tethyan; Speijer 1994). This event is referred to here as the Benthic Foraminiferal Extinction Event (BFE). In contrast to the sequential, stepwise extinction/turnover patterns (with punctuated accelerations) that deep sea faunas experienced in association with the late middle Eocene to Oligocene cooling, the terminal Paleocene event was relatively abrupt (several thousand years in duration). It occurred during the LPTM (Zachos and others 1993), and is represented by a stratigraphic interval of moderate to intense dissolution, condensation and/or single to multiple unconformities (Berggren and Aubry 1996; Aubry, chapter 3). The BFE coincided essentially with the $\delta^{13}C$ excursion spike, and exhibited a distinctly different signature than the pattern that occurred at the Eocene–Oligocene transition, some 15–20 my later, especially in the occurrence of a typical stressed, low-diversity and high-dominance fauna after the extinction.

A fauna typical of, and containing numerous elements from, the Late Cretaceous Period (denominated the *Stensionina beccariiformis* fauna because it was a characteristic component) was gradually "squeezed out" (exhibiting a upward shallowing) and eventually replaced by a *Nuttallides truempyi* dominated fauna (Tjalsma and Lohmann 1983). Pre-extinction faunas were cosmopolitan, taxonomically diverse, and dominated by thick-walled, heavily calcified calcareous, or agglutinated ("flysch-like"), taxa of the so-called Velasco-type fauna. Post-extinction faunas were, by comparison, characterized by smaller, thinner walled forms; most notable is the inter-site (i.e., geographic rather than depth dependent) difference in abundance and diversity patterns, particularly among buliminids (including *Tappanina selmensis*), *Nuttallides truempyi, Cibicidoides* spp. and abyssaminids.

Mid-bathyal agglutinated benthic faunas in the North Sea and Labrador Basin (and elsewhere in the world) also suffered a marked terminal Paleocene extinction; impoverished, low-diversity faunas characterize the earliest Eocene Epoch, and the sedimentary record points to dysoxic to anoxic conditions and associated strong carbonate corrosiveness. Comparable conditions, if at a lesser scale, have been observed at shallower depths (upper bathyal–outer neritic) in the eastern Tethys, and in sections in Spain, New Zealand and Japan.

Thomas (chapter 12) provides a comprehensive review of the complex and, as yet unresolved, data sets and tentatively suggests the following scenario. First, deep and intermediate waters of the ocean warmed rapidly, as a result of the rapid shift from high to low-latitude formation of WSBW, as suggested by several other investigators (e.g., Kennett and Stott 1991). Then, the increased temperatures of the deep waters led to dissociation and subsequent oxidation of isotopically light methane hydrates (Dickens and others 1995). This process released isotopically light carbon into the oceans, and caused the $\delta^{13}C$ anomaly, used up oxygen and aggravated the low-oxygen conditions caused by warming and sluggish circulation, and induced widespread $CaCO_3$-dissolution. Circulation changes and temperature increase led to expansion of the trophic resource continuum (TRC; Boersma et al., chapter 11). In this speculative scenario the benthic extinction was caused by a combination of lowered oxygen levels, increased temperatures, increased corrosivity and changing ocean productivity patterns.

Radiolaria

The sparsity of radiolarians and their generally poor preservation, coupled with a lack of well preserved and complete stratigraphic records preclude an accurate assessment of the nature of late Paleocene–early Eocene biotic changes in this group. Nevertheless, Sanfilippo and Nigrini (chapter 13) make a heroic effort at determining the stratigraphic range of some 70 taxa over a 10 million-year span (from the *Bekoma campechensis* Zone (= Zone P3) to the upper part of the *Buryella clinata* Zone (= Zones P8–9), based on analysis of 24 DSDP sites. The authors identify six radiolarian first appearance datum events (FADs) that are correlative with Zone CP9a (= Zone NP10) and thought to be reliable indicators of the Paleocene/Eocene Series boundary. Inasmuch as a formal taxonomic framework is still lacking for much of the early Paleogene radiolarian fauna, species/generic diversity trends cannot be rigorously described. A quantified morphologic diversity shows, however, a gradual increase in morphologic diversity during the Paleocene Epoch (following reduced diversity near the Cretaceous/Paleogene Period boundary) and until the early Eocene Epoch, by which time a maximum diversity pattern had been established that remained essentially stable through the remainder of the Cenozoic Era. No notable change in morphologic diversity has been observed to date at the Paleocene/Eocene Epoch boundary, although the recovery of more stratigraphically complete sections with well preserved radiolarian faunas may well alter this picture.

Dinoflagellates

Bujak and Brinkhuis (chapter 14) document the taxonomic and diversity changes that occurred in low-and mid-latitude dinocyst assemblages during the latest Paleocene Epoch (Thanetian Age; Biochron NP9). The changes affected primarily the *Apectodinium*-dominated assemblages (familiar from studies of NW Europe as well as from North Sea and Labrador Shelf petroleum exploration) and included suppression of several other dinocyst taxa at the expense of the diversification of early Eocene assemblages characterized by the (Eocene) Wezelliaceae.

The authors establish the succession of dinocyst stratigraphic events in North Sea subsurface stratigraphic sequences, as well as in NW European outcrops and SE England sequences, and then place these "stratigraphies" in an integrated magneto-bio- and chronostratigraphic framework. Comparative studies on low (Tethyan) and high northern (Spitsbergen, East Greenland, Norwegian Barents Sea, Canadian Beaufort Sea) latitudes are briefly reviewed, but lack of rigorous biostratigraphic control and incomplete sections/recovery preclude precise event correlations.

During the latest Paleocene (Biochron NP9) dinocyst assemblages changed at low-, mid- and high latitudes from relatively diverse assemblages with few *Apectodinium* to low-diversity assemblages dominated by *Apectodinium*. At mid-latitudes the *Apectodinium*-dominated assemblages were supplanted by increasingly diverse assemblages with few *Apectodinium*, whereas at low (Tethyan) latitudes *Apectodinium* continued in large numbers into mid-Biochron NP11 (close to the Chron C24r/n). The authors develop a model to explain the latest Paleocene *Apectodinium* dominance and subsequent geographic distribution/migration: the LPTM and nutrient availability enabled the migration of *Apectodinium* from Tethyan to mid- and high latitudes of both hemispheres (and concomitant suppression of other cooler water dinocysts); widespread transgression during the Ypresian and subsequent cooling pulses beginning in the early Lutetian of higher latitudes provided vacant niches for the (re)colonization by newly evolving Eocene forms.

This model is notable in its simplicity and ability to explain most of the observations pertaining to the latest Paleocene Epoch. Data on dinocysts from the North Sea, NW Europe and Italy (Bujak and Brinkhuis, personal communication 1997) indicate that conditions remained warm in higher northern latitudes until the early Lutetian (Chron C21 or possibly Chron C22). The dinocyst data therefore support the view that the short term LPTM event (which is part of a long-term increase in global temperatures that lasted well into the late early Eocene Epoch) resulted in a ~4–6° C increase in intermediate water temperatures at low latitudes (Bralower and others 1995) and ~8° C warming of high southern latitude surface waters (Kennett and Stott 1991). However, the fact that temperatures at low latitudes were similar to or lower than the present during the late Paleocene and early Eocene Epochs indicates that "cool" tropics have persisted through the Cenozoic Era, whereas high latitudes have undergone significant temperature variations. To date climate models have failed to produce the low equator to pole temperature gradients indicated by the isotopic data, and an increased heat transport from low- to high latitudes is difficult to achieve at low temperature gradients. Admittedly temperatures fell to near pre-LPTM values at high southern latitudes (Kennett and Stott 1991) but whether this was sufficient to account for the early Eocene immigration/diversification patterns in dinocysts delineated by Bujak and Brinkhuis (chapter 14) and presumably (although not explicitly) relied upon by them in their model remains uncertain. For instance, floras (with modern counterparts in SE Asia) were flourishing contemporaneously in the London–Hampshire Basin of SE England, widespread nummulitic deposits were forming in the Paris and Belgian Basins (during Biochrons NP11 and NP12) and alligators, varanid (monitor) lizards, dermopterans ("flying" "lemurs") and primates (paromomyids) lived on Ellesmere Island (Lat. 77–78° N); these data suggest warm temperatures extending well into the Arctic region and low equatorial–polar thermal gradients during the early Eocene Epoch.

Invertebrates

Ostracoda

A comparably dramatic and rapid turnover in deep sea ostracodes has been recently documented to have occurred at the same time as the BFE at Maud Rise (ODP Site 689) in the Southern Ocean (Steineck and Thomas 1996). A stratigraphically and taxonomically stable and persistent fauna of heavily calcified forms characterized by *Krithe* and a group of predominantly epifaunal taxa was suddenly replaced within 10 kyr by a taxonomically novel assemblage of thin-walled opportunistic and generalist forms (so-called "*Propontocypris*" fauna) considered by the authors to represent a no-analogue "disaster-fauna" comparable to modern assemblages associated with ephemeral, high-nutrient sites such as deep-sea vents. A second phase of recovery was reflected in a return ~25–40 kyr later of typical bathyal taxa characterized by calcium-poor carapaces, suggesting to the authors the presence at Site 689 of either a high flux of organic carbon or, more likely, the transport from low latitudes of a warm, low oxygen water mass comparable to the scenario suggested earlier by several authors to account for the dramatic turnover in the deep-sea benthic foraminiferal fauna.

Marine mollusks

In the Gulf Coastal Plain of North America the Paleocene/Eocene Series boundary lies within the Wilcox Group (Sabinian Stage) at the contact between the Tuscahoma and Bashi Formations. Dockery (chapter 15) documents major (97% and 94% respectively) extinction rates in molluscan faunas below and above the Wilcox Group coincident with Type 1 unconformities. A 69% extinction rate between the Tuscahoma and Bashi Formations is shown to occur at a (less prominent) Type 1 unconformity as well, together with a significant (molluscan) $^{87}Sr/^{86}Sr$ excursion. While interpretation of the Sr isotope record is not unequivocal, the author reviews the various lines of evidence and argues that the Sr ratios in the Gulf Coast Paleogene record reflect changes in ancient sea water, and that circulation patterns changed significantly at the Paleocene/Eocene Epoch boundary, thus affecting the larval distribution of mollusks as well (accounting for the flood of new taxa in the Bashi Formation).

The marine molluscan turnover at the Paleocene/Eocene Epoch boundary in the Gulf Coast is ascribed to the effects of sea-level lowering associated with the base of third-order cycle TE1.1, the creation of new environments following the subsequent early Eocene transgression and the introduction of new/novel taxic elements by means of larval transport entrained in changing oceanic currents. Indeed, the example cited by Dockery (chapter 15) of the appearance of Tethyan elements (*Volutilithes*) in the Hatchetigbee Formation of the northern Gulf Coastal Plain may reflect the major global warming event (the Late Paleocene Thermal Maximum, LPTM, of Zachos and others 1993) and concomitant latitudinal expansion of low latitude faunas that is seen in $\delta^{18}O$ and biotic records at this time (see chapters in this volume).

Terrestrial Biotic Response

During the late Paleocene and early Eocene Epochs, Europe, North America and Asia formed a boreally-connected landmass that constituted a single Holarctic paleobiogeographic province (e.g., McKenna 1983). Little is known about the terrestrial Paleocene–Eocene record in Africa, Australia and Antarctica. More is known concerning South America where the record indicates a high degree of endemism even though a dense volcanic archipelago (Greater Antillean arc) connected South and North America, at least intermittently, during the Paleocene Epoch. Best known and extensively discussed in several chapters in this volume is the Holarctic record.

Megaflora

Paleocene and Eocene terrestrial floras are well known from the Holarctic province, and Crane (1987), Manchester (1987) and Upchurch and Wolfe (1987) have presented broad reviews. Particularly well known are the floras from the northern Rocky Mountains and Great Plains of western North America. Here, deciduous plants (especially Juglandaceae and Betulaceae) underwent a great Paleocene diversification and came to dominate the flora, which indicates widespread temperate woodlands. In contrast, the early Eocene flora was quite different, corresponding to a megathermal vegetation dominated by families characteristic of extant megathermal rainforest, such as the Icacinaceae, Annonaceae, Lauraceae and Menispermaceae. This flora extended as far north and south as 50–60° latitude (Wolfe 1977, 1985), though by the middle Eocene Epoch its latitudinal spread was greatly reduced.

Wing (chapter 17; see also Wing et al. 1991) has dissected this change using the remarkable record in the Bighorn Basin, Wyoming. He argues that the change took place in two steps. The first, during the latest Clarkforkian–earliest Graybullian time, was a period of extended extinctions. The second, in the middle Wasatchian (Lysitean) North American Land Mammal Age (NALMA), represented the immigration into the Bighorn basin of megathermal taxa associated with a temperature maximum.

Collinson and Hooker (1987) attempted to correlate changes in vegetation and mammals that occurred in southeastern England across the Paleocene/Eocene Epoch boundary. In England, late Paleocene floras reflect open and disturbed environments with limited woodland and forested areas. Early Eocene (London Clay) floras indicate dense tropical forest. Early Eocene mammals included a low percentage of large ground-dwelling herbivores and arboreal browsers, as would be expected in a rain forest. However, late Paleocene mammals were mostly small ground dwellers. Collinson and Hooker (1987:288) concluded that "this incongruency suggests that these mammals [large ground dwellers] had not then [late Paleocene] invaded the herbivorous niche." However, a consideration of the North American late Paleocene mammal record, which is rich in large terrestrial herbivores (e.g., barylambdid pantodonts) and small arboreal herbivores (e.g., carpolestids and plesiadapids), reveals the incongruity of the results from an inadequate English fossil record.

Nonmarine mollusks

A comparable, but asynchronous turnover, occurred in nonmarine mollusks in the Bighorn Basin (NW Wyoming; Hartman and Roth, chapter 16). The history of early Paleogene nonmarine pelecypod and gastropod faunas is one of significant extinction at or near the Cretaceous/Paleogene Period boundary, increase in mesogastropod diversity during the late Torrejonian North American Land Mammal Age (NALMA; = late Danian Age) and Tiffanian NALMA (= Selandian–early Thanetian Age) and a major increase in terrestrial prosobranch and pulmonate families and concomitant diminution in aquatic prosobranchs during (and particularly in the middle of Biochron Cf2) the Clarkforkian NALMA (= late Thanetian Age). While the reduction of aquatic prosobranchs continued into, and remained a significant feature of, the Wasatchian NALMA, the authors note that increase in diversity among terrestrial taxa was the most significant aspect of the turnover in nonmarine mollusks and this event occurred during the middle Clarkforkian NALMA. Since the Clarkforkian NALMA was of relatively short duration (< 1 my, from Chron $C25n_{(o)}$ to a level within the early part of Chron C24r), the authors' suggestion that the molluscan taxa can serve to identify/delineate Clarkforkian biochrons in the absence of mammalian faunas in the Powder River Basin (SE Montana and NE Wyoming) seems to be well taken. While a definitive link has not been firmly established, it would appear reasonable to suggest that the decrease in aquatic prosobranchs and concomitant increase in diversity of terrestrial prosobranchs and pulmonates may reflect the increased aridity that characterized the latest Paleocene–earliest Eocene Epoch (see Rea, chapter 7, and Wing, chapter 17).

Vertebrates

Lower vertebrates

There is little evidence of significant evolutionary turnover in lower vertebrates (fishes, amphibians, reptiles) across the Paleocene/Eocene Epoch boundary. Particularly significant, however, is their late Paleocene–early Eocene boreal distribution. A mixture of aquatic, riparian and terrestrial lower vertebrates preserved in upper Paleocene-lower Eocene deposits on Ellesmere Island, is remarkably similar to broadly contemporaneous assemblages at much lower latitudes from North Dakota to New Mexico (e.g., Estes 1988). The Ellesmere Island assemblage includes bony fishes (bowfins and gars), salamanders, anguid and varanid lizards, turtles (including tortoises and trionychids) and the alligator *Allognathosuchus*. Extant relatives of these taxa have high temperature optima, and this suggests an equable climate with essentially frost-free winters on Ellesmere Island during Graybullian time (Estes and Hutchison 1980).

Fishes

Changes in the fish fauna across the Paleocene/Eocene Epoch boundary are difficult to evaluate because of insufficient record. In western North America, the record of Paleogene freshwater fishes is very fragmentary below the lower middle Eocene Green River Formation Lagerstätten (Grande 1980). The Green River fishes have their closest phylogenetic relationships with the Asian fish fauna, which suggests a shared ancestry of Mesozoic antiquity (Grande 1985, 1989).

The Chinese record of Paleogene fishes has been little studied, but Sychevskaya's (1986) review of the Paleogene fishes from Mongolia and the former USSR (FSU) supports Grande's arguments for a close affinity and shared ancestry of a trans-Pacific freshwater ichthyofauna. The Asian record suggests persistence of this ichthyofauna from the late Paleocene through early Oligocene Epoch without detectable, significant changes.

In Europe and North Africa, the ichthyofaunal record is mostly of marine selachians (e.g., Herman 1977; Ward 1980). Clearly little happened in selachian evolution—other than the appearance of a few batomorphs—at the Paleocene/Eocene Epoch boundary (Cappetta 1987; Cappetta et al. 1993).

Amphibians

The record of amphibians (frogs and salamanders) across the Paleocene/Eocene Epoch boundary is best in North America, less so in Europe and virtually nonexistent in Asia. As Estes (1981:101) wrote, "the fossil record of salamanders is so sparse and so geographically limited that it is difficult to make any extensive interpretation of the group at this time." All extant salamander families extend back at least as far as the Paleocene Epoch (Estes 1981:figure 27).

Turtles

Hutchison (chapter 18) reviews the upper Paleocene–lower Eocene record of turtles in North America (where the record is most diverse due to greater collecting effort and study), Western Europe and China–Mongolia. He calibrates this record to the Paleocene/Eocene Epoch boundary by referring to the Clarkforkian and Wasatchian LMAs of North America as the "boundary interval," placing the boundary in Europe between the Cernaysian and Dormaalian LMAs and in Asia between the Nongshanian and Bumbanian LMAs.

In North America, a major transition in the turtle fauna took place during the "boundary interval" with a rapid rise in diversity (linked to increased tropicality) and the immigration of Eurasian groups (emydids and testudinids). The European record suggests similar changes. Although it provides the earliest records of testudinids, emydids, platysternids and carettochelyids, the Asian record is too poor to document any change around the Paleocene/Eocene Epoch boundary. Extinction in North America and Europe of some archaic clades, such as the macrobaenids, indicates some modernization of the turtle fauna during the boundary interval.

Lizards and snakes

Almost all living families of lizards and snakes have a fossil record extending back to the Late Cretaceous Period (Estes 1983; Rage 1984). Yet, this record is so patchy that little inference about change across the Paleocene/Eocene Epoch boundary can be made. Iguanid, agamid and varanid lizards are found at sites located very far north (such as Ellesmere Island) during the early Eocene Epoch. Theses lizards indicate a single Holarctic province dominated by the heavily armored glyptosaurine anguids, which were relatively large omnivores. There seems to have been little interchange of lizards and snakes (indeed of any lower vertebrates) between North and South America during the late Paleocene–early Eocene interval. The main interchange had occurred during the Late Cretaceous Period or Paleocene Epoch (Bonaparte 1984; Estes and Báez 1985).

Crocodilians

The late Paleocene–early Eocene record of crocodilians is similar to that of turtles—most extensive in North America, more limited in Europe and too sparse to be of relevance in Asia (Berg 1969; Novacek et al. 1991; Sun et al. 1992). The North American record of the Bighorn Basin (Bartels 1983) and the San Juan Basin (Lucas 1992) has been examined in detail. Both studies document the continuous range of the most common alligators *Allognathosuchus* and *Leidyosuchus* across the Paleocene/Eocene Series boundary and show that only few genera, such as the horned crocodilian *Ceratosuchus*, were apparently restricted to the Paleocene Epoch. As Lucas (1992) noted, modern crocodilian feeding strategies—large predator, omnivore and piscivore—had evolved among eusuchians by the Late Cretaceous Period. *Allognathosuchus* had clearly migrated from North America to Europe by the early Eocene Epoch, and by middle Eocene time there was a rather homogeneous Euramerican crocodilian fauna.

Mammals

Mammalian paleontologists have long placed the Paleocene/Eocene Epoch boundary at a level of significant turnover in mammals that involved extinction and origination in the northern hemisphere resulting from a dispersal event and this independently from the stratigraphic position of the base of the Ypresian Stage (which defines the base of the Eocene Series) in the standard global chronostratigraphic scale (SGCS). Lucas (chapter 21) argues that the base of the Sparnacian deposits in western Europe, the bases of both the Clarkforkian and the Wasatchian Land Mammal Age (LMA) in North America and the base of the Bumbanian (= Lingchan) LMA in Asia are demonstrably older than the base of the Ypresian Stage and thus are of late Paleocene age. In Western Europe (England, Belgium, France), fossil mammals assigned to Reference Level MP7 are of late Paleocene age; the Paleocene/Eocene Series boundary is within Reference Levels MP 8–9 and is not associated with a detectable evolutionary turnover in mammals.

In North America, mammal-based correlation, particularly using coryphodontid pantodonts and plesiadapiforms, indicates that the Clarkforkian LMA correlates with Reference Level MP6 and with the temporal gap between Reference Levels MP6 and MP7, and that the Graybullian SLMA (early Wasatchian LMA) correlates with Reference Level MP8–9. Therefore, the Paleocene/Eocene Epoch boundary is within the Graybullian interval of Wasatchian time, a correlation supported by magnetostratigraphy, palynostratigraphy and the identification of the latest Paleocene carbon isotope excursion in lowest Graybullian strata in Wyoming.

Less precise mammal-based correlations with Asia suggest that the Paleocene/Eocene Epoch/Series boundary is within the Nongshanian LMA. It may coincide with or slightly precede the FAD of bona fide *Coryphodon* in sections in the Nemegt (Mongolia) and Turpan (Xinjiang Province, China) Basins.

The record of fossil mammals in western Europe across the Paleocene/Eocene Series boundary is particularly important because the marine stages (Thanetian and Ypresian Stages) used to define the boundary in the SGCS have their stratotypes there. It is possible to correlate directly the European mammal fossils to these stratotypes and thereby cross-correlate the boundary from marine to nonmarine strata (see below). Note that the most significant evolutionary turnovers in the Western European section occur between Reference Levels MP6 and MP7 and between Reference Levels MP8–9 and MP10. The two evolutionary turnovers readily delimit the Neustrian LMA. Within the stratigraphic interval between Reference Levels MP6 and MP7 (there are two levels within the Sables de Rilly/Berru, with slightly different faunas, and a higher level in the Marnes de Montchenot, although the time represented is probably quite short), perissodactyls, artiodactyls, chiropterans, euprimates, rodents and coryphodontid pantodonts appeared, and the diversity of multituberculates, "condylarths" and plesiadapiforms decreased. Note, however, that this "boundary" corresponds to a temporal gap between the Cernay (Reference Level MP6) fauna, which is a single point in Thanetian time, and the Reference Level MP7 fauna, and this gap certainly accentuates the apparent evolutionary turnover (compare with the North American record). At the boundary between Reference Levels MP8–9 and MP10, pantodonts and multituberculates disappeared and lophiodonts became dominant.

The beginning of the Wasatchian LMA represented a major evolutionary turnover in mammalian evolution (e.g., Krause and Maas 1990; Gunnell et al., 1993; Maas et al. 1995; Gunnell, chapter 19). Important FADs at the beginning of the Wasatchian Age included perissodactyls, artiodactyls, euprimates, hyaenodontid creodonts, didymoconids, didelphinine marsupials, several genera of insectivores, the miacid *Miacis* and the rodent *Microparamys*. This turnover provides the rationale behind placement of the Paleocene/Eocene Epoch boundary at the beginning of the Wasatchian LMA by many mammalian paleontologists.

Mammalian biostratigraphy provides robust correlations across the Paleocene/Eocene Series boundary in the nonmarine strata of Western Europe, North America and Asia. These correlations are consistent with magnetostratigraphy, radioisotopic dating and chemostratigraphy (carbon isotope excursion). The Paleocene/Eocene Series boundary (= base of the Ypresian Stage) is within the Neustrian (= "Sparnacian") LMA in Europe, within the Graybullian interval of the Wasatchian LMA in North America and in the younger part of the Nongshanian LMA in Asia. In Europe and North America, a major turnover event in mammalian evolution at the base of the Neustrian LMA and the Wasatchian LMA, respectively, predates the Paleocene/Eocene Series boundary. In Asia, the same turnover appears to postdate the Paleocene/Eocene Series boundary, but probably this turnover is more apparent than real due to the incompleteness of the Asian record.

Gunnell (chapter 19) discusses the remarkably rich mammalian record across the Paleocene/Eocene Series boundary in the northern Bighorn Basin, Wyoming. He locates the Paleocene/Eocene Series boundary at the Clarkforkian/Wasatchian LMA boundary, and argues that the "most fundamental change" in mammals occurred when Clarkforkian mammal assemblages dominated by Dinocerata, Mesonychia, "Notoungulata" and Plesiadapiformes were replaced by Wasatchian faunas dominated by Perissodactyla, Artiodactyla and Primates. He demonstrates that the earliest Wasatchian Age was a time of numerous FADs and low LADs. Standing species richness was high during the earliest Wasatchian Age and trophic groups were diverse and without dominance. Gunnell (chapter 19; also see Gunnell et al., 1993) argues from the "perspective of vertebrate paleontology" that the Clarkforkian/Wasatchian LMA boundary should be the location of the Paleocene/Eocene Series boundary.

Hooker (chapter 20) reviews a similar, synchronous major mammalian turnover event in Europe. He retains the original concept of the Sparnacian "Stage" as based upon the *Argiles à Lignites d'Epernay* in the eastern Paris Basin (= *Sables de Laon* = earliest Eocene Epoch–Ypresian Age) rather than the modified definition by Cavelier and Pomerol (1986), which excludes these strata from the concept of the Sparnacian Stage in favor of the demonstrably older *Argiles à Lignites du Soissonnais* to the NW (= Woolwich–Reading Beds [= mid-part of the Lambeth Group of the London Basin, above and exclusive of the Upnor Formation at the base of the Lambeth Group] = terminal Thanetian Age). He suggests, further, reinstatement of the term *Neustrian* for *Sparnacian* as a standard European land Mammal Age (ELMA), while at the same time observing that the so-called mammalian "ages" are biostratigraphic, rather than biochronologic, in connotation, based as they are primarily on multiple mammalian biozones or faunal associations. While this is certainly true in terms of structure and content, it is equally apparent that when a biozonal interval has been linked/calibrated to an independent chronology via paleomagnetic stratigraphy for example, that it assumes a temporal connotation as well, as in the sense of a biochron. This point has been emphasized in the case of marine biozones/biochrons in recent papers by Aubry et al. (1996) and Aubry (chapter 3).

Hooker reviews his recent attempts at placing the upper Paleocene–lower Eocene Reference Levels in a temporal sequence by means of parsimony analysis. Perhaps most pertinent in the context of this volume are the following points:

1. Reference Levels MP7 and MP8 have been replaced by 5 concurrent-range zones, denoted as P–E (for Paleocene–Eocene) I–V. The major mammalian turnover between the Cernaysian and Neustrian ELMA is centered on the transition from Reference Level MP6 to Zone P–E I.

2. This major turnover occurred at a level stratigraphically equivalent to a level within Zones NP9 and P5, the benthic foraminiferal extinction event (BFE) and the carbon isotope excursion event, and is correlative with the Clarkforkian/Wasatchian NALMA boundary in the Bighorn Basin.
3. The turnover event involved strong reduction/near elimination of multituberculates, "condylarths" (particularly the Arctocyonidae) and the "insectivore" family Adapisoriculidae. The immigration from North America included rodents, carnivores, pantodonts and apatotheres, as well as oxyaenid creodonts, amphilemurids, phenacodonts and the genus *Hypsodus*. Perissodactyls, artiodactyls, primates, hyaenodontids (creodonts) and the herpetotheriine "didelphids" (marsupials) immigrated into Europe at this time from centers of origin that remain obscure.
4. The scale of this turnover in mammalian faunas is greater in Europe than in North America and is ascribed, in part, to the earlier appearance in North America of coryphodontid pantodonts, rodents, *Hyopsodus*, oxyaenids, carnivores, phenacodonts and amphilemurids. Faunal turnover is even greater in Asia than in Europe between the Nongshanian and Bumbanian LMAs.
5. Hooker recommends placement of the Paleocene/Eocene GSSP at a level coincident with the major turnover in terrestrial mammalian faunas, which would have the virtue of providing a direct link between marine and terrestrial stratigraphies if this linkage can be unequivocally demonstrated via carbon isotope stratigraphy.

The Paleocene–Eocene transition demonstrates, as has been shown to be characteristic of many other epoch/series boundary intervals, that important perturbations in the integrated earth–atmosphere–hydrosphere circulatory system(s) are reflected by biotic turnovers in marine and terrestrial organisms that served as the basis for the fundamental chronostratigraphic subdivision of time and rock in the 19th century. But, whereas the Cretaceous–Paleogene "transition" was abrupt and the result of catastrophic (exogenic) causes—a classic example of the role of contingency in the evolution of life on Earth—the Paleocene–Eocene (and the Eocene–Oligocene) transitions were essentially endogenic in nature. They differ mainly in the fact that while the latter occurred during an interval of lowering global temperatures and concomitant initiation of glaciation on Antarctica—the transition from the "Hothouse World" to the "Doubthouse World"— the former took place over an interval of unprecedented global warming—in the midst of the "Hothouse World" mode.

Superimposed upon this overall warming trend was a brief, punctuated global warming, the Late Paleocene Thermal Maximum (LPTM) that resulted in dramatic changes in atmospheric circulation, global oceanic circulation and thermal stratification and concomitant extinction(s)/faunal turnovers in deep and (to a lesser extent) continental shelf benthic foraminifera and perturbations in planktonic ecosystems (but no dramatic turnovers in planktonic foraminifera faunas and calcareous nannoplankton and dinoflagellate floras). Significant turnovers occurred in terrestrial vertebrate faunas of NW Europe, North America and Asia with the replacement of (archaic) multituberculates, plesiadapiforms (primate relatives) and ungulates (the paraphyletic condylarths) by essentially modern mammalian groups such as the perissodactyls, artiodactyls, primates (s.s.), chiropterids (bats), hyaenodontids (creodonts) and rodents as well as the pantodont genus *Coryphodon*. While these faunal changes are believed to have occurred essentially contemporaneously, the disjunct nature of the terrestrial stratigraphic record precludes precise regional, temporal correlation and center(s) of origin of various groups remain(s) speculative.

The Paleocene/Eocene boundary lacks an internationally defined and accepted global boundary stratotype and point (GSSP). Its operational definition for the moment remains the base of the Ypresian Stage (base of the Ieper Clay Formation) in Belgium. The Paleocene–Eocene boundary interval is an approximately 2-my span between the top of the Thanetian Stage s.s. (latest Chron C25r) to the base of the Ypresian Stage (mid-Chron C24r). Various biotic and chemostratigraphic events characterize this interval, and their placement in a relative stratigraphic and chronologic framework has been the focus of efforts by members of IGCP Project 308 (Paleocene/Eocene Boundary Events in Space and Time) and members of the Working Group on the Paleocene/Eocene Boundary of the IUGS Subcommission on the Paleogene and this subject is treated at some length in several chapters in this volume. A consensus appears to be developing that the major $\delta^{13}C$ isotope spike that has now been found in marine and terrestrial carbonates and vertebrate bone enamel at stratigraphically correlative levels in lower Chronozone C24r would provide a unique signature/criterion for linking marine and terrestrial stratigraphies in the form of a Paleocene/Eocene boundary GSSP.

ACKNOWLEDGMENTS

We wish to thank the contributors to this volume; their cooperative spirit has rendered our task much easier as we passed through various iterations of manuscript editing (for stylistic, nomenclatural and terminological consistency), indexing and, finally, the preparation of this synthesis of their contributions. It has resulted in the timely appearance of a volume that clearly reflects our present understanding of the Paleocene–Eocene transition. While we have tried to represent the views expressed here by the authors to the best of our ability within the context of the main theme of this program: Paleocene/Eocene Boundary Events in Time and Space, the editors accept ultimate responsibility for any inadvertent oversights. This is Woods Hole Oceanographic Institution Contribution number 9320 and ISEM Contribution number 96–144.

REFERENCES

Aubry, M.-P., 1995. From chronology to stratigraphy: Interpreting the lower and middle Eocene stratigraphic record in the Atlantic Ocean. In Berggren, W. A., Kent, D. V., Aubry, M.-P. and Hardenbol, J. (eds.), *Geochronology Time Scales and Global Stratigraphic Correlation,* Society of Economic Paleontologists and Mineralogists Special Publication No 54, p. 213–274.

Aubry, M.-P., 1996. Towards an upper Paleocene-lower Eocene high resolution stratigraphy. In Aubry, M.-P. and Benjamini, C. (eds.), Paleocene/Eocene Boundary Events in Space and Time, *Israel Journal of Earth Sciences,* v. 44, p. 239–253.

Bartels, W. S., 1983. A transitional Paleocene–Eocene reptile fauna from the Bighorn basin, Wyoming. *Herpetologica,* 39, p. 359–374.

Berg, D. E., 1969. Characteristic crocodiles of the Paleogene in Europe. *Memoires du Bureau de Recherches géologiques et minières,* v. 69, p. 73–75.

Berggren W. A., Kent, D. V., Swisher, C. C., III and Aubry, M.-P., 1995. A revised Cenozoic geochronology and chronostratigraphy. In Berggren, W. A., Kent, D. V., Aubry, M.-P. and Hardenbol, J. (eds.), *Geochronology Time Scales and Global Stratigraphic Correlation,* Society of Economic Paleontologists and Mineralogists Special Publication No. 54, p. 129–212.

Berggren, W. A. and Prothero, D. R., 1992. Eocene-Oligocene climatic and biotic evolution: an overview. In Prothero, D. R. and Berggren, W. A. (eds.), *Eocene–Oligocene Climatic and Biotic Evolution,* Princeton University Press, Princeton, N.J., p. 1–28.

Bonaparte, J., 1984. El intercambio faunistico de vertebrados continentales entre America del Sur y del Norte al fines del Cretacico. *Memorias III Congreso Latinoamericano de Paleontologia,* p. 438–450.

Bralower, T. J., Zachos, J. C., Thomas, E., Parrow, M., Paull, C. K., Kelley, D. C., Premoli Silva, I., Sliter, W. V. and Lohmann, K. C., 1995. Late Paleocene to Eocene paleoceanography of the equatorial Pacific Ocean: Stable isotopes recorded at Ocean Drilling Program Site 865, Allison Guyot. *Paleoceanography,* v. 10, p. 841–865.

Cande, S. C. and Kent, D. V., 1992. A new geomagnetic polarity time scale for the Late Cretaceous and Cenozoic. *Journal of Geophysical Research,* v. 97(B10), p. 13917–13951.

Cande, S. C. and Kent, D. V., 1995. Revised calibration of the geomagnetic polarity time scale for the Late Cretaceous and Cenozoic. *Journal of Geophysical Research,* v. 100(B4), p. 6093–6095.

Cappetta, H., 1987. Mesozoic and Cenozoic Elasmobranchii, Chondrichthyes II. *Handbook of Paleoichthyology,* v. 3B, 193 pp.

Cappetta, H., Duffin, C. and Zidek, J., 1993. Chondrichthyes. In Benton, M. J. (ed.), *The Fossil Record* v. 2, Chapman & Hall, London, p. 593–609.

Cavelier, C. and Pomerol, C., 1986. Stratigraphy of the Paleogene. *Bulletin de la Société Géologique de France,* ser. 8, v. 2, no 2, p. 255–265.

Collinson, M. E. and Hooker, J. J., 1987. Vegetational and mammalian faunal changes in the Early Tertiary of southern England. In Friis, E. M., Chaloner, W. G. and Crane, P. R. (eds.), *The Origins of Angiosperms and their Biological Consequences,* Cambridge: Cambridge University Press, p. 259–304.

Corfield, R. M. and Cartlidge, R. M., 1992. Oceanographic and climatic implications of the Palaeocene carbon isotope maximum. *Terra Nova,* v. 4, p. 443–455.

Crane, P. R., 1987. Vegetational consequences of the angiosperm diversification. In Friis, E. M., Chaloner, W. G. and Crane, P. R. (eds.), *The Origins of Angiosperms and their Biological Consequences,* Cambridge: Cambridge University Press, p. 107–144.

Dickens, G. R., O'Neil, J. R., Rea, D. K., Owen, R. M., 1995. Dissociation of oceanic methane hydrate as a cause of the carbon isotope excursion at the end of the Paleocene. *Paleoceanography,* v. 10, p. 965–971.

Estes, R., 1981. *Gymnophonia, Caudata. Handbook of Paleoherpetology* Part 2, Stuttgart: Gustav Fischer Verlag, 115 pp.

Estes, R., 1983, *Sauria terrestria, Amphisbaenia. Handbook of Paleoherpetology,* Stuttgart: Gustav Fischer Verlag, part 10A, 249 pp.

Estes, R., 1988, Lower vertebrates from the Golden Valley Formation, early Eocene of North Dakota (U.S.A.). *Acta Zoologica Cracoviana,* 31, p. 541–562.

Estes, R. and Báez, A., 1985. Herpetofaunas of North and South America during the Late Cretaceous and Cenozoic: Evidence for interchange? In Stehli, F. G. and Webb, S. D. (eds.), *The Great American Biotic Interchange, Topics in Geobiology,* v. 4, Plenum Press, New York, p. 139–197.

Estes, R. and Hutchison, J. H., 1980. Eocene lower vertebrates from Ellesmere Island, Canadian Arctic Archipelago. *Palaeogeography, Palaeoclimatology, Palaeoecology,* v. 30, p. 325–347.

Grande, L., 1980, Paleontology of the Green River Formation with a review of the fish fauna: *The Geological Survey of Wyoming Bulletin* 63, 333 pp.

Grande, L., 1985, The use of paleontology in systematics and biogeography, and a time control refinement for historical biogeography: *Paleobiology,* 11, p. 234–243.

Grande, L., 1989, The Eocene Green River lake system, Fossil Lake, and the history of the North American fish fauna; in Flynn, J., ed., *Mesozoic/Cenozoic Vertebrate Paleontology: Classic Localities, Contemporary Approaches.* American Geophysical Union, Washington, D. C. [28th International Geological Congress Fieldtrip Guidebook T322], p. 18–28.

Hsü, K. J. and McKensie, J. A., 1985. A "Strangelove" ocean in the earliest Tertiary. in Sundquist, E. T. and Broeker, W. S. (eds.), *The Carbon Cycle and Atmospheric CO_2: Natural Variations Archean to Present,* Geophysical Monograph 32, p. 487–492.

Kennett, J. P. and Stott, L. D., 1991. Abrupt deep-sea warming, palaeoceanographic changes and benthic extinctions at the end of the Palaeocene. *Nature,* v. 353, p. 225–229.

Krause, D. W. and Maas, M. C., 1990. The biogeographic origins of late Paleocene–early Eocene mammalian immigrants to the Western Interior of North America. In Bown, T. M. and Rose, K. D. (eds.), *Dawn of the Age of Mammals in the Northern Part of the Rocky Mountain Interior, North America,* Geological Society of America Special Paper 243, p. 71–105.

Lu, G. and Keller, G., 1995a. Ecological stasis and saltation: Species richness change in planktic foraminifera during the late Paleocene to early Eocene, DSDP Site 577. *Palaeogeography, Palaeoclimatology, Palaeoecology,* v. 117, p. 211–227.

Lu, G. and Keller, G., 1995b. Planktic foraminiferal faunal turnovers in the subtropical Pacific during the late Paleocene to early Eocene. *Journal of Foraminiferal Research,* v. 25, p. 97–116.

Lucas, S. G., 1992, Cretaceous-Eocene crocodilians from the San Juan Basin, New Mexico. *New Mexico Geological Society, Guidebook* 43, p. 257–264.

Maas, M. C., Anthony, M. R. L., Gingerich, P. D., Gunnell, G. F. and Krause, D. W., 1995. Mammalian generic diversity and turnover in the late Paleocene and early Eocene of the Bighorn and Crazy Mountains Basins, Wyoming and Montana (USA). *Palaeogeography, Palaeoclimatology, Palaeoecology,* v. 115, p. 181–207.

McKenna, M. C., 1983. Holarctic landmass rearrangements, cosmic events, and Cenozoic terrestrial organisms. *Annals of the Missouri Botanical Garden,* v. 70, p. 459–489.

Miller, K. G., Janecek, T. R., Katz, M. E. and Keil, D. J., 1987. Abyssal circulation and benthic foraminiferal changes near the Paleocene/Eocene boundary. *Paleoceanography,* v. 2, p. 741–761.

Novacek, M. J., Ferrusquía-Villafranca, I., Flynn, J. J., Wyss, A. R. and Norell, M. A., 1991. Wasatchian (early Eocene) mammals and other vertebrates from Baja California, Mexico: The Lomas las Tetas de Cabra fauna. *Bulletin of the American Museum of Natural History,* 208, 88 pp.

Pak, D. K. and Miller, K. G., 1992. Paleocene to Eocene benthic foraminiferal isotopes and assemblages: Implications for deepwater circulation. *Paleoceanography,* v. 7, p. 405–422.

Prothero, D. W. and Berggren, W. A., 1992. *Eocene–Oligocene Climatic and Biotic Evolution.* Princeton: Princeton University Press, 568 pp.

Rage, J-C., 1984. Serpentes. *Handbook of Paleoherpetology,* Part 11, Stuttgart: Gustav Fischer Verlag, 80 pp.

Rage, J.-C. and Jaeger, J.-J., 1995. The sinking Indian raft: a response to Thewissen and McKenna. *Systematic Biology,* v. 44, p. 260–264.

Shackleton, N. J., Corfield, R. M. and Hall, M. A., 1985. Stable isotope data and the ontogeny of Paleocene planktonic foraminifera. *Journal of Foraminiferal Research,* v. 15, p. 321–336.

Speijer, R. P., 1994. Extinction and recovery patterns in benthic foraminiferal paleocommunities across the Cretaceous/Paleogene and Paleocene/Eocene boundaries. *Geologica Ultraiectina, Mededelingen van de Faculteit Aardwtensachappen Universiteit Utrecht,* v. 124, 191 pp.

Steineck, P. L. and Thomas, E., 1996. The latest Paleocene crisis in the deep sea: Ostracode succession at Maud Rise, Southern Ocean. *Geology,* v. 24, p. 583–586.

Sun, A., Li, J., Ye, X., Dong, Z. and Hou, L., 1992. *The Chinese fossil reptiles and their kins.* Beijing: Science Press, 260 pp.

Swisher, C. C. III and Knox, R. W. O'B., 1991. The age of the Paleocene/Eocene boundary: $^{40}Ar/^{39}Ar$ dating of the lower part of NP10, North Sea Basin and Denmark. IGCP Project 308 (Paleocene/Eocene boundary events), International Annual Meeting and Field Conference, 2–6 December 1991, Brussels, Abstracts with Program, p. 16.

Sychevskaya, E. K., 1986, Presnovodnaya Paleogenovaya ikhtiofauna SSSR i Mongolii [Freshwater Paleogene ichthyofauna of the USSR and Mongolia]: *Sovmestnaya Sovyetsko-Mongolskaya Paleontologicheskaya Ekspeditsiya Trudy,* v. 29, 157 pp.

Tjalsma, R. C. and Lohmann, G. P., 1983. Paleocene–Eocene bathyal and abyssal benthic foraminifera from the Atlantic Ocean. Micropaleontology Special Publication No. 4, 90 p.

Upchurch, G. P., Jr. and Wolfe, J. A., 1987. Mid-Cretaceous to early Tertiary vegetation and climate: Evidence from fossil leaves and woods. In Friis, E. M., Chaloner, W. G. and Crane, P. R. (eds.), *The Origins of Angiosperms and their Biological Consequences,* Cambridge: Cambridge University Press, p. 75–105.

Ward, D. J., in Hooker, J. J., Insole, A. N., Moody, R. T. J., Walker, C. A. and Ward, D. J., 1980. The distribution of sharks, rays and chimaeroids in the English Paleogene. *Tertiary Research,* v. 3, p. 13–19.

Wing, S. L., Bown, T. M. and Obradovich, J. D., 1991. Early Eocene biotic and climatic change in interior western North America. *Geology,* v. 19, p. 1189–1192.

Wolfe, J. A. 1977. Paleogene floras from the Gulf of Alaska region. U. S. Geological Survey Professional Paper 997, 108 pp.

Wolfe, J. A., 1985. Distribution of major vegetational types during the Tertiary. In Sundquist, E. T. and Broecker, W. S. (eds.), *The Carbon Cycle and Atmospheric CO_2: Natural Variations, Archean to Present.* Geophysical Monograph 32, Washington D. C: American Geophysical Union, p. 357–375.

Zachos, J. C., Lohmann, K. C., Walker, J. C. G. and Wise, S. W., 1993. Abrupt climate change and transient climates during the Paleogene: A marine perspective. *Journal of Geology,* v. 101, p. 191–213.

2

The Paleocene/Eocene Epoch/Series Boundary: Chronostratigraphic Framework and Estimated Geochronology

WILLIAM A. BERGGREN
MARIE-PIERRE AUBRY

ABSTRACT

We review the chronostratigraphic framework of the ~2-my Paleocene/Eocene boundary interval between the top of the Thanetian Stage (upper Paleocene Series; near the Chron C25n/r boundary) and the base of the Ypresian (lower Eocene Series; younger part of Chron C24r) from the historical perspective. The major global bio-chemo-tephra- and magnetostratigraphic events associated with the boundary interval are located in space and time with a view to providing criteria that may serve to denote and correlate the base of the Paleocene/Eocene Series boundary at an appropriate Global Stratotype Section and Point (GSSP). The Paleocene/Eocene Series boundary, as currently defined by the base of the Ieper Clay of the Belgian Basin (= the base of the London Clay Formation as recently redefined in the London Basin; and equated with the NP10a/b calcareous nannoplankton boundary) has an estimated age of 54.37 Ma. The base of the (recently redefined) Harwich Formation (including the Oldhaven Beds, Hales Clay) formerly included as the basal unit of the London Clay Formation and stratigraphically equivalent to the lower Phase 2a ash series (= 2.2; including the −17 Ash) has an estimated age of ~54.51 Ma. Estimated ages of several events suggested as appropriate to denote the Paleocene/Eocene Epoch/Series boundary include: 1) P5/P6 (planktonic foraminiferal) zonal boundary: 54.48 Ma; 2) NP9/NP10 (calcareous nannofossil) zonal boundary: 55 Ma; 3) Benthic Foraminiferal Extinction (BFE) in bathyal and abyssal (and to a lesser extent in neritic) environments: 55.5 Ma and 4) $\delta^{13}C$ isotope excursion/spike (CIE; in marine and terrestrial stratigraphies): 55.5 Ma. Choice of the latter criterion(a) would have the virtue of placing the Paleocene/Eocene Series boundary at a lithostratigraphic level where time and rock coincide in both marine and terrestrial sections and provide a unifying element for global correlation in these disjunct stratigraphies. A slightly different methodologic approach (which retains the original GPTS calibration of 55 Ma on the NP9/NP10 biochronal boundary, the original radioisotopic age determination on the −17 Ash [54.5 ± 0.05 Ma] and an extrapolated age estimate of the top of the ash series in Hole 550; see discussion below) to that used in Berggren and Aubry (1996) results in a radically different (briefer) interpretation of the duration (0.0521 my vs. 0.467 my) of that part of the ash-series represented at Site 550; the implications for NW European stratigraphy are briefly discussed.

Carbon isotope stratigraphy suggests that the Conglomérat de Meudon (generally assigned to European Reference Level [ERL] MP7 and considered by some vertebrate paleontologists to denote the base of the Sparnacian "Stage") is stratigraphically equivalent with the upper part of the Lambeth Group (Reading Bed 1) and a level low in Zone NP10 supporting a recent suggestion that it is actually referable to ERL MP8–9. The base of the Sparnacian "Stage" is slightly older than the base of the Wasatchian North American Land Mammal Age (NALMA) and the Conglomérat de Meudon is younger than the base of the Wasatchian NALMA.

Despite over a century of research the Paleocene/Eocene Series boundary remains without a rigorously defined Global Stratotype Section and Point (GSSP;

Cowie et al. 1986). The two series terms were defined in 1874 and 1833, respectively, and whereas the terms Thanetian and Ypresian have since come to serve as the upper and lower stage units, respectively, of these series, no precise limit(s) have been placed on their boundaries. IGCP Project 308 (Paleocene/Eocene Boundary Events in Space and Time) has had as its object the delineation of the various geologic events associated with the boundary interval that may serve as appropriate criteria for denoting and correlating a GSSP. While this project has now been officially terminated (1995), research continues on the subject of identifying events of global significance with the potential for global stratigraphic correlation of this boundary once it is established. Concurrently with the search for boundary events there are various studies being conducted in several countries of the world with a view to establishing a GSSP in a specific section that best meets the requirements of a GSSP.

In this chapter we review the Paleocene/Eocene boundary stratigraphic interval in its historical context, delineate the main bio-chemo- and magnetostratigraphic events that occur within the ~2-my span of this stratigraphic interval, present an estimated chronology of some 17 events between (and including) Chron $C25n_{(o)}$ and $C24n.3n_{(o)}$. For a recent, comprehensive examination of the stratigraphic setting of the lower Paleogene of NW Europe the reader is referred to Knox et al. (1996).

Chronostratigraphy

The marine stratigraphic succession developed in the sedimentary basins on the passive margins of the Baltic Plain surrounding the North Sea (the Anglo–Belgian–Paris–Danish–North German basin) has come to serve as the standard for mondial Paleogene chronostratigraphy (Berggren 1971; Pomerol 1973; Curry et al. 1978). Unfortunately exposures in this region are often poor and stratigraphic sequences are limited in both vertical and lateral extent and riddled with unconformities. Repeated transgressions and regressions have left a characteristic imprint of rapid lateral and vertical facies changes, and late Neogene glacial tectonism has added a coup de grâce in some areas (e.g., Denmark) in the form of physical dislocation of important parts of the stratigraphic record. Nevertheless a network of biostratigraphic and magnetostratigraphic studies (i. al., Aubry 1983, 1985, 1986; Aubry et al. 1986; Townsend and Hailwood 1985; Steurbaut 1988, 1990a, 1990b; Steurbaut and Nolf 1986; Ali and Jolley 1996; Ali et al. 1993, 1996; Ellison et al. 1996; Knox et al. 1990, 1994), coupled with petrologic and comparative studies on volcanoclastic deposits intercalated in the uppermost Paleocene–lowermost Eocene sedimentary sequence of northern Europe and the North Sea Basin (Knox and Morton 1988; Morton and Knox 1990; Knox 1996b; Ritchie and Hitchen 1996) is now providing a framework for an integrated Paleogene chronostratigraphy and an estimate of the position and numerical chronology of the Paleocene/Eocene Epoch boundary relative to the Geomagnetic Polarity Time Scale (GPTS).

Lyell's original three-fold (1833) and four-fold (1839) subdivision of the Tertiary included the Eocene Series as the lowest/oldest unit at the base and was based (in the marine realm) on the outcrops in the Paris and London basins. In actual fact his original concept of Eocene Series included the stratigraphic succession attributed today to a part of the lower Miocene Series, the Oligocene Series (Beyrich 1854, 1856; which Lyell never accepted in his subsequent modifications, e.g., 1852) and the remainder of the stratigraphic succession down to the top of the Chalk Series of Cuvier and Brongniart, 1810 (i.e., including the Paleocene Series of Schimper 1874, as modified by a generation of geologists and essentially accepted today down to the Cretaceous/Paleogene System boundary; figure 2.1). In his subsequent modification of the subdivision of the Tertiary Lyell (1852) divided the Eocene Series into an upper, middle and lower part, the latter corresponding essentially to what is today the Paleocene Series (but including the London Clay, as well as the underlying Thanet Sands in England and all rocks in the Paris Basin ascribed today to the Cuisian/Ypresian Stage down to and including the Dano-Montian chalks). The oldest Tertiary rocks in the London Basin are the Thanet Sands (including at their base the Bullhead Pebble Bed), which lie transgressively upon the (Campanian) Chalk and correspond approximately to Chron C26n (~58 Ma). Thus there are at least 7 my not represented by rock in the London Basin at the base of the Cenozoic Erathem (Paleogene System; and indeed considerably more if one includes the late Cretaceous part of the hiatus). In the Paris Basin the lower part of the Cenozoic (Paleogene) sequence is more complete and includes at its base the Dano-Montian chalks of Vigny–Montaineville–Laversines.

The term *Paleocene* was introduced by Schimper (1874) based on a study of floral samples from the Sables de Bracheux (NW Paris Basin), the Travertins de Sézanne (E Paris Basin) and the Lignites et grès du Soissonnais (NE Paris Basin), which he included in the Suessonian Stage of d'Orbigny (1852). This latter stage included, in addition to the Sables de Bracheux and the Argiles à lignites du Soissonnais the supradjacent Sables de Cuise (subsequently made the basis for the Cuisian Stage by Dollfus 1880). Dollfus (1880) also created the term Sparnacian Stage in the Paris Basin for

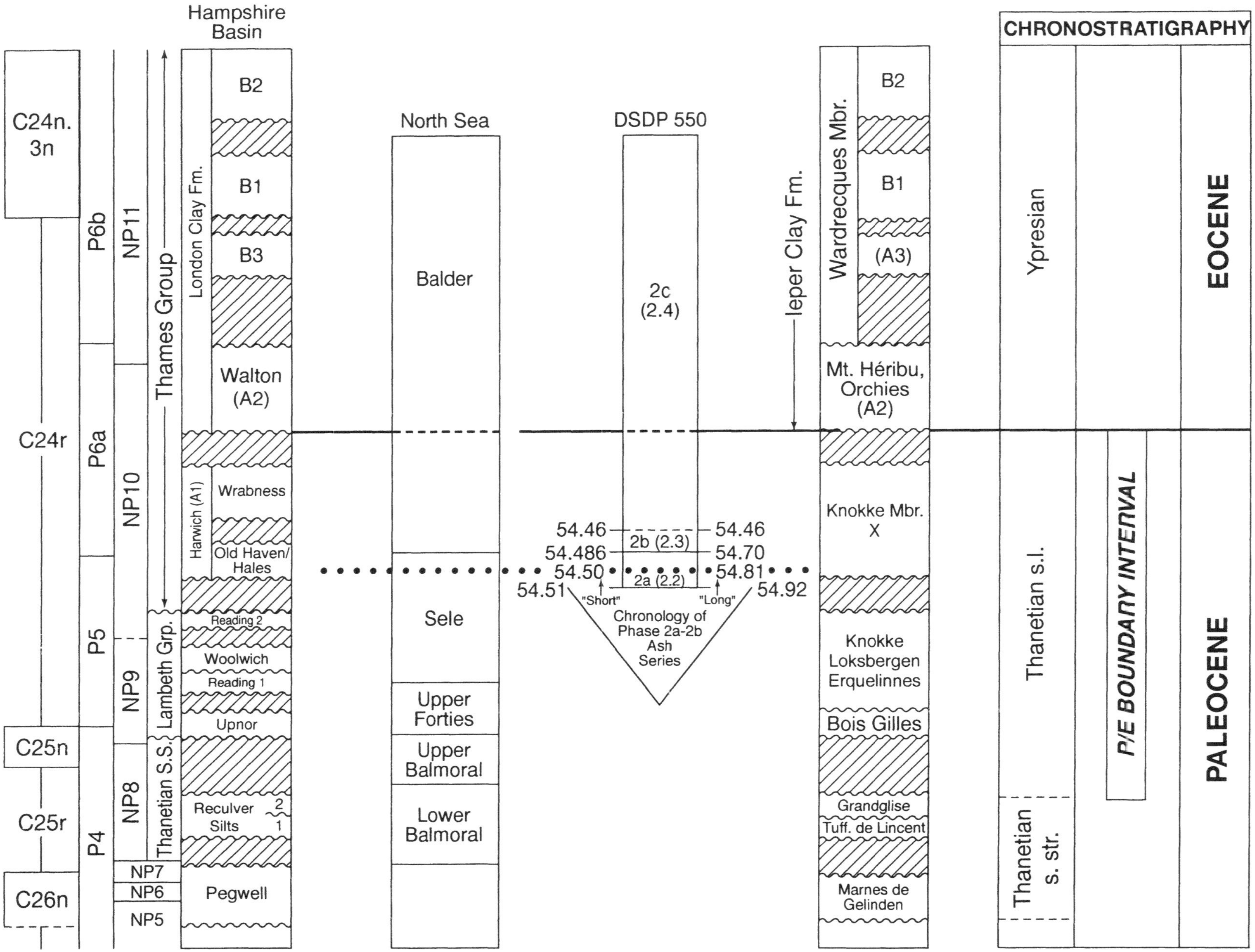

Figure 2.1 Regional sequence stratigraphic correlation framework of NW European and North Sea upper Paleocene—lower Eocene stratigraphic succession. Relative position of Chrons C25–C24 are placed in the margin only for purposes of clarifying approximate/relative placement of stratigraphic units/sequences in relation to some independent calibration/reference point(s). The distinctive −17 Ash is used as a major reference level for correlation. See text and Berggren and Aubry (1996) for further discussion.

the Argile plastique (near Paris) and the lignites de Soissons et d'Epernay. He subsequently added (1905) the fausses glaises at the top, the level just below the Sables d'Auteuil being considered equivalent to the Sables de Sinceny at the base of the Cuisian Stage. It is clear then that Schimper considered—and indeed restricted—his Paleocene Series to include rocks currently included in the Sparnacian Stage. However, Haug (1911) expanded Schimper's (1874) concept of the Paleocene Series in extending its upper boundary to include the Cuisian Stage (above) and its lower boundary to include the Montian Stage (below). In doing so he drew the Cretaceous/Tertiary System boundary between the Danian and Montian Stages, assigning the latter to the lowermost Tertiary Erathem (his Nummulitique; Paleogene System of Naumann 1858) and the former to the uppermost Cretaceous System. He stated that the Danian in Denmark is followed by beds that all authors consign to the Paleocene Series and that, "without doubt," correspond to the Montian of Belgium. However, he failed to explain why these beds correspond "without doubt" to the Montian; indeed Von Koenen (1885) had shown a quarter of a century earlier that they were probably correlative with the Thanet beds of England, and Brotzen (1948) and Haynes (1958) corroborated this subsequently in their foraminiferal studies. It was Haug's considerable influence that led to the general acceptance of the general dichotomy between Danian and Montian, and concomitant placement of the Cretaceous/Paleogene System boundary between these two units, and influenced most French and Belgian geologists well into the mid-part of the present century—to the unfortunate detriment of stratigraphy in Europe (see review in Berggren 1971) and the Former Soviet Union (FSU; see review in Yanshin 1953). The Danian Stage has resided safely, and appropriately, in the Paleogene System (Cenozoic Erathem) for the past thirty years.

Schimper was impressed by the fact that the Paleocene floras contained a number of plant elements now characteristic of the Northern Hemisphere and, in part, of the modern European flora, whereas in contrast the Cretaceous flora exhibited greater similarity with Southern hemisphere floras. At the same time it is well to recall that Schimper's original faunistic analysis was based on material from two restricted localities and whose stratigraphic position is not well documented. The original collections no longer exist and were in fact not found during a search of the collections at the University of Strasbourg in 1961. A further complicating fact is that we now know that the Lignites du Soissonnais (eastern Paris Basin), the basis of Dollfus' Sparnacian Stage (1880) are, in fact, younger (lower Eocene Series = Ypresian Stage) than the Argile plastique (western Paris Basin; Thanetian Stage; Laurain et al. 1983; Cavelier and Pomerol 1986; Pomerol 1989; Sinha et al. 1996; confirmed by more recent palynologic analysis, Hardenbol, personal communication 1995). This fact led Cavelier and Pomerol (1986) to recommend elimination of the (type) Sparnacian of the eastern Paris Basin from the Sparnacian s.l. and to restrict the term Sparnacian to the Argile plastique of the western Paris Basin. The fact that no definitive upper limit was originally set upon the Paleocene Series left open the question of where the boundary with the Eocene Series was to be set. The Thanet Sands, which crop out in Kent and Surrey (England) were described by Prestwich (1852) and the term *Thanetian Stage* was created by Renevier (1873) almost simultaneously with the Paleocene Series of Schimper (1874). Renevier (1897) believed that the Thanetian Stage represented only the lower part of the (Belgian Basin) Landenian Stage of Dumont (1839). There followed nearly a century of (often rancorous) debate on the stratigraphic correlation and affinities of the various lithic and stage units and the position of the Paleocene/Eocene Series boundary (see review by Berggren 1965, 1971; Pomerol 1973; Curry et al. 1978; additional information relating to Paleogene, and in particular Paleocene and Eocene, chronostratigraphy may be found in Vinken 1988 and Luterbacher 1992).

Our modern understanding of the Paleocene Series and the Paleocene/Eocene Series boundary stems from the work of von Koenen (1885) who, in presenting a monograph of the Paleocene molluscan faunas of the glauconitic clays and sands of Copenhagen, Denmark, redefined Schimper's Paleocene Series to include three subdivisions that correspond to units today equated with the Montian, Thanetian and Sparnacian s.l. stages. While Schimper had considered the Sables de Bracheux and Lignites du Soissonnais as Paleocene in age and the London Clay as Eocene in age, von Koenen placed the boundary between the Paleocene and Eocene Series between the Lignites du Soissonnais and Sables de Cuise (Paris Basin) and between the Woolwich–Reading Beds and the London Clay (London Basin). The Thanet Sands were first described by Prestwich (1852), but it was not until nearly a half century later that Burrows and Holland (1897) presented the first monographic treatment of the foraminiferal microfauna, updated, in turn, by the series of studies of Haynes (1955–1958) a half century later. Thus the marine molluscan faunas of the glauconites at Copenhagen have come to serve for many invertebrate paleontologists as the reference guide (for comparative purposes) for marine Paleocene faunas, whereas reference is more often made by marine micropaleontologists to the microfauna of the Thanet Sands as "typical" for the Paleocene strata.

Inasmuch as the overlying Woolwich and Reading Beds (of the recently defined Upnor Formation) are of predominantly lagoonal, fluviatile and marginal marine nature in England, as are their counterparts in the Paris Basin, the upper part of Schimper's Paleocene Series remained an enigma to invertebrate paleontologists. It should be recalled that the term *Thanetian* was introduced by Renevier (1873) and included the Thanet Sands of Prestwich (1852, 1854) and Woolwich–Reading Series of Prestwich (1854). Dollfus (1880) modified the concept of the Thanetian by creating the term Sparnacian for the Argile plastique and Lignites du Soissonnais (Paris Basin) to the older, Thanet Sands. It is in this latter sense that the term *Thanetian* has been used ever since, but with the recent recognition of the inadequacy of the Sparnacian as a chronostratigraphic term, the Woolwich–Reading Beds (Upnor Formation) now occupy a chronostratigraphic no-man's land with their ultimate residency seemingly dependent upon the choice of a GSSP for the Paleocene/Eocene Series boundary. (For a thorough discussion of the history of the term Thanetian Stage the reader is referred to Curry 1981.) However, because the terms *Thanetian* and *Sparnacian* were generally applied in the sense of facies development of rock units rather than as time-stratigraphic units, and because they gradually came to be accepted as the transgressive and regressive phase, respectively, of the first major sedimentary cycle of the Tertiary, the top of the Paleocene Series was drawn at the top of the Sparnacian Stage. It has remained there for the majority of invertebrate paleontologists, the Eocene Epoch generally considered to have begun with the transgression of the Ypresian sea, as represented in the type area.

Vertebrate paleontologists have cited the marked differences in the mammalian fauna of the Sparnacian lignites at Epernay and localities near Soissons (which are more closely related to Eocene faunas above) and the underlying Cernaysian fauna, which belongs to the Thanetian Stage, and have used this argument in placing the Paleocene/Eocene Series boundary between the Thanetian and Sparnacian Stages. Thus a gradual dichotomy in the determination of the Paleocene/Eocene Epoch/Series boundary has developed between marine/invertebrate and terrestrial vertebrate paleontologists over the past half century, which is at long last likely to see a unified and amicable resolution in the near future (Berggren and Aubry 1996).

With the exception of its acceptance in northern Europe (e.g., in Sweden and Denmark) and the former Soviet Union (FSU), the term Paleocene was little used until the mid-1930s (e.g., in England and Germany). It gained acceptance only in the 1970s in France (Pomerol 1973; [Gignoux 1955 still retained the term Eocene as encompassing rocks belonging to the Montian through Stampian Stages]). In the United States the U.S. Geological Survey officially adopted the term *Paleocene* in 1939 (in assigning the Martinez Group of California to "Paleocene"), thanks primarily to the proselytizing of W. D. Matthew, who recognized the peculiar, archaic nature of fossil mammal faunas in the montane-border basins of the Rocky Mountain province, which were decidedly older than those assigned to the Sparnacian Stage (Eocene Series) of western Europe. The use of the term *Paleocene* may be seen as a predominantly American phenomenon, thanks in great part to vertebrate paleontologists who proclaimed its virtues in postwar investigations; the term received respectability and relatively rapid acceptance in Europe and eventually worldwide within a decade of the conclusion of the war.

In epicontinental northwest Europe, the Paleocene Series is represented by the Danian, Selandian and Thanetian stages. (These stage names were approved by voting members of the Paleogene Subcommission of the International Union of Geological Scientists (IUGS) Commission on Stratigraphy (ICS) during the course of the 29th International Geological Congress, Washington, D.C., August, 1989; see also papers in Schmitz 1994.) Only the Thanetian Stage (upper Paleocene Series) and Ypresian Stage (lower Eocene Series) need concern us in the context of this chapter, whose main purpose is to locate the Paleocene/Eocene Epoch/Series boundary (and associated events) in space and time. The base of the Thanetian Stage (Bullhead Pebble Bed and Pegwell Marls and stratigraphic equivalents in outcrops and a borehole in northeast Kent (Knox et al. 1994) is within the calcareous nannofossil Zone NP6–NP7 interval and Magnetic Chron C26n, and the succeeding Thanet Sands are referable to Zone NP8 (see Aubry 1983, 1985, 1986). The lower part of the overlying Upnor Formation (the former Woolwich Bottom Bed) belongs to Zone NP9 (Siesser et al. 1987; Ellison et al. 1996) as does its counterpart in the Paris Basin, the Sables de Bracheux (Aubry 1983, 1985, 1986). The essentially non-calcareous, most extensive part of the upper Upnor Formation may belong to the basal Sparnacian, stratigraphically equivalent to mid-Zone NP9. The upper beds of the Lambeth Group (Reading Bed 2) may belong to upper Zone NP9 and/or basal Zone NP10. Indeed, carbon isotope data (see discussion below) from the (upper) Reading Bed 2 supports this suggestion. The lowest calcareous fossiliferous horizons in marine sediments representative of the Ypresian Stage have been assigned to Zone NP11 in the London Clay Formation of England (Aubry 1983, 1985, 1986), the Formation de Varengeville at Varengeville, Normandy, France (see Aubry 1983, 1985, 1986) and

Belgium (Steurbaut and Nolf 1986; Steurbaut 1988, 1990a, 1990b). Zone NP10 has not been recognized in northwest Europe owing to the absence of calcareous facies.[1] However, dinoflagellate biostratigraphy has enabled a correlation between calcareous and noncalcareous facies, which shows correlation of the lower London Clay Formation with Zone NP10; this has been discussed in detail elsewhere (Berggren et al. 1985; Aubry et al. 1996).

Where, then, does the Paleocene/Eocene Series boundary lie in the context of modern chronostratigraphy? Knox (1994) has made the point that the Paleogene stages of NW Europe are historically linked with rock units whose boundaries are unconformable, whereas standard/global stages (often redefined in areas remote from the historical stage stratotype) should (ideally) be defined in continuous sections. Inasmuch as the (newly (re)defined) boundaries of global stages may no longer correspond to the base of a historical stratotype, there is danger of creating a significantly different denotation for a stage, thereby compromising its historical context. In terms of the Paleocene Series, he notes that the bases of the Selandian, Thanetian and Ypresian stages, as originally defined, correspond to "non-sequences" and reflect renewed transgressions following intervals of lowered sea-level (regressions). Thus, the standard stage boundaries can be placed at stratigraphic levels corresponding to the time-interval denoted by the bounding unconformities of the contiguous rock units. With regard to the base of the Ypresian Stage (i.e., the Paleocene/Eocene Series boundary) it can be appropriately located at any stratigraphic level between the base of the Ieper Clay and the top of the Woolwich–Reading Beds (Lambeth Group of Ellison et al. 1994; Jolley 1996). Inasmuch as the Reading–Woolwich Beds have often been considered as "Sparnacian" (e.g., Cavelier and Pomerol 1986:256; Pomerol 1989:321; contra Curry et al. 1978), the boundary "window" could be extended to the top of the Thanet Beds s.s. (which would correspond stratigraphically to the Sequence Th-3/4 boundary of Hardenbol (1994:50; figure 2.1), the P4b/c zonal boundary of Berggren in Berggren et al. (1995) and a level near, but below/older than the Chron C25n/r boundary). In terms of the integrated magnetobiochronologic scale (IMBS; Berggren et al. 1995), this stratigraphic interval would span a ~2-my-long time interval between ~56.6–54.4 Ma (see figure 2.1 and discussion below).

It is within this stratigraphic/temporal interval that IGCP Project 308 has been investigating since 1989 the magneto-bio-chemo-stratigraphic events that may serve as appropriate criteria for denoting and correlating (not defining) the Paleocene/Eocene Epoch/Series boundary. It has been provisionally termed the "Paleocene/Eocene boundary interval." Numerous publications have resulted from the collaborative efforts of various geoscientists around the world. While a suitable GSSP awaits definition and characterization based on these investigations, an attempt has been made to estimate the geochronology of these major biochronal events/boundaries, chemostratigraphic events and litho- and chronostratigraphic boundaries (Berggren and Aubry 1996; Aubry et al. 1996). In the section below we present a brief introduction and review of the main events associated with the ~2-my-long interval in which the Paleocene/Eocene boundary GSSP will eventually be located.

Events

The Paleocene/Eocene boundary GSSP will eventually be located within a stratigraphic interval that spans ~2 my (between a level closely correlative with Chron $C25n_{(o)}$ and the younger part of Chron C24r). Within this time span a number of events have been identified over the past decade that are of general interest in terms of regional and global geology, paleoclimatology, and paleoceanography; indeed, some of them may well serve as criteria for denotation and correlation of the Paleocene/Eocene Series boundary once a GSSP is chosen. Some of these events are listed and briefly discussed below. We include three events (18–20) that lie stratigraphically above the base of the Ieper Clay Formation, by definition the base of the Ypresian Stage, and are thus irrelevant to the eventual location of a GSSP for the Paleocene/Eocene Series boundary. However, the chronology of the events in Chron C24r is based on estimates of the chronology of Chron C24n.3n and thus this "event" (20) and two biochronologic events (18, 19) in the later part of Chron C24r are included here for the sake of completeness. Their estimated geochronology is discussed in the next section. We also include lithostratigraphic events that define the base or top of the chronostratigraphic units of interest.

1. Top of the Thanet Sands. This level lies in Biochron NP8 and in upper Chronozone/late Chron C25r.
2. Chron C25n. The younger part of this chron is represented by a discrete interval with normal polarity in the lower part of the Upnor Formation in the Jubilee Line Borehole 404T, central London Core (Ellison et al., 1996).
3. NP8/NP9 (calcareous nannoplankton) biochronal boundary (= First Appearance Datum (FAD) of *Discoaster multiradiatus*).
4. P4/P5 (planktonic foraminifera) biochronal boundary (= Last Appearance Datum (LAD) of *Globanomalina pseudomenardii*).

5. Base (type) Sparnacian. Recent paleomagnetic studies indicate that the Argiles plastiques just below the Conglomérat de Meudon (near Paris) are of reversed polarity (Russell et al. 1993) and thus low in Chronozone C24r inasmuch as the $\delta^{13}C$ spike (see item 6 below) has been found several meters above the base of (exposed) Argiles plastiques in the Limay Quarry (western Paris Basin; Sinha and Stott 1993a, 1993b; Stott et al. 1996; Sinha et al. 1996). The base of the Sparnacian "Stage" is considered by European mammalian stratigraphers to denote the Paleocene/Eocene Series boundary in terrestrial sections (for a review of regional stratigraphic correlations in northwest European basins see Hooker 1991, 1996; chapter 20).

However, there is disagreement as to the (bio)stratigraphic position of the Conglomérat de Meudon, which has been regarded to lie at the base of the Argiles Plastiques Formation, a position that was recently questioned (Russell et al. 1989). Whereas the Meudon faunal assemblage has often been assigned to Reference Level MP7 (see Lucas, chapter 21), assignment to Reference Level MP8–9 has been proposed by Hooker (1991). Chemostratigraphic correlations may help resolve the disagreement(s). The carbon isotopic record from terrestrial organic matter from the Argiles Plastiques Formation shows a characteristic pattern of variations that parallels the late Paleocene trend in the $\delta^{13}C$ of marine carbonates (Sinha et al. 1996). Of particular interest in addition to the occurrence of the CIE within the Argiles plastiques bariolées is an interval of strongly positive values (~–20 to –22‰) that is recorded in several sections and, through cross correlations with the $\delta^{13}C$ marine record, has been proposed to correlate with the lowermost part of Zone NP10 (Sinha et al. 1996; Sinha, personal communication 1995). The Conglomérat de Meudon has yielded these strongly positive values, which suggests its correlation with basal Zone NP10 (Sinha, personal communication 1995). In addition, similar, strongly positive $\delta^{13}C$ values have also been measured on terrestrial organic matter from the upper part of the Lambeth Group (Reading 2 interval in Jubilee 404T Corehole, London Basin; Sinha, personal communication 1995), which indicates that the Conglomérat de Meudon is probably correlative with the upper part of the Woolwich–Reading Formation and with lowermost Zone NP10 (or uppermost Zone NP9). This would support Hooker's proposal of a younger age than Doormal (Reference Level MP7) for the Conglomérat de Meudon fauna. This also implies that a) the Conglomérat de Meudon does not constitute the base of the Argiles Plastiques Formation; b) it does not occur at the base of the Sparnacian "Stage"; c) the fauna of the Conglomérat de Meudon (= earliest Biochron NP10 or latest Biochron NP9) is somewhat younger than that of the basal Wasatchian (which is stratigraphically correlative with mid-Biochron NP9); d) the base of the Sparnacian Stage of the Paris Basin is slightly older than the base of the Wasatchian North American Land Mammal Age and only marginally younger than the Chron C24r/C25n boundary (see also Stott et al. 1996; Sinha et al. 1996).

6. $\delta^{13}C$ excursion/ spike: a highly significant 2.5 to 3.0‰ negative excursion in the isotope composition of total dissolved inorganic carbon first identified in the Subantarctic (Stott and Kennett 1990; Kennett and Stott 1991) has now been recognized in the global ocean as well as in marginally marine deposits (Bralower et al. 1995; Thomas and Shackleton 1996; Schmitz et al. 1996). Although associated with the NP9/NP10 zonal boundary in many records due to widespread unconformities (see Aubry, chapter 3) this geochemical event occurred in mid-Biochrons NP9 and P5. The excursion has also been recognized in terrestrial sections in North America associated with soil carbonates and basal Wasatchian mammalian faunas (Koch et al. 1992, 1995; Koch and Zachos 1993) and the "type" ("lower") Sparnacian soil carbonates of the Paris Basin (Sinha and Stott 1993a, 1993b; Sinha et al. 1996). It has recently been suggested that this $\delta^{13}C$ excursion and/or peak can be used to correlate distinct/discrete level(s) between disjunct marine and terrestrial stratigraphies owing to the transfer of ^{12}C between ocean and terrestrial carbon reservoirs via atmospheric CO_2 (Koch et al. 1992; Stott et al. 1996; Sinha et al. 1996). The Clarkforkian/ Wasatchian Land Mammal Age boundary is taken as the Paleocene/Eocene Series boundary by most terrestrial stratigraphers in North America.

7. Bathyal and abyssal benthic foraminiferal faunas underwent a major, global extinction/turnover (Tjalsma and Lohmann 1983; Miller et al. 1987; Thomas 1990a, 1990b; Kaiho 1988; Kaiho et al. 1991, 1996) and neritic–upper bathyal faunas experienced similar, but lesser extinctions coeval with temporary changes in faunal composition (Speijer 1994; Speijer et al. 1996). The event, termed the Benthic Foraminiferal Event (BFE) in Berggren and Aubry (1996), occurs at the same level and is coeval with the $\delta^{13}C$ excursion; the extinction has been linked to significant high latitude warming and formation of warm, salty intermediate and deep waters (WSBW) by evaporation in subtropical latitudes and concomitant lowering of oxygen levels in bottom waters (see further discussion in Thomas and Shackleton 1996, Thomas, chapter 12, and Kaiho et al. 1996).

8. A temporally brief but significant increase in the kaolinitic component of clay mineral assemblages (termed the KE here) occurred in association with the lower part of the $\delta^{13}C$ excursion and the BFE in Southern Ocean sites (Robert and Kennett 1992, 1994) and

reflects a temporary increase in perennial precipitation in East Antarctica and the development of (sub)tropical climates. Subsequent dominance during the $\delta^{13}C$ excursion by smectite indicates seasonal precipitation and enhanced chemical weathering. Increased atmospheric heat transport from low latitudes towards the pole(s) and a temporary alteration in atmospheric circulation from predominantly latitudinal to meridional characterized the latest Paleocene Epoch (Sloan and Thomas, chapter 9).

9. A dramatic decrease in grainsize of wind-blown eolian dust (temporally associated with the BFE) reflects a several fold reduction in wind intensity and concomitant reduction in the intensity of global atmospheric circulation (Miller et al. 1987; Hovan and Rea 1992). We shall refer to this event here as the (reduced) Eolian Dust Event (EDE). It has also been associated with an order of magnitude increase in the flux of seafloor derived hydrothermal materials thought to reflect regional tectonic events such as the East Greenland volcanic outpourings and the initiation of rifting between the European and North American plates (Norway and Greenland; Rea et al. 1990; see further discussion in Rea 1994 and chapter 7).

10. NP9/NP10 (calcareous nannoplankton) biochronal boundary (= FAD of *Tribrachiatus bramlettei*). The biozonal boundary is often used by calcareous nannoplankton specialists to denote the Paleocene/Eocene Series boundary.

11. The record of at least two phases of early Paleogene explosive volcanism in the NE Atlantic is found in the form of volcanoclastic deposits intercalated in marginal marine sediments of NW Europe (Knox and Morton 1983, 1988). Distal, attenuated representatives of the younger (Phase 2) ash series have been recorded over 1000 km to the SE in deep-sea sediments of the Goban Spur and Bay of Biscay (DSDP Holes 550 and 403, respectively; Knox 1985) in the NE Atlantic. The source of these pyroclastic deposits is thought to have been the so-called Hebridean–Thulean volcanic province of Scotland, the Faroes and East Greenland (Morton and Parsons 1988:table 7). The pyroclastic deposits of the (older) Phase 1 were restricted to volcanic activity in the British and Faroe–Greenland province (Knox and Morton 1988); those of the younger Phase 2 have been linked with the Faroe–Greenland igneous province, more specifically the proto-Greenland–Scotland Ridge (Morton and Knox 1990; Miller and Tucholke 1983; Berggren and Schnitker 1983).

Phase 2 pyroclastic deposits have been divided into 4 subphases (Knox and Morton 1983) and occur in the Sele (stratigraphic unit E of Knox et al. 1981) and Balder Formations of the (subsurface) North Sea and at equivalent stratigraphic levels in onshore and offshore boreholes, as well as in NE Atlantic drillsites (e.g., DSDP Sites 403, 550). In Denmark a remarkable series of almost 200 ashes corresponding to subphases 2a and 2b occur (partially) in a fine-grained diatomaceous unit known as the Mo Clay Formation and laterally equivalent facies known as the Fur and Ølst Formations. These ash beds are divided into a lower (negative; −1 to −39) series of mixed basaltic–rhyolitic composition (corresponding to phase 2a) and an upper (positive; +1 to +140) series of predominantly basaltic composition (corresponding to phase 2b). The lower part of the negative ash series has been correlated with the lower part of the *Apectodinium hyperacanthum* (dinoflagellate) Zone. The upper part of the negative series, beginning with the −19 Ash layer, and the positive ash series have been equated with the (?) local Acme Zone of *Deflandrea oebisfeldensis* in the upper part of the *A. hyperacanthum* Zone (Hansen 1979; Heilmann-Claussen 1982). The lowermost negative ashes have been (tenuously) correlated with the Sables d'Erquelinnes (Belgium) assigned to Zone NP9 (Heilmann-Clausen 1985).

Knox (1996a) has changed the annotation of the ash series from Knox and Morton (1983) in the following manner:

Knox and Morton (1983)	*Knox (1996a)*
2c	2.4
2b	2.3
2a	2.2 (upper negative series including −17 and −19 ashes)
	2.1 (lower negative series including −33 Ash)

The important distinction is the fact that the lower (negative) ash series consists of two subphases separated by an essentially ash-free interval that is documented in cores but not seen in outcrop(s).

Subphase 2.1 and 2.2 ashes occur extensively in the North Sea Basin (Sele Formation) and in onshore beds stratigraphically equivalent to the Woolwich–Reading Beds (2.1 series; Cox et al. 1985) as well as at deep sea drilling sites from Goban Spur (DSDP Site 550; probably only 2.2 series) and the Bay of Biscay (DSDP Sites 401 and 403). In Hole 550 Knox (1984) has shown that a series of over 40 ashes (correlated with the Phase 2 ashes of the North Sea area) are restricted to Zone NP10. The −17 and +19 ashes are seen to lie within the lower third of Zone NP10 and the mid-lower half of Chronozone C24r (see discussion in Berggren and Aubry 1996). A single trachytic ash has been identified in the Upnor Formation (Bottom Bed), which is correlative, in part, with the Chron C25n, and rhyolitic ashes have been identified in earliest Chron C24r (lower Zone

NP9) in DSDP Site 549 on the Goban Spur (NE Atlantic). Elsewhere the −17 Ash has been shown to lie within the *D. oebisfeldensis* Acme Zone, equivalent to a level in the upper part of the Sele Formation, within the palynofloral Assemblage Sequence Y4b–Y5 of the Orwell Member Unit B of the Hales Clay unit (of Jolley 1996) in East Anglia (formerly of the basal London Clay Formation, but now placed within the Harwich Formation by Ellison et al. 1994), and (probably) with a level close to (but above) the Woolwich–Reading/Oldhaven Beds boundary (Knox 1990). There is probably a significant unconformity between the Woolwich–Reading Beds and the base of the Harwich Formation, however, and another unconformity (with a ?shorter hiatus) between the Harwich Formation and the London Clay Formation (Knox 1990; see also Jolley 1996).

12. −17 Ash. This distinctive ash lies within the upper part (2.2) of the lower (negative) Phase 2.1–2.2 ash series and is located stratigraphically within the upper part of Zone P5 and lower Zone NP10, respectively (Swisher and Knox 1991; Berggren and Aubry 1996).

13. P5/P6 (planktonic foraminiferal) biochronal boundary (= LAD of *Morozovella velascoensis* and/or *M. acuta*). This event occurs within the range of the ash series in DSDP Hole 550 and is bracketed by the +19 (above) and −17 (below) ashes in Hole 550. The P5/P6 zonal boundary is commonly used by planktonic foraminiferal specialists to denote the Paleocene/Eocene Series boundary.

14. Phase 2.2 (negative)/2.2(positive) ash phase boundary.

15. Top (positive) Phase 2.3 ash series.

16. NP10a/b (calcareous nannoplankton) subbiochronal boundary (= FAD of *Tribrachiatus digitalis*). Zone NP10 has been recently subdivided into 4 subzones based on partial/total ranges of species of the *Tribrachiatus* lineage (Aubry 1996). This event occurs only slightly above the top of the ash series in DSDP Hole 550 and has been shown to be correlative with the basal transgression of the Ypresian Stage (Aubry 1996; see next item below).

17. Base Ypresian Stage. The base of the Ypresian Stage (Mt. Héribu Member of the Ieper Clay Formation) is stratigraphically correlative with the base of the London Clay Formation (Walton Member = Division A2 of King 1981) as redefined by Ellison et al. (1994). This level is correlative with (or only slightly subsequent to) the FAD of *Tribrachiatus digitalis* (as seen in DSDP Hole 550) and is only marginally subsequent to the contact between the ash series/carbonate succession (at ~385 m) in that hole. It corresponds closely to the termination of Phase 2.3 ashes, the Mt. Héribu Member (= base Ypresian Stage) and base of the London Clay Formation (Walton Member). The Paleocene/Eocene Series boundary is equated with the base of the Ypresian Stage here and by IGCP Project 308 until such time that a definitive GSSP is chosen to define the Paleocene/Eocene Series boundary.

18. P6a/b (planktonic foraminiferal) subbiochronal boundary (= FAD of *Morozovella formosa formosa* and/or *M. lensiformis*).

19. NP10/NP11 (calcareous nannoplankton) biochronal boundary (= LAD of *Tribrachiatus contortus*).

20. Chron C24n.3n.

Geochronology

Age determinations/estimates of the Paleocene/Eocene Epoch boundary have varied from 68 Ma (Holmes 1937) to 51.5 ± 1.5 Ma (Curry and Odin 1982; see also Menning 1989). The large differences in these numbers reflect a combination of improved technology and more precise determination of the chronostratigraphic level in question over the past 60 years. The Paleocene/Eocene Series boundary is currently defined by the base of the oldest/lowest stage of the Eocene Series—the Ypresian Stage (as represented by the Ieper Clay in the Belgium Basin, Belgium and its correlative—the London Clay Formation, as recently redefined in the London–Hampshire Basin, England). Several other (chrono)stratigraphic levels for the "Paleocene/ Eocene" Series boundary are also currently in use and/or vying for attention in ongoing attempts to reach a satisfactory determination of a GSSP for the Paleocene/Eocene Series boundary (see discussion above; Berggren and Aubry 1996; Aubry et al. 1996).

Recent geochronologic estimates of the Paleocene/Eocene boundary level(s) (Berggren and Aubry 1996) are based on the IMBS of Berggren et al. (1995), which is correlated, in turn, to the Geomagnetic Polarity Time Scale (Cande and Kent 1992, 1995). This latest incarnation of the GPTS incorporates several discrete components: 1) relative widths of the complete Upper Mesozoic and Cenozoic polarity intervals were reevaluated from magnetic anomaly profiles; 2) the geomagnetic polarity sequence was based predominantly upon data from the South Atlantic using a combination of finite rotation poles and averages of anomaly spacings from stacked profiles projected, in turn, onto a synthetic seafloor spreading flowline. Fine scale detail was derived, where necessary and possible, from magnetic polarity profiles on faster spreading ridges in the Pacific and Indian Oceans; 3) assuming that spreading on the South Atlantic ridge system varied smoothly over time rather than remaining constant, a time scale was generated by interpolating the ages of the polarity reversals using a cubic spline function fitted to 9 age calibration–anomaly distance tie-points (plus the zero-age ridge axis).

There are several salient points that bear repetition here and are fundamental for understanding how the age estimates have been derived for the levels variously used to characterize the Paleocene/Eocene Series boundary/ies:

1. The position of the (supposedly true) NP9/NP10 biozonal boundary in DSDP Hole 550 was used as a calibration point anchoring the lower part of the GPTS (Cande and Kent 1995). Its age was estimated based on extrapolation of age determinations on the +19 (now discredited) and −17 Ash of NW Europe (Denmark, London Basin, North Sea Basin).

2. Obradovich (in Wing et al. 1991; Berggren et al. 1992) obtained a bulk incremental heating $^{40}Ar/^{39}Ar$ plateau age on sanidine of 55.07 ± 0.16 Ma on the −17 Ash (Denmark), whereas Swisher (in Swisher and Knox 1991; see also Berggren et al. 1995) obtained a laser total fusion $^{40}Ar/^{39}Ar$ weighted mean (sanidine) age of 54.51 ± 0.05 Ma for the same ash and 54.0 ± 0.53 Ma on the stratigraphically higher +19 Ash (Denmark and offshore London Basin). An age of 55 Ma on the NP9/NP10 zonal boundary was used by Cande and Kent (1995) as calibration of the GPTS based on a simple sedimentation rate calculation in Swisher and Knox (1991) in Hole 550 in which the −17 and +19 ashes and the −17 Ash and the NP9/NP10 zonal boundary are each separated by 7 m.

3. The magnetostratigraphic position of the NP9/NP10 zonal boundary in Berggren et al. (1985) was used as the chronostratigraphic level for the calibration of the GPTS (Cande and Kent 1995); its position was estimated to lie at Chron C24r.0.66 (= 1221.20 km from the ridge axis in the composite section/profile of the South Atlantic).

4. Subsequent studies have shown that the calibration point (408 m in Hole 550) lies at an unconformity that separates mid-Zone NP9 from lower Zone NP10 (Aubry et al. 1996; Aubry, chapter 3). The implication is that the calibration point lies at a younger/higher level in Chron/Chronozone C24r than that used in the GPTS of Cande and Kent (1995).

5. The +19 Ash (at ~383 m in Hole 550) contains mineral inhomogeneities/impurities in the Danish and English material that was dated and its large error (± 0.5 my) renders it inappropriate for precise age determination. This fact, together with the recognition that the calibration level of 408 m in Hole 550 lies at an unconformity and that sediments below 400 m are strongly dissolved carbonates, renders simple sedimentation rate calculations based on the −17 and +19 ash dates for the (supposed) NP9/NP10 chronozonal boundary in Hole 550 unreliable. It can be shown that even if the sedimentation rate estimates (~1.4 cm/10^3 yr) used by Swisher and Knox (1991) are retained, the ~20-m-thick ash series at Hole 550 would correspond to deposition over a ~1.41-my duration (over half the duration of the ~2.5-my Chron C24r); in view of the fact that the Chron C24r interval is ~62 m thick at Hole 550 and that the interval from the top of the ash series (~385 m) to the base of Chron 24n. 3n (~360 m) is about 25 m, this implies that sedimentation rates of the carbonate sediments above the ash series would have been appreciably higher (> 5 cm/10^3 yr) than those of the ash series below (~1.4 cm/10^3 yr) to satisfy the age determinations on C24n.3n$_{(o)}$ (53.347 Ma; Cande and Kent 1995). This is both intuitively and demonstrably unlikely in view of the higher detrital component present in the ash series than in the carbonates (unless the detrital component were an "insoluble residue" and the carbonate preservation effectively controlled the rate of sedimentation, i.e., rate and final accumulation).

6. An alternative approach was used by Berggren and Aubry (1996) in deriving a chronology of events associated with the upper Paleocene–lower Eocene interval in Hole 550.

a. the chronology of the GPTS (Cande and Kent 1995) was used. This involves a certain amount of circular reasoning in that the precision of the GPTS is predicated on the precise placement of the 55 Ma calibration point (supposed NP9/NP10 chronozonal boundary) in Chron C24r as well as the linearity of the GPTS.

b. inasmuch as the −17 Ash has age estimates that differ by ~0.5 my and the fact that using the age determinations on the +19 and −17 ashes involves sedimentation rate extrapolation across/through detrital ash and carbonate sediments, a sedimentation rate (2.25 cm/10^3yr) extrapolation from Chron 24n.3n$_{(y)}$ (based on the chronology of Chron C24n.3n in the GPTS of Cande and Kent 1995) downwards to the top of the ash series at ~385 m was used to determine the relative magnetobiochronology of events in the younger part of Chrons C24r and C24n.3n (Berggren and Aubry 1996; Aubry et al. 1996). This yielded an age estimate of 54.46 Ma for the top of the ash series at 385 m in Hole 550.

c. a sedimentation rate of 4.28 cm/10^3 yr was used to estimate the duration of the ash series (~0.467 my) based on interpolation between the derived age of 54.46 Ma (top ash series at 385 m) and the 55 Ma age determination on the 408 m level used in the GPTS (Cande and Kent 1995). This yielded a relative biochronology of events within the ash series as well as an age estimate of 54.8 Ma for the −17 Ash (at ~400 m) in Hole 550, which is seen to lie essentially midway between the age determinations of 54.5 Ma and 55.0 Ma by Swisher and Obradovich, respectively. We acknowledge that this procedure compromises the point made above (6b) in that a constant sedimentation rate was used across the base of the ash (400 m) to the 408 m level (in a carbonate interval characterized by strong dissolution). This is required in order to remain consistent with the GPTS (Cande and Kent 1995), which is based on Hole 550 data. However, a significantly different/higher (and more realistic) rate of sedimentation was used in this iteration, which resulted in a more realistic estimate of the duration of the ash series vis à vis the duration of Chron C24r.

d. a relative magnetobiochronology of events in the early part of Chron C24r was derived by upward extrapolation of a sedimentation rate calculated on Chron C25n at Hole 690B using the GPTS (Cande and Kent 1995), inasmuch as the unconformity at 408 m, the questionable presence of the Chron C24r/C25n (at ~422–423 m) contact and strong dissolution in the stratigraphic interval below 408 m in Hole 550 precludes precise magnetobiochronology in early Chron C24r at that location.

With this background it is now possible to estimate the relative chronology of events associated with the stratigraphic interval spanning the Paleocene/Eocene Series boundary (Berggren and Aubry 1996). These estimates, it is emphasized, are preliminary and provisional and require substantial strengthening and verification/falsification via additional high-quality isotopic dating, integrated magnetobiostratigraphy, and a continuous stratigraphic section spanning Chron C24r (at least), which, up to the present, is demonstrably not the case. All age estimates used in this compilation are based on, and correlated to, the IMBS of Berggren et al. (1995) and based, unless otherwise discussed below, upon data in Berggren and Aubry (1996) and Aubry et al. (1996).

1. Top Thanet Beds (= top Thanetian Stage): 56.6 Ma (Chron C25r)

2. Chron C25n: 55.904–56.391 Ma

3. NP8/NP9 biochronal boundary: 56.2 Ma (Chron C25n)

4. P4/P5 biochronal boundary: 55.9 Ma (Chron $C25n_{(y)}$)

5. Base of the Sparnacian: estimated at ~55.8 Ma (Chron C24r)

6. $\delta^{13}C$ excursion/spike: apparently correlative with the base of the North American Wasatchian Mammal Age (Koch et al. 1992; Koch and Zachos 1993) it has an estimated age of 55.5 Ma (Chron C24r)

7. BFE (correlative with the $\delta^{13}C$ excursion/spike): ~55.5 Ma (Chron C24r)

8. KE (correlative with basal portion of $\delta^{13}C$ excursion/spike): ~55.5 Ma (Chron C24r)

9. EDE (correlative with the $\delta^{13}C$ excursion/spike): ~55.5 Ma (Chron C24r)

10. NP9/NP10 biochronal boundary: 55 Ma (Chron C24r, mid-late early part). The calibration point of 55 Ma in Hole 550 is located at the (supposed, but now known to be unconformable) NP9/NP10 biochronozonal boundary and the numerical age of Chron C24r is dependent upon this calibration, among others. In this chapter we have adopted the age of 55 Ma on the NP9/NP10 biochronal boundary as it was originally used/applied by Cande and Kent (1992, 1995; see Aubry, chapter 3 for detailed discussion).

11. Base ash series (in Hole DSDP 550): 54.92 Ma (Chron C24r)

12. −17 Ash: 54.8 Ma (Chron C24r). The −17 Ash is a component of the Phase 2a ash succession of NW Europe and the North Sea and occurs in Denmark and the eastern Atlantic. Basaltic ashes (Phase 2.3) occur in the Harwich Formation (Division A1) of East Anglia. The relationship between Phase 2.2 and Phase 2.3 ashes to the onshore London Basin lithic units now ascribed to the Harwich Formation (Hales Clay, Oldhaven Beds, among others) remains unclear, but it is clear that the age of the −17 Ash provides constraints on the age of the unconformity-bounded Harwich Formation. The estimated age of the −17 Ash is 54.8 Ma, ~0.5 my older than the (estimated) age of the base of the Ypresian Stage (as delineated by the transgressive base of the Ieper Clay and London Clay Formation and cross correlation with the age estimate of the termination of the Phase 2.3 ashes of 54.46 Ma (see below, item 15) just below the FAD of *Tribrachiatus digitalis*). The base of the London Clay Formation was previously estimated at 54.8 Ma based on inclusion of the Oldhaven/Hales Clay at the base of the London Clay (Berggren et al. 1995; Berggren and Aubry 1996). Thus, it should be noted that it is not the age of the "Paleocene/Eocene" Series boundary, as denoted by the base of the London Clay Formation, that has changed, but rather the definition of the lithostratigraphic boundary, that has necessitated a reestimate of the age of the stratigraphically higher/younger position of the "Paleocene/Eocene" Series boundary (see Ellison et al. 1994; Jolley 1996).

13. Phase 2.1 (negative)/2.2 ash boundary (in DSDP Hole 550): 54.5 Ma (Chron C24r)

14. P5/P6 biochronal boundary: 54.7 Ma (Chron C24r)

15. Top of the (positive) Phase 2.2 ash series: 54.46 Ma (Chron C24r)

16. NP10a/b Subbiochronal boundary: 54.37 Ma (Chron C24r)

17. Base Ypresian Stage (= base Ieper Clay Formation and base London Clay Formation): 54.37 Ma (Chron C24r)

18 P6a/b subbiochronal boundary: 54.0 Ma (Chron C24r)

19. NP10/NP11 biochronal boundary: 53.6 Ma (Chron C24r)

20. Chron C24n.3n: 52.903–53.347 Ma

The chronology of items 10–19 above has been derived (Berggren and Aubry 1996; Aubry et al. 1996) by using the 1) age of Chron $C24n.3n_{(o)}$ and 2) the age estimate of 55 Ma on the NP9/NP10 "chronozonal" boundary (from the GPTS of Cande and Kent 1995) plus 3) the age estimate on the top of the ash series—all in DSDP Hole 550—and deriving sedimentation rates of 2.25 cm/1000 yr and 4.28 cm/1000 yr, for the carbonate interval between the top of the ash series and the base of Chron $C24n.3n_{(o)}$ and the ash series itself, respectively (table 2.1). The age estimate on the NP9/NP10 "chronozonal" boundary at 408m (now known to lie at an unconformity/hiatus) was derived, in turn, by extrapolation from radioisotopic ages on the (now discredited +19 Ash) and the −17 Ash, which resulted in an anomalously low sedimentation rate within the ash series, extrapolated, in turn, through the ~3 m carbonate section between the base of the ash series (404.6 m) and the NP9/NP10 "chronozonal" boundary (408 m). The difficulties (pitfalls) of this approach were discussed at length in Berggren and Aubry (1996). An alternative (and scientifically more correct) approach to derive a chronology of the ash succession at Site 550 is discussed below (see also table 2.1).

Strict adherence to the method used in constructing the GPTS by Cande and Kent (1992, 1995) requires adherence to the calibration of 55 Ma at the NP9/NP10 biochronal (as opposed to biozonal) boundary. This (supposed) level (408 m in Site 550) lies at C24r.75 if the top of Chron C25n lies at ~422m (see discussion in Berggren and Aubry 1996). Cande and Kent (1992) moved the NP9/NP10 boundary to a position at C24r.0.66 (1221.20 km on the South Atlantic profile) and equated this level with the Paleocene/Eocene series boundary to be consistent with Berggren et al. (1985).

In our previous estimates of the Chron C24r chronology (Berggren and Aubry 1996; Aubry et al. 1996) we preferred to use the age of 55 Ma on the real level whose age had been estimated and is now known to correspond to the upper surface of the unconformity at 408 m in Hole 550. However, as explained in Aubry (chapter 3) although scientifically acceptable, this approach is not workable because it obliges us to accept the same age for two independent horizons, 1) a surface at 408 mbsf in Hole 550 (corresponding to a NP9/NP10 zonal contact) and 2) a level with a poorly constrained position within Chron C24r that corresponds to the NP9/NP10 chronozonal boundary, with the result that late Biochron NP9 and early Biochron NP10 overlap in the IMBS (Berggren et al. 1995). Aubry (chapter 3) proposes to retain the age of 55 Ma for the NP9/NP10 biochronal boundary, which is consistent with the GPTS, and to estimate the age of the surface at 408 mbsf in Hole 550 using a different methodology than used by Knox and Swisher (1991). This is done below.

If the radioisotopically determined age of the −17 Ash (54.5 Ma) in NW Europe, correlated to its presence at 400 m in DSDP Hole 550, is used in combination with the estimated age (54.46 Ma) of the top of the ash series (384.85 m) in Hole 550 it is then possible to estimate the age of the upper surface of the unconformity at

Table 2.1
"Long" (A) versus "short" (B) chronology of Phase 2a-2b ash series in DSDP Hole 550 based on various assumptions. See text, and discussion in BA (= Berggren and Aubry, 1996) and CK95 (= Cande and Kent, 1995) for further explanation. CP = calibration point; SR = sedimentation rates; T in Ma = Time in millions of years ago.

			A					B			
EVENT	DEPTH in meters	T in Ma	CP	SR in cm/1000 yr.	T in Ma	DURATION in m.y.			SR in cm/1000 yr.	CP	T in Ma
		CK 95	BA 96					This Work			
C24n.3n(y)	349	52.903									
C24n.3n(o)	360	53.347	(1)	2.25						(1)	–
Top 2b (+) Ash Ser.	384.85	–	(2)		54.46					(2)	–
+19 Ash	393	54.0±0.5			–		0.026				54.4815
2a/2b (+/-) Ash. Ser.	395	–		4.28	54.70	~0.46		~0.052	32.37		54.486
-17 Ash	400	54.5±0.05			54.81		0.024			(3)	–
Base 2a/2b (+/-) Ash Ser.	404.6	–			54.92						54.512
"NP9/10" boundary/ unconformity	408	55	(3)		–						

408 m (see table 2.1). This estimate is derived in two stages and is explained below:

1. Using the calibrations above a sedimentation rate of 32.37 cm/10^3 (= 0.3237m/10^3) is obtained within the 20 m Ash series (384.85 m–404.65 m).

a. Age of the base of the Ash series (404.65 m): 54.51 Ma
b. Age of the +19 Ash (393m): 54.481 Ma
c. Age of the 2a/2b (2.2/2.3) ash boundary: 54.4867 Ma
d. Age of the −17 Ash (400m): 54.50 Ma (calibration point)
e. Age of the base of the ash series (404.65m): 54.51 Ma
f. Duration (my) of ash series in Hole 550: 0.050 kyr (50,000 yr)
g. Duration (my) of upper (+) ash series: 26 kyr
h. Duration (my) of lower (−) ash series: 24 kyr

2. There are ~3.35 m between the base of the ash series (404.65 m) and the (upper surface of) the unconformity (408 m) in Hole 550. If we use/assume a sedimentation rate of 2.25 cm/10^3 yr (as used above the ash series in Hole 550 by Berggren and Aubry 1996) to derive a chronology of events within the younger part of Chron C24r we obtain the following estimates:

a. Age of the upper surface of the unconformity (408 m): 54.7 Ma
b. Duration of the NP10 hiatus (408 m): <0.3 my (54.7 Ma [Hole 550] to 55.0 Ma [NP9/NP10 zonal boundary; Hole 690])

While this latter estimate is arbitrary there is no alternative given the available information. However, this approach is consistent with Cande and Kent (1992, 1995) and allows estimates of events in the early part of Chron C24r and the younger part of Biochron NP9 and early part of Biochron NP10 (figure 2.2).

The 2.1–2.2 (−) and 2.3 (+) ash series present in Hole 550 are seen to have been of approximately equal duration (~0.024 my vs. 0.026 my, respectively), assuming the two series are completely represented. On the other hand, the presence of an unconformity at 408 m in Hole 550 results in a part of early Biochron NP10 and late Biochron NP9 being unrepresented. This suggests that only the upper part (2.2) of the Phase 2.1–2.2 (−) ashes are represented between 400 m and 405 m at Site 550 and that the remainder of the negative series of ashes are absent owing to the unconformity. This is supported by the fact that the base of the ash series in Hole 550 is within (lower) Zone NP10, whereas the ash series present in the Sele Formation in subsurface cores of the North Sea Basin are correlative with the Woolwich–Reading Beds of the Lambeth Group (= Zone NP9) in outcrop in SE England. The −17 and +19 ashes are seen to be separated by ~20 kyr (vs. 160 kyr using the approach in Berggren and Aubry 1996) an order of magnitude difference. It is interesting to note at the same time that these revised estimates on the duration of the ashes (~50,000 yr) are notably comparable to the estimates by Bonde (1974) who, in interpreting the ash layers in terms of annual (seasonal) varve deposits, estimated their duration at 60,000 yr.

Both the "long" and "short" chronologies depend upon the accuracy of the estimated age of the top of the ash series (at 385 m) based on extrapolation of the Chron C24n.3n sedimentation rate down to the top of the ash. Consider the following approaches to the problem:

1. Using the integrated sedimentation rate of 17.3 m/my from Chron $C24n.1n_{(y)}$ (342.6 m) to Chron $C24n.3n_{(o)}$ (362 m; Berggren and Aubry 1996) yields an estimated age of 54.78 Ma for the top of the ash series (385 m), some 0.32 my older than that obtained by using the rate of 22.5 m/my based on Chron C24n.3n alone (see table 2.1) and ~0.28 my older than the radioisotopic age obtained on the −17 Ash itself.

2. This may be a more appropriate rate to use inasmuch as it is possible that Subchrons C24n.2n and C24n.3n are concatenated over the stratigraphic interval of ~350–360 m in Hole 550 in which case the ~10-m interval would correspond to a duration of ~0.983 my rather than 0.444 my; the estimated rate of sedimentation would be ~10.2 m/my (rather than 22.5 m/my) and the top of the ash series (385 m) would have an estimated age of ~55.78 Ma, some 1.3 my older than the estimate based on assuming that the 10 m interval between ~350–360 m is Chron C24n.3n alone.

3. This value (55.78 Ma) is unlikely in view of the fact that the age of 54.5 Ma on the −17 Ash is almost 1 my younger than the estimated age of the top of the 2.3 ash series in this permutation. In short, both approaches suffer from the shortcoming of yielding slightly to significantly older ages for the top of the ash series than the radioisotopic age on the −17 Ash itself. Thus it would appear that the most appropriate calibration to use in estimating rates of sedimentation in the carbonate section above 385 m in Hole 550 is the ~10 m interval (349.62–359.65 m) interpreted as Chron C4n.3n (which yields an estimated rate of sedimentation of 22.5 m/my). Alternatively the ~17 m interval (342.6–359.65 m) representing Chron $C24n.1n_{(y)}$ to Chron $C24n.3n_{(o)}$ could be used with appropriate understanding of the limit(s) of extrapolation below this interval. The matter will have to rest there until such time that we discover a

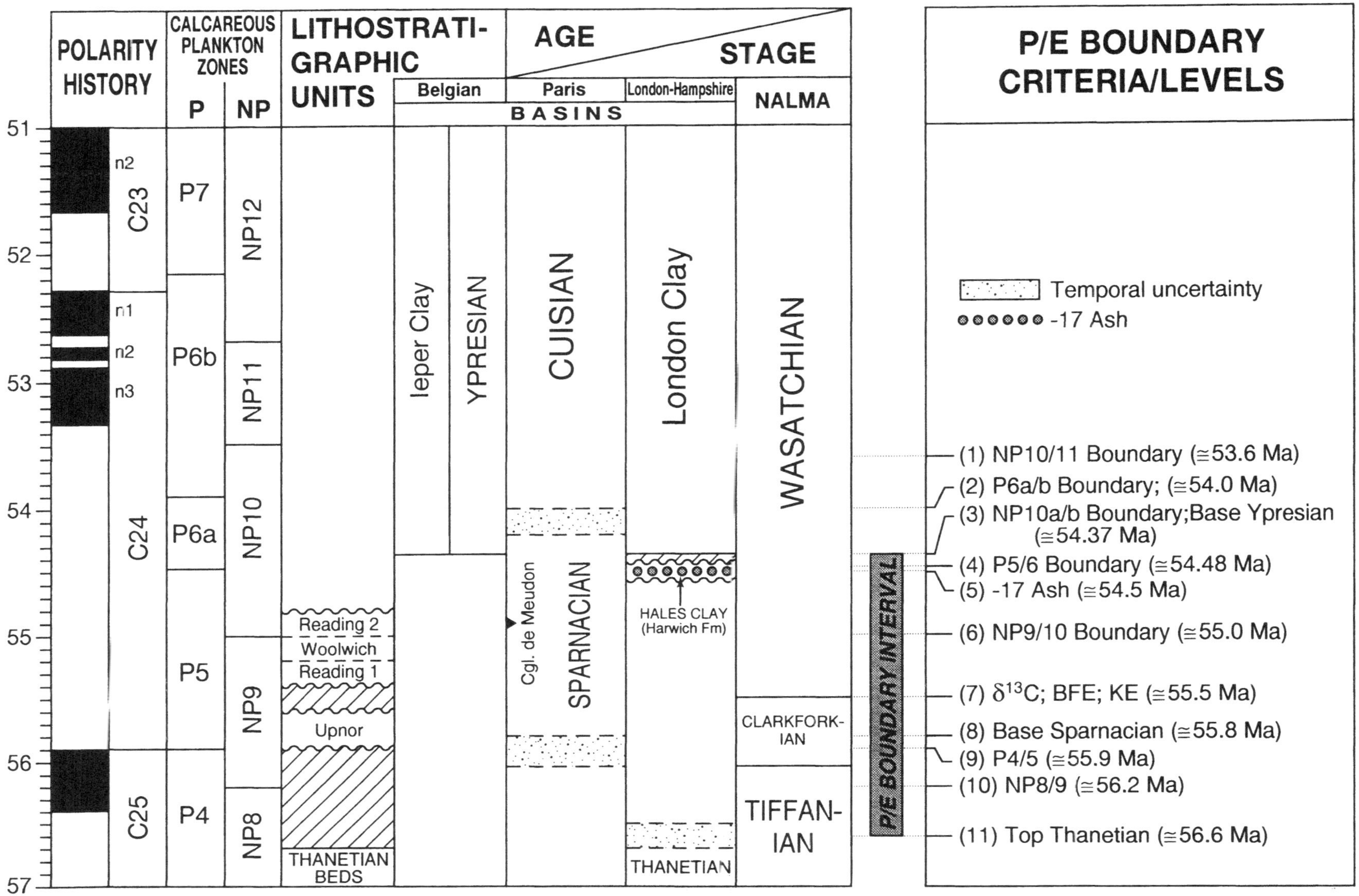

Figure 2.2 Relative stratigraphic position and estimated chronology of various events in the Paleocene–Eocene boundary interval, based on the "short" chronology as explained in the text. Some of these events have been used as criteria denoting the Paleocene/Eocene Epoch/Series boundary.

Note: Ma = mega-annum (millions of years ago); my = denotes million years (duration). The numbers 1 through 11 in the right hand column do not correspond to those given in the text.

complete magnetobiostratigraphic section spanning the Chron C24r interval (which is demonstrably not the case at present), or failing that, a complete and adequate composite section through this interval. Suffice to reiterate in conclusion that the ash series is probably incompletely represented at Site 550, the main part of the phase 2a (–) series is probably absent in the unconformity at 408 m with the result that various estimates on the duration of the ash beds present at Site 550 treat only a part of the total Phase 2a (2.1 and 2.2) and 2b (2.3) series.

The Paleocene Epoch/Series now rests on a firm conceptual framework; it/they encompasse(s) the Danian, Selandian, and Thanetian Ages/Stages. The Paleocene/Eocene Series boundary, however, awaits determination of a GSSP within the ~2-my time span between the top of the Thanetian Stage and the base of the Ypresian Stage. Over a dozen bio-, magneto-, tephra-, chemostratigraphic events within this interval provide a relatively high resolution stratigraphic subdivision. However, a precise chronology of these events requires a continuous magnetobiostratigraphic record of the 2.5-my-long Chron C24r or, at least, a demonstrably continuous composite section spanning this interval. While not favoring one criterion over another at this point, we note that the choice of a $\delta^{13}C$ excursion/spike in mid-Biochron NP9 (the base of which is correlative with the major benthic foraminiferal extinction event [BFE] and associated proxies for temporally correlative significant global atmospheric/climatic and oceanographic circulation events), which has been recorded in both marine and terrestrial stratigraphies, would have the virtue of unifying the Paleocene/Eocene Series boundary in these two disparate stratigraphies at a common level that is stratigraphically and temporally midway between the top of the Thanetian Stage and the base of the Ypresian Stage.

In order to place the stratigraphic succession (and related bio-magneto- and chemostratigraphic events) of the 2-my-long Paleocene–Eocene boundary interval in a chronologic framework for the purpose of chronologic (as opposed to stratigraphic) correlation, it is important to utilize an internally consistent time scale. We have pointed out the shortcomings of the current GPTS, and the IMBS derived from it. The IMBS, developed here is, however, consistent with the GPTS, and until such time that improvements are possible to these scales, we urge readers to use the IMBS presented here for the sake of consistency in correlation and comparison of marine and terrestrial stratigraphies.

ACKNOWLEDGMENTS

We thank R. W. O'B. Knox, R. D. Norris and D. V. Kent for thoughtful reviews of an earlier draft of this chapter and, in particular, D. V. Kent for lengthy discussions on problems relating to the chronologic aspects of this chapter and suggestions leading to the revisions to some of Chron C24r events as required by adherence to the (admitted shortcomings of) the GPTS of Cande and Kent (1992, 1995). Support of a part of the investigations of Berggren were provided by a consortium of oil companies (AMOCO, Chevron, EXXON and UNOCAL). This is Woods Hole Oceanographic Institution Contribution Number 9321 and ISEM Contribution Number 96145.

REFERENCES

Ali, J. R. and Jolley, D. W., 1996. Chronostratigraphic framework for the Thanetian and lower Ypresian deposits of southern England. In Knox, R. W. O'B., Corfield, R. M. and Dunay, R. E. (eds.), *Correlation of the Early Paleogene in Northwest Europe,* Geological Society Special Publication No. 101, p. 129–144.

Ali, J. R., Hailwood, E. A. and King, C., 1996. The "Oldhaven magnetozone" in East Anglia: a revised interpretation. In Knox, R. W. O'B., Corfield, R. M. and Dunay, R. E. (eds.), *Correlation of the Early Paleogene in Northwest Europe,* Geological Society Special Publication No. 101, p. 195–203.

Ali, J. R., King, C. and Hailwood, E. A., 1993. Magnetostratigraphic calibration of early Eocene depositional sequences in the southern North Sea Basin. In Hailwood, E. A. and Kidd, R. B. (eds.), *High Resolution Stratigraphy,* Geological Society Special Publication, No. 70, p. 99–125.

Aubry, M.-P., 1983. Biostratigraphie du Paléogène épicontinental de l'Europe du Nord-Ouest. Etude fondée sur les nannofossiles calcaires. *Documents des Laboratoires de Géologie, Lyon,* v. 89, 317p.

Aubry, M.-P., 1985. Northwestern European Paleogene magnetostratigraphy, biostratigraphy, and paleogeography: Calcareous nannofossil evidence. *Geology,* v. 13, p. 198–202.

Aubry, M.-P., 1986. Paleogene calcareous nannoplankton biostratigraphy of northwestern Europe. *Palaeogeography, Palaeoclimatology, Palaeoecology,* v. 55, p. 267–334.

Aubry, M.-P., 1996. Towards an upper Paleocene-lower Eocene high resolution stratigraphy. In Aubry, M.-P. and Benjamini, C. (eds.), Paleocene/Eocene Boundary Events in Space and Time, *Israel Journal of Earth Sciences,* v. 44, p. 239–253.

Aubry, M.-P., Berggren, W. A., Stott, L. and Sinha, A., 1996. The upper Paleocene–lower Eocene stratigraphic record and the Paleocene–Eocene boundary carbon isotope excursion: Implications for geochronology. In Knox, R. W. O'B., Corfield, R. M. and Dunay, R. E. (eds.), *Correlation of the early Paleogene in Northwest Europe,* Geological Society Special Publication No 101, p. 353–380.

Aubry, M.-P., Hailwood, E. A. and Townsend, H. A., 1986. Magnetic and calcareous-nannofossil stratigraphy of lower Palaeogene formations of the Hampshire and London basins. *Journal of the Geological Society, London,* v. 143, p. 729–735.

Berggren, W. A., 1965. Paleocene: A micropaleontologist's point of view. *Bulletin of the American Association of Petroleum Geologists,* v. 49, p. 1473–1484.

Berggren, W. A., 1971. Tertiary boundaries and correlations. In Funnell, B. M. and Riedel, W. R. (eds.), *The Micropaleontology of Oceans,* Cambridge: Cambridge University Press, p. 693–809.

Berggren, W. A. and Aubry, M.-P., 1996. A late Paleocene–early Eocene NW European and North Sea magnetobiochronological correlation network. In Knox, R. W. O'B., Corfield, R. M. and Dunay, R. E.

(eds.), *Correlation of the Early Paleogene in Northwest Europe,* Geological Society Special Publication No. 101, p. 309–352.

Berggren,W. A. and Schnitker, D., 1983. Cenozoic marine environments in the North Atlantic and Norwegian–Greenland Sea. In Bott, M. H. P., Saxov, S., Talwani, M. and Thiede, J. (eds.), *Structure and Development of the Greenland–Scotland Ridge. New Methods and Concepts,* NATO Conference Series, Series IV: *Marine Sciences,* v. 8, New York, London: Plenum Press, p. 495–548.

Berggren, W. A., Kent, D. V., Flynn, J .J. and Van Couvering, J. A., 1985. Cenozoic geochronology. *Geological Society of America Bulletin,* v. 96, p. 1407–1418.

Berggren, W. A., Kent, D. V., Obradovich, J. D. and Swisher, C. C., III, 1992. Towards a revised Paleogene geochronology. In Prothero, D. R. and Berggren, W. A. (eds.), *Eocene–Oligocene Climatic and Biotic Evolution,* Princeton University Press, Princeton, N.J., p. 29–45.

Berggren W. A., Kent, D. V., Swisher, C. C., III and Aubry, M.-P., 1995. A revised Cenozoic geochronology and chronostratigraphy. In Berggren, W. A., Kent, D. V., Aubry, M.-P. and Hardenbol, J. (eds.), *Geochronology Time Scales and Global Stratigraphic Correlation,* Society of Economic Paleontologists and Mineralogists Special Publication No. 54, p. 129–212.

Beyrich, E., 1854. *Über die Stellung des Hessischen Tertièrbildungen,* Berichte Verhandlungen kgl. preuss. Akademie Wissenschaft Berlin 1854, p. 640–666.

Beyrich, E., 1856. *Über den Zusammenhang des norddeutschen Tertièrbildungen,* Abhandlungen Akademie Wissenschaft Berlin 1856 (Physik Abhandlungen), p. 1–20.

Bonde, N., 1974. Palaeoenvironment as indicated by the "Mo-Clay Formation" (lowermost Eocene of Denmark). *Tertiary Times,* v. 2, p. 29–36.

Bralower, T. J., Zachos, J. C., Thomas, E., Parrow, M., Paull, C. K., Kelley, D. C., Premoli Silva, I., Sliter, W. V. and Lohmann, K. C., 1995. Late Paleocene to Eocene paleoceanography of the equatorial Pacific Ocean: Stable isotopes recorded at Ocean Drilling Program Site 865, Allison Guyot. *Paleoceanography,* v. 10, p. 841–865.

Brotzen, F., 1948. The Swedish Paleocene and its foraminiferal fauna. *Sveriges Geologiska Undersokelse,* series C, no. 493 (Crsbok 4, 1948), 140 p.

Burrows, H. and Holland, R., 1897. The foraminifera of the Thanet Beds of Pegwell Bay. *Proceedings of the Geologists' Association,* v. 15, p. 19–52.

Cande, S. C. and Kent, D. V., 1992. A new geomagnetic polarity time scale for the Late Cretaceous and Cenozoic. *Journal of Geophysical Research,* v. 97(B10), p. 13917–13951.

Cande, S. C. and Kent, D. V., 1995. Revised calibration of the geomagnetic polarity time scale for the Late Cretaceous and Cenozoic. *Journal of Geophysical Research,* v. 100(B4), p. 6093–6095.

Cavelier, C. and Pomerol, C., 1986. Stratigraphy of the Paleogene. *Bulletin de la Société Géologique de France,* ser. 8, v. 2, no 2, p. 255–265.

Cox, F. C., Hailwood, E. A., Harland, R., Hughes, M. J., Johnston, N. and Knox, R. W. O'B., 1985. Palaeocene sedimentation and stratigraphy in Norfolk, England. *Newsletters on Stratigraphy,* v. 14, p. 169–185.

Cowie, J. W., Ziegler, W., Boucot, A. J., Basset, M. G. and Remane, J., 1986. Guidelines and statutes of the International Commission on Stratigraphy (ICS). *Courier Forschungsinstitut Senckenbergiana,* v. 83, p. 1–14.

Curry, D., 1981. Thanetian. In Pomerol, C. (ed.), Stratotypes of Paleogene Stages, Bulletin d'Information des Géologues du Bassin de Paris, Mémoire hors série No. 2, p. 255–265.

Curry, D. and Odin, G. S., 1982. Dating of the Palaeogene. In Odin, G. S. (ed.), *Numerical Dating in Stratigraphy,* New York: John Wiley & Sons, Ltd., v. 1, p. 607–630.

Curry, D., Adams, C. G., Boulter, M. C., Dilley, F. C., Eames, F. E., Funnell, B. M. and Wells, M. K., 1978. A correlation of Tertiary rocks in the British Isles. Geological Society of London, Special Report No. 12, 72 pp.

Cuvier, G. and Brongniart, A., 1810. Essai sur la géographie minéralogique des environs de Paris. *Mémoire Classe Science, Mathématique, Physique, Institut impérial de France,* 278 p.

Dollfus, G. F., 1880. Essai sur l'étendue des terrains tertiaires dans le bassin anglo-parisien. *Bulletin de la Société géologique de Normandie,* v. 6, p. 584–605.

Dollfus, G. F., 1905. Critique de la classification de l'Eocène inférieur (lettre à M. Leriche). *Annales de la Société Géologique du Nord,* v. 34, p. 373–382.

D'Orbigny, A., 1852. *Cours élémentaire de paléontologie et géologie stratigraphiques.* vol. 2, Paris: Masson, 848 pp.

Dumont, A. H., 1839. Rapport sur les travaux de la carte géologique en 1839, avec une carte géologique des environs de Bruxelles. *Bulletin de l'Académie Royale de Belgique, Classe Sciences,* v. 6, p. 351–373.

Ellison, R. A., Ali, J. R., Hine, N. M. and Jolley, D. W., 1996. Recognition of Chron C25n in the upper Paleocene Upnor Formation of the London Basin, UK. In Knox, R. W. O'B., Corfield, R. M. and Dunay, R. E. (eds.), *Correlation of the Early Paleogene in Northwest Europe,* Geological Society Special Publication No 101, p. 185–193.

Ellison, R. A., Knox, R. W. O'B., Jolley, D. W. and King, C., 1994. A revision of the lithostratigraphical classification of the early Palaeogene strata of the London Basin and East Anglia. *Proceedings of the Geologists' Association,* v. 105, p. 187–197.

Gignoux, M., 1955. *Stratigraphic Geology,* San Francisco: W.H. Freeman and Co., 682 pp.

Hansen, J. M., 1979. Age of the Mo-Clay Formation. *Bulletin of the Geological Society of Denmark,* v. 27, p. 89–91.

Hardenbol, J. 1994. Sequence stratigraphic calibration of Paleocene and lower Eocene continental margin deposits in NW Europe and the US Gulf Coast with the oceanic chronostratigraphic record. *GFF, Meeting Proceedings* "Stratigraphy of the Paleocene," v. 116, p. 49–51.

Haug, E., 1911. *Traité de Géologie,* fasc. 3, *Les Périodes Géologiques,* Paris: p. 1397–2024.

Haynes, J., 1955. Pelagic foraminifera in the Thanet beds and the use of Thanetian as a stage name. *Micropaleontology,* v. 1, p. 189.

Haynes, J., 1956. Certain smaller British Paleocene Foraminifera. Part 1. Nonionidae, Chilostomellidae, Epistominidae, Discorbidae, Amphisteginidae, Globigerinidae, Globorotaliidae and Guembelinidae. *Contribution of the Cushman Foundation for Foraminiferal Research,* v. 7, p. 79–101.

Haynes, J., 1958. Certain British Paleocene Foraminifera. Part V. Distribution. *Contribution of the Cushman Foundation for Foraminiferal Research,* v. 9, p. 92–93.

Heilmann-Clausen, C., 1982. The Paleocene-Eocene boundary in Denmark. *Newsletters on Stratigraphy,* v. 11, p. 55–63.

Heilmann-Clausen, C., 1985. Dinoflagellate stratigraphy of the uppermost Danian to Ypresian in the Viborg 1 borehole, central Jylland, Denmark. *Danmarks Geologiske Undersrgelse,* Ser. A7, 69 pp.

Holmes, G. A., 1937. *The Age of the Earth.* London: Nelson, 263 pp.

Hooker, J. J., 1991. The sequence of mammals in the Thanetian and Ypresian of the London and Belgian Basins. Location of the Palaeocene–Eocene boundary. *Newsletters on Stratigraphy,* v. 25, p. 75–90.

Hooker, J. J., 1996. Mammalian biostratigraphy across the Palaeocene–Eocene boundary in the Paris, London and Belgian Basins. In Knox, R. W. O'B., Corfield, R. M. and Dunay, R. E. (eds.), *Correlation of the Early Paleogene in Northwest Europe,* Geological Society Special Publication, No. 101, p. 205–218.

Hovan, S. A. and Rea, D. K., 1992. Paleocene/Eocene boundary changes in atmospheric and oceanic circulation: A southern hemisphere record. *Geology*, v. 20, p. 15–18.

Jolley, D. W., 1996. The earliest Eocene sediments of eastern England: An ultra-high resolution palynological correlation. In Knox, R. W. O'B., Corfield, R. M. and Dunay, R. E. (eds.), *Correlation of the Early Paleogene in Northwest Europe*, Geological Society Special Publication No. 101, p. 219–254.

Kaiho, K., 1988. Uppermost Cretaceous to Paleogene bathyal benthic foraminiferal biostratigraphy of Japan and New Zealand: Latest Paleocene–middle Eocene benthic foraminiferal species turnover. *Revue de Paléobiologie*, Special Volume 2, p. 553–559.

Kaiho, K., 1991. Global changes of Paleogene aerobic/anaerobic benthic foraminifera and deep-sea circulation. *Palaeogeography, Palaeoclimatology, Palaeoecology*, v. 83, p. 65–86.

Kaiho, K., Arinobu, T., Ishiwara, R., Morgans, H. E. G., Okada, H., Takeda, N., Zhou, G., Kajiwara, Y., Matsumoto, R., Hirai, A., Niitsuma, N. and Wada, H., 1996. Latest Paleocene benthic foraminiferal extinction and environmental changes at Tawanui, New Zealand. *Paleoceanography*, v. 11, p. 447–465.

Kennett, J. P. and Stott, L. D., 1991. Abrupt deep-sea warming, palaeoceanographic changes and benthic extinctions at the end of the Palaeocene. *Nature*, v. 353, p. 225–229.

King, C., 1981. *The stratigraphy of the London Clay and associated deposits*. Tertiary Research Special Paper No. 6, Rotterdam: W. Backhuys, 158 pp.

Knox, R. W. O'B., 1984. Nannoplankton zonation and the Paleocene/Eocene boundary beds of NW Europe: An indirect correlation by means of volcanic ash layers. *Journal of the Geological Society, London*, v. 141, p. 993–999.

Knox, R. W. O'B, 1985. Stratigraphic significance of volcanic ash in Paleocene and Eocene sediments at Sites 549 and 550. In Graciansky, P. C. de, Poag, C. W., et al., Initial Reports of the Deep Sea Drilling Project, v. 80: Washington (U.S. Government Printing Office), p. 845–850.

Knox, R. W. O'B., 1990. Thanetian and early Ypresian chronostratigraphy in south-east England. *Tertiary Research*, v. 11, p. 57–64.

Knox, R. W. O'B., 1994. From regional to standard stage: implications for the historical Paleogene stratotypes of NW Europe. *GFF, Meeting Proceedings "Stratigraphy of the Paleocene,"* v. 116, p. 56–57.

Knox, R. W. O'B., 1996a. Correlation of the early Paleogene in northwest Europe: An overview. In Knox, R. W. O'B., Corfield, R. M. and Dunay, R. E. (eds.), *Correlation of the Early Paleogene in Northwest Europe*, Geological Society Special Publication No. 101, p.1–11.

Knox, R. W. O'B., 1996b. Tectonic controls on sequence development in the Paleocene and earliest Eocene of southeast England: Implications for North Sea stratigraphy. In Hesselbo, S. P. and Parkinson, D. N. (eds.), *Sequence Stratigraphy in British Geology*, Geological Society Special Publication No 103, p. 209–230.

Knox, R. W. O'B., Morton, A. C. and Harland, R. 1981. Stratigraphical relationships of Palaeocene Sands in the UK Sector of the Central North Sea. In Illing, L. V. and Hobson, G. D. (eds.), *Petroleum Geology of the Continental Shelf of North-West Europe*, Institute of Petroleum, Heyden and Son, London, p. 267–281.

Knox, R. W. O'B. and Morton, A. C., 1983. Stratigraphical distribution of early Palaeogene pyroclastic deposits in the North Sea Basin. *Proceedings of the Yorkshire Geological Society*, v. 44, p. 355–363.

Knox, R. W. O'B. and Morton, A. C., 1988. The record of early Tertiary N Atlantic volcanism in sediments of the North Sea Basin. In Morton, A. C. and Parson, L. M. (eds.), *Early Tertiary Volcanism and the Opening of the NE Atlantic*, Geological Society Special Publication No 39, p. 407–419.

Knox, R. W. O'B., Hine, N. and Ali, J. R., 1994. New information on the age and sequence stratigraphy of the type Thanetian of Southeast England. *Newsletters on Stratigraphy*, v. 30, p. 45–60.

Knox, R. W. O'B., Corfield, R. M. and Dunay, R. E., 1996. *Correlation of the Early Paleogene in Northwest Europe*, Geological Society Special Publication No 101, 480 pp.

Knox, R. W. O'B., Morigi, A. N., Ali, J. R., Hailwood, E. A., and Hallam, J. R., 1990. Early Paleogene stratigraphy of a cored borehole at Hales, Norfolk. *Proceedings of the Geologists' Association*, v. 101, p. 15–151.

Koch, P. L., Zachos, J. C., 1993. Chemostratigraphic correlation of marine and continental strata across the Paleocene/Eocene boundary. *Journal of Vertebrate Paleontology*, v. 13 (Supplement to No 3), p. 45A (Abstracts of papers, 53rd Annual Meeting of Society of Vertebrate Paleontology, Albuquerque, New Mexico).

Koch, P. L., Zachos, J. C. and Gingerich, P. D., 1992. Correlation between isotope records in marine and continental carbon reservoirs near the Palaeocene/Eocene boundary. *Nature*, v. 358, p. 319–322.

Koch, P. L., Zachos, J. C. and Dettman, D. L., 1995. Stable isotope stratigraphy and paleoclimatology of the Paleogene Bighorn Basin (Wyoming, USA). *Palaeogeography, Palaeoclimatology and Palaeoecology*, v. 115, p. 61–89.

Laurain, M., Barta, L., Bolin, C., Guernet, C., Gruas-Cavagnetto, C., Louis, P., Perreau, M., Riveline, J. and Thiry, M., 1983. Le sondage et la coupe du Mont Bernon à Epernay (Marne). Etude sédimentologique et paléontologique du stratotype du Sparnacien et de la série éocène. *Géologie de la France*, v. 3, p. 235–254.

Luterbacher, H. P. (ed.), 1992. Paleogene Stages and their Boundaries. *Neues Jahrbuch für Geologie und Paläontologie*, v. 186, 253 pp.

Lyell, C., 1833. *Principles of Geology, Being an Attempt to Explain the Former Changes of the Earth's Surface, by Reference to Causes Now in Operation*. London: J. Murray, v. 3, 398 pp. (Facsimile edition, Chicago, University of Chicago [published, 1991]).

Lyell, C., 1839. *Nouveaux éléments de géologie*, Pitois-Levranet, Paris, 648 pp.

Lyell, C., 1852. On the Tertiary strata of Belgium and French Flanders, Part II: The Lower Tertiary of Belgium. *Quarterly Journal Geological Society of London*, 8, p. 277–370, pls. 17–20.

Menning, M., 1989. A synopsis of numerical time scales. *Episodes*, v. 12, p. 3–5.

Miller, K. G., Janecek, T. R., Katz, M. E. and Keil, D. J., 1987. Abyssal circulation and benthic foraminiferal changes near the Paleocene/Eocene boundary. *Paleoceanography*, v. 2, p. 741–761.

Miller, K. G. and Tucholke, B. E., 1983. Development of Cenozoic abyssal circulation south of the Greenland–Scotland Ridge. In Bott, M. H. P., Saxov, S., Talwani, M. and Thiede, J. (eds.), *Structure and Development of the Greenland–Scotland Ridge*. New Methods and Concepts, NATO Conference Series, Series IV: *Marine Sciences*, v. 8, New York, London: Plenum Press, p. 549–589.

Morton, A. C. and Knox, R. W. O'B., 1990. Geochemistry of late Palaeocene and early Eocene tephras from the North Sea Basin. *Journal of the Geological Society, London*, v. 147, p. 425–437.

Morton, A. C. and Parsons, L. M. (eds.), 1988. *Early Tertiary Volcanism and the Opening of the NE Atlantic*. London: Geological Society Special Publication No 39, 477 pp.

Naumann, C. F., 1858. *Lehrbuch der Geognosie*. Leipzig: Englemann (2nd edition), 3, 576 pp.

Odin, G. S. (ed.), 1982. *Numerical Dating in Stratigraphy*. New York: John Wiley & Sons, Ltd., 1040 pp.

Pomerol, C., 1973. *Ere Cénozoique (Tertiaire et Quaternaire)*. Paris: DOIN editors, 269 pp.

Pomerol, C., 1989. Stratigraphy of the Palaeogene: Hiatuses and transitions. *Proceedings of the Geological Association*, v. 100, p. 313–324.

Prestwich, J., 1852. On the structure of the strata between the London Clay and the Chalk in the London and Hampshire Tertiary systems. Part II. *Quarterly Journal of the Geological Society of London,* v. 8, p. 235–264.

Prestwich, J., 1854. On the structure of the strata between the London Clay and the Chalk in the London and Hampshire Tertiary systems. Part III. *Quarterly Journal of the Geological Society of London,* v. 10, p. 75–154.

Rea, D. K., Zachos, J. C., Owen, R. M. and Gingerich, P. D., 1990. Global change at the Paleocene–Eocene boundary: climatic and evolutionary consequences of tectonic events. *Palaeogeography, Palaeoclimatology, Palaeoecology,* v. 79, p. 117–128.

Rea, D. K., 1994. The paleoclimatic record provided by eolian deposition in the deep sea: the geologic history of wind. *Reviews of Geophysics,* 32, p. 159–195.

Renevier, E., 1873. *Tableau des terrains sédimentaires (in 4o) plus un texte explicatif, Tab. 3.* Lausanne: Rouge and Dubois, 34 pp.

Renevier, E., 1897. *Chronologie géologique.* Congrès géologique international, IV Session, Zurich, 1894, p. 523–695.

Ritchie, J. D. and Hitchen, K., 1996. Early Paleogene offshore igneous activity to the northwest of the UK and its relationship to the North Atlantic Igneous Province. In Knox, R. W. O'B., Corfield, R. M. and Dunay, R. E. (eds.), *Correlation of the Early Paleogene in Northwest Europe,* Geological Society Special Publication No 101, p. 63–78.

Robert, C. and Kennett, J. P., 1992. Paleocene and Eocene kaolinite distribution in the South Atlantic and Southern Ocean: Antarctic climatic and paleoceanographic implications. *Marine Geology,* v. 103, p. 99–110.

Robert, C. and Kennett, J. P., 1994. Antarctic subtropical humid episode at the Paleocene–Eocene boundary, Clay mineral evidence. *Geology,* v. 22, p. 211–214.

Russell, D. E., Thiry, M. and Galoyer, A., 1989. Le conglomérat de Meudon: hier et aujourd'hui. 114e congrès national des Sociétés savantes, Paris, Géologie du Bassin parisien, p. 305–327.

Russell, D. E., Bignot, G., Galoyer, A., Guernet, C., Pomerol, C., Riveline, J., Sen, S., Thiry, M. and Tourenq, J., 1993. De la Craie à l'Argile Plastique: Un affleurement remarquable à Meudon près de Paris. *Bulletin d'Information des Géologues du Bassin de Paris,* v. 30, p. 3–9.

Schimper, W. P., 1874. *Traité de Paléontologie Végétale.* Paris: J. B. Baillière, v. 3, 896 pp.

Schmitz, B. (ed.), 1994. Meeting Proceedings "Stratigraphy of the Paleocene," *GFF,* v. 116, p. 39–67.

Schmitz, B., Speijer, R. P. and Aubry, M.-P., 1996. Latest Paleocene benthic extinction event on the southern Tethyan shelf (Egypt): Foraminiferal stable isotopic ($\delta^{13}C$, $\delta^{18}O$) records. *Geology,* v. 24, p. 347–350.

Siesser, W. G., Ward, D. J. and Lord, A. R., 1987. Calcareous nannoplankton biozonation of the Thanetian Stage (Palaeocene) in the type area. *Journal of Micropalaeontology,* v. 6, p. 85–102.

Sinha, A. and Stott, L. D., 1993a. Recognition of the Paleocene/Eocene boundary in the northwestern European stratotype sections on the basis of carbon isotope stratigraphy. *Journal of Vertebrate Paleontology,* v. 13 (Supplement to No 3), p. 57A (Abstracts of papers, 53rd Annual Meeting of Society of Vertebrate Paleontology, Albuquerque, New Mexico).

Sinha, A. and Stott, L. D., 1993b. Recognition of the Paleocene/Eocene boundary carbon isotope excursion in the Paris Basin, France. *Correlation of the Early Paleogene in Northwest Europe,* Geological Society of London, Programme and Abstracts (1–2 December 1993).

Sinha, A., Aubry, M.-P., Stott, L. D., Thiry, M. and Berggren, W. A., 1996. Chemostratigraphy of the "lower" Sparnacian deposits (Argiles plastiques bariolées) of the Paris Basin. In Aubry, M.-P. and Benjamini, C. (eds.), Paleocene/Eocene Boundary Events in Space and Time, *Israel Journal of Earth Sciences,* v. 44, p. 223–237.

Speijer, R. P., 1994. Extinction and recovery patterns in benthic foraminiferal paleocommunities across the Cretaceous/Paleogene and Paleocene/Eocene boundaries. *Geologica Ultraiectina, Mededelingen van de Faculteit Aardwtensachappen Universiteit Utrecht,* v. 124, 191 pp.

Speijer, R. P., Schmitz, B., Aubry, M.-P. and Charisi, S. D., 1996. The latest Paleocene benthic extinction event: Punctuated turnover in outer neritic foraminiferal faunas from Gebel Aweina, Egypt. In Aubry, M.-P. and Benjamini, C. (eds.), Paleocene/Eocene Boundary Events in Space and Time, *Israel Journal of Earth Sciences,* v. 44, p. 207–222.

Steurbaut, E., 1988. New early and middle Eocene calcareous-nannoplankton events and correlations in middle to high latitudes of the northern hemisphere. *Newsletters on Stratigraphy,* v. 18, p. 99–115.

Steurbaut, E., 1990a. Calcareous nannoplankton assemblages from the Tertiary in the Knokke borehole. In Laga, P. and Vandenberghe, N. (eds.), The Knokke Well (11E/138) with a Description of the Den Haan (22W/276) and Oostduinkerke (35E/142) Wells, *Mémoires pour servir à l'Explication des Cartes Géologiques et Minières de la Belgique,* v. 29, p. 47–62.

Steurbaut, E., 1990b. Ypresian calcareous nannoplankton biostratigraphy of the Belgian Basin. In Dupuis, C., DeConinck, J. and Steurbaut, E. (eds.), The Ypresian Stratotype, *Bulletin de la Société belge de Géologie,* 97–3/4 (1988), p. 251–285.

Steurbaut, E. and Nolf, D., 1986. Revision of Ypresian stratigraphy of Belgium and northeastern France. *Mededelinen van der Werkgroep voor Tertiare en Kwartair Geologie,* 23, p. 115–172.

Stott, L. D. and Kennett, J. P., 1990. Antarctic Paleogene planktonic foraminifer biostratigraphy: ODP Leg 113, Sites 689 and 690. In Barker, P. F., Kennett, J. P., et al., Proceedings of the Ocean Drilling Program, Scientific Results, v. 113: College Station, TX (Ocean Drilling Program), p. 549–569.

Stott, L. D., Sinha, A., Thiry, M., Aubry, M.-P. and Berggren, W. A., 1996. Global $\delta^{13}C$ changes across the Paleocene–Eocene boundary: Criteria for terrestrial-marine correlations. In Knox, R. W. O'B., Corfield, R. M. and Dunay, R. E. (eds.), *Correlation of the Early Paleogene in Northwest Europe,* Geological Society Special Publication No. 101, p. 381–399.

Swisher, C. C. III and Knox, R. W. O'B., 1991. The age of the Paleocene/Eocene boundary: $^{40}Ar/^{39}Ar$ dating of the lower part of NP10, North Sea Basin and Denmark. IGCP Project 308 (Paleocene/Eocene boundary events), International Annual Meeting and Field Conference, 2–6 December 1991, Brussels, Abstracts with Program, p. 16.

Thomas, E. 1990a. Late Cretaceous through Neogene deep-sea benthic foraminifers (Maud Rise, Weddell Sea, Antarctica). In Barker, P. F., Kennett, J. P., et al., Proceedings of the Ocean Drilling Program, Scientific Results., v. 113: College Station, TX (Ocean Drilling Program), p. 571–594.

Thomas, E., 1990b. Late Cretaceous–early Eocene mass extinctions in the deep sea. In Sharpton, V. L. and Ward, P. (eds.), *Global Catastrophes in Earth History: An interdisciplinary Conference on Impacts, Volcanism, and Mass Mortality,* Geological Society of America Special Paper No. 247, p. 481–495.

Thomas, E. and Shackleton, N. J., 1996. The Paleocene–Eocene benthic foraminiferal extinction and stable isotope anomalies. In Knox, R. W. O'B., Corfield, R. M. and Dunay, R. E. (eds.), *Correlation of the Early Paleogene in Northwest Europe,* Geological Society Special Publication No 101, p. 401–441.

Tjalsma, R. C. and Lohmann, G. P., 1983. Paleocene–Eocene bathyal and abyssal benthic foraminifera from the Atlantic Ocean. Micropaleontology Special Publication No. 4, 90 p.

Townsend, H. A. and Hailwood, E. A., 1985. Magnetostratigraphic correlation of Palaeogene sediments in the Hampshire and London Basins, southern UK. *Journal of the Geological Society, London,* v. 142, p. 957–982.

Vinken, R. (ed.), 1988. The Northwest European Tertiary Basin. Results of the International Geological Correlation Programme Project No. 124. *Geologisches Jahrbuch,* ser. A, v. 100, 508 pp.

Von Koenen, A., 1885. Über eine Palaocène Faune von Kopenhagen. *Abhandlungen Konglige Gesellschaft für Wissenschaft, Gottingen,* v. 32, 128 pp.

Wing, S. L., Bown, T. M. and Obradovich, J. D., 1991. Early Eocene biotic and climatic change in interior western North America. *Geology,* v. 19, p. 1189–1192.

Yanshin, A. L., 1953. Nekotorye voprosy stratigraficheskogo raschleneniya i stratigraficheskoi nomenklatury paleogenovykh otlozhenii (Some problems of stratigraphic subdivision and stratigraphic nomenclature of Paleogene deposits). *Geologiya Severnogo Priaraliya,* Moscow Society of Naturalists Press, Moscow, p. 180–243.

Addendum

It has come to our attention that owing to misprints at the printing stage of the paper by Berggren and Aubry (1996) several errors were incorporated in tables accompanying that paper. These are corrected here.

1. Table 4 (p. 336): column 4, 2nd row, mid-point value (346.61 m) of LAD of *T. contortus* (morphotype B) should read: 365.61 m.

2. Figure 13 (p. 340): Paris Basin column; Mammal locality Grauves (= Sables à Unios) given as MP8–9; should read (Reference Level) MP10.

3. Table 7 (p. 341): FAD *Morozovella lensiformis:* 53.40 Ma. Should read 54.0 Ma (this value is correctly given in Berggren et al. (1995: Table 9, p. 155, item 34).

Postscript

1. Since the completion of this manuscript, Zone NP10 has been identified in the Basement Bed of the London Clay outcropping in southern England, based on the occurrence of *Tribrachiatus bramlettei* and *T. digitalis* (Aubry, unpublished).

3

Stratigraphic (Dis)Continuity and Temporal Resolution of Geological Events in the Upper Paleocene–Lower Eocene Deep Sea Record

MARIE-PIERRE AUBRY

ABSTRACT

The rock record is the only support for reconstructing earth history. Although it is increasingly recognized that not all intervals of geological time are equally represented in sedimentary stratigraphic sections, it is commonly assumed that deep sea sections are essentially complete over long intervals (>10 my) of time. Consequently, stratigraphic records of isotopic and faunal changes are converted to temporal records mostly through linear interpolation/extrapolation using magnetic properties (reversals and susceptibility), sometimes complemented by characteristic isotopic signatures. With the help of theoretical cartoons I show how undeciphered unconformities in sections may blur the true patterns of temporal changes and how they affect temporal resolution in these sections. Temporal analysis of stratigraphic sections must be conducted to help distinguish between true events and pseudo-events.

The latest Paleocene–earliest Eocene (Magnetic Chron C24r) interval is one of the poorest represented in the Cenozoic deep sea record. Yet, most studies have ignored this fact and have assumed temporal continuity of upper Paleocene–lower Eocene sections. Based on two examples I discuss errors that incur from inappropriate conversion of a stratigraphic sequence into a temporal (numerical) record. In the first example, I examine the consequences of a shortcoming of the current Geomagnetic Polarity Time Scale (GPTS), which stems from the use of an age calibration tie-point on a level in DSDP Site 550, which was assumed to correspond to a biochronal boundary but in fact dates the upper surface of an unconformity. In the second example, I revise the stratigraphic and temporal significance of a 5.5 m thick upper Paleocene–lower Eocene section recovered from DSDP Site 577 and show that paleoceanographic reconstructions and biodiversity changes that have been inferred from high-resolution isotopic records and diversity patterns are illusory due to temporal misinterpretation of the section and incorrect (temporal) correlation with other sections.

The accuracy of geological interpretations is intimately dependent upon the correctness of the temporal interpretation of stratigraphic sections. For parts of the stratigraphic record that are riddled with unconformities, such as the upper Paleocene–lower Eocene interval, composite (virtual) reference sections provide the relative chronology of events and constitute the means of temporal correlations through stratigraphic analysis, not only of marine sections but also of sections in the continental record. Ultimately, the composite reference section assists in determining the best criterion(a)/event(s) to characterize epoch/series boundaries, and to select the most suitable Global Stratotype Section and Point (GSSP) to place the "golden spike."

Steady progress has been achieved in the last few years in delineating and correlating events that occurred during Chron C24r, the temporal interval that encompasses the Paleocene/Eocene Epoch boundary and has constituted a decisive turning point in the Cenozoic evolution of organisms. Events that are most critical in our attempt at identifying the best criteria with which to characterize the Paleocene/Eocene Epoch boundary and for providing a chronologic framework around it are discussed in Berggren and Aubry (chapter 2). Some provide the means for regional to global correlations (e.g., the carbon isotope excursion (CIE), the benthic foraminiferal extinction (BFE), magnetic reversals, and microfossil datums); others constitute the means for deriving a numerical chronology (i.e., the magnetic reversals, the volcanic ash series of the British Tertiary Igneous Province); still others (i.e., bases of the Ypresian and Sparnacian stages) supply information relevant to historical geology and ensure that (re)location of the boundary within Chron C24r will not violate historical and/or current chronostratigraphic usage. Numerous other events that are of no less importance for understanding the late Paleocene–early Eocene evolution of the planet but do not play as critical a role in stratigraphic correlations and in geochronology have also been described, and many are discussed in chapters of this volume. At the same time as events were described, mechanisms were proposed to explain the seemingly simultaneous high latitude warming, distinct sudden decrease in the isotopic composition of dissolved inorganic carbon in the oceans, sluggish atmospheric circulation, massive extinction among abyssal and bathyal foraminifera, and turnover among land mammals.

Documentation of the events in different geological contexts is still being refined as additional oceanic sections become available (e.g., ODP leg 179, Caribbean) and as increasingly sophisticated analyses are being performed (e.g., Thomson and Schmitz 1997), with the result that the inferred mechanisms for global change around the Paleocene/Eocene Epoch boundary can now be tested. By contrast, one area seems to have been generally neglected, that which concerns the correlation between sections themselves (as opposed to correlation between events in sections) and their temporal interpretation, and constitutes the very essence of stratigraphy.

The view that the stratigraphic record is essentially complete in basins and on margins has prevailed until the late seventies, although the occurrence of substantial stratigraphic gaps encompassing several biozones had long been recognized. In the late seventies and early eighties, the discontinuous nature of the sedimentary record on margins became recognized largely through the impetus of sequence stratigraphy. But it is only now that the equally discontinuous nature of the deep sea sedimentary record is being demonstrated (see Aubry 1995). In our search for a Paleocene/Eocene boundary stratotype, we have become aware of the highly incomplete nature of the sedimentary record corresponding to Chron C24r (see Berggren and Aubry 1996). Yet, many studies ignore the temporal offset in data that results from unconformities. Based on theoretical cartoons, I first illustrate difficulties and consequent shortcomings in identifying true events in the stratigraphic record. I then discuss two examples to show that stratigraphic correlation does not necessarily mean temporal correlation.

Identifying "True" Events in Stratigraphic Sections

I have shown (Aubry 1995) that a stratigraphic interval with normal polarity does not always represent a single magnetic chron but may result from the concatenation of two (or more) normal polarity intervals corresponding to different magnetochrons if one (or more) unconformity truncates the record, the same being true for intervals with reversed magnetic polarity. Determination of a magnetostratigraphic profile should not be based on numerical succession once a magnetozone has been identified, but on careful determination of individual reversals using biostratigraphic control whenever possible. The (incorrect) identifications of Chron C24n in ODP Hole 690B on Maud Rise and of Chrons C12 and C13 in the Chadronian terrestrial record of North America (see Aubry et al. 1996, and Prothero and Swisher 1992, respectively, for discussion) appropriately illustrate some of the more acute difficulties associated with magnetostratigraphy (see also Flynn and Tauxe, chapter 4). Likewise, anomalous patterns in the stratigraphic distribution of species may be produced either by diachrony or by unconformities that truncate the ranges of taxa (Aubry 1995).

Three geographically distant sections A, B, C, each representative of a different basin (A, B, C) are shown in figure 3.1. They are of different thickness but span the same stratigraphic interval as indicated by both magnetostratigraphy (from Magnetozone Mvn through Mxn) and biostratigraphy (not shown). The three sections record a sharp, abrupt turnover of some paleontologic group (of which distribution and abundance change are shown by dotted vertical column at right of each section). The different relationships between the paleontologic turnover and Magnetozone Mxn in sections A and C are seen as irrefutable evidence of diachrony. Classically, to achieve quantitative analysis, the sections are subdivided into intervals of equal duration based on the magnetostratigraphic reversals as shown in figure 3.1. The boundaries between Magnetozones Mvn and Mxr (level l1) and between Magnetozones Mxr and Mxn (level l6) are assigned ages T1 and T6 of the corresponding reversals between Chrons Cvn and Cxr and between Chrons Cxr and Cxn, respectively. Thus levels l2, l3, l4 and l5 are believed to be correlative between the three sections and to bear the same age, respectively, t2, t3, t4 and t5. Based on these stratigraphic correlations and temporal inferences, diachrony of the turnover is established between the three basins, the turnover occurring earlier in Basin B, slightly later in Basin A, and much later in Basin C.

The integrated magnetobiochronologic scale (IMBS; Berggren et al. 1985a, 1995) provides the means to test the validity of this interpretation. The same sections

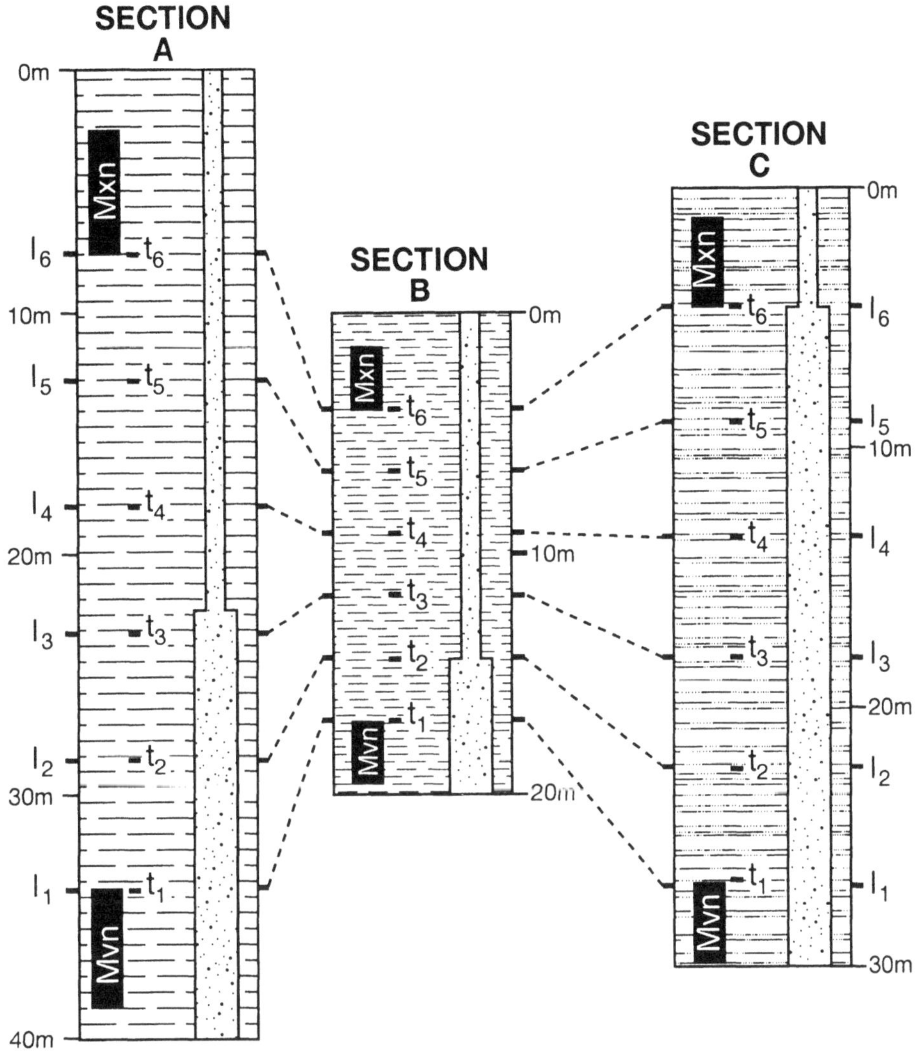

Figure 3.1 Stratigraphic correlations between three sections, A, B, C. Levels l2 to l5 have been determined to be correlative (and to bear ages t2 to t5, respectively) based on linear interpolation from the magnetic reversals that occur at levels l1 and l6. A paleontologic turnover occurs slightly above level l3 in section A, at level l2 in section B and at level l6 in section C. This indicates that the paleontologic turnover occurred slightly later than time t3 at the location of section A, at time t2 at the location of section B and at time t6 at the location of section C. This is seen to constitute an excellent example of diachrony, that lends itself to particularly interesting interpretations if the sections are located in different basins and/or at different latitudes.

correlated within a temporal framework are shown in figure 3.2. Each section has been submitted to a temporal interpretation based on a depth/time plot, following the methodology I have described and in which the two surfaces that compose any unconformity are dated (Aubry 1991, 1995). The supremacy of the IMBS over other time scales resides in the fact that it allows us to compare a (concrete) succession of stratigraphic events in a section to their temporal (virtual) succession as provided by the IMBS. The methodology has the same objective as graphic correlation methods but it is of more immediate application than the latter because only a few selected reference points (datums) are necessary. On the other hand, it is more readily applicable in land and cored-sections than in drilled sections when only cuttings are available. As clearly shown in figure 3.2 undeciphered unconformities may create a deceiving pattern of diachrony. Section A is temporally complete. Section B contains an unconformity at level l2 and section C at level l6. Whereas Section A records the turnover itself, Sections B and C do not. The turnover they exhibit is spurious and results from the juxtaposition of younger to much younger and older to much older records. In addition, it can be seen that a study of the differences between pre- and post-turnover horizons in the three sections would be rather indecisive if not

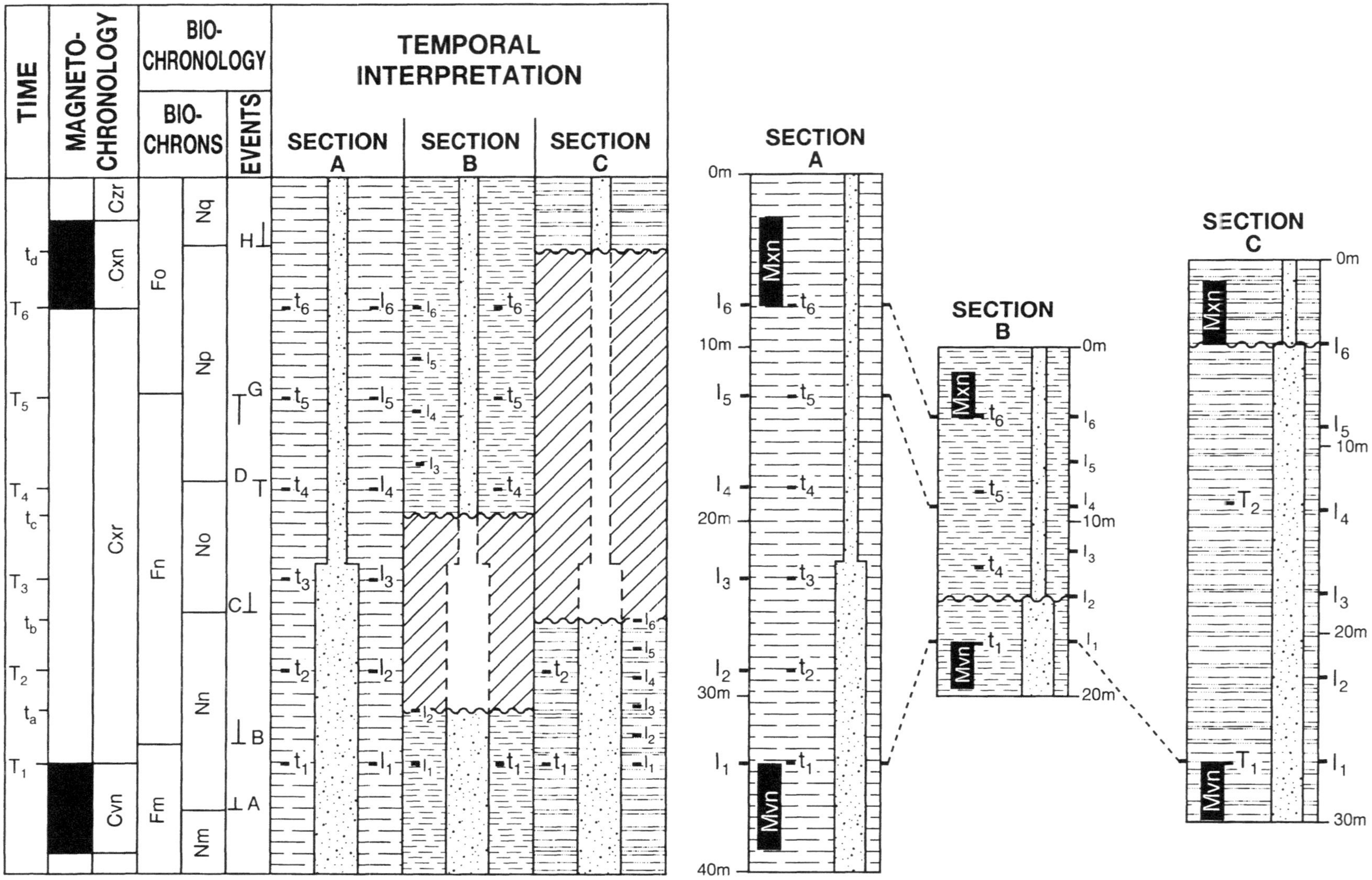

Figure 3.2 Re-interpretation of the significance of the paleontologic turnover seen in sections A, B and C, based on a rigorous temporal interpretation of the sections. The paleontologic event is seen only in section A, which is complete. Pseudoevents are seen in sections B and C where the record of turnover is truncated by unconformities. Comparison of pre- and post-turnover conditions can only be determined for pre-time ta and post time tc between sections A and B; and for pre-time tb and post time td between sections A and C. Note that times ta and tb on the one hand and times tc and td on the other are the ages of, respectively, the lower and upper surfaces of the unconformities in sections B and C. Note that sampling of levels l1 to l6 does not result in equal chronologic resolution for the three sections (contrary to what was expected following the methodology shown in figure 3.1). Chronologic resolution varies considerably between the three sections and is greater for section C. See text for further explanations.

meaningless because the levels l1 to l6 are now shown not to be correlative and to bear very different ages. Moreover, if levels l1 to l6 are those sampled in sections A, B and C, equal spacing between samples does not guarantee equal chronologic resolution in the three sections (compare figures 3.1 and 3.2). Whereas intervals of geological time are unrepresented in sections B and C, chronologic resolution for the intervals that are represented by them is higher than in Section C.

Paleontologic turnovers are not the only events that may be affected by incomplete stratigraphic records. The interpretation of isotopic events is as vulnerable to the presence of unconformities.

Two sections, A and D, from two different ocean basins, that span the same temporal interval are compared in figure 3.3. The isotopic records show the occurrence of a sharp excursion in both sections, at 20 m in Section A, at 10 m in Section D. Through interpolated magnetostratigraphy/magnetochronology (see above), it is established that the two excursions are correlative and thus reflect a geochemical event that occurred shortly before Time t4. The difference in amplitude of isotopic change is interpreted as reflecting differences in isotopic composition between the two basins. Temporal interpretation of the two sections (figure 3.4) reveals a different situation. Section A is complete and records the "true" geochemical change. But section D is incomplete and does not. It yields an unconformity at 10 m. The excursion simply results from the concatenation of two disjunct isotopic records (which in this case explains the amplitude difference in isotopic change). Isotopic correlations are favored as a

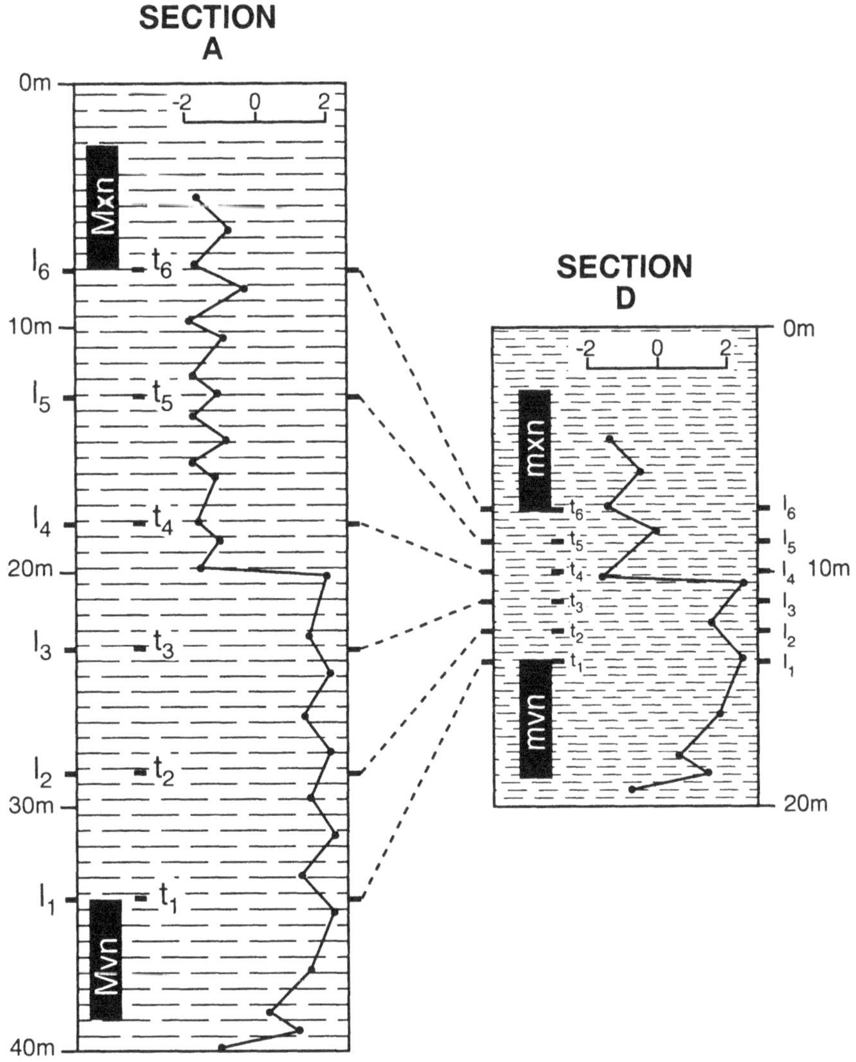

Figure 3.3 Stratigraphic correlations between 2 sections A and D, showing that the isotopic excursions at 20 m in section A and at 10 m in section D are correlative and thus correspond to a geochemical event that occurred slightly before time t4.

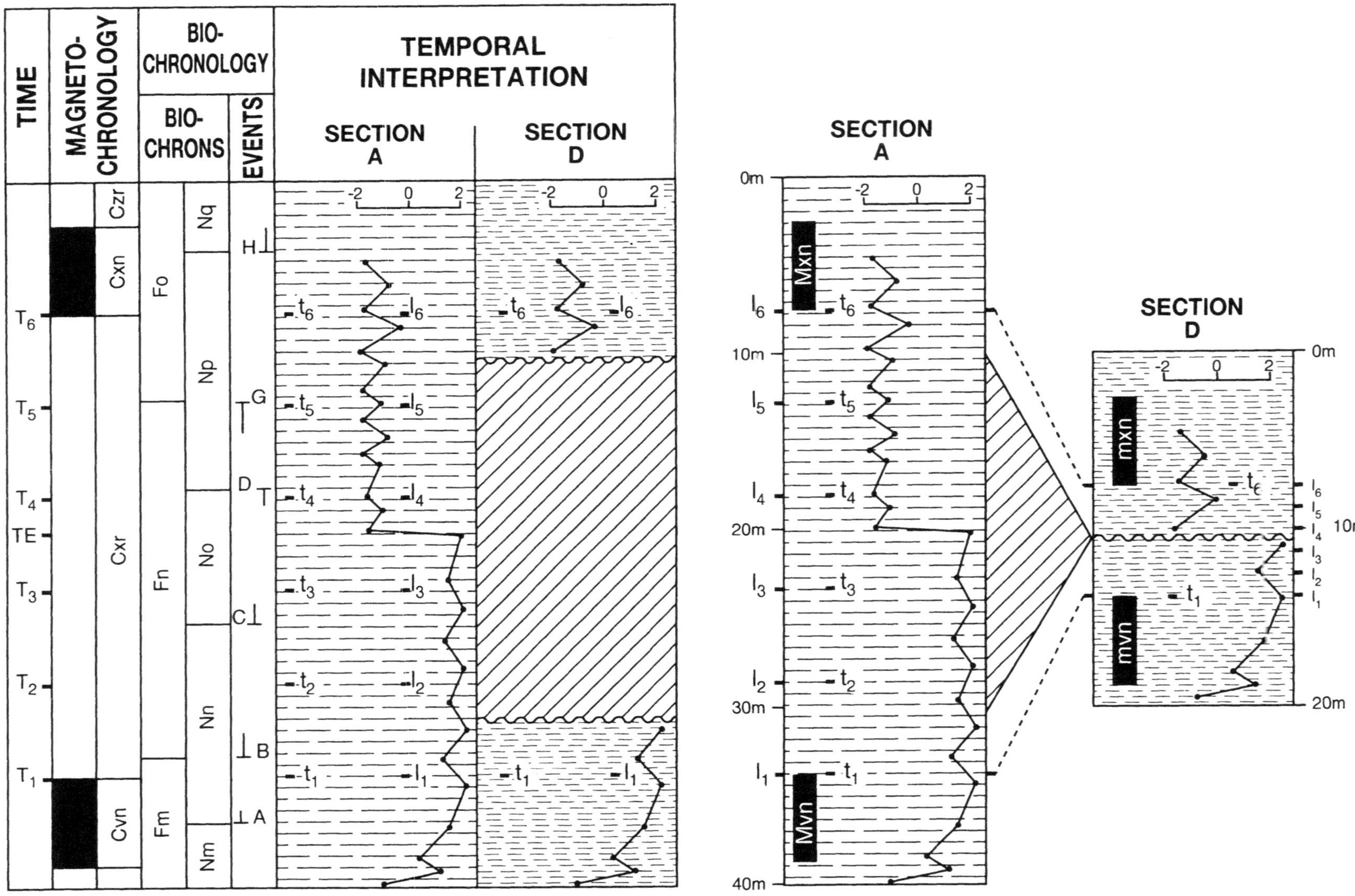

Figure 3.4 Re-interpretation of the correlation between sections A and D based on rigorous temporal interpretation of the sections. Section A registers the geochemical event itself, but the excursion in section D is an artifact that results from the concatenation of two disjunct parts of an isotopic record. In other words, the excursion in section A is a true event whereas the excursion in section D is a pseudoevent. See text for further explanation.

means of fine chronologic tuning because isotopic changes occur at a scale that corresponds to the mixing time (10^4 yr) of the ocean, which is geologically instantaneous. However, peak to peak correlations as currently practiced may result in miscorrelations if unconformities remain undeciphered (see Aubry et al. 1996).

As exemplified in figure 3.5, any stratigraphic event—e.g., paleontologic, geochemical, magnetic—that is recorded in a section must be shown to be the full expression of the historical (geological) event that it reflects before further determination is made. Stratigraphic Event "e_A" in Section A is the reliable expression of geological Event "E" that occurred at Time TE. But stratigraphic Event "e_F" in Section F is not, even though direct stratigraphic correlation between Sections A and F would point to the contrary. Event "e_F" is a pseudoevent. Events "e_A" and "e_F" may be the highest occurrence (HO) of a taxon in two sections. Only in Section A does the HO correspond to the last appearance datum (LAD) of the taxon. Events "e_A" and "e_F" may be the highest occurrence of a sequence of ash beds, or of a particular facies. In Section A, the event reflects the end of ash deposition. But in Section F (and eventually in the basin), the event reflects no more than the highest (youngest) recorded occurrence of the ash sequence. The point illustrated in figure 3.5 may seem trivial with regard to the chronologic resolution that is achieved for most of the stratigraphic record. However, the literature abounds with examples of miscorrelations due to the misinterpretation of the temporal significance of stratigraphic sections. Yet, correct interpretation of this temporal significance is critical if "we are to adequately comprehend the picture of the rock strata of the Earth as a whole, and to restore the history of *how, when* and *why* these strata came to be *what* and *where* they are today" (Hedberg 1976:v). This implies that the expression "stratigraphic correlations" must take on a new meaning. To correlate (in a stratigraphic sense) can no longer be "to show correspondence in character and/or in stratigraphic position (Salvador 1994:15, slightly modified from Hedberg 1976:14). Levels 11 and 16 in Sections A, B and C (figure 3.1), the isotopic shift in Sections A and D (figure 3.3) and Events e_A and e_F in Sections A and F, respectively (figure 3.5), do correlate in a stratigraphic sense. This does not mean that they are temporally correlative, and comparison of these levels/events is meaningless because they are asynchronous (see discussion above and below). Cause and effect relationships in earth history cannot be firmly established unless stratigraphy achieves truly temporal correlations. As we attempt to delineate the chronology of events for specific intervals, rigorous temporal analysis of stratigraphic sections becomes an absolute necessity, and stratigraphic correlation must mean temporal correlation and be based on such temporal analysis (see figures 3.2 and 3.4, which show which levels are correlative [i.e., have the same age] in Sections A, B, C [figure 3.2] and A, D [figure 3.4]).

Two additional comments are thus worth consideration here. First, the theoretical cartoons illustrated in figures 3.1 to 3.5 point to the fact that *neither* diachrony *nor* synchrony must be accepted as valid on the basis of stratigraphic correlations alone. *Both* diachrony *and* synchrony of levels/events must be demonstrated based on temporal analysis of stratigraphic sections. Second, I reiterate an earlier plea (see Aubry 1995) for a dual terminology for stratigraphic and temporal terms. Because we have come to understand that the stratigraphic record is incomplete in most, if not all, settings, and in order to avoid misinterpretation of the kind exemplified theoretically above, it is highly advisable to differentiate stratigraphic terms from temporal terms (see further discussion below).

The 55 Ma Calibration Tie-Point in the GPTS and the 408 mbsf Level in DSDP Site 550

The stratigraphic record is incomplete but not randomly incomplete. Sequence stratigraphy points to a global pattern of alternation of deposition and erosion on the continental shelves (Haq et al. 1988). Moore et al. (1978) showed that intervals of geological time are unevenly represented in ocean basins. I (1995) have demonstrated that this contrast is true at fine scale as well. The early Eocene, late Eocene and late Miocene, are epochs that are best represented in the stratigraphic record, with, commonly, complete sedimentary sections developed (e.g., Aubry 1993a, 1993b, 1995; Shackleton et al. 1995; Hilgen 1991). The latest early Eocene and earliest middle Eocene, the late middle Miocene and the early late Miocene are epochs that have a scanty stratigraphic record (e.g., Aubry 1993a, 1993b, 1995). Magnetic Chron C24r is among these temporal intervals that are poorly represented by incomplete sedimentary sections (see Berggren and Aubry 1996; see also Moore et al. 1978). Of the twenty marine land sections or so that have been investigated to date in various countries, only one, the Aweina section in Egypt (Schmitz et al. 1996), is thought to provide an essentially continuous record through Chron C24r (for logistic reasons, magnetostratigraphic investigations have not yet been possible on the section, but the isotopic and microfossil records are excellent).

Despite the incompleteness of the stratigraphic record of Chron C24r, an impressive numerical bio- and chemochronology has been established for it (table 3.1) with a resolution well below 0.1 my. However, though

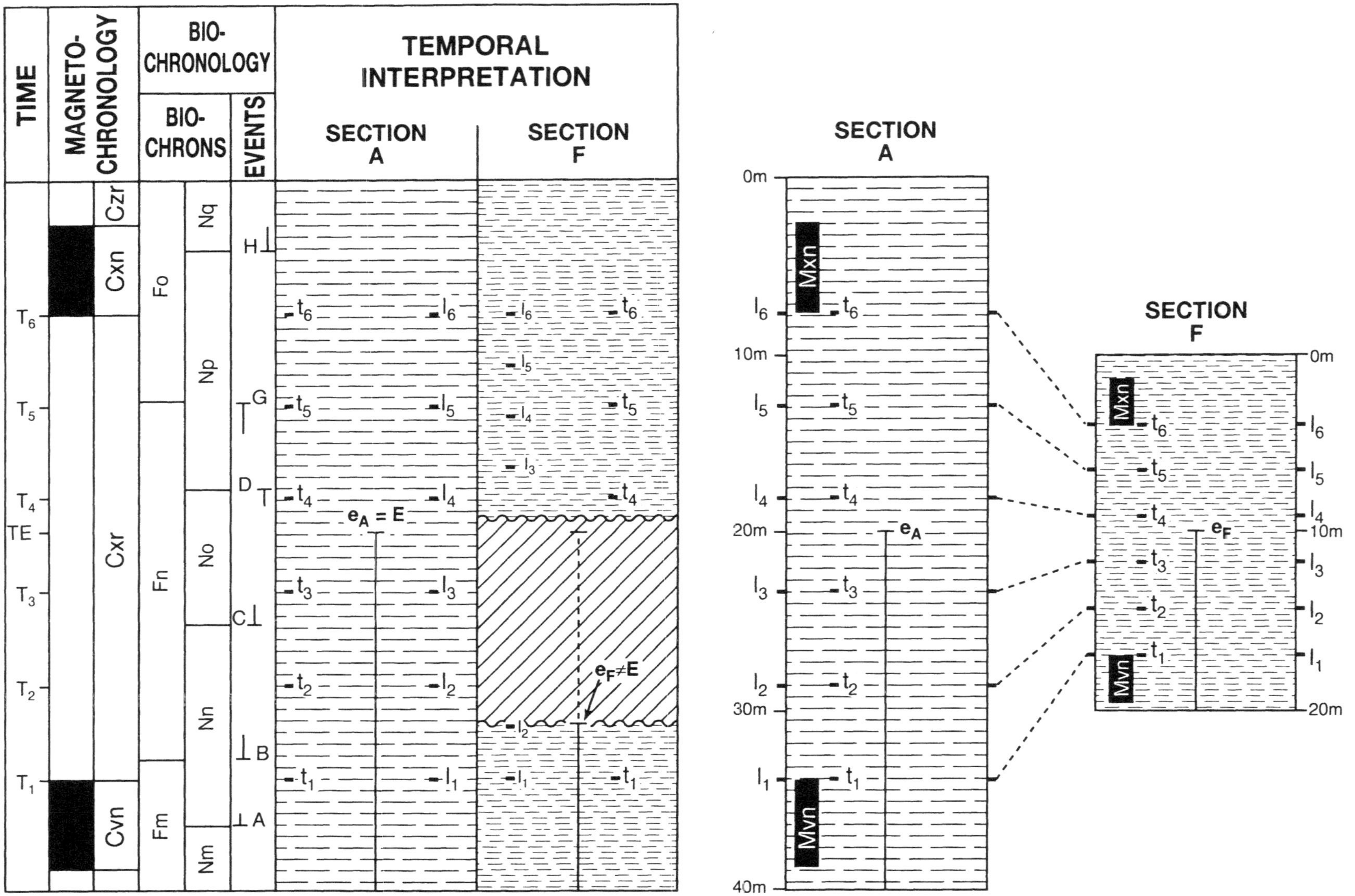

Figure 3.5 Straight stratigraphic correlations (right) indicate that the same geological event E is recorded in sections A (stratigraphic event e_A) and F (stratigraphic event e_F). Temporal interpretation of the sections (left) is critical in showing that only section A records event E. Stratigraphic event e_F is a pseudoevent that does not correspond to geological event E because of the stratigraphic gap. It can be seen that short duration events, such as abundance peaks of some taxon, which occurred between times T2 and TE, would be registered in section A but not in section F. Yet, without a temporal interpretation, the difference between abundance patterns in sections A and F would likely be interpreted in terms of biogeographic differentiation.

Table 3.1
Chronology of events in Chron C24r. Only events useful for global marine correlations are given

Events	Age (Ma)
Chron C24r/C23n.3n	53.347
LAD *Subbotina velascoensis*	53.53
LAD *Tribrachiatus contortus*	53.61
FAD *T. orthostylus*	53.64
LAD *Morozovella aequa*	53.65
LAD *T. bramlettei*	53.89
FAD *T. contortus*	53.93
FAD *M. lensiformis*	54.00
LAD *T. digitalis*	54.17
FAD *T. digitalis*	54.37
FAD *M. formosa gracilis*	54.69
FAD *Igorina broedermanni*	54.69
LAD *M. acuta*	54.72
FAD *M. marginodentata*	54.80
FAD *Acarinina wilcoxensis*	54.91
FAD *T. bramlettei*	55.00
FAD *Rhomboaster cuspis*	55.16
LAD *Fasciculithus tympaniformis*	55.33
FAD *Cruciplacolithus eodelus*	55.5
BFE/CIE	55.5
Chron C25n/C24r	55.904

For the chronology of other events see Berggren and Aubry, chapter 2. Data from Berggren and Aubry (1996), Berggren et al. (1995), and Cande and Kent (1995).

precise, this numerical chronology is inaccurate and will require revision once the exact position of the NP9/NP10 biochronal boundary within Chron C24r is determined (the sequence of events is unaffected by the problem discussed below, but the relative [distance between events] and numerical chronologies [duration in my between events] are). The nature of the problem has been discussed at length in Berggren and Aubry (1996, chapter 2) and Aubry et al. (1996). It is summarized here because this is a textbook example of how undeciphered unconformities falsify tools as essential as time scales in today's geological research and, also, because it provides a good illustration—if that is still necessary—of why periodic revisions of the time scale are required.

The need for improving the numerical chronology of events that occurred during Chron C24r stems from two missteps, first the dating of a series boundary through its approximated biozonal correlation, second, the incorrect identification of the biochronozonal nature of a biozonal contact.

The NP9/NP10 zonal boundary has often been (and regrettably still is) used to approximate the Paleocene/Eocene Series boundary. Yet, it has been shown that the lithostratigraphic horizon that defines the base of the Eocene Series (i.e., the base of the Ieper Clay, which characterizes the base of the Ypresian Stage and correlates with the base of the London Clay Formation sensu Ellison et al. 1994) lies well within Zone NP10 (Knox 1984; Aubry et al. 1989; Berggren and Aubry 1996, and references therein). Swisher and Knox (1991) estimated the age of a level they believed to correspond to the NP9/NP10 biochronozonal boundary in DSDP Hole 550 (figure 3.6) and presented it as the age of the Paleocene/Eocene Epoch/Series boundary. The confusion between definition and correlation that such a move created is exemplified by the age given to the ash series. The ash series in Hole 550 correlates with the lower part of Zone NP10 (Knox 1985), thus is early Eocene in age (following the proposal of Swisher and Knox). On the other hand, the ash series correlates with the Harwich Formation and (probably) part of the Woolwich–Reading formation of the London Basin, which are distinctly older than the London Clay Formation itself (e.g., Knox 1990). Thus it is of late Paleocene age in this interpretation. Obviously, the ash series is either Paleocene or Eocene in age. Current definition of the Paleocene/Eocene Epoch/Series boundary implies a Paleocene age. (Previous rules [see Hedberg 1976:24, 25] of the International Commission on Stratigraphy [ICS] dictated that the base of the Ypresian Stage [= base of the Ieper Clay] determines the base of the Eocene Series. Revised procedures for stratigraphic subdivision by ICS [Cowrie et al. 1986] have as a consequence that the lithostratigraphic horizon chosen to define the base of the Eocene Series [as accepted by ICS following proposal by the Working Group [WG] on the Paleocene/Eocene Epoch/Series boundary of the Subcommission on Paleogene Stratigraphy] will determine the base of the Ypresian Stage. Until the WG has selected the best criterion(a) to characterize the Paleocene/Eocene Epoch/Series boundary and, eventually, selected a GSSP, the older rule applies.)

Biozonal boundaries are often taken as biochronozonal boundaries (i.e., the biochronal boundary in temporal terms), just as normal successions of biozones are often taken as evidence of stratigraphic completeness (see discussion in Aubry 1995). The biochronozonal nature of biozonal boundaries must be established based on rigorous demonstration of stratigraphic completeness of the sections in which they occur (see above). There were probably no means of determining in the early eighties that the NP9/NP10 zonal boundary in Hole 550 lay at an unconformable contact at 408 mbsf, although the record of *Tribrachiatus contortus* shortly (2.4 m) above the boundary by Müller (1985) could have served to arouse suspicion as to the completeness of the section. Based on integrated chemo- ($\delta^{13}C$ stratigraphy) and biostratigraphy, we have demonstrated that the lithologic contact at 408 mbsf in Hole

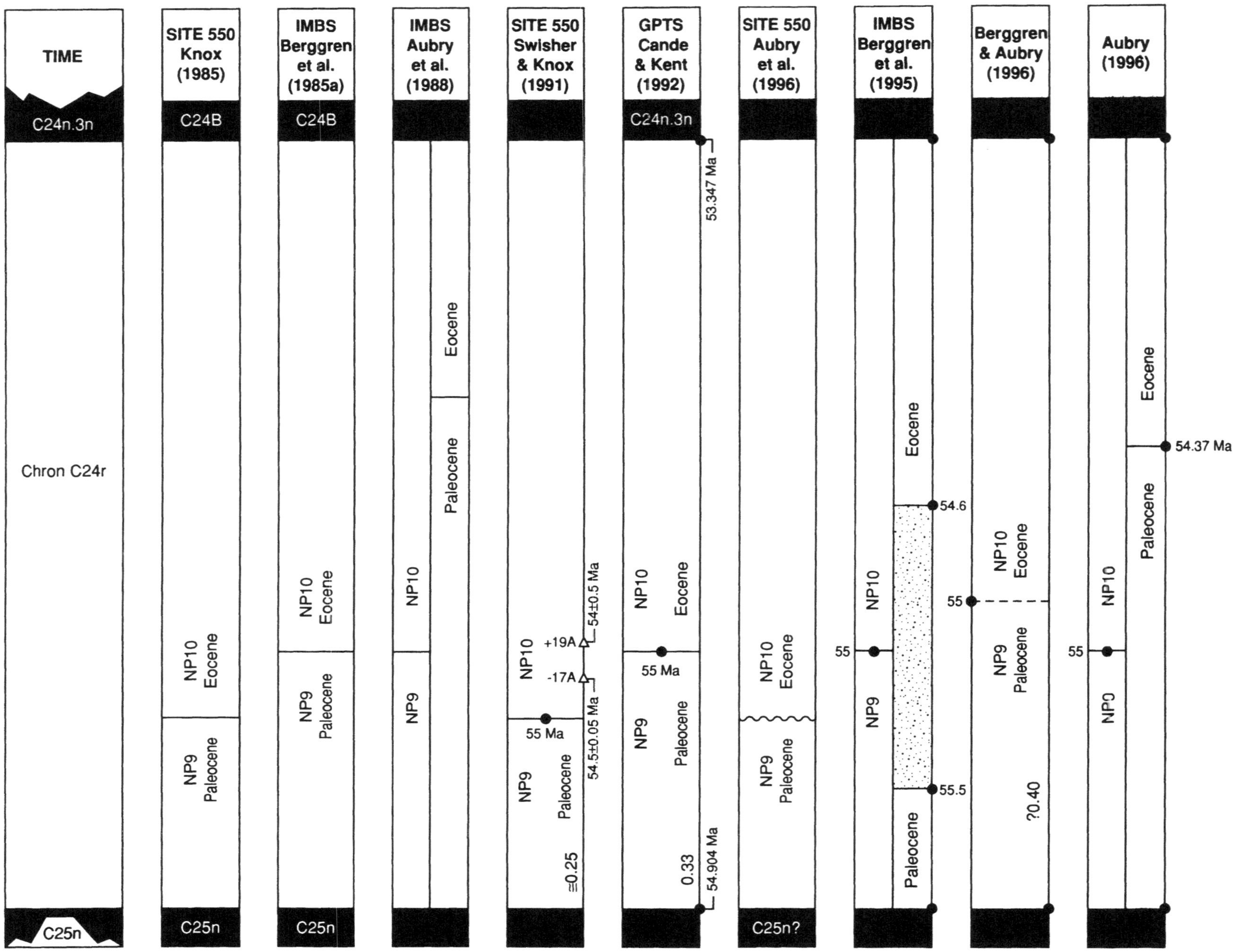

Figure 3.6 History of the relationships between the NP9/NP10 biochronal/zonal boundary, the Paleocene/Eocene Epoch/Series boundary and their location in Chron C24r.

550 is unconformable (Aubry et al. 1996). The implication for the GPTS in particular and the IMBS in general are discussed below.

The dating of the −17 Ash and the +19 Ash of the Danish Ash Series (= Mo Clay) in Denmark and their correlatives in southeast England (Obradovich in Wing et al. 1991; Swisher and Knox 1991; Berggren et al. 1992; Berggren et al. 1995) has constituted a notable progress for Paleogene geochronology and ended a long controversy as to the age of the Paleocene/Eocene Epoch/Series boundary. These $^{40}Ar/^{39}Ar$ ages on sanidine were the first to date upper Paleocene horizons with a well constrained magnetobiostratigraphic position through correlation with DSDP Site 550 on Goban Spur in the North Atlantic (Knox 1985; see discussion on the availability and reliance of radioisotopic ages on Paleocene to lower Eocene horizons in Berggren et al. 1985b). Consequently, they provided much needed age calibration tie-points between the ages of 49.5 and 84.00 Ma on the far distant Anomaly 21y and 34y, respectively, used (with 4 other tie-points) by Berggren et al. (1985b) to derive their magnetochronology. Cande and Kent (1992, 1995) used these ages on ashes indirectly by accepting a 55 Ma age estimate on the Paleocene/Eocene Epoch boundary as one of their nine calibration points for converting the sea floor anomaly spacing (distance in km) into time (chron duration in my).

Swisher and Knox (1991) estimated an age of 55 Ma for the NP9/NP10 zonal boundary in DSDP Hole 550, that they believed to correspond to the NP9/NP10 chronozonal boundary. DSDP Hole 550 on Goban Spur is a remarkable section that contains a thick (>60 m) sedimentary record representing Chron C24r (Townsend 1985) with a ~20-m-thick interval of ash-bearing calcareous nannofossil oozes (Knox 1985) restricted to the lower part of Zone NP10 and in which the characteristic −17 and +19 ash layers of the Fur Formation of Denmark were identified (Knox 1984). Total-fusion $^{40}Ar/^{39}Ar$ ages on sanidine from the −17 Ash and +19 Ash sampled in Denmark have yielded mean weighted ages of 54.5 ± 0.05 Ma and 54.0 ± 0.53 Ma, respectively (Swisher and Knox 1991). Using these two ages to derive a sedimentation rate between the two ashes in Hole 550 and extrapolating it to the NP9/NP10 zonal boundary in the hole, an age of 55 Ma was estimated for the NP9/NP10 chronozonal boundary. Because this boundary was used to approximate the Paleocene/Eocene Series boundary, these authors proposed an age of 55 Ma for the Paleocene/Eocene Epoch boundary.

Cande and Kent (1992) thus used a 55 Ma age calibration for the Paleocene/Eocene Epoch boundary tie-point in addition to 8 others. However, they astutely removed this point from the context of DSDP Hole 550 where the NP9/NP10 zonal boundary occurs ~75% of the way down in the reversed interval representing Chron C24r (figure 3.6). Instead, they used the position of the Paleocene/Eocene Epoch boundary estimated by Berggren et al. (1985b) at 2/3 of the duration of Chron C24r (= C24r(0.66) from the Chron C24n.3n/C24r reversal boundary (figure 3.6). They then determined a point on the sea floor with an age of 55 Ma located at a distance of 1221.20 Km from the ridge axis (figure 3.7) and, using the nine calibration points plus the zero-age ridge axis, applied a spline function to transform their metric composite polarity sequence into time. Thus, while the duration of Chron C24r is exclusively determined by the width between Anomaly 25_y and 24_o, the ages of the magnetic reversals C24n.3n/C24r and C24r/C25n are seen to be narrowly dependent upon placement of the 55 Ma tie-point. This is true for any other chron: the duration is determined by the distance on the sea floor between successive anomalies but the ages of the chron reversals would vary with different placement of the selected calibration tie-points between or in anomalies. Thus the fact that the NP9/NP10 zonal boundary in Hole 550 does not represent the chronozonal boundary jeopardizes the magnetochronology of Cande and Kent (1992, 1995) and more specifically the numerical chronology of Chron C24r-events. This is how.

The unconformity at 408 mbsf in Hole 550 corresponds to an interval on the sea floor of determined width that in turn measures the duration of the hiatus in Hole 550 (figure 3.8). We need then to add the corresponding amount of sediment between the 2 surfaces of the unconformity at 408 mbsf in order to obtain a complete stratigraphic record of Chron C24r (i.e., the chronozone corresponding to Chron C24r; figure 3.9). In doing so, we observe that the level dated at 408 mbsf corresponds to the upper surface of the unconformity and lies younger in Chron C24r than thought. If the age calibration point were placed at an older position in Chron C24r than it really is, the duration of the younger part of Chron C24r would be overestimated, whereas that of its older part would be underestimated. Cande and Kent's rejection of the position (~C24.75) of the NP9/NP10 biozonal boundary with respect to the thickness of the stratigraphic interval corresponding to Chron C24r in Hole 550 as indicative of the position of the NP9/NP10 biochronal boundary in Chron C24r in favor of the younger position (C24.66) of the NP9/NP10 biochronal boundary (then regarded as equivalent to the Paleocene/Eocene Epoch boundary) in Chron C24r in Berggren et al. (1985b; figure 3.6) was fortunate because this has compensated, at least partly, for the error introduced by Swisher and Knox (1991) on the position within Chron C24r of the stratigraphic horizon (408 mbsf) dated at 55 Ma in Hole 550. However, the position of the NP9/NP10 biochronal boundary relative to

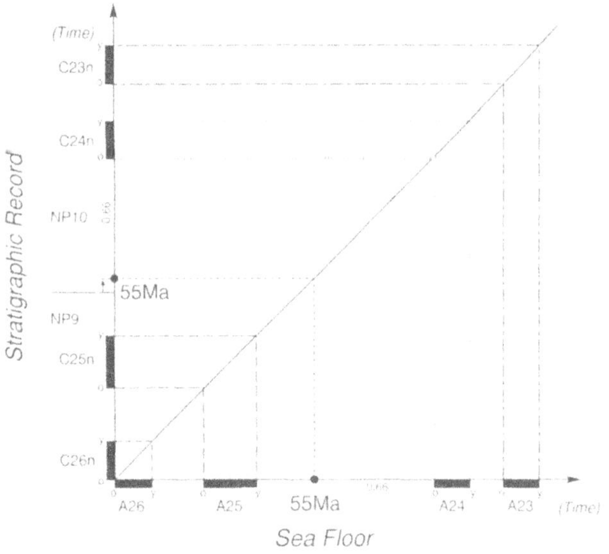

Figure 3.7 Conversion of the sea floor polarity sequence, expressed as metric distance between anomalies, into time in Cande and Kent (1992) involves the use of the Paleocene/Eocene Series boundary (as approximated by the NP9/NP10 chronozonal boundary [= the stratigraphic equivalent of the biochronal boundary]) as a calibration tie-point with an age of 55 Ma as determined from DSDP Hole 550. The position of the tie-point between Anomaly 25 and Anomaly 24 was determined at (0.66) from Anomaly 24o (i.e., 1221.20 km from the ridge-axis), based on the estimated position of the Paleocene/Eocene Series boundary 66% of the way down in a complete (virtual) section representing Chron C24r (Berggren et al. 1985b). In DSDP Hole 550, the boundary (thus the level dated at 55 Ma) occurs at ~75% of the way down in the reversed polarity interval. As shown here, Cande and Kent ignored the stratigraphic relationship in DSDP Hole 550, and readjusted the position of the boundary at Chron C24r(0.66).

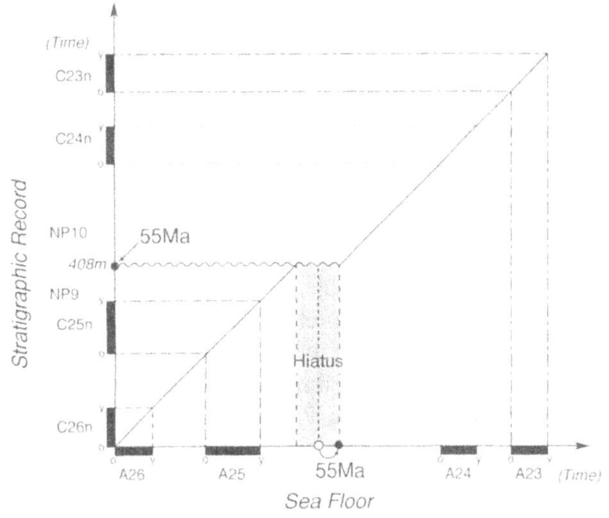

Figure 3.8 The unconformity at 408 mbsf in Hole 550 corresponds to a segment of the sea floor that measures the hiatus. The 55 Ma age calibration tie-point must therefore be moved to a position closer (thus younger) to Anomaly 24o.

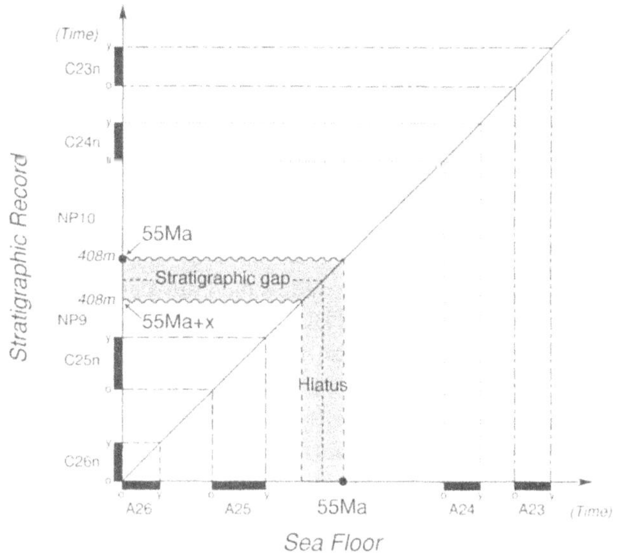

Figure 3.9 Through extrapolation of the sedimentation rates between the two ashes in Hole 550, Swisher and Knox (1991) dated in fact the upper surface of the unconformity at 408 mbsf. The lower surface has an age of 55 Ma + x my. The NP9/NP10 chronozonal boundary (dashed line) has an age comprised between 55 and 55 Ma + x my, and corresponds with a point on the sea floor of uncertain position (within the shaded interval). See text for further explanation.

the duration of Chron C24r remains undetermined and may be younger than accepted by Cande and Kent (1992; see discussion in Aubry et al. 1996; figure 3.6). Thus using the Cande and Kent (1995) GPTS, the relative (and thus numerical) chronology of events that occurred during late Chron C24r (i.e., during Biochron NP10) may be artificially stretched whereas that of events in early Chron C24r may be artificially compressed. However, despite the fact that we recognize that the ages of the magnetic reversals in the GPTS will likely need revision, revision would be inappropriate until a stratigraphic section that can be shown to be complete through Chron C24r provides the means to locate precisely the NP9/NP10 biochronal boundary in the chron.

It must be understood here that we are not concerned with the accuracy of the age estimates on the events in Chron C24r, but with their precision. Obviously, the relative chronology of events in Chron C24r must be correct if we are to derive meaningful rates of changes through the chron. On the other hand, once we accept the constraints of a GPTS, the accuracy of the age estimates is not as critical as their precision as long as internal consistency with the GPTS is preserved. The uncertainty in the age estimates determined for events in Chron C24r is certainly large in view of the uncertainty (>0.5 my) on the estimated age of the NP9/NP10 biochronal boundary. This age was derived through extrapolation of sedimentation rates between two ash layers. Although the −17 Ash yielded very precise $^{40}Ar/^{39}Ar$ ages, the uncertainty on its age is 0.5 my, reflecting interlaboratory age differences. Swisher and Knox (1991) reported an age of 54.50 ± 0.05 Ma on it whereas Obradovich (in Berggren et al. 1992) obtained an age of 55.07 ± 0.16 Ma. Similarly, the +19 Ash yields $^{40}Ar/^{39}Ar$ ages that differ by 0.5 my: 54.0 ± 0.5 (Swisher and Knox 1991) versus 54.5 ± 0.5 (Obradovich in Berggren and Aubry 1996). Should the radioisotopic ages obtained by Obradovich have been used to estimate the age of the NP9/NP10 zonal boundary in Hole 550 in a fashion similar to that employed by Swisher and Knox (1991), an age of 55.5 Ma rather than 55 Ma would have been obtained for it, and would have been used by Cande and Kent (1992) as the age calibration point to derive the numerical chronology in the GPTS. In addition, because it yields an uncertainty of 0.50 my, the radioisotopic age of the +19 Ash is indistinct from that of the −17 Ash. Thus, Swisher and Knox (1991) used sedimentation rates between the +19 and −17 ash layers that were poorly constrained by the radioisotopic age on the former ash. In all likelihood, the age of the NP9/NP10 zonal boundary in Hole 550 is younger than 55 Ma, owing to the anomalously low sedimentation rates used by Swisher and Knox (1991). We have independently estimated an age of 54.7 Ma for the NP9/NP10 contact at 408 mbsf in Hole 550 (see Berggren and Aubry, chapter 2). This would imply an uncertainty of 0.8 my on the age calibration used in Cande and Kent (1992). On the other hand the uncertainty in the position of the NP9/NP10 biochronal boundary is probably negligible in comparison to the uncertainty on the age calibration used. The uncertainty on the numerical chronology in Chron C24r and the problems related to the accuracy of the radioisotopic ages on the ashes are further discussed in Berggren and Aubry (1996).

Thus before proceeding with a revision of the GPTS two elements concerning the late Paleocene–early Eocene interval must be resolved, 1) the position of the NP9/NP10 biochronal boundary within Chron C24r in order to establish the correct relative (and ultimately numerical) chronology of events in the Chron, and more importantly, 2) the accuracy of radioisotopic ages in the interval in order to reduce the age uncertainty on any calibration point in Chron C24r used in the GPTS, on magnetic reversals, and on any other (e.g., bio, chemo, tephra) events in the interval and, ultimately, on the Paleocene/Eocene Epoch boundary.

More serious is the fact that the age of 55 Ma applies to two distinct levels: 1) the 408 mbsf-level in Hole 550 (Swisher and Knox 1991), which corresponds to a NP9/NP10 unconformable biozonal contact, and 2) the NP9/NP10 biochronozonal boundary (Cande and Kent 1992). The upper surface of the unconformity at 408 mbsf in Hole 550 is younger than the NP9/NP10 biochronozonal boundary but there is no possibility to determine the age difference between the two levels. The lower surface of the unconformity is older than the NP9/NP10 biochronozonal boundary as shown through chemobiostratigraphic correlations (Aubry et al. 1996; Stott et al. 1996) and has a undetermined age >55 Ma (55 Ma + x my; figure 3.9). The stratigraphic gap at 408 mbsf in Hole 550 includes the lower part of Zone NP10 and the upper part of Zone NP9. The use to calibrate the GPTS of an age of 55 Ma on a level believed to be of chronostratigraphic significance though corresponding to the upper surface of an unconformity has resulted in three surfaces being miscorrelated with the consequence that a substantial stratigraphic gap (in Hole 550) was collapsed to a single chronostratigraphic horizon (the NP9/NP10 chronozonal boundary). The resultant is a temporal interval (late Biochron NP9–early Biochron NP10) that cannot be discriminated using the current GPTS. This is particularly critical because it jeopardizes the temporal representation of fine scale records of variations such as that of isotopic change through the late part of Biochron NP9 and the early part of Biochron NP10 (figure 3.10; Aubry et al. 1996; Berggren and Aubry 1996; Stott et al. 1996). As a consequence of the use in the GPTS of an age of 55 Ma determined for an unconformable horizon rather than for a chronostratigraphically (= temporally) significant horizon, the later part of Biochron NP9 overlaps with the older part of Biochron NP10 in the IMBS of Berggren et al. (1995). This has also resulted in a difficult use of depth/duration plots to evaluate the completeness of stratigraphic sections through Chron C24r.

Theoretically, it should be possible to determine the age of the two surfaces of the unconformity through extrapolation of the sedimentation rates above and below it as shown in figure 3.7 using a magnetochronology different from that of Cande and Kent (1992). Considering that the duration of Chron C24r is almost the same in Berggren et al. (1985b) as in Cande and Kent (1992) this methodology would allow us to determine the duration of the hiatus at 408 mbsf. Recall that this is in fact how an age of 55 Ma was determined for the NP9/NP10 (chrono)zonal boundary in Hole 550 (extrapolating the sedimentation rate calculated between the −17 and +19 ash layers down to the zonal boundary). Because of the completeness of the lower Eocene sedimentary record in Hole 550 (Aubry 1995), it is possible to determine the age of the upper surface of the unconformity by extrapolating sedimentation rates using the magnetobiochronologic framework of Berggren et al. (1985b). Although possible, direct extrapolation is not without difficulties because of strong dissolution and differential sedimentation rates between pure calcareous nanno-

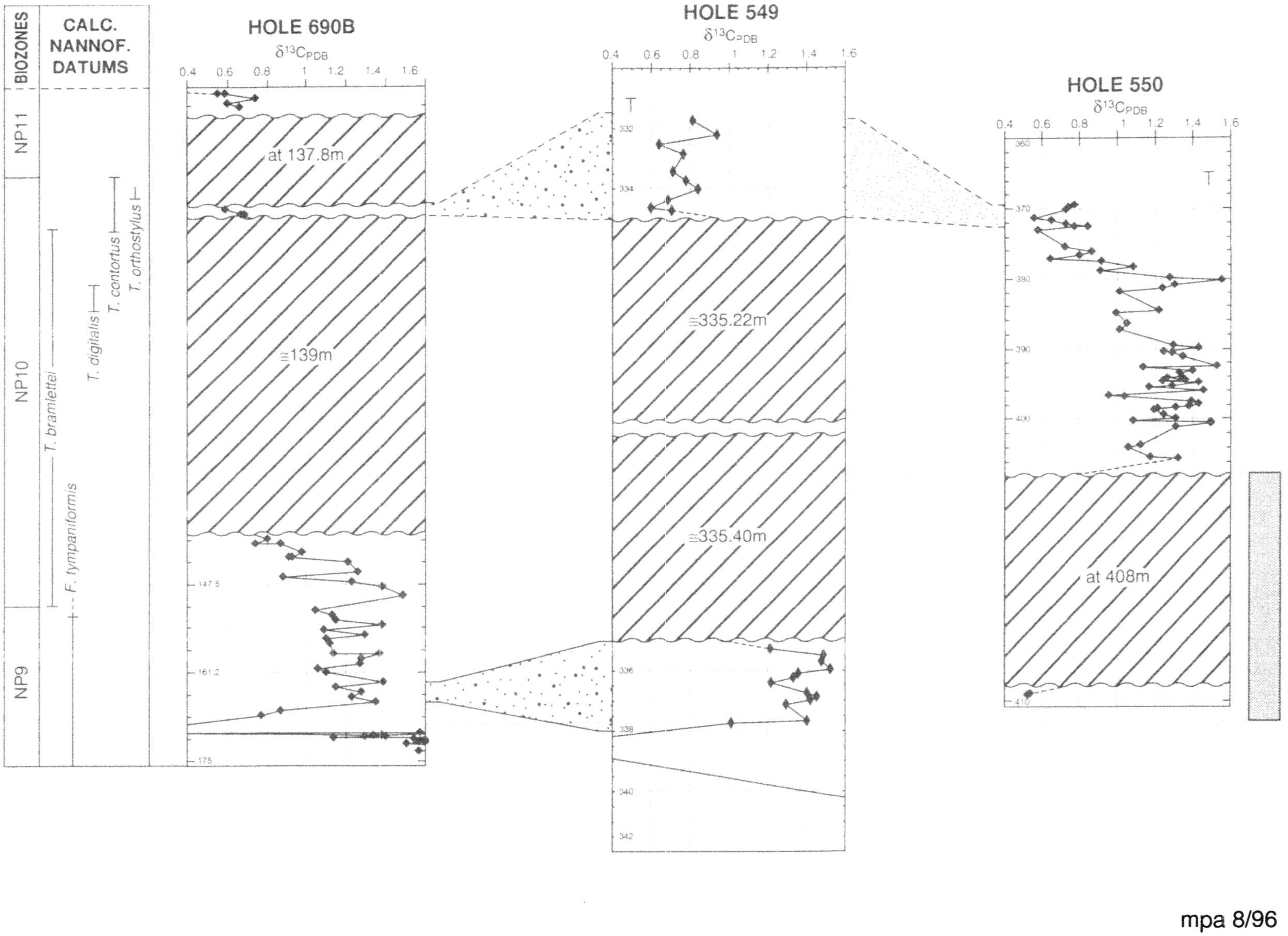

Figure 3.10 Temporal interval that is not accounted for by the GPTS of Cande and Kent (1992, 1995).
After Aubry et al. (1996:figure 9).

fossil oozes and ash-bearing calcareous nannofossil oozes (see discussion in Berggren and Aubry 1996 and chapter 2). On the other hand the presence of a series of unconformities below the lowest occurrence of *Discoaster multiradiatus* (= base Zone NP9) in Hole 550 (see Aubry et al. 1996; Berggren and Aubry 1996) precludes determining the age of the lower surface of the unconformity using this methodology. Nor is it possible to determine the time represented by the stratigraphic interval in Hole 690B correlative with the stratigraphic gap in Hole 550. A good record of Zone NP9 and Chron C25n is preserved in Hole 690B, but an unconformity at 137.5 m truncates Zone NP10 (see discussion in Aubry et al. 1996, and Berggren and Aubry 1996).

Thus, we are left with an unresolved situation until a section with a complete record of Chron C24r allows us to determine the position of the NP9/NP10 biochronal boundary in Chron C24r. Yet, the current situation is such that it hampers progress in temporal correlation between sections and in determining rates of changes during what can be seen as the most critical time in Cenozoic history. Also, this may lead to the proposal of alternative but incompatible solutions to derive a numerical chronology for Chron C24r. Yet, uniform use of a chronologic framework is essential to preserve integrity in comparing data from multiple sources in an attempt to resolve late Paleocene–early Eocene contingencies. Aubry et al. (1996) and Berggren and Aubry (1996) derived a chronology for the late and early parts of Chron C24r, using a different methodology for each interval. For the late part of the chron, they extrapolated the sedimentation rates through Chron C24n in Hole 550 down to the ash series and then interpolated between the ages of 54.5 ± 0.05 Ma on the −17 Ash and 55 Ma on the upper surface of the unconformity at 408 mbsf to estimate a numerical chronology for the intervening interval. For the early part of the chron, Aubry et al. (1996) used the sedimentation rates estimated at Site 690 during Chron C25n and extrapolated upwards to the NP9/NP10 zonal boundary (but did not use its postulated age of 55 Ma). They thus estimated an age of 55.52 Ma for the CIE and the BFE. In their calculations, Aubry et al. (1996) and Berggren et al. (1996) assigned the age of 55 Ma to a level dated by Swisher and Knox (1991) though recognizing the shortcomings of the methodology used by these authors (see in particular discussion in Berggren and Aubry 1996). Consequently, they eliminated the potential use of the age of 55 Ma on the NP9/NP10 (chrono)zonal boundary in Hole 690B to estimate the age of the CIE and BFE. Whereas this is probably the scientifically most sound approach, it allows no escape from the incompatible dual use of the age of 55 Ma. For instance, as pointed out above, the late Biochron NP9 overlaps with the early Biochron NP10 in the IMBS of Berggren et al. (1995) and, for this reason, variations of isotopic composition of organic matter through Chron C24r cannot be modeled satisfactorily (Sinha et al. unpublished). Thus, an arbitrary, temporary, workable solution is proposed here that preserves internal consistency for the already existing chronology of Chron C24r at the same time as it satisfies the contingencies of the current GPTS while only slightly violating the stratigraphic record in Hole 550. As explained above, the age of 55 Ma applies to two horizons. One, actual, is the upper surface of the unconformity at 408 mbsf in Hole 550. The other, virtual, is the NP9/NP10 chronozonal boundary (virtual, because it does not occur in Hole 550, and because its exact position in Chron C24r is unknown). Aubry et al. (1996) and Berggren and Aubry (1996) relied on observed data to establish an estimated chronology for Chron C24r. The alternative is to accept an age of 55 Ma for the NP9/NP10 chronozonal boundary as used in Cande and Kent (1992), and to estimate a (younger) age for the upper surface of the unconformity at 408 mbsf in Hole 550 (see Berggren and Aubry, chapter 2). This alternative is proposed here. It has several merits in addition to providing an escape from the dual use of the 55 Ma-age. First, this fulfills one of the contingencies of the GPTS. Second, it allows satisfactory discrimination of events that occurred during late Biochron NP9 and early Biochron NP10, and thus permits temporal representation of this interval (unlike the IMBS of Berggren et al. 1995). Third, it does not affect the estimated numerical chronology established for Chron C24r, except for events recorded in the intervening interval between the top of the ash series and 408 mbsf in Hole 550. Recall that as calculated in Aubry et al. (1996) and Berggren and Aubry (1996), internal consistency was preserved between the GPTS (Cande and Kent 1995) and this numerical chronology. The numerical chronology for the early part of Chron C24r was established independently of the 55 Ma-age on the NP9/NP10 zonal boundary in Hole 690B. Yet, serendipitously, interpolation between this age and the age of 55.904 Ma on the Chron C24r/C25n boundary results in an age estimate of 55.52 Ma for the BFE and CIE, exactly equal to the age determined through extrapolation of sedimentation rates for the interval representing Chron C25n in Hole 690B. This is serendipitous because there is no reason why the same age should be estimated through the two different methodologies, particularly when very different sedimentation rates are applied or implied. Aubry et al. (1996) applied a sedimentation rate of 2.14 cm/10^3 yr for the interval representing Chron C25n in Hole 690B. The constraints used here imply sedimentation rates of 3.96 cm/10^3 yr below the CIE and 3.8 cm/10^3 yr between the CIE and the NP9/NP10 zonal boundary (and this would contradict the suggestion that sedimentation rates increased after the CIE at high latitude Site 690; Thomas and Shackleton 1996). Although obviously good understanding of the interval representing early Chron C24r in Hole 690B is illusory, the agreement in age estimates for the CIE and BFE using either methodology means that acceptance of the age of 55 Ma for the NP9/NP10 chronozonal boundary does not necessitate revision of the estimated chronology for the early part of Chron C24r. Thus I propose to use an age of 55 Ma for the NP9/NP10 chronozonal boundary and an age of 54.7 Ma for the upper surface of the unconformity at 408 mbsf in Hole 550 (figure 3.11). This age of 54.7 Ma is derived through differential estimates of sedimentation rates through the calcareous oozes and ash series in the hole (see Berggren and Aubry, chapter 2).

In conclusion, the age of the magnetic reversals in the GPTS of Cande and Kent (1992, 1995) and the chronology of events that occurred during Chron C24r are constrained by the age of 55 Ma assigned to the NP9/NP10 biochronal boundary. Yet, this age of 55 Ma was estimated for a level within early Biochron NP10. In order to resolve the incompatible dual use of the same age for two distinct stratigraphic levels, it is proposed that the 55 Ma-age be used on the NP9/NP10 biochronal boundary to fit the constraints of the GPTS and to use an age of 54.7 Ma for the stratigraphic level that was originally dated at 55 Ma by Swisher and Knox (1991).

True Events and Pseudoevents: A Closer Look at the Upper Paleocene–Lower Eocene Stratigraphic Record at DSDP Site 577

Because the deep sea record is generally believed to be essentially complete over long intervals of time (>10 my), it is considered acceptable to date specific levels in stratigraphic sections through linear interpolation/extrapolation from magneto- and/or biostratigraphic events. Stratigraphic records are thus tuned to a particular ("high") chronologic resolution that merely corresponds to the distance between samples, close-spaced sampling being taken as synonymous with high resolution chronology. In this fashion, many stratigraphic records have been converted into temporal records that are believed to support comparison of various components of geological history between regions/basins. As I have explained above, transforming a stratigraphic record into a temporal one is not a simple matter (see figures 3.1 to 3.5) and incorrect temporal interpretation of sections may result in incorrect historical de-

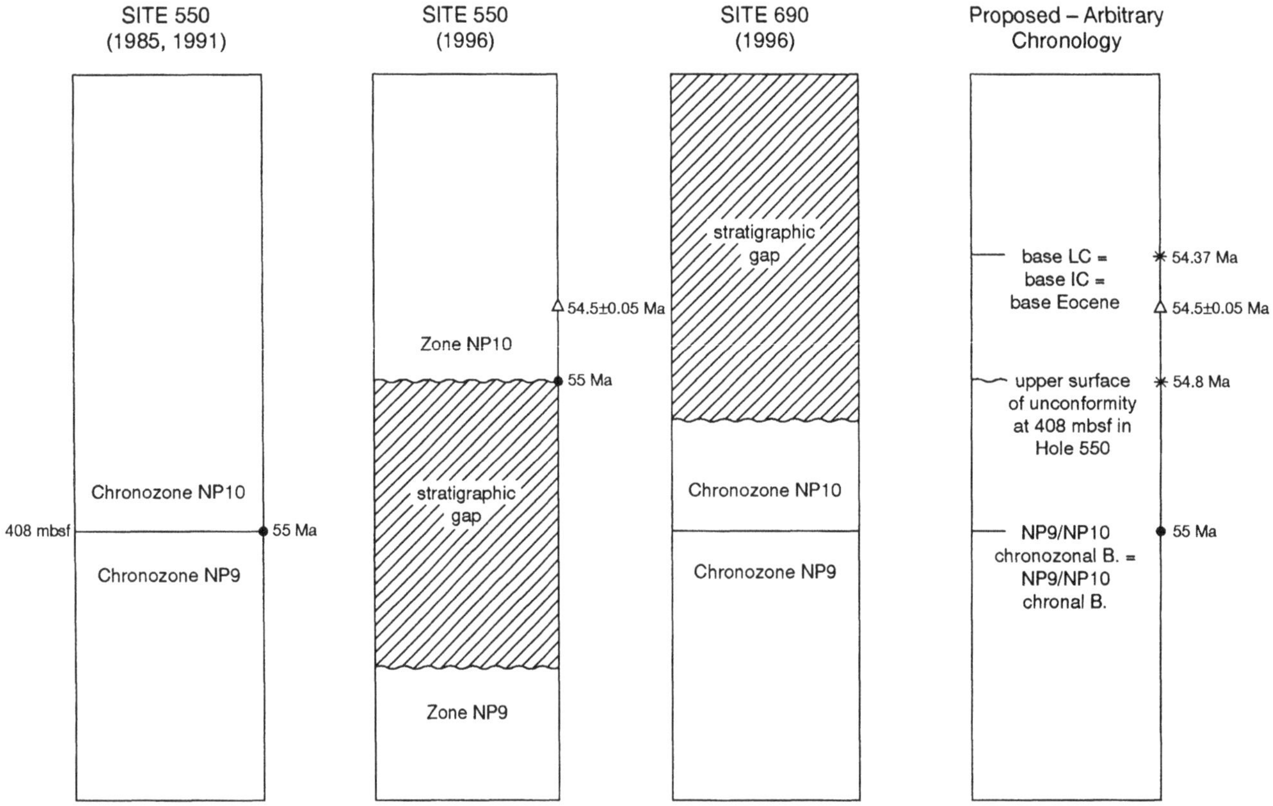

Figure 3.11 Proposed provisional and arbitrary solution to the problem created by the age assignment of 55 Ma to two discrete surfaces, the actual upper surface of an unconformity at 408 mbsf in Hole 550 and the virtual NP9/NP10 biochronozonal boundary

ductions. To illustrate further my contention that high resolution sampling does not imply high resolution chronology for a given interval, I shall compare opposite interpretations of the upper Paleocene–lower Eocene stratigraphic record recovered from DSDP Site 577 on Shatsky Rise, and discuss the implications for paleoceanographic interpretation. I have chosen this site because, as one of the few sites in the Pacific Ocean that provide a lower Paleogene stratigraphic record, it has been extensively used for paleontologic, biochronologic and isotopic purposes (e.g., Monechi et al. 1985; Shackleton et al. 1985; Backman 1986a, 1986b; Miller et al. 1987; Corfield and Cartlidge 1992; Pak and Miller 1992; Stott 1992 [immediately adjacent Hole 47.2]; Zachos et al. 1993; Berggren et al. 1995; Lu and Keller 1995a, 1995b; Wei and Zhong 1996; Charisi and Schmitz 1996). Despite its shortcomings (see below and Bralower et al. 1995) I shall show that the upper Paleocene–lower Eocene section recovered from this site can be rather tightly correlated with that recovered from DSDP Site 550 on Goban Spur, and discuss some resulting implications.

DSDP Site 577

DSDP Site 577 (32° 26.51' N; 157° 43.40' E) was drilled on the west side of the Shatsky Rise in 2675 m water depth (Heath et al. 1985a). A 5.50 m-thick section of nannofossil ooze representing Chron C24r (Bleil 1985; Monechi et al. 1985; figure 3.12) was recovered. It was subdivided by the lithologic contact at 82.8 mbsf (between Cores 9 and 10) between pale brown nannofossil ooze (Subunit IIA, above) and white and pale brown nannofossil ooze (Subunit IIB, below; Heath et al. 1985a). Recovery was good except at the bottom of Core 9 where a 1 m-long recovery gap occurs, and the sediment was only slightly deformed. Monechi (1985) suggested the presence of a stratigraphic gap in this section based on the absence of Subzone CP9a (in fact, the co-occurrence of *Tribrachiatus contortus* and *Discoaster diastypus* in one sample (9–5, 117–118 cm) indicates that a very thin interval is assignable to Zone CP9a). However, Monechi et al. (1985) apparently considered the section complete (contra their figure 2, p. 789) based on the study of additional closely spaced samples by Backman (in Monechi et al. 1985) who reported the highest occurrence of *Fasciculithus* spp. in Sample 10–1, 30 cm, the presence of *T. contortus* in 4 samples taken between 30 and 115 cm in Core 9, section 5 (presumably those studied by Shackleton et al. (1985) for isotopic analysis and taken at 30–31 cm, 60–61 cm, 90–91 cm, and 113–114 cm), as well as the presence of *T. bramlettei* in the lowermost of these four samples (i.e., between 113 and 114 cm). All subsequent studies on this 5.5 m-thick section have considered the sediment accumulation record to be complete through Chron C24r (see references above). Only recently have claims been made that the section is discontinuous. Wei and Zhong (1996) located an unconformity between 80.83 and 80.48 mbsf. Based on an integration of magneto-, bio-, and chemostratigraphic data available for the section and unpublished data on the range of planktonic foraminifera, Berggren and Aubry (1996) have indicated the presence of two unconformities, one between Cores 9 and 10, the other in Section 5 of Core 9, almost coincident with the boundary between the reversed and normal magnetozones.

Features of this record that are salient to our discussion are as follows:

1. Initial calcareous nannofossil stratigraphy (Monechi 1985; Monechi et al. 1985) indicated that the CP8/CP9 zonal boundary occurs between sections 5 and 6 of Core 9. Subzone CP9b was identified in Core 9–5.
2. The HO of *Morozovella velascoensis* is close to the Core 9/Core 10 break. As for other planktonic foraminiferal datums in the hole, there is disagreement between authors as to its exact location (table 3.2). The HO of *M. velascoensis* defines the P5/P6 zonal boundary and the LO of *Morozovella formosa formosa* defines the base of Subzone P6b of Berggren in Berggren et al. (1995). The LO of this latter taxon is also poorly determined (see table 3.2). Although differences in taxonomic concepts may partly explain the disagreement between authors, downhole contamination may have blurred the stratigraphic ranges of the taxa.
3. The highest occurrences (HO) of *Stensioina beccariiformis* and 14 other benthic foraminiferal taxa are located between samples 10–1, 30–32 cm and 10–1, 16–18 cm (between 83.15 and 82.97 mbsf; Pak and Miller 1992).
4. A 1‰ decrease in carbon isotopic values occurs between 83.15 and 82.97 mbsf, thus coincident with the decrease in diversity in the benthic foraminifera (Pak and Miller 1992).
5. A decrease in eolian grain size from 8.5 to 8.8 ϕ occurs between samples 10–1, 63–70 cm and 9–6, 53–60 cm (83.45 and 81.35 mbsf; Miller et al. 1987).

Biostratigraphic Revision

I have examined the calcareous nannofossil assemblages between 87.50 and 79.75 mbsf, using the samples that served for Pak and Miller's study (1992) plus additional samples (provided to me by Lu and Olsson) that provided the additional planktonic foraminiferal data in Berggren and Aubry (1996). I have not made counts but have searched the smear slides thoroughly

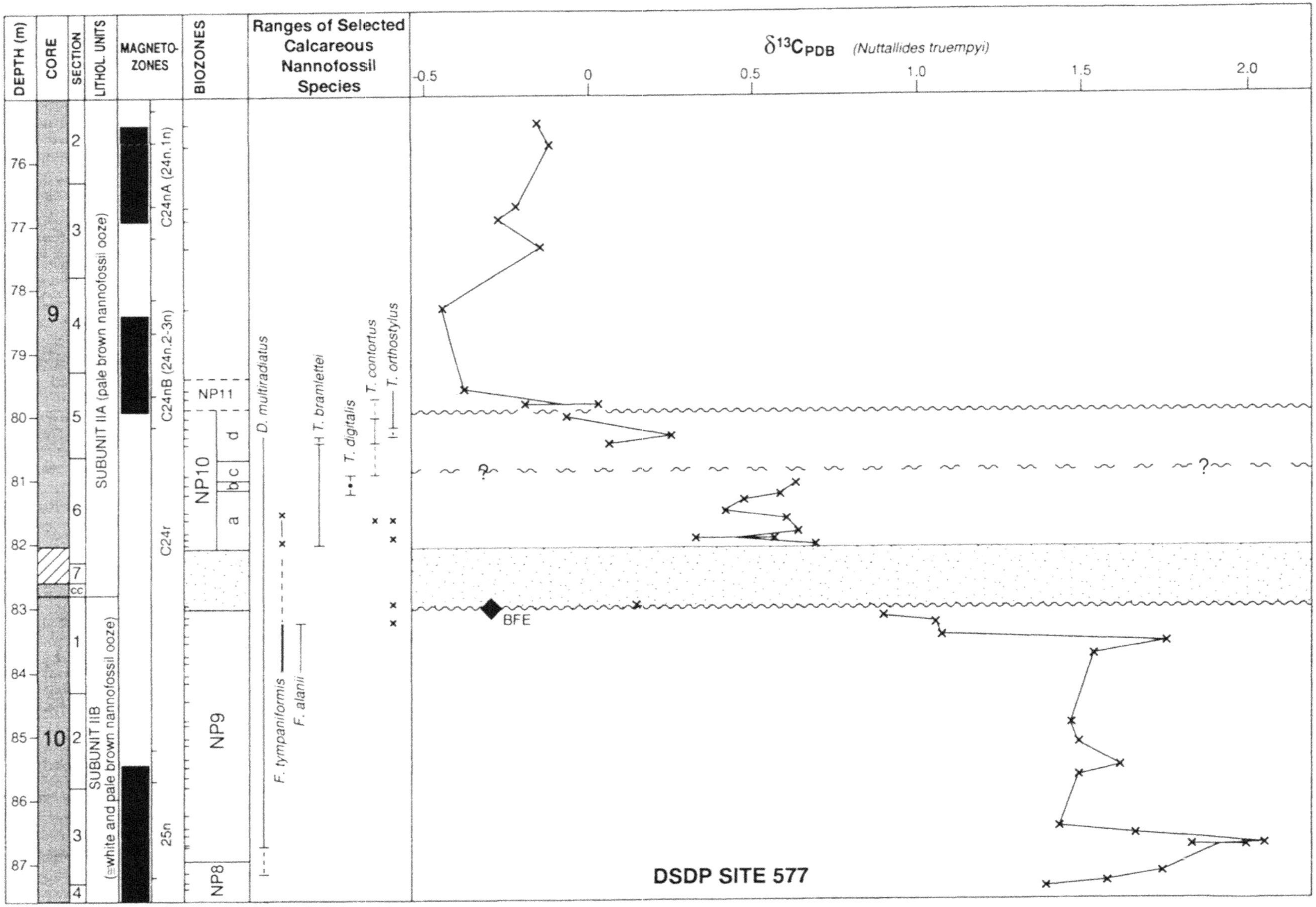

Figure 3.12 Stratigraphic interpretation of the upper Paleocene–lower Eocene interval recovered from DSDP Hole 577.

Lithology from Heath et al. (1985a); magnetostratigraphy from Bleil (1985); planktonic foraminiferal stratigraphy from Liu and Olsson (personal communication August 1995); isotope stratigraphy from Pak and Miller (1992).

Table 3.2
Main planktonic foraminiferal events in upper Core 10 and lower Core 9 in DSDP Hole 577 according to different authors

Datum	Miller et al. (1987)	Pak and Miller (1992)	Lu and Keller (1995)	Liu and Olsson (personal communication, 1995)
LO *M. aragonensis*	Between 78.33 and 76.93 mbsf (9-4, 53–60 and 9-3, 63.70 cm)	—	Given at 78.47 mbsf (9-4, 65–69 cm)	Between 79.60 and 79.17 mbsf (9-5, 26–31 and 9-4, 135–140 cm)
LO *M. lensiformis*	—	—	—	Between 79.60 and 79.17 mbsf (9-5, 26–31 and 9-4, 135–140 cm)
LO *M. formosa formosa*	Between 81.33 and 79.83 mbsf (9-6, 53–60 and 9-5, 53–60 cm)	—	Given at 83.13 mbsf (10-1, 33 cm)	Between 81.62 and 81.05 mbsf (9-6, 80-85 and 9-6, 24-29 cm)
HO *M. velascoensis*	Between 83.45 and 81.33 mbsf (10-1, 63–70 and 9-6, 53–60 cm)	Between 81.94 and 81.83 mbsf (9-6, 114–119 and 9-6, 103–105)	Given at 82.98 mbsf (10-1, 18 cm)	Between 83.15 and 81.95 mbsf (10-1, 30–35 and 9-6, 114–119 cm)
HO *G. pseudomenardii*	Between 86.43 and 84.83 mbsf (10-3, 63–70 and 10-2, 63–70 cm)	—	Given at 88.84 mbsf 10-4, 02–06 cm)	Between 88.71 and 88.66 mbsf (10-4, 139–142 and 10-4, 134–138 cm)

Precise position of samples analyzed not given in Lu and Keller (1995a). These authors indicate that the HO of *Globanomalina pseudomenardii* is difficult to delineate in the hole because of the rarity of the taxon in its upper range; the level given herein corresponds to the highest common occurrence, as delineated by Lu and Keller.

for representatives of the genus *Tribrachiatus*. In most slides, a 6.6 cm^2 area has been examined representing over 450 fields scanned at a magnification of x 950.

First, it must be emphasized that the well to moderately preserved assemblages of Core 10, Sections 1 to 3 contrast strongly with the poorly preserved assemblages in Core 9, Sections 5 and 6, where the large nannofossils are strongly overgrown and the smaller ones show strong dissolution artifacts. Second, downhole contamination (upper Miocene and Pliocene as well as lower Eocene elements) is pervasive. Backman (1986a) also remarked on significant amounts of downhole contamination in Hole 577, although in Core 11. Consequently, I have concentrated my efforts on determining the ranges of taxa that are critical for subdividing the NP9–NP10 zonal interval (*Fasciculithus* spp, *Tribrachiatus* spp.) and avoided the use of taxa that are useful for undisturbed sections but could have been displaced through downhole contamination (e.g., *Cruciplacolithus delus*). Contrary to Backman (1986b) who reported only one specimen of *T. bramlettei* in DSDP Site 577, and discouraged the use of this taxon as a zonal marker in general, I have found that this species has a continuous range and a consistent frequency throughout its range in Hole 577. Third, the upper 20 cm of Section 1 of Core 10 are strongly disturbed, as shown by the core photograph (Heath et al. 1985a:126). Thus, Sample 10–1, 16–18 cm contains a Zone NP11 calcareous nannofossil assemblage whereas Sample 10–1, 13–15 cm (82.95 mbsf) contains Pleistocene assemblages. Tooth pick samples taken at 10–1, 20 cm and 15 cm (83 and 82.95 mbsf, respectively) also contain Zone NP11 (?NP12) calcareous nannofossil assemblages. Disturbance is likely to explain the report by Lu and Keller (1995b) of the LO of *Morozovella formosa* at 83.13 mbsf, i.e., anomalously below (15 cm) the HO of *Morozovella velascoensis*. These two taxa are not known to overlap elsewhere.

The lowest occurrence of *D. multiradiatus* is between 87.15 and 86.70 mbsf, in agreement with Monechi (1985). The highest occurrence of representatives of the genus *Fasiculithus* is in Sample 10–1, 25 cm (83.05 mbsf) in agreement with Backman (in Monechi et al. 1985). Among others, *Fasciculithus alanii* is present in this sample. Only scarce representatives of *F. tympaniformis* occur in Core 9. *Tribrachiatus bramlettei* occurs from Sample 9–6, 123–126 cm (82.05 mbsf) to Sample 9–5, 116–119 cm (80.48 mbsf). *Tribrachiatus contortus* occurs from Sample 9–6, 116–119 cm to Sample 9–5, 72–75 cm (80.48 to 80.05 mbsf). Late morphotypes of the species occur at this latter level. Accordingly, the interval between 86.92 and 83.05 mbsf is assigned to Zone NP9 and that between 82.05 and 79.90 mbsf is assigned to Zone NP10 (figure 3.12). Because of the interval of non-recovery in Core 9 and the disturbed interval in upper Core 10, the NP9/NP10 zonal boundary cannot be located precisely between 83.05 and 82.05 mbsf. I place the NP10/NP11 zonal boundary between 80.05 and 79.75 mbsf. This is slightly lower than delineated by Backman (1986b) at ~79.45 mbsf. I suspect that this

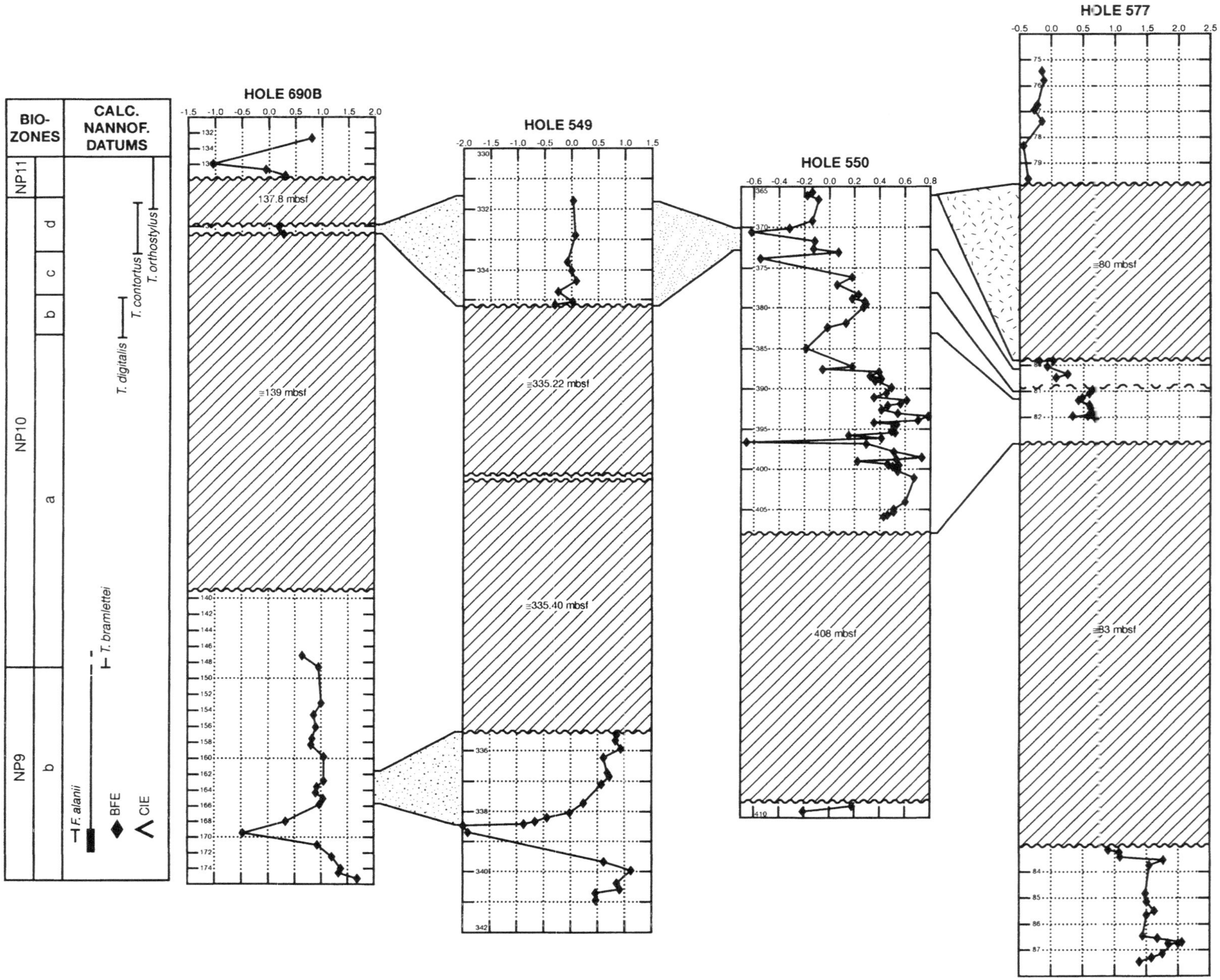

BIO-ZONES
CALC. NANNOF. DATUMS
NP11
NP10
NP9
d
c
b
a
b
T. orthostylus
T. contortus
T. digitalis
T. bramlettei
F. alanii
BFE
CIE
HOLE 690B
137.8 mbsf
≅139 mbsf
HOLE 549
≅335.22 mbsf
≅335.40 mbsf
HOLE 550
408 mbsf
HOLE 577
≅80 mbsf
≅83 mbsf

difference may result from slight shrinking of the core because there is a sharp contrast in composition between the assemblages at 80.05 and 79.75 mbsf and evidence for an unconformity at the NP10/NP11 zonal boundary (see below).

I have recently proposed subdivision of Zone NP10 into 4 subzones (Aubry 1996). All four are recognized in Hole 577 (figure 3.12). Subzones NP10b and d are total range subzones, defined by the range of *T. digitalis* and *T. contortus*, respectively. Subzones NP10a and c are interval zones, comprised respectively between the LO of *T. bramlettei* and the LO of *T. digitalis*, and between the HO of *T. digitalis* and the LO of *T. contortus*. The youngest Subzone NP10d, is well characterized in Hole 577 where the range of *T. contortus* extends between 80.48 and 80.05 mbsf. *Tribrachiatus digitalis* has long been included in *T. contortus* from which it differs in being essentially flat and in having bifurcated tips lying in the same plane. It differs from the somewhat comparable *T. orthostylus* in having long bifurcations and in being sturdier. Overgrown morphotypes of *T. digitalis*, with much shorter bifurcations but unquestionably distinct, occur in Sample 9–6, 24–29 cm (81.09 mbsf), thus assigned to Subzone NP10b. Sample 9–6, 13–16 cm (80.95 mbsf) is assigned to Subzone NP10c and the interval between Sample 9–6, 123–126 cm and 9–6, 43–45 cm (82.05 to 81.25 mbsf) to Subzone NP10a. It is important to note here that the same subzonal divisions are recognized in Hole 577 as in DSDP Hole 550 (Aubry et al. 1996; Aubry 1996).

Temporal Interpretation

The temporal interpretation of the interval between 87.50 and 79.75 mbsf in Hole 577 (figure 3.13) is extremely difficult to establish for various reasons. First, the recovery gap at the bottom of Core 9 and disturbance in the upper 25 cm of Core 10 prevent location of the NP9/NP10 zonal boundary (see above). Second, in addition to the recovery gap in Core 9 and disturbance in upper Core 10, sediment loss may have occurred at the core break. Sediment loss is commonly occurring during drilling of oozes (e.g., de Menocal et al. 1991) and multiple coring is required to construct complete composite stratigraphic sections (e.g., Raymo et al. 1989). In fact carbon isotope stratigraphy in Holes 577 and 577A has revealed substantial sediment loss at some core breaks in these holes (Shackleton et al. 1985) and permitted adjustment of depth between the two holes. However, no depth adjustment was made for the top of Core 10. Third, the sedimentation rates were extremely low, so that it is difficult to infer unconformities in the record. The interval for which sedimentation rates are best constrained is that assigned to Subzone NP10d (~80.72 to ~79.90 mbsf). There are several reasons to believe that the complete subbiochron is essentially represented because the same relationships between species of *Tribrachiatus* are seen as in Hole 550. *Tribrachiatus bramlettei* and *T. contortus* co-occur in the lowermost part of the subzone (at 80.48 mbsf). *Tribrachiatus orthostylus* and *T. contortus* co-occur in its upper part (between 80.19 and 80.05 mbsf). Late morphotypes of *T. contortus* occur at 80.05 mbsf. The average sedimentation rate is 0.25 cm/10^3 yr (table 3.3). Such a low sedimentation rate is not surprising if we consider the location of the hole, the low sedimentation rates at nearby sites (Heath et al. 1985b) as well as marked dissolution. Fourth, the inconsistent delineation of the planktonic foraminiferal datums by different authors (table 3.2) prevents their use in association with calcareous nannofossil datums to constrain the temporal interpretation of the section. For instance, the HO of *Morozovella velascoensis* would normally constitute a critical piece of evidence. The P5/P6a chronozonal boundary occurs in lower Chronozone NP10. Depending upon which report is followed (table 3.2), the HO of *M. velascoensis* in Hole 577 is located in Subzone NP10a (at ~82 mbsf; Pak and Miller 1992) or is correlative with the top of Zone NP9 (at 83.15 mbsf; Liu and Olsson, personal communication). This 1.15 m uncertainty in the position of the datum results in the following conundrum: If the HO of *M. velascoensis* is at 83.15 mbsf, it is likely that Zones NP9 and P5 are truncated at ~83 m, and the sediments not recovered in Section 6 of Core 9 belonged to Zone NP10. This would imply an unconformity between Cores 9 and 10 as marked by the

Figure 3.13 *(previous page)* Hybrid stratigraphic/temporal interpretation of the stratigraphic intervals representing Chron C24r in Holes 690B, 549, 550, and 577. A relative chronology is used for direct comparison with the interpretation of the records in Hole 690B, 550, and 549 in Aubry (1996:figure 9). The Chron C24r record at Hole 577 is correlated with those in Hole 690B, 549 and 550 previously described in Aubry et al. (1996). The record of $\delta^{13}C$ variations in *Nuttallides truempyi* in Holes 549 and 550 is from Stott et al. (1996), in Hole 690B from Kennett and Stott (1990) and in Hole 577 from Pak and Miller (1992). The interval between 82.05 and 79.75 mbsf in Hole 577 is represented as complete because it is not possible to confidently delineate unconformities. It is likely that short hiatuses are present. Yet, this interval essentially correlates with the stratigraphic interval recovered between 408 and 365.50 mbsf in Hole 550 although the section is twenty times thicker in the latter hole. This hybrid stratigraphic/ temporal interpretation permits reliable comparison of isotopic variations in different sections, and sound paleoceanographic interpretations. Because the sedimentation rates are considerably lower in Hole 577 than in the other holes, the hiatuses at ~83 m and ~80 mbsf are considerably enlarged. See text for further explanation.

Table 3.3

Comparative thicknesses of Biozone NP9 and Subzones NP10a to NP10d in DSDP Holes 577 and 550

Interval Site 577	Datums (mbsf)	Thickness (m)	Estimated Ages of Datums	Sedimentation Rates (cm/10^3 yr)	Thickness in Hole 550
NP10d	HO *T. contortus:* 80.05–79.75	Max: 1.20	LAD *T. contortus:* 53.61 Ma	Max: 0.37	7.25 m
	LO *T. contortus:* 80.95–80.48	Min: 0.43	FAD *T. contortus:* 53.93 Ma	Min: 0.13	
NP10c	LO *T. contortus:* 80.95–80.48	Max: 0.61	FAD *T. contortus:* 53.93 Ma	Max: 0.25	5.37 m
	HO *T. digitalis:* 81.09–80.95	Min: —	LAD *T. digitalis:* 54.17 Ma	Min: —	
NP10b	HO *T. digitalis:* 81.09–80.95	Max: 0.30	LAD *T. digitalis:* 54.17 Ma	Max: 0.15	5.05 m
	LO *T. digitalis:* 81.25–81.09	Min: —	FAD *T. digitalis:* 54.37 Ma	Min: —	
NP10a	LO *T. digitalis:* 81.25–81.09	Max: 2.06	FAD *T. digitalis:* 54.37 Ma	Max: 0.32	>24.72 m
	LO *T. bramlettei:* 82.05–83.15	Min: 0.80	FAD *T. bramlettei:* 55 Ma	Min: 0.12	
NP9	HO *S. beccariiformis:* 83.15–82.97	Max: 2.73	LAD *S. beccariiformis:* 55.5 Ma	Max: >0.68*	~14 m
	C24r/C25n reversal: 85.70–84.20	Min: 1.05	C24r/C25n reversal: 55.904 Ma	Min: 0.26	

Average thickness is given for DSDP Site 550, whereas maximum and minimum thickness are given for Site 577. Sedimentation rate decreased by half for Subzones NP10a to NP10d compared to Zone NP9.

*A maximum rate of sedimentation cannot be determined because of the unconformity at about 83 mbsf.

Thickness data in Hole 550 are from Aubry et al. 1996 and Aubry 1996). Estimated ages of datums are from Berggren et al. (1995).

lithologic contact. On the other hand, if the HO of *M. velascoensis* is at ~82 mbsf, the possibility exists that the section was complete (but poorly recovered and disturbed) between Cores 10 and 9. The problem is exacerbated by the low sedimentation rates at this site.

An anomalous contact between Cores 9 and 10 is indicated by the close juxtaposition of the HOs of *Fasciculithus tympaniformis* (abundant) and *F. alanii* at 83.05 mbsf and the LO of *T. bramlettei* at 82.05 mbsf as well as by a sharp change in calcareous nannofossil assemblage composition between these two levels. A decrease in abundance of *F. tympaniformis* (from common to rare) occurred in the late part of Biochron NP9 (after the BFE and the CIE) as inferred from the record in DSDP Site 549, ODP Site 690 and in the Aweina section (Egypt; Aubry et al. 1996; Aubry, unpublished data). Also, the LAD of *F. alanii* occurred in mid-Biochron NP10, close to the BFE (Aubry, unpublished data). This is also indicated by the occurrence of the 1‰ decrease in $\delta^{13}C$ that Pak and Miller (1992) interpreted as the CIE between 83.15 and 82.97 mbsf, and by the HO of *S. beccariiformis* also at 83.15 mbsf (but see below on the reliability of these events). This anomalous contact may be the result of an unconformity, or simply, of the combined effect of sediment loss at the break between Cores 9 and 10, the recovery gap in Core 9, and the disturbed interval in upper Core 10.

Although the anomalous contact at the Core 9/10 break may be a drilling artifact, I believe that an unconformity can be delineated between 83.05 mbsf and 82.03 mbsf (between Samples 10–1, 25 cm and 9–5, 123–126 cm). Previous work has shown that unconformities are often associated with lithologic boundaries (Aubry 1995), so that the lithologic contact at 82.8 mbsf, between Subunits II A and IIB can be seen as evidence of an unconformable (unrecovered) NP9/NP10 zonal contact. Thus I suggest that there is an actual stratigraphic gap in the section. However, sediment loss–recovery gap–sedimentary disturbance would result as well in an artificial stratigraphic gap (de Menocal et al. 1991), and both actual and virtual stratigraphic gaps would have the same consequences on the interpretation of the (recovered) upper Paleocene–lower Eocene section from Site 577. For reasons delineated above it is difficult if not impossible to determine which is the cause of the anomalous contact. Thus in the discussion below, I refer to an unconformity at the NP9/NP10 zonal boundary in Hole 577, aware that recoring at the site would be needed to establish firmly such unconformity.

Because of the ~25-cm-thick disturbed interval at the top of Core 10, I locate the unconformity at ~83 mbsf (figure 3.12). It is extremely difficult to estimate the duration of the associated hiatus. It is greater than 0.5 my because the later part of Biochron NP9 is not represented, but it is impossible to determine how much of Biochron NP10 is unrepresented.

An unconformity is likely present at the NP10/NP11 zonal boundary. This is indicated by 1) the close association between the zonal boundary and the normal polarity interval representing Chron C24n.3n (figure 3.12; see Berggren et al. 1995); 2) the rather abrupt change of morphotype morphologies in *T. orthostylus* between 79.75 and 80.05 mbsf, and 3) the sharp change in composition of the calcareous nannofossil assemblages between these levels.

Considering the extremely low sedimentation rates estimated for different subzonal intervals between 80 and 82 mbsf (table 3.3; the average rate for all of Zone NP10 is 0.14 cm/10^3 yr), it is likely that this record is truncated by a series of unconformities representing short hiatuses. Exceedingly low sedimentation rates (maximum 0.15 cm/10^3 yr; see table 3.3) for Subzone NP10b probably indicate that it is bounded by unconformities. As chemostratigraphy develops, it may become possible to use it to delineate thin stratigraphic gaps. Until then, the section is interpreted as being essentially complete and representing all of Biochron NP10 except its earliest part.

Implications for Paleoceanographic Reconstructions

Pak and Miller (1992) interpreted the 1‰ shift in the $\delta^{13}C$ record that occurs between 83.15 and 82.96 mbsf as correlative with the 3 to 4‰ transient excursion that occurs in mid Zone NP9 in Hole 690B on Maud Rise (Stott et al. 1990; Kennett and Stott 1990, 1991; Aubry et al. 1996; Stott et al. 1996). The record of the benthic foraminiferal extinction at the same level as the shift apparently reinforces the correlation. This correlation is, however, questionable. First, it is important to emphasize that this 1‰ shift does not occur in the uppermost part of Zone NP9 as shown by Wei (1994:figure 1) but between an *in situ* level (83.15 mbsf) in Zone NP9 and a level (82.97 mbsf) in a disturbed/displaced chunk of much younger sediments. The record of any event in Core 10 above ~83 m in Hole 577 is suspicious due to mixing in the upper ~25–30 cm of the Core (see above). Second, and more importantly, the presence of an unconformity above 83.15 mbsf implies that both the benthic and the isotopic record are truncated. Thus, the benthic extinction event and the $\delta^{13}C$ shift in DSDP Hole 577 are likely pseudo-events, not true events. The situation is equivalent to that described in figures 3.2 (for the benthic extinction) and 3.4 (for the isotopic excursion/shift, except that in figure 3.4, truncation of the isotopic record results in an excursion of greater amplitude than the true event). It is important to note that regardless of the suspected unconformity at ~83 mbsf, the recovery gap at the base of Core 9 and the disturbed interval in the upper part of Core 10 constitute an equivalent of a stratigraphic gap. This implies that 1) level 83.15 mbsf in Hole 577 is older than 55.5 Ma (estimate age of the CIE and BFE; Aubry et al. 1996; Berggren and Aubry 1996; note that Pak and Miller 1992 used the IMBS of Berggren et al. 1985b and estimated the CIE and BFE at 58 Ma), 2) that the shift in Hole 577 cannot be used as a chronologic tie-point in connection with the Chron C24n/C24r and C25n/r to estimate the ages of intermediate levels between the two magnetic reversals as done in Pak and Miller (1992), and 3) that the amplitude of the shift has no paleoceanographic significance. It also implies that the chronologic resolution between samples is different below and above the unconformity. It is ~25 kyr below the unconformity but >100 kyr above (compared to an assumed resolution of 13 kyr near the Paleocene/Eocene Series boundary in Pak and Miller 1992).

Pak and Miller (1992) further compared the carbon isotope records above the shift in Hole 577 with that above the excursion in Hole 690B. They remarked that the latest Paleocene to early Eocene carbon isotope values of the two oceans differed by 1.5‰, the Pacific values being lighter, determined that the southern Ocean waters were younger and more nutrient-depleted than the Pacific Ocean waters, and concluded that the Southern Ocean was a source of deep water during the early Eocene Epoch. They reached this conclusion by converting their chemostratigraphic record for Site 577 and that of Kennett and Stott (1990) for Site 690 into temporal records, using the magnetic reversals identified at both sites and the carbon isotope excursion/shift. Their tables 2a and c and their figure 10 show that, in particular, they compared the isotope stratigraphic records between 79.95 and 85.45 mbsf in Hole 577 and between 154.62 and 185.47 mbsf in Hole 690B. Aubry et al. (1996) have shown that the magnetic reversal pattern in the interval between 144.42 and 154.62 mbsf in Hole 690B is uninterpretable, and that the interval between 137.8 and 185.47 belongs to Zones NP9 and NP10, in agreement with Pospichal and Wise (1990). They delineated the NP9/NP10 zonal boundary at ~148.90 mbsf. They further showed that only the lower part of Zone NP10 is represented (see also Stott et al. 1996). I have shown above that in Hole 577 only the lower part of Zone NP9 is represented whereas Zone NP10 is essentially completely represented except for its lower part, although I concede that it is not possible to determine how much of the zone is missing. Thus by comparing carbon isotope records between 79.95 and 83.15 mbsf (level of the shift) in Hole 577 and between 154.82 and 170.60 mbsf (level of the excursion) in Hole 690B, Pak and Miller compared late Biochron NP9 values of the Southern Ocean with Biochron NP10 values of the Pacific Ocean (compare figures 3.14 and 3.15). Needless to say their conclusions are unsubstantiated. In fact, there is currently no section that provides a complete southern latitude record of Biochron NP10 (Site 690 provides a very incomplete representation of it), and there is no section that provides a Pacific Ocean record of late Biochron NP9 (Aubry, unpublished data).

On the other hand, the Biochron NP10 isotopic trends and values of the Pacific Ocean can be confidently compared with the Biochron NP10 isotopic trends and values of the NE Atlantic Ocean (figures 3.13, 3.14). As pointed out above, both Sites 550 and 577 provide a good stratigraphic record of Biochron NP10, although it is considerably better represented in DSDP Site 550 than in DSDP Site 577 because of the differences in sedimentation rates at the two sites. Also, the record at Site 550 is unquestionably complete whereas that at Site 577 likely contains small gaps. Lack of connecting lines between isotopic values for specific intervals in figures 3.10 and 3.12 indicates possible unconformities. The same general trend in the isotopic record of *Nuttallides truempyi* is seen in both sections, with an interval of values in Subzone NP10a ranging between 0.4 and 0.7‰,

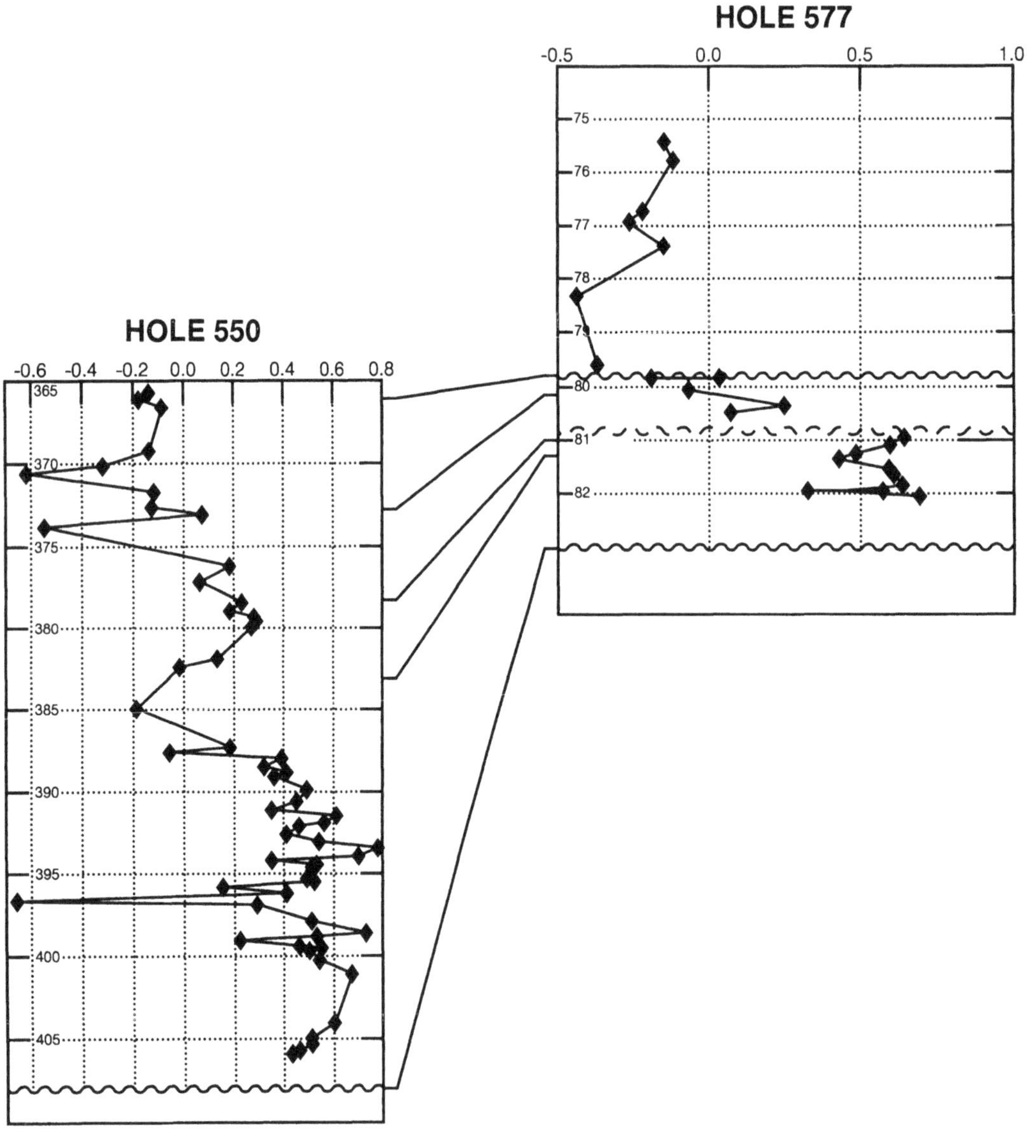

Figure 3.14 Detail of the chemostratigraphic correlations in the NP10 zonal interval in DSDP Holes 550 and 577.

Enlargement of part of figure 3.13.

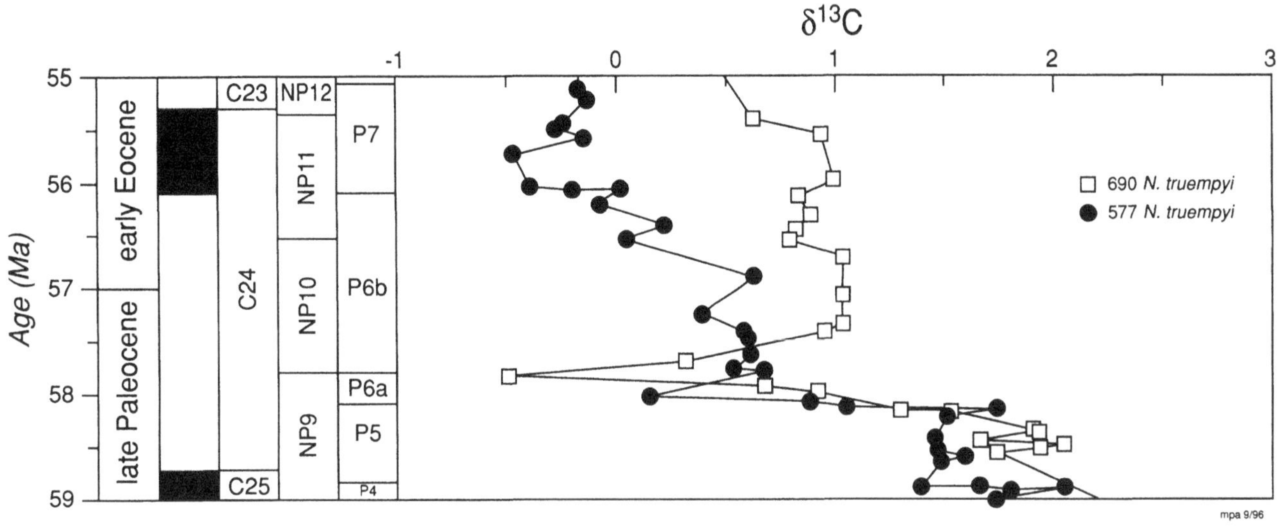

Figure 3.15 Comparison between late Paleocene–early Eocene $\delta^{13}C$ variations based on the upper Paleocene–lower Eocene stratigraphic sections recovered from ODP Hole 690B and DSDP Site 577. For this comparison, Pak and Miller (1992) converted the stratigraphic record into a temporal one using the magnetic reversals Chrons C24n/r and C24r/C25n and the isotope excursion/shift as temporal calibration points.

Compare with figure 3.11. Redrawn from Pak and Miller (1992). See text for further explanation.

succeeded by a shift towards more negative values in Subzone NP10d. Data for Site 577 are from Pak and Miller (1992) and for Site 550 from Stott et al. (1996). The isotopic values of 0.7‰ to 0. 4‰ on *Nuttallides truempyi* between 82.05 and 81.25 mbsf (Subzone NP10a) in Hole 577 compare well with those between 390 and 395 mbsf in the same Subzone in Hole 550. The isotopic values (0.07‰ and –0.07‰) at, respectively, 80.48 and 80.05 mbsf in Subzone NP10d in Hole 577 display similar range as the averages values of 0.12 to –0.13‰ obtained for the same subzone in Hole 550. Sample resolution in Hole 577 is insufficient to push further the comparison with Hole 550. This comparison between two records that are demonstrably time-equivalent shows that the NE Atlantic did not record values intermediate between the Southern Ocean and the Pacific as determined by Pak and Miller (1992) but similar values as in the Pacific. Site 550, with an early Eocene paleodepth of 3500 to 4000 m, and Site 577, with an early Eocene paleodepth of 1800–2100 m (Miller et al. 1987), were bathed by water masses of similar isotopic composition. In addition, until benthic foraminiferal isotopic records from a time-equivalent section from the Southern Ocean becomes available, it cannot be determined whether the Southern Ocean was a source of deep water.

Implications for Changes in Diversity Patterns Through the Paleocene/Eocene Epoch Boundary

Lu and Keller (1995a, 1995b) have used the sedimentary record of Site 577 to study early Paleogene planktonic foraminiferal evolution. They delineated periods of stasis and turnovers ("saltation" events), among which a sharp turnover event (their event T2) near the Paleocene/Eocene Epoch boundary "that nearly doubled species richness" (Lu and Keller 1995a:211). They (1995b) indicated that event T2 occurs at 83.03 mbsf, 5 cm below the HO of *Morozovella velascoensis* (at 82.98 mbsf) that delineates the P5/P6 zonal boundary. They further remarked that this turnover is coincident with the BFE and the CIE as delineated by Pak and Miller (1992) and associated with the onset of global warming and changes in circulation patterns.

I have indicated above that samples at 83.05 and 83.15 mbsf contain upper Paleocene Zone NP9 calcareous nannofossil assemblages whereas samples at 82.97 and 82.95 mbsf contain calcareous nannofossil assemblages indicative of younger (lower Eocene Zone NP11 [and NP12?] and Pleistocene) levels than those recovered in Section 6 of Core 9. Thus the strong disturbance of the upper 25 cm of sediment in Core 10 precludes any conclusive results from the analysis of samples in this thin interval. Turnover T2 of Lu and Keller (1995a, 1995b) falls exactly between the levels that are undisturbed (at 83.05 mbsf and below) and disturbed (82.97 mbsf and above; = the upper 25 cm of section 1 of Core 10). I have also indicated that downhole contamination is pervasive, even below Core 10–1, 30 cm (83.10 mbsf) and that disagreements between authors as to the ranges of planktonic taxa (see table 3.2) are likely to reflect, at least in part, downhole contamination. Finally, it is likely that an unconformity truncates Biozone NP9 at ~83 mbsf. Thus the remarkable turnover that Lu and Keller describe is unlikely to correspond to a true event, but is an artifact of core disturbance, downhole contamination and unrecognized sedimentary gap. The coincidence between the planktonic foraminiferal turnover, the BFE and the CIE in this hole is only superficial; just like the BFE and CIE in Hole 577, the planktonic foraminiferal turnover that Lu and Keller describe is not a true event but a pseudoevent.

Lu and Keller's study (1995a, 1995b) provides a good example that high resolution sampling (a sample every 5 cm in upper Core 10) does not necessarily equate with high chronologic resolution. Close sampling cannot correct for/eliminate the unconformity, the disturbed interval and the interval of no recovery. Lu and Keller converted their stratigraphic record of planktonic foraminiferal abundances and diversity into a temporal record using the magnetostratigraphic record of Bleil (1985) and the ages of the reversal boundaries of Chron C24BN/C24R and C24R/C25N in Berggren et al. (1985b). Chron C24r has an estimate duration of 2.5 my in Berggren et al. (1985b). If the 5.5 m-thick section in Hole 550 represents all of Chron C24r it was deposited at an average sedimentation rate of 0.22 cm/10^3 yr. Regardless of the unconformity at ~83 mbsf, there is a 20 to 25 cm disturbed interval that is not represented by sediment deposited during Chron C24r. In addition, there is at least a 75 cm recovery gap at the bottom of Core 9 (with the exception of ~12 cm of core catcher sediment). This means that at least up to 1 m of sediment that should have represented Chron C24r is missing in the hole. The disturbed interval represents ~91,000 to 114,000 yr; the recovery gap represents 285,000 to 340,000 yr. Thus, in addition to the hiatus associated with the unconformity at ~83 mbsf, as much as 450,000 yr may not be represented due to coring disturbances. Yet, this unrepresented interval is the critical one because it is in the vicinity of the late Paleocene BFE and CIE. It also implies that the temporal interval required to describe an eventual turnover in the planktonic foraminifera coincident with the BFE and CIE is unrepresented in Hole 577. Obviously, patterns of planktonic foraminiferal evolution must be established from complete (and continuously recovered) sections, criteria that Site 577 does not fulfill.

Implications for Correlations

Miller at al. (1987) used the record of Site 577 to constrain the timing of the decrease in eolian grain size that had been previously described from Core Ll44-GCP-3 and Site 576 (Janecek and Rea 1983; Janecek 1985). They reported that the decrease in grain size occurred between 10–1, 63–70 cm (83.45 mbsf) and 9–6, 53–60 cm (81.35 mbsf). Because the HO of *Morozovella velascoensis* is located between these two levels they concluded that the change in grain size occurred across the Paleocene/Eocene Epoch boundary (as determined by the LAD of *M. velascoensis*) rather than near the end of the Paleocene Epoch as indicated by ichthyolith stratigraphy (Doyle 1979; Doyle and Riedel 1985). It is clear from the discussion above that the record of change in grain size is far from being constrained in DSDP Hole 577. Level 83.45 mbsf lies in the lower part of Zone NP9 and has an age >55.5 Ma whereas Level 81.35 mbsf lies in the upper part of Subzone NP10a and has an age estimate of ~54.5 Ma.

It should be pointed out here that the record of *T. digitalis* at 81.09 mbsf in Hole 577 implies that this level is close in age to the base of the London Clay Formation as redefined by Ellison et al. (1994), which is correlative with the base of Ieper Clay (lithostratigraphic horizon that defines the base of the Ypresian Stage; see chapter 2). The base of the London Clay can be approximated by the LO of *T. digitalis* (Aubry 1996; see chapter 2). In Hole 577, the record of this species is restricted to a single level, and it is therefore not possible to determine whether it corresponds to the LO or HO of the species or to neither. However, *T. digitalis* had a very short life span (~200 kyr; Aubry et al. 1996). Considering the low sedimentation rates in the hole, it is safe to use it to approximate a level in the hole that correlates with the Ypresian transgression.

As the two examples above show, the accuracy of geological interpretations in reconstructing earth history depends on the correctness of the temporal interpretation of stratigraphic sections. Graphic correlation techniques (Shaw 1964; see also Mann and Lane 1995) and thickness/time plots (see Aubry 1995) are designed to evaluate the completeness of stratigraphic sections and delineate unconformities. Both point to the incompleteness of the stratigraphic record and to an allostratigraphic architecture specific to given basins (compare Carney and Pierce 1995, and Aubry 1993a), and find their optimal expression in the construction of composite reference sections (compare Aurisano et al. 1995 and Aubry et al. 1996).

Composite sections have long been constructed (cf. Wheeler 1958), but they are little used in the study of deep sea sediments older than Pliocene, presumably because of the assumed completeness of the deep sea record (see discussion in Aubry 1995). There are exceptions. For instance, when a paleontologic group has a scanty record, composite stratigraphic sections help in determining the full ranges of taxa (see chapter 13). The need for composite reference sections will increase as finer tools are used to subdivide the stratigraphic record, but it will increase unevenly, the need being greater for those intervals of time that are least well represented.

Because of multiple unconformities in the upper Paleocene–lower Eocene stratigraphic interval, the record of late Paleocene–early Eocene history must rely upon the construction of a composite reference section to which all other sections are compared. This virtual reference section yields the relative chronology of events that occurred during Magnetic Chron C24r (see chapter 2). The relative chronology is in turn transformed into a numerical chronology using the GPTS. We have seen that because of an unrecognized unconformity at 408 mbsf in Hole 550, the ages of the magnetic reversals C24n/r and C24r/C25n are inaccurate and that, more importantly, there is a stratigraphic interval that the current GPTS does permit to discriminate because the NP9/NP10 zonal boundary at 408 mbsf does not correspond to the NP9/NP10 biochronal (= temporal) boundary while both have an age estimate of 55 Ma. Thus, as explained, the current numerical chronology for the late Paleocene–early Eocene events is inaccurate while retaining a relatively high degree of precision. Yet, true temporal correlations are possible if the age of 55 Ma is arbitrarily assigned to the NP9/NP10 biochronal boundary while a new age estimate (54.7 Ma) is determined for the upper surface of the unconformity at 408 mbsf in Hole 550. I have shown that DSDP Site 577, one of the two sites used to infer late Paleocene–early Eocene paleoceanographic and biogeographic evolution of the Pacific Ocean, does not provide the continuous temporal record necessary to determine such histories. However, once its temporal significance is correctly established, the lower Eocene part of the section constitutes a reference for the Pacific Ocean to which equivalent lower Eocene sections (such as DSDP Site 550) can be compared. Through this example we have seen that paleoceanographic reconstructions are extremely sensitive to the temporal interpretations of sections.

The virtue of the (virtual) composite reference section is that it provides the means of temporal correlations through stratigraphic analysis, not only of marine sections but also of sections in the continental record. What stratigraphy attempts to achieve are temporal correlations, so that the contingent succession of events and their results in the chain-reaction of planetary changes can be determined. As we recognize that the stratigraphic record is incomplete, we must think of stratigraphic sections not as proxies for time, but as windows on specific intervals of time, and realize that stratigraphy is to time what fossils are to species. For this reason it is advisable to clearly differentiate between stratigraphic events and units (e.g., LO, HO, magnetozones, biozones, biochronozones, etc.) and temporal events and units (e.g., FAD, LAD, magnetochrons, biochrons, etc.) and use accordingly an appropriate terminology. As I have shown, not all stratigraphic events are true events; we have the means to distinguish between pseudo-events and true events, and only at the cost of some efforts can we progress in our efforts to understand the intricacies of Earth history. The critical role of biostratigraphy in this effort needs to be emphasized again in view of the common disbelief in biostratigraphic correlations. I have shown how calcareous nannofossil stratigraphy permits precise correlation between the upper Paleocene–lower Eocene sections recovered from DSDP Sites 550 and 577 (and with those from Sites 549, 690 (figure 3.13) and many other sites and outcrop sections (Aubry, unpublished data).

The benefit of the composite reference section is that it is predictive, and functions in two ways. First, it predicts events that have not yet been recorded in the stratigraphic record. For example, an additional carbon isotope excursion is believed/expected to have occurred close to the NP9/NP10 biochronal boundary (Aubry et al. 1996; Stott et al. 1996). Also, composite reference sections constitute the support for predicting and modeling temporal changes. For instance, late Paleocene–early Eocene changes in the isotopic composition of carbon in organic terrestrial matter have been derived through the record of changes of isotopic composition of dissolved inorganic carbon in the ocean (Stott et al. 1996; Sinha et al. 1996). Second, it helps in determining which temporal windows need to be documented and in which areas. As I have shown above, we now have a satisfactory documentation of the latest Paleocene–earliest Eocene (Biochron NP10) history of changes in isotopic composition of dissolved inorganic carbon in the eastern North Atlantic and Pacific Oceans, but almost no documentation of such history in the Southern Ocean. Conversely, we have a good documentation of the late Paleocene (late Biochron NP9) history of such changes in the Southern Ocean, but not in the Pacific and eastern North Atlantic Oceans. Planning for future deep sea drilling and for the sampling of outcrop sections should take into account the temporal coverage of sections currently available.

Finally, the composite reference section is critical for determining the (temporal) event best suited for characterizing epoch/series boundaries. Ideally this event can be identified in all settings (deep oceanic, shallow marine and continental) or else (demonstrably) contemporaneous events specific to each setting must ensure characterization of the same point in time in different settings. We have reached a point in history when we look for a specific interval in the stratigraphic record that records in a complete fashion the succession of events that have characterized a particular interval of time, and for a specific level that records a selected event that occurred at a specific point in time and chosen to characterize an epoch boundary. The Global Boundary Stratotype is this specific interval of the stratigraphic record that must best represent this selected interval of time, and the series boundary is the level that yields the event chosen to characterize the epoch boundary. The series boundary has thus become a subordinate of the epoch boundary. To the benefit of historical geology, it is becoming possible to avail ourselves of the classical approach in which a lithostratigraphic horizon determines a point in time. The advent of the GPTS and the IMBS and the simultaneous refinement in the construction of fine-scale composite reference sections will allow us to select a point in time that, in turn, determines a lithostratigraphic horizon. It now remains to choose the event that best characterizes the Paleocene/Eocene Epoch boundary, and subsequently to identify the lithologic horizon that will define the Paleocene/Eocene Series boundary in a suitable section. Together they will constitute the GSSP.

ACKNOWLEDGMENTS

I am grateful to Dorothy Pak and Chengie Liu for sharing their samples with me, to D. Pak for discussion and to C. Liu for providing unpublished data on the range of planktonic foraminiferal species in Hole 577, and to Dennis Kent, Robert Knox, Richard Norris and Ashish Sinha for thoughtful reviews. I am particularly indebted to Dennis Kent and William A. Berggren for discussion on the problems related to the current GPTS. This is ISEM Contribution No 96–143.

REFERENCES

Aubry, M-P., 1991. Sequence stratigraphy: Eustasy or tectonic imprint? *Journal of Geophysical Research,* v. 96(B4), p. 6641–6679.

Aubry, M.-P., 1993a. Neogene allostratigraphy and depositional history of the De Soto Canyon area, northern Gulf of Mexico. *Micropaleontology,* v. 39, p. 327–366.

Aubry, M.-P., 1993b. Calcareous nannofossil stratigraphy of the Neogene formations of eastern Jamaica. In Wright, R. M. and Robinson, E. (eds.), *Biostratigraphy of Jamaica,* Geological Society of America Memoir 182, p. 131–178.

Aubry, M.-P., 1995. From chronology to stratigraphy: Interpreting the lower and middle Eocene stratigraphic record in the Atlantic Ocean. In Berggren, W. A., Kent, D. V., Aubry, M.-P. and Hardenbol, J. (eds.), *Geochronology Time Scales and Global Stratigraphic Correlation,* Society of Economic Paleontologists and Mineralogists Special Publication No 54, p. 213–274.

Aubry, M.-P., 1996. Towards an upper Paleocene–lower Eocene high resolution stratigraphy. In Aubry, M.-P. and Benjamini, C. (eds.), Paleocene/Eocene Boundary Events in Space and Time, *Israel Journal of Earth Sciences,* v. 44, p. 239–253.

Aubry, M-P., Berggren, W. A., Kent, D. V., Flynn, J. J., Klitgord, K. D., Obradovich, J. D. and Prothero, D. W., 1988. Paleogene Geochronology: An integrated approach. *Paleoceanography,* v. 3, p. 707–742.

Aubry, M.-P., Berggren, W. A., Stott, L. and Sinha, A., 1996. The upper Paleocene–lower Eocene stratigraphic record and the Paleocene–Eocene boundary carbon isotope excursion: Implications for geochronology. In Knox, R. W. O'B., Corfield, R. M. and Dunay, R. E. (eds.), *Correlation of the early Paleogene in Northwest Europe,* Geological Society Special Publication No 101, p. 353–380.

Aurisano, R. W., Gamber, J. H., Lane, H. R., Loomis, E. C., ans Stein, J. A., 1995. Worldwide and local composite standards: Optimizing biostratigraphic data. In Mann, K. O. and Lane, H. R. (eds.), *Graphic Correlation,* Society of Economic Paleontologists and Mineralogists Special Publication No 53, p. 117–130.

Backman, 1986a. Late Paleocene to middle Eocene calcareous nannofossil biochronology from the Shatsky Rise, Walvis Ridge and Italy. *Palaeogeography, Palaeoclimatology, Palaeoecology,* v. 57, p. 43–59.

Backman, J., 1986b. Accumulation patterns of tertiary calcareous nannofossils around extinctions. *Geologische Rundschau,* v. 75, p. 185–196.

Berggren, W. A. and Aubry, M.-P., 1996. A late Paleocene–early Eocene NW European and North Sea magnetobiochronological correlation network. In Knox, R. W. O'B., Corfield, R. M. and Dunay, R. E. (eds.), *Correlation of the Early Paleogene in Northwest Europe,* Geological Society Special Publication No. 101, p. 309–352.

Berggren, W. A., Kent, D. V., Flynn, J .J. and Van Couvering, J. A., 1985a. Cenozoic geochronology. *Geological Society of America Bulletin,* v. 96, p. 1407–1418.

Berggren, W. A., Kent, D. V. and Flynn, J. J., 1985b. Paleogene geochronology and chronostratigraphy. In Snelling, N. J. (ed.), *The chronology of the Geological Record:* Geological Society Memoir No 10, Blackwell, Oxford, p. 141–195.

Berggren, W. A., Kent, D. V., Obradovich, J. D. and Swisher, C. C., III, 1992. Towards a revised Paleogene geochronology. In Prothero, D. R. and Berggren, W. A. (eds.), *Eocene–Oligocene Climatic and Biotic Evolution,* Princeton University Press, Princeton, N.J., p. 29–45.

Berggren W. A., Kent, D. V., Swisher, C. C., III and Aubry, M.-P., 1995. A revised Cenozoic geochronology and chronostratigraphy. In Berggren, W. A., Kent, D. V., Aubry, M.-P. and Hardenbol, J. (eds.), *Geochronology Time Scales and Global Stratigraphic Correlation,* Society of Economic Paleontologists and Mineralogists Special Publication No. 54, p. 129–212.

Bleil, U., 1985, The magnetostratigraphy of northwest Pacific sediments, Deep Sea Drilling Project Leg 86, in Heath, G. R., Burckle, L. H., et al., Initial Reports of the Deep Sea Drilling Project, v. 86: Washington (U.S. Government Printing Office), p. 441–458.

Bralower, T. J., Zachos, J. C., Thomas, E., Parrow, M., Paull, C. K., Kelley, D. C., Premoli Silva, I., Sliter, W. V. and Lohmann, K. C., 1995. Late Paleocene to Eocene paleoceanography of the equatorial Pacific Ocean: Stable isotopes recorded at Ocean Drilling Program Site 865, Allison Guyot. *Paleoceanography,* v. 10, p. 841–865.

Cande, S. C. and Kent, D. V., 1992. A new geomagnetic polarity time scale for the Late Cretaceous and Cenozoic. *Journal of Geophysical Research,* v. 97(B10), p. 13917–13951.

Cande, S. C. and Kent, D. V., 1995. Revised calibration of the geomagnetic polarity time scale for the Late Cretaceous and Cenozoic. *Journal of Geophysical Research,* v. 100(B4), p. 6093–6095.

Carney, J. L. and Pierce, R. W., 1995. Graphic correlation and composite standard databases as tool for the exploration biostratigrapher. In Mann, K. O. and Lane, H. R. (eds.), *Graphic Correlation,* Society of Economic Paleontologists and Mineralogists Special Publication No 53, p. 23–43.

Charisi, S. D. and Schmitz, B., 1996. Early Eocene palaeoceanography and palaeoclimatology of the eastern North Atlantic: Stable isotope results for DSDP Hole 550. In Knox, R. W. O'B., Corfield, R. M. and Dunay, R. E. (eds.), *Correlation of the Early Paleogene in Northwest Europe,* Geological Society Special Publication No 101, p. 457–472.

Corfield, R. M. and Cartlidge, J. E., 1992. Oceanographic and climatic implications of the Palaeocene carbon isotope maximum. *Terra Nova,* v. 4, p. 443–455.

Cowie, J. W., 1986. Guidelines for boundary stratotypes. *Episodes,* v. 9, p. 78–82.

de Menocal, P., Bloemendal, J. and King, J., 1991. A rock-magnetic record of monsoonal dust deposition of the Arabian Sea: Evidence for a shift in the mode of deposition at 2.4 Ma. In Prell, W. L., Niituma, N., et al., Proceedings of the Ocean Drilling Program, Scientific Results, v. 117: College Station (Ocean Drilling Program), p. 389–407.

Doyle, P. S. and Riedel, W. R., 1979. Cretaceous to Neogene ichthyoliths in a giant piston core from the Central North Pacific. *Micropaleontology,* v. 25, p. 337–364.

Doyle, P. S. and Riedel, W. R., 1985. Ichthyolith biostratigraphy of western North Pacific pelagic clays, Deep Sea Drilling Project Leg 86. In Heath, G. R., Burckle, L. H., et al., Initial Reports of the Deep Sea Drilling Project, v. 86: Washington (U.S. Government Printing Office), p. 349–366.

Ellison, R. A., Knox, R. W. O'B., Jolley, D. W. and King, C., 1994. A revision of the lithostratigraphical classification of the early Palaeogene strata of the London Basin and East Anglia. *Proceedings of the Geologists' Association,* v. 105, p. 187–197.

Haq, B. U., Hardenbol, J. and Vail, P. R., 1988. Mesozoic and Cenozoic chronostratigraphy and cycles of sea-level change. In Wilgus, C. K., Hastings, B. S., Posamentier, H., Van Wagoner, J., Ross, C. A. and Kendall, C. G. St. C. (eds.), *Sea-Level Changes: An Integrated Approach,* Society of Economic Paleontologists and Mineralogists Special Publication No 42, p. 71–108.

Heath, G. R., Burckle, L. H., et al., 1985a. Site 577. Initial Reports of the Deep Sea Drilling Project, v. 86: Washington (U.S. Government Printing Office), p. 91–137.

Heath, G. R., Rea, D. H. and Levi, S., 1985b. Paleomagnetism and accumulation rates of sediments at Sites 576 and 578, Deep Sea Drilling Project Leg 86, western North Pacific. In Heath, G. R., Burckle, L. H., et al., Initial Reports of the Deep Sea Drilling Project, v. 86: Washington (U.S. Government Printing Office), p. 459–502.

Hedberg, H. D. (ed.), 1976. *International stratigraphic guide. A guide to stratigraphic classification, terminology, and procedure.* New York: John Wiley and Sons, 200 pp.

Hilgen, F. J., 1991. Astronomical forcing and geochronological application of sedimentary cycles in the Mediterranean Pliocene–Pleistocene. *Geologica Ultraiectina, Mededelingen van de Faculteit Aardwtensachappen Universiteit Utrecht,* v. 93, 139 pp.

Janecek, T. R., 1985. Eolian sedimentation in the northwest Pacific Ocean: A preliminary examination of the data from Deep Sea Drilling Project Sites 576 and 578. In Heath, G. R., Burckle, L. H., et al., Initial Reports of the Deep Sea Drilling Project, v. 86: Washington (U.S. Government Printing Office), p. 589–603.

Janecek, T. R. and Rea, D. K., 1983. Eolian deposition in the northeast Pacific Ocean: Cenozoic history of atmospheric circulation. *Geological Society of America Bulletin,* v. 94, p. 730–738.

Kennett, J. P. and L. D. Stott, 1990. Proteus and Proto-Oceanus: Ancestral Paleogene oceans as revealed from Antarctic stable isotope results. In Barker, P. F., Kennett, J. P., et al., Proceedings of the Ocean Drilling Program, Scientific Results, v. 113: College Station, TX (Ocean Drilling Program), p. 865–880.

Kennett, J. P. and Stott, L. D., 1991. Abrupt deep-sea warming, palaeoceanographic changes and benthic extinctions at the end of the Palaeocene. *Nature,* v. 353, p. 225–229.

Knox, R. W. O'B., 1984. Nannoplankton zonation and the Paleocene/Eocene boundary beds of NW Europe: An indirect correlation by means of volcanic ash layers. *Journal of the Geological Society, London,* v. 141, p. 993–999.

Knox, R. W. O'B, 1985. Stratigraphic significance of volcanic ash in Paleocene and Eocene sediments at Sites 549 and 550. In Graciansky, P. C. de, Poag, C. W., et al., Initial Reports of the Deep Sea Drilling Project, v. 80: Washington (U.S. Government Printing Office), p. 845–850.

Knox, R. W. O'B., 1990. Thanetian and early Ypresian chronostratigraphy in south-east England. *Tertiary Research,* v. 11, p. 57–64.

Lu, G. and Keller, G., 1995a. Ecological stasis and saltation: Species richness change in planktic foraminifera during the late Paleocene to early Eocene, DSDP Site 577. *Palaeogeography, Palaeoclimatology, Palaeoecology,* v. 117, p. 211–227.

Lu, G. and Keller, G., 1995b. Planktic foraminiferal faunal turnovers in the subtropical Pacific during the late Paleocene to early Eocene. *Journal of Foraminiferal Research,* v. 25, p. 97–116.

Mann, K. O. and Lane, H. R., 1995. *Graphic Correlation.* Society of Economic Paleontologists and Mineralogists Special Publication No. 53, 263 p.

Miller, K. G., Janecek, T. R., Katz, M. E. and Keil, D. J., 1987. Abyssal circulation and benthic foraminiferal changes near the Paleocene/Eocene boundary. *Paleoceanography,* v. 2, p. 741–761.

Monechi, S., 1985. Campanian to Pleistocene calcareous nannofossil stratigraphy from the northwest Pacific Ocean, Deep Sea Drilling Project Leg 86. In Heath, G. R., Burckle, L. H., et al., Initial Reports of the Deep Sea Drilling Project, v. 86: Washington (U.S. Government Printing Office), p. 301–336.

Monechi, S., Bleil, U. and Backman, J., 1985. Magnetobiochronology of late Cretaceous–Paleogene and late Cenozoic pelagic sedimentary sequences from the Northwest Pacific (Deep Sea Drilling Project, Leg 86, Site 577). In Heath, G. R., Burckle, L. H., et al., Initial Reports of the Deep Sea Drilling Project, v. 86: Washington (U.S. Government Printing Office), p. 787–797.

Moore, T. C., Jr., van Andel, T. H., Sancetta, C. and Pisias, N., 1978. Cenozoic hiatuses in pelagic sediments. *Micropaleontology,* v. 24, p. 113–138.

Müller, C., 1985. Biostratigraphic and paleoenvironmental interpretation of the Goban Spur region based on a study of calcareous nannoplankton. In Graciansky, P. C., de, Poag, W. C., et al., Initial Reports of the Deep Sea Drilling Program, v. 80: Washington (U.S. Government Printing Office), p. 573–599.

Pak, D. K. and Miller, K. G., 1992. Paleocene to Eocene benthic foraminiferal isotopes and assemblages: Implications for deepwater circulation. *Paleoceanography,* v. 7, p. 405–422.

Pospichal, J. J. and Wise, S. W., Jr., 1990. Paleocene to middle Eocene calcareous nannofossils of ODP Sites 689 and 960, Maud Rise, Weddell Sea. In Barker, P. F., Kennett, J. P., et al., Proceedings of the Ocean Drilling Program, Scientific Results., v. 113: College Station, TX (Ocean Drilling Program), p. 613–666.

Prothero, D. R. and Swisher, C. C., III, 1992. Magnetostratigraphy and geochronology of the terrestrial Eocene–Oligocene transition in North America. In Prothero, D. R. and Berggren, W. A. (eds.), *Eocene–Oligocene Climatic and Biotic Evolution.* Princeton: Princeton University Press, p. 46–73.

Raymo, M. E., Ruddiman, W. F., Backman, J., Clement, B. M. and Martinson, D. G., 1989. Late Pliocene variation in northern hemisphere ice sheets and North Atlantic deep water circulation. *Paleoceanography,* v. 4, p. 413–446.

Salvador, A. (ed.), 1994. *International Stratigraphic Guide. A guide to stratigraphic classification, terminology, and procedure.* Boulder:

International Union of Geological Sciences and Geological Society of America, 214 pp.

Schmitz, B., Speijer, R. P. and Aubry, M.-P., 1996. Latest Paleocene benthic extinction event on the southern Tethyan shelf (Egypt): Foraminiferal stable isotopic ($\delta^{13}C$, $\delta^{18}O$) records. *Geology,* v. 24, p. 347–350.

Shackleton, N. J., Hall, M. A. and Bleil, U., 1985. Carbon-isotope stratigraphy, Site 577. In Heath, G. R., Burckle, L. H., et al., Initial Reports of the Deep Sea Drilling Project, v. 86: Washington (U.S. Government Printing Office), p. 503–511.

Shackleton, N. J., Baldauf, J. G., Flores, J.-A., Iwai, M., Moore, T. C., Jr., Raffi, I. and Vincent, E., 1995. Biostratigraphic summary for Leg 138. In Mayer, L. A., Janecek, T. R., Palmer-Judson, A. and van Andel, T. H., et al., Proceedings of the Ocean Drilling Program, Scientific Results, v. 138, College Station, TX (Ocean Drilling Program), p. 517–536.

Shaw, A. B., 1964. *Time in stratigraphy.* New York: McGraw-Hill, 365 pp.

Sinha, A., Aubry, M.-P., Stott, L. D., Thiry, M. and Berggren, W. A., 1996. Chemostratigraphy of the "lower" Sparnacian deposits (Argiles plastiques bariolées) of the Paris Basin. In Aubry, M.-P. and Benjamini, C. (eds.), Paleocene/Eocene Boundary Events in Space and Time, *Israel Journal of Earth Sciences,* v. 44, p. 223–237.

Stott, L. D., 1992. Higher temperatures and lower oceanic pCO_2: A climate enigma at the end of the Paleocene Epoch. *Paleoceanography,* v. 7, p. 395–404.

Stott, L. D., Kennett, J. P., Shackleton, N. J. and Corfield, R. C., 1990. The evolution of Antarctic surface waters during the Paleogene: Inferences from the stable isotopic composition of planktonic foraminifers, ODP Leg 113. In Barker, P. F., Kennett, J. P., et al., Proceedings of the Ocean Drilling Program, Scientific Results, v. 113, p. 849–863.

Stott, L. D., Sinha, A., Thiry, M., Aubry, M.-P. and Berggren, W. A., 1996. Global $\delta^{13}C$ changes across the Paleocene–Eocene boundary: Criteria for terrestrial–marine correlations. In Knox, R. W. O'B., Corfield, R. M. and Dunay, R. E. (eds.), *Correlation of the Early Paleogene in Northwest Europe,* Geological Society Special Publication No. 101, p. 381–399.

Swisher, C. C. III and Knox, R. W. O'B., 1991. The age of the Paleocene/Eocene boundary: $^{40}Ar/^{39}Ar$ dating of the lower part of NP10, North Sea Basin and Denmark. IGCP Project 308 (Paleocene/Eocene boundary events), International Annual Meeting and Field Conference, 2–6 December 1991, Brussels, Abstracts with Program, p. 16.

Thomas, E. and Shackleton, N. J., 1996. The Paleocene–Eocene benthic foraminiferal extinction and stable isotope anomalies. In Knox, R. W. O'B., Corfield, R. M. and Dunay, R. E. (eds.), *Correlation of the Early Paleogene in Northwest Europe,* Geological Society Special Publication No 101, p. 401–441.

Thompson, E., and Schmitz, B., 1997. Barium and the late Paleocene $\delta^{13}C$ maximum: Evidence of increased marine surface productivity. *Paleoceanography,* v. 12, p. 239–254.

Townsend, H. A., 1985. The paleomagnetism of sediments acquired from the Goban Spur on Deep Sea Drilling Project Leg 80. In Graciansky, P. C., de, Poag, W. C., et al., Initial Reports of the Deep Sea Drilling Program, v. 80: Washington (U.S. Government Printing Office), p. 389–414.

Wei, W., 1994. Chronology of the Wasatchian Land-Mammal Age (early Eocene); Magnetostratigraphic results from the McCullough Peaks section, northern Bighorn Basin, Wyoming: A discussion. *Journal of Geology,* v. 103, p. 461–463.

Wei, W. and Zhong, S., 1996. Taxonomy and magnetobiochronology of *Tribrachiatus* and *Rhomboaster*, two genera of calcareous nannofossils. *Journal of Paleontology,* v. 70, p. 7–22.

Wheeler, H. E., 1958, Time-Stratigraphy. *Bulletin of the American Association of Petroleum Geologists,* v. 42, p. 1047–1063.

Wing, S. L., Bown, T. M. and Obradovich, J. D., 1991. Early Eocene biotic and climatic change in interior western North America. *Geology,* v. 19, p. 1189–1192.

Zachos, J. C., Lohmann, K. C., Walker, J. C. G. and Wise, S. W., 1993. Abrupt climate change and transient climates during the Paleogene: A marine perspective. *Journal of Geology,* v. 101, p. 191–213.

4

Magnetostratigraphy of Upper Paleocene–Lower Eocene Marine and Terrestrial Sequences

JOHN J. FLYNN
LISA TAUXE

ABSTRACT
We have collected all the principal magnetostratigraphic data sets from both marine and terrestrial strata spanning the Paleocene/Eocene Series boundary. To assess the reliability of the data we use two simple measures (histogram analysis of polarity data; a jackknife parameter measuring dependence of a given polarity stratigraphy on sampling density). There are only sporadic magnetobiostratigraphic data, of questionable reliability, spanning the Paleocene/Eocene Series boundary in shallow marine/epicontinental sequences of northwestern Europe. There are several long marine sequences spanning the boundary, but there is poor consistency among the various ODP/DSDP sites, and the results are difficult to interpret. The most complete and reliable magnetostratigraphies from terrestrial sequences spanning the Paleocene/Eocene Series boundary appear to be those from the San Juan Basin, Ellesmere Island, and the Northern Bighorn Basin. The available magnetostratigraphic data clearly indicate that the Paleocene/Eocene Series boundary lies within Chron C24r. On the basis of poor documentation and suspected overprinting problems, we reject the term *Oldhaven event* (or *subchron*) used in reference to a short normal polarity interval within Chron C24r. However, there may be a different short normal polarity interval closely associated with the Paleocene/Eocene Epoch boundary, and possibly correlative with Cryptochron C24r.6 of Cande and Kent (1992); it may be represented by short normal polarity zones in at least one marine and two terrestrial sequences. If this short event is documented to represent a single magnetochron interval, it would subdivide the long reversed magnetozone (Chron C24r) within which the Paleocene/Eocene Series boundary has been shown to lie, and it would have great potential to improve local and global temporal correlations. Although much progress remains to be made, studies integrating radioisotopic dating, magnetostratigraphy, biostratigraphy, and stable isotope stratigraphy have greatly improved chronostratigraphic correlations and insights into the relative timing, and possible causes of biotic and physical events occurring near the Paleocene/Eocene Epoch boundary.

Understanding the timing, rates, and interrelationships of biologic and geologic events and processes requires a precise temporal framework. This framework permits comparison of records of these events and processes preserved in different kinds of sediments. Historically, it has been very difficult to tie together stratigraphic records from marine and terrestrial sequences to compare and contrast what happened in the oceans with what happened on continents, over geologic time. The primary age control in those sequences was biostratigraphy. Most organisms have mutually exclusive distributions in either marine or terrestrial environments (and the narrow transition interface between those environments). This leads to few areas in which fossiliferous marine and terrestrial strata interfinger and occur in direct superposition, thus making it extremely difficult to correlate from one to the other. Until very recently, independent radioisotopic age control has been very rare, and in most stratigraphic sequences still is. Therefore, developing a temporal framework for understanding regional events and processes, and correlating them globally across marine and terrestrial environments, has been done primarily with magnetostratigraphy.

The usual way of comparing magnetostratigraphic information is to convert some parameter related to the magnetic direction measured in rocks to a black (normal polarity) and white (reversed polarity) bar diagram.

However, this practice tends to disguise poor or ambiguous data, and actual spacing or coverage of samples within each polarity interval. Some magnetostratigraphies are accompanied by extensive rock and paleomagnetic evidence as to the reliability of the record. Arguments for reliability can include one or more of the following lines of evidence. The directions are derived from "clean" components of magnetization, uncontaminated by viscous or other secondary complications. The normal and reversed modes are antipodal and correspond to reasonable directions. In some cases, field tests such as the fold test or the conglomerate test have been performed. Other studies include no arguments for reliability, simply the directions, or in many cases, only the inclinations. The samples have been subjected to only cursory laboratory cleaning efforts, and the data quality is almost completely undocumented. Yet both sorts of studies are rendered into simple black and white patterns with no key to sampling density or data quality, and often used uncritically by the original investigators and/or later authors.

In this study, we employ two simple means of assessing the reliability of published magnetostratigraphic data sets. First, histograms of the parameter used to determine polarity (usually virtual geomagnetic pole [VGP] latitudes or inclinations), should show two distinct "humps" representing the two polarities. If polarities cannot be discriminated in the histogram plots, then the magnetostratigraphic patterns may not provide reliable polarity information. The second method is the jackknife parameter "J," defined by Tauxe and Gallet (1991). "J" is a quantitative assessment of the dependence of a given polarity stratigraphy on the sampling density. Sections with high "J" values have relatively robust magnetostratigraphies. In contrast, low "J's" indicate numerous polarity zones represented by only one or two sampling horizons.

The black and white bar/polarity diagrams yield age information when correlated with a standard time scale that tabulates the numerical ages of all major polarity reversals (referred to as the Geomagnetic Polarity Time Scale [GPTS]). This time scale is based on templates provided by marine magnetic anomalies, which are interpreted in terms of polarity sequences versus kilometers of ship track. Dated polarity reversals serve as calibration points, and the various versions of the GPTS stem from different selection of calibration points and/or different means of age interpolation between, or extrapolation from, calibration points. Like magnetic polarity stratigraphies, the GPTS is also presented as a black and white bar diagram, yet the original magnetic anomalies are not square waves. We believe that much may be lost in the common practice of "sanitizing" of magnetic anomaly and magnetostratigraphic patterns.

Thus, in this chapter, we collect all the principal magnetostratigraphic data sets, from both marine and terrestrial strata (figure 4.1), spanning the Paleocene/Eocene Series boundary. We rely on the actual directional records presented in the original publications, and the marine magnetic anomalies used for constructing the GPTS, for determining both magnetic polarity reversal sequences and correlations between the magnetostratigraphies and the GPTS. Consideration of chronostratigraphic and geochronologic correlations of both marine and terrestrial sequences has been further hampered by ambiguity and confusion surrounding the definition of the Paleocene/Eocene Series/Epoch boundary (see also chapter 2 and Berggren et al. 1995). Resolution of many of the controversies outlined below awaits unequivocal, formal definition of the boundary in IGCP Project 308.

Magnetostratigraphy: The Marine Record

In general, data from deep sea sediment cores are less reliable than data from outcrop sections. This is primarily because core data are usually based on single specimens, whereas the opportunity exists in outcrops to take multiple specimens and screen for within site reproducibility. Secondly, drill cores are usually unoriented (horizontally) and polarities are determined from inclination data only. At latitudes less than about 20°, scatter due to secular variation, core disturbance, and/or overprinting cannot be distinguished easily from polarity reversals, leading to possibly erroneous interpretations. When both declination and inclination are known, the data are often

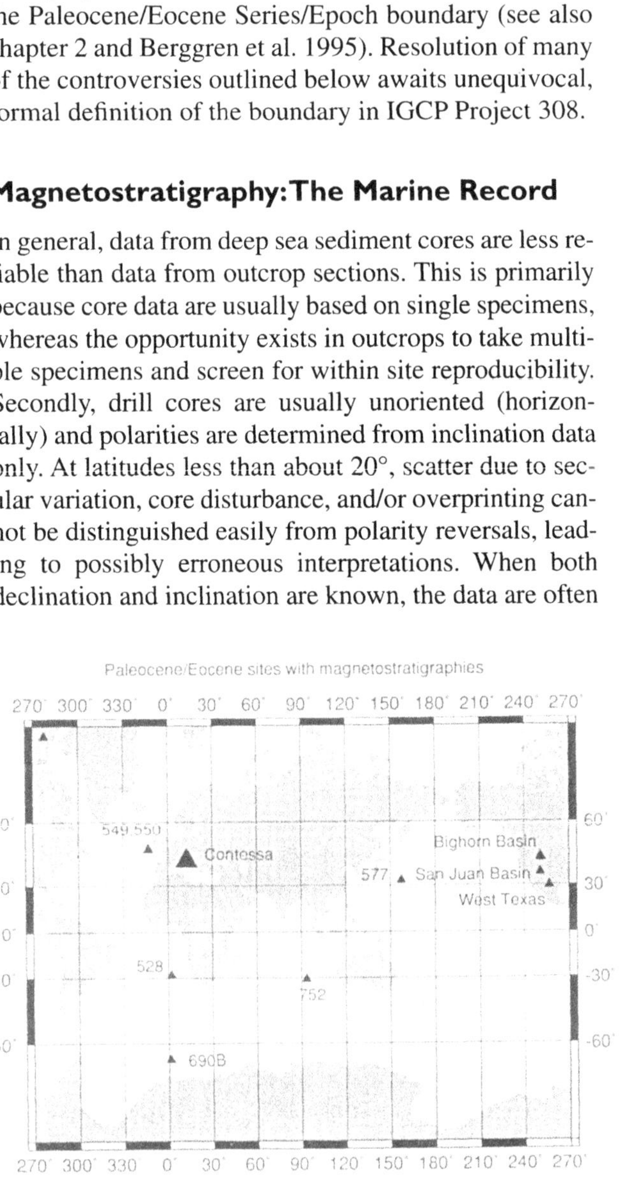

Figure 4.1 Map showing geographic locations of sections with magnetostratigraphies discussed in text and tables 4.1 and 4.2. The greater Bighorn Basin shown on the figure includes the Clark's Fork Basin, Northern Bighorn Basin, and Southern Bighorn Basin sections referred to in the text.

converted to virtual geomagnetic poles (VGPs), the latitudes of which serve to define polarity zones. VGP latitudes ideally cluster around the poles of ± 90°, hence are more easily distinguished in terms of polarity.

In decades of deep sea drilling, there are remarkably few records spanning the Paleocene/Eocene Series boundary. The most complete records published to date are shown in figure 4.2 (see also table 4.1). We have included one marine record from outcrop: the Contessa Highway Section (CHS) of Lowrie et al. (1982). Plotted to the side of each section are key fossil datums, the Cretaceous/Paleogene Period and the Paleocene/Eocene Series boundaries and the location of the "carbon excursion" in Aubry et al. (1996).

The Paleocene/Eocene Series boundary currently is defined by the base of the Ypresian Stage (the lowest stage of the Eocene). The Ypresian Stage is represented by the base (likely transgressive) of the Ieper Clay in the Belgium Basin, and the correlative base of the recently redefined London Clay Formation (base of the Walton Member [= Division A2 of King 1981]; in Ellison et al. 1994) in the London–Hampshire Basin (see Berggren et al. 1985b and chapter 2). Numerous litho-, chemo-, and bio-stratigraphic events have been suggested to correlate with, or approximate this boundary, including the NP9/NP10 zonal boundary. Berggren and Aubry (chapter 2) discuss seventeen of these events, and provide geochronologic estimates for the age of each. The NP9/NP10 zonal boundary is defined as the LO (lowest occurrence) of *Tribrachiatus bramlettei*. When *T. bramlettei* is absent, as it is in the Contessa Road Section, the HO (highest occurrence) of *Fasciculithus* spp. often is used to approximate this boundary (Aubry et al. 1986; but see Aubry et al. 1996, Berggren and Aubry 1996, and Wei and Zhong 1996). Here we use the HO of *Fasciculithus* spp. to identify the Paleocene/Eocene Series boundary in the Contessa Road Section. However, the HO of *Fasciculithus* probably lies well below the LO of *T. bramlettei* (Aubry et al. 1996), and therefore well below the Paleocene/Eocene Series boundary. Further, the lithostratigraphically defined Paleocene/Eocene Series boundary probably is even younger, lying higher within Zone NP10 (very close to the LO of *Tribrachiatus digitalis*, according to Aubry et al. 1996 and Berggren and Aubry 1996). Thus, the approximated Paleocene/Eocene Series boundary typically used in the Contessa Road Section probably lies below the actual Paleocene/Eocene Series boundary.

Table 4.1
Jackknife (Jackstrat; Tauxe and Gallet 1991) "J" values for marine magnetostratigraphic sections

Site/Section	Location	J
Contessa Road	Gubbio, Italy - (12.5° E, 43.5° N)	−0.23
690B	Deep South Atlantic (1° E, 65° S)	−0.35
577	Northwest Pacific (157.5° E, 32.5° N)	−0.39
528	South Atlantic (2° E, 28.5° S)	−0.68
550	North Atlantic (13.5° W, 48.5° N)	−0.57
549	North Atlantic (13° W, 49° N)	−0.51
752	Indian Ocean (93.5° E, 31° S)	−0.45

In the Contessa Road, DSDP Site 550, and ODP Site 752 sections, the Paleocene/Eocene Series boundary occurs in a robustly supported reversed polarity zone correlated to the reversed zone between Anomalies 24 and 25 (Chron C24r in the terminology of Cande and Kent 1992). However, both Wei and Zhong (1996) and Aubry et al. (1996) have suggested that there is a significant unconformity at the Paleocene/Eocene Series boundary in DSDP Site 550, and Aubry et al. (1996) also argued for a hiatus in DSDP Site 549. The characteristic nannofossil zones marking the Paleocene/Eocene Series boundary at lower latitudes appear to be much younger in ODP/DSDP Sites 690B and 528, occurring either in a younger, normal zone (ODP Site 690B) or at the even younger upper reversal boundary (DSDP Site 528). Traditional black and white polarity logs for the sections shown in figure 4.2 reveal little of the complexity of the primary data. Polarities are generally determined by the sign of the inclination (or VGP latitude), but this often is done in inconsistent ways. For example, polarity zones represented by one or a few samples often are ignored. Similarly, samples with shallow, but normal, inclinations are interpreted as indeterminate or of reversed polarity, while shallow, but reversed, inclinations are interpreted as questionably reversed.

To illustrate the difficulties in interpreting the data from some sections, we plot the inclinations or VGP latitudes from figure 4.2 as histograms in figure 4.3. Sites with clearly distinguishable normal and reversed data have two distinct "humps" in the histograms (see e.g., Contessa Road, ODP Site 690, DSDP Site 577, and ODP Site 752). Several sites have "smeared" inclination data (see DSDP Site 549 and to some extent DSDP Site 550). The polarities of such sites are very poorly constrained. On the basis of these plots, we can virtually exclude DSDP Site 549 from further discussion, and treat DSDP Site 550 with caution. Also, DSDP Site 528 has many low inclination values and the polarity of these samples is in doubt.

In addition to poor data quality, poor sample spacing often plays a role in determining the reliability of a

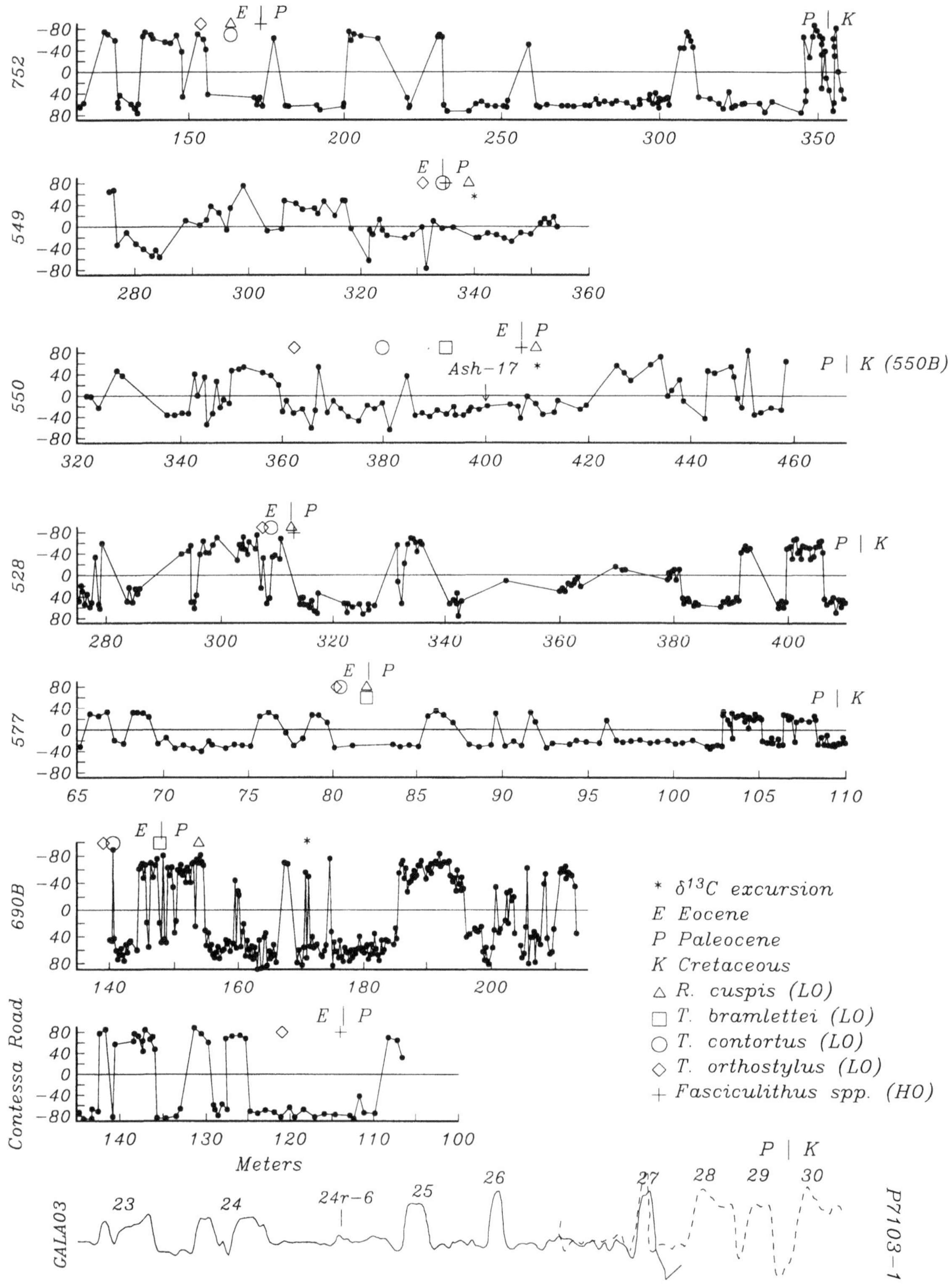

Figure 4.2 Summary of marine magnetobiostratigraphic constraints for the Paleocene/Eocene Series boundary. VGP latitudes, inclinations or positive anomalies indicating normal polarities plot up. GALA03: marine magnetic anomaly profiles from Cande and Kent (1992). Contessa Road: VGP latitudes from the Contessa Road section from Lowrie et al. (1982). 690B: inclination data of Spiess (1990), biostratigraphy of Wei and Zhong (1996). 577: inclination data of Bleil (1985), biostratigraphy of Wei and Zhong (1996). 528: inclination data of Shackleton et al. (1984), biostratigraphy of Wei and Zhong (1996). 752: inclination data of Gee et al. (1991), biostratigraphy of Wei and Zhong (1996), $\delta^{13}C$ data are from Aubry et al. (1996). 550: inclination data of Townsend (1985), biostratigraphy of Wei and Zhong (1996). 549: inclination data of Townsend (1985), biostratigraphy of Berggren and Aubry (1996).

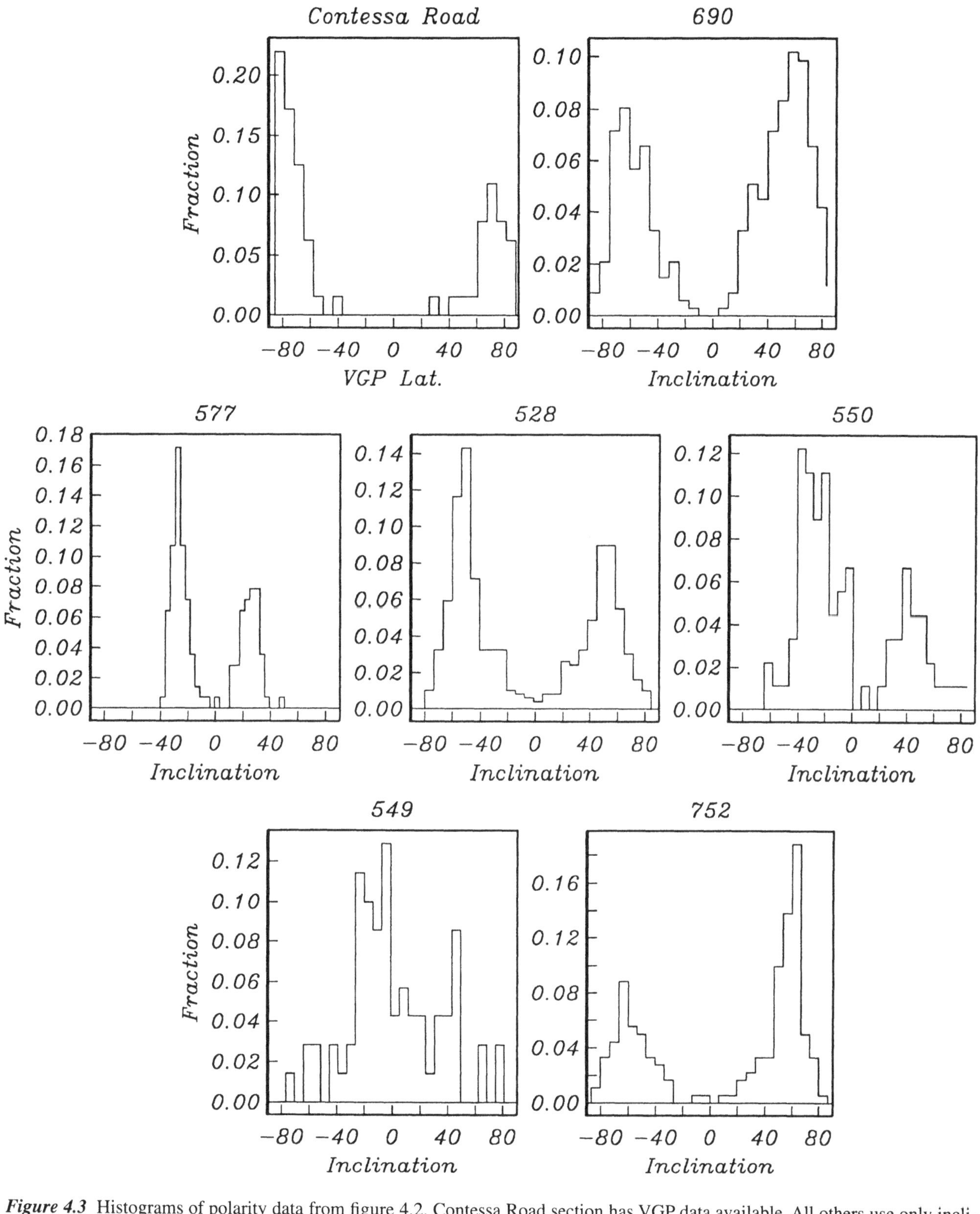

Figure 4.3 Histograms of polarity data from figure 4.2. Contessa Road section has VGP data available. All others use only inclination data for polarity determination. Most records show reasonable to good separation of the two polarity states, whereas data from DSDP Site 549 are smeared, preventing reliable polarity determination.

given polarity interpretation. In order to assess the importance of sampling density in the magnetostratigraphies discussed here, we use the jackknife proposed by Tauxe and Gallet (1991). The jackknife is a statistical technique (see Efron 1982) whereby a given parameter is calculated for a data set. In the stratigraphic jackknife, a Fortran program "jackstrat" (available from LT on request) counts the number of polarity reversals in a given section. Tauxe and Gallet (1991) defined a parameter, "J," as the slope relating the percentage of polarity zones (of a sampled section) remaining after random deletion of sites, increasing in proportion to a total of 20% of the initial collection. Thus "J" quantifies the dependence of a magnetostratigraphic record on the distribution of sampling sites. Based on simulations, they recommend values of "J" in the range of 0 to 0.5 for sections to be fairly robustly defined. In figure 4.4 we show typical "jackstrat" results by plotting the percentage of reversals remaining versus the

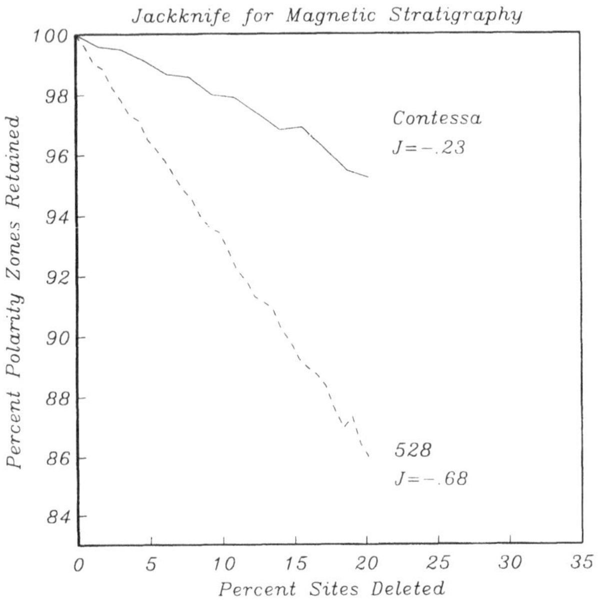

Figure 4.4 Examples of "jackstrat" jackknife results for two lower Paleogene magnetostratigraphic sections. Plotted is the percent of the polarity zones in the original section remaining as an increasing percentage of sites are randomly selected for deletion. Each point is the average of many samples with the same number of sites deleted. For example, when only one site is deleted, each site is deleted in turn (making a new jackknifed "pseudodata set," and the number of polarity units remaining are counted. The point plotted is the average number of sites remaining from all "pseudodata sets." When more than one site is to be deleted, sites are chosen at random for deletion from each "pseudodata set." The slope of the line is "J." The magnetostratigraphy of the Contessa Road section is fairly robust, with a "J" value of –0.23, whereas the magnetostratigraphy of DSDP Site 528 is rather fragile, with a "J" below –0.5.

percentage of sampling sites deleted. In table 4.1 we list "J" for the sections shown in figure 4.2.

Given the poor quality of the DSDP Site 528 inclination data and the poor score for "J," this section should also be treated with caution. Also, ODP Site 752 appears to be of high quality, but a relatively low score for "J" makes this section less than ideal.

In summary, the biostratigraphically-calibrated marine magnetostratigraphy of the Paleocene/Eocene Series boundary is disappointing. The Contessa Road and ODP Site 752 sections appear to give the most reliable results, indicating clearly that the Paleocene/Eocene Series boundary lies within Chron C24r. However, there is poor consistency among the various marine sequences examined. Some of the discrepancies could be due to hiatuses (e.g., in DSDP Sites 550 and 549; and likely in DSDP Site 577 and ODP Site 690, Aubry, personal communication 1996), poor quality data (DSDP Sites 550, 549, 528), or poor sampling density (ODP Site 752). However, none of these explanations alone appears to account for the large difference between the apparent boundary location in ODP Site 690 and the lower latitude sites such as DSDP Site 577. Perhaps the nannofossil datums marking the boundary are significantly time transgressive. Alternatively, the magnetic data from ODP Site 690 (and DSDP Site 528) could be overprinted. Overprinting alone, however, cannot account for the fact that the carbon excursion, which occurs near the Paleocene/Eocene Series boundary in DSDP Sites 550 and 549, lies some 40 m below it at ODP Site 690, in the lower third of a prominent reversed polarity interval likely to be Chron C24r. This discrepancy must be due to a combination of unconformities and overprinting, most likely in the ODP Site 690 record.

Magnetostratigraphy and Integrative Geochronology: The Terrestrial Record

North America

The North American stratigraphic record contains sequences that currently are likely to be the best terrestrial reference sections spanning the Paleocene/Eocene Series boundary (see figure 4.1 for geographic locations). In particular, deposition in the Bighorn and Clark's Fork Basins of northern Wyoming was relatively steady and accumulated at a high rate throughout the late Paleocene and early Eocene epochs. Lengthy magnetostratigraphies, biostratigraphies (including mammals, freshwater molluscs, plant macrofossils, and pollen), and geochemical stratigraphies have been developed by several working groups, for multiple sequences in several parts of the basin. Gunnell et al. (1993:142) considered the northern Bighorn Basin/Clark's Fork Basin stratigraphic sequence to represent "the most complete and best sampled record of vertebrate evolution from the late Paleocene through the early Eocene in the world," and that it "is the ideal continental reference section for recognizing the Paleocene–Eocene boundary." We agree that the greater Bighorn Basin sequences (Clark's Fork Basin, northern Bighorn Basin, southern Bighorn Basin), and particularly the Clark's Fork Basin/northern Bighorn Basin composite section because of the robust support of jackknife analysis for the Northern Bighorn Basin section (table 4.2), currently provide the best continental reference sections for recognizing the Paleocene/Eocene Epoch boundary and related events in the late Paleocene through early Eocene epochs. However, different environmental regimes and biotas possibly represented in the marine to mixed marine–terrestrial boundary stratotype sequences (containing the base of the Ypresian Stage) and fully terrestrial sequences, unique depositional and preservational biases in any sequence, slight temporal offset among the variety of observed events occurring during the latest Paleocene to earliest Eocene interval, possible asynchrony of these events in different areas (or any stratigraphic marker or unit), the fact that the Paleocene/Eocene Epoch boundary lies within a long interval of a single magnetic polarity (Chron C24r), and other problems make extremely high-precision temporal correlation difficult.

Berggren et al. (1985a) summarized the then-available magnetostratigraphic and biostratigraphic information, and temporal correlations, of Paleogene North American terrestrial sequences and the North American Land Mammal "Ages" (NALMAs). Their correlation charts showed that the Clarkforkian/Wasatchian NALMA boundary lay somewhere near or above the Paleocene/Eocene Epoch (and Series) boundary, within Chron C24r, but there was uncertainty in its precise position. This correlation was based on available data from Butler et al. (1981; Clark's Fork Basin/northern Bighorn Basin) and Rapp et al. (1983; West Texas) indicating that the Clarkforkian NALMA extended from the youngest part of Chron C25n to medial, or possibly late, Chron C24n and that the Wasatchian NALMA began within Chron C24r. Later results from the greater Bighorn Basin and Baja California, Mexico (Flynn et al. 1989) supported this correlation, but refined it to suggest that the Clarkforkian/Wasatchian NALMA boundary lay within the early to medial part of Chron C24r, possibly just below Cryptochron C24r.6.

Table 4.2

Jackknife (Jackstrat; Tauxe and Gallet 1991) "J "values for continental, terrestrial magnetostratigraphic sections

Section	Location	J
San Juan Basin	New Mexico (109° W, 37° N)	0.30
Ellesmere Island	Arctic Canada (82.7° W, 78.7° N)	0.34
West Texas	Texas (103° W, 29.5° N)	0.84
Clark's Fork Basin	Wyoming/Montana (109.2° W, 44.8° N)	0.82
Northern Bighorn Basin	Wyoming (108.6° W, 44.5° N)	0.33
Southern Bighorn Basin	Wyoming (108.6° W, 44.2° N)	1.00

We plot the relevant magnetostratigraphic profiles from North America in figure 4.5, with the reference anomaly pattern; "jackstrat" jackknifing analyses of robustness are presented in table 4.2.

The most recent synthetic summaries of Paleocene through Eocene North American mammalian biochronology and correlations are those of Archibald et al. (1987) and Krishtalka et al. (1987). Krause and Maas (1990) discussed hypotheses of mammalian dispersal events around the Paleocene/Eocene Epoch boundary. Below we detail available magnetostratigraphies, and correlation issues, from upper Paleocene to lower Eocene North American sections.

Baja California, Mexico

Lower Paleogene mammal-bearing terrestrial strata from the central Baja California peninsula were first reported by Morris (1966). Later investigations yielded a much more extensive mammalian fauna, and integrated detailed study of the biostratigraphy and magnetostratigraphy of intertonguing marine and terrestrial strata in this area (Flynn and Novacek 1984; Flynn et al. 1989; Novacek et al. 1991). These strata are clearly early Eocene in age, based on abundant gastropod and bivalve molluscs (restricted "Capay" North American Molluscan Stage [Givens 1974], correlative to the middle Ypresian Stage [middle lower Eocene Series]), a benthic foraminiferal assemblage (Penutian–Ulatisian Benthic Foraminiferal Stage), and the ostracode *Buntonia* (early Eocene first appearance). Both the marine assemblages and Wasatchian mammals occur within a reversed polarity interval, correlative with either Chron C23r or one of the two short reversed intervals in Chron C24n, indicating the strata are about 2.5–3 million years younger than the Paleocene/Eocene Series boundary.

West Texas

Rapp et al. (1983) sampled terrestrial strata in the Big Bend National Park area of West Texas that appear to span the Paleocene/Eocene Series boundary. Their magnetostratigraphy covered more than 160 m, from the uppermost Javelina Formation, through the entire Black Peaks Formation, to the lowermost part of the overlying Hannold Hill Formation. The West Texas magnetostratigraphy is not robustly supported by jack-

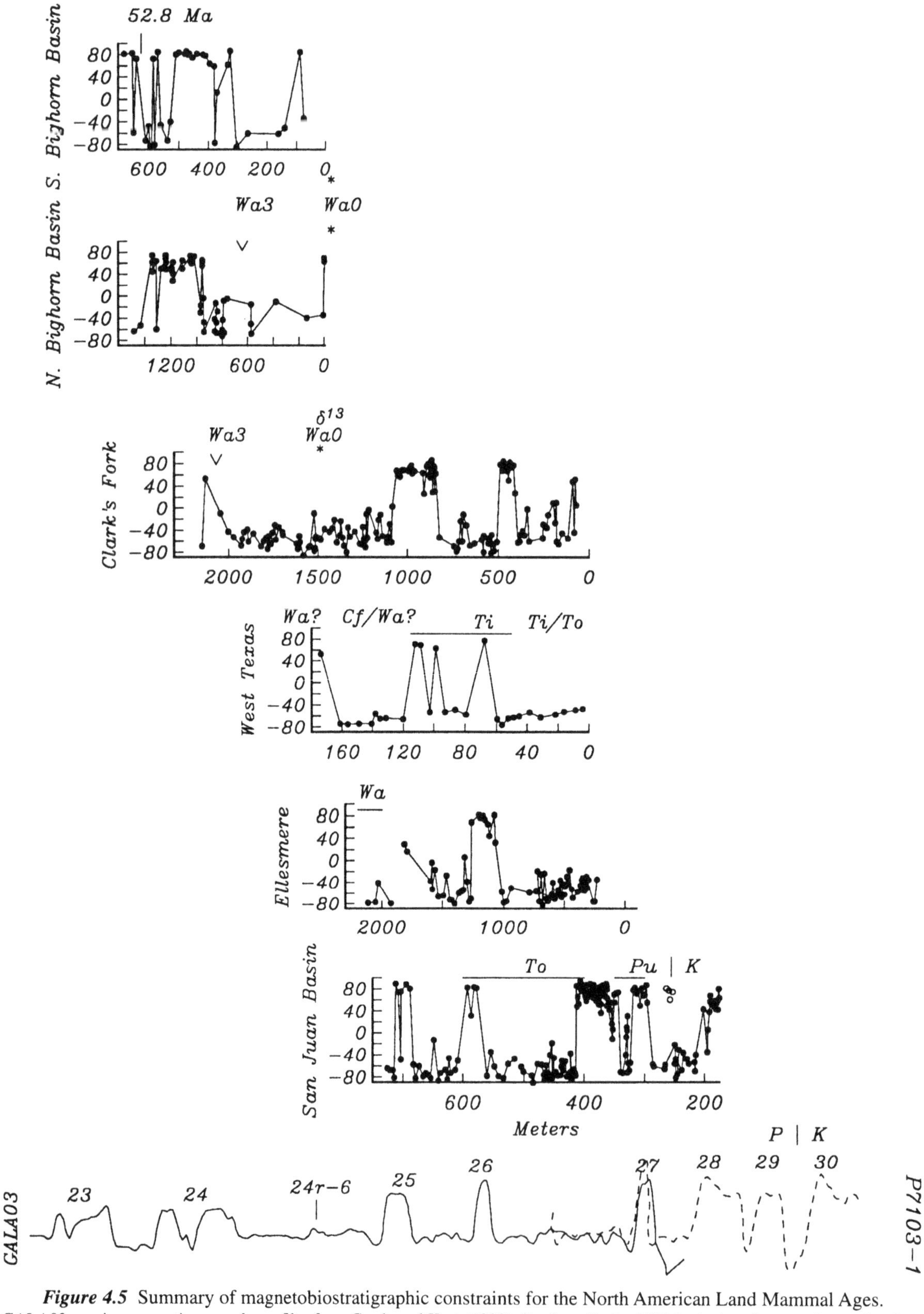

Figure 4.5 Summary of magnetobiostratigraphic constraints for the North American Land Mammal Ages. GALA03: marine magnetic anomaly profiles from Cande and Kent (1992). San Juan Basin: VGP Latitudes of composite section from Lindsay et al. (1981) and Taylor and Butler (1980); Butler and Lindsay (1985) suggested that the points shown as circles were overprinted. Ellesmere section: inclinations from the composite section of Tauxe and Clark (1987). West Texas: VGP latitudes from Rapp et al. (1983). Clark's Fork (Greater Northern Bighorn Basin): VGP latitudes from Big Sand Coulee and Polecat Bench sections of Butler et al. (1981). Northern Bighorn Basin: Inclinations from Clyde et al. (1994). Southern Bighorn Basin: VGP latitudes from composite section from Tauxe et al. (1994), δ^{13}C excursion from Koch et al. (1992).

knife analysis (table 4.2). This section contained superposed faunas assigned to the Torrejonian–Tiffanian (near the base of the section; undifferentiated mammal ages, because the fauna is poor), Tiffanian (represented by 4 superposed faunas), Clarkforkian (tentatively or questionably identified) and Wasatchian (above the sequence sampled) NALMAs, and yielded polarity zones correlated to Chrons C26r to C24n. The supposed Clarkforkian NALMA assemblage occurred in about the middle of a reversed polarity zone correlated to Chron C24r. Tauxe et al. (1994) implied that the faunal elements from this horizon might be better considered Wasatchian in age, as the Bighorn Basin and West Texas magnetostratigraphies (and location of the Clarkforkian/Wasatchian NALMA boundary within Chron C24r, and relative to the Paleocene/Eocene Series boundary) would be more consistent under this interpretation. It is important to realize that the West Texas "Clarkforkian assemblage" cited by Rapp et al. (1983; but see Schiebout 1995, Schiebout et al. 1987) is represented by sparse material of only two taxa *Barylambda* sp. (previously Tiffanian–Clarkforkian NALMAs in range) and *Hyracotherium angustidens* (previously Wasatchian NALMA in range). At that time *Barylambda* and *Hyracotherium* only were considered to co-occur in Baja California, in an assemblage that then was tentatively believed to be Clarkforkian in age (Morris 1966). However, the relevant taxa in Baja (Pantolamdidae, ?new genus; *Hyracotherium seekinsi*; see Novacek et al. 1991) differ generically and specifically, respectively, from the taxa present in the West Texas "Clarkforkian" "assemblage," and the Baja "Clarkforkian" assemblage is now known to be clearly Wasatchian in age (Flynn and Novacek 1984; Flynn et al. 1989; Novacek et al. 1991; and discussion above). The data from Baja, and elsewhere in North America, support the suggestion by Tauxe et al. (1994) that the West Texas "Clarkforkian" horizon might actually be Wasatchian in age, as *Hyracotherium* has not yet been documented to occur in strata older than Wasatchian, whereas a variety of pantodonts occur in the Wasatchian NALMA. The West Texas Wasatchian NALMA assemblage was in unsampled strata overlying a normal polarity zone correlated to Chron C24n. Thus, in the West Texas sequence the Paleocene/Eocene Series boundary (within Chron C24r) occurs somewhere in the "Clarkforkian" to Wasatchian NALMA interval, but determining its position relative to the NALMAs more precisely is limited by the small number of fossiliferous horizons and specimens.

New Mexico

The San Juan Basin of New Mexico contains classic mammalian faunas usually considered to be of Paleocene and Eocene age (see Lucas and Williamson 1993a). These include the "type faunas" for the Puercan and Torrejonian NALMAs (both early Paleocene in age, based on magnetostratigraphic correlations) preserved in the Nacimiento Formation, and the Wasatchian Almagre and Largo faunas of the San Jose Formation. Butler and Lindsay (1985) presented the most recent magnetostratigraphy spanning the upper Cretaceous–Paleocene rocks of the San Juan Basin. Identification of polarity intervals and correlation of this magnetostratigraphy had been quite complex and controversial (see discussions in Berggren et al. 1985b; Lucas and Williamson 1993a and references cited therein), but the currently accepted magnetostratigraphic correlations (Butler and Lindsay 1985) indicate that the Cenozoic strata preserved in the San Juan Basin span from Chron C29r through C25r. This magnetostratigraphy is robust, as indicated by the high "J" values (table 4.2). Interpretation of events around the Paleocene/Eocene Epoch boundary is difficult because there are no good early Wasatchian nor Clarkforkian faunas preserved in the basin, and no magnetostratigraphic sampling preserving Chron C24r (nor the older Chron C25n, nor any Eocene chrons younger than C24r). Correlation figures of Lucas and Williamson (1993a:figure 2) indicated a hiatus at or near the Paleocene/Eocene Series boundary, whereas those of Lucas and Williamson (1993b:figure 15) and Williamson and Lucas (1993:figure 2) indicated that deposition of the San Jose Formation may span the Clarkforkian–Wasatchian NALMAs (and thus presumably the Paleocene/Eocene Series boundary) relatively continuously. The correlation chart of Lucas and Williamson (1993b:figure 15) also suggested that the Blanco Basin Formation (Chama Basin), Galisteo Formation (Hagan-Cerrillos), and McRae and Love Ranch Formations (Caballo/Organ Mountains) also may span the Clarkforkian–Wasatchian interval and the Paleocene/Eocene Series boundary, although that interval in those areas is poorly fossiliferous and lacks magnetostratigraphies.

Ellesmere Island, Canada

Hickey et al. (1983) cited a preliminary magnetostratigraphy for Upper Cretaceous through lower Paleogene strata from Ellesmere Island. They suggested temporal diachroneity of up to 24 million years for Wasatchian mammal faunas, with high-latitude Ellesmere Island Wasatchian assemblages occurring in a zone correlated to Chron C25n (as does the putatively older Clarkforkian NALMA elsewhere; Butler et al. 1981; see also discussions in Flynn et al. 1984 and Berggren et al. 1985b), while temperate-latitude North American Wasatchian assemblages occur only within Chron C24r or younger strata. However, the magnetostratigraphic data and resultant chronostratigraphic conclusions were rejected by Kent et al. (1984) and Flynn et al. (1984), as there were serious problems with the primary paleomagnetic analyses (which were essentially unpublished except in abstract), the resultant magnetostratigraphy, and its correlation. Tauxe and Clark (1987) later resampled this sequence and performed detailed and thorough paleomagnetic analyses; the magnetostratigraphy is strongly supported in jackknife analyses (table 4.2). Their magnetostratigraphic results were consistent with studies from the Rocky Mountain region, as Ellesmere Island Wasatchian faunas occurred within a reversed polarity interval that probably correlated with Chron C24r.

Greater Bighorn Basin (BB, including the Clark's Fork Basin), Wyoming

There have been several magnetostratigraphic studies of the extremely important, fossiliferous, lower Paleogene strata of the Bighorn and Clark's Fork Basins, Wyoming. These strata probably represent the best terrestrial reference sequences in the world spanning the Paleocene/Eocene Series boundary. They were deposited rapidly and relatively continuously, represent a variety of terrestrial depositional environments and paleoclimatic conditions, contain abundant mammal and plant fossils (as well as fossils of other groups), have yielded at least one high-precision radioisotopic date, and have been analyzed extensively using magnetostratigraphic, sedimentologic and chemostratigraphic techniques. Butler et al. (1981; Clark's Fork Basin; Tiffanian–Wasatchian NALMAs [and Puercan–Torrejonian NALMAs, in strata difficult to interpret magnetostratigraphically]) presented the first magnetostratigraphy for these sequences, followed recently by the analyses of Clyde et al. (1994; northern Bighorn Basin; Wasatchian NALMA) and Tauxe et al. (1994; southern Bighorn Basin; Wasatchian NALMA). Unfortunately, suggestions for placement or recognition of the Paleocene/Eocene Series boundary have varied dramatically, ranging from well above the Clarkforkian/Wasatchian NALMA boundary (e.g., in the northern BB, at 150 m above [Koch et al. 1992] or 180 m above [Rea et al. 1990]), to about at the boundary (northern BB, Clyde et al. 1994; southern BB, Tauxe et al. 1994), to below the Clarkforkian/Wasatchian boundary (southern BB, Wing et al. 1991). Gunnell et al. (1993) suggested placing (or "recognizing") the Paleocene/Eocene Series boundary within the Bighorn Basin sequences at the Clarkforkian/Wasatchian NALMA boundary; although we agree that the Paleocene/Eocene Series boundary likely falls close to the Clarkforkian/Wasatchian NALMA boundary, we have methodological disagreements with this proposal (see below).

In the Clark's Fork Basin sequence (Butler et al. 1981) Clarkforkian NALMA assemblages occur in magnetozones correlated to Chrons C25n to C24r, while Wasatchian NALMA assemblages occur within the uppermost part of Chron C24r (as represented locally; as these horizons also represent the uppermost part of the section sampled magnetostratigraphically, Chron C24r may be present higher in the section in currently unsampled strata).

Rea et al. (1990:117) summarized a series of tectonic, climatic, and evolutionary events that they believed occurred within Chron C24r, at the Paleocene/Eocene Epoch boundary, including "a many-fold increase in seafloor hydrothermal activity, a global warming, a reduction in the intensity of atmospheric circulation, a conversion to salinity-driven deep ocean circulation, a marked lightening of oceanic $\delta^{13}C$ values, extinction and evolution of both benthic foraminifera and land mammals, and important plate-boundary reorganizations including the outpouring of the east Greenland volcanics and the initiation of the oceanic rift between Norway and Greenland." Rea et al. (1990) discussed changes in the mammalian fauna of the western interior of North America (Clark's Fork/Bighorn Basin) during Chron C24r (Butler et al. 1981). Body size of the largest Clarkforkian–Wasatchian mammal (*Coryphodon*) decreased rapidly and then increased again rapidly, which "parallels a change in isotopic values of C24r planktonic foraminifera in Antarctic surface waters, which corresponds to a change of 4–5° C," although there is no independent chronologic data presented suggesting precise temporal correlation of these two trends. A principal component analysis of family-level mammalian faunal composition indicated a brief and dramatic turnover at the Clarkforkian/Wasatchian NALMA boundary. This boundary is 200 meters below the body size decrease event in *Coryphodon* and about 150 m below a level indicated on their figure 6 as a "possible level of NP10/NP9; Paleocene–Eocene boundary," suggesting that there is complexity and asynchrony in the timing of these events.

Koch et al. (1992) similarly linked geochemical, biotic, and climatic events near the Paleocene/Eocene Epoch boundary within Chron C24r, recorded in terrestrial and marine environments. They noted that a sharp drop in $\delta^{13}C$ has been well documented in one marine sequence (ODP Site 690, see figure 4.1) and reported in several others; this isotopic shift is slightly older than the Paleocene/Eocene Series (and Epoch) boundary (if recognized to lie within calcareous nannoplankton Biochron NP10 and within planktonic foraminiferal Biochron P6b; Aubry et al. 1988). Their analyses of carbon isotopic ratios of mammalian teeth (*Phenacodus* enamel apatite) and paleosol carbonates in the upper Paleocene–lower Eocene sequence from the northern Bighorn Basin indicated a sharp drop in $\delta^{13}C$ values (decreases of 4.5–5.9‰) in both indicators precisely at the Clarkforkian/Wasatchian NALMA boundary (estimated to have occurred in less than 50,000 years; although not explicitly stated, statements in Clyde et al. 1994 suggest that this isotopic shift [Koch et al. 1992] occurs in Zone Wa0 in this Clark's Fork Basin/northern Bighorn Basin sequence).

Correlations of the Bighorn Basin terrestrial sequence (using a "correlation web" of magnetostratigraphic and mammalian biostratigraphic data from the northern Bighorn Basin, and mammalian biostratigraphic correlations to integrate palynofloral biostratigraphic and radioisotopic dating information from the southern Bighorn Basin) to marine records suggested that the age of the carbon isotopic excursions in both differed by less than 150,000 years, and could be synchronous given age uncertainties in the correlation techniques. The latest Paleocene marine carbon isotope excursion (believed to be controlled by climatic change and related modifications to oceanic circulation) is synchronous with significant extinctions in benthic foraminifera (Berggren and Aubry 1996) and, if temporally correlative, major changes in the mammalian fauna at the Clarkforkian/Wasatchian NALMA boundary. As they assumed that the marine and terrestrial carbon isotopic shift was synchronous and predated the Paleocene/Eocene Epoch boundary, Koch et al. (1992) concluded that the Paleocene/Eocene Epoch boundary would lie within the early Wasatchian NALMA, and therefore the major changes in the mammalian fauna at the Clarkforkian/Wasatchian NALMA boundary would not have coincided precisely with any changes occurring at the Paleocene/Eocene Epoch boundary. As summarized earlier, the exact relationship between the carbon isotopic shift and the Paleocene/Eocene Series boundary is not completely straightforward in the marine magnetostratigraphic records. The carbon isotopic shift occurs some 40 m below what has been considered to be the Paleocene/Eocene Series boundary (assuming it is closely approximated by the calcareous nannofossil NP9/NP10 zonal boundary) in ODP Site 690B, but there the boundary appears to be located within Chron C24n, at odds with other sections. However, the calcareous nannofossil NP9/NP10 zonal boundary likely is older (by several hundred thousand years) than the Paleocene/Eocene Series boundary as currently recognized (= base of the Ieper Clay), and Aubry et al. (1996) further argued that the polarity interval correlated to Chron C24n in ODP Site 690B has been misidentified. In ODP Sites 550 and 549, the carbon isotopic shift has been considered to occur just below the position of the boundary—but the boundary in those sections occurs at a significant hiatus, and Aubry et al. (1996) indicate that the excursion does not occur at all in ODP Site 550. Thus the synchroneity and exact timing of the carbon shift in both marine and terrestrial sequences is uncertain (but see also Berggren and Aubry 1996). It is heartening to note, however, that the carbon isotopic shift in all records appears to occur within the first half of chron C24r, and may be synchronous between marine and terrestrial sections.

Clyde et al. (1994) presented an extensive magnetostratigraphic analysis of Wasatchian strata in the northern Bighorn Basin, effectively extending the Clark's Fork Basin/northern Bighorn Basin sequence of Butler et al. (1981) to younger intervals. Their 1480 meter-long magnetostratigraphic section (37 magnetostratigraphic sites, 135 samples) is robustly supported by jackknife analysis (table 4.2). The section spanned from Wasatchian Zone Wa0 (earliest Sandcouleean Subage) through the base of Zone Wa7 (earliest Lostcabinian Subage), and was correlated from Chron C24r (samples from one magnetostratigraphic site at the base of this section suggest that a normal polarity event/zone [of ambiguous reality and uncertain correlation] may be present, although it has not been identified in the correlative parts of the Clark's Fork Basin/northern Bighorn Basin section) through C24n.1r, representing about 2.5 my. In detail, they proposed Wasatchian zonal correlations of: 1) Zone Wa0 through the middle of Zone Wa5 (equivalent to the beginning of the Sandcouleean Subage through most of the "Graybullian" Subage; or the base of the *Haplomylus–Ectocion* Range Zone through about the middle of the *Bunophorus* Interval Zone) within Chron C24r; 2) the middle of Zone Wa5 through about the middle of Zone Wa6 (equivalent to the latest Graybullian through early Lysitean Subages; or middle *Bunophorus* Interval Zone through lower *Heptodon* Range Zone) to Chron C24n.3n; 3) the remainder of Zone Wa6 (late Lysitean Subage; middle *Heptodon* Range Zone) to Chrons C24n.2–early C24n.1r (although the three intervals correlated to Chrons C24n.2r through C24n.1r are represented by only one or two magnetostratigraphic sites each, with some sites having only one sample of clearly defined polarity); and 4) the earliest part of Zone Wa7 (earliest Lostcabinian Subage; earliest part of the upper *Heptodon* Range Zone) to part of Chron C24n.1r.

Clyde et al. (1994) also used the magnetostratigraphic correlation of early Bridgerian age strata by Flynn (1986; preferred correlation to Chron C21n, but alternative correlation possibilities included Chron C22n alone or spanning Chrons C22n–C21n) to infer that the remainder of the Wasatchian NALMA (the remainders of Zone Wa7, the *Heptodon* Range Zone, and the Lostcabinian Subage) probably spanned an equivalent amount of time, from Chron C24n.1r to at least Chron C22r. We suggest that the remainders of Zone

Wa7 and the Lostcabinian Subage possibly may be as young as Chron C21n, if Zone Wa7 extends to the base of the Bridgerian NALMA and if the preferred magnetostratigraphic correlation of early Bridgerian age strata by Flynn (1986) is correct. Clyde et al. (1994) regarded the base of their section (base of Zone Wa0 [containing a carbon isotopic excursion, citing unpublished data of Koch and Zachos], or the Clarkforkian/Wasatchian NALMA boundary), as equivalent to the Paleocene/Eocene Series boundary, although they recognized the complexities and difficulties in precisely correlating the Paleocene/ Eocene Series boundary (as defined in a marine or an interfingering marine/coastal/terrestrial stratotype) to terrestrial sections.

Tauxe et al. (1994) analyzed a 700 m thick composite section in the southern Bighorn Basin (266 samples from 90 sites). This magnetostratigraphic section, weakly supported by jackknife analysis (table 4.2) probably spans the Paleocene/Eocene Series boundary, as: 1) there is a $^{40}Ar/^{39}Ar$ date of 52.8 ± 0.3 Ma near the top of the section (Wing et al. 1991), and 2) a floral datum (FAD of *Platycarya*; Wing 1984; Wing et al. 1991; Bown et al. 1994) considered to occur near the Paleocene/Eocene Series boundary has been identified just below (Wing et al. 1991) or about 100 m above the base of the section (Wing 1984; Bown et al. 1994). Although its precise location within the southern Bighorn Basin composite section is ambiguous (Tauxe et al. 1994), the *Platycarya* datum has been loosely correlated to the calcareous nannoplankton NP9/NP10 zonal boundary in marine strata of the Gulf Coast. Unfortunately, the confusion is exacerbated by the fact that the NP9/NP10 zonal boundary had been considered to occur at, or slightly below, the Paleocene/Eocene Series boundary (see Berggren et al. 1985b, and Aubry et al. 1986), but is now believed to be significantly older than the Paleocene/Eocene Series boundary (Aubry et al. 1988; Berggren et al. 1995). Using the radioisotopic date of 52.8 ± 0.3 Ma from within their section, and mammalian biostratigraphic correlation (Zone Wa0) to the Clark's Fork Basin/northern Bighorn Basin magnetostratigraphic section of Butler et al. (1981; Zone Wa0 lying within Chron C24r near the top of the section spanning from Chron C26r through C24r), Tauxe et al. (1994) correlated their magnetostratigraphy to an interval spanning from within Chron C24r to near the younger boundary of Chron C24n. They also tentatively identified Cryptochron C24r.6 (Cande and Kent 1992), calculated to be about 0.5 my younger than the Paleocene/Eocene Series boundary, about 100 m above the base of the section. Thus, the Paleocene/Eocene Series boundary remains difficult to correlate with certainty to the BB sequence; it probably lies just below the local base of the Clark's Fork Basin composite section sampled by Tauxe et al. (1994), near Zone Wa0 and the lower of the two possible alternative placements of the *Platycarya* FAD. Tauxe et al. (1994) noted that these results contradict the suggestion by Rea et al. (1990) that the Paleocene/Eocene Series boundary lies about 160 m above the Wa0 faunal level in the southern Bighorn Basin, and therefore, the observed decrease in *Coryphodon* body size near that horizon postdates the boundary. We further note here that it also would negate the temporal correlation proposed by Rea et al. (1990) between the $\delta^{18}O$ excursion, and inferred climatic/ oceanographic changes, occurring at about the Paleocene/Eocene Epoch boundary and the *Coryphodon* body size decrease. Their (Tauxe et al. 1994) correlation also suggested that Koch et al.'s (1992) temporal correlation of a $\delta^{13}C$ excursion at the Wa0 horizon in the Bighorn Basin and in deep marine sections is "likely to be fortuitous" rather than due to synchronous events, because either 1) the excursion falls within the lower third of Chron C24r in the terrestrial sequence but within the middle of Chron C24r in ODP Hole 690B (as noted by Tauxe et al. 1994), or 2) the carbon isotopic excursion lies in Zone Wa0, at or very near the likely position of the Paleocene/Eocene Series boundary in the terrestrial BB sequence, but significantly below the boundary in the marine sequence illustrated by Rea et al. (1990:figure 1) and other marine sequences spanning the boundary (see e.g., Aubry et al. 1996). Alternatively, if the *Platycarya* FAD does lie near the base of the BB sequence (not 100 m above the base) and is approximately equivalent to the calcareous nannofossil NP9/NP10 zonal boundary (and thus is several hundred thousand years older than the Paleocene/Eocene Series boundary, as currently recognized as the base of the Ieper Clay), *then* the carbon isotopic excursion might be synchronous in both the BB and marine sequences, and the excursion, Zone Wa0, and the Clarkforkian/ Wasatchian NALMA boundary all would be latest Paleocene in age (see also Lucas, chapter 21). If this alternative is correct, the *Coryphodon* body size decrease in the BB sequence may coincide with the Paleocene/ Eocene Series boundary (= base of the Ieper Clay), and the $\delta^{18}O$ excursion and hypothesized associated climatic/oceanographic changes (Rea et al. 1990), and the major faunal changes at the Clarkforkian/Wasatchian NALMA boundary would not be causally due to physical/biotic changes associated with the Paleocene/ Eocene Series boundary. This alternative would be contradicted if the correlation of the short normal magnetic polarity interval 100 m above the base of the BB sequence to Cryptochron C24r.6 (of Cande and Kent 1992) by Tauxe et al. (1994) is correct, and this cryptochron is 500,000 years younger than the Paleocene/ Eocene Epoch boundary.

As Tauxe et al. (1994) and Clyde et al. (1994) sampled similar stratigraphic intervals in different parts of the greater Bighorn Basin, and both studies were published at the same time and without the benefit of direct comparison, we take this opportunity to compare their results. Mammalian biostratigraphic data suggest that both studies sampled approximately the same temporal span, from at (Clyde et al., 1994) or just above (Tauxe et al. 1994) the base of the Wasatchian (Zone Wa0), and extending into Zone Wa7. The magnetostratigraphic results are similar, with both sections spanning from within Chron C24r (probably beginning at the end of the first third, based on data from all three Bighorn Basin magnetostratigraphic sequences) into Chron C24n, although the Tauxe et al. (1994) section extends to slightly younger intervals (ending in Chron C24n.1n, versus Chron C24n.2r in the Clyde et al. 1994 study). Interestingly, both studies identify a short normal polarity interval within Chron C24r (the brevity is clear in Tauxe et al. 1994, but is presumed in Clyde et al. 1994 because in that study the normal interval is represented only by a single horizon [with only 2 of 3 samples having positive inclinations] at the base of the section). Both studies tentatively correlated this normal zone with the "Oldhaven event/magnetozone" (Townsend and Hailwood 1985; but Ali 1988 and Ali et al. 1993 reject this normal as a secondary overprint, see discussion below), and Tauxe et al. (1994) also tentatively identified it as Cryptochron C24r.6 (which occurs above the correlated Paleocene/Eocene Series boundary in marine sections: Tauxe et al. 1994). In fact, however, actual magnetic data for the so-called "Oldhaven event" are published nowhere. Townsend and Hailwood (1985) and Ali et al. (1993) published black and white bar diagrams with black (normal polarity) zones associated with the "Oldhaven beds." On the basis of the poor documentation and suspected overprinting problems, we reject the term "Oldhaven event (or subchron)," but suggest that there may be a short normal polarity interval closely associated with the Paleocene/Eocene Series boundary. If the short normal polarity zone is the same in both Bighorn Basin sections, it has great potential for local to global temporal correlation via a synchronous horizon/zone, and for refining age estimates of events near the Paleocene/Eocene Series boundary by subdividing the very long reversed magnetochron (Chron C24r) in which it occurs.

Some mammalian biostratigraphic events appear to be synchronous, within the resolution of the magnetostratigraphic techniques, in the two Bighorn Basin sections, including the LO (lowest [regional stratigraphic] occurrence, Aubry, 1995) of *Lambdotherium* (base of Zone Wa7) in the Chron C24n.1r correlative, and the LO of *Heptodon* (base of Zone Wa6) in the Chron C24n.3n correlative. But other mammalian events have different relative stratigraphic and/or magnetostratigraphic occurrences in the two sections, including the HO (highest [regional stratigraphic] occurrence, Aubry 1995) of *Ectocion* and *Haplomylus* (the HO of *Ectocion* occurs slightly before the HO of *Haplomylus* in both sections, but both events occur earlier [within the Chron C24r correlative] in the northern Bighorn Basin than in the southern Bighorn Basin [within the Chron C24n.3n correlative]), and the LO of *Bunophorus* (which occurs in Chron C24r [above the *Ectocion* HO and at the *Haplomylus* HO] in the northern BB, but in Chron C24n.3n [and below the HOs of both *Ectocion* and *Haplomylus*] in the southern BB). These contradictory patterns suggest that some of the biostratigraphic events are: 1) poorly constrained in one or both areas, 2) asynchronous between both areas, or 3) that one or both magnetostratigraphic correlations are incorrect in part. We believe that both magnetostratigraphic correlations are likely to be largely correct, and that the apparently asynchronous mammalian LO/HO events are due to incomplete biostratigraphic or magnetostratigraphic sampling (e.g., the LO of *Platycarya* is difficult to establish clearly; lack of sampling horizons in the uppermost part of Chron C24r in the northern BB may have resulted in slightly higher placement of the base of Chron C24n.3n and apparent earlier LO of *Bunophorus* and HOs of *Ectocion* and *Haplomylus*) and/or slight temporal offset in their preserved or actual occurrences within the different parts of the BB. These apparent temporal offsets may seem minor, but they may significantly affect the ability to recognize the Wasatchian zonal subdivisions as synchronous, even within the same depositional basin (e.g., the base of Zone Wa5 [defined by the LO of *Bunophorus etsagicus*] appears to be high in Chron C24r in the northern BB, but well within Chron C24n.3n in the southern BB). Another point of interest is that even though both sequences span about the same temporal interval, average accumulation rate appears to be about twice as fast in the northern BB (1600 m in 2.5 my) than in the southern BB (800 m in 2.5 my). Yet accumulation rate appears relatively steady throughout the entire northern BB section, but increases dramatically in the upper half of the southern BB section (the approximately 1.7 my of Chron C24r preserved spans only 300 m, while the approx. 0.8 my of Chron C24n preserved spans 400 m).

Discussion

The most complete and reliable magnetostratigraphies from terrestrial sequences spanning the Paleocene/Eocene Series boundary appear to be those from the San Juan Basin, Ellesmere Island, and the Northern Bighorn

Basin, as those sequences all have high "J" values (between 0 and –0.35). In those terrestrial sequences, as in the boundary stratotype and correlated marine sections, the boundary clearly lies within a long reversed polarity interval correlated to Chron C24r.

There are short normal polarity zones at, or shortly above, the Wa0 Zone in both the Northern and Southern Bighorn Basin sections, as well as in the zone labeled Wa? in West Texas (see figure 4.5). Also, there is another short normal polarity zone in the Wa3 Zone of the Clark's Fork Basin. The lower part of the Southern Bighorn Basin section is poorly exposed, and there was some discussion about the exact placement of sites in the stratigraphic composite section of Bown et al. (1994). Thus it is possible that the normal zones (at or slightly above the Wa0 Zone) in the Northern and Southern Bighorn Basin sections are the same. The sporadic occurrence of poor mammalian faunas of the West Texas section precludes high resolution temporal correlation, but certainly permits the normal zone represented there to also be the same short normal polarity interval closely associated with the Paleocene/Eocene Series boundary. However, the normal polarity zone at the top of the Northern Bighorn Basin/Clark's Fork Basin sequence (exhibited in the Big Sand Coulee section of Butler et al. 1981) occurs in mammal Zone Wa3, and hence is unlikely to be the same normal polarity zone as that found in the other sections. Several deep sea sections show normal zones within Chron C24r, but the best sections do not (Contessa Road, DSDP Site 577; figure 4.2), perhaps because of low sampling density. As discussed above, if the short normal polarity zone is the same in all of these marine and terrestrial sections, it will provide a synchronous horizon/magnetozone to enhance greatly local-to-global temporal correlation.

In considering the difficulties inherent in correlating between marine and terrestrial sequences, over broad areas, Gunnell et al. (1993:140) rejected the use of biostratigraphic correlation for global geochronologic correlation, apparently because of presumptions that "the diachronous biotic reaction to geologic events, and the intra- and intercontinental time-transgressive nature of biotic reorganization make such biotic events unreliable indicators of temporal compatibility among geographically diverse samples." Gunnell et al. (1993:142) proposed that the "boundary between the Paleocene and Eocene in continental sediments be based on the most dramatic and obvious evidence of biotic reorganization," presumably either because unspecified dramatic changes at that boundary would lead to synchronous "reorganizations" in the biota worldwide, or that the difficulties in biostratigraphic correlation would require use of a "boundary" of convenience (but not necessarily synchrony) recognized at some obvious change in the biota in each particular sequence studied. Therefore, they "propose that the boundary between these two epochs continues to be placed at the Clarkforkian/Wasatchian boundary because of the dramatic reorganization of mammalian faunas that occurs there" (p. 137). We reject this approach, as any global chronostratigraphic/geochronologic boundary can be based *only* on correlation (as defined in the International Stratigraphic Guide 1994:15, "To correlate, in a stratigraphic sense, is to show correspondence in character and/or in stratigraphic position. There are different kinds of correlation [litho, bio, and chronocorrelation] depending on the feature to be emphasized" [Brackets added]; rather than Gunnell et al.'s 1993 unusual usage of "correlation"). Gunnell et al.'s use of an additional "recognition" criterion as an alternative to correlation is confusing, and probably is unnecessary. Parenthetically, if the Paleocene/Eocene Series boundary were to be "recognized" by dramatic reorganization of mammalian faunas, the faunal "reorganization" at the Clarkforkian/Wasatchian NALMA boundary has been suggested to be much less dramatic than that at the immediately preceding Tiffanian/Clarkforkian NALMA boundary (but see Gunnell, chapter 19).

The Paleocene/Eocene Series boundary first must be defined as the boundary between two chronostratigraphic units, in a boundary stratotype sequence, and then conceptualized as an instant in geologic time as a global geochronologic boundary between two geochronologic units (but see Aubry, chapter 3, for an alternative view). All available data must then be used to determine the temporal relation of various events and units (e.g., biostratigraphic, chemostratigraphic) to this boundary, and an attempt made to correlate chronostratigraphic data from Bighorn Basin strata and the boundary stratotype (as well as from other correlative sections) to identify the most likely position of the Paleocene/Eocene chronostratigraphic boundary, and its conceptually related geochronologic boundary and assigned numerical age, within the Bighorn Basin sequence. Of course it usually is more difficult to chronocorrelate between terrestrial (e.g., the Bighorn Basin) and marine/nearshore (e.g., the stratotypes for the Paleocene and Eocene Series) sequences than between different marine sequences. Additionally, the Paleocene/Eocene Epoch/Series boundary (in a conceptual, geochronologic sense, as well as a correlated chronostratigraphic horizon) may be represented by an unconformity in the Bighorn Basin sequence, and likely in most other stratigraphic sequences, and thus might never be recognized precisely. Although we agree with Gunnell et al. (1993) and others (e.g., Tauxe et al. 1994; Clyde et al. 1994) that the Paleocene/Eocene Series boundary may be correlated to and inferred to occur at about the Clarkforkian/Wasatchian NALMA boundary (and, if so, may coincide because of globally synchronous responses to major physical and biotic changes), as elaborated above, we differ methodologically in concluding that the boundary must be correlated from the stratotype (using all available chronostratigraphic information) and cannot be constrained or forced to fall precisely at that point simply because there is a "dramatic reorganization of the mammalian fauna." As mentioned above, and in agreement with the point of Gunnell et al. (1993) that global biostratigraphic correlation between marine and terrestrial strata is extremely difficult, magnetostratigraphy is very important for constraining attempted temporal correlation of the Paleocene/Eocene Epoch boundary.

Europe

European sequences are of the utmost significance to discussions of events and correlations near the Paleocene/Eocene Series boundary, as they contain the stratotype sequences historically used to define the boundary and relevant bracketing stages/ages. Unfortunately, these "greater North Sea basin" sequences were deposited in epicontinental seas (shallow to deeper shelf) and surrounding terrestrial environments, and therefore are riddled with minor to significant unconformities caused by sedimentary responses to changing sea levels and tectonic movements. Few radioisotopic dates are available from volcaniclastic sediments or primary volcanics in this sequence, although reliable $^{40}Ar/^{39}Ar$ dates of 54.5 Ma (Swisher and Knox 1991) and 55.0 Ma (J. Obradovich, USGS; cited in Berggren et al. 1995) have been produced on the "–17 Ash" in the North Sea, which had been considered close to the Paleocene/Eocene Series boundary (see discussion in Berggren et al. 1995). The "–17 Ash" previously was correlated to the "Oldhaven Beds," based on correlation of the "–17 Ash" to the Hales Clay, which had been presumed incorrectly to be correlative to the "Oldhaven Beds" (Cox et al. 1985). Thus, the "–17 Ash" (and the correlative Hales Clay [lower Harwich Formation]) is older than the "Oldhaven Beds"; further, changes in the lithostratigraphic definition of the base of the London Clay make all of these ("–17 Ash," Hales Clay, "Oldhaven beds") slightly older than the Paleocene/Eocene Series boundary. Odin (1982) summarized radioisotopic dates, largely K-Ar determinations on authigenic glauconites, available until that date. Although many glauconites appear to provide reasonable ages for associated sediments, the fact that they are authigenic minerals with variable and uncertain closure ages relative to the age of the deposits in which they are found, make it difficult to interpret the reliability and significance of any individual date derived from them (see discussions in Berggren et al. 1985; Aubry et al. 1988). Summaries of the biochronologic sequence of Paleogene mammalian faunas have been published by Hooker (1991, 1996, chapter 20). The sequence of Cernaysian–Neustrian European Land Mammal "Age" (ELMA) faunas appears to span the upper Paleocene through lower Eocene Series (Hooker, chapter 20). Faunas of these ELMAs have been assigned to the following biozones or MP "Reference Levels" ("Reference Level" in quotation marks, as these [sometimes also incorrectly referred to as "zones"] have not been defined yet as formal chronostratigraphic/geochronologic units), given in proposed superpositional sequence within basins: Paris Basin Cernay ("Reference Level" MP6), Meudon ("Reference Level" MP8–9; Zone PE II), Pourcy ("Reference Level" MP8–9; Zone PE III), Mutigny ("Reference Level" MP 8–9; Zone PE IV), Avenay ("Reference Level" MP8–9; Zone PE V), Grauves ("Reference Level" MP8–9); London–Hampshire Basins Suffolk Pebble Bed [reworked?]/Woolwich Shell Beds ("Reference Level" MP7; Zone PE I), Abbey Wood [Blackheath beds] ("Reference Level" MP8–9; Zone PE III), Swanscombe Member/Harwich Stone Band of London Clay Formation ("Reference Level" MP8–9; PE III), Herne Bay ("Reference Level" ?MP10); Belgian Basin Dormaal ("Reference Level" MP7; Zone PE I), Erquelinnes ("Reference Level" MP7; Zone PE I). However, isolated areal occurrences and bracketing sedimentary gaps still make it extremely difficult to determine precise superpositional relationships and temporal correlations of localities containing these faunas. Recent work (Hooker, chapter 20) suggests that the Paleocene/Eocene Series boundary occurs between "Reference Level" MP6 and Concurrent Range Zone PE1 (*Platychoerops georgei–Cymbalophus cuniculus* Biozone; therefore, within or somewhat older than the traditional concepts of "Reference Levels" MP8–9).

Townsend (1982) and Townsend and Hailwood (1985) presented magnetostratigraphic studies of lower Paleogene strata of the London–Hampshire Basins (see also Cox et al. 1985; Aubry 1985; Aubry et al. 1986), including analyses that suggested the presence of the widely discussed normal polarity interval of the "Oldhaven Formation" (probably latest Paleocene in age). Townsend and Hailwood (1985:981) proposed that the Whitecliff Bay section be adopted as "the UK Eocene magnetostratigraphic type-section"; although the lower part of the section is certainly very early Eocene in age, it is not clear whether upper Paleocene strata and the Paleocene/Eocene Series boundary are represented in the base of this section. Ali (1988), Ali et al. (1993), Ali and Hailwood (1995; also citing Ellison et al. "in press"), and Ali and Jolley (1996) present the most extensive magnetostratigraphic studies of terrestrial sequences from northwest Europe spanning the Paleocene/Eocene Series boundary, including strata from southern England, Belgium, northern France and the

Paris Basin, and Denmark. The following regional European summary is based primarily on the published results of Ali (1988), Ali et al. (1993), and Ali and Hailwood (1995), with consideration of the results of the other papers mentioned above.

In southern England the Thanet Sand Formation (= "Thanet Formation") spans Chron C26n (or possibly as old as Chron C26r, including the borehole sequences of the Ormesby Clay Formation, as a correlative of the Thanet Sand Formation) through Chron C25r (with Chron C26n well-represented in several borehole sequences). A hiatus follows in the southern England sequences, and Chron C25n had been widely believed to not be preserved. The overlying Reading and Woolwich formations (or lumped "Woolwich–Reading" formations) are of dominantly reversed polarity, with several short normal polarity events that may be correlatable regionally; although the base of the Woolwich–Reading Formation is marked by a significant unconformity, the unconformity had been considered to be of relatively short duration, and the reversed polarity interval was believed to probably represent a continuation of Chron C25r, with Chron 25n absent. Ali and Hailwood (1995; also citing Ellison et al. "in press"), and Ali and Jolley (1996) studied the Upnor Formation (Lambeth Group) in central London, and suggested that the lowermost of two distinct sedimentary packets in this formation is a small (a few square kilometers) erosional remnant, preserved below the upper sedimentary packet deposited during the main Upnor Formation transgression (early Chron C24r). This lowermost packet preserves a normal polarity interval correlated to Chron C25n (op cit.), in part because the sedimentary packet is correlative with Zone NP9 (Aubry, personal communication 1996). Therefore, the Woolwich and Reading formations are younger than had been thought (e.g., by Ali, 1988; entirely Chron C24r, rather than Chron C25r possibly extending into Chron C24r [with Chron C25n absent]). The Woolwich and Reading formations, together with the Harwich Formation (previously considered the Harwich Member, or "Division A1," of the London Clay Formation) and Walton Member (representing "Division A2") of the London Clay in East Anglia (as well as strata assigned to "Division A2" in Kent, the north and west London Basin, and the Hampshire Basin) are of reversed polarity, correlative with Chron C24r.

Ali (1988) provided evidence that the previously recognized (e.g., Townsend and Hailwood 1985) "Oldhaven" normal magnetozone in the upper Harwich Formation ("Harwich Member"; and stratigraphically correlative Oldhaven beds at Herne Bay) was caused by secondary overprints. This is consistent with the entirely reversed polarity interval in the correlative Harwich Formation interval of the Ormesby and Hales borehole sequences in Norfolk, England (Cox et al. 1985; although they suggested this might indicate a hiatus, and absence of the "Oldhaven" normal, in the Ormesby sequence). Ali (1988) identified a thin normal polarity interval at the base of the London Clay Formation ("London Clay Basement Bed," "Division ?A1") in Kent, which he suggested might represent a short normal polarity event within Chron C24r. Ali (1988:77) also thought that normal polarity intervals in the lower half of the Tilehurst Member ("Oldhaven Formation") at Harefield and Tilehurst (in the northwestern and western London Basin, respectively) and two normal polarity sites "in a similar unit" in the Formation de Varengeville in northern France represented, and supported the reality of, this event. Interestingly, Ali's (1988) magnetostratigraphic results indicated that the Tilehurst Member and/or the base of the dinoflagellate *Wetzeliella astra* Zone, often considered to lie at or very near the Paleocene/Eocene Series boundary, are time transgressive across southern England. Importantly, rejection of the reality of the "Oldhaven normal" magnetozone may cast further doubt on the proposal that the short normal polarity event in Chron C24r of the Bighorn Basin North American sequences is equivalent to the "Oldhaven normal"; and as discussed above, we reject the existence of an "Oldhaven event." However, an alternative possibility is that the thin normal event within Chron C24r in the Bighorn Basin sequence correlates instead with the other, slightly younger, normal polarity event that has been identified in Chron C24r in the Tilehurst Member ("Oldhaven Formation") and London Clay "Basement Bed" of southern England and Formation de Varengeville in northern France.

In the London (Sheppey) and Hampshire (Whitecliff Bay) Basins, the London Clay Formation spans from Chron C24r through early Chron C23n and is probably entirely of early Eocene (Ypresian) age. In the Hampshire Basin, the conformably overlying Wittering Formation spans from early Chron C23n through part of Chron C22r. Similarly, correlative sections of the Ieper Clay Formation in the Belgium Basin span from the very latest part of Chron C24r (at the base of Division B, with underlying Divisions A2 and A3 not sampled, and with an unconformity at the base of the formation) through very early Chron C22r. In detail, within southern England sections Chron C24n.2 spans from the base of Division B1 through about the middle of Division B2, Chron C24n.1 spans from about the middle of Division C1 through about the middle of Division C2, and Chron C23n spans the upper part of Division D to above the top of the London Clay Formation. In detail, within Belgium Basin sections Chron C24n.2 spans much of the lower half of Division B (the middle part of the Wardrecques Member), Chron C24n.1 spans most of

Division C2 (the lower part of the Baileul Member), and Chron C23n spans from low in the Aal Clay to the top of the Kortemark Silt of the Ieper Clay Formation.

In summary, the magnetostratigraphy of the upper Paleocene through lower Eocene terrestrial and near-shore marine sequences in epicontinental northwestern European basins has been intensively studied and is reasonably well understood. Many of the magnetozones through this interval are represented in those continental sequences, although evidence from biostratigraphy, sedimentology, and magnetostratigraphy make it clear that these sequences are riddled with unconformities that formed in response to eustatic (and relative) sea level changes. Unfortunately, although some of the classical northwest European Paleogene mammal-bearing localities (which form the basis for the "MP" mammalian biochronology/"reference level"/"zone" series, and the new formal biozones of Hooker, chapter 20) are present directly within this sequence, most are not. Therefore, it is very difficult to determine the exact temporal sequence of many isolated mammal-bearing localities and precise chronostratigraphic ranges of mammalian taxa, hindering temporal correlation of mammal faunas (and associated biotic and geologic events) between Europe and other continents. Confusion and debate over terminology, correlation, and age of lithostratigraphic, "biochronologic," and chronostratigraphic units has further complicated these correlations. This is true of the Paleocene Series, in general, and the "Sparnacian" and "Cuisian" Stages in particular (see discussions in Hardenbol and Berggren 1978; Berggren et al. 1985b). Strata used by Schimper (1874) to characterize the Paleocene Series/Epoch (formulated well after, and in a different manner than, the Eocene and other Cenozoic epochs of Lyell, 1833) actually are equivalent in age to units that will be considered early and late Paleocene in age, as well as to those assigned to the Eocene (both by Lyell and by working groups proposing stratotypes of all the Cenozoic ages and epochs). The Sparnacian Stage (s.l.) includes strata (or temporal correlatives) spanning much of this range, over more than 3 my (from at least Chron C24r through Chron C23r) and including mammalian faunas from strata assigned to "Reference Level" MP7 (e.g., Suffolk Pebble Bed or Erquelinnes) through possibly "Reference Level" MP10 (e.g., Herne Bay). The current, more restricted Sparnacian Stage (s.s.) ends at the beginning of the Cuisian/Ypresian Stage. It includes strata that are largely or entirely late Paleocene in age, including strata with mammalian faunas assigned to "Reference Level" MP7 (e.g., Suffolk Pebble Bed and Erquelinnes) and some, but not all, of those with mammalian faunas considered assignable to "Reference Levels" MP 8–9 (e.g., Abbey Wood [Blackheath Beds], Meudon, Pourcy).

South America

The South American lower Paleogene record is woefully constrained geochronologically. There are many fewer geochronologic studies integrating multiple correlation techniques than for the South American Neogene record, largely because of the paucity of radioisotopically datable horizons in known fossiliferous sequences. The Paleocene Riochican South American Land Mammal "Age" (SALMA; and superposed assemblages from older parts of the sequences, see below) from localities near the type sequence in Argentine Patagonia, including paleomagnetics (within Riochican SALMA, and older, strata), K-Ar dates (from within and below the marine Salamanca Formation underlying Peligran through Riochican strata), and Danian foraminifera (from the Salamanca Formation; alternatively considered upper Danian Stage [Bertels 1979; Marshall et al. 1981] or lower Danian Stage [Bonaparte et al. 1993, based on data in Toumarkine and Luterbacher 1985]). The Casamayoran SALMA traditionally has been considered to be early Eocene in age, based largely on the observation that Casamayoran faunas overlie, possibly conformably, late Paleocene Riochican faunas. However, there are no independent chronologic constraints currently available directly from Casamayoran strata, although work in progress by Wyss, Flynn, Charrier, and Swisher (unpublished) suggests that it may be possible to radioisotopically date (using $^{40}Ar/^{39}Ar$ techniques) Chilean sequences containing Casamayoran SALMA faunas.

Paleocene (Riochican SALMA and older units)

For many years, the Paleocene interval was represented by a single SALMA, the Riochican SALMA. Recent work suggests that there may be four, or more, distinct Paleocene SALMAs (see details in Flynn and Swisher 1995).

Short magnetostratigraphic sections of the Salamanca and Rio Chico formations at Punta Peligro and Cerro Redondo in Argentine Patagonia, representing the Paleocene Riochican SALMA, have been correlated to Chron C26r through Chron C25n (Marshall et al. 1981). These correlations would indicate that the Riochican is late Paleocene in age, and slightly older than the Paleocene/Eocene Series boundary (within Chron C24r). Bonaparte et al. (1993) disputed part of this correlation, especially for the basal stratigraphic interval sampling the "Banco Negro Inferior" of the Salamanca Formation (which contains the fauna they used to propose a new SALMA, the Peligran, older than the Riochican; fauna considered either Tiupampan SALMA [early Paleocene age] or possibly a distinct Peligran SALMA by Bond et al. 1994; tentatively considered Tiupampan

SALMA by Pascual et al. 1992a, 1992b, and Bond et al. 1994, and personal communication 1994), suggesting that while the majority of the longer sequence at Cerro Redondo (which includes the classical Riochican assemblages) might in fact span Chron C26r through Chron C25n, there may be problems with interpreting the basal part of that sequence and the very short section at Punta Peligro (both of which include the "Banco Negro Inferior").

The newly proposed Peligran SALMA (typified by the faunal assemblage from the "Banco Negro Inferior," Bonaparte et al. 1993) is considered broadly early to middle Paleocene in age, as the typifying fauna occurs in strata underlying (probably unconformably) Riochican faunas and overlying parts of the Salamanca Formation with Danian foraminifera. It overlies strata yielding low precision K-Ar dates of about 62.5–64 Ma, and occurs within a reversed polarity interval of uncertain correlation (correlated to Chron C26r by Marshall et al. 1981; part of which possibly may be Chron C27r instead, according to Bonaparte et al. 1993). Although Marshall et al. (1981) included the "Banco Negro Inferior" in the Rio Chico Formation, Bonaparte et al. (1993) pointed out that it is better assigned to the Salamanca Formation (Bond et al. 1994 and personal communication 1994 similarly call it the Hansen Member of the Salamanca Group, but without specifying a formation). Bonaparte et al. (1993) also suggested that there may be a missing normal polarity zone in the lower part of the Rio Chico Formation (as the stratigraphic thickness between the reversed horizons in the "Banco Negro Inferior" and the first reliable reversed horizon in the Rio Chico Formation is greater than the thickness of the two normal polarity intervals identified from the upper part of the section), and that there is an unconformity between the Rio Chico and Salamanca formations. Therefore, they conclude that there is no reason that the "Banco Negro Inferior" (and thus the Peligran SALMA) must be younger than Chron C27n (as proposed by Marshall et al. 1981) because there is an unconformity corresponding to a hiatus of unknown duration, there may be a missing normal polarity zone, the foraminiferal faunas from the underlying parts of the Salamanca Formation may be older than previously thought (e.g., as old as early Danian in age [Bonaparte et al. 1993, based on data in Toumarkine and Luterbacher 1985], rather than late Danian in age [Bertels 1975]), and the radioisotopic dates from the underlying Salamanca Formation suggest possible correlation of the magnetozones to older chrons if one correlates to newer GPTS (Cande and Kent 1992, 1995; although we would also point out that these old dates are poorly constrained and may not be useful for any precise correlation). Bonaparte et al. (1993) suggested instead that the Peligran falls within Chron C27r. Unfortunately, there are only preliminary studies of the fauna and poor geochronological constraint for the "Banco Negro Inferior," and a lack of independent geochronologic evidence from the Tiupampan SALMA in Bolivia. This makes it difficult to determine the temporal relationships between the Tiupampan and Peligran SALMAs and to decide if indeed the Peligran SALMA is younger than the Tiupampan SALMA (as suggested by Bonaparte et al. 1993) or the "Peligran" SALMA is temporally equivalent to the Tiupampan SALMA (as suggested tentatively by Pascual et al. 1992a, 1992b, and Bond et al. 1994 and personal communication 1994) and therefore invalid.

The same weak temporal constraints, and no direct magnetostratigraphic data, exist for the unnamed ("indeterminada"), post-Peligran and pre-Riochican, mammal age of Bonaparte et al. (1993).

Paleocene–Eocene marine intertongues in South American terrestrial sequences

Unfortunately, there are very few stratigraphic intervals or geographic areas from the Cenozoic record of South America in which fossiliferous marine strata intertongue with fossiliferous terrestrial sequences. The few marine sequences that do intertongue generally represent very shallow water environments that have not yielded good microplankton fossils. One of the most famous epicontinental marine sequences is the "Patagoniano" or "Patagonian Stage" (Simpson 1940; including "Formation Patagonienne" and "Etage superpatagoneen" of the "Formation Santacruzienne" of Ameghino 1900–1902; see Marshall et al. 1983), considered latest Oligocene or early Miocene in age, but now considered to represent at least three distinct units "ranging in age from Eocene to early Miocene" (Marshall et al. 1983:27, citing personal communication from Zinsmeister; see also Zinsmeister 1981). Two of the units considered "Patagoniano" include the "Estratos con *Monophoraster* y *Venericor*" (upper Eocene Series, molluscs: Camacho 1974; see also Zinsmeister 1981; Marshall et al. 1983), and San Julian Formation (lower [and middle?] Eocene Series, molluscs: Camacho 1974; but considered upper Eocene to lower Oligocene Series [upper Priabonian and lower Rupelian Stages] by Bertels 1970, 1975, based on foraminifera; see discussion in Zinsmeister 1981). The San Julian Formation is the marine unit most likely to be relevant to understanding events around the Paleocene/Eocene Epoch boundary; it has not yet been studied magnetostratigraphically, but it is underlain by strata containing Casamayoran SALMA faunas (traditionally considered "early Eocene" in age). The Riochican SALMA (and the older Itaboraian SALMA [or unnamed interval] and Peligran

SALMA from the same sequence, discussed above) is the other Paleocene–Eocene mammalian biochronologic interval constrained in age by associated marine units. All these units appear to be older than the Paleocene/Eocene Series boundary. In its type area, but not at the type section, the mammal-bearing Rio Chico Formation/Group is underlain conformably (Marshall et al. 1981, 1983) or unconformably (Bonaparte et al. 1993) by the marine Salamanca Formation/Group, which contains planktonic foraminifera indicative of late Danian (Bertels 1970, 1975, 1979) or early Danian (see Bonaparte et al. 1993) age, and the oyster *Odontogryphaea* is said to indicate a poorly resolved Thanetian or Ypresian age (Stenzel 1945).

Asia

No magnetostratigraphies are available yet for upper Paleocene–lower Eocene strata from Asia. However, J. Kappelman and W. Gose are developing a magnetostratigraphy for the mammal-bearing strata of the middle to upper Eocene Heti Formation, Yuanqu Basin, China (C. Beard, personal communication). Russell and Zhai (1987) provided the most recent summary of proposed temporal correlations of Asian Paleogene mammal-bearing strata. Their correlation charts indicated several sequences in China (including "Inner" Mongolia) and Mongolia that may span the Paleocene/Eocene Series boundary, and additional sequences from India and Pakistan that may be early Eocene in age, although none of these have yet produced a magnetostratigraphy.

Africa

The lower Paleogene terrestrial stratigraphic record of Africa has been poorly known, and only in the last 10–15 years have significant relevant mammalian faunas been reported from North Africa (see Gheerbrant 1989, 1990 and references cited therein). These faunas include those from the upper Paleocene record of Morocco (Adrar Mgorn locality, Ouarzazate Basin; Cappetta et al. 1978, 1987; Gheerbrant 1989, 1990), the lower Eocene strata of Tunisia (Chambi locality; Hartenberger et al. 1985), and the lower and upper Eocene sequences of Algeria (El Kohol locality [Mahboubi et al. 1984, 1986] and Bir El Ater locality [Coiffait et al. 1984], respectively). As in Asia, there are no magnetostratigraphies available yet for upper Paleocene–lower Eocene strata of Africa.

Australia

The Australian Paleogene terrestrial record is quite poor, and no magnetostratigraphies have been reported. The only Paleocene or Eocene vertebrate-bearing strata are freshwater clays near Murgon, southwestern Queensland, producing the Tingamarra Local Fauna (Godthelp et al. 1992; Archer et al. 1993; Hand et al. 1994). This fauna has been considered earliest Eocene in age because of a K-Ar date of 54.6 ± 0.5 Ma on poorly crystalline, authigenic illites (presumed to have crystallized for only a short time, and assuming little argon loss); an overlying basalt has been dated at 29.0 ± 0.2 Ma (Godthelp et al. 1992; Archer et al. 1993; Hand et al. 1994). The 54.6 Ma date is a minimum age estimate because it is on an authigenic illite, and if the Paleocene/Eocene Epoch boundary is considered to be about 55 Ma, the Tingamarra fauna may be either latest Paleocene or earliest Eocene in age and almost synchronous with the boundary. The fauna includes Australia's oldest placental mammals (a "condylarth-like" placental and a bat), and a diverse marsupial assemblage. The age of the Tingamarra fauna predates Australia's plate tectonic separation from Antarctica, and its components suggest biogeographic affinities to other Gondwanan continents continuing from the Mesozoic Era until at least the Paleocene/Eocene Epoch boundary.

Antarctica

To date, the only Paleogene terrestrial fossils from Antarctica have been reported from the upper Cretaceous–?upper Eocene sequence (e.g., Sadler 1988) on Seymour Island. This sequence is dominantly marine, but palynofloras have been reported for most of the sequence (e.g., Askin 1988, 1990; Case 1988). Within this sequence, the ?upper Eocene La Meseta Formation has yielded macrofloras (Case 1988) and a growing assemblage of fossil mammals (e.g., astrapothere; ?litoptern and toxodont "ungulates"; polydolopid, and possibly microbioteriid, marsupials; ?megalonychid tardigrade edentate; Woodburne and Zinsmeister 1982, 1984; Case et al. 1988; Bond et al. 1989; Carlini et al. 1990; Pascual and Ortiz Jaureguizar 1991; Hooker 1992; Marenssi et al. 1994). This fauna has been considered as probably late Eocene in age, based in part on the "stage-of-evolution" of the mammals, as well as poorly resolved temporal information from marine microfossil and macroinvertebrate biostratigraphies; there is no magnetostratigraphy for this sequence, and no reported radioisotopic dates. The mammal assemblages indicate a biogeographic association between South America and Antarctica at least until well into the Cenozoic Era. However, the generic-level endemism of the Antarctic taxa, relative to the coeval nearby faunas from Argentine Patagonia, suggests that significant landbridge connections between the two continents were disrupted prior to the mid-late Eocene Epoch.

Understanding the timing, rates, and interrelationships of paleobiologic and geologic events and processes requires a precise temporal framework in which to compare records of these events and processes preserved in different kinds of sediments. Magnetostratigraphy, particularly if associated with independent biostratigraphic data and/or radioisotopic dates, is essential for correlating between marine and terrestrial strata.

We have collected all the principal magnetostratigraphic data sets from both marine and terrestrial strata spanning the Paleocene/Eocene Series boundary. We rely on the actual directional records presented in the original publications, and the marine magnetic anomalies used for constructing the GPTS, for determining both magnetic polarity reversal sequences and correlations between the magnetostratigraphies and the GPTS.

We have employed two simple means of assessing the reliability of published magnetostratigraphic data sets. First, histograms of the parameter used to determine polarity, usually virtual geomagnetic pole (VGP) latitudes or inclinations, should show two distinct "humps" representing the two polarities. If polarities cannot be discriminated in the histogram plots, then the magnetostratigraphic patterns may not provide reliable polarity information. The second method is the jackknife parameter "J," a quantitative assessment of the dependence of a given polarity stratigraphy on the sampling density. Sections with high "J" values (0 to –0.5) are relatively robust, whereas low "J's" (high negative values, below –0.5) indicate numerous polarity zones represented by only one or two sampling horizons, and a greater probability that the resultant magnetostratigraphy is incomplete or incorrect.

The biostratigraphically-constrained marine magnetostratigraphy of the Paleocene/Eocene Series boundary interval is disappointing. The Contessa Road and ODP Site 752 sections appear to give the most reliable results, indicating clearly that the Paleocene/Eocene Series boundary lies within Chron C24r. However, there is poor consistency among the various marine sequences examined. Some of the discrepancies could be due to hiatuses (e.g., in DSDP Sites 550 and 549; and likely in DSDP Site 577 and ODP Site 690, Aubry personal communication 1996), poor quality data (Sites DSDP Sites 550, 549, and 528), or poor sampling density (ODP Site 752). However, none of these explanations appear to account fully for the large difference between the apparent boundary location in ODP Site 690 and the lower latitude sites such as DSDP Site 577. Perhaps the nannofossil datums marking the boundary are significantly time transgressive. Alternatively, the magnetic data from ODP Site 690 (and DSDP Site 528) could be overprinted. Overprinting alone, however, cannot account for the fact that the carbon isotopic excursion, which occurs near the Paleocene/Eocene Series boundary in DSDP Sites 550 (but see Aubry et al. 1996) and 549, has been considered to lie some 40 m below it at ODP Site 690 (presuming the boundary was closely approximated by the calcareous nannofossil NP9/NP10 zonal boundary), in the lower third of a prominent reversed polarity interval likely to be Chron C24r. This discrepancy must be due to a combination of unconformities and overprinting, most likely in the ODP Site 690 record. There currently is only sporadic magneto-biochronologic data, of questionable reliability, spanning the Paleocene/Eocene Series boundary in shallow marine sequences of northwestern Europe. Problems inherent in deriving and interpreting magnetostratigraphies in such unconformity-riddled settings will make it very difficult to ever derive long, continuous polarity sequences or to correlate the resultant magnetostratigraphies to the GPTS.

The most complete and reliable magnetostratigraphies from terrestrial sequences spanning the Paleocene/Eocene Series boundary appear to be those from the San Juan Basin, Ellesmere Island, and the Northern Bighorn Basin, as those sequences all have high "J" values (between 0 and –0.35). In those terrestrial sequences, as in several likely candidates for the boundary stratotype (to be defined soon by IGCP Project 308) and correlated marine sections, the boundary clearly lies within a long reversed polarity interval correlated to Chron C24r. Although we agree with Gunnell et al. (1993) and others (e.g., Tauxe et al. 1994; Clyde et al. 1994) that the Paleocene/Eocene Series boundary may be correlated and inferred to occur at about the Clarkforkian/Wasatchian North American Land Mammal "Age" boundary, and in fact possibly may coincide because of globally synchronous responses to major physical and biotic changes, we differ methodologically in concluding that the boundary ultimately must be correlated from the stratotype, and cannot be constrained or forced to fall precisely at that point simply because there is a "dramatic reorganization of the mammalian fauna" at the Clarkforkian/Wasatchian NALMA boundary. Within these North American terrestrial sections, the relative positions of the Clarkforkian/Wasatchian biochronologic boundary, the short normal polarity interval within Chron C24r (see below), and several proposed isotopic marker "events" are variable; the degree of synchrony (or amount of, and variability in, temporal offset) and relative ordering of these events are currently unresolved. Terrestrial sequences from Europe span this boundary, and lie within the boundary stratotype region, but have been directly sampled only sporadically for magnetostratigraphy. Terrestrial sequences of late Paleocene and/or early Eocene age occur in Africa, Asia, South America, Australia, Antarctica; however, magnetostratigraphies currently are available for only the lower-middle Paleocene strata of South America.

The available magnetostratigraphic data clearly indicate that the Paleocene/Eocene Series boundary lies within Chron C24r.

On the basis of the poor documentation and suspected overprinting problems, we reject the term *Oldhaven event* (or *subchron*), which has been used in reference to a short normal polarity interval within Chron C24r. However, we argue that there indeed may be a different short normal polarity interval closely associated with the Paleocene/Eocene Epoch boundary. If this short normal polarity zone, observed in at least one marine and two terrestrial sequences reflect the same magnetochron event, it has great potential for local to global temporal correlation via a synchronous horizon/zone, and for refining age estimates of events near the Paleocene/Eocene Epoch boundary by subdividing the very long reversed magnetochron (Chron C24r) in which it occurs.

There recently have been, and continue to be, extensive efforts to integrate magnetostratigraphic, marine and terrestrial biostratigraphic, radioisotopic dating, and marine and terrestrial stable isotopic information from sequences spanning the Paleocene/Eocene Series boundary. These include potential boundary and stage stratotypes, deep sea and shallow marine sections, and terrestrial sequences from many areas of the world. Integrating this great variety of stratigraphic data has provided improved chronostratigraphic correlations and insights into the relative timing, and possible causes of biotic and physical events occurring near the Paleocene/Eocene Epoch boundary. However, this chapter also summarizes many of the problems that still exist, including weakly supported magnetostratigraphic data in some areas and paucity of integrative chronostratigraphic studies in others, conflicts in relative positions of events near the boundary in different sequences, difficulties in correlating from marine to terrestrial sequences, and the fact that precise resolution of the position of the Paleocene/Eocene Series boundary relative to polarity reversal or chron boundaries is hindered by the fact that it lies within a long reversed polarity interval (although the possible presence of a globally recognizable short polarity chron near the boundary may alleviate some of the problem). Yet many intensive, integrative studies are just beginning, and there remains great potential to improve local and global correlations, and associated pattern and process analyses.

ACKNOWLEDGMENTS

The research was supported in part by NSF grants BSR8896178/8614133 and DEB9317943 to JF, and ACS 26498AC2 and EAR 93 15628 to LT. We thank Dennis Kent, Marie-Pierre Aubry, and Spencer Lucas for very thorough and thoughtful reviews.

REFERENCES

Ali, J. R., 1989. *Magnetostratigraphy of Early Palaeogene sediments from N.W. Europe.* Ph.D. thesis (unpublished), England, The University of Southampton, 277 pp.

Ali, J. R. and Hailwood, E. A., 1995. Magnetostratigraphy of upper Paleocene through lower middle Eocene strata of northwest Europe. In Berggren, W. A., Kent, D. V., Aubry, M.-P. and Hardenbol, J. (eds.), *Geochronology Time Scales and Global Stratigraphic Correlation,* Society of Economic Paleontologists and Mineralogists Special Publication No. 54, p. 275–279.

Ali, J. R. and Jolley, D. W., 1996. Chronostratigraphic framework for the Thanetian and lower Ypresian deposits of southern England. In Knox, R. W. O'B., Corfield, R. M. and Dunay, R. E. (eds.), *Correlation of the Early Paleogene in Northwest Europe,* Geological Society Special Publication No. 101, p. 129–144.

Ali, J. R., King, C. and Hailwood, E. A., 1993. Magnetostratigraphic calibration of early Eocene depositional sequences in the southern North Sea Basin. In Hailwood, E. A. and Kidd, R. B. (eds.), *High Resolution Stratigraphy,* Geological Society Special Publication, No. 70, p. 99–125.

Ameghino, F. 1900–1902. L'âge des formations sédimentaires de Patagonie. *Anales de la Sociedad Cientifica, Argentina,* v. 50, p. 109–130, 145–165, 209–229 (1900); v. 51, p. 20–39, 65–91 (1901); v. 52, p. 189–197, 244–250 (1901); v. 54, p. 161–180, 220–249, 283–342 (1902).

Archer, M., Godthelp, H. and Hand, S. J., 1993. Early Eocene marsupial from Australia. *Kaupia, Darmstädter Beiträge zur Naturgeschichte,* v. 3, p. 193–200.

Archibald, J. D., Clemens, W. A., Gingerich, P. H., Krause, D. W., Lindsay, E. H. and Rose, K. D., 1987. First North American land mammal ages of the Cenozoic Era. In Woodburne, M. O. (ed.), *Cenozoic Mammals of North America,* Berkeley: University of California Press, p. 24–76.

Askin, R. A., 1988. Campanian to Paleocene palynological succession of Seymour and adjacent islands, northeastern Antarctic Peninsula. In Feldmann, R. M. and Woodburne, M. O. (eds.), *Geology and Paleontology of Seymour Island, Antarctica,* Geological Society of America Memoir 169, p. 131–153.

Askin, R. A., 1990. Campanian to Paleocene spore and pollen assemblages of Seymour Island, Antarctica. *Review of Palaeobotany and Palynology,* v. 65, p. 105–113.

Aubry, M.-P., 1985. Northwestern European Paleogene magnetostratigraphy, biostratigraphy, and paleogeography: Calcareous nannofossil evidence. *Geology,* v. 13, p. 198–202.

Aubry, M.-P., 1995. From chronology to stratigraphy: Interpreting the lower and middle Eocene stratigraphic record in the Atlantic Ocean. In Berggren, W. A., Kent, D. V., Aubry, M.-P. and Hardenbol, J. (eds.), *Geochronology Time Scales and Global Stratigraphic Correlation,* Society of Economic Paleontologists and Mineralogists Special Publication No 54, p. 213–274.

Aubry, M.-P., Hailwood, E. A. and Townsend, H. A., 1986. Magnetic and calcareous-nannofossil stratigraphy of lower Palaeogene formations of the Hampshire and London basins. *Journal of the Geological Society, London,* v. 143, p. 729–735.

Aubry, M-P., Berggren, W. A., Kent, D. V., Flynn, J. J., Klitgord, K. D., Obradovich, J. D. and Prothero, D. W., 1988. Paleogene Geochronology: An integrated approach. *Paleoceanography,* v. 3, p. 707–742.

Aubry, M.-P., Berggren, W. A., Stott, L. and Sinha, A., 1996. The upper Paleocene–lower Eocene stratigraphic record and the Paleocene–Eocene boundary carbon isotope excursion: Implications for geochronology. In Knox, R. W. O'B., Corfield, R. M. and Dunay, R. E. (eds.), *Correlation of the early Paleogene in Northwest Europe,* Geological Society Special Publication No 101, p. 353–380.

Berggren, W. A. and Aubry, M.-P., 1996. A late Paleocene–early Eocene NW European and North Sea magnetobiochronological correlation network. In Knox, R. W. O'B., Corfield, R. M. and Dunay, R. E. (eds.), *Correlation of the Early Paleogene in Northwest Europe,* Geological Society Special Publication No. 101, p. 309–352.

Berggren, W. A., Kent, D. V., Flynn, J .J. and Van Couvering, J. A., 1985a. Cenozoic geochronology. *Geological Society of America Bulletin,* v. 96, p. 1407–1418.

Berggren, W. A., Kent, D. V. and Flynn, J. J., 1985b. Paleogene geochronology and chronostratigraphy. In Snelling, N. J. (ed.), *The Chronology of the Geological Record:* Geological Society Memoir No 10, Blackwell, Oxford, p. 141–195.

Berggren W. A., Kent, D. V., Swisher, C. C., III and Aubry, M.-P., 1995. A revised Cenozoic geochronology and chronostratigraphy. In Berggren, W. A., Kent, D. V., Aubry, M.-P. and Hardenbol, J. (eds.), *Geochronology Time Scales and Global Stratigraphic Correlation,* Society of Economic Paleontologists and Mineralogists Special Publication No. 54, p. 129–212.

Bertels, A., 1970. Sobre el "Piso Patagoniano" y la representacion de la epoca del Oligoceno en Patagonie austral, Republica Argentina. *Revista Asociación Geológica Argentina,* v. 25, p. 495–501.

Bertels, A., 1975. Bioestratigrafiá del Paleogeno en la Republica Argentina. *Revista Española de Micropaleontologiá,* v. 7, p. 429–450.

Bertels, A., 1979. Paleobiogeografiá de los foraminiferos del Cretácico superior y Cenozoico de América del Sur. *Ameghiniana,* v. 16, p. 273–356.

Bleil, U., 1985, The magnetostratigraphy of northwest Pacific sediments, Deep Sea Drilling Project Leg 86, in Heath, G. R., Burckle, L. H., et al., Initial Reports of the Deep Sea Drilling Project, v. 86: Washington (U.S. Government Printing Office), p. 441–458.

Bonaparte, J. F., Van Valen, L. M. and Kramartz, A., 1993. *La fauna local de Punta Peligro, Paleoceno Inferior, de la Province del Chubut, Patagonia, Argentina.* Chicago: Evolutionary Monographs, v. 14, 61 pp.

Bond, M., Carlini, A. A., Goin, F. J., Legarreta, L., Ortiz Jaureguizar, E., Pascual, R., Prado, J. L. and Uliana, M. A., 1994. Episodes in South American land mammal evolution and sedimentation: Testing their apparent concomitance in a Paleocene succession from Central Patagonia. VI Congreso Argentino de Paleontologiá y Bioestratigrafiá, Resumenes, p. 20.

Bond, M., Pascual, R., Reguero, M. A., Santillana, S. and Marenssi, S., 1989. Los primeros "ungulados" extinguidos sudamericanos de la Antártida. Ameghiniana, v. 26, p. 240.

Bown, T. M., Rose, K. D., Simons, E. L. and Wing, S. L., 1994. *Distribution and stratigraphic correlation of upper Paleocene and lower Eocene fossil mammal and plant localities of the Fort Union, Willwood and Tatman Formations, southern Bighorn Basin, Wyoming.* U.S. Geological Survey Professional Paper 1540, 103 pp.

Butler, R. F. and Lindsay, E. H., 1985. Mineralogy of magnetic minerals and revised magnetic polarity stratigraphy of continental sediments, San Juan Basin, New Mexico. *Journal of Geology,* v. 93, p. 535–554.

Butler, R. F., Gingerich, P. D. and Lindsay, E. H., 1981. Magnetic polarity stratigraphy and biostratigraphy of Paleocene and lower Eocene continental deposits, Clark's Fork Basin, Wyoming. *Journal of Geology,* v. 89, p. 299–316.

Camacho, H. H., 1974. Bioestratigrafiá de las formaciones marines del Eoceno y Oligoceno de la Patagonia. *Anales de la Academia [national] de Ciencias Exactas Fisicas y Naturales de Buenos Aires,* v. 26, p. 39–57.

Cande, S. C. and Kent, D. V., 1992. A new geomagnetic polarity time scale for the Late Cretaceous and Cenozoic. *Journal of Geophysical Research,* v. 97(B10), p. 13917–13951.

Cande, S. C. and Kent, D. V., 1995. Revised calibration of the geomagnetic polarity time scale for the Late Cretaceous and Cenozoic. *Journal of Geophysical Research,* v. 100(B4), p. 6093–6095.

Cappetta, H., Jaeger, J.-J., Sabatier, M., Sigé, B., Sudre, J. and Vianey-Liaud, M., 1978. Découverte dans le Paléocène du Maroc des plus anciens mammifères euthériens d'Afrique. *Géobios,* v. 11, p. 257–263.

Cappetta, H., Jaeger, J.-J., Sigé, B., Sudre, J. and Vianey-Liaud, M., 1987. Compléments et précisions biostratigraphiques sur la faune paléocène à mammifères et sèlaciens du Bassin d'Ouarzazate (Maroc). *Tertiary Research,* v. 8, p. 147–157.

Carlini, A. A., Pascual, R., Reguero, M. A., Scillato-Yané, G. J., Tonni, E. P. and Vizcaino, S. F., 1990. The first Paleogene land placental mammal from Antarctica; Its paleoclimatic and paleobiogeographical bearing. *Fourth International Congress on Systematics and Evolutionary Biology,* Abstracts, p. 325.

Case, J. A., 1988. Paleogene floras from Seymour Island, Antarctic Peninsula. In Feldmann, R. M. and Woodburne, M. O. (eds.), *Geology and Paleontology of Seymour Island, Antarctica,* Geological Society of America Memoir 169, p. 523–530.

Case, J. A., Woodburne, M. O. and Chaney, D. S., 1988. A new genus of polydolopid marsupial from Antarctica. In Feldmann, R. M. and Woodburne, M. O. (eds.), *Geology and Paleontology of Seymour Island, Antarctica,* Geological Society of America Memoir 169, p. 505–521.

Clyde, W. C., Stamatakos, J. and Gingerich, P. D., 1994. Chronology of the Wasatchian Land-Mammal Age (early Eocene): Magnetostratigraphic results from the McCullough Peaks Section, northern Bighorn Basin, Wyoming. *Journal of Geology,* v. 102, p. 367–377.

Coiffait, P. E., Coiffait, B., Jaeger, J. J. and Mahboubi, M., 1984. Un nouveau gisement d'âge éocène supérieur sur le versant sud des Nementcha (Algèrie orientale): Découverte des plus anciens rongeurs d'Afrique. *Comptes-Rendus de l'Académie des Sciences de Paris,* v. 299 (II, 13), p. 13,893–13,898.

Cox, F. C., Hailwood, E. A., Harland, R., Hughes, M. J., Johnston, N. and Knox, R. W. O'B., 1985. Palaeocene sedimentation and stratigraphy in Norfolk, England. *Newsletters on Stratigraphy,* v. 14, p. 169–185.

Efron, B., 1982. The jackknife, the bootstrap and other resampling plans. Society for Industrial and Applied Mathematics.

Ellison, R. A., Knox, R. W. O'B., Jolley, D. W. and King, C., 1994. A revision of the lithostratigraphical classification of the early Palaeogene strata of the London Basin and East Anglia. *Proceedings of the Geologists' Association,* v. 105, p. 187–197.

Ellison, R. A., Ali, J. R., Hine, N. M. and Jolley, D. W., 1996. Recognition of Chron C25n in the upper Paleocene Upnor Formation of the London Basin, UK. In Knox, R. W. O'B., Corfield, R. M. and Dunay, R. E. (eds.), *Correlation of the Early Paleogene in Northwest Europe,* Geological Society Special Publication No 101, p. 185–193.

Flynn, J. J., 1986. Correlation and geochronology of middle Eocene strata from the western United States. *Palaeogeography, Palaeoclimatology, Palaeoecology,* v. 55, p. 335–406.

Flynn, J. J. and Novacek, M. J., 1984. Early Eocene vertebrates from Baja California: Evidence for intracontinental age correlations. *Science,* v. 224, p. 151–153.

Flynn, J. J. and Swisher, C. C., III, 1995. Cenozoic South American Land Mammal Ages: Correlation to global geochronologies. In Berggren, W. A., Kent, D. V., Aubry, M.-P. and Hardenbol, J. (eds.), *Geochronology Time Scales and Global Stratigraphic Correlation,* Society of Economic Paleontologists and Mineralogists Special Publication No 54, p. 317–333.

Flynn, J. J., MacFadden, B. J. and McKenna, M. C., 1984. Land-Mammal Ages, faunal heterochrony, and temporal resolution in Cenozoic terrestrial sequences. *Journal of Geology,* v. 92, p. 687–705.

Flynn, J. J., Cipolletti, R. M. and Novacek, M. J., 1989. Chronology of early Eocene marine and terrestrial strata, Baja California, Mexico. *Bulletin of the Geological Society of America,* v. 101, p. 1182–1196.

Gee, J., Klootwijk, C. T. and Smith, G. M., 1991. Magnetostratigraphy of Paleogene and Upper Cretaceous sediments from Broken Ridge, eastern Indian Ocean. In Weissel, J., Peirce, J., Taylor, E., Alt, J., et al., Proceedings of the Ocean Drilling Program, Scientific Results, v. 121: College Station, TX (Ocean Drilling Program), p. 359–376.

Gheerbrant, E., 1989. *Les mammifères paléocènes du Bassin d'Ouarzazate (Maroc): Etude systématique, phylogénique, paléoécologique et palébiogéographique des plus anciens placentaires d'Afrique.* Mémoire des Sciences de la Terre [unpublished thesis], Paris, Université Pierre et Marie Curie, 473 pp.

Gheerbrant, E., 1990. On the early biogeographical history of the African placentals. *Historical Biology,* v. 4, p. 107–116.

Givens, C. R, 1974. Eocene molluscan biostratigraphy of the Pine Mountain area, Ventura County, California. *University of California Publication in Geological Sciences,* v. 109, 107 pp.

Godthelp, H., Archer, M., Cifelli, R., Hand, S. J. and Gilkeson, C. F., 1992. Earliest known Australian Tertiary mammal fauna. *Nature,* v. 356, p. 514–516.

Gunnell, G. F., Bartels, W. S. and Gingerich, P. D., 1993. Paleocene–Eocene boundary in continental North America: Biostratigraphy and Geochronology, northern Bighorn Basin, Wyoming. In Lucas, S. G. and Zidek, J. (eds.), *Vertebrate Paleontology in New Mexico,* New Mexico Museum of Natural History and Science, Bulletin 2, p. 137–144.

Hand, S., Novacek, M., Godthelp, H., and Archer, M., 1994. First Eocene bat from Australia. *Journal of Vertebrate Paleontology,* v. 14, p. 375–381.

Hardenbol, J. and Berggren, W. A., 1978. A new Paleogene numerical time scale. In Cohee, G. V., Glaessner, M. F. and Hedberg, H. D. (eds.), *The Geologic Time Scale,* Studies in Geology No. 6, American Association of Petroleum Geologists, p. 213–234.

Hartenberger, J.-L., Martinez, C., and Bensaid, A., 1985. Découverte de mammifères d'âge éocène inférieur en Tunisie centrale. *Comptes-Rendus de l'Académie des Sciences de Paris,* v. 301, p. 649–652.

Hickey, L. J., West, R. M., Dawson, M. R. and Choi, D. K., 1983. Arctic terrestrial biota: Paleomagnetic evidence of age disparity with mid-northern latitudes during the late Cretaceous and early Tertiary. *Science,* v. 221, p. 1153–1156.

Hooker, J. J., 1991. The sequence of mammals in the Thanetian and Ypresian of the London and Belgian Basins. Location of the Palaeocene–Eocene boundary. *Newsletters on Stratigraphy,* v. 25, p. 75–90.

Hooker, J. J., 1992. An additional record of placental mammal (Order Astrapotheria) from the Eocene of West Antarctica. *Antarctic Science,* v. 4, p. 107–108.

Hooker, J. J., 1996. Mammalian biostratigraphy across the Palaeocene–Eocene boundary in the Paris, London and Belgian Basins. In Knox, R. W. O'B., Corfield, R. M. and Dunay, R. E. (eds.), *Correlation of the Early Paleogene in Northwest Europe,* Geological Society Special Publication No. 101, p. 205–218.

Kent, D. V., McKenna, M. C., Opdyke, N. D., Flynn, J. J. and MacFadden, B. J., 1984. Arctic biostratigraphic heterochroneity. *Science,* v. 224, p. 173–174.

King, C., 1981. *The stratigraphy of the London Clay and associated deposits.* Tertiary Research Special Paper No. 6, Rotterdam: W. Backhuys, 158 pp.

Koch, P. L., Zachos, J. C. and Gingerich, P. D., 1992. Correlation between isotope records in marine and continental carbon reservoirs near the Palaeocene/Eocene boundary. *Nature,* v. 358, p. 319–322.

Krause, D. W. and Maas, M. C., 1990, The biogeographic origins of late Paleocene–early Eocene mammalian immigrants to the Western Interior of North America. In Rose, K. D. and Bown, T. M. (eds.), *Dawn of the Age of Mammals,* Geological Society of America Special Paper 243, p. 71–105.

Krishtalka, L., West, R. M., Black, C. C., Dawson, M. R., Flynn, J. J., Turnbull, W. D., Stucky, R. K., McKenna, M. C., Bown, T. M., Golz, D. J. and Lillegraven, J. A., 1987. Eocene (Wasatchian through Duschesnean) biochronology of North America. In Woodburne, M. O. (ed.), *Cenozoic Mammals of North America, Geochronology and Biostratigraphy,* Berkeley: University of California Press, p. 24–76.

Lindsay, E. H., Butler, R. F. and Johnson, N. M., 1981. Magnetic polarity zonation and biostratigraphy of Late Cretaceous and Paleocene continental deposits, San Juan Basin, New Mexico. *American Journal of Science,* v. 281, p. 390–435.

Lowrie, W., Alvarez, W., Napoleone, G., Perch-Nielsen, K., Premoli Silva, I. and Toumarkine, M., 1982. Paleogene magnetic stratigraphy in Umbrian pelagic carbonate rocks: The Contessa sections, Gubbio. *Geological Society of America Bulletin,* v. 93, p. 414–432.

Lucas, S. G. and Williamson, T. E., 1993a. Late Cretaceous–early Eocene vertebrate biostratigraphy and biochronology of the San Juan Basin, New Mexico. In Lucas, S. G. and Zidek, J. (eds.), *Vertebrate Paleontology in New Mexico,* New Mexico Museum of Natural History and Science, Bulletin 2, p. 93–104.

Lucas, S. G. and Williamson, T. E., 1993b. Eocene vertebrates and late Laramide stratigraphy of New Mexico. In Lucas, S. G. and Zidek, J. (eds.), *Vertebrate Paleontology in New Mexico,* New Mexico Museum of Natural History and Science, Bulletin 2, p. 145–158.

Lyell, C., 1833. *Principles of Geology, Being an Attempt to Explain the Former Changes of the Earth's Surface, by Reference to Causes Now in Operation.* London: J. Murray, v. 3, 398 pp. (Facsimile edition, Chicago, University of Chicago [published, 1991]).

Mahboubi, M., Ameur, R., Crochet, J.-Y. and Jaeger, J.-J., 1984. Earliest known proboscidean from early Eocene of north-west Africa. *Nature,* v. 308, p. 543–544.

Mahboubi, M., Ameur, R., Crochet, J.-Y., and Jaeger, J.-J., 1986. El Kohol (Saharan Atlas, Algeria): A new Eocene mammals locality in North Western Africa. *Palaeontographica,* v. 192, p. 15–49.

Marenssi, S. A., Reguero, M. A., Santillana, S. N. and Vizcaino, S. F., 1994. Review. Eocene land mammals from Seymour Island, Antarctica: Palaeobiogeographical implications. *Antarctic Science,* v. 6, p. 3–15.

Marshall, L. G., Butler, R. F., Drake, R. E. and Curtis, G. H., 1981. Calibration of the beginning of the age of mammals in Patagonia. *Science,* v. 212, p. 43–45.

Marshall, L. G., Hoffstetter, R. and Pascual, R., 1983. Mammals and stratigraphy: geochronology of the continental mammal-bearing Tertiary of South America. *Palaeovertebrata,* Mémoire extraordinaire, p. 1–93.

Morris, W. J. 1966. Fossil mammals from Baja California: New evidence on early Tertiary migrations. *Science,* v. 135, p. 1376–1378.

North American Stratigraphic Code. North American Commission on Stratigraphic Nomenclature, 1983. *Bulletin of the American Association of Petroleum Geologists,* v. 67, p. 841–875.

Novacek, M. J., Ferrusquía-Villafranca, I., Flynn, J. J., Wyss, A. R. and Norell, M. A., 1991. Wasatchian (early Eocene) mammals and other vertebrates from Baja California, Mexico: The Lomas las Tetas de Cabra fauna. *Bulletin of the American Museum of Natural History,* 208, 88 pp.

Odin, G. S. (ed.), 1982. *Numerical Dating in Stratigraphy.* New York: John Wiley & Sons, 1040 pp.

Pascual, R., Archer, M., Ortiz, J. E., Prado, J., Godthep, H. and Hand, S., 1992a. First discovery of monotremes in South America. *Nature,* v. 356, p. 704–706.

Pascual, R., Archer, M., Ortiz, J. E., Prado, J., Godthep, H. and Hand,

S., 1992b. The first non-Australian monotreme: An early Paleocene South American platypus (Monotremata, Ornithorhynchidae). In Augee, M. L. (ed.), *Platypus and Echidnas,* Sydney: The Royal Zoological Society of New South Wales, p. 2–15.

Pascual, R. and Ortiz Jaureguizar, E., 1991. El Ciclo Faunístico Cochabambiano (Paleoceno temprano): Su incidencia en la historia biogeográfica de los mamíferos sudamericanos. In Suárez-Soruco, R. (ed.), *Fósiles y Facies de Bolivia.* Santa Cruz: Revista Técnica de Yacimientos Petroliferos Fiscales Bolivianos, v. 12, p. 559–574.

Rapp, S. D., MacFadden, B. J. and Schiebout, J. A., 1983. Magnetic polarity stratigraphy in the early Tertiary Black Peaks Formation, Big Bend National Park, Texas. *Journal of Geology,* v. 91, p. 555–572.

Rea, D. K., Zachos, J. C., Owen, R. M. and Gingerich, P. D., 1990. Global change at the Paleocene–Eocene boundary: climatic and evolutionary consequences of tectonic events. *Palaeogeography, Palaeoclimatology, Palaeoecology,* v. 79, p. 117–128.

Russell, D. E. and Zhai, R.-J., 1987. The Paleogene of Asia: Mammals and stratigraphy. *Mémoire du Muséum National d'Histoire Naturelle,* Sér. C (*Sciences de la Terre*), v. 52, 488 pp.

Sadler, P. M., 1981. Sediment accumulation rates and the completeness of stratigraphic sections. *Geology,* v. 89, p. 569–584.

Salvador, A. (ed.), 1994. *International Stratigraphic Guide. A guide to stratigraphic classification, terminology, and procedure.* Boulder: International Union of Geological Sciences and Geological Society of America, 214 pp.

Schiebout, J. A., 1995. The Paleocene/Eocene transition on Tornillo Flat in Big Bend National Park, Texas. In Santucci, V. L. and McClelland, L. (eds.), *National Park Service Paleontological Research,* National Park Service Technical Report NPS/NRPO/NRTR 95/16, pp. 40–45.

Schiebout, J. A., Rigsby, C. A., Rapp, S. D., Hartnell, J. A. and Standhardt, B. R., 1987. Stratigraphy of the Cretaceous-Tertiary and Paleocene-Eocene transition rocks of Big Bend National Park, Texas. *Journal of Geology,* v. 95, pp. 359–371.

Schimper, W. P., 1874. *Traité de Paléontologie Végétale.* Paris: J. B. Baillière, v. 3, 896 pp.

Shackleton, N. J. and Members of the Shipboard Scientific Party, 1984. Accumulation rates in leg 74 sediments. In Moore, T. C., Jr., Rabinowitz, P. D., et al., Initial Reports of the Deep Sea Drilling Project, v. 74: Washington (U.S. Government Printing Office), p. 621–637.

Simpson, G. G., 1940. Review of the mammal-bearing Tertiary of South America. *Proceedings of the American Philosophical Society,* v. 83, p. 649–709.

Spiess, V., 1990. Cenozoic magnetostratigraphy of Leg 113 Drill Sites, Maud Rise, Weddell Sea, Antarctica. In Barker, P. F., Kennett, J. P., et al., Proceedings of the Ocean Drilling Program, v. 113: College Station, TX (Ocean Drilling Program), p. 261–315.

Stenzel, H. R., 1945. Stratigraphic significance of the Patagonian *Odontogryphaeas. Geological Society of America Bulletin,* v. 56, p. 1202.

Swisher, C. C. III and Knox, R. W. O'B., 1991. The age of the Paleocene/Eocene boundary: $^{40}Ar/^{39}Ar$ dating of the lower part of NP10, North Sea Basin and Denmark. IGCP Project 308 (Paleocene/Eocene boundary events), International Annual Meeting and Field Conference, 2–6 December 1991, Brussels, Abstracts with Program, p. 16.

Tauxe, L. and Clark, D. R., 1987. New paleomagnetic results from the Eureka Sound Group: Implications for the age of early Tertiary Arctic biota. *Geological Society of America Bulletin,* v. 99, p. 739–747.

Tauxe, L. and Gallet, Y., 1991. A jackknife for magnetostratigraphy. *Geophysical Research Letters,* v. 18, p. 1783–1786.

Tauxe, L., Gee, J., Gallet, Y., Pick, T. and Bown, T., 1994. Magnetostratigraphy of the Willwood Formation, Bighorn Basin, Wyoming: New constraints on the location of the Paleocene/Eocene boundary. *Earth and Planetary Science Letters,* v. 125, p. 159–172.

Taylor, L. H. and Butler, R. F., 1980. Magnetic-polarity stratigraphy of Torrejonian sediments, Nacimiento Formation, San Juan Basin, New Mexico. *American Journal of Science,* v. 280, p. 97–115.

Toumarkine, M. and Luterbacher, H. P., 1985. Paleocene and Eocene planktic foraminifera. In Bolli, H. M., Saunders, J. B. and Perch-Nielsen, K., (eds.), *Plankton Stratigraphy,* Cambridge: Cambridge University Press, p. 87–154.

Townsend, H. A., 1982. *Magnetostratigraphy of Early Palaeogene sediments from southern England.* Unpublished Ph.D. thesis, Southampton: University of Southampton, 249 pp.

Townsend, H. A., 1985. The paleomagnetism of sediments acquired from the Goban Spur on Deep Sea Drilling Project Leg 80. In Graciansky, P. C., de, Poag, W. C., et al., Initial Reports of the Deep Sea Drilling Program, v. 80: Washington (U.S. Government Printing Office), p. 389–414.

Townsend, H. A. and Hailwood, E. A., 1985. Magnetostratigraphic correlation of Palaeogene sediments in the Hampshire and London Basins, southern UK. *Journal of the Geological Society, London,* v. 142, p. 957–982.

Wei, W. and Zhong, S., 1996. Taxonomy and magnetobiochronology of *Tribrachiatus* and *Rhomboaster,* two genera of calcareous nannofossils. *Journal of Paleontology,* v. 70, p. 7–22.

Williamson, T. E. and Lucas, S. G., 1993. Paleocene vertebrate paleontology of the San Juan Basin, New Mexico. *New Mexico Museum of Natural History and Science Bulletin* 2, p. 105–135.

Wing, S. L., 1984. A new basis for recognizing the Paleocene/Eocene boundary in western interior North America. *Science,* v. 226, p. 439–441.

Wing, S. L., Bown, T. M. and Obradovich, J. D., 1991. Early Eocene biotic and climatic change in interior western North America. *Geology,* v. 19, p. 1189–1192.

Woodburne, M. O. and Zinsmeister, W. J., 1982. Fossil land mammal from Antarctica. *Science,* v. 218, p. 284–286.

Woodburne, M. O. and Zinsmeister, W. J., 1984. The first land mammal from Antarctica and its biogeographic implications. *Journal of Paleontology,* v. 58, p. 913–948.

Zinsmeister, W. J., 1981. Middle to late Eocene invertebrate fauna from the San Julian Formation at Punta Casamayor, Santa Cruz Province, Southern Argentina. *Journal of Paleontology,* v. 55, p. 1083–1102.

5

The Tectonic and Volcanic History of the North Atlantic Region During the Paleocene–Eocene Transition: Implications for NW European and Global Biotic Events

ROBERT W. O' B. KNOX

ABSTRACT
Northwest Europe experienced a long-term rise in sea level during much of the late Paleocene Epoch, but regional uplift associated with the Iceland plume caused a rapid and substantial fall in sea level in latest Paleocene times. The initial sea-level fall caused partial isolation of the North Sea and Faroe–Shetland basins, leading to the establishment of dysaerobic bottom-water conditions and a marked reduction in the abundance and diversity of benthic foraminifera. At the same time, the emergence of new land areas was marked by an increase in the proportion of terrestrial to marine palynomorphs. A combination of lava emplacement, sediment aggradation and uplift of local structural elements eventually caused complete isolation from oceanic waters, leading to the establishment of anoxic bottom-water conditions in deeper parts of the basins, as indicated by a change to laminated mudstone facies and the virtual elimination of benthic microfaunas. This closure also resulted in the completion of a land-bridge between North America and Europe, providing the pathway for the interchange of mammal faunas between the two regions. The tectonically induced sea-level fall was thus responsible for major environmentally related biotic events in marine and terrestrial environments in the NW European region. Evidence from palynomorph and clay mineral assemblages also indicates climatic warming at this time.

Important global events are also known to have taken place around the time of this phase of North Atlantic uplift, including negative carbon and oxygen isotope excursions and a widespread extinction of benthic foraminifera in the oceans. This temporal coincidence raises the possibility that these global changes were in some way related to a change in the pattern of oceanic circulation that resulted from closure of the connection between the Atlantic Ocean and northern high-latitude waters. However, this closure persisted long after the global isotope excursions and oceanic benthic extinction event, indicating that it cannot alone have been responsible for these rapid environmental changes.

It is similarly unlikely that either gas or dust emission associated with North Atlantic volcanism was responsible for short-term changes in global climate or ocean water chemistry at the time of the isotope excursions and oceanic benthic extinction. The latter events occurred during a much longer period of more or less continuous plateau lava extrusion and at a time of only minor pyroclastic activity.

Evidence for substantial regional tectonic activity in NW Europe during the latest Paleocene and earliest Eocene epochs, here referred to as the Paleocene–Eocene transition (P–E transition), is provided by widespread volcanism and major uplift of the North Atlantic rift margins. In the basinal, offshore lithofacies successions, these processes are reflected in the presence of abundant tephra layers and in the replacement of hemipelagic mudstone facies by submarine fan sands and prograding coastal plain deposits. In the marginal, onshore successions, they are reflected in discontinuous successions, and in a temporary change from marine to terrestrial facies; tephra layers are relatively scarce because of stratigraphic incompleteness and the poor preservation potential associated with marginal facies.

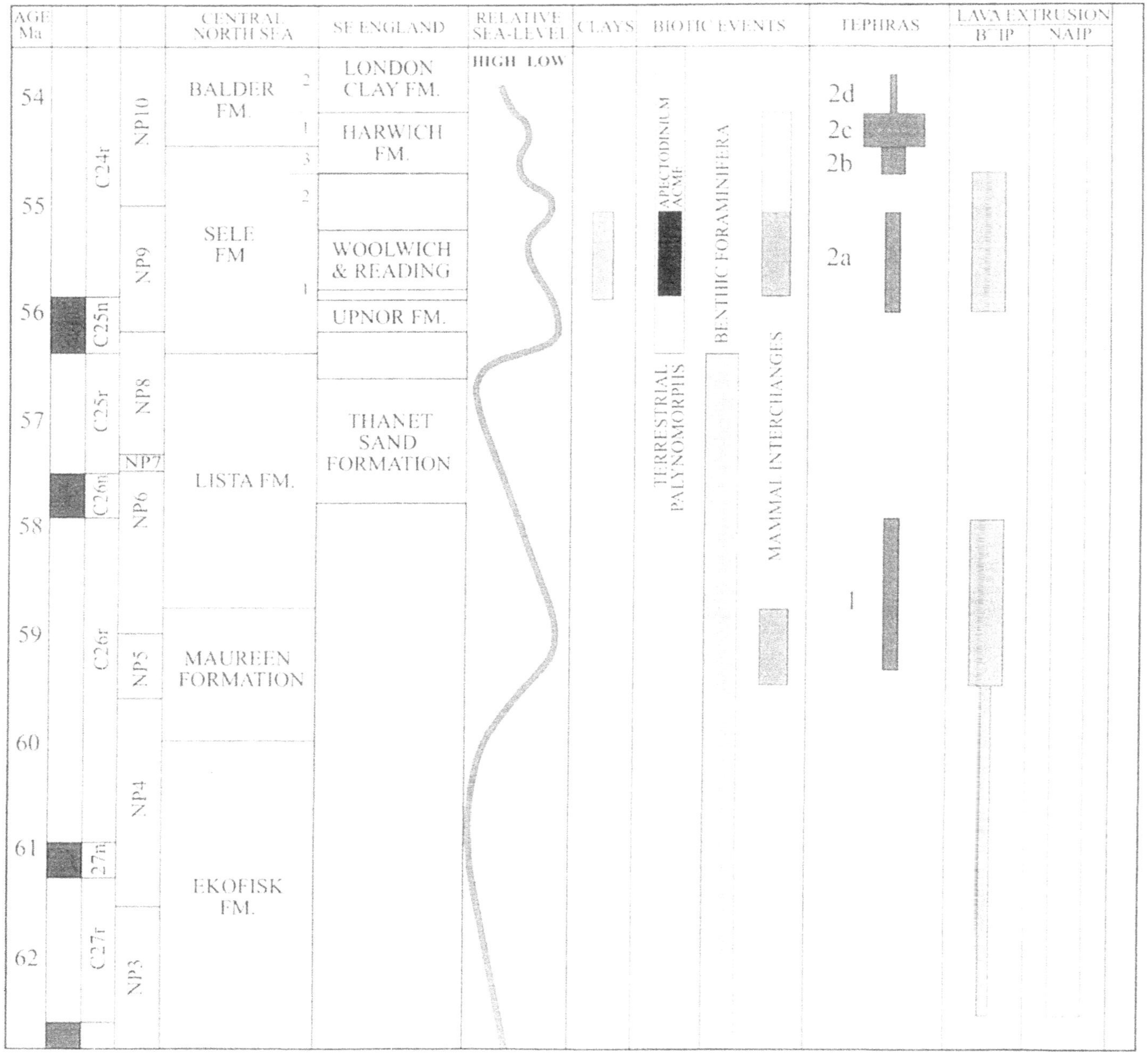
AGE Ma
CENTRAL NORTH SEA
SE ENGLAND
RELATIVE SEA-LEVEL
CLAYS
BIOTIC EVENTS
TEPHRAS
LAVA EXTRUSION
B' IP
NAIP
HIGH LOW
54
55
56
57
58
59
60
61
62
C24r
C25n
C25r
C26n
C26r
27n
C27r
NP10
NP9
NP8
NP7
NP6
NP5
NP4
NP3
BALDER FM.
SELE FM
LISTA FM.
MAUREEN FORMATION
EKOFISK FM.
LONDON CLAY FM.
HARWICH FM.
WOOLWICH & READING
UPNOR FM.
THANET SAND FORMATION
APECTODINIUM ACME
TERRESTRIAL PALYNOMORPHS
BENTHIC FORAMINIFERA
MAMMAL INTERCHANGES
2d
2c
2b
2a
1

The initial sea-level fall and the associated increase in pyroclastic volcanism in the North Atlantic region were broadly contemporaneous with important events recorded from elsewhere in the world, including a benthic foraminiferal extinction (BFE) event in the oceans, a global change to warmer and more humid climates, and a short-lived negative carbon isotope excursion (CIE) that has been recorded from both oceanic and terrestrial successions (see summary by Thomas and Shackleton 1996:402–404). This broad temporal coincidence raises the possibility that the North Atlantic crustal events may in some way have played a causative role. Thus uplift of the North Atlantic region, severing marine connection between the Atlantic Ocean and the Arctic region, could have affected the pattern of oceanic circulation, while gases and dust released during the volcanic activity could have affected climate and organic productivity. For such relationships to be properly assessed, however, requires an understanding of the relative timing of the various events. As a contribution to such an assessment, this chapter outlines the tectonic and volcanic history of the NE Atlantic region and the temporal relationship between these crustal processes and the major biotic events recorded from the P–E transition in NW Europe and beyond. The proposed temporal framework is summarized in figure 5.1.

Tectonism and Sea-Level Change

The mid to late Thanetian Epoch was a time of progressive sea-level rise in the North Atlantic region, leading to widespread transgression of the southern basin margin (Thanet Formation and equivalents) and to the deposition of fine-grained hemipelagic clays in basinal areas. These hemipelagic clays were deposited in fully marine conditions, with rich assemblages of dinoflagellate cysts and agglutinated foraminifera. They are represented by the Holmehus Formation of Denmark (see Heilmann-Clausen et al. 1985), the upper part of the Lista Formation in the Central North Sea (see Knox and Holloway 1992:65–67), and the upper part of the Lamba Formation in the Faroe–Shetland Basin (see Knox et al. 1997).

This episode of late Thanetian high sea-level stand was followed by a substantial fall in sea level, which is represented in southern marginal sections by a widespread unconformity at the top of the Thanet Formation and its equivalents. Evidence for exposure to meteoric circulation is provided by decalcification and by leaching of heavy mineral assemblages (Morton 1982) in the uppermost few meters of these deposits. The magnitude of the sea-level fall can be assessed from the superposition in Bradwell Borehole 217 (Knox et al. 1994) of pedogenetically altered marginal marine siltstone/sandstone (Upnor Formation) on fine-grained, homogeneous, waxy mudstones (Thanet Formation) with agglutinated benthic foraminiferal assemblages of outer neritic aspect, suggesting water depths of at least 50 m and possibly 100 m or more (C. King, personal communication 1996). The outer neritic biofacies is not restricted to the uppermost Thanet mudstones, however, since it has been recorded also in the underlying sequence (Sequence BT2, Knox 1994:figure 2, down to 64.9 m). Since the successive sequences are interpreted as representing a long term-rise in sea level, it is probable that water depths for the uppermost mudstones (sequence BT3) were at the deeper end of the outer neritic range. This interpretation is supported by the lithofacies characteristics, with the complete absence of sand or any bedding structure in the uppermost mudstones (sequence BT3) indicating deposition below even the strongest storm wave-base. Water depths of 100 m or more are therefore estimated for the uppermost Thanet mudstones in the Bradwell section. A comparable minimum figure may therefore be placed on the subsequent sea-level fall.

The duration of this sea-level fall is not easily determined. Although the onset of renewed sedimentation (around 56 Ma on the chronology of Berggren and Aubry, chapter 2) is clearly fixed by the presence of normal polarity sediments of (calcareous nannofossil) Zone NP9 age at the base of the Upnor Formation (see Ali and Jolley 1996; Ellison et al. 1996), the onset of the preceding sea-level fall cannot be dated in marginal sections because deposits of the regressive phase appear to have been removed by erosion. Such regressive de-

Figure 5.1 (previous page) Biotic events and phases of tectonism and volcanism in the North Atlantic region (temporal framework based on Knox 1996a).

1, 2 = upward reductions in abundance and diversity of benthic agglutinated foraminifera. The pyroclastic phases are incorrectly numbered and should read 1, 2.1, 2.2a (= 2.2), 2.2b (=2.3) and 3 (=2.4). Age range of the major phases of lava eruption based on Ritchie and Hitchen (1996) and Waagstein and Heilmann-Clausen (1993). BTIP = British Tertiary Igneous Province. Note that (1) the position of the NP9/NP10 zonal boundary in relation to the NW European successions is uncertain (see discussion in Berggren and Aubry, chapter 2) and (2) the timing of the terrestrial mammal migration event relative to the basinal events (kaolinite influx, *Apectodinium* acme, etc.) is conjectural. The horizontal pecked line represents the inferred level of the carbon and oxygen isotope shifts and the oceanic benthic extinction event. No direct correlation between the NW European sections and the oceanic isotope record has yet been established, although a negative carbon isotope excursion identified within the Argiles Plastiques bariolées in the Paris Basin (Stott et al. 1996) may represent the oceanic excursion. This would place the oceanic excursion within the basal part of the Reading Formation.

posits are certainly missing from the Bradwell succession, in which the Thanet Formation displays a progressive upward fining from sandstone to fine-grained mudstones of outer shelf facies (Knox et al. 1994). Here, the youngest datable mudstones are assigned to (calcareous nannofossil) Zone NTp14 of Varol (1989), equivalent to lower Zone NP8. The fine-grained, decalcified mudstones at the top of the succession cannot be dated, except that their consistently reversed polarity places them within Chron C25r. However, they clearly represent the culmination of the "Thanet" transgression, and probably represent relatively slow deposition. It is therefore reasonable to infer that they are no older than Zone NTp15 (upper mid Zone NP8) and thus probably extend upwards to near the top of Chron C25r (i.e., around 56.5 Ma).

These estimates indicate that while the transgression of the Thanet Formation extended from around 57.7 Ma (mid Chron C26n on the time-scale of Berggren et al. 1995) to around 56.5 Ma (end Chron C25r), with a duration of around 1.2 my, the subsequent regression, which was associated with a sea-level fall of some 100 m (see above) and led to widespread erosion and subaerial exposure of the Thanet Formation, appears to have taken place within about 0.5 my. Evidence for relatively rapid sea-level fall is also provided by the basinal succession of the central North Sea, in which the fine-grained, richly fossiliferous gray-green mudstones of the Lista Formation display a rapid upward transition to poorly fossiliferous, gray and increasingly laminated mudstones of the Sele Formation. This transition is marked by a sharp reduction in abundance and diversity of agglutinated benthic foraminifera (calcareous taxa are absent throughout) and by a dramatic increase in the proportion of terrestrial to marine palynomorphs (Thomas 1996). A comparable series of events is seen in the Danish succession, in which the Holmehus Formation (Lista Formation equivalent) is overlain by the ash-bearing Ølst Formation (Sele Formation equivalents; see Heilmann-Clausen et al. 1985), although the contact here is sharper and perhaps represents a discontinuity.

The rapidity and the magnitude of the sea-level fall, and the consequent asymmetry in the relative sea-level curve, is most easily explained as the result of tectonic uplift. This interpretation is supported by evidence for a change in basin configuration in southern England (Knox 1996a:figure 7) and by the association of the sea-level fall with the onset of renewed pyroclastic activity in the Greenland–Faroe province, corresponding to the basal tephras of Phase 2 of Knox and Morton (1988) and to the base of the Danish negative ash series.

The combination of regional uplift and renewed volcanism points to tectonic activity in the region of the North Atlantic rift zone, and, as would be expected, the effects of the regional uplift are most strongly expressed to the northwest of Scotland. Here, in the Faroe–Shetland Basin, the Lista–Thanet high sea-level stand is represented by mudstones of the upper part of the Lamba Formation (see Knox et al. 1997). As in the North Sea succession, these mudstones are characterized by relatively rich and diverse assemblages of agglutinated foraminifera and dinoflagellate cysts, and are overlain by mudstones with greatly reduced assemblages (basal Flett Formation). The effects of the regional uplift on facies distribution are more strongly expressed in the Faroe–Shetland Basin than in the North Sea Basin, with paralic facies extending across former deep-water facies within the axial part of the basin. This is probably a reflection of both greater proximity to the focus of the regional uplift and of smaller initial water depths in the basinal center due to higher rates of terrigenous influx. However, another factor may have been the superimposition on the regional uplift of more localized uplift to the southwest. Such uplift, which would probably have been related to changes in the pattern of uplift and subsidence elsewhere in Britain (see Knox 1996a:218), would account for the widespread non-sequence reported by Ebdon et al. (1995:figures 6, 10, base T40) from the southern part of the earlier Paleocene Faroe–Shetland Basin, which became an extension of the southern shelf area.

A major fall in sea-level thus affected all parts of the northeast Atlantic region at the beginning of the P–E transition, enhanced in some areas by uplift of local structural elements. It was followed by a rise in sea-level that led to deposition in southern marginal areas of transgressive sands (Upnor Formation and equivalents) followed by highstand paralic sediments (Woolwich and Reading formations and equivalents). Equivalent basinal sediments, which are characterized by an abundance of the dinoflagellate genus *Apectodinium*, include submarine-fan sands (e.g., Forties Sandstone Member) in more proximal settings, and laminated mudstones in more distal settings (including the Danish embayment). The subsequent record is incomplete in southern marginal areas, where, for example, the Woolwich and Reading formations of southern England are overlain with distinct angular unconformity by the largely sandy deposits of the Harwich Formation (see Knox 1996a). The northern offshore basins were the site of continued sedimentation, however, and a rise in relative sea-level led to renewed deposition in the southern part of the Faroe–Shetland Basin. At the same time, major progradation of coastal plain facies occurred in southern and central parts of the Faroe–Shetland Basin and along the western margin of the central and northern North Sea (Dornoch Formation). In the more distal parts of the North Sea, including the Danish embayment, laminated muds continued to accumulate (Sele Formation unit 2), but with lowered salinities, as indicated by the predominance of terrestrial palynomorphs over dinoflagellate cysts.

The progradational phase was terminated by the widespread transgression associated with Flett unit 3 in the Faroe–Shetland Basin, Sele unit 3 in the North Sea and the Harwich Formation in southern England. The basal deposits of this transgression are widely recognized offshore as a high-gamma mudstone overlain by the low-gamma Balder Formation unit 1, consisting of mudstone with abundant basaltic volcanic ash layers. A second major transgression, represented by the base of Balder unit 2 and the base of the London Clay Formation, marks the onset of substantial regional subsidence and the termination of large-scale pyroclastic eruptions as a result of drowning of the source volcanoes.

Volcanic Activity

The North Atlantic region was the site of intense volcanism in the mid to late Paleocene and earliest Eocene epochs, prior to and during the onset of sea-floor spreading between Scotland and Greenland. The volcanic activity appears to have been concentrated in two main periods, with maximum activity occurring around 62–58 Ma and 56–54 Ma (Ritchie and Hitchen 1996; see figure 5.1). Only the later period of activity is discussed in detail here, especially the activity that immediately preceded the onset of sea-floor spreading, represented by younger elements of the Rockall Trough and Faroe plateau lavas, by local volcanic centers in the northern part of the British Tertiary Igneous Province, and by associated tephra layers. Reference is also made to the later activity that took place during the earliest stages of sea-floor spreading, represented by the dipping reflector series and abundant basaltic tephras layers (see Waagstein and Heilmann-Clausen 1995).

The ages of the lavas and intrusives are mostly poorly constrained, because of the paucity of reliable radioisotopic dates. Fortunately, however, a few sections exist that provide some information on the ages of the extrusives in relation to the sedimentary succession. Thus distal representatives of the Faroe Plateau Basalt Group have been encountered in Well 205/9–1, in the Faroe–Shetland Basin, where they occur in the lower part of the Flett Formation. A comparable age assignment was inferred for the Faroe middle and upper basalt formations by Waagstein and Heilmann-Clausen (1995:figure 5). Jolley (1997) has demonstrated that the sediments that separate the Faroe lower and middle lava formations (the so-called "coal series") contain palynomorphs of upper Sele/Flett 1b to lowermost Sele/Flett unit 2a affinity. The 205/9–1 lavas, which extend from the upper part of Flett unit 1 into the lowest part of Flett unit 2 may therefore represent the earliest eruptions of the Faroe Middle Lava Formation.

Sedimentary intercalations within the lavas of the northern British Tertiary Igneous Province volcanic centers have not yet been dated by palynomorphs, but the upper limit of extrusive activity in the Erlend Volcanic center may be constrained by the occurrence in one well (Well 209/4–1A) of the characteristic mudstone-with-tuff facies of the Balder Formation above the lavas, underlain by a thin representative of Flett Formation unit 3 (= Sele Formation unit 3 of the North Sea). Waagstein and Heilmann-Clausen (1995) showed that the Balder Formation is widely distributed over the Faroe Upper Lava Formation in the offshore area to the east of the Faroe Islands.

By far the most detailed stratigraphic record of igneous activity in the North Atlantic region is provided by the occurrence of abundant tephra layers, principally in the mudstone-dominated basinal sections. The nature, occurrence and tectonic significance of these tephras layers have been described elsewhere (Knox and Morton 1983, 1988; Morton and Knox 1990; Knox 1994). The successive phases of pyroclastic activity were given the annotation Phase 1 and Phase 2 by Knox and Morton (1983), with Phase 2 being associated with the Sele and Balder formations of the North Sea. Phase 2 was subdivided into Subphases 2a (Sele Formation), 2b (Balder Formation unit 1) and 2c (Balder Formation unit 2). The Subphase 2a tephras thus equated with the "negative ash series" and the Subphase 2b ashes with the "positive ash series" of Denmark. This annotation was modified by Knox (1996a) to take account of new information from cored sections, which have shown that tephras are more or less confined to Sele units 1 and 3, with Sele unit 2 being devoid of significant tephra layers. A similar vertical distribution has been described from the negative series in the Harre Borehole, Denmark, where upper and lower groups of tephras are separated by about 13 meters of mudstone from which no tephras have been recorded (Nielsen et al. 1994). The Sele unit 1 tephras were assigned to Subphase 2.1, and the confluent Sele unit 3 and Balder unit 1 tephras to Subphases 2.2a and 2.2b, respectively (see figure 5.1).

Correlation of these tephra phases with the standard Paleogene chronostratigraphic scheme is hampered by the absence of standard zonal taxa in the associated mudstones. However, the occurrence of Phase 2 bentonites in DSDP Hole 550 (Knox 1984) does allow partial correlation with the planktonic foraminiferal and nannoplankton zonation, which has recently been revised by Aubry et al. (1996). These authors have also shown that the lower part of Zone NP10 is missing from this section, raising the possibility that the Phase 2 tephra succession may not be complete (see also chapter 2). Consideration of the relative positioning of the marker layer −17 Ash in relation to the entire tephra succession suggests that this may indeed be the case, since it is comparable to its relative position in other sections where only Phase 2.2 tephras are present (e.g., Borehole 81/46A, see Knox 1984:figure 4).

Biotic Events in Basinal Facies

One of the most clearly defined faunal changes within the interval under review is the sharp reduction in the abundance and diversity of benthic agglutinated foraminifera at the Lista/Sele formational boundary in the North Sea and at the equivalent, if disconformable, Holmehus/Ølst formational boundary in Denmark. Superficially, this appears comparable to the benthic foraminiferal extinction event (BFE) recorded from many oceanic sections (see Thomas and Shackleton 1996). However, the association of the North Sea benthic foraminiferal extinctions with a rapid change in facies from gray-green, heavily bioturbated mudstones of the Lista Formation to gray, pyritic, increasingly laminated mudstones of the Sele Formation strongly suggests control by local bottom-water oxygenation levels, with relatively well oxygenated conditions being replaced by dysaerobic conditions. It should also be noted that both the facies change and the loss of benthic foraminifera are not related to a single surface, but are rapidly transitional, so that the term *event* is not really appropriate. In addition, a further reduction in abundance and diversity occurs a little higher in the section at a change to more continuously laminated mudstone facies (see figure 5.1).

The other major feature of the Lista–Sele transition is the sharp influx of terrestrial palynomorphs reported by Thomas (1996). As discussed above, this is interpreted as reflecting a sharp increase in terrestrial run-off associated with a rapid fall in sea-level. A climatic change leading to increased rainfall may also have been involved, although this is not reflected in clay mineral assemblages, in which kaolinite, indicating warmer and/or more humid conditions, does not show a significant increase until higher in the section (Knox 1996b; Knox, in press; see also figure 5.1). A possible indication that surface-water conditions were also affected is indicated by an influx of diatoms (especially *Coscinodiscus*) at the Lista–Sele facies transition.

Further significant changes in the palynomorph assemblages occur higher up in the Sele Formation. The distribution of the genus *Apectodinium* is of particular interest as it displays a well defined acme occurrence in the middle and lower part of Sele unit 1. This acme occurrence is identified in all North Sea basinal successions and extends to southern marginal sections (e.g., Woolwich and Reading formations) as well as the Faroe–Shetland Basin. The most detailed account of this *Apectodinium* acme is that of Thomas (1996) in a study of a cored section from UK Central North Sea Well 22/10a-4. The abundance of *Apectodinium* in this section shows a broad correspondence to an increase in the proportion of kaolinite within the clay mineral assemblages (see figure 5.1), indicating that warming of the surface waters or increased fresh-water influx in response to climatic change may have been responsible for the change in the dinoflagellate cyst assemblages. The top of the longer-term *Apectodinium* acme was not encountered in the cored section, but it is known from extensive exploration drilling to occur close to the top of Sele unit 1.

An influx of new marine dinoflagellate taxa, accompanied in the Central North Sea by a change from laminated to non-laminated mudstone facies, takes place in the upper part of the Balder Formation (base of unit 2), indicating increased connection with oceanic waters to the west. This transgressive event, which equates to the base London Clay in southern England, led to the drowning of the major progradational complexes around Scotland and of part, at least of the Erlend volcanic center in the northwest. It marks the beginning of long-term thermal subsidence in the North Atlantic region that resulted from the onset of sea-floor spreading between Greenland and Scotland. Fully marine facies, comparable to those of the Thanet and Lista formations, do not recur until the top of the Balder Formation, which in the Central North Sea is marked by an upward change from gray, pyritic mudstone to variegated claystone with rich microfaunas and marine palynofloras.

Biotic Events in Marginal Facies

The marginal successions of the southern North Sea and Paris Basin regions play an important role in chronological calibration of the more continuous basinal successions because they include a more diverse biota and have yielded important information on magnetic polarity assignments.

Identification of true biotic events is hampered in the marginal successions by stratigraphic incompleteness. Nevertheless, the marginal successions are the only source of information on one of the most important faunal events of the P–E transition, namely the replacement of endemic northern European mammal faunas by more cosmopolitan faunas that had many features in common with those of North America (see chapter 20). This faunal turnover is recognized in the Paris Basin succession between marine sands (Sables de Bracheux and equivalents) and overlying continental to paralic deposits (Argiles Plastiques bariolées, Argiles à lignites du Soissonais, etc.). The Sables de Bracheux are believed to equate with the lower, calcareous unit of the Upnor Formation in the London Basin (see Ellison et al. 1996).

Correlation between these successions and those of the Central North Sea is provided primarily by the abundance of the dinoflagellate genus *Apectodinium* in the more marine facies of the Woolwich Formation and its Paris Basin equivalents. This indicates correlation with the middle to upper part of Sele unit 1. Further correlation is possible by inference only, whereby the major disconformity between the Thanet and the Upnor formations can be equated with the basal part of Sele unit 1, in which the palynomorph assemblages are dominated by terrestrial forms. On this correlation, the Upnor Formation would be broadly equivalent to the transition from non-laminated to laminated mudstone facies, and the base of the Argiles Plastiques bariolées, etc., would be close to the incoming of *Apectodinium* within the Central North Sea succession. These correlations are shown in figure 5.1, but it should be stressed that they are intuitive and that no means of direct correlation is currently available.

Interaction Between Crustal Events and Biotic Change

Volcanic Activity

Volcanic eruptions might be expected to affect the biosphere in two ways: by exhalation of gases into the atmosphere and by introduction of dust into the upper atmosphere during explosive eruptions. The emission of carbon dioxide in particular is likely to have contributed to climatic warming through the greenhouse effect, and volcanic activity in the North Atlantic region has therefore been considered as a possible contributor to the rapid climatic changes associated with the P–E transition (see Eldholm and Thomas 1993; Thomas and Shackleton 1996:402). Unfortunately, the volcanic and climatic records have been largely established in different regions, so that the relative timing of volcanic and climatic events can be determined only by indirect correlation, which at the present time is imprecise. Assessment of the influence of North Atlantic volcanism on biotic change must therefore rely on a relatively broad perspective.

It is necessary first to recognize that the volcanic activity consisted of both passive lava eruptions and explosive tephra eruptions. The former greatly exceeded the latter in volumetric terms, and may therefore be presumed to have made the greater contribution of gases to the atmosphere but a negligible contribution of dust. Interbedding of lavas and sediments at the distal eastward end of the Faroe Lava Group indicates that the greatest eastward extension of this lava field took place towards the end of the *Apectodinium* acme interval and continued beyond it, and was thus only partially coincident with the period of relatively warm and humid climatic conditions inferred from terrestrial palynomorphs (Schröder 1992) and clay mineral assemblages (Knox 1998) in contemporaneous marine successions of the central North Sea. Consideration of the timing of lava eruption on a regional scale also casts doubt on a direct genetic relationship between global warming and North Atlantic lava eruption, since there is no evidence for any exceptional activity at the time of the global events (figure 5.1), even allowing for uncertainty as to their precise location within the northwest European successions.

The pyroclastic record provides a firmer basis for assessment of the role of explosive eruptions in climatic and biotic change, although in this case it is likely that the effect would have been one of cooling rather than warming. The broad temporal equivalence of the phase 2.1 eruptions and the kaolinite influx and *Apectodinium* acme interval is thus likely to be coincidental. In any case, these eruptions were relatively infrequent and mostly of a small-scale nature, and their effects are likely to have been restricted to short-term changes in local weather conditions.

In conclusion, it seems unlikely that North Atlantic volcanism was responsible for short-term climatic warming during the Paleocene–Eocene transition. Although it seems that pyroclastic activity is unlikely to have caused any long-term effect on the global climate, the possibility of more local environmental effects should not be disregarded. In particular, it is notable that diatoms are at their most abundance in sediments associated with the Phase 2.3 ashes, indicating that productivity could have been enhanced by raised levels of dissolved silica resulting from the presence of abundant ash within the water column. However, substantial diatomite developments occur locally in the upper part of the Danish negative series, in which tephra layers are thin and sporadic, and it seems more likely that the high diatom productivity was the result of restricted circulation and water-column stratification (Mitlehner 1996) rather than high concentrations of dissolved volcanogenic silica.

Uplift and Subsidence

As discussed above, evidence from both basinal and marginal sections in NW Europe indicates that the long term rise in sea-level that characterized the early and mid Thanetian Age was abruptly terminated by a rapid, tectonically induced sea-level fall. This led to widespread exposure along the basin margins, and some former basinal areas became the sites of terrestrial and paralic sedimentation.

Uplift was particularly pronounced in the south of the Faroe–Shetland Basin, where sediments of Flett Formation unit 1 (equivalent to Sele Formation unit 1 of the North Sea) are missing (see Ebdon et al. 1995:figures 6, 10; Knox et al. 1997). This may reflect the existence of a broader area of uplift to the southwest. At about the same time, the Faroe lavas attained their greatest eastward extent, meeting up with paralic sediments prograding from the south and east.

A paleogeographic reconstruction for this period (figure 5.2B) shows that much of the North Atlantic region would have been subaerially exposed for all or part of the time, providing a pathway (land-bridge) for the interchange of mammal faunas between Europe and North America. Evidence from the Paris Basin indicates that this migration did not take place at the time of the profound sea-level fall represented by the Thanet/Upnor formation boundary (i.e., at the NP8/NP9 zonal boundary, which corresponds to an erosional contact = base Sables de Bracheux), but a short time later (top Upnor Formation, top Sables de Bracheux). This delay is best explained by paleogeographic control, in which a relatively deep-water Atlantic–Arctic seaway persisted along the Rockall Trough and Faroe–Shetland Basin during the initial uplift, but was eventually severed through a probable combination of continued lava emplacement coupled with more localized uplift to the west of Scotland. One possible site of such uplift is the Wyville–Thompson Ridge, which was proposed by McKenna (1983) and Roberts et al. (1983) as a probable component of the Greenland–Scotland land-bridge. However, although the Wyville–Thompson Ridge may in part represent accumulation of Paleocene lavas (Roberts et al. 1983) its current relief may be the result of later Cenozoic compressional tectonics (Boldreel and Andersen 1993). It is therefore more likely that a broader uplift was involved, possibly related to the contemporaneous eastward tilting of eastern Scotland described by Milton et al. (1990) and the more general structural reorganization in the British region inferred by Knox (1996a:218). Accumulation of lavas may have further served to bring the area above sea-level.

It should also be borne in mind that although faunal interchange between North America and Europe was clearly dependent on the establishment of the land-bridge, the location of the land-bridge necessitated migration via relatively high latitudes. It is therefore possible that the timing of faunal interchange was also controlled by climatic amelioration (see discussion in Hooker 1996:215). The recognition of the negative carbon isotope excursion in association with the earliest "cosmopolitan" faunas in North America (Koch et al. 1995) indicates a possible relationship between faunal migration and the short-lived climatic warming inferred from the isotopic signatures. However, as demonstrated by Hooker (chapter 20) faunal interchange was not restricted to this specific interval, indicating that a longer-term climatic control would have to be invoked.

The proposed correlation between the terrestrial and marine successions of the North Sea Basin (see figure 5.1) indicates that the mammal migration took place close to the transition from poorly laminated to well laminated mudstone within the Sele unit 1 mudstones of the Central North Sea. This facies change is consistent with isolation of the North Sea Basin as a result of the closure of both the Rockall Trough—Faroe–Shetland Basin seaway and the "proto-English Channel."

If the precise nature of the eastern part of the land-bridge is uncertain, that of the western part is even more so. Since basalt eruption in the East Greenland area was relatively localized at this time, the regional uplift may alone have been sufficient to destroy the Greenland–Rockall–Faroe seaway on account of its proximity to the plume center.

By establishing a land-bridge across the North Atlantic region, the rapid late Thanetian uplift and subsequent lava eruptions would have severed communication between the Atlantic Ocean and northern high latitude waters, with the possibility of significant effects on the global oceanic circulation pattern. The emergence of such a large land area may be expected to have had a major effect on the local climate, but any associated changes in the global climate are more likely to have resulted from changes in the pattern of global oceanic circulation.

How long the land-bridge persisted is not certain. Evidence from the Faroe–Shetland Basin (Flett unit 2) indicates that a long-term rise in relative sea level began within a few hundred thousand years of its establishment and that lava eruptions either ceased or were confined to areas closer to their source. Nevertheless, the sediment surface in central and southern parts of the Faroe–Shetland Basin lay close to sea-level during deposition of Flett unit 2 as indicated by paralic facies with sporadic soil horizons. Palynomorph data from the Faroe Islands succession indicates that after a period of quiescence (represented by the "Coal Series"), eruption of the Faroe plateau basalts resumed during deposition of the uppermost Flett unit 1 or Flett unit 2 (Jolley 1997), with lateritic flow tops indicating the continued

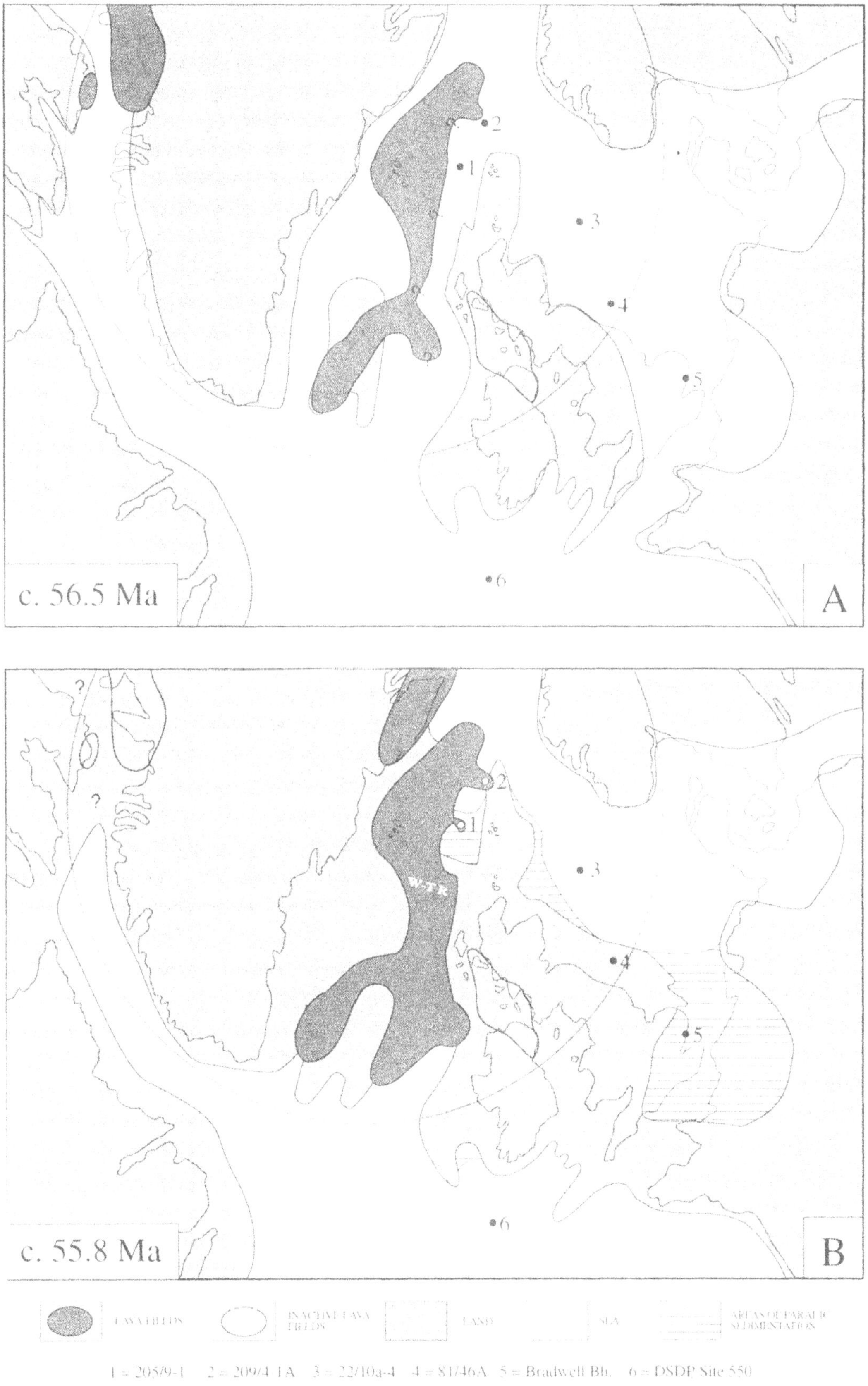
c. 56.5 Ma
A
c. 55.8 Ma
B
W-T R
LAVA FIELDS
INACTIVE LAVA FIELDS
LAND
SEA
AREAS OF PARALIC SEDIMENTATION
1 = 205/9-1 2 = 209/4-1A 3 = 22/10a-4 4 = 81/46A 5 = Bradwell Bh. 6 = DSDP Site 550

existence of a land surface to the west of the Faroe–Shetland Basin. The land-bridge is likely to have been substantially diminished in area during the widespread transgression that commenced during deposition of Flett unit 3 (equivalent to the Oldhaven (Harwich) transgression of southern England) and continued during deposition of the Balder Formation. This transgression appears to have led to widespread inundation of the Faroe lavas in eastern offshore areas (Waagstein and Heilmann-Clausen 1988). However, the frequency of pyroclastic eruption during deposition of the lower, tephra-rich, part of the Balder Formation indicates that the focus of the Greenland–Faroe volcanic province remained above sea-level until the major sea-level rise associated with Balder Formation unit 2 (equivalent to the basal London Clay Formation). Also evidence from the southern margin of the Faroe–Shetland Basin indicates that terrestrial environments existed in that region during deposition of the Balder Formation (see Knox et al. 1997), indicating that land persisted to the southwest. Maintenance of the land-bridge at this time, leading to continued closure of the Rockall Trough—Faroe–Shetland Basin seaway, would account for the persistence of bottom-water anoxia in the central North Sea Basin during deposition of the lower, tephra-rich part of the Balder Formation (see figure 5.1). Severance of the land-bridge is most likely to have taken place during the major transgression represented by the base of the London Clay Formation.

Uplift in the North Atlantic region, coupled with increased sediment influx and widespread lava eruption, is thought to have been responsible for severing the marine connection between the Atlantic Ocean and northern high-latitude waters. This led to the progressive isolation of the North Sea and Faroe–Shetland basins, reflected in profound changes in both lithofacies and biofacies. The emergence of the North Atlantic land area also provided a pathway or land-bridge for the interchange of mammal faunas between Europe and North America.

By severing marine connection between the Atlantic Ocean and northern waters, uplift of the North Atlantic region could have led to a significant change in the pattern of oceanic circulation. However, the period of emergence of the North Atlantic region persisted long after the end of the short-lived carbon isotope excursion, indicating that it is unlikely to have been solely responsible for the profound environmental changes recorded from both oceanic and terrestrial environments at the time of the negative carbon isotope shift. Volcanic activity similarly took place over a relatively long period, and there is no evidence for exceptional eruptive activity at the time of the oceanic benthic foraminiferal extinction event and negative isotope shift. It is therefore unlikely that North Atlantic volcanism was directly responsible for rapid global environmental change during the P–E transition.

ACKNOWLEDGMENTS

Reviews by Jeremy Hooker, David Jolley and Andrew Morton are gratefully acknowledged. Publication is with the approval of the Director, British Geological Survey (N.E.R.C.).

REFERENCES

Ali, J. R. and Jolley, D. W., 1996. Chronostratigraphic framework for the Thanetian and lower Ypresian deposits of southern England. In Knox, R. W. O'B., Corfield, R. M. and Dunay, R. E. (eds.), *Correlation of the Early Paleogene in Northwest Europe,* Geological Society Special Publication No. 101, p. 129–144.

Aubry, M.-P., Berggren, W. A., Stott, L. and Sinha, A., 1996. The upper Paleocene–lower Eocene stratigraphic record and the Paleocene–Eocene boundary carbon isotope excursion: Implications for geochronology. In Knox, R. W. O'B., Corfield, R. M. and Dunay, R. E. (eds.), *Correlation of the early Paleogene in Northwest Europe,* Geological Society Special Publication No 101, p. 353–380.

Berggren W. A., Kent, D. V., Swisher, C. C., III and Aubry, M.-P., 1995. A revised Cenozoic geochronology and chronostratigraphy. In Berggren, W. A., Kent, D. V., Aubry, M.-P. and Hardenbol, J. (eds.), *Geochronology Time Scales and Global Stratigraphic Correlation,* Society of Economic Paleontologists and Mineralogists Special Publication No. 54, p. 129–212.

Boldreel, L. O. and Andersen, M. S., 1993. Late Paleocene to Miocene compression in the Faeroe–Rockall area. In Parker, J. R. (ed.), *Petroleum Geology of Northwest Europe,* Geological Society, London, p. 1025–1034.

Ebdon, C., Granger, P. J., Johnson, H. D. and Evans, A. M., 1995. Early Tertiary evolution and sequence stratigraphy of the Faeroe–Shetland Basin: Implications for hydrocarbon prospectivity. in Scrutton, R. A., Stoker, M. S., Shimmield, G. B. and Tudhope, A. W.,(eds.), *The Tectonics, Sedimentation and Palaeoceanography*

Figure 5.2 *(previous page)* Paleogeographic reconstructions for the North Atlantic region.
Based on information from Waagstein (1988), Knott et al. (1993), Ritchie and Hitchen (1996) and Knox (1996a:figure 7). The inferred extent of the mantle plume is indicated by a circle. A. Around 56.5 Ma (late Chron C25r), at the maximum extent of late Paleocene marine transgression (upper Lista Formation, central North Sea; upper Lamba Formation, Faroe–Shetland Basin; upper Holmehus Clay Formation, Denmark; top Thanet Formation, Bradwell, SE England). B. Around 55.8 Ma (earliest Chron C24r), at a moment corresponding to the top of the Upnor Formation and Sables de Bracheux. Note that large-scale lava eruptions in West Greenland and the Hebridean region had probably ceased before 56.5 Ma. W–T R = location of present-day Wyville–Thompson Ridge.

of the North Atlantic Region, Geological Society Special Publication No. 90, p. 51–69.

Eldholm, O. and Thomas, E., 1993. Environmental impact of volcanic margin formation. *Earth and Planetary Science Letters,* v. 117, p. 319–329.

Ellison, R. A., Ali, J. R., Hine, N. M. and Jolley, D. W., 1996. Recognition of Chron C25n in the upper Paleocene Upnor Formation of the London Basin, UK. In Knox, R. W. O'B., Corfield, R. M. and Dunay, R. E. (eds.), *Correlation of the Early Paleogene in Northwest Europe,* Geological Society Special Publication No 101, p. 185–193.

Heilmann-Clausen, C., Nielsen, O. B. and Gersner, F., 1985. Lithostratigraphy and depositional environments in the Upper Paleocene and Eocene of Denmark. *Bulletin of the Geological Society of Denmark,* v. 33, p. 287–323.

Hooker, J. J., 1996. Mammalian biostratigraphy across the Palaeocene–Eocene boundary in the Paris, London and Belgian Basins. In Knox, R. W. O'B., Corfield, R. M. and Dunay, R. E. (eds.), *Correlation of the Early Paleogene in Northwest Europe,* Geological Society Special Publication, No. 101, p. 205–218.

Jolley, D. W., 1997. Palaeosurface palynofloras of the Skye lava field and the age of the British Tertiary volcanic province. In Widdowson, M. (ed.), *Palaeosurfaces: Recognition, Reconstruction and Palaeoenvironmental Interpretation,* Geological Society, London, Special Publication 120, p. 67–94.

Knott, S. D., Burchell, M. T., Jolley, E. J. and Fraser, A. J., 1993. Mesozoic to Cenozoic plate reconstructions of the North Atlantic and hydrocarbon plays of the Atlantic margins. In Parker, J. R. (ed.), *Petroleum Geology of Northwest Europe,* Geological Society, London, p. 953–974.

Knox, R. W. O'B., 1984. Nannoplankton zonation and the Paleocene/Eocene boundary beds of NW Europe: An indirect correlation by means of volcanic ash layers. *Journal of the Geological Society, London,* v. 141, p. 993–999.

Knox, R. W. O'B., 1994. Stratigraphical distribution of tephra layers in the Lower Paleogene of the southwestern margin of the North Sea Basin. *Bulletin de la Société Belge de Géologie,* v. 102 (for 1993), p. 159–164.

Knox, R. W. O'B., 1996a. Tectonic controls on sequence development in the Paleocene and earliest Eocene of southeast England: Implications for North Sea stratigraphy. In Hesselbo, S. P. and Parkinson, D. N. (eds.), *Sequence Stratigraphy in British Geology,* Geological Society Special Publication No 103, p. 209–230.

Knox, R. W. O'B., 1996b. Correlation of the early Paleogene in northwest Europe: An overview. In Knox, R. W. O'B., Corfield, R. M. and Dunay, R. E. (eds.), *Correlation of the Early Paleogene in Northwest Europe,* Geological Society Special Publication No. 101, p. 1–11.

Knox, R. W. O'B., 1998. Kaolinite influx within Paleocene/Eocene boundary strata of western Europe. *Newsletters on Stratigraphy,* v. 36, pp. 49–53.

Knox, R. W. O'B. and Holloway, S., 1992. Paleogene of the Central and Northern North Sea. In Knox, R. W. O'B. and Cordey, W. G. (eds.), *Lithostratigraphic nomenclature of the UK North Sea,* Nottingham: British Geological Survey, 133 pp.

Knox, R. W. O'B. and Morton, A. C., 1983. Stratigraphical distribution of early Palaeogene pyroclastic deposits in the North Sea Basin. *Proceedings of the Yorkshire Geological Society,* v. 44, p. 355–363.

Knox, R. W. O'B. and Morton, A. C., 1988. The record of early Tertiary N Atlantic volcanism in sediments of the North Sea Basin. In Morton, A. C. and Parson, L. M. (eds.), *Early Tertiary Volcanism and the Opening of the NE Atlantic,* Geological Society Special Publication No 39, p. 407–419.

Knox, R. W. O'B., Hine, N. and Ali, J. R., 1994. New information on the age and sequence stratigraphy of the type Thanetian of Southeast England. *Newsletters on Stratigraphy,* v. 30, p. 45–60.

Knox, R. W. O'B., Holloway, S., Kirby, G. A. and Baily, H., 1997. *Stratigraphic Nomenclature of the UK Northwest Margin. 2. Early Paleogene Lithostratigraphy and Sequence Stratigraphy.* Nottingham: British Geological Survey, 114 pp.

Koch, P. L., Zachos, J. C. and Dettman, D. L., 1995. Stable isotope stratigraphy and paleoclimatology of the Paleogene Bighorn Basin (Wyoming, USA). *Palaeogeography, Palaeoclimatology, Palaeoecology,* v. 115, p. 61–89.

McKenna, M. C., 1983. Cenozoic paleogeography of North Atlantic land bridges. In Bott, M. H. P., Saxov, S., Talwani, M. and Thiede, J. (eds.), *Structure and Development of the Greenland–Scotland Ridge. New Methods and Concepts,* NATO Conference Series, Series IV: Marine Sciences, v. 8, New York, London: Plenum Press, p. 351–399.

Milton, N. J., Bertram, G. T. and Vann, I. R., 1990. Early Palaeogene tectonics and sedimentation. In Hardman, R. F. P. and Brooks, J. (eds.), *Tectonic Events Responsible for Britain's Oil and Gas Reserves,* Geological Society Special Publication No. 55, p. 339–351.

Mitlehner, A., 1996. Palaeoenvironments in the North Sea Basin around the Paleocene–Eocene boundary: evidence from diatoms and other siliceous microfossils. In Knox, R. W. O'B., Corfield, R. M. and Dunay, R. E. (eds.), *Correlation of the Early Paleogene in Northwest Europe,* Geological Society Special Publication No 101, p. 255–273.

Morton, A. C., 1982. The provenance and diagenesis of Palaeocene sandstones of southeast England, as indicated by heavy mineral analysis. *Proceedings of the Geologists' Association,* v. 93, p. 263–274.

Morton, A. C. and Knox, R. W. O'B., 1990. Geochemistry of late Palaeocene and early Eocene tephras from the North Sea Basin. *Journal of the Geological Society, London,* v. 147, p. 425–437.

Nielsen, O. B., Friis, H. and Korsbech, U., 1994. Lithology and lithostratigraphy of the Harre borehole, Denmark. *Aarhus Geoscience,* v. 1, p. 5–14.

Ritchie, J. D. and Hitchen, K., 1996. Early Paleogene offshore igneous activity to the northwest of the UK and its relationship to the North Atlantic Igneous Province. In Knox, R. W. O'B., Corfield, R. M. and Dunay, R. E. (eds.), *Correlation of the Early Paleogene in Northwest Europe,* Geological Society Special Publication No 101, p. 63–78.

Roberts, D. G., Bott, M. H. P. and Uruski, C., 1983. Structure and origin of the Wyville–Thompson Ridge. In Bott, M. H. P., Saxov, S., Talwani, M. and Thiede, J. (eds.), *Structure and Development of the Greenland–Scotland Ridge.* New Methods and Concepts, NATO Conference Series, Series IV: Marine Sciences, v. 8, New York, London: Plenum Press, p. 133–158.

Schröder, T., 1992. A palynological zonation for the Paleocene of the North Sea Basin. *Journal of Micropalaeontology,* v. 11, p. 113–126.

Stott, L. D., Sinha, A., Thiry, M., Aubry, M.-P. and Berggren, W. A., 1996. Global $\delta^{13}C$ changes across the Paleocene–Eocene boundary: Criteria for terrestrial-marine correlations. In Knox, R. W. O'B., Corfield, R. M. and Dunay, R. E. (eds.), *Correlation of the Early Paleogene in Northwest Europe,* Geological Society Special Publication No. 101, p. 381–399.

Thomas, E. and Shackleton, N. J., 1996. The Paleocene–Eocene benthic foraminiferal extinction and stable isotope anomalies. In Knox, R. W. O'B., Corfield, R. M. and Dunay, R. E. (eds.), *Correlation of the Early Paleogene in Northwest Europe,* Geological Society Special Publication No 101, p. 401–441.

Thomas, J. E., 1996. The occurrence of the dinoflagellate cyst *Apectodinium* (Costa & Downie 1976) Lentin and Williams 1977 in the Moray and Montrose Groups (Danian to Thanetian) of the UK

Central North Sea. In Knox, R. W. O'B., Corfield, R. M. and Dunay, R. E. (eds.), *Correlation of the Early Paleogene in Northwest Europe,* Geological Society Special Publication No. 101, p. 115–120.

Varol, O., 1989. Palaeocene calcareous nannofossil biostratigraphy. In Crux, J. A. and van Heck, S. E. (eds.), *Nannofossils and their Applications,* Chichester: Ellis Horwood Ltd., p. 267–310.

Waagstein, R, 1988. Structure, composition and age of the Faeroe basalt plateau. In Morton, A. C. and Parson, L. M. (eds.), *Early Tertiary Volcanism and the Opening of the NE Atlantic,* Geological Society Special Publication No. 39, p. 225–238.

Waagstein, R. and Heilmann-Clausen, C., 1995. Petrography and biostratigraphy of Palaeogene volcaniclastic sediments dredged from the Faeroes shelf. In Scrutton, R. A., Stoker, M. S., Shimmield, G. B. and Tudhope, A. W. (eds.), *The Tectonics, Sedimentation and Palaeogeography of the North Atlantic Region,* Geological Society Special Publication No. 90, p. 179–197.

6

Climatic, Oceanographic, and Isotopic Consequences of the Paleocene India–Asia Collision

RICHARD A. BECK
ASHISH SINHA
DOUGLAS W. BURBANK
WILLIAM J. SERCOMBE
ASRAR M. KHAN

ABSTRACT

Available evidence from the India–Asia suture zone indicates that collision occurred during the Paleocene Epoch and may have triggered some of the late Paleocene–early Eocene biotic and climatic events recorded globally in marine and terrestrial strata. The rapid decrease in stable carbon and oxygen isotopic ratios shortly after collision suggests a fundamental link between continental collisions and the organic carbon sub-cycle. Destruction of the Neo-Tethyan margins as organic carbon sinks, as well as erosion of marine shelf strata in the early Himalaya, is inferred to have significantly decreased the apparent rate of organic carbon burial during the early Eocene Epoch. The India–Asia collision may also have affected the late Paleocene–early Eocene climate and oceans in ways not directly associated with the organic carbon sub-cycle. Those mechanisms may include a decrease in the flux of volcanic aerosols from the Trans-Himalayan volcanic Arc and an increase in the flux of warm saline deep water from collisional marine foreland basins to Antarctic seas as well as the construction of a physical barrier to circumequatorial ocean circulation.

The concept of the earth as a dynamic "geochemical factory" has become well established during the last three decades (Garrels and Mackenzie 1971). Meanwhile, the mechanical aspects of that factory, summarized as plate tectonic theory, have also become relatively well understood (Wilson 1965; Dewey and Bird 1970). These conceptual advances in combination with an increasingly precise geochronology have resulted in what has been referred to as "Earth system science." Although the Earth system view is not really new (Raymo 1991; Fleming 1992; Crowley and North 1991) data collected during the last few decades have made it possible to begin to formulate and test hypotheses regarding relationships between plate tectonics, evolution, oceanography and climate change (Fischer and Arthur 1977; Molnar and England 1990; Edmond 1992; Burbank 1992; Stille 1992; Kerrick and Caldeira 1993; Raymo 1994; Selverstone and Gutzler 1993; Beck et al. 1995c). Because of the complexity of the Earth system, it seems reasonable to examine extreme, long-term, episodes of global change and their relationship to major plate tectonic events in the hope that the signal-to-noise ratio will be high. We have examined possible links between the extraordinary global warmth of the late Paleocene–early Eocene epochs and the collision of India and Asia (figure 6.1) accordingly.

The late Paleocene–early Eocene (LP–EE, ~56 to ~51 Ma; Berggren et al. 1995) time interval represents one of the warmest episodes in the last 100 million years of Earth history (Crowley and North 1991; Wolfe and Poore 1982) and, with the possible exception of the mid to Late Cretaceous Period, of the entire Phanerozoic Eon (Frakes et al. 1992). However, data regarding

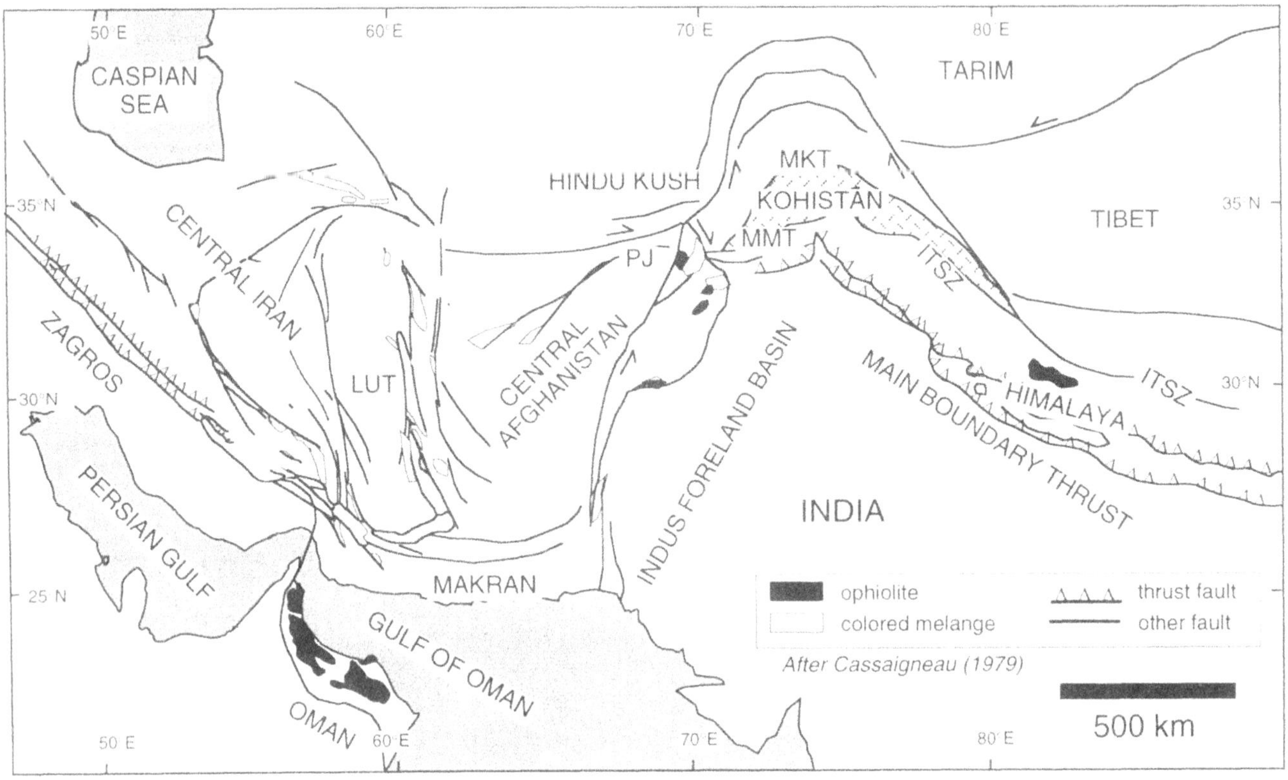

Figure 6.1 Neo-Tethyan sutures of Iran, Arabia, Afghanistan and Pakistan (after Cassaigneau 1979).
The sutures are the result of the northward motion of several Gondwanan fragments that have collided with the southern margin of Asia. Collision of India with Asia may have set the stage for the profound climatic, isotopic, and oceanographic changes that occurred during the late Paleocene and early Eocene epochs.
ITSZ = Indus–Tsangpo Suture Zone; MMT = Main Mantle Thrust; MKT = Main Karakorum Thrust (Shyok suture); PJ = Panjao suture.

many aspects of the LP–EE Earth system, including probable plant and animal physiology, the marine geochemical record, and contemporary tectonic events are more abundant and of greater precision than those available for the Cretaceous Period (Crowley and North 1991). Therefore, if one is interested in the long-term (multi-million year) record of climate change and possible mechanisms for extreme global warming, the LP–EE seems an ideal target for investigation.

This chapter examines hypotheses that relate extreme and well documented LP–EE long-term global warming (Douglas and Woodruff 1981; Shackleton 1985, 1987; Wolfe and Poore 1982) to contemporaneous plate reorganizations and mountain building in general and to the India–Asia collision in particular. Our discussion briefly reviews current concepts regarding the nature and timing of the India–Asia collision, long-term climate change, and the Cenozoic record of ocean isotopic chemistry, and concludes with a discussion of their possible relationships. Throughout the chapter we have adopted the premise that long-term climatic extrema are merely a corollary of plate tectonic theory (Garrels and McKenzie 1971; Fischer and Arthur 1977; Frakes et al. 1992).

Global Climate Change

Global climate change is somewhat analogous to water sloshing in a bucket. Whether or not the water stays in the bucket depends upon two things: the amount of water in the bucket and the vigor with which the bucket is shaken. If the shaking is gentle and the water level is low, the bucket may be shaken for long periods of time without spilling any of the water.

Similarly, the Earth's climate usually oscillates between relatively narrow limits to a regular but complex astronomical melody known as Milankovitch Cycles (Imbrie et al. 1993). That melody is driven by variations in the tilt of the Earth's axis, precession of its axis, and the eccentricity of its orbit about the sun and is played on a scale of tens to hundreds of thousands of years (Imbrie et al. 1993) unless it is near a biogeochemical threshold as during the Pleistocene Epoch. Several sections of pre-Pleistocene strata have been interpreted as a record of remarkably regular high-frequency climate change (e.g., Fischer et al. 1985). The magnitudes of these short-term climate changes are often significant but rarely extreme (Crowley and North 1991; Frakes et al. 1992). In other words, the water usually stays in the bucket.

However, occasionally and irregularly, the water sloshes out of the climate bucket and the Earth's mean surface temperature becomes extraordinarily cool (e.g., major ice ages at ~260 Ma and ~2 Ma) or extraordinarily warm (e.g., poleward expansion of tropical climates at ~250 Ma and ~55 Ma). What causes such extreme changes in the Earth's climate? Why do extreme climates occur on scales of millions to tens of millions of years and in some instances last for hundreds of thousands to millions of years?

Although several variables including long-term variations in solar irradiance, ocean circulation patterns, marine productivity and the position of the continents with respect to the poles are important in long-term climate change (Berggren and Hollister 1974, 1977; Barron and Washington 1984; Barron 1985; Wilson et al. 1986; Rea et al. 1990; Crowley and North 1991; Frakes et al. 1992; Eddy and Oescheger 1993) earth scientists have long suspected that the water in the climate bucket is carbon dioxide (CO_2) gas in the Earth's atmosphere (e.g., Broecker 1982; Barron and Washington 1985). Our atmospheric blanket interacts with short wavelength radiation (light) and long-wave length radiation (radiant heat) differently (e.g., Bolin et al. 1986). The result is an energy trap, known as the Greenhouse Effect. Without it, the Earth's surface temperature would be below the freezing point of water, and life as we know it would not exist (Cloud 1988; Crowley and North 1991; Erwin 1993).

Thus the Greenhouse Effect operates continuously but varies in magnitude. This magnitude depends upon the partial pressure of several greenhouse gases (GHGs) including water vapor, carbon dioxide, methane, chlorofluorocarbons and tropospheric ozone (Bolin et al. 1986). For several reasons it is suspected that the partial pressure of CO_2 (pCO_2) in the Earth's atmosphere largely determines the partial pressures of the other greenhouse gases and the overall effectiveness of the Greenhouse Effect, albeit indirectly (Bolin et al. 1986).

Many biologic and geologic factors affect pCO_2 and the other GHGs (the level of the water in the bucket; Barron and Washington 1985; Shackleton 1985; Vincent and Berger 1985). Variations in pCO_2 in the range of several ppm are demonstrably driven by seasonal changes in vegetation in the northern hemisphere (Keeling 1973) and the astronomical melody that modulates insolation as demonstrated by high-resolution ice-core stratigraphy (e.g., Barnola et al. 1987). The variations in pCO_2 that are apparently driven by such factors are relatively small, however, and are significant only if the background pCO_2 level is near a climatic threshold (the rim of the bucket) as it has been for the last 2 million years.

These observations and arguments lead us to search for mechanisms that may change the background concentration of CO_2 in the atmosphere on scales of millions of years. Such mechanisms may include the evolution of plants, volcanic eruptions, metamorphism of limestones (Kerrick and Caldeira 1993), periods of extensive limestone deposition/dissolution or perhaps the erosion or deposition of large amounts of organic carbon (C_{org}) during the rise of mountain ranges (Beck et al. 1995c; Faure et al. 1995). Mechanisms that affect climate through disturbances of the organic carbon cycle are of particular interest because, 1) they are chemically analogous to the combustion of fossil fuel, 2) they may be evaluated in light of the long-term record of the carbon isotopic ratio ($^{13}C/^{12}C$) in marine sediments, and 3) some of them are corollaries of plate tectonic theory. This chapter concentrates on the mechanisms by which continental collision and mountain building may affect the Earth system via changes in the flux of volcanic aerosols, changes in the rate of burial of C_{org}, and the oceanographic consequences of collisional foreland basin development. Indirect burial of C_{org} via the fertilization effect of weathering nutrient-rich rocks and the consumption of CO_2 by the weathering of calcsilicates have been discussed in the literature at some length (Vincent and Berger 1985; Flower and Kennett 1993; Raymo 1994; Föllmi 1995) and will be mentioned here only briefly.

Hypotheses Relating Mountain Building and Long-Term Global Climate Change

Mountain building may be responsible for global warming (Kerrick and Caldeira 1993; Beck et al. 1995c), as well as global cooling (Raymo et al. 1988; Vincent and Berger 1985; Selverstone and Gutzler 1993). Three tectono-climatic mechanisms for global cooling via long-term drawdown of atmospheric CO_2 due to the uplift of mountain ranges, have been proposed. They include: 1) chemical weathering of calcsilicates in igneous and metamorphic rocks exhumed by orogeny (Berner et al. 1983; Raymo et al. 1988), 2) enhanced burial of C_{org} due to an increase in the supply of limiting nutrients to the oceans during the chemical weathering of sedimentary rocks (Vincent and Berger 1985; Raymo 1994), and 3) multi-million-year storage of C_{org} in the mantle by subduction (Selverstone and Gutzler 1993).

Alternatively, three mechanisms for global warming via release or inefficient drawdown of CO_2 during mountain building have also been proposed. They include: 1) metamorphic volatilization of carbonate (Fyfe 1986; Craw 1990; Kerrick and Caldeira 1993), 2) the active exhumation of C_{org} (Beck et al. 1995c) or 3) a simple decrease in the flux of C_{org} to the lithosphere

(Shackleton 1985, 1987; Faure et al. 1995). Most of these hypotheses are not mutually exclusive. The character of any mountain range changes greatly throughout the orogenic process depending on starting conditions and tectonic environment. For example, a mountain range/foreland basin system could begin as a carbon source and end as a carbon sink or vice-versa. Similarly, volcanism may also be capable of a net warming (Caldeira and Rampino 1991) or net cooling effect (Bretherton 1986; McCormick 1995) depending on the relative importance of CO_2 and SO_2 production. Therefore each mechanism must be evaluated on a case-by-case basis in the light of available paleontologic, paleobotanic and isotopic constraints.

Evaluation of Tectono-Climatic Hypotheses

The abundance and variety of data for the LP–EE interval of global change allows us to evaluate the viability of several tectono-climatic hypotheses for that time interval. We begin with several well known observations regarding the LP–EE interval:

1. Paleobotanic and paleontological evidence such as the presence of subtropical to temperate flora and fauna in sub-polar regions of both hemispheres indicates that this was a time of exceptional global warmth (Wolfe and Poore 1982; Wolfe 1990; Estes and Hutchison 1980; Axelrod 1984; Boersma et al. 1987; Hallock et al. 1991; Aubry 1992). Mean annual temperatures (MAT) reconstructed from leaf-margin analysis increased from ~22° C to ~29° C (Wolfe and Upchurch 1987) during this interval. The overall mean global surface temperature is estimated to have been 2–4°C warmer than present (Barron 1987).
2. Oxygen isotopic ratios in calcareous shells of planktonic and benthic foraminifera declined dur-

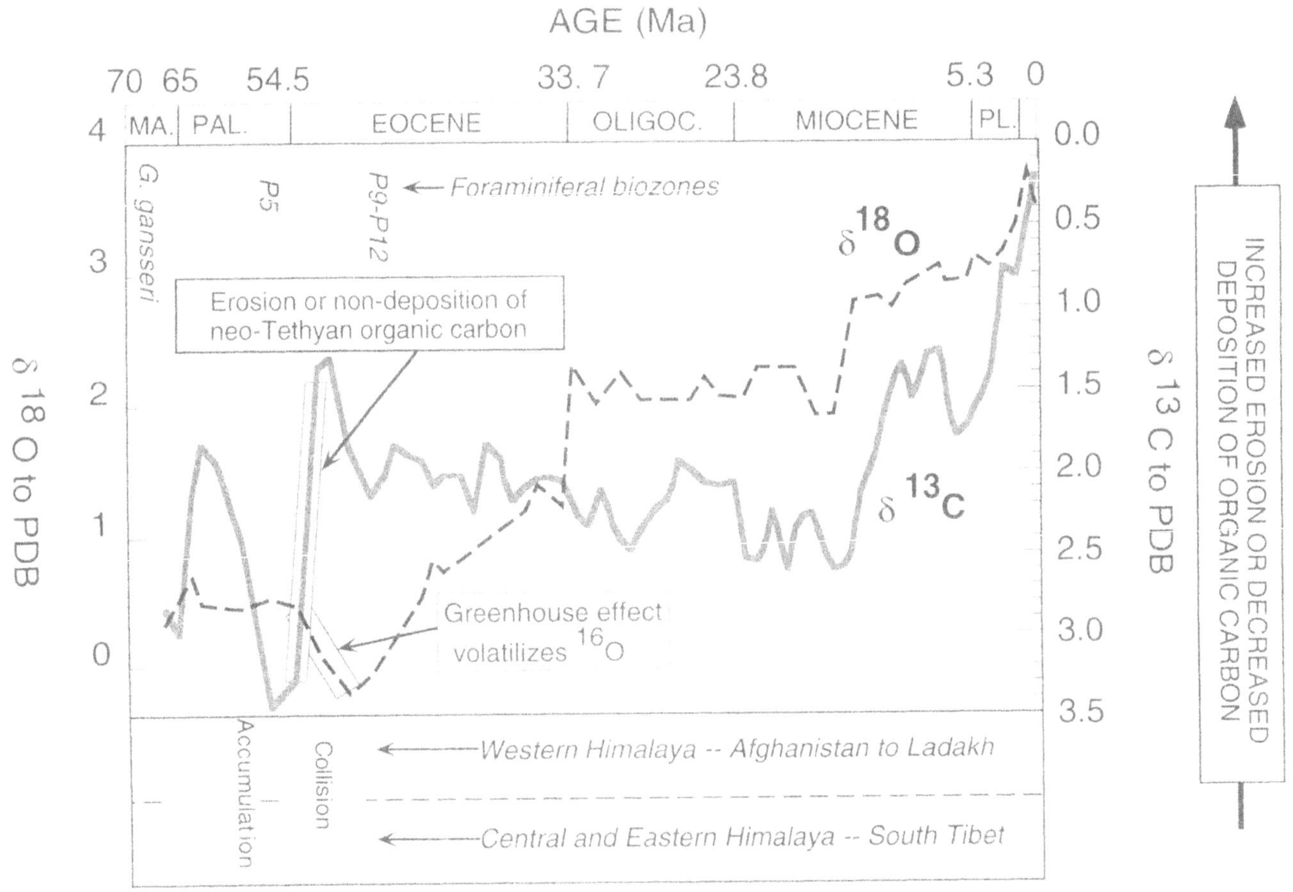

Figure 6.2 Values of $\delta^{18}O$ and $\delta^{13}C$ for marine Cenozoic carbonate strata (Shackleton 1987) vs. timing of Indian–Asian collision with consequent erosion of organic carbon-bearing north Indian slope, outer shelf, south Asian forearc and accretionary prism. Initial collision is determined as prior to the end of planktonic foraminiferal Biochron P5 (time scale of Berggren et al. 1995; Beck et al. 1995a, 1996).

MA. = Maastrichtian Age, PAL. = Paleocene Epoch, OLIGOC. = Oligocene Epoch, PL. = Pliocene Epoch.

ing the LP–EE interval (figure 6.2; e.g., Douglas and Woodruff 1981; Shackleton 1987). The oxygen isotopic record reveals that deep ocean temperatures warmed from 9° to 12° C between the late Paleocene and the early Eocene epochs, with mean annual sea surface temperatures (SST) exceeding 15° to 17° C in the Antarctic Ocean (Kennett and Stott 1990); however, there was no concomitant increase in tropical SSTs (Zachos et al. 1992) implying a greatly reduced latitudinal SST gradient.

3. Carbon isotopic ratios ($^{13}C/^{12}C$) of dissolved inorganic carbon in the ocean as recorded by benthic and planktonic foraminifera as well as in bulk marine carbonate ($\delta^{13}C_{carb}$), declined by as much as 3–4‰ during the LP–EE interval (figure 6.2; e.g., Shackleton 1987).
4. Strontium and osmium isotopic ratios in LP–EE marine carbonates were relatively non-radiogenic (figure 6.3; Edmond 1992; Rea 1992; Pegram 1992).

In light of the paleobiological and oxygen isotope evidence for LP–EE global warming, it is likely that tectonic–climatic mechanisms for global cooling via drawdown of CO_2 were not dominant. In making this

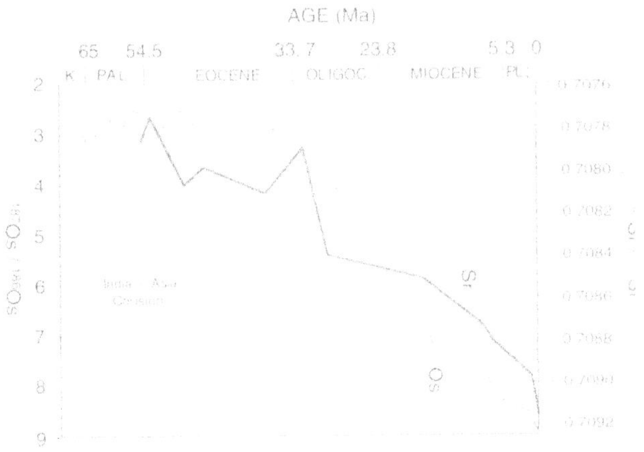

Figure 6.3 $^{187}Os/^{186}Os$ and $^{87}Sr/^{86}Sr$ values for marine Cenozoic strata vs. timing of the India–Asia collision.

Osmium data from Pegram (1992). Strontium data from Edmond (1992) and Rea (1992). The relatively non-radiogenic strontium and osmium isotopic ratios from Paleogene marine strata suggest that large areas of recently metamorphosed old crust were not being actively weathered during the LP–EE interval (Edmond 1992; W. K. Hart, personal communication 1995). However, initial collision occurred prior to the end of planktonic foraminiferal Biochron P5 (Beck et al. 1995a, 1996; Beck 1995). Therefore we infer that the lithology of eroding Himalayan rocks changed markedly between the Paleogene and Neogene periods and that the effect of the India–Asia collision on the biogeochemistry of the Earth changed accordingly (Beck et al. 1995c).

K. = Cretaceous Period, PAL. = Paleocene Epoch, OLIGOC. = Oligocene Epoch, PL. = Pliocene Epoch.

statement, we do not necessarily imply that the aforementioned link between drawdown of CO_2 and global cooling is correct; rather, we base our reasoning on available geochemical proxies.

For example, the non-radiogenic strontium and osmium isotopic ratios from Paleogene marine strata (Edmond 1992; Pegram 1992) suggest that large areas of recently metamorphosed old crust were not being actively weathered during the LP–EE (Edmond 1992; W. K. Hart, personal communication 1995). This implies that a large, well developed, collisional orogen exposing cratonal basement similar to the modern Himalaya and Tibet did not yet exist. Thus, this is at least consistent with the lack of evidence from the fossil record for cool temperatures and the lack of isotopic evidence for a large early Himalayan calcsilicate weathering sink for CO_2 during the Paleogene Period (Edmond 1992).

Similarly, the enhanced burial of C_{org} due to an increase in the supply of limiting nutrients to the ocean is unlikely, at least during the early Eocene warm interval. This is because enhanced burial of organic matter that is isotopically very depleted ($\delta^{13}C \sim -22.75‰$; Shackleton 1987) relative to total inorganic carbon in the ocean (~+2.25‰; Shackleton 1987) would have had the effect of increasing the $\delta^{13}C_{carb}$. This, however, is not consistent with the observed $\delta^{13}C_{carb}$ record that indicates decreasing $\delta^{13}C_{carb}$ values during the LP–EE interval.

In the other group of tectono-climatic mechanisms that invoke global warming due to release of CO_2 to the atmosphere, metamorphic volatilization (e.g., Kerrick and Caldeira 1993) of carbonates is not compatible with the marine carbon isotopic record. Declining $^{13}C/^{12}C$ ratios in contemporaneous marine carbonates ($\delta^{13}C_{carb}$; Shackleton 1987) imply that a large quantity of inorganic carbon was not volatilized. This is because the volatilization of large amount of carbonate ($\delta^{13}C$ ~+2.25‰) would have yielded ^{13}C enriched CO_2 that in turn would have pushed $\delta^{13}C_{carb}$ in the opposite direction relative to the observed record (Shackleton 1985, 1987; Beck et al. 1995c).

Indeed, the rapid decline of $\delta^{13}C_{carb}$ values during the latest Paleocene and early Eocene epochs suggests a decrease in the apparent rate of burial of C_{org} either through the loss of a lithospheric sink (Shackleton 1987; Beck et al. 1995b; Faure et al. 1995) or the active erosion and exhumation of C_{org} (Beck et al. 1995c). Either process would have decreased $\delta^{13}C_{carb}$.

Therefore the most important link between LP–EE global warming, decreases in $\delta^{13}C_{carb}$ and $\delta^{18}O_{carb}$, and plate tectonics must have involved the C_{org} cycle (Beck et al. 1995c). Accordingly, the remainder of this chapter discusses the loss of C_{org} sinks and the erosion of C_{org} during LP–EE plate reorganization and mountain building. We focus our attention on the collision of India and Asia and its climatic and isotopic consequences.

India–Asia Collision and Late Paleocene–Early Eocene Global Change

The most significant orogenic event of the LP–EE interval was the collision of India and Asia resulting in the initial erosion of the Himalaya (Garzanti et al. 1987; Bossart and Ottiger 1989; Beck et al. 1995a). The magnitude of this event and the material fluxes associated with it strongly suggest that it may have had a profound influence on the Earth's biogeochemistry and climate during the LP–EE interval (Raymo et al. 1988; Edmond 1992; Kerrick and Caldeira 1993; Beck et al. 1995b, 1995c).

Obviously, the timing of India–Asia collision is crucial to any hypothesis relating this tectonic event to the LP–EE biotic and climatic events. By "collision" we mean the time when and place where the crust of the Neo-Tethys ocean was completely subducted at some point along the convergent plate boundary and the Indian and Asian transitional crusts first came into contact (e.g., Beck et al. 1995a). Multiple lines of evidence suggest that the NW Indo-Pakistani craton began to collide with Asia during the Paleocene Epoch. They include:

1. The ophiolites of NW Pakistan had already been obducted onto the craton during the Late Cretaceous Period (Stonely 1974; Beck 1995; Beck et al. 1996) as opposed to the Paleocene Epoch (Allemann 1979). Therefore, intense and widespread post-Maastrichtian–pre-latest Paleocene deformation in the suture zone is interpreted as India–Asia collision.
2. The youngest colored mélange (accretionary prism) strata in eastern Afghanistan are middle Maastrichtian in age but structurally overlie the Khost Ophiolite (Cassaigneau 1979).
3. Laterites formed above the ophiolite and the overlying middle Maastrichtian colored mélange and before the deposition of unconformable shallow marine upper Paleocene limestone and shale (Cassaigneau 1979).
4. Syn-sedimentary extensional faulting along the western margin of Waziristan, west of the Waziristan and Khost Ophiolites, did not begin until the late Paleocene or early Eocene Epoch (Cassaigneau 1979).
5. Trans-Himalayan volcanism decreased during the Paleocene Epoch (Wolfart and Wittekindt 1980; Honnegger et al. 1982).
6. Rapid subsidence of the Kurram area during the early Paleocene Epoch (Beck et al. 1996) reflects a major tectonic load that post-dates ophiolite obduction by 15–20 my. This later and discrete event is interpreted as India–Asia collision.
7. Deposition of lower and lower upper Paleocene deep-to-open-marine strata in the foreland coincided with erosion above the deformed suture zone in NW Pakistan and eastern Afghanistan (Beck et al. 1995a, 1996).
8. Foreland strata at least as old as planktonic foraminiferal Biochron P3 and as young as Biochron P6 were thrust beneath the deformed belt before it was stratigraphically linked to the foreland by upper lower Eocene shallow marine strata (Davies 1926; Beck et al. 1995a, 1996). Deformation of Mesozoic and older Indo-Pakistani shelf strata in northern Pakistan also occurred before deposition of unconformable Paleocene strata (Yeats and Hussain 1987).
9. Biostratigraphic and petrologic analysis of clasts from turbidites in the bathyal to open marine lower Paleocene to upper Paleocene foreland fill sequence indicate that they were derived from a subaerially exposed ophiolite–radiolarite belt (Beck et al. 1995a, 1996).
10. A Paleocene foreland basin has previously been documented in northern Pakistan and India (Bossart and Ottiger 1989; Garzanti et al. 1987).
11. Rare "middle" Paleocene conglomerates with ultramafic and mafic clasts are known from the suture zone in easternmost Afghanistan (Bruggey 1973; Cassaigneau 1979).
12. Stratigraphic evidence for a Paleocene collision is consistent with the results of a number of recent radiometric studies of metamorphic rocks from the temporarily subducted northern margin of the Indo-Pakistani craton (Chamberlain et al. 1989; Spencer et al. 1993; Tonarini et al. 1993).

Given the evidence above we are reasonably confident that India–Asia collision began during the Paleocene Epoch in what is now the western India–Asia suture zone.

Once constraints are placed on the timing of initial India–Asia collision, we can relate the ensuing tectonic process to the Earth's biogeochemistry, oceanography and climate. Given the complexity of the Earth system any simple hypothesis for global change within the complex earth system is unlikely to be correct (Erwin 1993). It is difficult, however, to explore and evaluate all possible triggering events and their consequences simultaneously. Therefore, we limit this chapter to a discussion of the possible climatic and marine isotopic effects of the India–Asia collision with an emphasis on the organic carbon sub-cycle.

We consider several possible isotopic and climatic consequences of the India–Asia collision that may have led to the well documented multi-million-year LP–EE global warming and a rapid decline in $\delta^{13}C_{carb}$ and $\delta^{18}O_{carb}$ values in addition to the erosion of organic carbon from the early Himalaya (Beck et al. 1995b, 1995c). They include 1) the elimination of the north Indian passive margin as a C_{org} sink, 2) a decrease in the sulfate aerosol flux from the Trans-Himalayan Arc (Beck et al. 1995b), 3) an increase in the flux of low latitude warm saline bottom water from shallow restricted syn-collisional seaways to the deep Antarctic ocean, 4) faunal and floral interchange after allopatria (Jaeger et al. 1989), and 5) global plate reorganization (Molnar and Tapponnier 1975; Patriat and Achache 1984). In other words, the India–Asia collision may have had a multitude of effects on the earth system resulting in what Erwin (1993) called a "Murder on the Orient Express" scenario for marked global shifts in marine isotopic chemistry, ocean circulation, evolution and polar climate with multiple as yet unidentified and/or unquantified feedbacks.

Erosion of Organic Carbon

Beck et al. (1995c) suggested that the initial phase of the India–Asia collision may have actively exhumed C_{org} from the northern continental shelf of India. Such a mechanism is qualitatively and perhaps quantitatively compatible with the decrease in C_{org} burial during the LP–EE calculated by Shackleton (1985, 1987) whereas the volatilization of carbonate during metamorphism is not. However, there are many uncertainties in the estimates made by Beck et al. (1995c) regarding the rate at which C_{org} was exhumed.

First, little data regarding the total organic carbon (TOC) content of pre-collisional strata of the north Indian shelf are available. Although it may be possible to collect additional data from remnants of the inner and outer shelf in the central Himalaya (Thurow et al. 1992), the middle shelf is missing along most of the length of the Greater Himalaya (Brookfield 1993). Thus, we may never be able to accurately quantify the total amount of C_{org} eroded during the early India–Asia collision. Nonetheless, consideration of the loss of the pre-collisional Neo-Tethyan continental margins as organic carbon sinks suggests that our initial estimates for the net flux of C_{org} *eroded* from the early Himalaya may have been too high by as much as 50 percent.

Loss of the Pre-Collisional Neo-Tethyan Continental Margins as Organic Carbon Sinks

When considering the effects of any event on a biogeochemical cycle, one must consider the total imbalance in that cycle generated by the event (Garrels and Mackenzie 1971). Chemical budgets must include the destruction of sinks, as well as the construction of new sources. The LP–EE global warming, therefore, may reflect both the loss of C_{org} sinks as well as the creation of new sources. We point out that Beck et al. (1995c) considered only the creation of a C_{org} source rather than the total imbalance in the C_{org} sub-cycle due to the collision of India and Asia. They did not include the effect of the loss of the pre-collisional Neo-Tethyan active and passive margins as C_{org} sinks. Both aspects of the collision must be included when estimating the role of the India–Asia collision in the total apparent decrease in C_{org} burial during the LP–EE interval documented by Shackleton (1987).

Using the methods of Berner (1982), we conservatively estimate the decrease in C_{org} burial flux due merely to the loss of the Neo-Tethyan margins as C_{org} sinks as ~470 × 10^{15} moles/my for the ~5-my-long LP–EE interval (table 6.1). We have included estimates of the pre-collisional C_{org} sink for both margins, as well as two deltaic deposits, one on each of the pre-collisional margins.

Sedimentologic and petrographic evidence for a Cretaceous fluvio-deltaic system that drained the northern passive margin of the Indo-Pakistani craton is limited at best. However, a Paleocene–Eocene fluvio-deltaic system has been reported from northern India by Garzanti et al. (1987) and from northern Pakistan by Bossart and Ottiger (1989). Good evidence is available for river systems that drained the pre-collisional southern margin of Asia during the Late Cretaceous Period (Dürr 1993). Petrologic studies of Cretaceous active margin sediments from what is now southern Tibet indicate a source area at least as far north as the Bangong–Nujiang suture as well as in the Gangdese (eastern Trans-Himalayan) arc (Burg and Chen 1984; Dürr 1993).

The magnitudes of the deltaic C_{org} sinks are estimated from the modern sediment fluxes of large rivers draining active mountain belts and passive margins (Indus and Mississippi rivers) in combination with relatively low TOC values from syn-collisional strata at the western end of the "Himalayan" convergent boundary (tables 6.1, 6.2; Beck et al. 1995c). This estimate for the decrease in C_{org} burial due to the loss of the Neo-Tethyan sedimentary sink was calculated for the same time interval (~5 my) used to calculate the flux of C_{org} eroded from the early Himalaya. A decrease in the flux of organic carbon to the lithosphere of ~470 × 10^{15} moles/my would cause a shift of ~0.7‰ in oceanic $\delta^{13}C_{carb}$ values assuming Shackleton's (1987) parameters for the global LP–EE organic carbon cycle.

Both hypotheses, active tectonic exhumation and decreased burial of organic carbon, invoke increased at-

Table 6.1
Estimate of the magnitude of the Neo-Tethyan organic carbon sink eliminated by the India—Asia collision

Name of Organic Carbon Sink	Length (km)	Width (km)	Area (km^2)	Total Sediment Burial Flux (mg/cm^2/yr)	Total Sediment Burial Flux (g/yr)	Average Fraction Organic Carbon	Organic Carbon Burial Rate (Gt/my)	Preservation Factor After Burial	Organic Carbon Sequestration Rate (Gmoles/my)
Precollisional North Indian shelf	2500	5.00E + 02	1.25E + 06	12	1.5E + 14	0.015	2250	0.7	131
Precollisional South Asian margin	2500	2.00E + 02	5.00E + 05	25	125E + 14	0.01	1250	0.7	73
Precollisional delta systems					Suspended sediment transport Mt/yr	Average fraction organic carbon; Precollisional less post-collisional (0.8% less 0.2% TOC)			
Mississippi analog (from North Indian shelf)					315	0.006	1890	0.7	110
Indus analog (from MKT suture)					436	0.006	2616	0.7	152
Total									467

The decreased flux of organic carbon to the lithosphere due to the loss of the sink must be added to the flux of organic carbon added to the biosphere from eroding marine strata of the early Himalaya to determine the total apparent decrease in organic carbon burial caused by the collision of India and Asia. Data from Berner (1982).

mospheric CO_2 as the mechanism responsible for the LP–EE global warming on the my scale. While this scenario appears to be reasonable, it is important to remember that a link between pCO_2 and the LP–EE global warming has not yet been established conclusively. This is due to uncertain pCO_2 estimates and the current lack of knowledge regarding feedbacks in the LP–EE climate system. Further marine and non-marine stable isotope research may help evaluate the C_{org}-based tectono-climatic mechanisms discussed in this chapter.

Decreased Flux of Volcanic Aerosols from the Trans-Himalayan Arc

Because of the complexity of the earth system, it is important to consider tectono-climatic mechanisms other than those directly related to the carbon cycle. For ex-

Table 6.2
Modern rates of organic carbon burial for comparison with our estimate for organic carbon burial on pre-collisional Neo-Tethyan margins

	Organic Carbon Burial Rate (g/yr)	Preservation Factor	Net Organic Carbon Burial Rate (g/yr)
Deltaic-shelf sediments	1.3E + 14	0.8	1.0E + 14
Biogenous sediments in areas of high productivity	14E + 13	0.8	1.1E + 13
Shallow water carbonate sediments	7E + 12	0.8	5.6E + 12
Pelagic sediments in lower-productivity areas	6E + 12	0.8	4.8E + 12
Total			1.3E + 14

Data from Berner (1982).

ample, Trans-Himalayan volcanism and the production of sulfate aerosols (Beck et al. 1995b) may have had a net cooling effect upon the Paleocene earth. Sulfate and other volcanic aerosols are powerful cooling agents (Rampino and Self 1991). A single volcanic eruption can cool the Earth's atmosphere by 0.5–1° C for several years by reflecting a portion of incoming solar radiation (the "Pinatubo effect").

As India began to collide with Asia, the rate of convergence slowed and the Trans-Himalayan volcanic arc was gradually extinguished. We presume that the flux of aerosols from the volcanic arc to the atmosphere decreased as well. The expected decrease in the "Pinatubo effect" is conceptually equivalent to turning off the air conditioner. Indeed, relatively frequent eruptions (every few years) similar to those of El Chichon or Mt. Pinatubo have canceled the warming effect of even modern anthropogenic GHG emissions (Porter 1986; Rind 1993). The LP–EE global warming may, therefore, reflect decreased arc volcanism and a consequent relative increase in insolation as well as an enhanced Greenhouse Effect.

This hypothetical late Paleocene increase in global temperature due to a decrease in the "Pinatubo effect" is supported by existing radioisotopic dating. For example, volcanism in the Baluchistan–Afghan sector of the Trans-Himalayan arc declined markedly at 66 Ma with a few areas experiencing volcanism until 55 Ma (Debon et al. 1987; Wolfart and Wittekindt 1980). Radioisotopic cooling ages indicate that magmatism in the Trans-Himalayan Arc of what is now northern India and Tibet slowed between 60 and 50 Ma (Reuber et al. 1989).

Changes in Ocean Circulation Patterns

The most important paleogeographic implication of an early collision of India and Asia is that pre-collisional

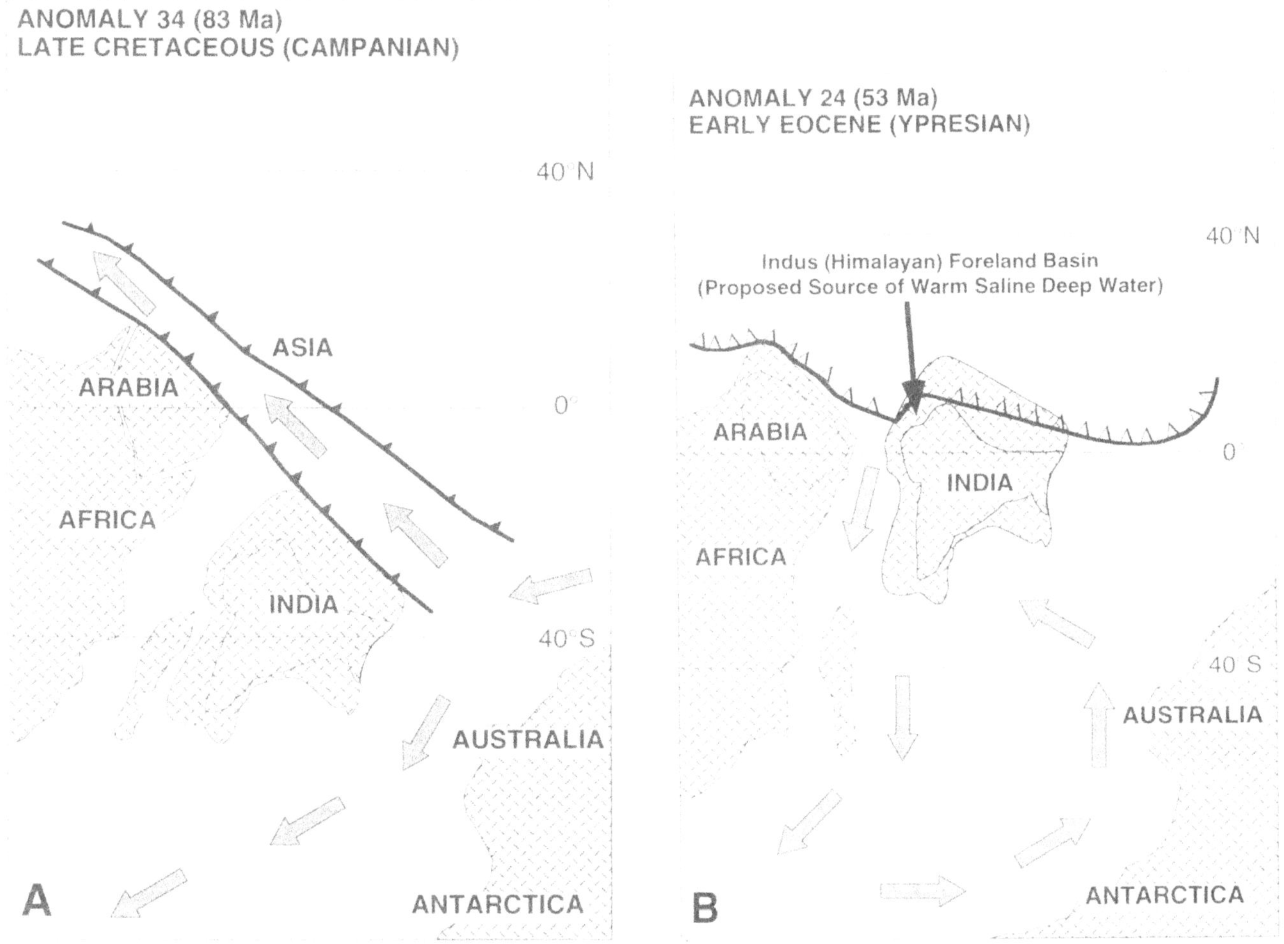

Figure 6.4 Ocean circulation patterns before (A, Late Cretaceous Period) and after (B, early Eocene Epoch) the India–Asia collision.

After Barron et al. (1981), modified according to Chen et al. (1993) and Beck et al. (1996). Cenozoic north–south tectonic barriers partitioned circum-equatorial (Tethyan) oceans into large equatorial–polar cells (Barron 1985). This may be one of several factors that increased the equatorial–polar heat flux and led to LP–EE warming of the deep Antarctic ocean (Barron 1985; Kennett and Stott 1990, 1991).

India extended farther to the north and/or pre-collisional Asia extended farther to the south relative to reconstructions based on an Eocene date for collision (Powell and Conaghan 1973; Besse et al. 1984; Besse and Courtillot 1988; Achache et al. 1983; Chen et al. 1993, Klootwijk and Bingham 1980; Treloar and Coward 1991; Klootwijk et al. 1992; Beck et al. 1995a, 1996; figure 6.4). Because our data are from the westernmost Himalaya, we cannot preclude the existence of early Eocene back-arc basins or small remnant oceanic basins similar to the modern Mediterranean Sea in areas to the east or south of our study area (e.g., Biju-Duval et al. 1976). Regardless, closure of the Neo-Tethys at any point would stop deep, through-going, equatorial ocean circulation between India and Asia and disrupt pre-existing chemical and heat flux patterns. Loss of this component of circulation influenced the eastern Atlantic Ocean as recorded by a change in marine $^{143}Nd/^{144}Nd$ values with respect to $^{87}Sr/^{86}Sr$ values (Barron et al. 1981; Barron 1985; Stille 1992).

While the deep Neo-Tethys was at least locally closed by collision during the Paleocene Epoch, a flexural marine foreland basin (Wells 1984) formed via tectonic loading in front of the early Himalaya as the range was underthrust by the northern Indo-Pakistani craton (Yeats and Hussain 1987; Gaetani and Garzanti 1991; Bossart and Ottiger 1989; Beck et al. 1993, 1995a, 1996). At least once during the Paleocene Epoch, erosion regionally beveled the allochthonous rocks of the suture zone before uppermost Paleocene–lowermost Eocene (tropical planktonic foraminiferal Biozones P5 and P6) shallow marine limestones were deposited above them (Davies 1926, 1940; Cizancourt 1938). The marine foreland basin gradually filled during the late early Eocene Epoch with evaporitic, lacustrine and fluvial sediments (Wells 1984; Pivnik and Sercombe 1993; Sercombe et al. 1994) before an early middle Eocene transgression drowned the foreland and resulted in the deposition of regionally extensive shallow marine limestones and shales (Davies 1926; Wells 1984; Meissner et al. 1974, 1975; Beck et al. 1995a).

Remnant Neo-Tethyan Seas and Marine Foreland Basins as Sources of Warm Saline Deep Water

Emphasis in this chapter is placed on the ~5-my-long LP–EE "greenhouse" interval and processes spanning millions of years. Significant insights into the LP–EE Earth-system are also available from the study of short-lived climatic–oceanographic events that span thousands to hundreds of thousands of years. One geologically brief event that may reveal information regarding possible "threshold" effects of the India–Asia collision is a very short episode of extraordinary global warming that occurred at the very beginning of the LP–EE interval (Kennett and Stott 1990, 1991) known as the Late Paleocene Thermal Maximum (LPTM, Zachos et al. 1992).

The LPTM that occurred at ~55.5 Ma (latest Paleocene Epoch; Aubry et al. 1996; Berggren and Aubry 1996) is characterized by several rapid, large and somewhat counter-intuitive events in the ocean. Remarkable observations regarding this event include: 1) rapid (<10 kyr) warming of deeper and surface Antarctic waters. At the peak of this warming event, the surface and deep water temperatures exceeded 22° and 17° C, respectively, with the bottom waters warming before surface waters (Kennett and Stott 1990, 1991); 2) benthic foraminifera suffered a massive extinction that eliminated 35–50% of species inhabiting bathyal to abyssal depths (Thomas 1990). In the Antarctic ocean, this extinction apparently proceeded from the bottom of the ocean upward (Kennett and Stott 1990, 1991); 3) the $\delta^{13}C$ gradient between the surface and deep ocean decreased dramatically within a few thousand years; 4) the $\delta^{18}O$ gradient between the deep and shallow parts of the ocean almost disappeared at higher southern latitudes (Kennett and Stott 1990, 1991) whereas the tropical oceans did not warm and may have cooled slightly (Douglas and Woodruff 1981; Crowley and North 1991; Zachos et al. 1992); 5) the grain-size of the eolian component of mid-ocean sediment decreased markedly during this event suggesting greatly reduced vigor of atmospheric circulation (Janecek and Rea 1983); and 6) the late Paleocene ocean may have entered an episode of halothermal circulation (Broecker et al. 1985; Kennett and Stott 1990, 1991) with a shift of the locus of bottom water formation from high polar latitudes to lower latitudes (Brass et al. 1982; Kennett and Stott 1991).

Kennett and Stott (1990) suggested that the changing late Paleocene configuration of the Neo-Tethys set the stage for the LPTM. They suspected that Tethyan shallow water platforms were the source of the Warm Saline Deep Water (WSDW) in the Paleogene Antarctic Ocean. Indeed, remnant seas similar to the modern Mediterranean existed between India and Asia until at least planktonic foraminiferal Biochron P5 (Bossart and Ottiger 1989; Garzanti et al. 1987; Beck et al. 1995a, 1995d). During Biochron P6 this narrow sea was divided by southward building Asian deltaic complexes (figure 6.5; Bossart and Ottiger 1989; Garzanti et al. 1987). Interruption of meridional circulation by the deltas created large restricted tropical seaways that would have served as excellent sources for large quantities of WSDW.

Stable oxygen and carbon isotope values from the tests of foraminifera in upper Paleocene and lower to middle Eocene strata on Maud Rise (Antarctica) indicate that decreased or reversed benthic to planktonic $\delta^{18}O$ gradients persisted throughout the LP–EE interval,

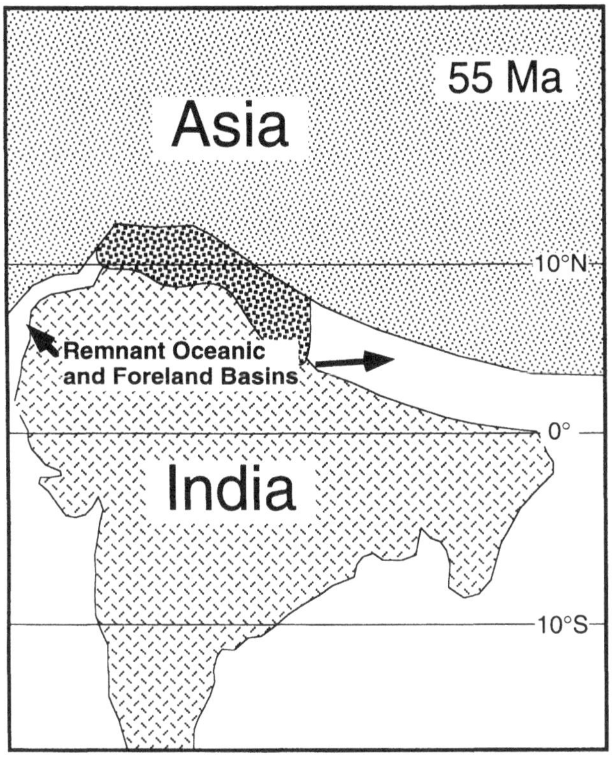

Figure 6.5 India–Asia collision resulted in the evaporitic concentration of seawater in remnant oceanic basins similar to the modern Mediterranean Sea and in flexural epicontinental collisional marine foreland basins (Yeats and Hussain 1987; Beck et al. 1995a, 1995d) similar to the modern Persian Gulf (Wells 1984; Pivnik and Wells 1996). Paleogene development of these restricted basins increased the production of Warm Saline Deep Water (WSDW) during the early Eocene Epoch and may have triggered the halothermal circulation pattern that characterized the early Eocene southern oceans (Kennett and Stott 1990).

Figure modified after Gaetani and Garzanti (1991).

from the late Paleocene to the middle Eocene Epoch, long after the LPTM (Kennett and Stott 1990). The data imply that Antarctic deeper waters remained warmer than surface waters throughout the early Eocene Epoch and that the halothermal ocean circulation established during the late Paleocene Epoch continued for almost 10 my, from ~56 Ma to ~46 Ma (time-scale of Berggren et al. 1995). We must search for a long-lived source of WSDW accordingly.

While Tethyan platforms in general might serve as sources of WSDW, eustatic sea-level falls such as those during the early and late Ypresian Age (Hallam 1992) would be expected to expose such platforms, periodically eliminating the source of WSDW. Therefore, the source of WSDW was probably a rapidly subsiding tropical marine basin with a floor that remained below sea-level. A good candidate would be a dynamically subsiding flexural marine foreland basin similar to the modern Persian Gulf in which the rate of subsidence induced by tectonic loading kept pace with the rate of sedimentation as well as with periodic eustatic sea-level falls.

The LP–EE Indus Basin was a mirror image of the modern Persian Gulf (Stoneley 1974; Wells 1984; Beck et al. 1995a; Pivnik and Wells 1996). Whereas the Persian Gulf owes its existence to an as yet incomplete collision of Arabia and Asia, the LP–EE Indus Basin formed in response to the collision of India and Asia (Beck et al. 1995a, 1996). Well exposed Paleogene strata along the western margin of the basin have been dated by foraminiferal biostratigraphy (Davies 1926; Meissner et al. 1975; Beck et al. 1995a) and indicate that a marine basin persisted throughout the late Paleocene and early Eocene epochs (Wells 1984; Beck 1995; Beck et al. 1995a) as the NW Indian continental margin was thrust beneath reactivated ophiolites and elements of the Iranian–Afghan tectonic collage (Asia; Beck et al. 1996). Subaqueous gypsum and halite deposition in the western arm of the flexural LP–EE Himalayan marine foreland (Indus Basin) imply high concentrations of salt within the seaway (Davies 1924; Gee 1945; Eames 1952; Wells 1984; Khan et al. 1986; Beck et al. 1995a; Pivnik and Wells 1996). The coincidence of the development of the hypersaline marine Indus foreland basin and the arrival of WSDW in Antarctica strongly suggests that the collision of India and Asia was at least partly responsible for the profound Paleogene oceanographic changes in the southern hemisphere documented by Kennett and Stott (1990).

We suggest that the India–Asia collision *may* have been the dominant driver of long-term (>1 my) LP–EE global change via the following mechanisms: 1) between ~60 and ~56 Ma (time-scale of Berggren et al. 1995), the collision may have passively warmed the earth by extinguishing at least some explosive Trans-Himalayan Arc volcanism thus decreasing the "Pinatubo effect." Meanwhile, stagnation and anoxia in Paleocene remnant oceanic basins similar to the modern Mediterranean and Black seas may have enhanced the preservation of marine C_{org} in sediments, depressed marine $\delta^{13}C_{carb}$ values, and enhanced global cooling via decreased pCO_2 (e.g., Calvert 1987); 2) from ~56 to ~51 Ma, $\delta^{13}C_{carb}$ values decreased, implying that the rate of net organic carbon burial decreased (Shackleton 1987; Zachos et al. 1992) suggesting that the collision may have led to the warming of the earth by exhuming organic carbon from the colliding continental margins and eliminating pre-existing C_{org} sinks, thus increasing the greenhouse effect (Beck et al. 1995, 1995c); 3) from ~51 to ~49 Ma, $\delta^{13}C_{carb}$ values increased, implying that the rate of net organic carbon burial increased (Shackleton 1987; Zachos et al. 1992) possibly due to C_{org} deposition in the Himalayan foreland basin (Wells 1984) and elsewhere; 4) from ~52 to ~48 Ma, $\delta^{18}O_{carb}$ and $\delta^{13}C_{carb}$ values were relatively stable and suggest that carbon storage in upper lower and middle Eocene Tethyan "nummulitic" carbonates (Kerrick and Caldeira 1993), thin organic-rich middle Eocene shales (probably less important than the Laramide coal deposits), and to a lesser degree subducted eclogitic rocks (Selverstone and Gutzler 1993) may have set the stage for Neogene global cooling.

Obviously some of these mechanisms overlapped in time and some acted in opposite directions with respect to the Earth's isotopic record and climate. However, it is the relative size of the associated chemical and heat fluxes that is important with respect to the Earth's long-term biogeochemistry and climate (Berner et al. 1983; Berner 1990; Edmond 1992; Raymo and Ruddiman 1992; Raymo 1994). Although the India–Asia collision was not the only event of the LP–EE interval, its timing, magnitude and numerous interactions with the rest of the extant lithosphere and biosphere, indicate that the hypotheses recounted above should be explored further in order to understand the long-term global context of the India–Asia collision.

ACKNOWLEDGMENTS

Funding for this study was provided by National Science Foundation and National Geographic Society grants to Burbank. Beck acknowledges support from Amoco, Fulbright Fellowships, Achievement Rewards for College Scientists Scholarships, the Phillip and Elaina Hampton Fund for International Initiatives and the Wayne D. Martin Fund for Field Geology. Comments by M.-P. Aubry, W. A. Berggren, R. Douglas, L. Stott and J. Kennett improved the manuscript significantly. W. K. Hart shared valuable insights with respect to osmium isotope systematics. N. Shafique and J. Burris provided valuable assistance in the field.

REFERENCES

Achache, J., Courtillot, V. and Besse, J., 1983. Paleomagnetic constraints on the late Cretaceous and Cenozoic tectonics of southeastern Asia. *Earth and Planetary Science Letters,* v. 63, p. 123–136.

Allemann, F., 1979. Time of Emplacement of the Zhob Valley Ophiolites and Bela Ophiolites, Baluchistan (Preliminary Report). In Farah, A. and DeJong, K. A. (eds.), *Geodynamics of Pakistan,* Quetta: Geological Survey of Pakistan, p. 215–242.

Aubry, M.-P., 1992. Late Paleogene calcareous nannoplankton evolution: A tale of climatic deterioration. In Prothero, D. R. and Berggren, W. A., *Eocene–Oligocene Climatic and Biotic Evolution,* Princeton University Press, Princeton, N.J., p. 272–309.

Aubry, M.-P., Berggren, W. A., Stott, L. and Sinha, A., 1996. The upper Paleocene–lower Eocene stratigraphic record and the Paleocene–Eocene boundary carbon isotope excursion: Implications for geochronology. In Knox, R. W. O'B., Corfield, R. M. and Dunay, R. E. (eds.), *Correlation of the early Paleogene in Northwest Europe,* Geological Society Special Publication No 101, p. 353–380.

Axlerod, D. I., 1984. An interpretation of Cretaceous and Tertiary biota in polar regions. *Palaeogeography, Palaeoclimatology, Palaeoecology,* v. 45, p. 105–147.

Barnola, J. M., Raynaud, D., Korotkevich, Y. S. and Lorius, C., 1987. Vostok ice core provides 160,000-year record of atmospheric CO_2. *Nature,* v. 329, p. 408–414.

Barron, E. J., 1985. Explanations of the Tertiary global cooling trend. *Palaeogeography, Palaeoclimatology, Palaeoecology,* v. 50, p. 45–61.

Barron, E. J., 1987. Eocene equator-to-pole surface ocean temperatures: A significant problem? *Paleoceanography,* v. 2, p. 729–739.

Barron, E. J. and Washington, W. M., 1984. The role of geographic variables in explaining paleoclimates: Results from Cretaceous climate model sensitivity studies. *Journal of Geophysical Research,* v. 89, p. 1267–1279.

Barron, E. J. and Washington, W. M., 1985. Warm Cretaceous climates: High atmospheric CO_2 as a plausible mechanism. In Sundquist, E. T. and Broecker, W. S. (eds.), *The Carbon Cycle and Atmospheric CO_2: Natural Variations Archean to Present,* Geophysical Monograph 32, Washington, D.C.: American Geophysical Union, p. 546–553.

Barron, E. J., Harrison, C. G. A., Sloan, J. L., II and Hay, W. W., 1981. Paleogeography, 180 million years ago to present. *Eclogae Geologicae Helvetiae,* v. 74, p. 443–470.

Beck, R. A., 1995. *Late Cretaceous ophiolite obduction and Paleocene India–Asia collision in the westernmost Himalaya.* Doctoral thesis, Los Angeles: University of Southern California, 550 pp.

Beck, R. A., Sercombe, W. J., and Burbank, D. W., 1993. Structure, stratigraphy and chronology of the Himalayan Indus–Tsangpo Suture Zone in NW Pakistan and eastern Afghanistan. *Geological Society of America, Abstracts with Programs,* v. 25, p. A-122.

Beck, R. A., Burbank, D. W., Sercombe, W. J., Riley, G. W., Barndt, J. K., Berry, J. R., Afzal, J., Khan, A. M., Jurgen, H., Metje, J., Cheema, A., Shafique, N. A., Lawrence, R. D. and Khan, M. A., 1995a. Stratigraphic evidence for an early collision between northwest India and Asia. *Nature,* v. 373, p. 55–58.

Beck, R. A., Burbank, D. W. and Sercombe, W. J., 1995b. Evidence for a very latest Cretaceous to Paleocene Indian–Eurasian collision: Paleogeographic, isotopic and climatic implications. In Spencer, D. A., Burg, J.-P., and Spencer-Cervato, C. (eds.), Proceedings of the 10th Himalaya–Karakoram–Tibet Workshop, Monte Verità Switzerland, *Mitteilugen aus dem Geologischen Institut der Eidgenössischen Technischen Hochschule und der Universität Zürich, Neue Folge,* No. 298., p. 4–8.

Beck, R. A., Burbank, D. W., Sercombe, W. J., Olson, T. L. and Khan, A. M., 1995c. Organic carbon exhumation and global warming during the early Himalayan collision. *Geology,* v. 23, p. 387–390.

Beck, R. A., Burbank, D. W. and Sercombe, W. J., 1995d. Late Cretaceous ophiolite obduction and Paleocene India–Asia collision in the westernmost Himalaya. *Geological Society of America Abstracts with Programs,* v. 27, p. A-456–A-456.

Beck, R. A., Burbank, D. W., Sercombe, W. J., Khan, A. M., and Lawrence, R. D., 1996. Late Cretaceous ophiolite obduction and Paleocene India–Asia collision in the westernmost Himalaya. Proceedings of the 10th Himalaya–Karakoram–Tibet Workshop, Monte Verità Switzerland, Special Volume, *Geodinamica Acta,* v. 9.

Berggren, W. A. and Hollister, C. D., 1974. Paleogeography, paleobiogeography and the history of circulation in the Atlantic Ocean. In Hay, W. W. (ed.), *Studies in Paleo-oceanography,* Society of Economic Paleontologists and Mineralogists Special Publication No. 20, p. 126–186.

Berggren, W. A. and Hollister, C. D., 1977. Plate tectonics and paleocirculation: commotion in the ocean. *Tectonophysics*, v. 38, p. 11–48.

Berggren W. A., Kent, D. V., Swisher, C. C., III and Aubry, M.-P., 1995. A revised Cenozoic geochronology and chronostratigraphy. In Berggren, W. A., Kent, D. V., Aubry, M.-P. and Hardenbol, J. (eds.), *Geochronology Time Scales and Global Stratigraphic Correlation*, Society of Economic Paleontologists and Mineralogists Special Publication No. 54, p. 129–212.

Berggren, W. A. and Aubry, M.-P., 1996. A late Paleocene–early Eocene NW European and North Sea magnetobiochronological correlation network. In Knox, R. W. O'B., Corfield, R. M. and Dunay, R. E. (eds.), *Correlation of the Early Paleogene in Northwest Europe*, Geological Society Special Publication No. 101, p. 309–352.

Berner, R. A., 1982. Burial of organic carbon and pyrite sulfur in the modern ocean: Its geochemical and environmental significance. *American Journal of Science*, v. 282, p. 451–473.

Berner, R. A., 1990. Atmospheric carbon dioxide levels over Phanerozoic time. *Science*, v. 249, p. 1382–1386.

Berner, R. A., Lasaga, A. C. and Garrells, R. M., 1983. The carbonate–silicate geochemical cycle and its effect on atmospheric carbon dioxide over the past 100 million years. *American Journal of Science*, v. 283, p. 641–683.

Besse, J. and Courtillot, V., 1988. Paleogeographic maps of the continents bordering the Indian Ocean since the Early Jurassic. *Journal of Geophysical Research*, v. 93(B10), p. 11,791–11,808.

Besse, J., Courtillot, V., Pozzi, J. P., Westphal, M. and Zhou, Y. X., 1984. Palaeomagnetic estimates of crustal shortening in the Himalayan thrusts and Zangbo suture. *Nature*, v. 311, p. 621–626.

Biju-Duval, B., Dercourt, J., and LePichon, X., 1976. From the Tethys Ocean to the Mediterranean Sea: A plate model of the evolution of the western Alpine system. In Biju-Duval, B. and Montadert, L. (eds.), *International Symposium on the Structural History of the Mediterranean Basins*, Split, Yugoslavia, Paris, Editions Techniq, p. 143–164.

Boersma, A., Premoli Silva, I. and Shackleton, N. J., 1987. Atlantic Eocene planktonic foraminiferal paleohydrographic indicators and stable isotope paleoceanography. *Paleoceanography*, v. 2, p. 287–331.

Bolin, B., Doos, B. R., Jager, J., Warrick, R. A., 1986. *SCOPE 29: The Greenhouse Effect, Climatic Change, and Ecosystems*. New York: John Wiley and Sons, 155 pp.

Bossart, P. and Ottiger, R., 1989. Rocks of the Murree Formation in northern Pakistan: Indicators of a descending foreland basin of late Paleocene to middle Eocene age. *Eclogae Geologicae Helvetiae*, v. 82, p. 133–165.

Brass, G. W., Southam, J. R. and Peterson, W. H., 1982. Warm saline bottom water in the ancient ocean. *Nature*, v. 296, p. 620–623.

Bretherton, F. P., 1986. The Oceans, climate, and technology. *Oceanus*, v. 29, p. 3–8.

Broecker, W. S., 1982. Glacial to Interglacial changes in Ocean Chemistry. *Progress in Oceanography*, v. 11, p. 151–197.

Broecker, W. S., Peteet, D. M. and Rind, D., 1985. Does the ocean-atmosphere system have more than one stable mode of operation? *Nature*, v. 315, p. 21–26.

Brookfield, M. E., 1993. The Himalayan passive margin from Precambrian to Cretaceous times. *Sedimentary Geology*, v. 84, p. 1–35.

Bruggey, J., Raufi, F. and Sickenberg, O., 1973. Mesozoikum und Alttertiar, in Nord-Paktia (SE-Afghanistan). *Geologisches Jahrbuch*, v. 3, p. 3–61.

Burbank, D. W., 1992. Causes of Himalayan uplift deduced from depositional patterns in the Ganges basin. *Nature*, v. 357, p. 680–683.

Burg, J. P. and Chen, G. M., 1984. Tectonics and structural zonation of southern Tibet, China. *Nature*, v. 311, p. 219–223.

Caldeira, K. and Rampino, M. R., 1991. The mid-Cretaceous super plume, carbon dioxide, and global warming. *Geophysical Research Letters*, v. 18, p. 987–990.

Calvert, S. E., 1987. Oceanographic controls on the accumulation of organic matter in marine sediments. In Brooks, J., and Fleet, A. J. (eds.), *Marine Petroleum Source Rocks*, Geological Society Special Publication No. 26, p. 137–153.

Cassaigneau, C., 1979. *Contribution à l'étude des sutures Inde-Eurasie: La zone de suture de Khost dans le Sud-Est de l'Afghanistan, l'obduction paléocène et la tectonique tertiaire.* Montpellier: Université des Sciences et Techniques du Languedoc, Doctoral Dissertation, 150 pp.

Chamberlain, C. P., Jan, M. Q. and Zeitler, P. K., 1989. *A petrologic record of the collision between the Kohistan Island-Arc and Indian Plate, northwest Himalaya.* Geological Society of America, Special Paper, v. 232, p. 23–32.

Chen, Y., Courtillot, V., Cogne, J.-P., Besse, J., Yang, Z. and Enkin, R., 1993. The configuration of Asia prior to the collision of India: Cretaceous Paleomagnetic constraints. *Journal of Geophysical Research*, v. 98 (B12), p. 21,927–21,941.

Cizancourt, H., 1938. Contribution à l'étude des faunes tertiaires de l'Afghanistan. I. Nummulites et Assilines du flysch de Gardez et du Khost, Afghanistan oriental. *Mémoires de la Société Géologique de France*, n.s. 17, v. 39, p. 5–28.

Cloud, P., 1988. *Oasis in Space. Earth History from the Beginning.* New York, London: W. W. Norton and Company, 508 pp.

Craw, D., 1990. Fluid evolution during uplift of the Annapurna Himal, central Nepal. *Lithos*, v. 24, p. 137–150.

Crowley, T. J. and North, G. R., 1991. *Paleoclimatology*. New York: Oxford University Press, 339 pp.

Davies, L. M., 1924. Notes on the geology of Kohat, with reference to the homotaxial position of the Salt Marl at Bahadur Khel. *Journal of the Asiatic Society of Bengal*, n.s., v. 20, p. 207–224.

Davies, L. M., 1926. The Ranikot beds at Thal (North-West Frontier Provinces of India). *Journal of the Geological Society of London*, v. 83, p. 260–290.

Davies, L. M., 1940. Geographical changes in northwest India during late Cretaceous and early Tertiary times. *Pacific Science Congress, 6, California, 1939*, v. 2, p. 483–501.

Debon, F., Afzali, N. H., Le Fort, P. and Sonet, J., 1987. Major intrusive stages in Afghanistan: Typology, age and geodynamic setting. *Geologisches Rundschau*, v. 76, p. 245–264.

Dewey, J. F. and Bird, J. M., 1970. Mountains belts and the new global tectonic. *Journal of Geophysical Research*, v. 75, p. 2625–2647.

Douglas, R. and Woodruff, F., 1981. Deep-sea benthic foraminifera. In Emiliani, C. (ed.), *The Sea*, New York: Wiley-Interscience, v. 7, p. 1233–1328.

Dürr, S. B., 1993. The mid- to Early Late Cretaceous Xigase forearc basin (South Tibet), Sedimentary Evolution and Provenance of Clastic Sediments. *Tubinger Geowissenschaftliche Arbeiten*, ser. A, v. 15, 119 pp.

Eames, F. E., 1952. A contribution to the study of the Eocene in Western Pakistan and western India. C. The description of the Scaphoda and Gastropoda from standard sections in the Rakhi Nala and Zinda Pir areas of the western Punjab and in the Kohat District. *Philosophical Transactions of the Royal Society of London*, series B, v. 236, p. 1–168.

Eddy, J. A. and Oescheger, H. 1993. The role of solar output variations. In Eddy, J. A., and Oescheger, H. (eds.), *Global Changes in the Perspective of the Past*, New York: John Wiley and Sons Ltd., p. 280–294.

Edmond, J. M., 1992. Himalayan tectonics, weathering processes, and the strontium isotope record in marine limestones. *Science*, v. 258, p. 1594–1597.

Erwin, D. H., 1993. *The Great Paleozoic Crisis: Life and death in the*

Permian. Critical Moments in Paleobiology and Earth History, New York: Columbia University Press, 328 pp.

Estes, R. and Hutchison, J. H., 1980. Eocene lower vertebrates from Ellesmere Island, Canadian Arctic Archipelago. *Palaeogeography, Palaeoclimatology, Palaeoecology,* v. 30, p. 325–347.

Faure, K., de Wit, M. J. and Willis, J. P., 1995. Late Permian global coal hiatus linked to ^{13}C-depleted CO_2 flux into the atmosphere during the final consolidation of Pangea. *Geology,* v. 23, p. 507–510.

Fischer, A. G. and Arthur, M. A., 1977. Secular variations in the pelagic realm. In Cook, H. E. and Enos, P. (eds.), *Deep Water Carbonate Environments,* Society of Economic Paleontologists and Mineralogists Special Publication No. 25, p. 19–50.

Fischer, A. G., Hebert, T. and Premoli Silva, I., 1985. Carbonate bedding cycles in Cretaceous pelagic and hemipelagic sequences. In Pratt, L. M., Kauffman, E. G. and Zelt, F. B. (eds.), *Fine-grained Deposits and Biofacies of the Cretaceous Western Interior Seaway: Evidence of Cyclic Sedimentary Processes,* Society of Economic Paleontologists and Mineralogists Field Trip Guidebook No. 4, 1985 Midyear Meeting, Golden, Colorado, p. 1–10.

Fleming, J. R., 1992. T. C. Chamberlin and H_2O climate feedbacks; a voice from the past. *EOS, Transactions of the American Geophysical Union,* v. 73, p. 505.

Flower, B. P. and Kennett, J. P., 1993. Relationships between Monterey Formation deposition and middle Miocene global cooling: Naples Beach section, California. *Geology,* v. 21, p. 877–880.

Föllmi, K. B., 1995. 160 m.y. record of marine sedimentary phosphorous burial: Coupling of climate and continental weathering under greenhouse and icehouse conditions. *Geology,* v. 23, p. 503–506.

Frakes, L. A., Francis, J. E. and Syktus, J. I., 1992. *Climate Modes of the Phanerozoic.* Cambridge: Cambridge University Press, 274 pp.

Fyfe, W. S., 1986. Fluids in deep continental crust. *Geodynamics Series, American Geophysical Union,* v. 14, p. 33–39.

Gaetani, M. and Garzanti, E., 1991. Multicyclic history of the northern India continental margin (Northwestern Himalaya). *American Association of Petroleum Geologists Bulletin,* v. 75, p. 1427–1446.

Garrels, R. M. and Mackenzie, F. T., 1971. *Evolution of Sedimentary Rocks.* New York: Norton, 397 pp.

Garzanti, E., Baud, A. and Mascle, G., 1987. Sedimentary record of the northward flight of India and its collision with Eurasia (Ladakh Himalaya, India). *Geodinamica Acta,* v. 1, p. 297–312.

Gee, E. R., 1945. The age of the Saline Series of the Punjab and Kohat. *Proceedings of the National Academy of Science of India,* v. 14, p. 269–312.

Hallam, A., 1992. *Phanerozoic Sea-Level Changes.* New York: Columbia University Press, 266 pp.

Hallock, P., Premoli Silva, I. and Boersma, A. 1991. Similarities between planktonic and larger foraminiferal evolutionary trends through Paleogene paleoceanographic changes. *Palaeogeography, Palaeoclimatology, Palaeoecology,* v. 83, p. 49–64.

Hamilton, W. B., 1988. Plate tectonics and island arcs. *Geological Society of America Bulletin,* v. 100, p. 1503–1527.

Honneger, K., Dietrich, V., Frank, W., Gansser, A., Thoni, M. and Trommsdorff, V., 1982. Magmatism and metamorphism in the Ladakh Himalayas (the Indus–Tsangpo suture zone). *Earth and Planetary Science Letters,* v. 60, p. 253–292.

Imbrie, J., Berger, A. and Shackleton, N. J., 1993. Role of orbital forcing: A two-million-year perspective. In Eddy, J. A. and Oescheger, H. (eds.), *Global Changes in the Perspective of the Past,* New York: John Wiley and Sons Ltd., p. 264–277.

Jaeger, J.-J., Courtillot, V. and Tapponnier, P., 1989. Paleontological view of the ages of the Deccan Traps, the Cretaceous/Tertiary boundary and the India–Asia collision. *Geology,* v. 17, p. 316–319.

Janecek, T. R. and Rea, D. K., 1983. Eolian deposition in the northeast Pacific Ocean: Cenozoic history of atmospheric circulation. *Geological Society of America Bulletin,* v. 94, p. 730–738.

Keeling, C. D., 1973. The carbon dioxide cycle: Reservoir models to depict the exchange of atmospheric carbon dioxide with the ocean and land plants. In Rasool, S. I. (ed.), *Chemistry of the Lower Atmosphere,* New York: Plenum Press, p. 251–329.

Kennett, J. P. and L. D. Stott, 1990. Proteus and Proto-Oceanus: Ancestral Paleogene oceans as revealed from Antarctic stable isotope results. In Barker, P. F., Kennett, J. P., et al., Proceedings of the Ocean Drilling Program, Scientific Results, v. 113: College Station, TX (Ocean Drilling Program), p. 865–880.

Kennett, J. P. and Stott, L. D., 1991. Abrupt deep-sea warming, palaeoceanographic changes and benthic extinctions at the end of the Palaeocene. *Nature,* v. 353, p. 225–229.

Kerrick, D. M. and Caldeira, K., 1993. Paleoatmospheric consequences of CO_2 released during early Cenozoic regional metamorphism in the Tethyan orogen. *Chemical Geology,* v. 108, p. 201–230.

Khan, M. A., Ahmed, R., Raza, H. A. and Kemal, A., 1986. Geology of Petroleum in Kohat–Potwar Depression, Pakistan. *American Association of Petroleum Geologists Bulletin,* v. 70, p. 396–414.

Klootwijk, C. T. and Bingham, D. K., 1980. The extent of greater India, III. Palaeomagnetic data from the Tibetan Sedimentary Series, Thakkhola region, Nepal Himalaya. *Earth and Planetary Science Letters,* v. 51, p. 381–405.

Klootwijk, C. T., Gee, J. S., Peirce, J. W., Smith, G. W. and McFadden, P. L., 1992. An early India–Asia contact: Paleomagnetic constraints from Ninetyeast Ridge, ODP Leg 121. *Geology,* v. 20, p. 395–398.

McCormick, M. P., 1995. Atmospheric Effects of the Mt. Pinatubo Eruption. *Nature,* v. 377, p. 399–404.

Meissner, C. R., Hussain, M., Rashid, M. A. and Sethi, U. B., 1975. *Geology of the Parachinar Quadrangle, Pakistan.* U.S. Geological Survey Professional Paper 716D, p. 1–34.

Meissner, C. R., Master, J. M., Rashid, M. A. and Hussain, M., 1974. *Stratigraphy of the Kohat Quadrangle, Pakistan.* U.S. Geological Survey Professional Paper 716B, p. 1–24.

Molnar, P. and England, P., 1990. Late Cenozoic uplift of mountain ranges and global climate change: Chicken or Egg? *Nature,* v. 346, p. 29–34.

Molnar, P. and Tapponnier, P., 1975. Cenozoic tectonics of Asia: Effects of a continental collision. *Science,* v. 189, p. 419–426.

Patriat, P. and Achache, J., 1984. India-Eurasia collision chronology has implications for crustal shortening and driving mechanism of plates. *Nature,* v. 311, p. 615–621.

Pegram, W. J., Krishnaswami, S., Ravizza, G. E. and Turekian, K. K., 1992. The record of sea water $^{187}Os/^{186}Os$ variation through the Cenozoic. *Earth and Planetary Science Letters,* v. 113, p. 569–576.

Pivnik, D. A. and Sercombe, W. J., 1993. Compression and transgression related deformation in the Kohat Plateau, NW Pakistan. In Treloar, P. J. and Searle, M. P. (eds.), *Himalayan Tectonics,* Geological Society Special Publication No. 74, p. 559–580.

Pivnik, D. A. and Wells, N. A., 1996. The transition from Tethys to the Himalaya as recorded in NW Pakistan. *Bulletin of the Geological Society of America,* v. 108, pp. 1295–1313.

Porter, S. C., 1986. Pattern and forcing of northern hemisphere glacier variations during the last millennium. *Quaternary Research,* v. 26, p. 27–48.

Powell, C. M. and Conaghan, P. J., 1973. Plate Tectonics and the Himalayas. *Earth and Planetary Science Letters,* v. 20, p. 5–12.

Rampino, M. R. and Self, S., 1991. Volcanic winter and accelerated glaciation following the Toba super-eruption. *Nature,* v. 359, p. 50–52.

Raymo, M. E., 1991. Geochemical evidence supporting T. C. Chamberlin's theory of glaciation. *Geology,* v. 19, p. 344–347.

Raymo, M. E., 1994. The Himalayas, organic carbon burial, and climate in the Miocene. *Paleoceanography,* v. 9, p. 399–404.

Raymo, M. E. and Ruddiman, W. F., 1992. Tectonic forcing of the late Cenozoic climate. *Nature,* v. 359, p. 117–122.

Raymo, M. E., Ruddiman, W. F. and Froelich, P. N., 1988. Influence of late Cenozoic mountain building on ocean geochemical cycles. *Geology,* v. 16, p. 649–653.

Rea, D. K., 1992. Delivery of Himalayan sediment to the northern Indian Ocean and its relation to global climate, sea level, uplift and seawater strontium. In Duncan, R. A., Rea, D. K., Kidd, R. B., von Rad, U. and Weissel, J. K. (eds.), *Synthesis of Results from Scientific Drilling in the Indian Ocean,* Geophysical Monograph 70, Washington, D.C.: American Geophysical Union, p. 387–402.

Rea, D. K., Zachos, J. C., Owen, R. M. and Gingerich, P. D., 1990. Global change at the Paleocene-Eocene boundary: climatic and evolutionary consequences of tectonic events. *Palaeogeography, Palaeoclimatology, Palaeoecology,* v. 79, p. 117–128.

Reuber, I., Montigny, R., Thuizat, R. and Heitz, A., 1989. K-Ar ages of ophiolites and arc volcanics of the Indus suture zone: Clues on the early evolution of the Neo-Tethys. *Eclogae Geologicae Helvetiae,* v. 82, p. 699–715.

Rind, D., 1993. How will future climate changes differ from the past? In Eddy, J. A. and Oescheger, H. (eds.), *Global Changes in the Perspective of the Past,* New York: John Wiley and Sons Ltd., p. 39–49.

Selverstone, J. and Gutzler, D. S., 1993. Post-125 Ma carbon storage associated with continent-continent collision. *Geology,* v. 21, p. 885–888.

Sercombe, W. J., Stratton, M. A., Albertin, M., Wilson, W. P., Roth, B. L., Nieuwenhuise, R. E., Pivnik, D. A. and Beck, R. A., 1994. Wrench faulting in the Northern Pakistan Foreland Region. *The Leading Edge,* v. 11, p. 1107–1110.

Shackleton, N. J., 1985. Oceanic carbon isotope constraints on oxygen and carbon dioxide in the atmosphere. In Sundquist, E. T. and Broecker, W. S. (eds.), *The Carbon Cycle and Atmospheric CO_2: Natural variations Archean to Present,* Geophysical Monograph 32, Washington, D.C.: American Geophysical Union, p. 412–417.

Shackleton, N. J., 1987. The carbon isotope record of the Cenozoic: History of organic carbon burial and of oxygen in the ocean and atmosphere. In Brooks, J. and Fleet, A. J. (eds.), *Marine Petroleum Source Rocks,* Geological Society Special Publication No 26, p. 423–434.

Spencer, D. A., 1993. What was the timing of the Himalayan continent-continent collision (45, 50, 55, 60, or 65 Ma)? Constraints from eclogites. *Geological Society of America, Abstracts with Programs,* v. 25, p. 121.

Stille, P., 1992. Nd-Sr isotope evidence for dramatic changes of paleocurrents in the Atlantic Ocean during the past 80 m.y. *Geology,* v. 20, p. 387–390.

Stoneley, R., 1974. Evolution of the Continental Margins bounding a former Southern Tethys. In Burk, C.A. and Drake, C.A. (eds.), *The Geology of Continental Margins,* New York: Springer-Verlag, p. 889–903.

Sundquist, E. T. and Broecker, W. S. (eds.), 1985. *The Carbon Cycle and Atmospheric CO_2: Natural Variations Archean to Present.* Geophysical Monograph 32, Washington, D.C.: American Geophysical Union, 627 pp.

Thomas, E., 1990. Late Cretaceous–early Eocene mass extinctions in the deep sea. In Sharpton, V. L. and Ward, P. (eds.), *Global Catastrophes in Earth History: An interdisciplinary Conference on Impacts, Volcanism, and Mass Mortality,* Geological Society of America Special Paper No. 247, p. 481–495.

Thurow, J., Brumsack, H.-J., Littke, R., Meyers, P. and Rullkotter, J., 1992. The Cenomanian/Turonian boundary event in the Indian Ocean: A key to understand the global picture. In Duncan, R. A., Rea, D. K., Kidd, R. B., von Rad, U. and Weissel, J. K. (eds.), *Synthesis of Results from Scientific Drilling in the Indian Ocean,* Geophysical Monograph 70, Washington, D.C.: American Geophysical Union, p. 253–273.

Tonarini, S., Villa, I. M., Oberli, F., Meier, M., Spencer, D. A., Pogante, U. and Ramsay, J. G., 1993. Eocene age of eclogite metamorphism in Pakistan Himalaya: Implications for India–Eurasia collision. *Terra Nova,* v. 5, p. 13–20.

Treloar, P. J. and Coward, M. P., 1991. Indian Plate motion and shape: Constraints on the geometry of the Himalayan orogen. *Tectonophysics,* v. 191, p. 189–198.

Treloar, P. J. and Izatt, C. N., 1993. Tectonics of the Himalayan collision between the Indian Plate and the Afghan Block: A synthesis. In Treloar, P. J. and Searle, M. P. (eds.), *Himalayan Tectonics, Geological Society of London,* v. 74, p. 69–87.

Vincent, E. and Berger, W. H., 1985. Carbon dioxide and polar cooling in the Miocene: The Monterey hypothesis. In Sundquist, E. T. and Broecker, W. S. (eds.), *The Carbon Cycle and Atmospheric CO_2: Natural Variations Archean to Present,* Geophysical Monograph 32, Washington D. C.: American Geophysical Union, p. 455–468.

Wells, N. A., 1984. *Marine and continental sedimentation in the early Cenozoic Kohat basin and adjacent northern Indo-Pakistan.* Doctoral Dissertation, University of Michigan, U.S.A. 465 pp.

Wilson, J. T., 1965. A new class of faults and their bearing on continental drift. *Nature,* v. 207, p. 343–347.

Wilson, R. C., Hudson, H. S. and Frohlich, C., 1986. Long-term downward trend in total solar irradiance. *Science,* v. 28, p. 1114–1117.

Wolfart, R. and Wittekindt, H., 1980. *Geologie von Afghanistan.* Berlin: Gebrüder Borntrager, 500 pp.

Wolfe, J. A., 1990. Palaeobotanical evidence for a marked temperature increase following the Cretaceous/Tertiary boundary. *Nature,* v. 343, p. 153–156.

Wolfe, J. A. and Poore, R. Z., 1982. Tertiary marine and non-marine climatic trends. In *Climate in Earth History, Studies in Geophysics,* Washington, D.C.: National Academy Press, p. 154–158.

Wolfe, J. A. and Upchurch, G. R., 1987. North American non-marine climates and vegetation during the Late Cretaceous. *Palaeogeography, Palaeoclimatology, Palaeoecology,* v. 61, p. 33–77.

Yeats, R. S. and Hussain, A., 1987. Timing of structural events in the Himalayan foothills of northwestern Pakistan. *Geological Society of America Bulletin,* v. 99, p. 161–176.

Zachos, J. C., Rea, D. K., Seto, K., Nomura, R. and Niitsuma, N.,1992. Paleogene and early Neogene deep water paleoceanography of the Indian Ocean as determined from benthic foraminifer stable carbon and oxygen isotope records. In Duncan, R. A., Rea, D. K., Kidd, R. B., von Rad, U. and Weissel, J. K. (eds.), *Synthesis of Results from Scientific Drilling in the Indian Ocean,* Geophysical Monograph 70, Washington, D.C.: American Geophysical Union, p. 351–385.

7

Changes in Atmospheric Circulation During the Latest Paleocene and Earliest Eocene Epochs and Some Implications for the Global Climate Regime

DAVID K. REA

ABSTRACT
Examination of eolian sediments from deep-sea cores has revealed evidence for a significant reduction in the intensity of atmospheric circulation that occurred over a span of 1–2 my in the temporal vicinity of the Paleocene/Eocene Epoch boundary. Early Eocene winds were much more sluggish than the surprisingly strong winds of the latest Cretaceous Period and Paleocene Epoch, and implicate a significant reduction in the planetary temperature gradient. In a sequence of longer-term changes that occurred over approximately 3 my beginning in the late Paleocene Epoch (from Biochron P4 to P6), the onset of ocean deep-water warming (based on $\delta^{18}O$ data) preceded the beginning of changes in the partitioning of carbon between ocean and sedimentary reservoirs (based on $\delta^{13}C$ data), which preceded the start of a pronounced reduction in wind intensity, which preceded the benthic foraminifera extinction. These four events were spaced roughly one-half to one million years apart. Explanations for this rather drawn-out series of changes, particularly considering the order in which they occur, are not yet fully satisfactory. The introduction to the ocean–atmosphere system of another greenhouse gas, methane, may be more important than previously considered.

Environmental changes in the latest Paleocene and earliest Eocene epochs have long been of interest to paleontologists because of the oceanic benthic faunal turnover among foraminifera (Tjalsma and Lohmann 1983) and the terrestrial evolutionary radiation of mammals (Gingerich 1980). In the early 1980s two new data sets resulted in a broadening of the interest in these events to include paleoceanographers and paleoclimatologists. First, Shackleton and co-workers (Shackleton and Boersma 1981; Shackleton and Hall 1984; Shackleton et al. 1984) clearly defined the large changes in $\delta^{13}C$ and $\delta^{18}O$ that occurred in conjunction with the paleontologic event in the deep sea. These changes include an upcore lightening of oxygen isotopic values by more than 1‰ and of carbon isotopic values by more than 2‰. Secondly, Janecek and Rea (1983), in a study of eolian deposition in the central North Pacific, found evidence for a several-fold reduction in the intensity of atmospheric circulation near the time of the Paleocene/Eocene Epoch boundary, such that the early Eocene winds were much more sluggish than those of the late Paleocene Epoch. This work, based on central North Pacific Core LL44-GPC3, was confirmed by finding evidence for similar changes at northwest Pacific Deep-Sea Drilling Project (DSDP) Site 576 (Janecek 1985; Rea et al. 1985). Other changes that may have influenced the global environment also occurred at or near the time of the Paleocene/Eocene Epoch boundary, such as the voluminous eruptions of the east Greenland volcanics and the opening of the North Atlantic between Greenland and Norway (White and McKenzie 1989), changes in plate motions and ridge-transform geometry (Rona and Richardson 1978; Rea and Dixon 1983; Aubry et al. 1988), and an increase in the amount of sea-floor hydrothermal activity (Owen and Rea 1985; Rea et al. 1990). Rea et al. (1990) have suggested that these global environmental events, which occurred over a time span of a few million years, were a direct result of the tectonic events that began in the latest Paleocene Epoch.

Processes of Eolian Transport and Deposition

In the pelagic ocean, far from continents, the terrigenous mineral component of deep sea sediments is eolian dust that has been transported long distances from its continental source region. Dust grains accumulate slowly and, in the absence of other sedimentary components, form the pelagic (red) clay deposits that characterize much of the deep Pacific and Indian oceans. The mass accumulation rate of eolian sediment reflects the dust supply to the ocean, a direct function of the aridity of the eolian source region. Furthermore, the grain size of dust is a measure of the intensity of the transporting winds such that stronger winds transport coarser grains. These two parameters, the flux and size of eolian grains, vary independently. Thus it is incorrect to equate stronger winds with more dust when stronger winds mean only coarser grains. Mineralogy and geochemistry of the dust grains give information regarding provenance. Rea (1994) provided a review of the paleoclimatic record interpreted from eolian sediments in the pelagic ocean and the reader is referred to that paper for details of methodology and interpretation.

Changes in Atmospheric Circulation at the End of the Paleocene Epoch

The eolian grain size data from central North Pacific piston Core LL44-GPC3 (30.3° N, 157.8° W; Janecek and Rea 1983) and northwest Pacific DSDP Site 576 (32.4° N, 164.3° E; Janecek 1985; figure 7.1) were determined from pelagic clay sediments characterized by low stratigraphic resolution. A third study, by Miller et al. (1987) on samples from DSDP Site 577 (32.4° N, 157.7° E), examined an upper Paleocene to lower Eocene section from the shallower Shatsky Rise where the sediments were carbonate oozes and chalks. The data from the three North Pacific cores all show a significant reduction in eolian grainsize that, based on the stratigraphy outlined by Miller et al. (1987) at Site 577, occurs very close to the Paleocene/Eocene Epoch boundary, although nannofossil stratigraphy suggests a short hiatus at the NP9/NP10 zonal contact at this site (Monechi 1985; Berggren and Aubry 1996). The Miller et al. (1987) study is particularly valuable because it was the first to link biostratigraphic, magnetostratigraphic, isotopic, and eolian records of environmental change in the same site near the Paleocene/Eocene Epoch boundary. The magnitude of the grain size change at the paleolocation of piston Core LL44-GPC3 is more than twice that found at Sites 576 and 577, which are 3500 to 4000 km farther west. In latest Paleocene time these sites lay along a paleolatitude of about 15° north beneath the tradewinds. Late Paleocene winds are interpreted on the basis of the eolian grainsize to have been similar in intensity to those of the Pleistocene Epoch, and the winds of the early Eocene Epoch to have been quite weak (figure 7.1; Janecek and Rea 1983; Rea et al. 1985). The indication of vigorous atmospheric circulation in the latest Cretaceous Period and Paleocene Epoch was an unexpected result of these eolian grainsize investigations. Atmospheric circulation intensity remained low for most of the Eocene Epoch and began to intensify (based on coarser median grainsize) in the early part of the Oligocene Epoch (figure 7.1).

This demonstration and verification from the same ocean of a large reduction in the intensity of global wind near the time of the Paleocene/Eocene Epoch boundary holds important consequences for interpretation of the overall global change of that time. As such it became important to try to verify this change from another ocean and, hopefully, another hemisphere. To this end we undertook studies of the eolian record in the

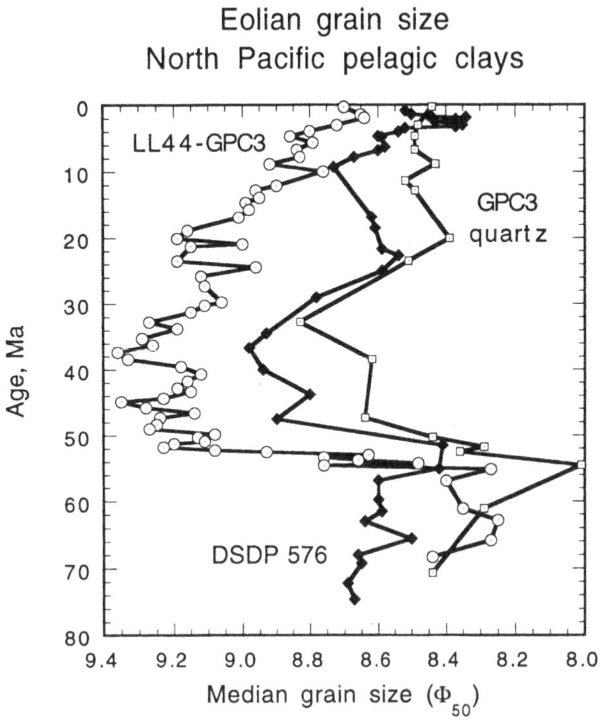

Figure 7.1 Eolian grain size of North Pacific pelagic clays. Open circles indicate data from giant piston Core LL44-GPC3 (Janecek and Rea 1983) that show a reduction in grainsize from about 8.3Φ (3.2 mm) to 9.2Φ (1.7 mm) near the time of the Paleocene/Eocene Epoch boundary. Open squares indicate grain size data on quartz grains extracted from the sediment of the giant piston core. Solid diamonds indicate data from DSDP Site 576 (Janecek 1985) showing a grain size reduction from 8.4Φ (3.0 mm) to 8.9Φ (2.1 mm) near the time of the Paleocene/Eocene Epoch boundary. Ages on this plot are based on ichthyolith stratigraphy.

southern Indian Ocean at DSDP Site 215 (8.1° S, 86.8° E; paleolatitude 47° S; Hovan and Rea 1992) and in the South Atlantic at Walvis Ridge (DSDP Sites 525, 527 and 528, paleolatitude 30°S). We found that the mineral component of deep sea sediment at the Walvis Ridge is dominated by hemipelagic and not eolian grains (Rea and Hovan 1995), leaving only Site 215 in the Indian Ocean as the one well-studied, southern-hemisphere eolian record spanning the Paleocene/Eocene Series boundary (Hovan and Rea 1992).

DSDP Site 215, located on the sea floor west of Ninetyeast Ridge, contains a 45-meter long section of nannofossil ooze across the Paleocene/Eocene Series boundary; a short hiatus may occur at the boundary itself (W. A. Berggren, personal communication 1995). Based on the recently revised Paleogene chronology (Berggren et al. 1995), linear sedimentation rates through this section are between 0.8 and 1.0 cm/kyr, so the temporal resolution is adequate. Hovan and Rea (1992) reported on the carbon and oxygen isotopic values of benthic foraminifera and on the grain size of the eolian dust extracted from these oozes. Foraminiferal assemblage data and the horizon of the benthic foraminiferal extinctions were determined by McGowran (1974). Results of the work at Site 215 (figure 7.2) show the first documentation of the southern hemisphere expression of the reduction in dust grain size associated with Paleocene/Eocene Epoch boundary climatic transition. Comparison of all the data show that the long term lightening in oxygen isotopic values begun first, followed by the onset of the large change in carbon isotopic values. Both of these isotopic trends were established before the trend toward reduced dust grain size began. The benthic foraminifera extinction event was the last step of the Paleocene/Eocene Epoch boundary sequence of events. The same ordering of events, beginning with the onset of $\delta^{18}O$ changes and ending with the benthic foraminiferal extinctions occurs in Site 577 in the northwestern Pacific (Miller et al. 1987). The entire series of changes, continuing into the early Eocene Epoch, may span 4 my or more (Zachos et al. 1993; Corfield 1994).

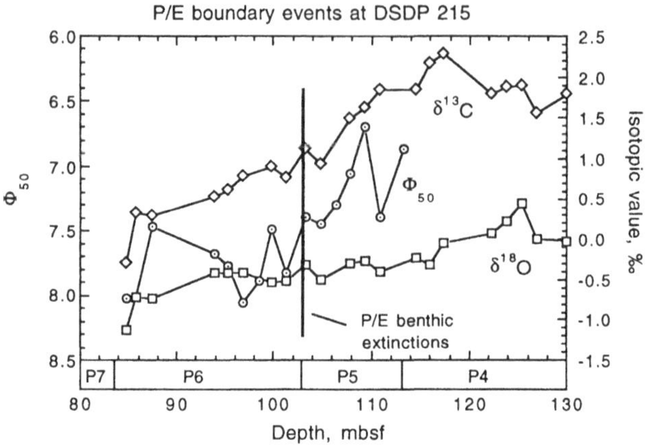

Figure 7.2 Events of the Paleocene–Eocene transition at DSDP Site 215 in the southern Indian Ocean (Hovan and Rea 1992).

Open squares are the $\delta^{18}O$ values of benthic foraminifera, open diamonds show the $\delta^{13}C$ values of benthic foraminifera, and open circles show the eolian grain size data. At this site the 1‰ reduction in benthic $\delta^{18}O$ begins at 126 mbsf, the 2.5‰ lightening in $\delta^{13}C$ values starts at 117 mbsf, the 1Φ reduction in eolian grainsize starts at 109 mbsf, and the benthic extinction event is at 103 mbsf. Planktic foraminiferal zonation along the bottom is based on the work of McGowran (1974) and transposed to the chronology of Berggren et al. (1995). Any explanation of the events of the latest Paleocene to early Eocene transition must include the reason for this sequence of events, which appears to be the same everywhere.

Discussion

Warm, high-latitude environments of the early Eocene Epoch have been known for a long time (Frakes 1979). The reduction in wind intensity at the beginning of the Eocene Epoch as interpreted from the eolian grainsize data implies a coeval significant reduction in the pole-to-equator temperature gradient. Zachos et al. (1994) provided the details of this change in planetary temperature gradient, showing that in the latest Paleocene and earliest Eocene epochs the high latitudes warmed about 6° C while the tropics remained at the same temperature, or perhaps even cooled. Sea-surface temperature gradients between the equator and 70° latitude were reduced from 17° C in the latest Paleocene Epoch to 10° C in the early Eocene Epoch (Zachos et al. 1994; Corfield 1994).

The sequence of events that together give evidence of pronounced environmental changes across the Paleocene/Eocene Epoch boundary are found in data from the southern mid-latitude Indian Ocean (Hovan and Rea 1992; figure 7.2), the northern low-latitude Pacific Ocean (Miller et al. 1987), and at the Walvis Ridge in the mid-latitude South Atlantic (J. Zachos and D. Rea, unpublished data; Thomas and Shackleton 1996). The first change was the warming of ocean deep waters that started during the early to middle part of Biochron P4, presumably because of a warming at the latitudes of deep water formation. Perhaps 800,000 years later, during the later part of Biochron P4, the global carbon system began to change with $\delta^{13}C$ values eventually becoming 2‰ to 3‰ lighter. This $\delta^{13}C$ shift has been interpreted to mean reintroduction of organic carbon to the ocean from sedimentary storage reservoirs either on continents or in the ocean. It is only after this event, another 800,000 years later, in Biochron P5, that a slowing of (mid- and low-latitude) atmospheric circulation occurred as indicated by the reduction in eolian grain size. The benthic foraminiferal extinction event followed another 600,000 years later, perhaps longer, depending on the duration of any hiatus in the stratigraphic record at Site 215 at the P5/P6 zonal boundary.

A rapid, large negative excursion in oxygen and carbon isotopic values, often associated with a relatively clay-rich layer, occurs in the uppermost Paleocene oozes at the same stratigraphic horizon as the benthic foraminiferal extinctions (Kennett and Stott 1991; Thomas and Shackleton 1996). This brief event, which may have had a duration of up to a few hundred thousand years, is superimposed on the much longer-term changes outlined above. Carbon isotopic data from terrestrial sequences have linked this terminal Paleocene event with the sudden evolutionary radiation of mammals (Koch et al. 1992) providing a rare direct tie between continental and marine evolutionary changes.

An ultimate tectonic cause for the long-term changes that occurred near the Paleocene/Eocene Epoch boundary has been proposed (Owen and Rea 1985; Rea et al. 1990; Hovan and Rea 1992; Rea 1994; Thomas and Shackleton 1996). The scenario begins with the massive extrusive events of the North Atlantic region culminating in the Norwegian–Greenland rifting, along with associated sea-floor hydrothermal activity adding CO_2 to the atmosphere.

What's Going On Here, Anyway?

Any explanation of the sequence of events that characterize the global changes at the Paleocene/Eocene Epoch boundary must account for each of four major phenomena: warming, especially at high latitudes, preceding the change in carbon partitioning between the oceanic and other reservoirs, preceding the reduction in the intensity of atmospheric circulation, preceding both oceanic and terrestrial biotic changes. There is as of yet no single scenario explaining these phenomena and their sequencing that paleoceanographers agree upon. The 3-my time span required for all of these events and especially their spacing in time poses a particularly difficult problem of interpretation. Most changes in climate occur on time scales similar to 1) the mixing time of the oceans, up to a few thousand years and 2) the periodicities of orbital variability, up to one hundred thousand, occasionally four hundred thousand, years; or they occur on scales of 1–2 my, often directly associated with tectonic events. The sequence of events described above, separated in time by many hundred of thousands of years implies changes that occur on scales longer than those associated with orbital variability and shorter than those related to most tectonic events. What follows are offered as some partial explanations.

A warming trend, evidenced in the benthic $\delta^{18}O$ data, began in the late Paleocene Epoch. This warming was much more pronounced at high latitudes and resulted in warmer ocean deep waters and a reduced planetary temperature gradient (Zachos et al. 1994; Corfield 1994). Most workers ascribe some portion of this latest Paleocene to early Eocene warming to enhanced atmospheric CO_2 concentrations resulting from late Paleocene volcanism/tectonism. The problem with relying only on CO_2 to change the pole-to-equator temperature gradient is that enhanced CO_2 does not warm high latitudes dramatically more than low latitudes (Sloan and Rea 1995) and therefore has little effect on the planetary temperature gradient. Some additional process must be called upon to either keep the equatorial region relatively cool or to keep the polar regions relatively warm. Most attention has been paid to the idea that enhanced ocean deep water circulation, perhaps carrying warm salty bottom water from the subtropics or the Tethys to high latitudes, may solve this heat transport discrepancy (Barron 1987; Covey and Barron 1988). However, a recent calculation of the amount of water of appropriate temperature needed to be transported poleward suggests that 80 times the outflow of the modern Mediterranean is required (Sloan et al. 1995), an unlikely scenario. Furthermore, ocean heat transport, as a sole causal factor, also does not explain the apparent warming of continental interiors during the early Eocene Epoch (Wing and Greenwood 1994).

We then are left with the quandary that, based on the evidence of declining zonal wind speeds, neither the atmosphere nor the wind-driven surface circulation is capable of carrying the heat necessary to warm the poles, and the deep ocean may not be capable of that amount of heat transport either (Sloan et al. 1995). At this point it is appropriate to consider other greenhouse gasses, especially methane. Sloan et al. (1992) have suggested that methane released from wetlands in the course of the latest Paleocene warming would both serve as an additional greenhouse gas and, more importantly, act to retard winter cooling at the poles because of its effect on the nature of polar clouds. Oxidation of methane is a source of stratospheric water, resulting in polar stratospheric clouds that trap outgoing long-wavelength radiation. The polar stratospheric clouds retard winter cooling of the earth's surface and may result in significant high-latitude warming (Sloan et al. 1992).

There is an additional source of methane besides early Eocene wetlands that needs to be considered, the great reservoir of gas in sub-seafloor methane hydrates that underlie all continental margins. Kvenvolden (1993) has suggested that destabilization of gas hydrates during an episode of ocean warming would release large quantities of methane into the ocean-atmosphere system. The $\delta^{13}C$ values of methane in gas hydrates ranges from –40‰ to –70‰ (Kvenvolden 1993). Recently, Dickens and Quinby-Hunt (1994) have determined the pressure–temperature stability fields for methane and water and defined the pressure–temperature conditions of hydrate formation in seawater. A quick determination of the amount of methane that would be released by a deep ocean warming of 5° C, from 6 to 11° C, shows that the entire $\delta^{13}C$ shift can be accounted for by the release of only a few percent of the potentially available methane (Dickens et al. 1995). Methane exolution from sub-seafloor hydrates also provides an explanation for the shift to light carbon isotopic values of the oceanic reservoir in response to the global warming associated with the latest Paleocene events. Incorporating the release of methane from the sub-seafloor hydrates into the sequence of events near the Paleocene/Eocene Epoch boundary provides considerable assistance in explaining both the extreme nature of the carbon isotopic shift without requiring unusual continental weathering, and the maintenance of unusually warm polar regions via the polar cloud mechanism proposed by Sloan et al. (1992).

The long term $\delta^{13}C$ shift of the latest Paleocene and earliest Eocene epochs began roughly 800,000 years after the oxygen isotope values began to change, perhaps an indication of how long it took for the intermediate and deep waters to become warm enough to affect the methane hydrate horizons. The change in eolian grain size began another 800,000 years or so after the onset of the $\delta^{13}C$ change. Slowing of the winds must be in response to reduction of the planetary temperature gradient, and thus may be an indication of when methane-related cloudiness became important in polar regions. Finally, the altered abyssal circulation patterns (perhaps even including the formation of warm salty bottom water in the Tethys) coupled with reduced sea-surface productivity and declining organic carbon rain resulting from more sluggish sea surface circulation, may have caused enough stress in the abyssal environment to result in the benthic extinction event.

The few combined records of atmospheric and oceanic change near the Paleocene/Eocene Epoch boundary have been studied at a relatively low temporal resolution. To fully evaluate the scenario of climate changes proposed here likely requires both new records and that the available records be sampled at a detail necessary to define changes certainly at orbital timescales and probably at the timescales of ocean mixing.

A significant reduction in the intensity of zonal winds followed changes in global temperature and the carbon system in the late Paleocene and early Eocene epochs. Any explanation for the several events that occurred successively through this interval must account for both the timing and sequence of those events. For instance, the beginning of the change in atmospheric circulation follows the beginning of the change in the carbon isotopic record by almost a million years, so may be the result of change in the carbon system, but not its cause. The application of computer models to the study of global change has provided several new and useful insights to the behavior of the late Paleocene and early Eocene climate system and pointed out that other greenhouse gases besides CO_2 may have important roles to play (Sloan et al. 1992, 1995; Sloan and Rea 1995). Methane may be particularly important because a vast reservoir of isotopically very light methane from hydrates is available to any newly warmed ocean (Kvenvolden 1993; Dickens et al. 1995).

Events of the Paleocene–Eocene transition, each separated in time by one-half to one million years, were:

1. Late Paleocene tectonism/volcanism/hydrothermal activity, which led to
2. A CO_2-induced global warming, which led to
3. Methane release from sub-seafloor gas hydrates, which led to
4. Relatively warmer poles, reduction of the planetary temperature gradient, and weak winds, which led to
5. Altered abyssal circulation rates and pathways, low sea-surface productivity, and benthic extinctions.

ACKNOWLEDGMENTS

In the past years I have had fruitful discussions about the nature of global change near the time of the Paleocene/Eocene Epoch boundary with a number of people including: Lisa Sloan, Jim Zachos, Lowell Stott, Ted Moore, Jim Walker, and Jerry Dickens. Research on the events of the late Paleocene and early Eocene epochs done at the University of Michigan was supported in part by National Science Foundation grant ATM9014184. This chapter was reviewed by Marie-Pierre Aubry and Ted Moore, who provided thoughtful comments.

REFERENCES

Aubry, M-P., Berggren, W. A., Kent, D. V., Flynn, J. J., Klitgord, K. D., Obradovich, J. D., Prothero, D. W., 1988. Paleogene Geochronology: An integrated approach. *Paleoceanography,* v. 3, p. 707–742.

Barron, E. J., 1987. Eocene equator-to-pole surface ocean temperatures: A significant problem? *Paleoceanography,* v. 2, p. 729–739.

Berggren, W. A. and Aubry, M.-P., 1996. A late Paleocene–early Eocene NW European and North Sea magnetobiochronological correlation network. In Knox, R. W. O'B., Corfield, R. M. and Dunay, R. E. (eds.), *Correlation of the Early Paleogene in Northwest Europe,* Geological Society Special Publication No. 101, p. 309–352.

Berggren W. A., Kent, D. V., Swisher, C. C., III and Aubry, M.-P., 1995. A revised Cenozoic geochronology and chronostratigraphy. In Berggren, W. A., Kent, D. V., Aubry, M.-P. and Hardenbol, J. (eds.), *Geochronology Time Scales and Global Stratigraphic Correlation,* Society of Economic Paleontologists and Mineralogists Special Publication No. 54, p. 129–212.

Corfield, R. M., 1994. Palaeocene oceans and climate: An isotopic perspective. *Earth Science Reviews,* v. 37, p. 225–252.

Covey, C. and Barron, E., 1988. The role of ocean heat transport in climatic change. *Earth Science Reviews,* v. 24, p. 429–445.

Dickens, G. R. and Quinby-Hunt, M. S., 1994. Methane hydrate stability in seawater. *Geophysical Research Letters,* v. 21, p. 2115–2118.

Dickens, G. R., O'Neil, J. R., Rea, D. K., Owen, R. M., 1995. Dissociation of oceanic methane hydrate as a cause of the carbon isotope excursion at the end of the Paleocene. *Paleoceanography,* v. 10, p. 965–971.

Frakes, L. A., 1979. *Climates Throughout Geologic Time.* New York: Elsevier Scientific Pub. Co., 310 pp.

Gingerich, P. D., 1980. Evolutionary patterns in early Cenozoic mammals. *Annual Review of Earth and Planetary Sciences,* v. 8, p. 407–424.

Hovan, S. A. and Rea, D. K., 1992. Paleocene/Eocene boundary changes in atmospheric and oceanic circulation: A southern hemisphere record. *Geology,* v. 20, p. 15–18.

Janecek, T. R., 1985. Eolian sedimentation in the northwest Pacific Ocean: A preliminary examination of the data from Deep Sea Drilling Project Sites 576 and 578. In Heath, G. R., Burckle, L. H., et al., Initial Reports of the Deep Sea Drilling Project, v. 86: Washington (U.S. Government Printing Office), p. 589–603.

Janecek, T. R. and Rea, D. K., 1983. Eolian deposition in the northeast Pacific Ocean: Cenozoic history of atmospheric circulation. *Geological Society of America Bulletin,* v. 94, p. 730–738.

Kennett, J. P. and Stott, L. D., 1991. Abrupt deep-sea warming, palaeoceanographic changes and benthic extinctions at the end of the Palaeocene. *Nature,* v. 353, p. 225–229.

Koch, P. L., Zachos, J. C. and Gingerich, P. D., 1992. Correlation between isotope records in marine and continental carbon reservoirs near the Palaeocene/Eocene boundary. *Nature,* v. 358, p. 319–322.

Kvenvolden, K. A., 1993. Gas hydrates: geological perspective and global change. *Reviews of Geophysics,* v. 31, p. 173–187.

McGowran, B., 1974. Foraminifera. In von der Borch, C. C., Sclater, J. G., et al., Initial Reports of the Deep Sea Drilling Project, v. 22: Washington (U.S. Government Printing Office), p. 609–627.

Miller, K. G., Janecek, T. R., Katz, M. E. and Keil, D. J., 1987. Abyssal circulation and benthic foraminiferal changes near the Paleocene/Eocene boundary. *Paleoceanography,* v. 2, p. 741–761.

Monechi, S., 1985. Campanian to Pleistocene calcareous nannofossil stratigraphy from the northwest Pacific Ocean, Deep Sea Drilling Project Leg 86. In Heath, G. R., Burckle, L. H., et al., Initial Reports of the Deep Sea Drilling Project, v. 86: Washington (U.S. Government Printing Office), p. 301–336.

Owen, R. M. and Rea, D. K., 1985. Sea-floor hydrothermal activity links climate to tectonics: The Eocene carbon dioxide greenhouse. *Science,* v. 227, p. 166–169.

Rea, D. K., 1994. The paleoclimatic record provided by eolian deposition in the deep sea: the geologic history of wind. *Reviews of Geophysics,* v. 32, p. 159–195.

Rea, D. K. and Dixon, J. M., 1983. Late Cretaceous and Paleogene tectonic evolution of the North Pacific Ocean. *Earth and Planetary Science Letters,* v. 65, p. 145–166.

Rea, D. K. and Hovan, S. A., 1995. Grain size distribution and depositional processes of the mineral component of abyssal sediments: Lessons from the North Pacific. *Paleoceanography,* v. 10, p. 251–258.

Rea, D. K., Leinen, M. and Janecek, T. R., 1985. Geologic approach to the long-term history of atmospheric circulation. *Science,* v. 227, p. 721–725.

Rea, D. K., Zachos, J. C., Owen, R. M. and Gingerich, P. D., 1990. Global change at the Paleocene–Eocene boundary: climatic and evolutionary consequences of tectonic events. *Palaeogeography, Palaeoclimatology, Palaeoecology,* v. 79, p. 117–128.

Rona, P. A. and Richardson, E. S., 1978. Early Cenozoic global plate reorganization. *Earth and Planetary Science Letters,* v. 40, p. 1–11.

Shackleton, N. J. and Boersma, A., 1981. The climate of the Eocene ocean. *Journal of the Geological Society of London,* v. 138, p. 153–157.

Shackleton, N. J. and Hall, M. A., 1984. Carbon isotope data from Leg 74 sediments. In Moore, T. C., Jr., Rabinowitz, P. D., et al., Initial Reports of the Deep Sea Drilling Project, v. 74: Washington (U.S. Government Printing Office), p. 613–619.

Shackleton, N. J., Hall, M. A. and Boersma, A., 1984. Oxygen and carbon isotope data from Leg 74 foraminifers. In Moore, T. C., Jr., Rabinowitz, P. D., et al., Initial Reports of the Deep Sea Drilling Project, v. 74: Washington (U.S. Government Printing Office), p. 599–612.

Sloan, L. C. and Rea, D. K., 1995. Atmospheric carbon dioxide and early Eocene climate: A general circulation modeling sensitivity study. *Palaeogeography, Palaeoclimatology, Palaeoecology,* v. 119, p. 275–292.

Sloan, L. C., Walker, J. C. G. and Moore, T. C. Jr., 1995. Possible role of oceanic heat transport in early Eocene climate. *Paleoceanography,* v. 10, p. 347–356.

Sloan, L. C., Walker, J. C. G., Moore, T. C. Jr., Rea, D. K. and Zachos, J. C., 1992. Possible methane-induced polar warming in the early Eocene. *Nature,* v. 357, p. 320–322.

Thomas, E. and Shackleton, N. J., 1996. The Paleocene–Eocene benthic foraminiferal extinction and stable isotope anomalies. In Knox, R. W. O'B., Corfield, R. M. and Dunay, R. E. (eds.), *Correlation of the Early Paleogene in Northwest Europe,* Geological Society Special Publication No 101, p. 401–441.

Tjalsma, R. C. and Lohmann, G. P., 1983. *Paleocene–Eocene bathyal and abyssal benthic foraminifera from the Atlantic Ocean.* Micropaleontology Special Publication No. 4, 90 pp.

White, R. and McKenzie, D., 1989. Magmatism at rift zones: The generation of volcanic continental margins and flood basalts. *Journal of Geophysical Research,* v. 94(B6), p. 7685–7729.

Wing, S. L. and Greenwood, D. R., 1994. Fossils and fossil climate: The case for equable continental interiors in the Eocene. In Allen, J. R. L., Hoskins, B. J., Sellwood, B. W., Spicer, R. A. and Valdes, P. J. (eds.), *Palaeoclimates and their Modeling with Special Reference to the Mesozoic Era,* London: Chapman and Hall, p. 35–44.

Zachos, J. C., Lohmann, K. C., Walker, J. C. G. and Wise, S. W., 1993. Abrupt climate change and transient climates during the Paleogene: A marine perspective. *Journal of Geology,* v. 101, p. 191–213.

Zachos, J. C., Stott, L. D. and Lohmann, K. C., 1994. Evolution of early Cenozoic marine temperatures. *Paleoceanography,* v. 9, p. 353–387.

8

The Oxygen and Carbon Isotopic Context of the Paleocene/Eocene Epoch Boundary

RICHARD M. CORFIELD
RICHARD D. NORRIS

ABSTRACT

The Paleocene Period is characterized by the most isotopically positive $\delta^{13}C$ values in Cenozoic marine carbonates. Although the contribution of productivity to the Paleocene $\delta^{13}C$ record has been confirmed by recent elemental abundance studies, the amplitude of the $\delta^{13}C$ increase varies according to the material in which it is measured (i.e., planktonic foraminifera or bulk sediment) implying that the $\delta^{13}C$ record of surface produced-carbonate should be interpreted with caution. The difference between the bulk $\delta^{13}C$ curve and the surface dwelling planktonic $\delta^{13}C$ curve may be attributable to the presence of photosymbionts in surface dwelling planktonic foraminifera. Additionally, the two major genera of the surface dwelling planktonic foraminifera of Paleocene time (*Morozovella* and *Acarinina*) yield subtly different $\delta^{13}C$ values implying that one or both fractionated carbon isotopes out of equilibrium with ambient sea water. The $\delta^{13}C_{benthic}$ curve probably reflects the rate of organic carbon burial in the Paleocene ocean, which, although of lesser amplitude, changed broadly in synchrony with the changes in the $\delta^{13}C_{planktonic}$ and the $\delta^{13}C_{bulk}$ curves.

$\delta^{13}C$ measurements of benthic foraminifera show that, during the Late Cretaceous Period and early Paleocene Epoch, deep ocean waters were formed in the temperate North Atlantic as well as in the high southern latitudes. As Paleocene time progressed the Southern Ocean around Antarctica assumed progressively more importance as the region for the production of deep waters.

Superimposed on the broad increase and subsequent decrease of the Paleocene $\delta^{13}C$ maximum is a transient negative excursion below the Paleocene/Eocene Series boundary (calcareous nannofossil Zone NP9). This has been linked to a climatic anomaly found in coeval $\delta^{18}O$ records as well as a short term change in the prevailing mode of deep water circulation to the production of saline deep waters in the low latitudes, as well as a severe benthic foraminiferal extinction event.

$\delta^{18}O$ data show an overall decrease from the Late Cretaceous Period into the early Eocene Epoch interrupted by an increase between 64 and 57 Ma. This is interpreted as an overall warming trend with a superimposed cooling phase in the early to late Paleocene Epoch.

The Paleocene Epoch remains an enigma in the study of Cenozoic climates and oceans. It is generally considered to have been a time of global warmth and reduced latitudinal temperature gradients compared with the present day (e.g., Zachos et al. 1993) and to have had experienced comparatively elevated levels of atmospheric CO_2 that may have been at least partly responsible for the inferred high global temperatures (Berner 1990; Arthur et al. 1991; Freeman and Hayes 1992).

One of the most intriguing features of the study of Paleocene climates and oceans is the pattern of carbon isotope change following the extinctions at the Cretaceous/Paleogene Period boundary (figure 8.1). $\delta^{13}C$ values increased to between 3 and 4‰ (in bulk and whole-rock samples) by late Paleocene time and then decreased again to values similar to those that prevailed at the Cretaceous/Paleogene Period boundary (although these more negative values persisted longer than across the Cretaceous/Paleogene Period boundary). This pattern has been interpreted (Shackleton et al. 1985a; Corfield and Cartlidge 1992; Corfield 1994) as the result of an increase in global ocean productivity followed by a decline of similar magnitude.

Although this pattern of carbon isotope change is

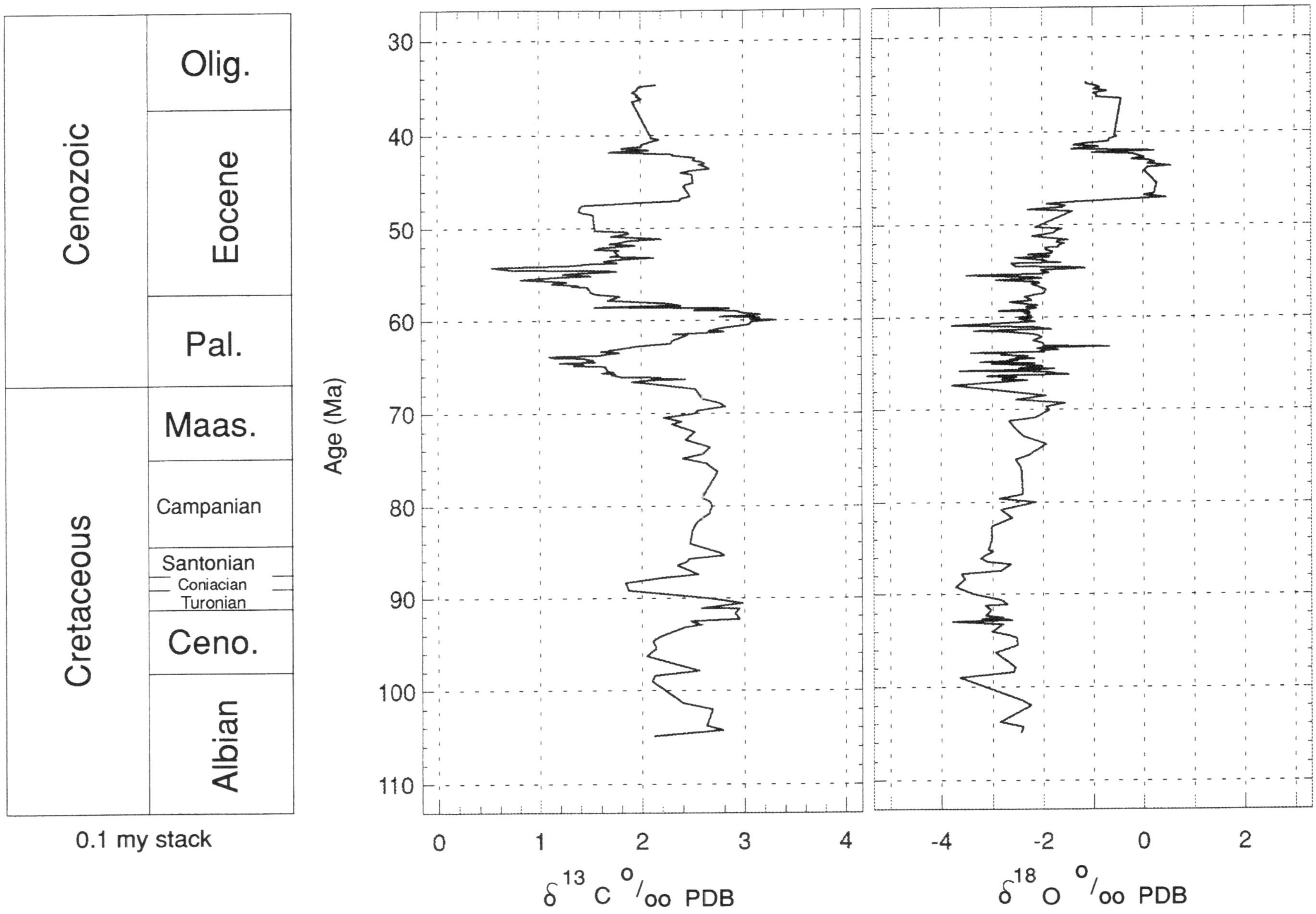

Figure 8.1 General ^{13}C curve for the Cenozoic deposits of the Bottaccione Gorge section (Umbrian Apennines) showing the magnitude of the Paleocene carbon isotope maximum. Modified from Corfield et al. (1991). Note that these data are shown as plotted originally on the time scale of Berggren et al. (1985).

found in all forms of marine carbonate routinely measured (planktonic and benthonic foraminifera as well as bulk sediment) and has been found in cores from the deep ocean (Shackleton et al. 1985a; Stott et al. 1990; Corfield and Cartlidge 1992; Zachos et al. 1992) as well as from outcrops (Corfield et al. 1991; Schmitz et al. 1996b) the absolute values of the $\delta^{13}C$ signal from different locations vary for a variety of reasons. For example, planktonic foraminifera of the genera *Morozovella* and *Acarinina* (plate 8.1) routinely exhibit heavier $\delta^{13}C$ values than deep dwelling foraminifera (*Subbotina* and *Globanomalina*; plate 8.1) because they inhabited the photic zone where maximum extraction of ^{12}C by photosynthesis into organic carbon occurred. Similarly benthic foraminifera exhibit the lightest $\delta^{13}C$ values because they were unaffected by photosynthetic effects and consequent local enrichment in ^{13}C. As such they are useful monitors of the average $\delta^{13}C$ composition of the world ocean. The differences in $\delta^{13}C$ (the carbon isotope gradient; $\Delta\delta^{13}C$) between foraminifera as well as between foraminifera and bulk carbonate values can be useful in deciphering subtleties in the oceanography and climatology of the Paleocene Epoch.

Despite a decade of research, the ultimate cause(s) of the Paleocene increase in $\delta^{13}C$ and the transient $\delta^{13}C$ excursion near the Paleocene/Eocene Epoch boundary remains largely unknown. However, recent work shows that efforts are being made to identify the forcing phenomena for both these events. These are discussed below.

Vertical Carbon Isotope Gradients in the Paleocene and Early Eocene Epochs: Productivity Versus Carbon Burial

Figure 8.2 shows the compilation of carbon isotope records of Corfield and Cartlidge (1992) calculated using the new time scale of Berggren et al. (1995). Surface, sub-thermocline and benthic foraminifera show the now familiar increase and subsequent decrease that define the Paleocene carbon isotope maximum. However, the magnitude of the long-term carbon isotope trend in benthic foraminifera is of significantly reduced amplitude when compared to the intermediate amplitude of the trend as measured in the sub-thermocline planktonic *Subbotina* and the maximum amplitude when measured using the surface dwelling planktonic *Morozovella* or *Acarinina*. Corfield and Cartlidge (1992) have interpreted the $\delta^{13}C_{benthic}$ curve as reflecting an increase in the rate of burial of organic (i.e., isotopically light) carbon during the Paleocene Epoch. The increasing gradient between the $\delta^{13}C_{Morozovella}$ and $\delta^{13}C_{benthic}$ curves was interpreted by Corfield and Cartlidge (1992) as evidence that surface waters were registering an increase in productivity (i.e., the rate of organic carbon production) during the increase in $\delta^{13}C$ values until late Paleocene time and then as $\delta^{13}C$ values decreased into the early Eocene Epoch, productivity decreased. The increase in benthic $\delta^{13}C$ values, together with differences in the amplitude of the $\delta^{13}C$ signal between the different depth groups, suggest that the Paleocene carbon isotope maximum is a composite of both an internal carbon-isotope fractionation effect as well as an external carbon-isotope fractionation effect (see Corfield and Cartlidge 1992 for more extensive discussion). Further evidence for the role of an external carbon isotope fractionation mechanism during the Paleocene Epoch is that correlative short-period $\delta^{13}C$ excursions at the Paleocene/Eocene Epoch boundary have been located in terrestrial carbonates (soil nodules and herbivore tooth enamel) from the Bighorn Basin in the continental USA and the deep sea (Zachos et al. 1991; Koch et al. 1992).

It is clear therefore that pivotal to the original hypothesis by Shackleton et al. (1985a) and Corfield and Cartlidge (1992) was the assumption that the $\delta^{13}C$ values derived from isotopic analyses of foraminifera were truly a high-fidelity reflection of Paleocene ocean and atmosphere $\delta^{13}C$ composition. However, more recent studies (D'Hondt et al. 1994; Norris 1996) have presented compelling evidence that the surface dwelling planktonic foraminifera of the Paleocene Epoch may have harbored photosymbionts in analogous manner to that found in modern surface dwelling planktonic foraminifera. Such symbiosis typically enriches the metabolic carbon pool in ^{13}C as the light isotope is incorporated into protoplasm by carbon fixation. If this inference is correct, then the clear implication is that estimates of Paleocene surface water $\delta^{13}C$ based on surface dwelling planktonic foraminifera may be too high. Norris (1996) has shown that members of the Paleocene morozovellids probably developed photosymbiosis as they evolved from the *Praemurica uncinata* plexus of morphotypes at the same time as they acquired the characteristic muricate wall texture of the group.

In order to test the effect that photosymbiosis may have had on estimates of Paleocene sea surface $\delta^{13}C$ and productivity change, we have added the bulk $\delta^{13}C$ measurements of Shackleton et al. (1985b) to the existing foraminiferal data set of Corfield and Cartlidge (1992) from DSDP Site 577 and present this comparison in figure 8.3. To reduce scatter we have stacked the data set into 0.1 million year increments of modeled time. This figure shows several intriguing features many of which can be explained by the photosymbiosis hypothesis:

1. Interval between 64.5–61.5 Ma. The isotopic records for bulk carbonate, *Subbotina* and *Morozovella* overlap. This suggests either that there was

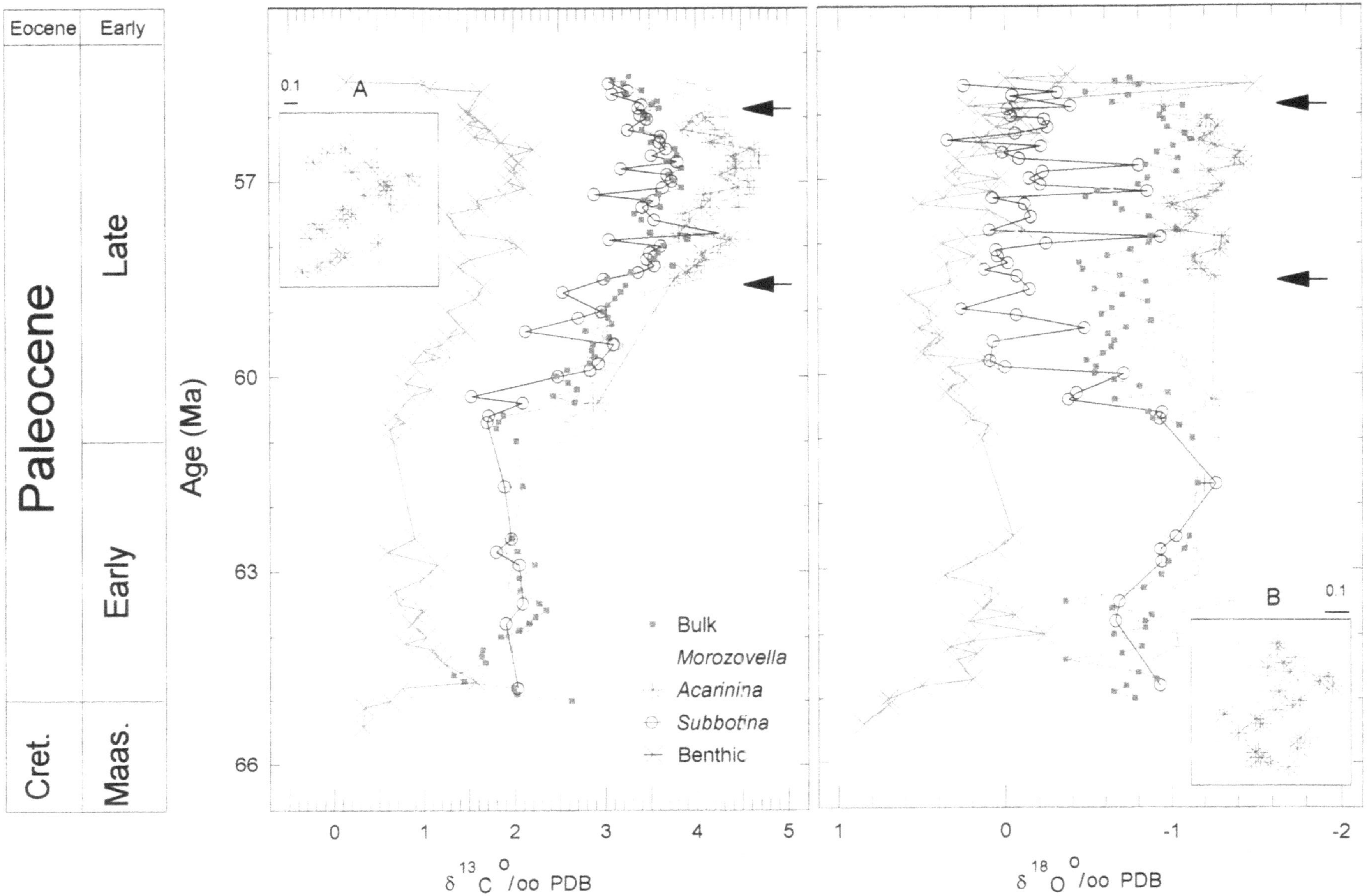

Figure 8.4 Oxygen and carbon isotope stratigraphies of bulk values, *Morozovella*, *Acarinina*, *Subbotina* and benthic foraminifera in the Paleocene sediments of DSDP Hole 577. Panels A and B highlight the non-equilibrium fractionation of *Morozovella* or *Acarinina*.
Time scale is that of Berggren et al. (1995). Modified from data in Corfield and Cartlidge (1992).

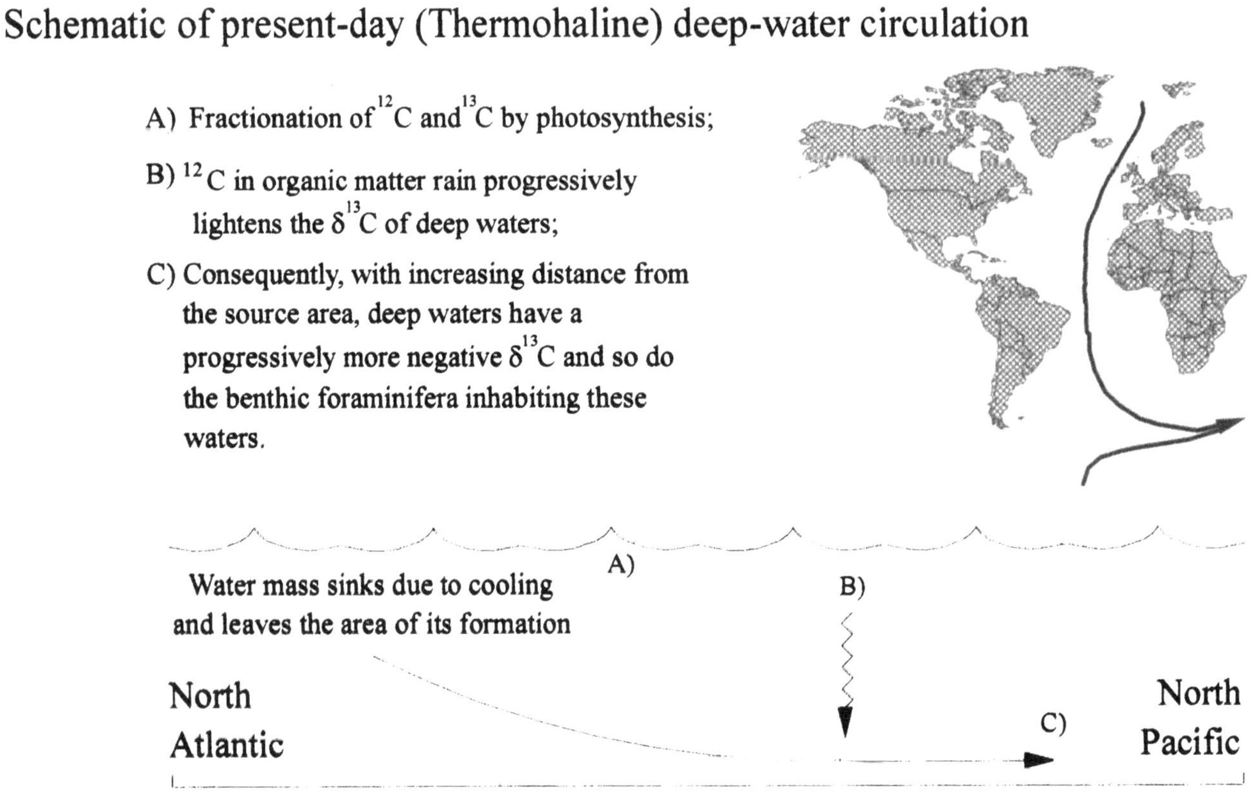

Figure 8.6 The application of benthic carbon isotope variability in tracing paleo–deep water circulation.

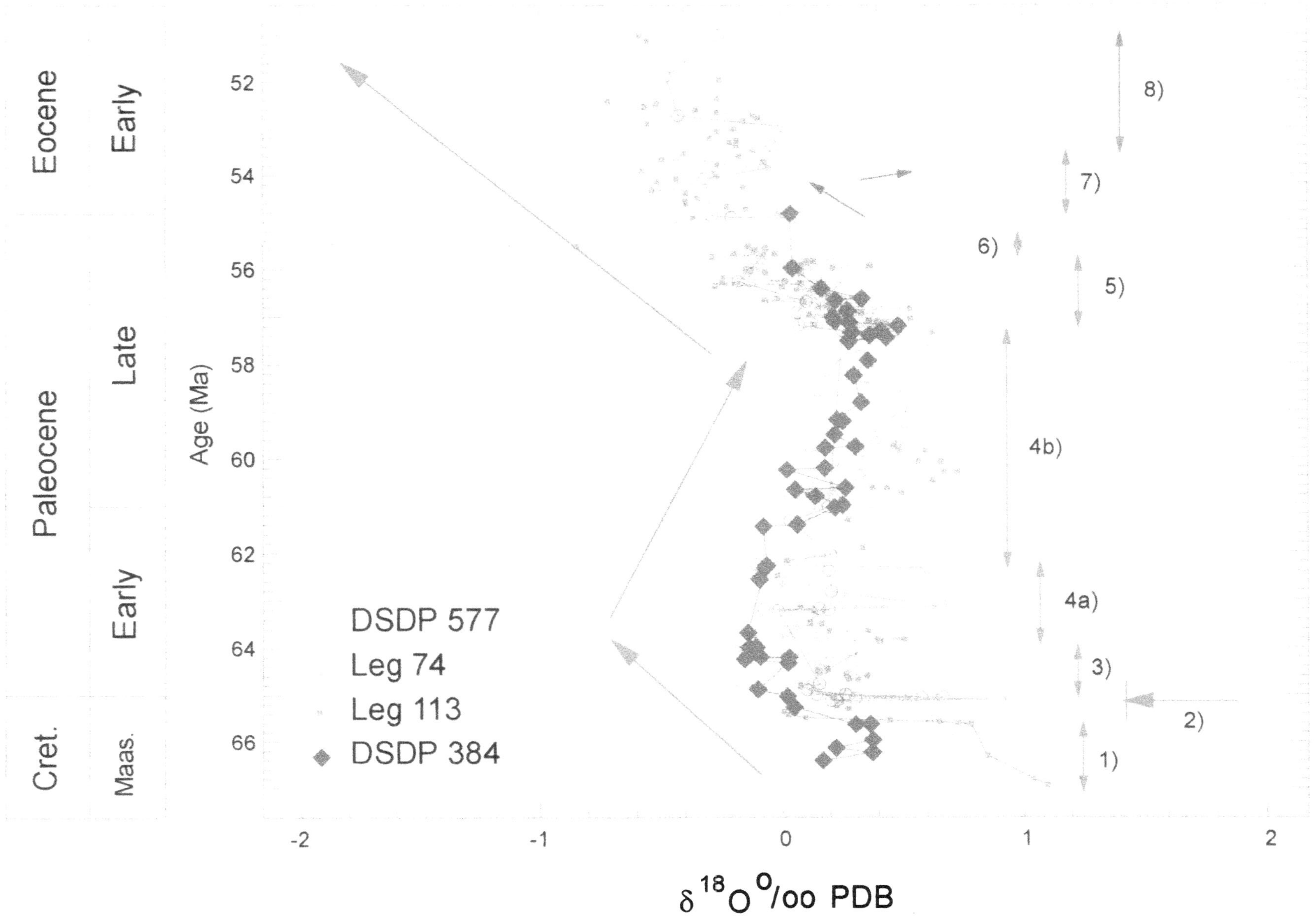

Figure 8.7 Comparative benthic $\delta^{18}O$ from several sites in the Paleocene world ocean plotted on the time scale of Berggren et al. (1995). Modified from data in Corfield and Norris (1996).

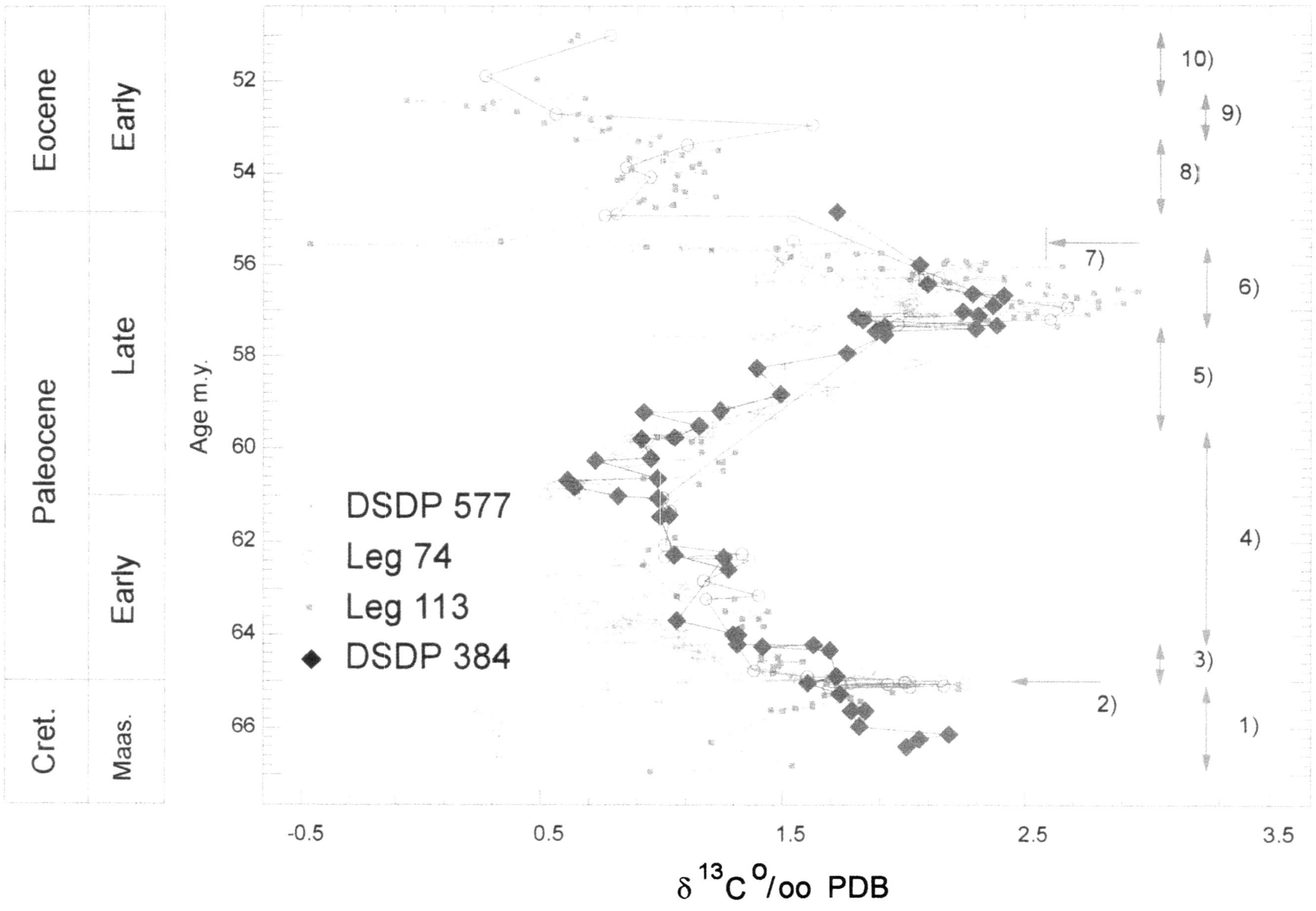

Figure 8.8 Comparative benthic $\delta^{13}C$ from several sites in the Paleocene world ocean plotted on the time scale of Berggren et al. (1995). Modified from data in Corfield and Norris (1996).

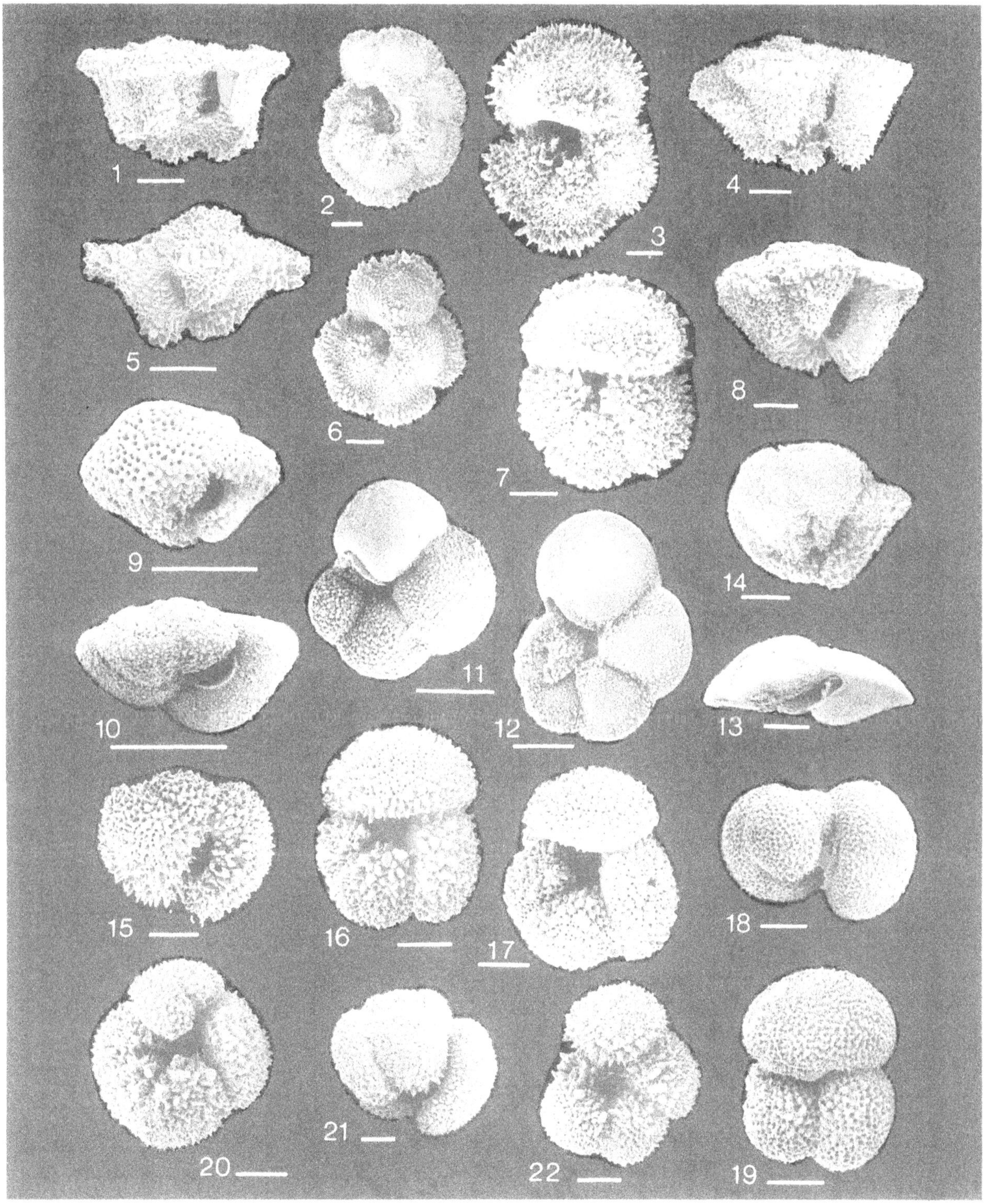

Plate 8.1 Axial and umbilical views of common late Paleocene and early Eocene planktonic foraminiferal species.

All scale bars = 100 μm. 1, 2: *Morozovella velascoensis* (late Paleocene–early Eocene species) from DSDP Hole 213-16-1, 104–106 cm. 3, 4, 8: *Morozovella subbotinae* (latest Paleocene–earliest Eocene species). 3, 4: from DSDP Hole 213-16-1, 104–106 cm; 8: from ODP Site 758A-28X-1, 50–52 cm. 5, 6: *Morozovella marginodentata* (early Eocene species) from DSDP Hole 213-15-6, 94–96 cm. 7: *Morozovella aequa* (latest Paleocene and early Eocene species) from ODP Site 758A-28X-5, 50–52 cm. 9: *Igorina tadjikistanensis* (late Paleocene species) from DSDP Site 465, 3-3, 98–100 cm. 10, 11: *Globanomalina australiformis* (high latitude marker species for the Paleocene/Eocene Series/Epoch boundary) from ODP Site 747C-2R-4, 90–92 cm. 12, 13: *Globanomalina pseudomenardii* (its stratigraphic range defines Zone P4). 12: from DSDP Site 465-4-1, 62–64 cm; 13: from ODP Site 758A-29X-1, 50–52 cm. 14: *Morozovella lensiformis* (early Eocene species) from DSDP Hole 213-15-4, 64–66 cm. 15, 16: *Acarinina coalingensis* (late Paleocene–early Eocene species) from DSDP Hole 20C-6-4, 17–19 cm. 17, 21, 22: *Acarinina soldadoensis* (late Paleocene and early Eocene species). 17: from DSDP Hole 20C-6-4, 17–19 cm; 21, 22: from DSDP Hole 213-16-1, 104–106 cm. 18, 19: *Subbotina velascoensis* (late Paleocene and early Eocene species) from DSDP Site 465-4-2, 60–62 cm. 20: *Acarinina primitiva* (latest Paleocene–early Eocene species) from DSDP Site 465-4-4, 62–64 cm.

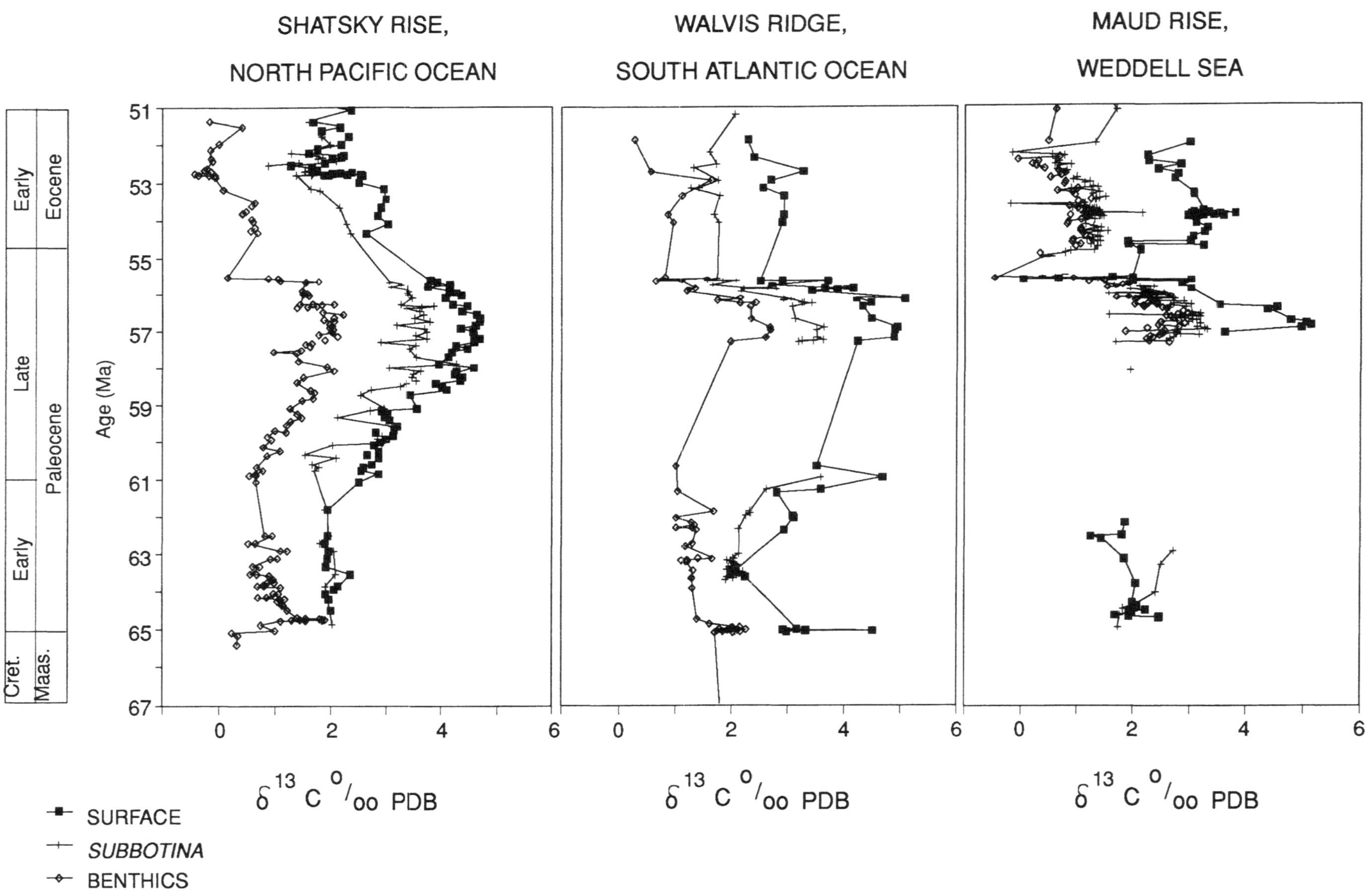

Figure 8.2 $\delta^{13}C$ curves from DSDP Site 577, the Leg 74 sites and the Leg 113 sites showing differing amplitudes of $\delta^{13}C$ increase and subsequent decrease. Note that the surface monitor at Shatsky Rise and the Walvis Ridge are species of the genus *Morozovella*, while at the Maud Rise sites, species of the genus *Acarinina* are used.

Modified from data in Corfield and Cartlidge (1992). These data are shown using the time scale of Berggren et al. (1995).

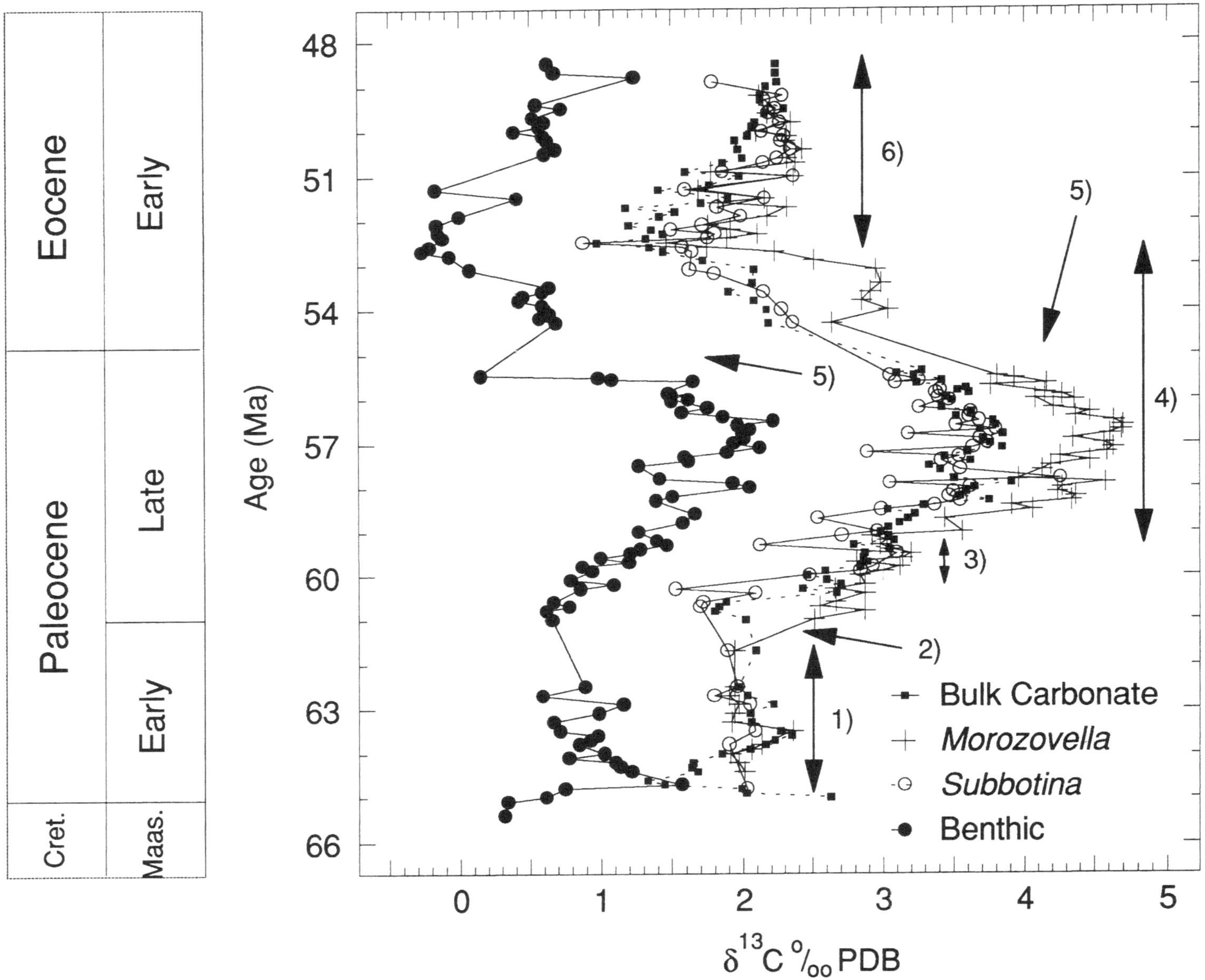

Figure 8.3 Comparative $\delta^{13}C$ signals from surface dwelling *Morozovella*, deeper dwelling *Subbotina*, benthic foraminifera and bulk carbonate from DSDP Hole 577.

Time scale is that of Berggren et al. (1995). Modified from data in Corfield and Cartlidge (1992).

a very small vertical carbon isotope gradient in the upper 500 meters of the water column, or that the early morozovellid morphotypes (*P. praecursoria*, *P. uncinata*, *P. trinidadensis*) had not yet developed photosymbionts as suggested by Norris (1996).

2. Interval between 61.5–60.5 Ma. The $\delta^{13}C$ signature of *Morozovella* became rapidly heavier than the records for bulk carbonate and *Subbotina*. This occurred simultaneously with the rapid diversification of this lineage into progressively more muricate forms, coupled with the trend toward angulo-conicality, and was probably due to the acquisition of photosymbionts by members of the genus *Morozovella*.
3. Interval between 60.2–59.1 Ma. For a short interval the $\delta^{13}C$ signature of bulk carbonate and *Subbotina* increased to the same values as those of *Morozovella*. For a 1.1-my-long interval all three curves are congruent.
4. Interval between 59.1–52.2 Ma. Maximum Paleocene $\delta^{13}C$ values were reached during this interval. *Morozovella* $\delta^{13}C$ values remained more positive than bulk carbonate and *Subbotina* values until 56.5 Ma when they began to converge toward bulk and *Subbotina* values. At the same time as maximum $\delta^{13}C$ values were achieved the vertical carbon isotope gradient as measured between bulk carbonate and benthic foraminifera was also at a maximum. Values then decreased to a minimum at 52.5 Ma. If the bulk record is interpreted as a more reliable indicator of sea surface $\delta^{13}C$ than measurements based on foraminifera, then the increasing difference between the bulk and benthic record suggests that ocean productivity was at a maximum during this interval as previously interpreted by Corfield and Cartlidge (1992). The decrease in the difference between *Morozovella* and the bulk/*Subbotina* record starting at ~56.4 Ma may reflect the decreasing vertical carbon isotope gradient noted by $\delta^{13}C_{bulk-benthic}$ comparisons but may also reflect the fact that the species used to generate the *Morozovella* measurements were changed near this horizon from those predominantly based on *M. velascoensis* to those based on *M. subbotinae*. At the present time however, it is not known whether these species may have had different non-equilibrium fractionation effects.
5. At 55.5 Ma. A pronounced negative $\delta^{13}C$ excursion occurred in nannofossil Biochron NP9 just prior to the Paleocene/Eocene Epoch boundary (as defined by the P5/P6 zonal boundary). This event is pronounced in both the deep sea and terrestrial carbon isotope records and is considered to have been of very short duration (<100,000 yr; Thomas and Shackleton 1996). It is discussed in more detail below.
6. Interval between 52.2–49 Ma. Following the decline in all monitors of ocean $\delta^{13}C$ to the minimum at 52.2 Ma in the early Eocene Epoch, $\delta^{13}C$ values partially recovered but the vertical carbon isotope gradient between *Morozovella* vs. bulk carbonate/*Subbotina* had disappeared. This may be due to the predominant use of *M. aragonensis* for this interval to monitor morozovellid $\delta^{13}C$. It is yet to be established whether this species may have had different non-equilibrium fractionation characteristics from *M. subbotinae*. The oxygen isotope data from the same interval (figure 8.4; see color insert) continue to suggest habitat stratification between *Morozovella* and *Subbotina*.

We conclude that much further work needs to be done on the carbon isotopic characteristics of the *Morozovella* lineage at the species level to establish whether $\delta^{13}C$ variations are due to the use of different clades or whether they really reflect Paleocene sea surface $\delta^{13}C$. If, as we now suspect, significant carbon isotopic differences exist between different species of the *Morozovella* clade, then all existing sea-surface $\delta^{13}C$ records will need to be re-evaluated. At the very least, it is probable that a correction will have to be applied to all measurements based on *Morozovella* to correct for non-equilibrium fractionation effects and to properly estimate Paleocene (and Eocene) sea-surface $\delta^{13}C$.

The other important muricate planktonic foraminiferal genus of the Paleocene Epoch is *Acarinina*. Figure 8.4 (see color insert) shows $\delta^{13}C$ data from both *Morozovella* and *Acarinina* in DSDP Site 577. It is clear from this figure that *Acarinina* measurements yield lower $\delta^{13}C$ values than contemporaneous *Morozovella* measurements. This is a consistent trend where measurements of both genera occur, and is generally greater than the 0.1‰ uncertainty generally associated with carbonate oxygen and carbon isotope ratio measurements. Oxygen isotope measurements from the same interval however, suggest that species of *Acarinina* are slightly *lighter* than *Morozovella* although the difference is close to the analytical uncertainty of stable isotope analyses. These subtle differences cannot be ascribed to slight differences in depth habitat because lighter $\delta^{13}C$ as in *Acarinina* implies a deeper water habitat whereas lighter $\delta^{18}O$ implies a shallower habitat. Neither can these slight differences be ascribed to the use of different size fractions because the differences are maintained for both the "D" (250–300 μm) and the "C" (300–355 μm) fraction (Corfield and Norris, unpublished data). We conclude therefore that these offsets between *Acarinina* or *Morozovella* are due to non-equilibrium fractionation, probably due to local metabolic enrichment of ^{13}C in *Morozovella*. Panels A and B on Figure 8.4 (see color insert) highlight these differences.

Bralower et al. (1995) also noted similar isotopic differences between species of *Morozovella* and *Acarinina* in the upper Paleocene and lower Eocene sediments of ODP Hole 865, and concluded that the variance was attributable to minor differences in depth habitat rather than variation in the season of growth or non-equilibrium fractionation. However, their data clearly show that, as at DSDP Hole 577, Paleocene species of *Acarinina* (*A. mckannai* and *A. soldadoensis*) are systematically lighter for *both* $\delta^{13}C$ and $\delta^{18}O$ than contemporaneous morozovellids (*M. velascoensis* and *M. subbotinae*). This apparent inconsistency in depth ranking on the basis of different isotope systems tends to imply that the differences are *not* ascribable to habitat differences unless there was some form of pan-Pacific inversion of either the vertical oxygen and carbon isotope gradient during the late Paleocene and the early Eocene epochs. They also noted that Eocene acarininids tend to show heavier $\delta^{18}O$ values than contemporaneous morozovellids (i.e., the reverse situation in the Paleocene record) a pattern that is inconsistent with Pearson et al.'s assertion (1993) that acarininids were surface dwellers during Eocene time.

The possibility that surface dwelling planktonic foraminifera may have fractionated carbon isotopes out of equilibrium with surrounding seawater raises questions about whether our estimates of vertical $\Delta\delta^{13}C$ (and therefore the role of productivity in causing the Paleocene carbon isotope maximum) are reliable. To test this, figure 8.5 shows several estimates of vertical carbon isotope gradient change during Paleocene time at DSDP Hole 577 based on estimation of surface water $\delta^{13}C$ using both *Morozovella* and bulk sediment measurements. It is clear that $\Delta\delta^{13}C$ increases with increasing surface $\delta^{13}C$ (as gauged using either *Morozovella* or bulk measurements) between any combination of surface to deep water monitors. This provides strong support for the assertion that despite the probable photosymbiosis of Paleocene muricate planktonic foraminifera, productivity did indeed increase and then decreased during Paleocene time as asserted by Shackleton et al. (1985a) and Corfield and Cartlidge (1992). The vertical carbon isotope gradient between deep dwelling planktonic and benthic foraminifera, as predicted, did not change.

An important recent test of the contribution of productivity to the Paleocene $\delta^{13}C$ signal is that of Thompson and Schmitz (1997). They measured Ba* elemental abundance in bulk sediment of Paleocene age and compared it with the $\delta^{13}C$ composition of foraminifera from the same samples. Their reasoning was based on the fact that the flux of $BaSO_4$ to the sea floor is considered to be proportional to sea-surface productivity (Revelle 1944, 1955; Goldberg and Arrhenius 1958; Dehairs et al. 1980, 1990; Dymond 1992) and hence sedimentary barium/barite has strong potential as a paleoproductivity indicator independent of $\delta^{13}C$. They found that the Ba* signal was highly correlated with the $\delta^{13}C$ signal measured in bulk sediment, *Morozovella*, *Subbotina* and benthic foraminifera. This therefore provides independent verification that productivity did indeed increase until the latest Paleocene Epoch and thereafter decreased into the early Eocene Epoch.

From the foregoing discussion we conclude that significant variability in $\delta^{13}C$ (and possibly $\delta^{18}O$) composition existed at both the genus and species level in muricate planktonic foraminifera from the Paleocene Epoch at least through the late Eocene Epoch. It is clear that, as a matter of urgency, future work needs to concentrate on unraveling details of the isotopic ranking of these genera and species to assess their suitability as paleoceanographic proxies.

Horizontal Carbon Isotope Gradients: Deep Water Circulation

Carbon isotope measurements in benthic foraminifera have been widely used to trace deep water circulation. This is because the $\delta^{13}C$ composition of benthic foraminifera reflects the isotopic composition of surrounding water at the time of calcification (assuming fractionation in equilibrium with surrounding seawater). As deep waters leave the area of their formation they become isotopically more negative due to the addition of organic material that progressively adds ^{12}C to the deep water mass. Hence, comparison of sites of similar paleodepth in different areas can be used to suggest the location of deep water sources (e.g., Curry and Lohmann 1982, 1983; Duplessy et al. 1984; Shackleton et al. 1984; Miller and Fairbanks 1985; Woodruff and Savin 1989) and the pathway taken by deep waters through the world ocean. Figure 8.6 illustrates the application of $\delta^{13}C$ benthic analysis in tracing deep water circulation.

Applications of this technique to Paleogene sediments have been focused on the suggestions (e.g., Chamberlin 1906; Brass et al. 1982; Bralower and Thierstein 1984; Kennett and Stott 1990, 1991; Zachos et al. 1993) that deep waters may have been formed by evaporative processes at low latitudes inducing salinity driven (halothermal) circulation, as opposed to the temperature driven (thermohaline) circulation characteristic of the late Neogene Period.

Until now the evidence for the sources of Paleocene

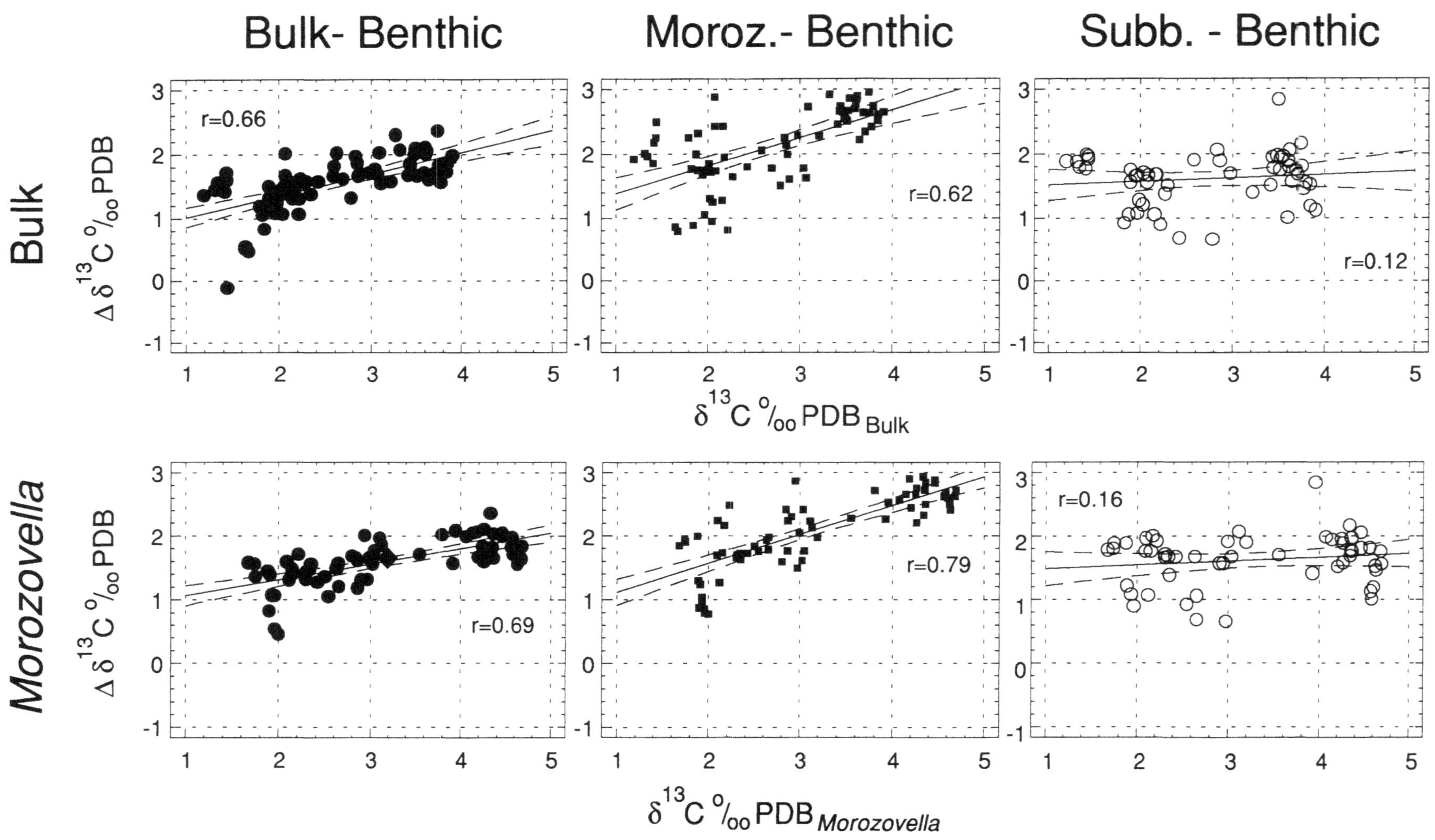

Figure 8.5 Vertical carbon isotope gradients during Paleocene time as measured between *Morozovella*–benthic foraminifera, bulk carbonate–benthic foraminifera and *Subbotina*–benthic foraminifera.

Modified from data in Corfield and Cartlidge (1992).

deep waters has been very incomplete and suggested that the most ^{13}C enriched waters were generally found in the vicinity of Antarctica (Corfield and Cartlidge 1992; Pak and Miller 1992; Thomas 1992). This in turn implied a predominantly southern ocean source for deep waters during the Paleocene Epoch. The most recent global compilation is that of Corfield and Norris (1996) who combined data from DSDP Site 384 with existing data sets from other oceans. They discovered that the temperate North Atlantic was apparently an important source of deep waters during the Late Cretaceous Period and the early Paleocene Epoch. They suggested on the basis of the available evidence that deep water production in the Late Cretaceous to early Eocene interval could be subdivided into several stages, which are summarized below.

The Late Cretaceous Period 67–65 Ma ($\delta^{13}C$ stage 1 and $\delta^{18}O$ stage 1)

The youngest (isotopically lighter) waters are identified in the western sub-tropical Atlantic suggesting that the temperate northern latitudes were important producer of deep water. The $\delta^{18}O$ record (figure 8.7; see color insert) indicates that the temperature of deep waters at DSDP Site 384 was 2° C higher than either high southern latitude or Pacific deep waters at this time (assuming uniformity of deep water salinities). Furthermore, it is likely that DSDP Site 384 experienced some ^{18}O enrichment, if this site, as suggested, was adjacent to a locus of paleo-NADW formation; consequently the estimated paleotemperatures for this site would be too low.

The Cretaceous/Paleogene Period boundary interval; $\delta^{13}C$ stage 2 and $\delta^{18}O$ stage 2)

Below the Cretaceous/Paleogene System boundary $\delta^{13}C$ values in DSDP Hole 577, the Maud Rise holes (ODP Leg 113), the Walvis Ridge holes (DSDP Leg 74) and ODP Hole 738 increase and converge. This change is seen at DSDP Site 384, which suggests that it reflects a temporary reorganization of deep water circulation. This convergence implies that the high southern latitudes became the dominant source of deep waters at this time. A coincident decrease in $\delta^{18}O$ between these sites briefly eliminated the thermal contrast between the ocean basins.

The aftermath of the Cretaceous/Paleogene Period boundary; 65–64 Ma ($\delta^{13}C$ stage 3 and $\delta^{18}O$ stage 3)

Values rapidly decrease in all sites except at DSDP Site 384, which also fails to register the $\delta^{13}C$ increase immediately below the Cretaceous/Paleogene Series boundary. Hence, DSDP Site 384 attains the heaviest lowermost Paleocene $\delta^{13}C$ values suggesting a North Atlantic deep water source. $\delta^{18}O$ values from DSDP 384 are once again more negative than all but the Indian Ocean sites, suggesting that paleo-NADW was being produced in or near the sub-tropical North Atlantic.

The early to late Paleocene Epoch; 64–57.4 Ma ($\delta^{13}C$ stages 4–5, $\delta^{18}O$ stages 4–5)

During $\delta^{13}C$ stage 4, between 64.2 and 59.5 Ma, interbasinal $\Delta\delta^{13}C$ was relatively low. This implies a comparatively weak south to north aging gradient in the Atlantic. Apparently, the Southern Ocean was a source area for deep waters and the sub-tropical North Atlantic was at least adjacent to a deep water source area (such as Tethys). $\delta^{18}O$ data indicate two stages in ocean thermal evolution during this time; stage 4a when the North Atlantic was warmer than all other sites; and stage 4b when the high southern latitudes became cooler than the equatorial Pacific. $\delta^{18}O$ stage 4b overlaps with $\delta^{13}C$ stage 5. $\delta^{13}C$ stage 5 (between 59.6 and 57.4 Ma) is a transitional interval. Although data from the Leg 113 holes are sparse, they appear to trend toward more positive values than those of DSDP Hole 384. Additionally, the Atlantic and Pacific sites became more positive than the Indian Ocean sites. In general latitudinal $\delta^{13}C$ gradients increased during stage 5 suggesting an increase in the importance of the Southern Ocean as the dominant source of deep waters and a decrease in the importance of other (i.e., the low latitude) source areas.

The prelude to the end of the Paleocene Epoch; 57.4–55.4 Ma ($\delta^{13}C$ stage 6 and $\delta^{18}O$ stage 5)

The most positive $\delta^{13}C$ values occur in the high southern latitude Leg 113 holes. The south Atlantic shows only slightly more negative values implying that by the latest Paleocene Epoch the predominant source of deep water production had become the region around Antarctica. North Atlantic values are intermediate between Southern Ocean/South Atlantic and Pacific $\delta^{13}C$ ratios, which is consistent with an area receiving waters generated elsewhere via a smaller ocean basin with less aging potential than the Indo-Pacific and not producing deep waters of its own. $\delta^{18}O$ stage 5 marks the beginning of the ^{18}O depletion characteristic of the late Paleocene and early Eocene epochs (e.g., Miller et al. 1987; Corfield and Cartlidge 1992). Values from the Leg 113 holes converged with those of other sites for the same interval. Hence, it appears that all ocean basins were filled with waters of apparently similar temperature supporting the idea of a single deep water source during this interval. Superimposed on the (Pacific) DSDP Hole

577 trend were at least two short negative excursions in $\delta^{13}C$ and $\delta^{18}O$ at 57.6 and 56.4 Ma; these may represent brief periods when warm, nutrient-rich waters entered the Pacific perhaps through renewed deep water production in low-latitude basins.

The Paleocene/Eocene Epoch boundary interval; 55.4–55 Ma ($\delta^{13}C$ stage 7 and $\delta^{18}O$ stage 6)

This event has been associated with a temporary change from thermohaline to halothermal mode of deep water circulation, which caused profound benthic foraminiferal extinction in the marine realm (Thomas 1991, 1992; Thomas and Shackleton 1996). It may also have been coupled with a transient climate change (Stott 1992; Zachos et al. 1993; Stott et al. 1996). Figure 8.8 (see color insert) shows that the Leg 113 holes register negative $\delta^{13}C$ excursions superimposed on the long term late Paleocene–early Eocene decline. This was coeval with a negative excursion in the $\delta^{18}O$ record shown in figure 8.6 (see color insert). The Paleocene/Eocene boundary interval is discussed in more detail below.

The aftermath of the Paleocene/Eocene Epoch boundary; 54.8–53.4 Ma ($\delta^{13}C$ stage 8 and $\delta^{18}O$ stage 7)

Following the negative P/E boundary interval, $\delta^{13}C$ values in the Leg 113 holes increase and are again similar to values from the Leg 74 holes as in $\delta^{13}C$ stage 6. A steep latitudinal $\delta^{13}C$ gradient once again existed between Atlantic and Indo-Pacific sites suggesting a predominantly Antarctic source for deep waters. However, our record from DSDP Site 384 does not extend into these relatively young horizons, preventing any further deductions concerning Tethyan deep water production. During this stage $\delta^{13}C$ values temporarily plateaued from the overall long term late Paleocene to early Eocene decline. The $\delta^{18}O$ record suggests a temperature increase followed at the end of $\delta^{18}O$ stage 7 by a transient temperature decline (the trajectories of $\delta^{18}O$ change are marked with arrows on figure 8.7; see color insert). The temperature increase phase of $\delta^{18}O$ stage 7 is characterized by shallow latitudinal $\delta^{18}O$ gradients suggesting homogenous ocean temperatures, consistent with the idea of a single deep water source. The temperature decline at the end of $\delta^{18}O$ stage 7 marked the beginning of an increase in latitudinal $\delta^{18}O$ gradients with DSDP 577 values becoming progressively more negative than Leg 113 values.

The earliest Eocene Epoch; Part 1; 53.4–51 Ma ($\delta^{13}C$ stage 9 and $\delta^{18}O$ stage 8)

The late Paleocene to early Eocene $\delta^{13}C$ decline resumed. Although the latitudinal $\delta^{13}C$ gradient remained the same, all three localities areas show a pronounced negative trend suggesting that the Southern Ocean was still the major producer of deep waters. $\delta^{18}O$ values in DSDP Hole 577 continued to diverge from those of the Leg 113 holes.

The earliest Eocene Epoch; Part 2; 53.4–51 Ma ($\delta^{13}C$ stage 10 and $\delta^{18}O$ stage 8)

$\delta^{13}C$ stage 10 was the start of the post late Paleocene–early Eocene $\delta^{13}C$ recovery. $\delta^{18}O$ values in DSDP 577 continued to decline, whereas those in the Leg 113 and Leg 74 holes did not.

Charisi and Schmitz (1996) have noted that, during the early Eocene Epoch, the eastern North Atlantic (DSDP Holes 401 and 550) yielded relatively negative $\delta^{13}C$ and $\delta^{18}O$ values. They interpret this to indicate that this region was dominated by a single source of deep waters probably produced in the low latitudes. Since the data-set from DSDP Site 384 does not extend into this interval we cannot compare the eastern and western North Atlantic, but it is possible that this region was once again a net producer of deep waters as it was in the Late Cretaceous Period and early Paleocene Epoch. However, on the basis of existing $\delta^{13}C$ comparisons, it is unlikely that this region was as important an area of deep water production as the high southern latitudes by early Eocene time. Detailed $\delta^{13}C$ comparisons from several additional sites in the Atlantic region will be needed to verify and assess the importance of paleo-NADW production during the early Eocene Epoch.

Schmitz et al. (1996a) have investigated the deep water circulation of the North Sea during the early Eocene Epoch. They found that the North Sea was episodically closed and probably dysaerobic but that this condition alternated with intervals of more open exchange with the North Atlantic depending on regional sea level variations and the height of sub-surface sills. Planktonic foraminifera $\delta^{18}O$ indicate episodic significant input of fresh water. Taken together, their findings suggest that the early Eocene North Sea was fundamentally a semi-isolated arm of the North Atlantic.

Schmitz et al. (1996b) have examined sections in Egypt to test whether this region of Tethys could have produced Warm Saline Deep Waters (WSDW) during the latest Paleocene Epoch. Their $\delta^{13}C$ comparisons with the high southern latitude ODP Hole 690B show that the southern Tethyan Shelf area did not produce WSDW during this interval. They speculated that WSDW may have been produced elsewhere in the Tethyan Seaway. An alternative explanation is that, at times during the Late Cretaceous Period to early Eocene Epoch, the sub-tropical North Atlantic was not merely proximal to a Tethyan deep water source but was itself the source area (Corfield and Norris 1996). It is probable that deep water circulation in the Paleogene Ocean resulted from a balance between production in the Southern Ocean and the low latitude North Atlantic. The flux of deep waters from these two source areas may have varied during the early Paleogene Period. Hence the late Paleocene "event" needs only reflect a reduction in the importance of Southern Ocean production, which allowed the North Atlantic source region to temporarily dominate global deep waters, warming them and thereby causing the extinction of many benthic foraminiferal taxa.

Latest Paleocene Extinction and Isotopic Events

As the foregoing discussion has shown the late Paleocene world was evolving towards a predominantly thermohaline mode of deep water production in the region around Antarctica; also, surface waters were undergoing a marked decline in productivity, which may have been associated with an increase in atmospheric CO_2 and concomitant warming of the high latitudes, which resulted in a warming of deeper waters. The Paleocene/Eocene Epoch boundary interval was characterized by a warming event of about 4° C in the deep ocean (with maximum temperatures in excess of 15° C), surface waters of the high latitudes warmed as well by 5–6° C reaching peak temperatures in excess of 20° C.

Although much has been written recently about the sequence of events that characterize the Paleocene/Eocene Epoch boundary interval (e.g., numerous references in Knox et al. 1996), the formal identification of a "Global Stratigraphic Section and Point" (GSSP) has yet to occur (Berggren and Aubry 1996 and chapter 2). It is clear however that this horizon (where by definition rock and time are congruent) will be placed at a level bracketed by the base of the Ieper Clay (at ~54.37 Ma, Aubry 1996) and the pronounced, short-term $\delta^{13}C$ late Paleocene excursion, at ~55.5 Ma, first identified by Kennett and Stott (1991) and now recognized widely in deep sea sediments and on land. In terms of microfossil stratigraphy the horizon will be within calcareous nannofossil Biozone NP9 and mid planktonic foraminiferal Biozone P5 (Berggren and Aubry 1996).

The cause of the transient ocean/climate anomaly at (or near) the Paleocene/Eocene Epoch boundary is still unclear. Essentially two categories of theory can be invoked to explain it: 1) forcing by an external medium; or 2) forcing by some inherent feature of the system. In the former case there is no compelling evidence for phenomena such as the bolide impact implicated in the cause of Cretaceous/Paleogene Period boundary extinctions. In the latter case, Zachos et al. (1993) have speculated that this "Late Paleocene Thermal Maximum" may have been caused by some form of feedback effect inherent to changing ocean/climate systems. Crowley and North (1988) have noted that gradually changing climatic boundary conditions can lead to abrupt responses when threshold levels with more than one equilibrium state are reached. Clearly such systems are chaotic and not predictable by existing climate models. However, there is abundant evidence to suggest that the late Paleocene to early Eocene Epoch was a time of changing boundary conditions and this could indeed be the cause of the transient anomaly at (or near) the Paleocene/Eocene Epoch boundary.

Thomas and Shackleton (1996) have speculated that the cause was rooted in a complex combination of volcanogenic emission of CO_2, changing oceanic circulation patterns and surface water productivity, and changes in the relative size of the oceanic and atmospheric carbon reservoirs. A plausible single-cause explanation for the transient $\delta^{13}C$ component of the latest Paleocene event has been proposed by Dickens et al. (1995). These authors suggested that dissociation of sedimentary methane hydrate was caused by the increase in bottom water temperatures at this time and that this would have been capable of decreasing the $\delta^{13}C$ of oceanic CO_3– observed in foraminiferal carbonates by 2–3‰. They further suggested that the temperature "trigger" necessary for this dissociation might have come from volcanic exhalation and subsequent Greenhouse warming as suggested by Thomas and Shackleton (1996). One aspect of the Paleocene–Eocene transition that does seem to enjoy general agreement, however, is the suggestion that the negative $\delta^{13}C$ excursion in benthic foraminifera indicates a temporary change from deep water production in the high latitudes to production in the low latitudes (Corfield and Cartlidge 1992; Corfield and Norris 1996; Thomas and Shackleton 1996). Such a radical change in deep water temperatures was the most likely cause of the benthic foraminiferal extinctions in Biochron NP9. A recent suggestion (Aubry et al. 1996) is that there may have been more than one $\delta^{13}C$ excursion during the latest Paleocene Epoch. It is inferred on the basis of detailed analysis of upper Paleocene and lower Eocene sections that, in addition to the documented carbon isotope excursion in mid-Biochron NP9, there was an as yet undocumented additional excursion in early Biochron NP10. It is clear that despite much theorizing, the ultimate causes of the paleoceanographic change associated with the Paleocene carbon isotope maximum as well as the events around the Paleocene/Eocene Epoch boundary remain unknown.

ACKNOWLEDGMENTS

We thank Julie Cartlidge for generating many of the measurements on which this review is based. We also thank Marie-Pierre Aubry and Bill Berggren for inviting us to contribute to this volume and for their reviews of an earlier version of this manuscript. We are particularly grateful for the robust criticisms of Birger Schmitz, which materially improved the final product. We dedicate this contribution to the memory of the late Professor Robert Kidd, whose contributions to the Deep Sea Drilling Project and the Ocean Drilling Program helped foster the study of past climatic and oceanographic change.

REFERENCES

Arthur, M. A., Hinga, K. R., Pilson, M. E. Q., Whitaker, E. and Allard, D., 1991. Estimates of pCO_2 for the last 120 m.y. based on the $\delta^{13}C$ of marine phytoplankton organic matter. *Eos*, p. 72, 166.

Aubry, M.-P., Berggren, W. A., Stott, L. and Sinha, A., 1996. The upper Paleocene–lower Eocene stratigraphic record and the Paleocene–Eocene boundary carbon isotope excursion: Implications for geochronology. In Knox, R. W. O'B., Corfield, R. M. and Dunay, R. E. (eds.), *Correlation of the early Paleogene in Northwest Europe*, Geological Society Special Publication No 101, p. 353–380.

Berggren, W. A., Kent, D. V. and Flynn, J. J., 1985. Paleogene geochronology and chronostratigraphy. In Snelling, N. J. (ed.), *The chronology of the Geological Record:* Geological Society Memoir No 10, Blackwell, Oxford, p. 141–195.

Berggren W. A., Kent, D. V., Swisher, C. C., III and Aubry, M.-P., 1995. A revised Cenozoic geochronology and chronostratigraphy. In Berggren, W. A., Kent, D. V., Aubry, M.-P. and Hardenbol, J. (eds.), *Geochronology Time Scales and Global Stratigraphic Correlation*, Society of Economic Paleontologists and Mineralogists Special Publication No. 54, p. 129–212.

Berggren, W. A. and Aubry, M.-P., 1996. A late Paleocene–early Eocene NW European and North Sea magnetobiochronological correlation network. In Knox, R. W. O'B., Corfield, R. M. and Dunay, R. E. (eds.), *Correlation of the Early Paleogene in Northwest Europe*, Geological Society Special Publication No. 101, p. 309–352.

Berner, R. A., 1990. Atmospheric carbon dioxide levels over Phanerozoic time. *Science*, v. 249, p. 1382–1386.

Bralower, T. J. and Thierstein, H. R., 1984. Low productivity and slow deep-water circulation in mid-Cretaceous oceans. *Geology*, v. 12, p. 614–618.

Bralower, T. J., Zachos, J. C., Thomas, E., Parrow, M., Paull, C. K., Kelley, D. C., Premoli Silva, I., Sliter, W. V. and Lohmann, K. C., 1995. Late Paleocene to Eocene paleoceanography of the equatorial Pacific Ocean: Stable isotopes recorded at Ocean Drilling Program Site 865, Allison Guyot. *Paleoceanography*, v. 10, p. 841–865.

Brass, G. W., Southam, J. R. and Peterson, W. H., 1982. Warm saline bottom water in the ancient ocean. *Nature*, v. 296, p. 620–623.

Chamberlin, T. C., 1906. On a possible reversal of deep-sea circulation and its influence on geologic climates. *Journal of Geology*, v. 14, p. 363–373.

Charisi, S. D. and Schmitz, B., 1996. Early Eocene palaeoceanography and palaeoclimatology of the eastern North Atlantic: Stable isotope results for DSDP Hole 550. In Knox, R. W. O'B., Corfield, R. M. and Dunay, R. E. (eds.), *Correlation of the Early Paleogene in Northwest Europe*, Geological Society Special Publication No 101, p. 457–472.

Corfield, R. M., Cartlidge, J. E., Premoli Silva I. and Housley, R. A., 1991. Oxygen and carbon isotope stratigraphy of the Palaeogene and Cretaceous limestones in the Bottaccione Gorge and the Contessa Highway sections, Umbria, Italy. *Terra Nova*, v. 3, p. 414–422.

Corfield, R. M., 1994. Palaeocene oceans and climate: An isotopic perspective. *Earth Science Reviews*, v. 37, p. 225–252.

Corfield, R. M. and Cartlidge, J. E., 1992. Oceanographic and climatic implications of the Palaeocene carbon isotope maximum. *Terra Nova*, v. 4, p. 443–455.

Corfield, R. M. and Norris, R. D., 1996. Deep water circulation in the Paleocene Ocean. In Knox, R. W. O'B., Corfield, R. M. and Dunay, R. E. (eds.), *Correlation of the Early Paleogene in Northwest Europe*, Geological Society Special Publication No 101, p. 443–456.

Crowley, T. J. and North, G. R., 1991. *Paleoclimatology. Oxford Monographs in Geology and Geophysics*, Oxford: Oxford University Press, v. 16, 339 pp.

Curry, W. B. and Lohmann, G. P., 1982. Carbon isotopic changes in benthic foraminifera from the western South Atlantic: Reconstruction of glacial abyssal circulation patterns. *Quaternary Research*, v. 18, p. 218–235.

Curry, W. B. and Lohmann, G. P., 1983. Reduced advection into Atlantic Ocean deep eastern basins during last glaciation maximum. *Nature*, v. 306, p. 577–580.

D'Hondt, S., Zachos, J. C. and Schultz, G., 1994. Stable isotopic signals and photosymbiosis in the late Paleocene planktic foraminifera. *Paleobiology*, v. 20, p. 391–406.

Dehairs, F., Chesselet, R. and Jedwab, J., 1980. Discrete suspended particles of barite and the barium cycle in the open ocean. *Earth and Planetary Science Letters*, v. 49, p. 528–550.

Dehairs, F., Goeyens, L., Stroobants, N., Bernard, P., Goyet, C., Poisson, A. and Chesselet, R., 1990. On suspended barite and the oxygen minimum in the Southern Ocean. *Global Biogeochemical Cycles*, 4, p. 85–102.

Dickens, G. R., O'Neil, J. R., Rea, D. K., Owen, R. M., 1995. Dissociation of oceanic methane hydrate as a cause of the carbon isotope excursion at the end of the Paleocene. *Paleoceanography*, v. 10, p. 965–971.

Duplessy, J-C., Shackleton, N. J., Matthews, R. K., Prell, W., Ruddiman, W. F., Caralp, M. and Hendy, C. H., 1984. ^{13}C record of benthic foraminifera in the last interglacial ocean: Implication for the carbon cycle and the global deep water circulation. *Quaternary Research*, v. 21, p. 225–243.

Dymond, J., 1992. Geochemistry of Nazca Plate surface sediments: An evaluation of hydrothermal, biogenic, detrital and hydrothermal sources. In Kulm, L. D., et al (eds.), *Nazca Plate: Crustal formation and Andean Convergence*, Geological Society of America Memoir, 1981, p. 133–174.

Freeman, K. H. and Hayes, J. M., 1992. Fractionation of carbon isotopes by phytoplankton and estimates of ancient CO_2 levels. *Global Biogeochemical Cycles*, v. 6, p. 185–198.

Goldberg, E. D. and Arrhenius, G. O. S., 1958. Chemistry of Pacific pelagic sediments. *Geochimica et Cosmochimica Acta*, 13, p. 153–212.

Kennett, J. P. and L. D. Stott, 1990. Proteus and Proto-Oceanus: Ancestral Paleogene oceans as revealed from Antarctic stable isotope results. In Barker, P. F., Kennett, J. P., et al., Proceedings of the Ocean Drilling Program, Scientific Results, v. 113: College Station, TX (Ocean Drilling Program), p. 865–880.

Kennett, J. P. and Stott, L. D., 1991. Abrupt deep-sea warming, palaeoceanographic changes and benthic extinctions at the end of the Palaeocene. *Nature*, v. 353, p. 225–229.

Knox, R. W. O'B., Corfield, R. M. and Dunay, R. E., 1996. *Correlation of the Early Paleogene in Northwest Europe*, Geological Society Special Publication No 101, 480 pp.

Koch, P. L., Zachos, J. C. and Gingerich, P. D., 1992. Correlation between isotope records in marine and continental carbon reservoirs near the Palaeocene/Eocene boundary. *Nature*, v. 358, p. 319–322.

Miller, K. G. and Fairbanks, R. G., 1985. Oligocene to Miocene benthic foraminifera and abyssal circulation changes. In Sundquist, E. T.

and Broecker, W. S. (eds.), *The Carbon Cycle and Atmospheric CO_2: Natural Variations Archean to Present.* Geophysical Monograph 32, Washington, D.C.: American Geophysical Union, p. 469–486.

Miller, K. G., Janecek, T. R., Katz, M. E. and Keil, D. J., 1987. Abyssal circulation and benthic foraminiferal changes near the Paleocene/Eocene boundary. *Paleoceanography,* v. 2, p. 741–761.

Norris, R. D., 1996. Symbiosis as an evolutionary innovation in the radiation of Paleocene planktonic foraminifera. *Paleobiology,* v. 22, p. 461–480.

Pak, D. K. and Miller, K. G., 1992. Paleocene to Eocene benthic foraminiferal isotopes and assemblages: Implications for deepwater circulation. *Paleoceanography,* v. 7, p. 405–422.

Pearson, P. N., Shackleton, N. J. and Hall, M. A., 1993. Stable isotope paleoecology of middle Eocene planktonic foraminifera and multispecies isotope stratigraphy, DSDP Site 523, South Atlantic. *Journal of Foraminiferal Research,* v. 23, p. 123–140.

Revelle, R., 1944. *Marine bottom samples collected in the Pacific by the Carnegie on its seventh cruise.* Carnegie Institute of Washington Special Publication No. 556, 180 pp.

Revelle, R., Bramlette, M., Arrhenius, G. and Goldberg, E. D., 1955. *Pelagic sediments of the Pacific.* Geological Society of America Special Paper No. 62, p. 221–235.

Schmitz, B., Heilmann-Clausen, C., King, C., Steurbaut, E., Andreasson, F. P., Corfield, R. M. and Cartlidge, J. E., 1996a. Stable isotope and biotic evolution in the North Sea during the early Eocene: The Albaek Hoved section, Denmark. In Knox, R. W. O'B., Corfield, R. M. and Dunay, R. E. (eds.), *Correlation of the Early Paleogene of North West Europe,* Geological Society Special Publication No. 101, p. 275–306.

Schmitz, B., Speijer, R. P. and Aubry, M.-P., 1996b. Latest Paleocene benthic extinction event on the southern Tethyan shelf (Egypt): Foraminiferal stable isotopic ($\delta^{13}C$, $\delta^{18}O$) records. *Geology,* v. 24, p. 347–350.

Shackleton, N. J., Imbrie, J. and Hall, M. A., 1984. Oxygen and carbon isotope record of East Pacific core V19–30: Implications for the formation of deep water in the late Pleistocene North Atlantic. *Earth and Planetary Science Letters,* v. 65, p. 233–244.

Shackleton, N. J., Corfield, R. M. and Hall, M. A., 1985a. Stable isotope data and the ontogeny of Paleocene planktonic foraminifera. *Journal of Foraminiferal Research,* v. 15, p. 321–336.

Shackleton, N. J., Hall, M. A. and Bleil, U., 1985b. Carbon-isotope stratigraphy, Site 577. In Heath, G. R., Burckle, L. H., et al., Initial Reports of the Deep Sea Drilling Project, v. 86: Washington (U.S. Government Printing Office), p. 503–511.

Stott, L. D., 1992. Higher temperatures and lower oceanic pCO_2: A climate enigma at the end of the Paleocene Epoch. *Paleoceanography,* v. 7, p. 395–404.

Stott, L. D., Kennett, J. P., Shackleton, N. J. and Corfield, R. C., 1990. The evolution of Antarctic surface waters during the Paleogene: Inferences from the stable isotopic composition of planktonic foraminifers, ODP Leg 113. In Barker, P. F., Kennett, J. P., et al., Proceedings of the Ocean Drilling Program, Scientific Results, v. 113, p. 849–863.

Stott, L. D., Sinha, A., Thiry, M., Aubry, M.-P. and Berggren, W. A., 1996. Global $\delta^{13}C$ changes across the Paleocene/Eocene boundary: Criteria for terrestrial–marine correlations. In Knox, R. W. O'B., Corfield, R. M. and Dunay, R. E. (eds.), *Correlation of the Early Paleogene in Northwest Europe,* Geological Society Special Publication No. 101, p. 381–399.

Thomas, E., 1991. The latest Paleocene mass extinction of deep sea benthic foraminifera: Result of global climate change. *Geological Society of America, Abstracts with Programs,* v. 23, p. A141.

Thomas, E., 1992. Middle Eocene–Late Oligocene bathyal benthic foraminifera (Weddell Sea): faunal changes and implications for ocean circulation. In Prothero, D. A. and Berggren, W. A. (eds.), *Eocene–Oligocene Climatic and Biotic Evolution,* Princeton University Press, Princeton, p. 245–271.

Thomas, E. and Shackleton, N. J., 1996. The Paleocene–Eocene benthic foraminiferal extinction and stable isotope anomalies. In Knox, R. W. O'B., Corfield, R. M. and Dunay, R. E. (eds.), *Correlation of the Early Paleogene in Northwest Europe,* Geological Society Special Publication No 101, p. 401–441.

Thompson, E. and Schmitz, B., 1997. Barium and the late Paleocene $\delta^{13}C$ maximum: Evidence of increased marine surface productivity. *Paleoceanography,* v. 12, p. 239–254.

Woodruff, F. and Savin S. M., 1989. Miocene deepwater oceanography. *Paleoceanography,* v. 4, p. 87–140.

Zachos, J. C., Gingerich, P. D. and Koch, P. L., 1991. Marine and terrestrial carbon isotope records across the Paleocene/Eocene boundary: Application to reconstructing paleoclimates. *Geological Society of America, Abstracts with Programs,* v. 23, p. A-142.

Zachos, J. C., Rea, D. K., Seto, K., Nomura, R. and Niitsuma, N.,1992. Paleogene and early Neogene deep water paleoceanography of the Indian Ocean as determined from benthic foraminifer stable carbon and oxygen isotope records. In Duncan, R. A., Rea, D. K., Kidd, R. B., von Rad, U. and Weissel, J. K. (eds.), *Synthesis of Results from Scientific Drilling in the Indian Ocean,* Geophysical Monograph 70, Washington, D.C.: American Geophysical Union, p. 351–385.

Zachos, J. C., Lohmann, K. C., Walker, J. C. G. and Wise, S. W., 1993. Abrupt climate change and transient climates during the Paleogene: A marine perspective. *Journal of Geology,* v. 101, p. 191–213.

9

Global Climate of the Late Paleocene Epoch: Modeling the Circumstances Associated with a Climatic "Event"

LISA CIRBUS SLOAN
ELLEN THOMAS

ABSTRACT

In the latest Paleocene Epoch the world warmed during an unusual and rapid event that lasted for a short period of time, 10^4–10^5 years. According to various types of evidence, this Late Paleocene Thermal Maximum (LPTM) was characterized by extreme high latitude warming of surface ocean waters, global warming of deep ocean waters, a strong negative excursion of carbon isotopic records in marine and terrestrial environments, increased rates of species turnover and species richness in land mammals and in oceanic surface-dwellers such as planktonic foraminifera and coccolithophorids, and widespread deep sea benthic foraminiferal extinctions. All of this occurred against the background of a longer interval of time during which the global climate was warmer than it is today. Estimates of LPTM sea surface temperatures (SSTs) for two seasonal extremes were used to drive climate model experiments with the National Center for Atmospheric Research GENESIS climate model. The goal of the modeling study was to produce climatic conditions that might have occurred in equilibrium with the ocean surface temperatures that have been reconstructed for the LPTM. The LPTM model results were compared to results from the same model representing a "background" climate. Definition of more complete global climatic conditions for the LPTM may provide a basis for understanding how and why the LPTM came about. Major conclusions of the comparison of model results for the LPTM and the "background climate" are the following: (1) global mean surface temperature for the LPTM case was more than 2° C warmer than temperatures for the "background climate" case; (2) for LPTM model results, continental warming occurred primarily polewards of 40° latitude, and winter warming of land masses generally exceeded summer warming magnitude; (3) the LPTM climate produced by the model had a more even latitudinal distribution of precipitation and had more precipitation occurring over land surfaces than for the "background climate" case; (4) LPTM wind velocities exhibited substantial decreases relative to the "background climate." This decrease included an expanded equatorial doldrum zone during the LPTM; (5) precipitation-minus-evaporation patterns over the oceans, which influence surface ocean salinity distributions, responded to the imposed LPTM SSTs, but the model results are not unequivocally in support of the hypothesis that formation of deep and intermediate ocean waters switched to subtropical latitudes during the latest Paleocene Epoch. In fact, model results imply that the "background climate" may have led to conditions conducive to tropical and subtropical deep water formation, and the surface salinity patterns that occur with the subsequent LPTM SSTs would not sustain such deep water formation at mid latitudes; (6) the absence of polar ice caps during the late Paleocene and early Eocene epochs makes the pronounced warming at high latitudes during the short-lived LPTM extremely difficult to explain. Had it existed at the time, polar ice would have provided a mechanism of warming via ice–albedo feedbacks and melt-back of sea ice, but the absence of this mechanism to amplify high latitude warming leaves few other likely mechanisms to explain the warming. Increased atmospheric greenhouse gas concentrations would not preferentially warm the high latitudes. Other possible mechanisms include radiatively warming clouds, increased atmospheric transport of heat to high latitudes, and increased oceanic heat transport contributions to high latitudes.

The early Cenozoic Era was characterized by warm, generally ice-free conditions (e.g., Wise et al. 1992; Zachos et al. 1993) and low vertical and latitudinal temperature gradients in the oceans (e.g., Shackleton 1986). The late Paleocene and early Eocene epochs were the warmest epochs of the Cenozoic Era (e.g., Savin 1977; Frakes 1979; Shackleton 1986; Miller et al. 1987a; Adams et al. 1990; McGowran 1991; Zachos et al. 1993, 1994; Wolfe 1994). During the late Paleocene and early Eocene epochs, warm-water pelagic organisms penetrated into the high latitudes of the world's oceans (e.g., Haq et al. 1977; Boersma et al. 1987; Stott and Kennett 1990; McGowran 1991; Aubry 1992). On land at this time, crocodiles, turtles and other thermophilic biota lived at high northern latitudes (e.g., Estes and Hutchison 1980; McKenna 1980; Markwick 1994). Vegetation and soil types in these regions likewise indicated mild temperatures, even for the winter seasons (Kemp 1978; Nilsen and Kerr 1978; Wolfe 1978, 1985, 1994; Parra et al. 1986; Schmidt 1991; Basinger et al. 1994; Manum 1994). Latitudinal sea surface temperature (SST) gradients derived from oxygen isotopic records of planktonic foraminifera appear to have been much reduced relative to the present gradient, with high latitude values that were much warmer than, and tropical values that were similar to or slightly cooler than, current values (e.g., Shackleton and Boersma 1981; Stott et al. 1990; Zachos et al. 1994; Bralower et al. 1995a). During this time atmospheric pCO_2 concentration may have been higher than it presently is (e.g., Cerling 1991; Berner 1991, 1994; Freeman and Hayes 1992), but enhanced greenhouse gas concentrations do not appear to explain completely the shallow latitudinal temperature gradients reconstructed for this time (Sloan et al. 1995; Sloan and Rea 1995).

Against the background of this warm world, an abrupt, short-lived and extreme warming event occurred (Kennett and Stott 1991), called the late Paleocene Thermal Maximum (LPTM; Zachos et al. 1993). This event was associated with a strong negative excursion in carbon isotopes in the marine record (Kennett and Stott 1991; Lu and Keller 1993; Bralower et al. 1995a, 1995b; Thomas and Shackleton 1996) and the continental record (Koch et al. 1992, 1995; Stott et al. 1996). The exact timing of this brief event is difficult to establish because the geomagnetic polarity time scale (GPTS), as well as the correlation between zonations based on different microfossil groups for the upper Paleocene–lower Eocene Series, are in a state of flux (e.g., Cande and Kent 1992; 1995; Berggren et al. 1985, 1995; Berggren and Aubry 1996; Aubry et al. 1986, 1988, 1996). The LPTM occurred within the early, reversed part of paleomagnetic Chron C24, with numerical age varying from 56.14–58.64 Ma in Berggren et al. (1985) and Aubry et al. (1988), to 53.250–55.981 Ma (Cande and Kent 1992), to 53.347–55.904 Ma (Cande and Kent 1995). The LPTM occurred before the evolutionary appearance (FAD) of *Tribrachiatus bramlettei*, within calcareous nannofossil Biochron NP9, and before the last appearance (LAD) of the planktonic foraminifer *M. velascoensis* within planktonic foraminiferal Biochron P5 (Berggren et al. 1995). These relationships place the event in the latest Paleocene Epoch (Berggren and Aubry 1996), with the numerical age of 58 Ma in the integrated magnetobiochronologic scale (IMBS) of Berggren et al. (1985) and Aubry et al. (1988), 55.7 Ma in the GPTS of Cande and Kent (1992), and 55.5 Ma in the chronology of Berggren and Aubry (1996) and Aubry et al. (1996).

Oceanic microfossil data and oxygen isotope data indicate the existence of a shallow pole-to-equator ocean surface temperature gradient in the late Paleocene through early Eocene Epoch (e.g., McGowran 1991; Zachos et al. 1994). However, this gradient decreased even more during the latest Paleocene Epoch (Bralower et al. 1995a, 1995b; Corfield 1995; hereafter referred to as the LPTM). During the LPTM the high latitudes and the deep waters of the oceans warmed by 4–6° C over less than 10,000 years, and remained at high temperatures for 50,000 to a few hundred thousand years (Kennett and Stott 1991; Lu and Keller 1993; Thomas and Shackleton 1996).

The kaolinite content of clay mineral assemblages in sediments, indicating high humidity and intensive chemical weathering, was generally high during the early Eocene Epoch (Robert and Maillot 1990; Robert and Chamley 1991; Robert and Kennett 1992), but it peaked during the LPTM on the Antarctic continent (Robert and Kennett 1994), in the North Sea area (Knox, personal communication 1993), and on the New Jersey continental shelf (Gibson et al. 1993). Dust concentrations in oceanic sediments dropped precipitously at the LPTM, suggesting a rapid decrease in wind strength (Janecek and Rea 1983; Miller et al. 1987b; Rea et al. 1990; Rea 1994).

Correlation between oceanic and terrestrial sediments during the LPTM is difficult, but land-mammal faunas apparently showed major evolutionary changes, including the introduction of many new taxa, at the time of the short-lived negative excursion in carbon isotope values in land sections (Rea et al. 1990; Hooker 1991; Koch et al. 1992, 1995; Berggren and Aubry 1996;

Maas et al. 1995; Stott et al. 1996; Wing et al. 1995). In reference to the North American land-mammal ages (NALMA) the LPTM, as recognized by the short-lived carbon isotope excursion, occurred at the Clarkforkian/Wasatchian Age boundary (Koch et al. 1992, 1995).

Concurrent with the short interval of extreme warming, deep-sea benthic foraminifera underwent their most extensive extinctions since the end of the Cenomanian Age (Schnitker 1979; Tjalsma and Lohmann 1983; Miller et al. 1987b, 1992; Thomas 1990a, 1990b, 1992; Kaiho 1991, 1994; Katz and Miller 1991; Nomura 1991; Pak and Miller 1992; Mackensen and Berggren 1992; Bolli et al. 1994; Speijer 1994; Thomas and Shackleton 1996). Benthic faunal patterns indicate that many regions in the global ocean suffered relatively low-oxygen conditions during this interval.

A short-term, large (2‰) excursion to low carbon isotopic values in surface and deep oceans as well as in the atmosphere was coeval with the abrupt warming (Kennett and Stott 1991; Lu and Keller 1992; Koch et al. 1992; Pak and Miller 1992; Seto et al. 1991; Stott et al. 1996; Thomas and Shackleton 1996). This carbon isotope excursion occurred contemporaneously with increased dissolution of carbonate at many locations in the deep sea, suggesting that the concentration of total dissolved inorganic carbon in the deep oceans increased. The carbon isotopic excursion has proven to be very challenging to explain because of its large amplitude and short duration, and the absence of a coeval mass extinction in the terrestrial biosphere that could have caused transfer of enough isotopically light carbon into the oceans and atmosphere (Thomas and Shackleton 1996). The short duration of the event makes it improbable that erosion of organic-rich matter from the continental shelf areas during a sea-level fall could have been the cause.

The LPTM as a climatic event was concurrent with a major upheaval in the global carbon cycle, but which was cause and which was effect is difficult to determine. The available data are not unequivocal, but high-latitude warming appears to have started slightly before the carbon isotope excursion, and before the increased dissolution of carbonate sediments in the deep sea (Kennett and Stott 1991; Thomas 1992; Thomas and Shackleton 1996).

The causes of this extreme and rapid episode of climatic change are an enigma. Paleoceanographers have speculated that some factor, possibly related to increased emissions of CO_2 or other greenhouse gases into the atmosphere, caused high-latitude warming above and beyond the already high temperatures of the late Paleocene Epoch (Rea et al. 1990; Sloan et al. 1992; Eldholm and Thomas 1993; Kaiho and Saito 1994). The additional warming could have led to a temporary cessation of, or strong decrease in volume of, intermediate to deep water formation at high latitudes, and an increase in the contribution of deep water from subtropical regions (e.g., Kennett and Stott 1991; Pak and Miller 1992, 1995; Sloan et al. 1992; Thomas 1992; Zachos et al. 1993). This rapid change in deep water source areas could have caused the benthic foraminiferal extinction, either directly, by decreasing oxygen concentration in the warmer waters, or indirectly, by changing the location of upwelling and high primary productivity in the oceans (Thomas and Shackleton 1996).

The oceanic circulation dynamics necessary for this chain of events, including the mechanism that would bring the system back to the regime of high latitude deep water formation after several tens of thousands of years, are unknown at the current time (e.g., Sloan et al. 1995). Additionally, explaining the carbon isotopic record through the LPTM remains a challenge, in terms of explaining the reason for the shift to more negative values, and the return to pre-LPTM conditions (e.g., Koch et al. 1995; Thomas and Shackleton 1996).

A further challenge exists in explaining the initial high latitude warming that may have set off this hypothesized chain of events. The signature of greenhouse gas warming that is predicted from climate models is not consistent with the reconstructed temperature gradient for the LPTM. Increased atmospheric concentration of greenhouse gases would cause warming at all latitudes

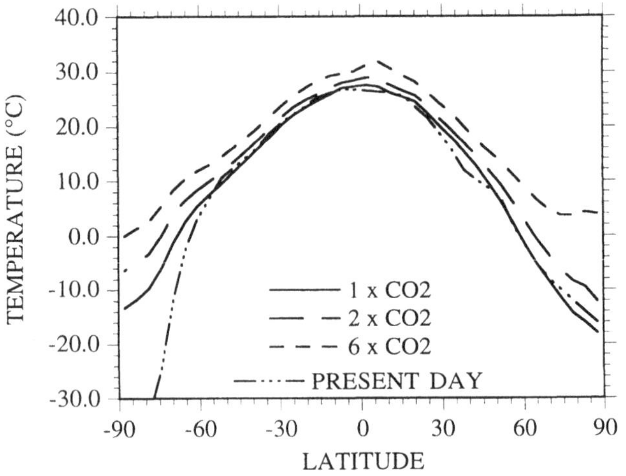

Figure 9.1 Latitudinally averaged, mean annual surface temperatures produced by the NCAR GENESIS general circulation model for the present day and for three early Eocene model cases. Each of the Eocene cases had Eocene geography, no continental ice, and the model employed a mixed layer ocean.

Atmospheric pCO_2 concentration specified in the Eocene cases was 300 ppm (1×), 560 ppm (2×) and 1680 ppm (6×). See text and Sloan and Rea (1995) for further explanation. Present-day model results taken from Pollard and Thompson (1994).

(figure 9.1; see Sloan and Rea 1995 for discussion). Some factor other than greenhouse gases may be required to generate the additional warming only for high latitude regions. For example, high latitude, radiatively warming clouds, caused by increased atmospheric methane concentrations and the subsequent conversion of methane to stratospheric water vapor and then ice clouds, have been suggested as an explanation for preferential warming of high latitudes (e.g., Sloan et al. 1992). However, the timing and magnitude of this mechanism is uncertain, and there is little geological evidence to support or refute this idea.

Climate Model Application: Rationale

Little is known about the behavior of the climate system during the latest Paleocene Epoch, because evidence of such a short-lived climatic event can be documented only when sedimentation rates are relatively high. Sections in which the event has been described in detail are therefore rare, especially because many sections have not been recovered completely, and/or have hiatuses that occur at least partially across the interval of interest (Aubry et al. 1996).

In order to address the question of what caused the global oceanic and climatic changes that are documented in the geologic record for the LPTM, a more complete view of what the climate system was like during that time is needed. A general circulation model (GCM) is a useful tool to apply in order to create a conceptual model of the global climate conditions that might have occurred in conjunction with the LPTM SSTs, including the exceptionally shallow pole-to-equator SST gradient. Using a GCM to estimate the consequences of an extremely shallow temperature gradient in terms of climatic parameters such as precipitation, evaporation, surface moisture, temperature, and winds, can provide information necessary for understanding how such warming occurred in the latest Paleocene Epoch. One can then evaluate whether the climatic consequences of the SSTs as predicted by the model are in agreement with relevant, existing paleoclimatic data. The model-produced climate estimates also can be used to predict regional climatic conditions in areas for which no data presently exist, and in the future, when more data become available, those model predictions also can be tested.

Data Manipulation and Sea Surface Temperature Derivation

A modeling study was carried out in which global climatic conditions that might have occurred in equilibrium with LPTM SSTs were produced with a GCM. The

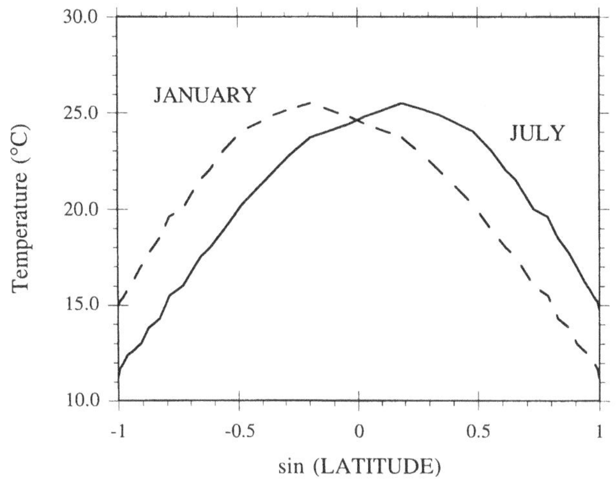

Figure 9.2 Sea surface temperature gradients prescribed in LPTM model cases.

LPTM SSTs, with warm high latitude temperatures and an extremely shallow pole-to-equator SST gradient, were reconstructed from marine oxygen isotopic records (figure 9.2). "Climatic conditions" of interest here refer to atmospheric and continental temperatures, precipitation, evaporation, net moisture at land surfaces, moisture fluxes at the ocean surface, and wind velocities. Rather than try to address why the LPTM conditions occurred, in this study we first ask what conditions most likely occurred in response to, or in conjunction with, the SSTs. From that point, future studies may address the question of how such climatic conditions might have evolved.

The paleoclimatic data coverage for conditions during the extreme LPTM is currently very sparse. As a consequence, several assumptions were made to maximize the constraints that could be placed upon the SST record (table 9.1). At sites for which records were available, we used the oxygen isotopic composition of keeled planktonic foraminifera (morozovellids), because these species have oxygen and carbon isotope signatures of surface dwellers (e.g., Boersma et al. 1987; Shackleton et al. 1985; Corfield and Shackleton 1988; Corfield and Cartlidge 1992a; Corfield 1993; Pearson et al. 1993; d'Hondt et al. 1994; Bralower et al. 1995a, 1995b). Such species are usually absent from high latitude sites, but they appeared just before and during the LPTM at (paleo)latitudes up to 65° S (Stott and Kennett 1990; Stott et al. 1990; Lu and Keller 1993; Thomas and Shackleton 1996). The keeled species *Morozovella aequa* was present in a few samples only, and we averaged the data for those samples. The occurrence of morozovellids in polar regions makes it possible to compare oxygen isotopically derived temperatures for specimens of one surface-dwelling genus from an equa-

Table 9.1
Data used for derivation of SST estimates

Paleolatitude	Location, Site	Warmest Temperature (°C)	Coldest Temperature (°C)
70° N	Ellesmere Island[1]	12	15
46° N	Bay of Biscay, 549, 550[2]		19
22° N	Shatsky Rise, 577[3]		25
2° N	Allison Guyot, 865[4]		25
32° S	Indian Ocean, 762[5]		18
35–38° S	Walvis Ridge, 525, 527[6]		18
60° S	Kerguelen Plateau, 738[7]		16
64–65° S	Weddell Sea, 689, 690[8]	14	15–17

[1]Temperature estimates from oxygen isotope analyses of scaphopod molluscs, Ellesmere Island; McKenna 1980.
[2]Bay of Biscay, Sites 550, 549; Stott et al. 1996; Aubry et al. 1996.
[3]Corfield and Cartlidge 1992a, 1992b.
[4]Bralower et al. 1995a, 1995b.
[5]Unpublished data, Thomas and Shackleton; Thomas et al. 1992.
[6]Shackleton et al. 1984; Thomas and Shackleton 1996.
[7]Lu and Keller 1993.
[8]Stott et al. 1990; Thomas and Shackleton 1996.
The equation by Erez and Luz (1983) was used, and the salinity correction of Zachos et al. (1994) was applied.

torial site (ODP Site 865, at latitude 2° N, Bralower et al. 1995a, 1995b) and from polar sites (ODP Sites 689 and 690, at latitude 64–65° S, Stott et al. 1990; Kennett and Stott 1991; Thomas and Shackleton 1996).

We assumed that at high latitudes the growth of keeled planktonic foraminifera would occur dominantly during the summer months, when light would enable surface-dwelling species with symbionts to grow. Therefore we assumed that the temperature derived from analysis of morozovellid species was typical for the warmest months of the year for the polar sites. We also assumed that temperatures derived from oxygen isotopic data of deep-dwelling planktonic foraminifera (subbotinids) and of benthic foraminifera would give us a temperature close to what surface waters were in the coldest season of the year at these high latitudes, where vertical temperature gradients are minimal. Where data on bulk carbonate were available, we compared temperatures derived from bulk carbonate data with those derived from planktonic foraminifera. For the oxygen isotopic data representing the peak warming, temperatures derived from bulk data were within 2° C of temperatures derived from subbotinid foraminiferal data. For lower latitude sites, where seasonality would be expected to be less, we assumed that the dominant growth period of keeled planktonic foraminifera would be in the warmer months, and we did not make an estimate for the colder months. No data were available for extreme northern high latitudes, and instead, oxygen isotope data from scaphopod molluscs from the lower Eocene deposits from Ellesmere Island (McKenna 1980) were used. It was assumed that those temperatures were applicable because deep-ocean temperatures, as well as carbon isotope signatures, in the earliest Eocene Epoch were similar to these during the LPTM (e.g., Zachos et al. 1993; Corfield and Cartlidge 1992b; Corfield 1995).

A salinity correction (after Zachos et al. 1994) was used on the isotopic data. Even with the correction, surface water temperatures in the equatorial region were between 24 and 26° C, clearly lower than present values, but in agreement with estimates in other records (Shackleton and Boersma 1981; Zachos et al. 1994; Bralower et al. 1995b). These low temperature estimates for tropical regions cannot be fully reconciled with data on the distribution of coastal thermophilic biota such as corals and mangroves (Adams et al. 1990), and data on land floras at similar latitudes (e.g., Wolfe 1994).

The paleotemperature interpretations were used to produce two SST gradients, one for January (as coldest northern hemisphere and warmest southern hemisphere month) and one for July (opposite conditions). Because there was very little tropical data control, the tropical data points were fixed to have a tropical maximum value of 26° C (figure 9.2). These temperature gradients were specified in two experiments with the GENESIS GCM (described below).

Model Description and Experimental Design

The general circulation model used in this work is known as "GENESIS" (Global ENvironmental and Ecological Simulation of Interactive Systems; version 1.02A). The model was developed at the National Center for Atmospheric Research (NCAR) by D. Pollard and S. Thompson (Thompson and Pollard 1994, 1995;

Pollard and Schulz 1994). GENESIS is based on a version of the Community Climate Model (CCM v.1) atmospheric general circulation model, with major modifications to facilitate paleoclimate investigations. The atmospheric model is coupled to several other systems, including a land surface model, and submodels for soil and snow. The modeled atmosphere consists of 12 vertical layers encompassing the distance up to the lower stratosphere.

Modifications to CCM (v.1) for GENESIS include a diurnal cycle and improved cloud parameterizations more similar to those found in version 2 of the CCM (Thompson and Pollard 1994, 1995). Additional changes are documented in Thompson and Pollard (1994, 1995). The land surface model linked to GENESIS explicitly models the physical effects of vegetation upon surface and boundary layer dynamics. The model describes surface vegetation at all land points, with a total of 12 vegetation categories recognized. Vegetation attributes, including albedo, are determined by the assigned vegetation type, and the influence of vegetation upon climate processes takes place via albedo feedbacks, interception of rain and snow, and radiative and turbulent effects of the vegetation upon soil, snow, and surface winds.

GENESIS contains a soil submodel that consists of 6 vertical layers extending from the surface to 4.25 meters depth. Heat is diffused linearly and moisture is diffused nonlinearly within the soil. Water or ice within soil layers is computed by the model, based on temperature and moisture content of each soil level. Surface runoff occurs if precipitation minus evaporation exceeds the maximum possible infiltration rate of all soil layers, and drainage occurs from the bottom soil layer (Pollard and Schulz 1994). The snow submodel has 3 layers and is calculated at each snow covered point, whether snow overlays soil, ice, or sea ice.

LPTM Model Case

The horizontal resolution of both the atmospheric GCM and the land surface component used in this study is approximately 4.5° latitude by 7.5° longitude (resulting in a global grid of 48 longitudinal cells and 40 latitudinal cells). The model performs up to standards with other models at this resolution in the simulation of present day climate (Thompson and Pollard 1994, 1995). Additionally, GENESIS is used in many paleoclimate studies (e.g., Crowley et al. 1993; Barron et al. 1993; Otto-Bliesner 1993; Pollard and Schulz 1994; Sloan 1994; Sloan and Rea 1995). Detailed descriptions of the model and its performance under present day conditions can be found in Thompson and Pollard (1994, 1995).

The ocean component of GENESIS as used for the LPTM cases consists of an ocean surface with temperatures that are provided to the model; the ocean surface temperatures do not change during the course of model calculations. The model produces atmospheric and terrestrial climatic conditions that are, by definition, forced to be in equilibrium with the specified SSTs. This means that temperatures in the atmosphere over both land and ocean, ground surface temperature, and all other atmospheric conditions, adjust in response to the SSTs. The theory behind this technique is that atmospheric and terrestrial climate responds much more rapidly to forcing than the ocean system does, and so these results represent a climate that would evolve on time scales shorter than that on which the ocean can respond ($<10^3$ years).

Boundary conditions for the LPTM cases were based upon geologic reconstructions of the surface of the Earth during the early Eocene Epoch (late Chron C24r), and were taken from another study (Sloan and Rea 1995). At the spatial resolution employed by the model, these conditions are very similar to late Paleocene geography. The global land/sea distribution and continental topography were based upon an early Eocene paleogeographic compilation (Scotese and Golonka 1992), with minor adjustments made to this based upon recent ODP results (for example, the positions of Australia and India; Sloan and Rea 1995). Global vegetation distribution was based upon limited paleobotanical information for the Paleocene and early Eocene epochs (e.g., Wolfe 1985), and on latitudinal zonation trends of flora. The greatest difference from present conditions was the specification of deciduous forests at high latitudes (e.g., Kemp 1978; Schweitzer 1980), and replacement of all desert areas by tropical or subtropical ground cover. Intermediate values for both soil texture and color were uniformly specified for all continental grid points. SSTs were kept constant at each latitude because surface ocean temperature structure at that time is unknown (after Sloan and Barron 1990). Land masses were described to be free of continental ice, atmospheric pCO_2 was set at 340 ppm (approximate present concentration), and orbital parameters were set at present configurations. Each model case, January and July, was run for 300 days. The final 120 days of model results were averaged together and are presented here.

"Background" Model Case and General Conditions for Comparison

A previous study investigated the climate of the early Eocene Epoch and the sensitivity of that climate to atmospheric pCO_2 concentration (Sloan and Rea 1995). Model experiment cases from that study have the same boundary conditions as listed above, with the exception of atmospheric pCO_2 concentration and the prescribed SSTs. The Sloan and Rea pCO_2 sensitivity study (1995) used a model version that included a mixed-layer ocean. Operationally, this meant that SSTs, as well as land temperatures and all atmospheric conditions were calculated by the model. Results from a 2 × CO_2 (doubled preindustrial level; 560 ppm) early Eocene case are used here for comparison to the LPTM model results. The early Eocene 2 × CO_2 results compare reasonably well with early Eocene paleoclimate data (see Sloan and Rea 1995 for details). Results from this case are the final three years of January averaged together, and final 3 years of July averaged together, from a 14-year model execution. These early Eocene results will be referred to as the "background climate" throughout the remainder of this chapter.

The purpose of using the "background climate" model results in a comparison to the LPTM model case results is to gauge the change from a warm, ice-free climate with a SST gradient less steep than present (but not as shallow as that interpreted for the LPTM), to the climate of the LPTM (compare figure 9.1 and 9.2). Both global climate scenarios are ice-free, but they are also fundamentally different in character. It is important to compare these cases, rather than compare the LPTM results to present day climate, if the cause of the LPTM is to be found.

Even though one set of experiments used prescribed SSTs and one set used computed SSTs, comparing the results is reasonable because ocean surface temperature results and full mean annual results, where the two experiments are fundamentally different, are not considered. Atmospheric conditions and continental conditions between the two studies depend upon the total climate forcing. The general character of the respective pole-to-equator SST gradients between the model LPTM and Eocene cases will substantially influence the nature of the results.

Model Results

Continental Surface Temperature

Despite maintaining polar ocean surface temperatures at values greater than 10° C in the winter hemisphere (figure 9.3), continental interiors experienced freezing conditions in both winter hemispheres of the LPTM cases (figure 9.4). In January large regions of North America and Asia experienced freezing conditions, and the interior of Antarctica experienced freezing temperatures in July. No continental regions froze in the summer hemisphere. Maximum continental surface temperatures were generally not higher than 30° C in either season. Between January and July, the seasonal continental interior temperature amplitude in North America reached ~40° C, and was ~30–35° C for the Antarctic interior.

In comparison to the "background climate" case results, land temperatures are higher in the LPTM case at latitudes poleward of ~35° in both January and July (figure 9.4). The temperature difference is greater in the winter hemisphere than in summer between the cases, because LPTM winters are warmer due to the warmer prescribed SSTs of that case relative to the SSTs of the "background climate" case (figure 9.4). In the tropics, some land areas have lower surface temperatures in the LPTM cases than the "background climate" cases, but with temperature differences ranging between 1 and 8° C (figure 9.5). Since all cases contain the same terrestrial boundary conditions, this temperature change must be caused by the surrounding ocean surface temperatures, or atmospheric feedbacks (including clouds), or both factors. For both the January and July cases, the LPTM global mean surface temperatures are 2.3° C warmer than the corresponding temperatures for the "background climate."

Precipitation

The January and July latitudinal distribution of precipitation over both land and oceans in the LPTM cases is more evenly distributed across latitudes than in the "background climate" cases (figure 9.6), although the global mean total for each month is similar. Warmed by the high latitude SSTs in the LPTM case, air moving to high latitude regions transports more moisture than in the "background climate" cases. Tropical maximum precipitation rates for the LPTM case are comparable to those produced for the "background climate" case, and reflect the influence of slightly cooler tropical SST values in the LPTM cases. For both January and July, land surfaces receive more precipitation in the LPTM case than in the "background climate" cases (figure 9.6). In January, the east coast margins of both North America and Eurasia are wetter in the LPTM case than in the "background climate" cases, as are large regions of South America and Africa (not shown). Small regions of the eastern coast of Antarctica and central Australia receive slightly less mean monthly precipitation in January in the LPTM case relative to the "background climate" case. In July, southern and eastern Asia receives far more precipitation in the LPTM case than for the "background climate" cases, while central Asia receives less precipitation. Northern Africa and southern regions of North America receive more precipitation in the LPTM July case relative to the "background climate" case (not shown).

Available Moisture

To estimate the amount of net moisture that is available from the balance of precipitation and evaporation at land surfaces, model results for surface soil moisture (from 0 to 5 cm depth below surface) were calculated. In

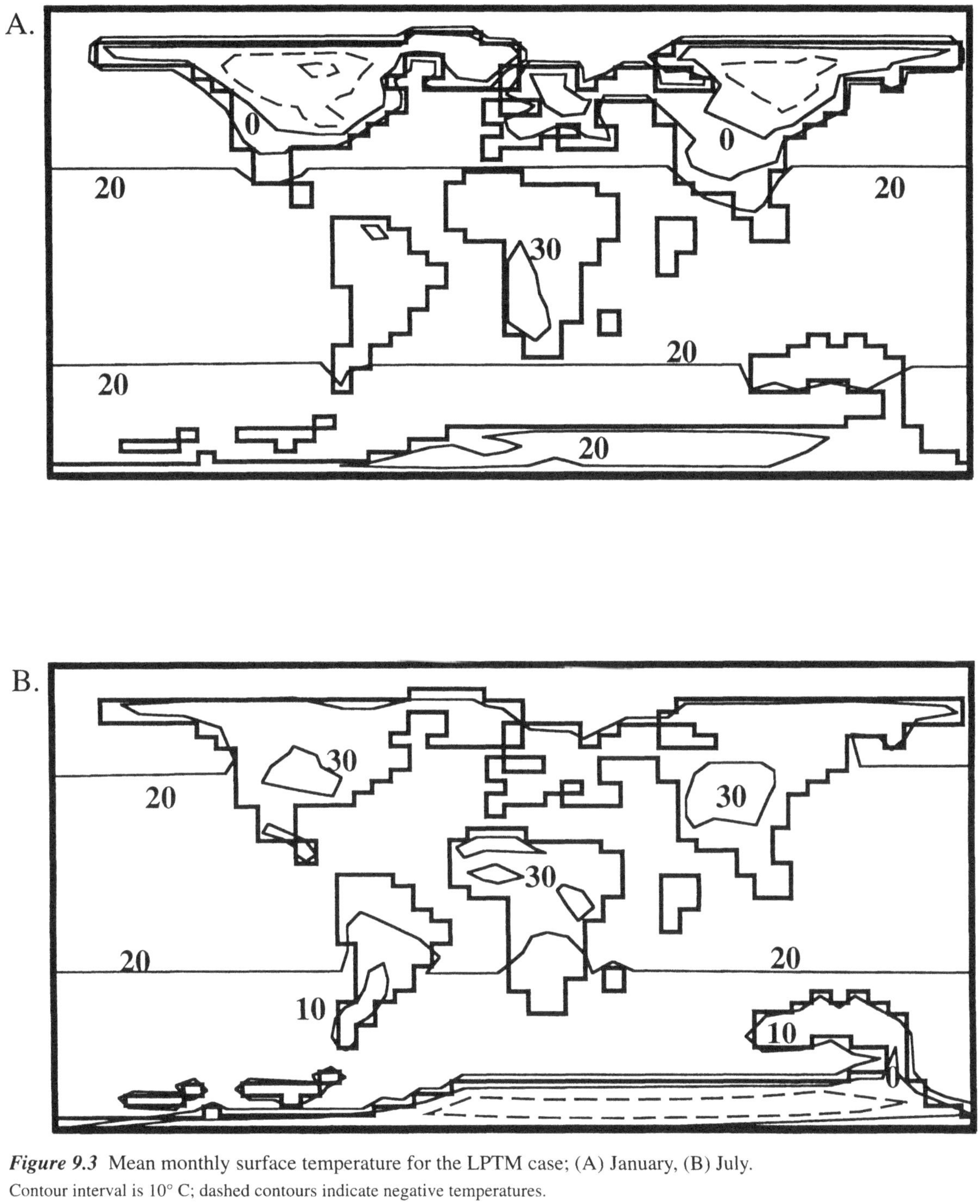

Figure 9.3 Mean monthly surface temperature for the LPTM case; (A) January, (B) July. Contour interval is 10° C; dashed contours indicate negative temperatures.

the winter hemispheres poleward of ~40° latitude, there is greater available terrestrial moisture in the LPTM world than in the "background climate" case (figure 9.7). Summer hemispheres have drier soils in the LPTM case between latitudes 30° and 90°, and wetter soils in the tropics, relative to the "background climate" case.

Over the oceans available moisture is represented by precipitation-minus-evaporation balance at the ocean surface. This quantity, to first order, provides some estimate of relative surface salinity with more negative fluxes indicating higher salinity, and more positive fluxes indicating lower salinity. For both January and July, the global, total moisture flux over the oceans is slightly more negative in the LPTM case relative to the "background climate" results, suggesting the potential for more saline surface ocean conditions in the LPTM case. However, mean latitudinal negative extreme fluxes are more pronounced in the "background cli-

A.

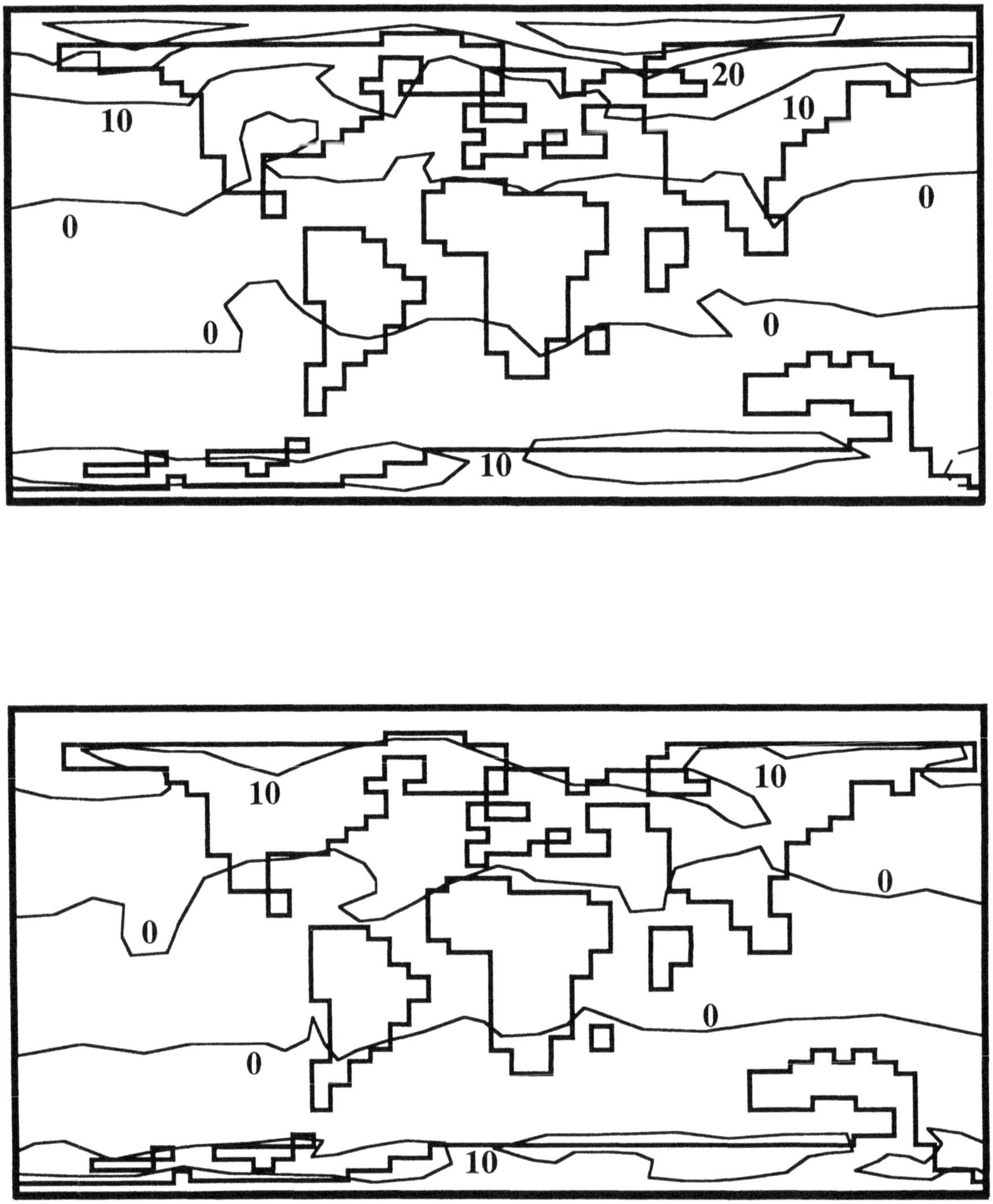

B.

Figure 9.4 Mean monthly surface temperature difference, LPTM case minus "background climate" case; (A) January, (B) July.

Contour interval is 10° C.

mate" cases relative to the LPTM cases (figure 9.8). Both the early Eocene and the LPTM cases show large subtropical negative moisture fluxes in areas, but the locations of the minimum values shift substantially between the cases (figure 9.9). In both seasons, regional variations in surface moisture fluxes are greatest in the Pacific Ocean basin. For January and July conditions there are larger tropical and subtropical areas of evaporation in the "background climate" cases than the LPTM cases (figure 9.9). These results could have implications for hypothesized processes that address the formation and intensity of deep water formation at mid and low latitudes (see discussion, below).

Zonal Winds

In both January and July, zonal winds (east–west winds) are much weaker in the LPTM cases than for the corre-

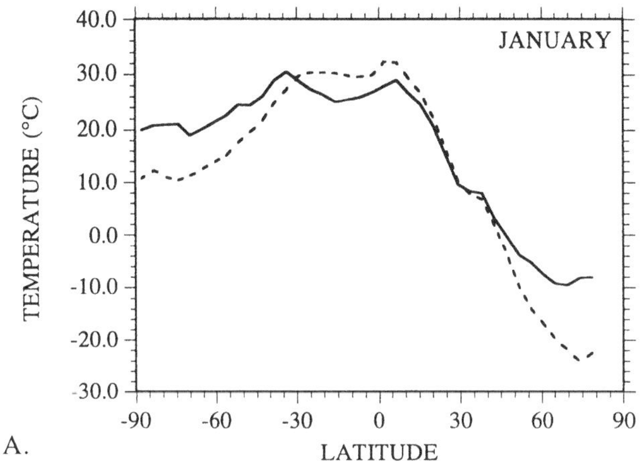

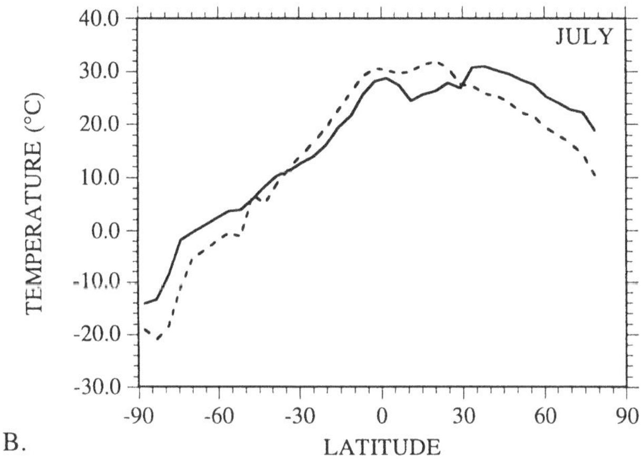

Figure 9.5 Mean monthly surface temperature, averaged over land only, for each model latitude.

Solid lines are the LPTM results, dashed lines the "background climate" results; (A) January, (B) July.

sponding "background climate" cases (figure 9.10). The "background climate" results themselves are substantially less than results calculated by the same model for the present day (Sloan and Rea 1995), and so this result represents an extreme departure from present day conditions. Winter maximum jet stream velocities decrease by nearly 50% over the "background climate" cases, and instead of the weak westerlies in the equatorial region that are seen in present day and "background climate" model results, in both LPTM cases there is a large latitudinal expanse within which wind velocities decrease to nearly zero m/s velocity (figure 9.10). The greatly reduced wind velocities occur in response to the very shallow latitudinal temperature gradient that exists in the LPTM cases, caused largely by the imposed SSTs. If this result is correct, such atmospheric conditions should have had major effects upon equatorial upwelling processes occurring during LPTM time.

Discussion

Validation of Model Results

The model results suggest possible conditions that might have existed around the time of the LPTM "event." In comparison to what might have been the "background climate" at the time, the LPTM climate, driven by the specified SSTs, may have had several unique characteristics. The annual temperature range may have been 5–10° C less than that of the "background climate," but continental interiors still would have retained a substantial annual temperature cycle under those conditions (~35° maximum for continental interior regions at mid-latitudes). Perhaps paleobotanical data could evaluate the temperature amplitude prediction (e.g., Wolfe 1985). Also, if a large lake or more inland seas existed at the time (e.g., remnants of the Cannonball Seaway), the temperature range would have been even less than predicted in this study (e.g., Kutzbach and Ziegler 1993; Sloan 1994).

According to model results, continental minimum temperatures rose above those of the "background climate" case during the LPTM event, especially in high latitude regions. This result primarily is caused by the prescribed warm winter SSTs, and the lack of ice and associated high albedo in the summer. Thus, this response is more likely an effect and not a cause of the LPTM event. Most of the continental thermal response to the warm SSTs is limited to regions poleward of ~35° latitude. If model results for the LPTM are accurate, the continental paleoclimate record should indicate warming conditions on land during this "event." There is some support for this conclusion from the work of Koch et al. (1995), but this is not definitive. Also driven by the imposed SSTs, the LPTM model results suggest that there would have been a slight cooling of tropical land areas during this time, relative to the "background climate." This might also be detectable in the geologic record, but low-latitude, fossil-bearing land sections are rare, especially those sections with high-resolution stratigraphic data. Additionally, in many cases estimates of the duration of the deposition of sediments containing thermophilic biota are insufficient as are estimates of the continuity of deposition (e.g., the deposits containing crocodiles and other biota on Ellesmere Island; Estes and Hutchison 1980; McKenna 1980; Markwick 1994). Possibly studies of continental carbon-isotope records of herbivore teeth or soil concretions might help to establish such a detailed stratigraphy, similar to data for the American and European mid-latitude sections that have been derived (Koch et al. 1992, 1995; Stott et al. 1996).

There is a large precipitation response to the LPTM SST gradient, with a far more even latitudinal distribu-

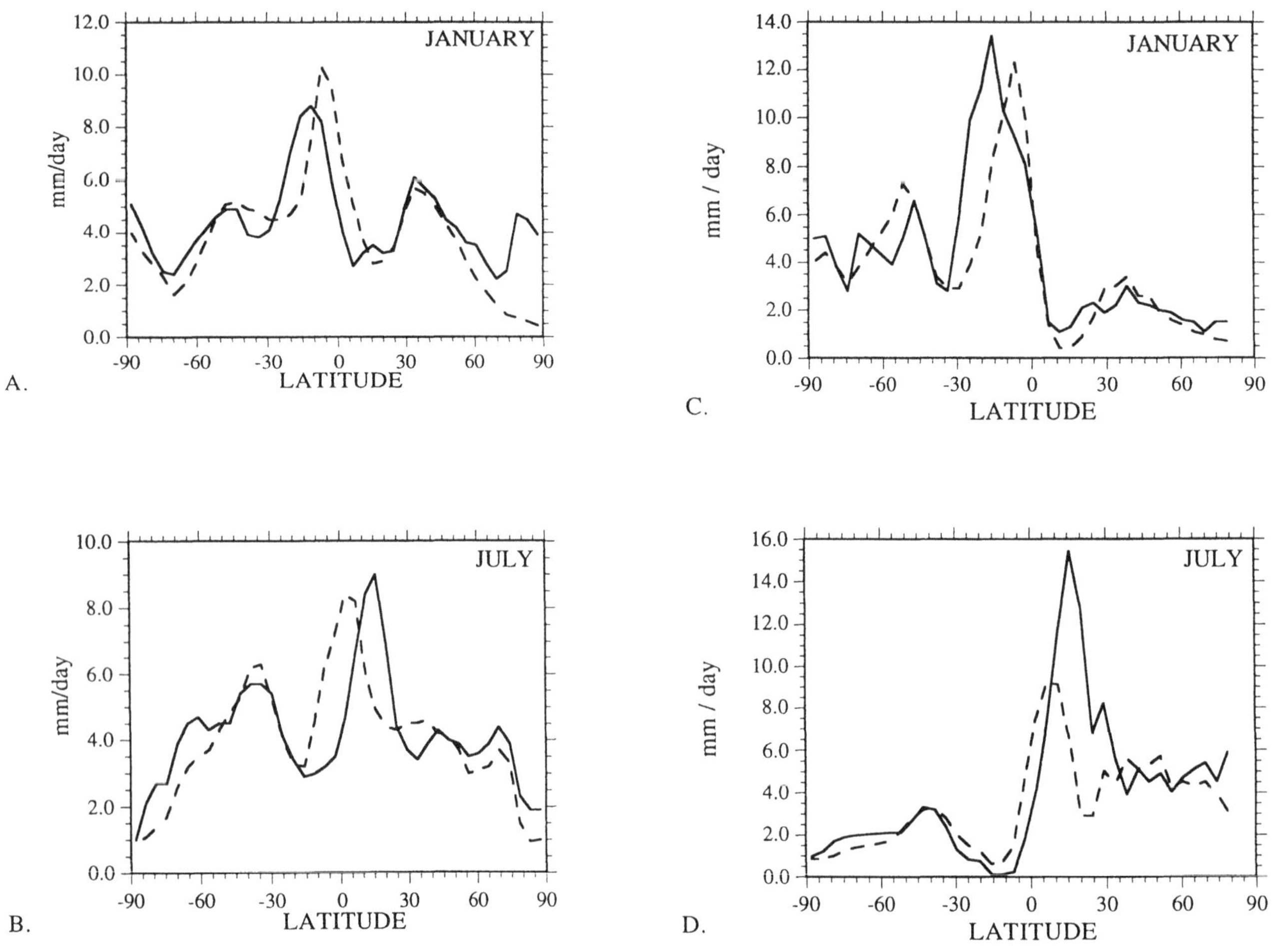

Figure 9.6 Mean monthly precipitation for the LPTM cases (solid line) and "background climate" cases (dashed line); (A) values over land and oceans for January; (B) values over land and oceans for July; (C) values over land only for January; (D) values over land only for July.

tion of precipitation under LPTM conditions than calculated for the "background climate" cases. This latitudinal distribution is also more even than for the present day (not shown). According to model results, land surfaces would have received more precipitation in both January and July under the LPTM conditions than during the "background climate" cases. The precipitation and evaporation distributions that occur in the LPTM cases produced much wetter soils at mid- and high latitude regions of the winter hemisphere, and drier soil moisture conditions in the summer hemisphere (relative to the background climate results). Thus, winter conditions effectively became warmer and wetter, while summer conditions, which were already warm and wet, experienced a small decrease in available moisture relative to the "background climate" results.

The LPTM model results of precipitation and soil moisture are in agreement with sedimentological data. Clay mineralogy studies in oceanic cores at high southern latitudes show an increase in kaolinite and in smectite in sediments deposited during the LPTM, while illite decreased in relative abundance (Robert and Kennett 1994). According to Robert and Kennett (1994), the clay mineral abundances probably reflect conditions on the Antarctic continent. Moreover, the trends in clay mineral associations parallel the trends in oxygen isotopic values from benthic foraminifera. Kaolinite increased in abundance sharply during the period in which the high-latitude surface waters warmed to the LPTM peak; such a strong increase in kaolinite probably meant summer temperatures in excess of 15° C, and year-round high precipitation. During the peak LPTM, kaolinite decreased relative to smectite, suggesting a warm climate, intensive chemical weathering, but an alternation of humid and dry seasons (Robert and Chamley 1991; Robert and Kennett 1994). Clay mineral assemblages on the New Jersey continental margin showed increased abundance of kaolinite (Gibson et al. 1993); however, this record does not exist at sufficiently high resolution to determine the exact position of the LPTM.

Model results suggesting high humidity during the LPTM agree with inferences based on vegetation

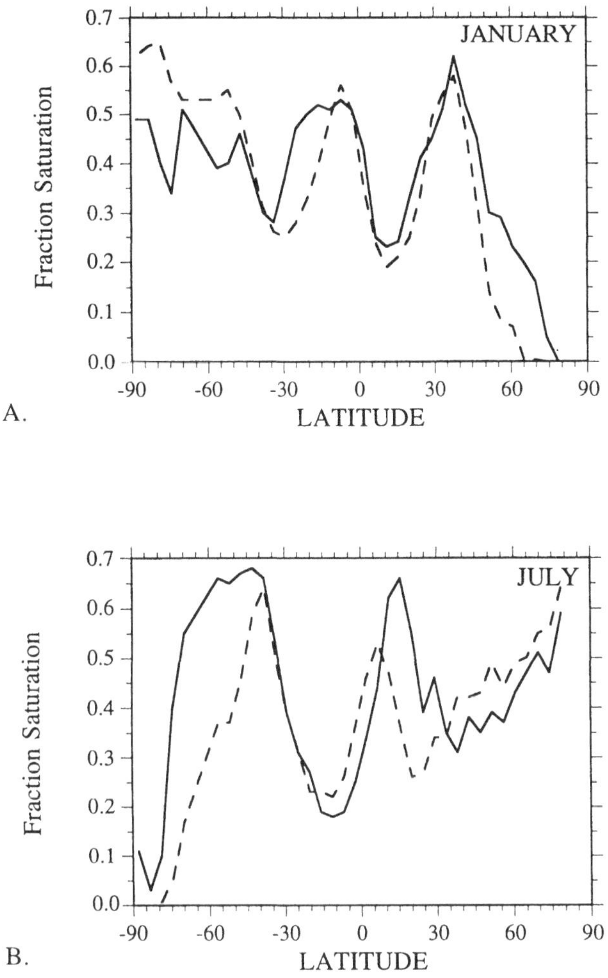

Figure 9.7 Mean monthly soil moisture content (fraction saturation) for the top 5 cm of the soil column, averaged over land only, for each model latitude.

Solid lines are the LPTM results, dashed lines the "background climate" results; (A) January, (B) July.

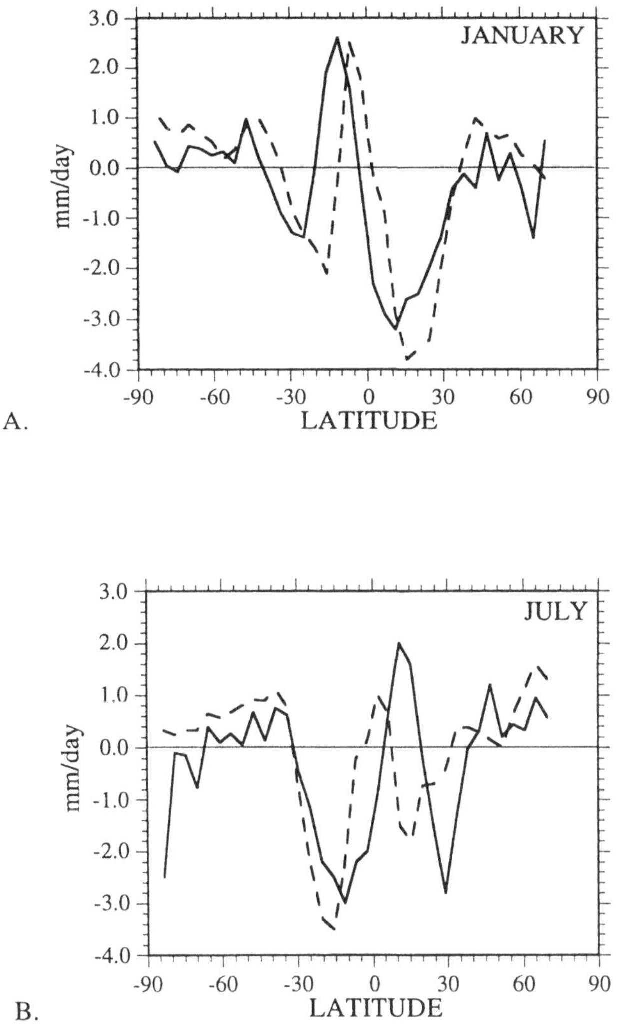

Figure 9.8 Mean monthly precipitation minus evaporation, over the oceans only, for the LPTM case (solid line) and the "background climate" case (dashed line); (A) January, (B) July.

around the North Sea region and along the volcanic region in the North Atlantic (the Brito–Arctic Province; paleolatitudes 50–60° N; Boulter and Manum 1989). Pollen studies from material of LPTM age in the North Sea region also suggest high humidity and temperature (Schroeder 1992).

Model results of available moisture fluxes (precipitation minus evaporation) at the ocean surface indicate changes in the locations of maximum and minimum fluxes (which generally correspond to regions of relatively high or low salinity, respectively) in response to the imposed SSTs. For any given location, the LPTM minimum and maximum values are no greater than those calculated by the model for the "background climate" cases, but the locations of the maxima and minima change between the cases. Such changes might affect the potential for marginal basins, such as the western Tethys, to be able to function as source regions for warm and saline water masses that could sink to intermediate levels in the oceans. Overall, there is a slightly more negative global mean value over oceans for the LPTM cases than for the "background climate" cases, which could imply a tendency towards surface waters of higher salinity than in the "background climate" case. However, the distribution of moisture flux (figure 9.9) shows that the "background climate" case has greater mean negative flux in tropical and subtropical regions, which may indicate conditions that led to the LPTM. The LPTM surface ocean moisture flux results do not show this subtropical precipitation-minus-evaporation minima, while the "background climate" does, indirectly supporting the idea that the LPTM ocean circulation with a large deep water contribution from the subtropics was not sustained for long on a

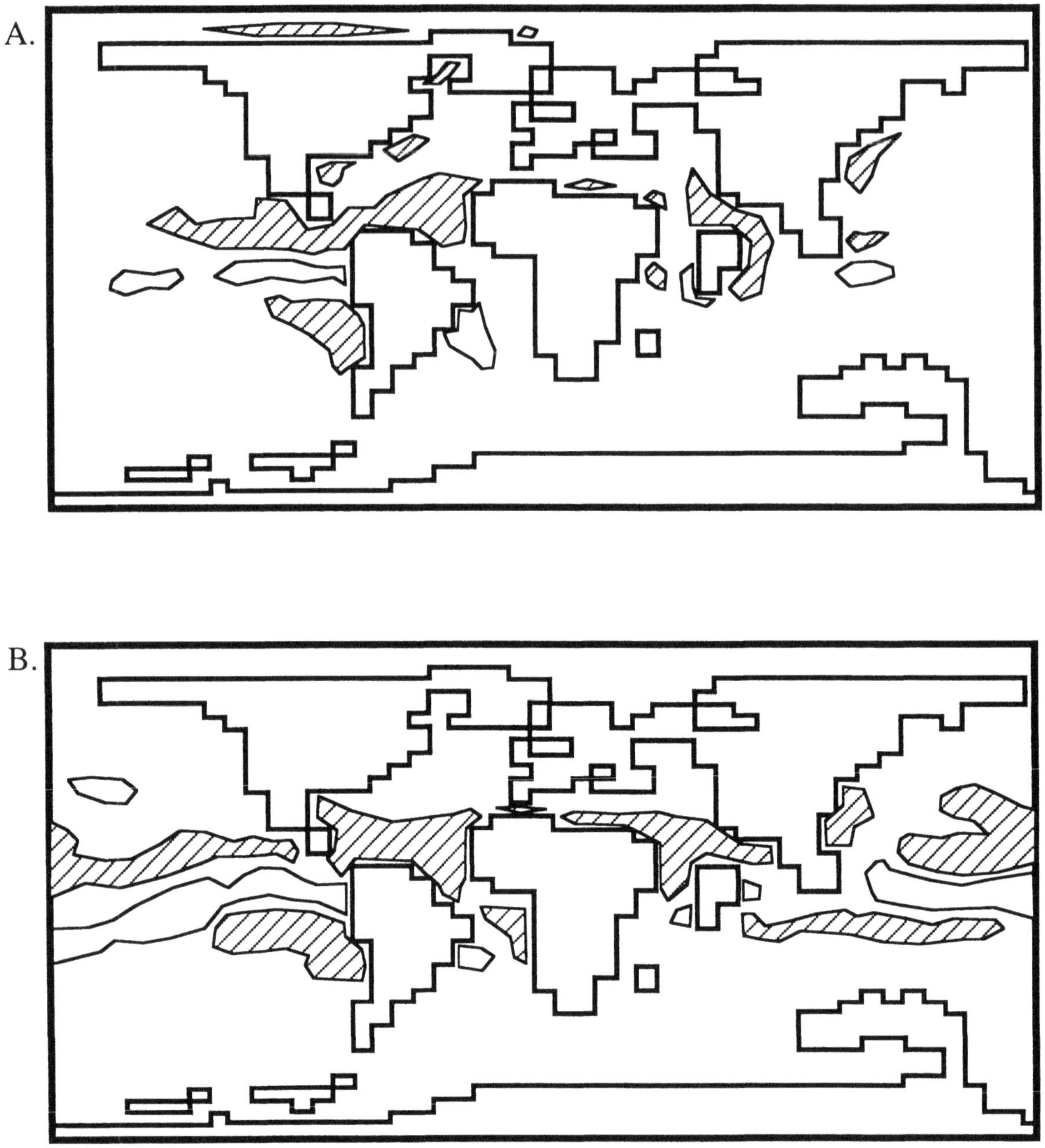

Figure 9.9 Mean monthly precipitation minus evaporation; (A) LPTM January; (B) "background climate" case January; (C) LPTM July; (D) "background climate" case July.

Hatched closed contours indicate areas with moisture flux less than or equal to –3 mm/day (net evaporation); unfilled closed contours indicate areas with moisture flux equal to or greater than +3 mm/day (net precipitation).

geologic scale. The distributions of surface ocean moisture fluxes (figure 9.9) show greatest regional variations in the Pacific basin. Bralower et al. (1995b) suggested that Pacific equatorial waters might have been a source of deep water in the latest Paleocene Epoch, and the model results are consistent with that idea.

Model results show that zonal atmospheric circulation intensity during the LPTM decreased by 25–50% compared to the corresponding "background climate" velocities. The "background climate" zonal wind velocities themselves have been shown to have been much weaker than model-produced present day velocities (Sloan and Rea 1995), and so this result represents an extreme change for the LPTM relative to other time intervals. Zonal westerly winds in the equatorial region decrease to nearly zero velocity in the LPTM cases (figure 9.10). A decrease in wind speed at the time of the late Paleocene benthic foraminiferal extinction is clearly indicated by data reflecting atmospheric dust transport (Miller et al. 1987b; Hovan and Rea 1992;

C.

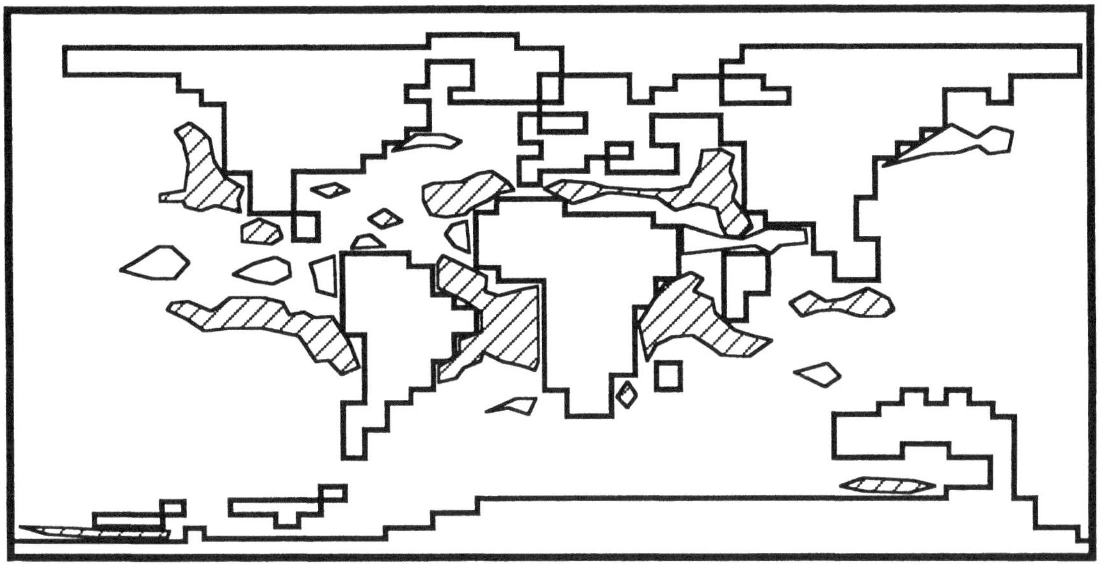

D.

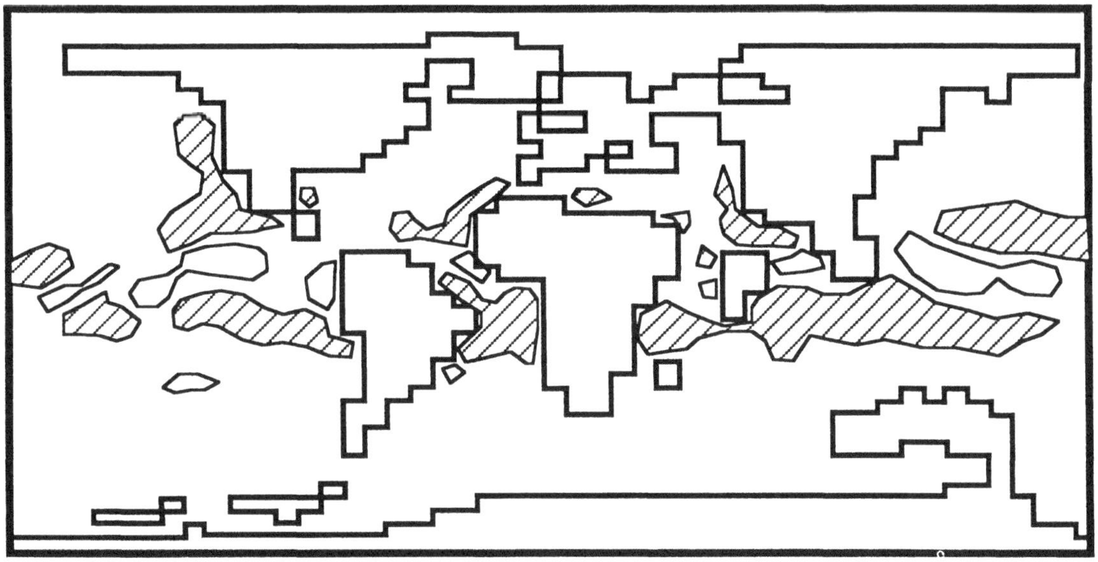

Rea et al. 1990; Rea 1994). The available data on dust concentrations, however, are not available at a sufficiently high temporal resolution and therefore it cannot be determined whether or not the precipitous drop in wind velocity occurred exactly during the LPTM. In addition, the data on atmospheric dust transport suggest that such transport remained low during the early Eocene Epoch. Additional high–temporal resolution data on dust transport would be very useful to obtain more information on wind patterns during the LPTM and at its close, especially in the tropics, since the change predicted by the model is so extreme relative to the "background climate" results.

Decreasing wind velocities could be seen as a cause of less intense upwelling in many areas of the oceans, including the southern margin of Tethys. Decreased upwelling, and thus decreased nutrient supply, to the surface waters may have caused an overall decrease in surface primary productivity in the late Paleocene Epoch (e.g., Shackleton 1987; Shackleton et al. 1985; Corfield and Shackleton 1988; Rea et al. 1990; Corfield and Cartlidge 1992b; Corfield 1993). It cannot be determined how pronounced this decrease in nutrients was during the LPTM. At equatorial Pacific ODP Site 865, however, the already low surface productivity declined even more during the LPTM (Kelly et al. 1996). Decreased wind velocities and decreased upwelling might have been one of the factors causing less vigorous ventilation of the oceans and widespread low-oxygen conditions at middle to bathyal depths along the southern

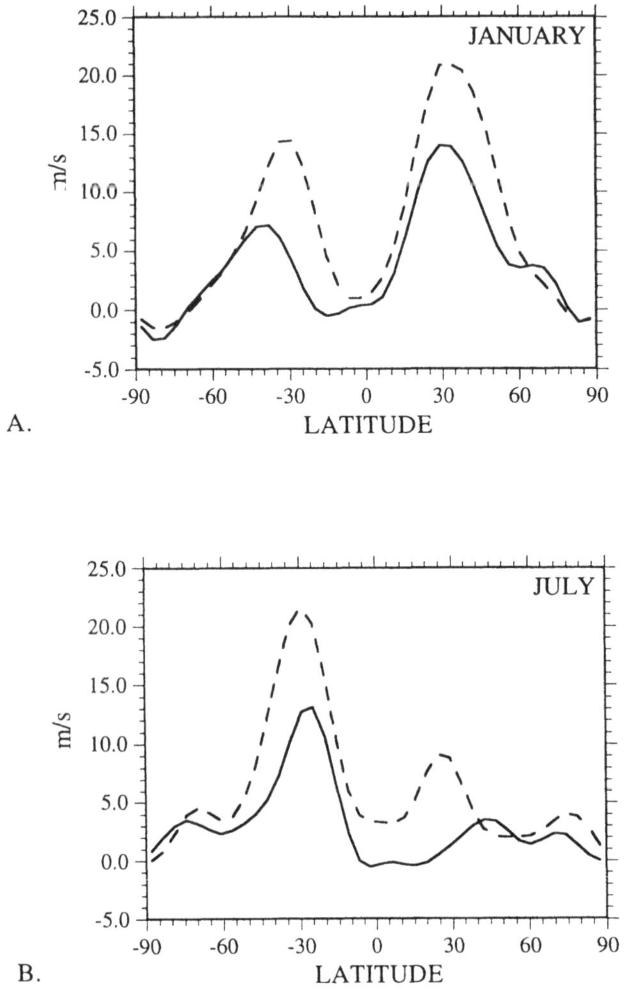

Figure 9.10 Mean monthly zonal wind velocity at 500 mb for the LPTM case (solid line) and "background climate" case (dashed line); (A) January; (B) July.

Tethys Ocean margins from Egypt to Spain (Speijer 1994). However, the reduced winds were more likely to have been a response to the LPTM latitudinal temperature gradient, rather than have caused the LPTM.

Possible Causes of the LPTM

The LPTM model results, compared to model results for a generally representative "background climate," suggest that it is extremely difficult to bring about the conditions recorded for the LPTM. Causing extreme warming at high latitudes—in excess of the warm conditions that existed during the latest Paleocene Epoch—and especially without any ice-albedo feedbacks, presents a significant challenge. The late Paleocene and the early Eocene epochs are thought to have been ice-free. In the current climate state, high-latitude ice feedbacks are one of the most powerful ways to warm high latitudes, given some perturbation of the climate state (e.g., Washington and Meehl 1989). As climate warms, the sea ice extent at high latitudes is reduced, there is greater exchange of heat between the oceans and the atmosphere, leading to winter warming, reduced albedo, and additional absorption of sunlight during the summer months. Warming the climate also results in the melting of land ice, thus lowering the albedo and allowing more absorption of solar radiation at the surface, and therefore contributing positively to additional warming.

In the absence of ice, effective methods of increasing high latitude temperatures are more difficult to identify, especially when concomitant tropical warming is prohibited by the data (table 9.1; figure 9.2). If the tropics were also permitted to warm, greenhouse gases, which would warm all latitudes (figure 9.1), might be a possible cause for the LPTM (e.g., Eldholm and Thomas 1993; Sloan and Rea 1995). Lack of evidence for tropical warming during the LPTM (Bralower et al. 1995a) makes this factor a less likely cause, however. High latitude, polar stratospheric ice clouds may have helped to increase winter temperatures (e.g., Sloan et al. 1992), but that mechanism is not likely to have caused the 4–6° C warming indicated by the paleoceanographic data.

Poleward oceanic heat transport is another possible mechanism to explain the LPTM event, even though that process may be unable to generate sufficient warming (Sloan et al. 1995). However, poleward oceanic heat transport is a solution that is supported indirectly by the model results presented here. The ocean surface available moisture flux results (precipitation minus evaporation; figures 9.8, 9.9) suggest that, under the "background climate" conditions, an increased contribution of deep water formation from tropical and subtropical latitudes, especially in the Pacific, might have taken place; the distribution of moisture flux (figure 9.9) shows that the "background climate" case has greater mean negative flux in tropical and subtropical regions. Such conditions might have led to a small degree of high latitude warming (Sloan and Thomas 1995), leading to increased high-latitude precipitation, and then increased high latitude ocean surface water freshening. This would hinder high latitude deep water production, and cause an increase in the relative contribution of deep water from tropical and/or subtropical latitudes. The "background climate" case might more readily have lead to increased deep water formation, bringing about the LPTM conditions. The LPTM surface ocean moisture flux results show a smaller area of the subtropical salinity maximum than in the background cases, and in fact, results from the LPTM cases show more of a high latitude negative moisture flux than in the "background climate" cases. The LPTM surface water salinity distributions might have tended to return the system to one of dominant deep water formation at high latitudes. These results indirectly support the suggestion that the LPTM ocean circulation, with a large deep water contribution from the tropics and subtropics, was not sustained for long on a geological scale. It is important to note, however, that this idea requires further testing with additional modeling experiments.

This conclusion is in agreement with the observation that the LPTM event was an extremely rare event, and the only time of the Late Cretaceous Period and the Cenozoic Era when deep-sea benthic microfaunas apparently underwent global, synchronous, rapid extinction (Thomas 1992; Kaiho 1994). Likewise, the short-term, extreme excursion to negative values of $\delta^{13}C$ in the whole ocean was the only event of that severity that has been detected to date in the Late Cretaceous Period and Cenozoic Era. We suggest that such an extraordinary global climate event, with extreme high latitude warming resulting in very low temperature gradients, can occur only as the result of a complex chain of events. As demonstrated by the short duration of the LPTM event, such conditions may not be able to exist for prolonged time because of innate instabilities. It could be speculated that short, LPTM-like events occurred not once, but several times during the earliest Eocene Epoch (Aubry et al. 1996), leading to the overall very warm early Eocene climate. Such multiple events would be difficult to detect with certainty, because very high-resolution isotope records would be needed, and benthic foraminiferal faunal records would not be useful to detect later events; these organisms would not be expected to undergo another major extinction since sensitive taxa were already eliminated during the first event.

The LPTM was characterized by extremely high temperatures at high latitudes, even compared with the warm background climate of the late Paleocene and early Eocene epochs. Using available data to reconstruct and then specify SSTs (and ocean surface temperature gradient) for model cases for January and July, climatic conditions for the LPTM were produced. Major conclusions of the comparison between model results for the LPTM and for the "background climate" are the following: (1) global mean surface temperature for the LPTM cases are more than 2° C warmer than temperatures produced by the same model for the "background climate" cases; (2) for LPTM model results, continental warming caused by the imposed SSTs primarily occurs polewards of ~40° latitude, and winter warming of landmasses generally exceeds summer warming; (3) the climate produced by the model for the LPTM has a more even latitudinal distribution of precipitation and has more precipitation occurring over land than in the "background climate" cases; (4) LPTM zonal wind velocities show a substantial decrease relative to the "background climate." This decrease includes an expanded equatorial doldrum zone during the LPTM, with implications for vastly reduced upwelling occurring under these conditions; (5) precipitation-minus-evaporation patterns over the oceans, which influence surface ocean salinity distributions, respond to the imposed LPTM SSTs, but the model results are not unequivocally in support of the hypothesis that formation of deep and intermediate ocean waters switched to subtropical latitudes during the latest Paleocene Epoch. In fact, it appears that the "background climate" case might more readily lead to increased deep water formation in subtropical and tropical regions, bringing about the LPTM conditions. The LPTM surface water salinity conditions might have tended to return the system to one of dominant deep water formation at high latitudes; (6) the absence of polar ice caps during the late Paleocene and early Eocene epochs makes the pronounced warming at high latitudes during the short-lived LPTM more difficult to explain. Increased atmospheric greenhouse gas concentrations would have caused warming at all latitudes and would not preferentially warm the high latitudes.

The model results presented here provide some ideas regarding environmental conditions that may have existed during the LPTM warm event, and of most value are conditions that can be tested with evidence from the geologic record. Terrestrial records to evaluate model predictions of warmer minimum temperatures, slightly greater mean annual temperatures, reduced annual temperature seasonality, increased winter season terrestrial moisture availability, and reduced wind strength would be the most meaningful evidence to use. Evaluating these conditions requires more marine and terrestrial paleoclimatic evidence for the LPTM interval obtained from studies conducted at higher temporal resolution than is currently available. Land sections in which such high resolution records can be studied are, unfortunately, not common, but data could be derived from high resolution studies of clay minerals similar to the one by Robert and Kennett (1994), from more localities, and from more high resolution studies of dust records (e.g., Rea 1994). In addition, evidence for changes in paleoproductivity could be used to infer changes in upwelling regimes in the equatorial region at LPTM time (Kelly et al. 1996), as well as at high latitudes (Thomas and Shackleton 1996), and evidence of upwelling regimes could be used to evaluate model results of drastic wind changes in the equatorial region at LPTM time.

ACKNOWLEDGMENTS

We thank M.-P. Aubry, D. Pak, M. Katz, P. Fawcett, and T. Bralower for helpful reviews, and D. Pollard for helpful discussions. Sloan gratefully acknowledges the support of NSF (EAR-9312014). Computing was carried out at the National Center for Atmospheric Research, which is supported by the National Science Foundation.

REFERENCES

Adams, C. G., Lee, D. E. and Rosen, B. R., 1990. Conflicting isotope and biotic evidence for tropical sea-surface temperatures during the Tertiary. *Palaeogeography, Palaeoclimatology, Palaeoecology,* v. 77, p. 289–313.

Aubry, M.-P., 1992. Paleogene calcareous nannofossils from the Kerguelen Plateau, Leg 120. In Wise, S. W., Jr., Schlich, R., et al., Proceedings of the Ocean Drilling Program, Scientific Results, v. 120: College Station, TX (Ocean Drilling Program), p. 471–491.

Aubry, M.-P., Hailwood, E. A. and Townsend, H. A., 1986. Magnetic and calcareous-nannofossil stratigraphy of lower Palaeogene formations of the Hampshire and London basins. *Journal of the Geological Society, London,* v. 143, p. 729–735.

Aubry, M-P., Berggren, W. A., Kent, D. V., Flynn, J. J., Klitgord, K. D., Obradovich, J. D. and Prothero, D. W., 1988. Paleogene Geochronology: An integrated approach. *Paleoceanography,* v. 3, p. 707–742.

Aubry, M.-P., Berggren, W. A., Stott, L. and Sinha, A., 1996. The upper Paleocene–lower Eocene stratigraphic record and the Paleocene–Eocene boundary carbon isotope excursion: Implications for geochronology. In Knox, R. W. O'B., Corfield, R. M. and Dunay, R. E. (eds.), *Correlation of the early Paleogene in Northwest Europe,* Geological Society Special Publication No 101, p. 353–380.

Barron, E. J., Peterson, W. H., Pollard, D. and Thompson, S., 1993. Past climate and the role of ocean heat transport: Model simulations for the Cretaceous. *Paleoceanography,* v. 8, p. 785–798.

Basinger, J. R., Greenwood, D. R., and Sweda, T., 1994. Early Tertiary vegetation of Arctic Canada and its relevance to paleoclimatic interpretation. In Boulter, M. C. and Fisher, H. C. (eds.), *Cenozoic Plants and Climates of the Arctic,* NATO ASI Series, 127, Dordrecht: Springer-Verlag, p. 175–198.

Berggren, W. A. and Aubry, M.-P., 1996. A late Paleocene–early Eocene NW European and North Sea magnetobiochronological correlation network. In Knox, R. W. O'B., Corfield, R. M. and Dunay, R. E. (eds.), *Correlation of the Early Paleogene in Northwest Europe,* Geological Society Special Publication No. 101, p. 309–352.

Berggren, W. A., Kent, D. V., Flynn, J. J. and Van Couvering, J. A., 1985a. Cenozoic geochronology. *Geological Society of America Bulletin,* v. 96, p. 1407–1418.

Berggren W. A., Kent, D. V., Swisher, C. C., III and Aubry, M.-P., 1995. A revised Cenozoic geochronology and chronostratigraphy. In Berggren, W. A., Kent, D. V., Aubry, M.-P. and Hardenbol, J. (eds.), *Geochronology Time Scales and Global Stratigraphic Correlation,* Society of Economic Paleontologists and Mineralogists Special Publication No. 54, p. 129–212.

Berner, R. A., 1991. A model for atmospheric CO_2 over Phanerozoic time. *American Journal of Science,* v. 291, p. 339–376.

Berner, R. A., 1994. 3GEOCARB II: A revised model of atmospheric CO_2 over Phanerozoic time. *American Journal of Science,* v. 294, p. 56–91.

Boersma, A., Premoli Silva, I. and Shackleton, N. J., 1987. Atlantic Eocene planktonic foraminiferal paleohydrographic indicators and stable isotope paleoceanography. *Paleoceanography,* v. 2, p. 287–331.

Bolli, H. M., Beckmann, J.-P. and Saunders, J. B., 1994. *Benthic foraminiferal biostratigraphy of the south Caribbean region.* Cambridge: Cambridge University Press, 408 pp.

Boulter, M. C. and Manum, S. B., 1989. The Brito–Arctic Igneous Province Flora around the Paleocene/Eocene boundary. In Eldhom, O., Thiede, J., Taylor, E., et al., Proceedings of the Ocean Drilling Program, Scientific Results, v. 104: College Station, TX (Ocean Drilling Program), p. 663–680.

Bralower, T. J., Zachos, J. C., Thomas, E., Parrow, M., Paull, C. K., Kelley, D. C., Premoli Silva, I., Sliter, W. V. and Lohmann, K. C., 1995a. Late Paleocene to Eocene paleoceanography of the equatorial Pacific Ocean: Stable isotopes recorded at Ocean Drilling Program Site 865, Allison Guyot. *Paleoceanography,* v. 10, p. 841–865.

Bralower, T. J., Parrow, M., Thomas, E. and Zachos, J. C., 1995b. Data Report: Stable isotopic stratigraphy of the Paleogene pelagic cap at Site 865, Allison Guyot. In Winterer, E. L., Sager, W. W., Firth, J. V. and Sinton, J. M. (eds.), Proceedings of the Ocean Drilling Program, Scientific Results, v. 143: College Station, TX (Ocean Drilling Program), p. 581–586.

Cande, S. C. and Kent, D. V., 1992. A new geomagnetic polarity time scale for the Late Cretaceous and Cenozoic. *Journal of Geophysical Research,* v. 97(B10), p. 13917–13951.

Cande, S. C. and Kent, D. V., 1995. Revised calibration of the geomagnetic polarity time scale for the Late Cretaceous and Cenozoic. *Journal of Geophysical Research,* v. 100(B4), p. 6093–6095.

Cerling, T. E., 1991. Carbon dioxide in the atmosphere: Evidence from Cenozoic and Mesozoic paleosols. *American Journal of Science,* v. 291, p. 377–400.

Corfield, R. M., 1993. Depth habitats and the Palaeocene radiation of the planktonic foraminifera monitored using oxygen and carbon isotopes. In Lees, D. and Edwards, D. (eds.), *Evolutionary Patterns and Processes,* Linnean Society of London, p. 59–70.

———., 1994. Palaeocene oceans and climate: An isotopic perspective. *Earth Science Reviews,* v. 37, p. 225–252.

Corfield, R. M. and Cartlidge, J. E., 1992a. Isotopic evidence for the depth stratification of fossil and recent Globigerina; a review. *Historical Biology,* v. 5, p. 37–63.

Corfield, R. M. and Cartlidge, J. E., 1992b. Oceanographic and climatic implications of the Palaeocene carbon isotope maximum. *Terra Nova,* v. 4, p. 443–455.

Corfield, R. M. and Shackleton, N. J., 1988. Productivity change as a control on planktonic foraminiferal evolution after the Cretaceous/Tertiary boundary. *Historical Biology,* v. 1, p. 323–343.

Crowley, T. J., Yip, K.-J. and Baum, S. K., 1993. Milankovitch cycles and Carboniferous climate. *Geophysical Research Letters,* v. 20, p. 1175–1178.

D'Hondt, S., Zachos, J. C. and Schultz, G., 1994. Stable isotopic signals and photosymbiosis in the late Paleocene planktic foraminifera. *Paleobiology,* v. 20, p. 391–406.

Eldholm, O. and Thomas, E., 1993. Environmental impact of volcanic margin formation. *Earth and Planetary Science Letters,* v. 117, p. 319–329.

Erez, J. and Luz, B., 1983. Experimental paleotemperature equation for planktonic foraminifera. *Geochimica et Cosmochimica Acta,* v. 47, p. 1025–1031.

Estes, R. and Hutchison, J. H., 1980. Eocene lower vertebrates from Ellesmere Island, Canadian Arctic Archipelago. *Palaeogeography, Palaeoclimatology, Palaeoecology,* v. 30, p. 325–347.

Frakes, L. A., 1979. *Climates Throughout Geologic Time.* New York: Elsevier Scientific Pub. Co., 310 pp.

Freeman, K. H. and Hayes, J. M., 1992. Fractionation of carbon isotopes by phytoplankton and estimates of ancient CO_2 levels. *Global Biogeochemical Cycles,* v. 6, p. 185–198.

Gibson, T. G., Bybell, L. M. and Owens, J. P., 1993. Latest Paleocene lithologic and biotic events in neritic deposits of southwestern New Jersey. *Paleoceanography,* v. 8, p. 495–514.

Haq, B. U., Premoli Silva, I. and Lohmann, G. P., 1977. Calcareous plankton paleobiogeographic evidence for major climatic fluctuations in the early Cenozoic Atlantic Ocean. *Journal of Geophysical Research,* v. 82, p. 3861–3876.

Hooker, J. J., 1991. The sequence of mammals in the Thanetian and Ypresian of the London and Belgian Basins. Location of the Palaeocene–Eocene boundary. *Newsletters on Stratigraphy,* v. 25, p. 75–90.

Hovan, S. A. and Rea, D. K., 1992. Paleocene/Eocene boundary changes in atmospheric and oceanic circulation: A southern hemisphere record. *Geology,* v. 20, p. 15–18.

Janecek, T. R. and Rea, D. K., 1983. Eolian deposition in the northeast Pacific Ocean: Cenozoic history of atmospheric circulation. *Geological Society of America Bulletin,* v. 94, p. 730–738.

Kaiho, K., 1991. Global changes of Paleogene aerobic/anaerobic benthic foraminifera and deep-sea circulation. *Palaeogeography, Palaeoclimatology, Palaeoecology,* v. 83, p. 65–86.

Kaiho, K., 1994. Planktonic and benthic foraminiferal extinction events during the last 100 m.y. *Palaeogeography, Palaeoclimatology, Palaeoecology,* v. 111, p. 45–71.

Kaiho, K. and Saito, S., 1994. Oceanic crust production and climate during the last 100 Myr. *Terra Nova,* v. 6, p. 376–384.

Katz, M. E. and Miller, K. G., 1991. Early Paleogene benthic foraminiferal assemblages and stable isotopes in the Southern Ocean. In Cieselski, P. F., Kristoffersen, Y., et al., Proceedings of the Ocean Drilling Program, Scientific Results, v. 114: College Station, TX (Ocean Drilling Program), p. 481–512.

Kelly, D. C., Bralower, T. J., Zachos, J. C., Premoli Silva, I. and Thomas, E., 1996, Rapid diversification of planktonic foraminifera in the tropical Pacific (ODP Site 865) during the late Paleocene Thermal Maximum. *Geology,* v. 24, p. 423–426.

Kemp, E. M., 1978. Tertiary climatic evolution and vegetation history in the Southeast Indian Ocean region. *Palaeogeography, Palaeoclimatology, Palaeoecology,* v. 24, p. 169–208.

Kennett, J. P. and Stott, L. D., 1991. Abrupt deep-sea warming, palaeoceanographic changes and benthic extinctions at the end of the Palaeocene. *Nature,* v. 353, p. 225–229.

Koch, P. L., Zachos, J. C. and Gingerich, P. D., 1992. Correlation between isotope records in marine and continental carbon reservoirs near the Palaeocene/Eocene boundary. *Nature,* v. 358, p. 319–322.

Koch, P. L., Zachos, J. C. and Dettman, D. L., 1995. Stable isotope stratigraphy and paleoclimatology of the Paleogene Bighorn Basin (Wyoming, USA). *Palaeogeography, Palaeoclimatology and Palaeoecology,* v. 115, p. 61–89.

Kutzbach, J. E. and Ziegler, A. M., 1993. Simulation of late Permian climate and biomes with an atmospheric/ocean model: Comparisons with observations. *Royal Society of London Philosophical Transactions,* ser. B, v. 341, p. 327–340.

Lu, G. and Keller, G., 1993. The Paleocene–Eocene transition in the Antarctic Indian Ocean: Inference from planktic foraminifera. *Marine Micropaleontology,* v. 21, p. 101–142.

Maas, M. C., Anthony, M. R. L., Gingerich, P. D., Gunnell, G. F. and Krause, D. W., 1995. Mammalian generic diversity and turnover in the late Paleocene and early Eocene of the Bighorn and Crazy Mountains Basins, Wyoming and Montana (USA). *Palaeogeography, Palaeoclimatology, Palaeoecology,* v. 115, p. 181–207.

Mackensen, A. and Berggren, W. A., 1992. Paleogene benthic foraminifers from the Southern Indian Ocean (Kerguelen Plateau): Biostratigraphy and paleoecology. In Wise, S. W., Jr., Schlich, R. et al., Proceedings of the Ocean Drilling Program, Scientific Results, v. 120: College Station, TX (Ocean Drilling Program), p. 603–630.

Manum, S. B., 1994. The Palaeogene flora of Spitsbergen. In Boulter, M. C. and Fisher, H. C. (eds.), *Cenozoic Plants and Climates of the Arctic,* NATO-ASI Series, 127, Dordrecht: Springer-Verlag, p. 215–222.

Markwick, P. J., 1994. "Equability," continentality, and Tertiary "climate": the crocodilian perspective. *Geology,* v. 22, p. 613–616.

McGowran, B., 1991. Evolution and Environment in the early Paleogene. *Memoirs of the Geological Society of India,* v. 20, p. 21–53.

McKenna, M. C. 1980. Eocene paleolatitude, climate, and mammals of Ellsemere Island. *Palaeogeography, Palaeoclimatology, Palaeoecology,* v. 30, p. 349–362.

Miller, K. G., Fairbanks, R. G. and Mountain, G. S., 1987a. Tertiary oxygen isotope synthesis, sea level history, and continental margin erosion. *Paleoceanography,* v. 2, p. 1–19.

Miller, K. G., Janecek, T. R., Katz, M. E. and Keil, D. J., 1987b. Abyssal circulation and benthic foraminiferal changes near the Paleocene/Eocene boundary. *Paleoceanography,* v. 2, p. 741–761.

Miller, K. G., Katz, M. E. and Berggren, W. A., 1992. Cenozoic deep-sea benthic foraminifera: A tale of three turnovers. *Studies in Benthic Foraminifera, BENTHOS '90,* Tokai University Press, Sendai, p. 67–75.

Nilsen, T. H. and Kerr, D. R., 1978. Paleoclimatic and paleogeographic implications of a lower Tertiary laterite (latosol) on the Iceland–Faeroe Ridge, North Atlantic region. *Geological Magazine,* v. 115, p. 153–236.

Nomura, R., 1991. Paleoceanography of Upper Maestrichtian to Eocene benthic foraminiferal assemblages at Sites 752, 753 and 754, eastern Indian Ocean. In Weissel, J., Pierce, J., Taylor, E., Alt, J., et al., Proceedings of the Ocean Drilling Program, Scientific Results, v. 121: College Station, TX (Ocean Drilling Program), p. 3–29.

Otto-Bliesner, B. L., 1993. Tropical mountains and coal formation: A climate model study of the Westphalian (306 Ma). *Geophysical Research Letters,* v. 20, p. 1947–1950.

Pak, D. K. and Miller, K. G., 1992. Paleocene to Eocene benthic foraminiferal isotopes and assemblages: Implications for deepwater circulation. *Paleoceanography,* v. 7, p. 405–422.

Pak, D. K. and Miller, K. G., 1995. Isotopic and faunal record of Paleogene deep-water transitions in the North Pacific. In Rea, D. K., Basov, I. A., Scholl, D. W. and Allan, J. F. (eds.), Proceedings of .the Ocean Drilling Program, Scientific Results, v. 145: College Station, TX (Ocean Drilling Program), p. 265–281.

Parra, M., Puechmaille, C., Dumon, J. C., Delmont, P. and Ferragne, A., 1986. Geochemistry of Tertiary alterite clay phases on the Iceland–Faeroe Ridge (Northeast Atlantic), Leg 38, Site 336. *Chemical Geology,* v. 54, p. 165–176.

Pearson, P. N., Shackleton, N. J. and Hall, M. A., 1996. Stable isotope paleoecology of middle Eocene planktonic foraminifera and multi-species isotope stratigraphy, DSDP Site 523, South Atlantic. *Journal of Foraminiferal Research,* v. 23, p. 123–140.

Pollard, D. and Thompson, S. L., 1994. Sea-ice dynamics and CO_2 sensitivity in a global climate model. *Atmosphere–Ocean,* v. 32, p. 449–467.

Pollard, D. and Schulz, M., 1994. A model for the potential locations of Triassic evaporite basins driven by paleoclimatic GCM simulations. *Global and Planetary Change,* v. 9, p. 233–249.

Rea, D. K., 1994. The paleoclimatic record provided by eolian deposition in the deep sea: the geologic history of wind. *Reviews of Geophysics,* 32, p. 159–195.

Rea, D. K., Zachos, J. C., Owen, R. M. and Gingerich, P. D., 1990. Global change at the Paleocene–Eocene boundary: climatic and evolutionary consequences of tectonic events. *Palaeogeography, Palaeoclimatology, Palaeoecology,* v. 79, p. 117–128.

Robert, C. and Chamley, H., 1991. Development of early Eocene warm climates, as inferred from clay mineral variations in oceanic sediments. *Palaeogeography, Palaeoclimatology, Palaeoecology,* v. 89, p. 315–331.

Robert, C. and Kennett, J. P., 1992. Paleocene and Eocene kaolinite distribution in the South Atlantic and Southern Ocean: Antarctic climatic and paleoceanographic implications. *Marine Geology,* v. 103, p. 99–110.

Robert, C. and Kennett, J. P., 1994. Antarctic subtropical humid episode at the Paleocene–Eocene boundary, Clay mineral evidence. *Geology,* v. 22, p. 211–214.

Robert, C. and Maillot, H., 1990. Paleoenvironments in the Weddell Sea area and Antarctic climates, as deduced from clay mineral associations and geochemical data ODP Leg 113. In Barker, P. F., Kennett, J. P., et al., Proceedings of the Ocean Drilling Program, Scientific Results, v. 113: College Station, TX (Ocean Drilling Program), p. 51–70.

Savin, S. M., 1977. The history of earth's surface temperature during the past 100 million years. *Annual Review of Earth and Planetary Sciences,* v. 5, p. 319–355.

Schmidt, K.-H., 1991. Tertiary palaeoclimatic history of the southeastern Colorado Plateau. *Palaeogeography, Palaeoclimatology, Palaeoecology,* v. 86, p. 283–296.

Schnitker, D., 1979. Cenozoic deep water benthic foraminifers, Bay of Biscay. In Montadert, L., Roberts, D. G., et al., Initial Reports of the Deep Sea Drilling Project, v. 48: Washington (U.S. Government Printing Office), p. 377–413.

Schröder, T., 1992. A palynological zonation for the Paleocene of the North Sea Basin. *Journal of Micropalaeontology,* v. 11, p. 113–126.

Schweitzer, H.-J., 1980. Environment and climate in the early Tertiary of Spitsbergen. *Palaeogeography, Palaeoclimatology, Palaeoecology,* v. 30, p. 297–311.

Scotese, C. R. and Golonka, J., 1992. *PALEOMAP Paleogeographic atlas.* PALEOMAP Progress Report no. 20, Arlington, University of Texas, 43 pp.

Seto, K., Nomura, R. and Niitsuma, N., 1991. Data report: Oxygen and carbon isotope records of the Upper Maestrichtian to lower Eocene benthic foraminifers at Site 752 in the Eastern Indian Ocean. In Weissel, J., Pierce, J., Taylor, E., Alt, J., et al., Proceedings of the Ocean Drilling Program, Scientific Results, v. 121: College Station, TX (Ocean Drilling Program), p. 885–889.

Shackleton, N. J., 1986. Paleogene stable isotope events. *Palaeogeography, Palaeoclimatology Palaeoecology,* v. 57, p. 91–102.

Shackleton, N. J., 1987. The carbon isotope record of the Cenozoic: History of organic carbon burial and of oxygen in the ocean and atmosphere. In Brooks, J. and Fleet, A. J. (eds.), *Marine Petroleum Source Rocks,* Geological Society Special Publication No 26, p. 423–434.

Shackleton, N. J. and Boersma, A., 1981. The climate of the Eocene ocean. *Journal of the Geological Society of London,* v. 138, p. 153–157.

Shackleton, N. J., Corfield, R. M. and Hall, M. A., 1985. Stable isotope data and the ontogeny of Paleocene planktonic foraminifera. *Journal of Foraminiferal Research,* v. 15, p. 321–336.

Sloan, L. C., 1994. Equable climates during the early Eocene: Significance of regional paleogeography for North American climate. *Geology,* v. 22, p. 881–884.

Sloan, L. C. and Barron, E. J., 1990. "Equable" climates during Earth history? *Geology,* v. 18, p. 489–492.

Sloan, L. C. and Rea, D. K., 1995. Atmospheric carbon dioxide and early Eocene climate: A general circulation modeling sensitivity study. *Palaeogeography, Palaeoclimatology, Palaeoecology,* 119, p. 275–292.

Sloan, L.C. and Thomas, E., 1995. Paleocene/Eocene global climate reconstructed from sea surface temperature data and climate model computations. International Conference on Paleoceanography V, Halifax, Nova Scotia.

Sloan, L. C., Walker, J. C. G. and Moore, T. C. Jr., 1995. Possible role of oceanic heat transport in early Eocene climate. *Paleoceanography,* v. 10, p. 347–356.

Sloan, L. C., Walker, J. C. G., Moore, T. C. Jr., Rea, D. K. and Zachos, J. C., 1992. Possible methane-induced polar warming in the early Eocene. *Nature,* v. 357, p. 320–322.

Speijer, R. P., 1994. Extinction and recovery patterns in benthic foraminiferal paleocommunities across the Cretaceous/Paleogene and Paleocene/Eocene boundaries. *Geologica Ultraiectina, Mededelingen van de Faculteit Aardwtensachappen Universiteit Utrecht,* v. 124, 191 pp.

Stott, L. D., Sinha, A., Thiry, M., Aubry, M.-P. and Berggren, W. A., 1996. Global $\delta^{13}C$ changes across the Paleocene–Eocene boundary: Criteria for terrestrial–marine correlations. In Knox, R. W. O'B., Corfield, R. M. and Dunay, R. E. (eds.), *Correlation of the Early Paleogene in Northwest Europe,* Geological Society Special Publication No. 101, p. 381–399.

Stott, L. D., Kennett, J. P., Shackleton, N. J. and Corfield, R. C., 1990. The evolution of Antarctic surface waters during the Paleogene: Inferences from the stable isotopic composition of planktonic foraminifers, ODP Leg 113. In Barker, P. F., Kennett, J. P., et al., Proceedings of the Ocean Drilling Program, Scientific Results, v. 113, p. 849–863.

Stott, L. D. and Kennett, J. P., 1990. Antarctic Paleogene planktonic foraminifer biostratigraphy: ODP Leg 113, Sites 689 and 690. In Barker, P. F., Kennett, J. P., et al., Proceedings of the Ocean Drilling Program, Scientific Results, v. 113: College Station, TX (Ocean Drilling Program), p. 549–569.

Thomas, E. 1990a. Late Cretaceous through Neogene deep-sea benthic foraminifers (Maud Rise, Weddell Sea, Antarctica). In Barker, P. F., Kennett, J. P., et al., Proceedings of the Ocean Drilling Program, Scientific Results., v. 113: College Station, TX (Ocean Drilling Program), p. 571–594.

Thomas, E., 1990b. Late Cretaceous–early Eocene mass extinctions in the deep sea. In Sharpton, V. L. and Ward, P. (eds.), *Global Catastrophes in Earth History: An Interdisciplinary Conference on Impacts, Volcanism, and Mass Mortality,* Geological Society of America Special Paper No. 247, p. 481–495.

Thomas, E., 1992. Cenozoic deep sea circulation: Evidence from deep sea benthic foraminifera. In Kennett, J. P. and Warnke, D. (eds.), *The Antarctic Paleoenvironment: A Perspective on Global Change,* American Geophysical Union Antarctic Research Series No. 56, p. 141–165.

Thomas, E., Shackleton, N. J. and Hall, M. A., 1992. Data Report: Carbon isotope stratigraphy of Paleogene bulk sediments, Hole 762C (Exmouth Plateau, eastern Indian Ocean). In von Rad, U., Haq, B. U., et al., Proceedings of the Ocean Drilling Program, Scientific Results, v. 122: College Station, TX (Ocean Drilling Program), p. 897–901.

Thomas, E. and Shackleton, N. J., 1996. The Paleocene–Eocene benthic foraminiferal extinction and stable isotope anomalies. In Knox, R. W. O'B., Corfield, R. M. and Dunay, R. E. (eds.), *Correlation of the Early Paleogene in Northwest Europe,* Geological Society Special Publication No 101, p. 401–441.

Thompson, S. L. and Pollard, D., 1994. Sea–ice dynamics and CO_2 sensitivity in a global climate model. *Atmospheres: Oceans,* v. 32, p. 449–467.

Thompson, S. L. and Pollard, D., 1995. A global climate model (GENESIS) with a land-surface-transfer scheme (LSX). Part 1. Present-day climate. *Journal of Climate,* v. 8, p. 732–761.

Tjalsma, R. C. and Lohmann, G. P., 1983. *Paleocene–Eocene bathyal and abyssal benthic foraminifera from the Atlantic Ocean.* Micropaleontology Special Publication No. 4, 90 pp.

Washington, W. M. and Meehl, G. A., 1989. Climate sensitivity due to increased CO_2: Experiments with a coupled atmosphere and ocean general circulation model. *Climate Dynamics,* v. 4, p. 1–38.

Wing, S. L., Alroy, J. and Hickey, L. J, 1995. Plant and mammal diversity in the Paleocene to early Eocene of the Bighorn Basin. *Palaeogeography, Palaeoclimatology, Palaeoecology,* v. 115, p. 117–155.

Wise, S. W., Jr., Breza, J. R., Harwood, D. M., Wei, W. and Zachos, J. C., 1992. Paleogene glacial history of Antarctica in light of Leg 120 drilling results. In Wise, S. W., Jr., Schlich, R., et al., Proceedings of the Ocean Drilling Program, Scientific Results, v. 120: College Station, TX (Ocean Drilling Program), p. 1001–1030.

Wolfe, J. A., 1978. A paleobotanical interpretation of Tertiary climates in the northern hemisphere. *American Scientist,* v. 66, p. 694–703.

Wolfe, J. A., 1985. Distribution of major vegetational types during the Tertiary. In Sundquist, E. T. and Broecker, W. S. (eds.), *The Carbon Cycle and Atmospheric CO_2: Natural Variations, Archean to Present.* Geophysical Monograph 32, Washington D. C: American Geophysical Union, p. 357–375.

Wolfe, J. A., 1994. Tertiary climatic changes at middle latitudes of western North America. *Palaeogeography, Palaeoclimatology, Palaeoecology,* v. 108, p. 195–205.

Zachos, J. C., Lohmann, K. C., Walker, J. C. G. and Wise, S. W., 1993. Abrupt climate change and transient climates during the Paleogene: A marine perspective. *Journal of Geology,* v. 101, p. 191–213.

Zachos, J. C., Stott, L. D. and Lohmann, K. C., 1994. Evolution of early Cenozoic marine temperatures. *Paleoceanography,* v. 9, p. 353–387.

10

Early Paleogene Calcareous Nannoplankton Evolution: A Tale of Climatic Amelioration

MARIE-PIERRE AUBRY

ABSTRACT

Temporal distribution of the Cenozoic calcareous nannofossil genera is analyzed to determine the modes and rates of evolution of the Cenozoic coccolithophorids. The analysis shows that diversification did not occur at a regular pace. 50% of the Cenozoic genera evolved during the first 15 my (Paleocene and early Eocene epochs) of the Cenozoic Era, which corresponds to an average rate of 3.3 genera/my compared to 1 genus/my for the middle Eocene to Pleistocene interval (Recent not included because many Recent taxa are unknown from the fossil record). In addition over 70% of the Cenozoic families evolved during the same interval. Generic diversification during the Paleocene Epoch occurred in three successive steps, in the earliest Paleocene (65–63.8 Ma), the earliest late Paleocene (60.6–59.4 Ma) and close to the Paleocene/Eocene Epoch boundary, during a 1-my-long interval that is referred to as the Paleocene/Eocene boundary interval (P/E BI). Although unrecognized for various reasons that are discussed, the P/E BI turnover was profound and involved both extinctions and new evolutionary appearances. The phylogenetic relationships between Paleocene genera and Cretaceous ancestors are discussed, and it is shown that Cretaceous relict species gave rise to Cenozoic genera first in the earliest Paleocene, next during the P/E BI. Paleocene diversification of the coccolithophorids is tentatively explained in terms of progressive increase in oligotrophy of the photic zone as a result of co-evolution in the coccolithophorids and the photosymbiont-bearing planktonic foraminifera. This resulted in increased late Paleocene productivity as indicated by carbon isotopic and barium records. Decrease in productivity following the Paleocene carbon maximum was probably related to well established oligotrophic conditions that intensified during the early Eocene Epoch. Unexpectedly, the P/E BI turnover, which began during the late Paleocene thermal maximum (LPTM) and at the time of the extremely negative carbon excursion (CIE), appears to have been selective, as shown through comparison between the dominance and success histories of the genera that evolved during the Paleocene Epoch. This enigmatic selectivity is thought to be related to the indeterminate event responsible for the LPTM and CIE and possibly affected biomineralization in the coccolithophorids.

In contrast to the latest Cretaceous Period when marine calcareous planktonic communities (calcareous nannoplankton and planktonic foraminifera) underwent a radical turnover (e.g., Berggren 1962a, 1962b; Bolli 1957a, 1957b; Bramlette and Martini 1964; Loeblich and Tappan 1957a, 1957b; Olsson 1960; Percival and Fisher 1977; Perch-Nielsen 1969, 1981a; Tappan 1979), the latest Paleocene Epoch is essentially seen as a time of extensive turnover among deep-water calcareous benthic communities (bathyal and abyssal benthic foraminifera and ostracods; e.g., Tjalsma and Lohmann 1983; Thomas 1992; Kaiho 1994a, 1994b; Steineck and Thomas 1996; Thomas and Gooday 1996). While strong emphasis is being placed on the global synchrony of the late Paleocene benthic foraminiferal extinction event (BFE) and the short term (~100 kyr) 3 to 4‰ negative excursion in the carbon isotopic composition of the ocean recorded on Maud Rise (Stott et al. 1990; Kennett and Stott 1991) and on its paleoceanographic significance (e.g., Miller et al. 1987a; Pak and Miller 1992; Thomas and Shackleton 1996), the latest Paleocene–earliest Eocene evolution of marine planktonic communities has received little attention, probably because of lack of striking event(s). The commonly accepted picture is that extinctions in the deep ocean benthic communities occurred against a background of increasing diversity in the marine calcareous planktonic groups that culminated in the early Eocene Epoch (Berggren 1969; Tappan and Loeblich 1973; Haq 1973; Zachos et al. 1993). Yet, there are indications that latest Paleocene changes also occurred in the calcareous plankton. Well documented incursion(s) of tropical species into northern and southern high latitude areas indicate(s) latest Paleocene (Biochron P5 of Berggren and Miller 1988) poleward extension of the subtropical water masses (Haq et al. 1977; Gradstein and Srivastava 1980; Boersma and Premoli Silva 1983; Aubry 1992a; Berggren 1992; Pospichal and Wise 1990; Boersma et al., chapter 11). Also, shifts in calcareous nannoplankton and planktonic foraminiferal species dominance, rapid diversification in planktonic foraminifera associated with extreme morphologic variability, and maximum diversity of the tropical water-indicator genus *Discoaster* have been reported associated with the late Paleocene thermal maximum (LPTM, Zachos et al. 1993) at sites in the equatorial Pacific, subantarctic Indian Ocean and East Antarctica (Kelly et al. 1996; Lu and Keller 1993, 1995a, 1995b; Pospichal and Wise 1990).

In this chapter, I first explore the significance of the latest Paleocene crisis on the Cenozoic phylogenetic history of the calcareous nannoplankton. I then discuss the modalities of the diversification of the group through the Paleocene Epoch. Because of problems in correlating sections due to the frequency of unconformities in upper Paleocene–lower Eocene sections (see Aubry, chapter 3), the biogeography of the calcareous nannoplankton is not discussed but the main biogeographic features are mentioned.

Chronologic Framework

The chronologic framework in this study is that of Berggren et al. (1995) with slight modifications around the Paleocene/Eocene Epoch boundary. Unlike in Cande and Kent (1992, 1995) and for reasons explained in Aubry (chapter 3), the Paleocene/Eocene Epoch boundary and the NP9/NP10 biochronal boundary are dissociated. An age of 55 Ma for the NP9/NP10 biochronal boundary, used in Cande and Kent (1992, 1995) as an age calibration tie-point, is accepted here, whereas an age of 54.37 Ma is assigned to the Paleocene/Eocene Epoch boundary (as corresponding to the Ypresian transgression). It must be emphasized that the numerical chronology of events in Biochrons NP9 and NP10 needs revision due to the likely mislocation of the NP9/NP10 biochronal boundary within Chron C24r in Cande and Kent (1992, 1995; see detailed discussions of the problem in Aubry et al. 1996, Berggren and Aubry 1996; see also Aubry, chapter 3; Berggren and Aubry, chapter 2). Revision requires the identification of a marine stratigraphic section representing the entire Chron C24r, with good magnetostratigraphy, biostratigraphy and isotope stratigraphy. Using the current GPTS, the chronology of events is compressed in Biochron NP9 and stretched in Biochron NP10. Even though the difference involved may be only on the order of a few hundred thousand years, it alters substantially rates calculated between events in this interval.

In referring to the time scale of Berggren et al. (1995), the Paleocene Epoch is 10.63 my long, from 65 Ma, the radioisotopic age of the Cretaceous/Paleogene Period boundary (Dalrymple et al. 1993; Swisher et al. 1993; see discussion in Berggren et al. 1995), to 54.37 Ma, the estimated age of the base of the London Clay and Ieper Clay Formations (Aubry 1996). Depending upon the recommendation of the Working Group on the Paleocene/Eocene Boundary of the International Subcommission on Paleogene Stratigraphy and following the revised guidelines of the International Commission on Stratigraphy (ICS; Cowie et al. 1986), the Paleocene/Eocene Series boundary may be relocated at a lithostratigraphic level easier to correlate worldwide than the base of the London Clay (Berggren and Aubry, chapter 2). If the carbon isotope excursion that has been documented in both the marine and terrestrial records (e.g., Stott et al. 1990; Kennett and Stott 1991; Schmitz et al. 1996; Koch et al. 1992, 1995; Stott et al. 1996; Sinha et al. 1996) and has an age estimate of 55.5 Ma (Aubry et al. 1996; Berggren and Aubry 1996) is chosen to characterize the Paleocene/Eocene Epoch/Series boundary, the duration of the Paleocene Epoch will be reduced to 9.5 my. Several authors have prematurely accepted the carbon isotope excursion (or the BFE) as the new criterion delineating the boundary. In this chapter, the Paleocene/Eocene Series boundary is placed at the base of the oldest stage of the (Eocene) series in agreement with the recommendation of the ICS (Hedberg 1976), i.e., at the base of the Ieper Formation. The 1.13-my duration between the carbon isotope excursion and the Ieper clay transgression is the younger part of the interval which is under close scrutiny for determining the event(s) best suited to characterize the Paleocene/Eocene Epoch boundary and is often referred to as the "Paleocene/Eocene boundary interval." This appellation is regrettable in as much as a boundary is an instant between two epochs or a point (or a horizon) between two series. What is meant by Paleocene/Eocene boundary interval is the temporal/stratigraphic interval under consideration for redefining/relocating the Paleocene/Eocene Epoch/Series boundary. Stratigraphically, it corresponds to the interval comprised between the top of the Thanet Beds (Thanetian Stage) and the base of the Ieper Clay/London Clay Formations (Ypresian Stage). Temporally, it is ~2 my long and comprises the late part of

Chron C25r, Chron C25n, and the greater part of Chron C24r (see Berggren and Aubry, chapter 2). With this in mind, I will use "P/E BI" below to designate the 1.13-my temporal span that corresponds to the late/younger part of the Paleocene/Eocene boundary interval.

The P/E BI encompasses late Biochron NP9 and early Biochron NP10 (calcareous nannoplankton), late Biochron P5 and earliest Biochron P6a (planktonic foraminifera). It is essentially bracketed by the last appearance datum (LAD) of *Fasciculithus alanii* (Aubry, unpublished) and the first appearance datum (FAD) of *Tribrachiatus digitalis*. At southern high latitude, its beginning is marked by the FAD of *Globanomalina autraliformis*. The P/E BI falls within the ~8.5-my-long late Paleocene to early Eocene warming trend, and its beginning corresponds to the late Paleocene thermal maximum (LPTM; Zachos et al. 1993).

It is in this 1.13-my-long P/E BI, which currently represents the latest Paleocene Epoch, that many innovations occurred in the calcareous nannoplankton: the origins of several modern lineages are to be found in it.

Taxonomic Framework: Diversity Analysis

A sound taxonomic framework is a prerequisite to meaningful analysis of diversity changes through time. The least controversial level in calcareous nannoplankton taxonomy currently is that of the genus. This view differs from that recently expressed by Bown et al. (1991) who regard the species "as the most useful taxonomic unit" (op. cit., p. 281). I have shown (Aubry 1989a) that somewhat paradoxically, the paleontologic concept of genus in the coccolithophorids is more "natural" than that of species. This latter is quite artificial due to a variety of factors such as dimorphism, dithecatism, production of dissimilar coccoliths during different phases of the life cycle, and poorly known morphologic variability linked either to environmental factors and/or to genetic differentiation (see Bollmann 1997). On the other hand, rigorous generic definition and differentiation can be achieved based on the structural characteristics of the bulk of the calcareous nannofossils referred to as "heterococcoliths."

The fundamental building block of a heterococcolith is a "cycle" of strongly modified rhombohedrons of calcite called "elements." Cycles differ by the shape, size and crystallographic orientation of their elements and heterococcoliths differ by the number, shape and orientation of their cycles. This construction confers heterococcoliths a tremendous potential for morphologic diversification associated with structural innovations, which, indeed, has been vastly explored by the coccolithophorids throughout the Mesozoic and Cenozoic eras. It is on this basis that phylogenetic reconstructions in the group can be soundly established. Thus, diversification in the coccolithophorids constitutes an excellent monitor of extinction and radiation events in relation to past global changes that is probably far superior to that of other microfossil groups, in particular the planktonic foraminifera whose basic building block, a spherical shell, confers a somewhat reduced diversification potential to the group.

Discussion below on calcareous nannoplankton evolution during the Cenozoic Era is thus mainly based on the temporal range of genera. However, the analysis of generic diversity is complemented by an indication of the diversity achieved in each genus (expressed as species richness), as a means of determining its evolutionary dominance. This is also complemented by groupings of genera into families, intended to show when the main innovations in the calcareous nannoplankton took place.

Change in specific diversity through the Cenozoic Era is not reassessed in this work. There is no doubt that the total number of species increased through the Paleocene Epoch, but it is not clear when maximum species richness was reached. Different databases produce different results. Haq (1973) and Bown et al. (1991) showed maximum species richness in the early Eocene Epoch, closely followed by high values in the middle Eocene Epoch. Constructed on a compilation of Perch-Nielsen's work (1985a, 1985b) Knoll's histogram (1989: figure 2) shows that maximum diversity occurred during the latest Paleocene Epoch.

Generic Taxonomy

The generic taxonomy in this study is that of Aubry (1984, 1988a, 1989b, 1990, 1998) in which genera are rigorously defined based on coccolith/nannolith structure rather than on morphology (see discussion in Aubry 1988b, 1989a). For genera that are not treated in these works, the taxonomic framework in Perch-Nielsen (1985a) has been used complemented by my own unpublished interpretations. Diversity analysis at the generic level is extremely dependent upon the criteria used for generic determination. For instance, based on structural criteria, I place all cribriliths (a group of coccoliths) into a single genus, *Pontosphaera*. Based on morphologic criteria, Perch-Nielsen (1985a) separated them into two genera, *Pontosphaera* and *Transversopontis*. Until recently, up to 5 generic names (*Crassapontosphaera*, *Discolithina*, *Koczya*, *Pontosphaera*, *Transversopontis*) have been used to designate the coccoliths that are now assigned to *Pontosphaera*. Also, different views are currently expressed regarding the relationships between *Rhomboaster* and *Tribrachiatus* (discrete or synonymous taxa), and their respective specific contents (Bybell and Self-Trail 1995; Angori and Monechi 1996; Aubry 1996; Wei and Zhong 1996). I believe that, although very closely related, the two genera are distinct, their nannoliths exhibiting different symmetries.

Occasionally, I have united generic names where criteria for generic differentiation remain indecisive. This is the case for several controversial genera included in the Family Coccolithaceae Poche (*Coccolithus–Ericsonia* and *Cruciplacolithus–Birkelundia*), and in the Family Noelarhabdaceae Jerkovic (*Reticulofenestra, Dictyococcites, Cyclicargolithus* and *Cribrocentrum*). I believe that this grouping has only minor impact on the diversity curves discussed below, but recognized that some adjustments may be necessary once the taxonomy in these families is sorted out.

Specific Taxonomy and Diversity

The number of species assigned to each genus should not be regarded as absolute. It does not correspond to the total number of species validly described and assigned to it, which would be meaningless. Rather, it corresponds to the number of species currently assigned to the genus once the superfluous (synonymous) forms have been eliminated. This is very important because in genera such as *Pontosphaera* and *Coccolithus–Ericsonia*, the number of sound taxa is about half the total number of forms that have been ascribed to them. The synonymy is to be found in Aubry (1984, 1988a, 1989b, 1990, 1998, and unpublished). As species variability is very poorly known, early morphotypes of a taxon have sometimes been elevated to specific rank. On the other hand there are distinct morphotypes that have not yet been described but deserve a species status. Whereas the splitting of species may be useful for biostratigraphic purposes, it artificially increases species diversity. Thus the number of species given herein should be regarded as informative and not absolute. Nevertheless, it is unlikely that any number of species in a genus should be doubled or halved (unless the genus currently includes only 1 or 2 species), and although approximate, the numbers yield significant information with regard to diversification in the genera.

Family Taxonomy

Published taxonomic frameworks at the family level show large disparities. Comparison between the three synthetic frameworks currently available reveals disagreements both as to the authenticity of families and to their generic content (appendices 10.1, 10.2). Thus a genus may be placed by different authors in what appears to be unrelated families.

Because it incorporates improved understanding of the role of structure in calcareous nannofossil taxonomy, Perch-Nielsen's framework (1985a) has represented a major improvement in the classification of the Cenozoic calcareous nannofossil taxa. Although this has resulted in greater segregation of Mesozoic and Cenozoic families than previously admitted, some uneasiness in establishing a comprehensive family framework transpires from the assignment of genera that occurred both in the Mesozoic and Cenozoic Eras to different families depending on the considered era (see appendix 10.2, for instance).

A new framework is being developed (Aubry 1989a, 1990, 1998, and in progress; appendix 10.3). As genera are based on structural characteristics, families are likewise established based on structural homology. It would be beyond the scope of this chapter to discuss this framework, but a brief overview is given in appendix 10.4. To avoid overlap in generic content of Mesozoic and Cenozoic families as in Perch-Nielsen (1985a, 1985b), it would appear necessary to separate Cenozoic and Mesozoic taxa in different families except when no structural change has occurred across the Cretaceous/Paleogene boundary as in a few families (e.g., Braarudosphaeraceae, Calciosoleniaceae, and Goniolithaceae). This approach is easily justified using the example illustrated in appendix 10.5.

The classification that is being developed and is outlined here contributes to enhance the differences between the Mesozoic and the Cenozoic coccolithophorids. As classification is currently under revision, several family names are used herein informally pending documentation of the proposed amendments.

Diversity Analysis

In addition to sound taxonomy, one must be concerned with how representative fossil assemblages are of real diversity at any one time. Holococcolith-bearing coccolithophorids, which are extremely diverse in the modern calcareous nannoplankton (e.g., Kleijne 1991) show little diversity throughout the Cenozoic Era. The group is likely underrepresented in fossil assemblages, possibly due to its sensitivity to dissolution (Roth and Thierstein 1972). Modern genera that are distinguished based on coccolith dimorphism may not be known from the fossil record if one of the coccolith types is not preserved. For instance the two extant genera *Acanthoica* and *Anacanthoica* differ from each other by dimorphism in the latter. The sacculiform rhabdoliths that are common to both are known from upper Miocene sediments (Aubry 1998) but the tiny styliform rhabdoliths that characterize *Acanthoica* are unlikely to be preserved or recognized in smear slide preparations, which jeopardizes the possibility of determining a fossil record for the genus.

Diversity in a family may be increased considerably with the serendipitous finding of new genera in particularly well preserved sediments. For instance, Paleogene diversity in the Family Rhabdosphaeraceae almost doubled with the description of new genera of tiny rhabdoliths exceptionally well preserved in upper Eocene and lower Oligocene sediments from Australia (Shafik 1989; and see Aubry 1998). Many isolated monospecific genera have been described, and, no doubt, more will be. Thus total diversity may never be known. However, our concern here is not total diversity. It is mainly to analyze the timing of the origination of genera that contributed most to the Cenozoic calcareous nannoplankton. Rare, isolated monospecific genera provide little additional information about global patterns of diversification. They merely confirm these patterns.

Determination of the timing of the evolution of genera and their temporal ranges remains a challenge. The quality of the data is extremely uneven, being very precise for some, speculative for others. Whereas for many of the highly diversified genera the FAD and LAD of their first and last representatives, respectively, have been well calibrated using magnetostratigraphy, those of other genera are tentative. This is particularly true for genera with few species known from few locations. To determine the temporal range of taxa other than those discussed in Berggren et al. (1995), I have used the synthetic overview of Perch-Nielsen (1985a) on the stratigraphic range of taxa and complemented it with an evaluation of data in subsequent biostratigraphic studies from deep sea as well as land sections (e.g., Theodoridis 1984; Aubry 1992a; Pospichal and Wise 1990, Bralower and Mutterlose 1995; Bybell and Self-Trail 1995; Aubry et al. 1996; Angori and Monechi 1996 among many others). Greater emphasis has been placed on determining as precisely as possible the temporal range of species assigned to genera that are common to both the Mesozoic and the Cenozoic eras (see appendix 10.6) and that of Paleocene and early Eocene taxa. The same age has been given to FADs and LADs that occurred closely, but whose succession is not firmly known. Whereas this has no consequence for the purpose of this study, it results in artificially synchronous clusters of evolutionary events. It must be understood that this synchrony is artificial. Yet, it indicates that there were short intervals within which accelerated evolution occurred. For valid mono- or bi-specific genera described without precise information as to the range of their representatives, and whose species have been reported only once or twice (e.g., *Aspidorhabdus*), I have estimated the age of the type material using either the zonal assignment given by the author(s) or the general composition of the assemblages.

With all the caution expressed above, diversification through the Cenozoic is discussed below with emphasis on the late Paleocene–early Eocene interval.

Early Cenozoic Calcareous Nannofossil Evolution and the P/E BI

To correctly evaluate the significance of the extinctions that occurred among the calcareous nannoplankton species near the Eocene/Oligocene Epoch boundary and to determine their cause, it proved necessary to examine evolutionary patterns much before and after the boundary, i.e., from the early middle Eocene through the early Oligocene interval, i.e., over a 20-my span (Aubry 1992b). Similarly to determine any significant event in the evolution of the Cenozoic calcareous nannoplankton during the P/E BI, it is necessary to examine its evolutionary patterns in a broader context. These are examined below at different scales.

Cenozoic Diversity and Diversification

As presented here, the Cenozoic calcareous nannoplankton comprise 91 genera (exclusive of Recent taxa and about 12 poorly documented isolated genera; figure 10.1, color insert) although, because of the grouping of some genera for the purpose of this chapter, a total number of 95 genera may be more appropriate. Seventy-five of these are divided among 25 families, the remaining sixteen genera being considered *incertae sedis* (figure 10.2; appendix 10.3). Not all 91 genera are strictly Cenozoic genera. Thirteen genera are common to both the Mesozoic and Cenozoic eras (figure 10.1). They are referred to herein as "relict genera" as opposed to (78) Cenozoic genera.

Generic diversification did not occur at a regular pace through the Cenozoic Era (figure 10.3; appendix 10.7). Fifty percent of the Cenozoic genera evolved during the first 15 my of the Paleogene Period (i.e., from the earliest Paleocene Epoch to the mid early Eocene Epoch), with 32% of them during the Paleocene Epoch alone (i.e., in 10 my). This corresponds to an average rate of diversification of 3.3 genera/my through the Paleocene–early Eocene Epoch versus an average rate of 1 genus/my through the remaining of the Cenozoic Era. There is thus a large contrast between early Paleogene high generic diversification and subsequent slow diversification. Averaged rates over long intervals are, however, somewhat misleading. No genus evolved during the early Oligocene Epoch, probably as a result of oceanic eutrophication as a response to global cooling (Boersma et al. 1987; Hallock et al. 1991; Aubry 1992b). In contrast, accelerated generic diversification occurred during the P/E BI (see below).

Of the 18 strictly Cenozoic families, 13 evolved during the first 15 my of the Cenozoic Era (figure 10.2; see color insert). The proportion (72%) of families that had evolved before 50 Ma may be high if we consider that 66% of the incertae sedis taxa evolved after 50 Ma. Taxa such as *Minylitha* and *Florisphaera*, *Oolithothus*, and *Hayaster* are likely to be closely related, which are potential representatives of discrete families. Nevertheless, it reflects the high rate of family diversification (1.1 family/my) between 65 and 50 Ma.

Paleocene Diversity

The Paleocene calcareous nannoplankton comprises two groups of taxa. Relict members of the Cretaceous nannoplankton constitute the first group. The other consists of new genera, which evolved from Cretaceous taxa but with radical structural differences. I will refer to them below as Cenozoic genera.

The relict Cretaceous nannoplankton genera

The striking contrast between Maestrichtian and Danian calcareous nannofossil assemblages was documented over three decades ago (Bramlette and Martini 1964). However, regardless of the number of taxa that became extinct at the Cretaceous/Paleogene Period boundary, and of the unsolved puzzle regarding whether Maestrichtian species survived into the earliest Paleocene Epoch (Perch-Nielsen et al. 1982) or Cretaceous coccoliths were reworked into the lowermost Paleocene sediments (Thiertein and Okada 1979; Thierstein 1981), it is now well established that a number of taxa that were rare during the Maestrichtian Age became common during the early Paleocene Epoch (Perch-Nielsen 1969, 1972; Percival and Fischer 1977; Pospichal and Wise 1990; Erhendorfer and Aubry 1992). These are the persistent species of Percival and Fischer (1977) and the survivors of Perch-Nielsen (1969, 1979). They belong to 13 genera that thus remained unaffected by the K/P extinction event (figure 10.1; see color insert). Perch-Nielsen (1981a:table 2) proposed that 18 Cretaceous genera either persisted into or had direct descendants in the Paleocene Epoch. Differences in interpreting phylogenetic relationships between Paleocene genera in Perch-Nielsen (1981a, 1985a) and herein explain this numerical difference. It is beyond the scope of this chapter to discuss these differences, but I provide a short documentation of the relationships that I propose (figure 10.4; appendix 10.8). By way of explanation I can cite for instance the fact that I propose that *Ellipsolithus* evolved from *Biscutum* rather than from *Repagalum* as suggested by Perch-Nielsen (1981a).

The relatively large number (13) of relict genera appears disproportionate with regard to their contribution to the Paleocene calcareous nannoplankton, both in term of numbers of species and specimens (see below). On the other hand, these 13 genera had tremendous importance in the evolutionary history of the Cenozoic calcareous nannoplankton that evolved from them.

The Cenozoic genera

At least 24 genera evolved during the Paleocene Epoch, among which 7 evolved during the P/E BI, and 7 are restricted to the Paleocene Epoch (figures 10.1, 10.5; see color insert). Generic diversification did not occur regularly through the epoch, but in three successive steps (table 10.1). The first step was in the earliest Paleocene Epoch, essentially during Biochrons NP1 and NP2 (65 to 63.8 Ma) when 8 genera evolved. The next step was during late Biochron NP4–early Biochron NP5 (60.6 to 59.4 Ma) when 4 additional genera evolved and at least one relict genus became extinct. The third step was during the P/E BI when 4 relict and 5 Cenozoic genera became extinct and 8 new genera evolved, one of them with a short life span in the interval. In addition, among the Cenozoic genera, *Heliolithus* became extinct and *Zygrhablithus* evolved shortly before the P/E BI (figures 10.1, 10.5; see color insert). Thus the P/E BI corresponds to an unsuspected sharp turnover in the calcareous nannoplankton. There may be some contention as to the number of genera that evolved then. First, *Helicosphaera* sp. was described from upper Paleocene sediments of the Atlantic Ocean (Perch-Nielsen 1977), but this represents an isolated occurrence and the genus was not firmly established until Biochron NP12. Second, contrary to the more commonly accepted view (see discussion in Aubry 1989a, 1990) I have earlier proposed that *Pontosphaera* might have evolved during the Maestrichtian Age and had become firmly established during the late Paleocene Epoch. Even if this proposition should prove correct, there would be an occurrence gap of over 10 my between the evolution of *Pontosphaera* and its radiation. This implies that whichever suggestion is correct regarding the timing of its origination, the P/E BI was an important time in the radiation of the genus. The same is true for *Zygodiscus*, which is usually considered to have evolved during the Paleocene Epoch, but may have been represented in the Maestrichtian Age (Aubry 1989a).

Discussion

Until now, no sharp turnover near the Paleocene/Eocene Epoch boundary has been proposed for the calcareous nannoplankton. This is probably due to a combination

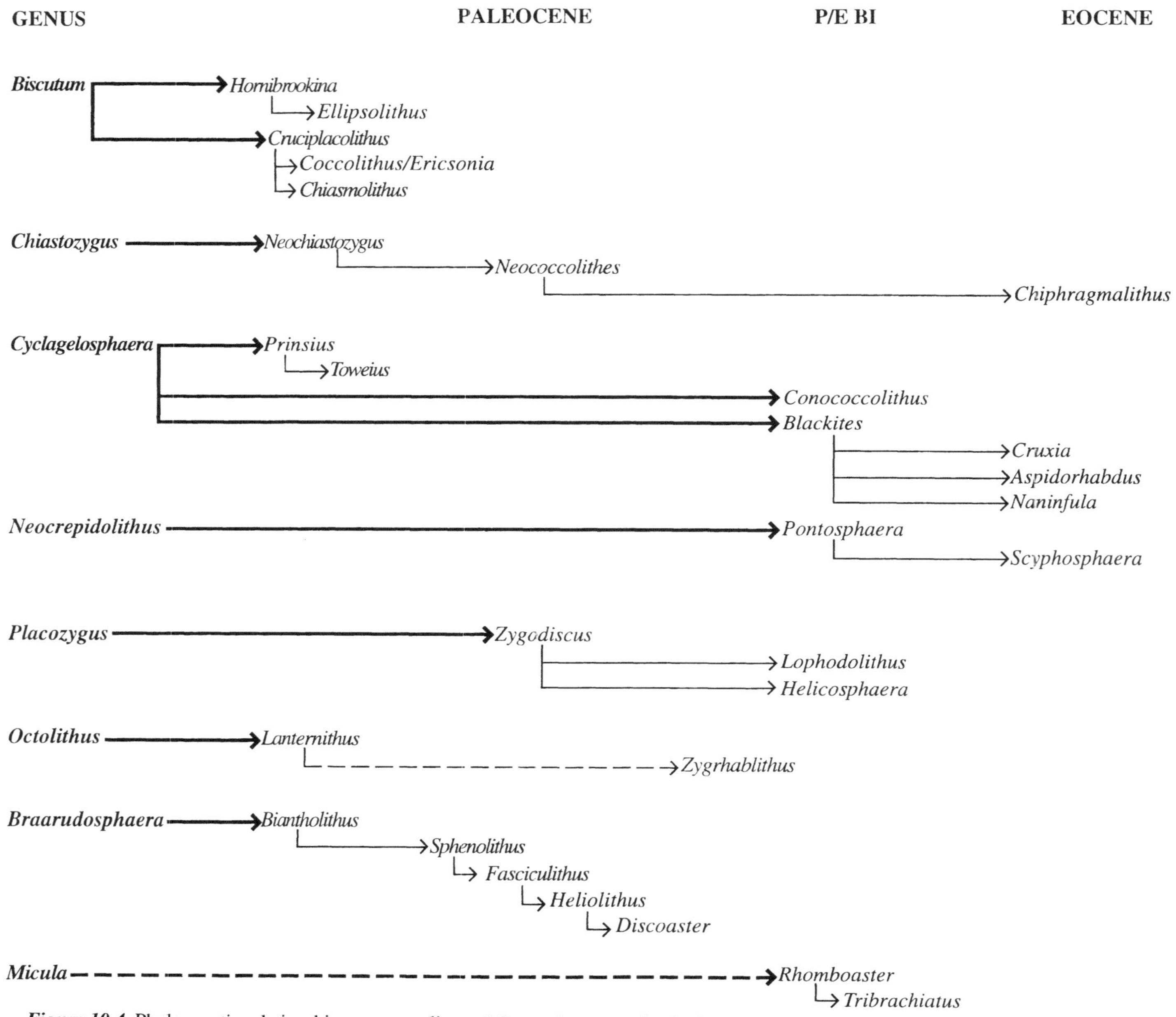

Figure 10.4 Phylogenetic relationships among relict and Cenozoic genera for the Paleocene and early Eocene epochs. Relict genera are in bold characters. Phylogenetic links between relict genera and Cenozoic genera indicated by bold arrows. Regular arrows indicate phylogenetic relationships between Cenozoic genera. See appendix 10.7 for further explanation.

Table 10.1
Paleocene calcareous nannoplankton generic diversification

Evolutionary Steps	Genus		Evolutionary Event	
Earliest Paleocene diversification (65 to 63.8 Ma)	New Paleocene Genera:	*Biantholithus*	FAD	
		Prinsius	FAD	
		Hornnibrookina	FAD	
		Neochiastozygus	FAD	
		Lanternithus	FAD	
		Cruciplacolithus	FAD	
		Chiasmolithus	FAD	
		Ericsonia/Coccolithus	FAD	
Mid-Paleocene diversification (60.6 to 59.4 Ma)	Relictual genus:	*Chiastozygus*		LAD
	New Paleocene genera:	*Fasciculithus*	FAD	
		Zygodiscus	FAD	
		Toweius	FAD	
		Neococcolithes	FAD	
P/E BI turnover (55.5 to 54.37 Ma)	Relictual genera:	*Placozygus*		LAD
		Neocrepidolithus		LAD
		Cyclagelosphaera		LAD
		Biscutum		LAD
	New Paleocene genera:	*Biantholithus*		LAD
		Prinsius		LAD
		Hornibrookina		LAD
		Fasciculithus		LAD
		Zygodiscus		LAD
		Conococcolithus	FAD	LAD
		Rhomboaster	FAD	
		Tribrachiatus	FAD	
		Lophodolithus	FAD	
		Blackites	FAD	
		Pontosphaera	FAD	
		Helicosphaera	FAD	
		Clausicoccus	FAD	

of reasons. First, the turnover is seen at the generic level, and representatives of many of the genera involved (e.g., *Biantholithus*, *Neocrepidolithus*, *Cyclagelosphaera*, *Conococcolithus*), particularly those that became extinct, are usually scarce in assemblages. Second, several of these genera (e.g., *Biantholithus*, *Neocrepidolithus*, *Placozygus*, *Cyclagelosphaera*), were represented by few species. Third, representatives of several genera (e.g., *Biantholithus*, *Hornibrookina*, *Zygodiscus*, *Lophodolithus*, *Pontosphaera*) are mostly common in deposits from epicontinental settings that are sensitive to sea-level changes and, thus 1) do not provide expanded sections of continuously occurring planktonic microfossils, and 2) offer sections that are riddled with unconformities. Fourth, upper Paleocene–lower Eocene deep sea sections also yield stratigraphic gaps (Aubry et al. 1996, chapter 2; Berggren and Aubry 1996), and, in most sections through oceanic sediments, there is no striking change in specific diversity and species abundance. This is in strong contrast, for instance, with the remarkable change in calcareous nannoplankton composition across the Eocene/Oligocene Epoch boundary, particularly at low and mid-latitudes (see Aubry 1992b), resulting from the extinction of the rosette-shaped discoasters. Somehow, the high species diversity of the late Paleocene–early Eocene calcareous nannoplankton has concealed a remarkable generic turnover.

Significance of the P/E BI Turnover

Cenozoic calcareous nannoplankton evolution

The P/E BI event was the most pronounced and decisive Cenozoic turnover in the history of the Cenozoic calcareous nannoplankton (figures 10.1, 10.2; see color insert). Unlike the turnover associated with the Eocene/Oligocene Epoch boundary, which spread over several my, the P/E BI turnover occurred over a short time (~1.1 my). Most importantly, it gave rise to some of the longest ranging (figures 10.1, 10.2; see color insert) and most prolific (figures 10.2, 10.3; see color insert) Cenozoic genera, some of them still extant. In addition, the P/E BI straddles a 5-my-long interval (Biochron NP9 to Biochron NP12; 57.5 to ~52.5 Ma) during which genera whose species constituted the bulk of late Paleogene and Neogene assemblages evolved (figures 10.1–10.3; see color insert). *Discoaster* (>130 spp.), *Pontosphaera* (>60 spp.), *Helicosphaera* (48 spp.), and *Scyphosphaera* (52 spp.) evolved then, shortly followed by *Reticulofenestra* (>40 spp.).

The importance of the P/E BI turnover is however best seen in the context of Paleocene diversification.

Paleocene diversification

Few studies have dealt with the phylogenetic relationships among Cenozoic calcareous nannoplankton genera. Perch-Nielsen (1981a, 1981b) was the first author to designate ancestor–descendant relationships between Cretaceous and Paleocene genera. Some of the relationships that she proposed are now well established but others are contentious, and I have constructed a different phylogenetic history (compare table 2 in Perch-Nielsen (1981a) with figure 10.4 herein). Establishing phylogenetic links between calcareous nannoplankton genera is often hampered by parallel and convergent evolution (Aubry 1988b, 1989a, 1990). The phylogenetic relationships given below are thus those established by Perch-Nielsen (1981a, 1981b), Aubry (1989a, 1990, 1998, and unpublished) and Romein (1979). It is noteworthy that the Paleocene diversification pattern described below is essentially independent of the phylogenetic history chosen. This is seen in the plot of taxonomic richness through the Late Cretaceous and Paleogene Periods based on data compiled from Perch-Nielsen (1985a, 1985b) in Knoll (1989:figure 3). Although at a coarser scale, the plot shows the same main Paleocene–early Eocene features as described below. This reflects the fact that, although there may be disagreement as to which relict genus was ancestral to a particular Cenozoic genus, there is little disagreement as to the timing of the evolution of the Cenozoic genera.

Paleocene generic diversification involved two processes. One was the evolution of Cenozoic genera from Cretaceous ancestors. The other was the evolution of additional Cenozoic genera from Paleocene ancestors. It is unquestionable that relict genera were at the core of Paleocene (and in fact Cenozoic) diversification.

As indicated above, 13 relict genera were part of the Paleocene calcareous nannoplankton. In term of species diversity and frequency, they contributed very little to it. First, no more than 2 species per relict genus spanned the Cretaceous/Paleogene Period boundary. Second, only six of the relict species increased in frequency immediately following the boundary so as to have contributed significantly to the bulk of the earliest Paleocene nannoplankton (table 10.2). Third, with the exception of *Micrantholithus*, the relict genera did not diversify or diversified little during the Paleocene Epoch. Lastly, also with the exception of *Micrantholithus*, their species were common only in the earliest Danian Age; they became quickly dominated by the newly evolved truly Cenozoic taxa (table 10.3; Pospichal and Wise 1990; Ehrendorfer and Aubry 1992). Contrary to what happened at the generic level, specific diversification during the Paleocene Epoch was mostly the fact of the Cenozoic genera with limited contribution from the relict genera.

Despite their modest share in the composition of the Paleocene calcareous nannoplankton, some of the relict genera came to play a tremendous role in the Paleocene repopulation of the world ocean photic zone. Not all relict genera were involved: *Goniolithus*, *Lapideacassis*, *Micrantholithus*, *Scampanella* and *Scapholithus* do not seem to have been at the origin of any Cenozoic genus during the Paleocene Epoch. In contrast, *Braarudosphaera*, *Biscutum*, *Chiastozygus*, *Cyclagelosphaera*, *Markalius*, *Neocrepidolithus*, *Octolithus*, and *Placozygus* are at the origin of at least 10 Paleocene lineages (figure 10.4; appendix 10.8). In addition, *Micula*, a genus usually not regarded as relict is thought to be at the origin of the *Rhomboaster–Tribrachiatus* lineage (Romein 1979). Paleocene diversification through the relict genera occurred in three waves (figures 10.1, 10.2, 10.5; see color insert). The first was in the earliest Paleocene Epoch (65 to 62.2 Ma) when *Hornibrookina*, *Cruciplacolithus* and subsequently *Ellipsolithus* evolved from *Biscutum*, *Neochiastozygus* from *Chiastozygus*, *Prinsius* from *Cyclagelosphaera*, *Biantholithus* possibly from *Braarudosphaera*, *Lanternithus* from *Octolithus*. The second wave occurred in the mid-Paleocene Epoch (60.6 to 59.4 Ma) when *Zygodiscus* evolved from *Placozygus*, and *Sphenolithus* and *Fasciculithus* from *Biantholithus*. The third wave took place during the P/E BI when *Blackites* and *Conococcolithus* evolved from *Cyclagelosphaera*, *Pontosphaera* from *Neocrepidolithus* and *Rhomboaster* possibly from *Micula*.

The mid-Paleocene wave was minor and differed from the other two in involving more earlier Cenozoic (4) than relict genera (1). However, while there is a clear mid-Paleocene increase in generic diversification, the role played by the relict genera in the evolution of *Sphenolithus* and *Fasciculithus* is uncertain. In contrast to Perch-Nielsen (1981a) who suggested *Cyclagelosphaera* and *Markalius* as ancestral to *Sphenolithus* and *Fasciculithus*, respectively, I suggest *Biantholithus* as a common ancestor (figure 10.4).

Based on the discussion above, the distribution of generic originations during the Paleocene Epoch is as follows (figures 10.4, 10.5):

Table 10.2
Persistent genera and their Cenozoic species

Persistent Genera	Species Spanning the K/P Boundary	Paleocene Species
Biscutum	*B. castrorum**	*B. romeini*
		B. parvulum
Braarudosphaera	*B. bigelowii**	*B. alta*
	B. turbinea	
Chiastozygus	—	*C. ultimus*
Cyclagelosphaera	*C. margereli**	*C. alta*
		C. reinhardtii
Goniolithus	*G. flukigeri*	—
Lapideacassis	—	*L. blackii*
		L. trispina
		L. multispina
Markalius	*M. inversus**	*M. apertus*
Micrantholithus	—	*M. aequalis*
		M. bramlettei
		M. concinnus
		M. crenulatus
		M. entaster
		M. fornicatus
		M. inaequalis
		M. pinguis
		M. truncus
Neocrepidolithus	*N. neocrassus**	*N. cruciatus*
		N. fossus
		N. dirimosus
		N. biskayae
		N. bukryi
Octolithus	*O. multiplus*	—
Placozygus	*P. sigmoides**	—
Scampanella	*S. magnifica*	*S. asymmetrica*
		S. bispinosa
		S. wisei
Scapholithus	*S. fossilis*	*S. apertus*
		S. rhombiformis
		S. solidus

* indicates species that became notably abundant during the earliest Paleocene Epoch.

1. In the earliest Paleocene Epoch, 7 Cenozoic genera evolved from 5 relict genera. One Cenozoic genus gave rise to 2 younger Cenozoic genera.
2. In the mid-Paleocene Epoch, 5 additional Cenozoic genera evolved, 4 from older Cenozoic genera, one from a relict genus.
3. During the P/E BI, 3 relict genera and 2 older Cenozoic genera gave rise to 7 other Cenozoic genera.

Of the 24 Cenozoic genera that evolved during the Paleocene Epoch, 21 were produced during these three waves. The 3 remaining genera evolved during the late Paleocene Epoch from older Cenozoic genera. *Heliolithus* likely evolved from *Fasciculithus* at ~58.4 Ma; *Discoaster* evolved from *Heliolithus* at ~57.5 Ma; *Zygrhablithus* evolved shortly before the P/E BI from a holococcolith-bearing genus, possibly from *Lanternithus*. It cannot be overemphasized that over half of the Cenozoic genera that evolved during the Paleocene Epoch evolved from 6 relict genera, represented by a tiny number (1 to 4) of Paleocene species. Instead, Paleocene specific diversity increased through the radiation of the Cenozoic genera. The modes of generic and specific diversifications during the Paleocene Epoch are in striking contrast.

Paleocene extinctions

Whereas only 2 relict genera became extinct in the course of the Paleocene Epoch, a total of 12 genera

Table 10.3
Generic diversity of different epochs

Epoch	No. of Cenozoic Genera	No. of Relict Genera
Pleistocene	18	2
late Pliocene	18	2
early Pliocene	21	2
late Miocene	23	2
middle Miocene	18	3
early Miocene	18	3
late Oligocene	19	5
early Oligocene	27	5
late Eocene	34	5
middle Eocene	35	6
early Eocene	31	6
P/E BI	22	9
late Paleocene	16	13
early Paleocene	9	13

The Pleistocene count does not include Recent taxa (that would increase diversity considerably because of the high number of Recent [holococcolith-bearing] Calyptrosphaeraceae). Poorly documented taxa (e.g., *Vermiculithina*, *Gongylis*, *Nannoturba*) are not included.

It is worth noting that, actually, maximum generic diversity was not reached during the early Eocene but during the middle and late Eocene Epoch as ecotonal communities developed in response to progressive ocean mixing (see discussion in Aubry 1992b).

became extinct shortly before and during the P/E BI. One relict genus and one Cenozoic genus became extinct just before the P/E BI. Four relict and 6 Cenozoic genera became extinct during the P/E BI. This means that by the end of the Paleocene Epoch, over half of the relict species and almost a third of the Cenozoic genera had become extinct. In addition, all the relict genera that gave rise to Cenozoic genera during the Paleocene Epoch had become extinct. Following the P/E BI, none of the remaining relict genera contributed to fundamental generic differentiation as occurred before. Pentaliths (a type of coccolith) of the middle Eocene genus *Pemma* are very close to pentaliths of *Micrantholithus*. Pentaliths of *Pentaster* are reminiscent of those of *Braarudosphaera*.

Although additional analysis is required, it appears as a first approximation that the Cenozoic genera that became extinct shared a common character. They were robust, massive, strongly calcified (e.g., *Fasciculithus*, *Heliolithus*, *Zygodiscus*, *Neocrepidolithus*), and in this respect were reminiscent of many Cretaceous genera (e.g., *Micula*, *Cylindralithus*).

Paleocene–Early Eocene Nannofossil Diversification: Evidence of Climatic Amelioration

Because of their role as primary producers in the photic zone of the world ocean, changes in evolutionary and biogeographic patterns in the calcareous nannoplankton provide critical evidence for understanding global evolution of the biosphere and the feedback relationships between it and the physical environment. As for other planktonic groups, long-term changes in diversity in the calcareous nannoplankton are now seen to be linked to changes in productivity resulting from climatically induced decrease or increase in ocean mixing. Based on a comparison with evolutionary patterns in the planktonic foraminifera (Boersma et al. 1987; Hallock et al. 1991), I have shown (Aubry 1992b) that both decreasing diversity and increasing latitudinal biogeographic differentiation among the calcareous nannoplankton during the late Paleogene Period (Chron C22n to Chron C6Cr) reflected increased ocean mixing as a response to global cooling, which eventually led to glaciation (see reviews in Miller et al. 1987b and Miller 1992). In contrast, I propose below that increasing diversity and decreasing latitudinal biogeographic differentiation during the early Paleogene Period (Chron C29r to Chron C22r) were linked to increased stratification in the photic zone as a result of progressive warming. However, there was a relay in this trend. It was smooth and brief, and lasted the duration of the P/E BI. It consisted in a rapid shift among the calcareous nannoplankton genera that contributed to increasing diversification; and it resulted in the establishment of what may be called the modern calcareous nannoplankton.

Paleocene–Early Eocene Diversification: A Byproduct of Warming

The early Eocene has long been recognized as the warmest epoch of the Cenozoic Era (see review in Frakes 1979). Benthic foraminiferal isotope records, the best available tool currently for determining long-term global environmental changes (e.g., Zachos et al. 1993, 1994; Corfield 1994), have helped to delineate the timing, modalities and amplitude of the early Cenozoic warming. Compilation of benthic foraminiferal isotope records from various latitudes and oceans show that Paleocene warming began in mid Chron C26r at ~60 Ma in reference to the time scale of Berggren et al. (1995). Warming appears to have been progressive until latest Chron C23n (~50.5 Ma) except for an abrupt short term (~10 kyr) event first identified at Maud Rise Site 690 (Kennett and Stott 1991) and referred to as the "latest Paleocene thermal maximum" (LPTM, Zachos et al. 1993). According to Zachos et al. (1994) deep sea temperatures increased ~4° C, from 7° to 10° C in the late Paleocene to 10° to 14° C in the early Eocene. During the LPTM the deep ocean warmed by 4° C and the sea surface temperatures (SST) at high latitudes increased by 5 to 6° C (Zachos et al. 1993). Temperature began to decrease again during Chron C22r.

It appears that early Paleogene calcareous nannoplankton diversification largely followed this warming trend. First, I have shown that generic diversification during the Paleocene–early Eocene epochs (65 to 50 Ma) occurred at a rate three times greater than later (figure 10.3): 50% of the Cenozoic genera had evolved by the mid early Eocene Epoch (i.e., before 50 Ma). Second, I have indicated that most Cenozoic families evolved during the first 15 my of the Cenozoic, which corresponds to a rate of 1.1 family/my in striking contrast to a rate of 0.1 family/my after 50 Ma. Third, as described above, Paleocene diversification proceeded in three waves. The earliest Paleocene wave can be seen as the immediate consequence of the end Maestrichtian mass extinction, which made essentially vacant a huge ecologic reservoir. The second wave occurred in mid Chron C26r and, as a first approximation its timing appears essentially synchronous with the beginning of warming, although direct comparison of calcareous nannofossil distribution and stable isotopic values in several Paleocene stratigraphic sections is required to ascertain a link between earliest warming and generic diversification. The third wave occurred during the P/E BI, which began with the LPTM. Lastly, successive radiations in the genera *Fasciculithus*, *Heliolithus*, *Neochiastozygus* and *Discoaster* (Romein 1979; Perch-Nielsen 1981a, 1981b) paralleled the late Paleocene warming trend.

The link between diversification and rising temperature is, however, indirect. Nutrient availability is recognized as the main factor that controls diversification (Valentine 1973; Dodd and Stanton 1981; Hutchinson 1978; Tappan and Loeblich 1973; Tappan 1986; Martin 1996). Warming results in decreased intensity of oceanic circulation, which leads to the expansion of oligotrophic areas (see Boersma et al., chapter 11).

Hallock (1987) has discussed how spatial heterogeneity in the distribution of trophic resources, which is referred to as trophic resource continuum (TRC), controls the diversity of pelagic biotas. Trophic resources are defined as including "both nutrients available to primary producers and the consequent organic carbon available to consumers" (op. cit., p. 458). Trophic resources thus vary with factors that control the diversity and abundance of the primary producers. In addition to nutrients (nitrogen, phosphorus, silica, iron and other trace elements; see Hallock 1987), environmental stability (which favors specialization) and water transparency (depth of light penetration critical for photosynthesis) constitute parameters whose variations through time result in increased or decreased diversity among primary producers.

Coccolithophorids are intriguing by their adaptation to extremely low nutrient availability (Kilham and Soltau Kilham 1980) and their maximum diversity today is seen in the oligotrophic tropical oceanic gyres where the vertical distribution of species produces a discrete stratification of the photic zone (Okada and Honjo 1973; see discussion and additional references in Aubry 1992b), indicating that species are strongly specialized. Inasmuch as greater biologic diversity in the modern ocean means finer stratification in the photic zone, it seems reasonable to infer that past diversity changes in the calcareous nannoplankton reflect changes in the stratification of the photic zone. If that is correct, times of diversification were times when stratification in the photic zone increased, whereas times of extinction were times of decreased stratification.

Hallock (1987) has developed models to explain diversity and stratification in the euphotic benthic environment. Because species adaptation is an intricate interplay between several parameters (e.g., nutrient, light, temperature), the models are based on the assumption that all species have the same ecologic requirements except for light. These models show that 1) the number of species that share the euphotic benthic environment depends on its eu-, meso- or oligotrophic (i.e., unclear, clear, very clear waters) nature, 2) species diversity increases with decreasing nutrient availability (i.e., increasing water transparency) as specialization towards narrower light intensity ranges increases, and 3) diversification (i.e., specialization) occurs as a means of reducing competition between species. As emphasized by Hallock, they also show that species that are unspecialized towards light are best adapted to eutrophic environments, and that highly specialized taxa are likely to be eliminated through loss of water transparency. These models that represent unlikely simple interactions between species and between species and their environments, and were designed for benthic species, have the merit of being generally applicable to photosynthetic primary producer, and in particular they may be applied to the calcareous nannoplankton. For instance, the modern species *Emiliania huxleyi* is a nutrient-rich indicator (see Brand 1994). It is dominant in eutrophic high latitude waters, but occurs also at any depth at any latitude (Okada and Honjo 1973) and, as shown by McIntyre and Bé (1967) its abundance pattern around Bermuda (latitude 40–45° N) shows little seasonality, except for its greater abundance in the winter months when light intensity is lower.

Hallock's models also help to predict the dynamics of diversification of the coccolithophorids (adapted to extremely oligotrophic conditions) through time. However, this is a first approximation since species structure in an oligotrophic ocean is poorly understood (Venrick 1982, 1990). When the photic zone of the ocean was primary eutrophic, diversity must have been very low. Evolution towards meso- and subsequently oligotrophy would have resulted in increased diversification with extremely specialized taxa evolving during maximum oligotrophy.

Paleocene diversification can adequately be explained along these lines. The earliest Paleocene nannoplankton were of extremely low diversity, implying an eutrophic to mesotrophic oceanic photic zone. Two species first dominated: *Braarudosphaera bigelowii* mostly at low latitudes and *Thoracosphaera* spp. preferentially at high latitudes (Haq and Lohmann 1976). They were soon replaced by newly evolved species that were tiny to small and successively became predominant, among which *Hornibrookina edwardsii*, and species of the genus *Prinsius* (Pospichal and Wise 1990; Ehrendorfer and Aubry 1992). Although taxa evolved during the early Paleocene Epoch, the nannoplankton remained of low diversity until mid-Chron C26r. Radiation occurred in only a few genera, suggesting minimal specialization. This implies that the photic zone remained essentially mesotrophic through the early Paleocene Epoch with little stratification in the distribution of taxa. In contrast the late Paleocene and early Eocene were times of accelerated generic diversification and radiation, suggesting decreasing nutrient availability, resulting in increasing water transparency, and consequently increasing stratification in species distribution in the photic zone (i.e., greater specialization).

Two datasets confirm the global trend described above. The Paleocene and early Eocene biogeographic patterns exhibited by the planktonic foraminifera point to early Paleocene mesotrophic conditions and late Paleocene–early Eocene strong stratification. The carbon isotopic record supports low productivity during the early Paleocene and increasing species stratification during the late Paleocene–early Eocene. As described by Boersma and Premoli Silva (1991:39) "Earliest Paleocene planktonic foraminifera were cosmopolitan, low in diversity, small in size, generalized in their shapes. . . . Polar and equatorial faunas were typified by large numbers of fertility indices such as the elongate heterohelicids. . . . Maximum biserial heterohelicid diversity of the Paleogene was, in fact, characteristic of the early Paleogene," and "fertility-related biserial heterohelicids . . . decreased in abundance near the end of Zone P1 coeval with the first maximum of tropical indices at both lower and middle latitudes" (op. cit., p. 38). Morozovellids and acarininids indicate a well stratified late Paleocene and early Eocene oceanic photic zone (Boersma et al. 1987; Boersma and Premoli Silva 1988, 1991; Premoli Silva and Boersma 1988). As for the calcareous nannoplankton, a change occurred in mid Chron C26r, with increasing diversification through the late Paleocene reflecting increasing stratification.

Preferential removal of ^{12}C in the photic zone through plankton photosynthetic activity results in a surface-to-depth vertical gradient. Changes in this gradient through time are reconstructed from the measurements of the $\delta^{13}C$ of benthic and deep- and surface-dwelling planktonic foraminifera and used to monitor changes in productivity.

Paleocene to early Eocene carbon isotope records show several successive phases. In the earliest Paleocene Epoch (late Chron C29r–early chron C29n) the surface-to-deep water gradient collapsed, implying a major reduction in productivity as confirmed by barium accumulation rates (Zachos et al. 1989; see review in Corfield 1994). This was followed (late Chron C29n) by a progressive reestablishment of a surface-to-deep water gradient, indicating a progressive increase in productivity (Zachos et al. 1989). Isotopic values then remained constant or increased slightly until earliest Chron C26r (Zachos et al. 1989; Corfield and Cartlidge 1992). A strong and rapid increase in $\delta^{13}C$ at all depths occurred from early Chron C26r to mid Chron C25n to reach a late Paleocene maximum (Shackleton et al. 1985; Shackleton 1986; Corfield and Cartlidge 1992; Zachos et al. 1993). This long-term late Paleocene trend was coupled with an increasing difference in the surface-to-depth gradient (Corfield and Cartlidge 1992). This has been interpreted to reflect both increasing productivity and increasing depth-related specialization of the planktonic foraminifera during the late Paleocene Epoch (Shackleton et al. 1985; Corfield and Cartlidge 1992). Increased productivity through the Paleocene Epoch is supported by a strong increase in barium records (Thompson and Schmitz 1997). However, the amplitude of the productivity increase as measured on tests of *Morozovella* and *Acarinina* may be overestimated due to photosymbiotic activity in these genera (D'Hondt and Zachos 1994). From late Chron C25n to latest Chron C24r carbon isotopic values decreased and remained low through Chron C23 (Corfield and Cartlidge 1992; Zachos et al. 1993; Charisi and Schmitz 1996). Opinions differ as to the significance of this long-term decrease, which has been interpreted as reflecting decreasing productivity linked to the evolution of planktonic foraminifera adapted to oligotrophic conditions, reduced productivity due to decreased ocean circulation as a result of weak atmospheric circulation (Rea et al. 1985), input of ^{12}C through continental erosion (Rea et al. 1990; Beck et al. 1995), or input of ^{12}C from North Atlantic volcanism (Eldholm and Thomas 1993). Superimposed on this trend, there was at least one carbon isotope negative anomaly. It was recognized in both the marine and terrestrial records (Kennett and Stott 1991; Koch et al. 1992; Stott et al. 1996; Sinha et al. 1996), and was synchronous with the LPTM. This excursion has not yet received a satisfactory explanation (see discussion in Thomas and Shackleton 1996) although hypothetical massive warming-induced release of methane from oceanic hydrates is currently seen as the best explanation for ^{12}C enrichment during the LPTM (Dickens et al. 1995).

This carbon fractionation history correlates well with the history of calcareous nannoplankton diversification. Extremely low productivity during the earliest Paleocene Epoch is well supported by extremely low calcareous nannoplankton diversity. Progressive early Paleocene reestablishment of a surface-to-depth carbon isotope gradient correlates with slow and low diversification. The long-term late Paleocene increase in productivity parallels the history of generic diversification and radiation (figure 10.6). The late Paleocene carbon maximum was reached just prior to the evolution of *Discoaster multiradiatus* and the diversification and dominance of the discoasters as a group. If this group is correctly interpreted as adapted to oligotrophy (see Aubry 1992b), its late Paleocene–early Eocene (mid-Chron C25n to Chron C23) flourishing can be seen as supportive of the long-term decrease in productivity inferred from carbon isotopic records. Finally, the enigmatic short-term carbon isotope excursion occurred just at the beginning of the P/E BI.

Paleocene Calcareous Nannofossil Evolution: Increased Oligotrophy and Coevolution with Symbiont-Bearing Planktonic Foraminifera

As Hsü and McKenzie (1985) pointed out, eutrophic conditions prevailed in the earliest Paleocene ocean, nutrients being concentrated in oceanic waters as a result of mass mortality and low productivity. In this "Strangelove Ocean," only a few r-mode opportunistic, photosynthetic taxa bloomed. The calcareous nannofossil species that evolved in this early Paleocene ocean were thus adapted to, or tolerant of, nutrient-rich waters, as were the species that survived the Cretaceous/Paleogene Period mass extinction (e.g., species in the Cenozoic genera *Prinsius*, *Biantholithus*, *Hornibrookina*, but also in the relict genera, in particular in *Biscutum*, *Cyclagelopshaera*, *Neocrepidolithus*, *Placozygus*). The greater frequency of these taxa in shallow epicontinental than in oceanic Paleocene deposits supports well this interpretation (compare for instance assemblage composition in NW Europe [Aubry 1986] and New Jersey [Bybell and Self-Trail 1995] Paleocene deposits with that in eastern Pacific [Monechi 1985] and western Atlantic oceanic oozes [Okada and Thierstein 1979]). As productivity resumed in the early Paleocene Epoch, nutrient concentration progressively decreased and calcareous nannoplankton diversity progressively increased. However, species (in *Cruciplacolithus*, *Chiasmolithus*, *Ericsonia*) probably remained relatively unspecialized and tolerant of mesotrophic conditions.

The first critical moment in calcareous nannoplankton diversification was in early Chron C26r, when temperature began to rise. As generic diversification increased and as radiation occurred, productivity increased. However, increased productivity occurred not only in the calcareous nannoplankton but also, simultaneously, in the planktonic foraminifera through the development of photosymbiosis (Kelly et al. 1996; Norris 1997) conferring on them an acquired adaptation to oligotrophic environments. Because photosymbiont-bearing planktonic foraminifera seem to derive their energy primarily from their symbionts who they provide with concentrated nutrients (Jorgensen et al. 1985), they may be seen as primary producers rather than low-level consumers. In this case, coccolithophorids and photosymbiont-bearing planktonic foraminifera may have competed with each other in the photic zone, resulting in high specialization in both groups to alleviate competition (see above).

Two lines of evidence support this view. Whereas no significant Paleocene radiation occurred among asymbiontic planktonic foraminifera (*Globanomalina*, *Parasubbotina*, and *Subbotina*) radiations occurred among the symbiont-bearing planktonic foraminifera (in particular in *Morozovella* and *Acarinina*; Norris 1997; Olsson et al. 1998). Second, diversification in this latter group paralleled diversification in the calcareous nannoplankton. The development of symbiosis in the planktonic foraminifera occurred at ~61.5 Ma, i.e., in mid-Chron C26r and radiation in *Morozovella* started at ~61 Ma (Norris 1997), slightly before the second diversification event in the calcareous nannoplankton (60.6 to 59.4 Ma). Norris (1997) remarked that because of their timing, evolution of photosymbiosis appears to explain diversification in the morozovellid plexus, although correlation between the two may have been "mere coincidence" (op. cit., p. 472). I propose here that indeed the acquisition of photosymbionts by the morozovellid–acarininid plexus may have led to radiation through competition between this plexus and the calcareous nannoplankton. This competition was, at least in part, the triggering mechanism for diversification in both groups. This coupled diversification must have contributed to a rapid depletion of the nutrients in the photic zone and consequently, the evolution of more and more specialized (calcareous nannoplankton and planktonic foraminiferal) species. It is possible that the evolution of photosymbiosis in the planktonic foraminifera contributed to accelerated recovery of the coccolithophorids. Today, planktonic foraminifera establish symbiosis with chrysophytes, diatoms, and dinoflagellates (Margulis and Schwartz 1988; Hemleben et al. 1989). Most species of these marine protists have requirements for high to relatively high nutrient levels (Tappan 1980; Margulis and Schwartz 1988). If the morozovellid– acarininid plexus established symbiosis with either group, it would have contributed to a more

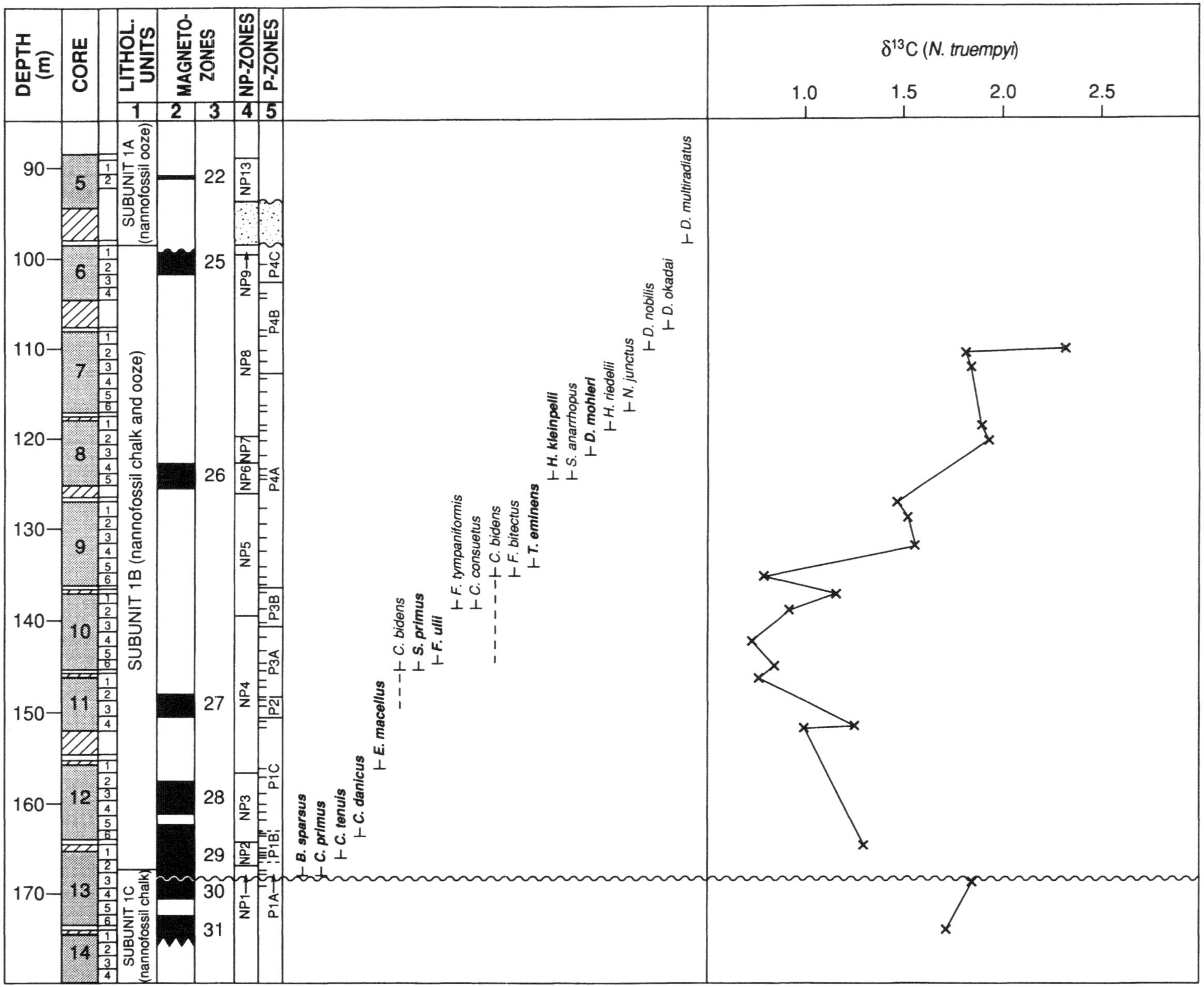

Figure 10.6 Correlation between increasing calcareous nannoplankton diversity and the rapid increase in $\delta^{13}C$ values as registered in the Paleocene section recovered from North Atlantic DSDP Site 384. Magnetostratigraphy from M. Von Fossein (unpublished); planktonic foraminiferal stratigraphy from W. A. Berggren and R. Norris (unpublished); carbon isotopic data from R. Norris (unpublished).

rapid and much greater utilization of the nutrients than the coccolithophorids alone would have. There are good reasons to believe that the Late Cretaceous coccolithophorids were adapted to oligotrophy; thus recovery of the calcareous nannoplankton to high diversity levels following the "Strangelove Ocean" episode could only be accelerated if there was an external means to assist them in reestablishing oligotrophic conditions. The photosymbiont-bearing foraminifera may have contributed to this, as strongly supported by the fact that the mid Paleocene diversification event in the calcareous nannoplankton occurred ~1 my after the establishment of photosymbiosis in the planktonic foraminifera and ~0.5 my after the first radiation in the genus *Morozovella* (see Norris 1997).

As a whole, the world ocean thus changed from nutrient-rich in the earliest Danian Age to less and less nutrient-rich during the early Paleocene Epoch, and then to more and more oligotrophic during the late Paleocene Epoch. Oligotrophic conditions probably prevailed by Chron C25n. It is interesting to compare the duration of each phase in this transition. The "Strangelove Ocean" conditions persisted less than 0.4 my (Zachos et al. 1989; duration adjusted to the time scale of Berggren et al. 1995). Normal oceanic productivity resumed in ~0.4 my and productivity remained constant or slightly increased for the next 5 my (until early Chron C26r). In contrast the Paleocene carbon maximum was reached in less than 3 my. Joint exploitation of the nutrients by two rapidly diversifying groups may have resulted in the swift transition from an "opaque" nutrient-rich ocean with a thin photic zone to a transparent-water nutrient-depleted ocean with a well developed photic zone. During this transition, species evolved towards increasing adaptation to oligotrophy. Environmental conditions were certainly favorable to rapid diversification and evolution of K-mode specialists. As latitudinal (equator to high southern) sea-surface and vertical temperature gradients were very low in the Paleocene Epoch (Corfield 1994; Zachos et al. 1993, 1994), oceanic circulation was slow and the ocean poorly mixed, with the result that oligotrophic to very oligotrophic areas could develop. Thus, joint photosynthetic activity in an expanding (thickening) photic zone by the planktonic foraminifera and the calcareous nannoplankton with little contribution from the siliceous microplankton (diatoms and radiolarians; see Fourtanier 1991; Sanfilippo et al. 1985) and co-evolution between the two resulted in the high productivity registered by the long term increase in $\delta^{13}C$ values.

It is not possible to determine with confidence the habitat and life strategies of the calcareous nannoplankton species. Isotopic analyses have helped to determine how Paleogene planktonic foraminiferal species stratified in the water column (Shackleton et al. 1985; Boersma et al. 1987; Boersma and Premoli Silva 1988; Corfield and Cartlidge 1992; Norris 1997). Although carbon and oxygen isotope analyses on individual cultured species and on Pleistocene sediments have yielded promising results (Dudley et al. 1986; Paull and Thierstein 1987; Dudley and Nelson 1988; see review in Steinmetz 1994) it is not yet possible to isolate individual coccoliths for isotopic analysis. Biogeographic patterns may help determine the habitat preferences and life strategy of at least some coccolithophorid species as done in Aubry (1992b). There has been little new information on quantitative composition of calcareous nannofossil assemblages in lower Paleogene sediments since the work of Haq and Lohmann (1976). These authors and Haq et al. (1977) described a number of assemblages and showed their cross latitudinal migrations as a function of temperature evolution. Because of the small late Paleocene equator to high latitude surface temperature gradients species distributions were rather broad and little or no biogeographic diachrony occurred (see discussion in Berggren et al. 1995), so that assemblages from different latitudes/regions are difficult to characterize. The main differences between assemblages from different latitudes concern the relative abundance of different taxa. Because the geographic coverage in these studies was limited, and because the time slices considered may have been much broader than thought (considering the new relationships that have been established between stratigraphic sections and geological time [see Aubry, chapter 3]), Paleocene biogeographic patterns need to be reevaluated. For these reasons, Paleocene evolution of the TRC, largely inferred from biogeographic patterns is not discussed here. The reader is referred to Boersma et al. (chapter 11). I only point out that early biogeographic studies showed the extension of low latitude assemblages towards high latitude during the late Paleocene–early Eocene epochs; and such extension has been confirmed by drilling in the Southern Ocean (Pospichal and Wise 1990; Aubry 1992a), implying expanded oligotrophy.

Although it is likely that each Paleocene structural innovation (resulting in a new genus) represents an adaptation to a particular depth range in the photic zone, this range is extremely difficult to determine. If correct, the interpretation given above concerning the modes of diversification in the coccolithophorids implies that the genera and species that evolved earliest were more tolerant to changing conditions and able to modify their depth range as new taxa evolved. In contrast genera and species that evolved later, i.e., when oligotrophy prevailed, were more specialized and less adaptable to changing environmental conditions and selection pressure. As indicated above, early species of *Prinsius* and *Hornibrookina* whose coccoliths constitute up to 50% of lowermost Paleocene southern high latitude assemblages (e.g., Pospichal and Wise 1990; Erhendorfer and Aubry 1992) were r-mode selected taxa. They became extinct relatively early in the Paleocene Epoch, at or before the Chron C26r diversification event (see Perch-Nielsen 1985a:figure 58). Later species in these genera were common either at high southern latitudes or in epicontinental seas. At the other end of the spectrum are species that evolved when oligotrophic conditions were well established. Discoasters evolved prior to the late Paleocene carbon maximum but became abundant only when productivity, as inferred from carbon isotopic data, begun to decrease. I have suggested earlier (Aubry 1992b) that most discoasters were indices of oligotrophic conditions. The radiation in the genus during Chron C24r supports this interpretation. I have also suggested that discoasters were deep-water species adapted to low light intensity. The holococcolith *Zygrhablithus bijugatus* is also an oligotrophic indicator. It is extremely abundant in lower Eocene assemblages (e.g., DSDP Site 577). Kleijne (1991) observed that the frequency of holococcolith-bearing species in today's photic zone is related to oligotrophic waters. Species common in upper Paleocene and lower Eocene assemblages with discoasters and *Z. bijugatus* were likely adapted to oligotrophic conditions. This includes *Ericsonia subpertusa* and *E. formosa* (see also Aubry 1992b). Species in the *Rhomboaster–Tribrachiatus* lineage were probably very specialized with regard to oligotrophy. The evolution and radiation of *Fasciculithus*, *Heliolithus*, *Neochiastozygus*, *Zygodiscus*, genera that are characteristic of the Paleocene Epoch was certainly induced by the rapidly changing fertility of the photic zone. However, while their species actively contributed to the restructuring of the ocean photic zone, it is only possible to speculate on their habitats and life strategies.

The Turnover at the P/E BI: An Enigma

While oligotrophic conditions prevailed in the late Paleocene (early Chron C24r) ocean, an enigmatic turnover occurred that changed the course of evolution.

I have described above a ~1-my-long late Paleocene turnover in the calcareous nannoplankton. It involved the extinction of 4 relict and 5 Cenozoic genera and the evolution of 8 new genera. This turnover event occurred during the long-term carbon isotopic decrease that may, at least in part, reflect the lower productivity associated with late Paleocene–early Eocene oligotrophy. It began with the LPTM and at the time of the global carbon isotope excursion (CIE). The turnover may have thus been triggered by the sudden warming although warming apparently affected only the high latitudes (Zachos et al. 1994) and thus poorly account for a profound turnover. The turnover was more likely triggered by the as yet unknown condition/event that the CIE reflects. The difficulty to explain the CIE is explored by Sloan and Thomas (chapter 9). The calcareous nannoplankton do not shed light on its origin, but emphasize the enigma.

An unexpected selectivity, which cannot be explained by a simple cause, accompanied the P/E BI turnover. It is seen in the dominance and success histories of the Paleocene coccolithophorids. Dominance and success are here taken as defined by Wilson (1992). Success is "best defined as the longevity of a species and all its descendants" (op. cit., p. 129). The longevity of a genus (eventually a family) is thus a measure of success. Dominance "is both an ecological and evolutionary concept . . . measured by the relative abundance of the species group in comparison with other, related groups, and by the relative impact it has on the life around it" (op. cit., p. 129). The frequency of species in Paleocene assemblages is thus a measure of their dominance. Selectivity occurred as follows.

First, only the relict genera that became ancestral to Cenozoic genera became extinct during the P/E BI. The only exception is *Braarudosphaera* if this genus was ancestral to *Biantholithus* as proposed here. It is to note however that the two species of *Biantholithus* are always rare in Paleocene deposits and the genus itself became extinct during the P/E BI. Second (with the exception of *Biantholithus*) the Cenozoic genera that became extinct are those that radiated and/or became dominant during the Paleocene Epoch (table 10.4). The genera *Fasciculithus*, *Heliolithus*, *Hornibrookina*, *Prinsius*, which radiated and/or dominated at one time or another during the Paleocene Epoch became extinct during the P/E BI or shortly before (*Heliolithus*). The Cenozoic genera that did not radiate but produced only a few Paleocene species (*Neococcolithes*, *Chiasmolithus*, *Ericsonia/Coccolithus*, *Sphenolithus*, *Toweius*) radiated after the P/E BI, even if one of their species had been abundant at any time during the Paleocene Epoch. Third, strikingly, the genera that radiated during the Paleocene Epoch and became extinct during the P/E BI correspond to heavily calcified, massive structures (*Fasciculithus*, *Heliolithus*) or to coccoliths whose margins consists of relatively thick elements radially arranged or poorly imbricated (e.g., *Hornibrookina*, *Prinsius*), and reminiscent of the margin of the Late Cretaceous coccoliths. In contrast, the genera that did not become extinct during the P/E BI mostly correspond to coccoliths with a margin formed of strongly imbricated elements as

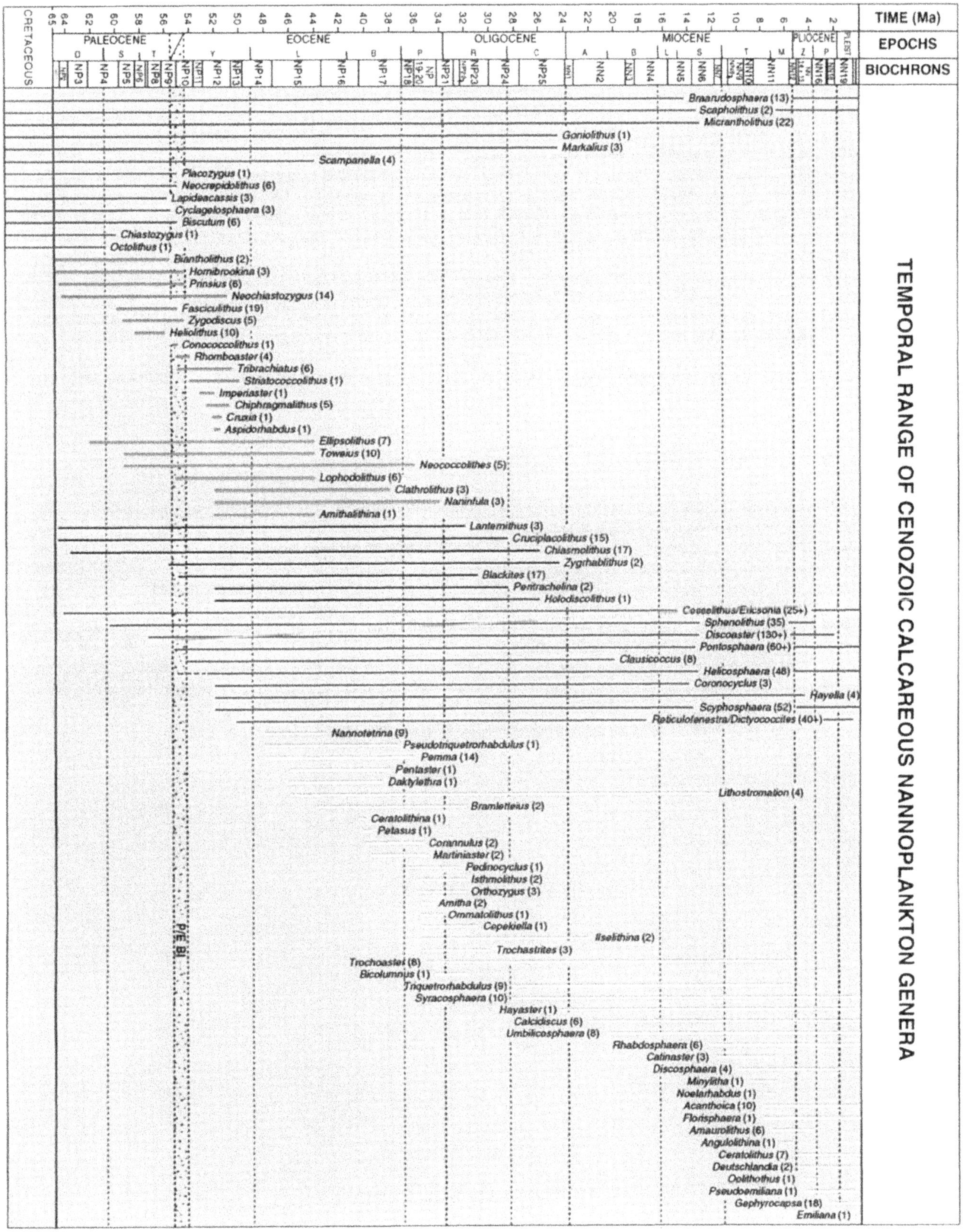

Figure 10.1 *(see caption next page)*

Figure 10.1 Temporal distribution of Cenozoic calcareous nannoplankton genera.

Temporal framework of Berggren et al. (1995). Taxonomic framework of Aubry (1984, 1988b, 1989b, 1990, 1998) and of Aubry (unpublished) and Perch-Nielsen (1985a, 1985b). Monospecific genera that are poorly documented, in particular *Quinquerhabdus* (= *Zygrhablithus*?), *Discoturbella, Neoangulolithina, Calciopilleus, Nannocorbis, Vermiculithina, Discolobatus, Gongylis, Pistillum, Nannoturba, Perforocalcinella, Tansinius* are not included. A number of dubious species have been ignored even though their stratigraphic distribution would modify the temporal distribution of the genera they have been assigned to. This is the case for instance of *Isthmolithus unipons*, described from the lower Eocene marls of the Lodo Formation (Bramlette and Sullivan 1961). If proven to be a representative of the genus *Isthmolithus*, currently regarded to be restricted to the late Eocene and early Oligocene epochs, this species would extend the distribution of the genus back to the early Eocene Epoch. Similarly, *Cycloperfolithus carlae*, described from middle Miocene deposits of the Paratethys (Lehotayova and Priewalder 1978) is regarded here as a representative of the genus *Calcidiscus*. If the genus *Cycloperfolithus* should not be considered superfluous, it would constitute an additional genus with a FAD in the middle Miocene Epoch. Extant species that have no fossil record (e.g., *Cyrtosphaera, Halopappus, Umbellosphaera, Anacanthoica, Algirosphaera, Lohmannosphaera*, and many others) are not included. The artificial genus *Semihololithus* (Aubry 1988) is rejected. *Semihololithus kerabyi* is included in *Zygrhablithus* whereas *S. biscayae* is included in *Lanternithus*.

Color code. Purple: Relict Genera: Genera that evolved during the Cretaceous Period and persisted into the Cenozoic Era. Red: Genera that were restricted to the Paleocene and early Eocene epochs. Orange: Genera that evolved during the Paleocene or early Eocene Epoch and became extinct in the middle or late Eocene Epoch. Green: Genera that evolved during the Paleocene or early Eocene Epoch and became extinct during the Oligocene Epoch. Blue: Genera that evolved during the Paleocene or early Eocene Epoch and became extinct during the Neogene Period or are still represented. Pink: other genera.

Genera are arranged in order of appearance. Numbers following the genus name indicates the total number of species in the genus. See text for further explanation.

Figure 10.2 *(following page)* Temporal distribution and generic composition of Cenozoic calcareous nannoplankton families.

Temporal framework of Berggren et al. (1995). Each single column represents a single genus. Coalescent columns show the change in generic composition of families through time. Families separated by even spaces, except when very closely related (e.g., Biantholithaceae to Discoasteraceae). Unrelated *incertae sedis* genera separated by single column-wide spaces. Formal family names in boldface; informal family names in lightface. Taxonomic framework given in table 10.3. Column arrangement in the figure corresponds to the sequence of families and genera in the table.

Color code refers to number of species per genus: As in Figure 1, relict species in purple. Monospecific genera: marine blue. Genera with 2 to 5 species: sky blue; with 6 to 10 species: aqua; with 11 to 15 species: green; with 16 to 20 species: yellow green; with 21 to 40 species: yellow; with 40 to 60 species: orange; with 61 to 80: vermilion; over 80 species: carmine.

Due to the lumping of several genera (*Cyclicargolithus, Dictyococcites* and *Reticulofenestra*), the Family Noelarhadaceae is somewhat misrepresented in this figure. This has little bearing on our discussion however, first because *Cyclicargolithus* and *Dictyococcites* evolved in the middle Eocene Epoch; second because the number of species shown for *Reticulofenestra* s.l. is the sum of species in the three genera.

Inferred phylogenetic relationships between relict genera/families shown by oblique dashed lines.

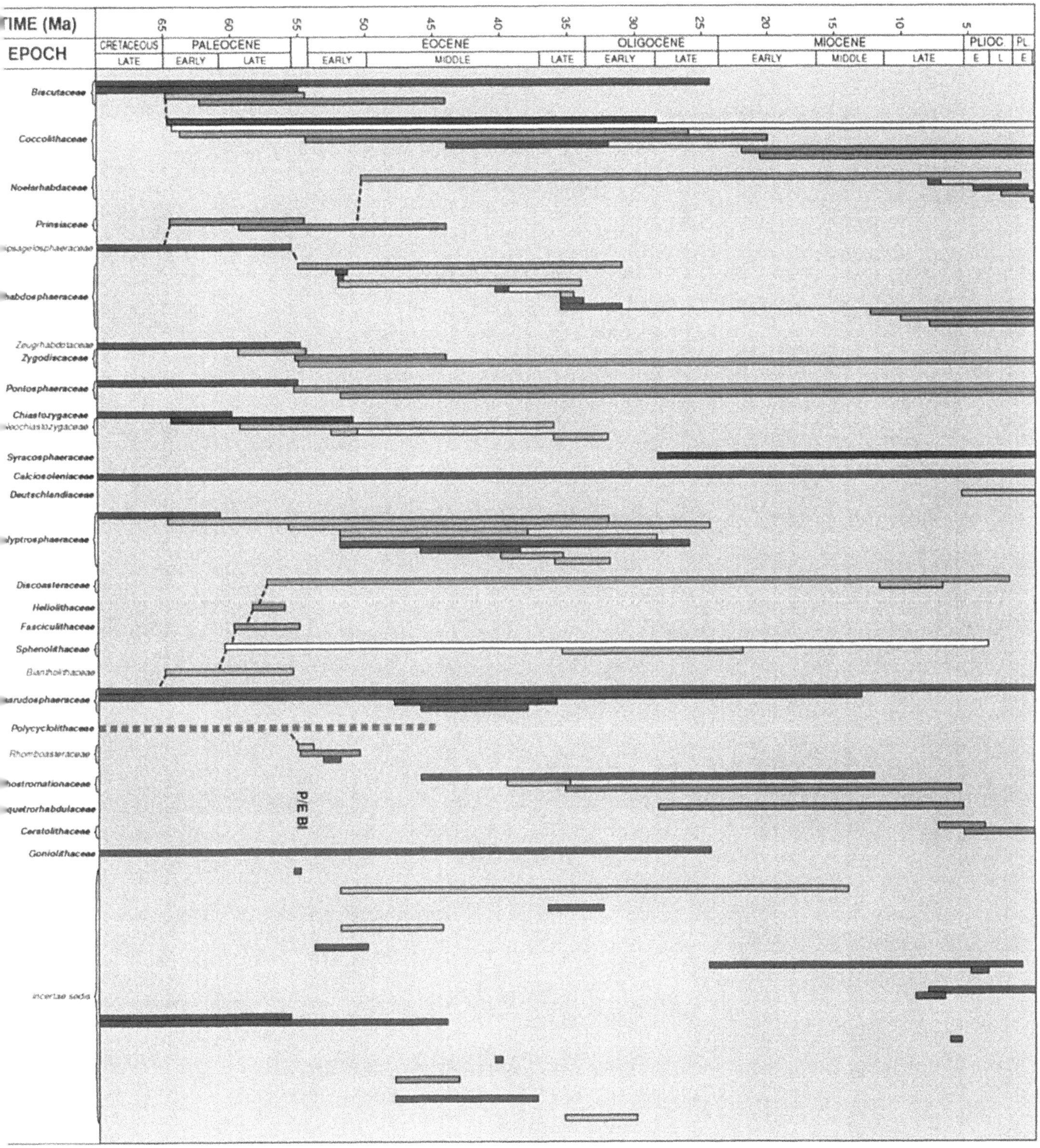

TIME (Ma)
65
60
55
50
45
40
35
30
25
20
15
10
5
EPOCH
CRETACEOUS
LATE
PALEOCENE
EARLY
LATE
EOCENE
EARLY
MIDDLE
LATE
OLIGOCENE
EARLY
LATE
MIOCENE
EARLY
MIDDLE
LATE
PLIOC.
E
L
PL
E
Biscutaceae
Coccolithaceae
Noelarhabdaceae
Prinsiaceae
Zeugrhabdotaceae
Zygodiscaceae
Pontosphaeraceae
Chiastozygaceae
Syracosphaeraceae
Calciosoleniaceae
Deutschlandiaceae
Discoasteraceae
Heliolithaceae
Fasciculithaceae
Sphenolithaceae
Polycyclolithaceae
Rhomboasteraceae
Ceratolithaceae
Goniolithaceae
P/E BI
incertae sedis

Figure 10.2

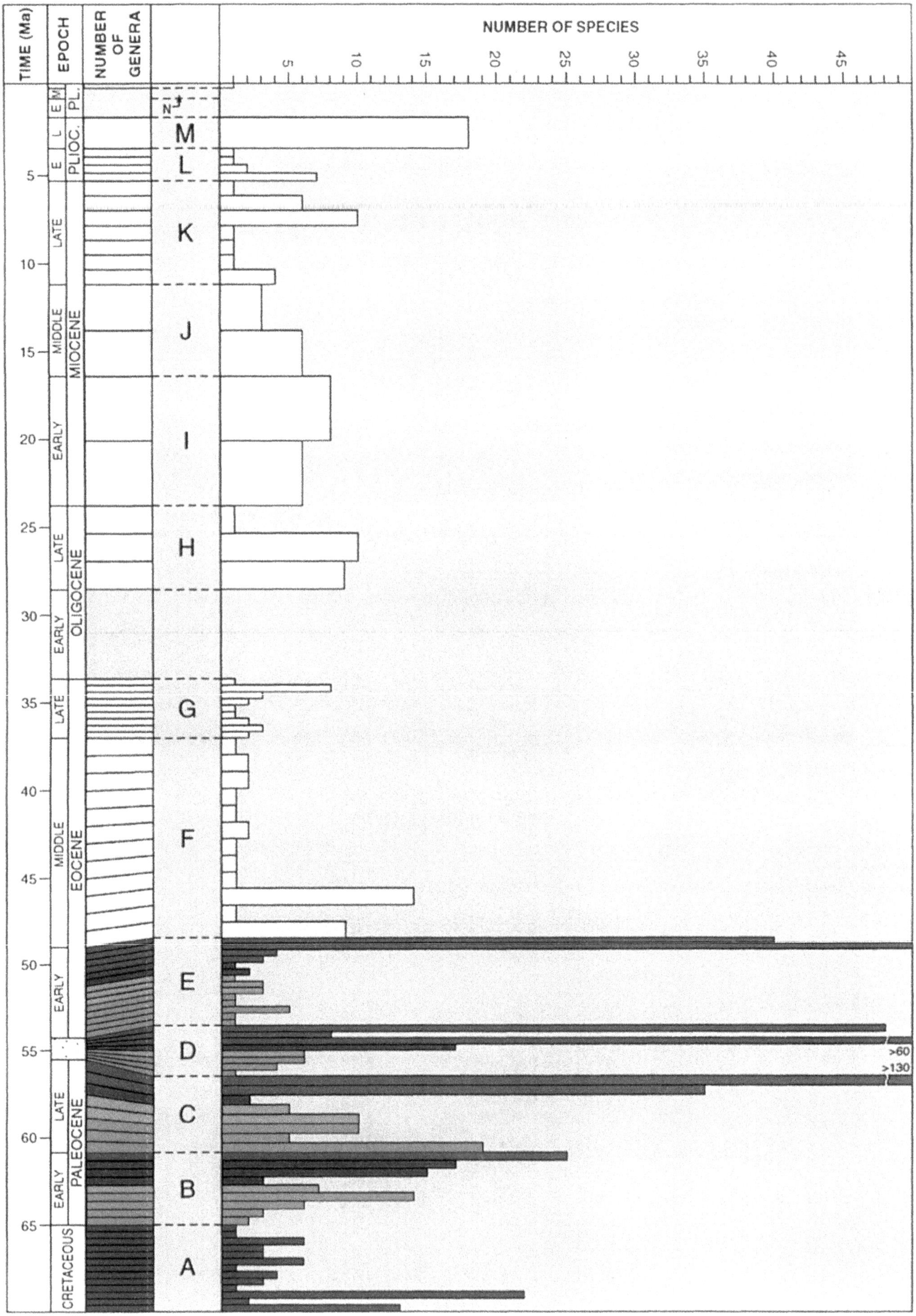

TIME (Ma)
EPOCH
NUMBER OF GENERA
NUMBER OF SPECIES
5
10
15
20
25
30
35
40
45
50
55
60
65
PL.
PLIOC.
MIOCENE
OLIGOCENE
EOCENE
PALEOCENE
CRETACEOUS
E
L
M
EARLY
MIDDLE
LATE
N
M
L
K
J
I
H
G
F
E
D
C
B
A
>60
>130

Table 10.4
Comparative radiation history and abundance patterns of genera that evolved during the Paleocene Epoch

Genus	Remarks	Dominance/Success
Hornibrookina	Never radiated (three species) but dominant in terms of specimens	High early Paleocene dominance
Prinsius	Radiated and dominant at all latitudes in early Paleocene; abundant until Biochron NP10 at high latitudes	High early Paleocene dominance
Neochiastozygus	Radiated during the Paleocene Epoch; common to abundant in oceanic and epicontinental deposits	Moderate dominance
Fasciculithus	Paleocene radiations, and dominant in oceanic and epicontinental assemblages	High dominance throughout the late Paleocene, until the P/E BI
Heliolithus	Paleocene radiation, and abundant to dominant in oceanic and epicontinental deposits	High late Paleocene dominance until shortly before the P/E BI
Toweius	Abundant at mid- and low latitudes during the late Paleocene Epoch in oceanic and epicontinental areas	Moderate success
Cruciplacolithus	Dominant to common at all latitudes during the Paleocene Epoch; secondary role during the Eocene Epoch	Success
Chiasmolithus	Common at all latitudes during the Paleocene Epoch but radiated after the P/E BI	High success
Coccolithus/Ericsonia	Common during the Paleocene Epoch but radiated after the P/E BI	High success
Sphenolithus	Common during the Paleocene Epoch but radiated after the P/E BI	High success
Discoaster	Evolved in the latest Paleocene Epoch but radiated after the P/E BI	High success

found today in the extant coccolithophorids, as for example in *Coccolithus pelagicus* and *Helicosphaera carteri*. The placoliths of *Chiasmolithus*, *Cruciplacolithus*, *Coccolithus/Ericsonia* have such a margin, and thus have a very modern look. *Toweius* is also among this group. Interestingly, most of these genera were among the first to evolve.

Thus, the enigmatic nature of the P/E BI turnover can be summarized as follows. A general trend towards increasing diversity in the calcareous nannoplankton occurred throughout the late Paleocene Epoch and, through co-evolution with the photosymbiont-bearing planktonic foraminifera, this led to prevailing late Paleocene oligotrophy. Further diversification in both groups resulted in very oligotrophic conditions in the early Eocene ocean. Superimposed on this diversification trend, a turnover occurred in the nannoplankton during the P/E BI. The genera that were dominant in the late Paleocene Epoch (until early Chron C24r), both in term of species richness and specimen abundance, became extinct. In contrast, genera that were common in terms of specimen abundance but had little diversified, did not become extinct, and most radiated during the early Eocene Epoch.

Extinctions were selective at the Cretaceous/Paleogene Period boundary in the sense that only tiny forms with simple structures survived. The species with complex structures were likely very specialized and became extinct. This selectivity is easily explained in the context of an ecosystem being destroyed by a catastrophic event. Only unspecialized forms had a chance for survival. Extinctions were also selective around the Eocene/Oligocene Epoch boundary, but this selectivity is easily explained by the progressive eutrophication of the Eocene ocean due to increased mixing as a result of global cooling.

The situation at the P/E BI is completely different. The ecosystem was strongly affected by an unidentified event as reflected by anomalous carbon and oxygen isotopic signatures. This event struck when conditions were close to maximal for the coccolithophorids, i.e., when oligotrophy was well established.

Several of the genera that radiated during the early Eocene Epoch (e.g., *Chiasmolithus*, *Coccolithus/Ericsonia*) evolved in the nutrient-rich early Paleocene ocean, which means that their Paleocene species may have been more adaptable to changing ecologic conditions than species of genera that evolved during the late Paleocene and were more specialized (see Hallock 1987 and discussion above). However, if anything, the ocean was evolving towards greater oligotrophic conditions throughout Chron C24r, and in this context, the extinction of the genera that had evolved under oligotrophic conditions is difficult to understand. The most puzzling question concerns the selectivity in extinction that operated at the level of coccolith structure. As indicated above, the taxa that became extinct have structural features similar to those of Mesozoic coccoliths. In particular, the shields of the placoliths of *Biscutum*, *Hornibrookina*, and *Prinsius*, consist of elements that are weakly imbricated, as in the placoliths of Mesozoic genera (e.g., *Watznaueria*). A new feature in Cenozoic placoliths (*Coccolithus/Ericsonia*, *Chiasmolithus*, and *Cruciplacolithus*) was the strong imbrication of the elements of their shields. Strongly imbricated elements were not the apanage of the Cenozoic placoliths. For instance, a change in the manner elements imbricate also occurred in the evolution of the Cenozoic genus *Zygodiscus* from the ancestral relict genus *Placozygus* as well as in the evolution of the Cenozoic genus *Neochiastozygus* from the relict genus *Chiastozygus*. The likelihood that the orientation of the elements of the coccoliths had something to do with the selectivity during the P/E BI turnover is strongly supported by the two examples that follow. The genera *Prinsius* and *Toweius* are regarded as closely related (Romein 1979; Perch-Nielsen 1981b; this chapter). The shields of the placoliths in *Prinsius* consist of poorly overlapping elements, as in Mesozoic placoliths. The shields of the placoliths of *Toweius* consist of strongly overlapping elements as in *Chiasmolithus*. *Prinsius* became extinct during the P/E BI; *Toweius* did not. Among the short-lived late Paleocene genera, those with radially arranged elements (e.g., *Fasciculithus*, *Heliolithus*) left no close descendant; those with strongly imbricated elements (e.g., *Zygodiscus*) did. There is no Eocene *Fasciculithus*-like or *Heliolithus*-like nannolith. On another hand, *Lophodolithus nascens* evolved from *Zygodiscus plectopons* through anagenesis (Romein 1979). The Paleocene genus *Zygodiscus* became extinct but had an immediate Eocene descendant, *Lophodolithus*. The only two exceptions to the selectivity described here are two successful genera, *Sphenolithus* and *Discoaster*. In both genera, the elements are radially arranged. However, their coccoliths are unlike any that evolved during the Mesozoic Era.

If coccolith basic structure (the orientation of the elements in discrete cycles of the heterococcoliths) is at the core of the selectivity as emphasized here, the cause for selectivity must be found at an intimate biologic level, heterococcolith secretion being an extremely sophisticated internal biologic process that involves an organic matrix (see review in Pienaar 1994). This organic matrix determines "the form and shape of the individual $CaCO_3$ elements that are deposited" (Pienaar 1994:30). In as much as the role of coccoliths remains largely unknown (see review in Young 1994), it is difficult to imagine which ecologic factor would have had a direct selective effect on biocalcification. However, it is likely that the enigmatic event known from anomalous isotopic signatures strongly affected biomineralization in the coccolithophorids.

Epilogue: Although analyses of species diversity changes in the calcareous nannoplankton did not reveal any significant event close to the Paleocene/Eocene Epoch boundary, a very important turnover is delineated at the generic and family levels. This turnover was not as abrupt as in the benthic foraminifera (Thomas, chapter 12) but lasted ~1 my and began at the time of the LPTM and the CIE. The P/E BI turnover was not as sudden as the extinction event at the Cretaceous/Paleogene Period boundary, but it was much more rapid than the Eocene/Oligocene Epoch boundary turnover, which spread over several millions of years. Whereas the greater part of the coccolithophorids were eradicated by the catastrophic Cretaceous/Paleogene boundary event, the P/E BI turnover was selective in a manner that is difficult to explain. Recovery of the calcareous nannoplankton was initially slow in the early Paleocene and then accelerated as co-evolution with the photosymbiont-bearing planktonic foraminifera led to a vertical expansion of the photic zone. The P/E BI event did not interrupt the recovery, but modified the course of evolution by causing the extinction of the genera that radiated during the Paleocene Epoch, and by favoring the post Paleocene radiation of other genera that were less prominent during the Paleocene Epoch. Most of these genera and those that evolved during the P/E BI became successful (long-ranging) and had several post-Paleocene radiations. Some are still represented in the modern ocean. The P/E BI thus appears as a turning point in the Cenozoic calcareous nannoplankton evolution, the time when the modern calcareous nannoplankton became established. The P/E BI was also a turning point in the evolution of the benthic foraminifera (Thomas 1992, chapter 12), the terrestrial mammals (see Gunnell, chapter 19), and many other groups (see several contributions in this volume). The biotic events at the Paleocene/Eocene Epoch boundary thus heralds the commencement of our modern world.

ACKNOWLEDGMENTS

I am grateful to W. A. Berggren, L. Beaufort, T. Bralower and R. D. Norris for their thoughtful reviews of the manuscript. Discussions with W. A. Berggren, A. Boersma and P. Hallock have helped to clarify the concepts expressed in this chapter. This is ISEM Contribution No 96-167.

REFERENCES

Angori, E. and Monechi, S., 1996. High-resolution calcareous nannofossil biostratigraphy across the Paleocene/Eocene boundary at Caravaca (southern Spain). In Aubry, M.-P. and Benjamini, C. (eds.), Paleocene/Eocene Boundary Events in Space and Time, *Israel Journal of Earth Sciences*, v. 44, p. 197–206.

Aubry, M-P., 1984. *Handbook of Paleogene Calcareous Nannofossils,*

v. 1. New York: Micropaleontology Press, American Museum of Natural History, 263 pp.

Aubry, M.-P., 1986. Paleogene calcareous nannoplankton biostratigraphy of northwestern Europe. *Palaeogeography, Palaeoclimatology, Palaeoecology,* v. 55, p. 267–334.

Aubry, M.-P., 1988a. *Handbook of Paleogene Calcareous Nannofossils,* v. 2. New York: Micropaleontology Press, American Museum of Natural History, 279 pp.

Aubry, M.-P., 1988b. Phylogeny of the Cenozoic calcareous nannoplankton genus Helicosphaera. *Paleobiology,* v. 14, p. 64–80.

Aubry, M.-P., 1989a. Phylogenetically based calcareous nannofossil taxonomy: Implications for the interpretation of geological events. In Crux, J. A. and Van Heck, S. E. (eds.), *Nannofossils and their applications: Proceedings of the INA Conference, London 1987,* Chichester: Ellis Horwood, p. 21–40.

Aubry M-P., 1989b. *Handbook of Paleogene Calcareous Nannofossils,* v. 3. New York, Micropaleontology Press, American Museum of Natural History, 279 pp.

Aubry M-P., 1990. *Handbook of Paleogene Calcareous Nannofossils,* v. 4. New York: Micropaleontology Press, American Museum of Natural History, 381 pp.

Aubry, M.-P., 1992a. Paleogene calcareous nannofossils from the Kerguelen Plateau, Leg 120. In Wise, S. W., Jr., Schlich, R., et al., Proceedings of the Ocean Drilling Program, Scientific Results, v. 120: College Station, TX (Ocean Drilling Program), p. 471–491.

Aubry, M.-P., 1992b. Late Paleogene calcareous nannoplankton evolution: A tale of climatic deterioration. In Prothero, D. R. and Berggren, W. A., *Eocene–Oligocene Climatic and Biotic Evolution,* Princeton: Princeton University Press, p. 272–309.

Aubry, M.-P., 1996. Towards an upper Paleocene–lower Eocene high resolution stratigraphy. In Aubry, M.-P. and Benjamini, C. (eds.), Paleocene/Eocene Boundary Events in Space and Time, *Israel Journal of Earth Sciences,* v. 44, p. 239–253.

Aubry M-P., 1998. *Handbook of Paleogene Calcareous Nannofossils,* v. 5. New York: Micropaleontology Press, American Museum of Natural History, in press.

Aubry, M.-P. and Bonnemaison, M., 1997. NannoStrat: Cenozoic calcareous nannofossils. New York: Micropaleontology Press, American Museum of Natural History. (Database, in binary code.)

Aubry, M.-P., Berggren, W. A., Stott, L. and Sinha, A., 1996. The upper Paleocene–lower Eocene stratigraphic record and the Paleocene–Eocene boundary carbon isotope excursion: Implications for geochronology. In Knox, R. W. O'B., Corfield, R. M. and Dunay, R. E. (eds.), *Correlation of the early Paleogene in Northwest Europe,* Geological Society Special Publication No 101, p. 353–380.

Beck, R. A., Burbank, D. W., Sercombe, W. J., Olson, T. L. and Khan, A. M., 1995. Organic carbon exhumation and global warming during the early Himalayan collision. *Geology,* v. 23, p. 387–390.

Berggren, W. A., 1962a. Some planktonic foraminifera from the Maestrichtian and type Danian stages of southern Scandinavia. *Stockholm Contributions in Geology,* v. 9, 106 pp.

Berggren, W. A., 1962b. Stratigraphic and taxonomic–phylogenetic studies of upper Cretaceous and Paleogene planktonic foraminifera. *Stockholm Contributions in Geology,* v. 9, p. 107–128.

Berggren, W. A., 1969. Rates of evolution in some Cenozoic planktonic foraminifera. *Micropaleontology,* v. 15, p. 351–365.

Berggren, W. A., 1992. Paleogene planktonic foraminifer magnetobiostratigraphy of the southern Kerguelen Plateau (Sites 747–749). In Wise, S. W., Jr., Schlich, R., et al., Proceedings of the Ocean Drilling Program, Scientific Results, v. 120: College Station, TX (Ocean Drilling Program), p. 551–568.

Berggren, W. A. and Aubry, M.-P., 1996. A late Paleocene–early Eocene NW European and North Sea magnetobiochronological correlation network. In Knox, R. W. O'B., Corfield, R. M. and Dunay, R. E. (eds.), *Correlation of the Early Paleogene in Northwest Europe,* Geological Society Special Publication No. 101, p. 309–352.

Berggren, W. A. and Miller, K. G., 1988. Paleogene tropical planktonic foraminiferal biostratigraphy and magnetobiochronology. *Micropaleontology,* v. 34, p. 362–380.

Berggren W. A., Kent, D. V., Swisher, C. C., III and Aubry, M.-P., 1995. A revised Cenozoic geochronology and chronostratigraphy. In Berggren, W. A., Kent, D. V., Aubry, M.-P. and Hardenbol, J. (eds.), *Geochronology Time Scales and Global Stratigraphic Correlation,* Society of Economic Paleontologists and Mineralogists Special Publication No. 54, p. 129–212.

Boersma, A. and Premoli Silva, I., 1983. Paleocene planktonic foraminiferal biogeography and the paleoceanography of the Atlantic Ocean. *Micropaleontology,* v. 29, p. 355–381.

Boersma, A. and Premoli Silva, I., 1988. Boundary conditions of Atlantic Eocene oxygen minimum zones. *Rivista Italiana Paleontologia e Stratigrafia,* v. 93, p. 479–506.

Boersma, A. and Premoli Silva, I., 1991. Distribution of Paleogene planktonic foraminifera: analogies with the Recent? *Palaeogeography, Palaeoclimatology, Palaeoecology,* v. 83, p. 29–48.

Boersma, A., Premoli Silva, I. and Shackleton, N. J., 1987. Atlantic Eocene planktonic foraminiferal paleohydrographic indicators and stable isotope paleoceanography. *Paleoceanography,* v. 2, p. 287–331.

Bolli, H. M., 1957a. The genera *Praeglobotruncana, Rotalipora, Globotruncana,* and *Abathomphalus* in the upper Cretaceous of Trinidad, B.W.I. *U.S. National Museum Bulletin* 215, p. 61–60.

Bolli, H. M., 1957b. The genera *Globigerina* and *Globorotalia* in the Paleocene–lower Eocene Lizard springs Formation of Trinidad, B.W.I. *U.S. National Museum Bulletin* 215, p. 61–81.

Bollmann, J., 1997. Morphology and biogeography of *Gephyrocapsa* coccoliths in Holocene sediments. *Marine Micropaleontology,* v. 29, p. 319–350.

Bonnemaison, M., 1997. "CompuStrat" database manager, version 1 for Mac and PC (software and manual). New York: Micropaleontology Press, American Museum of Natural History. 2 Mb; 150 pp.

Bonnemaison, M. and Aubry, M.-P., 1997. NannoStrat: Mesozoic calcareous nannofossils. New York: Micropaleontology Press, American Museum of Natural History. (Database, in binary code.)

Bown, P. R., Burnett, J. A. and Gallagher, L. T., 1991. Critical events in the evolutionary history of calcareous nannoplankton. *Historical Biology,* v. 5, p. 279–290.

Bralower, T. J. and Mutterlose, J., 1995. Calcareous nannofossil biostratigraphy of Site 865, Allison Guyot, central Pacific Ocean: A tropical Paleogene reference section. In Winterer, E. L., Sager, W. W., Firth, J. V. and Sinton, J. M. (eds.), Proceedings of the Ocean Drilling Program, Scientific Results, v. 143: College Station, TX (Ocean Drilling Program), p. 31–74.

Bramlette M. N. and Martini, E., 1964. The great change in calcareous nannoplankton fossils between the Maestrichtian and Danian. *Micropaleontology,* v. 10, p. 291–322.

Bramlette, M. N. and Sullivan, F. R., 1961. Coccolithophorids and related nannoplankton of the early Tertiary in California. *Micropaleontology,* v. 7, p. 129–188.

Bramlette, M. N. and Wilcoxon, J. A., 1967. Middle Tertiary calcareous nannoplankton of the Cipero section, Trinidad, W. I., *Tulane Studies in Geology,* v. 5, 93–131.

Brand, L., 1994. Physiological ecology of marine coccolithophores. In Winter, A. and Siesser, W. G. (eds.), *Coccolithophores,* Cambridge: Cambridge University Press, p. 39–49.

Bukry, D., 1969. Upper Cretaceous coccoliths from Texas and Europe. *The University of Kansas Paleontological Contributions,* v. 51, 79 pp.

Bybell, L. M. and Self-Trail, J. M., 1995. *Evolutionary, biostratigraphic, and taxonomic study of calcareous nannofossils from a continuous Paleocene–Eocene boundary section in New Jersey.* U.S. Geological Survey Professional Paper 1554, 36 p.

Cande, S. C. and Kent, D. V., 1992. A new geomagnetic polarity time scale for the Late Cretaceous and Cenozoic. *Journal of Geophysical Research,* v. 97(B10), p. 13917–13951.

Cande, S. C. and Kent, D. V., 1995. Revised calibration of the geomagnetic polarity time scale for the Late Cretaceous and Cenozoic. *Journal of Geophysical Research,* v. 100(B4), p. 6093–6095.

Charisi, S. D. and Schmidtz, 1996. Early Eocene palaeoceanography and palaeoclimatology of the eastern North Atlantic: Stable isotope results for DSDP Hole 550. In Knox, R. W. O'B., Corfield, R. M. and Dunay, R. E. (eds.), *Correlation of the early Paleogene in Northwest Europe,* Geological Society Special Publication No 101, p. 457–472.

Corfield, R. M., 1994. Palaeocene oceans and climate: An isotopic perspective. *Earth-Science Reviews,* v. 37, p. 225–252.

Corfield, R. M. and Cartlidge, R. M., 1992. Oceanographic and climatic implications of the Palaeocene carbon isotope maximum. *Terra Nova,* v. 4, p. 443–455.

Cowie, J. W., Ziegler, W., Boucot, A. J., Basset, M. G. and Remane, J., 1986. Guidelines and statutes of the International Commission on Stratigraphy (ICS). *Courier Forschungsinstitut Senckenbergiana,* v. 83, p. 1–14.

Dalrymple, G. B., Izett, G. A., Snee, L. W. and Obradovich, J. D., 1993. $^{40}Ar/^{39}Ar$ age spectra and total fusion-ages of tektites from Cretaceous–Tertiary boundary sedimentary rocks in the Beloq Formation, Haiti. *United States Geological Bulletin,* v. 2065, p. 1–20.

D'Hondt, S. and Zachos, J. C., 1994. Stable isotopic signals and photosymbiosis in late Paleocene planktic foraminifera. *Paleobiology,* v. 20, p. 391–406.

Dickens, G. R., O'Neil, J. R., Rea, D. K. and Owen, R. M., 1995. Dissociation of oceanic methane hydrate as a cause of the carbon isotope excursion at the end of the Paleocene. *Paleoceanography,* v. 10, p. 965–971.

Dodd, J. R. and Stanton, R. J., Jr., 1981. *Paleoecology, Concept and Applications.* New York: John Wiley, 559 pp.

Dudley, W. C. and Nelson, C. S., 1988. The $\delta^{13}C$ content of calcareous nannofossils as an indicator of Quaternary paleoproductivity in the southwest Pacific region. *New Zealand Journal of Geology and Geophysics,* v. 31, p. 111–116.

Dudley, W. C., Blackwalder, P., Brand, L. and Duplessy, J.-C., 1986. Stable isotopic composition of coccoliths. *Marine Micropaleontology,* v. 10, p. 1–8.

Ehrendorfer, T. and Aubry, M.-P., 1992. Calcareous nannoplankton changes across the Cretaceous/Paleocene boundary in the southern Indian Ocean (Site 750). In Wise, S. W., Jr., Schlich, R., et al., Proceedings of the Ocean Drilling Program, Scientific Results, v. 120: College Station, TX (Ocean Drilling Program), p. 451–470.

Eldholm, O. and Thomas, E., 1993. Environmental impact of volcanic margin formation. *Earth and Planetary Science Letters,* v. 117, p. 319–329.

Fourtanier, E., 1991. Paleocene and Eocene diatom biostratigraphy and taxonomy of eastern Indian Ocean sites. in Weissel, J., Pierce, J., Taylor, E., Alt, J. et al., Proceedings of the Ocean Drilling Program, Scientific Results, v. 121: College Station, TX (Ocean Drilling Program), p. 171–187.

Frakes, L. A., 1979. *Climates throughout geological time.* Amsterdam: Elsevier, 310 pp.

Gradstein, F. and Srivastava, S., 1980. Aspects of Cenozoic stratigraphy and paleoceanography of the Labrador Sea and Baffin Bay. *Palaeogeography, Palaeoclimatology, Palaeoecology,* v. 30, p. 261–295.

Hallock, P., 1987. Fluctuations in the trophic resource continuum: A factor in global diversity cycles? *Paleoceanography,* v. 2, p. 457–471.

Hallock, P., Premoli Silva, I. and Boersma, A. 1991. Similarities between planktonic and larger foraminiferal evolutionary trends through Paleogene paleoceanographic changes. *Palaeogeography, Palaeoclimatology, Palaeoecology,* v. 83, p. 49–64.

Haq, B. U., 1973. Transgressions, climatic change and the diversity of calcareous nannoplankton. *Marine Geology,* v. 15, p. M25–M30.

Haq, B. U. and Lohmann, G. P., 1976. Early Cenozoic calcareous nannoplankton biogeography of the Atlantic Ocean. *Marine Micropaleontology,* v. 1, p. 119–194.

Haq, B. U., Premoli Silva, I., and Lohmann, G. P., 1977. Calcareous plankton paleobiogeographic evidence for major climatic fluctuations in the early Cenozoic Atlantic Ocean. *Journal of Geophysical Research,* v. 82(27), p. 3861–3876.

Hay, W. W., 1977. Calcareous nannofossils. In Ramsay, A. T. S (ed.), *Oceanic Micropaleontology,* New York: Academic Press, v. 2, p. 1055–1200.

Hedberg, H. D. (ed.), 1976. *International stratigraphic guide. A guide to stratigraphic classification, terminology, and procedure.* New York: John Wiley and Sons, 200 p.

Hemleben, C., Spindler, M. and Anderson, O. R., 1989. *Modern Planktonic Foraminifera.* New York: Springer-Verlag, 363 pp.

Hsü, K. J. and McKenzie, J. A., 1985. A "Strangelove" ocean in the earliest Tertiary. in Sundquist, E. T. and Broeker, W. S. (eds.), *The Carbon Cycle and Atmospheric CO_2: Natural Variations Archean to Present,* Geophysical Monograph 32, p. 487–492.

Hutchinson, G. E., 1978. *An Introduction to Population Ecology.* New Haven: Yale University Press, 260 pp.

Jorgensen, B. B., Erez, J., Revsbech, N. P. and Cohen, Y. P., 1985. Symbiotic photosynthesis in the planktonic foraminifera, *Globigerinoides sacculifer* (Brady), studied with microelectrodes. *Limnology and Oceanography,* v. 30, p. 1253–1267.

Kaiho, K., 1994a. Benthic foraminiferal dissolved-oxygen index and dissolved-oxygen levels in the modern ocean. *Geology,* v. 22, p. 719–722

Kaiho, K., 1994b. Planktonic and benthic foraminiferal extinction events during the last 100 m.y. *Palaeogeography, Palaeoclimatology, Palaeoecology,* v. 111, p. 45–71.

Kelly, D. C., Bralower, T. J., Zachos, J. C., Premoli Silva, I. and Thomas, E., 1996, Rapid diversification of planktonic foraminifera in the tropical Pacific (ODP Site 865) during the late Paleocene Thermal Maximum. *Geology,* v. 24, p. 423–426.

Kennett, J. P. and Stott, L. D., 1991. Abrupt deep-sea warming, palaeoceanographic changes and benthic extinctions at the end of the Palaeocene. *Nature,* v. 353, p. 225–229.

Kilham, P. and Soltau Kilham, S., 1980. The evolutionary ecology of phytoplankton. In Morris, I. (ed.), *The Physiological Ecology of Phytoplankton,* Studies in Ecology, 7, p. 493–570.

Kleijne, A., 1991. Holococcolithophorids from the Indian Ocean, Red Sea, Mediterranean Sea and North Atlantic Ocean. *Marine Micropaleontology,* 17:1–96.

Knoll, A. H., 1989. Evolution and extinction in the marine realm: Some constraints imposed by phytoplankton. *Philosophical Transactions of the Royal Society of London,* v. B325, p. 279–290.

Koch, P. L., Zachos, J. C. and Dettman, D. L., 1995. Stable isotope stratigraphy and paleoclimatology of the Paleogene Bighorn Basin (Wyoming, USA). *Palaeogeography, Palaeoclimatology and Palaeoecology,* v. 115, p. 61–89.

Koch, P. L., Zachos, J. C. and Gingerich, P. D., 1992. Correlation between isotope records in marine and continental carbon reservoirs near the Palaeocene/Eocene boundary. *Nature,* v. 358, p. 319–322.

Lehotayova, R. and Priewalder, H., 1978. *Cycloperfolithus,* eine neue Nannofossil-Gattung aus dem Badenian der Zentralen Paratethys. In Papp, A., Cicha, I., Senes, J., et al. (eds.), *Chronostratigraphie und Neostratotypen, M4–Badenian Chronostratigraphie und Neostratotypen Miozaen der zentralen Paratethys,* Bratislava: Veda Verl. Slow. Akad. Wiss., v. 6, p. 486–489.

Loeblich, A. R., Jr. and Tappan, H., 1957a. Planktonic foraminifera of Paleocene and early Eocene age from the Gulf and Atlantic coastal plains. *U.S. National Museum Bulletin* 215, p. 173–198.

Loeblich, A. R., Jr. and Tappan, H., 1957b. Correlation of the Gulf and Atlantic coastal plain Paleocene and lower Eocene formations by means of planktonic foraminifera. *Journal of Paleontology,* v. 31, p. 1109–1137.

Lu, G. and Keller, G., 1993. The Paleocene–Eocene transition in the Antarctic Indian Ocean: Inference from planktic foraminifera. *Marine Micropaleontology,* v. 21, p. 101–142.

Lu, G. and Keller, G., 1995a. Ecological stasis and saltation: Species richness change in planktic foraminifera during the late Paleocene to early Eocene, DSDP Site 577. *Palaeogeography, Palaeoclimatology, Palaeoecology,* v. 117, p. 211–227.

Lu, G. and Keller, G., 1995b. Planktic foraminiferal faunal turnovers in the subtropical Pacific during the late Paleocene to early Eocene. *Journal of Foraminiferal Research,* v. 25, p. 97–116.

Margulis, L. and Schwartz, K. V., 1988. *Five Kingdoms. An Illustrated Guide to the Phyla of Life on Earth.* New York: Freeman and Company, 376 pp.

Martin, R. E., 1996. Secular increase in nutrient levels through the Phanerozoic: Implications for productivity, biomass, and diversity. *Palaios,* v. 11, p. 209–219.

McIntyre, A., and Bé, A. W. H., 1967. Modern coccolithophoridae of the Atlantic Ocean: I. Placoliths and cyrtoliths. *Deep Sea Research,* v. 14, p. 561–597.

Miller, K. G., 1992. Middle Eocene to Oligocene stable isotopes, climates and deep-water history: The terminal Eocene event? In Prothero, D. R. and Berggren, W. A., *Eocene–Oligocene Climatic and Biotic Evolution,* Princeton: Princeton University Press, p. 160–177.

Miller, K. G., Janecek, T. R., Katz, M. E. and Keil, D. J., 1987a. Abyssal circulation and benthic foraminiferal changes near the Paleocene/Eocene boundary. *Paleoceanography,* v. 2, p. 741–761.

Miller, K. G., Fairbanks, R. G. and Mountain, G. S., 1987b. Tertiary oxygen isotope synthesis, sea level history, and continental margin erosion. *Paleoceanography,* v. 2, p. 1–19.

Monechi, S., 1985. Campanian to Pleistocene calcareous nannofossil stratigraphy from the northwest Pacific Ocean, Deep Sea Drilling Project Leg 86. In Heath, G. R., Burckle, L. H., et al., Initial Reports of the Deep Sea Drilling Project, v. 86: Washington (U.S. Government Printing Office), p. 301–336.

Norris, R. D., 1997. Symbiosis as an evolutionary innovation in the radiation of Paleocene planktic foraminifera. *Paleobiology,* v. 22, p. 461–480.

Okada, H., and Honjo, S., 1973. The distribution of oceanic coccolithophorids in the Pacific. *Deep Sea Research,* v. 20, p. 355–374.

Okada, H. and Thierstein, H. R., 1979. Calcareous nannoplankton: leg 43, Deep Sea Drilling Project. In Tucholke, B. E., Vogt, P. R. et al., Initial Report of the Deep Sea Drilling Project, v. 43: Washington (U.S. government Printing Office), p. 507–573.

Olsson, R., 1960. Foraminifera of latest Cretaceous and earliest Tertiary age of the New Jersey coastal plain. *Journal of Paleontology,* v. 34, p. 1–58.

Olsson, R. K., Hemleben, C., Berggren, W. A. and Huber, B. T. (eds.), 1998. *Atlas of Paleocene Planktonic Foraminifera,* Washington: Smithsonian Contributions to Paleobiology, in press.

Pak, D. K. and Miller, K. G., 1992. Paleocene to Eocene benthic foraminiferal isotopes and assemblages: Implications for deepwater circulation. *Paleoceanography,* v. 7, p. 405–422.

Paull, C. K. and Thierstein, H. P., 1987. Stable isotope fractionation among particles in Quaternary coccolith-size deep-sea sediments. *Paleoceanography,* v. 2, p. 423–429.

Perch-Nielsen, K., 1968. Der Feinbau und die Klassification der Coccolithen aus dem Maestrichtian von Dänemark. *Kong. Danske Vidensk. Selsk., Biol. Skr.* v. 16, 96 pp.

Perch-Nielsen, K., 1969. Die Coccolithen einiger Dänischer Maastrichtien- und Danienlokalitäten. *Bulletin of the Geological Society of Denmark,* v. 19, p. 51–68.

Perch-Nielsen, K., 1972. Les nannofossiles calcaires de la limite Crétacé–Tertiaire. *Mémoire du Bureau de Recherches Géologiques et Minières,* v. 77, p. 181–188.

Perch-Nielsen, K., 1977. Albian to Pleistocene calcareous nannofossils from the western South Atlantic, DSDP Leg 39. In Supko, P. R., Perch-Nielsen, K., et al., Initial Reports of the Deep Sea Drilling Project, v. 39. Washington, D.C. (U.S. Government Printing Office), p. 699–823.

Perch-Nielsen, K., 1979. Calcareous nannofossil zonation at the Cretaceous/Tertiary boundary in Denmark. In Birkelund, T., and Bromley, R. G. (eds.), *Cretaceous–Tertiary boundary events, I. The Maastrichtian and Danian of Denmark,* Copenhagen: University of Copenhagen, p. 115–135.

Perch-Nielsen, K., 1981a. Les coccolithes du Paléocène près de El Kef, Tunisie, et leurs ancLtres. Cahiers de Micropaléontologie, v. 3, p. 7–23.

Perch-Nielsen, K., 1981b. New Maastrichtian and Paleocene calcareous nannofossils from Africa, Denmark, the USA and the Atlantic, and some Paleocene lineages. *Eclogae Geologicae Helvetiae,* v. 74, p. 31–863.

Perch-Nielsen, K., 1985a. Cenozoic calcareous nannofossils. In Bolli, H. M., Saunders, J. B. and Perch-Nielsen, K., *Plankton Stratigraphy,* Cambridge: Cambridge University Press, p. 427–554.

Perch-Nielsen, K., 1985b. Mesozoic calcareous nannofossils. In Bolli, H. M., Saunders, J. B. and Perch-Nielsen, K., *Plankton Stratigraphy,* Cambridge: Cambridge University Press, p.329–426.

Perch-Nielsen, K., McKensie, J. and He, Q., 1982. Biostratigraphy and isotope stratigraphy and the "catastrophic" extinction of calcareous nannoplankton at the Cretaceous/Tertiary boundary. In Silver, L. T. and Schultz, P. H., Geological *Implications of Impacts of Large Asteroids and Comets on the Earth,* Geological Society of America Special Paper 190, p. 353–371.

Percival, F. S., and Fischer, A. G., 1977. Changes in calcareous nannoplankton in the Cretaceous–Tertiary biotic crisis at Zumaya, Spain. *Evolutionary Theory,* v. 2, p. 1–35.

Pienaar, R. N., 1994. Ultrastructure and calcification of coccolithophores. In Winter, A. and Siesser, W. G. (eds.), *Coccolithophores,* Cambridge: Cambridge University Press, p. 13–37.

Pospichal, J. J. and Wise, S. W., Jr., 1990. Paleocene to middle Eocene calcareous nannofossils of ODP Sites 689 and 690, Maud Rise, Weddell Sea. In Barker, P. F., Kennett, J. P., et al., Proceedings of the Ocean Drilling Program, Scientific Results, v. 113: College Station, TX (Ocean Drilling Program), p. 613–638.

Premoli Silva, I. and Boersma, A., 1988. Atlantic Eocene planktonic foraminiferal historical biogeography and paleohydrographic indices. *Palaeogeography, Palaeoclimatology, Palaeoecology,* v. 67, p. 315–356.

Rea, D. K., Leinen, M. and Janecek, T. R., 1985. Geologic approach to the long-term history of atmospheric circulation. *Science,* 227, p. 721–725.

Rea, D. K., Zachos, J. C., Owen, R. M. and Gingerich, P. D., 1990. Global change at the Paleocene–Eocene boundary: climatic and evolutionary consequences of tectonic events. *Palaeogeography, Palaeoclimatology, Palaeoecology,* v. 79, p. 117–128.

Romein, A. J. T., 1979. Lineages in early Paleogene calcareous nannoplankton. *Utrecht Micropaleontological Bulletin,* v. 22, 231 pp.

Romein, A. J. T., Willems, H. and Mai, H., 1996. Calcareous nannoplankton of the Geulhemmerberg K/T boundary section, Maastrichtian type area, the Netherlands. *Geologie en Mijnbown,* v. 75, p. 231–238.

Roth, H. P. and Thierstein, H., 1972. Calcareous nannoplankton: Leg 14 of the Deep Sea Drilling Project. In Hayes, D. E., Pimm, A. C. et al.,

Initial Report of the Deep Sea Drilling Project, v. 14: Washington (U.S. government Printing Office), p. 421–494.

Sanfilippo, A., Westberg-Smith, M. J. and Riedel, W. A., 1985. Cenozoic radiolaria. In Bolli, H. M., Saunders, J. B. and Perch-Nielsen, K., *Plankton Stratigraphy,* Cambridge: Cambridge University Press, p. 631–712.

Schmitz, B., Speijer, R. P. and Aubry, M.-P., 1996. Latest Paleocene benthic extinction event on the southern Tethyan shelf (Egypt): Foraminiferal stable isotopic ($\delta^{13}C$, $\delta^{18}O$) records. *Geology,* v. 24, p. 347–350.

Shackleton, N. J., 1986. Paleogene stable isotope events. *Palaeogeography, Palaeoclimatology, Palaeoecology,* v. 57, p. 91–102.

Shackleton, N. J., Corfield, R. M. and Hall, M. A., 1985. Stable isotope data and the ontogeny of Paleocene planktonic foraminifera. *Journal of Foraminiferal Research,* v. 15, p. 321–336.

Shafik, S., 1989. Some new calcareous nannofossils from upper Eocene and lower Oligocene sediments in the Otway Basin, southeastern Australia. *Alcheringa,* 13:69–83.

Sinha, A., Aubry, M.-P., Stott, L. D., Thiry, M. and Berggren, W. A., 1996. Chemostratigraphy of the "lower" Sparnacian deposits (Argiles plastiques bariolées) of the Paris Basin. In Aubry, M.-P. and Benjamini, C. (eds.), Paleocene/Eocene Boundary Events in Space and Time, *Israel Journal of Earth Sciences,* v. 44, p. 223–237.

Steineck, P. L. and Thomas, E., 1996. The latest Paleocene crisis in the deep sea: Ostracode succession at Maud Rise, Southern Ocean. *Geology,* v. 24, p. 583–586.

Steinmetz, J. C., 1994. Stable isotopes in modern coccolithophores. In Winter, A. and Siesser, W. G. (eds.), *Coccolithophores,* Cambridge: Cambridge University Press, p. 219–229.

Stott, L. D., Kennett, J. P., Shackleton, N. J. and Corfield, R. C., 1990. The evolution of Antarctic surface waters during the Paleogene: Inferences from the stable isotopic composition of planktonic foraminifers, ODP Leg 113. In Barker, P. F., Kennett, J. P., et al., Proceedings of the Ocean Drilling Program, Scientific Results, v. 113, p. 849–863.

Stott, L. D., Sinha, A., Thiry, M., Aubry, M.-P. and Berggren, W. A., 1996. Global $\delta^{13}C$ changes across the Paleocene–Eocene boundary: Criteria for terrestrial–marine correlations. In Knox, R. W. O'B., Corfield, R. M. and Dunay, R. E. (eds.), *Correlation of the Early Paleogene in Northwest Europe,* Geological Society Special Publication No. 101, p. 381–399.

Swisher, C. C. III, Dingus, L. and Butler, R. F., 1993. $^{40}Ar/^{39}Ar$ dating and magnetostratigraphic correlation of the terrestrial Cretaceous–Paleogene boundary and Puercan mammal Age, Hell Creek–Tullock formations, eastern Montana. *Canadian Journal of Earth Sciences,* v. 30, p. 1981–1996.

Tappan, H., 1979. Protistan evolution and extinction at the Cretaceous/Tertiary boundary. In Birkelund, T., and Bromley, R. G. (eds.), *Cretaceous–Tertiary boundary events, II. Proceedings,* Copenhagen: University of Copenhagen, p. 13–21.

Tappan, H. 1980. *The Paleobiology of Plant Protists.* San Francisco: Freeman and Company, 1028 pp.

Tappan, H. and Loeblich, A. R., Jr., 1973. Evolution of the oceanic plankton. *Earth-Science Reviews,* v. 9, p. 207–240.

Tappan, H., 1986. Phytoplankton: Below the salt at the global table. *Journal of Paleontology,* v. 60, p. 545–554.

Theodoridis, S., 1984. Calcareous nannofossil biozonation of the Miocene and revision of the helicoliths and discoasters. *Utrecht Micropaleontological Bulletin,* v. 32, 271 pp.

Thierstein, H., 1981. *Late Cretaceous nannoplankton and the change at the Cretaceous–Tertiary boundary.* Society of Economic Paleontologists and Mineralogists Special Publication No. 32, p. 355–394.

Thierstein, H. and Okada, H., 1979. The Cretaceous/Tertiary boundary event in the North Atlantic. In Tucholke, B. E., Vogt, P. R., et al., Initial Reports of the Deep Sea Drilling Project, v. 43: Washington (U.S. Government Printing Office), p. 601–616.

Thomas, E., 1992. Cenozoic deep sea circulation: Evidence from deep sea benthic foraminifera. In Kennett, J. P. and Warnke, D. (eds.), *The Antarctic Paleoenvironment: A Perspective on Global Change,* American Geophysical Union Antarctic Research Series No. 56, p. 141–165.

Thomas, E. and Gooday, A. J., 1996. Deep-sea benthic foraminifera: tracers for changes in Cenozoic oceanic productivity. *Geology,* v., p. 355–358.

Thomas, E. and Shackleton, N. J., 1996. The Paleocene–Eocene benthic foraminiferal extinction and stable isotope anomalies. In Knox, R. W. O'B., Corfield, R. M. and Dunay, R. E. (eds.), *Correlation of the Early Paleogene in Northwest Europe,* Geological Society Special Publication No 101, p. 401–441.

Thompson, E. and Schmitz, B., 1997. Barium and the late Paleocene $\delta^{13}C$ maximum: Evidence of increased marine surface productivity. *Paleoceanography,* v. 12, p. 239–254.

Tjalsma, R. C. and Lohmann, G. P., 1983. *Paleocene–Eocene bathyal and abyssal benthic foraminifera from the Atlantic Ocean.* Micropaleontology Special Publication No. 4, 90 p.

Valentine, J. M., 1973. *Evolutionary Ecology of the Marine Biosphere.* Englewood Cliffs: Prentice-Hall, 511 pp.

Venrick, E. L., 1982. Phytoplankton in an oligotrophic ocean: Observations and questions. *Ecological Monographs,* v. 52, p. 129–154.

Venrick, E. L., 1990. Phytoplankton in an oligotrophic ocean: Phytoplankton in an oligotrophic ocean: Species structure and interannual variability. *Ecology,* v. 71, p. 1547–1563.

Wei, W. and Zhong, S., 1996. Taxonomy and magnetobiochronology of *Tribrachiatus* and *Rhomboaster*, two genera of calcareous nannofossils. *Journal of Paleontology,* v. 70, p. 7–22.

Wilson, E. O., 1992. *The Diversity of Life.* Cambridge: The Belknap Press of Harvard University Series, 429 pp.

Young, J., 1994. Functions of coccoliths. In Winter, A. and Siesser, W. G. (eds.), *Coccolithophores,* Cambridge: Cambridge University Press, p. 63–82.

Zachos, J. C., Arthur, M. A. and Dean, W. E., 1989. Geochemical evidence for suppression of pelagic marine productivity at the Cretaceous/Tertiary boundary. *Nature,* v. 337, p. 61–64.

Zachos, J. C., Lohmann, K. C., Walker, J. C. G. and Wise, S. W., 1993. Abrupt climate change and transient climates during the Paleogene: A marine perspective. *Journal of Geology,* v. 101, p. 191–213.

Zachos, J. C., Stott, L. D. and Lohmann, K. C., 1994. Evolution of early Cenozoic marine temperatures. *Paleoceanography,* v. 9, p. 353–387.

Appendix 10.1

Comparison of the synthetic classifications (families) for the Class Coccolithophyceae proposed by Hay (1977), Tappan (1980) and Perch-Nielsen (1985a, 1985b)

Family	Hay (1977)	Tappan (1980)	Perch-Nielsen (1985a,b)
ACANTHOICACEAE Hay, 1977	- ***Acanthoica*** Lohmann, 1903 - **Anacanthoica* Deflandre, 1952	- ***Acanthoica*** Lohmann, 1903 - **Anacanthoica* Deflandre, 1952	—
ACTINOZYGACEAE Rood et al., 1971 Family regarded synonymous with Rhagodiscaceae Hay, 1977 in Tapan (1980)	- *Actinozygus* Gartner, 1968 - *Angulofenestrellithus* Bukry, 1969 - *Diadozygus* Rood et al., 1971 - *Diadorhombus* Worsley, 1971 - *Nodosella* Prins ex Rood et al., 1973 - *Pontilithus* Gartner, 1968 - *Rhabdolithina* Reinhardt, 1967 - *Truncatoscaphus* Rood et al., 1971	—	—
AHMUELLERELLACEAE Reinhardt, 1965	- *Ahmuellerella* Reinhardt, 1964 - *Crucirhabdus* Prins ex Rood, Hay & Barnard, 1973 - *Staurolithites* Caratini, 1960 (= *Zygostephanos* Hoffmann, 1970) - *Vekshinella* Loeblich & Tappan, 1963 (= *Ephippium* Vekshina, 1959; *Vagalapilla* Bukry, 1969; *Staurorhabdus* Noël, 1972)	- *Ahmuellerella* Reinhardt, 1964 - *Crucirhabdus* Prins ex Rood, Hay & Barnard, 1973 - *Helicolithus* Noël, 1970 - ***Placozygus**** Hoffmann, 1970 - *Staurolithites* Caratini, 1960 (= *Zygostephanos* Hoffmann, 1970) - *Vekshinella* Loeblich & Tappan, 1963 (= *Ephippium*Vekshina, 1959; *Vagalapilla* Bukry, 1969; *Staurorhabdus* Noël, 1972)	- *Ahmuellerella* Reinhardt, 1964 - *Bukrylithus* Black, 1971 - *Crucirhabdus* Prins ex Rood, Hay & Barnard, 1973 - *Haslingfieldia* Black, 1973 - *Heteromarginatus* Bukry, 1969 - *Mitrolithus* Deflandre, 1954 - *Monomarginatus* Wind & Wise, 1977 - *Pontilithus* Gartner, 1968 - *Saeptella* Goy, 1979 ex Goy, 1981 - *Staurolithites* Caratini, 1960 - *Staurorhabdus* Noël, 1973 - *Tubirhabdus* Prins, 1969 ex Rood et al., 1973 - *Vagalapilla* Bukry, 1969 - *Vekshinella* Loeblich & Tappan, 1963) - *Vacherauvillius* Goy, 1979 ex Goy, 1981
ARKHANGELSKIELLACEAE Bukry, 1969	- *Arkhangelskiella* Vekshina, 1959 - *Aspidolithus* Noël, 1969 - *Broinsonia* Bukry, 1969 - *Gartnerago* Bukry, 1969 - *Kamptnerius* Deflandre, 1959 - *Laffittius* Noël, 1969 - *Prolatipatella* Gartner, 1968	- *Acaenolithus* Black, 1973 - *Arkhangelskiella* Vekshina, 1959 - *Aspidolithus* Noël, 1969 - *Broinsonia* Bukry, 1969 - *Cribricatillus* Black, 1973 - *Crucicribrum* Black, 1973 - *Gartnerago* Bukry, 1969 - *Kamptnerius* Deflandre, 1959 - *Laffittius* Noël, 1969	- *Acaenolithus* Black, 1973 - *Arkhangelskiella* Vekshina, 1959 - *Aspidolithus* Noël, 1969 - *Broinsonia* Bukry, 1969 - *Cribricatillus* Black, 1973 - *Crucicribrum* Black, 1973 - *Diloma* Wind & Cepek, 1979 - *Gartnerago* Bukry, 1969 (= *Laffittius* Noël, 1969)

Appendix 10.1 *(continued)*

Family	Hay (1977)	Tappan (1980)	Perch-Nielsen (1985a,b)
		- *Prolatipatella* Gartner, 1968	- *Kamptnerius* Deflandre, 1959 - *Misceomarginatus* Wind & Wise, 1977 - *Thiersteinia* Wise & Watkins, 1983
BISCUTACEAE Black, 1971	—	- *Bennocyclus* Zweili, 1974 - *Bidiscus* Bukry, 1969 - ***Biscutum**** Black, 1959 - *Discorhabdus* Noël, 1965 - *Noellithina* Grün & Zweili, 1974 - *Palaeopontosphaera* Noël, 1965 - *Seribiscutum* Filewicz, Wind & Wise, 1977 - *Similicoronolithus* Bukry, 1969	- ***Biscutum**** Black, 1959 - *Discorhabdus* Noël, 1965 - *Noellithina* Grün & Zweili, 1974 - ***Seribiscutum*** Filewicz, Wind & Wise, 1977
BRAARUDOSPHAERACEAE Deflandre, 1947	- ***Braarudosphaera**** Deflandre, 1947 (= Eodiscoaster Martini, 1961) - ***Micrantholithus**** Deflandre, 1950 - ***Pemma*** Klumpp, 1953 - ***Quinquerhabdus***** Bukry & Bramlette, 1971 - *Rucinolithus* Stover, 1966 - ***Vermiculithina***** Bukry & Percival, 1971	- *Astrionis* Wilcoxon, 1972 - ***Biantholithus*** Bramlette & Martini, 1964 - ***Braarudosphaera**** Deflandre, 1947 (= Eodiscoaster Martini, 1961) - *Hexalithus* Gardet, 1955 - *Hexangolithus* Bukry, 1969 - ***Micrantholithus**** Deflandre, 1950 - *Octolithites* Caratini, 1963 - ***Pemma*** Klumpp, 1953 - ***Pentaster*** Bybell & Gartner, 1972 - ***Quinquerhabdus***** Bukry & Bramlette, 1971 - ***Vermiculithina***** Bukry & Percival, 1971	- ***Braarudosphaera**** Deflandre, 1947 (= Eodiscoaster Martini, 1961) - *Bukryaster* Prins, 1971 - ***Micrantholithus**** Deflandre, 1950 - ***Pemma*** Klumpp, 1953 - ***Pentaster*** Bybell & Gartner, 1972 - *Trapezopentus* Wind & Cepek, 1979
CALCIOSOLENIACEAE Kamptner, 1937	- **Acanthosolenia* Bernard, 1939 - *Anoplosolenia* Deflandre, 1952 - **Calciopappus* Gaarder & Ramsfjell, 1954 - **Calciosolenia* Gran, 1912 - *Dictyolithus* Gorka, 1957 - **Navisolenia* Lecal, 1965 - ***Scapholithus**** Deflandre, 1954	- **Acanthosolenia* Bernard, 1939 - *Anoplosolenia* Deflandre, 1952 - **Calciopappus* Gaarder & Ramsfjell, 1954 - **Calciosolenia* Gran, 1912 - *Dictyolithus* Gorka, 1957 - **Navisolenia* Lecal, 1965 - ***Scapholithus**** Deflandre, 1954	- *Anoplosolenia* Deflandre, 1952 - *Dictyolithus* Gorka, 1957 - ***Scapholithus**** Deflandre, 1954

Appendix 10.1 _(continued)_

Family	Hay (1977)	Tappan (1980)	Perch-Nielsen (1985a,b)
CALYCULACEAE Noël, 1973	—	—	- _Calyculus_ Noël, 1973 - _Catillus_ Goy, 1979 ex Goy, 1981 - _Incerniculum_ Goy, 1979 ex Goy, 1981 - _Vikosphaera_ Goy, 1979 ex Goy, 1981
CALYPTROSPHAERACEAE Boudreaux & Hay, 1969	- *_Anthosphaera_ Kamptner, 1937 - _Calyptrolithus_ Kamptner, 1948 - _Calyptrosphaera_ Lohmann, 1902 - _Cyclocalyptra_ Kamptner, 1958 - *_Lacrymasphaera_ Lecal, 1965 - _Petalosphaera_ Lohmann, 1913 - _Sphaerocalyptra_ Deflandre, 1952	- _Acuturris_ Wind & Wise, 1977 - *_Balaniger_ Thomsen & Oates, 1978 - _Calyptrosphaera_ Lohmann, 1902 (= _Calyptrolithus_ Kamptner ex Deflandre, 1952 - __*Clathrolithus*__ Deflandre, 1954 - _Corisphaera_ Kamptner, 1936 - _Cyclocalyptra_ Kamptner, 1936 - __*Daktylethra*__ Gartner, 1969 - __*Discoturbella*__** Roth, 1970 - *_Helladosphaera_ Kamptner, 1936 (= *_Anthosphaera_ Kamptner, 1936) - __*Holodiscolithus*__ Roth, 1970 - *_Homozogosphaera_ Deflandre, 1952 - *_Lacrymasphaera_ Lecal, 1965 - __*Lanternithus*__ Stradner, 1962 - _Lucianorhabdus_ Deflandre, 1959 (= _Eurhabdus_ Reinhardt, 1965) - __*Naninfula*__ Perch-Nielsen, 1968 - _Okkolithus_ Wind & Wise, 1977 - _Orastrum_ Wind & Wise, 1977 - _Ottavianus_ Risatti, 1973 - _Periphyllopora_ Kamptner, 1936 - __*Peritrachelina*__ Deflandre, 1952 - _Petalosphaera_ Lohmann, 1913 - _Phanulithus_ Wind & Wise, 1977 - _Pharus_ Wind & Wise, 1977 - _Polycladolithus_ Deflandre, 1954 - _Pyxolithus_ Deflandre, 1954 - _Russelia_ Risatti, 1973 - __*Scampanella*__* Forchheimer & Stradner, 1973 - _Semihololithus_ Perch-Nielsen, 1971 - _Sphaerocalyptra_ Deflandre, 1952 - __*Trochoaster*__ Klumpp, 1953 - *_Turrispira_ Manton, Sutherland & Oates, 1976	- _Acuturris_ Wind & Wise, 1977 - _Anfractus_ Medd, 1979 - _Athenagalea_ Hattner & Wise, 1980 - _Calculites_ Prins & Sissingh, 1977 (= _Phanulithus_ Wind & Wise, 1977) - __*Clathrolithus*__ Deflandre, 1954 - __*Corannulus*__ Stradner, 1962 - __*Daktylethra*__ Gartner, 1969 - __*Holodiscolithus*__ Roth, 1970 - _Isocrystallithus_ Verbeek, 1976 - __*Lanternithus*__ Stradner, 1962 - _Lucianorhabdus_ Deflandre, 1959 - _Metadoga_ Wind & Cepek, 1979 - _Multipartis_ Risatti, 1973 - _Munarinus_ Risatti, 1973 - __*Octolithus*__* Romein, 1979 - _Okkolithus_ Wind & Wise, 1977 - _Orastrum_ Wind & Wise, 1977 - __*Orthozygus*__ Bramlette & Wilcoxon, 1967 - _Ottavianus_ Risatti, 1973 - __*Peritrachelina*__ Deflandre, 1952 - _Pharus_ Wind & Wise, 1977 - __*Polycladolithus*__ Deflandre, 1954 - __*Quinquerhabdus*__ Bukry & Bramlette, 1969 - _Ramsaya_ Risatti, 1973 - _Russellia_ Risatti, 1973 - _Semihololithus_ Perch-Nielsen, 1971 - _Tetralithus_ Gardet, 1955 - __*Zygrhablithus*__ Deflandre, 1959 (= _Pseudozygrhablithus_ Haq, 1971)

Appendix 10.1 (*continued*)

Family	Hay (1977)	Tappan (1980)	Perch-Nielsen (1985a,b)
		- *Zygosphaera* Kamptner, 1936 (= ***Orthozygus*** Bramlette & Wilcoxon, 1967) - ***Zygrhablithus*** Deflandre, 1959 (= *Orthorhabdus* Bramlette & Wilcoxon, 1967)	
CERATOLITHACEAE Norris, 1965	- ***Angulolithina*** Bukry, 1973 - *Ceratolithina* Martini, 1967 - *Ceratolithoides* Bramlette & Martini, 1964 - ***Ceratolithus*** Kamptner, 1950	- ***Amaurolithus*** Gartner & Bukry, 1975 - ***Angulolithina*** Bukry, 1973 - ***Ceratolithus*** Kamptner, 1950	- ***Amaurolithus*** Gartner & Bukry, 1965 - ***Angulolithina*** Bukry, 1973 - ***Ceratolithus*** Kamptner, 1950
CHIASTOZYGACEAE Rood et al., 1973 Considered synonym of Zygodiscaceae in Tappan, 1980	- ***Chiastozygus**** Gartner, 1968 - ***Chiphragmalithus*** Bramlette & Sullivan, 1961 - ***Isthmolithus*** Deflandre, 1954 - ***Nannotetrina*** Achutan & Stradner, 1967 - ***Neococcolithes*** Sujkowski, 1931 (= *Zygolithus* Kamptner ex Matthes, 1956; *Indumentalithus* Vekshina, 1959) - *Tegumentum* Roth & Thierstein, 1972	—	- ***Chiastozygus**** Gartner, 1968 - *Helicolithus* Noël, 1970 - *Tegumentum* Thierstein, 1972
COCCOLITHACEAE Poche, 1913 (Coccolithaceae Kamptner, 1928, emended Hay & Mohler, 1967 in Hay, 1977)	- ***Biscutum**** Black, 1959 - ***Coccolithus*** Schwarz, 1894 (= *Cyatholithus* Huxley, 1868; *Coccosphaera* Wallich, 1877; *Cyathosphaera* Haeckel, 1894) - ***Conococcolithus*** Hay & Mohler, 1967 - ***Corannulus*** Stradner, 1962 (= *Diademopetra* Hay, Mohler & Wade, 1966) - ***Coronocyclus*** Hay, Mohler & Wade, 1966 - *Cribrocentrum* Perch-Nielsen, 1972 - ***Cyclicargolithus*** Bukry, 1971 - *Cyclococcolithina* Wilcoxon, 1970	- *Birkelundia* Perch-Nielsen, 1971 - ***Calcidiscus*** Kamptner, 1950 (= *Cyclococcolithus* Kamptner, 1958; *Tiarolithus* Kamptner, 1958; *Cycloplacolithus* Kamptner, 1963; *Cycloplacolithella* Haq 1968; *Cyclococcolithina* Wilcoxon, 1970) - ***Chiasmolithus*** Hay, Mohler & Wade, 1966 - ***Coccolithus*** Schwarz, 1894 (= *Cyatholithus* Huxley, 1868; *Coccosphaera* Wallich, 1877; *Coccolithophora* Lohmann, 1902; **Crystallolithus* Gaarder & Markali, 1956)	- *Birkelundia* Perch-Nielsen, 1971 - ***Bramletteius*** Gartner, 1969 - ***Calcidiscus*** Kamptner, 1950 - *Campylosphaera* Kamptner, 1963 - ***Chiasmolithus*** Hay, Mohler & Wade, 1966 - ***Clausicoccus*** Prins, 1979 - ***Coccolithus*** Schwarz, 1894 - ***Conococcolithus*** Hay & Mohler, 1967 - ***Coronocyclus*** Hay, Mohler & Wade, 1966 - ***Cruciplacolithus*** Hay & Mohler, 1967 - ***Cyclagelosphaera**** Noël, 1965 - *Cycloperfolithus* Lehotayova & Priewalder, 1978

Appendix 10.1 (*continued*)

Family	Hay (1977)	Tappan (1980)	Perch-Nielsen (1985a,b)
COCCOLITHACEAE (*continued*)	(= *Cyclococcolithus* Kamptner, 1954; ***Calcidiscus*** Kamptner, 1952; *Tiarolithus* Kamptner, 1958) - *Cyclolithella* Loeblich & Tappan, 1963 (= *Cyclolithus* Kamptner ex Deflandre, 1952) - *Heyneckia* Lohman, 1913 - ***Ilselithina*** Stradner, 1966 (= *Hayella* Roth, 1969) - ***Markalius**** Bramlette & Martini, 1964 - **Neosphaera* Lecal-Schlauder, 1950 - ***Pedinocyclus*** Bukry & Bramlette, 1971 - ***Striatococcolithus*** Bukry, 1971 - *Tergestiella* Kamptner, 1941 - ***Umbilicosphaera*** Lohmann, 1902 (= *Craspedolithus* Kamptner, 1963)	- ***Conococcolithus*** Hay & Mohler, 1967 - ***Corannulus*** Stradner, 1962 (= *Diademopetera* Hay, Mohler & Wade, 1966) - ***Coronocyclus*** Hay, Mohler, & Wade, 1966 - *Cribrocentrum* Perch-Nielsen, 1972 - *Cricolithus* Kamptner, 1958 - ***Cruciplacolithus*** Hay & Mohler, 1967 (= *Crucilithus* Stradner, 1968) - ***Cyclicargolithus*** Bukry, 1971 - *Heyneckia* Lohmann, 1913 - ***Hornibrookina*** Edwards, 1973 - ***Ilselithina*** Stradner, 1966 (= *Hayella* Roth, 1969) - ***Markalius*** Bramlette & Martini, 1964 - **Neosphaera Lecal-Schlauder, 1950* - ***Oolithothus*** Reinhardt, 1968 - ***Pedinocyclus*** Bukry & Bramlette, 1971 - *Rhabdolekiskus* Hill, 1976 - ***Striatococcolithus*** Bukry, 1971 - *Tergestiella* Kamptner, 1941 - ***Umbilicosphaera*** Lohmann, 1902 (= *Craspedolithus* Kamptner, 1963)	- ***Ericsonia*** Black, 1964 - ***Hayella*** Gartner, 1969 - ***Ilselithina*** Stradner, 1966 - ***Markalius**** Bramlette & Martini, 1964 -**Neosphaera* Lecal-Schlauder, 1950 - ***Pedinocyclus*** Bukry & Bramlette, 1971 - ***Striatococcolithus*** Bukry, 1971 - ***Umbilicosphaera*** Lohmann, 1902
CREPIDOLITHACEAE Black, 1972	—	- *Carinolithus* Prins, 1974 - *Crepidolithus* Noël, 1965 - *Mitosia* Worsley, 1971 - *Parhabdolithus* Deflandre, 1952	- *Calcicalathina* Thierstein, 1971 - *Carinolithus* Prins, 1974 - *Conusphaera* Trejo, 1969 - *Eoconusphaera* Jafar, 1983 - *Crepidolithus* Noël, 1965 - *Diazomatolithus* Noël, 1965 - *Millbrookia* Medd, 1979 - *Parhabdolithus* Deflandre, 1952 - *Proculithus* Medd, 1979

Appendix 10.1 *(continued)*

Family	Hay (1977)	Tappan (1980)	Perch-Nielsen (1985a,b)
DEFLANDRIACEAE Black, 1968 (in Tappan, 1980) = PREDICOSPHAERACEAE Rood, Hay & Barnard, 1971 (in Hay, 1977; Perch-Nielsen, 1985b)	- *Predicosphaera* Vekshina, 1959	- *Deflandrius* Bramlette & Martini, 1964 (= *Predicosphaera* Vekshina, 1959)	- *Predicosphaera* Vekshina, 1959 (= *Deflandrius* Bramlette & Martini, 1964)
DEUTSCHLANDIACEAE Kamptner, 1928 (in Hay, 1977; Perch-Nielsen, 1985b) = HALOPAPPACEAE Kamptner, 1928 (in Tappan, 1980)	- ***Deutschlandia*** Lohmann, 1912 - ***Florisphaera*** Okada & Honjo, 1973 - **Thorosphaera* Ostenfeld, 1910	—	- ***Deutschlandia*** Lohmann, 1912
DISCOASTERACEAE Tan Sin Hok, 1927	- *Agalmastoaster* Klumpp, 1953 - ***Catinaster*** Martini & Bramlette, 1963 - *Clavodiscoaster* Prins, 1971 - ***Discoaster*** Tan Sin Hok, 1927 - *Discoasteroides* Bramlette & Sullivan, 1961 - *Gyrodiscoaster* Stradner, 1961 - ***Hayaster*** Bukry, 1973 - *Heliodiscoaster* Tan Sin Hok, 1927 - *Hemidiscoaster* Tan Sin Hok, 1927 - *Radiodiscoaster* Prins, 1971 - *Truncodiscoaster* Prins, 1971 - *Turbodiscoaster* Prins, 1971	- *Agalmastoaster* Klumpp, 1953 - ***Catinaster*** Martini & Bramlette, 1963 - *Clavodiscoaster* Prins, 1971 - ***Discoaster*** Tan Sin Hok, 1927 - *Discoasteroides* Bramlette & Sullivan, 1961 - *Gyrodiscoaster* Stradner, 1961 - ***Hayaster*** Bukry, 1973 - *Heliodiscoaster* Tan Sin Hok, 1927 - *Hemidiscoaster* Tan Sin Hok, 1927 - *Radiodiscoaster* Prins, 1971 - ***Tribrachiatus*** Shamrai, 1963 - *Truncodiscoaster* Prins, 1971 - *Turbodiscoaster* Prins, 1971	- ***Catinaster*** Martini & Bramlette, 1963 - ***Discoaster*** Tan Sin Hok, 1927 (= *Agalmastoaster* Klumpp, 1953; *Clavodiscoaster* Prins, 1971; *Gyrodiscoaster* Stradner, 1961; *Heliodiscoaster* Tan Sin Hok, 1927; *Hemidiscoaster* Tan Sin Hok, 1927; *Radiodiscoaster* Prins, 1971; *Truncodiscoaster* Prins, 1971; *Turbodiscoaster* Prins, 1971) - *Discoasteroides* Bramlette & Sullivan, 1961
EIFFELLITHACEAE Reinhardt, 1965	- *Eiffellithus* Reinhardt, 1965 (= *Clinorhabdus* Stover, 1966)	- *Eiffellithus* Reinhardt, 1965 (= *Clinorhabdus* Stover, 1966)	- *Eiffellithus* Reinhardt, 1965 (= *Clinorhabdus* Stover, 1966)
ELLIPSAGELOSPHAERACEAE Noël, 1965	- *Bidiscus* Bukry, 1968 - ***Cyclagelosphaera**** Noël, 1965 - *Esgia* Worsley, 1971 - *Flabellites* Thierstein, 1973 - *Helenia* Worsley, 1971	- *Caterella* Black, 1971 - *Coptolithus* Black, 1973 - ***Cyclagelosphaera**** Noël, 1965 - *Diazomatolithus* Noël, 1965 - *Ellipsagelosphaera* Noël, 1965	- *Ansulasphaera* Grün & Zweili, 1980 - *Calolithus* Noël, 1965 - *Caterella* Black, 1971 - *Coptolithus* Black, 1973 - ***Cyclagelosphaera**** Noël, 1965 - *Ellipsagelosphaera* Noël, 1965

Appendix 10.1 *(continued)*

Family	Hay (1977)	Tappan (1980)	Perch-Nielsen (1985a,b)
	(= *Cruciellipsis* Thierstein, 1971) - *Similicoronilithus* Bukry, 1969 - *Stenoradiatus* Worsley, 1971 - *Watznaueria* Reinhardt, 1964 (= *Colvillea* Black, 1964; *Actinosphaera* Noël, 1965; *Calolithus* Noël, 1965; *Ellipsagelosphaera* Noël, 1965; *Maslovella* Tappan & Loeblich, 1966)	- *Esgia* Worsley, 1971 - *Lotharingius* Noël, 1972 (= *Striatomarginis* Prins ex Rood, Hay & Barnard, 1973) - *Manivitella* Thierstein, 1971 - *Margolatus* Forchheimer, 1972 - *Sphenoradiatus* Worsley, 1971 - *Tretosestrum* Wilcoxon, 1972 - *Tubodiscus* Thierstein, 1973 - *Watznaueria* Reinhardt, 1964 (= *Colvillea* Black, 1964; *Actinosphaera* Noël, 1965; *Calolithus* Noël, 1965; *Maslovella* Tappan & Loeblich, 1966)	- *Haqius* Roth, 1978 - *Lotharingius* Noël, 1972 - *Manivitella* Thierstein, 1971 - *Margolatus* Forchheimer, 1972 - ***Markalius**** Bramlette & Martini, 1964 - *Tretosestrum* Wilcoxon, 1972 - *Tubodiscus* Thierstein, 1973 - *Watznaueria* Reinhard, 1964 (= *Actinosphaera* Noël, 1965; *Colvillea* Black, 1964; *Esgia* Worsley, 1971; *Maslovella* Tappan & Loeblich, 1966)
FASCICULITHACEAE Hay & Mohler, 1967	- *Bomolithus* Roth, 1973 - ***Fasciculithus*** Bramlette & Sullivan, 1961	- *Bomolithus* Roth, 1973 - ***Fasciculithus*** Bramlette & Sullivan, 1961	- ***Fasciculithus*** Bramlette & Sullivan, 1961
GEPHYROCAPSACEAE Black, 1971	- ***Cruciplacolithus*** Hay & Mohler, 1967 - ***Chiasmolithus*** Hay, Mohler & Wade, 1966 - *Crenalithus* Roth, 1973 - [?] ***Dictyococcites*** Black, 1967 - ***Emiliania*** Hay & Mohler, 1967 - ***Gephyrocapsa*** Kamptner, 1943 - ***Pseudoemiliania*** Gartner, 1969 - ***Reticulofenestra*** Hay, Mohler & Wade, 1966 (= *Apertapetra* Hay, Mohler & Wade, 1966; [?] ***Dictyococcites*** Black, 1967)	- *Crenalithus* Roth, 1973 - **Dicrateria* Parke, 1949 - ***Emiliania*** Hay & Mohler, 1967 (= *Ellipsoplacolithus* Kamptner, 1963) - ***Gephyrocapsa*** Kamptner, 1943 - *Imantonia* Reynolds, 1974 - ***Reticulofenestra*** Hay, Mohler & Wade, 1966 (= Apertapetra Hay, Mohler & Wade, 1966; ***Dictyococcites*** Black, 1967; *Stradnerius* Haq, 1968) - *Tremalithus* Kamptner ex Deflandre, 1952	—
GONIOLITHACEAE Deflandre, 1957	- ***Goniolithus*** Deflandre, 1957	- *Calyculus* Noël, 1972 - *Carinolithus* Prins, 1974 - ***Goniolithus*** Deflandre, 1957	- ***Goniolithus*** Deflandre, 1957

Appendix 10.1 *(continued)*

Family	Hay (1977)	Tappan (1980)	Perch-Nielsen (1985a,b)
HALOPAPPACEAE Kamptner, 1928	- **Bernardosphaera* Lecal-Schlauder, 1951 - *Calcioconus* Schiller, 1925 - **Halopappus* Lohmann, 1912	- *Calcioconus* Schiller, 1916 - ***Deutschlandia*** Lohmann, 1912 (= ***Florisphaera*** Okada & Honjo, 1973 - **Halopappus* Lohmann, 1912 - **Michaelsarsiae* Gran, 1912 - **Ophiaster* Gran, 1912 (= **Bernardosphaera* Lecal, 1951) - **Pappomonas* Manton & Oates, 1975 - **Papposphaera* Tangen, 1970 - **Thalassopappus* Kamptner, 1941 - **Thorosphaera* Ostenfeld, 1910	—
HELICOSPHAERACEAE Black, 1971	—	- ***Helicosphaera*** Kamptner, 1954 (= <u>*Helicopontosphaera*</u> Hay & Mohler, 1967)	- ***Helicosphaera*** Kamptner, 1954 (?= <u>*Helicopontosphaera*</u> Hay & Mohler, 1967)
HELIOLITHACEAE Hay & Mohler, 1967	- ***Heliolithus*** Bramlette & Sullivan, 1961	- ***Heliolithus*** Bramlette & Sullivan, 1961	- <u>*Bomolithus*</u> Roth, 1973 - ***Heliolithus*** Bramlette & Sullivan, 1961
HYMENOMONADACEAE Senn, 1900 (in Tappan, 1980) = HYMENOMONACEAE Hay, 1977	- **Coccochrysis* Conrad, 1926 - **Cricosphaera* Braarud, 1960 - **Heimiella* Lohmann, 1913 - **Hymenomonas* Stein, 1878	- **Apistonema* Pascher, 1925 - **Calciarcus* Manton, Sutherland & Oates, 1977 - **Chrysonema* Anand, 1937 - **Coccochrysis* Conrad, 1926 - **Cricosphaera* Braarud, 1960 - **Heimiella* Lohmann, 1913 - **Hymenomonas* Stein, 1878 (= **Pleurochrysis* Pringsheim, 1955) - **Wigwamma* Manton, Sutherland & Oates, 1977	—
LITHOSTROMATIONACEAE Deflandre, 1959	- ***Lithostromation*** Deflandre, 1942 - ***Martiniaster*** Loeblich & Tappan, 1963 (= <u>*Coronaster*</u> Martini, 1961) - ***Trochoaster*** Klumpp, 1953	- ***Lithostromation*** Deflandre, 1942 - ***Martiniaster*** Loeblich & Tappan, 1963 (= <u>*Coronaster*</u> Martini, 1961) - ***Trochoaster*** Klumpp, 1953	- ***Lithostromation*** Deflandre, 1942 (= ***Trochoaster*** Klumpp, 1953) - ***Martiniaster*** Loeblich & Tappan, 1963

Appendix 10.1 *(continued)*

Family	Hay (1977)	Tappan (1980)	Perch-Nielsen (1985a,b)
ISOCHRYSIDACEAE Parke, 1949 (in Tappan, 1980) = MARTHASTERITACEAE Hay, 1977	- *Assipetra* Roth, 1973 - *Marthasterites* Deflandre, 1959 - *Micular Vekshina, 1959* (= <u>*Nannotetraster*</u> Martini & Stradner, 1960) - ***Rhomboaster*** Bramlette & Sullivan, 1961 - ***Trochastrites*** Stradner, 1961	- *Chrysotila* Anand, 1937 (= *Ruttneria* Geitler, 1942) - Isochrysis Parke, 1949 - *Marthasterites* Deflandre, 1959 - *Tetralithus* Gardet, 1955	—
MICHELSARSIOIDEAE Lecal-Schlauder, 1951	- **Michelsarsia* Gran, 1912 - **Ophiaster* Gran, 1912 - **Thalassopappus* Kamptner, 1941	—	—
MICRORHABDULACEAE Deflandre, 1963	- *Lithraphidites* Deflandre, 1963 - *Microrhabdulinus* Deflandre, 1963 - *Microrhabduloides* Deflandre, 1963 - *Microrhabdulus* Deflandre, 1959	- *Lithraphidites* Deflandre, 1963 - *Microrhabdulinus* Deflandre, 1963 - *Microrhabduloides* Deflandre, 1963 - *Microrhabdulus* Deflandre, 1959 - *Pseudolithraphidites* Keupp, 1976	- *Lithraphidites* Deflandre, 1963 - *Microrhabdulus* Deflandre, 1959 - *Pseudolithraphidites* Keupp, 1976 - *Pseudomicula* Perch-Nielsen, 1978
NANNOCONACEAE Deflandre, 1959	- *Nannoconus* Kamptner, 1931 - *Polycostella* Thierstein, 1971	- *Conusphaera* Trejo, 1969 (= *Cretaturbella* Roth, 1971) - *Nannoconus* Kamptner, 1931 - *Polycostella* Thierstein, 1971	- *Nannoconus* Kamptner, 1931 (= *Brachiolithus* Noël, 1959 ex Loeblich & Tappan, 1963)
OCHROSPHAERACEAE Schussnig, 1930	—	- **Ochrosphaera* Schussnig, 1930	—
PAVLOVACEAE Green, 1976	—	- **Corcontochrysis* Kalina, 1970 - **Diacronema* Prauser, 1958 - **Exanthemachrysis* Lepailleur, 1970 - **Pavlova* Butcher, 1952	—
PERCIVALIACEAE Black, 1973	—	- *Percivalia* Bukry, 1969	—
PHAEOCYSTACEAE Lagerheim, 1893	—	- **Phaeoystis* Lagerheim, 1893	—
PODORHABDACEAE Noël, 1965	- *Bipodorhabdus* Noël, 1965 - *Cretadiscus* Gartner, 1968 - *Cretarhabdus* Bramlette & Martini, 1964	- *Allemanites* Grün, 1975 - *Axopodorhabdus* Wind & Wise, 1977 - *Bipodorhabdus* Noël, 1965 - *Boletuvelum* Wind & Wise, 1977	- *Allemanites* Grün, 1975 - *Axopodorhabdus* Wind & Wise, 1977 - *Bipodorhabdus* Noël, 1965 - *Boletuvelum* Wind & Wise, 1977

Appendix 10.1 *(continued)*

Family	Hay (1977)	Tappan (1980)	Perch-Nielsen (1985a,b)
	(= *Stradneria* Reinhardt, 1964) - *Cribrosphaera* Arkhangelski, 1912 (= *Favocentrum* Black, 1964) - *Diazomatolithus* Noël, 1965 - *Discorhabdus* Noël, 1965 - *Dodekapodorhabdus* Perch-Nielsen, 1968 - *Ethmorhabdus* Noël, 1965 - *Heteromarginatus* Bukry, 1961 - *Hexapodorhabdus* Noël, 1965 - *Manivitella* Thierstein, 1971 - *Nephrolithus* Gorka, 1957 - *Octopodorhabdus* Noël, 1965 - *Paleopontosphaera* Noël, 1965 - *Percivalia* Bukry, 1969 - *Podorhabdus* Noël, 1965 - *Polypodorhabdus* Noël, 1965 - *Sollasites* Black, 1967 (= *Costacentrum* Bukry, 1969) - *Striatomarginis* Prins ex Rood et al., 1973 - *Tubodiscus* Thierstein, 1973	- *Cleistorhabdus* Black, 1972 - *Cretadiscus* Gartner, 1968 - *Cretarhabdella* Black, 1971 - *Cretarhabdus* Martini & Bramlette, 1964 (= *Stradneria* Reinhardt, 1965; *Heterorhabdus* Noël, 1970; *Helenea* Worsley, 1971) - *Cribrosphaerella* Deflandre, 1952 (= *Cribrosphaera* Arkhangelski, 1912) - *Cruciellipsis* Thierstein, 1972 - *Dodekapodorhabdus* Perch-Nielsen, 1968 - *Ethmorhabdus* Noël, 1965 - *Flabellites* Thierstein, 1973 - *Favocentrum* Black, 1964 - *Gephyrorhabdus* Hill, 1976 - *Grantarhabdus* Black, 1971 - *Hemipodorhabdus* Black, 1971 - *Hexapodorhabdus* Noël, 1965 - *Microstaurus* Black, 1971 - *Miravetesina* Grün, 1975 - *Nephrolithus* Gorka, 1957 - *Octocyclus* Black, 1972 - *Octopodorhabdus* Noël, 1965 - *Perissocyclus* Black, 1971 - *Podorhabdus* Noël, 1965 - *Polypodorhabdus* Noël, 1965 - *Retacapsa* Black, 1971 - *Teichorhabdus* Wind & Wise, 1977 - *Tetrapodorhabdus* Black, 1971	- *Cleistorhabdus* Black, 1972 - *Cretarhabdus* Martini & Bramlette, 1964 (= *Cretarhabdella* Black, 1971) - *Cribrocorona* Perch-Nielsen, 1973 - *Cribrosphaerella* Deflandre, 1952 (= *Cretadiscus* Gartner, 1968; *Cribrosphaera* Arkhangelski, 1912; *Favocentrum* Black, 1964) - *Cruciellipsis* Thierstein, 1972 - *Dekapodorhabdus* Medd, 1979 - *Dodekapodorhabdus* Perch-Nielsen, 1968 - *Ethmorhabdus* Noël, 1965 - *Flabellites* Thierstein, 1973 - *Gephyrorhabdus* Hill, 1976 - *Grantarhabdus* Black, 1971 - *Helenea* Worsley, 1979 - *Heterorhabdus* Noël, 1970 - *Hexapodorhabdus* Noël, 1965 - *Microstaurus* Black, 1971 - *Miravetesina* Grün, 1975 - *Nephrolithus* Gorka, 1957 - *Octocyclus* Black, 1972 - *Octopodorhabdus* Noël, 1965 - *Perissocyclus* Black, 1971 - *Podorhabdus* Noël, 1965 (= *Hemipodorhabdus* Black, 1971) - *Polypodorhabdus* Noël, 1965 - *Retacapsa* Black, 1971 - *Speetonia* Black, 1971 - *Stradneria* Reinhardt, 1964 - *Teichorhabdus* Wind & Wise, 1977 - *Tetrapodorhabdus* Black, 1971
POLYCYCLOLITHACEAE Forchheimer, 1972	—	- *Bukryaster* Prins, 1971 - *Cylindralithus* Bramlette & Martini, 1964 - *Eprolithus* Stover, 1966	- *Assipetra* Roth, 1973 - *Eprolithus* Stover, 1966 (= *Polycyclolithus* Forchheimer, 1968) - *Hayesites* Manivit, 1971

Appendix 10.1 *(continued)*

Family	Hay (1977)	Tappan (1980)	Perch-Nielsen (1985a,b)
		- *Hayesites* Manivit, 1971 - *Lapideacasis* Black, 1971 - *Lithastrinus* Stradner, 1962 - *Polycostella* Thierstein, 1971 - *Polycyclolithus* Forchheimer, 1968 - *Radiolithus* Stover, 1966 - *Rhombogyrus* Black, 1973 - *Rucinolithus* Stover, 1966	- *Hexalithus* Gardet, 1955 - *Lithastrinus* Stradner, 1962 - *Micula* Vekshina, 1959 - *Pervilithus* Crux, 1981 - *Polycostella* Thierstein, 1971 - *Quadrum* Prins & Perch-Nielsen, 1977 (= *Uniplanarius* Hattner & Wise, 1980) - *Radiolithus* Stover, 1966 (= *Rhombogyrus* Black, 1973) - *Rucinolithus* Stover, 1966
PONTOSPHAERACEAE Lemmermann, 1908	- **Algirosphaera* Schlauder, 1945 - *Crassapontosphaera* Boudreaux & Hay, 1969 - *Discolithina* Loeblich & Tappan, 1963 - *Koczyia Boudreaux & Hay, 1969* - ***Lophodolithus*** Deflandre, 1954 - *Helicopontosphaera* Hay & Mohler, 1967 [vel ***Helicosphaera***] - **Lohmannosphaera* Schiller, 1913 - ***Pontosphaera*** Lohmann, 1902 - ***Scyphosphaera*** Lohmann, 1902 - *Transversopontis* Hay, Mohler & Wade, 1966	- **Algirosphaera* Schlauder, 1945 - *Crassapontosphaera* Boudreaux & Hay, 1969 - *Discolithina* Loeblich & Tappan, 1963 (= *Discolithus* Huxley, 1868) - *Koczya* Boudreaux & Hay, 1969 - ***Lophodolithus*** Deflandre, 1954 - *Pontolithina* Black, 1968 - ***Pontosphaera*** Lohmann, 1902 - *Transversopontis* Hay, Mohler & Wade, 1966	- ***Calciopilleus***** Müller, 1974 - ***Neocrepidolithus**** Romein, 1979 - ***Nannocorbis***** Müller, 1974 - ***Pontosphaera*** Lohmann, 1902 (= *Crassapontosphaera* Boudreaux & Hay, 1969; *Discolithina* Loeblich & Tappan, 1963; *Discolithus* Huxley, 1868; *Koczya* Boudreaux & Hay, 1969) - ***Scyphosphaera*** Lohmann, 1902 - *Transversopontis* Hay, Mohler & Wade, 1966
PRINSIACEAE Hay & Mohler, 1967	- ***Bramletteius*** Gartner, 1969 - *Campylosphaera* Kamptner, 1963 - ***Ericsonia*** Black, 1964 - ***Ellipsolithus*** Sullivan, 1964 - ***Prinsius*** Hay & Mohler, 1967 - ***Toweius*** Hay & Mohler, 1967	- ***Bramletteius*** Gartner, 1969 - *Campylosphaera* Kamptner, 1963 - ***Ellipsolithus*** Sullivan, 1964 - ***Ericsonia*** Black, 1964 - ***Prinsius*** Hay & Mohler, 1967 - *Repagalum* Forchheimer, 1972 - ***Toweius*** Hay & Mohler, 1967	- ***Biscutum**** Black, 1959 - ***Cribrocentrum*** Perch-Nielsen, 1971 - ***Cyclicargolithus*** Bukry, 1971 - ***Dictyococcites*** Black, 1967 (= *Stradnerius* Haq, 1968 - ***Emiliania*** Hay & Mohler, 1967 - ***Gephyrocapsa*** Kamptner, 1943 - ***Hornibrookina*** Edwards, 1973 - ***Noelarhabdus*** Jerkovic, 1970 - ***Prinsius*** Hay & Mohler, 1967 - ***Pseudoemiliania*** Gartner, 1969

Appendix 10.1 *(continued)*

Family	Hay (1977)	Tappan (1980)	Perch-Nielsen (1985a,b)
			- ***Reticulofenestra*** Hay, Mohler & Wade, 1966 (= *Apertapetra* Hay, Mohler & Wade, 1966) - ***Toweius*** Hay & Mohler, 1967
PRYMNESIACEAE Conrad, 1926	—	- **Chrysochromulina* Lackey, 1939 (= **Chrysocampanula* Fournier, 1971) - **Corymbellus* Green, 1976 - **Platychrysis* Geitler, 1930 - **Prymnesium* Massart, 1920	—
RHABDOSPHAERACEAE Lemmermann, 1980	- ***Aspidorhabdus*** Hay & Towe, 1962 - ***Blackites*** Hay & Towe, 1962 - ***Discosphaera*** Haeckel, 1894 - **Lecalia* Loeblich & Tappan, 1963 (= *Rhabdocyclus* Lecal, 1969) - **Palusphaera* Lecal, 1967 - *Pyrocyclus* Hay & Towe, 1962 - *Rhabdolithus* Kamptner ex Deflandre, 1952 - ***Rhabdosphaera*** Haeckel, 1894 (= *Rhabdolithes* Schmidt, 1870; *Clavosphaera* Lecal-Schlauder, 1951) - **Umbellosphaera* Paasche, 1955	- ***Aspidorhabdus*** Hay & Towe, 1962 - ***Blackites*** Hay & Towe, 1962 - ***Discosphaera*** Haeckel, 1894 - *Lecalia* Loeblich & Tappan, 1963 (= *Rhabdocyclus* Lecal, 1960) - ***Naninfula*** Perch-Nielsen, 1968 - **Palusphaera* Lecal, 1965 - ***Petasus*** Perch-Nielsen, 1971 - *Pyrocyclus* Hay & Towe, 1962 - *Rhabdolithus* Kamptner ex Deflandre, 1952 - ***Rhabdosphaera*** Hoekel 1894 (= *Rhabdolithes*)Schmidt, 1870; *Clavosphaera* Lecal-Schlauder, 1951; ***Aspidorhabdus*** Hay & Towe, 1962) - **Umbellosphaera* Paasche, 1955	- ***Blackites*** Hay & Towe, 1962 - ***Discosphaera*** Haeckel, 1894 - ***Rhabdosphaera*** Haeckel, 1894
RHAGODISCACEAE Hay, 1977 = Actinozygaceae; Alfordiaceae	- *Reinhardites* Perch-Nielsen, 1968 - *Rhagodiscus* Reinhardt, 1967	- *Reinhardites* Perch-Nielsen, 1968 - *Rhagodiscus* Reinhardt, 1967 (= *Actinozygus* Gartner, 1968; *Alfordia* Black, 1975; *Rhabdolithina* Reinhardt, 1967; *Viminites* Black, 1975)	- *Rhagodiscus* Reinhardt, 1967 (= *Actinozygus* Gartner, 1968; *Alfordia* Black, 1975; *Rhabdolithina* Reinhart, 1967; *Viminites* Black, 1975)

Appendix 10.1 *(continued)*

Family	Hay (1977)	Tappan (1980)	Perch-Nielsen (1985a,b)
SCYPHOSPHAERACEAE Jafar, 1975	—	- ***Calciopilleus***** Müller, 1974 - **Lohmannosphaera* Schiller, 1913 - ***Nannocorbis***** Müller, 1974 - ***Scyphosphaera*** Lohmann, 1902	—
SOLLASITACEAE Black, 1971	—	- *Gaarderella* Black, 1973 - *Heteromarginatus* Bukry, 1961 - *Sollasites* Black, 1967 (= *Costacentrum* Bukry, 1969)	- *Gaarderella* Black, 1973 - *Sollasites* Black, 1967 (= *Costacentrum* Bukry, 1969)
SPEETONIACEAE Black, 1971	—	- *Speetonia* Black, 1971	—
SPHENOLITHACEAE Deflandre, 1952	- <u>*Furcatolithus*</u> Martini, 1965 - ***Sphenolithus*** Deflandre, 1952 (= <u>*Nannoturbella*</u> Bronnimann & Stradner, 1960)	- *Assipetra* Roth, 1973 - *Calculites* Prints & Sissingh, 1977 - *Ceratolithina* Martini, 1967 - *Ceratolithoides* Bramlette & Martini, 1964 - <u>*Furcatolithus*</u> Martini, 1965 - *Micula* Vekshina, 1959 (= <u>*Nannotetraster*</u> Martini & Stradner, 1960) - *Quadrum* Prins & Perch-Nielsen, 1977 - ***Rhomboaster*** Bramlette & Sullivan, 1961 - ***Sphenolithus*** Deflandre, 1952 (= <u>*Nannoturbella*</u> Bronnimann & Stradner, 1960) - ***Trochoastrites*** Stradner, 1961	-***Sphenolithus*** Deflandre, 1952
STEPHANOLITHIONACEAE Black, 1968	- *Bukryaster* Prins, 1971 - *Calcicalathina* Thierstein, 1971 - *Carinolithus* Prins, 1969 - *Corollithion* Stradner, 1961 - *Crepidolithus* Noël, 1965 - *Cylindralithus* Bramlette & Martini, 1964 - *Eprolithus* Stover, 1966 - *Hayesites* Manivit, 1971 - *Lithastrinus* Stradner, 1962	- *Anguslofenestrellithus* Bukry, 1969 - *Corollithion* Stradner, 1961 - *Diadorhombus* Worsley, 1971 - *Nodosella* Prins ex Rood, Hay & Barnard, 1973 - *Pontilithus* Gartner, 1968 - *Rhombolithion* Black, 1973 - *Rotelapillus* Noël, 1972 - *Stephanolithion* Deflandre, 1939 - *Stradnerlithus* Black, 1971	- *Corollithion* Stradner, 1961 - *Cylindralithus* Bramlette & Martini, 1964 - *Diadorhombus* Worsley, 1971 - *Nodosella* Prins ex Rood, Hay & Barnard, 1973 - *Rhombolithion* Black, 1973 - *Rotelapillus* Noël, 1972 - *Stephanolithion* Deflandre, 1939 - *Stoverius* Perch-Nielsen, 1984

Appendix 10.1 *(continued)*

Family	Hay (1977)	Tappan (1980)	Perch-Nielsen (1985a,b)
	- *Mitosia* Worsley, 1971 - *Polycyclotithus* Forchheimer, 1968 - *Radiolithus* Stover, 1966 - *Stephanolithion* Deflandre, 1939	(= *Diadozygus* Rood, Barnard & Hay, 1971) - *Truncatoscaphus* Rood, Barnard & Hay, 1971	- *Stradnerlithus* Black, 1971 (= *Diadozygus* Rood, Barnard & Hay, 1971) - *Thurmannolithion* Grün & Zweili, 1980 - *Truncatoscaphus* Rood, Barnard & Hay, 1971
SYRACOSPHAERACEAE Lemmerman, 1908	- ***Cepekiella*** Roth, 1970 - *Syracolithus* Kamptner, 1941 ex Deflandre, 1952 - ***Syracosphaera*** Lohmann, 1902 (= *Eusyracosphaera* Lohmann, 1902; *Syracorhabdus* Lecal, 1965)	- **Alisphaera* Heimdal, 1973 - **Anthosphaera* Kamptner, 1936 - ****Caneosphaera*** Gaarder, 1977 - ***Cepekiella*** Roth, 1970 - ***Coronosphaera*** Gaarder, 1977 - *Syracolithus* Kamptner, 1941 - ***Syracosphaera*** Lohmann, 1902 (= *Eusyracosphaera* Kamptner, 1941; *Syracorhabdus* Lecal & Bernheim, 1960)	- ***Caneosphaera*** Gaarder, 1977 - ***Coronosphaera*** Gaarder, 1977 - ***Syracosphaera*** Lohmann, 1902
TRIQUETRORHABDACEAE Lipps, 1969	- ***Triquetrorhabdulus*** Martini, 1965	- ***Pseudotriquetrorhabdulus*** Wise & Constans, 1976 - ***Triquetrorhabdulus*** Martini, 1965	- *Orthorhabdus* Bramlette & Wilcoxon, 1967 - ***Pseudotriquetrorhabdulus*** Wise & Constans, 1976 - ***Triquetrorhabdulus*** Martini, 1965
ZYGODISCACEAE Hay & Mohler, 1967	- *Amphizygus* Bukry, 1969 - *Eurhabdus* Reinhardt, 1965 - *Glaukolithus* Reinhardt, 1964 - *Lucianorhabdus* Deflandre, 1959 - *Mitrolithus* Deflandre & Fert, 1954 (= *Alvearium* Black, 1967) - *Parhabdolithus* Deflandre, 1952 - *Tranolithus* Stover, 1966 - *Tubirhabdus* Prins ex Rood et al., 1973 - *Zeugrhabdotus* Reinhardt, 1965 - ***Zygodiscus*** Bramlette & Sullivan, 1961	- *Amphizygus* Bukry, 1969 - *Barringtonella* Black, 1973 - *Calcicalathina* Thierstein, 1971 - ***Chiastozygus**** Gartner, 1968 - ***Chiphragmalithus*** Bramlette & Sullivan, 1961 - *Cyclolithella* Loeblich & Tappan, 1963 (= *Cyclolithus* Kamptner ex Deflandre, 1952) - *Glaukolithus* Reinhardt, 1964 - *Haslingfieldia* Black, 1973 - *Heteromarginatus* Bukry, 1969	- *Barringtonella* Black, 1973 - ***Chiastozygus**** Gartner, 1968 - ***Chiphragmalithus*** Bramlette & Sullivan, 1961 - *Glaukolithus* Reinhardt, 1964 - ***Isthmolithus*** Deflandre, 1954 - ***Lophodolithus*** Deflandre, 1954 - *Loxolithus* Noël, 1965 - ***Neochiastozygus*** Perch-Nielsen, 1971 - ***Neococcolithes*** Sujkowski, 1931 (= *Zygolithus* Kamptner ex Matthes, 1956; *Indumentalithus* Vekshina, 1958;

Appendix 10.1 (*continued*)

Family	Hay (1977)	Tappan (1980)	Perch-Nielsen (1985a,b)
		- ***Isthmolithus*** Deflandre, 1954 - *Loxolithus* Noël, 1965 - *Misceomarginatus* Wind & Wise, 1977 - *Mitrolithus* Deflandre & Fert, 1954 (= *Alvearium* Black, 1967) - *Monomarginatus* Wind & Wise, 1977 - ***Nannotetrina*** Achutan & Stradner, 1967 - ***Neochiastozygus*** Perch-Nielsen, 1971 - ***Neococcolithes*** Sujkowski, 1931 (= <u>*Zygolithus*</u> Kamptner ex Matthes, 1956; <u>*Indumentalithus*</u> Vekshina, 1958; <u>*Heliorthus*</u> Bronnimann & Stradner, 1960) - *Tegumentum* Thierstein, 1972 - *Tranolithus* Stover, 1966 - *Tubirhabdus* Prins ex Rood, Hay & Barnard, 1973 - *Zeugrhabdotus* Reinhardt, 1965 - ***Zygodiscus*** Bramlette & Sullivan, 1961 - *Zygolithites* Black, 1972	<u>*Heliorthus*</u> Bronnimann & Stradner, 1960) - ***Placozygus**** Hoffmann, 1970 - *Reinhardites* Perch-Nielsen, 1968 - *Tranolithus* Stover, 1966 - *Zeugrhabdotus* Reinhardt, 1965 (= *Zygolithites* Black, 1972) - ***Zygodiscus*** Bramlette & Sullivan, 1961
ZYGOSPHAERACEAE Hay, 1977	- *Corisphaera* Kamptner, 1937 - *Helladosphaera* Kamptner, 1937 - *Homozygosphaera* Deflandre, 1952 - *Periphyllophora* Kamptner, 1937 - *Zygosphaera* Kamptner, 1937	—	—
ZYGRHABLITHACEAE Hay, 1977	- ***Clathrolithus*** Deflandre, 1954 - ***Daktylethra*** Gartner, 1969 - ***Discoturbella***** Roth, 1970 - ***Holodiscolithus*** Roth, 1970 - ***Lanternithus*** Stradner, 1962 - ***Naninfula*** Perch-Nielsen, 1968 - ***Orthozygus*** Bramlette & Wilcoxon, 1967 - ***Peritrachelina*** Deflandre, 1952 - <u>*Polycladolithus*</u> Deflandre, 1954 - *Pyxolithus* Deflandre, 1954 - ***Zygrhablithus*** Deflandre, 1959	—	—

Appendix 10.1 (*continued*)

Family	Hay (1977)	Tappan (1980)	Perch-Nielsen (1985a,b)
GENERA *INCERTAE SEDIS*	- *Annulithus* Rood et al., 1973 - *Bucculinus* Noël, 1957 - *Conusphaera* Trejo, 1969 - ***Ellipsodiscoaster***** Boudreaux & Hay, 1969 - ***Hayella*** Gartner, 1969 - *Hexalithus* Gardet, 1955 - ***Gongylis***** Hoffmann, 1970 - <u>Guttilithion</u> Stradner, 1962 - ***Minylitha*** Bukry, 1973 - *Noelites* Caratini, 1960 - *Octolithites* Caratini, 1960 - ***Oolithotus*** Reinhardt, 1968 - <u>*Orthohabdus*</u> Bramlette & Wilcoxon, 1967 - *Tetralithus* Gardet, 1955	—	- *Annulithus* Rood et al., 1973 - ***Biantholithus*** Bramlette & Martini, 1964 - *Ceratolithina* Martini, 1967 - *Ceratolithoides* Bramlette & Martini, 1964 - ***Ellipsolithus*** Sullivan, 1964 - ***Florisphaera*** Okada & Honjo, 1973 - ***Hayaster*** Bukry, 1973 - ***Imperiaster*** Martini, 1970 - *Laguncula* Black, 1971 - ***Lapideacassis**** Black, 1971 - *Liliasterites* Stradner & Steinmetz, 1984 - *Lithastrinus* Stradner, 1962 - *Marthasterites* Deflandre, 1959 - ***Minylitha*** Bukry, 1973 - ***Nannotetrina*** Achutan & Stradner, 1969 - ***Nannoturba***** Müller, 1974 - ***Oolithotus*** Reinhardt, 1968 - *Percivalia* Bukry, 1969 - *Petrarhabdus* Wind & Wise, 1977 - *Prolatipatella* Gartner, 1968 - *Repagalum* Forchheimer, 1972 - ***Rhomboaster*** Bramlette & Sullivan, 1961 - ***Scampanella**** Forchheimer & Stradner, 1973 - *Tortolithus* Crux, 1982 - ***Tribrachiatus*** Shamrai, 1963 - ***Trochastrites*** Stradner 1961 - **Umbellosphaera* Paasche, 1955

Note the differences in generic content of any single family and the different acceptance of families between the three authors. For instance, regarding the Cenozoic Coccolithophyceae only Tappan (1980) recognizes the monospecific Family Scyphosphaeraceae Jafar 1975 while contrary to Tappan (1980) and Perch-Nielsen (1985a), Hay (1977) does not accept the monospecific Family Helicosphaeraceae Black 1971 (see also table 10.2). Note also that several families include Mesozoic and Cenozoic genera in Hay (1977) and Tappan (1980) whereas there is almost complete segregation in Perch-Nielsen (1985a, 1985b). On the other hand, this latter author assigned to distinct families Mesozoic and Cenozoic representatives of some relictual genera (e.g., *Biscutum*, *Chiastozygus*, *Cyclgelosphaera*) attesting to the difficulty in constructing a sound taxonomic framework.

Complete taxonomic nomenclature and references from NannoStrat databases (Aubry and Bonnemaison 1997; Bonnemaison and Aubry 1997) using the CompuStrat database manager (Bonnemaison 1997).

Boldface: valid Cenozoic genera; bold underlined: Cenozoic genera that have been lumped with others; bold followed by *: genera that are represented in both the Mesozoic and the Cenozoic eras; bold followed by **: Cenozoic genera of poorly known structure with sporadic occurrences that are omitted in this study.

Underlined regular characters: superfluous Cenozoic generic names; names preceded by *: Recent genera; all other regular characters: Mesozoic taxa (no indication of current validity).

Appendix 10.2
Classification of 13 Cenozoic genera into families as proposed in synthetic works and herein

<table>
<tr><th>Genus</th><th>Hay (1977)</th><th>Tappan (1980)</th><th>Perch-Nielsen (1985a, 1985b)</th><th>Aubry (1989a, 1997, and in progress)</th></tr>
<tr><td>Chiastozygus</td><td rowspan="5">Chiastozygaceae</td><td rowspan="5">Zygodiscaceae</td><td>Chiastozygaceae/Zygodiscaceae</td><td>Chiastozygaceae</td></tr>
<tr><td>Chiphragmalithus</td><td rowspan="7">Zygodiscaceae</td><td rowspan="4">Neochiastozygaceae</td></tr>
<tr><td>Isthmolithus</td></tr>
<tr><td>Neochiastozygus</td></tr>
<tr><td>Neococcolithes</td></tr>
<tr><td>Placozygus</td><td>—</td><td>Ahmuellerellaceae</td><td>Zeugrhabdotaceae</td></tr>
<tr><td>Zygodiscus</td><td>Zygodiscaceae</td><td>Zygodiscaceae</td><td rowspan="3">Zygodiscaceae</td></tr>
<tr><td>Lophodolithus</td><td rowspan="4">Pontosphaeraceae</td><td>Pontosphaeraceae</td></tr>
<tr><td>Helicosphaera</td><td>Helicosphaeraceae</td><td>Helicosphaeraceae</td></tr>
<tr><td>Pontosphaera</td><td>Pontosphaeraceae</td><td rowspan="3">Pontosphaeraceae</td><td rowspan="3">Pontosphaeraceae</td></tr>
<tr><td>Scyphosphaera</td><td>Scyphosphaeraceae</td></tr>
<tr><td>Neocrepidolithus</td><td>—</td><td>—</td></tr>
<tr><td>Algirosphaera</td><td>Pontosphaeraceae</td><td>Pontosphaeraceae</td><td>—</td><td>Rhabdosphaeraceae</td></tr>
</table>

Note the large differences between authors. Cenozoic representatives of the genus *Chiastozygus* have been assigned to the Family Chiastozygaceae, whereas its Cenozoic species has been assigned to the Family Zygodiscaceae. The Recent genus *Algirosphaera* is included here to emphasize the difference between early and new classifications.

Appendix 10.3
Taxonomic framework (family level) used in this chapter

Biscutaceae Black(1971):
Biscutum
Markalius
Hornibrookina
Ellipsolithus

Coccolithaceae Poche (1913):
Cruciplacolithus (Birkelundia, Campylosphaera)
Coccolithus/Ericsonia
Chiasmolithus
Clausicoccus
Bramletteius
Calcidiscus
Umbilicosphaera

Noelarhabdaceae Jerkovic (1970):
Reticulofenestra (Cribrocentrum, Cyclicargolithus, Dictyococcites)
Noelarhabdus
Pseudoemiliania
Gephyrocapsa
Emiliania

Prinsiaceae Hay and Mohler (1967)
Prinsius
Toweius

Ellipsagelosphaeraceae Noël (1965)
Cyclagelosphaera

Rhabdophaeraceae Lemmermann (1908)
Blackites
Cruxia
Aspidorhabdus
Naninfula
Petasus
Amitha
Ommatolithus
Cepekiella

Appendix 10.3 *(continued)*

	Rhabdosphaera
	Discosphaera
	Acanthoica
Zeugrhabdotaceae	*Placozygus*
Zygodiscaceae Hay and Mohler (1967)	*Zygodiscus*
	Lophodolithus
	Helicosphaera
Pontosphaeraceae Lemmermann (1908)	*Neocrepidolithus*
	Pontosphaera
	Scyphosphaera
Chiastozygaceae Rood, Hay and Barnard (1973)	*Chiastozygus*
Neochiastozygaceae inf. name	*Neochiastozygus*
	Neococcolithes
	Chiphragmalithus
	Isthmolithus
Syracosphaeraceae Lemmermann (1908)	*Syracosphaera* s.l.
Calciosoleniaceae Kamptner (1927)	*Scapholithus*
Deutschlandiaceae Kamptner (1928)	*Deutschlandia*
Calyptrosphaeraceae Boudreaux and Hay (1969)	*Octolithus*
	Lanternithus
	Zygrhablithus
	Clathrolithus
	Peritrachelina
	Holodiscolithus
	Daktylethra
	Corannulus
	Orthozygus
Discoasteraceae Tan (1927)	*Discoaster*
	Catinaster
Heliolithaceae Hay and Mohler (1967)	*Heliolithus*
Fasciculithaceae Hay and Mohler (1967)	*Fasciculithus*
Sphenolithaceae Deflandre (1952)	*Sphenolithus*
	Ilselithina
Biantholithaceae	*Biantholithus*
Braarudosphaeraceae Deflandre (1947)	*Braarudosphaera*
	Micrantholithus
	Pemma
	Pentaster
Polycyclolithaceae Forchheimer (1972)	*Micula*
Rhomboasteraceae	*Rhomboaster*
	Tribrachiatus
	Imperiaster
Lithostromationaceae Deflandre (1959)	*Lithostromation*
	Martiniaster
	Trochoaster

Appendix 10.3 *(continued)*

Triquetrorhabdulaceae Lipps (1969)	*Triquetrorhabdulus*
Ceratolithaceae Norris (1965)	*Amaurolithus* *Ceratolithus*
Goniolithaceae Deflandre (1957)	*Goniolithus*
Incertae sedis	*Conococcolithus* *Coronocyclus* *Pedinocyclus* *Hayella* *Striatococcolithus* *Hayaster* *Oolithothus* *Florisphaera* *Minylitha* *Lapideacassis* *Scampanella* *Angulolithina* *Ceratolithina*? *Nannotetrina* *Pseudotriquetrorhabdulus* *Trochastrites*

Recent genera with no fossil record are excluded. Formal family names in boldface. The family names such as *Biantholithaceae* and *Rhomboasteraceae* are coined here for an informal use. Families that belong to the coccolithophorids (in which cells produce coccoliths) are given first. Families represented by nannoliths (minute particles always associated with coccoliths in sediments that show structures unknown among living coccolithophorids) are given next. A number of genera are difficult to assign to well-established families, and phylogenetic links with other genera seem remote. They are placed among *incertae sedis*. The first 5 genera listed in alphabetical order, are represented by coccoliths. The last 5 genera, also listed by alphabetic order, are represented by nannoliths. Three groups of two genera in between are distinguished to reflect close phylogenetic relationships within each group.
The order in which families are given follows that in figure 10.2. Overview of the families is given in appendix 10.1.

Appendix 10.4

Family Biscutaceae

Rarely used for Cenozoic genera. There is much disagreement as to its generic content. *Markalius* is often assigned to the same family as *Cruciplacolithus*, *Chiasmolithus* and *Ericsonia*, either Coccolithaceae (Perch-Nielsen 1985a) or Ellipsagelosphaeraceae (Romein 1979). Generic assignment herein is based on structural similarity, which is well indicated by the optical behavior in cross-polarized light of the coccoliths in the four genera.

Family Coccolithaceae

Generally a well established family. However, several of the genera assigned to it by Perch-Nielsen (1985a) are here placed in the Family Biscutaceae.

Family Noelarhabdaceae

There is much disagreements as to the generic content of this family, which often overlaps with the Prinsiaceae. Based on structural grounds, *Prinsius* and *Toweius* are excluded from the family contrary to Perch-Nielsen (1985a) and Romein (1979).

Family Prinsiaceae

See Noelarhabdaceae.

Family Ellipsagelosphaeraceae

This is a well-established, essentially Mesozoic family.

Family Rhabdosphaeraceae

This is a well-defined, strictly Cenozoic family. Its generic content is described in Aubry (1998).

Family Zygodiscaceae

This is a strictly Cenozoic family, unlike proposed by Hay (1977), Tappan (1980), Perch-Nielsen (1985a, 1985b), and Aubry (1989a). The genus *Placozygus* is seen as the likely ancestor of the Cenozoic genus *Zygodiscus*, and both genera have been assigned to the Family Zygodiscaceae. However, there are sufficient structural differences between the two genera, particularly when compared to the *Zygodiscus–Lophodolithus–Helicosphaera* trio of closely related genera to warrant separation of *Placozygus* and *Zygodiscus* into separate families.

Family Zeugrhabdotaceae

Informally used here to include the Mesozoic genera that have been assigned to the Family Zygodiscaceae. *Placozygus* is the only genus in this family that is discussed herein and was represented during the Cenozoic Era. The family name is based on the name of its oldest genus.

Family Pontosphaeraceae

The generic content is that given in Aubry (1989a).

Family Chiastozygaceae

This is a Mesozoic family. The imbrication of the elements of the proximal (outer) cycle in the margin is clockwise and thus opposite to that of the coccoliths in the Cenozoic genera *Neochiastozygus*, *Neococcolithes* and *Chiphragmalithus*. Only one Cenozoic representative (*Chiastozygus ultimus*) has been described (Perch-Nielsen 1981b), but assignment of this form to *Chiastozygus* is questionable if we consider that the imbrication of the proximal (outer) cycle of the margin is counterclockwise, opposite to that of zygoliths of *Chiastozygus* and all other Mesozoic genera assigned to the Family Chiastozygaceae by Perch-Nielsen (1985b). Generic differentiation based on the single- or double-cycle composition of the margin as proposed by Perch-Nielsen (1981a, 1981b; 1985a, 1985b) is not workable. Unlike Perch-Nielsen's description of the margin of *Neococcolithes* and *Chiastozygus* as consisting of a single cycle in opposition to that in *Neochiastozygus*, the margin in the three genera is formed of two cycles, the distal (or inner) one being little developed in zygoliths of *Chiastozygus* and *Neococcolithes* (see Aubry 1998).

Family Neochiastozygaceae

Informally proposed here to include the Cenozoic zygoliths: *Neochiastozygus*, *Neococcolithes*, *Chiphragmalithus*, and tentatively *Isthmolithus* (see text for further explanation).

Family Syracosphaeraceae

A well-established Cenozoic (Neogene) family.

Family Calciosoleniaceae

See Perch-Nielsen (1985a, 1985b)

Family Deutschlandiaceae

See Perch-Nielsen (1985a, 1985b).

Family Calyptrosphaeraceae

This family includes holococcoliths that are not known to represent a stage in the complex life cycle of coccolithophoridae species.

Family Discoasteraceae

See Romein (1979) and Perch-Nielsen (1985a).

Family Heliolithaceae

See Romein (1979) and Perch-Nielsen (1985a).

Family Fasciculithaceae

See Romein (1979) and Perch-Nielsen (1985a).

Family Sphenolithaceae

A well-established Cenozoic family, although there are disagreements as to its generic content.

Family Biantholithaceae

Informally used here to emphasize the original structure of *Biantholithus*.

Family Braarudosphaeraceae

See Perch-Nielsen (1985a).

Family Polycyclolithaceae

This family may or may not be represented in the Cenozoic Era. See Perch-Nielsen (1985b).

Family Rhomboasteraceae

Informally used here to include the *Rhomboaster–Tribrachiatus* group, in which the nannoliths consist of a single crystal with a rhombohedral or axial symmetry.

Family Lithostromationaceae

See Perch-Nielsen (1985a).

Family Triquetrorhabdulaceae

See Perch-Nielsen (1985a).

Family Ceratolithaceae

See Perch-Nielsen (1985a) and Aubry (1988a).

Family Goniolithaceae

See Perch-Nielsen (1985a).

Incertae Sedis

Incertae sedis genera differ considerably between authors (see appendix 10.1) because of the different phylogenetic relationships proposed by different authors.

Appendix 10.5

It is thought that the genus *Chiastozygus* gave rise to the genus *Neochiastozygus* in the early Paleocene Epoch. Both genera have been placed in the same family, either Chiastozygaceae by Hay (1977) and Aubry (1989), or Zygodiscaceae by Tappan (1980). Perch-Nielsen (1985a, 1985b) ambiguously assigned Mesozoic species of *Chiastozygus* to the Family Chiastozygaceae while assigning its Cenozoic representatives (together with *Neochiastozygus*) to the Family Zygodiscaceae. Evolution of *Neochiastozygus* from *Chiastozygus* did not involve the development of an additional marginal cycle as claimed by Perch-Nielsen (1981a, 1985b) but has resulted in the re-orientation of the elements of the proximal (outer) cycle of the margin (see terminology in Aubry 1998). Whereas imbrication of the elements is clockwise in *Chiastozygus* it is anticlockwise in its three Cenozoic descendants (*Neochiastozygus*, *Neococcolithes*, *Chiphragmalithus*). Re-orientation of the elements of the proximal cycle of the margin constitutes a Paleocene innovation that justifies the erection of a new family, Neochiastozygaceae, as informally used here.

For reasons discussed in Aubry (1989a), the Cenozoic zygoliths should not be placed in the family Zygodiscaceae (see also appendix 10.4).

Appendix 10.6

Of critical importance is the determination of the range of species assigned to genera that are common both to the Mesozoic and the Cenozoic Eras. The following is a case to the point. It is generally considered that *Cyclagelosphaera margereli* is the only species of this genus that spanned the Cretaceous/Paleogene Period boundary, and that 2 additional species (*C. reinhardtii* and *C. alta*) evolved during the Paleocene Epoch. However, Bybell and Self-Trail have recently introduced a new combination, *Cyclagelosphaera prima* (Bukry 1969) Bybell and Self Trail 1995, and introduced synonymy between *C. prima* and *C. reinhardtii* (Perch-Nielsen 1968) Romein 1979, implying that two species of the genus *Cyclagelosphaera* spanned the Cretaceous/Paleogene Period boundary and that only one species evolved during the Paleocene Epoch. Bukry (1969) described *Similicoronolithus primus* from transmission electron microscope analysis of a carbon replica of the distal view of a Campanian placolith that he mistook for a proximal view. *Similicoronolithus primus* shows the structural characteristics of *Cyclagelosphaera*, thus justifying the recombination *C. prima* by Bybell and Self-Trail (1995). However, whereas they also show the structural characteristics of *Cyclagelosphaera*, the specimens illustrated by these authors from upper Paleocene deposits of the Solomon Islands and New Jersey Borehole GL 915 do not show the morphological characteristics of *C. prima*. The crest on their distal shield, which marks the boundary between the margin and the central area, is located at half distance between the center of the placolith and its periphery whereas it is at ~2/3 inwards in *C. prima* (compare Bybell and Self-Trail 1995:plate 5, figures 1–3 with Bukry 1969:plate 9, figure 11); their central structure does not fill the central area as it does in this latter species; and their shields form a higher cone than in *C. prima*. Thus these upper Paleocene coccoliths do not represent *C. prima*, nor do they belong to *C. reinhardtii* as implied by the synonymy given by Bybell and Self-Trail, notwithstanding the fact that the name *reinhardtii* introduced in 1968 (*Markalius reinhardtii* Perch-Nielsen 1968) has priority over the name *prima*. Based on comparison with the illustrations given by Perch-Nielsen (1979:plate 3, figures 3–7) they belong to *C. alta* or represent a new form closely related to it (I point out that the photomicrographs given of *C. prima* by Bybell and Self-Trail 1995:plate 32, figures 19 and 20 are not of the same species as illustrated in plate 3; figure 19 corresponds to *C. margereli* and figure 30 represents *C. reinhardtii*).

I have questioned the validity of the combination *Chiastozygus ultimus* Perch-Nielsen 1981a (Aubry 1998; appendix 10.4). However, for the purpose of this chapter, occurrence of the genus *Chiastozygus* during the Paleocene Epoch is accepted, following Perch-Nielsen (1981a, 1981b, 1985a, 1985b).

Appendix 10.7

Database used to establish figure 10.3

A		B		C	
Braarudosphaera	12	*Biantholithus*	2	*Fasciculithus*	19
Scapholithus	2	*Hornibrookina*	3	*Zygodiscus*	5
Micrantholithus	22	*Prinsius*	6	*Heliolithus*	10
Goniolithus	1	*Neochiastozygus*	14	*Toweius*	10
Markalius	3	*Ellipsolithus*	7	*Neococcolithes*	5
Scampanella	4	*Lanternithus*	3	*Zygrhablithus*	2
Placozygus	1	*Cruciplacolithus*	15	*Sphenolithus*	35
Neocrepidolithus	6	*Chiasmolithus*	17	*Discoaster*	>130
Lapideacassis	3	*Coccolithus/Ericsonia*	>25		
Cyclagelosphaera	3				
Biscutum	6				
Chiastozygus	1				
Octolithus	1				

D		E		F	
Conococcolithus	1	*Striatococcolithus*	1	*Nannotetrina*	9
Rhomboaster	4	*Imperiaster*	1	*Pseudotriquetrorhabdulus*	1
Tribrachiatus	6	*Chiphragmalithus*	5	*Pemma*	14
Lophodolithus	6	*Cruxia*	1	*Pentaster*	1
Blackites	17	*Aspidorhabdus*	1	*Daktylethra*	1
Pontosphaera	>60	*Clathrolithus*	3	*Lithostromation*	1
Clausicoccus	>8	*Naninfula*	3	*Bramletteius*	2
Helicosphaera	48	*Amithalithina*	1	*Ceratolithina?*	1
		Peritrachelina	2	*Petasus*	1
		Holodiscolithus	1	*Corannulus*	2
		Coronocyclus	3	*Martiniaster*	2
		Hayella	4	*Pedinocyclus*	1
		Scyphosphaera	52		
		Reticulofenestra s.l.	>40		

G		H		I	
Isthmolithus	2	*Triquetrorhabdulus*	9	*Calcidiscus*	6
Orthozygus	3	*Syracosphaera* s.l.	10	*Umbilicosphaera*	8
Amitha	2	*Hayaster*	1		
Ommatolithus	1				
Cepekiella	1				
Ilselithina	2				
Trochastrites	3				
Trochoaster	8				
Bicolumnus	1				

J		K		L	
Rhabdosphaera	6	*Discosphaera*	4	*Ceratolithus*	7
Catinaster	3	*Minylitha*	1	*Deustchlandia*	2
		Noelarhabdus	1	*Oolithothus*	1
		Acanthoica	10	*Pseudoemiliania*	1
		Florisphaera	1		
		Amaurolithus	6		
		Angulolithina	1		

M		N	
Gephyrocapsa	>18	*Emiliania*	1

Appendix 10.8

Convergent and parallel evolution often hinder the establishment of firm phylogenetic relationships among calcareous nannoplankton genera. The links proposed in figure 10.4 are based on the following.

Biscutum–Hornibrookina–Ellipsolithus lineage(s): as herein, Perch-Nielsen (1981a) confidently proposed that *Hornibrookina* evolved from *Biscutum*, but tentatively proposed *Repagalum* as ancestral to *Ellipsolithus*. I have observed intermediate forms between *Biscutum* and *Ellipsolithus* in the lower Paleocene sediments recovered from the Southern Indian Ocean (ODP Site 750).

Biscutum–Cruciplacolithus–Coccolithus/Ericsonia–Chiasmolithus lineage: The *Cruciplacolithus–Chiasmolithus* lineage has been well established (Romein 1979). No ancestor has been confidently proposed for *Cruciplacolithus*. I tentatively suggest *Biscutum* herein.

Chiastozygus–Neochiastozygus–Neococcolithes–Chiphragmalithus lineage: see Perch-Nielsen (1981a,1981b).

Cyclagelopshaera–Prinsius–Toweius lineage. This is a tentative proposition. Perch-Nielsen (1981a) proposed that *Prinsius* evolved from *Biscutum*. The *Prinsius–Toweius* lineage was discussed by Romein (1979).

Cyclagelosphaera–Blackites–Cruxia–Aspidorhabdus–Naninfula lineage: Perch-Nielsen (1979) remarked on similarities between the base of the rhabdolith *Blackites vitreus* and *Cyclagelosphaera alta*. I (1998) have demonstrated close structural relationships between the last four genera, and between *Cyclagelosphaera* and *Blackites*. The distal shield in *Cyclagelosphaera* likely evolved to form the third cycle of the base in *Blackites*.

Neocrepidolithus–Pontosphaera–Scyphosphaera lineage: see Aubry 1988b, 1989a.

Octolithus–Lanternithus–Zygrhablithus lineage: see Perch-Nielsen (1981a). The Family Calyptrolithaceae gathers all the holococcoliths (coccoliths formed of simple rhombs and without differentiated cycles). Thus the family may be polyphyletic.

Placozygus–Zygodiscus–Lophodolithus–Helicosphaera lineage(s): see Aubry 1988b, 1989a.

Braarudosphaera–Biantholithus–Sphenolithus–Fasciculithus–Heliolithus–Discoaster. The *Fasciculithus–Heliolithus–Discoaster* lineage is well established (see Romein 1979; Perch-Nielsen 1981a). Perch-Nielsen (1981a) proposed *Markalius* as ancestral to *Fasciculithus* and *Cyclagelosphaera* to *Sphenolithus*. Close structural relationships between *Sphenolithus* and *Fasciculithus* were demonstrated in Aubry (1989b) and are exemplified by photographs in Bybell and Self-Trail (1995; compare plate 17, figures 1a–5b with plate 24, figures 8a–10b). Close structural relationship between *Sphenolithus* and *Biantholithus* is also shown by photographs in Bybell and Self-Trail (1995; compare plate 1, figures 1–8 with SEM photographs of sphenoliths in Aubry 1989b) and in Romein et al. (1996:plate 2, figure 16). I propose that the lower cycle of elements in *Biantholithus* evolved to form the basal cycle (column) in *Sphenolithus* whereas the upper cycle evolved into the lateral and distal cycles. The origin of *Biantholithus* is problematic. Based on a combination of characters such as basic construction and extinction pattern I suggest that it evolved from *Braarudosphaera*.

Micula–Rhomboaster–Tribrachiatus lineage: see Romein (1979).

Appendix 10.7 Taxonomic framework of Aubry (1984, 1988, 1989b, 1990, 1998) and of Aubry (unpublished) and Perch-Nielsen (1985).

Genera grouped in order of temporal evolution. A: Relict genera. B to N: Genera that appeared during the early Paleocene Epoch (B), during the late Paleocene Epoch (C), during the P/E BI (D), during the early Eocene Epoch (E), during the middle Eocene Epoch (F), during the late Eocene Epoch (G), during the late Oligocene Epoch (H), during the early Miocene Epoch (I), during the middle Miocene Epoch (J), during the late Miocene Epoch (K), during the early Pliocene Epoch (L), during the late Pliocene Epoch (M), and during the late Pleistocene Epoch (N). Note that no genus evolved during the early Oligocene Epoch.

The number of species representatives of each genus is given in facing columns.

11

Trophic Models for the Well-Mixed and Poorly Mixed Warm Oceans Across the Paleocene/Eocene Epoch Boundary

ANNE BOERSMA
ISABELLA PREMOLI SILVA
PAMELA HALLOCK

ABSTRACT

Models of the trophic resource continuum (TRC) predict aspects of nutrient cycling and oceanic circulation across the Paleocene/Eocene Epoch boundary. The abundance, diversity, and areal extent of oligotrophic planktonic foraminifera combined with the presence, but geographic compartmentalization of eutrophic planktonic foraminifera indicate the expansion of the TRC from late Paleocene to early Eocene time. Because development of oligotrophic habitats is maximal under poorly mixed conditions, we infer decreased intensity of oceanic mixing through the thermocline across the Paleocene/Eocene Epoch boundary.

The Trophic Resource Continuum (TRC) in euphotic waters of the world's oceans is the spectrum of conditions from the richest runoff and upwelling areas to the most nutrient deficient subtropical seas. That spectral range has likely varied through geologic time as more efficient phytoplankton taxa evolved, as oceanic circulation rates changed, and in response to ocean-mixing events. How fluctuations in the TRC control diversity of euphotic habitats in the oceans provides an hypothesis by which to interpret global taxonomic diversity cycles, taxonomically selective extinctions, changes in species dominance, and morphologic evolution (Hallock 1987).

The purpose of this chapter is to examine TRC models for changes from a relatively well mixed to a poorly mixed warm ocean. We chose these scenarios because isotopic, thermal gradient, surface paleotemperature, and biotic studies all indicate that the ocean was uniformly warm and circulation was sluggish in the early Eocene Epoch (Boersma et al. 1979; Tjalsma and Lohman 1983; Shackleton et al. 1985; Barrera and Huber 1991; Kaiho 1991; Miller et al. 1987). Models, formulated from the modern situation and the ecologic principles describing the modern situation, allow us to infer from biotic patterns the changing physical oceanography across the Paleocene/Eocene Epoch boundary.

The Trophic Resource Continuum in Modern Oceans

The term *resource continuum* refers to the spectrum of possible resource levels in an environment (figure 11.1). In the modern ocean, fertility, as indicated by primary productivity levels often in excess of 200 g C/m$^{2\text{-yr}}$ and trophic resources as indicated by sea-surface chlorophyll concentrations of more than 1 mg Chl/m^3, are at least seasonally high in regions of meridional upwelling, such as the eastern tropical Pacific, the eastern tropical Atlantic, the Arabian Sea, and also in ocean waters near large rivers such as the Mississippi or those of southeast Asia. Minimal primary productivity, less than 35 g C/m$^{2\text{-yr}}$, and trophic resources, less than 0.1 mg Chl/m^3, are characteristic of subtropical gyres. The greatest difference in the modern ocean is between the eastern tropical Pacific upwelling and the north-central Pacific gyre center. Higher latitude oceans and equatorial divergences tend to be intermediate in productivity (35–200 g C/m$^{2\text{-yr}}$) and in trophic resources (0.1–1 mg Chl/m^3; Hallock et al. 1991).

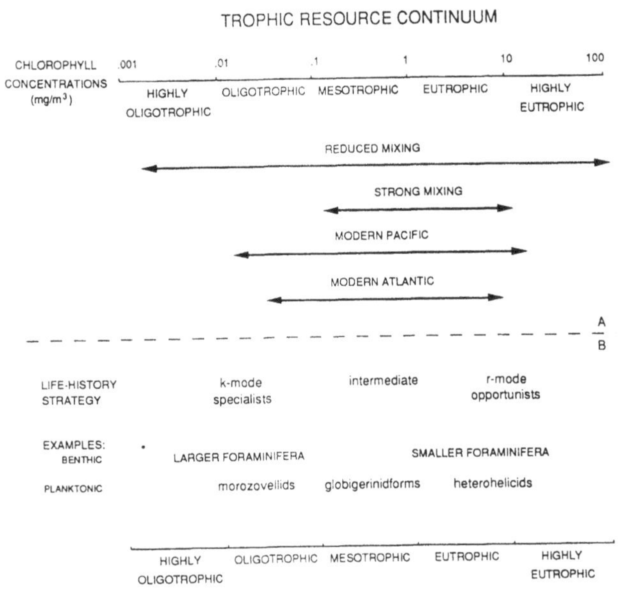

Figure 11.1 The Trophic Resource Continuum (TRC) in the modern ocean (Hallock 1987). Chlorophyll concentrations (mg/m^3) in the euphotic zone differentiate more and less nutrient-rich zones along the continuum of possible resources. Four situations—the modern Atlantic and Pacific continua and two hypothetical scenarios—are plotted against the modern TRC. Model for the Paleocene–Eocene Trophic Resource Continuum is adapted from Hallock et al. (1991). Paleocene–Eocene k-mode specialists, such as the larger foraminifera and the muricate-walled morozovellid planktonic foraminifera, were adapted to stable, low nutrient environments, whereas r-mode opportunists, such as the microperforate chiloguembelinids, were adapted to resource-rich, low-stability habitats. The smaller benthics and honeycomb-walled globigerinidform subbotinids were intermediate between these two extremes (Premoli Silva and Boersma 1988).

As shown in figure 11.2, the global range of chlorophyll levels in surface waters is one indicator of the breadth of the TRC. Highest sea surface chlorophyll levels are associated with meridional upwelling, seasonal blooms at high latitudes, and fluvial sources. Equatorial divergences have intermediate chlorophyll levels, while lowest levels are measured in the subtropical gyres. In one way, chlorophyll levels in sea surface waters can be thought of as "potential primary productivity" (Hallock et al. 1991).

Trophic Resources and Ocean Temperature

Trophic resources are physiologically interrelated with temperatures because in ectotherms, rates of chemical reactions associated with metabolism increase as temperature increases. For many physiological processes, the rate of reactions doubles for every 10° C of temperature increase (Hallock 1987). Theoretically it takes twice as much energy for a foraminifer to live at 11° C as at 1° C, four times as much energy to live at 21° C, and 8 times as much energy to live at 31° C. Thus, a rate of nutrient flux that might support a tropical oligotrophic community is effectively doubled in the cool subtropics, quadrupled in temperate waters, and 8-fold more effective in polar waters. Likewise, upwelling or runoff that doubles nutrient flux is metabolically similar to a 10° C drop in temperature; and a four-fold increase in nutrient flux is energetically equivalent to a 20° C drop in temperature (Hallock et al. 1991).

The case of *Globigerina bulloides* is instructive. This species, long considered a subpolar index (Kennett 1977) is typically abundant in meridional upwelling zones in the tropics. In the southern Caribbean upwelling where *G. bulloides* constitutes a majority of the microfauna, water temperatures are depressed only a few degrees from those of tropical surface waters (de Miro 1971). While *G. bulloides* is not inhabiting cold waters in this upwelling, the increase in nutrient flux in the upwelling is energetically equivalent to a temperature drop and thus *G. bulloides* can support its metabolic needs at these higher temperatures.

This reasoning could be used to explain the incursions of tropical acarininid and morozovellid species into the Southern Ocean during early Paleogene warmings described by Nocchi et al. (1991). The nutrient flux at warm poles is sufficient only to support an oligotrophic community in near surface waters.

The distributions of a number of other species have been related to nutrient flux rather than to temperature (Bé and Toderlund 1971; Hudson 1977; Reynolds and Thunnell 1985, 1986; Ottens 1991). *Globigerinoides sacculifer*, long considered a warm, tropical surface-dwelling species reaches maximum abundances in the northeast Atlantic near 40° N Latitude, even in April. Its distributions are more closely correlated with nutrient concentrations and oligotrophy than with temperatures. Highest concentrations of *Globigerinoides ruber* follow maximum temperature, but minimal phosphate values in surface waters. The cosmopolitan species *Globigerinita glutinata* also tracks phosphate concentrations rather than temperature. *Globorotalia inflata* shows no temperature preference since in the northeast Atlantic it lives at cooler temperatures in April than in August. Its maximum abundance is found in isothermal waters with highest phosphate concentrations (Ottens 1991).

The decreased diversity of many organisms in polar and subpolar waters has long been attributed to low temperatures. Another key feature of subpolar waters is, however, their variability (Hallock 1987). Steep seasonal changes in temperature, salinity, nutrients, concentration and oxygen, together with an ephemeral thermocline and fluctuating apposition of watermasses

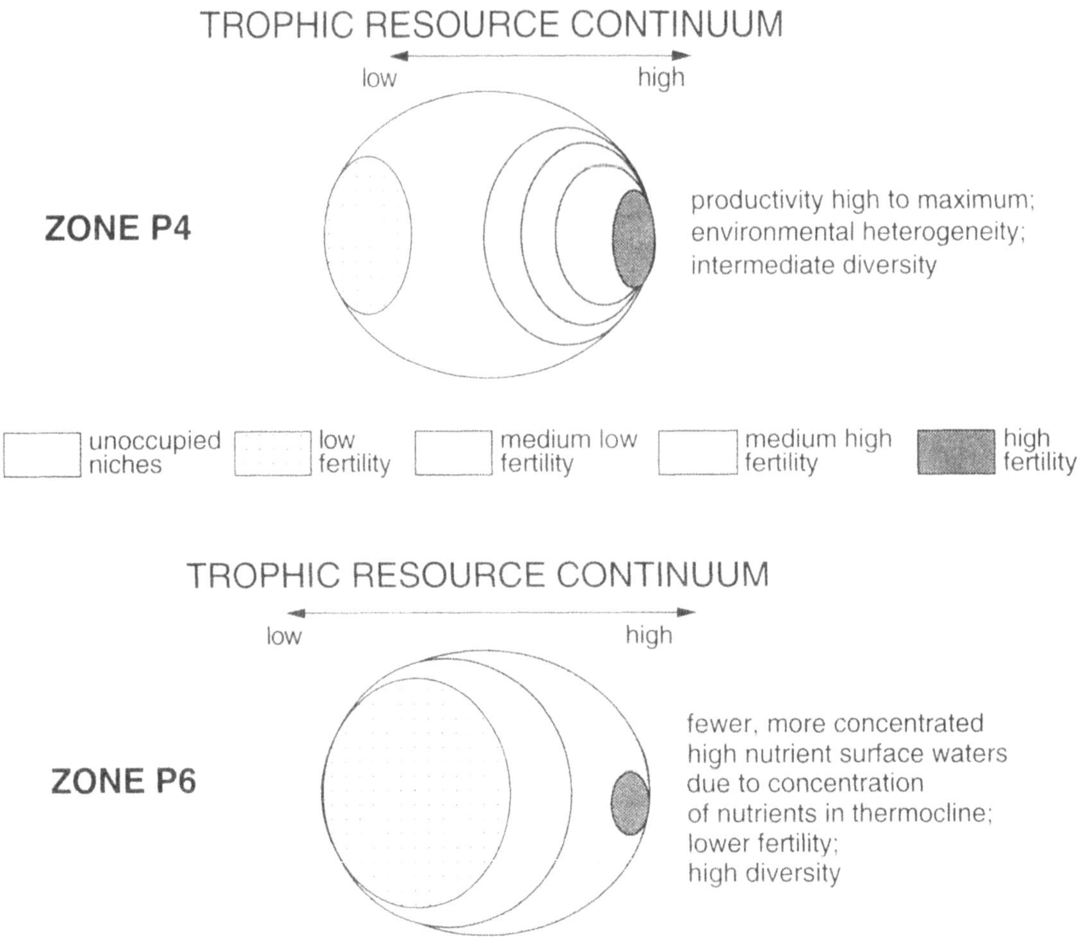

Figure 11.2 Fertility along the Trophic Resource Continuum in a well stirred ocean and in a warmer ocean with weaker thermohaline circulation. In the well stirred case (Biochron P4) with a strong thermohaline circulation, there is a broad TRC with high environmental heterogeneity. Niches of medium fertility (mesotrophic) predominate, but highly fertile and low fertility realms develop at the poles and tropics, respectively. In the uniformly warm ocean (Biochron P6) the TRC is even broader, from the very low fertility niches of the tropics to the very high fertility niches at the base of the thermocline in upwellings and in offshore upwellings. The variety of niches and the numbers of mesotrophic habits are fewer.

in strong frontal zones, all characterize these waters. Polar and subpolar waters are also considered supersaturated with nutrients. In such conditions *Globigerina quinqueloba* predominates during the North Atlantic April blooms (Ottens 1991). *Globigerina quinqueloba* is not a cool-water preferring species, being found in greater abundance in subtropical waters. Neither this species nor *G. bulloides* is located at or correlates with the deep chlorophyll maximum or the base of the thermocline. The abundant occurrence of *G. quinqueloba* together with *G. bulloides* in subpolar regions, then, cannot be attributed strictly to tolerance of low temperature or development of the deep chlorophyll maximum, but appears to be linked to its tolerance of ecologic variability and preference for a high nutrient habitat and the food sources therein (Bé and Toderlund 1971; Ottens 1991).

Oligotrophic conditions expand with global climatic amelioration for at least two reasons. First, rates of thermohaline circulation likely decline (the sluggish oceans of Fischer and Arthur 1977), reducing the rate of nutrient flux from subsurface waters to surface waters (Bralower and Thierstein 1984). But even if nutrient flux to sea surface waters does not decline, the expansion of warm-water habitats increases metabolic demand for nutrients and thereby expands the geographic range of oligotrophic habitats. Thus, the high latitude occurrences of the morozovellids may indicate the combined effects of warmer sea surface conditions and effectively lower nutrient supplies.

Hallock et al. (1991) described how trophic resources can explain the relatively poor correlation between temperature changes and biotic reorganization in the Paleogene Epoch. For example oxygen isotope values at the

Eocene/Oligocene Epoch boundary reflect a major drop in bottom water temperatures, while cooling of subtropical surface waters was much less dramatic (Poore and Matthews 1984). Yet biotic reorganization in surface waters (Gradstein and Srivastava 1980; Premoli Silva and Boersma 1988) was far more pronounced than changes among deep sea benthic foraminifera at the same time (Corliss 1981). This apparent discrepancy can be reconciled using the argument of increasing nutrient flux to surface waters. A minor drop in surface water temperatures in mid to lower latitude seas can result from upwelling or deep mixing that increases trophic resources several fold, promotes proliferation of "cooler water," mesotrophic to subeuphotic taxa, and excludes taxa indicative of warm, oligotrophic waters.

Strong Stirring Model

A well-stirred ocean provides relatively few warm water, euphotic habitats along the TRC (table 11.1; figure 11.2) whereas cool water or subeuphotic habitats may be more common. The range of nutrient fluxes into surface waters should narrow if ocean mixing rates increase (Fischer and Arthur 1977). This is because upward mixing through the thermocline redistributes nutrients, thus reducing or eliminating low nutrient surface water habitats. Processes that increase mixing also accelerate advection of surface waters from areas of nutrient input (terrestrial runoff, meridional and equatorial upwelling) to more remote regions. Marine areas rich in resources tend to be unstable because the processes that deliver nutrients to sea surface waters tend to be seasonal (Valentine 1973).

Abundant inorganic nutrients support concentrated standing stocks of rapidly growing and reproducing primary producers (Rhyther 1969; Hallock et al. 1991). The most nutrient-rich waters are characterized by primary production of large, chain-forming diatoms that are fed upon by consumers such as anchovies and krill, which in turn are fed upon by marine mammals, birds, and tuna. Food chains are short, simple, and the greatest diversities seem to be among the carnivores, which are specialized to these high food supplies. As nutrient flux declines, size of primary producers generally declines (depending on temperature) and planktonic foraminifera and coccolithophorids become increasingly important links in increasingly complex food webs (Ziveri et al. 1995). Among consumers, abundant food supplies typically favor a generalist feeding strategy with rapid growth to a relatively small maturation size and a high potential for population size increase (r-selection; Longhurst 1967; Valentine 1973). As nutrient flux declines with distance from source, primary producers become scarcer and smaller, while planktonic foraminifera tend to increase in size and complexity as they specialize to the limited resources.

In a well stirred ocean (table 11.1) both carbonate and siliceous organisms are producing sediments. Siliceous production tracks upwelling regions especially during upwelling maxima, while widespread "mesotrophic" conditions promote relatively high rates of carbonate production by the coccolithophorids (Ziveri et al. 1995).

Table 11.1
Summary of observations on modern higher and lower fertility conditions, with associated thermal and biotic features

	Modern Ocean	
	More Oligotrophic	More Eutrophic
Food chains	More complex	Shorter
Diversity	Higher	Lower
Primary producers	Nannoplankton	Large diatoms
Life strategies	Slower growth	Rapid growth, opportunism
Thermophilic indices	Increase	Decrease
Evolution	Oligotrophs	—
Sea level	Higher	Lower
Wind strength	Decreased	Increased
Deep circulation	Decreased vertical mixing	Vigorous
Global gradients	Expanded	Contracted
Euphotic zone upwelling	Low, warm	Higher low, cool
Nutrient input	Broad, deep	Shoaler

More oligotrophic conditions favor longer food webs, higher diversity, larger numbers of specialists and oligotrophs, greater diversity of the oligotrophic habitat, mesotrophic production, and warmer mixed-layer temperatures in spring and fall in the middle latitudes. Under more eutrophic conditions, food webs are shorter, diversity is less, and primary production is higher. More opportunistic species proliferate and oligotrophs are not favored or exist in lower numbers in reduced numbers of habitats as the warm water sphere of the euphotic zone contracts. Tropical warm sphere indices decrease in abundance whereas biotic indices for the chlorophyll maximum increase in number.

Reduced Stirring Model

A poorly stirred, or sluggish ocean, should provide a wide range of euphotic habitats along the TRC (table 11.1; figure 11.2), ranging from more constricted but richer upwelling zones (Fischer and Arthur 1977) to more nutrient-poor subtropical gyres. Advection declines and nutrients tend to be concentrated both geographically and vertically at the top of the thermocline, in upwellings and during seasons of upwellings, and near fluvial sources. A warm, poorly stirred ocean may be more stratified because neither vertical nor advective circulation breaks down picnal barriers and hence it may be more stable (Valentine 1973).

Nutrient-poor, stable habitats provide great potential for specialization, with the limited resources finely subdivided. Tiny nano- and picoplankton support low densities of specialized consumers in long, complex food webs (Ryther 1969). Algal symbiosis is common (Longhurst 1967a; Valentine 1973). The ability to feed and recycle metabolic nutrients to support primary productivity by symbionts provides one means of concentrating and efficiently utilizing scarce resources (Hallock 1987).

Paleocene–Eocene Planktonic Foraminiferal Trophic Index Groups

Evidence from habitat, morphology, and geography indicates that the early Paleogene oligotrophs were the morozovellids while the more eutrophic planktonics were the chiloguembelinids (Boersma and Premoli Silva 1983, 1991; Hallock et al. 1991). Morozovellids were usually large sized, morphologically specialized forms that predominated in the tropics. Their ranges, evolution and diversity changed through time, and their morphologic complexities paralleled those of other oligotrophic groups of the euphotic zone, such as the larger benthic foraminifera, the alveolinids, assilinids, and nummulitids (Hallock et al. 1991). By contrast, analogy with modern distribution patterns, opportunistic behavior, small size, and high densities indicates that the chiloguembelinids occupied more eutrophic habitats (Boersma and Premoli Silva 1988, 1991; Resig 1990).

Preliminary evidence indicates that wall structural types reflect a more eutrophic versus oligotrophic habitat preference among Paleocene–Eocene planktonic foraminifera. Muricate, non-spinose walls characterize the oligotrophic indices, whereas the microperforate wall is associated with the more eutrophic habitat (Boersma and Premoli Silva 1991). Living representatives of the microperforate chiloguembelinids inhabit low latitude eutrophic areas such as the Indian Ocean upwelling zone and are known to feed upon small diatoms (Kroon and Ganssen 1989; Kroon and Nederbragt 1990). Paleogene age chiloguembelinid–diatom oozes from fertile equatorial currents have been found on the Ceara Rise in the equatorial Atlantic.

Members of the genus *Acarinina* secreted muricate walls. While the morozovellids and acarininids registered similar near surface oxygen isotope ratios, the acarininids usually registered enriched carbon isotope ratios (Boersma et al. 1979; Shackleton et al. 1985). The acarininids characteristically ranged into and diversified at higher latitudes than the large morozovellids, and we suspect that they occupied a more mesotrophic surface water habitat than the morozovellids and may have possessed photosymbionts (Shackleton et al. 1985; Spero et al. 1991; D'Hondt 1994).

If these different groups represented more versus less oligotrophic conditions in euphotic waters of the Paleocene–Eocene ocean, their diversities, sizes, relative abundances, and distribution patterns should then depict, by analogy to the above models, the late Paleocene to early Eocene extent of the TRC.

Muricate Morozovellids and Microperforate Chiloguembelinids Across the Paleocene/Eocene Epoch Boundary

From latest Paleocene to earliest Eocene time (contrasting Biochrons P4–P5 to Biochrons P6b–P7; Berggren et al. 1985) there was an increase in abundance and areal extent of the muricate foraminifera (figure 11.3). That is, the large muricate forms (Groups 7 and 16) expanded at low and mid latitudes and extended into the Antarctic. At the same time the microperforate chiloguembelinids (Group 18) were rare to absent in all but higher latitudes (Boersma and Premoli Silva 1983, 1988; Nocchi et al. 1991). This can be interpreted as an increase in the areal extent of oligotrophs coeval with the survival and compartmentalization of eutrophs in Subantarctic and Antarctic current systems.

Increase in global planktonic foraminiferal species richness was caused by increased numbers of muricate planktonics, but a decrease in microperforate chiloguembelinid species (table 11.2). The lower number of chiloguembelinid species paralleled their reduced geographic ranges (figure 11.3) and minimal abundances outside cool temperate areas and the Southern Ocean.

Chiloguembelinids can be subdivided into two morphologic types (figure 11.4) indicative of warmer conditions and tropical communities versus cooler conditions and/or upwelling zones. The warm type (called flaring herein) is represented by the low latitude group, *Chiloguembelina midwayensis* Cushman s.l., while the temperate species, *Chiloguembelina virgata* Khalilov and *Chiloguembelina gradata* Khalilov exemplify the elongate, parallel-sided morphology. While all chiloguembelinids including both the flaring (warm) and elon-

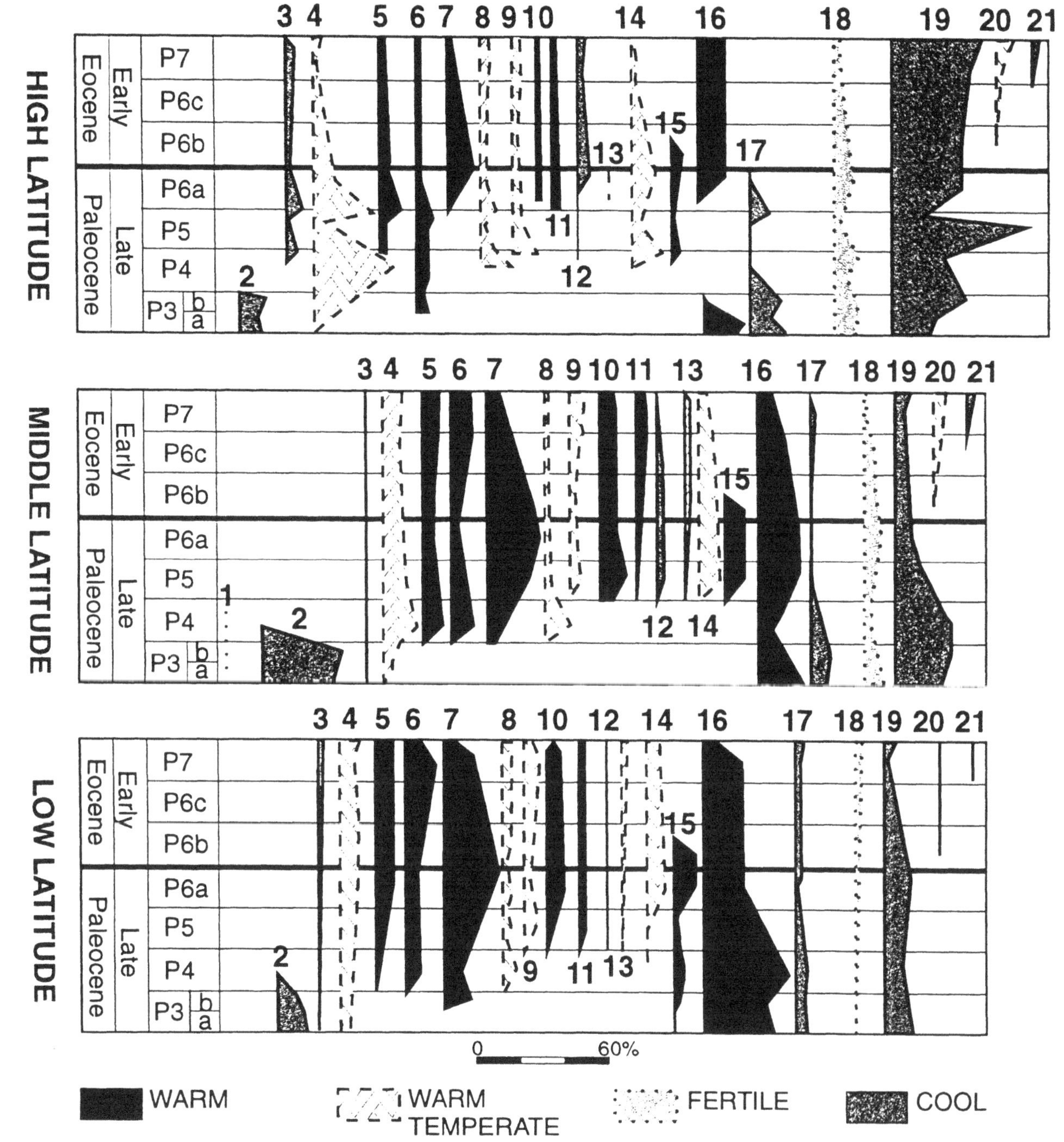

Figure 11.3 Abundances of planktonic foraminiferal groups and genera through the late Paleocene and earliest Eocene epochs at low, middle and higher latitude sites (adapted from Boersma and Premoli Silva 1983, 1991). In the late Paleocene Epoch muricate forms dominated at low and middle latitudes; they then also extended into high latitudes in the early Eocene Epoch. They were the most abundant forms at all sites outside the Southern Ocean in the early Eocene Epoch. Chiloguembelinids, never abundant, were compartmentalized together with the honeycomb-walled subbotinids in higher latitudes by early Eocene time.

Fertile group: 1: *Guembelitria*; 18: *Chiloguembelina*. Cool group: 2: *S. pseudobulloides*; 3: *Tenuitella*; 12: *Globorotaloides*; 17: *Globanomalina* (*Planorotalites* of Premoli Silva and Boersma 1988); 19: low-spired subbotinids; 21: *Catapsydrax*. Warm temperate group: 4: finely muricate acarininids (smooth walled acarininids of Boersma and Premoli Silva 1988); 8: globigerinid-like acarininids; 9: small acarininids; 13: *Turborotalia pseudoampliapertura* group; 14: biconvex morozovellids; 20: *Pseudohastigerina*. Warm group: 5: *Igorina* (spinose acarininids of Premoli Silva and Boersma 1988); 6: *A. spinuloinflata* lineage; 7: weakly muricate acarininids; 10: subquadrate acarininids; 11: *Morozovella broedermanni*; 15: small morozovellids; 16: large morozovellids.

Table 11.2
Global diversity (species richness) of all planktonic foraminiferal genera from late Paleocene into early Eocene times

Zones	Muricates	*Morozovella*	"*Igorina*"	*Acarinina*	Subbotinids	*Planorotalites*	Flaring chiloguembelinids	Elongate chiloguembelinids	*Globorotaloides*	Finely perforates	*Catapsydrax*	*Globigerinatheka*	Others	Species richness
P6b-P7	33	11	4	18	3	2	2	1	2	2	1	2	2	50
P6a	27	8	4	15	3	2	3	1	1	2	1	0	2	43
P5	22	5	4	13	2	2	8	3	1	0	0	0	5	44
P4b	24	11	5	8	2	4	6	3	1	0	0	0	0	40
P4a	24	13	4	7	5	5	6	3	1	0	0	0	0	44

Total muricate diversity (shown in gray) including the genera *Igorina*, *Acarinina*, and *Morozovella*, made up nearly 50% of total diversity in the late Paleocene Epoch and over 60% in the early Eocene Epoch. Total chiloguembelinid diversity (shown in hachured lines) decreased by 66% from late Paleocene to early Eocene time (data from Boersma and Premoli Silva 1983, 1988, 1991).

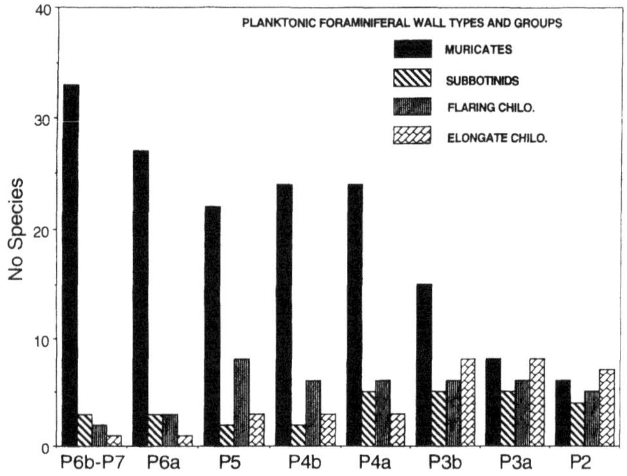

Figure 11.4 Diversity changes among planktonic foraminifera from early Paleocene Biochron P2 to early Eocene Biochrons P6b–P7.

gate (cool) morphologies decreased across the Paleocene/Eocene Epoch Boundary, the flaring group persisted longest and was more diverse in the earliest Eocene Epoch when the elongate group was restricted to the southernmost Indian Ocean. We interpret this to mean that the cool, nutrient-rich habitat preferred by the elongate group was nearly eliminated across the Paleocene/Eocene Epoch boundary outside the Southern Ocean.

In summary, in the late Paleocene planktonic foraminiferal communities were lower in diversity (species richness), global planktonic foraminiferal diversity was lower, the muricate oligotrophs were diverse (but less so than later), and the tropical morozovellids were more common in the tropics but did not extend into higher latitudes. By contrast, the eutrophic chiloguembelinids were more widespread and not confined to high latitudes and the elongate morphology was more common and represented by more species. By early Eocene time, however, the oligotrophs diversified, dominated low latitude communities, increased in abundance at middle latitudes, and entered Antarctica. Early Eocene planktonic faunas of the Southern Ocean began to resemble those of the late Paleocene time in the western subtropical gyres. Global species richness was higher because of diversification among the muricates. Chiloguembelinids were rare outside the Indian Ocean sector of the Southern Ocean, which was a refugium for the elongate morphology.

TRC Model Predictions for the Paleocene/Eocene Epoch Boundary Based on Planktonic Foraminiferal Communities

If trophic resource partitioning and community structure of the Paleogene were analogous to those of the modern ocean, then our observations on planktonic foraminifera allow us to describe changes in the TRC across the Paleocene/Eocene Epoch boundary (table 11.3). From a description of the TRC we can surmise some features of oceanic circulation and nutrient cycling.

Late Paleocene planktonic foraminiferal communities indicate an intermediate TRC (table 11.4). Diversities of oligotrophs were intermediate, not as low as in the early Paleocene or middle Eocene Epoch, but significantly lower than in the early Eocene Epoch. Global planktonic foraminiferal diversities (species richness) were also intermediate, neither as low as in the early Paleocene Epoch nor as high as in the early or middle Eocene Epoch. Diversities of the eutrophic chiloguembelinids were also intermediate, neither as high as in the early Paleocene Epoch nor as low as in the early Eocene Epoch (Boersma and Premoli Silva 1983, 1991). Eutrophs were widespread though not abundant, while oligotrophs were concentrated in the tropics and subtropics (see figure 11.3).

The predominance of oligotrophs, their geographic expansion, and their increased diversification contributing to an overall increase in global diversity, all indicate maximum development of warm, low nutrient surface waters during early Eocene times (see figure 11.3). The low numbers and clustering of chiloguembelinid eutrophs in the Southern Ocean indicate a decrease or restriction of subsurface, higher nutrient habitats. The foraminifera thus depict a broad TRC and the expanded areal extent of the oligotrophic habitat (figure 11.3).

We conclude that the late Paleocene ocean conforms to the moderately well stirred model in a warm ocean (figure 11.5). The model indicates that there was a higher thermal contrast between deep and surface waters, that nutrient cycling through the oceans and through the euphotic zone was efficient, and that fertility was high in the late Paleocene Epoch (table 11.4). Both calcareous and siliceous production including that of the diatoms should have been high during this time.

The early Eocene Epoch conforms to the poorly stirred ocean model (figure 11.5). With low thermal contrasts, fertility was reduced, nutrients were more concentrated both geographically and near the base of the thermocline, and fertility in many regions should have been insufficient to support the short, diatom-based food chains of previous times (table 11.4).

Vertical expansion of the oligotrophic habitat and the mixed layer is represented by the high numbers of morozovellids. Because oxygen isotope studies indicate that morozovellids inhabited the warmest water layers of the Paleocene–Eocene ocean, we can use their high abundances to predict a deep mixed layer and a stable, deep thermocline in tropical–subtropical areas, similar to that of 18° C subtropical-mode water in summer today (de Miro 1971; Ravelo et al. 1990; Boersma and Premoli Silva, 1991). This broad, deep mixed layer is

Table 11.3
Observations on late Paleocene–early Eocene planktonic foraminiferal communities and derivative predictions about the TRC through the Paleocene–Eocene transition

Zones P4/P5	Zones P6B/P7
Observations	
Intermediate foraminiferal species richness	High foraminiferal species richness
Intermediate abundance of oligotrophs	Maximum abundance of oligotrophs
Intermediate species richness of oligotrophs	Maximum species richness of oligotrophs
Predictions from Models	
Higher meridional and vertical thermal contrasts	Lower thermal contrasts
Higher fertility and siliceous production	Lower fertility
Efficient nutrient recycling	Nutrient concentration in thermocline
High siliceous and calcareous production	Higher calcareous production

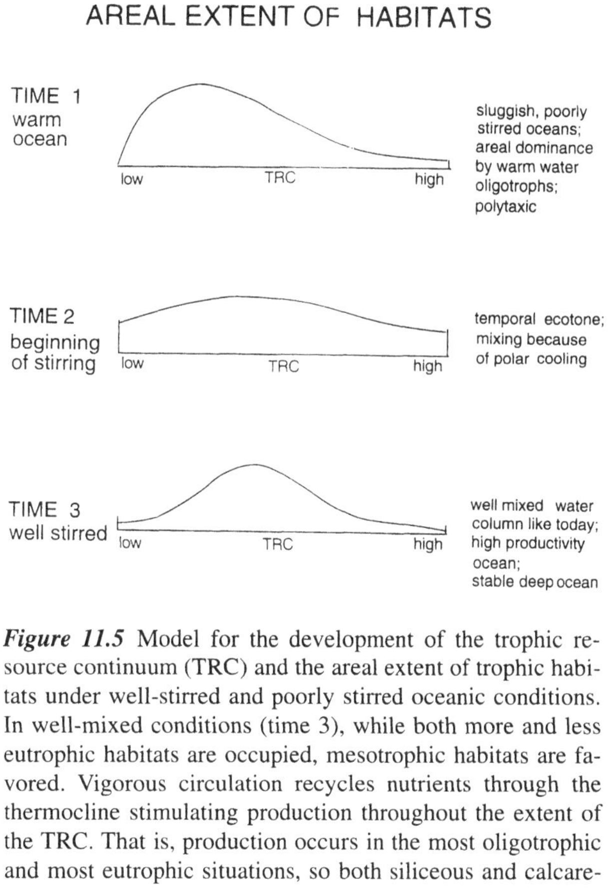

Figure 11.5 Model for the development of the trophic resource continuum (TRC) and the areal extent of trophic habitats under well-stirred and poorly stirred oceanic conditions. In well-mixed conditions (time 3), while both more and less eutrophic habitats are occupied, mesotrophic habitats are favored. Vigorous circulation recycles nutrients through the thermocline stimulating production throughout the extent of the TRC. That is, production occurs in the most oligotrophic and most eutrophic situations, so both siliceous and calcareous primary producers are supported. Diversity is high, but not maximal due to the only moderately developed oligotrophic realm.

In the poorly mixed scenario (time 1) the areal extent of oligotrophic conditions increases markedly, producing higher abundances and diversities of oligotrophs. Due to decreased cycling through the thermocline and in warm upwellings, both fertility and production are lower. Eutrophic habitats develop only where nutrients are concentrated, in the lower thermocline, in upwelling regions, and possibly over poorly oxygenated or ventilated shelves and basins.

Table 11.4
Summary of observations on late Paleocene to early Eocene planktonic foraminifera and predictions for the late Paleocene (Biochrons P4–P5) versus early Eocene (Biochrons P6–P7) oceans inferred from the TRC models in figures 11.2 and 11.5

	Early Eocene	Late Paleocene
Food chain	Nannoplankton based	Diatom based
Species richness	50 species	40 species
Productivity	Lower sedimentation rates	High sedimentation rates
Life strategies	More specialized forms	Fewer
Thermal indices	Tropical species in Antarctic	Subtropical species in Antarctic
Evolution	Oligotrophs diversify	All diversify
Sea level	High	Rising
Winds	Decreased	Increased
Deep circulation	Warmer (13° C)	Cooler (10° C)
Surface water temperature	Antarctic 16° C	Antarctic 11° C
Global gradients	Decreased	Increased
Euphotic zone	More oligotrophs	Fewer oligotrophs
Upwelling	Warmer, more rads	Cooler, more diatoms

The marked expansion of oligotrophs and compartmentalization of eutrophs implies the transition from a well-mixed to a poorly mixed ocean from late Paleocene to early Eocene times.

consistent with the interpretation of a warm, poorly mixed euphotic zone.

Decreased oceanic circulation intensity associated with diminished global thermal gradients from the late Paleocene into the early Eocene Epoch should be indicated by stable isotope data (table 11.4). And this is the case; both the meridional temperature gradient and the zonal wind intensities decreased from the late Paleocene Epoch into the warmer periods of the Eocene Epoch (Rea et al. 1990; Miller et al. 1987; Stott et al. 1990; Barrera and Huber 1991; Zachos et al. 1994). A sluggish circulation model for the warm periods of the early Eocene Epoch is corroborated by benthic foraminiferal studies (i.e., Tjalsma and Lohmann 1983; Katz and Miller 1991; Kennett and Stott 1991; Thomas 1992).

Planktonic foraminiferal population structure indicates the transition from a warm, relatively well stirred to a warmer, sluggish, poorly stirred global ocean across the Paleocene/Eocene Epoch boundary. The relatively well mixed late Paleocene euphotic zone supported a range from moderately eutrophic to moderately oligotrophic habitats. The broad meridional paleotemperature gradient of this time indicates active oceanic circulation and mixing. By contrast, the broader early Eocene TRC was dominated by oligotrophs, indicating a deep, poorly mixed euphotic zone. The reduced early Eocene meridional temperature gradient and decreased zonal wind intensities reflect sluggish oceanic and atmospheric circulation.

ACKNOWLEDGMENTS

The Deep Sea Drilling Project and Ocean Drilling Program kindly supplied the numerous samples used in this study. We thank Marie-Pierre Aubry, D. Corselli, and Brian Huber for reviewing the manuscript. This chapter is a contribution towards IGCP Project 308, Paleocene/Eocene boundary events in space and time. We thank Marie-Pierre Aubry for inviting our participation in IGCP Project 308.

REFERENCES

Barrera, E. and Huber, B. T., 1991. Paleogene and early Neogene oceanography of the southern Indian Ocean: Leg 119 foraminifer stable isotope results. In Barron, J., Larsen, B., et al., Proceedings of the Ocean Drilling Program, Scientific Results, v. 119: College Station, TX (Ocean Drilling Program), p. 693–717.

Bé, A. W. H. and Tolderlund, D. S., 1971. Distribution and ecology of living planktonic foraminifera in surface waters of the Atlantic and Indian Oceans. In Funnell, B. M. and Riedel, W. R. (eds.), *The Micropaleontology of Oceans,* Cambridge: Cambridge University Press, p. 105–149.

Berggren, W. A., Kent, D. V., Flynn, J. J. and Van Couvering, J. A., 1985. Cenozoic geochronology. *Geological Society of America Bulletin,* v. 96, p. 1407–1418.

Boersma, A. and Premoli Silva, I., 1983. Paleocene planktonic foraminiferal biogeography and the paleoceanography of the Atlantic Ocean. *Micropaleontology,* v. 29, p. 355–381.

Boersma, A. and Premoli Silva, I., 1988. Boundary conditions of Atlantic Eocene oxygen minimum zones. *Rivista Italiana Paleontologica i Stratigraphia,* v. 93, p. 479–506.

Boersma, A. and Premoli Silva, I., 1991. Distribution of Paleogene planktonic foraminifera: analogies with the Recent? *Palaeogeography, Palaeoclimatology, Palaeoecology,* v. 83, p. 29–48.

Boersma, A., Shackleton, N., Hall, M., Given, Q., 1979. Carbon and oxygen isotope records at DSDP Site 384 (North Atlantic) and some Paleocene paleotemperatures and carbon isotope variations in the Atlantic Ocean. In Tucholke, B. E., Vogt, P. R., et al., Proceedings of the Deep Sea Drilling Project, 43: Washington (U.S. Government Printing Office), p. 695–717.

Bralower, T. J. and Thierstein, H. R., 1984. Low productivity and slow deep-water circulation in mid-Cretaceous oceans. *Geology,* v. 12, p. 614–618.

Corliss, B. H., 1981. Deep-sea benthonic foraminiferal faunal turnover near the Eocene/Oligocene Boundary. *Marine Micropaleontology,* v. 6, p. 367–384.

D'Hondt, S., Zachos, J. C. and Schultz, G., 1994. Stable isotopic signals and photosymbiosis in the late Paleocene planktic foraminifera. *Paleobiology,* v. 20, p. 391–406.

Fischer, A. and Arthur, M., 1977. *Secular variations in the pelagic realm.* Society of Economic Paleontologists and Mineralogists Special Publication No. 25, p.19–50.

Gradstein, F. and Srivastava, S., 1980. Aspects of Cenozoic stratigraphy and paleoceanography of the Labrador Sea and Baffin Bay. *Palaeogeography, Palaeoclimatology, Palaeoecology,* 30, p. 261–295.

Hallock, P., 1987. Fluctuations in the trophic resource continuum: a factor in global diversity cycles? *Paleoceanography,* v. 2, p. 457–471.

Hallock, P., Premoli Silva, I. and Boersma, A. 1991. Similarities between planktonic and larger foraminiferal evolutionary trends through Paleogene paleoceanographic changes. *Palaeogeography, Palaeoclimatology, Palaeoecology,* v. 83, p. 49–64.

Hutson, W. H., 1977. Variations in planktonic foraminiferal assemblages along north–south transects in the Indian Ocean. *Marine Micropaleontology,* v. 2, p. 47–66.

Kaiho, K., 1991. Global changes of Paleogene aerobic/anaerobic benthic foraminifera and deep-sea circulation. *Palaeogeography, Palaeoclimatology, Palaeoecology,* v. 83, p. 65–86.

Katz, M. E. and Miller, K. G., 1991. Early Paleogene benthic foraminiferal assemblages and stable isotopes in the Southern Ocean. In Cieselski, P. F., Kristoffersen, Y., et al., Proceedings of the Ocean Drilling Program, Scientific Results, v. 114: College Station, TX (Ocean Drilling Program), p. 481–512.

Kennett, J. P., 1977. Cenozoic evolution of Antarctic glaciation, the circum-Antarctic Ocean, and their impact on global paleoceanography. *Journal of Geophysical Research,* v. 82, p. 3843–3860.

Kennett, J. P. and Stott, L. D., 1991. Abrupt deep-sea warming, palaeoceanographic changes and benthic extinctions at the end of the Palaeocene. *Nature,* v. 353, p. 225–229.

Kroon, D. and Ganssen, G., 1989. Northern Indian Ocean upwelling cells and the stable isotope composition of living planktonic foraminifers. *Deep-Sea Research,* v. 36, p. 1219–1236.

Kroon, D. and Nederbragt, A. J., 1990. Ecology and paleoecology of triserial planktic foraminifera. *Marine Micropaleontology,* v. 16, p. 25–38.

Longhurst, A., 1967. Diversity and trophic structure of zooplankton communities in the California Current. *Deep-Sea Research,* v. 14, p. 393–408.

Miller, K. G., Janecek, T. R., Katz, M. E. and Keil, D. J., 1987. Abyssal circulation and benthic foraminiferal changes near the Paleocene/Eocene boundary. *Paleoceanography,* v. 2, p. 741–761.

Miro, M. D. de., 1971. Los foraminiferos planctonicos vivos y sedimentados del margen continental de Venezuela (resumen). *Acta Geologia Hispanica,* v. 6, p. 102–108.

Nocchi, M., Amici, E. and Premoli Silva, I., 1991. Planktonic foraminiferal biostratigraphy and paleoenvironmental interpretation of Paleogene faunas from the Subantarctic transect, Leg 114. In Cieselski, P. F., Kristoffersen, Y., et al., Proceedings of the Ocean Drilling Program, Scientific Results, v. 114: College Station, TX (Ocean Drilling Program), p. 233–279.

Ottens, J. J., 1991. Planktic foraminifera as North Atlantic water mass indicators. *Oceanologica Acta,* v. 14, p. 123–140.

Poore, R. Z. and Mathews, R. K., 1984. Oxygen isotope ranking of late Eocene and Oligocene planktonic foraminifers: Implications for Oligocene sea-surface temperature and global ice-volume. *Marine Micropaleontology,* v. 9, p. 111–134.

Premoli Silva, I. and Boersma, 1988. Atlantic Eocene planktonic foraminiferal historical biogeography and paleohydrographic indices. *Palaeogeography, Palaeoclimatology, Palaeoecology,* v. 67, p. 315–356.

Ravelo, A. C., Fairbanks, R. G. and Philander, S. G. H., 1990. Reconstructing tropical Atlantic hydrography using planktonic foraminifera and an ocean model. *Paleoceanography,* v. 5, p. 409–431.

Rea, D. K., Zachos, J. C., Owen, R. M. and Gingerich, P. D., 1990. Global change at the Paleocene–Eocene boundary: climatic and evolutionary consequences of tectonic events. *Palaeogeography, Palaeoclimatology, Palaeoecology,* v. 79, p. 117–128.

Resig, J. M., 1989. Stratigraphic distribution of late Neogene species of the planktonic foraminifer *Streptochilus* in the Indo-Pacific. *Micropaleontology,* v. 35, p. 49–62.

Reynolds, L. and Thunnel, R. C., 1985. Seasonal succession of planktonic foraminifera in the subpolar North Pacific. *Journal of Foraminiferal Research,* v. 15, p. 282–301.

Reynolds, L. A. and Thunnel, R. C., 1986. Seasonal production and morphologic variation of *Neogloboquadrina pachyderma* (Ehrenberg) in the Northeast Pacific. *Micropaleontology,* v. 32, p. 1–18.

Rhyther, J., 1969. Phytosynthesis and fish production in the sea. *Science,* 166, p. 72–76.

Shackleton, N. J., Corfield, R. M. and Hall, M. A., 1985. Stable isotope data and the ontogeny of Paleocene planktonic foraminifera. *Journal of Foraminiferal Research,* v. 15, p. 321–336.

Spero, H. J., Lerche, I. and Williams, D. F., 1991. Opening the carbon isotope "vital effect" black box, 2, Quantitative model for interpreting foraminiferal carbon isotope data. *Paleoceanography,* v. 6, p. 639–655.

Stott, L. D., Kennett, J. P., Shackleton, N. J. and Corfield, R. C., 1990. The evolution of Antarctic surface waters during the Paleogene: Inferences from the stable isotopic composition of planktonic foraminifers, ODP Leg 113. In Barker, P. F., Kennett, J. P., et al., Proceedings of the Ocean Drilling Program, Scientific Results, v. 113, p. 849–863.

Thomas, E., 1992. Cenozoic deep sea circulation: Evidence from deep sea benthic foraminifera. In Kennett, J. P. and Warnke, D. (eds.), *The Antarctic Paleoenvironment: A Perspective on Global Change,* American Geophysical Union Antarctic Research Series No. 56, p. 141–165.

Tjalsma, R. C. and Lohmann, G. P., 1983. *Paleocene–Eocene bathyal and abyssal benthic foraminifera from the Atlantic Ocean.* Micropaleontology Special Publication No. 4, 90 pp.

Valentine, J., 1973. *Evolutionary Paleoecology of the Marine Biosphere.* Englewood Cliffs: Prentice Hall, 511 pp.

Zachos, J. C., Stott, L. D. and Lohmann, K. C., 1994. Evolution of early Cenozoic marine temperatures. *Paleoceanography,* v. 9, p. 353–387.

Ziveri, P., Thunnel, R. C. and Rio, D., 1995. Export production of coccolithophores in an upwelling region: Results from San Pedro Basin, Southern California Borderlands. *Marine Micropaleontology,* v. 24, p. 335–358.

12

Biogeography of the Late Paleocene Benthic Foraminiferal Extinction

ELLEN THOMAS

ABSTRACT

During the Late Paleocene Thermal Maximum (LPTM) benthic foraminifera at middle bathyal and greater depths suffered extinction of 30–50% of species during a few thousand years. Extinction was less severe at neritic to upper bathyal depths, where temporary changes in faunal composition prevailed. Preextinction deep-sea faunas were cosmopolitan and diverse, and contained heavily calcified species. Immediate postextinction faunas were more variable geographically, exhibited low diversity, and were dominated by thin-walled calcareous or agglutinated taxa, possibly because $CaCO_3$ dissolution increased globally from neritic to abyssal depths just before the extinction. These assemblages were dominated either by long-lived taxa such as *Nuttallides truempyi* or by buliminid taxa, the latter accompanied by agglutinants in some areas.

Faunas dominated by *N. truempyi* were common in the South Atlantic and at lower bathyal through upper abyssal depth in the Indian Ocean, and might indicate oligotrophic conditions as well as increased corrosiveness. Buliminid-dominated faunas might indicate high rates of deposition of organic matter or low-oxygen conditions. Such faunas were common globally along continental margins, and locally co-occurred with sedimentologic or planktonic faunal indicators of high productivity. In the bathyal central Pacific, however, buliminid-dominated faunas co-occurred with planktonic faunas suggesting oligotrophy, and they could reflect low-oxygen conditions resulting from sluggish ocean circulation, oxidation of dissociated methane hydrates, or warming of bathyal–abyssal waters caused by a change in deep-sea circulation. Alternatively, they could indicate that the fraction of organic matter reaching the seafloor increased as a result of decreased oceanic oxygenation. The latest Paleocene benthic extinction thus was complex, and factors such as changes in deep-sea circulation, increased $CaCO_3$ corrosiveness, increased temperatures, decreased oxygenation, and changes in the patterns of high productivity may have contributed to its severity.

Environmental Setting

The late Paleocene and early Eocene were the warmest epochs of the Cenozoic Era; polar ice caps were almost certainly absent, and shallow seas covered large parts of all continents. During these epochs, major changes occurred in global climate and in the carbon cycle on scales of millions to thousands of years. Long-term warming of the deep oceans started in the mid-Paleocene Epoch (Miller et al. 1987a; Shackleton 1987; Kennett and Stott 1990; Zachos et al. 1992, 1993; Seto 1995). Warm-water pelagic marine organisms penetrated to polar latitudes by the end of the Paleocene Epoch (Stott and Kennett 1990; Premoli Silva and Boersma 1984; Boersma et al. 1987; Aubry 1992; Berggren 1992). Thermophilic vertebrates occurred in the Arctic (Estes and Hutchison 1980; McKenna 1980; Markwick 1994), with vegetation and soil types indicating warm climates (Kemp 1978; Nilsen and Kerr 1978; Wolfe 1994; Basinger et al. 1994; Manum 1994). Clay mineral associations in oceanic sediments at high to midlatitudes indicate high humidity and intense chemical weathering (Antarctic, Robert and Kennett 1992; New Jersey margin, Gibson et al. 1993; New Zealand, Kaiho et al. 1996). Oxygen isotope data suggest that latitudinal temperature gradients were very shallow (Shackleton and Boersma 1981; Stott et al. 1990; Barrera and Huber 1991; Zachos et al. 1994; Bralower et al. 1995a, 1995b; Lu and Keller 1995b).

Rates of species origination and diversity were high for terrestrial vertebrates and flora (Hooker 1991; Collinson 1983; Rea et al. 1990; Wing et al. 1991, 1995; Maas et al. 1995), as well as for pelagic oceanic organisms (Romein 1979; Boersma et al. 1987; Boersma and Premoli Silva 1991; Corfield and Shackleton 1988; McGowran 1991; Aubry 1992; Berggren 1992; Corfield 1993; Kelly et al. 1996). The late Paleocene and earliest Eocene epochs thus constitute a time of major innovation of the biosphere (Briggs 1995).

The unusual warmth of these epochs has been commonly explained by high atmospheric pCO_2 concentrations, probably caused by plate tectonic processes such as massive volcanism in the North Atlantic Volcanic Province during initial opening of the North Atlantic, decarbonation of limestone or oxidation of organic-rich sediments during the beginning of the India–Asia continental collision, and high hydrothermal activity along midoceanic ridges (see Thomas and Shackleton 1996 for a review). Proxy data for atmospheric pCO_2 levels tentatively support higher values than today (Cerling 1991; Freeman and Hayes 1992), although this remains debatable (Stott 1992). Climate modeling indicates that at such CO_2 levels tropical temperatures would be much higher than deduced from oxygen isotope data, and mechanisms for highly increased heat transport from low to high latitudes at the low latitudinal temperature gradients remain unexplained (Sloan et al. 1995; Sloan and Rea 1995).

In the latest Paleocene Epoch the world warmed even more for a short time (Kennett and Stott 1991; Pak and Miller 1992; Thomas and Shackleton 1996), which has been named the Late Paleocene Thermal Maximum (LPTM; Zachos et al. 1993). This event occurred within paleomagnetic Chron C24r, in nannofossil Biochron NP9 (= Biochron CP8), and before the last appearance datum (LAD) of the planktonic foraminifer *Morozovella velascoensis* (Aubry et al. 1996; Berggren and Aubry 1996). Oxygen isotope data indicate that intermediate to deep ocean waters globally warmed by 4 to 6° C over less than a few thousand years (Kennett and Stott 1991; Thomas and Shackleton 1996), but tropical to subtropical surface to neritic temperatures remained constant (Stott 1992; Bralower et al. 1995a, 1995b; Lu and Keller 1995b; Schmitz et al. 1996). This rapid warming of the deep ocean has been explained by a change in deep- to intermediate-water circulation, during which the dominant sources of deep and intermediate waters shifted from high to subtropical latitudes (Kennett and Stott 1991; Pak and Miller 1992; Thomas and Shackleton 1996).

Coeval with this short-term warming was a shift in carbon isotope values by about −2‰ in surface and deep waters of all oceans (Kennett and Stott 1991; Pak and Miller 1992; Lu and Keller 1993, 1995a, 1995b; Canudo et al. 1995; Aubry et al. 1996; Thomas and Shackleton 1996; Schmitz et al. 1996), superimposed on a long-term decrease in values that started in the middle Paleocene Epoch (Shackleton and Hall 1984; Shackleton 1986, 1987; Corfield et al. 1991; Corfield and Cartlidge 1992; Corfield 1995). An isotope excursion similar in magnitude and estimated duration occurred in the tooth enamel of herbivores and in carbonate concretions on land (Koch et al. 1992, 1995; Stott et al. 1996), and in terrestrial-derived organic matter in a New Zealand marine section (Kaiho et al. 1996). The whole atmospheric–oceanic reservoir thus underwent a rapid, negative shift in carbon isotope values during the LPTM, the duration of which has been estimated at between 50 kyr (Thomas and Shackleton 1996) and 200 kyr (Kennett and Stott 1991).

Mass balance equations show that this carbon isotope excursion was so large that it probably could not have been caused by transfer of terrestrial biomass into the ocean–atmosphere system or by eruption of volcanogenic CO_2; it was so rapid that it probably could not have been caused by a change in deposition or erosion rates of carbon in carbonate as compared to carbon in organic matter (see Thomas and Shackleton 1996, for a review). This leaves few explanations open, and recently it has been speculated that this unusual, negative carbon isotope excursion could have been caused by massive dissociation of (isotopically extremely light) oceanic methane hydrates as a result of the deep-ocean warming (Dickens et al. 1995; Kaiho et al. 1996).

Benthic Foraminifera

The latest Paleocene benthic foraminiferal extinction has been described from land sections and at oceanic drill sites, at depths ranging from outer neritic to abyssal (appendices 12.1 and 12.2; figure 12.1). This extinction was rapid; it affected benthic foraminiferal faunas catastrophically worldwide, and it occurred during or close to the beginning of the LPTM and the negative excursion in carbon isotopes, just before the Paleocene/Eocene Epoch boundary as most commonly defined (Thomas 1989, 1990b; Stott and Kennett 1990; Nomura 1991 in combination with Seto et al. 1991; Pak and Miller 1992; Bralower et al. 1995a, 1995b; Canudo et al. 1995 in combination with Ortiz 1995; Aubry et al. 1996; Thomas and Shackleton 1996; Kaiho et al. 1996).

This major break in deep-sea benthic foraminiferal faunas occurred between the Paleocene and Eocene epochs if viewed at coarse timescales, whereas planktonic foraminifera underwent a major extinction at the end of the Cretaceous Period (Cushman 1946; Beckmann 1960; von Hillebrandt 1962). After the late Paleocene

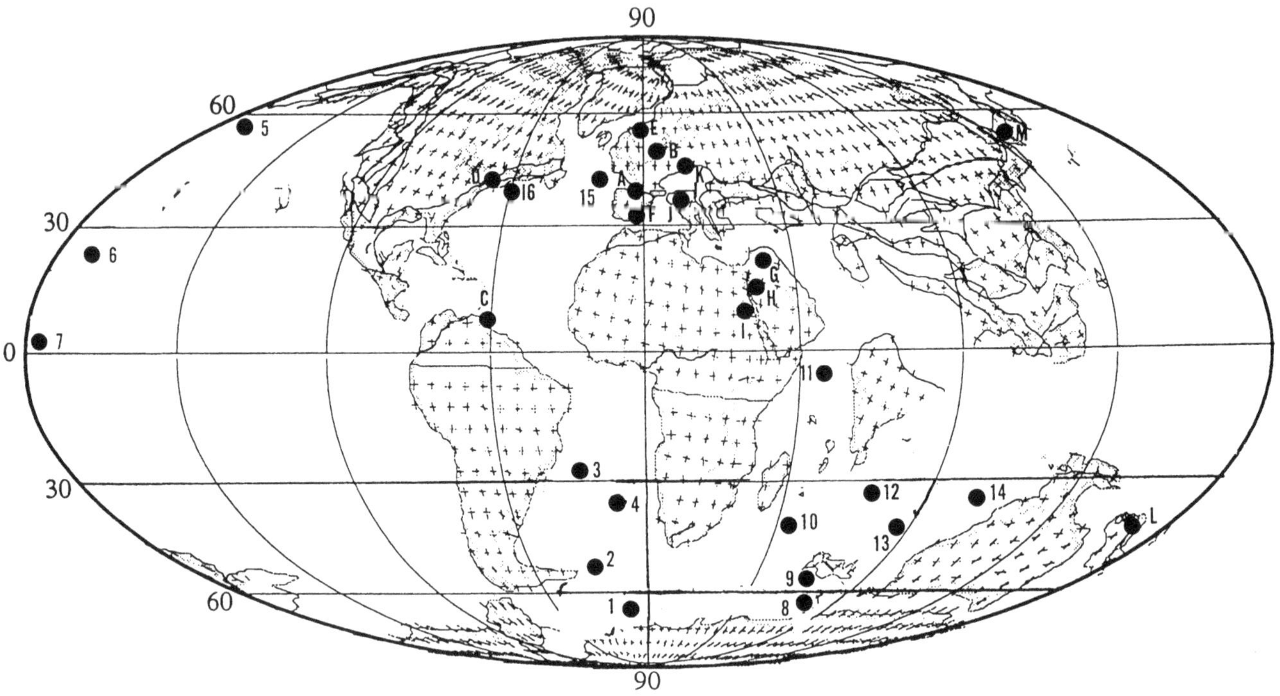

Figure 12.1 Location of drilling sites and land sections as described in appendices 12.1 and 12.2; numbers (drill sites) and letters (land sections) identify locations in the appendices.
The paleogeography of the continents in the late Paleocene Epoch is after Zachos et al. (1994).

benthic extinction faunas were predominantly of low diversity throughout the bathyal and abyssal regions of the oceans, whereas coeval planktonic organisms showed high rates of species turnover, high rates of species origination, and high diversity (Shackleton et al. 1985; Boersma et al. 1987; Corfield and Shackleton 1988; Boersma and Premoli Silva 1991; Hallock et al. 1991; Aubry 1992; Corfield and Cartlidge 1991; Corfield 1993; Lu and Keller 1993, 1995a, 1995b; Kelly et al. 1996). Major extinctions of planktonic and benthic organisms in the oceans thus appear to have been decoupled (e.g., Thomas 1990b; Kaiho 1994b).

The rapidity of the extinction makes it difficult to compare immediate postextinction faunas at different locations, especially because many sections contain unconformities in the interval straddling the extinction (Aubry et al. 1996). Only for high-resolution data sets can we know that postextinction faunas were indeed coeval, because we can compare faunas that co-occurred with the short-term negative carbon isotope event. Faunal comparisons may be more difficult because of differences in taxonomic concepts of different authors and because of the difference in size fraction being studied, varying between 63 and 150 μm in most cases (appendices 12.1 and 12.2).

High-resolution faunal data and carbon isotope analyses are available from Ocean Drilling Program (ODP) Sites 689 and 690 (Weddell Sea; Thomas and Shackleton 1996), South Atlantic (Walvis Ridge) Deep Sea Drilling Project (DSDP) Sites 525 and 527 (Thomas and Shackleton 1996), and equatorial Pacific ODP Site 865 (Bralower et al. 1995a, 1995b; Thomas unpublished data). Detailed data are also available for the Zumaya and Caravaca land sections in Spain (Canudo et al. 1995; Ortiz 1995) and for the Tawanui section in New Zealand (Kaiho et al. 1993, 1996). The short-term carbon isotope anomaly has been recognized in neritic Egyptian sections (Schmitz et al. 1996), but benthic foraminiferal data have not yet been published at high resolution (Speijer 1994; Speijer et al. 1996a, 1996b). Isotope records for ODP Site 738 registered the $\delta^{13}C$ excursion (Lu and Keller 1993), but only limited data on the benthic faunas are available (Nomura and Kennett, personal communication 1995; Thomas, unpublished data). Faunal data for equatorial Pacific DSDP Site 577 and Bay of Biscay DSDP Site 401 (Pak and Miller 1992) probably do not cover the time interval directly after the extinction because the magnitude of the $\delta^{13}C$ anomaly is much smaller at these sites than at others, suggesting that its minimum value (and thus the details of the benthic foraminiferal extinction) occurred between samples.

Most authors concluded that a change in deep-water circulation played a major role in the extinction, that most of the deep to intermediate waters of all oceans formed at least for some time at subtropical latitudes, and that decreased levels of deep-sea oxygenation resulting from the increased temperature of these waters were a causal factor (e.g., Miller et al. 1987b; Thomas 1989, 1990b, 1992; Katz and Miller 1991; Pak and Miller 1992, 1995; Kaiho 1991, 1994b; Nomura 1991; Coccioni et al. 1994; Ortiz 1995; Speijer et al. 1996b). Others, however, suggested that the patterns of faunal extinction vary geographically and that their complexity cannot be explained by decreased oxygenation only (e.g., Nomura 1991; Coccioni et al. 1994; Thomas and Shackleton 1996).

In this chapter, I first explore postextinction benthic foraminiferal faunal patterns following the extinction using high-resolution data from the sites mentioned above. Then I use literature data (figure 12.1, appendices 12.1 and 12.2) to try to evaluate oceanic environmental changes between late Paleocene and early Eocene times, to determine the biogeographic pattern of the extinction, and to speculate on its possible causes.

Methods

Sample Preparation and Stratigraphic Framework for High-Resolution Study

In this chapter new data are presented from ODP Site 865, and a few additional samples from ODP Sites 689 and 690 and DSDP Site 527 (figures 12.2 and 12.3), as compared to data in Thomas and Shackleton (1996). Samples for benthic foraminiferal faunal and isotope analysis were taken from DSDP Site 527 (Walvis Ridge, South Atlantic), ODP Sites 689 and 690 (Maud Rise, Weddell Sea), and ODP Site 865 (equatorial Pacific). They were dried overnight at 50° C and weighed, then soaked overnight in distilled water. Most samples disaggregated readily and could be washed over a 63-μm screen. Benthic foraminifera for faunal analysis were picked from the >63 μm size fraction, following Thomas (1990a). All specimens were picked and mounted in cardboard slides. All samples contained sufficient specimens for analysis (>250). Taxonomy is as in Thomas (1990a) and Thomas and Shackleton (1996), and largely follows van Morkhoven et al. (1986). The species richness was used to calculate species richness in a sample of 100 specimens using rarefaction (Sanders 1968).

I use the integrated magnetobiochronologic framework of Berggren et al. (1995), which is based on the geomagnetic polarity timescale of Cande and Kent (1992) and modified by Cande and Kent (1995). For ODP Site 690, I use the biostratigraphic interpretation of Aubry et al. (1996), giving an age of 55.5 Ma to the benthic extinction event and 55 Ma to the NP9/NP10 biochronal boundary. I derive numerical ages for ODP Site 690 assuming constant sedimentation rates between the benthic extinction and the NP9/NP10 zonal boundary, and between the benthic extinction and the top of Chron C25. For samples from DSDP Sites 525 and 527 and ODP Sites 689 and 865 I assign numerical ages according to Berggren et al. (1995) to datum levels in Thomas et al. (1990), Shackleton et al. (1984), and Bralower et al. (1995b). Finally, I revise the numerical ages at these four sites until the benthic foraminiferal carbon isotope curves agree with that for ODP Site 690 (Thomas and Shackleton 1996) in order to correct optimally for varying sedimentation rates, especially for the dissolution zone at Sites 525 and 527 and for the interval with very low sedimentation rates just above the extinction at Site 865.

Selection of Literature Data Sources

Appendices 12.1 (DSDP and ODP sites) and 12.2 (land sections) list sites for which data are available that straddle the extinction level; hence only a few sites from Tjalsma and Lohmann (1983) are included (figure 12.1). Data on sections in the Norwegian–Greenland Sea and Labrador Sea were not included because in many of the sections sedimentation probably started after the extinction event (Hulsbos et al. 1989; Kaminski et al. 1989). Data on agglutinated faunas were mostly not included, even though they suggest that changes occurred in agglutinated faunas at the end of the Paleocene Epoch (e.g., Geroch and Novak 1984; Kuhnt and Kaminski 1990; Kaminski et al. 1996), because exact data and precise correlations are difficult in the absence of carbonate and thus of isotope data. Data from the North Sea region are included because detailed stratigraphic data are available (Gradstein et al. 1994).

Few papers on neritic faunas are included because benthic faunas at these depths are more geographically variable and planktonic zonation is commonly difficult; therefore, it is not possible to determine whether observed faunal change can indeed be correlated to the late Paleocene extinction in the deep sea. For instance, papers on neritic to brackish water Paleogene faunas from the northern North Atlantic were not included (Berggren 1974a; Murray 1989). Data on neritic sections in Egypt (Speijer 1994; Speijer et al. 1996a,b) were included because detailed carbon isotope data are available (Charisi and Schmitz 1995; Schmitz et al. 1996). Data from Spain (Molina et al. 1992; Ortiz and McDougall 1991) were tentatively included because planktonic foraminiferal data suggest that benthic extinctions were at least close in time to the extinction in the deep sea. Data from coastal New Jersey were tentatively

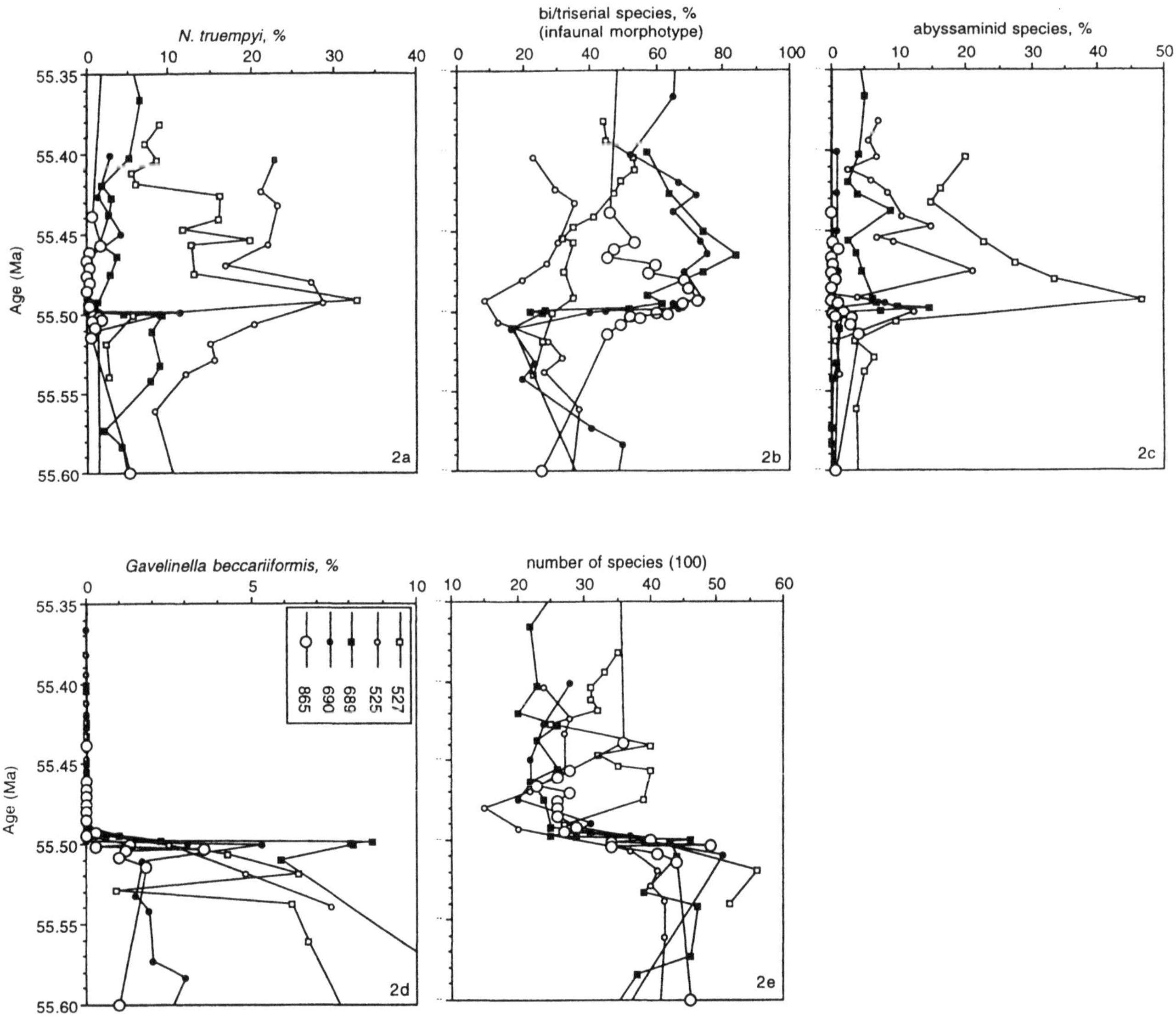

Figure 12.2 Comparison of the species richness and relative abundance of significant taxa at the time of the late Paleocene benthic foraminiferal extinction. See text for derivation of the age model.

Data for DSDP Sites 525 and 527 and ODP Sites 689 and 690 after Thomas and Shackleton (1996), with the addition of a few data points for Sites 527, 689, and 690; data for ODP Site 865 have not yet been published.

included (Olsson and Wise 1987; Gibson et al. 1993; Gibson and Bybell 1994) because the clay mineral record at these sites may be a means of correlation to ODP Site 690 (Robert and Kennett 1994).

Results

High-Resolution Studies

At ODP Sites 689 (paleodepth 1100 m) and 690 (1900 m), DSDP Sites 525 (1600 m) and 527 (3400 m), and ODP Site 865 (1300–1500 m) benthic foraminiferal diversity dropped precipitously at the extinction level, with 35 to 50% of species becoming extinct (figures 12.2 and 12.3). This percentage is a minimum estimate because several nodosariid species are rare, so their ranges cannot be determined with accuracy. Some nodosariids are absent in the interval corresponding to the LPTM, but reappear higher in the section at varying levels throughout the lower Eocene section.

At both Walvis Ridge sites the extinction occurs just above the base of a level with severe dissolution, although benthic foraminifera were well preserved; planktonic foraminifera show severe dissolution and fragmentation (Thomas and Shackleton 1996). No lowering of the $CaCO_3$ percentage of the sediment occurred at Sites 689 and 865, but faunas at these sites show more etching and are less well preserved just above the

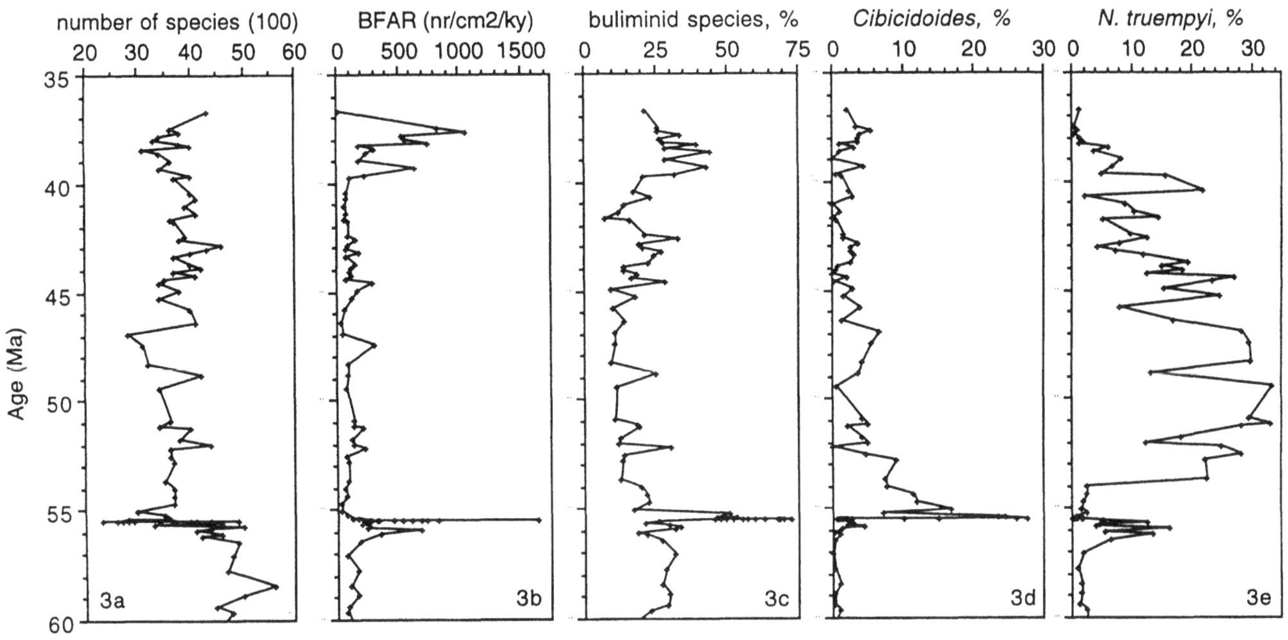

Figure 12.3 (a) Species richness at 100 specimens, (b) benthic foraminiferal accumulation rates (BFAR), and relative abundances of (c) buliminid taxa, (d) *Cibicidoides* spp., and (e) *Nuttallides truempyi* at equatorial Pacific Site 865.
Numerical ages according to Cande and Kent (1995); age model after Bralower et al. (1995a, 1995b).

extinction level. At Site 690 similar slight dissolution occurs in the faunas, and $CaCO_3$ values decrease just below the extinction level (O'Connell 1990; Thomas 1992).

At all sites, *Gavelinella beccariiformis* has its uppermost occurrence at the extinction level of several other taxa (see below). Our data confirm the observation by Katz and Miller (1991) that this species is most common at higher-latitude sites; it never reaches abundances higher than about 5% at equatorial Pacific Site 865. Other species that have their highest occurrence at this level include several large agglutinated species with calcareous cement (*Dorothia oxycona*, *Tritaxia havanensis*, and *Tritaxia paleocenica*) and calcareous species such as *Aragonia velascoensis*, *Bolivinoides delicatulus*, *Bulimina midwayensis*, *Bulimina thanetenis*, *Neoflabellina reticulata*, *Neoeponides hillebrandti*, *Neoeponides lunata*, and *Pullenia coryelli*.

Immediately above the extinction level, specimens are smaller and have thinner walls than below, even specimens of species that straddle the extinction interval (e.g., *Oridorsalis umbonatus* and *Nuttallides truempyi*). At ODP Sites 689 and 865 the change to smaller and thinner-walled taxa occurs a few centimeters below the highest occurrence of *G. beccariiformis*, possibly because of bioturbation. *Tappanina selmensis* and *Aragonia aragonensis* are at all sites more common above the extinction level than below (as also observed by Boersma 1984b; Speijer 1994), but their relative abundances vary strongly between sites. *Tappanina selmensis* reaches up to 40% at the Maud Rise sites, up to 15% at the Walvis Ridge sites, and up to 25% at Pacific Site 865. At all sites, peak occurrences of *T. selmensis* occur directly above the extinction level, whereas those of *A. aragonensis* occur at the top of the LPTM interval.

Despite these similarities, postextinction faunas are most notable for the major differences between sites, especially in the abundance patterns of buliminid species, *Nuttallides truempyi*, *Oridorsalis umbonatus*, *Cibicidoides* spp., and abyssaminid species (Thomas and Shackleton 1996; appendix 12.1). The postextinction faunas differ much more from site to site than the preextinction faunas, which are comparable to Velasco-type faunas at all five sites (see also Berggren and Aubert 1975; Kaiho 1988). These postextinction faunal differences cannot be completely attributed to the presence of stratigraphic gaps and comparison of noncoeval faunas, because faunas at all five sites were coeval with the short-term carbonate isotope excursion. Data were collected by one researcher, so there is homogeneity in the database with regard to taxonomic concepts and size fraction. Postextinction faunas differ more by geographic region than by depth; faunas from the two Walvis Ridge sites and the two Maud Rise sites are quite similar despite large differences in depth.

At the Walvis Ridge sites, postextinction faunas are dominated by *N. truempyi*, *O. umbonatus*, and abyssaminid species; the latter are especially abundant at the deepest site (figure 12.2). Abyssaminids (*Abyssamina poagi*, *Abyssamina quadrata*, and *Clinapertina planispira*) increase in relative abundance postextinction at Maud Rise, but much less than at the Walvis Ridge sites. At Pacific Site 865 these species have very low abundances throughout, with a slight further decrease at the extinction level. *Nuttallides truempyi* declines strongly in relative abundance at and just above the extinction level at the Maud Rise sites as well as at Site 865, in strong contrast with the situation in the South Atlantic. At the Walvis Ridge sites the postextinction *N. truempyi* specimens are smaller and have thinner walls than preextinction specimens, but are much more numerous than in preextinction faunas.

At the extinction level the relative abundance of the buliminid species group decreases at Site 527, does not change at Site 525, but increases strongly at both Maud Rise sites and Site 865 (figure 12.2). Above the extinction level this group is dominated by small species such as *T. selmensis*, *Bolivinoides* sp. cf. *B. decorata,* and *Bulimina simplex*. At the Maud Rise sites *Siphogenerinoides brevispinosa* is common (up to 40%) across the extinction level, at the Walvis Ridge sites it is much less common (up to 15%), and at Site 865 its abundance is similar to that at Walvis Ridge, but the species has its highest occurrence at the extinction level. *Bulimina semicostata* is abundant at Pacific Site 865 just above the extinction level, then decreases in abundance but remains present throughout the lower and middle Eocene Series. At the Maud Rise sites this species appears only in Zone NP12 (51–52 Ma, Berggren et al. 1995). In the lowermost postextinction specimens of *B. semicostata* at Site 865 the costae on the lower part of the test, typical for this species, are unusually thin and the specimens are small; possibly the species originated at this time from a noncostate, trihedral *Bulimina* species.

At Site 865 the predominance of buliminid species appears to have lasted for a shorter time than at the Maud Rise sites. At this site the interval of buliminid dominance is only 15 cm thick, representing about 10 kyr in our age model, whereas it is about 500 cm thick, representing about 100 kyr, at the Maud Rise sites. Above the buliminid-dominated interval at this site faunas are dominated by *Cibicidoides* spp. up to a level in Zone NP11 (zonal assignment in Bralower et al. 1995b; 52.85–53.61 Ma in Berggren et al. 1995; figure 12.3), where the relative abundance of *N. truempyi* increases to 12–25%. At none of the other sites is there an interval dominated by *Cibicidoides* species (up to 30%). The taxonomy of these *Cibicidoides* species is still under investigation; they resemble large, flat shallow-water European species such as *C. proprius* (Brotzen 1948).

A detailed record of planktonic foraminifera and isotopes is available from southern Indian Ocean ODP Site 738 (Lu and Keller 1993), but the benthic foraminiferal record is limited. Immediate postextinction faunas are very poorly preserved because the extinction occurs within a layer with very strong dissolution. I observed very common small, buliminid species (including *T. selmensis* and *B. simplex*) just above the extinction level, suggesting that the faunal patterns at this site resemble those at Maud Rise.

Detailed data on faunas with age control by the LPTM carbon isotope excursion are available for the Zumaya and Caravaca sections in Spain (Canudo et al. 1995; Ortiz 1995; appendix 12.2). In both sections strong dissolution occurs at the extinction level, as indicated by the presence of dark gray, laminated shales. Faunas just above the extinction level of *G. beccariiformis* are dominated by the agglutinated species *Haplophragmoides retrosepta*. At Zumaya, above the dissolution interval there is a thin interval with common *Bulimina tuxpamensis* (*Bulimina* spp. Assemblage); higher *N. truempyi* is dominant. At Caravaca *N. truempyi* is not as common as at Zumaya, and it is accompanied by *Anomalinoides capitatus* and *Oridorsalis umbonatus*. The *Bulimina* spp. Assemblage may have persisted somewhat longer at the somewhat shallower Caravaca section than at Zumaya, but for this part of the section age control is not as precise, and short-term changes in sedimentation rates may obscure the pattern.

In the Tawanui Section (New Zealand; Kaiho et al. 1993, 1996; appendix 12.2) the benthic extinction occurs in laminated, dark gray, organic carbon-rich shales. Age control is provided by recognition of the LPTM $\delta^{13}C$ excursion in terrestrial organic matter. Within a 4-cm interval directly above the highest occurrence of *G. beccariiformis* faunas are dominated by *Bulimina tuxpamensis* and *Praeglobobulimina pupoides*. Above the *Bulimina* spp. Assemblage there is an ~80-cm-thick interval with very strong dissolution, in which faunas are dominated by agglutinant taxa (*Rhabdammina* spp., *Conotrochammina depressa*, and *Reophax regularis*). Higher in the section faunas are more diverse, with common *Cibicidoides* spp., *Anomalinoides* spp., and *B. tuxpamensis*.

We can conclude from these high-resolution studies that postextinction faunas are characterized by low diversity, and that geographic differences cannot be attributed to insufficient age control, differing taxonomic concepts, or differing size fractions studied. With this in mind, we can try to evaluate the lower-resolution data available in the literature.

Literature Data

Beckmann (1960) was the first author to recognize the importance of the benthic foraminiferal extinction at the end of the Paleocene Epoch, based on his study of the bathyal Lizard Springs Formation (Trinidad). His data (as given in Bolli et al. 1994) show a very strong increase in highest occurrences as well as a decrease in lowest occurrences of benthic foraminifera in the upper Paleocene *Morozovella velascoensis* Zone, close to the Paleocene/Eocene Series boundary (Bolli et al. 1994). Benthic foraminiferal faunas in the overlying zone have a much lower diversity. There are more last appearances of agglutinated, nodosariid, and rotaliid taxa than of buliminid taxa (figure 12.4). Another early recognition of the importance of a late Paleocene benthic foraminiferal extinction event was by von Hillebrandt (1962), who noticed a strong drop in diversity of benthic foraminiferal faunas toward the top of his *M. velascoensis* Zone (Zone F) in the Reichenhall–Salzburg Basin (figure 12.5). In these sections, however, rotaliid taxa appeared to be the more common survivors.

The importance of the benthic foraminiferal extinction was recognized by many later authors (appendices 12.1 and 12.2) in studies of land sections as well as of DSDP and ODP sites in all oceans. High-resolution studies demonstrate that the extinction was a very short-term event, so comparison of faunal events between sites is problematic for data that are so disparate in chronologic resolution, age control, taxonomic concept, and size fraction studied, as presented in the appendices. However, some overall conclusions can be drawn from these data, especially in combination with the high-resolution data.

In the first place, the extinction was severe at middle bathyal through abyssal depth, with extinction of species ranging between 30 and 50%. At upper bathyal through neritic depths the extinctions were much less severe, but faunal composition changes were highly significant. These compositional changes were temporary and many species returned to the sites (Speijer 1994; Gibson and Bybell 1994).

Second, preextinction Paleocene faunas were remarkably cosmopolitan, extending over a wide depth range (Brotzen 1948; Berggren 1974a, 1974b; Berggren and Aubert 1975). The latter authors distinguished a continental shelf fauna (Midway-type) and a lower continental slope and abyssal plain fauna (Velasco-type),

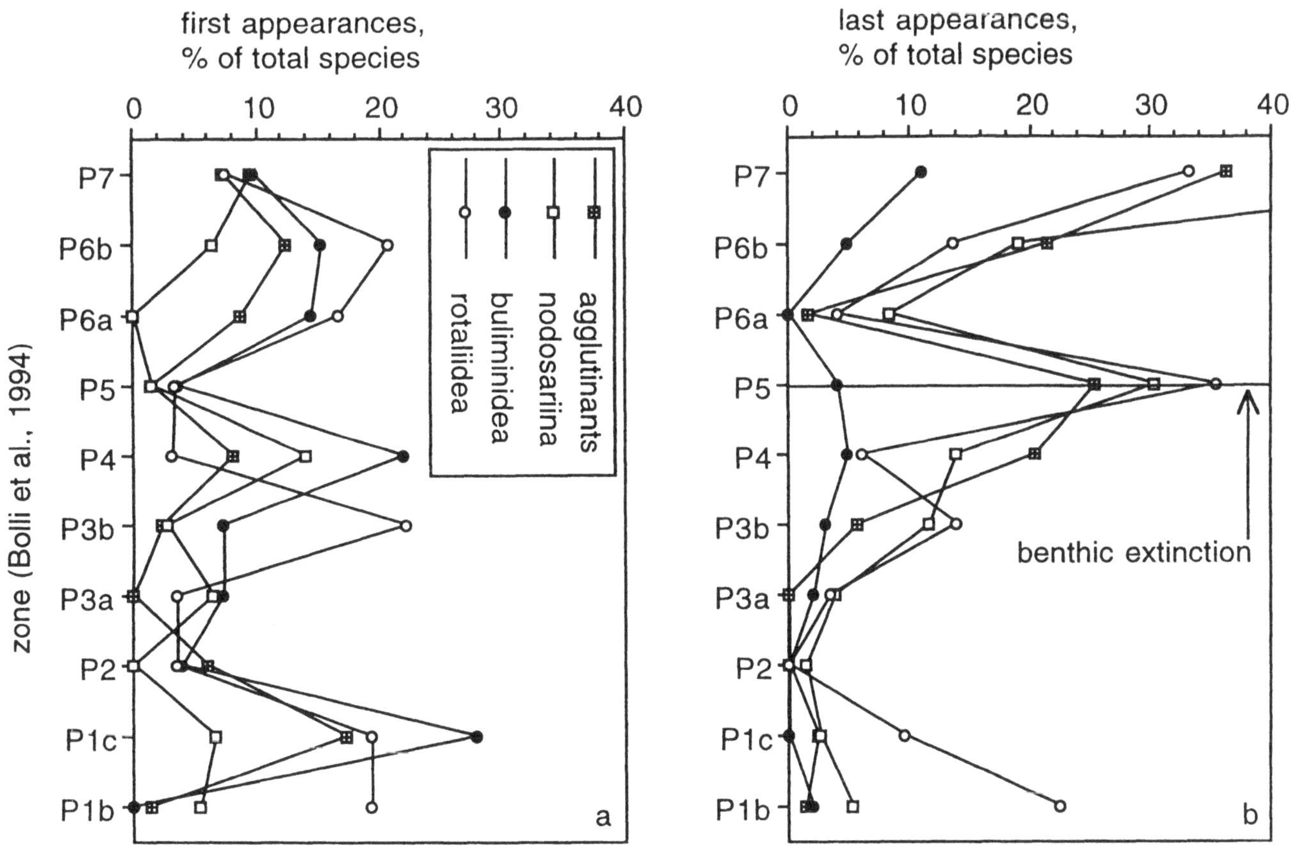

Figure 12.4 (a) Lowest and (b) highest occurrences of species belonging to the four most important groups of taxa as a percentage of the total number of species in the Lizard Spring Formation, Trinidad (Beckmann 1960; Bolli et al. 1994). The horizontal line shows the level of the benthic extinction.

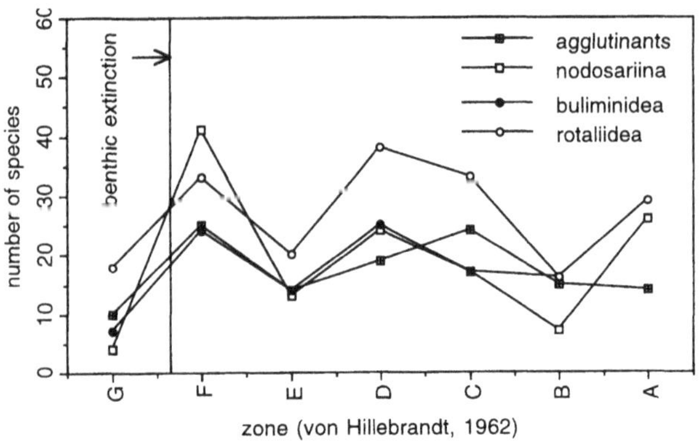

Figure 12.5 Numbers of species of the four most important groups of species in the Reichenhall–Salzburg Basin (von Hillebrandt 1962). The vertical line shows the level of the benthic extinction.

although several species (e.g., *Bulimina midwayensis* and *Coryphostoma midwayensis*) occur in both faunas (e.g., Kaiho 1988, 1991; Nomura 1991; Kaiho et al. 1993; Speijer 1994; Coccioni et al. 1994). The Midway fauna typically contains more *Cibicidoides* and especially *Anomalinoides* species, whereas the Velasco fauna has *A. velascoensis*, *Gyroidinoides* spp., *N. truempyi*, and agglutinant taxa such as *Dorothia oxycona*, *Gaudryina laevigata*, *Tritaxia havanensis*, and *Tritaxia paleocenica*. The common species *N. truempyi* generally is considered typical for the Velasco fauna (below about 500 m). *Tappanina selmensis* occurs from the shelf (about 100 m) down to abyssal depths, but is rare in bathyal–abyssal faunas below the level of the extinction. Both Midway and Velasco faunas are characterized by extremely high diversities, the occurrence of many rare taxa (especially nodosariids), and thick-walled calcareous taxa.

Third, there were geographic differences in faunal assemblages in spite of the cosmopolitan occurrence of many species. *Gavelinella beccariiformis*, for instance, appears to be more common at high-latitude than low-latitude sites (Katz and Miller 1991), although it was also common in the Tethys (Speijer 1994). *Aragonia velascoensis* is extremely rare at both Maud Rise sites, whereas *S. brevispinosa* is abundant at these sites, but shows large fluctuations in relative abundance (Thomas 1990a, 1990b). The geographic differences have not been well evaluated; the availability of more oceanic drill sites in depth transects would help in separating depth from geographic effects, and in distinguishing Paleocene biogeographic provinces, possibly similar to those in the latest Cretaceous Period (Widmark 1995).

Fourth, this high degree of faunal cosmopolitanism ended with the late Paleocene extinction (Kaiho 1988; Thomas 1992; Thomas and Shackleton 1996). This apparent difference in provinciality between late Paleocene and early Eocene faunas might be thought to result at least partly from the lack of a monograph on the early Eocene benthic foraminiferal taxonomy comparable to that of Berggren and Aubert (1975) for the late Paleocene Epoch, so that there are greater differences in taxonomic concepts between authors. This cannot be the full explanation, however, because greater differences between early Eocene faunas from different locations have been observed in faunas studied by one investigator (Tjalsma and Lohmann 1983; Kaiho 1988; Katz and Miller 1991).

Finally, postextinction faunas were of low diversity and high dominance, in stark contrast with the highly diverse, low-dominance preextinction faunas. The postextinction community thus has a structure typical for faunas affected by a major environmental perturbance (e.g., Pearson and Rosenberg 1978). In these perturbed communities the species *T. selmensis* and *A. aragonensis* are common to abundant everywhere below shelf depths (appendices 12.1 and 12.2; Boersma 1984a).

Otherwise, few faunal observations are valid on a worldwide basis. The postextinction pattern as seen by Tjalsma and Lohmann (1983) and Katz and Miller (1991), i.e., dominance by *N. truempyi* with survivor species such as *O. umbonatus*, appears to be typical for the South Atlantic and possibly Gulf of Mexico; abyssaminid taxa are common below about 2500 m in this type of assemblage. Thomas (1990a, 1990b) suggested that this association might be seen only in studies using the large size fraction (>150 μm) because many of the buliminid taxa are small. High-resolution data in Thomas and Shackleton (1996) and this chapter (figure 12.2) show that this is incorrect, and that faunas are dominated by *N. truempyi* and abyssaminids even though the >63-μm fraction is studied, as was also observed by Müller-Merz and Oberhänsli (1991).

Patterns in the North Atlantic are more variable. Most North Atlantic data are from the Goban Spur–Bay of Biscay Sites, and postextinction faunas vary from buliminid dominated to *N. truempyi* dominated. Possibly, the two types of faunas can be differentiated by depth: Buliminid-dominated faunas appear to occur at middle bathyal and shallower bathyal depths (DSDP Sites 548, 549; Reynolds 1992; Boltovskoy et al. 1992), whereas *N. truempyi*–dominated faunas occur at lower bathyal through abyssal sites (Schnitker 1979; Pak and Miller 1992; appendix 12.1). All these sites, however, show increased dissolution and occurrence of unconformities in the lower Eocene stratigraphic record, which makes faunal comparisons more difficult (Aubry et al. 1996).

Bathyal land sections close to the Bay of Biscay (Coccioni et al. 1995; Ortiz 1995) show buliminid-dominated postextinction faunas, although the pattern is complicated by the occurrence of agglutinated faunas in an interval of severe dissolution. In New Jersey margin DSDP Site 605 *N. truempyi*–dominated faunas occur above the extinction level (Hulsbos 1987; Thomas unpublished data). Shelf faunas (Olsson and Wise 1987; Gibson et al. 1993) in the region show common *Epistominella*, *Pulsiphonina*, and *Turrilina*, and cannot easily be compared to the deeper faunas; coastal sections generally have unconformities at the extinction level (Gibson and Bybell 1994). The postextinction faunas from Trinidad are buliminid dominated (Beckmann 1960; Bolli et al. 1994).

In conclusion, we can speculate that the northern Atlantic abyssal to lower bathyal faunas were similar to South Atlantic faunas (*N. truempyi*–dominated), whereas they were buliminid dominated in the upper- to mid-bathyal reaches. We cannot be certain, however, that we are indeed comparing coeval faunas because dissolution and unconformities are common.

The bathyal faunas in the North Sea (depths 750–1000 m) were dominated by noncalcareous agglutinated taxa through the late Paleocene and early Eocene epochs (Charnock and Jones 1990; Gradstein et al. 1994). These agglutinated taxa underwent a period of severe extinction close to the Paleocene/Eocene Epoch boundary, after which extremely impoverished, low-diversity, agglutinant faunas remained (e.g., Charnock and Jones 1990; King 1989; Gradstein et al. 1994). Direct postextinction sediments indicate strong $CaCO_3$ corrosiveness and dysoxic to anoxic conditions (Charnock and Jones 1990; Schroeder 1992). At somewhat shallower levels (100–500 m) calcareous taxa (including *G. beccariiformis*) were present during the Paleocene Epoch (King 1989). These faunas were replaced by extremely low-diversity agglutinated faunas similar to those from deeper regions. The timing of the disappearance of the calcareous taxa is difficult to establish because of the biostratigraphic complexity of the North Sea region (Gradstein et al. 1994; Berggren and Aubry 1996). The local North Sea disappearance of *G. beccariiformis* probably did not predate the global extinction of *G. beccariiformis* at the LPTM by very much time because it can be correlated to the lower part of dinoflagellate Zone D5 (King 1989), which has been correlated to just below the middle part of nannofossil Zone NP9. Palynological studies indicate the occurrence of a peak warm period somewhere in dinoflagellate Zone D5 (thus Zone NP9; Schroeder 1992), which might possibly be correlated to the LPTM in mid-Biochron NP9 (Aubry et al. 1996).

Tethyan faunas are known from land sections only (appendix 12.2). Sections in the eastern Tethys (Speijer 1994) from a depth range of 500–1000 m show postextinction faunas with common *N. truempyi*, *Anomalinoides* species, and *Cibicidoides*. Shallower faunas (100–300 m) contain more buliminids, but on the shelf *Cibicidoides* spp. and especially *Anomalinoides aegyptiacus* dominate. Sections in Spain (Ortiz 1995) and Italy (Braga et al. 1975; Di Napoli et al. 1970; 500–1500 m depth) show strongly increased percentages of buliminids above the extinction level, including *Bulimina semicostata*. Faunas from the Reichenhall–Salzburg Basin (von Hillebrandt 1962; 200–800 m) show increased abundance of rotaliid taxa, but detailed stratigraphy is not available from these sections. We can conclude tentatively that western Tethys faunas were similar to North Atlantic faunas from the same depth.

In the Indian Ocean, no simple faunal pattern can be extracted, particularly because many DSDP and ODP sites have records with poor recovery and unconformities. Faunas at some lower bathyal–abyssal sites (appendix 12.1) seem to resemble South Atlantic faunas (*N. truempyi*–dominated), but abyssaminids are much less common or absent, possibly because no lower abyssal sites are available. Other taxa in the postextinction faunas are mostly long-term survivors such as *O. umbonatus*, although *B. semicostata* may be present above the extinction level. Faunas at upper and middle bathyal sites (appendix 12.1) have very different postextinction faunas, dominated by *Lenticulina* species (Mackensen and Berggren 1992) or by *Anomalinoides* species (similar to faunas in the eastern Tethys; Speijer 1994). Such faunas dominated by *Anomalinoides* species have been interpreted as indicating low-oxygen conditions or high productivity (Speijer et al. 1996b; Nomura and Kennett, personal communication 1995). At all these sites, however, core recovery was poor and the existence of the LPTM $\delta^{13}C$ excursion has not been fully documented. High-latitude ODP Site 738 (1350 m), however, where the carbon isotope excursion is present (Lu and Keller 1993), has the buliminid-dominated pattern of the Maud Rise sites.

Data on the Pacific Ocean are available from few sites only at bathyal depths. Northern Pacific ODP Site 883 (Pak and Miller 1995; 1000–2000 m) has a poor upper Paleocene record and an unconformity, but its postextinction faunas appear to be of the *N. truempyi*–dominated type. Faunas at DSDP Site 577 (Pak and Miller 1992; Kaiho in press; 1800–2100 m) have common *Bulimina semicostata* above the extinction level, although buliminid percentages are much lower than at Site 865. At this site, however, a 0.5-m section in the critical interval was not recovered (between Cores 9 and 10 in Hole 577), and the section probably contains unconformities (Aubry, personal communication 1995). Sites 577 and 865 (described above) thus have buliminid-dominated faunas. Land sections in Japan and New Zealand (500–1500 m; Kaiho 1988; Kaiho et al. 1993,

1996) show buliminid-dominated faunas, although complicated by dissolution intervals with agglutinated taxa.

Data on the neritic faunas do not indicate a relationship between the benthic extinction (thus the LPTM) and changes in sea level. The extinction event occurred during the second order supercycle TA2 of Haq et al. (1987), and might have been coeval with part of short-term third-order cycle 2.3, in mid-nannofossil Biochron NP9. Speijer (1994) and Speijer et al. (1996b) did not observe significant sea-level effects in neritic sections in Egypt. European and North American sections have many unconformities in their upper Paleocene–lower Eocene stratigraphic record (e.g., Olsson and Wise 1992; Gibson and Bybell 1994; Berggren and Aubry 1996), suggesting that sea level was relatively low at the time. In the northern Spanish sections the benthic extinction event has been correlated to a lowstand (Pujalte et al. 1995). It has been proposed, however, that sea-level trends in the North Atlantic reflected the development and collapse of the Iceland Plume under the North Atlantic Volcanic Province, and thus were regional rather than global (Nadin and Kusznir 1995).

Discussion and Speculation on $CaCO_3$ Dissolution, Productivity, and O_2 Levels

$CaCO_3$ Dissolution

The global data set on the benthic foraminiferal extinction event is incomplete and requires high-resolution studies to describe in detail the events that occurred during the LPTM. Data are needed from sites located on depth transects in order to discriminate between depth and geographic effects. Most information is from sites at lower bathyal depths (figure 12.6a), with few data from abyssal depths. Geographically, most data are from the North and South Atlantic oceans. Coverage in the Pacific Ocean is extremely poor, data being available only from middle to lower bathyal depths (figure 12.6b). Coverage is slightly better for the Indian Ocean, but recovery is poor at many sites. In addition, taxonomic concepts for the postextinction faunas, specifically for *Bulimina* spp., *Cibicidoides* spp., and *Anomalinoides* spp., needs to be uniformized between researchers before global patterns can be established with any degree of confidence.

Combining lithological and paleontological data in this incomplete set, however, suggests that a remarkable global episode of carbonate dissolution occurred at least partially coeval with the LPTM (figure 12.7). In the North Sea region calcareous species occurred only in the shallowest neritic areas of the time (King 1989; Gradstein et al. 1994). This severe dissolution has commonly been attributed to local factors such as basin restriction, which certainly may have contributed, but the severe dissolution occurred synchronously with worldwide dissolution. In the eastern Tethys (Egypt), Speijer (1994) observed dissolution layers over a wide depth range, Benjamini (1992) reported such layers from the north central Negev (Israel), and Berggren and Aubry (1996) from the Khieu River section in Georgia (NW Caucasus). Similar dissolution intervals occur in land sections in Spain (Tethys and Atlantic sides; Coccioni et al. 1994; Canudo et al. 1995).

Dissolution is more difficult to trace in DSDP and ODP sites than in land sections because poor recovery is a problem. A thin layer (<50 cm) of sediment may not be recovered, and therefore many sites are listed as "no data" in figure 12.7 and appendix 12.1. We cannot be sure of the full geographic and bathymetric extent of carbonate dissolution, but it was extensive in and around the Atlantic Ocean, as well as at least in parts of the Indian Ocean, along the Tethys margin, and along the Pacific margins. We have no data on the Pacific deep basin floors, which were probably below the CCD during most of the Paleocene and Eocene epochs (e.g., as shown by red clays in Core LL44-GPC 3, Rea 1994). We cannot be sure about the exact timing of the initiation of this increased dissolution, but it may have started slightly before the LPTM (Thomas 1992; Thomas and Shackleton 1996; Kaiho et al. 1996).

The occurrence of such widespread $CaCO_3$ dissolution during a period of warming is unexpected because rising temperatures alone would result in decreased solubility of calcite (e.g., Broecker and Peng 1984). Therefore, the occurrence of widespread dissolution strongly indicates massive addition of carbon to the ocean–atmosphere reservoir, and the pattern of dissolution (more severe around the Atlantic, less in the Pacific) resembles that predicted to result from massive burning of fossil fuel and deforestation (e.g., Walker and Kasting 1992). The occurrence of the extremely negative carbon isotope anomaly in this interval of dissolution suggests that the added mass of carbon must have been isotopically unusually light. The amount of biosphere-derived organic carbon needed to cause such a large isotopic anomaly would have to be reflected in widespread destruction of land biomass, which does not agree with paleontological evidence (Thomas and Shackleton 1996). Therefore, widespread carbonate dissolution associated with the $\delta^{13}C$ excursion supports the speculation that dissociation of methane hydrates (and their oxidation of methane to CO_2) may have been involved (Dickens et al. 1995; Kaiho et al. 1996). This speculation implies that the temperature increase of the deep waters preceded the carbon isotope excursion slightly, in agreement with Thomas and Shackleton (1996).

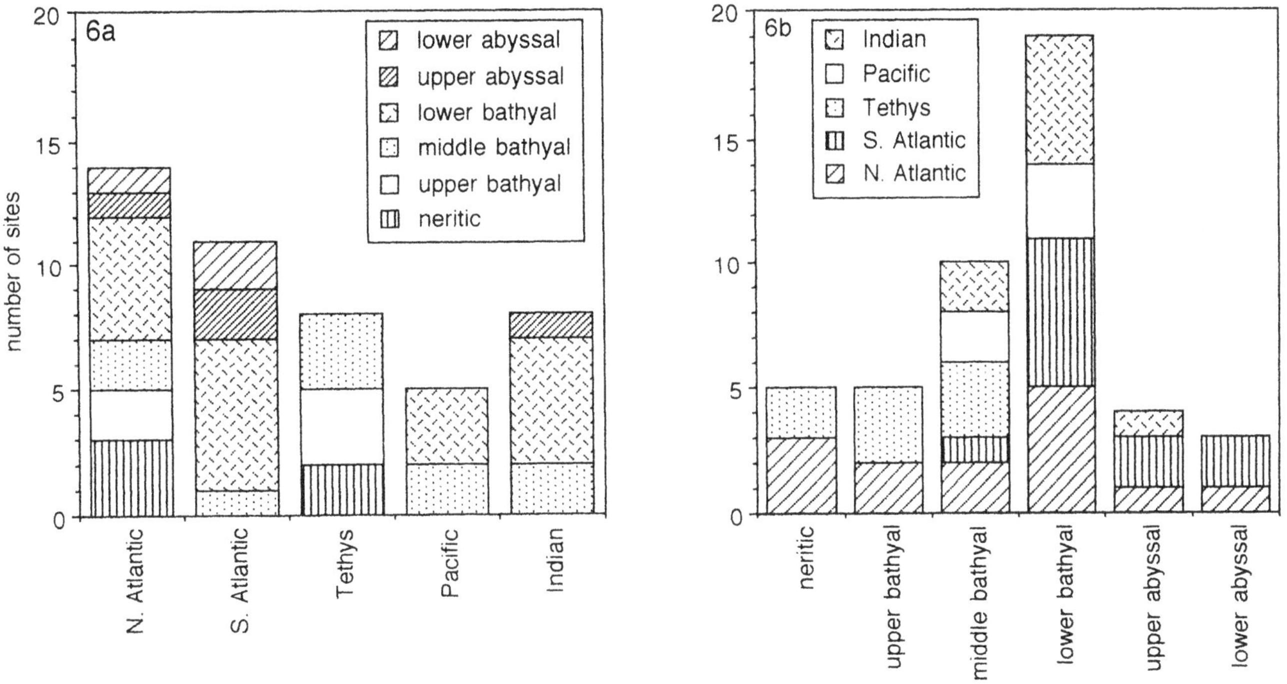

Figure 12.6 Distribution of sites listed in appendices 12.1 and 12.2 according to (a) depth and (b) geographic location.
The depth categories are after van Morkhoven et al. (1986): neritic, 0–200 m; upper bathyal, 200–600 m; middle bathyal, 600–1000 m; lower bathyal, 1000–2000 m; upper abyssal, 2000–3000 m; lower abyssal, >3000 m. Plotted was the midpoint of the paleodepth range for each site, as estimated by the authors of the original papers and given in appendices 12.1 and 12.2.

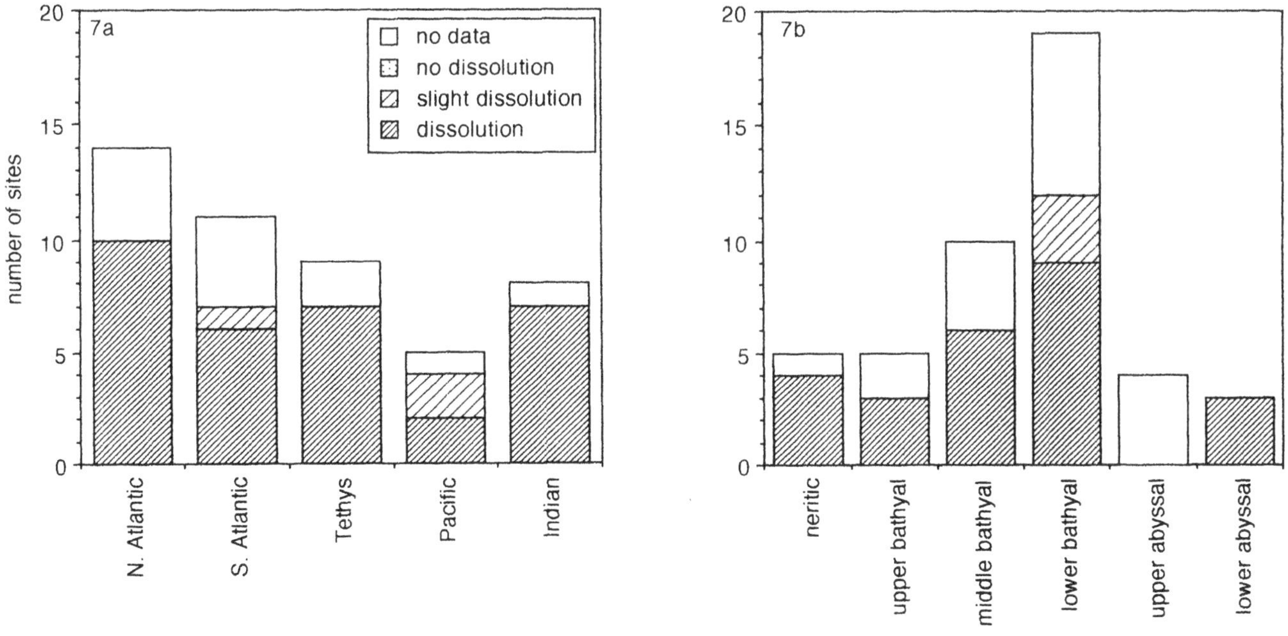

Figure 12.7 Occurrence of dissolution across the interval of the benthic foraminiferal extinction by (a) depth and (b) geographic location; sites listed in appendices 12.1 and 12.2, depth categories as in figure 12.6.
Dissolution: decrease in $CaCO_3$ content of the sediment by 30% or more; occurrence of black or dark gray, laminated shales. Slight dissolution: increased dissolution and etching of microfaunas, but no clear sedimentary evidence.

Such an increase of the ocean–atmosphere carbon reservoir is in contradiction with Stott's interpretation (1992) of data on the carbon isotopic composition of organic matter in foraminiferal tests, although such data cannot be easily interpreted (Goericke and Fry 1994; Hinga et al. 1994). Stott's data (1992) might be explained by many factors, such as species-specific effects, diagenetic effects, effects of varying temperature, or salinity. More problematic for the methane hydrate hypothesis might be the precise timing and sequence of events during the LPTM; dissolution appears to have preceded the $\delta^{13}C$ anomaly.

Dissolved Oxygen Levels and Productivity

What caused the benthic foraminiferal extinction, and what can explain the biogeographic patterns observed? This is a difficult question to answer because there is no agreement on the interpretation of Recent benthic faunas (see e.g., Gooday 1994, Schnitker 1994, for reviews). There is no modern analog for faunas dominated by *N. truempyi* because this species became extinct in the late Eocene Epoch. Its modern relative, *Nuttallides umbonifera*, is common to abundant in $CaCO_3$-corrosive waters, and specifically in Antarctic bottom water (e.g., Bremer and Lohmann 1982; Schmiedl 1995). Others, however, have described *N. umbonifera* as a typically oligotrophic species (Loubere 1991, 1994; Gooday 1993, 1994). Possibly, the waters in which *N. umbonifera* is common are $CaCO_3$ corrosive because of the presence of abundant dissolved CO_2 derived from oxidation of organic matter, leaving little food for benthic faunas, thus combining oligotrophy and corrosiveness. Some support for the view that *N. truempyi* may have been an oligotrophic indicator may be seen in its Eocene distribution pattern: This species had its highest abundance during the middle Eocene Epoch (e.g., figure 12.3; Kaminski et al. 1989; Berggren and Miller 1989; Miller et al. 1992). It then declined in abundance and became extinct during the late Eocene Epoch, at a time when global productivity may have increased considerably (e.g., Hallock 1987; Hallock et al. 1991; Aubry 1992; Brasier 1995; Thomas and Gooday 1996).

In the Recent oceans, low-diversity faunas dominated by small, thin-walled specimens belonging to the buliminid group commonly occur in low-oxygen environments (e.g., Boltovskoy et al. 1991; Sen Gupta and Machain-Castillo 1994; Kaiho 1994a). Such environments usually occur below a well-developed oxygen minimum zone, where oxygen levels are low because of enhanced productivity and oxidation of organic matter, and where $CaCO_3$ dissolution may be severe because of the high concentrations of dissolved CO_2 resulting from this oxidation (e.g., Mullins et al. 1985; Hermelin and Shimmield 1990; Levin et al. 1991; Schmiedl 1995; Wishner et al. 1995). There is considerable discussion whether such benthic faunas reflect the low oxygen levels or the high food supplies (e.g., Corliss et al. 1986; Linke and Lutze 1993; Miao and Thunell 1993; Loubere 1994; Rathburn and Corliss 1994; Jorissen et al. 1995). Recent species of *Bulimina* are particularly common in areas with high rates of deposition of nondegraded organic matter (Corliss and Chen 1988; Caralp 1989; Corliss 1991; Mackensen et al. 1993). Additional factors of importance might be the lack of metazoan predators under low-oxygen conditions (Wishner et al. 1995), and some thin-walled taxa have been described as being opportunistic taxa rather than directly linked to high productivity or low oxygen (Alve 1994; Gooday 1994).

We thus cannot determine from the benthic faunas alone whether the buliminid-dominated postextinction faunas primarily reflect low-oxygen conditions, high-productivity conditions, stressed conditions, or high $CaCO_3$ corrosiveness. Similarly, we cannot determine whether the *N. truempyi*–dominated fauna reflects the occurrence of $CaCO_3$-corrosive waters or low productivity. We need to consider evidence from co-occurring planktonic organisms and sedimentary and isotopic evidence in order to evaluate the most probable causes for the extinction, and compare data on dissolution with those on benthic faunas (figures 12.7 and 12.8).

It appears probable that the occurrence of small and thin-walled benthic foraminifera following the benthic extinction primarily reflects increased $CaCO_3$ corrosiveness of deep waters because both dissolution and thin-walled tests occurred at almost all locations (figure 12.7), and the association of thin-walled tests with corrosive waters is well documented (Boltovskoy et al. 1991). Proxies for high productivity or lower dissolved oxygen levels are not as common as for dissolution, and thus neither can be seen as the dominant factor controlling the occurrence of the thin-walled taxa (figure 12.8). In addition, ostracodes, which are complex metazoans, became thin-walled during the LPTM at ODP Site 689 at the same time as the benthic foraminifera (protistans) did. Ostracodes have been shown to form such "degraded" tests in hydrochemical regimes unfavorable to biomineralization, i.e., in $CaCO_3$-corrosive waters (Steineck and Thomas 1996).

Increased corrosiveness thus probably played a role in the extinction. It has been hypothesized that the benthic foraminiferal extinction was largely caused by decreased levels of dissolved oxygen (Thomas 1989), and a parallel has been established with the Cretaceous anoxic oceanic events, although of lesser intensity (Kaiho 1994b). The LPTM benthic extinction occurred

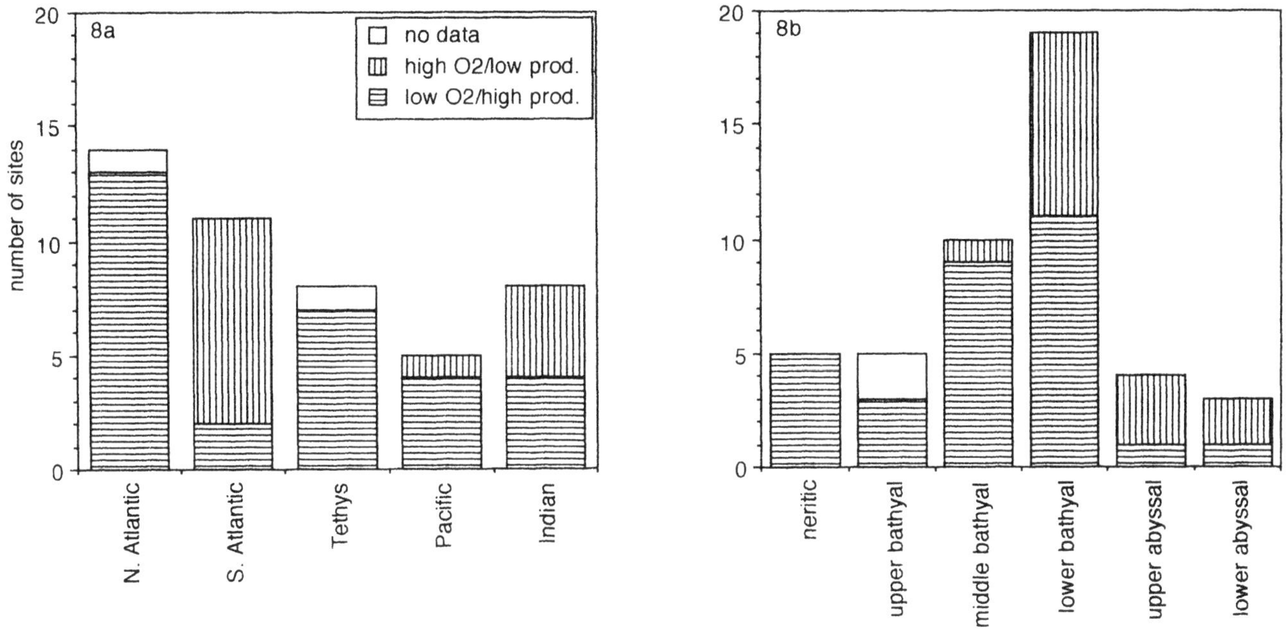

Figure 12.8 Occurrence of benthic faunas suggesting either low-oxygen conditions or increased productivity by (a) depth and (b) geographic location; sites listed in appendices 12.1 and 12.2, depth categories as in figure 12.6.

during rapid warming of the deep waters, possibly caused by a switch to dominant formation at subtropical latitudes of warm and saline deep to intermediate waters (e.g., Thomas 1989; Kennett and Stott 1991; Pak and Miller 1992, 1995). Warmer deep waters can be predicted to have a lower oxygen content because of the strong decrease in solubility of oxygen at higher temperatures (Thomas 1989, 1990a; Kaiho 1991). If the speculation about dissociation of methane hydrates is correct (Dickens et al. 1995; Kaiho et al. 1996), oxidation of these compounds would have used large amounts of oxygen and thus have contributed to the creation of low-oxygen environments. There is no agreement between the distribution of dissolution and that of low-oxygen conditions and/or high productivity with regard to depth and paleogeography (figures 12.7b and 12.8b).

The buliminid-dominated faunas, however, did not occur globally. Faunas indicating either high productivity or low oxygen contents were most common in the North Atlantic over a wide range of depths (with possible exception of abyssal sites), along Tethys, along the Antarctic continent, in the central Pacific, and at upper to middle bathyal depths in the Indian Ocean (figure 12.8). Faunas dominated by *N. truempyi* were common in the South Atlantic, and occurred in the northernmost Pacific and at middle to lower bathyal depths in the Indian Ocean. It is difficult to interpret such patterns because of the small number of localities to compare: The North Atlantic DSDP sites, for instance, are all close to the continents (Bay of Biscay, New Jersey margin), whereas the South Atlantic sites are mainly along oceanic ridges and rise (figures 12.1 and 12.6; appendix 12.1). We thus cannot determine whether the differences in distribution patterns between the North and the South Atlantic are true geographic differences between deep waters in these basins, or the differences between open ocean and margin. The lack of indications that South Atlantic waters were especially low in oxygen content, however, may suggest that warm and saline waters did not dominantly flow from subtropical regions of Tethys into the Atlantic. If Tethys functioned as a source area, the more probable region of outflow was the eastern Tethys, in agreement with carbon isotope data (Seto 1995). On the other hand, agglutinated benthic foraminiferal assemblages do support a Tethys–Atlantic connection (Kaminski et al. 1996).

It remains that faunal patterns were not globally the same, and, consequently, decreased oxygen content as a result of global deep-water warming does not fully explain the faunal changes and the benthic extinction. Thomas and Shackleton (1996) used carbon isotope data to interpret the postextinction, buliminid-dominated faunas at the Maud Rise sites as indicative of a high rate of supply of organic matter to the seafloor. This hypothesis appears to be in contradiction with the view that the Paleocene long-term carbon isotope record reflects a decrease in productivity (e.g., Shackleton et al. 1985; Shackleton 1986; Corfield and Shackleton 1988; Corfield and Cartlidge 1992; Corfield 1995), a view that is in agreement with biogeographical patterns of planktonic microfossils (e.g., Hallock 1987; Boersma et al. 1987; Hallock et al. 1991; Ottens and Nederbragt 1992; Brasier 1995). Theoretically, one would expect extremely low productivity during the LPTM, when temperature gradients were shallow, so that wind velocities over the equatorial regions were very low and upwelling and nutrient supply to the surface waters were almost null (chapter 9). In addition, benthic foraminifera have higher metabolic rates at higher temperatures and increased temperatures thus lead to oligotrophy.

Thomas and Shackleton (1996), however, argued that global productivity might have declined, but that the carbon isotope records at the Maud Rise sites suggested increased local upwelling, and thus possibly enhanced local productivity. Along other continental margins productivity might have also increased during the LPTM, possibly caused by changed patterns of upwelling as a result of changing patterns of deep-intermediate water circulation. High abundances of chiloguembelinid planktonic foraminifera in lowermost Eocene sediments in the Bay of Biscay and along the New Jersey margin have been interpreted as indicating increased local productivity and an expanded oxygen minimum zone (Saint-Marc 1991a, 1991b; Pardo et al. 1997). Productivity may also have increased locally in the eastern Tethys, where typically a "flood of radiolarians" occurs in the dissolution layer (Benjamini 1992; Berggren and Aubry 1996). At Indian Ocean Sites 738, 752, and 762 the abundance of chiloguembelinid planktonic foraminifera also increases above the extinction level (Nomura and Kennett, personal communication 1995). In other eastern Tethys sections, however, benthic faunas suggest low-oxygen conditions as well as increased productivity (Speijer et al. 1996b).

Local increased productivity during the LPTM in spite of the decreased temperature gradients could have resulted from an increased influx of nutrients from the continents. An increased abundance of kaolinite in the sediments deposited during the LPTM indicates enhanced precipitation and chemical weathering around the world (e.g., Robert and Kennett 1994; Gibson et al. 1993; Kaiho et al. 1996), in agreement with climate modeling (chapter 9). The higher rainfall could have led to increased runoff and nutrient supply to coastal waters; the increased influx of fresh water could also have increased water stratification and caused local shelf or slope anoxia (Malone 1991; van der Zwaan and Jorissen 1991).

Increased productivity cannot be seen as the cause of all occurrences of buliminid-dominated faunas: at equatorial Pacific ODP Site 865 planktonic foraminiferal faunas and coccolithophorids indicative of extreme oligotrophy occur in the same samples as buliminid-dominated benthic faunas (Kelly et al. 1996). At such sites the faunas probably reflect low-oxygen conditions in the water column. Such low-oxygen waters in the bathyal open Pacific may have resulted from the high temperatures (e.g., Thomas 1990a, Kennett and Stott 1991), their formation possibly exacerbated by sluggish circulation due to the extremely low temperature gradients. It may also be possible, however, that the buliminid-dominated faunas reflect the input of a larger proportion of organic matter on the seafloor under lower oxygen conditions, as supported by the strong increase in benthic foraminiferal accumulation rates at the level of extinction at Site 865 (figure 12.3; Herguera and Berger 1991).

The peak abundance in buliminid species at ODP Site 865 is just below a peak abundance of *Cibicidoides* species that was not observed at other sites. The *Cibicidoides* specimens resemble the Recent species *C. wuellerstorfi*, which lives on elevated supports above the sediment–water interface and is most common in regions with active bottom currents (Linke and Lutze 1993). If the earliest Eocene *Cibicidoides* had a similar habitat, its high abundance might indicate increased bottom current activity and winnowing, as also indicated by increased percentages of material in the >63-μm size fraction, low accumulation rates, and intensive reworking of planktonic foraminifera (Kelly et al. 1996).

The widespread occurrence of the *N. truempyi*–abyssaminid faunas can be tentatively explained as reflecting increased corrosiveness of bottom waters, as shown by the occurrence of widespread $CaCO_3$ dissolution (figure 12.6). These faunas, however, might also reflect an overall decline in open-ocean productivity during the LPTM. The pattern observed by Tjalsma and Lohmann (1983), that *N. truempyi*–dominated faunas migrated to shallower depths in the oceans during the late Paleocene Epoch, could be explained by the hypothesis that organic matter derived from the surface waters became exhausted at shallower and shallower levels in the oceans because of the declining productivity, as also suggested by agglutinated microfaunas (Kaminski et al. 1996). This pattern was not global, and was not observed, for instance, at Falkland Plateau Sites 698–702 (Katz and Miller 1991).

The question remains open concerning the significance of faunas containing common to abundant *Tappanina selmensis*, a species that occurs in postextinction faunas worldwide over a large depth range but was common only at neritic to upper bathyal depths before the extinction. Although it might be seen as a part of the buliminid-dominated faunas, this is not a satisfactory explanation because the species occurs in *N. truempyi*–

dominated assemblages as well. At ODP Site 689 it is most abundant together with an ostracode fauna dominated by taxa related to the present-day propontocyprids (Steineck and Thomas 1996). Propontocyprids are opportunistic forms that live dominantly in neritic environments. They are confined in the deep sea to wood-based and hydrothermal vent communities that form in transient, food-rich environments. I therefore suggest that *T. selmensis*, originally described from shallow-water regions, was likewise an opportunistic taxon rather than a low-oxygen indicator, similar to modern taxa such as *Stainforthia fusiformis* (Alve 1994).

In conclusion, I suggest that the latest Paleocene benthic foraminiferal extinction was a complex phenomenon. Factors that contributed to the extinction may have been the global increase in $CaCO_3$ corrosiveness and the global decrease in oxygenation of neritic through abyssal waters as a result of increased temperatures. The average oceanic productivity may have declined, but sluggish circulation and low oxygenation could have led to the sequestering of organic matter in the deep sea. The upwelling of such nutrient-enriched deep waters, possibly in combination with an increase in nutrient-carrying runoff along continental margins, may have caused increased local productivity, leading to the expansion of the trophic resource continuum, i.e., the occurrence of highly oligotrophic as well as highly eutrophic conditions (Hallock 1987; Hallock et al. 1991). Benthic faunas can then be seen as reflecting both extremes, with the buliminid-dominated faunas at the eutrophic end, the *N. truempyi*–dominated faunas at the oligotrophic end, with patterns complicated by the occurrence of opportunistic taxa such as *Tappanina selmensis*.

During the LPTM, benthic foraminiferal faunas at middle bathyal and greater depths exhibited high rates of extinction (30 to 50% of species), whereas at upper bathyal through neritic depths extinction was less severe and faunas showed significant but temporary changes in species composition.

After the extinction deep-sea benthic foraminiferal faunas were typical for perturbed communities (low-diversity, high-dominance). Strong $CaCO_3$ dissolution occurred worldwide, resulting in thin-walled faunas; this dissolution might have resulted from the addition of carbon to the atmosphere–ocean system through dissociation of methane hydrates.

Two widely occurring postextinction assemblages at bathyal through abyssal depths are *N. truempyi*–dominated and buliminid–dominated. In regions where both assemblages occurred, the *N. truempyi*–dominated assemblage lived deeper (lower bathyal–abyssal).

Productivity may have decreased globally during the LPTM, but this effect could have been counteracted locally by lower oxygen levels in the water column, resulting in delivery of a larger fraction of organic matter to the seafloor. Productivity may have increased along continental margins as a result of increased continental runoff and upwelling of nutrient-enriched deep waters.

Speculatively, *N. truempyi*–dominated faunas may be seen as indicators of highly oligotrophic conditions on the seafloor, and buliminid-dominated faunas as indicators of eutrophic conditions. The occurrence of both faunas after the extinction suggests an expansion of the trophic resource continuum.

ACKNOWLEDGMENTS

This chapter is a contribution to IGCP Project 308. The research was partially funded by NSF grant EAR 9405784 to Jim Zachos and Ellen Thomas. I thank Marie-Pierre Aubry, Murlene Clark, Rodolfo Coccioni, Andy Gooday, and Mimi Katz for their helpful and constructive reviews, and Robert Speijer, Birger Schmitz, and Kunio Kaiho for sharing preprints and reprints of their work.

REFERENCES

Alve, E., 1994. Opportunistic features of the foraminifer *Stainforthia fusiformis* (Williamson): Evidence from Frierfjord, Norway. *Journal of Micropalaeontology,* v. 13, p. 24.

Aubry, M.-P., 1992. Paleogene calcareous nannofossils from the Kerguelen Plateau, Leg 120. In Wise, S. W., Jr., Schlich, R., et al., Proceedings of the Ocean Drilling Program, Scientific Results, v. 120: College Station, TX (Ocean Drilling Program), p. 471–491.

Aubry, M.-P., Berggren, W. A., Stott, L. and Sinha, A., 1996. The upper Paleocene–lower Eocene stratigraphic record and the Paleocene–Eocene boundary carbon isotope excursion: Implications for geochronology. In Knox, R. W. O'B., Corfield, R. M. and Dunay, R. E., (eds.), *Correlation of the early Paleogene in Northwest Europe,* Geological Society Special Publication No 101, p. 353–380.

Barrera, E. and Huber, B. T., 1991. Paleogene and early Neogene oceanography of the southern Indian Ocean: Leg 119 foraminifer stable isotope results. In Barron, J., Larsen, B., et al., Proceedings of the Ocean Drilling Program, Scientific Results, v. 119: College Station, TX (Ocean Drilling Program), p. 693–717.

Basinger, J. R., Greenwood, D. R., and Sweda, T., 1994. Early Tertiary vegetation of Arctic Canada and its relevance to paleoclimatic interpretation. In Boulter, M. C. and Fisher, H. C., (eds.), *Cenozoic Plants and Climates of the Arctic,* NATO ASI Series, 127, Dordrecht: Springer-Verlag, p. 175–198.

Beckmann, J.-P., 1960. Distribution of benthic foraminifera at the Cretaceous–Tertiary boundary of Trinidad (West Indies). Report 21st session, International Geological Congress, Norden, Copenhagen, v. 5, p. 57–69.

Benjamini, C., 1992. The Paleocene–Eocene boundary in Israel: a candidate for the boundary stratotype. *Neues Jahrbuch für Geologie und Paläontologie Abhandlungen,* v. 186, p. 49–61.

Berggren, W. A., 1974a. Late Paleocene–Early Eocene benthonic foraminiferal biostratigraphy and paleoecology of Rockall bank. *Micropaleontology,* v. 20, p. 426–448.

Berggren, W. A., 1974b. Paleocene benthonic foraminiferal biostratigraphy, biogeography, and paleoecology of Libya and Mali. *Micropaleontology,* v. 20, p. 449–465

Berggren, W. A., 1992, Paleogene planktonic foraminifer magnetobiostratigraphy of the southern Kerguelen Plateau (Sites 747–749). In Wise, S. W., Jr., Schlich, R., et al., Proceedings of the Ocean Drilling Program, Scientific Results, v. 120: College Station, TX (Ocean Drilling Program), p. 551–568.

Berggren, W. A. and Aubert, J., 1975. Paleocene benthonic foraminiferal biostratigraphy, paleobiogeography and paleoecology of Atlantic–Tethyan regions: Midway type fauna. *Palaeogeography, Palaeoecology, Palaeoclimatology,* v. 18, p. 73–192.

Berggren, W. A. and Aubry, M.-P., 1996. A late Paleocene–early Eocene NW European and North Sea magnetobiochronological correlation network. In Knox, R. W. O'B., Corfield, R. M. and Dunay, R. E., (eds.), *Correlation of the Early Paleogene in Northwest Europe,* Geological Society Special Publication No. 101, p. 309-352.

Berggren W. A., Kent, D. V., Swisher, C. C., III and Aubry, M.-P., 1995. A revised Cenozoic geochronology and chronostratigraphy. In Berggren, W. A., Kent, D. V., Aubry, M.-P. and Hardenbol, J., (eds.), *Geochronology Time Scales and Global Stratigraphic Correlation,* Society of Economic Paleontologists and Mineralogists Special Publication No. 54, p. 129–212.

Berggren, W. A. and Miller, K. G., 1989. Cenozoic bathyal and abyssal calcareous benthic foraminiferal zonation. *Micropaleontology,* v. 35, p. 308–320.

Boersma, A., 1984a. Cretaceous–Tertiary planktonic foraminifers from the southeastern Atlantic, Walvis Ridge area, Deep Sea Drilling Project Leg 74. In Moore, T. C., Jr., Rabinowitz, P. D., et al., Proceedings of the Deep Sea Drilling Project, v. 74: Washington (U.S. Government Printing Office), p. 501–523.

Boersma, A., 1984b. Oligocene and other Tertiary benthic foraminifera from a depth traverse down Walvis Ridge, Deep Sea Drilling Project Leg 75. In Hay, W. W., Sibuet, J. C., et al., Proceedings of the Deep Sea Drilling Project, v. 75: Washington (U.S. Government Printing Office), p. 1273–1300.

Boersma, A. and Premoli Silva, I., 1991. Distribution of Paleogene planktonic foraminifera: analogies with the Recent? *Palaeogeography, Palaeoclimatology, Palaeoecology,* v. 83, p. 29–48.

Boersma, A., Premoli Silva, I. and Shackleton, N. J., 1987. Atlantic Eocene planktonic foraminiferal paleohydrographic indicators and stable isotope paleoceanography. *Paleoceanography,* v. 2, p. 287–331.

Bolli, H. M., Beckmann, J.-P. and Saunders, J. B., 1994. *Benthic foraminiferal biostratigraphy of the south Caribbean region.* Cambridge: Cambridge University Press, 408 pp.

Boltovskoy, E. and Boltovskoy, D., 1989, Paleocene–Pleistocene benthic foraminiferal evidence of major paleoceanographic events in the eastern South Atlantic (DSDP Site 525, Walvis Ridge), *Marine Micropaleontology,* v. 14, p. 283–316.

Boltovskoy, E., Scott, D. B. and Medioli, F. S., 1991. Morphological variations of benthic foraminiferal tests in response to changes in ecological parameters: A review. *Journal of Paleontology,* v. 65, p. 175–185.

Boltovskoy, E., Watanabe, S., Totah, V. I. and Vera Ocampo, J., 1992. Cenozoic benthic bathyal foraminifers of DSDP Site 548 (North Atlantic). *Micropaleontology,* v. 38, 183–207.

Boltovskoy, E., Watanabe, S., Totah, V. and Vera Ocampo, J., 1995. Benthic foraminifers from DSDP Site 516 (upper Maestrichitan–Quaternary, South Atlantic). *Revista Española de Micropaleontologia,* v. 27, p. 111–139.

Braga, G., De Biase, R., Grunig, A. and Proto Decima, F., 1975. Foraminiferi bentonici del Paleocene ed Eocene della Sezione di Possagno. In Bolli, H. M., (ed.), *Monographia Micropaleontologica sul Paleocene e l'Eocene di Possagno, Provincia di Trevisio, Italia,* Schweizerische Paläontologische Abhandlungen, v. 97, p. 85–198.

Bralower, T. J., Parrow, M., Thomas, E. and Zachos, J. C., 1995b. Data Report: Stable isotopic stratigraphy of the Paleogene pelagic cap at Site 865, Allison Guyot. In Winterer, E. L., Sager, W. W., Firth, J. V. and Sinton, J. M., (eds.), Proceedings of the Ocean Drilling Program, Scientific Results, v. 143: College Station, TX (Ocean Drilling Program), p. 581–586.

Bralower, T. J. , Zachos, J. C., Thomas, E., Parrow, M., Paull, C. K., Kelley, D. C., Premoli Silva, I., Sliter, W. V. and Lohmann, K. C., 1995. Late Paleocene to Eocene paleoceanography of the equatorial Pacific Ocean: Stable isotopes recorded at Ocean Drilling Program Site 865, Allison Guyot. *Paleoceanography,* v. 10, p. 841–865.

Brasier, M. D., 1995. *Fossil indicators of nutrient levels. 2. Evolution and extinction in relation to oligotrophy.* Geological Society Special Publication No 83, p. 133–150.

Bremer, M. L. and Lohmann, G. P., 1982. Evidence for primary control of the distribution of certain Atlantic Ocean benthonic foraminifera by degree of carbonate saturation. *Deep-Sea Research,* v. 29, p. 987–998.

Briggs, J. C., 1995. *Global Biogeography.* Amsterdam: Elsevier, 749 pp.

Broecker, W. S. and Peng, T.-H., 1984. The climate–chemistry connection. *American Geophysical Union Geophysical Monographs,* v. 29, p. 327–336.

Brotzen, F., 1948. *The Swedish Paleocene and its foraminiferal fauna.* Sveriges Geologiska Undersokelse, series C, no. 493 (ürsbok 4, 1948), 140 pp.

Cande, S. C. and Kent, D. V., 1992. A new geomagnetic polarity time scale for the Late Cretaceous and Cenozoic. *Journal of Geophysical Research,* v. 97(B10), p. 13,917–13,951.

Cande, S. C. and Kent, D. V., 1995. Revised calibration of the geomagnetic polarity time scale for the Late Cretaceous and Cenozoic. *Journal of Geophysical Research,* v. 100(B4), p. 6093–6095.

Canudo, J. I., Keller, G., Molina, E. and Ortiz, N., 1995. Planktic foraminiferal turnover and $\delta^{13}C$ isotopes across the Paleocene-Eocene transition at Caravaca and Zumaya, Spain. *Palaeogeography, Palaeoclimatology, Palaeoecology,* v. 114, p. 75–100.

Caralp, M. H., 1989. Abundance of *Bulimina exilis* and *Melonis barleeanum*: Relationship to the quality of marine organic matter. *Geo-Marine Letters,* v. 9, p. 37–43.

Cerling, T. E., 1991. Carbon dioxide in the atmosphere: Evidence from Cenozoic and Mesozoic paleosols. *American Journal of Science,* v. 291, p. 377–400.

Charisi, S. D. and Schmitz, B., 1995. Stable ($\delta^{13}C$, $\delta^{18}O$) and strontium ($^{87}Sr/^{86}Sr$) isotopes through the Paleocene at Gebel Aweina, eastern Tethys region. *Palaeogeography, Palaeoclimatology, Palaeoecology,* v. 116, p. 103–129.

Charnock, M. A. and Jones, R. W., 1990. Agglutinated foraminifera from the Palaeogene of the North Sea. In Hemleben, C., Kaminski, M. A., Kuhnt, W. and Scott, D. B., (eds.), *Paleoecology, Biostratigraphy, Paleoceanography and Taxonomy of Agglutinated Foraminifera,* NATO ASI Series, Series C: Mathematical and Physical Sciences, v. 327, Dordrecht: Kluwer Academic, p. 129–263.

Clark, M. W. and Wright, R. C., 1984. Paleogene abyssal benthic foraminifers from the Cape and Angola Basins, South Atlantic Ocean. In Hsü, K. J., LaBrecque, J. L., et al., Initial Reports of the Deep Sea Drilling Project, v. 73: Washington (U.S. Government Printing Office), p. 459–480.

Coccioni, R., Di Leo, R., Galeotti, S. and Monechi, S., 1994. Integrated biostratigraphy and benthic foraminiferal faunal turnover across the Paleocene-Eocene boundary at Trabuaka Pass section, Northern Spain. *Palaeopelagos,* v. 4, p. 87–100.

Collinson, M. E., 1983. *Fossil Plants of the London Clay.* Palaeontological Association, Field Guides to Fossils, v. 1 (Palaeontological Association, London), 121 pp.

Corfield, R. M., 1993. Depth habitats and the Palaeocene radiation of the planktonic foraminifera monitored using oxygen and carbon isotopes. In Lees, D. and Edwards, D. (eds.), *Evolutionary Patterns and Processes,* Linnean Society of London, p. 59–70.

Corfield, R. M., 1995. Palaeocene oceans and climate: an isotopic perspective. *Earth Science Review,* v. 37, p. 225–252.

Corfield, R. M., and Cartlidge, J. E., 1991. Isotopic evidence for the depth stratification of fossil and recent Globigerinina; a review. *Historical Biology,* v. 5, p. 37–63.

Corfield, R. M. and Cartlidge, J. E., 1992. Oceanographic and climatic implications of the Palaeocene carbon isotope maximum. *Terra Nova,* v. 4, p. 443–455.

Corfield, R. M. and Shackleton, N. J., 1988. Productivity change as a control on planktonic foraminiferal evolution after the Cretaceous/Tertiary boundary. *Historical Biology,* v. 1, p. 323–343.

Corfield, R. M., Cartlidge, J. E., Premoli Silva I. and Housley, R. A., 1991. Oxygen and carbon isotope stratigraphy of the Palaeogene and Cretaceous limestones in the Bottaccione Gorge and the Contessa Highway sections, Umbria, Italy. *Terra Nova,* v. 3, p. 414–422.

Corliss, B. H., 1991. Morphology and microhabitat preferences of benthic foraminifera from the northwest Atlantic Ocean. *Marine Micropaleontology,* v. 17, p. 195–236.

Corliss, B. H. and Chen, C., 1988. Morphotype patterns of Norwegian Sea deep-sea benthic foraminifera and ecological implications. *Geology,* v. 16, p. 716–719.

Corliss, B. H., Martinson, D. G. and Keffer, T., 1986. Late Quaternary deep-ocean circulation. *Geological Society of America Bulletin,* v. 97, p. 1106–1121.

Cushman, J. A., 1946. *Upper Cretaceous foraminifera of the Gulf Coastal Region of the United States and adjacent areas.* U.S. Geological Survey Professional Paper 206, 241 pp.

Dailey, D. H., 1983. Late Cretaceous and Paleocene benthic foraminifers from Deep Sea Drilling Project Site 516, Rio Grande Rise, western South Atlantic Ocean. In Barker, P. F., Carlson, R. L., Johnson, D. A., et al., Initial Reports of the Deep Sea Drilling Project, v. 72: Washington (U.S. Government Printing Office), p. 757–782.

Dickens, G. R., O'Neil, J. R., Rea, D. K., Owen, R. M., 1995. Dissociation of oceanic methane hydrate as a cause of the carbon isotope excursion at the end of the Paleocene. *Paleoceanography,* v. 10, p. 965–971.

Di Napoli Alliata, E., Proto-Decima, F. and Pelegrini, G. B., 1970. Studio Geologico, Stratigrafico, e Micropaleontologico dei Dintorni di Belluno. *Memorie Societa Geologica Italiana,* v. 9, p. 1–28

Estes, R. and Hutchison, J. H., 1980. Eocene lower vertebrates from Ellesmere Island, Canadian Arctic Archipelago. *Palaeogeography, Palaeoclimatology, Palaeoecology,* v. 30, p. 325–347.

Freeman, K. H. and Hayes, J. M., 1992. Fractionation of carbon isotopes by phytoplankton and estimates of ancient CO_2 levels. *Global Biogeochemical Cycles,* v. 6, p. 185–198.

Geroch, S. and Novak, W., 1984. Proposal of zonation for the late Tithonian–late Eocene, based upon arenaceous foraminifera from the outer Carpathians, Poland. BENTHOS `83, Second International Symposium on Benthic Foraminifera, (Pau, France, April 1983), p. 225–239.

Gibson, T. G. and Bybell, L. M., 1994. Sedimentary patterns across the Paleocene-Eocene boundary in the Atlantic and Gulf coastal plains of the United States. *Bulletin de la Société Belge de Géologie,* v. 103, p. 237 265.

Gibson, T. G., Bybell, L. M. and Owens, J. P., 1993. Latest Paleocene lithologic and biotic events in neritic deposits of southwestern New Jersey. *Paleoceanography,* v. 8, p. 495–514.

Goericke, R. and Fry, B., 1994. Variations of marine plankton $\delta^{13}C$ with latitude, temperature, and dissolved CO_2 in the world ocean. *Global Biogeochemical Cycles,* 8, 85–90

Gooday, A. J., 1993. Deep-sea benthic foraminiferal species which exploit phytodetritus: Characteristic features and controls on distribution. *Marine Micropaleontology,* v. 22, p. 187–205.

Gooday, A. J., 1994. The biology of deep-sea foraminifera: A review of some advances and their applications in paleoceanography. *Palaios,* v. 9, p. 14–31.

Gradstein, F. M., Kaminski, M. A., Berggren, W. A., Kristiansen, I. L. and D'Ioro, M. A., 1994. Cenozoic biostratigraphy of the North Sea and Labrador Shelf. *Micropaleontology,* v. 40 (Suppl.), 152 pp.

Hallock, P., 1987. Fluctuations in the trophic resource continuum: a factor in global diversity cycles? *Paleooceanography,* v. 2, p. 457–471.

Hallock, P., Premoli Silva, I. and Boersma, A. 1991. Similarities between planktonic and larger foraminiferal evolutionary trends through Paleogene paleoceanographic changes. *Palaeogeography, Palaeoclimatology, Palaeoecology,* v. 83, p. 49–64.

Haq, B. U., Hardenbol, J. and Vail, P. R., 1987. Chronology of fluctuating sea levels since the Triassic. *Science,* v. 235, p. 1156–1167.

Herguera, J. C. and Berger, W. H., 1991. Paleoproductivity from benthic foraminiferal abundance: Glacial to postglacial changes in the west-equatorial Pacific. *Geology,* v. 19, p. 1173–1176.

Hermelin, J. O. R. and Shimmield, G. B., 1990. The importance of the oxygen minimum zone and sediment geochemistry in the distribution of Recent benthic foraminifera in the Northwest Indian Ocean. *Marine Geology,* v. 91, p. 1–29.

Hinga, K. R., Arthur, M. A., Pilson, M. E. Q. and Whitaker, D., 1994. Carbon isotope fractionation by marine phytoplankton in culture: the effects of CO2 concentration, pH, temperature, and species. *Global Biogeochemical Cycles,* v. 8, p. 91–102.

Von Hillebrandt, A., 1962. Das Palaeozaen und seine Foraminiferenfauna im Becken von Reichenhall und Salzburg. Bayerische Akademie der Wissenschaften, Mathematisch-Naturwissenschaftliche Klasse, Abhandlungen, Neue Folge, v. 108, 182 pp.

Von Hillebrandt, A., 1965. Foraminiferen Stratigraphie im Alttertiaer von Zumaya (Provinz Guipozcoa, NW-Spanien) und ein Vergleich mit anderen Tethys-Gebieten. Bayerische Akademie der Wissenschaften, Mathematisch-Naturwissenschaftliche Klasse, Abhandlungen, Neue Folge, v. 123, 62 pp.

Hooker, J. J., 1991. The sequence of mammals in the Thanetian and Ypresian of the London and Belgian Basins. Location of the Palaeocene–Eocene boundary. *Newsletters on Stratigraphy,* v. 25, p. 75–90.

Hornibrook, N. De B., Brazier, R. C. and Strong, C. P., 1989. Manual of New Zealand Permian to Pleistocene foraminiferal biostratigraphy. *New Zealand Geological Survey Paleontological Bulletin* No 56, 198 pp.

Hovan, S. A. and Rea, D. K., 1992. Paleocene/Eocene boundary changes in atmospheric and oceanic circulation: A southern hemisphere record. *Geology,* v. 20, p. 15–18.

Hsü, K. J., McKenzie, J. A., Oberhänsli, H., Weissert, H. and Wright, R. C., 1984. South Atlantic Cenozoic Paleoceanography. In Hsü, K. J., LaBrecque, J. L., et al., Initial Reports of the Deep Sea Drilling Project, v. 73: Washington (U.S. Government Office), p. 771–785.

Hulsbos, R. E., 1987. Eocene benthic foraminifers from the upper continental rise off New Jersey, Deep Sea Drilling Project Site 605. In van Hinte, J. E., Wise, S. W., Jr., et al., Initial Reports of the Deep Sea Drilling Project, v. 93: Washington (U.S. Government Office), p. 525–538.

Hulsbos, R. E., Kroon, D., Jansen, H. S. M. and van Hinte, J. E., 1989. Lower Eocene benthic foraminifera and paleoenvironment of the outer Vøring Plateau, Norwegian Sea (DSDP Site 338). *Micropaleontology,* v. 35, p. 256–273.

Jorissen, F. J., de Stigter, H. C. and Widmark, J. G. V., 1995. A conceptual model explaining benthic foraminiferal microhabitats. *Marine Micropaleontology,* v. 26, p. 3–15.

Kaiho, K., 1988. Uppermost Cretaceous to Paleogene bathyal benthic foraminiferal biostratigraphy of Japan and New Zealand: Latest Paleocene–middle Eocene benthic foraminiferal species turnover. *Revue de Paléobiologie,* Special Volume 2, p. 553–559.

Kaiho, K., 1991. Global changes of Paleogene aerobic/anaerobic benthic foraminifera and deep-sea circulation. *Palaeogeography, Palaeoclimatology, Palaeoecology,* v. 83, p. 65–86.

Kaiho, K., 1994a. Benthic foraminiferal dissolved-oxygen index and dissolved-oxygen levels in the modern ocean. *Geology,* v. 22, p. 719–722.

Kaiho, K., 1994b. Planktonic and benthic foraminiferal extinction events during the last 100 m.y. *Palaeogeography, Palaeoclimatology, Palaeoecology,* v. 111, p. 45–71.

Kaiho, K., in press. Dissolved oxygen changes in the deep-sea during the last 100 million years based on benthic foraminifera. *Palaeogeography, Palaeoclimatology, Palaeoecology.*

Kaiho, K., Morgans, H. E. G. and Okada, H., 1993. Faunal turnover of intermediate-water benthic foraminifera during the Paleogene in New Zealand. *Marine Micropaleontology,* v. 21, p. 51–86.

Kaiho, K., Arinobu, T., Ishiwara, R., Morgans, H. E. G., Okada, H., Takeda, N., Zhou, G., Kajiwara, Y., Matsumoto, R., Hirai, A., Niitsuma, N. and Wada, H., 1996. Latest Paleocene benthic foraminiferal extinction and environmental changes at Tawanui, New Zealand. *Paleoceanography,* v. 11, p. 447–465.

Kaminski, M. A., Gradstein, F. M. and Berggren, W. A., 1989. Paleogene benthic foraminifer biostratigraphy and paleontology at Site 647, Southern Labrador Sea. In Srivasta, S. P., Arthur, M., Clement, B., et al., Proceedings of the Ocean Drilling Program, Scientific Results, v. 105: College Station, TX (Ocean Drilling Program), p. 705–730.

Kaminski, M. A., Kuhnt, W. and Radley, J. D., 1996. Palaeocene–Eocene deep water agglutinated foraminifera from the Numidian flysch (Rif, northern Morocco): their significance for the palaeoceanography of the Gibraltar gateway. *Journal of Micropaleontology,* v. 15, p. 1–19.

Katz, M. E. and Miller, K. G., 1991. Early Paleogene benthic foraminiferal assemblages and stable isotopes in the Southern Ocean. In Cieselski, P. F., Kristoffersen, Y., et al., Proceedings of the Ocean Drilling Program, Scientific Results, v. 114: College Station, TX (Ocean Drilling Program), p. 481–512.

Kelly, D. C., Bralower, T. J., Zachos, J. C., Premoli Silva, I. and Thomas, E., 1996, Rapid diversification of planktonic foraminifera in the tropical Pacific (ODP Site 865) during the late Paleocene Thermal Maximum. *Geology,* v. 24, p. 423–426.

Kemp, E. M., 1978. Tertiary climatic evolution and vegetation history in the Southeast Indian Ocean region. *Palaeogeography, Palaeoclimatology, Palaeoecology,* v. 24, p. 169–208.

Kennett, J. P. and L. D. Stott, 1990. Proteus and Proto-Oceanus: Ancestral Paleogene oceans as revealed from Antarctic stable isotope results. In Barker, P. F., Kennett, J. P., et al., Proceedings of the Ocean Drilling Program, Scientific Results, v. 113: College Station, TX (Ocean Drilling Program), p. 865–880.

Kennett, J. P. and Stott, L. D., 1991. Abrupt deep-sea warming, palaeoceanographic changes and benthic extinctions at the end of the Palaeocene. *Nature,* v. 353, p. 225–229.

King, C., 1989. Cenozoic of the North Sea. In D. G., Jenkins and J. W. Murray, (eds.), *Stratigraphical Atlas of Fossil Foraminifera,* (2nd ed.), British Micropaleontological Society Series, Chichester: Ellis Horwood Limited, p. 418–489.

Koch, P. L., Zachos, J. C. and Gingerich, P. D., 1992. Correlation between isotope records in marine and continental carbon reservoirs near the Palaeocene/Eocene boundary. *Nature,* v. 358, p. 319–322.

Koch, P. L., Zachos, J. C. and Dettman, D. L., 1995. Stable isotope stratigraphy and paleoclimatology of the Paleogene Bighorn Basin (Wyoming, USA). *Palaeogeography, Palaeoclimatology and Palaeoecology,* v. 115, p. 61–89.

Kuhnt, W. and Kaminski, M. A., 1990. Paleoecology of Late Cretaceous to Paleocene deep-water agglutinated foraminifera from the North Atlantic and western Tethys. In Hemleben, C., Kaminski, M. A., Kuhnt, W. and Scott, D. B., (eds.), *Paleoecology, Biostratigraphy, Paleoceanography and Taxonomy of Agglutinated Foraminifera,* NATO ASI Series, Series C: Mathematical and Physical Sciences, v. 327, Dordrecht: Kluwer Academic Publishers, p. 433–505.

Le Calvez, Y., 1970. Contribution à l'étude des Foraminifères paléogènes du Bassin de Paris. Cahiers de Paléontologie, 92 pp.

Levin, L. A., Huggett, C. L. and Wishner, K. F., 1991. Control of deep-sea benthic community structure by oxygen and organic-matter gradients in the eastern Pacific Ocean. *Journal of Marine Research,* v. 49, p. 763–800.

Linke, P. and Lutze, G. F., 1993. Microhabitat preferences of benthic foraminifera—a static concept or a dynamic adaptation to optimize food acquisition? *Marine Micropaleontology,* v. 20, p. 215–234.

Loubere, P., 1991. Deep-sea benthic foraminiferal assemblage response to a surface ocean productivity gradient: A test. *Paleoceanography,* v. 6, p. 193–204.

Loubere, P., 1994. Quantitative estimation of surface ocean productivity and bottom water oxygen concentration using benthic foraminifera. *Paleoceanography,* v. 9, p. 723–737.

Lu, G. and Keller, G., 1993. The Paleocene–Eocene transition in the Antarctic Indian Ocean: Inference from planktic foraminifera. *Marine Micropaleontology,* v. 21, p. 101–142.

Lu, G. and Keller, G., 1995a. Planktic foraminiferal faunal turnovers in the subtropical Pacific during the late Paleocene to early Eocene. *Journal of Foraminiferal Research,* v. 25, p. 97–116.

Lu, G. and Keller, G., 1995b. Ecological stasis and saltation: Species richness change in planktic foraminifera during the late Paleocene to early Eocene, DSDP Site 577. *Palaeogeography, Palaeoclimatology, Palaeoecology,* v. 117, p. 211–227.

Maas, M. C., Anthony, M. R. L., Gingerich, P. D., Gunnell, G. F. and Krause, D. W., 1995. Mammalian generic diversity and turnover in the late Paleocene and early Eocene of the Bighorn and Crazy Mountains Basins, Wyoming and Montana (USA). *Palaeogeography, Palaeoclimatology, Palaeoecology,* v. 115, p. 181–207.

Mackensen, A. and Berggren, W. A., 1992. Paleogene benthic foraminifers from the Southern Indian Ocean (Kerguelen Plateau): Biostratigraphy and paleoecology. In Wise, S. W., Jr., Schlich, R. et al., Proceedings of the Ocean Drilling Program, Scientific Results, v. 120: College Station, TX (Ocean Drilling Program), p. 603–630.

Mackensen, A., Fütterer, D. K., Grobe, H. and Schmiedl, G., 1993. Benthic foraminiferal assemblages from the eastern South Atlantic Polar Front region between 35° and 75° S: Distribution, ecology and fossilization potential. *Marine Micropaleontology,* v. 22, p. 33–69.

Malone, T. C., 1991. *River flow, phytoplankton production and oxygen depletion in Chesapeake Bay.* Geological Society Special Publication No. 58, p. 83–93.

Manum, S. B., 1994. The Palaeogene flora of Spitsbergen. In Boulter, M. C. and Fisher, H. C., (eds.), *Cenozoic Plants and Climates of the Arctic,* NATO-ASI Series, 127, Dordrecht: Springer-Verlag, p. 215–222.

Markwick, P. J., 1994. "Equability," continentality, and Tertiary "climate": the crocodilian perspective. *Geology,* v. 22, p. 613–616.

McGowran, B., 1974. Foraminifera. In von der Borch, C. C., Sclater, J. G., et al., Initial Reports of the Deep Sea Drilling Project, v. 22: Washington (U.S. Government Printing Office), p. 609–627.

McGowran, B., 1991. Evolution and environment in the early Paleogene. *Memoir of the Geological Society of India,* v. 20, p. 21–53.

McKenna, M. C., 1980. Eocene paleolatitude, climate, and mammals of Ellesmere Island. *Palaeogeography, Palaeoclimatology, Palaeoecology,* v. 30, p. 349–362.

Miao, Q. and Thunell, R. C., 1993. Recent deep-sea benthic foraminiferal distributions in the South China and Sulu Seas. *Marine Micropaleontology,* v. 22, p. 1–32.

Miller, K. G., Katz, M. E. and Berggren, W. A., 1992. Cenozoic deep-sea benthic foraminifera: A tale of three turnovers. *Studies in Benthic Foraminifera,* BENTHOS '90, Tokai University Press, Sendai, p. 67–75

Miller, K. G., Fairbanks, R. G. and Mountain, G. S., 1987a. Tertiary oxygen isotope synthesis, sea level history, and continental margin erosion. *Paleoceanography,* v. 2, p. 1–19.

Miller, K. G., Janecek, T. R., Katz, M. E. and Keil, D. J., 1987b. Abyssal circulation and benthic foraminiferal changes near the Paleocene/Eocene boundary. *Paleoceanography,* v. 2, p. 741–761.

Molina, E., Canudo, J. J., Martínez-Ruiz, F. and Ortiz, N., 1994. Integrated stratigraphy across the Paleocene/Eocene boundary at Caravaca, southern Spain. *Eclogae geologica Helvetiae,* v. 87, p. 47–61.

Molina, E., Canudo, J. J., Guernet, C., McDougall, K., Ortiz, N., Pascual, J. O., Pares, J. O., Samso, J. M., Serra-Kiel, J. and Tosquella, J., 1992. The stratotypic Ilerdian revisited: Integrated stratigraphy across the Paleocene/Eocene boundary. *Revue de Micropaléontologie,* v. 35, p. 143–156.

Müller-Merz, E. and Oberhänsli, H., 1991. Eocene bathyal and abyssal benthic foraminifera from a South Atlantic transect at 20°–30° S. *Palaeogeography, Palaeoclimatology, Palaeoecology,* v. 83, p. 117–171.

Mullins, H. T., Thompson. J. B., McDougall, K. and Vercoutere, T. L., 1985. Oxygen-minimum zone edge effects: Evidence from the central California coastal upwelling system. *Geology,* v. 13, p. 491–494.

Murray, J. W., 1989. Paleogene and Neogene benthic foraminifers from Rockall Plateau. In Roberts, D. G., Schnitker, D., et al., Initial Reports of the Deep Sea Drilling Project, v. 81: Washington (U.S. Government Printing Office), p. 503–534.

Nadin, P. A. and Kusznir, N. J., 1995. Palaeocene uplift and Eocene subsidence in the northern North Sea basin from 2D forward and reverse stratigraphic modeling. *Journal of the Geological Society, London,* v. 152, p. 833–848.

Nilsen, T. H. and Kerr, D. R., 1978. Paleoclimatic and paleogeographic implications of a lower Tertiary laterite (latosol) on the Iceland-Faeroe Ridge, North Atlantic region. *Geological Magazine,* v. 115, p. 153–236.

Nomura, R., 1991. Paleoceanography of Upper Maestrichtian to Eocene benthic foraminiferal assemblages at Sites 752, 753 and 754, eastern Indian Ocean. In Weissel, J., Pierce, J., Taylor, E., Alt, J., et al., Proceedings of the Ocean Drilling Program, Scientific Results, v. 121: College Station, TX (Ocean Drilling Program), p. 3–29.

O'Connell, S. B., 1990. Variations in Upper Cretaceous and Cenozoic calcium carbonate percentages, Maud Rise, Weddell Sea, Antarctica. In Barker, P. F., Kennett, J. P., et al., Proceedings of the Ocean Drilling Program, Scientific Results, v. 113: College Station, TX (Ocean Drilling Program), p. 971–984.

Olsson, R. K. and Wise, S. W., Jr., 1987. *Upper Paleocene to middle Eocene depositional sequences and hiatuses in the New Jersey Atlantic margin.* Cushman Foundation for Foraminiferal Research, Special Publication 24, p. 99–112.

Orue Extebarria, X., Appelliniz, E., Baceta, J. I., Coccioni, R., Monechi, S., Nunez-Betelli, K., Pares, J. M., Jaume, Payros A., Pujalte, V., Sanso, J. and Serra-Kiel, J., in press. Prospective Paleocene–Eocene boundary stratotype sections in the deep water Basque basins, western Pyrenees, Spain: biostratigraphy and magnetostratigraphy of the Trabakua pass and Ermua sections. *Micropaleontology.*

Ortiz, N., 1995. Differential patterns of benthic foraminiferal extinctions near the Paleocene/Eocene boundary in the North Atlantic and the western Tethys. *Marine Micropaleontology,* v. 26, p. 341–359.

Ortiz, N. and McDougall, K., 1991. Biostratigraphy and paleoecology of the smaller benthic foraminifers from the European Paleogene marine Ilerdian Stage (NE Spain). *Geological Society of America, Abstracts with Programs,* v. 23, p. A37.

Ottens, J. J. and Nederbragt, A. J., 1992. Planktic foraminiferal diversity as indicator of ocean environments. *Marine Micropaleontology,* v. 19, p. 13–28.

Pak, D. K. and Miller, K. G., 1992. Paleocene to Eocene benthic foraminiferal isotopes and assemblages: Implications for deepwater circulation. *Paleoceanography,* v. 7, p. 405–422.

Pak, D. K. and Miller, K. G., 1995. Isotopic and faunal record of Paleogene deep-water transitions in the North Pacific. In Rea, D. K., Basov, I. A., Scholl, D. W. and Allan, J. F., (eds.), Proceedings of the Ocean Drilling Program, Scientific Results, v. 145: College Station, TX (Ocean Drilling Program), p. 265–281.

Pardo, A., Keller, G., Molina, E. and Canudo, J. I., 1997. Planktic foraminiferal turnover across the Paleocene–Eocene transition at DSDP Site 401, Bay of Biscay, North Atlantic. *Marine Micropaleontology,* v. 29, p. 129–158.

Parker, W. C., Clark, M. W., Wright, R. C. and Clark, R. K., 1984. Population dynamics, Paleogene abyssal benthic foraminifers, eastern South Atlantic. In Hsü, K. J., LaBrecque, J. L., et al., Initial Reports of the Deep Sea Drilling Project, v. 73: Washington (U.S. Government Printing Office), p. 481–486.

Pearson, T. H. and Rosenberg, R., 1978. Macrobenthic succession in relation to organic enrichment and pollution of the marine environment. *Oceanography and Marine Biology Annual Review,* v. 16, p. 229–311.

Premoli Silva, I. and Boersma, 1984. Atlantic Eocene planktonic foraminiferal historical biogeography and paleohydrographic indices. *Palaeogeography, Palaeoclimatology, Palaeoecology,* v. 67, p. 315–356.

Pujalte, V., Baceta, J.-I., Dinarès-Turell, J., Orue-Extebarria, X., Parès, J.-M. and Payros, A., 1995. Biostratigraphic and magnetostratigraphic intercalibration of latest Cretaceous and Paleocene depositional sequences from the deep-water Basque Basin, western Pyrenees, Spain. *Earth and Planetary Science Letters,* v. 136, p. 17–30.

Rathburn, A. E. and Corliss, B. H., 1994. The ecology of living (stained) deep-sea benthic foraminifera from the Sulu Sea. *Paleoceanography,* v. 9, p. 87–150.

Rea, D. K., 1994. The paleoclimatic record provided by eolian deposition in the deep sea: the geologic history of wind. *Reviews of Geophysics,* 32, p. 159–195.

Rea, D. K., Zachos, J. C., Owen, R. M. and Gingerich, P. D., 1990. Global change at the Paleocene–Eocene boundary: climatic and evolutionary consequences of tectonic events. *Palaeogeography, Palaeoclimatology, Palaeoecology,* v. 79, p. 117–128.

Reynolds, L. E., 1992. *Faunal change in benthic foraminifera at the Paleocene/Eocene boundary, DSDP Site 549.* Master of Science Thesis, Orono: University of Maine, 130 pp.

Robert, C. and Kennett, J. P., 1992. Paleocene and Eocene kaolinite distribution in the South Atlantic and Southern Ocean: Antarctic climatic and paleoceanographc implications. *Marine Geology,* v. 103, p. 99–110.

Robert, C. and Kennett, J. P., 1994. Antarctic subtropical humid episode at the Paleocene–Eocene boundary, Clay mineral evidence. *Geology,* v. 22, p. 211–214.

Romein, A. J. T., 1979. Lineages in early Paleogene calcareous nannoplankton. *Utrecht Micropaleontological Bulletin,* v. 22, 231 pp.

Saint-Marc, P., 1987. Biostratigraphic and paleoenvironmental study of Paleocene benthic and planktonic foraminifers, Site 605, Deep Sea Drilling Project Leg 93. In van Hinté, J. E., Wise, S. W., Jr., et al., Initial Reports of the Deep Sea Drilling Project, v. 93: Washington (U.S. Government Printing Office), p. 539–547.

Saint-Marc, P., 1991a. Le Paléocène et le passage Paléocène–Eocène sur la bordure septentrionale du Golfe de Biscaye (Atlantique Nord, Site DSDP 401, Leg 48). Biostratigraphie et paléocéanographie. *Bulletin de la Société Géologique de France,* v. 162, p. 1125–1132.

Saint Marc, 1991b. Données stratigraphiques et paléocéanographiques sur les dépôts du Paléocène supérieur de la partie occidentale de l'Atlantique Nord (Fosse de Baltimore, Site DSDP 605). *Palaeogeography, Palaeoclimatology, Palaeoecology,* v. 85, p. 335–350.

Sanders, H. L., 1968. Marine benthic diversity: A comparative study. *American Naturalist,* v. 102, p. 243–282.

Schmiedl, G., 1995. Late Quaternary benthic foraminiferal assemblages from the eastern South Atlantic: Reconstruction of deep water circulation and productivity changes. *Reports on Polar Research, Alfred Wegener Institute for Polar and Marine Research,* v. 160, 207 pp.

Schmitz, B., Speijer, R. P. and Aubry, M.-P., 1996. Latest Paleocene benthic extinction event on the southern Tethyan shelf (Egypt): Foraminiferal stable isotopic ($\delta^{13}C$, $\delta^{18}O$) records. *Geology,* v. 24, p. 347–350.

Schnitker, D., 1979. Cenozoic deep water benthic foraminifers, Bay of Biscay. In Montadert, L., Roberts, D. G., et al., Initial Reports of the Deep Sea Drilling Project, v. 48: Washington (U.S. Government Printing Office), p. 377–413.

Schnitker, D., 1994. Deep-sea benthic foraminifers: Food and bottom water masses. In Zahn, R., Pedersen, T. F., Kaminski, M. A. and Labeyrie, L., (eds.), *Carbon Cycling in the Glacial Ocean: Constraints on the Ocean's Role in Global Change,* New York: Springer Verlag, p. 539–554.

Schröder, T., 1992. A palynological zonation for the Paleocene of the North Sea Basin. *Journal of Micropalaeontology,* v. 11, p. 113–126.

Sen Gupta, B. K. and Machain-Castillo, M. L., 1994. Benthic foraminifera in oxygen-poor habitats. *Marine Micropaleontology,* v. 20, p. 183–201.

Seto, 1995. Carbon and oxygen isotopic paleoceanography of the Indian and South Atlantic Oceans: paleoclimate and paleo-ocean circulation. *Journal of Science of Hiroshima University,* Ser. C, v. 10, p. 393–485.

Seto, K., Nomura, R. and Niitsuma, N., 1991. Data report: Oxygen and carbon isotope records of the Upper Maestrichtian to lower Eocene benthic foraminifers at Site 752 in the Eastern Indian Ocean. In Weissel, J., Pierce, J., Taylor, E., Alt, J., et al., Proceedings of the Ocean Drilling Program, Scientific Results, v. 121: College Station, TX (Ocean Drilling Program), p. 885–889.

Shackleton, N. J., 1986. Paleogene stable isotope events. *Palaeogeography, Palaeoclimatology Palaeoecology,* v. 57, p. 91–102.

Shackleton, N. J., 1987. The carbon isotope record of the Cenozoic: History of organic carbon burial and of oxygen in the ocean and atmosphere. In Brooks, J. and Fleet, A. J., (eds.), *Marine Petroleum Source Rocks,* Geological Society Special Publication No 26, p. 423–434.

Shackleton, N. J. and Boersma, A., 1981. The climate of the Eocene ocean. *Journal of the Geological Society of London,* v. 138, p. 153–157.

Shackleton, N. J. and Hall, M. A., 1984. Carbon isotope stratigraphy of bulk sediments, ODP Sites 689 and 690. In Barker, P. F., Kennett, J. P., et al., Proceedings of the Ocean Drilling Program, Scientific Results, v. 113: College Station, TX (Ocean Drilling Program), p. 985–989.

Shackleton, N. J. and Members of the Shipboard Scientific Party, 1984. Accumulation rates in leg 74 sediments. In Moore, T. C., Jr., Rabinowitz, P. D., et al., Initial Reports of the Deep Sea Drilling Project, v. 74: Washington (U.S. Government Printing Office), p. 621–637.

Shackleton, N. J., Corfield, R. M. and Hall, M. A., 1985. Stable isotope data and the ontogeny of Paleocene planktonic foraminifera. *Journal of Foraminiferal Research,* v. 15, p. 321–336.

Schlich, R., et al., 1974. Site 245. In Simpson, E. S. W., Schlich, R., et al., Initial Reports of the Deep Sea Drilling Project, v. 25: Washington (U.S. Government Printing Office), p. 187–236.

Sloan, L. C. and Rea, D. K., 1995. Atmospheric carbon dioxide and early Eocene climate: A general circulation modeling sensitivity study. *Palaeogeography, Palaeoclimatology, Palaeoecology,* 119, p. 275–292.

Sloan, L. C., Walker, J. C. G. and Moore, T. C. Jr., 1995. Possible role of oceanic heat transport in early Eocene climate. *Paleoceanography,* v. 10, p. 347–356.

Speijer, R. P., 1994. Extinction and recovery patterns in benthic foraminiferal paleocommunities across the Cretaceous/Paleogene and Paleocene/Eocene boundaries. *Geologica Ultraiectina, Mededelingen van de Faculteit Aardwetenschappen Universiteit Utrecht,* v. 124, 191 pp.

Speijer, R. P., van der Zwaan, G. J. and Schmitz, B., 1996a. The impact of Paleocene/Eocene boundary events on middle neritic benthic foraminiferal assemblages from Egypt. *Marine Micropaleontology,* v. 28, p. 99–132.

Speijer, R. P., Schmitz, B., Aubry, M.-P. and Charisi, S. D., 1996b. The latest Paleocene benthic extinction event: Punctuated turnover in outer neritic foraminiferal faunas from Gebel Aweina, Egypt. In Aubry, M.-P. and Benjamini, C., (eds.), Paleocene/Eocene Boundary Events in Space and Time, *Israel Journal of Earth Sciences,* v. 44, p. 207–222.

Steineck, P. L. and Thomas, E., 1996. The latest Paleocene crisis in the deep sea: Ostracode succession at Maud Rise, Southern Ocean. *Geology,* v. 24, p. 583–586.

Stott, L. D., 1992. Higher temperatures and lower oceanic pCO_2: A climate enigma at the end of the Paleocene Epoch. *Paleoceanography,* v. 7, p. 395–404.

Stott, L. D. and Kennett, J. P., 1990. Antarctic Paleogene planktonic foraminifer biostratigraphy: ODP Leg 113, Sites 689 and 690. In Barker, P. F., Kennett, J. P., et al., Proceedings of the Ocean Drilling Program, Scientific Results, v. 113: College Station, TX (Ocean Drilling Program), p. 549–569.

Stott, L. D., Sinha, A., Thiry, M., Aubry, M.-P. and Berggren, W. A., 1996. Global $\delta^{13}C$ changes across the Paleocene–Eocene boundary: Criteria for terrestrial–marine correlations. In Knox, R. W. O'B., Corfield, R. M. and Dunay, R. E., (eds.), *Correlation of the Early Paleogene in Northwest Europe,* Geological Society Special Publication No. 101, p. 381–399.

Stott, L. D., Kennett, J. P., Shackleton, N. J. and Corfield, R. C., 1990. The evolution of Antarctic surface waters during the Paleogene: Inferences from the stable isotopic composition of planktonic foraminifers, ODP Leg 113. In Barker, P. F., Kennett, J. P., et al., Proceedings of the Ocean Drilling Program, Scientific Results, v. 113, p. 849–863.

Thomas, E., 1989. Development of Cenozoic deep-sea benthic foraminiferal faunas in Antarctic waters. in Crame, J. A., (ed.), *Origins and Evolution of the Antarctic Biota,* Geological Society Special Publication No 47, p. 283–296.

Thomas, E. 1990a. Late Cretaceous through Neogene deep-sea benthic foraminifers (Maud Rise, Weddell Sea, Antarctica). In Barker, P. F., Kennett, J. P., et al., Proceedings of the Ocean Drilling Program, Scientific Results., v. 113: College Station, TX (Ocean Drilling Program), p. 571–594.

Thomas, E., 1990b. Late Cretaceous–early Eocene mass extinctions in the deep sea. In Sharpton, V. L. and Ward, P., (eds.), *Global Catastrophes in Earth History: An interdisciplinary Conference on Impacts, Volcanism, and Mass Mortality,* Geological Society of America Special Paper No. 247, p. 481–495.

Thomas, E., 1992. Cenozoic deep sea circulation: Evidence from deep sea benthic foraminifera. In Kennett, J. P. and Warnke, D., (eds.), *The Antarctic Paleoenvironment: A Perspective on Global Change,* American Geophysical Union Antarctic Research Series No. 56, p. 141–165.

Thomas, E., and Gooday, A. J., 1996. Deep-sea benthic foraminifera: tracers for changes in Cenozoic oceanic productivity. *Geology,* p. 355–358.

Thomas, E. and Shackleton, N. J., 1996. The Paleocene–Eocene benthic foraminiferal extinction and stable isotope anomalies. In Knox, R. W. O'B., Corfield, R. M. and Dunay, R. E., (eds.), *Correlation of the Early Paleogene in Northwest Europe,* Geological Society Special Publication No 101, p. 401–441.

Thomas, E., Barrera, E., Hamilton, N., Huber, B. T., Kennett, J. P., O'-Connell, S. B., Pospichal, J. J., Spiess, V., Stott, L. D., Wei, W. and Wise, S. W., Jr., 1990. Upper Cretaceous–Paleogene stratigraphy of Sites 689 and 690, Maud Rise (Antarctica). In Barker, P. F., Kennett, J. P., et al., Proceedings of the Ocean Drilling Program, Scientific Results, v. 113: College Station, TX (Ocean Drilling Program), p. 901–914.

Thomas, E., Shackleton, N. J. and Hall, M. A., 1992. Data Report: Carbon isotope stratigraphy of Paleogene bulk sediments, Hole 762C (Exmouth Plateau, eastern Indian Ocean). In von Rad, U., Haq, B. U., et al., Proceedings of the Ocean Drilling Program, Scientific Results, v. 122: College Station, TX (Ocean Drilling Program), p. 897–901.

Tjalsma, R. C., 1976. Cenozoic foraminifera from the South Atlantic, DSDP Leg 36. In Barker, P. F., Dalziel, I. W. D., Initial Reports of the Deep Sea Drilling Project, v. 36: Washington (U.S. Government Printing Office), p. 493–517.

Tjalsma, R. C., 1983. Eocene to Miocene benthic foraminifers from Deep Sea Drilling Project Site 516, Rio Grande Rise. In Barker, P. F., Carlson, R. L., Johnson, D. A., et al., Initial Reports of the Deep Sea Drilling Project, v. 72: Washington (U.S. Government Printing Office), p. 731–755.

Tjalsma, R. C. and Lohmann, G. P., 1983. *Paleocene–Eocene bathyal and abyssal benthic foraminifera from the Atlantic Ocean.* Micropaleontology Special Publication No. 4, 90 pp.

Van Morkhoven, F. P. C. M., Berggren, W. A. and Edwards, A. S., 1986. *Cenozoic cosmopolitan deep-water benthic foraminifera.* Bulletin des Centres de Recherches Exploration-Production Elf-Aquitaine, Memoir 11, 421 pp.

Van der Zwaan, G. J. and Jorissen, F. J., 1991. Biofacial patterns in river-induced shelf anoxia. *Journal of the Geological Society,* v. 58, p. 65–82.

Vincent, E. and Brun, L., 1974. Paleocene and early Eocene microfacies, benthonic foraminifera, and paleobathymetry of Deep Sea Drilling Project Sites 236 and 237, Western Indian Ocean. In Fisher, R. L., Bunce, E. T., et al., Initial Reports of the Deep Sea Drilling Project, v. 24: Washington (U.S. Government Printing Office), p. 859–885.

Von Hillebrandt, A., 1962. Das Palaeozaen und seine Foraminiferenfauna im Becken von Reichenhall und Salzburg. Bayerische Akademie der Wissenschaften, Mathematisch-Naturwissenschaftliche Klasse, Abhandlungen, Neue Folge, v. 108, 182 pp.

Walker, J. C. G. and Kasting, J. F., 1992. Effects of fuel and forest conservation on future levels of atmospheric carbon dioxide. *Palaeogeography, Palaeoclimatology, Palaeoecology,* v. 97, p. 151–189.

Widmark, J. G. V., 1995. Multiple deep-water sources and trophic regimes in the latest Cretaceous deep-sea: Evidence from benthic foraminifera. *Marine Micropaleontology,* v. 26, p. 361–384.

Wing, S. L., Alroy, J. and Hickey, L. J., 1995. Plant and mammal diversity in the Paleocene to early Eocene of the Bighorn Basin. *Palaeogeography, Palaeoclimatology, Palaeoecology,* v. 115, p. 117–155.

Wing, S. L., Bown, T. M. and Obradovich, J. D., 1991. Early Eocene biotic and climatic change in interior western North America. *Geology,* v. 19, p. 1189–1192.

Wishner, K. F., Ashjian, C. J., Gelfman, C., Gowing, M. M., Kann, L., Levin, L. A., Mullineaux, L. S. and Saltzmann, J., 1995. Pelagic and benthic ecology of the lower interface of the Eastern Tropical Pacific oxygen minimum zone. *Deep-Sea Research,* v. 42, p. 93–115.

Wolfe, J. A., 1994. Alaskan Palaeogene Climates as inferred from the CLAMP database. In: Boulter, M. C., and Fisher, H. C., eds., *Cenozoic Plants and Climate of the Arctic,* NATO ASI Series, 127, Frankfurt: Springer-Verlag, p. 223–238.

Zachos, J. C., Rea, D. K., Seto, K., Nomura, R. and Niitsuma, N.,1992. Paleogene and early Neogene deep water paleoceanography of the Indian Ocean as determined from benthic foraminifer stable carbon and oxygen isotope records. In Duncan, R. A., Rea, D. K., Kidd, R. B., von Rad, U. and Weissel, J. K., (eds.), *Synthesis of Results from Scientific Drilling in the Indian Ocean,* Geophysical Monograph 70, Washington, D. C.: American Geophysical Union, p. 351–385.

Zachos, J. C., Lohmann, K. C., Walker, J. C. G. and Wise, S. W., 1993. Abrupt climate change and transient climates during the Paleogene: A marine perspective. *Journal of Geology,* v. 101, p. 191–213.

Zachos, J. C., Stott, L. D. and Lohmann, K. C., 1994. Evolution of early Cenozoic marine temperatures. *Paleoceanography,* v. 9, p. 353–387.

Appendix 12.1

	Site	Location	Reference	Depth	Resolution	LPTM $\delta^{13}C$	Size	Recovery/ Unconformity
South Atlantic								
1.1	ODP 689	Weddell Sea	Thomas 1990a; Thomas and Shackleton 1996	1100	High	Yes	>63	Medium unconformity
1.2	ODP 690	Weddell Sea	Thomas 1990a; Thomas and Shackleton 1996	1900	High	Yes	>63	Good
2.1	ODP 698	Falklands Plateau	Katz and Miller 1991	800–900	Low	No	>150	Poor
2.2	ODP 702	Falklands Plateau	Katz and Miller, 1991	1700–1800	Low	No	>150	Poor
2.3	ODP 700	Falklands Plateau	Katz and Miller 1991	2250–2600	Low	No	>150	Poor
2.4	DSDP 329	Falklands Plateau	Tjalsma 1977; Tjalsma and Lohmann 1983	1740–1745	Low	No	>150	Medium
3.1	DSDP 20C	Mid-Atlantic Ridge, west	Tjalsma and Lohmann 1983; Müller-Merz and Oberhänsli 1991	2275–3000	Low	No	>150, >63	Poor
3.2	DSDP 516	Rio Grande Rise	Dailey 1983; Tjalsma 1983; Boltovskoy et al. 1995	1000–2000	Low	No	>150, >150, >63	Good
4.1	DSDP 524	Walvis Ridge	Clark and Wright 1984; Hsu et al. 1984; Parker et al. 1984	3500–4000	Low	No	>150	Poor
4.2	DSDP 525	Walvis Ridge,	Thomas and Shackleton 1996; Boersma 1984; Boltovskoy and Boltovskoy 1989	1600	High	Yes	>63	Poor/LPTM in core
4.3	DSDP 527	Walvis Ridge,	Thomas and Shackleton 1996; Boersma 1984	3400	High	Yes	>63	Poor/LPTM in core
Pacific Ocean								
5	ODP 883	North Pacific	Pak and Miller 1995	1000–2000	Low	No	>150	Poor/ unconformity
6	DSDP 577	Subtropical North Pacific	Miller et al. 1987b; Pak and Miller 1992; Kaiho in press	1800–2100	High/ medium	Partim	>150	Good/gap unconformity
7	ODP 865	Equatorial Pacific	Bralower et al. 1995a,b; Thomas, unpublished data, this chapter	1300–1500	High	Yes	>63	Good/ unconformity?

Appendix 12.1

Pre-Extinction Taxa	Postextinction/ Short-Term	Postextinction/ Long-Term	Low O2/ High Productivity	Dissolution
G. beccariiformis, B. midwayensis, N. truempyi, B. thanetensis, Neoeponides spp., *S. brevispinosa*	*T. selmensis, B. simplex, A. aragonensis, S. brevispinosa*	Buliminids (*B. semicostata*), *S. eleganta, N. truempyi, O. umbonatus*	Yes	Slight
G. beccariiformis, R. carpentierae, B. thanetensis, N. truempyi	*T. selmensis, B. simplex,* abyssaminids, *S. brevispinosa*	Buliminids (*B. semicostata*), *S. eleganta, N. truempyi*	Yes	Yes
G. beccariiformis, Cibicidoides spp., *P. coryelli, buliminids*		*Cibicidoides* spp., *N. truempyi, O. umbonatus*	No?	Yes
G. beccariiformis, Gyroidinoides spp., *Cibicidoides* spp.		*N. truempyi,* buliminids, *Lenticulina* spp., *O. umbonatus*	No?	?
G. beccariiformis, P. coryelli, A. velascoensis, Neoeponides spp.		Abyssaminids, *Cibicidoides* spp., buliminids, *N. truempyi*	No?	?
G. beccariiformis, P. coryelli, agglutinants, *Gyroidinoides* spp.	Abyssaminids, *N. truempyi, A. aragonensis*	*N. truempyi, O. umbonatus, T. selmensis, A. aragonensis, B. semicostata*	No?	?
G. beccariiformis, buliminids, *Cibicidoides* spp., agglutinants	Abyssaminids, *N. truempyi, A. aragonensis*	Abyssaminids, *N. truempyi, A. aragonensis,* buliminids	No?	?
G. beccariiformis, P. coryelli, B. velascoensis, S. brevispinosa, B. delicatulus, N. truempyi	?	*N. truempyi, O. umbonatus, A. aragonensis, P. prolixa*	No?	Yes
G. beccariiformis, Neoeponides spp., agglutinants	?	*N. truempyi,* abyssaminids, *B. trihedra, T. selmensis*	Yes	Yes?
G. beccariiformis, N. truempyi, agglutinants, *Neoeponides* spp., *Cibicidoides* spp.	*N. truempyi,* abyssaminids, *T. selmensis, A. aragonensis*	*N. truempyi,* buliminids, *Cibicidoides* spp.	No	Yes
G. beccariiformis, A. velascoensis, N. truempyi, B. thanetensis	*N. truempyi,* abyssaminids, *T. selmensis*	*N. truempyi,* buliminids, *Cibicidoides* spp.	No	Yes
G. beccariiformis, B. midwayensis, agglutinants, *P. coryelli*	?	*N. truempyi, A. aragonensis,* abyssaminids	No?	?
G. beccariiformis, P. coryelli, A. velascoensis, B. midwayensis	? *T. selmensis, A. aragonensis, Neoeponides* spp.	*N. truempyi,* buliminids	Yes	Slight
G. beccariiformis, A. velascoensis, A. semicribratus, agglutinants	*T. selmensis, B. semicostata*	*Cibicidoides* spp. *N. truempyi,* buliminids (*B. semicostata*)	Yes	Slight

Appendix 12.1 *(continued)*

	Site	Location	Reference	Depth	Resolution	LPTM $\delta^{13}C$	Size	Recovery/ Unconformity
Indian Ocean								
8.1	ODP 747	Kerguelen	Mackensen and Berggren 1992	2000–3000	Low	No	>150	good/ unconformity
8.2	ODP 748	Kerguelen	Mackensen and Berggren 1992	600–2000	Low	No	>150	Poor
9	ODP 738	Kerguelen	Lu and Keller 1993; Nomura and Kennett personal communication; Thomas unpublished data	1350	High	Yes	>150, >63	poor/ LPTM in core
10	DSDP 245	South Madagascar Basin	Sigal 1974	600–1000	Low	No	>150	Poor
11	DSDP 237	Mascaren Plateau	Vincent and Brün 1974	600–2500	Low	No	>150	Poor
12	DSDP 215	Ninetyeast Ridge	McGowran 1974; Hovan and Rea 1992	600–2500	Low	No	>150	Good/LPTM not in core?
13	ODP 752	Kerguelen	Nomura 1991; Nomura and Kennett personal communication	500–1000	High/ medium	Partim	>150	Poor
14	ODP 762	Eastern Indian Ocean	Nomura and Kennett personal communication; Thomas unpublished data; Thomas et al. 1992	1000–1500	High/ medium	Partim	>150, >63	Poor/ disturbed
North Atlantic								
15.1	DSDP 549	Bay of Biscay/ Goban Spur	Reynolds 1992; Berggren and Aubry 1996	600–1500	Low	Yes/no benthics	>150	Good/ unconformity
15.2	DSDP 401	Bay of Biscay	Pak and Miller 1992; Schnitker 1979; Saint Marc 1991a, 1991b; Pardo et al. in press	1800	High/ medium	Partim/ no/ no yes	>150	Good/ unconformity
15.3	DSDP 400A	Bay of Biscay	Schnitker 1979	3100–4200	Low	No	>150	Poor
15.4	DSDP 548	Goban Spur/ Bay of Biscay	Boltovskoy et al. 1992	600–2000	Medium	No	>63	good/post
16	DSDP 605	New Jersey margin	Hulsbos 1987; Saint Marc 1987; 1991b; Thomas, unpublished data	1800–2300	Low/high	No	>150, >63	Good/ unconformity

Appendix 12.1 *(continued)*

Pre-Extinction Taxa	Postextinction/ Short-Term	Postextinction/ Long-Term	Low O2/ High Productivity	Dissolution
G. beccariiformis, Cibicidoides spp., *P. coryelli, N. truempyi*	?	*N. truempyi, Cibicidoides* spp., *O. umbonatus*	No?	?
G. beccariiformis, buliminids	?	*Stilostomella* spp., *Lenticulina* spp.	No?	?
G. beccariiformis, N. truempyi, B. thanetensis, agglutinands	*T. selmensis,* buliminids	*N. truempyi,* buliminids, *Cibicidoides* spp.	Yes	Yes
G. beccariiformis, A. velascoensis, A. welleri, B. thanetensis		? *T. selmensis, A. aragonensis, N. truempyi*	Yes?	?
G. beccariiformis, N. truempyi		*N. truempyi, Anomalinoides* spp.	No?	?
G. beccariiformis, N. truempyi, agglutinants		*T. selmensis, Aragonia* spp.	Yes?	?
G. beccariiformis, P. coryelli, Cibicidoides spp., *N. truempyi,* buliminids		*N. truempyi, Anomalinoides danicus, O. umbonatus*	Yes?	?
G. beccariiformis, N. truempyi, Cibicidoides spp., *B. thanetensis, B. delicatulus*	*N. truempyi, B. semicostata, Cibicidoides* spp., *T. selmensis*	*N. truempyi,* buliminids, *Cibicidoides* spp., *A. aragonensis, O. umbonatus*	No	?
G. beccariiformis, agglutinants		Buliminids, *T. selmensis*	Yes?	Yes
G. beccariiformis, C. hyphalus, C. midwayensis, B. delicatulus, Neoeponides spp., *Gyroidinoides*	? *N. truemypyi,* abyssaminids, buliminids (*B. semicostata*)	*N. truempyi,* abyssaminids, buliminids, *Cibicidoides* spp.	Yes	Yes
Cibicidoides spp., *N. truempyi,* abyssaminids, *Gyroidinoides* spp.	?	*Cibicidoides* spp., *Aragonia aragonensis, Cibicidoides* spp.	Yes	Yes
No data	*T. selmensis, S. brevispinosa, A. aragonensis, B. semicostata*	*N. truempyi,* buliminids, *Cibicidoides* spp.	Yes	Yes
G. beccariiformis, Cibicidoides spp., *A. velascoensis, B. midwayensis, P. coryelli,* agglutinants	? Probably unconformity	*N. truempyi, Cibicidoides* spp., abyssaminids, *T. selmensis,* buliminids (*B. semicosta*)	Yes	Yes?

Appendix 12.2

	Country	Section/ Formation	References	Depth	Resolution	LPTM $\delta^{13}C$	Size	Lithology
North Atlantic								
A.1	Spain	Zumaya (Atlantic)	Ortiz 1995; Canudo et al. 1995; Molina et al. 1994; von Hillebrandt 1965	600–1500	High	Yes	>63	Marl, calcarenite turbidites
A.2	Spain	Trabakua Pass, Ermua, west Pyrenees	Coccioni et al. 1994; Orue-Extebarria et al. in press	800–1000	High	No	>100	Limestones/marls
A.3	Spain	Tremp/Campo (Ilerdian stratotype)	Molina et al. 1992; Ortiz and McDougall 1991; von Hillebrandt 1965	?50–500	Low/ medium	No	>150	Marls
B	France	Paris Basin	Le Calvez 1970	10–200	Low	No	>150	Clays/silts
C	Trinidad	Lizard Springs Formation	Beckmann 1960; Bolli et al. 1994	500–1500	Low	No	>150	Marls
D.1	USA	New Jersey; Vincentown Manasquan Formations	Gibson et al. 1993; Gibson and Bybell 1995	50–200	Low	No	>63	Clayey sands/ clays
D.2	USA	New Jersey; Vincentown Manasquan Formations	Olsson and Wise 1987	50–250	Low	No	>150	Silty sands/clays
E.1	NW Europe	North Sea region	King 1989	100–500	Low	No	>150	Sand/clays
E.2	NW Europe	North Sea region	Gradstein et al. 1994	750–1000	Low	No	>150	Sand/clays
Tethys								
F	Spain	Caravaca (Tethys)	Ortiz 1995; Canudo et al. 1995	500–1000	High	Yes	>63	Marls, limestone
G	Israel	Nahal Avdat (Negev)	Speijer 1994	500–700	Low	No	>125	Marls/limestone
H	Egypt	Abu Rudeis = Wadi Nukhl (Sinai)	Speijer 1994	500–700	Low/ medium	No	>125	Marl/limestone
I.1	Egypt	Gebel Aweina	Speijer 1994; Speijer et al. 1996a; Charisi et al. 1995; Schmitz et al. 1996	100–150	Low/ medium	Yes	>125	Marl/limestone
I.2	Egypt	Bir Inglisi = Gebel Duwi	Speijer 1994; Speijer et al. 1996b; Schmitz et al. 1996	50–200	Low/ medium	No/ Yes	>125	Marl/limestone
J.1	Italy	Belluno	Di Napoli et al. 1970	500–1500	Low	No	>150	Marls/scaglias

Appendix 12.2

Pre-Extinction Taxa	Postextinction/ Short-Term	Low O_2/ Postextinction/ Long-Term	High Productivity	Dissolution
G. beccariiformis, Cibicidoides, buliminids, *P. coryelli*	First agglutinants, *O. umbonatus,* then *B. tuxpamensis*	*N. truempyi,* abyssaminids, *O. umbonatus,* buliminids	Yes	Yes
G. beccariiformis, N. truempyi, agglutinants, buliminids, *Anomalinoides welleri*	Rare agglutinants, then small buliminids, *N. truempyi*	*A. dissonata, N. truempyi,* buliminids, *O. umbonatus*	Yes	Yes
A. rubiginosus, Bulimina trigonalis, P. coryelli, agglutinants		*Cibicidoides* spp., *Uvigerina* . spp., *Hanzawaia* spp	?	?
Protelphidium, Pulsiphonina, Cibicidoides, buliminids		Buliminids, *Cibicides suc-, cedens, Bolivina* spp.	Yes?	?
A. velascoensis, B. midwayensis, buliminids, *Cibicidoides* spp.		*N. truempyi, T. selmensis, B. semicostata* (buliminids)	Yes?	?
Gavelinella spp., *Cibicidoides* spp., *Anomalinoides* spp., *Uvigerina*	*T. selmensis, Epistominella* sp.A A., buliminids, *Pseudouviger-ina* spp.	*T. selmensis, Pulsiphonina prima, Anomalinoides* spp., *Epistominella*	Yes	Yes
Gavelinella spp., *Anomalinoides* spp., *Cibicidoides* spp.	Agglutinants, *T. selmensis*	*Epistominella* sp., *T. robertsi, N. truempyi*	Yes	Yes
G. beccariiformis, B. midwayensis, Bulimina trigonalis, Cibicidoides	Low-diversity agglutinants	*T. brevispira, Gaudryina hilter-manni,* buliminids	Yes	Yes
Trochammina ruthyenmurrayi, Reticulophragmium paupera, assemblage (agglutinant)	Low-diversity agglutinants	*T. robertsi, R. amplectens, R. intermedia* assemblage (agglutinant)	Yes	Yes
G. beccariiformis, A. velascoensis, P. coryelli, Gyroidinoides spp.	Rare *Haplophragmoides* (barren), *T. selmensis,* buliminids, *A. aragonensis*	Buliminids, *N. truempyi, Anomalinoides* spp.	Yes	Yes
G. beccariiformis, agglutinants, *P. coryelli,* buliminids, *Cibicidoides* spp.		*N. truempyi, Cibicidoides* spp.	Yes?	Yes
G. beccariiformis, agglutinants, *P. coryelli,* buliminids, *Cibicidoides* spp.	(?) *Anomalinoides* spp., vagin-ulinids, *Stainforthia* spp., buliminids	*T. truempyi, Cibicidoides* spp., *T. selmensis,* buliminids	Yes	Yes
Cibicidoides spp., lagenids, *C. mid-wayensis, A. avnimelechi*	(?) *B. callahani, Cibicidoides* spp., *A. aragonensis, P. wilcoxensis*	Buliminids, *Cibicidoides* spp., *Anomalinoides* spp.	Yes	Yes
Anomalinoides, spp., *Valvulineria* spp., *Loxostomoides applinae*	*Anomalinoides aegyptaicus, Lenticulina* spp., *L. applinae*	*Valvulineria scrobiculata, Cibicidoides* spp., *Anomalin-oides* spp., *B. farafraensis*	*Yes*	*Yes*
Agglutinants, N. truempyi, G. bec-cariiformis, buliminids, *A. velasco-ensis, P. coryelli*	? *B. semicostata, B. trinitatensis*	*N. truempyi, Anomalinoides* spp., buliminids	Yes?	?

Appendix 12.2 *(continued)*

	Country	Section/ Formation	References	Depth	Resolution	LPTM $\delta^{13}C$	Size	Lithology
J2	Italy	Possagno	Braga et al. 1975	500–1500	Low	No	>150	Marls
K	Austria	Reichenhall–Salzburg Basin	von Hillebrandt 1962	200–800	Low	No	>150	Silts/marls
Pacific								
L	New Zealand	Tawanui	Hornibrook et al. 1989; Kaiho et al. 1993, in press	500–1500	High	No	>63	Siltstone, silty limestone
M	Japan	Hokkaido, Tokachi district	Kaiho 1988	500–1000	Low	No	>63	Silty marls

Appendix 12.2 *(continued)*

Pre-Extinction Taxa	Postextinction/ Short-Term	Low O_2/ Postextinction/ Long-Term	High Productivity	Dissolution
G. beccariiformis, N. truempyi, agglutinants, buliminids, *A. velascoensis, P. coryelli*	*N. truempyi, O. umbonatus*	*N. truempyi, B. semicostata,* buliminids	Yes?	Yes
Agglutinants, buliminids, nodosariids, lenticulinids		Agglutinants, *Anomalinoides rubiginosus*	?	?
G. beccariiformis, agglutinants, *C. midwayensis, A. rubiginosus*	First: *B. tuxpamensis,* agglutinants Second: agglutinants only	*Cibicidoides* spp., *O. umbonatus,* agglutinants	Yes	Yes
A. welleri, agglutinants, *B. midwayensis*		Agglutinants, buliminids	Yes	Yes

13

Upper Paleocene–Lower Eocene Deep-Sea Radiolarian Stratigraphy and the Paleocene/Eocene Series Boundary

ANNIKA SANFILIPPO
CATHERINE NIGRINI

ABSTRACT
Twenty-four DSDP sites with radiolarian-bearing upper Paleocene and/or lower Eocene sediment have been reexamined. None of the investigated sediment sequences contains the Paleocene/Eocene Series boundary. There are frequent coring gaps and/or poor recovery, and the radiolarian fauna is usually sparse and poorly preserved. Nevertheless, we have established the stratigraphic sequence of 70 lowest and highest occurrences in a 10-my interval spanning the boundary from the Paleocene *Bekoma campechensis* Zone to the upper part of the lower Eocene *Buryella clinata* Zone, and have related them to the calcareous nannofossil zonation. Using current nannofossil and foraminifera criteria to approximate the Paleocene/Eocene Series boundary, we have determined that there are six reliable, easily recognizable, and potentially useful radiolarian lowest occurrences that approximate the Paleocene/Eocene Series boundary: *Calocycloma castum*, *Theocotylissa auctor*, *Lamptonium fabaeforme fabaeforme*, *Podocyrtis* (*Podocyrtis*) *papalis*, *Giraffospyris lata*, and *Phormocyrtis turgida*.

Recent interest in the precise definition of Paleogene epoch boundaries has prompted us to examine the radiolarian fauna from the Paleocene/Eocene Series boundary interval in deep-sea sediments. In this chapter we present a summary of currently available radiolarian-bearing sediments found in the open ocean in a stratigraphic interval from the Paleocene *Bekoma campechensis* Zone to the upper part of the lower Eocene *Buryella clinata* Zone.

The latest Paleocene–earliest Eocene Epoch is a critical interval in Earth history. Warming at high latitudes (Stott and Kennett 1990; Kennett and Stott 1991) and changes in oceanic circulation associated with brief production of deep water at low latitudes (Miller et al. 1987; Kennett and Stott 1991; Thomas 1990a, 1990b, 1992) concomitant with major turnovers in benthic foraminifera (Tjalsma and Lohmann 1983; Thomas 1990a, 1990b) are events that occurred abruptly. The effects on the radiolarian fauna have not yet been investigated.

Magnetobiostratigraphic correlations are useful in the determination or recognition of chronostratigraphic boundaries. However, the Paleocene/Eocene Epoch boundary falls within a long magnetic reversal, Chron 24r, with an estimated duration of 2.5 my (Berggren et al. 1985) or 2.556 my (Cande and Kent 1995), thus making this magnetostratigraphic interval of limited use in precise placement of the serial boundary. In addition, calcareous nannofossils are absent at and/or near the boundary in many deep-sea sections.

Deep-Sea Radiolarian Occurrences in the Upper Paleocene to Lower Eocene Series

Occurrences

Present knowledge of the stratigraphic ranges of Paleocene and early Eocene radiolarians is limited because radiolarian-bearing sediment sequences of this age are rare. Although there are a number of papers describing either Paleocene or Eocene radiolarians from land-based sections, we are not aware of any land-based sections that are continuous across the Paleocene/Eocene Series boundary.

Land-based sections

It is not the purpose of this chapter to give a complete account of all land-based Paleocene and lower Eocene sections. Some noteworthy occurrences are listed below.

Paleocene North America, in Missouri (Frizzell and Middour 1951) and in California (Foreman 1968); Cuba (Florez Abin 1983; Florez Abin and Fernandez 1985; Sanfilippo and Riedel 1976); France (Riedel and Sanfilippo 1971); and New Zealand (Hollis 1993). Paleocene radiolarians have also been reported from the former Soviet Union (FSU) by Borisenko (1958, 1960), Lipman (1972), Tochilina (1975), and Khokhlova (1994). In Japan, early Paleogene radiolarians have been reported by Ueda et al. (1993), Nanayama (1992), and Suzuki (1992).

Lower Eocene Caribbean (summarized in Sanfilippo and Riedel 1976), England (rare and fragmented radiolarians from the London Clay accompanied by common and well-preserved foraminifera and diatoms; Shrubsole 1889), Germany (Brandt 1935), FSU (Borisenko 1960; Tochilina 1975; Kozlova 1984), and Japan (Suyari and Yamasaki 1987).

Deep-sea sections

Open-ocean radiolarian occurrences in Paleocene and lower Eocene sediments are limited to the continental margins (Palmer 1987), isolated deep-sea sites in the western Atlantic (Nigrini 1970; Petrushevskaya and Kozlova 1972; Weaver and Dinkelman 1978; Johnson 1983; Nishimura 1987, 1992; Palmer 1987), off Spain and northwest Africa (Sanfilippo and Riedel 1979; Westberg et al. 1980), the Gulf of Mexico and Caribbean (Foreman 1973; Sanfilippo and Riedel 1973, 1976; Riedel and Sanfilippo 1973), the tropical Pacific (Riedel and Sanfilippo 1971; Foreman 1975; Moore 1973), the Tasman Sea (Dumitrica 1973) and scattered localities in the Indian Ocean (Sanfilippo and Riedel 1974; Blome 1992).

Diversity

It is clearly desirable to investigate the extent to which radiolarian diversity shows trends that are similar to or different from other shelled microplankton. Because the total number of radiolarian species in a sample at any given time is rather large (approximately 150) and a high proportion of taxa are still unclassified, diversity cannot be discussed on the basis of numbers of species, or even genera, found at various geological periods. Using deep-sea material, Riedel and Sanfilippo (1981) attempted to quantify radiolarian diversity. To avoid the problem of dealing with a group within which species could not be satisfactorily delineated, they applied a procedure whereby morphological diversity was quantified. The general trend showed that, after a period of low values near the Cretaceous/Paleogene Period boundary, morphological diversity in radiolarians increased gradually through the Paleocene Epoch and, by the early Eocene Epoch, had achieved a level that has been more of less maintained throughout the Cenozoic Era. Studies by Hollis (1993) on land-based sections from New Zealand confirmed this general trend. Strong et al. (1995), using the same material as Hollis (1993), documented rapid faunal turnover as evident from progressive disappearances of several key species and the appearance of many middle to late Eocene species during the early middle Eocene Epoch. No marked change in diversity at the Paleocene/Eocene Epoch boundary has been recorded to date. Detailed documentation of radiolarian diversity over this time interval awaits the recovery of stratigraphically complete sections containing well-preserved radiolarians and additional taxonomic studies of the entire faunal assemblage.

Radiolarian Zonation for the Paleocene–Lower Eocene Series

A detailed zonation for the Eocene through the Pleistocene stratigraphic interval in the tropics was proposed by Riedel and Sanfilippo (1971, 1978) and improved upon for the upper Eocene Series by Saunders et al. (1985). Foreman (1973) defined four radiolarian zones from the uppermost Paleocene to the lower middle Eocene interval based on material from the Gulf of Mexico. Listed from oldest to youngest, they are the *Bekoma bidartensis*, *Buryella clinata*, *Phormocyrtis striata striata*, and *Theocotyle cryptocephala* Zones. The stratigraphic interval between the *B. bidartensis* Zone and the Cretaceous/Paleogene System boundary had been referred to as the "unzoned interval" by Sanfilippo et al. (1985) until Nishimura (1987) described a new upper Paleocene zone, the *Bekoma campechensis* Zone, just below the *B. bidartensis* Zone. It is approximately equivalent to nannofossil Zones CP3–CP6 and part of Zone CP7 of Okada and Bukry (1980). Subsequent biostratigraphic work by Nishimura (1987, 1992) in the northwest Atlantic resulted in a threefold subdivision of the *B. campechensis* Zone (figure 13.1).

Zonal schemes derived from land-based sections have been proposed for most of the Paleocene and lowermost Eocene stratigraphic record of the Volga Region by Kozlova (1983) and for the lower Paleocene sequence in New Zealand by Hollis (1993). In the case of the Russian sequences, the chronostratigraphic assign-

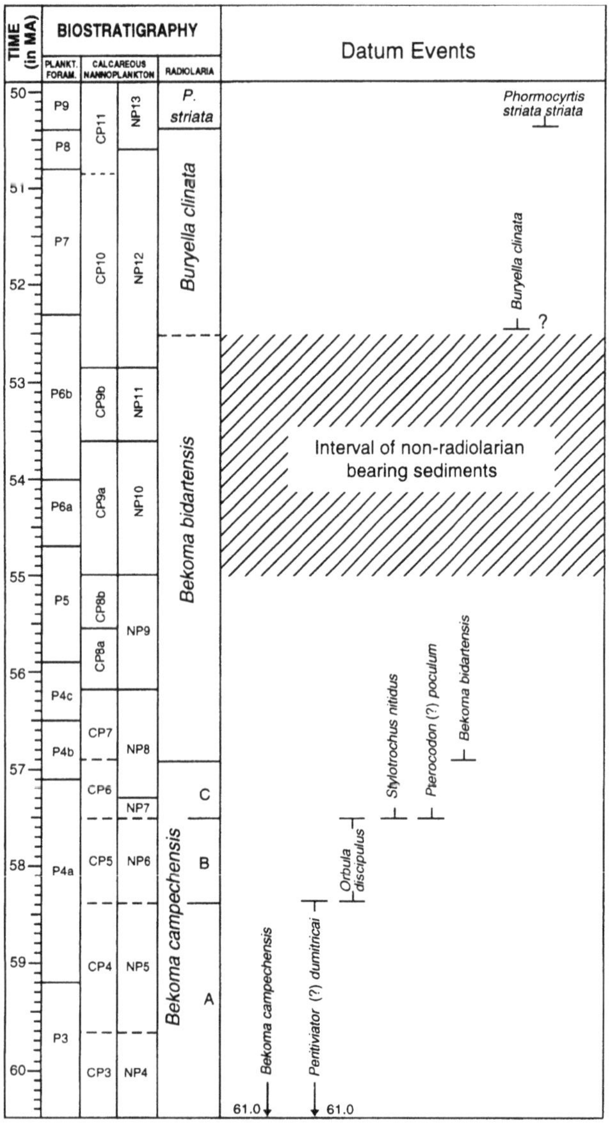

Figure 13.1 Chronology of planktonic foraminiferal, calcareous nannoplankton, and radiolarian zonations for the upper Paleocene and lower Eocene stratigraphic record.
Foraminiferal and calcareous nannoplankton zonations and magnetostratigraphy from Berggren et al. (1995). Radiolarian zonation and position of datum events from Sanfilippo et al. (1985), Nishimura (1992), and Sanfilippo and Nigrini (unpublished). Subzone A = *Peritiviator (?) dumitricai*, B = *Orbula discipulus*, C = *Stylotrochus nitidus–Pterocodon (?) poculum*

ments are doubtful because of the lack of associated calcareous microfossils.

The interval critical to this investigation includes the upper part of the *B. campechensis* Zone, the *B. bidartensis*, and *Buryella clinata* Zones. We have not applied Nishimura's subzones in the *B. campechensis* Zone because the zonal markers could not be recognized in most of our material.

Morphotypic limits have been applied throughout this investigation. We have not applied evolutionary transitions because the short and often discontinuous sediment sequences available for this study are unsuitable for taking advantage of the increased stratigraphic resolution associated with evolutionary transitions. As a result, we are using the lowest morphotypic occurrences of *B. clinata* and *B. bidartensis* to approximate the base of their respective zones, rather than the formal zonal boundary definition, which is based on evolutionary transitions.

Method and Material Studied

The Paleocene/Eocene Series boundary (as defined lithostratigraphically by the base of the London Clay Formation; see Berggren and Aubry, chapter 2) is considered herein to be in the middle part of nannofossil Zone NP10 (= Zone CP9a) and in the middle part of planktonic foraminiferal Zone P6a (Aubry et al. 1996). In terms of radiolarian stratigraphy it is within the *B. bidartensis* Zone. Because none of the sections currently being considered as potential boundary stratotypes contain Radiolaria, a two-step correlation process, using co-occurring calcareous microfossils and radiolarians in deep-sea sediments, must be used to determine the placement of the boundary in terms of radiolarian stratigraphy.

With this in mind, we have reexamined all available deep-sea sections (DSDP/ODP Legs 1–135) in the interval around the boundary. Our survey has revealed that only 26 sites contain material from the boundary interval. All but 12 of the DSDP sites are either too poorly preserved or unsuitable for stratigraphic analysis (figure 13.2). ODP Sites 752 and 761 may contain a workable fauna and they are the only sites for which there is a paleomagnetic record over the appropriate interval, but samples are currently unavailable to the authors. Fourtanier (1991) stated that the Paleocene/Eocene Series boundary was recovered for the first time in diatom-bearing sediments at Site 752. No radiolarian study was made for this site, but the site chapter (Peirce et al. 1989) indicates that they are present. The 12 sites we have reexamined contain a sufficient number of well-preserved to moderately well-preserved radiolarians to be useful in the construction of a list of radiolarian "events" (morphotypic lowest and highest occurrences) that can be correlated to nannofossil and foraminiferal zones.

In all our tables and diagrams we have equated the top of the recovered sediment with the top of the cored interval, as is the current practice of ODP. If only a core catcher is recovered, then it is considered to come from

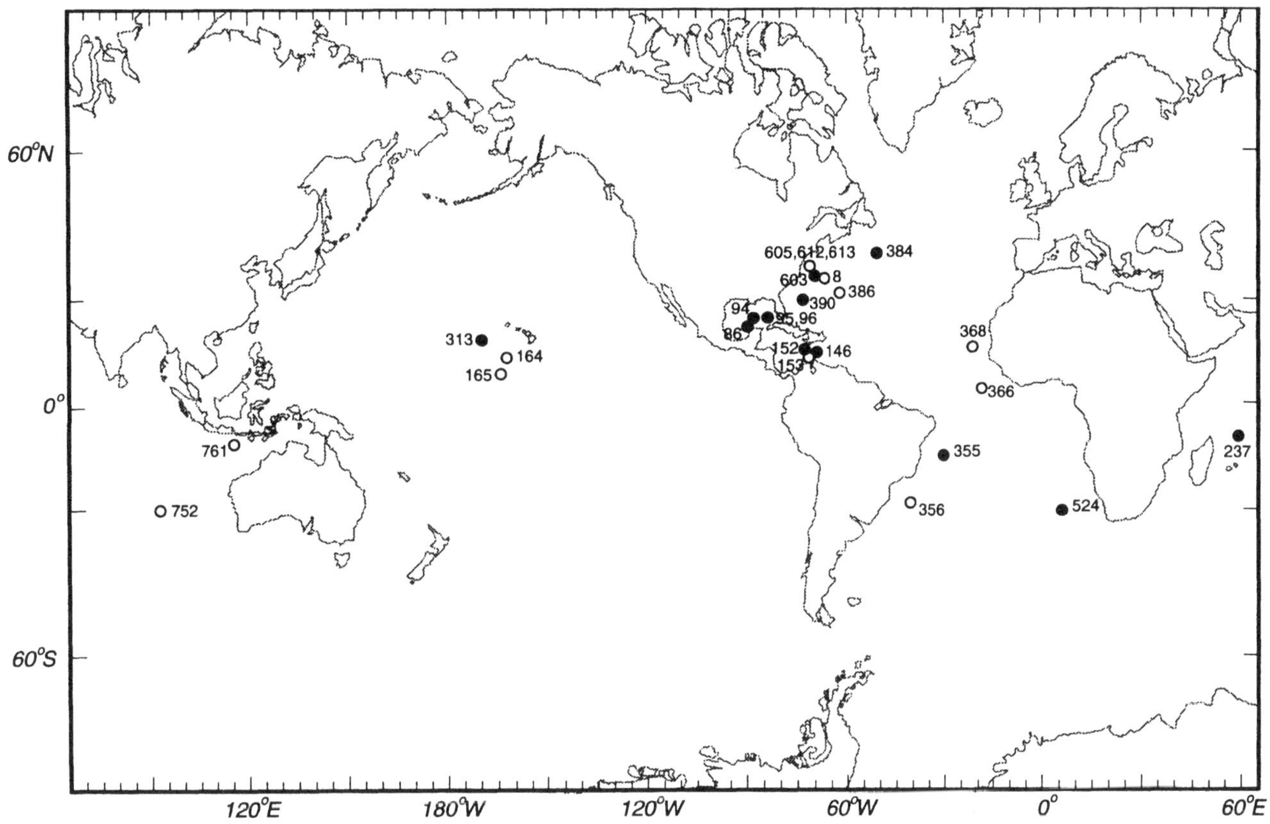

Figure 13.2 Location of DSDP/ODP Sites (through Leg 135) containing upper Paleocene and/or lower Eocene radiolarian-bearing sediments.
Open circles represent sites in which material is too poorly preserved for a meaningful analysis. Filled circles indicate the sites reexamined herein.

the top of the cored interval. In early DSDP legs if only a core catcher was recovered, it was equated with the bottom of the cored interval. Hence, some of our depths (meters below sea floor), particularly for Leg 10, are not the same as those published in the original reports.

Description of Sites (tables 13.1–13.5)

Site 10-86 (Gulf of Mexico, 22°53' N, 90°58' W, 1481 m)

Original biostratigraphic data (Sanfilippo and Riedel 1973; Foreman 1973) indicate that the upper Eocene sequence is unconformable with an unzoned Paleocene interval in sample 7-cc and throughout Core 8. Radiolarians are common and well preserved. Bukry (1973) determined that sample 7-cc is in the *Discoaster mohleri* Zone (Zone CP6) and all of Core 8 is within the *Heliolithus kleinpelli* Zone (Zone CP5).

Although the site was, for the most part, discontinuously cored, Cores 7 and 8 are contiguous. Our reexamination indicates that sample 7-cc and Core 8 belong to the *B. campechensis* Zone of Nishimura (1992) and, although there are no radiolarian lowest or highest occurrences, this zonal assignment agrees with Bukry's zonal determinations.

Site 10-94 (Gulf of Mexico, 24°32' N, 88°28' W, 1793 m)

Sanfilippo and Riedel (1973) and Foreman (1973) indicated that samples 29-cc to 31-cc lie within the *B. clinata* Zone and sample 32-cc within the *B. bidartensis* Zone. Nannofossil data from Bukry (1973) place Cores 29 and 30 in Zone CP11. Radiolarians are abundant and well preserved throughout the studied interval.

Coring in this interval was discontinuous and recovery was poor (Cores 29 (0.3 m), 30 (2.5 m), 31 (0.1 m), 32 (0.1 m)).

Site 10-96 (Gulf of Mexico, 23°45' N, 85°46' W, 3439 m)

Original biostratigraphic data (Sanfilippo and Riedel 1973; Foreman 1973) indicate that Core 3 is in the *B. bidartensis* Zone and samples 4-cc and 5-cc are within an unzoned Paleocene interval. Our reexamination of the radiolarian assemblage suggests that Cores 4 and 5 should also be placed in the *B. bidartensis* Zone even though the zonal marker is absent.

Radiolarians are abundant and well preserved in Core 3, but decline in abundance and preservation in Cores 4 and 5. Nannofossil analysis of this site (Bukry 1973:391) indicates that Cores 3 to 5 contain an in situ Paleocene assemblage belonging to Subzone CP8b.

Coring was discontinuous except for Cores 4 and 5, which were continuously cored. The average rate of deposition for the Paleocene interval is low (0.6 cm/10^3 yr) presumably because of the presence of a number of disconformities. Recovery in Core 3 was good (9.3 m), whereas Core 4 recovered only a catcher sample and Core 5 recovered 1 m of nannochalk.

Site 15-146 (Caribbean, 15°07' N, 69°23' W, 3949 m)

Both original (Riedel and Sanfilippo 1973) and revised (this chapter) biostratigraphic analysis place Core 4 in the *B. clinata* Zone, Core 7 in the *B. bidartensis* Zone. Core 10, previously placed in the "unzoned" Paleocene interval, is now thought to belong to the *B. campechensis* Zone. Nannofossil data (Bukry 1973) place Core 7 in Zone CP8b. Because radiolarian abundances vary from rare to common and preservation is always poor to moderate, we have been unable to give zonal assignments to Cores 3, 5, 6, 8, and 9.

Coring of Cenozoic material at this site was discontinuous and recovery was consistently poor. Core 4 recovered only the core catcher, whereas Core 5 recovered 1.5 m and Cores 6 and 7 each recovered 1 m. Cores 8, 9, and 10 recovered 2 m each.

Site 15-152 (Caribbean, 13°53' N, 74°36' W, 3899 m)

Biostratigraphic data from Riedel and Sanfilippo (1973) indicate that samples 1-cc to 2-4, 61–63 cm, belong to the *B. bidartensis* Zone and no zonal determination could be made for samples from 2–4, 128–130 cm to 4–3, 103–106 cm. Samples 4-cc through 9-cc belong to the "unzoned" part of the Paleocene. Our reexamination (this chapter) places Cores 1 through 3 in the *B. bidartensis* Zone and Cores 5 through 9 in the *B. campechensis* Zone of Nishimura (1992). The boundary between these zones lies within Core 4, but cannot be placed precisely because the zonal marker species is absent.

Nannofossil data (Hay and Beaudry 1973) place samples 1-1, 99–100 cm to 4-2, 78–79 cm in Zone CP8; samples 4-3, 100–101 cm to 6-2, 64–65 cm belong to Zone CP5; samples 6-5, 108–109 cm to 8-1, 127–128 cm belong to Zone CP4; samples 9-1, 82–83 cm to 9-1, 138–140 cm belong to Zone CP2. Additional information from Bukry (1973) indicates that Cores 2 and 3 belong to Zone CP8b.

Foraminiferal data (Premoli Silva and Bolli 1973), using the highest occurrence of *Globorotalia* (now *Morozovella*) *velascoensis*, indicate that the Paleocene/Eocene Series boundary is between samples 152-3-5, 48–50 cm and 152-3-5, 68–70 cm. This is in conflict with the nannofossil data, which indicates that Cores 1 through 9 at Site 152 are Paleocene in age.

Coring over this interval was continuous; recovery was good in Cores 2 (7.5 m), 3 (7.5 m), 4 (4.5 m), 6 (8 m), and 7 (8.5 m), and poor in Cores 1 (2 m), 5 (core catcher only), 8 (1 m), 9 (1 m), and 10 (core catcher only).

Site 24-237 (Tropical Indian Ocean, 07°05' S, 58°08' E, 1640 m)

Original biostratigraphic data (Sanfilippo and Riedel 1974) place samples from Cores 36-1 and 37-1 in the *B. clinata* Zone and samples from Cores 38-1, 41-1, and 42-1 in the *B. bidartensis* Zone. Our revision places Core 38-1 in the *B. clinata* Zone and, consequently, the boundary with the underlying zone is within the unrecovered section. Radiolarians are common, but only moderately well preserved. The *Phormocyrtis striata exquisita* to *P. striata striata* transition used to define the base of the *P. striata striata* Zone is overshadowed by other evidence indicating that sample 36-1 should be assigned to the *B. clinata* Zone. However, if this assignment is wrong, and the sample should be assigned to the *P. striata striata* Zone, it would be more consistent with the nannofossil assignment of Zone CP12. The recovered material does not permit a more precise determination.

Both Roth (1974) and Bukry (1974) place Core 36-1 in Zone CP12, but the radiolarian data indicate equivalence to Zone CP10 or Zone CP11. However, there is low diversity in the nannofossil assemblage and the fauna is apparently mixed (Aubry, personal communication 1992). Additional nannofossil data from Roth (1974) and Bukry (1974) place Core 37-1 in Zone CP10, Core 37-2 in Zone CP9, Core 38-1 in Zone CP9 or possibly lower Zone CP10, Core 41-2 in Zone CP8a, and Core 42-1 in Zone CP6.

Coring was continuous, but recovery was consistently poor: Cores 36 (1.3 m), 37 (1.9 m), 38 (1.5 m), 39 and 40 (no recovery), 41 (2.2 m), and 42 (1.3 m).

Site 32-313 (Tropical Pacific, Mid-Pacific Mountains, 20°11' N, 170°57' W, 3848 m)

According to Foreman (1973) the base of the *B. clinata* Zone is defined by the lowest morphotypic occurrence of *B. clinata*. In 1975, Foreman emended the definition of the base of the zone to the evolutionary transition of *Pterocodon(?) anteclinata* to *B. clinata*.

The *B. clinata/B. bidartensis* boundary shown on Foreman's table 8 (1975) does not reflect her emended definition. For our reexamination of this and other sites we have chosen to retain the original definition of the zonal boundary. Thus, Core 9 and the upper part of Core 12 are in the *B. clinata* Zone and the rest of Core 12 and Core 13 are in the *B. bidartensis* Zone. Although radiolarians are abundant and well preserved in the studied interval at Site 313, the evolutionary transition from *P.(?) anteclinata* to *B. clinata* (between Cores 9 and 12) cannot be determined because Cores 10 and 11 are barren.

The nannofossil zonation, determined by Bukry (1975), places samples from Cores 12-1 to 13-1 in Zone CP10 and Core 13-2 in Zone CP9. Correlation with foraminiferal data suggests that Core 13-2 is in Zone CP9b. However, Wei (personal communication 1995) assigns samples 313-13-1, 57–59 cm and 313-13-6, 139–141 cm to nannofossil Zone CP10 based on the presence of *Discoaster lodoensis*, *D. saipanensis*, and *Tribrachiatus orthostylus* and the absence of *Discoaster multiradiatus*.

Coring was continuous; recovery was poor in Cores 9 (0.9 m), 10 (0.1 m), and 11 (0.1 m) and good in Cores 12 (6.5 m) and 13 (9.5 m).

Site 39-355 (Atlantic Ocean, Brazil Basin, 15°43' S, 30°36' W, 4886 m)

Original radiolarian data (Dinkelman in Supko, Perch-Nielsen et al. 1977 in site chapter) places the interval from Core 12-1 to 15-cc in the *B. clinata* Zone. Calcareous microfossils are absent from this interval. Our reexamination places samples from Core 12-2 to Core 13-3, where radiolarians are common, but only poorly to moderately well preserved, in the *B. clinata* Zone. The fauna below this level is too poor to support a zonal determination.

Coring was continuous, but recovery was usually less than 65%.

Site 43-384 (Northwest Atlantic Ocean, 40°22' N, 51°40' W, 3909 m)

There is no radiolarian chapter in the DSDP reports for this site. However, a detailed stratigraphy was subsequently presented by Nishimura (1992). In her report, samples from Core 6-1 to Core 7-2 belong to the *B. bidartensis* Zone; samples from Core 7-3 to Core 10-1 belong to the *B. campechensis* Zone, which is subdivided into three subzones: *Stylotrochus nitidus–Pterocodon (?) poculum* Subzone from Core 7-3 to Core 8-3, *Orbula discipulus* Subzone from Core 8-4 to 8-cc, and *Peritiviator (?) dumitricai* Subzone from Core 9-1 to Core 11-2. The sample from Core 11-3 is in an unzoned interval. In general, our reexamination confirms Nishimura's zonal assignments, but we place the *B. bidartensis/B. campechensis* zonal boundary between samples 7-6, 62–66 cm and 7-6, 119–125 cm. Radiolarians are abundant and well preserved throughout.

There is a complete Paleocene sequence of abundant and well-preserved nannofossils (Okada and Thierstein 1979) as follows: Core 6-1, 14–15 cm = Zone CP8b; Core 6-1, 39–40 cm to 6-1, 109–110 cm = Zone CP8a; Core 6-1, 149–150 cm to 7-5, 59–60 cm = Zone CP7; Core 7-6, 49–50 cm to 8-3, 124–125 cm = Zone CP6; Core 8-4, 49–50 cm to 8-cc = Zone CP5; Core 9-1, 4–5 cm to 10-2, 119–120 cm = Zone CP4; Core 10-3, 49–50 cm to 12-2, 51–52 cm = Zone CP3.

Coring was continuous with good recovery.

Hole 44-390A (Atlantic Ocean, Blake Plateau, 30°09' N, 76°07' W, 2670 m)

Original biostratigraphic data (Weaver and Dinkelman 1978) indicate that samples from Core 6-6 to Core 7-4 belong to the *B. clinata* Zone and sample 7-cc belongs to the *B. bidartensis* Zone. Our reexamination of samples from Core 6-6 and Core 7-4 shows that these samples probably belong to the *P. striata striata* Zone (equivalent to Zone CP12), which is inconsistent with the nannofossil assignment.

The nannofossil data (Bukry 1978) place a sample from Core 6-6 in Zone CP11 and samples from Core 7-2 to Core 7-4 in Zone CP10.

Cores 6 and 7 were continuously cored with good recovery.

Site 524 (Atlantic Ocean, Walvis Ridge, 29°29' S, 03°31' W, 4796 m)

There was no radiolarian report for this site. Our examination of this site shows that Cores 4 and 5 are in the *B. bidartensis* Zone. The radiolarian fauna is common and well to moderately well preserved.

Percival (1984) reports that Cores 4-3 to 5-1 are in Zone CP8 and Core 5-5 is in Zone CP7.

Coring was continuous and recovery for Cores 4 (5.1 m) and Core 5 (7.5 m) was good.

Hole 93-603B (Atlantic Ocean, New Jersey Margin, 35°30' N, 70°02' W, 4633 m)

Nishimura (1987) places the core catcher samples from Cores 18 and 19 in the *B. bidartensis* Zone and the samples from Core 20-1 to 21-cc in the *B. campechensis* Zone. Nannofossils are absent. The fauna is abundant in

Table 13.1
Occurrence charts for DSDP Sites 86, 94, and 96

EPOCH	Nannofossil Zones (Bukry 1973)	Radiolarian Zones (this chapter)	LEG 10 SITE 86	mbsf	ABUNDANCE	PRESERVATION	*Amphicraspedum murrayanum*	*Amphicraspedum prolixum*	*Bekoma bidartensis*	*Bekoma campechensis*	*Buryella clinata*	*Buryella pentadica*	*Buryella tetradica*	*Calocycloma castum*	*Dictyoprora urceolus*	*Giraffospyris lata*	*Lamptonium fab. chaunothorax*	*Lamptonium fab. constrictum*	*Lamptonium fab. fabaeforme*	*Lamptonium pennatum*	*Lamptonium sanfilippoae*
PALEOCENE	CP6	*Bekoma campechensis*	86-7-CC	504.5	C	MG	–	–	–	C	–	C	C	–	–	–	–		–	C	–
	CP5		86-8-2,3-5	509.2	R	M	–	–	–				C		–	–					
			86-8-2,48-50	509.7	C	MG	–	–	–	R		C	C		–	–					
			86-8-2,114-116	510.3	C	MG	–	–	–	R	–	F	C	–	–	–	–		–	R	–
			86-8-3,2-4	510.7	C	MG	–	–		F	–	F	C	–	–	–	–		–	R	–
			86-8-3,44-46	511.1	C	MG	–	–		C		C	C		–	–				R	
			86-8-3,114-116	511.8	C	MG	–	–			–	C	C	–	–	–	–		–	VF	–
			86-8-4,1-2	512.2	C	MG	–	–	–	C		C	C		–	–	–		–		
			86-8-4,46-48	512.7	C	MG	–	–		C	–	C	C	–	–	–				F	–
			86-8-4,119-121	513.4	C	MG	–	–					C		–	–					
			86-8-CC	513.5	C	MG	–	–		C	–	C	C	–	–	–	–		–	F	–

LEG 10 SITE 94

EPOCH	Nannofossil Zones	Radiolarian Zones	Sample	mbsf	ABUNDANCE	PRESERVATION	*A. murrayanum*	*A. prolixum*	*B. bidartensis*	*B. campechensis*	*B. clinata*	*B. pentadica*	*B. tetradica*	*C. castum*	*D. urceolus*	*G. lata*	*L. fab. chaunothorax*	*L. fab. constrictum*	*L. fab. fabaeforme*	*L. pennatum*	*L. sanfilippoae*
EARLY EOCENE	CP 11	*Buryella clinata*	94-29-CC	551.3	A	G	F	R	–	–	C	–	–	F	C	–	R	R	F	–	+
			94-30-1,33-35	570.8	C	G	F	R	–	–	C	–	+	F	C	R	R	R	F	R	+
			94-30-1,80-82	571.3	A	G	F	R	–	–	C	–	+	F	C	R	R	–	F	R	+
			94-30-2,44-46	572.4	A	G	R	R	–	–	C	–	R	F	C	R	R	–	F	R	+
			94-30-2,114-116	573.1	A	G	R	R	–	–	C	–	R	MR	C	R	F	–	F	R	R
			94-30-CC	573.5	A	G	R	F	–	–	C	–	F	F	C	+	R	–	F	R	+
	?	*B.*	94-31-CC	589.1	A	G	R	F	+	–	C	–	F	R	F	R	R	–	F	R	+
	?	*bidar*	94-32-CC	609.1	A	G	R	F	R	–	–	–	C	–	–	R	–	–	F	R	–

LEG 10 SITE 96

EPOCH	Nannofossil Zones	Radiolarian Zones	Sample	mbsf	ABUNDANCE	PRESERVATION	*A. murrayanum*	*A. prolixum*	*B. bidartensis*	*B. campechensis*	*B. clinata*	*B. pentadica*	*B. tetradica*	*C. castum*	*D. urceolus*	*G. lata*	*L. fab. chaunothorax*	*L. fab. constrictum*	*L. fab. fabaeforme*	*L. pennatum*	*L. sanfilippoae*
PALEOCENE	CP 8b	*Bekoma bidartensis*	96-3-1,41-43	301.4	A	G	+	R	R	–	–	–	C	–	–	MR	–		F	R	–
			96-3-1,114-121 MRC	302.2	A	G	R	R	R	–		–	C	–		VF	–	–	VF	R	–
			96-3-3,42-44	304.4	C	G	MR	R	R	–		–	C				–		VF	MR	
			96-3-4,42-44	305.9	A	G	R	R	R			–	C	–		F		–		R	–
			96-3-5,42-44	307.4	A	G			F	–		–	C	–		F	–	–	F	R	–
			96-3-6,42-44	308.9	A	G	F	R		–		–	C	–		F	–		R	F	–
			96-3-6,93-100 MRC	309.5	A	G		R	R			–	C	–		F	–		R	F	–
			96-3-CC	310.3	A	G	R	R	R	–		–	C	–	–	F		–	R	R	–
			96-4-CC	321.6	F	M	F	F	–			–	C			R			–	R	–
			96-5-1,114-116	329.9	R	G	X	X	X	X		–	F						–	–	
			96-5-CC	330.0	F	M	F	R	X	X		X	C			X			–	–	

Species abundances are C (common) = >10%, F (few) = 1–10%, VF (very few) and R (rare) = 0.1–1%, MR (moderately rare) and VR (very rare) = 0.01–0.1%, + = less than 0.01% of the total assemblage, – = searched for but not found, ? = doubtful identification, ! = reworked or contaminant, X = absence of a species that, according to other data, is expected to be present (absence is inferred to be due to poor preservation combined with low abundance). Overall abundance is denoted as B = barren, VR = very rare, R = rare, F = few, C = common, A = abundant; preservation is shown as P = poor, M = moderate, G = good. MRC indicates samples from the DSDP/ODP Micropaleontological Reference Collection, West Coast Repository, Scripps Institution of Oceanography, La Jolla, California. Dashed lines indicate uncertainty in the zonal assignment.

Table 13.1 *(continued)*
Occurrence charts for DSDP Sites 86, 94, and 96

Lithochytris archaea	*Lophocyrtis biaurita*	*Lychnocanoma auxilla*	*Phormocyrtis cubensis*	*Phormocyrtis striata exquisita*	*Phormocyrtis striata striata*	*Phormocyrtis turgida*	*Podocyrtis acalles*	*Podocyrtis papalis*	*Pterocodon ampla*	*Pterocodon anteclinata*	*Pterocodon poculum*	*Pterocodon tenellus*	*Spongatractus balbis*	*Spongatractus pachystylus*	*Spongodiscus cruciferus*	*Spongodiscus phrix*	*Spongodiscus quartus quartus*	*Stylotrochus alveatus*	*Theocorys acroria*	*Theocorys phyzella*	*Theocotyle nigriniae*	*Theocotylissa alpha*	*Theocotylissa auctor*	*Theocotylissa ficus*	*Thyrsocyrtis hirsuta*	*Thyrsocyrtis tarsipes*
		–	–	C	–	–	–	–	–		–	–			–	–	–	R	–	–	–	–	–		–	–
				C	–		–	–								–	–								–	
				C	–		–	–								–	–								–	
		–	–	C	–	–	–	–	–		–	–			–	–	–	R	–	–	–	–	–			–
		–	–	C	–	–	–	–	–		–	–			–	–	–	R	–	–	–	–	–			–
				C			–	–								–	–	F								
		–		C		–	–	–	–			–				–	–	MR	–	–	–	–	–			–
		–		C		–	–	–	–			–				–	–	MR	–	–			–			–
			–	C			–	–							–	–	–	F			–	–				
				C			–	–								–	–									
		–		C		–	–	–	–			–	–			–	–	F	–	–	–	–	–			–
F	R	–	–	F	R	–	–	F		–	–		R	–	R	–	–	–	R		F	–	–	F	C	–
F	R	–	R	C	F	–	–	R		–	–			–		–	R	–	+		F	–	–	F	C	–
F	R	–	R	C	F	–	–	F	–	–	–	–	R	–		–	R	–	+	–	F	–	–	F	C	–
F	R	–	R	C	F	–	–	F	–	–	–	–		–		–	R	–	R	–	F	R	–	F	C	–
C	+	–	R	F	R	–	–	F	–	–	–	–	+	–	R	–	R	–	+	–	F	R	–	C	F	–
F	F	–		F	R	–	–	F	–	–	–	–	R		R	–	R	–	R	–	F	R	R	C	F	–
R	R	R	R	C	R	R	–	F	F	R	–	R	–		R	–	R	–	R	MR	R	R	F	–	R	+
–	–	R	F	C	R	F	–	F	C	R	+	R	–		R	–	F	–	R	F	–	–	C	–	–	+
+		F	+	C	R	C	–	F	F	R	–	VR	–	–	+	–	R	–	R	F	–	–	C		–	MR
R	–	VF	R	C	R	VF	–	F	F	R			–	–		–	R	–	R	F	–	–	R	–	–	VF
–		VF	R	C		F	–	F	F	R	–	R	–	–		–	MR	–	R	F		–	MR		–	VF
–	–	R	R	C	R	F	–	F	F	–	–	VF	–	–	R	–	R	–	R	F	–	–	C	–	–	VF
–		R	F	C		C	–	F	C	+	–	R	–	–	R	–		–	VR	F	–	–	C	–	–	C
–	–	F	F	C	R	F	–	F	F	R	–	F	–	–		–	F	–	R	F	–	–	C	–	–	F
		F	R	C	R	VF	–	F	F	R		F				–	F	–	R	F	–	–	C		–	F
	–	R	R	C	R	C	–	F	F	–	–	F			R	–	F	VR	R	F	–	–	C	–	–	R
		R	+	C	–	R		R	R			R					F	R	VR	R		–	–		–	R
		R	+	R	–	X		–	R			X			F		X	X	–	X		–	X			–
		–	+	C	X	–		–				X			F		R	VR	–	–		X	X			–

Table 13.2
Occurrence chart for DSDP Site 152

EPOCH	Nannofossil Zones (Bukry 1973); Hay & Beaudry 1973)	Radiolarian Zones (this chapter)	LEG 15 SITE 152	mbsf	ABUNDANCE	PRESERVATION	*Amphicraspedum murrayanum*	*Amphicraspedum prolixum*	*Bekoma bidartensis*	*Bekoma campechensis*	*Buryella clinata*	*Buryella pentadica*	*Buryella tetradica*	*Calocycloma castum*	*Dictyoprora urceolus*	*Giraffospyris lata*	*Lamptonium fab. chaunothorax*	*Lamptonium fab. constrictum*	*Lamptonium fab. fabaeforme*	*Lamptonium pennatum*	*Lamptonium sanfilippoae*
PALEOCENE	CP8b	*Bekoma bidartensis*	152-1-1,91-93	153.9	A	G	R	F	–	–	–	–	+	–	–	F	R	–	R	F	–
			152-1-1,130-137 MRC	154.3	A	G		F		–	–		R	–		VF			R	R	
			152-1-2,80-85	155.3	A	G	F	F	–	–	–	–	R	–	–	F	MR	–	R	R	–
			152-1-CC	155.3	C	G							R				R		VF	F	–
			152-2-3,99-106 MRC	166.0	F	M		R					R			+	X		X	X	
			152-2-4,61-63	167.1	F	M		+					R			R			+	+	
			152-2-4,128-130	167.8	F	M		X	–	–	–	–	R			R			X	–	
			152-2-5,49-51	168.5	R	M							R			R				+	
			152-2-5,127-129	169.3	R	P															
			152-2-CC	169.5	R	M					X	X	+						X	X	
			152-3-3,126-133 MRC	176.3	R	M							F								
			152-3-4, 132-134	177.8	F	M	X		–	–	–		F			R			X	X	
			152-3-5,15-17	178.2	R	M		X				–	F			+					–
			152-3-5,133-135	179.3	A	G	R	R	–	–	–	–	F	–		R	R		F	F	–
			152-3-CC	179.5	C	G			–	–		–				R				VF	
			152-4-2,30-37 MRC	183.8	A	G							F		–	R	–		R	F	
			152-4-2,85-87	184.4	C	G	+	+	–	–	–	–	F		–	R	–		R	R	
	CP5		152-4-3,103-106	186.0	R	M						–				+			X	X	
			152-4-CC	186.4	R	M	X	+			–	–	F			+			–	+	
		Bekoma campechensis	152-5-CC	192.0	R	P						R	+							+	
			152-6-2,26-33 MRC	202.8	R	P						R	R								
	CP4		152-6-5,92-94	207.9	F	M	X	X		–		F	F			X			–	–	
			152-6-5,144-146	208.4	F	M				–		R	F			X			–	X	
			152-7-3,126-129	215.3	F	M				+		R									
			152-7-4,21-23	215.7	F	M		X		X		F	C						–	+	
			152-7-4,109-116 MRC	216.6	F	M															
			152-7-4,145-147	217.0	F	M		X		+		R	F						X	X	
			152-7-CC	217.0	F	M				?		F	F							+	
			152-8-1,80-82	220.8	F	M				–		F	F						–	R	
			152-8-1,125-127	221.3	F	M				–		F	F						–	+	
			152-8-CC	221.5	C	G				–		F	F						–	R	
	CP2		152-9-1,83-85	229.8	C	G				–		MR	F							VF	
			152-9-1,141-143	230.4	A	G				–		R	F							VF	
			152-9-CC	230.5	R	P						R	R								

For explanation of symbols, see table 13.1

Table 13.2 *(continued)*
Occurrence chart for DSDP Site 152

Lithochytris archaea	*Lophocyrtis biaurita*	*Lychnocanoma auxilla*	*Phormocyrtis cubensis*	*Phormocyrtis striata exquisita*	*Phormocyrtis striata striata*	*Phormocyrtis turgida*	*Podocyrtis acalles*	*Podocyrtis papalis*	*Pterocodon ampla*	*Pterocodon anteclinata*	*Pterocodon poculum*	*Pterocodon tenellus*	*Spongatractus balbis*	*Spongatractus pachystylus*	*Spongodiscus cruciferus*	*Spongodiscus phrix*	*Spongodiscus quartus quartus*	*Stylotrochus alveatus*	*Theocorys acroria*	*Theocorys phyzella*	*Theocotyle nigriniae*	*Theocotylissa alpha*	*Theocotylissa auctor*	*Theocotylissa ficus*	*Thyrsocyrtis hirsuta*	*Thyrsocyrtis tarsipes*
F	–	F	VF	F	R	R	–	C	R	R	VF	R	–	–	+	–	R	–	R	R	–	F	R	–	–	R
F			R		R	VF		F	VF	MR	VF	R								R	–	R	R	–	–	+
–	–	R	–	F	R	F	–	C	VF	R	VF	R	–	–	+	–	F	–	R	R	R!	R	F	–	–	R
F		R	MR	F		R		C	R	R		R					R			X	–	R	R			R
X		X	–	R	R	+		R	R	X	X	X						+	X	X	X	R	+		X	X
		+	–	R		+		R	R		+	+							+			X	+			
		X	–	R	–	+		R	R		+						X	X	+	+			+			+
		+	–	R		R		R	R		–							+		–			–			+
			–																							
			–	R		+			+											X			X			
			–																	X			X		X	X
		R	–	R	–	+		R	+			X					–	+	X	–		X	–			X
		MR	–	R		+		F	+		?							R					R			
		F	+	C	–	R		F	R		R	R					R	R	R	+	–	–	R		–	–
		F	VF	F		MR			F		R	R					R	R	X	+			VR		–	
–		F	F	F		MR		F	MR		–	R					R		–	+			R		–	
–		F		F	–	R		F	+		–	R					R	–	–	R			–		–	+
			MR			X		+	R		X									X						
		R	+	F		R		R	+		?	X					+		–	–			?			X
			–			X		R			+									+						
			–					–																		
		–	–	R		–		–	–			X					–	+		–			–			–
		X	–	R		–		–	–			–					–	R					–			–
			–						–										–	–						
		X	–	R		–		–	–			–					–	+	–				–			–
								–																		
		X		R		–		–	–								X	+	–							
				R		–		–	–									R	–	–						
				F		–		–	–			–						R		–						
				F		–		–	–			–						R		–						
				F		–		–	–			–						R								
				VF		–		–	–			–														
						–		–	–			–						+								
						–		–	–																	

Table 13.3
Occurrence charts for DSDP Sites 146, 237, and 313

EPOCH	Nannofossil Zones (1)	Radiolarian Zones (this chapter)	LEG 15 SITE 146	mbsf	ABUNDANCE	PRESERVATION	*Amphicraspedum murrayanum*	*Amphicraspedum prolixum*	*Bekoma bidartensis*	*Bekoma campechensis*	*Buryella clinata*	*Buryella pentadica*	*Buryella tetradica*	*Calocycloma castum*	*Dictyoprora urceolus*	*Giraffospyris lata*	*Lamptonium fab. chaunothorax*	*Lamptonium fab. constrictum*	*Lamptonium fab. fabaeforme*	*Lamptonium pennatum*	*Lamptonium sanfilippoae*
EOCENE			146-3-CC	406.0	F	M					-										
		B. clinata	146-4-CC	413.0	C	M			–	–	R		–								
	CP9		146-5 Top	422.0	C	P					?										
			146-5 Bottom	425.0	R	P															
			146-6-1,117-122	432.2	F	M	X	X			X		R			X			X	X	
PALEOCENE	CP8b	*Bekoma bidartensis*	146-6-CC	432.5	F	P	+	+	X	X			X						X	X	
			146-7-1,35-40	440.4	C	M		+	–	–		–	F	–		R			+	+	
			146-7-1,83-86	440.8	C	G	R	X	–	–	–	–	C	–		–	X		X	R	
			146-7-CC	441.5	R	P	X	X	X	X	–	R	+			X			X	X	
			146-8-1,124-128	450.3	R	P															
			146-8-2,115-120	451.7	R	P															
			146-8-CC	452.0	F	P															
			146-9 Top	458.0	R	M															
			146-9-2,15-18	459.7	R	P															
			146-9-2,100-102	460.5	C	P						F	F								
			146-9 Bottom	461.0	B	–															
		Bekoma campechensis	146-10-1,91-98	467.9	C	P															
			146-10-2,25-28	468.8	C	M							C								
			146-10-2,85-90	469.4	A	M			–	–			C							F	
			146-10-2,138-142	469.9	C	M			–	–			C							F	
			146-10-CC	470.0	F	P			–	–			X							X	

EPOCH	(2)	Radiolarian Zones	LEG 24 SITE 237	mbsf	ABUNDANCE	PRESERVATION	*A. murrayanum*	*A. prolixum*	*B. bidartensis*	*B. campechensis*	*B. clinata*	*B. pentadica*	*B. tetradica*	*C. castum*	*D. urceolus*	*G. lata*	*L. fab. chaunothorax*	*L. fab. constrictum*	*L. fab. fabaeforme*	*L. pennatum*	*L. sanfilippoae*
EOCENE	CP12	*Buryella clinata*	237-36-1, ca. 130	330.8	C	G	F	–	–	–	F	–	+	MR	R	R	+	–	+	–	–
	CP10		237-37-1, ca. 133	340.3	C	M	F	+	–	–	R	–	R	MR	R	+	+	+	R	–	–
	CP9		237-38-1, ca. 90	345.9	C	M	R	+	–	–	R	–	R	MR	+	R	+	–	+	–	–
PAL	CP8a	*Bekoma bidartensis*	237-41-1, ca. 70	383.7	C	M	R	–	+	+	–	–	F	–	–	–	–	–	–	R	–
	CP6		237-42-1, ca. 72	393.2	F	M	R		–	+	–	–	F	–	–	–	–	–	–	R	–

EPOCH	(3)	Radiolarian Zones	LEG 32 SITE 313	mbsf	ABUNDANCE	PRESERVATION	*A. murrayanum*	*A. prolixum*	*B. bidartensis*	*B. campechensis*	*B. clinata*	*B. pentadica*	*B. tetradica*	*C. castum*	*D. urceolus*	*G. lata*	*L. fab. chaunothorax*	*L. fab. constrictum*	*L. fab. fabaeforme*	*L. pennatum*	*L. sanfilippoae*
EOCENE	CP11	*Buryella clinata*	313-9-1,134-136	187.9	A	G	F	F	–	–	C	–	–	F	C	–	R	R	R	–	–
			313-9-1,144-150 MRC	188.0	A	G	F	R	–	–	C	–	–	F	C	–	F	R	F	–	–
	CP10		313-12-1,144-150 MRC	206.5	A	G	F	R	R	–	R	–	F	R	R	F	R	–	R	R	+
			313-12-2,34-36	206.8	A	G	R	R	R	–	VR	–	C	R	R	F	R	–	R	R	+
		Bekoma bidartensis	313-12-3,58-64 MRC	208.6	A	G	+	R	R	–	–	–	VF	R	R	C	R	–	F	R	–
			313-12-5,129-131	212.3	A	G	R	VR	R	–	–	–	F	F	–	C	R	–	F	R	–
			313-13-1,57-59	214.6	A	G	+	R	R	–		–	F	VF	–	C	R	–	R	–	–
			313-13-1,70-76 MRC	214.7	A	G	–	R	R	–		–	C	R	–	C	–	–	R	+	–
	CP9/CP10		313-13-5,64-70 MRC	220.7	A	G	+	R	R	–		–	F	R	–	C	–	–	F	R	+
			313-13-6,139-141	222.9	A	G	+	R	R	–		–	C	R	–	C	VR	–	F	R	+

For explanation of symbols, see table 13.1
[1]Hay & Beaudry, 1973; Bukry, 1973
[2]Bukry, 1974; Wei, pers. commun., 1995
[3]Bukry, 1975; Wei, pers. commun., 1995

Table 13.3 *(continued)*
Occurrence charts for DSDP Sites 146, 237, and 313

Lithochytris archaea	*Lophocyrtis biaurita*	*Lychnocanoma auxilla*	*Phormocyrtis cubensis*	*Phormocyrtis striata exquisita*	*Phormocyrtis striata striata*	*Phormocyrtis turgida*	*Podocyrtis acalles*	*Podocyrtis papalis*	*Pterocodon ampla*	*Pterocodon anteclinata*	*Pterocodon poculum*	*Pterocodon tenellus*	*Spongatractus balbis*	*Spongatractus pachystylus*	*Spongodiscus cruciferus*	*Spongodiscus phrix*	*Spongodiscus quartus quartus*	*Stylotrochus alveatus*	*Theocorys acroria*	*Theocorys phyzella*	*Theocotyle nigriniae*	*Theocotylissa alpha*	*Theocotylissa auctor*	*Theocotylissa ficus*	*Thyrsocyrtis hirsuta*	*Thyrsocyrtis tarsipes*
								+																		
				–	+			+																		
		X		R	X	+		X	X								X	X	X	X						X
		+		+		+		+	R						VR				X							
		R		F	R	+		R	R						VR		+	R	R	R						+
		–		F	–	–		+	–			–			F		R	R	X	+			–			X
		X		R	–	X		X	X			–					X	+	X	+						X
												X														
				R		–												+								
				–		–												R								
				–		–											X	R								
				–		–												X								
R	R	–	–	+	F	–	–	R	–	–	–	–			–	–	–	–	–	+	R	+	+	R	R	
R	+	–	–	+	R	–	–	R	–	–	–	–			–	–	+	–	–	+	+	+	+	+	+	
R	–	–	–	R	R	–	–	R	–	–	–	R			–	–	R	–	–	R	+	VF	+	–	R	
–	–	R	+	F	+	–	–	–	–	X	–	X			R	–	–	R	–	–	–	–	–	–	–	
–	–	R	+	F	–		–	–	–	X	–	X			R	–	–	R	–	–	–	–	–	–	–	
R	F	–	–	R	F	–	F	F	–		–	–	R	R	–	F	–	–	F	–	C	–	–	C	F	–
R	+	–	–	F	F	–	F	F	–	–	–	–	R	R	–	R	–	–	F	–	F	–	–	C	F	–
R	–	–	R	C	R	R	–	F	–	F	–	R	–	–	–	–	R	–	R	F	R	F	–	–	R	–
R	–	–	R	C	VR	R	–	F	–	F	–	F	–	–	–	–	R	–	R	F	F	F	–	–	R	VR
R	–	–	R	C	VR	F	–	F	–	F	–	F	–	–	–	–	R	–	R	F	R	F	–	–	R	–
R	–	–	R	C	R	F	–	F	!	F	–	F	–	–	–	–	R	R!	R	F	F	F	–	–	R	–
F	–	–	R	C	R	R	–	C	–	VF	–	F	–	–	–	–	R	–	R	R	R	VF	–	–	R	–
F	–	–	R	C	+	R	–	F	–	VF	–	F	–	–	–		R	–	R	R	R	+	–	–	R	+
MR	–	–	R	C	–	F	–	F	–	R	–	F	–	–	–		R	–	–	R	R	F	–		R	–
MR	–	–	R	C	R	F	–	F	–	R	–	F	–	–	+		R	–	R	R	R	F	–		R	–

Table 13.4
Occurrence charts for DSDP Sites 355, 384, and 390A

EPOCH	Nannofossil Zones (1)	Radiolarian Zones (this chapter)	LEG 39 SITE 355	mbsf	ABUNDANCE	PRESERVATION	*Amphicraspedum murrayanum*	*Amphicraspedum prolixum*	*Bekoma bidartensis*	*Bekoma campechensis*	*Buryella clinata*	*Buryella pentadica*	*Buryella tetradica*	*Calocycloma castum*	*Dictyoprora urceolus*	*Giraffospyris lata*	*Lamptonium fab. chaunothorax*	*Lamptonium fab. constrictum*	*Lamptonium fab. fabaeforme*	*Lamptonium pennatum*	*Lamptonium sanfilippoae*
EOCENE	CP13	*P. striata*	355-10-2,24-27	339.8	C	M/P	VR				F		–	MR			R	+	R		R
	CP11	*B. clinata*	355-11-2,100-103	350.0	A	M	VR				F		–	R			MR	MR	R		R
			355-12-2,75-77	359.3	C	M	VR		–		F		VR	R		–	VR	–	R		VR
			355-12-5,50-53	363.5	F	P	R				F		VR	R			VR	–	VR		VR
			355-13-2,52-55	368.5	C	P	VR		X		R		MR	VR		X		–			–
			355-13-3,120-123	370.7	F	VP	R				R		VR				X		X		X

EPOCH	Nannofossil Zones (2)	Radiolarian Zones (this chapter)	LEG 43 SITE 384	mbsf	ABUNDANCE	PRESERVATION	*Amphicraspedum murrayanum*	*Amphicraspedum prolixum*	*Bekoma bidartensis*	*Bekoma campechensis*	*Buryella clinata*	*Buryella pentadica*	*Buryella tetradica*	*Calocycloma castum*	*Dictyoprora urceolus*	*Giraffospyris lata*	*Lamptonium fab. chaunothorax*	*Lamptonium fab. constrictum*	*Lamptonium fab. fabaeforme*	*Lamptonium pennatum*	*Lamptonium sanfilippoae*
EOC	CP12	*P. striata*	384-5-3,93-99	92.9	C	G	–	–	–	–	MR	–	–	–	R	–	–	MR	–	–	–
			384-5-3,138-143	93.4	VR	–															
			384-5-3,142-149 MRC	93.5	F	MG	–	–	–	–	R	–	–	R	R	–	–	–	R	–	–
PALEOCENE	CP8a	*B. clinata*	384-6-1,74-80	99.4	A	G	+	+	F	VF	–	–	C	–	–	–	–	–	–	VF	–
	CP7		384-6-3,83-89	102.5	A	G	+	R	F	F	–	–	C	–	–	–	–	–	–	F	–
			384-6-3,102-109 MRC	102.7	A	G	+	R	F	+	–	–	C	–	–	–	–	–	–	+	–
			384-6-4,89-94	104.0	A	G	R	R	F	R	–	–	C	–	–	–	–	–	–	F	–
			384-7-2,13-18	109.7	A	G	R	–	F	R	–	–	C	–	–	–	–	–	–	F	–
			384-7-2,52-57	110.0	A	G	+	–	F	R	–	–	C	–	–	–	–	–	–	F	–
			384-7-4,91-96	113.4	C	MG	+	–	VR	VF	–	–	C	–	–	–	–	–	–		–
	CP6		384-7-6,62-66	116.1	C	G	+	–	VR	F	–	VF	C	–	–	–	–	–	–	MR	–
		Bekoma campechensis	384-7-6,119-125	116.7	C	G	R	–	–	F	–	VF	C	–	–	–	–	–	–	MR	–
			384-8-2,45-50	119.3	C	G	–	–	–	MR		VF	F				–	–	–	MR	
			384-8-2,108-113	119.9	A	G	–	–	–	C		F	F				–	–	–	MR	
			384-8-2,114-121 MRC	120.0	A	G	–		–	MR		F	MR				–	–	–	VF	
	CP5		384-8-4,52-59	122.4	A	P	–		–								–	–	–		
			384-8-4,65-71	122.5	B	–		–									–	–	–		
			384-8-5,108-113	124.4	C	M	–		–	F		F	C				–	–	–	R	
			384-8-5,113-120 MRC	124.5	F	M						+					–	–	–		
	CP4		384-9-1,125-130	128.2	F	M		–				R	?				–	–	–		
			384-9-2,102-107	129.4	A	M	–	–	–	F		–	C				–	–	–	+	
			384-9-3,84-90	130.8	C	G	–		–	F		F	F				–	–	–	+	
			384-9-4,103-110 MRC	132.5	A	G			–	F		R	R				–	–	–	+	
			384-9-5,56-61	133.5	A	G				F		MR	R				–	–	–	R	
			384-9-5,69-75	133.6	A	MG				F							–	–	–	R	
			384-10-1,90-95	137.3	A	G				F		MR					–	–	–	MR	
			384-10-1,115-120	137.6	C	M				R		R	R				–	–	–	R	

EPOCH	Nannofossil Zones (3)	Radiolarian Zones (this chapter)	LEG 44 HOLE 390A	mbsf	ABUNDANCE	PRESERVATION	*Amphicraspedum murrayanum*	*Amphicraspedum prolixum*	*Bekoma bidartensis*	*Bekoma campechensis*	*Buryella clinata*	*Buryella pentadica*	*Buryella tetradica*	*Calocycloma castum*	*Dictyoprora urceolus*	*Giraffospyris lata*	*Lamptonium fab. chaunothorax*	*Lamptonium fab. constrictum*	*Lamptonium fab. fabaeforme*	*Lamptonium pennatum*	*Lamptonium sanfilippoae*
EOC.	?	*P. striata*	390A-6-1,50-57	57.5	A	G	–	F	–	–	F	–	–	MR	R	–	–	R	R	–	–
	CP11		390A-6-6,50-57	65.0	A	G	+	R	–	–	F	–	–	MR	R	–	–	R	R	–	–
	CP10		390A-7-4,100-107	72.0	A	G	VR	R	–	–	VF	–	VF	MR	R	MR	R	–	R	R	+

For explanation of symbols, see table 13.1

[1]Perch-Nielsen, 1977

[2]Okada and Thierstein, 1979

[3]Bukry, 1978

Table 13.4 *(continued)*
Occurrence charts for DSDP Sites 355, 384, and 390A

Lithochytris archaea	*Lophocyrtis biaurita*	*Lychnocanoma auxilla*	*Phormocyrtis cubensis*	*Phormocyrtis striata exquisita*	*Phormocyrtis striata striata*	*Phormocyrtis turgida*	*Podocyrtis acalles*	*Podocyrtis papalis*	*Pterocodon ampla*	*Pterocodon anteclinata*	*Pterocodon poculum*	*Pterocodon tenellus*	*Spongatractus balbis*	*Spongatractus pachystylus*	*Spongodiscus cruciferus*	*Spongodiscus phrix*	*Spongodiscus quartus quartus*	*Stylotrochus alveatus*	*Theocorys acroria*	*Theocorys phyzella*	*Theocotyle nigriniae*	*Theocotylissa alpha*	*Theocotylissa auctor*	*Theocotylissa ficus*	*Thyrsocyrtis hirsuta*	*Thyrsocyrtis tarsipes*
R	X	–	–	R	R			F					R	VR	X				+	–	R	–	–	R	MR	X
R	X	–	–	R	R			F					R	–	–				R	–	X	–	–	VF	MR	–
R		–	VR	MR	VF			F	–		–	–	R	–	VR				R	–	R	–	–	R	R	–
R	VR	–	VR	R	R			F				–	–	–	X		+		R	X	VR	–	–	R	R	–
X	VR	X		R	VR			R	X		X	–	–	–					X	X	VR	VR	X	R	VR	X
X	–	X		R				R				–		–					X	VR	X	VR	X	X	X	X
–	F	–	–	–	VF	–	R	F	–	–	–	–	–			R	–	–	VR	–	R	–	–	MR	VF	–
–	F	–	–	–	R	–	R	F	–	–	–	–	+	–	–	R	–	–	R	–	R	–	+	–	MR	–
–	–	R	–	R	R	–	–	–	–	–	MR	VR	–	–	R	–	+	F		–	–	–	–	–	–	–
–	–	R	–		R	–	–	–	–	–	MR	VR	–	–	MR	–	–	F	–	–	–	–	–	–	–	–
–	–	R	–	F	R	–	–	–	–	–	MR	VR	–	–		–	–		–	–	–	–	–	–	–	–
–	–	R	–	F	R	–	–	–	–	–	MR	VR	–	–	MR	–	–	F	–	–	–	–	–	–	–	–
–	–	VR	–		R	–	–	–	–	–	R	VR	–	–	F	–	–	F	–	–	–	–	–	–	–	–
–	–	VR	–		R	–	–	–	–	–	R	VR	–	–	R	–	–		–	–	–	–	–	–	–	–
–	–	VR	–	C	R	–	–	–	–	–	R		–	–	–	–	–		–	–	–	–	–	–	–	–
–	–	VR	–	F	R	–	–	–	–	–	R	R	–	–	–	–	–	R	–	–	–	–	–	–	–	–
–	–	VR		F	–	–	–	–	–	–	R	+	–	–	–	–	–	VF	–	–	–	–	–	–	–	–
		VR	–	F	–	–	–	–	–		R	+	–		–	–	–		–	–	–	–	–	–	–	–
		VR	–	F	–	–		–	–		R	–	–		–			F	–	–	–	–	–	–		
		–	–	VF	–	–		–	–		–	–	–		–				–	–	–	–	–	–		
		–							–		–	–	–						–	–	–	–	–	–		
																				–						
		–	–	F					–		–	–				–		R		–						
		–		R					–		–									–						
		–		X																						
				R							–	–						–								
				R		–					–	–						R								–
		–	–	VR		–		–	–			–							–	–			–			–
			–	R		–					–							R								–
			–	R		–												R								–
			–	–		–			–							–										–
				–												–		R								
–	R	–	–	–	F	–	VF	C	–	–	–	–	R	R	R	R	–	–	R	–	F	–	–	C	F	–
–	R	–	–	–	F	–	VF	C	–	X	–	–	+	R	R	R	–	–	R	–	F	–	F	F	F	–
R	VF	+	R	R	F	–	–	F	–	X	–	–	–	–	R	–	R	–	R	VF	VF	R	VF	VF	F	+

Table 13.5
Occurrence charts for DSDP Site 524 and ODP Hole 603B

EPOCH	Nannofossil Zones (Percival 1984)	Radiolarian Zones (this chapter)	LEG 73 SITE 524	mbsf	ABUNDANCE	PRESERVATION	*Amphicraspedum murrayanum*	*Amphicraspedum prolixum*	*Bekoma bidartensis*	*Bekoma campechensis*	*Buryella clinata*	*Buryella pentadica*	*Buryella tetradica*	*Calocycloma castum*	*Dictyoprora urceolus*	*Giraffospyris lata*	*Lamptonium fab. chaunothorax*	*Lamptonium fab. constrictum*	*Lamptonium fab. fabaeforme*	*Lamptonium pennatum*	*Lamptonium sanfilippoae*
PALEOCENE	CP8	*B. bidartensis*	524-4-3,88-94 MRC	51.4	A	G	–	–	+	+	–	–	C	–	–	–	–	–	–	R	–
			524-4-3,98-103	51.5	F	M	–	–	–	–	–	–	F	–	–	–	–	–	–	–	–
			524-4-4,40-45	52.4	A	G	–	–	VF	?	–	–	C	–	–	–	–	–	–	R	–
			524-5-1,68-75 MRC	57.7	A	G	–	–	R	–	–	–	C	–	–	–	–	–	–	+	–
			524-5-1,97-102	58.0	A	G	–	–	F	–	–	–	F	–	–	–	–	–	–	R	–
			524-5-2,68-73	59.2	C	G	–	–	–	–	–	–	F	–	–	–	–	–	–	–	–
			524-5-4,121-126	62.7	R	M	–	–	+	–	–	–	F	–	–	–	–	–	–	–	–
	CP7		524-5-5,114-119	64.2	F	M	–	–	–	–	–	–	F	–	–	–	–	–	–	–	–

EPOCH	Nannofossil Zones	Radiolarian Zones	LEG 93 HOLE 603B	mbsf	ABUNDANCE	PRESERVATION	*Amphicraspedum murrayanum*	*Amphicraspedum prolixum*	*Bekoma bidartensis*	*Bekoma campechensis*	*Buryella clinata*	*Buryella pentadica*	*Buryella tetradica*	*Calocycloma castum*	*Dictyoprora urceolus*	*Giraffospyris lata*	*Lamptonium fab. chaunothorax*	*Lamptonium fab. constrictum*	*Lamptonium fab. fabaeforme*	*Lamptonium pennatum*	*Lamptonium sanfilippoae*
EOCENE	Barren	*P. striata*	603B-17-1,91-92	976.0	F	VP								+							
			603B-17-3,118-126 MRC	979.3	C	P							R	+						X	
		B. cl.	603B-17-4,20-22	979.8	A	P	R	R	R		R		R	+		+	R			–	
		Bekoma bidartensis	603B-18-1,7-9	984.6	A	VP		+	+	–			R					–			
			603B-18-2,134-136	987.4	C	VP		+	R		X		R								
PALEOCENE			603B-18-3,105-107	988.7	A	VP	F		+				R								
			603B-19-1,12-14	993.7	A	VP	R	R	R	–			F							+	
			603B-19-1,117-124 MRC	994.8	A	VP	–			–			F								
			603B-19-2,130-132	996.4	A	VP			F	?			F								
			603B-19-3,117-125 MRC	997.8	A	P			R	–			R								
		Bekoma campechensis	603B-20-1,72-74	1003.3	A	VP			+	R			F							+	
			603B-20-1,120-127 MRC	1003.8	A	VP				R			R								
			603B-20-2,35-37	1004.5	A	VP	+			?			+								
			603B-21-1,46-48	1012.1	F	VP			+	+			R								
			603B-21-3,122-124	1015.8	R	VP															
			603B-22-1,109-111	1021.7	R	VP						R	+								
			603B-22-1,132-138 MRC	1021.9	F	P				+		F	F								

For explanation of symbols, see table 13.1

Table 13.5 *(continued)*
Occurrence chart for DSDP Site 524 and ODP Hole 603B

Lithochytris archaea	*Lophocyrtis biaurita*	*Lynchnocanoma auxilla*	*Phormocytis cubensis*	*Phormocytis striata exquisita*	*Phormocyrtis striata striata*	*Phormocyrtis turgida*	*Podocyrtis acalles*	*Podocyrtis papalis*	*Pterocodon ampla*	*Pterocodon anteclinata*	*Pterocodon poculum*	*Pterocodon tenellus*	*Spongatractus balbis*	*Spongatractus pachystylus*	*Spongodiscus cruciferus*	*Spongodiscus phrix*	*Spongodiscus quartus quartus*	*Stylotrochus alveatus*	*Theocorys acroria*	*Theocorys phyzella*	*Theocotyle nigriniae*	*Theocotylissa alpha*	*Theocotylissa auctor*	*Theocotylissa ficus*	*Thyrsocyrtis hirsuta*	*Thyrsocyrtis tarsipes*
–	–	VF	–	F	R	–	–	–	–	–	R	–	–	–	–	–	–	R	–	–	–	–	–	–	–	–
–	–	R	–	F	R	–	–	–	–	–	MR	–	–	–	–	–	–	MR	–	–	–	–	–	–	–	–
–	+	F	R	F	F	–	–	–	–	–	F	+	–	–	–	–	–	R	–	–	–	–	–	–	–	–
–	+	MR	–	F	F	–	–	–	–	–	R	–	–	–	–	–	–	R	–	+	–	–	–	–	–	–
–	–	MR	–	VF	VF	–	–	–	–	–	R	–	–	–	–	–	–	MR	–	–	–	–	–	–	–	–
–	–	MR	–	R	R	–	–	–	–	–	R	–	–	–	–	–	–	R	–	–	–	–	–	–	–	–
–	–	R	–	F	+	–	–	–	–	–	R	–	–	–	–	–	–	+	–	–	–	–	–	–	–	–
–	–	R	–	MR	VR	–	–	–	–	–	R	–	–	–	–	–	–	+	–	–	–	–	–	–	–	–

Lithochytris archaea	*Lophocyrtis biaurita*	*Lynchnocanoma auxilla*	*Phormocytis cubensis*	*Phormocytis striata exquisita*	*Phormocyrtis striata striata*	*Phormocyrtis turgida*	*Podocyrtis acalles*	*Podocyrtis papalis*	*Pterocodon ampla*	*Pterocodon anteclinata*	*Pterocodon poculum*	*Pterocodon tenellus*	*Spongatractus balbis*	*Spongatractus pachystylus*	*Spongodiscus cruciferus*	*Spongodiscus phrix*	*Spongodiscus quartus quartus*	*Stylotrochus alveatus*	*Theocorys acroria*	*Theocorys phyzella*	*Theocotyle nigriniae*	*Theocotylissa alpha*	*Theocotylissa auctor*	*Theocotylissa ficus*	*Thyrsocyrtis hirsuta*	*Thyrsocyrtis tarsipes*
					R			+																		
					R	+		R					+							+			+			
			R	+	R	–		R									R	–		R		R	–			
			R	R		R														R						
			+	R		R																				
				R																R						
			+	R														–								
					+	+		R																		
			?	+	+	R																				
				+		R																	+			
				R														–		–						
						+												–								
																		–		–						
				R																						

Table 13.6
List of radiolarian events in terms of meters below seafloor

Event No.	RADIOLARIAN EVENTS	Leg 10 Site 86		Leg 10 Site 94		Leg 10 Site 96		Leg 15 Site 152		Leg 24 Site 237	
1	T *Calocycloma*			above						above	
	castum			551.3	G					330.8	G
2	T *Spongodiscus*										
	phrix										
3	T *Lamptonium*			above		above		above		above	
	fab. fabaeforme			551.3	G	301.4	G	153.9	G	330.8	M
4	T *Amphicraspedum*			above		above		above		**330.8**	
	prolixum			551.3	G	301.4	G	153.9	G	**340.3**	P
5	T *Amphicraspedum*			above		above		above		above	
	murrayanum			551.3	G	301.4	M	153.9	M	330.8	G
6	T *Lithochytris*			above		above				above	
	archaea			551.3	G	301.4	M			330.8	G
7	T *Phormocyrtis*	above		above		above		above		above	
	striata exquisita	504.5	G	551.3	G	301.4	G	153.9	G	330.8	P
8	T *Lamptonium*			above							
	sanfilippoae			551.3	M						
9	T *Giraffospyris*			551.3		above		above		above	
	lata			570.8	G	301.4	G	153.9	G	330.8	M
10	B *Spongatractus*										
	pachystylus										
11	T *Buryella*	above		551.3		above		above		above	
	tetradica	504.5	G	570.8	G	301.4	G	153.9	M	330.8	M
12	B *Podocyrtis*										
	acalles										
13	T *Spongodiscus*			551.3		above		above		330.8	
	quartus quartus			570.8	G	301.4	G	153.9	G	340.3	M
14	B *Lamptonium*			**570.8**							
	fab. constrictum			**571.3**	G						
15	T *Spongodiscus*			**above**		above				345.9	
	cruciferus			**551.3**	G	301.4	M			383.7	M
16	T *Phormocyrtis*			551.3		above				345.9	
	cubensis			570.8	G	301.4	M			383.7	P
17	T *Lamptonium*	above		551.3		above		above		345.9	
	pennatum	504.5	G	570.8	G	301.4	G	153.9	G	383.7	M
18	T *Theocotylissa*			571.3				above		**above**	
	alpha			572.4	G			153.0	G	**330.8**	P
19	T *Pterocodon*			573.5		above		above			
	ampla			589.1	G	301.4	G	153.9	G		
20	B *Theocotylissa*			573.5						**340.3**	
	ficus			589.1	G					**345.9**	M
21	T *Theocorys (?)*			573.5		above		above		**above**	
	phyzella			589.1	G	301.4	G	153.9	G	**330.8**	M
22	T *Theocotylissa*			**573.1**		above		above		**above**	
	auctor			**573.5**	G	301.4	G	153.9	G	**330.8**	P
23	B *Spongartractus*			573.5							
	balbis			589.1	M						
24	T *Phormocyrtis*			573.5		above		above			
	turgida			589.1	G	301.4	G	153.9	G		
25	T *Pterocodon (?)*			573.5		above		above		**340.3**	
	tenellus			589.1	G	301.4	M	153.9	G	**345.9**	M

T = morphotypic high highest occurrence, B = morphotypic lowest occurrence. The quality of each event is described as G = good, M = moderate, and P = poor. Bold type indicates out-of-sequence events. See text for further explanation of symbols.

Table 13.6
List of radiolarian events in terms of meters below seafloor

Leg 32 Site 313		Leg 39 Site 355		Leg 43 Site 384		Leg 44 Hole 390A		Leg 73 Site 524		Leg 93 Hole 603B	
above		above				above					
187.9	G	339.8	G			57.5	G				
above						above					
187.9	G					57.5	G				
above		above				above					
187.9	G	339.8	G			57.5	G				
above				92.9		above				above	
187.9	G			99.4	M	57.5	G			984.6	M
above		above		93.5		57.5				above	
187.9	G	339.8	P	99.4	M	65.0	P			984.6	G
above		above				65.0					
187.9	G	339.8	G			72.0	M				
above		above		93.5		65.0		above		above	
187.9	G	339.8	G	99.4	G	72.0	G	51.4	G	984.6	G
188.0		above				65.0					
206.5	P	339.8	G			72.0	P				
188.0						65.0					
206.5	G					72.0	G				
188.0		339.8				65.0					
206.5	G	350.0	P			72.0	G				
188.0		350.0		93.5		65.0		above		above	
206.5	G	359.3	M	99.4	G	72.0	G	51.4	G	984.6	G
188.0				93.5		65.0					
206.5	G			99.4	G	72.0	G				
188.0						65.0					
206.5	G					72.0	G				
188.0		350.0		92.9		65.0					
206.5	G	359.3	G	99.4	M	72.0	G				
220.7		350.0		93.5		**above**					
222.0	P	359.3	P	99.4	G	**57.5**	G				
188.0		350.0				65.0		**51.5**		above	
206.5	G	359.3	P			72.0	M	**52.4**	P	984.6	M
188.0				93.5		65.0		above		**988.7**	
206.5	G			99.4	G	72.0	M	51.4	P	**993.7**	P
188.0		**363.5**				65.0					
206.5	G	**368.5**	P			72.0	M				
188.0		**below**		92.9		**below**				above	
206.5	G	**368.5**	P	99.4	G	**72.0**	G			976.0	M
188.0		359.3				65.0				976.0	
206.5	G	370.7	P			72.0	G			979.3	P
						57.5				976.0	
						65.0	G			979.3	P
188.0		359.3				65.0				979.3	
206.5	G	363.5	M			72.0	P			979.8	P
188.0										above	
206.5	G									984.6	M
188.0				93.5							
206.5	G			99.4	M						

Table 13.6 *(continued)*
List of radiolarian events in terms of meters below seafloor

Event No.	RADIOLARIAN EVENTS	Leg 10 Site 86		Leg 10 Site 94		Leg 10 Site 96		Leg 15 Site 152		Leg 24 Site 237	
26	T *Bekoma*			573.5		above				345.9	
	bidartensis			589.1	G	301.4	G			383.7	P
27	B *Spongodiscus*										
	phrix										
28	T *Thyrsocyrtis*			573.5		above		above			
	tarsipes			589.1	M	301.4	G	153.9	G		
29	T *Pterocodon (?)*			573.5		above					
	anteclinata			589.1	M	301.4	G				
30	B *Lophocyrtis*			**589.1**						**340.3**	
	biaurita			**609.1**	G					**345.9**	M
31	T *Lychnocanoma*			573.5		above		above		345.9	
	auxilla			589.1	G	301.4	G	153.9	G	383.7	M
32	B *Buryella*			589.1						345.9	
	clinata			609.1	G					383.7	G
33	B *Dictyoprora*			589.1						345.9	
	urceolus			609.1	G					383.7	M
34	T *Pterocodon (?)*			589.1				above			
	poculum			609.1	P			153.9	G		
35	T *Stylotrochus*	above				**309.5**		155.3		345.9	
	alveatus	504.5	G			**310.3**	M	168.5	P	383.7	M
36	B *Lithochytris*			589.1		302.2		below		345.9	
	archaea			609.1	G	304.4	M	155.3	M	383.7	G
37	B *Lamptonium*			589.1				below		345.9	
	fab. chaunothorax			609.1	G			155.3	M	383.7	M
38	B *Lamptonium*			589.1							
	sanfilippoae			609.1	M						
39	B *Theocotyle*			589.1						345.9	
	nigriniae			609.1	G					383.7	M
40	B *Thyrsocyrtis*			589.1						345.9	
	hirsuta			609.1	M					383.7	G
41	B *Theocotylissa*			589.1				166.0		345.9	
	alpha			609.1	G			167.1	M	383.7	G
42	B *Calocycloma*			589.1						345.9	
	castum			609.1	G					383.7	G
43	B *Pterocodon (?)*			below		309.5		**155.3**			
	anteclinata			609.1	M	310.3	M	**166.0**	M		
44	B *Theocotylissa*			below		310.3		183.8		345.9	
	auctor			609.1	G	321.6	G	184.4	M	383.7	M
45	B *Lamptonium*			below		310.3		184.4		345.9	
	fab. fabaeforme			609.1	G	321.6	G	186.0	M	383.7	M
46	B *Podocyrtis*			below		321.6		192.0		345.9	
	papalis			609.1	G	330.1	G	202.8	G	383.7	G
47	B *Theocorys*			below				**179.3**			
	acroria			609.1	M			**183.8**	P		
48	B *Giraffospyris*			below		below		186.4		345.9	
	lata			609.1	G	321.6	G	192.0	M	383.7	M
49	B *Theocorys (?)*			below		below		**184.4**		345.9	
	phyzella			609.1	G	321.6	G	**186.0**	P	383.7	M
50	B *Thyrsocyrtis*			below		below		**184.4**			
	tarsipes			609.1	P	321.6	G	**207.9**	P		
51	B *Pterocodon*			below		below		186.4			
	ampla			609.1	G	330.1	G	192.0	P		

Table 13.6 *(continued)*
List of radiolarian events in terms of meters below seafloor

Leg 32 Site 313		Leg 39 Site 355		Leg 43 Site 384		Leg 44 Hole 390A		Leg 73 Site 524		Leg 93 Hole 603B	
188.0				93.5				above		above	
206.5	G			99.4	G			51.4	P	984.6	M
188.0				93.5		65.0					
206.5	G			99.4	G	72.0	M				
						65.0					
						72.0	P				
188.0						below					
206.5	G					72.0	P				
188.0		368.5		93.5		below					
206.5	M	370.7	P	99.4	G	72.0	G				
				93.5		65.0		above			
				99.4	G	72.0	P	51.4	G		
206.8		below		93.5		below				979.8	
208.6	M	370.7	G	99.4	G	72.0	G			984.6	P
208.6				93.5		below					
212.3	G			99.4	G	72.0	G				
				93.5				above			
				99.4	G			51.4	G		
				93.5				above			
				99.4	G			51.4	G		
below		below				below				**above**	
222.0	G	363.5	G			72.0	M			**976.0**	P
below		below				below					
222.0	P	363.5	M			72.0	M				
below		363.5				below					
222.0	P	368.5	P			72.0	P				
below		below		93.5		below					
222.0	G	368.5	P	99.4	G	72.0	G				
below		below		93.5		below				above	
222.0	G	368.5	P	99.4	G	72.0	G			976.0	P
below		below				below				979.8	
222.0	G	370.7	P			72.0	M			984.6	P
below		below		93.5		below				979.8	
222.0	G	368.5	P	99.4	M	72.0	G			984.6	M
below											
222.0	G										
				93.5		below					
				99.4	P	72.0	G				
below		below		93.5		below				below	
222.0	G	363.5	M	99.4	M	72.0	G			979.8	P
below		below		93.5		below				994.8	
222.0	G	370.7	M	99.4	G	72.0	G			996.4	P
below				93.5		below					
222.0	M			99.4	M	72.0	G				
below						below					
222.0	G					72.0	G				
below		below				below				988.7	
222.0	G	370.7	P			72.0	G			993.7	M
						below					
						72.0	P				

Table 13.6 *(continued)*
List of radiolarian events in terms of meters below seafloor

Event No.	RADIOLARIAN EVENTS	Leg 10 Site 86		Leg 10 Site 94		Leg 10 Site 96		Leg 15 Site 152		Leg 24 Site 237	
52	T *Bekoma*	above						**208.4**		345.9	
	campechensis	504.5	G		G			**215.3**	P	383.7	P
53	B *Spongodiscus*			below		below		186.4		345.9	
	quartus quartus			609.1		330.0	G	192.0	P	383.7	M
54	B *Amphicraspedum*					below		below		345.9	
	prolixum				G	330.0	G	186.4	P	383.9	P
55	B *Spongodiscus*			below		below					
	cruciferus			609.1		330.0	G				
56	T *Buryella*	above						186.4			
	pentadica	504.5	G		M			192.0	M		
57	B *Phormocyrtis*			below		below		**166.0**		383.7	
	striata striata			609.1		310.3	G	**167.8**	P	393.2	P
58	B *Bekoma*			below		below				383.7	
	bidartensis			609.1	G	310.3	G			393.2	P
59	B *Phormocyrtis*			below		below		186.4		below	
	cubensis			609.1	G	310.3	G	192.0	P	393.2	P
60	B *Amphicraspedum*			below		below		**184.4**		below	
	murrayanum			609.1	G	330.0	G	**186.4**	P	393.2	G
61	B *Pterocodon (?)*			below		below		**184.4**			
	tenellus			609.1	G	321.6	G	**186.4**	M		
62	B *Lychnocanoma*			below		below		186.4		below	
	auxilla			609.1	G	330.1	G	207.9	P	393.2	G
63	B *Pterocodon (?)*			below				192.0			
	poculum			609.1	G			221.3	P		
64	B *Phormocyrtis*			below		below		**186.4**			
	turgida			609.1	P	321.6	G	**192.0**	M		
65	B *Phormocyrtis*	below		below		below		below		below	
	striata exquisita	513.5	G	609.1	G	330.0	G	230.5	G	393.2	G
66	B *Buryella*	below		below		below		below		below	
	tetradica	513.5	G	609.1	G	330.0	G	230.5	G	393.2	G
67	B *Bekoma*	below						below		below	
	campechensis	513.5	G		G			217.0	P	393.2	P
68	B *Stylotrochus*	below						below		below	
	alveatus	513.5	G					221.5	M	393.2	G
69	B *Buryella*	below						below			
	pentadica	513.5	G					230.5	G		
70	B *Lamptonium*	below		below		below		below		below	
	pennatum	513.5	G	609.1	G	321.6	G	230.4	G	393.2	G

Table 13.6 *(continued)*
List of radiolarian events in terms of meters below seafloor

Leg 32 Site 313		Leg 39 Site 355		Leg 43 Site 384		Leg 44 Hole 390A		Leg 73 Site 524		Leg 93 Hole 603B	
				93.5						997.8	
				99.4	G					1003.3	M
below				99.4		below					
222.0	G			102.5	P	72.0	G				
below				104.0		below				993.7	
222.0	G			109.7	M	72.0	G			994.8	P
		below		110.0		below					
		359.3	P	113.4	G	72.0	M				
				113.4						**1015.8**	
				116.1	G					**1021.7**	M
below		below		116.1		below		below			
222.0	M	368.5	M	116.7	M	72.0	G	64.2	P		
below				116.1				below		**1012.1**	
222.0	G			116.7	M			62.7		**1015.8**	P
below		below				below			P	**993.7**	
222.0	G	363.5	P			72.0	M			**994.8**	P
below		below		116.7		below				**1004.5**	
222.0	P	370.7	M	119.3	M	72.0	P			**1012.1**	P
below				119.3							
222.0	G			119.9	M						
				119.9		below		below			
				120.0	M	72.0	P	64.2	G		
				119.9				below			
				120.0	M			64.2	G		
below										**1003.8**	
222.0	G									**1004.5**	P
below		below		133.6		below		below		below	
222.0	G	370.7	M	137.3	G	72.0	M	64.2	G	1021.9	G
below		below		below		below		below		below	
222.0	G	370.7	M	137.6	G	72.0	G	64.2	G	1021.9	G
				below				below		below	
				137.6	G			64.2	P	1021.9	P
				below							
				137.6	G						
				below				below		below	
				137.6	G			64.2	P	1021.9	G
below				below		below		**58.0**		**1003.3**	
222.0	M			137.6	G	72.0	M	**59.2**	P	**1003.8**	P

Cores 18 to 20, but rare in Cores 21 and 22. Coring was continuous with good recovery.

Additional DSDP Sites (8, 95, 153, 164, 165, 356, 366, 368, 386, 605, 612, and 613) also contain Eocene to Paleocene material but the core recovery and/or radiolarian content and preservation are too poor for meaningful analysis.

Results

Events List

Table 13.6 details 70 events (lowest and highest morphotypic occurrences) in the Paleocene to lower Eocene stratigraphic interval arranged chronologically from youngest to oldest. Additional radiolarian events can be found, but are not, as yet, well documented. The meters below seafloor for the pair of samples between which an event takes place are given for each event in each hole. Thus, this list serves to correlate in detail from hole to hole. If one assumes that events are globally synchronous, they should be in the same order at each site and the list has been constructed so as to have the least number of out-of-sequence events. The reliability of each event has been labeled as good, moderate, or poor. Reliability criteria are based on the abundance of the taxon, the quality of preservation of the sample, and the certainty of identification. Because coring was often discontinuous and recovery was poor, clustered events within a particular core or site probably reflect poor recovery or discontinuous coring rather than a hiatus in sedimentation.

Composite Sections and Fence Diagram

In order to understand the sparse and often discontinuous record provided by radiolarian-bearing sediments, we have constructed, using the events list, three composite sections that span a 10-my interval across the Paleocene/Eocene Series boundary (nannofossil Zones CP4 to CP11). Each composite section consists of two or more individual sites (figure 13.3). The most complete composite section (Sites 94, 313, 96, and 384) is shown in the center column. The other two composite sections include sequences with less stratigraphic coverage, and thus a higher proportion of lowest and highest occurrences cluster in the unrecovered intervals.

The key to the numbered lowest and highest occurrences can be found in the events list (table 13.6). Events included within boxes cluster in the unrecovered intervals; there is no significance to the order of events within each box. The lowest morphotypic occurrences of *Buryella clinata* (#32) and *Bekoma bidartensis* (#58) are considered herein to mark the bases of their respective zones and are shown in bold on the diagram. The lowest occurrence of *B. clinata* (#32) is in the interval between Core 237-38 and Core 524-4, in the uppermost part of Core 313-12, and in the interval between Core 355-13 and Core 152-1. The lowest occurrence of *B. bidartensis* (#58) is between Core 524-5 and Core 86-7, within section 6 of Core 384-7, but in Site 152, due to poor preservation, it is absent.

Because of the many unrecovered intervals in our material, the correlation of events between cores shows few crossing lines. Those events that are out of sequence can be explained in the following ways. (1) Because of the scarcity of the species and/or poor preservation we think that the highest occurrence of *Theocotylissa alpha* (#18) in Site 355 is too old, the lowest occurrence of *Theocorys acroria* (#47) in Site 152 is too young, and the highest occurrence of *B. campechensis* (#52) in Site 152 is too old. (2) The lowest occurrence of *Podocyrtis papalis* (#46) appears to be too old in Site 152. However, it is only its presence in the core catcher sample from Core 5 that is inconsistent with its lowest occurrence in other sites. Because that sample is described as being "bioturbated . . . with conchoidal and blocky fractures" (Edgar et al. 1973:345), we suspect that the sample is mixed.

Placement of the Paleocene/Eocene Series Boundary

Currently the Paleocene/Eocene Series boundary corresponds to the base of the Ypresian Stage or the base of the London Clay (see discussion in Berggren and Aubry 1996 and chapter 2) and lies in the middle part of planktonic foraminifera Zone P6a. The upper boundary of Zone P5, which is marked by the highest occurrence of *Morozovella velascoensis*, is estimated at 54.7 Ma (Berggren et al. 1995). On the basis of calcareous nannofossil stratigraphy, the Paleocene/Eocene Series boundary is currently considered to be in the middle part of nannofossil Zone NP10 (= Zone CP9a) at the lowest occurrence of *Tribrachiatus digitalis* (= *T. contortus* morphotype A) and is estimated to be at 54.37 Ma (Aubry 1996). Diatom data for the upper Paleocene to lower Eocene stratigraphic interval in the open ocean is limited, and diatom-bearing sediments that may span the boundary have been described only in ODP Site 752 (see Fourtanier 1991 for a review of available data).

In our most complete composite section (Sites 94, 313, 96, and 384), the published nannofossil data show that Cores 96-3 to 96-5 are Paleocene deposits (Zone CP8b), whereas according to the published foraminifera data, Core 96-3 is Eocene in age (*Globorotalia rex* Zone or Zone P6) and Cores 96-4 and 96-5 are Paleocene in age. In our second most complete composite section

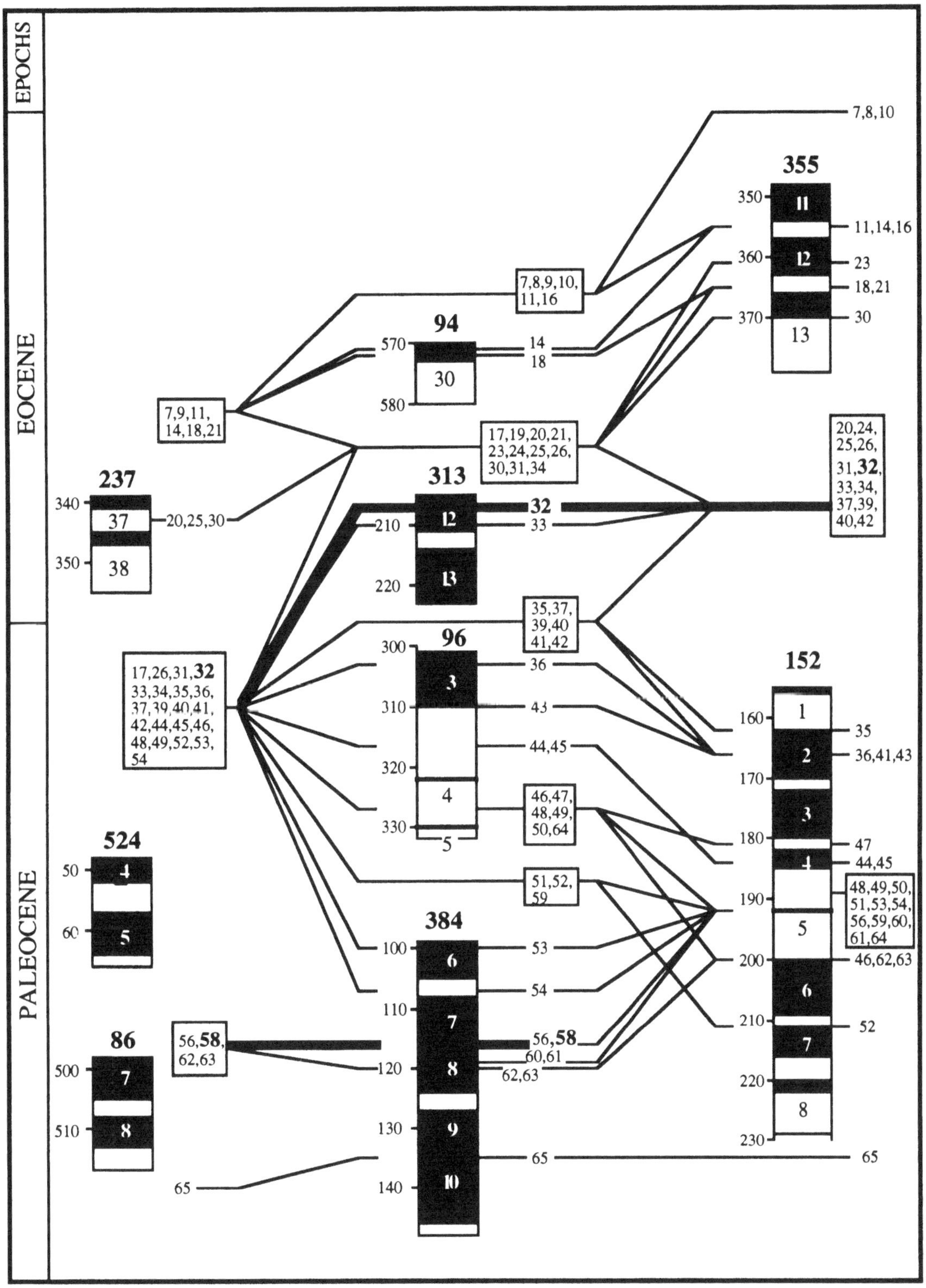

Figure 13.3 Fence diagram showing three composite sections in the Paleocene/Eocene Series boundary interval.

Numbers above each core are DSDP site designations. Numbers within the columns indicate core sections, with core recovery indicated in black. Numbers to the left of each column indicate meters below seafloor. Numbers to the right of each column indicate radiolarian events (see table 13.6 for event associated with each number). The zonal marker events #32 (B *Buryella clinata*) and #58 (B *Bekoma bidartensis*) are shown in bold.

(Sites 355 and 152) the published nannofossil data place the Paleocene/Eocene Series boundary above Core 152-1, but according to the published foraminifera data the boundary is within Core 152-3. These discrepancies will require reexamination by nannofossil and foraminifera experts. However, our radiolarian data favor the nannofossil interpretation.

Unless a direct correlation can be established, positioning of radiolarian events relative to the Paleocene/Eocene Series boundary stratotype and/or the geomagnetic polarity timescale is most easily accomplished by first linking the events to nannofossil zones. In figure 13.4, using data from our most complete composite section, we show the position of the radiolarian zones relative to the nannofossil zones. This composite section includes Core 94-30 (part of the *Buryella clinata* Zone, equivalent to nannofossil Zone CP11) and Cores 313-12 and 313-13 (uppermost part of the *B. bidartensis* Zone, equivalent to nannofossil Zone CP10), but there is some disagreement about the zonal assignment for samples below Core 13-2 (see discussion above), Cores 96-3 to 96-5 (*B. bidartensis* Zone, equivalent to nannofossil Zone CP8b), and Cores 384-6 through 384-10 (*B. bidartensis* and *B. campechensis* Zones, equivalent to nannofossil Zones CP8a through CP4). There are no radiolarian-bearing sediments which can be unequivocally correlated with Zone CP9 in this composite section.

However, it has been determined that the following radiolarian events occur in the unrecovered interval corresponding to nannofossil Zone CP9: the highest occurrence of *Stylotrochus alveatus* and the lowest occurrences of *Calocyclas castum*, *Lamptonium fabaeforme chaunothorax*, *Theocotyle nigriniae*, *Thyrsocyrtis hirsuta*, and *Theocotylissa alpha*.

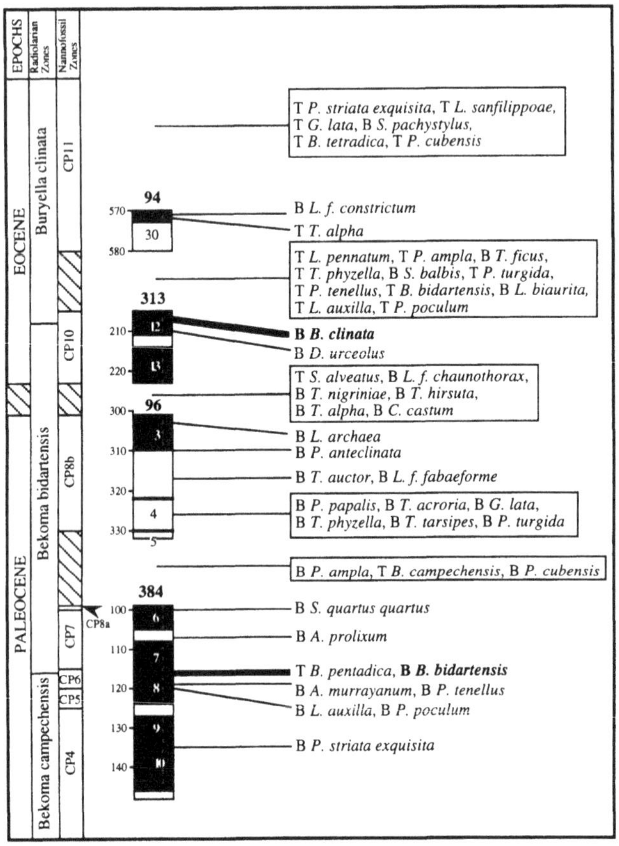

Figure 13.4 Summary diagram showing position of selected radiolarian events in a composite section. The positions of the radiolarian zones in the left-hand column are based on our reexamination of the sedimentary sequences across the Paleocene/Eocene Series boundary in DSDP Sites 94, 313, 96, and 384 (see tables 13.1–13.5).

Hachured intervals indicate uncertainty in limits of zonal or epoch assignment. Numbers to the left of each column indicate meters below seafloor. Numbers within the columns indicate core sections with core recovery shown in black. T = highest occurrence; B = lowest occurrence; the order of events within boxes is not significant. Bold lines and type indicate the position of zonal marker species.

It is the mandate of the Working Group on the Paleocene/Eocene Boundary of the Subcommission on Paleogene Stratigraphy to determine a stratotype for the Paleocene/Eocene Series boundary. The boundary is considered herein to be in the middle part of nannofossil Zone NP10 (= Zone CP9a; Berggren and Aubry 1996; Aubry 1996), in the middle part of foraminiferal Zone P6a, and within the *Bekoma bidartensis* radiolarian Zone. Because radiolarians are absent from all sections currently being considered as potential boundary stratotypes, a two-step correlation process, using co-occurring calcareous microfossils and radiolarians in deep-sea sediments, has been used to determine the placement of the boundary in terms of radiolarian events. We have reexamined all deep-sea sections (DSDP/ODP Legs 1–135) currently available to us, which yield an upper Paleocene to lower Eocene stratigraphic record. In general, the radiolarian fauna is sparse and poorly preserved and documentation of the ranges of radiolarian species is hampered by frequent coring gaps. In addition, none of the radiolarian-bearing sediments reexamined herein belong unequivocally to nannofossil Zone CP9 and, therefore, it is not possible to place the boundary precisely in terms of radiolarians.

However, by piecing together discrete sections from several drilling sites, we have been able to correlate 70 radiolarian events (morphotypic lowest and highest occurrences) with calcareous nannofossil zones on either side of the boundary. Some events are clustered in the unrecovered intervals, but others can be constrained within a site or core so as to provide a reproducible sequence of events. From this we are able to determine the position of numerous radiolarian events relative to calcareous nannofossil zones.

Because the Paleocene/Eocene Series boundary lies

in the middle part of nannofossil Subzone CP9a, the following radiolarian events are at or near the boundary: lowest morphotypic occurrences of *Lamptonium fabaeforme chaunothorax* (#37), *Theocotyle nigriniae* (#39), *Thyrsocyrtis hirsuta* (#40), *Theocotylissa alpha* (#41), *Calocycloma castum* (#42), and *Lithochytris archaea* (#36) and the highest morphotypic occurrence of *Stylotrochus alveatus* (#35).

Published foraminifera data for the cores considered herein place the boundary in a slightly lower position. If that position proves to be closer to the formal boundary, then the following radiolarian events are at or near the boundary: lowest morphotypic occurrences of *Pterocodon* (?) *anteclinata* (#43), *Theocotylissa auctor* (#44), *Lamptonium fabaeforme fabaeforme* (#45), *Podocyrtis papalis* (#46), *Theocorys acroria* (#47), *Giraffospyris lata* (#48), *Theocorys* (?) *phyzella* (#49), *Thyrsocyrtis tarsipes* (#50), and *Phormocyrtis turgida* (#64).

Of these 16 events, recognition of the lowest morphotypic occurrences of *L. archaea*, *P.* (?) *anteclinata*, *T. acroria*, *T.* (?) *phyzella*, and *T. tarsipes* is dependent upon good preservation. The morphotypic limits of both *L. fabaeforme chaunothorax* and *S. alveatus* are poorly defined; because these two species appear to be paleoclimatically sensitive, they are not reliable as universal stratigraphic markers. *Theocotyle nigriniae* and *T. alpha* are members of two rapidly evolving lineages and their identification depends upon precise and consistent application of their morphotypic limits. Because the lowest occurrence of *T. hirsuta* in all three composite sections falls within an unrecovered interval, we are unable to evaluate its usefulness as a marker species. The lowest morphotypic occurrences of *C. castum*, *T. auctor*, *L. fabaeforme fabaeforme*, *P. papalis*, *G. lata*, and possibly *P. turgida* are all easily recognizable events because these species are distinctive, robust, and relatively abundant. Any of these six events could be used as a reliable radiolarian marker to approximate the Paleocene/Eocene Series boundary.

The cohesive order of radiolarian events described herein and their correlation with described nannofossil zones provide a foundation for future placement of the boundary in terms of radiolarians when (and if) the appropriate material becomes available.

ACKNOWLEDGMENTS

We would like to thank Marie-Pierre Aubry for inviting us to participate in IGCP Project 308 (Paleocene/Eocene Boundary Events in Space and Time), for providing us with some useful discussions, and for her careful editing of the manuscript. John Barron and William Riedel kindly reviewed the manuscript and made numerous constructive comments. This work was supported in part by NSF grant number OCE91-23712 (to AS). The bibliography was prepared using RadRefs, a comprehensive computer-based library of all radiolarian literature. This is a contribution toward IGCP Project 308.

REFERENCES

Aubry, M.-P., 1996. Towards an upper Paleocene–lower Eocene high resolution stratigraphy. In Aubry, M.-P. and Benjamini, C., (eds.), Paleocene/Eocene Boundary Events in Space and Time, *Israel Journal of Earth Sciences*, v. 44, p. 239–253.

Aubry, M.-P., Berggren, W. A., Stott, L. and Sinha, A., 1996. The upper Paleocene–lower Eocene stratigraphic record and the Paleocene–Eocene boundary carbon isotope excursion: Implications for geochronology. In Knox, R. W. O'B., Corfield, R. M. and Dunay, R. E., (eds.), *Correlation of the early Paleogene in Northwest Europe*, Geological Society Special Publication No 101, p. 353–380.

Berggren, W. A. and Aubry, M.-P., 1996. A late Paleocene–early Eocene NW European and North Sea magnetobiochronological correlation network. In Knox, R. W. O'B., Corfield, R. M. and Dunay, R. E., (eds.), *Correlation of the Early Paleogene in Northwest Europe*, Geological Society Special Publication No. 101, p. 309–352.

Berggren, W. A., Kent, D. V. and Van Couvering, J. A., 1985. Neogene geochronology and chronostratigraphy. In N. J. Snelling (ed.), *The chronology of the Geological Record:* Geological Society Memoir No 10, Blackwell, Oxford, p. 211–260.

Berggren W. A., Kent, D. V., Swisher, C. C., III and Aubry, M.-P., 1995. A revised Cenozoic geochronology and chronostratigraphy. In Berggren, W. A., Kent, D. V., Aubry, M.-P. and Hardenbol, J., (eds.), *Geochronology Time Scales and Global Stratigraphic Correlation*, Society of Economic Paleontologists and Mineralogists Special Publication No. 54, p. 129–212.

Blome, C. D., 1992. Radiolarians from Leg 122, Exmouth and Wombat Plateaus, Indian Ocean. In von Rad, U., Haq, B. U., et al., Proceedings of the Ocean Drilling Program, Scientific Results, v. 122: College Station, TX, (Ocean Drilling Program), p. 633–652.

Borisenko, N. N., 1958. Radiolyarii paleotsena zapadnoi Kubani [Paleocene Radiolaria of Western Kubanj]. *Trudy Vsesoyuznyi Neftegazovyi Nauchno-Issledovalelskii Institut (VNII), Krasnodarskii Filial*, v. 17, p. 81–100.

Borisenko, N. N., 1960. Novye radiolyarii iz paleotsenovykh otlozhenii Kubani [New radiolarians from the Paleocene deposits of the Kubanj]. *Trudy Vsesoyuznyi Neftegazovyi Nauchno-Issledovalelskii Institut (VNII), Krasnodarskii Filial*, v. 4, p. 199–207.

Brandt, R., 1935. Die Mikropalaeontologie des Heiligenhafener, Kieseltones (OberEozän) Radiolarien; Systematik. In Wetzel, E. O., (ed.), *Jahresbericht des Niedersächsischen geologischen Vereins*, v. 27, p. 48–59.

Bukry, D., 1973. Low-latitude coccolith biostratigraphic zonation. In Edgar, N. T., Saunders, J. B., et al., Initial Reports of the Deep Sea Drilling Project, v. 15: Washington (U.S. Government Printing Office), p. 685–703.

Bukry, D., 1974. Coccolith zonation of cores from the western Indian Ocean and the Gulf of Aden, Deep Sea Drilling Project Leg 24. In Fisher, R. L., Bunce, E. T., et al., Initial Reports of the Deep Sea Drilling Project, v. 24: Washington (U.S. Government Printing Office), p. 995–996.

Bukry, D., 1975. Coccolith and silicoflagellate stratigraphy, northwestern Pacific Ocean, Deep Sea Drilling Project Leg 32. In Larson, R. L., Moberly, R., et al., Initial Reports of the Deep Sea Drilling

Project, v. 32: Washington (U.S. Government Printing Office), p. 683–701.

Bukry, D., 1978. Cenozoic coccolith and silicoflagellate stratigraphy, offshore northwest Africa, Deep Sea Drilling Project Leg 41. In Lancelot, Y., Seibold, E., et al., Initial Reports of the Deep Sea Drilling Project, v. 41: Washington (U.S. Government Printing Office), p. 689–707.

Cande, S. C. and Kent, D. V., 1995. Revised calibration of the geomagnetic polarity time scale for the Late Cretaceous and Cenozoic. *Journal of Geophysical Research,* v. 100(B4), p. 6093–6095.

Clark, B. L. and Campbell, A. S., 1942. *Eocene radiolarian faunas from the Monte Diablo area, California.* Geological Society of America, Special Paper 39, 112 pp.

Dumitrica, P., 1973. Paleocene Radiolaria, DSDP Leg 21. In Burns, R. E., Andrews, J. E., et al., Initial Reports of the Deep Sea Drilling Project, v. 21: Washington (U.S. Government Printing Office), p. 787–817.

Edgar, N. T., Saunders, J. B., et al., 1973. Site 152. Initial Reports of the Deep Sea Drilling Project, v. 15: Washington (U.S. Government Printing Office), p. 331–365.

Ehrenberg, C. G., 1847. Über die mikroskopischen kieselschaligen Polycystinen als mächtige Gebirgsmasse von Barbados und über das Verhältniss der aus mehr als 300 neuen Arten bestehenden ganz eigenthümlichen Formengruppe jener Felsmasse zu den jetzt lebenden Thieren und zur Kreidebildung. Eine neue Anregung zur Erforschung des Erdlebens. *Königliche Preussische Akademie der Wissenschaften zu Berlin,* Bericht, Jahre 1847, p. 40–60.

Ehrenberg, C. G., 1854. *Mikrogeologie. Das Erden und Felsen schaffende Wirken des unsichtbar kleinen selbstständigen Lebens auf der Erde.* Leipzig: Leopold Voss, p. xxviii + 374 p., Atlas, 31 p., 41 pl.

Ehrenberg, C. G., 1873. Grössere Felsproben des Polycystinen-Mergels von Barbados mit weiteren Erläuterungen. *Königliche Preussische Akademie der Wissenschaften zu Berlin,* Monatsberichte, Jahre 1873, p. 213–263.

Ehrenberg, C. G., 1875. Fortsetzung der mikrogeologischen Studien als Gesammt-Uebersicht der mikroskopischen Paläontologie gleichartig analysirter Gebirgsarten der Erde, mit specieller Rücksicht auf den Polycystinen-Mergel von Barbados. *Königliche Akademie der Wissenschaften zu Berlin, Abhandlungen,* Jahre 1875, 225 pp.

Florez Abin, E., 1983. Radiolarios de algunas formaciones del Cretacico Paleogeno Inferior de Cuba occidental (Radiolarians from Cretaceous and lower Paleogene formations in western Cuba). *Ciencias de la Tierra y del Espacio,* v. 7, p. 3–36.

Florez Abin, E. and Fernandez, G., 1985. Acereca de la bioestratigrafia del limite entre los depositos del paleoceno–eoceno inferior (formaciones Alkazar y Capdevila) en la provincia Ciudad Habana [Biostratigraphy of the limit between Paleocene–Lower Eocene sediments (Alkazar and Capdevila formations) in Havana City Province]. *Revista Tecnologica,* v. 15, Ser. Geologia, p. 8–18.

Foreman, H. P., 1968. Upper Maestrichtian Radiolaria of California. The Palaeontological Association, *Special Papers in Palaeontology,* v. 3, p. iv + 1–82.

Foreman, H. P., 1973. Radiolaria of Leg 10 with systematics and ranges for the families Amphipyndacidae, Artostrobiidae, and Theoperidae. In Worzel, J. L., Bryant, W., et al., Initial Reports of the Deep Sea Drilling Project, v. 10: Washington (U.S. Government Printing Office), p. 407–474.

Foreman, H. P., 1975. Radiolaria from the North Pacific, Deep Sea Drilling Project, Leg 32. In Larson, R. L., Moberly, R., et al., Initial Reports of the Deep Sea Drilling Project, v. 32: Washington (U.S. Government Printing Office), p. 579–676.

Fourtanier, E., 1991. Paleocene and Eocene diatom biostratigraphy and taxonomy of eastern Indian Ocean Site 752. In Weissel, J., Peirce, J., Taylor, E., Alt, J., et al., Proceedings of the Ocean Drilling Program, Scientific Results, v. 121: College Station, TX (Ocean Drilling Program), p. 171–187.

Frizzell, D. L. and Middour, E. S., 1951. Paleocene Radiolaria from southeastern Missouri. *Bulletin of the University of Missouri School of Mines and Metallurgy, Technical Series,* v. 77, 41 pp.

Goll, R. M., 1969. Classification and phylogeny of Cenozoic Trissocyclidae (Radiolaria) in the Pacific and Caribbean basins. Part II. *Journal of Paleontology,* v. 43, p. 322–339.

Haeckel, E., 1887. Report on the Radiolaria collected by H.M.S. Challenger during the years 1873–1876: Report on the Scientific Results of the Voyage of the H.M.S. Challenger. *Zoology,* v. 18, 2 parts, Atlas, clxxxviii + 1803 pp., 140 pl.

Hay, W. W. and Beaudry, F. M., 1973. Calcareous nannofossils: Leg 15. In Edgar, N. T., Saunders, J. B., et al., Initial Reports of the Deep Sea Drilling Project, v. 15: Washington (U.S. Government Printing Office), p. 625–683.

Hollis, C. J., 1993. Latest Cretaceous to Late Paleocene radiolarian biostratigraphy: A new zonation from the New Zealand region. *Marine Micropaleontology,* v. 21, p. 295–327.

Johnson, D. A., 1983. Cenozoic radiolarians from the Brazil Basin and Rio Grande Rise. In Barker, P. F., Carlson, R. L., Johnson, D. A., et al., Initial Reports of the Deep Sea Drilling Project, v. 72: Washington (U.S. Government Printing Office), p. 783–791.

Kennett, J. P. and Stott, L. D., 1991. Abrupt deep-sea warming, palaeoceanographic changes and benthic extinctions at the end of the Palaeocene. *Nature,* v. 353, p. 225–229.

Khokhlova, I. E., 1994. Paleogenovye radiolyarii massiva trodos (KIPR) [Paleogene Radiolaria from the Trodos (KIPR) massif]. Stratigrafiya. *Geologicheskaya korrelyatsiya,* v. 2, p.78–86.

Kozlova, G. E., 1983. Rasprostraneniye radiolyariyevykh zon Atlantiki v paleogene Povolzh'ya [Distribution of radiolarian zones of the Atlantic in the Paleogene of the Volga region]. *Akademiya Nauk SSSR, Seriya Geologicheskaya [Proceedings of the USSR Academy of Sciences, Geological Series],* v. 3, p. 46–51.

Kozlova, G. E., 1984. Zonalnoe podrazdelenie borealnogo paleogena po radiolariyam [Zonal subdivision of the boreal Paleogene by radiolarians]. In Petrushevskaya, M. G. and Stepanjants, S. D., Morfologiya, ekologiya i evolutsiya Radiolyarii. Materialy IV simpoziuma evropeiskikh radiolyaristov EURORAD IV [Morphology, ecology, and evolution of radiolarians. Material from the IV symposium of European radiolarists EURORAD IV]. Akademiya Nauk SSSR, Zoologicheskii Institut, Leningrad, USSR, p. 196–210.

Kozlova, G. E. and Gorbovets, A. N., 1966. Radiolyarii verkhnemelovykh i verkhneeotsenovykh otlozhenii Zapadno-Sibirskoi nizmennosti [Radiolarians of the Upper Cretaceous and Upper Eocene deposits of the West Siberian Lowland]. *Trudy vsesoyuznogo neftyanogo nauchno-issledovatelskogo geologorazvedochnogo instituta* (VNIGRI) [*Proceedings of the All Union Petroleum Scientific Research Institute for Geological Survey* (VNIGRI)], v. 248, 159 pp.

Krasheninnikov, V. A., 1960. Nekotorye Radiolyarii Nizhnego i Srednego Eotsena Zapadnogo Predkavkazya [Some radiolarians of the lower and middle Eocene of the Western Pre-Caucasus]. *Mineralogicko-Geologicka i Okhrana Nedr SSSR Vsesoyuznogo Nauchno-Issledovatelskogo Geologorazved Neftyanogo Instituta,* v. 16, p. 271–308.

Lipman, R. K., 1972. Paleogenovye radiolyarii SSSR [Paleogene Radiolarians of the USSR]. In Ministerstvo Vysshego i Srednego Spetsialnogo Obrazovaniya Rsfsr Leningradskii Ordena Lenina i Ordena Trudovogo Krasnogo Znameni Gornyi Institut G. B. Plekhanova, Dissertatsii na soiskanie uchenoi stepeni Doktora Geologo-Mineralogicheskikh. Leningrad, USSR: Nauk, 55 pp.

Miller, K. G., Janecek, T. R., Katz, M. E. and Keil, D. J., 1987. Abyssal circulation and benthic foraminiferal changes near the Paleocene /Eocene boundary. *Paleoceanography,* v. 2, p. 741–761.

Moore, T. C., Jr., 1973. Radiolaria from Leg 17 of the Deep Sea Drilling Project. In Winterer, E. L., Ewing, J. I., et al., Initial Reports of the Deep Sea Drilling Project, v. 17: Washington (U.S. Government Printing Office), p. 797–869.

Nanayama, F., 1992. Stratigraphy and facies of the Paleocene Nakanogawa Group in the southern part of central Hokkaido, Japan. *Journal of the Geological Society of Japan (Chishitsugaku Zasshi)*, v. 98, p. 1041–1059.

Nigrini, C., 1970. Radiolarian assemblages in the north Pacific and their application to a study of Quaternary sediments in Core V20–130. In Hays, J. D., *Geological Investigations of the North Pacific,* Geological Society of America Memoir 126, p. 139–183.

Nigrini, C., 1977. Tropical Cenozoic Artostrobiidae (Radiolaria). *Micropaleontology,* v. 23, p. 241–269.

Nishimura, A., 1987. Cenozoic Radiolaria in the western North Atlantic, Site 603, Leg 93 of the Deep Sea Drilling Project. In van Hinte, J. E., Wise, S. W., Jr., et al., Initial Reports of the Deep Sea Drilling Project, v. 93: Washington (U.S. Government Printing Office), p. 713–737.

Nishimura, A., 1992. Paleocene radiolarian biostratigraphy in the northwest Atlantic at Site 384, Leg 43, of the Deep Sea Drilling Project. *Micropaleontology,* v. 38, p. 317–362.

Okada, H., and Bukry, D., 1980. Supplementary modification and introduction of code numbers to the low-latitude coccolith biostratigraphic sonation. *Marine Micropaleontology,* v. 5, p. 321–325.

Okada, H. and Thierstein, H. R., 1979. Calcareous nannoplankton—Leg 43, Deep Sea Drilling Project. In Tucholke, B. E., Vogt, P. R., et al., Initial Reports of the Deep Sea Drilling Project, v. 43: Washington (U.S. Government Printing Office), p. 507–573.

Palmer, A. A., 1987, Cenozoic radiolarians from Deep Sea Drilling Project Sites 612 and 613 (Leg 95, New Jersey Transect) and Atlantic Slope Project Site ASP 15. In Poag, C. W., Watts, A. B., et al., Initial Reports of the Deep Sea Drilling Project, v. 95: Washington (U.S. Government Printing Office), p. 339–357.

Peirce, J., Weissel, J., et al., 1989. Proceedings of the Ocean Drilling Program, Initial Reports, v.121: College Station, TX (Ocean Drilling Program), 1000 pp.

Perch-Nielsen, K., 1977. Albian to Pleistocene calcareous nannofossils from the western South Atlantic, DSDP Leg 39. In Supko, P. R., Perch-Nielsen, K., et al., Initial Reports of the Deep Sea Drilling Project, v. 39: Washington (U.S. Government Printing Office), p. 699–823.

Percival, S. F., Jr., 1984. Late Cretaceous to Pleistocene calcareous nannofossils from the South Atlantic, Deep Sea Drilling Project Leg 73. In Hsü, K. J., LaBreque, J. L., et al., Initial Reports of the Deep Sea Drilling Project, v. 73: Washington (U.S. Government Printing Office), p. 391–424.

Petrushevskaya, M. G. and Kozlova, G. E., 1972. Radiolaria: Leg 14, Deep Sea Drilling Project. In Hayes, D. E., Pimm, A. C., et al., Initial Reports of the Deep Sea Drilling Project, v. 14: Washington (U.S. Government Printing Office), p. 495–648.

Premoli Silva, I. and Bolli, H. M., 1973. Late Cretaceous to Eocene planktonic foraminifera and stratigraphy of Leg 15 sites in the Caribbean Sea. In Edgar, N. T., Saunders, J. B., et al., Initial Reports of the Deep Sea Drilling Project, v. 15: Washington (U.S. Government Printing Office), p. 499–547.

Riedel, W. R. and Sanfilippo, A., 1970. Radiolaria, Leg 4, Deep Sea Drilling Project. In Bader, R. G., et al., Initial Reports of the Deep Sea Drilling Project, v. 4: Washington (U.S. Government Printing Office), p. 503–575.

Riedel, W. R. and Sanfilippo, A., 1971. Cenozoic Radiolaria from the western tropical Pacific, Leg 7. In Winterer, E. L., Riedel, W. R., et al., Initial Reports of the Deep Sea Drilling Project, v. 7: Washington (U.S. Government Printing Office), p. 1529–1672.

Riedel, W. R. and Sanfilippo, A., 1973. Cenozoic Radiolaria from the Caribbean, Deep Sea Drilling Project, Leg 15. In Edgar, N. T., Saunders, J. B., et al., Initial Reports of the Deep Sea Drilling Project, v. 15: Washington (U.S. Government Printing Office), p. 705–751.

Riedel, W. R. and Sanfilippo, A., 1978. Stratigraphy and evolution of tropical Cenozoic radiolarians. *Micropaleontology,* v. 24, p. 61–96.

Riedel, W. R. and Sanfilippo, A., 1981. Evolution and diversity of form. In Simpson, T. L. and Volcani, B. E., (eds.), *Silicon and Siliceous Structures in Biological Systems, Radiolaria,* New York: Springer-Verlag, p. 323–346.

Roth, P. H., 1974. Calcareous nannofossils from the northwestern Indian Ocean, Leg 24, Deep Sea Drilling Project. In Fisher, R. L., Bunce, E. T., et al., Initial Reports of the Deep Sea Drilling Project, v. 24: Washington (U.S. Government Printing Office), p. 969–994.

Sanfilippo, A. and Riedel, W. R., 1973. Cenozoic Radiolaria (exclusive of theoperids, artostrobiids and amphipyndacids) from the Gulf of Mexico, Leg 10. In Worzel, J. L., Bryant, W., et al., Initial Reports of the Deep Sea Drilling Project, v. 10: Washington (U.S. Government Printing Office), p. 475–611.

Sanfilippo, A. and Riedel, W. R., 1974. Radiolaria from the west-central Indian Ocean and Gulf of Aden, DSDP Leg 24. In Fisher, R. L., Bunce, E. T., et al., Initial Reports of the Deep Sea Drilling Project, v. 24: Washington (U.S. Government Printing Office), p. 997–1035.

Sanfilippo, A., and Riedel, W. R., 1976. Radiolarian occurrences in the Caribbean Region. Publication de la VII conférence géologique des Caraibes du 30 juin au 12 juillet 1974 (VII Conférence Géologique des Caraibes, Cayenne, Département français de la Guyane), p. 145–168.

Sanfilippo, A. and Riedel, W. R., 1979. Radiolaria from the northeastern Atlantic Ocean DSDP Leg 48. In Montadert, L., Roberts, D. G., et al., Initial Reports of the Deep Sea Drilling Project, v. 48: Washington (U.S. Government Printing Office), p. 493–511.

Sanfilippo, A. and Riedel, W. R., 1982. Revision of the radiolarian genera *Theocotyle*, *Theocotylissa* and *Thyrsocyrtis*. *Micropaleontology,* v. 28 p. 170–188.

Sanfilippo, A. and Riedel, W. R., 1992. The origin and evolution of Pterocorythidae (Radiolaria): A Cenozoic phylogenetic study. *Micropaleontology,* v. 38, p. 1–36.

Sanfilippo, A., Westberg-Smith, M. J. and Riedel, W. R., 1985. Cenozoic Radiolaria. In Bolli, H. M., Saunders, J. B. and Perch-Nielsen, K., *Plankton Stratigraphy,* Cambridge: Cambridge University Press, p. 631–712.

Saunders, J. B., Bernoulli, D., Müller-Merz, E., Oberhänsli, H., Perch-Nielsen, K., Riedel, W. R., Sanfilippo, A. and Torrini, R., Jr., 1985. Stratigraphy of the late middle Eocene to early Oligocene in the Bath Cliff Section, Barbados, West Indies. *Micropaleontology,* v. 30, p. 390–425.

Shrubsole, W. H., 1889. Notes on the Radiolaria of the London Clay. *Quarterly Journal of the Geological Society of London,* v. 45, p. 121–124.

Stott, L. D. and Kennett, J. P., 1990. Antarctic Paleogene planktonic foraminifer biostratigraphy: ODP Leg 113, Sites 689 and 690. In Barker, P. F., Kennett, J. P., et al., Proceedings of the Ocean Drilling Program, Scientific Results, v. 113: College Station, TX (Ocean Drilling Program), p. 549–569.

Strong, C. P., Hollis, C. J. and Wilson, G. J., 1995. Foraminifera, radiolarian and dinoflagellate biostratigraphy of Late Cretaceous to Middle Eocene pelagic sediments (Muzzle Group), Mead Stream, Marlborough, New Zealand. *New Zealand Journal of Geology and Geophysics,* v. 38, p. 165–206.

Supko, P. R., Perch-Nielsen, K., et al., 1977. Site 355: Brazil Basin. In Supko, P. R., Perch-Nielsen, K., et al., Initial Reports of the Deep Sea Drilling Project, v. 39: Washington (U.S. Government Printing Office), p. 101–140.

Suyari, K. and Yamasaki, T., 1987. Boundary between the north and south Shimanto Subbelts in Tokushima Prefecture. *Journal of Science, University of Tokushima,* v. 20, p. 37–46.

Suzuki, H., 1992. Discovery of Eocene Radiolaria from the Otonashigawa Belt of the Shimanto Superbelt in the Kii Peninsula, Southwest Japan. *Earth Science,* v. 47, p. 75–79.

Thomas, E., 1990a. Late Cretaceous–early Eocene mass extinctions in the deep sea. In Sharpton, V. L. and Ward, P., (eds.), *Global Catastrophes in Earth History: An interdisciplinary Conference on Impacts, Volcanism, and Mass Mortality,* Geological Society of America Special Paper No. 247, p. 481–495.

Thomas, E., 1990b. Late Cretaceous through Neogene deep-sea benthic foraminifers (Maud Rise, Weddell Sea, Antarctica). In Barker, P. F., Kennett, J. P., et al., Proceedings of the Ocean Drilling Program, Scientific Results., v. 113: College Station, TX (Ocean Drilling Program), p. 571–594.

Thomas, E., 1992. Cenozoic deep sea circulation: Evidence from deep sea benthic foraminifera. In Kennett, J. P. and Warnke, D., (eds.), *The Antarctic Paleoenvironment: A Perspective on Global Change,* American Geophysical Union Antarctic Research Series No. 56, p. 141–165.

Tjalsma, R. C. and Lohmann, G. P., 1983. *Paleocene–Eocene bathyal and abyssal benthic foraminifera from the Atlantic Ocean.* Micropaleontology Special Publication No. 4, 90 pp.

Tochilina, S. V., 1975. Biostratigraficheskaya kharakteristika paleotsen–eotsenovykh otlozhenii Voronezhskoi anteklizy (Biostratigraphic characteristics of Paleocene–Eocene deposits of the Voronezh antecline). In Zhamoida, A. I. (eds.), Sistematika i stratigraficheskoe znachenie radiolyarii (Systematics and stratigraphic importance of Radiolaria). Leningrad, USSR: Trudy vsesoyuznogo nauchno-issledovatelskogo Geologicheskogo Instituta (VSEGEI) (Publication of the All-Union Institute of Geology), n. s., v. 226, p. 78–84.

Ueda, H., Kawamura, M. and Iwata, K., 1993. Occurrence of the Paleocene radiolarian fossils from the Idon'nappu Belt, central part of Hokkaido, northern Japan. *Journal of the Geological Society of Japan (Chishitsugaku Zasshi),* v. 99, p. 565–568.

Weaver, F. M. and Dinkelman, M. G., 1978. Cenozoic radiolarians from the Blake Plateau and the Blake–Bahama Basin, DSDP Leg 44. In Benson, W. E., Sheridan, R. E., et al., Initial Reports of the Deep Sea Drilling Project, v. 44: Washington (U.S. Government Printing Office), p. 865–885.

Westberg, M. J., Sanfilippo, A. and Riedel, W. R., 1980. Radiolarians from the Moroccan Basin, Deep Sea Drilling Project Leg 50. In Lancelot, Y., Winterer, E. L., et al., Initial Reports of the Deep Sea Drilling Project, v. 50: Washington (U.S. Government Printing Office), p. 429–437.

Appendix 13.1

Species List and Taxonomic Notes

Amphicraspedum murrayanum Haeckel 1887: 523, plate 44, figure 10; Sanfilippo and Riedel 1973:524, plate 10, figures 3–6, plate 28, figure 1 [plate 13.1, figure 1].

Amphicraspedum prolixum Sanfilippo and Riedel 1973:524, plate 10, figures 7–11, plate 28, figures 3 and 4.

Bekoma bidartensis Riedel and Sanfilippo 1971: 1592, plate 7, figures 1–7; Foreman 1973:432, plate 3, figures 20–21, plate 10, figure 6; Sanfilippo et al. 1985:667, figures 13.1a and 13.1b; Nishimura 1992:332, plate 5, figures 8 and 9 [plate 13.1, figures 2 and 3].

Bekoma campechensis Foreman 1973:432, plate 3, figure 24, plate 10, figures 1–2 and 4; Nishimura 1992:332, plate 5, figures 1–4, plate 13, figure 1 [plate 13.1, figure 4].

Buryella clinata Foreman 1973:433, plate 8, figures 1–3, plate 9, figure 19. Foreman 1975:620, plate 9, figures 35–36; Sanfilippo et al. 1985:668, figures 14.1a and 14.1b [plate 13.1, figures 5 and 6].

Buryella pentadica Foreman 1973:433, plate 8, figure 8, plate 9, figures 15–16 [plate 13.1, figure 7].

Buryella tetradica Foreman 1973:433, plate 8, figures 4–5, plate 9, figures 13–14; Sanfilippo et al. 1985:668, figures 14.3a and 14.3b [plate 13.1, figures 8 and 9].

Calocycloma castum (Haeckel). *Calocyclas casta* Haeckel 1887:1384, plate 73, figure 10. *Calocycloma castum* (Haeckel) Foreman 1973:434, plate 1, figures 7, 9, and 10; Sanfilippo et al. 1985:669, figure 14.4 [plate 13.1, figure 10].

Dictyoprora urceolus (Haeckel). *Dictyocephalus urceolus* Haeckel 1887:1305; *Dictyoprora urceolus* (Haeckel) Nigrini 1977:251, plate 4, figures 9, 10 [plate 13.1, figure 11].

Giraffospyris lata Goll 1969:334, plate 58, figures 22, 24–26; Sanfilippo and Riedel 1973:529, plate 18, figures 3–7, plate 33, figure 4 [plate 13.1, figures 12 and 13].

Lamptonium fabaeforme chaunothorax Riedel and Sanfilippo. *Lamptonium (?) fabaeforme (?) chaunothorax* Riedel and Sanfilippo 1970:524, plate 5, figures 8–9; Sanfilippo et al. 1985:673, figure 18.3 [plate 13.1, figure 14].

Lamptonium fabaeforme constrictum Riedel and Sanfilippo. *Lamptonium (?) fabaeforme (?) constrictum* Riedel and Sanfilippo 1970:523, plate 5, figure 7; Sanfilippo et al. 1985:674, figure 18.4 [plate 13.1, figures 15 and 16].

The early forms encountered in the material used for this study have the cephalis enclosed in the base of a very long, robust, bladed horn, as is typical for this species, but the base of the narrow part of the thorax is not as markedly constricted as in later forms.

Lamptonium fabaeforme fabaeforme (Krasheninnikov). *[?] Cyrtocalpis fabaeformis* Krasheninnikov 1960:296, plate 3, figure 11. *Lamptonium (?) fabaeforme fabaeforme* (Krasheninnikov) Riedel and Sanfilippo 1970:523, plate 5, figure 6; Foreman 1973:436, plate 6, figures 6–9; Sanfilippo et al. 1985:674, figure 18.2 [plate 13.1, figure 17].

Lamptonium pennatum Foreman 1973:436, plate 6, figures 3–5, plate 11, figure 13; Sanfilippo et al. 1985:674, figure 18.1 [plate 13.1, figure 19].

Early forms have been observed with a closed, inverted, caplike third segment, and with wings as long as the maximum width of the thorax.

Lamptonium sanfilippoae Foreman 1973:436, plate 6, figures 15–16, plate 11, figures 16–17; Sanfilippo et al. 1985:674, figure 18.5 [plate 13.1, figure 18].

Lithochytris archaea Riedel and Sanfilippo 1970, plate 9, figure 8; 1973:1594, plate 7, figure 13; Foreman 1973:436, plate 2, figures 4 and 5 [plate 13.1, figure 20].

Lophocyrtis biaurita (Ehrenberg). *Eucyrtidium biauritum* Ehrenberg 1873:226; 1875:70, plate 10, figures 7 and 8. *Lophocyrtis biaurita* (Ehrenberg) Haeckel 1887:1411; Foreman 1973:442, plate 8, figures 23–26.

Lychnocanoma auxilla Foreman 1973:437, plate 2, figure 6; plate 11, figures 1 and 2 [plate 13.1, figure 21].

Phormocyrtis cubensis (Riedel and Sanfilippo). *Eucyrtidium cubense* Riedel and Sanfilippo 1971:1594, plate 7, figures 10–11. *Phormocyrtis cubensis* (Riedel and Sanfilippo) Foreman 1973:438, plate 7, figures 11–12, 14.

Phormocyrtis striata exquisita (Kozlova). *Podocyrtis exquisita* Kozlova in Kozlova and Gorbovet 1966:106, plate 17, figure 2. *Phormocyrtis striata exquisita* (Kozlova) Foreman 1973:438, plate 7, figures 1–4, 7–8, plate 12, figure 5 (with synonymy); Sanfilippo et al. 1985:678, figures 20.2a–20.2c [plate 13.2, figure 1].

Phormocyrtis striata striata Brandt. *Phormocyrtis striata* Brandt 1935:55, plate 9, figure 12; Riedel and Sanfilippo 1970:532, plate 10, figure 7; Foreman 1973:438, plate 7, figures 5–6, 9; Sanfilippo et al. 1985:679, figures 21.1a and 21.1b [plate 13.2, figure 2].

Phormocyrtis turgida (Krasheninnikov). *Lithocampe turgida* Krasheninnikov 1960:301, plate 3, figure 17. *Phormocyrtis turgida* (Krasheninnikov) Foreman 1973:438, plate 7, figure 10; plate 12, figure 6 [plate 13.2, figure 3].

Podocyrtis (Lampterium) acalles Sanfilippo and Riedel 1992:12, plate 3, figures 2–5 [plate 13.2, figure 4].

Podocyrtis (Podocyrtis) papalis Ehrenberg. *Podocyrtis papalis* Ehrenberg 1847:55, figure 2; 1854, plate 36, figure 23; 1873:251; Riedel and Sanfilippo 1970:533, plate 11, figure 1; Sanfilippo and Riedel 1973:531, plate 20, figures 11–14, plate 36, figures 2–3, 5–6; Sanfilippo and Riedel 1992, plate 3, figures 9 and 10 [plate 13.2, figure 5].

Pterocodon ampla (Brandt). *Theocyrtis ampla* Brandt 1935:56, plate 9, figures 13–15. *Pterocodon (?) ampla* (Brandt) Foreman 1973:438, plate 5, figures 3–5. *Pterocodon ampla* (Brandt) Sanfilippo et al. 1985:680, figures 20.3a and 20.3b [plate 13.2, figure 6].

Pterocodon (?) anteclinata Foreman 1975:621, plate 9, figures 32–34 [plate 13.2, figure 12].

Pterocodon (?) poculum Nishimura 1992:350, plate 8, figures 1–3; plate 13, figure 13 [plate 13.2, figure 11].

Pterocodon (?) tenellus Foreman 1973:439, plate 5, figure 7; plate 12, figure 4 [plate 13.2, figure 13].

Spongatractus balbis Sanfilippo and Riedel 1973:518, plate 2, figures 1–3, plate 25, figures 1–2; Sanfilippo et al. 1985:652, figure 6.1.

Spongatractus pachystylus (Ehrenberg). *Spongosphaera pachystyla* Ehrenberg 1873:256; 1875, plate 26, figure 3. *Spongatractus pachystylus* (Ehrenberg) Sanfilippo and Riedel 1973:519, plate 2, figures 4–6, plate 25, figure 3 (with synonymy); Sanfilippo et al. 1985:652, figure 6.2.

Spongodiscus cruciferus (Clark and Campbell). *Spongasteriscus cruciferus* Clark and Campbell 1942:50, plate 1, figures 1–6, 8, 10, 11, 16–18. *Spongodiscus cruciferus* (Clark and Campbell) Sanfilippo and Riedel 1973:524, plate 11, figures 14–17, plate 28, figures 10 and 11 [plate 13.2, figure 8].

Spongodiscus phrix Sanfilippo and Riedel 1973:525, plate 12, figures 1 and 2; plate 29, figure 2. [plate 13.2, figure 7].

Spongodiscus quartus quartus (Borisenko). *Staurodictya quartus* Borisenko 1958:96, plate 2, figure 5. *Spongodiscus quartus quartus* in Sanfilippo and Riedel 1973:525, plate 12, figures 6 and 7; plate 29, figures 5 and 6 [plate 13.2, figure 10].

Stylotrochus alveatus Sanfilippo and Riedel 1973:525, plate 13, figures 4, 5; plate 30, figures 3 and 4 [plate 13.2, figure 9].

Theocorys acroria Foreman 1973:439, plate 5, figures 11–13; plate 12, figure 2 [plate 13.2, figures 14 and 15].

Theocorys (?) phyzella Foreman 1973:440, plate 5, figure 8; plate 12, figure 1 [plate 13.2, figure 16].

Theocotyle nigriniae Riedel and Sanfilippo. *Theocotyle cryptocephala (?) nigriniae* Riedel and Sanfilippo 1970:525, plate 6, figure 5 (non 6); *Theocotyle nigriniae* Riedel and Sanfilippo, Sanfilippo and Riedel 1982:178, plate 2, figures 1–3 (with synonymy); Sanfilippo et al. 1985:685, figures 25.1a and 25.1b [plate 13.2, figures 17 and 18].

Theocotylissa alpha Foreman. *Theocotyle (Theocotylissa) alpha* Foreman 1973:441, plate 4, figures 13–15 (non 14), plate 12, figure 16; Foreman 1975:621; *Theocotylissa alpha* Foreman, Sanfilippo and Riedel 1982:179, plate 2, figures 16–17 (with synonymy); Sanfilippo et al. 1985:686, figure 25.6 [plate 13.2, figure 19].

Theocotylissa auctor Foreman. *Theocotyle (Theocotylissa) auctor* Foreman 1973:441, plate 4, figures 8–10, plate 12, figure 13. *Theocotylissa auctor* Foreman, Sanfilippo and Riedel 1982:180, plate 2, figures 14 and 15; Sanfilippo et al. 1985:686, figures 25.5a and 25.5b [plate 13.2, figure 20].

Theocotylissa ficus (Ehrenberg). *Eucyrtidium ficus* Ehrenberg 1873:228; 1875, plate 11, figure 19. *Theocotylissa ficus* (Ehrenberg) Sanfilippo and Riedel 1982:180, plate 2, figures 19–20 (with synonymy); Sanfilippo et al. 1985:686, figures 25.7a and 25.7b [plate 13.2, figure 21].

Early forms are difficult to distinguish from the ancestral species *Theocotylissa alpha* because their general outline, shape, and pore pattern are very similar. Typical forms of *T. alpha* terminate in a thick, poreless peristome with three short, triangular feet. However, at the time of the evolutionary transition to *T. ficus*, the three feet are commonly expressed only as three shallow undulations in the peristome that are difficult to distinguish from the wavy peristome in the earliest morphotypes of *T. ficus*.

Thyrsocyrtis (Thyrsocyrtis) *hirsuta* (Krasheninnikov). *Podocyrtis hirsutus* Krasheninnikov 1960:300, plate 3, figure 16. *Thyrsocyrtis (Thyrsocyrtis) hirsuta* (Krasheninnikov) Sanfilippo and Riedel 1982:173, plate 1, figures 3–4 (with synonymy); Sanfilippo et al. 1985:687, figure 26.2 [plate 13.2, figure 24].

Thyrsocyrtis (Thyrsocyrtis) tarsipes Foreman. *Thyrsocyrtis tarsipes* Foreman 1973:442, plate 3, figure 9, plate 12, figure 14; *Thyrsocyrtis (Thyrsocyrtis) tarsipes* Foreman, Sanfilippo and Riedel 1982:174, plate 1, figures 1 and 2; Sanfilippo et al. 1985:689, figure 26 [plate 13.2, figures 22 and 23].

Plate 13.1 *(following page)* On this and plate 13.2, a sequence of numbers in the form 2-10-9-3, 66–68 cm indicates DSDP leg, site, core, section, and centimeter interval. Cs.2, for example, is a slide designation and H38/4 is an England Finder position for a specimen. Sample designations from the DSDP/ODP Micropaleontological Reference Collection are followed by the abbreviation MRC and an England Finder position. Scale bar for figure 1 and figures 2–21, as marked on fig. 1 and fig. 2 respectively, equal 150 μm. 1: *Amphicraspedum murrayanum*, 2-10-9-3, 66–68 cm Cs.2 H38/4. 2: *Bekoma bidartensis*, 43-384-6-1, 74–80 cm Cs.1 H35/3. 3: *Bekoma bidartensis*, 32-313-12-2, 34–36 cm Cs.1 R23/1. 4: *Bekoma campechensis*, 43-384-7-4, 91–96 cm Cs.1 S36/2. 5: *Buryella clinata*, late form 10-94-30-1, 33–35 cm Ph.C V17/4. 6: *Buryella clinata*, early form 2-10-9-3, 66–68 cm Sl.1 H9/1. 7: *Buryella pentadica*, 10-86-8-CC Ph.1 O41/3. 8: *Buryella tetradica*, 10-96-3-4, 42–44 cm Ph.1 X37/0. 9: *Buryella tetradica*, 43-384-6-1, 64–80 cm Ph.1 U8/3. 10: *Calocycloma castum*, 2-10-9-3, 125–127 cm Cs.2 V44/4. 11: *Dictyoprora urceolus*, 10-94-30-1, 33–35 cm Ph.C V17/4. 12: *Giraffospyris lata*, late form 32-313-12-2, 34–36 cm Ph.2 Q49/0. 13: *Giraffospyris lata*, early form 10-96-3-4, 42–44 cm Ph.1 K47/0. 14: *Lamptonium fabaeforme chaunothorax*, 2-10-9-3, 66–68 cm Cs.3 P44/4. 15: *Lamptonium fabaeforme constrictum*, late form 10-94-24-2, 42–44 cm Cs.2 N45/4. 16: *Lamptonium fabaeforme constrictum*, early form 44-390A-6-1, 50–57 cm MRC R14/4. 17: *Lamptonium fabaeforme fabaeforme*, 32-313-12-2, 34–36 cm Cs.1 M50/1. 18: *Lamptonium sanfilippoae*, 2-10-9-3, 125–127 cm Cs.2 C26/4. 19: *Lamptonium pennatum*, 10-96-3-4, 42–44 cm Cs.1 V31/0. 20: *Lithochytris archaea*, 2-10-9-3, 125–127 cm Cs.2 N50/0. 21: *Lychnocanoma auxilla*, late form 15-152-1-CC Sl.3 F22/3

Plate 13.2 *(following page)* Scale bar for figures 1–6 and 11–24, as marked on figure 24, and scale bar for figures 7–10, as marked on figure 7 respectively, equal 150 μm. 1: *Phormocyrtis striata exquisita*, 10-96-3-4, 42–44 cm Cs.1 N33/0. 2: *Phormocyrtis striata striata*, 2-10-9-3, 66–68 cm Cs.2 B21/1. 3: *Phormocyrtis turgida*, 10-96-3-4, 42–44 cm Ph.1 X43/1. 4: *Podocyrtis (Lampterium) acalles*, 44-390A-6-1, 50–57 cm MRC R26/0. 5: *Podocyrtis (Podocyrtis) papalis*, 10-96-3-4, 42–44 cm Ph.1 V51/0. 6: *Pterocodon ampla*, 10-96-3-4, 42–44 cm Ph.1 L37/2. 7: *Spongodiscus phrix*, 2-10-9-3, 66–68 cm Cs.3 K38/0. 8: *Spongodiscus cruciferus*, 10-96-3-4, 42–44 cm Cs.1 B13/2. 9: *Stylotrochus alveatus*, 43-384-6-3, 83–89 cm Cs.1 V30/4. 10: *Spongodiscus quartus quartus*, 10-96-3-4, 42–44 cm Cs.1 D21/0. 11: *Pterocodon (?) poculum*, 43-384-6-1, 74–80 cm Ph.1 W9/1. 12: *Pterocodon (?) anteclinata*, 10-96-3-4, 42–44 cm Ph.1 Q37/4. 13: *Pterocodon (?) tenellus*, 32-313-12-2, 34–36 cm Ph.2 J22/3. 14: *Theocorys acroria*, 10-96-3-4, 42–44 cm Cs.1 U16/1. 15: *Theocorys acroria*, 2-10-9-3, 66–68 cm Sl.1 R32/1. 16: *Theocorys (?) phyzella*, 10-96-3-4, 42–44 cm Ph.1 D47/0. 17: *Theocotyle nigriniae*, late form 10-94-30-1, 33–35 cm Ph.C H13/3. 18: *Theocotyle nigriniae*, early form 2-10-9-3, 125–127 cm Cs.2 O44/1. 19: *Theocotylissa alpha*, 32-313-12-2, 34–36 cm Ph.2 Q25/4. 20: *Theocotylissa auctor*, 10-96-3-1, 41–43 cm Cs.A Q22/0. 21: *Theocotylissa ficus*, early morphotype, 2-10-9-3, 66–68 cm Cs.2 G41/0. 22: *Thyrsocyrtis (Thyrsocyrtis) tarsipes*, 10-96-3-4, 42–44 cm Ph.1 A43/0. 23: *Thyrsocyrtis (Thyrsocyrtis) tarsipes*, same slide as 22, C20/0. 24: *Thyrsocyrtis (Thyrsocyrtis) hirsuta*, 2-10-9-3, 66–68 cm Cs.2 L33/1.

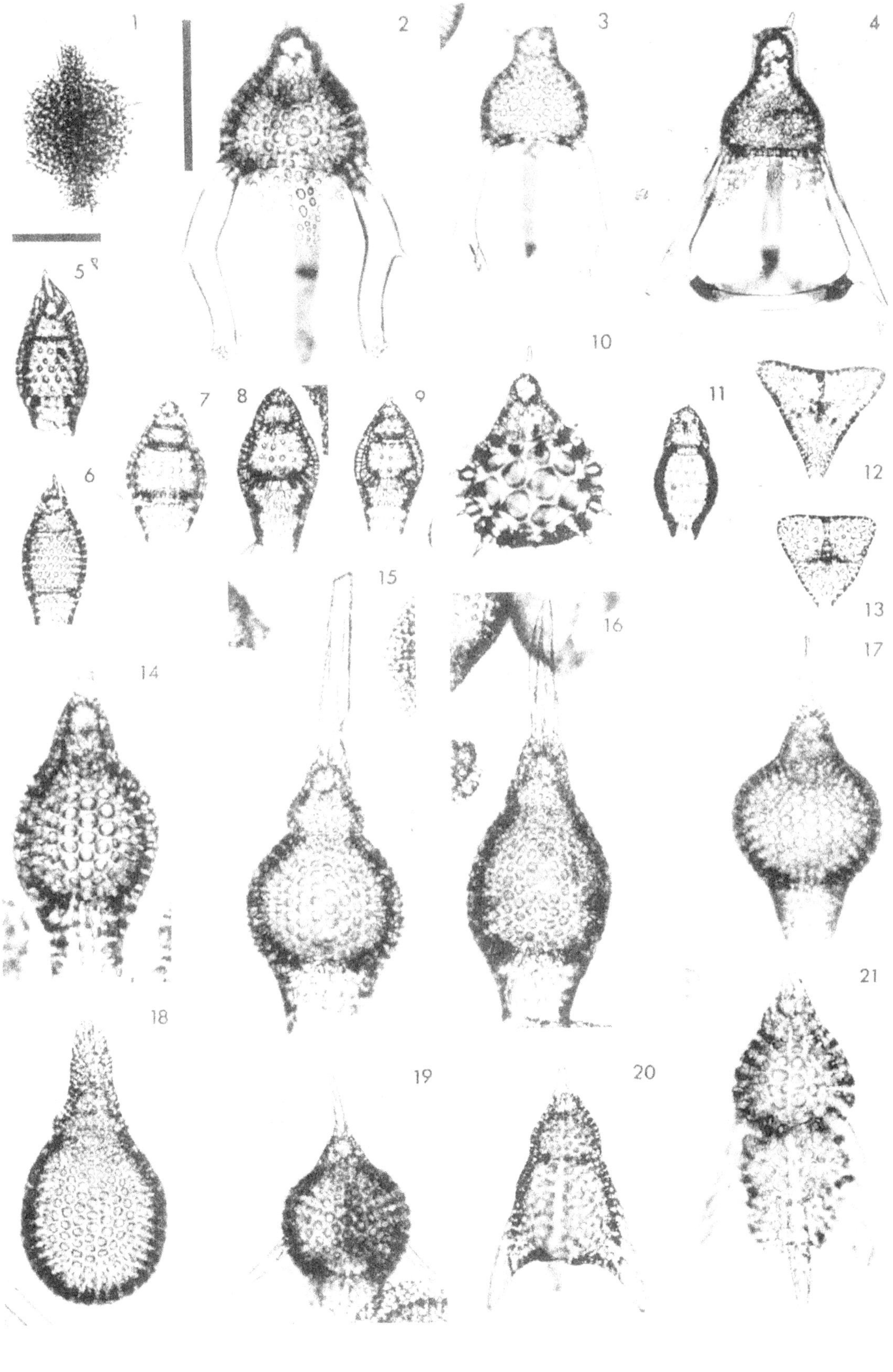
1
2
3
4
5
6
7
8
9
10
11
12
13
14
15
16
17
18
19
20
21

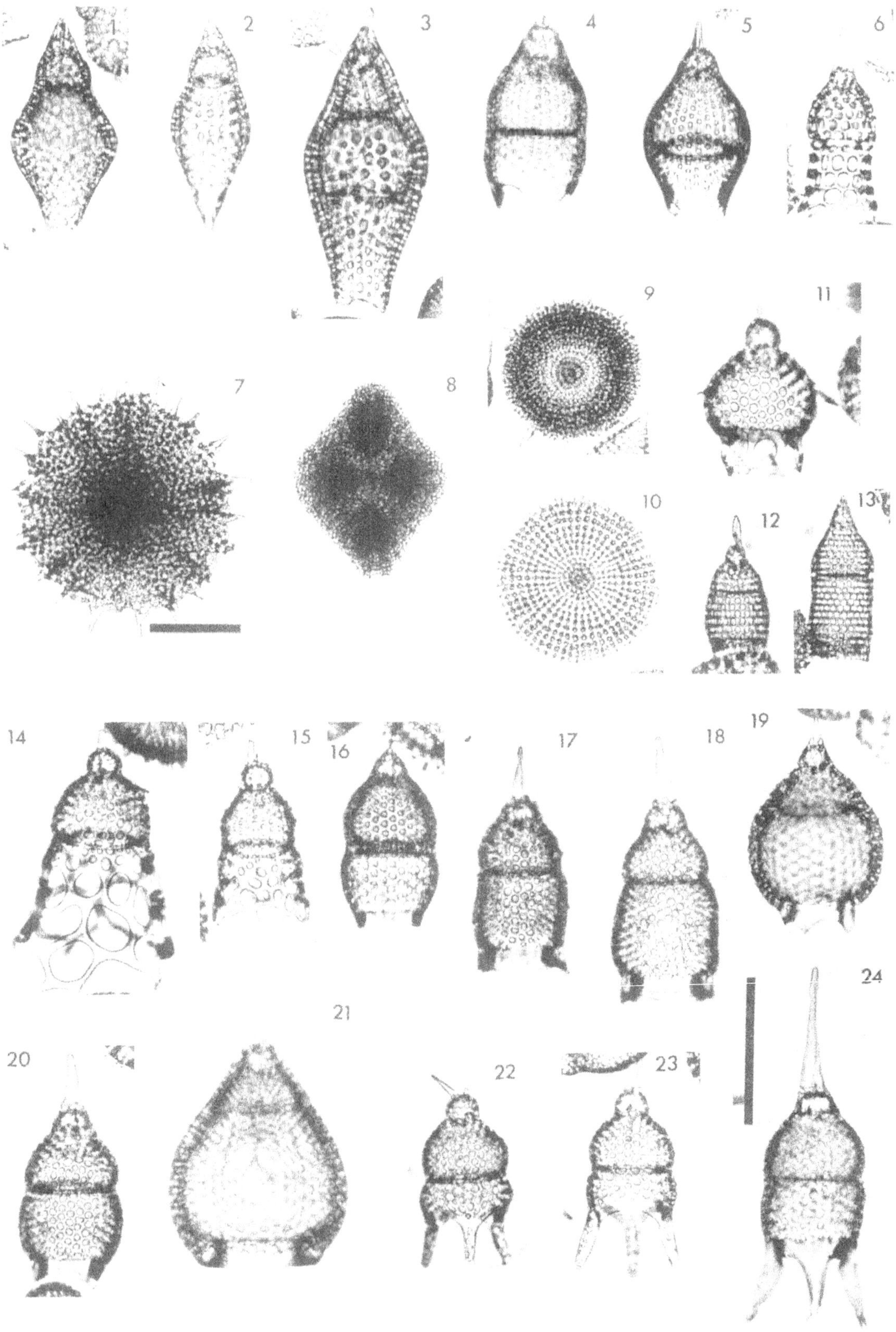
1
2
3
4
5
6
7
8
9
10
11
12
13
14
15
16
17
18
19
20
21
22
23
24

14

Global Warming and Dinocyst Changes Across the Paleocene/Eocene Epoch Boundary

JONATHAN P. BUJAK
HENK BRINKHUIS

ABSTRACT

Dinocyst assemblages changed abruptly during latest Thanetian mid-Biochron NP9 in low and midlatitudes from relatively diverse populations with few *Apectodinium* to low-diversity populations dominated by *Apectodinium*. A shift to *Apectodinium* dominance also occurred during the latest Paleocene Epoch in high northern latitudes, but precise age assignment is hampered by poor correlation to chronostratigraphy. The *Apectodinium*-dominated assemblages were replaced in midlatitudes during late Biochron NP9 by increasingly diverse dinocyst populations with few *Apectodinium*, but abundant *Apectodinium* persisted until mid-Biochron NP11 in lower latitudes. A model is outlined that relates the *Apectodinium* dominance to the late Thanetian carbon and oxygen isotope excursions. Late Thanetian warming and changes in nutrient availability enabled *Apectodinium* to migrate from the Tethys into mid- and high latitudes in both hemispheres, while many cooler water dinocysts were suppressed. Subsequent temperature fall and retreat of the Tethyan dinoflagellates from mid- and high latitudes, combined with widespread marine transgression, provided vacant niches that were colonized by newly evolving Eocene species.

Recent investigations have documented a rapid drop in carbon ($\delta^{13}C$) and oxygen ($\delta^{18}O$) isotopes during the latest Paleocene Epoch in areas including the Paris Basin (Stott et al. 1996), the Weddell Sea (Kennett and Stott 1991), and South Atlantic DSDP Site 525 (Thomas and Shackleton 1993). Kennett and Stott referred to this event as the carbon and oxygen isotope excursion, and suggested that it represented "rapid global warming and oceanographic changes that caused one of the largest deep sea benthic extinctions of the past 90 million years. For a few thousand years, oceanic circulation underwent fundamental changes producing a transient state that, although brief, had long term effects on environmental and biotic evolution" (1991:225).

The warming event resulted in bottom-water anoxia, benthic extinction, and reduction of the latitudinal gradient of surface waters to approximately 10° C between the equator and poles (Thomas and Shackleton 1993). Increased kaolinite abundance reflects high weathering rates associated with an unusually warm climate with high temperatures and precipitation, and a large influx of organic matter to sediments resulting from either high productivity or low oxygenation of bottom waters (Thomas and Shackleton 1993).

Recent evidence from accurately dated Thanetian–Ypresian sections indicates that major changes in dinoflagellate cyst (dinocyst) assemblages occurred in low to midlatitudes of the Northern Hemisphere at the same time as the warming event. These changes primarily involved the development of *Apectodinium*-dominated assemblages and suppression of many other dinocyst taxa. Similar population shifts also occurred at high northern latitudes and in the Southern Hemisphere near the end of the Paleocene Epoch, but correlation of these sections to the chronostratigraphic framework is less well established than in northern low to midlatitudes. The *Apectodinium*-dominated populations were succeeded by increasingly diverse early Eocene (Ypresian) dinocyst assemblages, including characteristic Eocene taxa such as *Wetzeliella*.

This chapter first documents the changes in dinocyst assemblages across the Paleocene–Eocene Epoch transition at various latitudes. The timing and nature of these changes are then summarized, and a model is outlined to explain them. We suggest that increased surface water temperatures, possibly associated with nutrient availability, permitted the migration of *Apectodinium* into mid- and high latitudes, resulting in a nearly worldwide *Apectodinium* Acme. Subsequent cooling of surface waters and retreat of the Tethyan dinoflagellates, combined with widespread marine transgression, provided vacant niches that were colonized by newly evolving Eocene species.

Lower Eocene strata of midlatitude localities such as those in northwestern Europe are therefore characterized by a particularly high number of new taxa, whereas fewer new species occurred in low-latitude localities such as the Mediterranean region. Many of the midlatitude taxa showed considerable diversification during the early and middle Eocene epochs, resulting in the highest species diversity known from the fossil dinocyst record (Bujak and Williams 1979), and permitting exceptionally high-resolution dinocyst biostratigraphy.

Author citations for all dinocyst taxa discussed in the text are given in Lentin and Williams's index (1993) to fossil dinocysts.

Fossil Dinoflagellate Cysts

Dinoflagellates are unicellular organisms that have a free-swimming motile biflagellate stage in their life cycle. They are one of the most important living groups of phytoplankton, forming a crucial link in the aquatic food chain and the earth's oxygen supply. They are also the notorious cause of red tides and paralytic shellfish poisoning, resulting from blooms of certain species containing toxins.

The life cycle of modern dinoflagellates is complex, and includes fossilizable cyst formation in approximately 10% of living species. The motile stage is quickly destroyed by bacterial action, but the cyst is composed of resistant organic material similar to sporopollenin, so that organic-walled dinocysts are typically seen in preparations that also contain the transported spores and pollen of plants.

Unequivocal cysts of fossil dinoflagellates have a record beginning in the Late Triassic Period, when they probably developed their characteristic cingular-sulcal (dinokont) biflagellar arrangement (Bujak and Williams 1981). However, dinoflagellates probably evolved as a separate group in the Precambrian Era, with pre-Triassic dinoflagellates having an anterior (desmokont) biflagellar arrangement, so that their cysts would be indistinguishable from those of other groups of desmokont algae.

Most modern dinoflagellates that form fossilizable cysts are marine and photosynthetic, but certain groups, including the Protoperidiniaceae, are either exclusively holozoic or combine photosynthetic and holozoic characteristics. These holozoic dinoflagellates feed upon organic detritus or ingest phytoplankton such as diatoms, so that blooms of these organisms are intimately associated. Similar relationships can be inferred throughout the Neogene Period, based on observed associations of protoperidiniacean-dominated dinocyst assemblages with diatom-rich sediments.

Protoperidiniacean dinocysts are rare in Paleogene sediments, where dinoflagellates with holozoic capability may be represented by several extinct genera such as *Apectodinium*, *Cerodinium*, and *Deflandrea* (Bujak and Davies 1983). These genera were probably related to the Protoperidiniaceae via the Late Cretaceous Tethyan genus *Senegalinium* (Lentin and Williams 1980), and are associated with diatom-rich assemblages in the North Sea, as discussed below.

Paleocene–Eocene Dinoflagellate Cyst Stratigraphy

Following an early period emphasizing morphology and taxonomy, dinocyst research moved into a phase of biozonal description that began in the mid-1970s. Most Paleogene zonations were initially established using the onshore sequences of western Europe and emphasized species of the Wetzeliellaceae due to their relatively large size and distinctive morphology (e.g., Caro 1973; Costa and Downie 1976; Jan du Chêne 1977; Chateauneuf and Gruas-Cavagnetto 1978). These are shown in figure 14.1, which indicates the original chronostratigraphy suggested by the authors of the zonal schemes. Subsequent schemes proposed for upper Paleocene–lower Eocene deposits of onshore northwest Europe, such as those of Bujak et al. (1980) and Powell et al. (1996), were based on the total dinocyst assemblage, and permitted worldwide correlation based on the presence of cosmopolitan taxa.

Many zonal schemes that have been proposed are based on sections with poorly constrained chronostratigraphy, or represent generalized syntheses of existing schemes (e.g., Costa and Manum 1988; Powell 1992). In this respect, the study by Powell et al. (1996) is particularly important and relevant to the characterization of the Paleocene/Eocene Series boundary because it is based on detailed analysis of Thanetian to lower Ypresian strata from southeast England, including the type Thanetian sections. Their results provide first-order chronostratigraphic calibration of Thanetian and lower Ypresian dinocyst events based on magnetostratigraphic and nannofossil investigations on the same sections.

During the 1980s and 1990s, as North Sea ex-

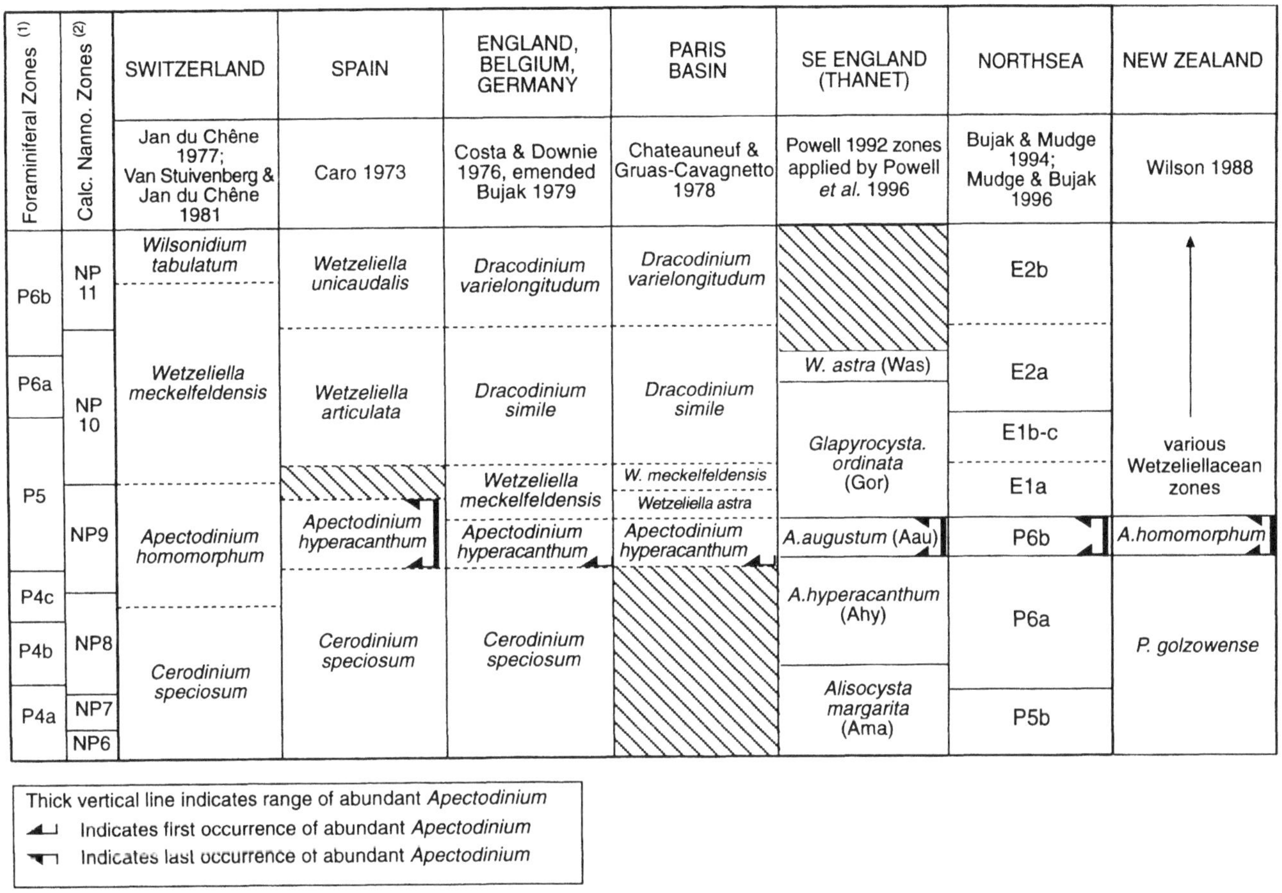

Figure 14.1 Various worldwide dinocyst zonal schemes, showing their emphasis on *Apectodinium* abundance approximately during Biochron NP9. Correlations of the dinocyst and NP/P zones are based on the original authors' assignments.
(1) After Berggren (1994); (2) after Hardenbol (1994).

ploratory data became less confidential, a succession of Paleogene dinocyst zonal schemes were proposed for the United Kingdom (U.K.) and Norwegian offshore areas (e.g., Powell 1988; Gradstein et al. 1992; Mudge and Copestake 1992a, 1992b; Schröder 1992). These schemes became progressively refined, with the latest including 14 zones and 33 subzones for the Paleocene–Eocene interval (Bujak and Mudge 1994; Mudge and Bujak 1996). Despite these latest improvements, chronostratigraphic calibration of the North Sea dinocyst zones remains largely based on correlation of dinocyst events in various onshore sections because planktonic foraminifera and calcareous nannofossils are either absent or restricted in their occurrence due to water depth limitations in the upper Paleocene–lower Eocene deposits.

No low-latitude dinocyst zonal schemes have been proposed for sections spanning the Paleocene/Eocene Series boundary, and only a few of the published papers on low-latitude dinocysts provide some correlation with calcareous nannofossil and planktonic foraminiferal zonal schemes. In most cases, these calibrations are based on isolated calcareous nannoplankton (NP) or planktonic foraminiferal (P) zonal determinations. Although of limited value, they provide some information on low-latitude areas such as Nigeria (Jan du Chêne and Adediran 1984) and Pakistan (Köthe 1988).

Only recently has a detailed succession of dinocyst events been documented for the Tunisian El Kef section (Brinkhuis et al. 1994). This section is significant because calcareous nannofossil and planktonic foraminiferal determinations have been made on the same section, permitting first-order correlations. The dinocyst successions of North Africa and northwest Europe can thus be compared within a rigid chronostratigraphic framework, as discussed below.

Published information on Thanetian to Ypresian dinocysts at northern high latitudes is scarce and has little chronostratigraphic control, but it includes relevant data on Spitsbergen (Manum and Throndsen 1986) and East Greenland (Soper et al. 1976a, 1976b). Previously unpublished observations are also presented herein for the Barents Sea and Beaufort Sea in order to construct a low to high transect of upper Paleocene to lower Eocene sections.

Although Paleocene–Eocene dinocyst records in the Southern Hemisphere are confined mainly to Argentina (Archangelsky 1969a, 1969b), Australia (Cookson and Eisenack 1967; Partridge 1976), and New Zealand (Wilson 1988), these provide insight into the global pattern of dinocyst changes close to the Paleocene/Eocene Epoch boundary, and aid attempts to understand these changes within a global paleoceanographic and paleoclimatic context.

Dinoflagellate Cyst Stratigraphy in the Thanetian and Ypresian Deposits of Southern England

Upper Paleocene and lower Eocene deposits in the type area of the Thanetian Stage, southeast England, are now relatively well understood in terms of magnetostratigraphy, sedimentology, and calcareous nannofossil stratigraphy (e.g., Townsend and Hailwood 1985; Aubry 1985, 1986, 1994; Aubry et al. 1986; Siesser et al. 1987; Ali 1994; Knox 1994a, 1994b; Knox et al. 1994).

Powell et al. (1996) presented an upper Paleocene to lower Eocene dinocyst biostratigraphy and sequence stratigraphy of key sections from southeast England, including the type Thanetian Stage (figure 14.2). Their analysis of the sections at Pegwell Bay, Herne Bay, Lower Upnor, and Wrabness resulted in the delineation of a succession of dinocyst events and permits their calibration with recently published data on magneto- and nannofossil stratigraphy from the area (e.g., Ali 1994; Knox et al. 1994). These calibrations provide the basis for detailed chronostratigraphic correlation across the entire North Sea Basin, as dinocysts are the principal biostratigraphic correlation tool in the area.

Powell et al. (1996) recognized eight third-order cycles (sensu Haq et al. 1987) in the Thanetian to lower Ypresian succession (figure 14.2). Most of these cycles correlate with those plotted by Hardenbol (1994), except that Powell et al. (1996) recognized five Thanetian cycles rather than the six of Hardenbol (1994), plus an additional lowermost Ypresian cycle to those plotted by Hardenbol (1994).

Figure 14.2 shows the succession of major dinocyst events observed by Powell et al. (1996) for the Thanetian and lower Ypresian interval of southeast England. Also shown is their calibration with the chronostratigraphy of Berggren (1994) and Hardenbol (1994), the benthic foraminiferal extinction event, the carbon and oxygen isotope excursion of Kennett and Stott (1991), the U.K. dinocyst zones of Powell (1992), and the North Sea dinocyst zones and bioevents of Bujak and Mudge (1994) and Mudge and Bujak (1996). A comprehensive discussion of the dinocyst succession in southeast England, and the basis for chronostratigraphic calibration of the dinocyst bioevents, is given in Powell et al. (1996), but the salient features of this succession are summarized below.

The highest abundance of *Areoligera gippingensis* occurs near the base of the Pegwell Marls in the Thanet Sand Formation, just above strata assigned to calcareous nannofossil Zone NP6 by Knox et al. (1994). This horizon was assigned to Chron C26n by Knox et al. (1994). Powell et al. (1996) placed this part of the section in their sequence Tht-1, which they equated with sequence Th-1 of Hardenbol (1994).

The highest occurrence of *Alisocysta margarita* occurs slightly above the base of the Reculver Silts in the Thanet Sand Formation. This horizon was assigned to Chron C25r by Ali (1994) and Knox et al. (1994). Powell et al. (1996) placed this horizon in their sequence Tht-2, which they equated with sequence Th-2 of Hardenbol (1994). The highest occurrence of *A. margarita* coincides in the section with the lowest consistent occurrence of *Deflandrea oebisfeldensis.*

The lowest occurrence of *Apectodinium* spp. observed by Powell et al. (1996) occurs within the Reculver Silts of the Thanet Sand Formation, at the base of the section exposed at Herne Bay. These strata were assigned to Zone NP8 by Aubry (1986) and Knox et al. (1994), and to Chron C25r by Townsend and Hailwood (1985) and Ali (1994). Powell et al. (1996) placed this horizon near the base of their sequence Tht-3, which they equated with sequence Th-3 of Hardenbol (1994). This occurrence is isolated stratigraphically from younger *Apectodinium* in the section and is designated as "*Apectodinium* spp. (Tht-3)" in figure 14.2. All *Apectodinium* present in the sections examined by Powell et al. (1996) from southeast England are representatives of the *A. homomorphum–hyperacanthum* plexus, which has a rather conservative morphology compared to other species of *Apectodinium.*

A second sporadic occurrence of *Apectodinium* spp. occurs in the Woolwich Bottom Beds of the Upnor Formation, assigned to Zone NP9 by Siesser et al. (1987) and Ali (1994) and probably representing Chron C25n (Ali 1994). Powell et al. (1996) placed this in the lower part of their sequence Tht-4, which they equated with sequence Th-4 of Hardenbol (1994). This occurrence is isolated stratigraphically from other *Apectodinium* occurrences in the sections and is designated as "*Apectodinium* spp. (Tht-4)" in figure 14.2.

The lowest occurrence of the acme of *Apectodinium* spp. is located at the base of the Woolwich Shell Beds (WSB) of the Woolwich Formation, in strata assigned to Chron C24r by Ali (1994). This occurs near the base of sequence Tht-5 of Powell et al. (1996), which they equated with sequence Th-5 of Hardenbol (1994). According to Hardenbol (1994), this correlates with the carbon isotope excursion ($\delta^{13}C$) and oxygen isotope excursion ($\delta^{18}O$) of Kennett and Stott (1991), and the benthic foraminiferal extinction event (BFE), shown in figure 14.2.

The highest occurrence of the acme of *Apectodinium* spp. occurs at the top of the Woolwich Formation, in strata assigned to Chron C24r by Ali (1994). This horizon is just below the base of sequence Ypr-1 of Powell et al. (1996), which they equated with sequence Th-6 of Hardenbol (1994). The highest occurrence of the *Apectodinium* spp. acme in the section coincides with the lowest occurrence of common leiospheres of indeterminate affinity assigned to *Leiosphaeridia* spp. by Powell et al. (1996).

Leiosphaeridia spp. have an acme throughout most of the Oldhaven Beds of the Harwich Formation, assigned to Chron C24r by Ali (1994). No calcareous nannofossil determinations have been made for the Oldhaven Beds, but second-order correlation with DSDP Hole 550 indicates assignment to Zone NP10 (Berggren and Aubry 1996). Powell et al. (1996) placed this part of the section in their sequence Ypr-2, which they equated with sequence Yp-1 of Hardenbol (1994). The base of this acme coincides with the lowest occurrence of *Deflandrea phosphoritica* at the base of the Oldhaven Beds and the base of sequence Ypr-2 of Powell et al. (1996).

The highest occurrence of common *Leiosphaeridia* spp. coincides with the highest occurrence of *Phelodinium magnificum* in the Wrabness Member of the Harwich Formation, assigned to Chron C24r by Ali (1994). Powell et al. (1996) placed this horizon in their sequence Ypr-2, which they equated with sequence Yp-1 of Hardenbol (1994).

The lowest occurrence of common *D. oebisfeldensis* is located near the base of the Wrabness Member of the Harwich Formation. Powell et al. (1996) placed this horizon in their sequence Ypr-2, which they equated with sequence Yp-1 of Hardenbol (1994).

The highest observed occurrence of common *D. oebisfeldensis* is near the top of the Wrabness Member. This occurs in sequence Ypr-3 of Powell et al. (1996), which they equated with sequence Yp-2 of Hardenbol (1994).

Upper Paleocene–Lower Eocene Dinoflagellate Cyst Stratigraphy in the North Sea Basin

The North Sea Basin has been the subject of intense petroleum exploration activity over the past 20 years, with more than 1000 wells now being released from confidentiality. This extensive data base has resulted in the publication of a series of biostratigraphic schemes based primarily on highest occurrence and abundance events. The emphasis on highest events results from the predominance of ditch cuttings samples and attendant downhole caving contamination in many exploration wells, obscuring the lowest occurrence and abundance events. However, due to the availability of core samples in some wells, it is possible to document these lowest occurrence events and hence construct the stratigraphic ranges and abundances of taxa in the North Sea (figure 14.3).

Most North Sea dinocyst zonal schemes were erected during the past decade, although some preliminary schemes were proposed in the late 1970s and early 1980s (e.g., Ioakim 1978; Heritier et al. 1979, 1981; Knox et al. 1981). These schemes were reviewed by Powell (1988), who also erected four dinocyst zones for the upper Paleocene to lowermost Eocene interval and recommended that the highest North Sea occurrence of *Apectodinium augustum* be selected to characterize the top of the Paleocene Series in the central North Sea. Other North Sea dinocyst zonations include Gradstein et al. (1992), Schröder (1992), and Wills and Peattie (1990), with the latter two schemes including several upper Thanetian subzones based on the lowest and highest occurrences or acmes of *Apectodinium* species.

Bujak and Mudge (1994) and Mudge and Bujak (1996) erected a Paleocene–Eocene dinocyst zonation for the central and northern North Sea based on data from more than 150 wells. They defined 14 zones and 33 subzones based exclusively on a combination of highest occurrence and abundance events. Their scheme has particularly high resolution in the Ypresian to Lutetian (lower middle Eocene) stratigraphic record due to combination of high diversity and rapid species turnover. Bujak and Mudge (1994) and Mudge and Bujak (1996) were able to recognize most of their zonal bioevents in onshore northwest European localities fringing the North Sea Basin, providing a broad chronostratigraphic framework for their scheme from the Thanetian Stage to the Bartonian Stage.

Mudge and Bujak (1994, 1996) also distinguished nine stratigraphic sequences in the Paleocene–Eocene stratigraphic record of the North Sea (figures 14.4 and 14.5), and demonstrated that the individual sequence boundaries had a consistent relationship with the dinocyst zones across the whole of the central and northern North Sea Basin. Their sequences were based on electric log data, with each sequence boundary being marked by a high gamma log peak marking a thin condensed claystone horizon.

Bujak and Mudge (1994) and Mudge and Bujak (1996) calibrated their Paleocene–Eocene dinocyst zones and stratigraphic sequences with planktonic and benthic foraminiferal, diatom, and radiolarian bioevents. The distribution of these microfossils was

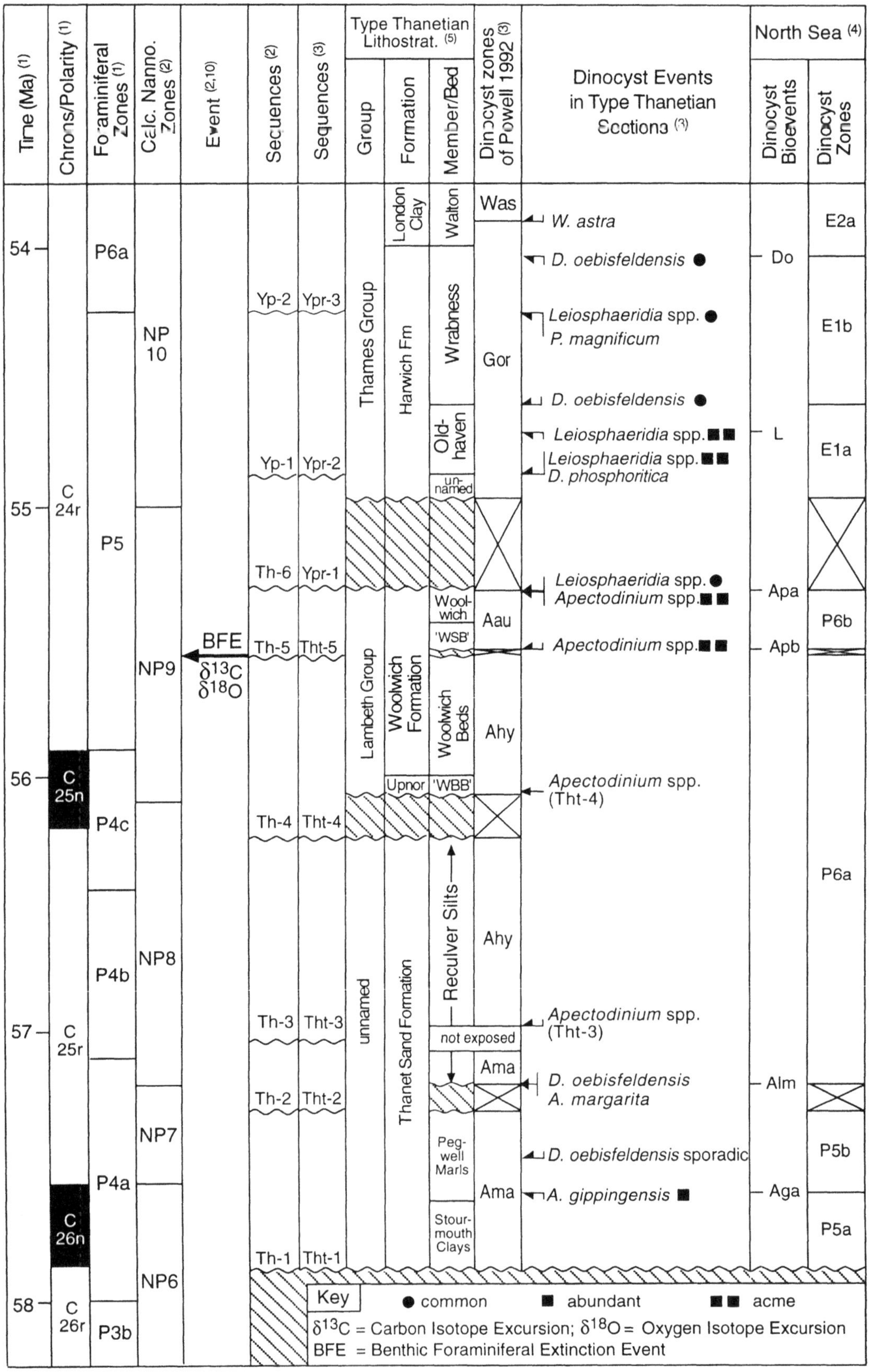
Time (Ma) (1)
Chrons/Polarity (1)
Foraminiferal Zones (1)
Calc. Nanno. Zones (2)
Event (2,10)
Sequences (2)
Sequences (3)
Type Thanetian Lithostrat. (5)
Group
Formation
Member/Bed
Dinocyst zones of Powell 1992 (3)
Dinocyst Events in Type Thanetian Sections (3)
North Sea (4)
Dinocyst Bioevents
Dinocyst Zones
54
55
56
57
58
C 24r
C 25n
C 25r
C 26n
C 26r
P6a
P5
P4c
P4b
P4a
P3b
NP 10
NP9
NP8
NP7
NP6
BFE
δ13C δ18O
Yp-2
Ypr-3
Yp-1
Ypr-2
Th-6
Ypr-1
Th-5
Tht-5
Th-4
Tht-4
Th-3
Tht-3
Th-2
Tht-2
Th-1
Tht-1
Thames Group
Lambeth Group
unnamed
London Clay
Harwich Fm
Woolwich Formation
Upnor
Thanet Sand Formation
Walton
Wrabness
Old-haven
un-named
Wool-wich
'WSB'
Woolwich Beds
'WBB'
Reculver Silts
not exposed
Peg-well Marls
Stour-mouth Clays
Was
Gor
Aau
Ahy
Ama
W. astra
D. oebisfeldensis ●
Leiosphaeridia spp. ●
P. magnificum
D. oebisfeldensis ●
Leiosphaeridia spp. ■■
Leiosphaeridia spp. ■■
D. phosphoritica
Leiosphaeridia spp. ●
Apectodinium spp. ■■
Apectodinium spp. ■■
Apectodinium spp. (Tht-4)
Apectodinium spp. (Tht-3)
D. oebisfeldensis
A. margarita
D. oebisfeldensis sporadic
A. gippingensis ■
Do
L
Apa
Apb
Alm
Aga
E2a
E1b
E1a
P6b
P6a
P5b
P5a
Key
● common
■ abundant
■■ acme
δ13C = Carbon Isotope Excursion; δ18O = Oxygen Isotope Excursion
BFE = Benthic Foraminiferal Extinction Event

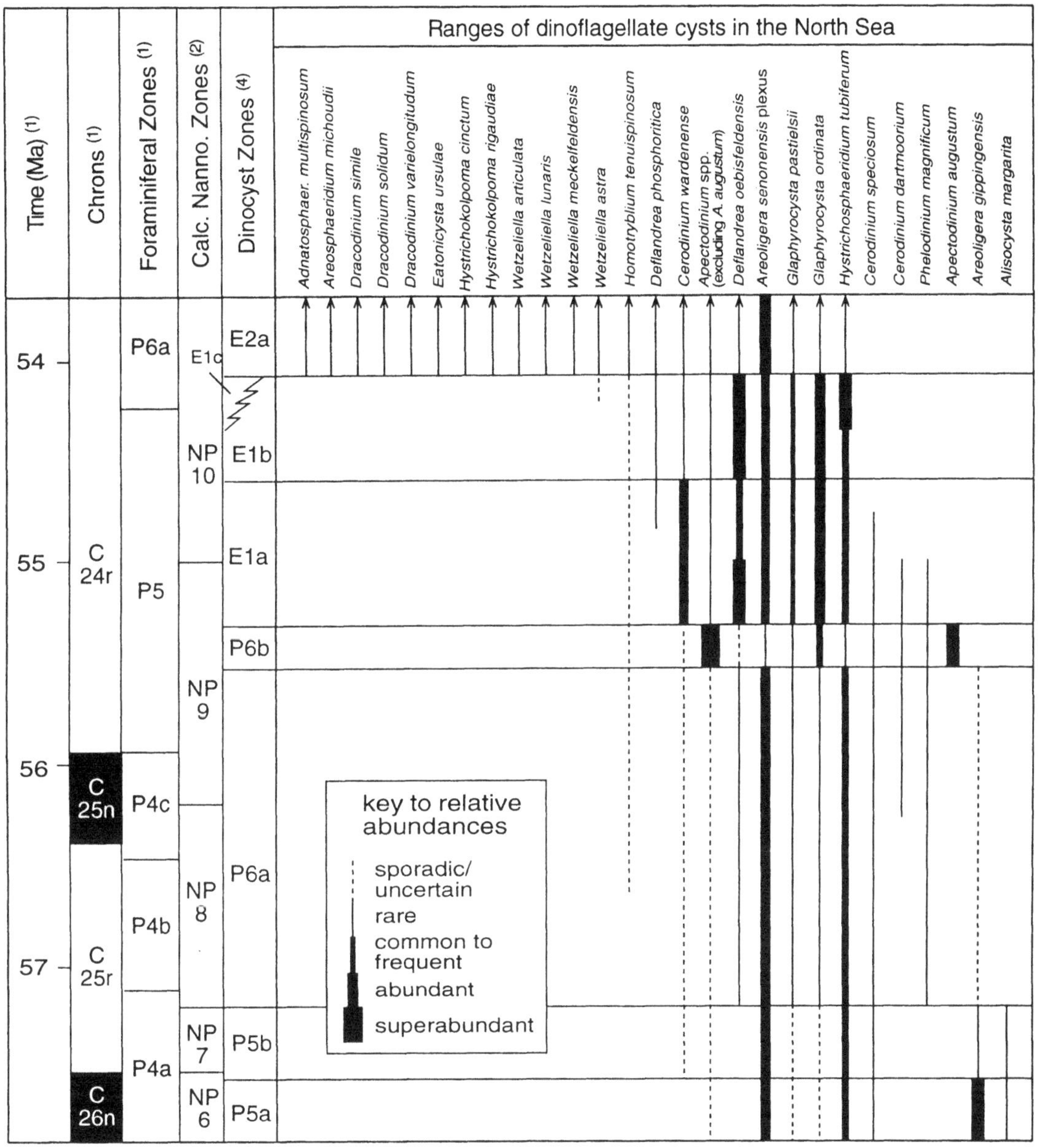

Figure 14.3 Dinocyst ranges across the Paleocene/Eocene Series boundary interval in the North Sea Basin.
(1) After Berggren (1994); (2) after Hardenbol (1994); (4) after Bujak and Mudge (1994) and Mudge and Bujak (1996).

strongly influenced by the prevailing paleoceanography of the North Sea marine basin, which became increasingly restricted during the latest Paleocene and earliest Eocene epochs, resulting in highly endemic, low-diversity assemblages of diatoms and agglutinated benthic foraminifera (figure 14.4).

The offshore North Sea dinocyst succession is very similar to that documented by Powell et al. (1996) from the Thanetian to lower Ypresian succession exposed in southeast England (figure 14.2), permitting calibration of the North Sea dinocyst events to magneto- and calcareous nannofossil stratigraphy. It is thus possible to correlate the North Sea upper Paleocene–lower Eocene stratigraphic record to the global chronostratigraphic framework, as shown in figure 14.4. Although chronostratigraphic correlation of older North Sea Paleocene

Figure 14.2 (left) Dinocyst succession in the lower Ypresian and type Thanetian sections of southeast England, and its relationship to chronostratigraphy and North Sea stratigraphy.
(1) After Berggren (1994); (2) after Hardenbol (1994); (3) after Powell et al. (1996); (4) after Bujak and Mudge (1994) and Mudge and Bujak (1996); (5) after Knox et al. (1994); (10) after Kennett and Stott (1991).

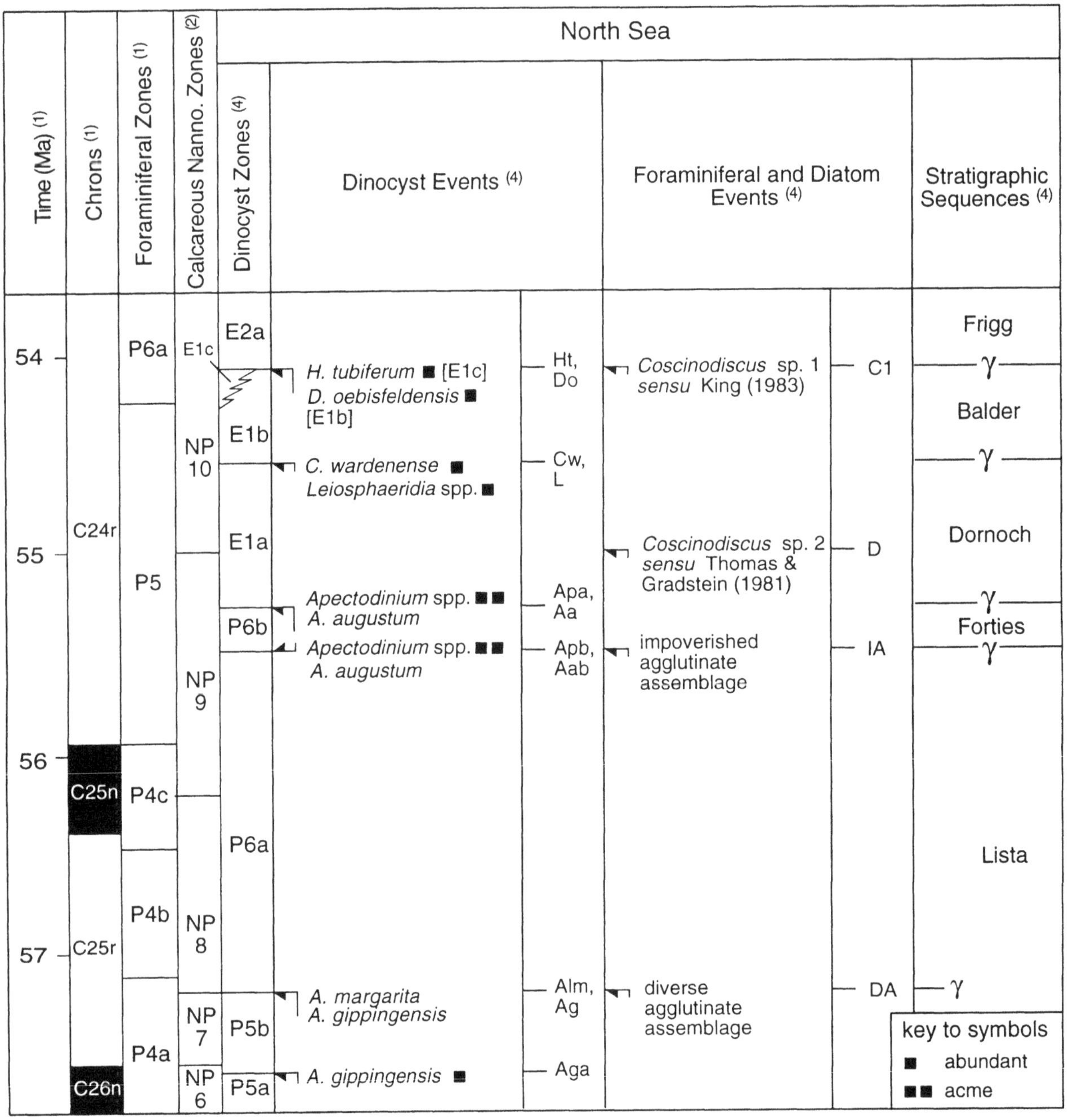

Figure 14.4 Chronostratigraphic calibration of North Sea microfossil events and stratigraphic sequences across the Paleocene/Eocene Series boundary interval.
(1) After Berggren (1994); (2) after Hardenbol (1994); (4) after Bujak and Mudge (1994) and Mudge and Bujak (1996).

deposits is less certain, Mudge and Bujak (1996) suggested the tentative correlations shown in figure 14.5.

The following discussion summarizes the Thanetian to lower Ypresian North Sea microfossil succession and chronostratigraphy (figures 14.4 and 14.5), based primarily on correlations with the Thanetian and Ypresian deposits of southeast England (figure 14.2). The North Sea lithostratigraphy and stratigraphic sequences of Mudge and Bujak (1994, 1996) are also shown in figure 14.5.

The highest abundance of the dinocyst *Areoligera gippingensis* (event Aga) defines the top of Subzone P5a of Mudge and Bujak (1996). Comparison with the type Thanetian section indicates correlation with uppermost Zone NP6, with Chron C26n, and with a level within sequence Tht-1 of Powell et al. (1996).

The highest occurrence of the dinocyst *Alisocysta margarita* (event Alm) defines the top of Subzone P5b of Mudge and Bujak (1996), coincident with the highest Paleocene occurrence of diverse agglutinated benthic foraminifera (event DA), including species such as *Spiro-*

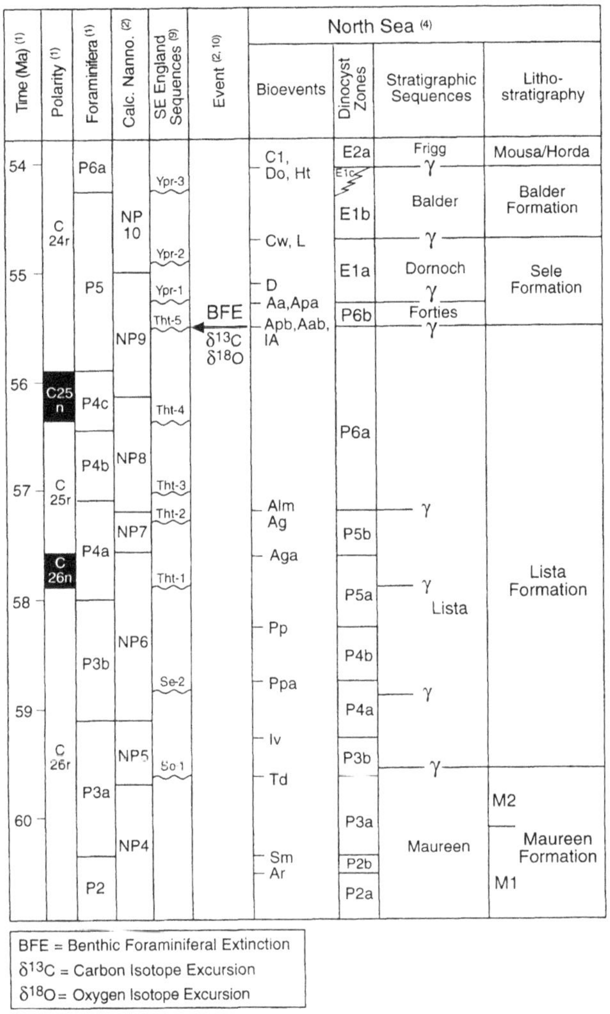

Figure 14.5 Selandian to lower Ypresian sequence stratigraphy and biostratigraphy of the North Sea Basin.
(1) After Berggren (1994); (2) after Hardenbol (1994); (4) after Bujak and Mudge (1994) and Mudge and Bujak (1996); (9) after Hardenbol (1994) for Selandian sequences and Powell et al. (1996) for Thanetian–Ypresian sequences; (10) after Kennett and Stott (1991).

plectammina spectabilis, and a high-gamma subsequence boundary within the Lista stratigraphic sequence (Mudge and Bujak 1996). Comparison with the type Thanetian section indicates correlation with Chron C25r, and a level within sequence Tht-2 of Powell et al. (1996).

The top of Subzone P5b is also characterized by the highest occurrence of the dinocyst *Areoligera gippingensis.* This species was not recorded by Powell et al. (1996) from the type Thanetian section above its highest abundance near the base of the Pegwell Marls (figure 14.2). Its absence from the middle and upper parts of the Pegwell Marls, which contain *Alisocysta margarita,* may be due to paleoenvironmental restriction. The lowest occurrence of the acme of *Apectodinium* spp. (event Apb) and the lowest occurrence of *Apectodinium augustum* (event Aab) define the top of Subzone P6a of Mudge and Bujak (1996). These events coincide with the highest Paleocene occurrence of impoverished agglutinated benthic foraminifera (event IA), including such taxa as *Bathysiphon* and *Haplophragmoides,* with the lowest occurrence of diatom-dominated assemblages, and with the high-gamma boundary between the Lista and Forties stratigraphic sequences of Mudge and Bujak (1996). Comparison with the type Thanetian section indicates correlation with Chron C24r, and a level near the base of sequence Tht-5 of Powell et al. (1996).

Dinocyst event Apb was thought to represent the earliest occurrence of *Apectodinium* in the North Sea, but Thomas (1993) recently recorded sporadic older occurrences of *Apectodinium* from a cored section in the U.K. central North Sea well 22/10a-4. She observed its lowest occurrence in the uppermost part of unit M1 of the Maureen Formation, just above the highest occurrence of *Alisocysta reticulata* (event Ar in figure 14.5), and a second occurrence in unit M2 of the Maureen Formation. The age of these two North Sea events is currently uncertain, but it is probable that the lower event occurs in nannofossil Zone NP4 and that the upper event occurs in Zone NP4 or lower Zone NP5.

The highest occurrence of the acme of *Apectodinium* spp. (event Apa) and the highest occurrence of *Apectodinium augustum* (event Aa) define the top of Subzone P6b of Bujak and Mudge (1994), coincident with the high-gamma boundary between the Forties and Dornoch stratigraphic sequences of Mudge and Bujak (1996). Comparison with the type Thanetian section indicates correlation with Chron C24r and a level just below the base of sequence Ypr-1 of Powell et al. (1996).

The North Sea acme of *Apectodinium* includes a variety of morphotypes that have been distinguished by various authors into more than 10 species based primarily on horn development. Several of these species form an intergrading plexus with conservative horn development that can be assigned to the *A. homomorphum/ hyperacanthum* plexus. These relatively simple forms are the only morphotypes known to predate the *Apectodinium* acme.

An acme of the probable algal cyst *Leiosphaeridia* spp. characterizes the North Sea *Leiosphaeridia* biofacies of Bujak and Mudge (1994). Comparison with southeast England indicates correlation with Chron C24r, and a level within sequence Ypr-2 of Powell et al. (1996). Second-order correlation with DSDP Hole 550 indicates assignment to Zone NP10 (Berggren and Aubry 1996). The *Leiosphaeridia* biofacies occurs within North Sea Subzone E1a of Bujak and Mudge (1994), which has an upper boundary that is defined by the highest abundance of the dinocyst *Cerodinium wardenense,* close to the Dornoch/Balder sequence boundary (Mudge and

Bujak 1994). *Cerodinium wardenense* is rare in the onshore Thanetian sections examined by Powell et al. (1996), so that correlation with North Sea Subzone E1a is tentatively based on the acme of *Leiosphaeridia* spp. in the Oldhaven Beds of southeast England.

The highest abundance of the dinocyst *Deflandrea oebisfeldensis* (event Do) defines the top of Subzone E1b of Bujak and Mudge (1994) and coincides with the highest occurrence of the diatom *Coscinodiscus* sp. 1 sensu King (1983; event C1) and of the diatom-dominated microfaunal assemblages, and with the high-gamma boundary between the Balder and Frigg stratigraphic sequences of Mudge and Bujak (1996). Comparison with southeast England indicates correlation with Zone NP10, Chron C24r, and a level within sequence Ypr-3 of Powell et al. (1996).

The top of North Sea Zone E1 is dominated by the dinocyst *Hystrichosphaeridium tubiferum* and contains few *D. oebisfeldensis* in wells drilled near the southeastern margin of the Shetland Platform. Bujak and Mudge (1994) interpreted these to be marginal marine assemblages synchronous with the upper part of Subzone E1b, and designated the *H. tubiferum*–dominated assemblages as Subzone E1c.

The base of the overlying North Sea Zone E2 corresponds to the base of the Frigg stratigraphic sequence of Mudge and Bujak (1994), and is characterized by the first occurrences of typical Eocene dinocyst taxa (figure 14.3). These taxa indicate correlation with the London Clay Formation, based primarily on the work of G. L. Williams, published in Davey et al. (1966) and Bujak et al. (1980), so that the North Sea appearance of these dinocysts at the base of the Frigg sequence corresponds to the London Clay transgression.

Upper Thanetian and lower Ypresian dinocyst assemblages have also been described from the shallow marine strata of Belgium (De Coninck 1975) and Denmark (Heilmann-Clausen 1985), and lower Ypresian assemblages have been described from northwest Germany (Heilmann-Clausen and Costa 1989). In all three areas, the examined sections were from boreholes with little or no data on magnetostratigraphy or calcareous nannofossil stratigraphy.

The dinocyst successions in the three sections are very similar to those from the North Sea and southeast England, and include an acme of *Apectodinium*, including common *Apectodinium augustum*, in the Negative Series of the Ølst Formation in the Viborg 1 borehole of Denmark (Heilmann-Clausen 1985), and abundant *Apectodinium homomorphum/hyperacanthum*, recorded as *Wetzeliella* spp., at the top of the Sables d'Ostende-ter-Streep in the Kallo borehole of Belgium (De Coninck 1975). In the Wursterheide borehole in northwest Germany, the lowest sample examined for dinocysts contains abundant *Deflandrea oebisfeldensis* (Heilmann-Clausen and Costa 1989), indicating correlation with North Sea Subzone E1b of Bujak and Mudge (1994).

Dinoflagellate Cyst Stratigraphy in Low-Latitude Regions

Only limited information is available on Paleocene and Eocene dinocyst distribution at low latitudes. Most studies are derived from Africa (Williams 1978; Jan du Chêne and Adediran 1984; Jan du Chêne 1988), Pakistan and India (e.g., Varma and Dangwal 1964; Van Erve et al. 1980; Sarka and Singh 1988; Köthe 1988), and the Mediterranean area (e.g., Caro 1973; Jan du Chêne 1977; Stampfi et al. 1978; Ediger 1981; Brinkhuis and Zachariasse 1988; Michoux 1988; El-Beialy 1988). In addition, Damassa (1979a, 1979b) and Drugg (1967) documented Paleocene and Eocene dinocysts from California, and Brinkhuis et al. (1994) recently presented a summary of stratigraphically important bioevents (foraminifera, calcareous nannofossils, and dinocysts) in the Paleocene part of the El Kef section in northwest Tunisia.

Very few of these studies actually cover the Paleocene–Eocene transition. Among them are investigations by Jan du Chêne and Adediran (1984) on Nigeria and Köthe (1988) on Pakistan sections. However, neither study has a reliable chronostratigraphic control. No detailed studies of dinocyst ranges across the Paleocene/Eocene Series boundary are available from the Mediterranean, except for Caro's (1973) analysis of the Campo section in the Tremp–Grauss Basin of Spain. Combining all of the published information with unpublished data from Tunisian and Italian sections (Brinkhuis unpublished data), a succession of apparently correlative lower-latitude dinocyst events around the Paleocene/Eocene Series boundary can be recognized (figure 14.6).

The earliest definite occurrence of *Apectodinium* is in the Tunisian El Kef section, where it occurs in strata directly assigned to planktonic foraminiferal Zone P3a at the base of the Selandian Stage. In this section, it is continuously present in low numbers up to the base of planktonic foraminiferal Zone P5 (figure 14.6). The *Apectodinium* acme has its lowest occurrence at the base of Zone P5 and extends into sediments assigned to the middle part of calcareous nannofossil Zone NP11 in the Jebel Boudabous section of central Tunisia (Brinkhuis and Driever, personal observation). *Apectodinium* occurs in low numbers above this horizon up to the middle Eocene strata.

The *Apectodinium* acme also occurs in uppermost Paleocene–lowermost Eocene strata in the Bolca section of northeast Italy (H. Brinkhuis, personal observation).

All specimens of *Apectodinium* observed from the Paleocene and Eocene sections of Elf Kef belong to the

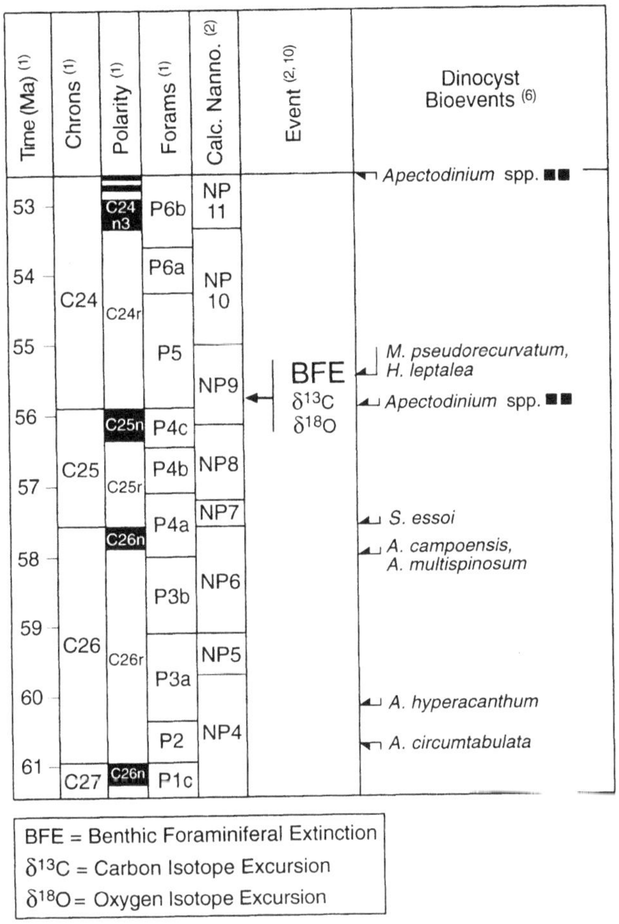

Figure 14.6 Major dinocyst events and chronostratigraphy in the El Kef section, Tunisia.

(1) After Berggren (1994); (2) after Hardenbol (1994); (6) after Brinkhuis (personal observation) and Brinkhuis et al. (1994); (10) after Kennett and Stott (1991).

A. homomorphum/hyperacanthum plexus, with extreme morphotypes such as *Apectodinium augustum* being absent. The environments represented by these sections were shelfal marine during the Paleocene Epoch, shallowing upward into typical *Nummulites*-platform facies during the Eocene Epoch.

Dinoflagellate Cyst Stratigraphy in Northern High Latitudes

Manum and Throndsen (1986) and Soper et al. (1976a, 1976b) documented dinocyst assemblages in upper Paleocene–lower Eocene deposits from Spitsbergen and East Greenland, respectively, but no independent chronostratigraphy is currently available for these deposits. In both areas, strata associated with either ash layers or basaltic material contain dinocyst assemblages dominated by *Apectodinium* to which a latest Thanetian age was assigned based on dinocyst evidence. Slightly younger assemblages are more diverse and include *Wetzeliella meckelfeldensis* in both areas, indicating an Ypresian age. Manum and Throndsen (1986) and Soper et al. (1976a, 1976b) noted strong similarities with northwest European dinocyst assemblages, and assigned the Thanetian assemblages to the *Apectodinium hyperacanthum* Zone, and the Ypresian assemblages to the *Wetzeliella meckelfeldensis* Zone of Costa and Downie (1976), shown in figure 14.1.

No other information has been published on dinocyst successions across the Paleocene/Eocene Series boundary in high latitudes, but unpublished data from exploration wells drilled in the Canadian Beaufort Sea (J. Bujak, personal observation) and the Norwegian Barents Sea (H. Brinkhuis, personal observation) indicate that *Apectodinium* became the dominant dinocyst for a relatively brief period near the Paleocene/Eocene Epoch boundary in these two areas. Chronostratigraphic control in these wells is based primarily on dinocysts, with the *Apectodinium* dominance preceding the earliest occurrence of typical Ypresian dinocysts, including *Wetzeliella.* This suggests that the *Apectodinium* dominance was approximately coeval with the *Apectodinium* acme recorded from midlatitude areas.

Southern Hemisphere Dinoflagellate Cyst Stratigraphy

The primary source of Southern Hemisphere data close to the Paleocene/Eocene Series boundary with independent chronostratigraphic control is from New Zealand (figure 14.7). Wilson (1988) examined the dinocyst succession from the middle to upper Paleocene Teurian and the uppermost Paleocene to lower Eocene Waipawan Stages. He noted an abrupt change in dinocyst assemblages close to the base of the Waipawan Stage from relatively diverse populations assigned to the upper Teurian *Palaeocystodinium golzowense* dinocyst Zone, to populations dominated by *Apectodinium* in the lower part of the Waipawan Stage, which Wilson (1984, 1988) designated the *Apectodinium homomorphum* Zone. These are succeeded in the middle and upper parts of the Waipawan Stage by assemblages assigned to the lower Eocene *Wetzeliella spinulosa* Zone and *Dracodinium waipawaense* Zone, respectively.

Calcareous nannofossil and planktonic foraminiferal zonal assignments have been determined for most of the section. The *A. homomorphum* Zone is equivalent to the *Globigerina triloculinoides* and lower part of the *Globanomalina wilcoxensis* foraminiferal Zones (probable mid-Zone NP9 to lower Zone NP10 equivalents). Edwards (in Grant-Taylor and Hornibrook 1976) correlated these zones with the *Rhomboaster cuspis, Marthasterites tribrachiatus,* and lower part of the *Chiasmolithus grandis* calcareous nannofossil Zones.

Partridge (1976) also distinguished an upper Pale-

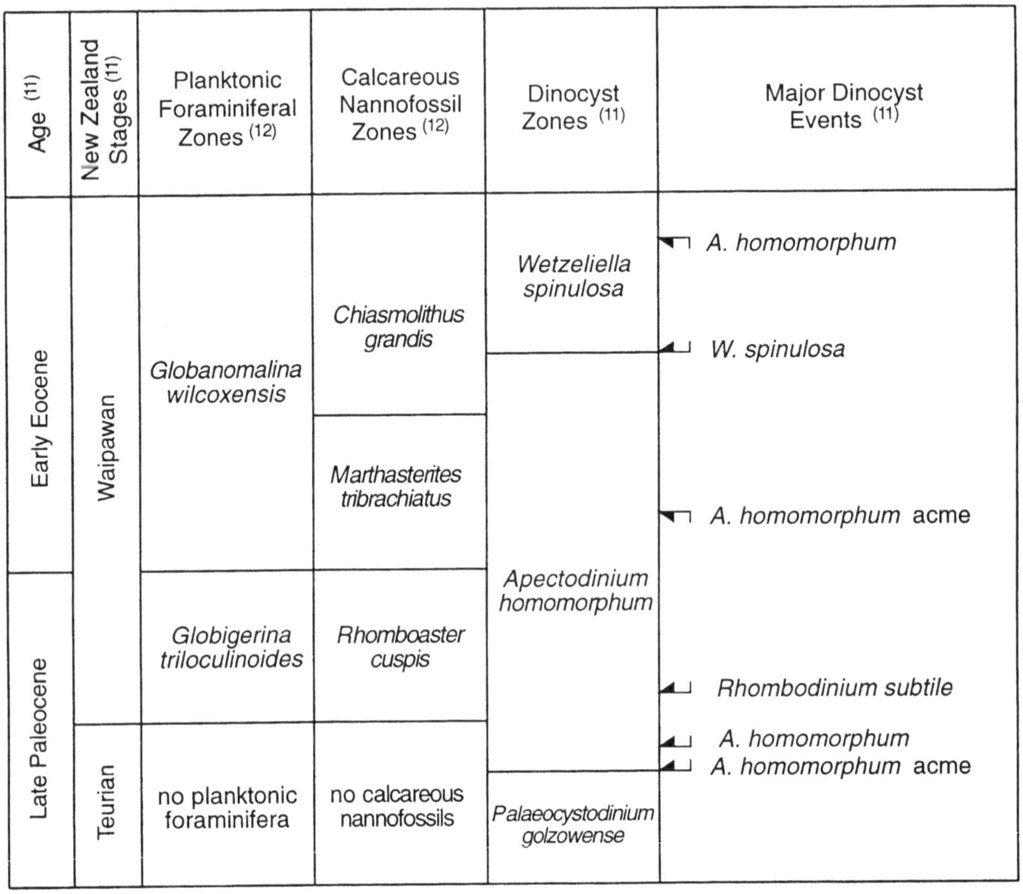

Figure 14.7 Major dinocyst events and zonal framework in the type Waipawan section at Hawkes Bay, New Zealand.
(11) After Wilson (1988); (12) after Edwards in Grant-Taylor and Hornibrook (1976).

ocene *Apectodinium homomorphum* Zone, lowermost Eocene *Apectodinium hyperacanthum* Zone, and a lower Eocene *Dracodinium waipawaense* Zone in the Gippsland Basin of southeast Australia. He calibrated these with planktonic foraminiferal zones, upper Zone P4 to Zone P5, Zone P6 and Zone P7, respectively, based on his unpublished correlations with the Otway Basin, Australia, and New Zealand. No details of these correlations are available, and the *A. hyperacanthum* Zone may be upper Paleocene, rather than lower Eocene, based on Cookson and Eisenack's (1967) work on the Paleocene Rivernook Bed, Victoria. Regardless of this, it appears that the genus *Apectodinium* first became abundant during the latest Paleocene in both southeast Australia and New Zealand.

Worldwide *Apectodinium* Distribution Through Time

From the above discussion, several observations can be made regarding the Paleocene and early Eocene distribution of *Apectodinium* and other dinocyst taxa.

With the exception of a questionable Maastrichtian occurrence in the subsurface of India (Mehrotra and Sarjeant 1987; Garg and Jain 1995), the earliest known occurrence of *Apectodinium* is in the Tunisian El Kef section in strata assigned to planktonic foraminiferal Zone P3a at the base of the Selandian Stage. The genus is continuously present in the section in low numbers up to the base of planktonic foraminiferal Zone P5, and has an acme from the base of this zone up to the middle part of nannofossil Zone NP11, above which it occurs rarely up to middle Eocene strata. All specimens of *Apectodinium* observed from the Paleocene and Eocene sections of Elf Kef belong to the *A. homomorphum/hyperacanthum* plexus, which is the most conservative morphotype of the genus.

In northwest Europe, the earliest record of *Apectodinium* is a stratigraphically isolated occurrence in the U.K. North Sea well 22/10a-4 in the lower part of the Selandian Maureen Formation, tentatively assigned to nannofossil Zone NP4. It has at least one other isolated North Sea occurrence higher in the Maureen Formation, in strata that probably belong to either nannofossil Zone

NP4 or lower Zone NP5. Both of these North Sea Selandian occurrences represent specimens of the *A. homomorphum/hyperacanthum* plexus.

Apectodinium has at least two stratigraphically isolated occurrences in the type Thanetian sections of southeast England, first in the Reculver Silts of the Thanet Sand Formation (lower nannofossil Zone NP8, Chron C25r, sequence Tht-3), and next in the lower part of the Woolwich Formation (lower nannofossil Zone NP9, Chron C25n, sequence Tht-4). Both occurrences represent the *A. homomorphum/hyperacanthum* plexus.

In the type Thanetian sections, the *Apectodinium* acme occurs in the upper part of the Woolwich Formation, within sequence Tht-5 of Powell et al. (1996), equivalent to sequence Th-5 of Hardenbol (1994). Data on carbon isotope stratigraphy of these sections have not yet been published, but second-order correlations with DSDP Hole 550 (Berggren and Aubry 1996) suggest that the initiation of this acme coincides with the carbon isotope excursion of Kennett and Stott (1991). This supports Hardenbol's correlation (1994) of the carbon isotope excursion and the benthic foraminiferal extinction event with his sequence Th-5 in the upper part of the Woolwich Formation. The work of Kennett and Stott (1991) and others indicates that the carbon isotope excursion is associated with the oxygen isotope excursion, which reflected increased bottom-water temperatures at higher latitudes.

In southeast England, the *Apectodinium* acme includes *A. homomorphum/hyperacanthum, A. parvum,* and *A. paniculatum. Apectodinium augustum* does not occur in the type Thanetian sections.

An apparently coeval acme of *Apectodinium*, including *A. augustum,* occurs in Belgium and Denmark, and extends across the North Sea Basin. In the North Sea Basin, its onset is coincident with the last Paleocene occurrence of agglutinated benthic foraminifera, and a change to assemblages dominated by pyritized diatoms. At the same time, the kerogen changes to predominantly pyritized sapropelic material, reflecting bottom-water anoxia.

Poorly dated but potentially coeval acmes of *Apectodinium* also occur in upper Paleocene strata as far north as Spitsbergen, the Barents Sea, and the Beaufort Sea. *Apectodinium* is absent or extremely rare in the Eocene strata of these areas.

In the Southern Hemisphere, the first accurately dated occurrence of *Apectodinium* is in the uppermost Paleocene (mid–Zone NP9 equivalent) deposits of New Zealand. In this area, the genus *Apectodinium* was abundant from approximately mid–Biochron NP9 to early Biochron NP10. An *Apectodinium* abundance in the Gippsland Basin of southeast Australia is also of probable latest Paleocene to questionable earliest Eocene age.

The *Apectodinium* acme assemblages are succeeded in the offshore North Sea Basin by lower Ypresian dinocyst assemblages with abundant *Deflandrea oebisfeldensis* and *Cerodinium wardenense.* Pyritized diatoms are still dominant, and pyritized sapropelic kerogen is common, but reduced compared to strata containing the *Apectodinium* Acme.

In southeast England, close to the southern shoreline of the North Sea Basin, shallow marine assemblages overlying sediments containing the acme of *Apectodinium* contain abundant *Deflandrea oebisfeldensis* in low-diversity dinocyst assemblages including *Glaphyrocysta, Hystrichosphaeridium,* and *Spiniferites. Apectodinium* is absent or rare. Similar dinocyst assemblages occur in the lower Ypresian deposits of Belgium, Denmark, and northwest Germany and also contain few or no *Apectodinium.* Brackish to freshwater lower Ypresian sediments from southeast England contain abundant leiospheres and no dinocysts.

A change to diverse dinocyst and planktonic to benthic foraminiferal assemblages occurs across northwest Europe and the North Sea in strata that are probably coeval with the basal London Clay Formation, close to the NP10/NP11 zonal boundary. *Apectodinium* occurs consistently above but in low numbers up to the Lutetian strata, above which it is sporadic and probably has its last in situ occurrence in the Bartonian deposits.

The *Apectodinium* Migration Model

Based on these observations, a model is outlined below that integrates data on late Paleocene–early Eocene dinocyst distributions from low to high latitudes with recently published information on worldwide oceanographic changes that occurred close to the Paleocene/Eocene Epoch boundary (figure 14.8).

The dinocyst genus *Apectodinium* first evolved near the beginning of the Selandian Age at low latitudes, and was produced by a motile dinoflagellate that may have been wholly or partially holozoic, having the ability to feed upon phytoplankton and organic detritus. The genus became a minor component of the plankton associations during the Selandian and most of the Thanetian ages in the Tethyan realm. It periodically migrated into midlatitudes, including the North Sea Basin, during periods of warmer water conditions. *Apectodinium* showed little morphological diversification during this time, and all morphotypes can be assigned to the *A. homomorphum/hyperacanthum* plexus. There is no evidence suggesting that these *Apectodinium* were not fully marine, either in the Tethys or in higher latitudes.

Near the end of the Thanetian Age, in mid–Biochron NP9, global warming occurred (Stott and Kennett 1990; Kennett and Stott 1991), and was associated with re-

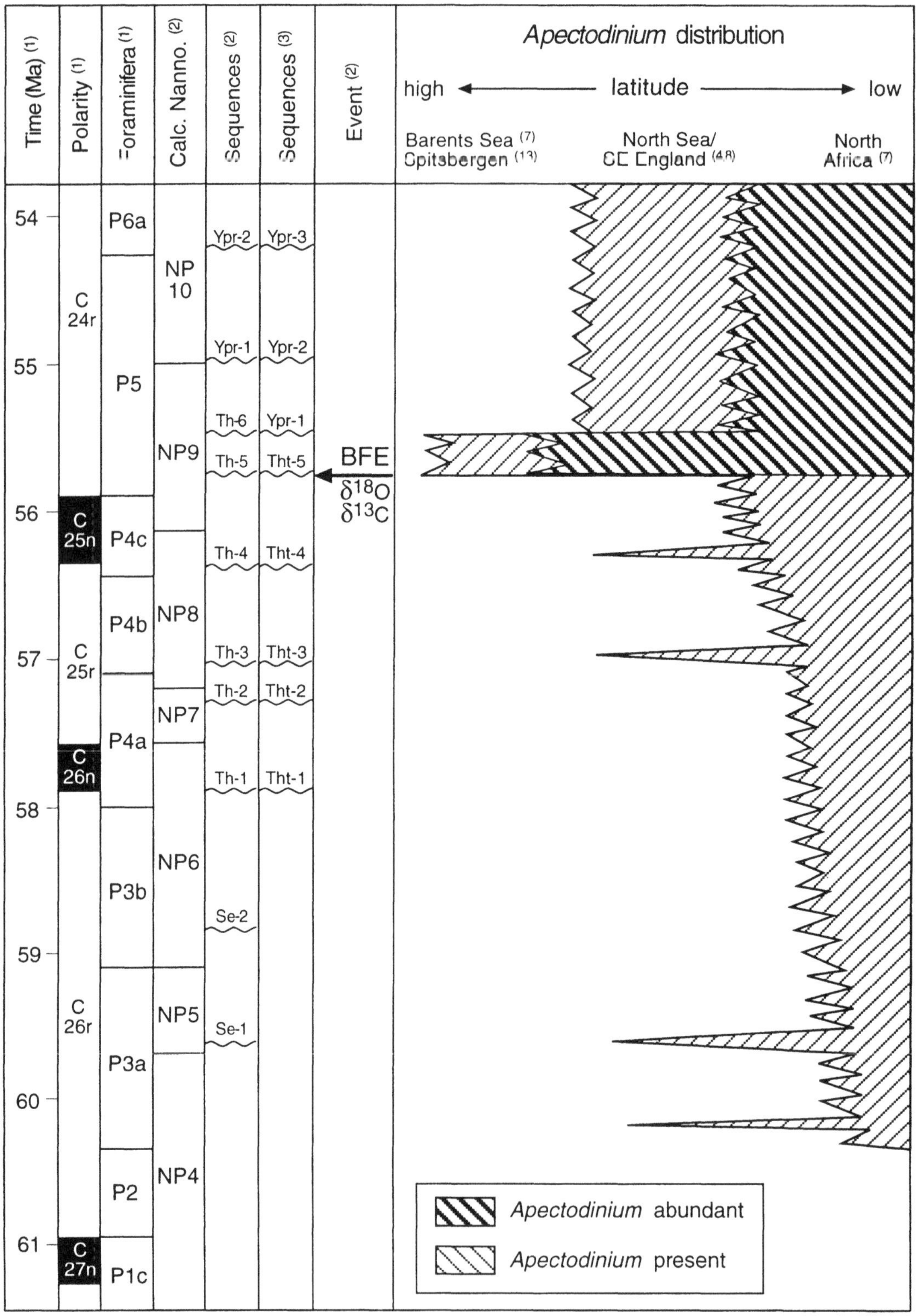

Figure 14.8 North Hemisphere transect of *Apectodinium* distribution and abundance through the Selandian to lower Ypresian stages.

(1) After Berggren (1994); (2) after Hardenbol (1994); (3) after Powell et al. (1996); (4) after Bujak and Mudge (1994) and Mudge and Bujak (1996); (7) J. P. Bujak and H. Brinkhuis (personal observation); (8) after Thomas (1993); (13) after Soper et al. (1976a, 1976b).

duced atmospheric circulation (Rea et al. 1990), changes in oceanic circulation, and the brief production of deep water at low latitudes (Miller et al. 1987; Kennett and Stott 1991; Thomas 1990a, 1990b, 1993). Benthic foraminifera underwent major extinction (Tjalsma and Lohman 1983; Thomas 1990a, 1990b) associated with bottom-water anoxia.

At the same time, the temperatures increased abruptly in mid- to high latitudes, resulting in a strongly reduced temperature gradient between the equator and poles (Thomas and Shackleton 1993). Associated shifts in nutrient concentrations may have resulted from the expanded oligotrophy of euphotic waters in the open ocean and margins, similar to those of the early Eocene oligotrophic expansion (Boersma et al. 1987; Hallock et al. 1991).

It is difficult to separate the role of temperature and nutrient concentrations, but the warming events and associated oceanographic changes were optimum for *Apectodinium* and permitted its widespread migration into high latitudes at least as far north as the Barents Sea and Beaufort Sea and into the Southern Hemisphere at least as far south as Australia and New Zealand.

The late Thanetian event also resulted in the suppression of many dinocyst taxa in mid- and high latitudes, and *Apectodinium* became the dominant dinocyst genus. Its successful exploitation of these previously cooler water locations enabled it to undergo rapid morphological diversification, resulting in several distinct taxa. The dinocyst assemblages in mid- to high latitudes thus changed from predominantly photosynthetic assemblages to populations dominated by *Apectodinium* species that may have been wholly or partially holozoic.

In low latitudes, the late Thanetian event had less impact on the overall dinocyst assemblages than in higher latitudes. An increased abundance of *Apectodinium* occurred, but the dinocyst assemblages retained most of their earlier species. Low-latitude *Apectodinium* also showed less species diversification than in higher latitudes, probably due to fewer low-latitude opportunities for the exploitation of vacant niches.

During the latest Thanetian Age, the North Sea Basin was characterized by a codominance of Coscinodiscid diatoms and *Apectodinium* species (figure 14.9). It is possible that diatoms were the primary food source of the *Apectodinium* motile stage, so that diatom blooms would have been the triggering mechanism for *Apectodinium* blooms. This is similar to the relationship between modern marine diatoms and many holozoic dinoflagellates (F. G. R. Taylor, personal communication 1994). Bottom-water anoxia extended across the North Sea Basin, so that benthic organisms were unavailable to provide an essential stage for recycling organic material back into shallower water. Sediments therefore accumulated pyritized sapropelic kerogen plus the pyritized remains of diatoms and dinocysts.

The onset of diatom–*Apectodinium* codominance and the last occurrence of North Sea Paleocene benthic foraminifera coincided with the Lista/Forties sequence boundary (figure 14.9). Preceding this, the basin had been the subject of increasing restriction during the later phase of deposition of the Lista sequence, so that the foraminiferal assemblages were already becoming impoverished. This restriction reached its maximum during the latest Thanetian and earliest Ypresian deposition of the Forties, Dornoch, and Balder stratigraphic sequences, with the basin having marine connection only to the north and east. Nevertheless, surface waters continued to be fully marine across much of the basin and *Apectodinium* diversification included the probable offshore marine species *A. augustum,* which is not known from nearshore marine sections, including those at the southern end of the basin in southeast England.

Termination of the *Apectodinium* acme at a time equivalent to the Forties/Dornoch sequence boundary in the North Sea was accompanied by a change to dinocyst assemblages containing abundant holozoic *Deflandrea oebisfeldensis* and *Cerodinium wardenense* (figure 14.9). Coscinodiscid diatom dominance continued due to their tolerance of cooler surface water conditions, and bottom-water anoxia still restricted the benthic faunas. The *Deflandrea-Cerodinium*–dominated assemblages were therefore a niche replacement of the earlier warmer-water *Apectodinium* assemblages, and probably had the same relationship to diatom blooms that persisted during deposition of the entire Dornoch and Balder sequences.

The *Apectodinium* acme terminated in the North Sea and southeast England close to the Thanetian/Ypresian Age boundary (= Paleocene/Eocene Epoch boundary; figure 14.2). Assemblages in the lowermost Ypresian deposits of southeast England are dominated by the cysts of probable photosynthetic dinoflagellates, such as *Glaphyrocysta* and *Hystrichosphaeridium* in marine sediments, or leiospheres in brackish to freshwater sediments. *Apectodinium* species are rare or absent in the lower Ypresian deposits.

The late Thanetian to early Ypresian North Sea anoxic regime was terminated by widespread marine transgression that flooded the intertidal lowland swamps at the beginning of Frigg sequence deposition (figure 14.9). *Taxodium* pollen became strongly reduced, and the basin was opened and flushed sufficiently to end the stratified, anoxic bottom-water regime. The shift to aerobic bottom-water conditions resulted in recolonization by benthic foraminifera, with surface-water populations changing to planktonic foraminiferal microfaunas.

At the same time, a succession of Eocene dinocyst

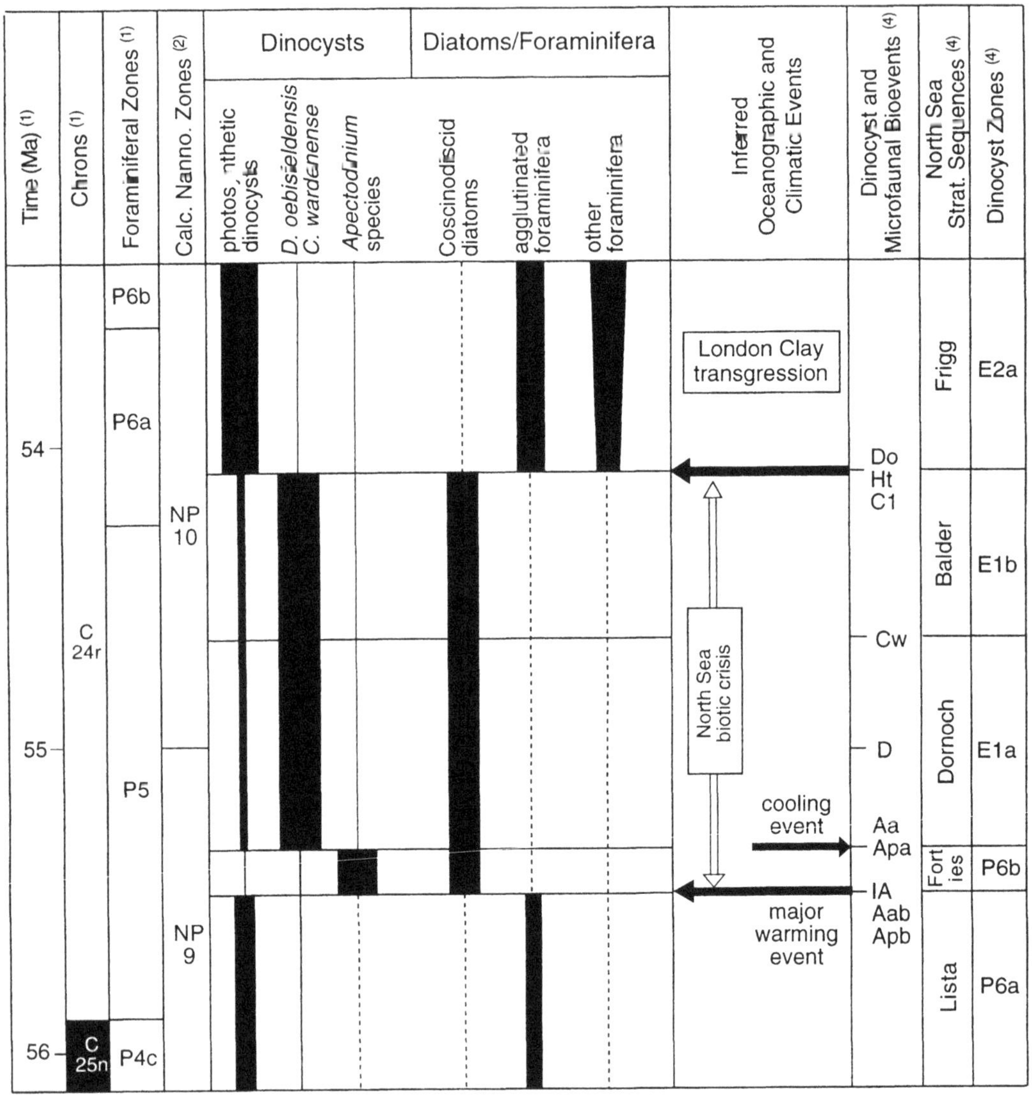

Figure 14.9 North Sea biotic changes across the Paleocene/Eocene Epoch boundary.
(1) After Berggren (1994); (2) after Hardenbol (1994); (4) after Bujak and Mudge (1994) and Mudge and Bujak (1996).

species evolved during a widespread transgressive phase, which resulted in the deposition of the lower part of the Frigg stratigraphic sequence in the North Sea Basin and the London Clay Formation in southern England. These dinocysts comprised mostly taxa that were probably derived from photosynthetic dinoflagellates, with only rare *Apectodinium*. In contrast, the genus was common at low latitudes until the end of planktonic foraminiferal Biochron P6 and mid-calcareous nannofossil Biochron NP11, reflecting warm-water conditions.

ACKNOWLEDGMENTS

The authors are grateful to G. L. Eaton and A. J. Powell for their constructive comments during preparation of the manuscript. We are also grateful to J. W. Zachariasse and B. Driever for providing unpublished foraminiferal and nannofossil data on Tunisian Paleocene and Eocene sections, to G. J. Wilson for providing unpublished data on New Zealand, and to L. I. Costa for providing unpublished data on East Greenland and the Rockall Plateau. This publication is the Netherlands Research School for Sedimentary Geology (NSG) Contribution No. 970172.

REFERENCES

Ali, J. R., 1994. Magnetostratigraphy of the upper Paleocene and lowermost Eocene of SE England. GFF, Meeting Proceedings "Stratigraphy of the Paleocene," v. 116, p. 41–42.

Archangelsky, S. 1969a. Sobre el paleomicroplancton del Terciaro inferior de Rio Tunbro, Provincia de Santa Cruz. *Ameghiana,* v. 5, p. 406–416.

Archangelsky, S. 1969b. Estudio del paleomicroplancton de la Formación Ris Tunbro (Eoceno), Provincia de Santa Cruz. *Ameghiana,* v. 6, p. 1–18.

Aubry, M.-P., 1985. Northwestern European Paleogene magnetostratigraphy, biostratigraphy, and paleogeography: Calcareous nannofossil evidence. *Geology,* v. 13, p. 198–202.

Aubry, M.-P., 1986. Paleogene calcareous nannoplankton biostratigraphy of northwestern Europe. *Palaeogeography, Palaeoclimatology, Palaeoecology*, v. 55, p. 267–334.

Aubry, M.-P. 1994. The Thanetian Stage in NW Europe and its significance in terms of global events. GFF, Meeting Proceedings "Stratigraphy of the Paleocene," v. 116, p. 43–44.

Aubry, M.-P., Hailwood, E. A. and Townsend, H. A., 1986. Magnetic and calcareous-nannofossil stratigraphy of lower Palaeogene formations of the Hampshire and London basins. *Journal of the Geological Society*, London, v. 143, p. 729–735.

Berggren, W. A. 1994. In defense of the Selandian Age/Stage. GFF, Meeting Proceedings "Stratigraphy of the Paleocene," v. 116, p. 44–46.

Berggren, W. A. and Aubry, M.-P., 1996. A late Paleocene–early Eocene NW European and North Sea magnetobiochronological correlation network. In Knox, R. W. O'B., Corfield, R. M. and Dunay, R. E., (eds.), *Correlation of the Early Paleogene in Northwest Europe,* Geological Society Special Publication No. 101, p. 309–352.

Boersma, A., Premoli Silva, I. and Shackleton, N. J., 1987. Atlantic Eocene planktonic foraminiferal paleohydrographic indicators and stable isotope paleoceanography. *Paleoceanography,* v. 2, p. 287–331.

Brinkhuis, H. and Zachariasse, W. J., 1988. Dinoflagellate cysts, sea level changes and planktonic foraminifers across the Cretaceous–Tertiary boundary at El Haria, Northwest Tunisia. *Marine Micropaleontology*, v. 13, p. 153–191.

Brinkhuis, H., Romein, A. J. T., Smit, J. and Zachariasse, J.-W., 1994. Danian–Selandian dinoflagellate cysts from lower latitudes with special reference to the El Kef section, NW Tunisia. GFF, Meeting Proceedings "Stratigraphy of the Paleocene," v. 116, p. 46–48.

Bujak, J. P., 1979. Proposed phylogeny of the dinoflagellates *Rhombodinium and Gochtodinium. Micropaleontology,* v. 25, p. 308–324.

Bujak, J. P. and Davies, E. H., 1981. *Modern and fossil Peridiniineae.* American Association of Stratigraphic Palynologists Contribution Series no. 13, 203 pp.

Bujak, J. P. and Davies, E. H., 1983. *Modern and fossil Peridiniineae.* American Association of Stratigraphic Palynologists Contribution Series no. 13, 203 pp.

Bujak, J. P., Downie, C., Eaton, G. L. and Williams, G. L., 1980. Dinoflagellate cysts and acritarchs from the Eocene of southern England. *Special Papers in Palaeontology,* v. 24, p. 1–100.

Bujak, J. P. and Mudge, D., 1994. A high-resolution North Sea Eocene dinocyst zonation. *Journal of the Geological Society of London,* v. 151, p. 449–462.

Bujak, J. P. and Williams, G. L., 1979. Dinoflagellate diversity through time. *Marine Micropaleontology*, v. 4, p. 1–12.

Bujak, J. P. and Williams, G. L., 1981. The evolution of dinoflagellates. *Canadian Journal of Botany*, v. 59, p. 2077–2087.

Caro, Y., 1973. Contribution à la connaissance des dinoflagellés du Paléocène–Eocène inférieur des Pyrénées espagnoles. *Revista Española de Micropaleontologia,* v. 5, p. 329–372.

Chateauneuf, J. J. and Gruas-Cavagnetto, C., 1978. Les zones de Wetzeliellaceae (Dinophyceae) du Bassin de Paris. *Bulletin du Bureau de Recherches Géologiques et Minières,* série 4, v. 2, p. 59–93.

Cookson, I. C. and Eisenack, A., 1967. Some microplankton from the Paleocene Rivernook Bed, Victoria. *Proceedings of the Royal Society of Victoria,* v. 80, p. 247–257.

Costa, L. and Downie, C., 1976. The distribution of the dinoflagellate Wetzeliella in the Palaeogene of north-western Europe. *Palaeontology,* v. 19, p. 591–614.

Costa, L. and Manum, S. B., 1988. The description of the interregional zonation of the Paleogene (D1–D15) and the Miocene (D16–D20). In Vinken, R., compiler, *The Northwest European Tertiary Basin. Results of the International Geological Correlation Programme Project No. 124, Geologisches Jahrbuch, ser. A,* v. 100, p. 321–330.

Damassa, S. P., 1979a. Eocene dinoflagellates from the Coastal Belt of the Franciscan Complex, northern California. *Journal of Paleontology,* v. 53, p. 815–840.

Damassa, S. P., 1979b. Danian dinoflagellates from the Franciscan Complex, Mendocino County, California. *Palynology*, v. 3, p. 191–207.

Davey, R. J., Downie, C., Sarjeant, W. A. S. and Williams, G. L., 1966. Studies on Mesozoic and Caenozoic dinoflagellate cysts. *Bulletin of the British Museum (Natural History) Geology,* Supplement 3, p. 1–248.

De Coninck, J., 1976. *Microfossiles à paroi organique de l'Yprésien du Bassin belge.* Service Géologique de Belgique, Professional Paper 12, 151 pp.

Drugg, W. S., 1967. Palynology of the Upper Moreno Formation (Late Cretaceous–Paleocene) Escarpado Canyon, California. *Palaeontographica,* Part B, v. 120, p. 1–71.

Ediger, V. S. ,1981. Some Tertiary dinoflagellate cysts from Thrace Basin, Turkey, and their usage in Palaeoenvironmental analyses. *Revista Española de Micropaleontología,* v. 13, p. 89–103.

El-Beialy, S. Y., 1988. The dinocyst biostratigraphy of the Upper Eocene subsurface sediments, West Nile Delta, Egypt. *Newsletters on Stratigraphy,* v. 119, p. 131–1141.

Garg, R. Khowaja-Ateequzzaman and Jain, K. P. 1995. Occurrences of the marker dinoflagellate cyst *Apectodinium* in Narsapur Well-1, Krishna-Godavari Basin, India. *Palaeobotanist* v. 42, p. 363–371.

Gradstein, F. M., Kristiansen, I. L., Loemo, L. and Kaminski, M. A., 1992. Cenozoic foraminiferal and dinoflagellate cyst biostratigraphy of the central North Sea. *Micropaleontology,* v. 38, p. 101–137.

Grant-Taylor, T. L. and Hornibrook, N. de B., 1976. *Late Mesozoic to Cenozoic stratigraphy, North Island (Eastern Part).* 25th International Geological Congress, Sydney, New Zealand Excursion Guide No. 54A.

Hallock, P., Premoli Silva, I. and Boersma, A. 1991. Similarities between planktonic and larger foraminiferal evolutionary trends through Paleogene paleoceanographic changes. *Palaeogeography, Palaeoclimatology, Palaeoecology*, v. 83, p. 49–64.

Haq, B. U., Hardenbol, J. and Vail, P. R., 1987. Chronology of fluctuating sea levels since the Triassic. *Science,* v. 235, p. 1156–1167.

Hardenbol, J. 1994. Sequence stratigraphic calibration of Paleocene and lower Eocene continental margin deposits in NW Europe and the US Gulf Coast with the oceanic chronostratigraphic record. GFF, Meeting Proceedings "Stratigraphy of the Paleocene," v. 116, p. 49–51.

Heilmann-Clausen, C., 1985. Dinoflagellate stratigraphy of the uppermost Danian to Ypresian in the Viborg 1 borehole, central Jylland, Denmark. *Danmarks Geologiske Undersogelse,* ser. A, v. 7, p. 1–69.

Heilmann-Clausen, C. and Costa, L. I. 1989. Dinoflagellate zonation of the uppermost Paleocene? to Lower Miocene in the Wursterheide research well, NW Germany. *Geologisches Jahrbuch,* v. A111, p. 431–521.

Heritier, F. E., Lossel, P. and Wathne, E. 1979. Frigg Field: Large submarine-fan trap in Lower Eocene rocks of North Sea Viking Graben. *American Association of Petroleum Geologists Bulletin,* v. 63, p. 1999–2020.

Heritier, F. E., Lossel, P. and Wathne, E. 1981. The Frigg Gas Field. In Illing, L. V. and Hobson, G. D., (eds.), *Petroleum Geology of the Continental Shelf of North-West Europe,* London: Heyden and Son, p. 380–391.

Ioakim, C., 1978. Étude comparative des dinoflagellés du Tertiaire inférieur de la Mer du Labrador et de la Mer du Nord. Thèse de Doctorat (unpublished), Paris: Université Pierre et Marie Curie (Paris VI), 205 pp.

Jan du Chêne, R. E., 1977. Palynostratigraphie (Maastrichtien–Éocène inférieur) des Flyschs du Schlieren (Canton d'Obwald, Suisse centrale). *Revue de Micropaléontologie*, 20, p. 147–156.

Jan du Chêne, R. E., 1988. Étude systématique des kystes de dinoflagellés de la Formation des Madeleines (Danien du Sénégal). *C.N.R.S.*, v. 2, p. 147–174.

Jan du Chêne, R. E. and Adediran, S. A., 1984. Late Paleocene to Early Eocene dinoflagellates from Nigeria. *C.N.R.S.*, v. 3, p. 3–38.

Kennett, J. P. and Stott, L. D., 1991. Abrupt deep-sea warming, palaeoceanographic changes and benthic extinctions at the end of the Palaeocene. *Nature*, v. 353, p. 225–229.

King, C., 1983. Cainozoic micropaleontological biostratigraphy of the North Sea. Institute of Geological Sciences, Report 82/7, p. 1–40.

Knox, R. W. O'B., 1994a. The age and regional context of the Thanetian stratotype sections in SE England. GFF, Meeting Proceedings "Stratigraphy of the Paleocene," v. 116, p. 55–56.

Knox, R. W. O'B., 1994b. From regional stage to standard stage: implications for the historical Paleogene stratotypes of NW Europe. GFF, Meeting Proceedings "Stratigraphy of the Paleocene," v. 116, p. 56–57.

Knox, R. W. O'B., Hine, N. and Ali, J. R., 1994. New information on the age and sequence stratigraphy of the type Thanetian of Southeast England. *Newsletters on Stratigraphy*, v. 30, p. 45–60.

Knox, R. W. O'B, Morton, A. C. and Harland, R. 1981. Stratigraphical relationships of Palaeocene Sands in the UK Sector of the Central North Sea. In Illing, L. V. and Hobson, G. D., (eds.), *Petroleum Geology of the Continental Shelf of North-West Europe*, Institute of Petroleum, Heyden and Son, London, p. 267–281.

Köthe, A., 1988. Biostratigraphy of the Surghar Range, Salt Range, Sulaiman Range and the Kohat area, Pakistan, according to Jurassic through Paleogene calcareous nannofossils and Paleogene dinoflagellates. *Geologisches Jahrbuch*, v. B71, p. 3–87.

Lentin, J. K. and Williams, G. L. 1980. Dinoflagellate provincialism with emphasis on Campanian peridiniaceans. *American Association of Stratigraphic Palynologists Contribution Series* No. 7, p. 1–47.

Lentin, J. K, and Williams, G. L., 1993. *Fossil dinoflagellates: index to genera and species 1993 edition*. American Association of Stratigraphic Palynologists Contribution Series No. 28, 856 pp.

Manum, S. B. and Throndsen, T., 1986. Age of Tertiary formations in Spitsbergen. *Polar Research*, n. s., v. 4, p. 103–131.

Mehrotra, N. C. and Sarjeant, W. A. S., 1987. Late Cretaceous to Early Tertiary dinoflagellate cysts from Narasapur Well-1, Godavari-krishna Basin, south India. *Geobios*, v. 20, p. 149–191.

Michoux, D., 1988. Dinoflagellate cysts of the Wetzeliella complex from Eocene sediments of the Aquitaine Basin, southwestern France. *Palynology*, v. 12, p. 11–37.

Miller, K. G., Janecek, T. R., Katz, M. E. and Keil, D. J., 1987. Abyssal circulation and benthic foraminiferal changes near the Paleocene/Eocene boundary. *Paleoceanography*, v. 2, p. 741–761.

Mudge, D. C. and Bujak, J. P., 1994. Eocene stratigraphy of the North Sea Basin. *Marine and Petroleum Geology*, v. 11, p. 166–181.

Mudge, D. C. and Bujak, J. P. 1996. Palaeocene biostratigraphy and sequence stratigraphy of the UK central North Sea. *Marine and Petroleum Geology*, v. 13, p. 295–312.

Mudge, D. C. and Copestake, P., 1992a. Revised Lower Palaeogene lithostratigraphy for the Outer Moray Firth, North Sea. *Marine and Petroleum Geology*, v. 9, p. 53–69.

Mudge, D. C. and Copestake, P., 1992b. Lower Palaeogene stratigraphy of the northern North Sea. *Marine and Petroleum Geology*, v. 9, p. 287–301.

Partridge, A. D., 1976. The geological expression of eustasy in the Early Tertiary of the Gippsland Basin. *APEA Journal*, v. 16, p. 73–79.

Powell, A. J., 1988. A modified dinoflagellate biozonation for the latest Palaeocene and earliest Eocene sediments from the central North Sea. *Review of Palaeobotany and Palynology*, v. 56, p. 327–344.

Powell, A. J., 1992. Dinoflagellate cysts of the Tertiary System. In Powell, A. J., (ed.), *A Stratigraphic Index of Dinoflagellate Cysts*, London: Chapman and Hall, p. 155–251.

Powell, A. J., Brinkhuis, H. and Bujak, J. P., 1996. Upper Paleocene–lower Eocene dinoflagellate cyst sequence biostratigraphy of southeast England. In Knox, R. W. O'B., Corfield, R. M. and Dunay, R. E., (eds.), *Correlation of the Early Paleogene in Northwest Europe*, Geological Society Special Publication No. 101, p. 145–183.

Rahul, Khowaja-Ateequzzaman G. and Jain, K. P., 1995. Occurrence of the marker dinoflagellate cyst *Apectodinium* in Narsapur Well-1, Krishna-Godavari Basin, India. *Palaeobotanist*, v. 42. p. 363–371.

Rea, D. K., Zachos, J. C., Owen, R. M. and Gingerich, P. D., 1990. Global change at the Paleocene–Eocene boundary: climatic and evolutionary consequences of tectonic events. *Palaeogeography, Palaeoclimatology, Palaeoecology*, v. 79, p. 117–128.

Sarkar, S. and Singh, H. P., 1988. Palynological investigation of the Subathu Formation (Eocene) in the Banethi-Bagthan area of Himachal Pradesh, India. *Palaeontographica*, ser. B, v. 209, p. 29–109.

Schröder, T., 1992. A palynological zonation for the Paleocene of the North Sea Basin. *Journal of Micropalaeontology*, v. 11, p. 113–126.

Siesser, W. G., Ward, D. J. and Lord, A. R., 1987. Calcareous nannoplankton biozonation of the Thanetian Stage (Palaeocene) in the type area. *Journal of Micropalaeontology*, v. 6, p. 85–102.

Sinha, A. and Stott, L. D., 1993b. Recognition of the Paleocene/Eocene boundary carbon isotope excursion in the Paris Basin, France. *Correlation of the Early Paleogene in Northwest Europe*, Geological Society of London, Programme and Abstracts (1–2 December 1993).

Soper, N. J., Downie, C., Higgins, A. C. and Costa, L. I., 1976a. Biostratigraphic ages of Tertiary basalts on the East Greenland continental margin and their relationship to plate separation in the northeast Atlantic. *Earth and Planetary Science Letters*, v. 32, p. 149–157.

Soper, N. J., Higgins, A. C. Downie, C., Matthews, D. W. and Brown, P. E., 1976b. Late Cretaceous–early Tertiary stratigraphy of the Kangerdlugssuaq, east Greenland. *Journal of the Geological Society of London*, v. 132, p. 85–102.

Stampfi, G., Jan du Chêne, R. E. and Heub, R., 1978. Géologie et micropaléontologie (Nummulites et palynologie) de la formation éocène de Ziarat, Elbourz ourental (Iran). *Rivista Italiana Paleontologia*, v. 84, p. 383–402.

Stott, L. D. and Kennett, J. P., 1990. Antarctic Paleogene planktonic foraminifer biostratigraphy: ODP Leg 113, Sites 689 and 690. In Barker, P. F., Kennett, J. P., et al., Proceedings of the Ocean Drilling Program, Scientific Results, v. 113: College Station, TX (Ocean Drilling Program), p. 549–569.

Stott, L. D., Sinha, A., Thiry, M., Aubry, M.-P. and Berggren, W. A. 1996. Global $\delta^{13}C$ changes across the Paleocene–Eocene boundary: Criteria for terrestrial–marine correlations. In Knox, R. W. O'B., Corfield, R. M. and Dunay, R. E. (eds.), *Correlation of the Early Paleogene in Northwest Europe*, Geological Society Special Publication No. 101, p. 381–399.

Thomas, E., 1990a. Late Cretaceous–early Eocene mass extinctions in the deep sea. In Sharpton, V. L. and Ward, P., (eds.), *Global Catastrophes in Earth History: An interdisciplinary Conference on Impacts, Volcanism, and Mass Mortality*, Geological Society of America Special Paper No. 247, p. 481–495.

Thomas, E. 1990b. Late Cretaceous through Neogene deep-sea benthic foraminifers (Maud Rise, Weddell Sea, Antarctica). In Barker, P. F., Kennett, J. P., et al., Proceedings of the Ocean Drilling Program, Scientific Results., v. 113: College Station, TX (Ocean Drilling Program), p. 571–594.

Thomas, E., 1992. Cenozoic deep sea circulation: Evidence from deep sea benthic foraminifera. In Kennett, J. P. and Warnke, D., (eds.), *The Antarctic Paleoenvironment: A Perspective on Global Change,* American Geophysical Union Antarctic Research Series No. 56, p. 141–165.

Thomas, E. and Shackleton, N. J., 1993. The Paleocene benthic foraminiferal extinction: timing, duration and association with stable isotope anomalies. *Correlation of the Early Paleogene in Northwest Europe,* Geological Society of London, Programme and Abstracts (1–2 December 1993).

Thomas, F. C. and Gradstein, F. M. 1981. Tertiary subsurface correlations using pyritized diatoms, offshore Eastern Canada. *Current Research Part B, Geological Survey of Canada,* Paper 81-1B, p. 17–23.

Thomas, J. E., 1993. The occurrence of the dinoflagellate cyst *Apectodinium* in a cored section through the Montrose and basal Sele groups (Paleocene) of the Central North Sea. *Correlation of the Early Paleogene in Northwest Europe,* Geological Society of London, Programme and Abstracts (1–2 December 1993).

Tjalsma, R. C. and Lohmann, G. P., 1983. *Paleocene–Eocene bathyal and abyssal benthic foraminifera from the Atlantic Ocean.* Micropaleontology Special Publication No. 4, 90 pp.

Townsend, H. A. and Hailwood, E. A., 1985. Magnetostratigraphic correlation of Palaeogene sediments in the Hampshire and London Basins, southern UK. *Journal of the Geological Society*, London, v. 142, p. 957–982.

Van Erve, A. W., Visscher, H. and Gupta, V. J., 1980. An Eocene phytoplankton assemblage from the Subathu Formation of the Simla Himalayas, India. *Recent Researches in Geology,* New Delhi: Hindustan Publ. Co., v. 6, p. 614–630.

Van Stuijvenberg, J. and Jan du ChLne, R, E., 1981. Nouvelles observations stratigraphiques dans le massif des Voirons. Bulletin du Bureau de Recherches *Géologiques et Minières,* ser. 2., Section 1, No. 1, p. 3–9.

Varma, C. P. and Dangwal, A. K., 1964. Tertiary hystrichosphaerids from India. *Micropaleontology,* v. 10, p. 53–71.

Williams, G. L., 1978. Palynological biostratigraphy, Deep Sea Drilling Project Sites 367 and 370. In Lancelot, Y., Seibold, E., et al., Initial Reports of the Deep Sea Drilling Project, v. 41: Washington (U.S. Government Printing Office), p. 783–815 (supplement to volumes XXXVIII, XXXIX, XL and XLI).

Wills, J. M. and Peattie, D. K., 1990. The Forties Field and the evolution of a reservoir management strategy. *North Sea Oil and Gas Reservoirs:* II, Graham and Trotman, p. 1–23.

Wilson, G. J. 1984. A new Paleocene dinoflagellate cyst from the Chatham Islands, New Zealand. *New Zealand Journal of Botany*, v. 22, p. 545–547.

Wilson, G. J., 1988. Paleocene and Eocene dinoflagellate cysts from Waipawa, Hawkes Bay, New Zealand. *New Zealand Geological Survey Paleontological Bulletin* 57, 96 pp.

15

Molluscan Faunas Across the Paleocene/Eocene Series Boundary in the North American Gulf Coastal Plain

DAVID T. DOCKERY III

ABSTRACT

The Paleocene/Eocene Series boundary occurs within the Wilcox Group of the U.S. Gulf Coastal Plain at the contact of the Tuscahoma and Bashi formations. Molluscan extinction rates of 97% and 94%, respectively, below and above the Wilcox Group coincide with high-relief Type 1 unconformities overlain by lowstand channel sands. A lesser extinction rate of 69% occurs between the Tuscahoma and Bashi formations and coincides with both a Type 1 unconformity (though not as prominent as those above and below the Wilcox Group) and a significant deviation in the $^{87}Sr/^{86}Sr$ seawater curve as measured from outcrop sections. The Paleocene/Eocene boundary extinction rate in the Gulf Coastal Plain is subordinate to those bracketing the Wilcox Group and is slightly higher than the background extinction rate of 64%.

This chapter documents molluscan faunal turnovers in the Paleogene section, with particular attention to the Paleocene–Eocene interval, of the northern Gulf Coastal Plain of the southeastern United States and explores causative mechanisms. The North American Gulf Coast Paleogene contains some of the world's best preserved and most diverse marine molluscan faunas. These faunas are rivaled in diversity only by those of the Paris Basin in Europe. The highest molluscan diversities for both Europe and North America are in the middle Eocene Series, peaking in the Lutetian formations of Europe and a little later in the Bartonian deposits of North America. In North America, 580 molluscan taxa are recorded from the Cook Mountain Formation (of Bartonian Age) over a four-state area from Texas to Alabama. Although this number is impressive, it is dwarfed by the report of Dolin et al. (1980) of 773 molluscan taxa of similar age from a modest borrow pit in Auversian sands at Baron, France. The upper Paleocene and lower Eocene stratigraphic record of the North American Gulf Coastal Plain has a comparable diversity to that of equivalent European sections. This is partially due to better preservation in the Gulf Coast section and to the low diversity of the marginal marine Sparnacian deposits of the Paris Basin.

Figures 15.1 and 15.2 show the number of molluscan species recorded respectively from Paleogene groups/stages of the northern Gulf and those of the Paris Basin. In each figure, turnovers are high at the group/stage boundaries. These boundaries correspond to those of stratigraphic sequences, suggesting that extinctions may be related to the lowering of sea level on the continental shelves and/or long-duration lowstand-related hiatuses. Within stratigraphic sequences, the highest molluscan diversities occur in transgressive marine units and the lowest in lowstand and regressive highstand units. Faunal turnovers are generally high between successive formations. This can be attributed in part to the fact that most species are uncommon, and many are known from only one formation. For this reason, molluscan turnovers at the Paleocene/Eocene Epoch/Series boundary must be studied in context with those of the Paleogene section in general.

Paleogene Sequences and Molluscan Diversities

Perhaps the best Paleocene–Eocene molluscan record in the western hemisphere is in the Wilcox Group

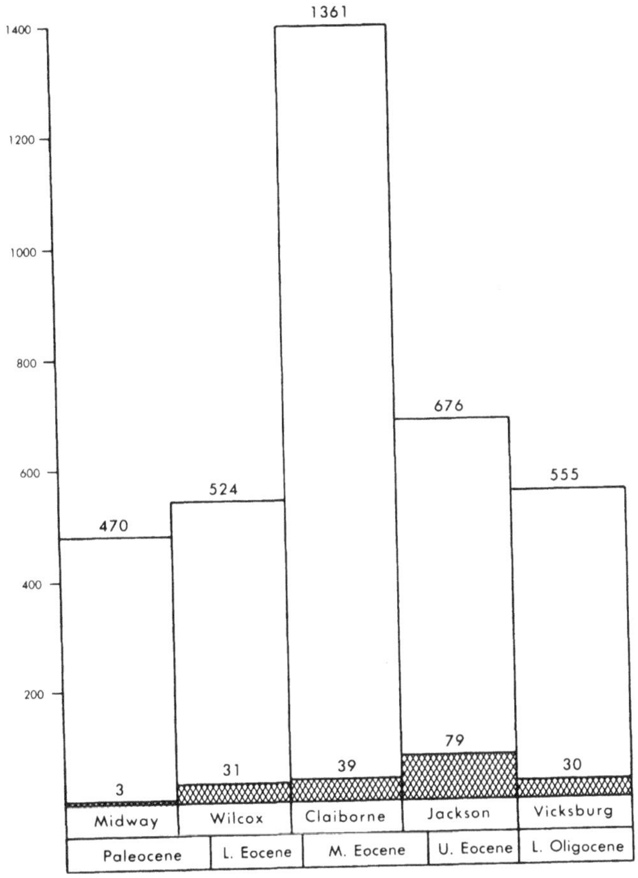

Figure 15.1 Bar graph showing the number of molluscan taxa (including species, subspecies, and unnamed forms) reported from the Paleogene groups of the southeastern United States.
Shaded segment indicates the number of species that do not have FADs within the group. From Dockery (1984).

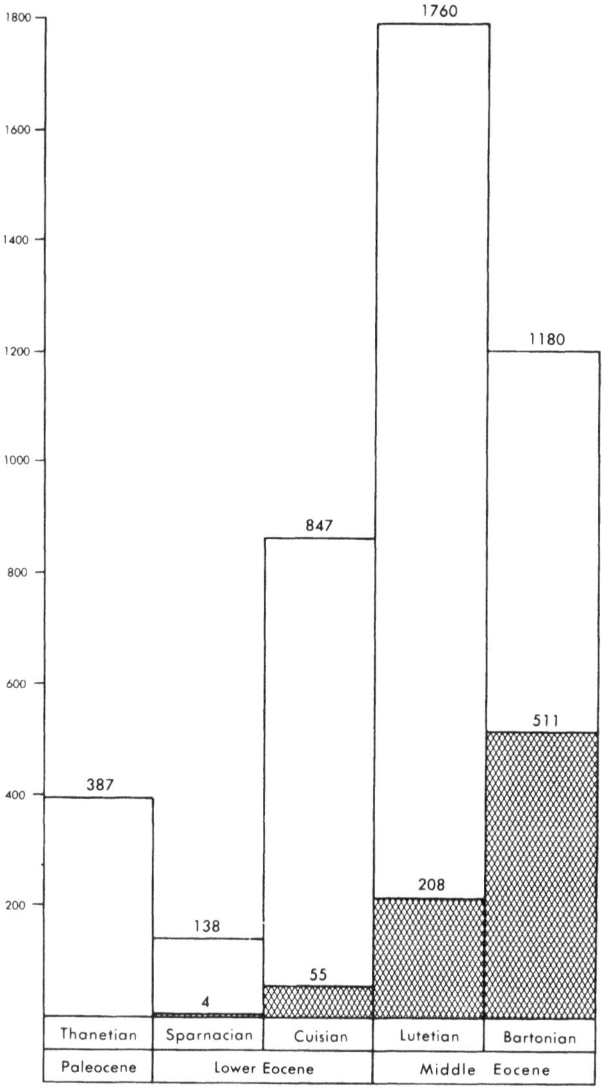

Figure 15.2 Bar graph showing the number of molluscan taxa (including species, subspecies, and unnamed forms) reported by Cossmann and Pissarro (1904–1913) from the Paleocene and Eocene deposits of the Paris Basin, France.
Shaded segments indicate the number of species that do not have FADs within the stage. From Dockery (1984).

(= Sabine Stage) of Alabama. Here, the Paleocene/Eocene Series boundary, as placed by Mancini and Tew (1991a), falls within the Wilcox Group at the Tuscahoma/Bashi Formation contact and approximates the NP9/NP10 zonal boundary (Martini 1971). Placement of the Paleocene/Eocene Series boundary worldwide is under discussion, and the boundary may be lowered to some point within the sparsely fossiliferous beds of the upper Tuscahoma Formation. Such a move could change the age assignment of the vertebrate-bearing Tuscahoma "T4 sand" discussed below from late Paleocene to early Eocene, but would probably not reach as low as the mollusk-bearing Bells Landing Member, and thus would have no bearing on the following discussions on molluscan faunal turnovers across the Paleocene/Eocene Epoch boundary.

Late Paleocene mollusks within the Wilcox Group are known largely from the transgressive "*Ostrea*" *thirsae* beds of the Nanafalia Formation and from three relatively thin, transgressive marine units in the Tuscahoma Formation, which in ascending order are the Bear Creek marl, Greggs Landing Marl, and Bells Landing Marl members. Each of these transgressive units is overlain by a nonfossiliferous or sparsely fossiliferous highstand regressive unit. Mancini and Tew (1991a) placed the Nanafalia Formation and lower Tuscahoma Formation within their sequence TP2.1 (figure 15.3). They recognized the Greggs Landing and Bells Landing Marl members as the transgressive deposits of separate sequences, TP2.3 and TP2.4, respectively. Lower Eocene marine sediments of the Bashi and overlying Hatchetigbee formations were placed in sequence TE1.1.

The basal, transgressive, fossiliferous marine sands

Super-cycle	Cycle	Relative Changes in Coastal Onlap (Landward – Seaward)	Sequence Components	Lithostratigraphy	Group	Planktonic Foraminiferal Zonation	European Stage	Series/ Epoch
Tb	TE1.1		Highstand Condensed Section Transgressive	Hatchetigbee Bashi	Wilcox	*M. subbotinae* I.Z.	Ypresian	Eocene
	TP2.3		Highstand Condensed Section Transgressive	upper Tuscahoma Bells Landing		*M. velascoensis* I. Z.	Selandian	Paleocene
	TP2.2		Highstand Condensed Section Transgressive Lowstand	middle Tuscahoma Greggs Landing middle Tuscahoma				
	TP2.1		Highstand Condensed Section Transgressive Incised Valley	lower Tuscahoma Grampian Hills "Ostrea thirsae" beds Gravel Creek		*P. pseudomenardii* R.Z.		

Figure 15.3 Sequence stratigraphy of the Wilcox Group in the north-central Gulf Coastal Plain. Modified from Mancini and Tew (1991a:figure 5).

of the Bashi Formation overlie the Tuscahoma Formation in Alabama and in most Mississippi localities. This contact in the northern Gulf is recognized as a Type 1 unconformity by Baum and Vail (1988) and Mancini and Tew (1991a), even though lowstand deposits were not identified. Ingram (1991) recognized basal Bashi lowstand deposits in a much-studied outcrop behind the Red Hot Truck Stop in Meridian, Mississippi. Here a 15-cm-thick, glauconitic, fossiliferous (teeth and bones but no shells) estuarine sand at the top of the Tuscahoma Formation is unconformably overlain by 3 m of Eocene sands and leaf-bearing clays. These latter sediments are truncated and burrowed beneath the overlying fossiliferous (with shells), transgressive sands of the Bashi marine section.

Dynocysts and pollen in the upper marine sand, informally named the T4 sand, in the Tuscahoma Formation at the Red Hot Truck Stop locality indicate a late Paleocene age (Lucy Edwards and Norman Frederiksen, personal communication 1991). Pollen in the overlying leaf-bearing clays of the Bashi lowstand section indicate an early Eocene age (Norman Frederiksen, personal communication 1991). An important feature of this section (figure 15.4) is the presence of land mammal bones and teeth mixed with shark and bony fish teeth in both the T4 sand and the transgressive marine sands of the Bashi Formation. Both the T4 and Bashi faunas contain index taxa of the Wasatchian North American Land Mammal Age (NALMA). One omomyid primate from the Bashi marine sand was published by Beard and Tabrum (1990). Two additional taxa from the Bashi Sand at Gallagher Creek in Meridian were identified by Beard and Tabrum as a *Hyracotherium* right lower jaw preserving the last lower premolar P4 and a creodont lower jaw and molar M3. The latter was noted to be close in size and morphology to *Prototomus deimos* Gingerich and Deutsch 1989. Unfortunately, the *Hyracotherium* and *Prototomus* specimens were lost by a well-known shipping company in the return shipment (even with tracking systems as many as 1 in 10,000 items are lost). All three mammal fossils known from the Bashi Formation have first occurrences in the Watsatchian NALMA. It is evident from their fine state of preservation, relatively larger size (than related Tuscahoma taxa), and attachment to jaw fragments that they have not been reworked from the Tuscahoma Formation below.

The T4 sand has a concentration of vertebrate remains overlying a diastem at its base and has been the subject of several excavations by the Carnegie Museum of Natural History staff. Mammal teeth in this horizon are isolated and not attached to jaw fragments. Over 23 land mammal taxa have been found, all of which remain to be published (Beard, et al. 1995). Mammal remains in the Tuscahoma and Bashi deposits at Meridian provide direct evidence for correlations between land mammal ages of the western interior with the marine stages of the northern Gulf Coastal Plain. They extend the Wasatchian NALMA below the Paleocene/Eocene Series boundary, as it is presently recognized, into the upper Paleocene Series. Other vertebrate fossils in the T4 sand include abundant shark teeth (Case 1994b), bony fish teeth (Case 1986, 1994a), vertebrae from three species of marine/estuarine snakes (Holman et al. 1990), turtle bones, and rare lizard and bird remains.

Molluscan faunas characteristic of the Bashi type locality in Alabama are present in the Bashi marine section at the Red Hot Truck Stop locality. However, here

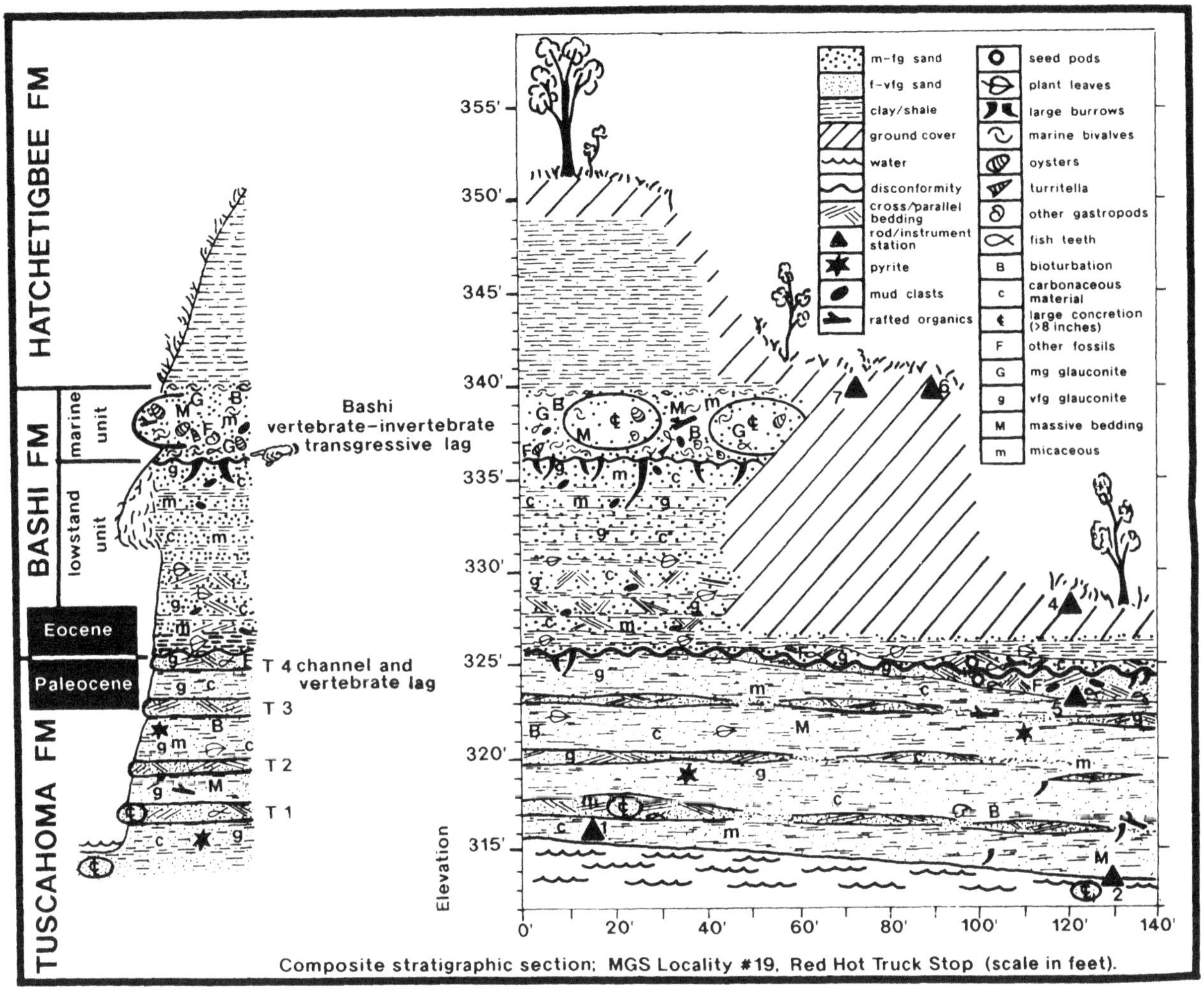

Figure 15.4 The Tuscahoma–Bashi section at the Red Hot Truck Stop locality in Meridian, Mississippi. From Ingram (1991).

the fauna is less diverse due to breakage of delicate species in a high-energy nearshore environment. Only molds of molluscan shells are present in the Tuscahoma Formation of Mississippi. A distinctive late Paleocene guide fossil to the Nanafalia and Tuscahoma formations, *Turritella praecincta* Conrad, was found as an external mold in the Greggs Landing Marl equivalent at Alamucha Creek southeast of Meridian, Mississippi (Ingram, personal communication 1994), but most Tuscahoma molds cannot be identified at specific level.

Moldic molluscan faunas in the Tuscahoma Formation extend as far north as Kemper County, Mississippi. Farther north, marine sands grade into deltaic facies containing lignites. The Bashi and overlying Hatchetigbee formations are cut out below the regional Wilcox/Claiborne Group unconformity about 10 miles north of Meridian so that the full extent of the Bashi marine transgression is unknown.

Molluscan Strontium Isotope Variation Across the Paleocene/Eocene Epoch Boundary

The Paleocene/Eocene Epoch boundary was a time of global tectonic events and restructuring of deep-sea currents (Rea et al. 1990). This boundary is marked by a lightening of oceanic $\delta^{13}C$ and a major turnover in benthic foraminifera (Koch et al. 1992). Rea et al. (1990) correlated these events with the initiation of the oceanic rift between Norway and Greenland, the outpouring of the east Greenland volcanics, and a manifold increase in sea-floor hydrothermal activity. They hypothesized that this period of oceanic volcanism flooded the atmosphere with CO_2, causing warming at high latitudes and increased evaporation at low latitudes, and resulting in a change from temperature-driven to salinity-driven, deep-water circulation. A change in oceanic currents may have influenced benthic molluscan faunas as well as benthic foraminiferal assemblages.

The strontium-isotopic composition of seawater is strongly influenced by variations in the input of hydrothermal strontium. Denison et al. (1993) noted anomalous low $^{87}Sr/^{86}Sr$ ratios in the lower Eocene Bashi Formation of the North American Gulf Coastal Plain. However, Charisi and Schmitz (1995) found invariable $^{87}Sr/^{86}Sr$ ratios in the Paleocene/Eocene boundary section at Gebel Aweina, Egypt, and at DSDP Sites 366 and 216. Based on these conflicting reports, the Bashi may contain an oceanic isotopic record not resolved in other sections, or possibly has low $^{87}Sr/^{86}Sr$ ratios due to diagenesis. Denison et al. (1993) argue against the latter.

Temporal variations in the $^{87}Sr/^{86}Sr$ isotopic ratio of seawater are produced by changes in the relative contribution of strontium from various isotopically distinct sources within the earth's crust (Denison et al. 1993). Old sialic rocks from continental interiors have high strontium ratios averaging 0.720. Mafic rocks from plate margins and ocean basins have low ratios averaging 0.704, and marine carbonates and sulfates average 0.708. Because all Phanerozoic carbonates are confined within the narrow range of 0.7068–0.7091, contributions from sialic rocks play the most important role in temporal variations of the strontium ratio. Thus episodes of cratonic exposure and erosion factor heavily in fluctuations of the strontium isotopic variation line for seawater. Such episodes may correlate to sea-level lowstands and the displacement of molluscan faunas inhabiting shallow marine environments.

Burke et al. (1982) produced a seawater curve for Cenozoic strontium isotopes based exclusively on samples from the Deep Sea Drilling Project. DePaolo and Ingram (1985) and Goldsmith et al. (1994) published a data set based on both DSDP samples and fossil shell material from the U.S. and Europe. Strontium ratio values from the non-DSDP samples fell below the best seawater line of Burke et al. (1982). Ratios from land-based samples suggest that the original curve of Burke et al. (1982) may lie above the true Paleogene seawater curve. To investigate this possibility, Denison et al. (1993) analyzed strontium ratios for 230 Paleogene rock and shell samples from 83 sites in Alabama, Mississippi, and Louisiana. Many of these were from classic Gulf Coast Paleogene sections including those for the Tuscahoma and Bashi formations. Multiple analyses of samples from these units showed very little scatter, and in most cases, precise agreement was obtained from geographically separated collection sites.

Burke et al. (1982) and Denison et al. (1993) used an assumed value of 0.71014 for the $^{87}Sr/^{86}Sr$ of the NBS987 strontium standard. Because of mass spectrometer differences, results from all laboratories must be corrected to a standard value in order to be directly compared. This can be eliminated by calculating the difference between the unknown and modern seawater (Δsw). Based on more than 100 measurements of modern shells, the value for seawater used here is 0.709073.

$$\Delta sw = (^{87}Sr/^{86}Sr \text{ unknown} - {}^{87}Sr/^{86}Sr \text{ seawater}) \times 10^5$$

Denison et al. (1993) obtained a weighted mean value of $^{87}Sr/^{86}Sr = 0.707722 \pm 7$ (Δsw −135.1) from four samples of the Bells Landing Member of the Tuscahoma Formation. This value fits smoothly with older Paleocene units sampled (figure 15.5).

Some of the best lower Eocene Bashi samples analyzed by Denison et al. (1993) included fossil oysters from three localities in Mississippi, one of which was the Red Hot Truck Stop at Meridian. The weighted mean $^{87}Sr/^{86}Sr$ of 0.707594 ± 5 (Δsw −147.9) was substantially lower than that of the underlying Tuscahoma Formation. It is also much lower than that of the overlying Tallahatta Formation. This abrupt shift in the strontium isotope variation line across the Paleocene/Eocene Series boundary was not seen in the Burke et al. (1982) or Hess et al. (1986) data derived from DSDP sites, although it falls within the envelope of data scatter of Hess et al. An important sample supporting the Bashi weighted mean, but not published by Denison et al. (1993), is from the overlying Hatchetigbee Formation at its type locality on the Tombigbee River in Alabama. Here the formation contains several marine layers, the highest containing a shell hash about 15 m below the upper contact (Thomas Gibson, personal communication 1994). Calcitic shells from this layer had a weighted mean $^{87}Sr/^{86}Sr$ of 0.707571 ± 10 (Δsw −150.2).

The strontium isotopic composition of freshwater runoff at the time when the Bashi Formation was deposited cannot be determined from the available data. However, a reasonable range can be estimated using analyses of modern rivers. Goldstein and Jacobsen (1987) showed that interior drainage in the central United States ranges from 0.7085 to 0.7112 (Δsw −57 to +213). All values are more radiogenic than those on the Bashi mollusks (Δsw −148). Therefore, diagenesis in contact with fresh water should cause the $^{87}Sr/^{86}Sr$ to rise. The measurements on the Bashi and underlying Tuscahoma shells show just the opposite, a significant drop in the $^{87}Sr/^{86}Sr$ (figure 15.5). Denison et al. (1993) attributed the reason for this decline to a low strontium ratio in the ancient seawater rather than to freshwater contamination. During deposition of the Bashi and Hatchetigbee formations, increased oceanic volcanism

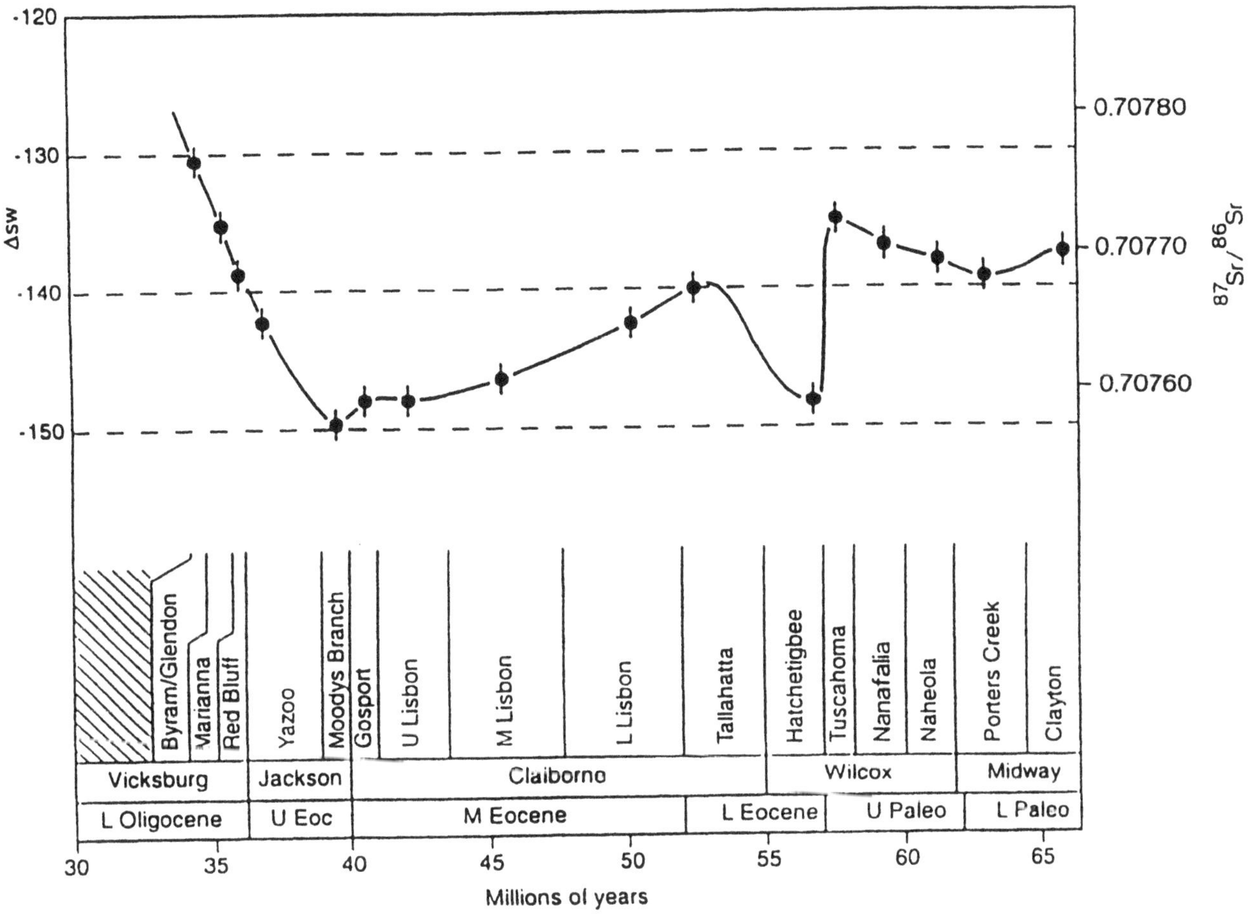

Figure 15.5 Plot of the weighted mean seawater $^{87}Sr/^{86}Sr$ for Paleogene formations sampled from the central Gulf Coastal Plain versus age.

The line represents the best estimate of the curve of seawater $^{87}Sr/^{86}Sr$, based on outcrop data. The bar through individual points reflects the uncertainty associated with the weighted mean. The Bashi and Hatchetigbee formations are grouped together under Hatchetigbee in the stratigraphic column. From Denison et al. (1993).

from basaltic sources was lowering the strontium ratio of seawater to that determined for the Bashi mollusks.

Denison et al. (1993) argued that diagenetic alteration cannot explain the low ratio strontium of Bashi mollusks from widely separated localities. Fluviodeltaic sediments found above and below the Bashi Formation are like those bracketing the Bells Landing Member of the Tuscahoma Formation, which has a higher strontium ratio. It is unlikely that freshwater strontium from ancient fluvial depositional systems (if it by some means had a low ratio) would contribute to the strontium ratio of Bashi mollusks and not those of the Bells Landing Member as well. The shell hash layer near the top of the Hatchetigbee Formation at its type locality shows no evidence of alteration and has a ratio within the error limits of the Bashi Formation. Bryant et al. (1995) believed that, based on modern analogs with salinities greater than 10 ppt, the $^{87}Sr/^{86}Sr$ ratios of marine beds in the Gulf Coast Paleogene should reflect those of the ancient seawater. Denison et al. (1993) concluded that the Tuscahoma/Bashi contact records a major shift in seawater $^{87}Sr/^{86}Sr$ ratios.

Schmitz et al. (1991) found the $^{87}Sr/^{86}Sr$ ratios in skeletal fish debris to be useful as a measure of the paleosalinity of fossil-fish habitats. Analyses by Birger Schmitz (personal communication 1995), in an ongoing work on the $^{87}Sr/^{86}Sr$ ratios of vertebrate fossils in the T4 sand of the Tuscahoma Formation and the Bashi Formation at the Red Hot Truck Stop locality, suggest that these fossils were buried in a freshwater deposit, where they acquired meteoric strontium-isotopic signals during early diagenesis and were then reworked into the fully marine Bashi Formation (Schmitz et al. 1997). Denison (personal communication 1994) believes the more solid shells of his study to be less susceptible to such diagenesis than would be porous bones and teeth.

Whether or not the strontium isotopes of fossil mollusks in the Bashi Formation provide an accurate record of ancient seawater chemistry, the global factors previously mentioned strongly suggest a change in oceanic circulation patterns at the Paleocene/Eocene Epoch boundary. If these changes included the circulation of surface currents as well, the dispersal of molluscan larvae would have been affected. Changes in surface currents can introduce new molluscan taxa from distant shallow marine habitats and facilitate faunal turnovers. A flood of such new taxa appear in the Bashi Formation, as discussed below.

Database for Paleocene–Eocene Boundary Molluscan Faunas

A meaningful study of Paleocene and Eocene mollusks in the Gulf Coastal Plain is largely possible due to the catalog of Palmer and Brann (1965, 1966). This catalog summarizes a large volume of literature on over 2800 species. It was completed under the scrutiny of the senior author, then the foremost expert on Paleocene and Eocene mollusks, and places in synonymy a large number of redundant taxa. Additional data used here were obtained from the monographs of Toulmin (1977) and Dockery (1977, 1980) and from various taxonomic papers in *Tulane Studies in Geology* and *Mississippi Geology.*

A total of 2816 Paleocene and Eocene molluscan taxa from both the Gulf and Atlantic coastal plains is contained within the Mississippi Office of Geology database file PALSTRAT (Dockery 1985, 1986). Of these, 2261 (or 80%) occur within four adjoining states: Texas, Louisiana, Mississippi, and Alabama. Taxa counted include species, subspecies, variations, and unnamed forms that Palmer and Brann (1965, 1966) did not find to be synonyms of other species. The use of forms not formally named as species increases taxon diversity to more realistic levels for tropical open marine environments and increases the resolution of biostratigraphic control by breaking down broadly defined long-ranging species.

Figure 15.6 is a biostratigraphic correlation of Paleogene formations between the four states mentioned above. PALSTRAT and a database file on Oligocene mollusks (OLISTRAT) were used for the molluscan diversity graph of the composite northern Gulf Paleogene section in figure 15.7. Shaded areas in the diversity bars indicate the number of species known only from that horizon. This number is given as a percentage of the total fauna in the right column. Transgressive marine units having high diversities include the Clayton, Hatchetigbee (including the Bashi), Weches, Cook Mountain, Gosport, Moodys Branch, and Mint Spring intervals. Other transgressive units including the Nanafalia, Tallahatta, and Chickasawhay intervals have low diversities due to poor preservation or dissolution in silicic or carbonate lithologies.

The FAD percentage column in figure 15.8 shows the percentage of taxa in each interval having their first appearance datum in that interval. The LAD percentage column shows the percentage of taxa with last appearance data. The difference of these percentages is shown in a difference column. At the base of each group, FADs outnumber LADs. The balance of these values shifts in successive intervals until the LADs outnumber the FADs at the top of each group. A stippled curve to the right of the difference column illustrates the punctuated increase of LADs, which are concentrated below unconformities at group boundaries. A sea-level column at the right is derived from a composite section of the northern Gulf Paleogene. Major unconformities that mark sequence boundaries overlie the concentrated LADs noted above.

The sea-level column of figure 15.8 agrees with the Paleogene sequence stratigraphy chart of Mancini and Tew (1991a) with the exception of the Jackson/Vicksburg Group boundary, which they and Baum and Vail (1988) consider to be a condensed section rather than a sequence boundary. This point, which is argued by Dockery (1990, 1992), Coleman and Galloway (1990), Combes (1993), and Pasley and Hazel (1995), has considerable bearing on the interpretation of extinction events. If classic Gulf Coast stratigraphy is followed, as it is here, a regional Type 1 unconformity is present at the base of each group. Thus, there is a strong correlation between sea-level lowstands and extinctions. If the Jackson–Vicksburg hiatus is attributed to a condensed section representing a time with no sediment input rather than an unconformity, then alternative explanations such as bolide impacts (Hansen 1987) must be found. Opposing views on this issue are given by Dockery and Hansen (1987, 1988).

Gaskell (1991) found extinction patterns in northern Gulf Paleogene benthic foraminifera similar to those shown for mollusks in figure 15.8, the only difference being that based on the FAD and LAD trends she records (figure 2 in her report), the top of the Claiborne cycle occurs at the base of the Yegua Formation–Gosport Sand interval rather than at the base of the Moodys Branch Formation. This apparent discrepancy in molluscan and foraminiferal data is due in part to the nearshore depositional environment of the Yegua Formation and Gosport Sand. This environment favored a high molluscan diversity in the Gosport Sand but a lower benthic foraminiferal diversity than in the deeper water Cook Mountain interval. As a result, many foraminiferal taxa have LADs in the Cook Mountain Formation, whereas the mollusks do not.

To best determine molluscan faunal changes across

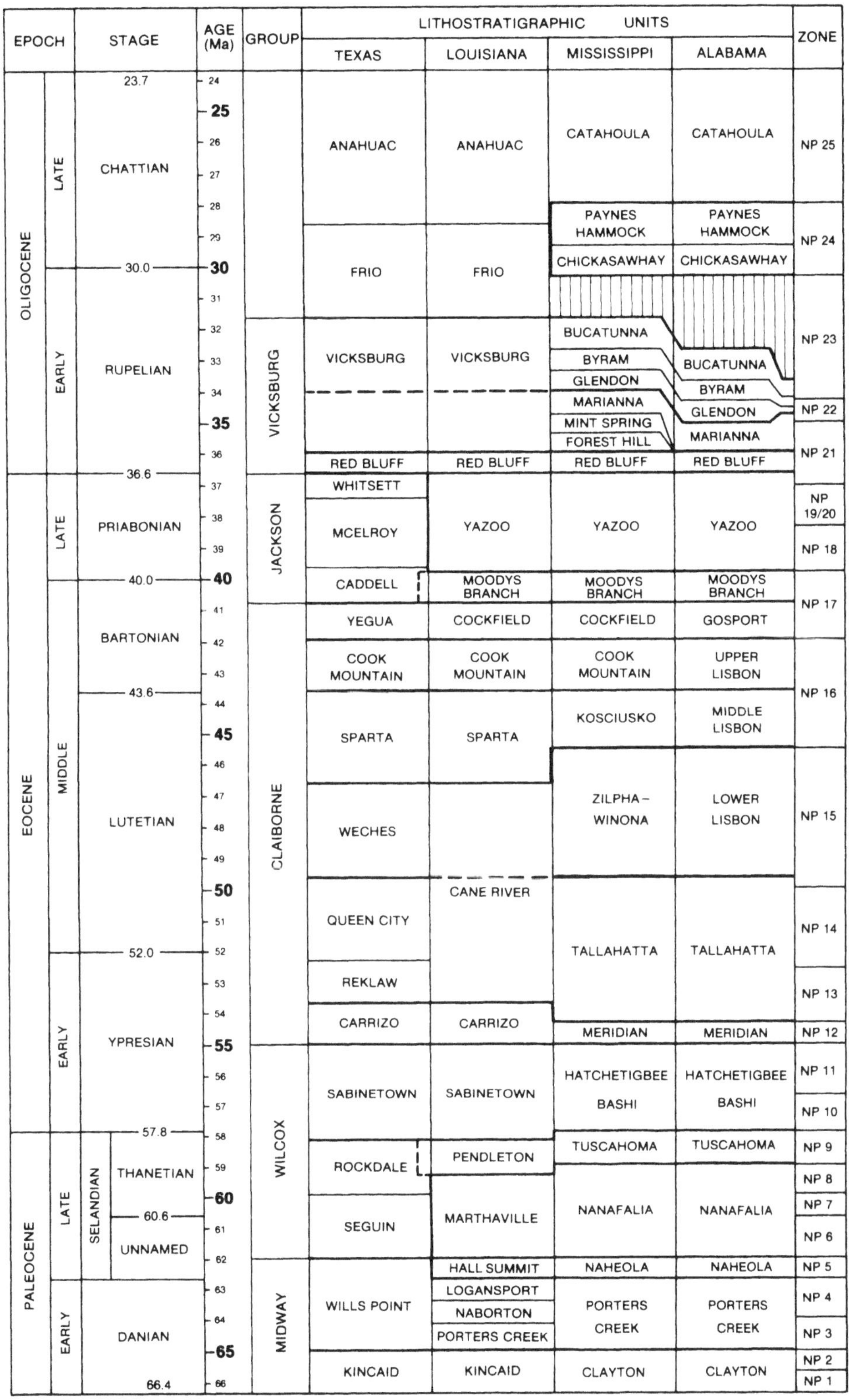

Figure 15.6 Correlation chart of Paleogene formations in Texas, Louisiana, Mississippi, and Alabama.

Heavy lines separate stratigraphic horizons given in figures 15.6 and 15.7. The timescale and synchronism of calcareous nannoplankton zonation follow Berggren et al. (1985). From Dockery (1986).

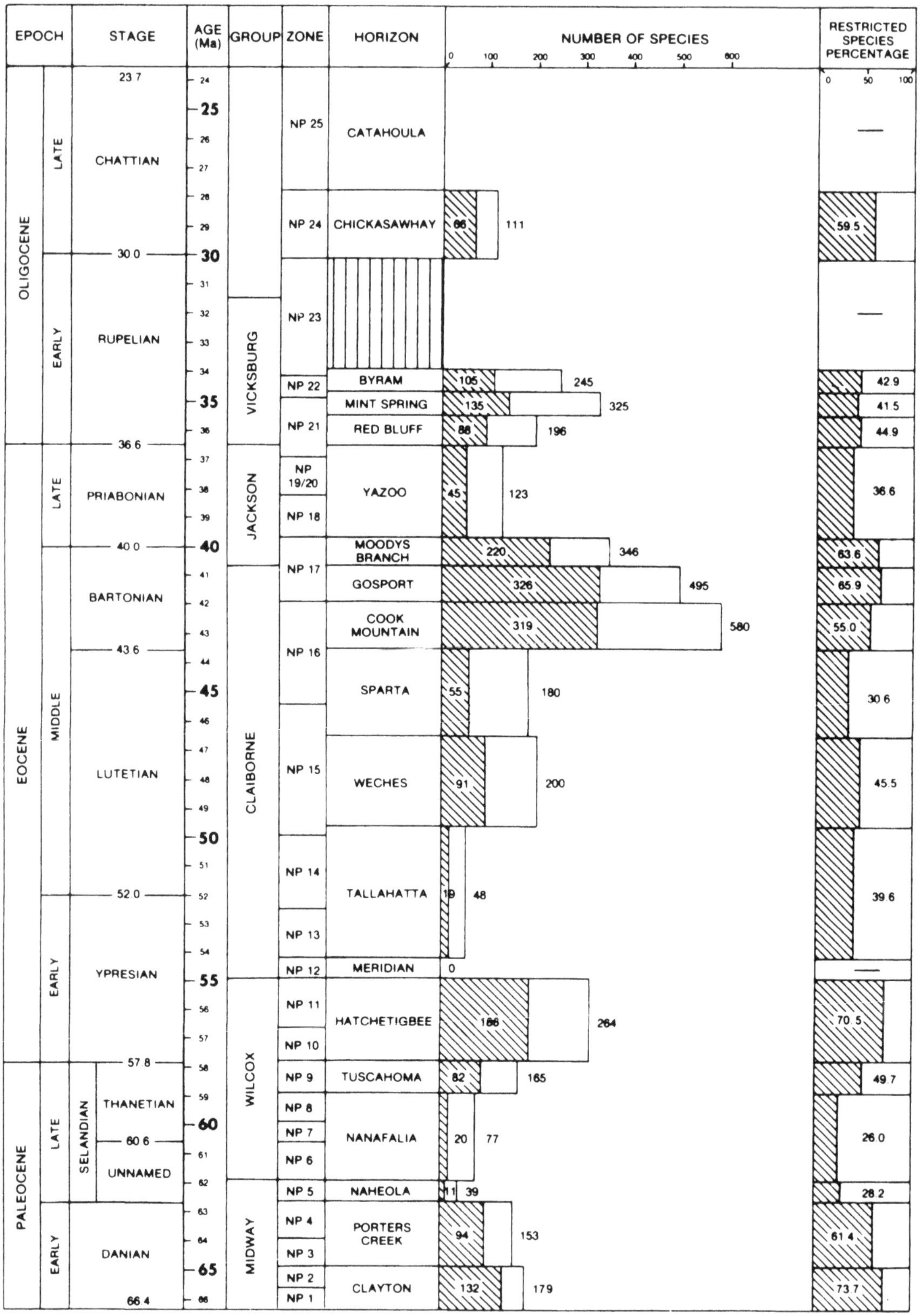

Figure 15.7 Composite section for the Paleogene record of the northern Gulf with bar graph showing the number of molluscan taxa (including species, subspecies, variations, and unnamed forms) recorded for each horizon.

Shaded areas of the bar graph indicate the number of species restricted to each horizon. The percentage of the fauna restricted to each horizon is given in the right-hand column. From Dockery (1986).

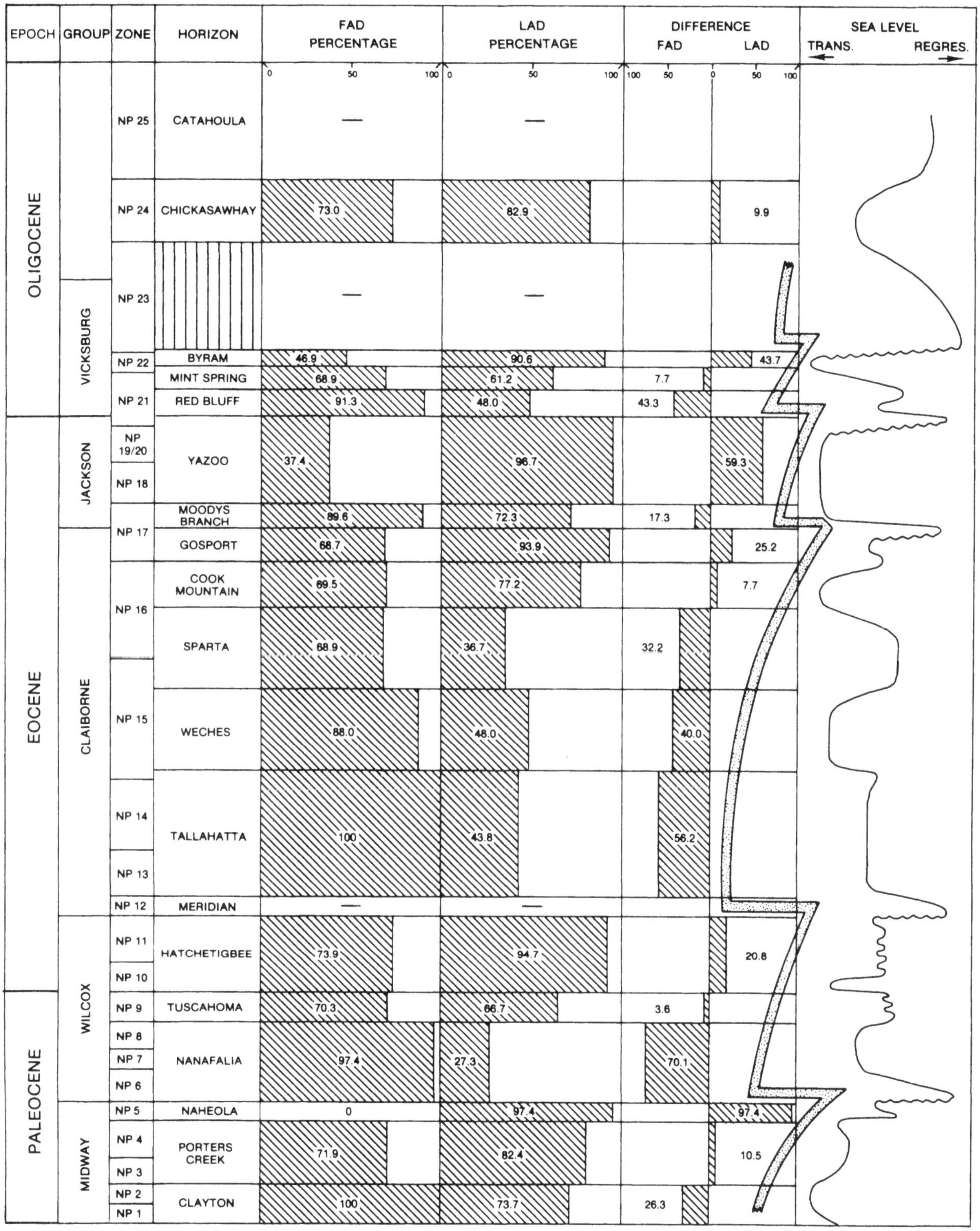

Figure 15.8 Composite Paleogene section with percentages of molluscan taxa at each horizon having FAD, LAD, and the difference between the FAD and LAD percentages given in adjacent columns.

A stippled curve follows trends indicated in the difference columns. The right-hand column contains a sea-level curve for the northern Gulf showing transgressive and regressive cycles. From Dockery (1986).

the Paleocene/Eocene Epoch/Series boundary, the Wilcox Group in the north-central Gulf Coastal Plain was analyzed on a bed-by-bed basis. Appendix 15.1 lists 368 molluscan taxa from the Wilcox Group of Alabama and Mississippi. These are arranged according to their first order of appearance and by class with species in alphabetical order. Eight stratigraphic horizons are recognized. These are the Midway Group (M), Nanafalia Formation (NA), members of the Tuscahoma Formation including the Bear Creek marl (BC), Greggs Landing Marl (GL), and Bells Landing Marl (BL), Bashi Formation (BA), Hatchetigbee Formation (HA), and Claiborne Group (C).

Molluscan distribution data from appendix 15.1 are displayed in the bar graphs of figure 15.9. To illustrate the completeness of the molluscan record across the Paleocene/Eocene Series boundary, a stratigraphic section of the Wilcox Group as encountered in the OSM No. 2 Wahalak corehole (Mancini 1981), 8 km southwest of Butler in Choctaw County, Alabama, is shown at left. Here the Wilcox Group is almost 280 m thick. Fossiliferous beds from which molluscan occurrences have been published are shaded and named to the right, with the exceptions of unnamed fossiliferous beds of the Hatchetigbee Formation. Collectively, these beds comprise a small fraction of the overall Wilcox section. There is also a large interval of nonfossiliferous strata between the uppermost Paleocene molluscan horizon of the Tuscahoma Formation, the Bells Landing Marl Member, and the lower Eocene Bashi Formation. The T4 sand of the Red Hot Truck Stop section in Meridian bridges this gap, but lacks calcareous fossils.

The earliest Paleocene mollusks of the Wilcox Group are in the "*Ostrea*" *thirsae* beds of the Nanafalia Formation. Most of the molluscan taxa known for this unit are from the Nanafalia type locality at Nanafalia Landing on the Tombigbee River in Marengo County and Becks Mill on the Pea River near the Highway 231 bridge in Dale County, Alabama. At most other localities, only calcitic oyster shells are preserved. The Nanafalia Formation contains 64 molluscan taxa; 61 or 95.3% of them have their first occurrences in the formation.

The upper Paleocene Tuscahoma Formation is 154 m thick in the OSM No. 2 Wahalak core. Of this thickness, only a 2-m-thick interval in the Greggs Landing Marl Member and, 10 m higher, a 4-m-thick interval in the Bells Landing Marl Member account for all the molluscan taxa known from this formation. The Bear Creek marl is an informal name for a 0.31-m-thick, fossiliferous marl on Bear Creek that lies 30 m below the Greggs Landing Marl (Mancini and Tew 1991b). All of the 19 molluscan taxa recorded from it are also present in the Greggs Landing Marl Member. Toulmin (1977) is the source for most of the Bear Creek molluscan data, although he mistakenly placed these occurrences in the Greggs Landing Member at his locality AWi-22 (the Bear Creek marl had not been recognized at the time of his publication).

The Greggs Landing Marl contains the most diverse late Paleocene molluscan fauna of the Wilcox Group. Unfortunately, most of the species published from this unit are from the type locality, which is now below low river level on the Alabama River Waterway. In fact, the Bear Creek marl may now be one of the best horizons for collecting molluscan species once known only from Greggs Landing Member. Figure 15.9 in the right-hand column shows a reversal in dominance from FAD to LAD taxa between the Greggs and Bells Landing members, with the Bells Landing Member having more species with previous occurrences than new ones.

A flood of new species appear in the lower Eocene Bashi Formation, increasing diversity to 218 taxa, the highest of the Wilcox Group. However, 133 of these are known only from the Bashi Formation so that LAD taxa have a slight edge of 2.7% over FAD taxa. The large number of species recorded for the Bashi Formation is partially due to the excellent exposures of this unit that were once available along the Tombigbee River. Woods Bluff, as shown in the artist drawing in figure 15.10, was one such locality. Today it is below the waters of the Tombigbee Waterway. Based on current exposures, the diversity of the early Bashi collections cannot be replicated.

The molluscan fauna in the Hatchetigbee Formation is known largely from the formation type locality at Hatchetigbee Bluff on the Tombigbee River. Fortunately this locality is still collectable. Here the section contains several fossiliferous marine beds interlayered within a predominantly nonfossiliferous sequence. These beds are dominated by a few taxa, most notably planicostate venericard bivalves, many of which appear to have been buried in life position.

A shell-hash layer high in the type section contains the formation's greatest diversity. A corresponding 0.6-m-thick shelly layer was encountered in USGS Core Hole C-226 2 miles west of Hatchetigbee Bluff (Thomas G. Gibson, personal communication 1994). This layer was 13.7 m below the base of the Tallahatta Formation and contained calcareous nannoplankton species, which, in the type section, occur only within the upper shell bed.

Of the 78 molluscan taxa in the Hatchetigbee Formation, 73 (or 93.6%) represent last occurrences. It is arguable that some of these taxa might turn up in moldic faunas of the overlying Tallahatta Formation. A recent monograph by Garvie (1996) on the mollusks of the Reklaw Formation in Texas, a Tallahatta Formation equivalent, sheds new light on the overlooked fossilifer-

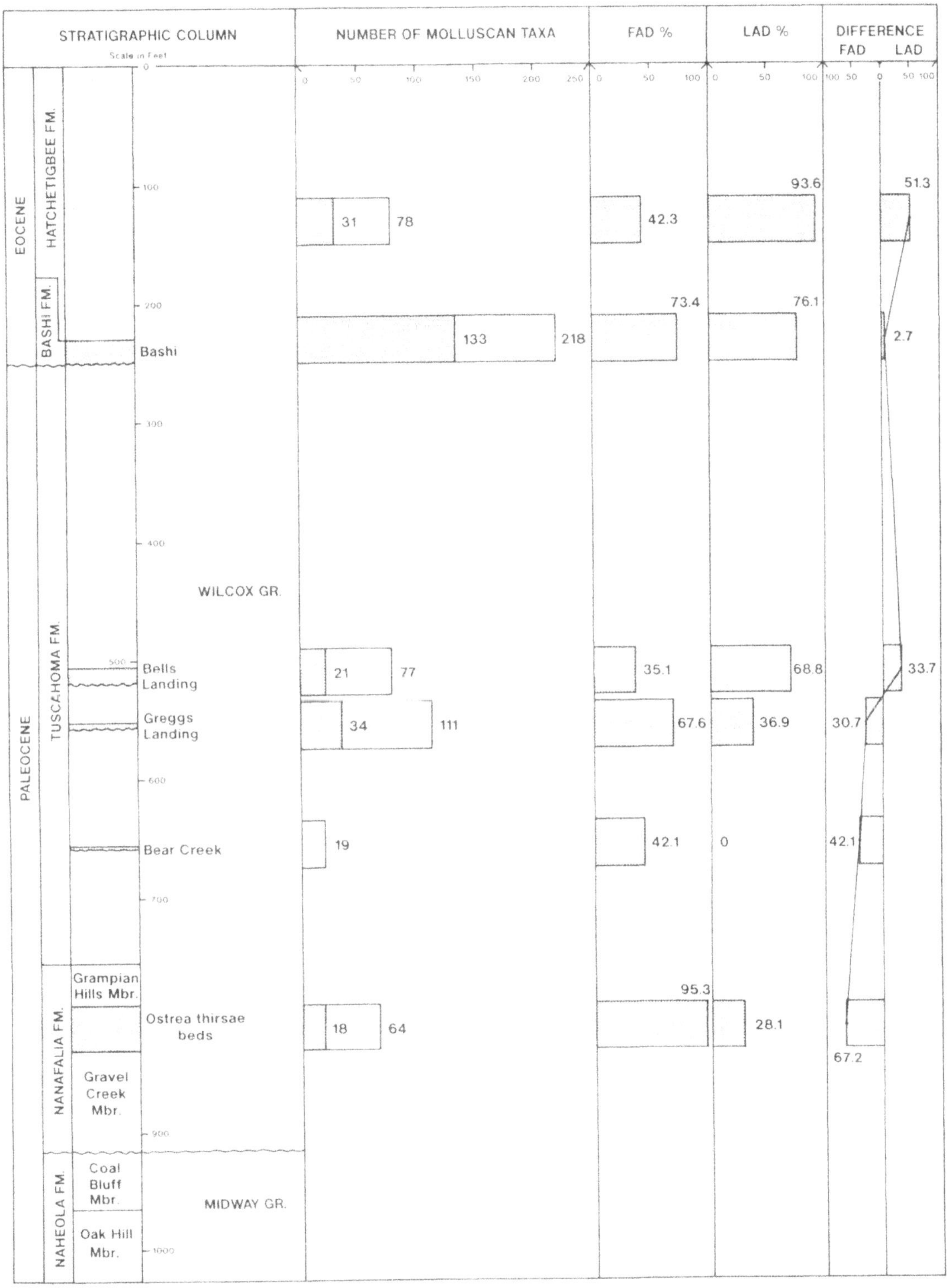

Figure 15.9 The Wilcox section as encountered in the OSM No. 2 Wahalak corehole (Mancini 1981), 5 miles southwest of Butler, Choctaw County, Alabama.

Shaded intervals of section are those from which molluscan occurrence data were obtained (occurrence data shown here do not include the new species in figures 15.11–15.16). Bar graphs to the right opposite the shaded intervals show (1) the number of molluscan taxa per interval as listed in appendix 15.1; shaded segments indicate number of species restricted to the interval; (2) percentages of molluscan taxa with FAD in the interval; (3) percentage of molluscan taxa with LAD in the interval; and (4) the difference between the FAD and LAD percentages.

Figure 15.10 The Bashi Formation at Woods Bluff on the Tombigbee River in Alabama. The hummocky terrain in the foreground contains boulder-sized concretions characteristic of the fossiliferous Bashi sands. Such broad and fresh exposures account for the diverse collections of early Eocene mollusks known from this site. No comparable sites for the Bashi Formation exist today. Smith and Johnson (1887).

ous lenses of this interval. Garvie recognizes 191 molluscan taxa, including 131 new species and subspecies, 35 previously named taxa, and 25 unnamed species. Of the 35 published taxa, 27 occur in overlying Claiborne strata, 3 occur in underlying Wilcox units, 1 occurs in both groups, and 4 are restricted to the Reklaw Formation. Garvie's diverse Reklaw fauna does not significantly increase the occurrences of Wilcox molluscan taxa in the Claiborne Group.

Extinctions, Faunal Turnovers, and Current-Driven Migrations

One aspect of Gulf Coast molluscan faunas that must be considered when studying extinction events is the large number of species restricted to a single stratigraphic horizon. Of the 368 forms listed from the north-central Gulf in appendix 15.1, 237 (or 64%) are restricted to one horizon. Therefore the loss of 64% of a fauna between stratigraphic horizons is an average extinction rate and not an accelerated extinction event. Major extinction events are noted across regional unconformities (= sequence boundaries) bounding the Wilcox Group. Only 3 Wilcox molluscan taxa are known from the Midway Group below and only 15 from the Claiborne Group above. These numbers translate into 99% new occurrences within the Wilcox Group and a 96% cumulative extinction rate at the group's upper boundary. Only one of 38 molluscan taxa recorded from the Naheola Formation of the upper Midway Group occurs in the Wilcox Group. This equates with an extinction rate of 97% below the basal Wilcox unconformity.

The right-hand column of figure 15.9 shows a steady gain in the percentage of taxa having their LAD over taxa having their FAD in successive molluscan horizons. The only deviation in this trend is the Bashi Formation, which overlies the Paleocene/Eocene series boundary. Here taxa having their LAD have only a slight increase over those having their FAD. This may be categorized as a second-order faunal turnover occurring within the Wilcox Group, a turnover produced by the formation's high diversity and abundance of taxa having their FAD, and by extinctions of Paleocene mollusks.

Important Paleocene and Eocene molluscan groups (see figures 15.11–15.16) that are ubiquitous and sensi-

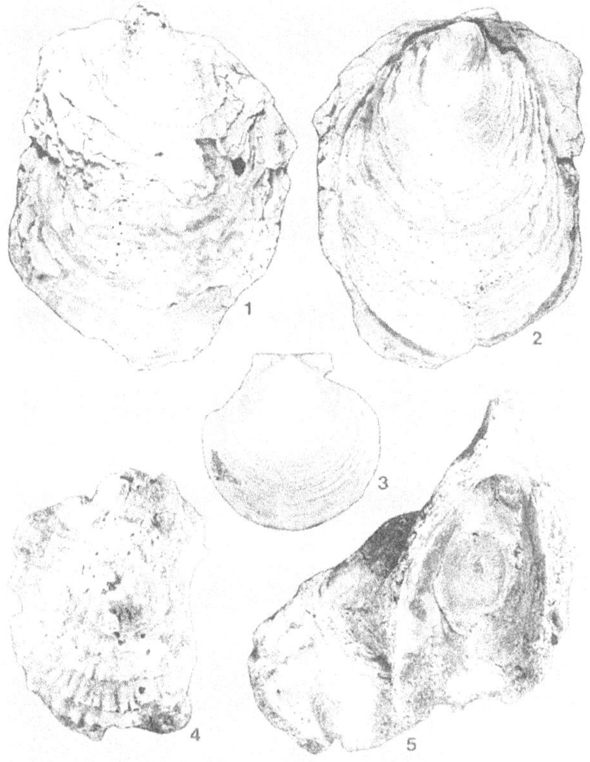

Figure 15.11 1–2: *Ostrea sinuosa* Rogers and Rogers 1837, Bells Landing Member of the Tuscahoma Formation, Bells Landing, Alabama River, Monroe County, Alabama, Mississippi Geological Survey (MGS) specimen 1755. 1: Left valve, height 170 mm, width 140 mm. 2: Right valve, height 153 mm, width 113 mm. 3: *Eburneopecten corneoides* (Harris 1919), left valve, height 41.8 mm, width 42.0 mm, upper Hatchetigbee Formation, Hatchetigbee Bluff, Tombigbee River, Washington County, Alabama, MGS specimen 1756. 4: *Ostrea brevifronta* Dockery 1980, left valve, Bashi Formation, behind the Red Hot Truck Stop parking lot, Meridian, Mississippi, Mississippi Geological Survey (MGS) locality 19, MGS specimen 1757, height 105 mm, width 78 mm. 5: *Crassostrea* sp., Bashi Formation, MGS locality 19, MGS specimen 1758, two left valves fused together, height of upper valve (incomplete) 130 mm, width 67 mm.

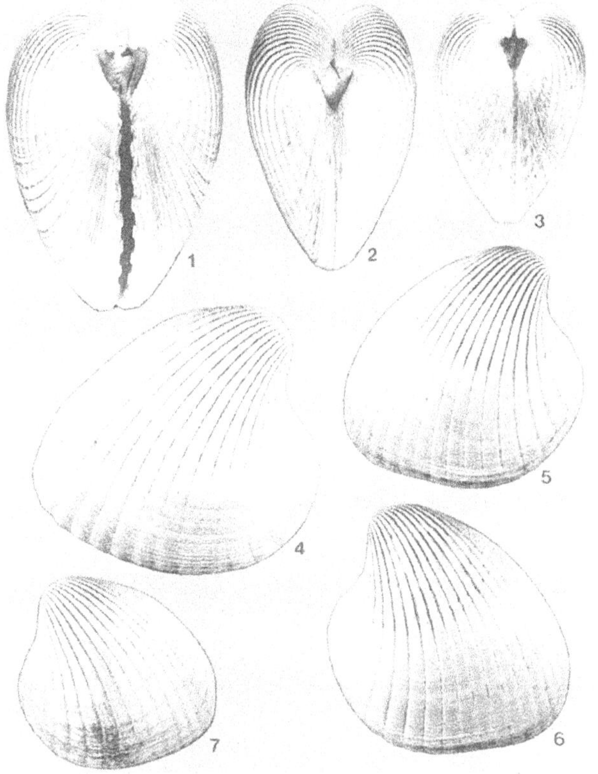

Figure 15.12 1, 4: *Venericardia (Venericor) aposmithii* Gardner and Bowles 1939, Bells Landing Member of the Tuscahoma Formation, Bells Landing, Alabama River, Monroe County, Alabama, MGS specimen 1759. 1: Anterior view of both valves, height 111.2 mm, width 109.0 mm, thickness of both valves 75 mm. 4: Right valve. 2, 5, 6: *Venericardia (Venericor) bashiplata* Gardner and Bowles 1939, Bashi Formation, Gallagher Creek, Meridian, Mississippi, MGS locality 125, MGS specimen 1760. 2: Anterior view of both valves, height 83.5 mm, width 87.0 mm, thickness of both valves 54.0 mm. 5: Right valve. 6: Left valve. 3, 7: *Venericardia (Venericor) hatcheplata* Gardner and Bowles, 1939 Hatchetigbee Formation, Hatchetigbee Bluff (from the venericard beds near water level), Tombigbee River, Washington County, Alabama, MGS specimen 1761. 3: Anterior view of both valves. 7: Left valve.

tive to environmental disruptions, and thus serve well as guide fossils, are the oysters (figure 15.11), planicostate venericard bivalves of the subgenus *Venericor* (figures 15.12 and 15.13), and turritellid gastropods (figure 15.14). All these groups suffered species extinctions at the Paleocene/Eocene Epoch boundary. The long-ranging and common turritellids *Turritella mortoni postmortoni* Harris, *T. praecincta* Conrad, *T. multilira* Whitfield, and *T. eurynome* Whitfield were extinct after this boundary (Allmon 1996) as were the large oyster *Ostrea sinuosa* Rogers and Rogers and the planicostate venericards *Venericardia (Venericor) aposmithi* Gardner and Bowles and *V. (V.) nanaplata* Gardner and Bowles. Whereas new species of oysters and planicostate venericards are common in the Bashi Formation, the turritellids are poorly represented. Turritellids in the Bashi and Hatchetigbee formations are diminutive in comparison to those of the Tuscahoma Formation and are not a dominant element of the fauna, with the exception of some local beds. A similar decrease in abundance and diversity in the turritellids across the Eocene/Oligocene Epoch boundary was noted by Allmon and Dockery (1992). The late Eocene turritellid–planicostate venericard, Tethyan, molluscan fauna was

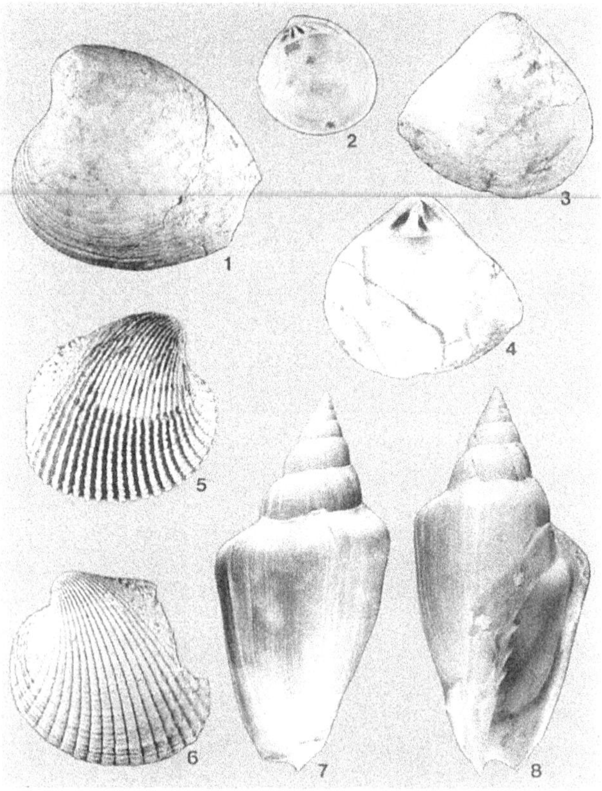

Figure 15.13 1: *Venericardia* (*Leuroactis*) *horatiana* Gardner 1936, Bashi Formation, behind B&B Truck Sales, Meridian, Mississippi, MGS locality 134, MGS specimen 1762, height 85 mm, width 95 mm. 2: *Pelecyora hatchetigbeensis* (Aldrich 1886), upper Hatchetigbee Formation, Hatchetigbee Bluff, Tombigbee River, Washington County, Alabama, MGS specimen 1761, height 18.7 mm, width 20.3 mm. 3, 4: *Crassatella tumidula* Whitfield 1865, Bells Landing Member of the Tuscahoma Formation, Bells Landing, Alabama River, Monroe County, Alabama, MGS specimen 1762, height 38.7 mm, width 42.1 mm. 5: *Venericardia gregiana* Dall 1903, Bear Creek marl member of the Tuscahoma Formation, Bear Creek, Wilcox County, Alabama, MGS specimen 1763, height 55 mm, width 53 mm. 6: *Venericardia* (*Venericor*) n. sp., Bear Creek Marl Member of the Tuscahoma Formation, Bear Creek, Wilcox County, Alabama, MGS specimen 1764, height 76.5 mm, width (incomplete) 76.3 mm. 7, 8: *"Voluta" newcombiana* Whitfield 1865, Bells Landing Member of the Tuscahoma Formation, Bells Landing, Alabama River, Monroe County, Alabama, MGS specimen 1765, height 109 mm, width 51 mm.

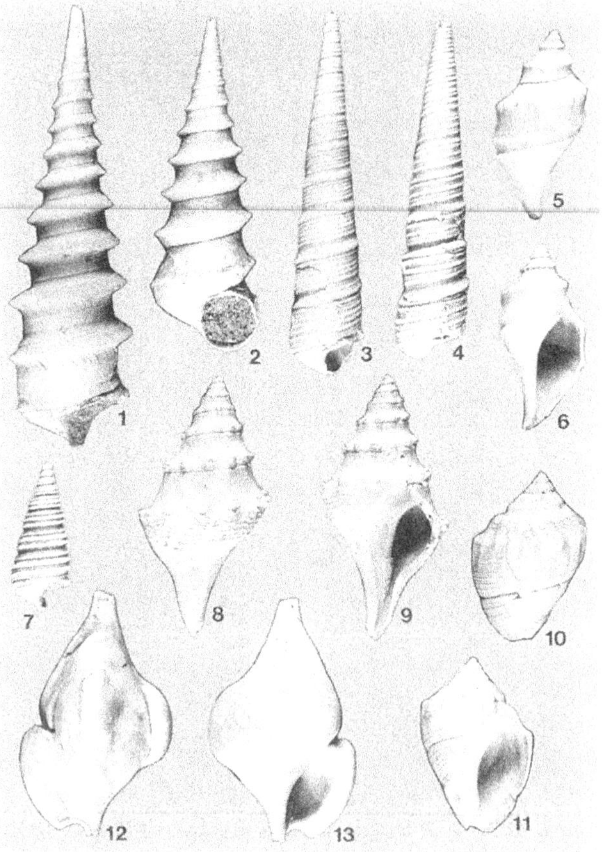

Figure 15.14 All specimens from the Bells Landing Member of the Tuscahoma Formation, Bells Landing, Alabama River, Monroe County, Alabama. 1: *Turritella praecincta* Conrad 1864, MGS specimen 1766, height 99 mm, width 28.5 mm. 2: *Turritella mortoni postmortoni* Harris 1894, MGS specimen 1767, height 67.1 mm, width 22.4 mm. 3, 4: *Turritella eurynome* Whitfield 1865. 3: *T. bellifera* of Aldrich, 1885 (synonym of *T. eurynome* according to Allmon 1996), MGS specimen 1768, height 67 mm, width 13.2 mm; .4, *T. eurynome* Whitfield 1865, s.s., MGS specimen 1769, height 64 mm, width 14 mm. 5, 6: *Priscofiscus* (*Fulguroficus*) *juvenis* (Whitfield 1865), MGS specimen 1770, height 23.7 mm, width 11.9 mm. 7: *Mesalia alabamiensis tetradeirus* Stenzel and Turner 1940, MGS specimen 1771, height 19.3 mm, width 8.2 mm. 8, 9: *Levifusus? pagodiformis* (Heilprin 1880), MGS specimen 1772, height 41.9 mm, width 21.4 mm. 10, 11: *Pseudoliva tuberculifera* Conrad 1860, MGS specimen 1773, height 26.9 mm, width 17.6 mm. 12, 13: *Calyptraphorus trinodiferus* Conrad 1858, MGS specimen 1774, height (incomplete) 55.0 mm, width 35.0 mm.

replaced by an Oligocene fauna more akin to that of the Miocene Gulf and Recent Caribbean (Ray Givens, personal communication 1995).

Various mechanisms have been proposed that could produce large extinctions among shallow shelf marine invertebrates. Among these, lowering of sea level and bolide impacts are perhaps the best accepted. Palmer (1979) noted that the "rhythm of the Paleocene–Eocene seas" could be illustrated in the fossils as well as the sediments. Changing sea level best explains the molluscan faunal turnovers occurring in the Paleocene and lower Eocene deposits of the Gulf Coastal Plain. One adaptation to changing sea level is the thickened parietal callus and strongly callous spires of common gas-

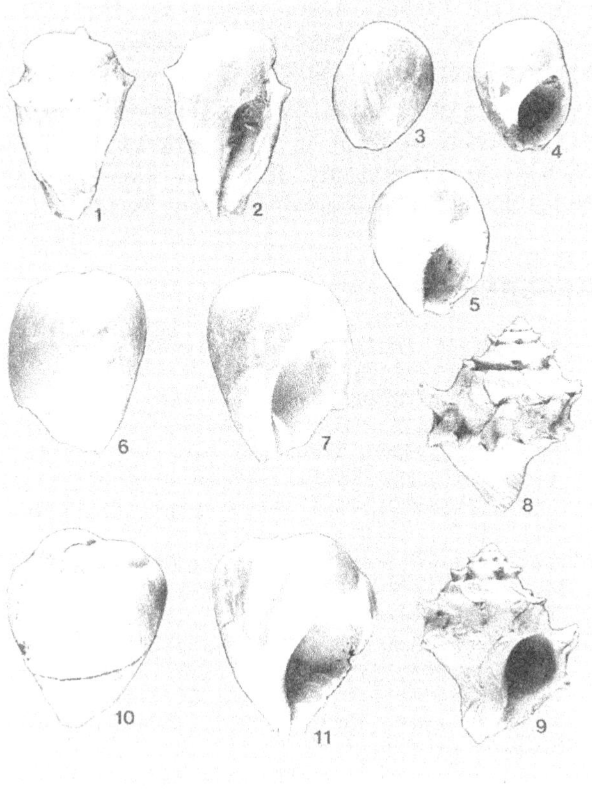

Figure 15.15 All specimens except for 6 and 7 are from the lower venericard beds of the Hatchetigbee Formation at Hatchetigbee Bluff on the Tombigbee River, Washington County, Alabama. 1, 2: *Athleta tuomeyi* Conrad 1853, MGS specimen 1775, height 43.5 mm, width 30.0 mm. 3–5: *Bullia calluspira* Dockery 1980. 3, 4: MGS specimen 1776, height 29.3 mm, width 23.6 mm. 5: MGS specimen 1777, height 29.4 mm, width 24.5 mm. 6, 7: *Pseudoliva* n. sp., Bells Landing Member of the Tuscahoma Formation, Bells Landing, Alabama River, Monroe County, Alabama, MGS specimen 1778, height 48.2 mm, width 38.2 mm. This species has a strongly callous spire like *P. santander*, but differs in its polished shell. 8, 9: *Cornulina minax compressa* Dockery 1980, MGS specimen 1779, height 53.4 mm, width (including spines) 45.6 mm. 10, 11: *Pseudoliva santander* Gardner 1945, MGS specimen 1780, height 45.6 mm, width 37.4 mm.

tropod taxa in the shallow-water facies of the Bashi and Hatchetigbee formations, as shown in figure 15.15. Dockery (1980) suggested the thickened shells provided protection from increased predation and served as ballast in high-energy, nearshore environments.

At well-recognized sequence boundaries and associated regional unconformities, including those at the base of the Wilcox Supercycle Tb and Claiborne Supercycle Tc, the molluscan extinction rate is over 90%. In the Gulf Coast, the Paleocene/Eocene Series contact occurs at the base of a third-order cycle TE1.1, and the extinction rate is 69%. While this is greater that the average rate, it is much lower than those at the previously mentioned second-order boundaries.

Extinctions are only half the equation involving Paleogene faunal turnovers in the northern Gulf Coastal Plain. The second half concerns the origin of new molluscan faunas, which occur in transgressive marine units overlying Type 1 unconformities. After a fall in sea level and the loss of shallow shelf habitats, there is no guarantee that native faunas will be reinstated in the transgressive systems tract of a subsequent sea-level rise. Many times it seems that native molluscan faunas are supplanted by others with unrelated species. This may be explained by the introduction of larvae from distant areas riding on oceanic currents. A modified ocean circulation at the Paleocene/Eocene Epoch boundary is a likely source of new species in the lower Eocene Bashi Formation. Evidence for such current-driven faunal migrations is found in the lower Eocene of the North American West Coast provincial molluscan "Capay Stage" with the first appearance of several Tethyan molluscan taxa (Squires 1987). Among the early Eocene newcomers found in the northern Gulf Coastal Plain, Givens (1989) noted *Volutilithes clarae* (Harris) of the Hatchetigbee Formation (figure 15.16, numbers 14 and 15) to be the first record of the Tethyan genus *Volutilithes* in the Gulf Coast.

Accelerated molluscan extinction rates, including that at the Paleocene/Eocene Epoch boundary, are associated with Type 1 unconformities overlain by lowstand sands. This may indicate that marine regression exposing shelf habitats is a cause for molluscan extinctions or perhaps the extinctions only appear sudden due to a "truncation effect" at the unconformity (Jablonski 1986). The abrupt decline in the $^{87}Sr/^{86}Sr$ across the Paleocene/Eocene unconformity in the central Gulf Coastal Plain coincides with the initiation of oceanic rifting between Norway and Greenland and an increase in seafloor hydrothermal activity. Global tectonic events at this boundary may have altered sea level and oceanic circulation, which adversely affected benthic molluscan communities. A change in oceanic currents could have introduced larvae of allochthonous molluscan taxa into the northern Gulf. Such currents could explain the high number of taxa having their FAD in the Bashi Formation. Changing habitats and competition from new arrivals stressed the indigenous Paleocene Gulf molluscan assemblage, producing the faunal turnover at the Paleocene/Eocene Epoch boundary.

ACKNOWLEDGMENTS

The writer gratefully acknowledges the reviews and helpful comments of Warren D. Allmon, Paleontological

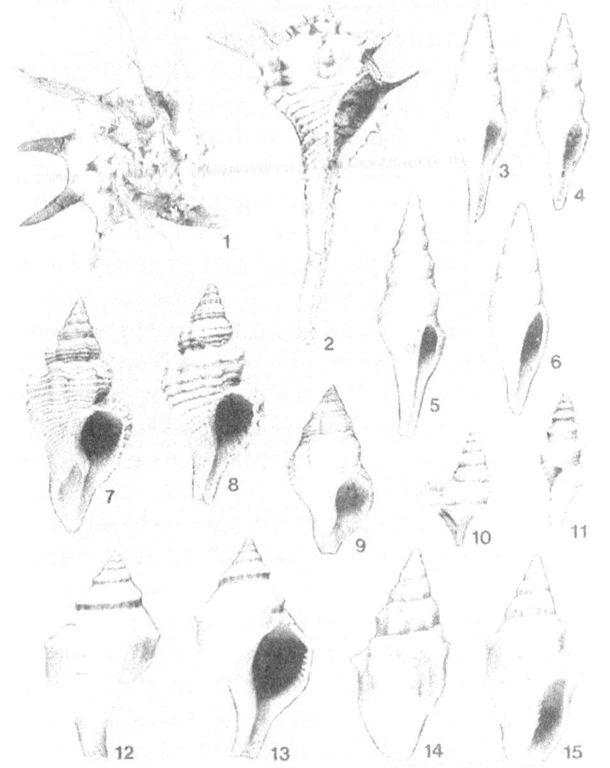

Figure 15.16 1, 2: *Sceptrum* sp., Matthews Landing Member of the Porters Creek Formation (Paleocene), Alabama River below Dixons Creek, Wilcox County, Alabama, MGS specimen 1781, height 102.5 mm, width (including spines) 69.0 mm. A closely related species *S. perulum* (Aldrich 1886) occurs in the Bashi Formation. 3, 5: *Surcula nasuta* (Whitfield 1865), Bells Landing Member of the Tuscahoma Formation, Bells Landing, Alabama River, Monroe County, Alabama. 3: Un-noded variety, MGS specimen 1782, height 38.8 mm, width 10.4 mm. 5: Noded variety, MGS specimen 1783, height 42.0 mm, width 12.3 mm. 4: *Orthosurcula? langdoni* (Aldrich 1895), upper Hatchetigbee Formation, Hatchetigbee Bluff, Tombigbee River, Washington County, Alabama, MGS specimen 1784, height 27.0 mm, width 7.4 mm. 6: Cf. *Sullivania* n. sp., Bells Landing Member of the Tuscahoma Formation, Bells Landing, Alabama River, Monroe County, Alabama, MGS specimen 1785, height 34.0 mm, width 9.8 mm. 7: *Latirus* n. sp., Bells Landing Member of the Tuscahoma Formation, Bells Landing, Alabama River, Monroe County, Alabama, MGS specimen 1786, height 36.8 mm, width 17.4 mm. 8: *Lirofusus subtenuis* (Heilprin 1881), Bells Landing Member of the Tuscahoma Formation, Bells Landing, Alabama River, Monroe County, Alabama, MGS specimen 1787, height 35.0 mm, width 16.9 mm. 9: *"Siphonalia"* sp. B, Hatchetigbee Formation, Hatchetigbee Bluff, Tombigbee River, Washington County, Alabama, MGS specimen 1788, height 25.0 mm, width 13.0 mm. 10: *Aporrhais aldrichi* (Cossmann 1904), Bear Creek Marl Member of the Tuscahoma Formation, Bear Creek, Wilcox County, Alabama, MGS specimen 1789, height (incomplete) 16.4 mm, width (incomplete) 10.4 mm. 11: *Cochlespira greggi* (Harris 1937), Bells Landing Member of the Tuscahoma Formation, Bells Landing, Alabama River, Monroe County, Alabama, MGS specimen 1790, height 20.8 mm, width 7.3 mm. 12, 13: *Levifusus* sp. J, upper Hatchetigbee Formation, Hatchetigbee Bluff, Tombigbee River, Washington County, Alabama, MGS specimen 1791, height 42.4 mm, width 23.7 mm. 14, 15: *Volutilithes clarae* (Harris 1897), upper Hatchetigbee Formation, Hatchetigbee Bluff, Tombigbee River, Washington County, Alabama, MGS specimen 1792, height 41.6 mm, width 20.0 mm.

Research Institution, Ithaca, New York; Marie-Pierre Aubry, Institut des Sciences de l'Evolution, Université Montpellier II, France; Michael B. E. Bograd, Mississippi Office of Geology, Jackson; Rodger E. Denison, the University of Texas at Dallas; Charles R. Givens, Nicholls State University, Thibodaux, Louisiana; and Birger Schmitz, University of Göteborg, Sweden.

REFERENCES

Aldrich, T. H., 1886. Part I. Preliminary report on the Tertiary fossils in Alabama and Mississippi, P. 15–60, pl. 1–6. In Smith, E. A., Geological Survey of Alabama, Bulletin No. 1, 85 pp.

Allmon, W. D., 1996. Systematics and evolution of Cenozoic American Turritellidae (Mollusca: Gastropoda). I. Paleocene and Eocene coastal plain species related to "*Turritella mortoni* Conrad" and "*Turritella humerosa* Conrad." *Palaeontographica Americana*, no. 59, 134 pp.

Allmon, W. D. and Dockery, D. T., III, 1992. A Turritelline gastropod-dominated bed in the Byram Formation (Oligocene) of Mississippi. *Mississippi Geology*, v. 13, p. 29–35.

Baum, G. R. and Vail, P. R., 1988. Sequence stratigraphic concepts applied to Paleogene outcrops, Gulf and Atlantic Basins. In Wilgus, C. K., et al., (eds.), *Sea-Level Changes: An Integrated Approach, Society of Economic Paleontologists and Mineralogists* Special Publication No. 42, p. 309–327.

Beard, K. C., Dawson, M. R. and Tabrum, A. R. 1995. First diverse land mammal fauna from the early Cenozoic of the southeastern United States: The early Wasatchian Red Hot Local Fauna, Lauderdale County, Mississippi. *Geological Society of America, 1995 Annual Meeting, Abstracts with Programs*, v. 27, p. 453.

Beard, K. C. and Tabrum, A. R., 1990. The first early Eocene mammal from eastern North America: An omomyid primate from the Bashi Formation, Lauderdale County, Mississippi. *Mississippi Geology*, v. 11, p. 1–6.

Berggren, W. A., Kent, D. V., Flynn, J. J. and Van Couvering, J. A., 1985. Cenozoic geochronology. *Geological Society of America Bulletin*, v. 96, p. 1407–1418.

Bryant, J. D., Jones, D. S. and Mueller, P. A., 1995. Influence of freshwater flux on $^{87}Sr/^{86}Sr$ chronostratigraphy in marginal marine environments and dating of vertebrate and invertebrate faunas. *Journal of Paleontology*, v. 69, p. 1–6.

Burke, W. H., Denison, R. E., Hetherington, E. A., Koepnick, R. B., Nelson, H. F. and Otto, J. B., 1982. Variation of seawater $^{87}Sr/^{86}Sr$ throughout Phanerozoic time. Geology, v. 10, p. 516–519.

Case, G. R., 1986, The bony fishes (teleosts) of the Tuscahoma and Bashi formations, early Eocene, Meridian, Lauderdale County, Mississippi. *Mississippi Geology,* v. 6, p. 6–8.

Case, G. R., 1994a. Fossil fish remains from the late Paleocene Tuscahoma and early Eocene Bashi formations of Meridian, Lauderdale County, Mississippi. Part I. Selachians. *Sonder-Abdruck aus Palaeontographica Beitrage zur Naturgeschichte der Vorzeit, Pal. A,* v. 230, p. 97–138.

Case, G. R., 1994b. Ibid, Part II. *Teleosteans,* p. 139–153.

Charisi, S. D. and Schmitz, B., 1995. Stable ($\delta^{13}C$, $\delta^{18}O$) and strontium ($^{87}Sr/^{86}Sr$) isotopes through the Paleocene at Gebel Aweina, eastern Tethys region. *Palaeogeography, Palaeoclimatology, Palaeoecology,* v. 116, p. 103–129.

Coleman, J. and Galloway, W. E., 1990. Sequence stratigraphic analyses of the lower Oligocene Vicksburg Formation of Texas. *Sequence Stratigraphy as an Exploration Tool, Concepts and Practices in the Gulf Coast, Gulf Coast Section,* Society of Economic Paleontologists and Mineralogists Foundation, Eleventh Annual Research Conference, p. 99–112.

Combes, J. M., 1993. The Vicksburg Formation of Texas: depositional systems distribution, sequence stratigraphy, and petroleum geology. *American Association of Petroleum Geologists Bulletin,* v. 77, p. 1942–1970.

Cossmann, M. and Pissarro, G., 1904–1913. *Iconographie Complète des Coquilles Fossiles de l'Eocène des Environs de Paris,* v. 1 (1904–1906), *Pélécypodes,* pl. 1–45; v. 2 (1910–1913), *Scaphopodes, Gastropodes, Brachiopodes, Céphalopodes, and supplément,* pl. 1–65, privately published, Paris.

Denison, R. E., Koepnick, R. B., Fletcher, A., Dahl, D. A. and Baker, M. C., 1993. Reevaluation of early Oligocene, Eocene, and Paleocene seawater strontium isotope ratios using outcrop samples from the U.S. Gulf Coast. *Paleoceanography,* v. 8, p. 101–126.

DePaolo, D. J. and Ingram, B. L., 1985. High-resolution stratigraphy with strontium isotopes. *Science,* v. 227, p. 938–941.

Dockery, D. T., III, 1977. Mollusca of the Moodys Branch Formation, Mississippi. *Mississippi Geological Survey Bulletin* 120, 212 pp.

Dockery, D. T., III, 1980. The invertebrate macropaleontology of the Clarke County, Mississippi, area. *Mississippi Bureau of Geology Bulletin* 122, 387 pp.

Dockery, D. T., III, 1984. Crisis events for Paleogene molluscan faunas in the southeastern United States. *Mississippi Geology,* v. 5, p. 1–7.

Dockery, D. T., III, 1985. PALSTRAT: A biostratigraphic computer program for Paleocene and Eocene mollusks. *Mississippi Geology,* v. 6, p. 18–19.

Dockery, D. T., III, 1986. Punctuated succession of Paleogene mollusks in the northern Gulf Coastal Plain. *Palaios,* v. 1, p. 582–589.

Dockery, D. T., III, 1990. The Eocene–Oligocene boundary in the northern Gulf: A sequence boundary. *Sequence Stratigraphy as an Exploration Tool, Concepts and Practices in the Gulf Coast, Gulf Coast Section,* Society of Economic Paleontologists and Mineralogists Foundation, Eleventh Annual Research Conference, p. 141–150.

Dockery, D. T., III, 1992. The Eocene–Oligocene boundary in the northern Gulf Coastal Plain: A Type 1 sequence boundary. *Geological Society of America, 26th Annual South-Central Section,* Abstracts with Programs, v. 24, p. 9.

Dockery, D. T., III and Hansen, T. A., 1987. Eocene–Oligocene molluscan extinctions: Comment and reply. *Palaios,* v. 2, p. 620–622.

Dockery, D. T., III and Hansen, T. A., 1988. Molluscan extinction rates in question. (Comment and reply). *Nature,* v. 331, p. 123.

Dolin, C., Dolin, L. and LeRenard, V., 1980. Inventaire systématique des mollusques de l'Auversien et faciès charrié de Baron (Oise), et remarques paléontologiques. *Bulletin d'Information des Géologues du Bassin de Paris,* v. 17, p. 26–48.

Garvie, C. L., 1996. The molluscan macrofauna of the Recklaw Formation, Marquez Member (Eocene: lower Claibornian), in Texas. *Bulletins of American Paleontology,* v. 111, 177 pp.

Gaskell, B. A., 1991. Extinction patterns in Paleogene benthic foraminiferal faunas: Relationship to climate and sea level. *Palaios,* v. 6, p. 2–16.

Gingerich, P. D. and Deutsch, H. A., 1989. Systematics and evolution of early Eocene hyaenodontidae (Mammalia, Creodonta) in the Clark's Fork Basin, Wyoming. *Contributions from the Museum of Paleontology, University of Michigan,* v. 27, p. 327–391.

Givens, C. R., 1989. First record of the Tethyan genus *Volutilithes* (Gastropoda: Volutidae) in the Paleogene of the Gulf Coastal Plain, with a discussion of Tethyan molluscan assemblages in the *Gulf Coastal Plain and Florida. Journal of Paleontology,* v. 63, p. 852–856.

Goldsmith, N. F., Martinell, J., Demarcq, G., Bohn-Havas, M. and Dockery, D. T., III, 1994. Sr-isotopic calibration of Cenozoic Bivalvia and Early Miocene migrations: Eurasian carnivores to Africa (the Hazeva Formation, Israel) and African gazelles and proboscidia to Ipolytarnoc, Hungary. *Newsletter on Stratigraphy,* v. 31, p. 167–183.

Goldstein, S. J. and Jacobsen, S. B., 1987. The Nd and Sr isotopic systematics of river-water dissolved material: Implications for the sources of Nd and Sr in seawater. *Earth and Planetary Science Letters,* v. 66, p. 245–272.

Hansen, T. A., 1987. Extinction of Late Eocene to Oligocene molluscs: Relationship to shelf area, temperature changes, and impact events. *Palaios,* v. 2, p. 69–75.

Harris, G. D., 1919. Pelecypods of the St. Maurice and Claiborne stages. *Bulletins of American Paleontology,* v. 6, 268 pp.

Harris, G. D., 1937. Turrid illustrations. *Paleontographica Americana,* v. 2, 122 pp.

Heilprin, A., 1880. On some new species of Eocene Mollusca from the southern United States. United States National Museum, Proceedings, v. 3, p. 149–152.

Heilprin, A., 1881. On some new lower Eocene Mollusca from Clarke Co., Ala., with some points as to the stratigraphical positions of the beds containing them. *Academy of Natural Sciences of Philadelphia Proceedings,* v. 32, p. 364–375.

Hess, J., Bender, M. L. and Schilling, J. G., 1986. Evolution of the ratio of strontium-87 to strontium-86 in seawater from Cretaceous to Present. *Science,* v. 231, p. 979–984.

Holman, J. A., Dockery, D. T., III and Case, G. R., 1990. Paleogene snakes of Mississippi. *Mississippi Geology,* v. 11, p. 1–12.

Ingram, S. L., 1991. The Tuscahoma–Bashi section at Meridian, Mississippi: First notice of lowstand deposits above the Paleocene-Eocene TP2/TE1 sequence boundary. *Mississippi Geology,* v. 11, p. 9–14.

Jablonski, D., 1986. Causes and consequences of mass extinctions: A comparative approach. In Elliot, D. K., (ed.), *Dynamics of Extinction,* Toronto: John Wiley & Sons, p. 183–229.

Koch, P. L., Zachos, J. C. and Gingerich, P. D., 1992. Correlation between isotope records in marine and continental carbon reservoirs near the Palaeocene/Eocene boundary. *Nature,* v. 358, p. 319–322.

Mancini, E. A., 1981. Lithostratigraphy and biostratigraphy of Paleocene subsurface strata in southwest Alabama. *Transactions of the Gulf Coast Association of Geological Societies,* v. 31, p. 359–367.

Mancini, E. A. and Tew, B. H., 1991a. Relationships of Paleogene stage and planktonic foraminiferal zone boundaries to lithostratigraphic and allostratigraphic contacts in the eastern Gulf Coastal Plain. *Journal of Foraminiferal Research,* v. 21, p. 48–66.

Mancini, E. A. and Tew, B. H., 1991b. Tertiary sequence stratigraphy and biostratigraphy of southwestern Alabama. *International Field Symposium on the Paleocene/Eocene Boundary of the Gulf and Atlantic Coastal Plains, USA: Alabama Gulf Coastal Plain Fieldtrip Guidebook,* October 26–29, 1991, 51 pp. plus additional stops.

Martini, E., 1971. Standard Tertiary and Quaternary calcareous nannoplankton zonation. In Farinacci, A., (ed.), *Proceedings of the Second Planktonic Conference:* Roma, Italy, *Tecnoscienza,* p. 739–785.

Palmer, K. V. W., 1979. Rhythm of the Paleocene–Eocene seas of the Central Gulf Coast, as defined by fossils and sediments. *Tulane Studies in Geology and Paleontology,* v. 15, p. 69–74.

Palmer, K. V. W. and Brann, D. C., 1965. Catalogue of the Paleocene and Eocene Mollusca of the Southern and Eastern United States. Part I. Pelecypoda, Amphineura, Pteropoda, Scaphopoda, and Cephalopoda. *Bulletin of American Paleontology,* v. 48, 466 pp.

Palmer, K. V. W. and Brann, D. C., 1966. *Ibid. Part II.* Gastropoda. Ibid, p. 447–1057.

Pasley, M. A. and Hazel, J. E., 1995. Revised sequence stratigraphic interpretation of the Eocene–Oligocene boundary interval, Mississippi and Alabama, Gulf Coast Basin, U.S.A. *Journal of Sedimentary Research,* v. B65, p. 160–169.

Rea, D. K., Zachos, J. C., Owen, R. M. and Gingerich, P. D., 1990. Global change at the Paleocene–Eocene boundary: climatic and evolutionary consequences of tectonic events. *Palaeogeography, Palaeoclimatology, Palaeoecology,* v. 79, p. 117–128.

Schmitz, B., Åberg, G., Werdelin, L., Forey, P. and Bendix-Almgreen, S. E., 1991. $^{87}Sr/^{86}Sr$, Na, F, Sr, and La in skeletal fish debris as a measure of the paleosalinity of fossil-fish habitats. *Geological Society of America Bulletin,* v. 103, p. 786–794.

Schmitz, B., Ingram, S. L., Dockery, D. T. III, and Åberg, G., 1997. Testing $^{87}Sr/^{86}Sr$ as a paleosalinity indicator on mixed marine, brackish-water and terrestrial vertebrate skeletal apatite in late Paleocene–early Eocene near-coastal sediments. Mississippi, *Chemical Geology,* v. 140, p. 275–287.

Smith, E. A. and Johnson, L. C., 1887. Tertiary and Cretaceous strata of the Tuscaloosa, Tombigbee, and Alabama Rivers. *U.S. Geological Survey, Bulletin,* v. 43, 189 pp.

Squires, R. L., 1987. Eocene molluscan paleontology of the Whitaker Peak area, Los Angeles and Ventura Counties, California. *Natural History Museum of Los Angeles County, Contributions in Science,* no. 388, 93 pp.

Toulmin, L. D., 1977. *Stratigraphic Distribution of Paleocene and Eocene Fossils in the Eastern Gulf Coast Region.* Geological Survey of Alabama, Monograph 13, 602 pp.

Whitfield, R. P., 1865. Descriptions of new species of Eocene fossils. *American Journal of Conchology,* v. 1, p. 259–268.

Appendix 15.1

Wilcox Group–Sabinian Stage molluscan taxon ranges in the Late Paleocene Nanafalia (NA) Formation, Bear Creek (BC), Greggs Landing (GL), and Bells Landing (BL) marl members of the Tuscahoma Formation, and the Early Eocene Bashi (BA) and Hatchetigbee (HA) formations. Those taxa with ranges extending into the Early-Late Paleocene Midway or Early-Middle Eocene Claiborne groups are checked under M and C respectively. Type horizons are indicated in bold. New species illustrated in this report and not found in the published record are accompanied by their figure numbers.

	Range						
Taxon	Paleocene				Eocene		
Bivalvia with FAD in NA or earlier							
Acanthocardia tuomeyi (Aldrich)	**NA**						
Atrina sp. C	NA						
Barbatia ludoviciana (Harris)	NA						C
Bornia prima (Aldrich)	NA				**BA**		
Codakia? sp.	NA				BA	HA	
Corbula subengonata Dall	NA		**GL**	BL	BA	HA	
Crassatella halei Harris	NA		**GL**		BA		
C. tumidula Whitfield	NA	BC	**GL**	BL			
Diplondonta hopinsensis Clark	NA				BA	HA	
Glycymeris idonea (Conrad) subsp. B	NA	BC	GL	BL	BA	HA	
Lithorhadia aldrichiana (Harris)	NA				**BA**		
Martesia recurva Aldrich	**NA**						
Nuculana corpulentoides (Aldrich)	**NA**						
Odontogryphaea thirsae (Gabb)	**NA**						
O. thirsae (Gabb) subsp.	NA						
Ostrea arrosis Aldrich	**NA**						
O. sinuosa Rogers and Rogers	NA		GL	BL			
Pholas alatoidea Aldrich	NA		**GL**	BL			
Pitar nuttalliopsis fulvus (Harris)	NA			BL	**BA**		
P. nuttalliopsis greggi (Harris)	NA		GL	**BL**			
Tellina bellsiana Aldrich	NA		**GL**	BL			
T. greggi Harris	NA	BC	**GL**				

Appendix 15.1 *(continued)*

	Range							
Taxon		Paleocene			Eocene			
T. subtriangularis Aldrich		NA			BL	BA	**HA**	
veneicardia aposmithi Gardner and Bowles		NA	BC	GL	**BL**			
V. greggiana Dall		NA	BC	**GL**	BL	BA		
V. nanaplata Gardner and Bowles		**NA**		GL				
V. pilsbryi Stewart		**NA**		GL	BL			
Yoldia pharcida (Dall)		NA		GL	BL	**BA**		
Gastropoda with FAD in NA or earlier								
Athleta tuomeyi Conrad		NA	BC	GL	**BL**	BA	HA	
Bullia calluspira Dockery		NA				**BA**	HA	
Calyptraea aperta (Solander in Grander)		NA		GL	BL	BA	HA	C
Calyptraphorus trinodiferus Conrad	M	NA	BC	GL	BL	BA		
Capulus expansus (Whitfield)		NA			**BL**	BA	HA	C
Caricella podagrina (Dall)		NA			**BL**			
Clavithites sp. A		NA				BA		
Cornulina philadelphica (Harris)		NA					**HA**	
C. minas compressa Dockery		NA		GL		**BA**	HA	
Diododra sp. A		NA						
Diodora sp. B		NA						
Elimia trigemmata (Conrad)	M	NA				**BA**		
Euspira marylandica (Conrad) var. A		NA		GL		BA	HA	
Exilia sp.		NA						
"Fusus" sp.		NA						
Hermisurcula silicata Harris		NA	BC	**GL**				
Latirus tortilis nanafalius Harris		**NA**						
Levifusus supraplanus Harris		NA		**GL**	BL			
Levifusus sp. J		NA			BL	BA	HA	
Mesalia alabamiensis (Whitfield		NA		**GL**	BL			
M. biplicata Bowles		**NA**						
Cf. *Natica aperta* Whitfield		NA		**GL**	BL			
Orthosurcula indenta (Harris)		**NA**						
Pleurotomella whitfieldi (Aldrich)	**M**			GL				
Polinices sp.		NA				BA		
Popenoeum scalina Heilprin		NA				**BA**		
Pseudoliva nanafaliaensis Aldrich		**NA**						
P. santander Gardner		NA	BC	GL	BL	BA	HA	
Pseudoliva sp. G		NA						
Sinum fiski Le Blanc		NA		GL	BL	BA		
Cf. *Sullivania tombigbeensis* (Aldrich)		NA				**BA**		
Surcula nasuta (Whitfield)		NA	BC	GL	**BL**			
Turritella mortoni postmortoni Harris		NA	BC	GL	**BL**			
T. multilira Whitfield		**NA**						
T. praecincta Conrad		NA	BC	GL	BL			
Turritella sp. B		NA						
"Voluta" newcombiana (Whitfield	NA				**BL**			
Bivalvia with FAD in BC								
Ostrea sp.			BC			BA	HA	
V. nanaplata nanna Gardner and Bowles			BC	**GL**	BL	BA		
V. nanaplata sp. (Figure 13.6)			BC					
Gastropoda with FAD in BC								
Aporrhais aldrichi (Cossmann)			BC	**GL**				
Caricella demissa Conrad "var."			BC	GL				

Appendix 15.1 *(continued)*

Taxon	Range					
	Paleocene	Eocene				
Levifusus? pagodiformis (Heilprin)	BC	GL	BL	**BA**	HA	
Priscoficus juvenis (Whitfield)	BC	**GL**	BL	BA	HA	
Pseudoliva tuberculifera Conrad	BC	GL	**BL**			
Turritella eurynome Whitfield	BC	GL	**BL**			
Bivalvia with FAD in GL						
Barbatia cuculoides (Conrad) var.		GL		BA	HA	
Basterotia prima Aldrich		**GL**	BL			
Chlamys greggi Harris		**GL**	BL			
Coralliophaga prima (Harris)		GL		**BA**		
Cucullaea gigantea Conrad "var."		GL	BL			
Ervilia lignitica Aldrich		GL	**BL**			
Gari harrisi (Aldrich)		**GL**				
G. smithi (Aldrich)		**GL**				
Litorhadia elongatoidea (Aldrich) "var." B		GL				
Miltha greggi (Harris)		**GL**				
Nemocardium harrisi (Dall)		GL	BL	**BA**		
Panopea alabama (Harris)		**GL**	BL			
Pholas aldrichi de Gregorio		**GL**	BL			
Pitar nuttalliopsis (Heilprin)		GL	BL	**BA**		
Saccella robusta (Aldrich)		GL		**BA**		
Tellina semirotunda Aldrich		**GL**	BL			
T. temperata Aldrich		**GL**				
T. trumani Harris		**GL**				
Vokesula aldrichi (Meyer)		GL		**BA**	HA	
Scaphopoda with FAD in GL						
Cadulus sp. D		GL				
D. multannulatum Aldrich		**GL**				
Gastropoda with FAD in GL						
Agaronia sp. B		GL		BA		
Ampullella recurva (Aldrich) "var."		GL		BA	HA	
Architectonica greggi (Harris)		**GL**				
A. intusa (Harris)		**GL**				
Astyris subfraxa Harris		GL		**BA**		
Atys robustoides Aldrich		**GL**				
Bonellitia graciloides (Aldrich)		**GL**				
B. graciloides bella (Aldrich)		**GL**				
Bullia tuomeyi (Aldrich)		**GL**				
Cantharus elegantissimus (Aldrich)		**GL**		BA		
Caricella pyruloides (Conrad)		GL				C
Cerithiopsis greggiensis Aldrich		**GL**				
Cochlespira greggi (Harris)		**GL**	BL	BA		
Coronia childreni (Lea) var. A		GL		BA		C
Crenilabium cossmanni (Aldrich)		**GL**			HA	
Cf. *Cyllene bellana* Harris		**GL**	**BL**	BA		
Diodora alabama (Harris)		**GL**				
Eocypraea eosmithi (Aldrich)		**GL**		BA	HA	
Epitonium exquisitum (Aldrich)		**GL**				
E. munistriatum Aldrich		**GL**		BA		
Fulgurofusus rugatus (Aldrich)		**GL**				
Fusitoma sipha (Aldrich)		**GL**				
Glyptostyla bacula (Aldrich)		GL	**BL**			

Appendix 15.1 *(continued)*

	Range			
Taxon	Paleocene	Eocene		
Lacunaria alabamiensis (Whitfield)	GL	**BL**		
Lirofusus subtenuis (Heilprin)	GL	BL	**BA**	HA
Lyrosurcula sylvaerupis (Harris)	GL		**BA**	HA
Mazzalina tuomeyi (Aldrich)	**GL**	**BL**		
Melanopsis anita Aldrich	**GL**			
Mesalia alabamiensis tetradeiras (Stenze and Turner)	GL	**BL**		
Narona greggi (Harris)	**GL**	BL		
Nassarius exilis (Conrad)	GL		BA	HA
Odontopolys sublevis (Haris)	**GL**			
Odostomia insignifica (Aldrich)	**GL**			
Palaeorhaphis pergracilis (Aldrich)	**GL**			
Perissolax eocensis (Aldrich)	**GL**			
"Plurotoma" capax (Whitfield)	GL	**BL**		
P. triserialis (Whitfield)	GL			
Cf. *Pseudoneptunea harrisi* (Aldrich)	**GL**			
Pyramidella obtusoides Aldrich	**GL**			
Scaphander alabamensis Aldrich	**GL**			
Sceptrum perulum (Aldrich)	GL		**BA**	
Serpulorbis sp. B	GL			
Sinum sp. A	GL		BA	
Solariorbis sp. B	GL	BL		
Strepsidura heilprini (de Gregorio)	GL	BL	**BA**	
Cf. *Sullivania roscoei* (Harris)	**GL**			
Surculoma? servatoidea (Aldrich)	**GL**		BA	
Sveltia gilberti (Aldrich)	**GL**			
Tenuiacteon pertenuis Aldrich	**GL**			
Turboniolla clinensis Aldrich	**GL**			
T. tuscahomensis Aldrich	**GL**			
"Turris" sp.	GL			
Bivalvia with FAD in BL				
Corbula concha Aldrich		**BL**	BA	
Donax acutangula Aldrich		**BL**		
Dosiniopsis lenticularis (Rogers and Rogers)		BL		
Martesia elongata Aldrich		**BL**		
Montacuta bicuspidata Aldrich		**BL**		
Mysella minuta (Aldrich)		**BL**		
Ostrea intermedoides Aldrich		**BL**		
Panopea bellsensis Aldrich		**BL**		
Saxicavella alabamensis Aldrich		**BL**		
Semele langdoniana Aldrich		**BL**		
S. monroensis Aldrich		**BL**		
Venericardia sp. G		BL	BA	
Gastropoda with FAD in BL				
Acteocina leai (Aldrich)		**BL**		
Architectonica bellensis (Harris)		**BL**		
Buccinanops ellipticum (Whitfield)		**BL**		
Cantharus johnsoni (Aldrich)		**BL**		
Galeodea koureos Gardner		**BL**	BA	
Lacunaria erecta (Whitfield)		**BL**		
Latirus n. sp. (Figure 16.7)		BL		
Levifusus bellanus (Harris)		**BL**		

Appendix 15.1 *(continued)*

Taxon	Range				
	Paleocene	Eocene			
Mazzalina plena (ALdrich)		**BL**			
Odostomia bartschi Aldrich		**BL**			
Pseudoliva n. sp. (Figures 15.6–15.7)		BL			
Pyramidella anita (Aldrich)		**BL**	**BA**		
Ringicula butleriana lignitifera (Aldrich)		**BL**	BA		
Simnia regularoidea (Aldrich)		**BL**			
"Siphonalia" sp. B		BL		HA	
Solariorbis subangulatus smithi (Aldrich)		**BL**			
Cf. *Sullivania* n. sp. (Figure 16.6)		BL			
Turbonilla sp. A		BL			
Bivalvia with FAD in BA					
Amusium cf. *A. squamulum* (Lamarck)			BA		C
Andontia ozarkana (Harris)			BA	**HA**	C
Anomia navicelloides Aldrich			**BA**		
Arca hatchetigbeensis Harris			BA	**HA**	
Atrina sp. D			BA		
Barbatia lignitifera Aldrich			**BA**		
Brachidontes alabamensis (Aldrich)			**BA**		
Callista sylvaerupis (Harris)			**BA**	HA	
Chlamys choctavensis (Aldrich)			**BA**		
Crassostrea n. sp. (Figure 11.5)			BA		
Cuspidaria prima (Aldrich)			**BA**		
Diplodonta sp. C			BA		
Felaniella sp.			BA		
Gari ozarkana (Harris)			**BA**		
Gastrochaena striatula Aldrich			**BA**		
Glyptoactis alticostata (Conrad) "var." A			BA		
Haliris granuloides Aldrich			**BA**		
Lepton vaughani Aldrich			**BA**		
Lirodiscus smithvillensis (Harris) "var."			**BA**		
Litorhadia acala (Dall)			**BA**		
L. elongatoidea (Aldrich)			**BA**		
Lucina pomilia smithi Meyer			BA		C
"Lucina" sylvaerupis Harris			**BA**		
Lucina ulrichi Harris			**BA**		
Cf. *Miodontiscus timothii* Palmer and Brann			**BA**		
Cf. *Miodontiscus* sp.			BA		
Montacuta herberti Aldrich			**BA**		
Myrtea? astartiformis (Aldrich)			**BA**		
Nucula sp. C			BA		
Nuculana marieana (Aldrich)			**BA**		
Ostrea sp. I			BA		
O. brevifronta Dockery			**BA**		
Pachecoa microcancellata (Barry)			**BA**	HA	
Periploma butlerianum Aldrich			**BA**		
Physoida clarkeana (Aldrich)			**BA**		
Pitar nuttalliopsis heilprini (Palmer)			**BA**		
Plicatula filamentosa Conrad "var." A			BA	HA	
Pteria sp. B			BA		
Pycnodonta sylvaerupis Harris			**BA**		
Scintilla clarkeana Aldrich			**BA**		
Solemya alabamensis Harris			**BA**		

Appendix 15.1 *(continued)*

	Range			
Taxon	Paleocene	Eocene		
Spisula parilis bistriata (Harris)		**BA**		
Sportella oblonga (Aldrich)		**BA**		
Tellina cynoglossa Dall		**BA**		
T. semipapyria Aldrich		**BA**		
Vernericardia bashiplata Gardner and Bowles		**BA**		
V. gulielmi Gardner and Bowles		**BA**		
V. horatiana Gardner		BA	**HA**	
V. turneri Gardner and Bowles		BA	**HA**	
Scaphopoda with FAD in BA				
Cadulus sp. E		BA		
Dentalium microstria Heilprin		**BA**	HA	
D. sylvaerupis (Harris)		**BA**		
Cephalopoda with FAD in BA				
"Nautilus" sp.		BA		
Gastropoda with FAD in BA				
Abderospira sabina Palmer		**BA**		
Acteon pomilius multannulatus Aldrich		**BA**		
Ancilla staminea maternae Palmer		**BA**		
Architectonica cupola (Heilprin)		**BA**		
A. huppertzi (Harris) "var."		BA		
A. sabinia Palmer		**BA**		
A. sylvaerupis (Harris)		**BA**		
Athleta clarae (Harris)		BA	**HA**	
Balcis wheeleri (Palmer)		**BA**		
Bittium anita (Aldrich)		BA		
Bonellitia silvaerupis (Harris)		**BA**		
Buccitriton sp. A		BA		C
Cancelrana lanceolata (Aldrich)		**BA**		
Caricella heilprini Dall		**BA**		
Caricella sp. D		BA		
Cerithiella fluviatilis (Aldrich)		**BA**		
C. regularoides (Aldrich)		**BA**		
C. terebropsis (Harris)		**BA**		
Cerithiella sp. A		BA		
"Cerithium" tombigbeense Aldrich		**BA**		
Cochlespirella nana (Lea)		BA		C
Conomitra traceyi Harris		**BA**		
Coronia mediavia (Harris) "var."		BA		
C. mediavia equiseta (Harris)		BA	HA	
Cf. *Coronia nodoidea* (Aldrich)		**BA**		
Cf. *Cryptochorda clarkensis* (Aldrich)		**BA**		
Cyclostremiscus dalli (Harris)		**BA**		
C. sylvaerupis (Harris)		**BA**		
Diodora unilineata (Aldrich)		**BA**		
Diodora sp. A		BA		
Dolicholatirus? interstriatus (Heilprin)		**BA**		
D? interstriatus (Heilprin) subsp.		BA		
D.? tombigbeensis (Aldrich)		**BA**		
Eopleurotoma bimoniata (Aldrich)		**BA**		
E. cainei (Harris)		**BA**		

Appendix 15.1 *(continued)*

Taxon	Range			
	Paleocene	Eocene		
E. veatchi (Harris)		**BA**		
Eosinica elevata (Aldrich)		**BA**		
Eosurcula tuomeyi (Aldrich)		**BA**		
Epitonium multiliniferum Aldrich		**BA**		
E. subacutum Aldrich		**BA**		
Eulima cainei Harris		**BA**		
E. sylvaerupsis (Harris)		**BA**		
Euspira leana (Palmer)		BA		C
E. sabina (Palmer)		**BA**	HA	
Euthria dubia (Aldrich)		**BA**		
Falsifusus ottonis (Aldrich)		**BA**		
Galeodea dubia (Aldrich)		**BA**		C
Gegania sp.		BA		
Hipponix sylvaerupis Harris		**BA**		
Laevibuccinum harrisi Aldrich		**BA**		
L. lineatum Heilprin		**BA**		
Levifusus prepagoda Palmer		**BA**		
Levifusus sp. F		BA		
Limacina choctavensis (Aldrich)		**BA**		
L. elongatoidea (Aldrich)		**BA**		
Mathilda elongatoides Aldrich		**BA**		
M. leana Aldrich		**BA**		
M. singularis Aldrich		**BA**		
Metula sylvaerupis Harris		**BA**	HA	
Microdrillia infans (Meyer)? "var."		BA		
M. rostratula Casey		**BA**		
M. turriculata (Aldrich)		**BA**		
M. sp. B		BA		
Mitrella erecta (Aldrich)		**BA**		
M. mississippiensis meyeri Palmer		**BA**		
"Natica" gilberti Palmer		**BA**		
Niso sp. A		BA		
Orthosurcula? langdoni (Aldrich)		BA	**HA**	
Penion gracilis (Aldrich)		**BA**		
P. imbricatulus (Harris)		**BA**		
Philine alabamensis Aldrich		**BA**		
"Pleurotoma" moniliata Helprin		**BA**	HA	
"P." sigma (Harris)		**BA**		
"Pleurotoma" sp. A		BA		
Poirieria woodsensis Vokes		**BA**		
Pyramidella sp.		BA		
Ranella tuomeyi Aldrich		**BA**	HA	
Raphitoma georgei (Harris)		**BA**		
R. specus (Aldrich)		**BA**		
Retusa sylvaerupis (harris)		**BA**		
Ringicula butleriana Aldrich		**BA**		
Rudiscala harrisi Palmer		**BA**		
Scaphander ligniticus Aldrich		**BA**		
Scaphella sp. D		BA		
Sceptrum sp. B		BA		
Sigatica clarkeana Aldrich		**BA**	HA	
Sinum declive (Conrad)		BA		C
Solariella sylvaerupis Harris		**BA**	HA	

Appendix 15.1 *(continued)*

	Range			
Taxon	Paleocene	Eocene		
Solariella sp.		BA	HA	
Solariorbis liniferus Aldrich		**BA**		
Sullivania exilloides (Aldrich)		**BA**		
Surculites engonatus (Heilprin)		**BA**		
Surculoma? nebulosa (Harris)		**BA**		
Terebrifusus n. sp. A		BA		
Tornatellaea bella Conrad		**BA**	HA	
Trigonostoma marieanum (Aldrich)		**BA**		
T. sabinetownense Le Blanc		BA		
Trypanotoma carlottae (Harris)		**BA**		
Trypanotoma sp. A		BA		
Turbonilla harrisi Aldrich		**BA**		
T. sabina Palmer		**BA**		
T. tardiusculus Aldrich		**BA**		
Turritella gilberti Bowles		**BA**	HA	
T. rina Palmer var.		BA		
Umbraculum sylvaerupis (Harris)		**BA**		
Vitrinella aldrichi (Harris)		**BA**		
Volvariella aldrichi Cossmann		**BA**		
Xenophora sp. A		BA		
Bivalvia with FAd in HA				
Acanthocardia hatchetigbeensis (Aldrich)			**HA**	
Anomia sp. A			HA	
Corbicula? cornelliana Harris			**HA**	
Corbula milium Dall			**HA**	
Crassostrea sp. A			HA	
Crassostrea sp. C			HA	
Cucullaria ozarkensis Aldrich			**HA**	
Eburneopecten corneoides (Harris)			**HA**	
Lima ozarkana Harris			**HA**	
Ostrea sp. E			HA	
Pelecyora hatchetigbeensis (Aldrich)			**HA**	
Solenlisbonensis (Aldrich)			**HA**	
Spondylus sp. B			HA	
Venericardia hatcheplata Gardner and Bowles			**HA**	
Yoldia sp. A			HA	
Gastropoda with FAD in HA				
Architectonica leana (Dall)			HA	**C**
Bathytoma marieana Aldrich			**HA**	
Bulliopsis choctavensis (Aldrich)			**HA**	
Cerithiella delicatula (Aldrich)			**HA**	
"Chrysodomus" striatus (Aldrich)			**HA**	
Cornulina hatchetigbeensis (Aldrich)			**HA**	
Cyclostremiscus n. sp.			HA	
Mesorhytis? hatchetigbeensis (Aldrich)			**HA**	
Mitrella alabamensis (Aldrich)			**HA**	
Neritina unidenta Aldrich			**HA**	
Olivella semilignitica Aldrich			**HA**	
"Potamides" fulvarupis Harris			**HA**	
"Pseudomalaxis" verrilli (Aldrich)			**HA**	
Pyramidella mitchelliana (Aldrich)			**HA**	

Appendix 15.1 *(continued)*

	Range	
Taxon	Paleocene	Eocene
Scobinella sculpturata Aldrich		**HA**
Trophon caudatoides Aldrich		**HA**
Turbonilla agrestis Aldrich		**HA**
Volutilithes clarae (Harris)		HA

Upper Paleocene Nanafalia (NA) Formation, bear Creek (BC), Greggs Landing (GL), and Bells Landing (BL) marl members of the Tuscahoma Formation, and the lower Eocene Bashi (BA) and Hatchetigbee (HA) formations. Taxa with ranges extending into the lower–upper Paleocene Midway or lower–Middle Eocene Claiborne groups are checked under M and C respectively. New species illustrated in this chapter and not found in the published record are accompanied by their figure numbers.

16

Late Paleocene and Early Eocene Nonmarine Molluscan Faunal Change in the Bighorn Basin, Northwestern Wyoming and South-Central Montana

JOSEPH H. HARTMAN
BARRY ROTH

ABSTRACT
The nonmarine molluscan fauna in the Western Interior of the United States underwent large-scale changes during the time represented by the Laramide Orogeny. This period began with the loss of many bivalve and gastropod taxa near the Cretaceous/Paleogene Period boundary, followed by an increase in mesogastropod diversity in the late Torrejonian and Tiffanian North American Land Mammal Ages (NALMA), then a reduction in such diversity into the Clarkforkian NALMA, which is associated with a dramatic increase in the representation of terrestrial prosobranch and pulmonate families prior to the beginning of the Wasatchian NALMA. The latter portion of this record is well documented in the northern portion of the Bighorn Basin in Wyoming and Montana. The Clarkforkian/Wasatchian NALMA boundary, equated by some with the Paleocene/Eocene Epoch/Series boundary on the basis of the pattern of mammalian evolution and other evidence in this area, is delimited by aquatic and terrestrial mollusks, but the faunal change is less dramatic in the Bighorn Basin than changes associated with the increase in terrestrial taxa during the middle Clarkforkian NALMA. However, certain nonmarine molluscan taxa appear to be good indicators of an early Wasatchian Age, whether they occur in the San Juan Basin of New Mexico or on the Wasatch Plateau of Utah. The present study shows that molluscan taxa appear capable of indicating Clarkforkian biochrons in the absence of mammalian fossils in the Powder River Basin of southeastern Montana and northeastern Wyoming. The approximate stratigraphic horizon corresponding to the boundary between the lower and middle parts of Clarkforkian NALMA is near the level of the Anderson coal bed, while Wasatchian Age strata occur just above the Arvada coal bed in the main body of the "Wasatch" Formation. The boundary is documented by both the loss and introduction of different species of mesogastropods and aquatic pulmonates, as well as a few species of unionid and pisidiid bivalves. The Clarkforkian/Wasatchian NALMA boundary in the Willwood Formation of the Clark's Fork Basin can be determined on the basis of these taxa and also by a well-differentiated and diverse terrestrial snail fauna.

The mammalian-based Clarkforkian/Wasatchian North American Land Mammal Age (NALMA) boundary has been interpreted to represent the Paleocene/Eocene Series boundary in North America (Gingerich 1989a; Clyde et al. 1994). The beginning of the Wasatchian NALMA can be shown, generally, to be well delimited by the record of nonmarine mollusks in the Western Interior of the United States. Recognition of the Clarkforkian/Wasatchian NALMA boundary has not been documented beyond generalized biochronologic interpretations, differentiating between distinctive Paleocene and Eocene faunas (Hartman 1992, 1993a, 1993b). Without question, major changes in nonmarine molluscan faunal composition occurred during the Paleogene Period. But precise interpretation of the timing of faunal events in well-controlled stratigraphic settings has been lacking. In particular, there has not been the systematic and biochronologic foundation to establish, in detail, the molluscan species assemblages through the Clarkforkian biochrons. This chapter provides a basis upon which to begin to understand the nonmarine molluscan composition of latest Tiffanian, Clarkforkian, and early Wasatchian biochrons through interpretation of taxa and their stratigraphic occurrence in the northern portion of the Bighorn Basin, including the Clark's Fork Basin, Wyoming, and the Red Lodge–Belfry area, Montana. The molluscan record in the type area of the Clarkforkian mammal stage is the logical area in which to establish the foundation to interpret the temporal occurrence of probable Clarkforkian mollusks elsewhere in western North America.

The delineation of the Clarkforkian/Wasatchian NALMA boundary in nonmarine sequences elsewhere in North America has been an important topic of discussion (Wing 1984; Gunnell et al. 1993), as reflected in the presentations given at "The Paleocene/Eocene Boundary" symposium held at the 1993 annual meeting of the Society of Vertebrate Paleontology (SVP 1993). Biostratigraphic studies on nonmarine mollusks, at present, cannot address the correlation of this boundary with the European record. Thus, although convincing arguments have been made that the Clarkforkian/Wasatchian NALMA and Paleocene/Eocene Epoch boundaries are equivalent, all figures and discussions will specifically refer to the record of nonmarine mollusks in terms of mammalian biochronology through the interval likely to contain the Paleocene/Eocene Epoch boundary.

Notes

The abbreviations for the NALMAs of the Paleocene and Eocene Series used in this chapter are, from oldest to youngest, Pu (Puercan NALMA), To (Torrejonian NALMA), Ti (Tiffanian NALMA), Cf (Clarkforkian NALMA), and Wa (Wasatchian NALMA). These ages are subdivided, by number, into several biochrons, as shown in subsequent illustrations. These figures are placed on a metric scale of Gingerich (personal communication 1994, 1995). The thin stratigraphic intervals that cannot be assigned to a mammalian biochron because of lack of mammalian data are shown as gray-shaded units.

All fossil localities mentioned in this chapter are referenced by a unique locality number (e.g., L0136) in a database system maintained by Hartman. The appended Locality Register presents a brief description of each locality mentioned in this chapter.

The identification of mollusks was divided along environmental lines. Aquatic mollusks were identified by Hartman and terrestrial mollusks by Roth. The identifications of the original authors are cited, except where noted, in the following review of the contributions of earlier workers. Systematic or nomenclatural revisions introduced in this report are noted in tables 16.1 and 16.2 and Comments on Taxa.

The institutional abbreviations for collections mentioned in this report are as follows: American Museum of Natural History Fossil Invertebrate collection (AMNH-FI), Academy of Natural Sciences of Philadelphia Malacology collection (ANSP-MA), Princeton University Invertebrate collection (PU-INV; see note below), University of Michigan Museum of Paleontology collections (UMMP), University of Michigan Sand Coulee (Bighorn Basin) vertebrate locality number (UM SC), U.S. Geological Survey Mesozoic (Mz) and Cenozoic (Cz) and unnumbered collections in Denver (USGS-D), University of Colorado Museum collections (UCM), and U.S. National Museum Invertebrate Paleontology collections (USNM-I). Most of the identifications were made of specimens collected by the field crews of John H. Hanley (USGS) and Philip D. Gingerich (UMMP). Other evaluated specimens, where available, were reported by Cockerell and Henderson (1912), Cockerell (1914a, 1914b, 1915a, 1915b), and Russell (1931b). With the loss of its paleontological museum, the invertebrate paleontology collections of Princeton University have been transferred to the Yale University Peabody Museum of Natural History.

General Geography and Stratigraphy

Geography

The Bighorn Basin is located primarily in Park, Big Horn, and Hot Springs Counties in northwestern Wyoming and in Carbon County in south-central Montana (figure 16.1). Although this chapter covers the entire basin in an evaluation of the historic record of molluscan collection and taxonomy, the emphasis is on the northern Bighorn Basin, sometimes referred to as the Clark's Fork Basin (figure 16.2), located just east of Yellowstone National Park. The biostratigraphic record of mollusks can be documented in detail in this area as a result of the extensive studies on the temporal and stratigraphic distribution of mammals conducted throughout most of this century.

Stratigraphy

Jepsen (1940) and Gingerich et al. (1980) reviewed the early nomenclatural history of the Paleocene strata in the Bighorn Basin and adjacent areas. Simpson (1937:21) argued that the use of the name "Fort Union"

Table 16.1
Nonmarine Molluscan Fauna of the latest Tiffanian (Ti5), Clarkforkian (Cf1–Cf3), and Wasatchian (Wa0–Wa3) NALMAs in the Bighorn Basin, Wyoming–Montana

Phylum Mollusca
Class Bivalvia
Subclass Heteroconchia
Order Unionoida
  Unionidae Fleming 1828
    Plesielliptio cf. *P. priscus* (Meek and Hayden 1856d)
    Plesielliptio silberlingi (Russell 1934b)
    Plesielliptio cf. *P. silberlingi*
    Plesielliptio wasatchensis (Cockerell 1914a)
    Plesielliptio sp. form 4
    "*Unio*" *didymictidis* Cockerell 1915b
    "*Unio*" *eomargaron* Cockerell 1915b (*nomen dubium*)
    "*Unio*" *sinopae* Cockerell 1915b
    "*Unio wasatchensis*" (Cockerell 1914a)*
    "*Unio*" sp. form 1
    "*Unio*" sp. form 2
    "*Unio*" sp. undet.
    "*Unio*" sp. indet.
Order Veneroida
  Pisidiidae Gray 1857
    Sphaerium cf. *S. recticardinale* Meek and Hayden 1860a
    Sphaerium sp. form 1
    Sphaerium sp. form 2
    Sphaerium? sp. form 3
    Sphaerium? sp. form 5
    Eupera? sp. form 4
    Eupera sp. form 7
    Eupera? sp. form 8
    Genus undet. sp. form 6
    Genus and sp. indet.
Class Gastropoda
Subclass Prosobranchia
  Helicinidae Latreille 1825
    Eohipptychia eohippina (Cockerell 1915)
    "*Helicina*" *evanstonensis* (White 1878)
    "*Helicina*" cf. *H. evanstonensis* (White 1878)
    cf. *Tozerpina* sp. undet.
    "proserpinid" sp. form A
    "proserpinid" sp. form B
    "proserpinid" sp. form C
    Helicinidae gen. and sp. undet.
  Hendersoniidae Baker 1925
    Grangerella phenacodorum (Cockerell 1914b)
    Grangerella sinclairi (Cockerell and Henderson 1912)
    Grangerella cf. *G. sinclairi* (Cockerell and Henderson 1912)
    Grangerella sp. form D
    Grangerella cf. *G.* sp. form D
    Grangerella sp. undet.
    Protoboysia complicata Cockerell 1914b
    Hendersoniidae gen. and sp. undet. (operculum)
  Valvatidae Gray 1840
    Valvata cf. *V. bicincta* Whiteaves 1885
    Valvata cv. *V. subumbilicata* (Meek and Hayden 1856)
    Valvata sp. 1
    Valvata sp. undet.
    Valvata sp. indet.
    Valvata? sp. undet.
  Viviparidae Gray 1847
    Viviparus formosus? Meek 1876
    Viviparus retusus (Meek and Hayden 1856)
    Viviparus aff. *V. retusus* (Meek and Hayden 1856)
    Viviparus wyomingensis Meek 1871a
    Viviparus sp. form H
    Viviparus cf. sp. form H
    Viviparus sp. undet.
    Viviparus sp. undet.
    Campeloma nebrascense nebrascense (Meek and Hayden 1860b)
  Bithyniidae Troschel 1857
    Reesidella nana (White 1886)
    Reesidella nana? (White 1886) (unassociated opercula)
  Hydrobiidae Troschel 1857
    Hydrobia anthonyi (Meek and Hayden 1856)
    Hydrobia sp. form 1
    Hydrobia sp. form 2
    Hydrobia aff. *H.* sp. form 2
    Hydrobia sp. undet.
    Hydrobia sp. undet.
    Clenchiella cf. *C. planospiralis* (Yen 1946)
  Pleuroceridae Fischer 1885
    Lioplacodes nebrascensis producta (White 1882)
    Lioplacodes nebrascensis nebrascensis (Meek and Hayden 1856)
    Lipplacodes nebrascensis subsp. indet. (Meek and Hayden 1856)
    Lioplacodes tenuicarinata (Meek and Hayden 1857)
    Lioplacodes sp. form 1
    Lioplacodes sp. form 2
    Lioplacodes sp. undet.
    Lioplacodes sp. undet.
    Elimia sp. form 1
    Elimia sp. form 2
    Elimia sp. form 3
    Elimia sp. form 4
    Elimia sp. indet.
  incertae sedis
    New Genus A *limneaformis* (Meek and Hayden 1856)
Subclass Pulmonata
Order Archaeopulmonata
  Ellobiidae Adams 1855
    Pleurolimnaea cf. *P. tenuicosta* (Meek and Hayden 1856)
    Pleurolimnaea sp. indet.
    Pleurolimnaea? sp. indet.
Order Basommatophora
  Acroloxidae Thiele 1931
    Acroloxus radiatulus Whiteaves 1885
    Acroloxus cf. *A. radiatulus* Whiteaves 1885
    Acroloxus minutus (Meek and Hayden 1856)
    Acroloxus sp. indet.
    Acroloxus? sp. indet.
  Physidae Fiztinger 1833
    Physa bridgerensis Meek 1873

(continued)

Table 16.1 *(continued)*

Physa sp. form 1
Physa sp. form 2
Physa sp. form 3
Physa sp. form 4
Physa sp. form 5
Physa sp. form 6
Physa sp. form 7
Physa sp. undet.
Physa sp. indet
Planorbidae Rafinesque 1815
Biomphalaria storchi (Russell 1931b)
Planorbella sp. form 1
Planorbinae gen. and sp. indet.
Planorbidae gen. and sp. undet.
Planorbidae? gen. and sp. indet.
Order Stylommatophora Schmidt 1855
Stylommatophora gen. and sp. undet.
Pupillidae Turton 1831
cf. *Albertanella* sp. undet.
Pupillidae gen. and sp. undet.
?Pupillidae gen. and sp. undet.
Haplotrematidae Baker 1925
Ancotrema sp. form E
Haplotrema sp. form F
Haplotrema sp. form G
Haplotrema cf. *H.* sp. form G
Urocoptidae Pilsbry and Vanatta 1898
Holospira sp. undet.
Bulimulidae Tryon 1867
cf. *Oreoconus* sp. indet.
Helicodiscidae Baker 1927
Helicodiscus? sp. form H
Discidae Thiele 1931
Discus ralstonensis (Cockerell 1914a)
Discus cf. *D. ralstonensis* (Cockerell 1914a)
Discus sp. form I
?Discidae gen. and sp. undet.
Oreohelicidae Pilsbry 1939
Radiocentrum grangeri (Cockerell and Henderson 1912)
Radiocentrum cf. *R. grangeri* (Cockerell and Henderson 1912)
Radiocentrum sp. form J
New Genus S *spatiosa* (Meek and Hayden 1862)
New Genus S cf. *S. spatiosa* (Meek and Hayden 1862)
New Genus S sp. undet.
cf. New Genus S sp. undet. 1
cf. New Genus S sp. undet. 2
"*Gastrodonta*" *coryphodontis* Cockerell 1914a (not *Gastrodonta*)
"*Gastrodonta*" cf. "*G.*" *coryphodontis* Cockerell 1914a (not *Gastrodonta*)
"*Eucalodium*" *eophilum* Cockerell 1915a (not *Eucalodium*)
"*Eucalodium*" cf. "*E.*" *eophilum* Cockerell 1915a (not *Eucalodium*)
New Genus L sp. form K
?Zonitidae Möerch 1864
?Zonitidae gen. and sp. undet.
Polygyridae Pilsbry 1895
Neohelix? sp. form M
Thysanophoridae Pilsbry 1926
Microphysula cf. *M. ingersolli* (Bland 1875)
Microphysula oxyaenae (Cockerell 1914b)
Microphysula cf. *M. oxyaenae* (Cockerell 1914b)
Microphysula sinoparum (Cockerell 1914b)
Megomphicidae Baker 1930
Polygyrella cf. *P. parvula* (Whiteaves 1885)
?Bradybaenidae Pilsbry 1939
New Genus P *jepseni* (Russell 1931)
Helminthoglyptidae Pilsbry 1939
Glypterpes veternior (Cockerell 1915a)[#]
Glypterpes cf. *G. veternior* (Cockerell 1915a)[#]
Noyo? sp. form N
Noyo? cf. *N.*? sp. form N

*Paratype specimen not comparable to holotype or other morphotype in present collections.
[#]Not a subspecies of *Polygyra veternior* (Cockerell) as reported by Russell (1931b).

should be limited to "group" status and include all of the Paleocene strata in the northwestern Great Plains and that "for more precise work it will, in any case, be necessary to define and use more local names for particular formations and members." Following this logic, Jepsen (1940) named the Polecat Bench Formation, which included all of the Paleocene strata in the vicinity of Polecat Bench north of Powell, Park County, Wyoming, to avoid the implications that would result from correlation of these strata with the type Fort Union in the vicinity of the Missouri–Yellowstone River confluence. In contrast, Stow (1938, 1946) correlated the previously named Tullock, Lebo, and Tongue River formations with rocks of the Bighorn Basin on the basis of mineralogy. Stow (1946) also synonymized the formation names proposed by Simpson (1937) in the Crazy Mountain field (Bear = Tullock, Melville = Tongue River), thereby eliminating the local names established by vertebrate paleontologists in the Crazy Mountains and Bighorn basins (see Hartman and Krause 1993). The name "Polecat Bench Formation" was subsequently widely used by paleontologists and others (e.g., Gingerich et al. 1980), but such usage, although recognized informally by the USGS (Keroher 1966; Swanson et al. 1981; but not in GNULEX 1992) was not subsequently followed. Recent studies have abandoned the use of the Polecat Bench Formation for the more widely used Fort Union Formation (Gingerich 1989a). Hickey

Table 16.2
Holotype Specimens Recorded in the Bighorn Basin

Taxon	Age	Specimen	Number	Locality	Author
complicata, Protoboysia	Wa2–Wa3	AMNH-FI	22255	L3822	Cockerell (1914b)
coryphodontis, Gastrodonata	Wa2–Wa3	AMNH-FI	22361	L3822	Cockerell (1914a)
didymictidis, Unio	Wa3–Wa4	AMNH-FI	24939	L4283	Cockerell (1915b)
eomargaron, Unio	Cf3	AMNH-FI	24938	L3952a	Cockerell (1915b)
eohippina, Pleurodonte	Wa1	AMNH-FI	22357	L3952c	Cockerell (1915a)
eophilum, Eucalodium	Cf3	AMNH-FI	22257	L3820	Cockerell (1915a)
grangeri, Oreohelix	Cf3–Wa0	AMNH-FI	Unknown	L4282a	Cockerell and Henderson (1912)
jepseni, Viviparus	Wa2–Wa3	PU-INV	20034	L2316	Russell (1931b)
megarche, Oreohelix (1)	Wa3–Wa4	AMNH-FI	Unknown	L4284	Cockerell and Henderson (1912)
megastoma, Grangerella (2)	Wa1	AMNH-FI	22256	L3952c	Cockerell (1915a)
oxyaenae, Microphysula	Wa2–Wa3	AMNH-FI	22986	L3822	Cockerell (1914b)
phenacodorum, Boysia	Wa2–Wa3	AMNH-FI	22389	L3822	Cockerell (1914b)
ralstonensis, Pyramidula	Wa1	AMNH-FI	22356	L3821	Cockerell (1914a)
reticulata, nebrascensis, Campeloma (3)	Ti3–Ti4	PU-INV	20009	L0134	Russell (1931b)
sinclairi, evanstonensis, Gastrodonata?	Cf3–Wa0	AMNH-FI	Unknown	L4110	Cockerell and Henderson (1912)
sinopae, Unio	Wa3–Wa4	AMNH-FI	24940	L4283	Cockerell (1915b)
sinoparum, Vitrea	Wa2–Wa3	AMNH-FI	22987	L3822	Cockerell (1914b)
storchi, Planorbis	Wa3–Wa4	PU-INV	20042	L2311	Russell (1931b)
veternior, veterna, Helix	Wa2–Wa3	AMNH-FI	22359	L3822	Cockerell (1915a)
wasatchensis, Unio	Cf–Wa	AMNH-FI	22363	L4282b	Cockerell (1914a)

*Taxon name follows that of original author. See table 16.4 for revised identication relevant to this chapter.
(1) *Oreohelix megarche* synonymized with New Genus S *spatiosa*.
(2) *Grangerella megastoma* tentatively synonymized with *Grangerella sinclairi*.
(3) *Campeloma nebrascensis reticulata* synonymized with *Campeloma nebrascense nebrascense*.

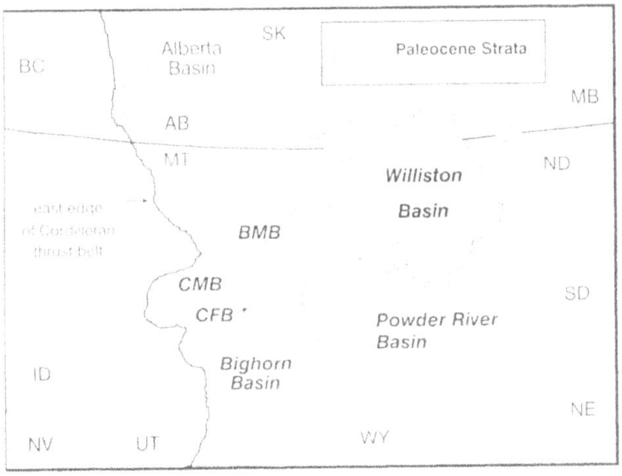

Figure 16.1 Basins of the Northern Great Plains.
Rocks of Paleocene Age delimit basins discussed in this chapter (from state and provincial geologic maps; see Bayer 1983). Abbreviations: Bull Mountain Basin (BMB), Clark's Fork Basin (CFB), Crazy Mountains Basin (CMB), and state and provincial postal codes.

(1980) mapped the Fort Union Formation by environments and designated informal environmental members (e.g., fluvial member). We evaluated all of the Montana localities in terms of Hickey's members, as well as the mammalian stages as portrayed in map view by Gingerich (1989b), and found that the biostratigraphy of nonmarine molluscan faunas was consistent with their sedimentological and temporal presentations.

The collections of nonmarine mollusks made up through the period of Jepsen's later efforts are, for the most part, stratigraphically uncontrolled. The absence of stratigraphic control applies most specifically to nearly all the collections made in the northern Bighorn Basin in Montana. Only those that can now be tied to a specific horizon, such as a coal bed in the coal-mining area south of the Red Lodge area, can, in some cases, be placed in an overall litho- or chronostratigraphic framework. This includes the majority of the occurrences of Zones Ti5 and Cf1 aquatic mollusks. Thus the biostratigraphic framework for the mollusks collected prior to the studies

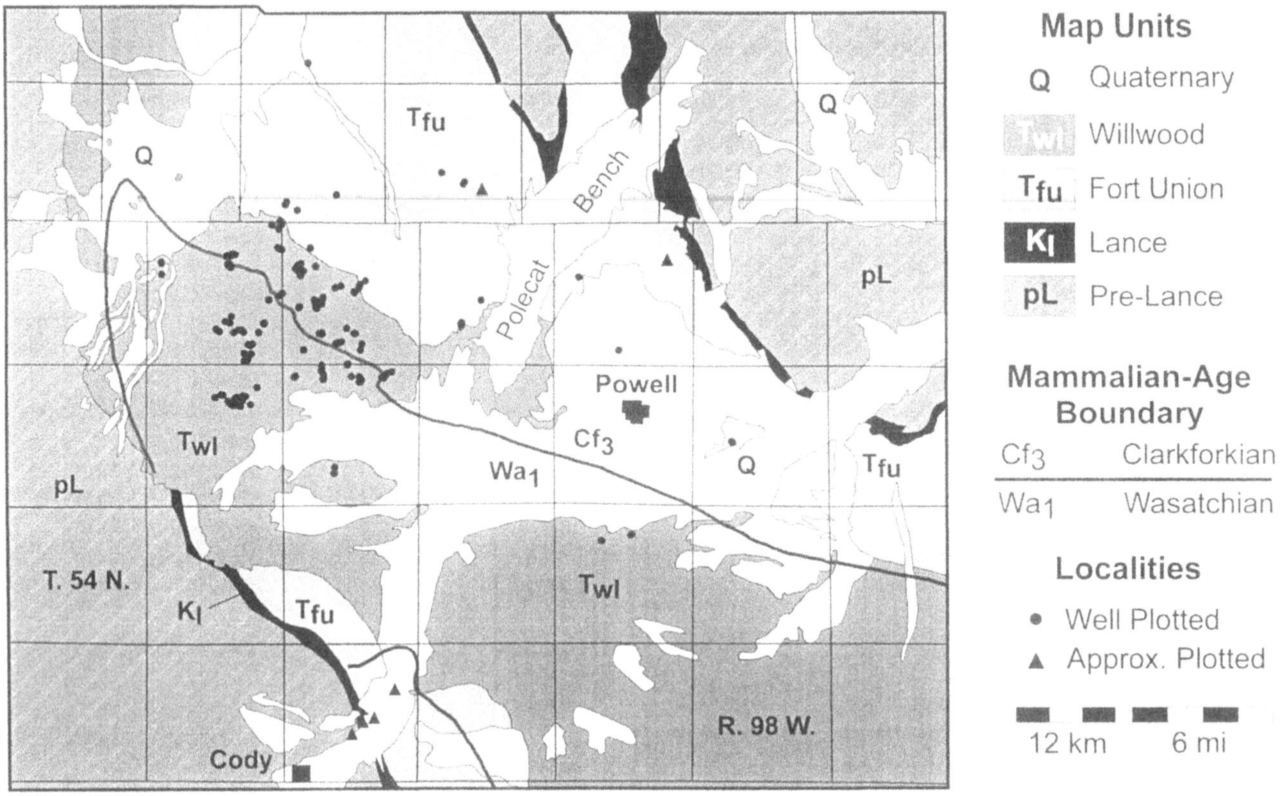

Figure 16.2 Geologic map of the Clark's Fork Basin, Wyoming.
Compiled and modified after the geology of Love and Christiansen (1985) and the Cf3/Wa1 isochron of Gingerich (1989a). Nonmarine molluscan localities are shown as filled dots and triangles. The currently available state geologic map of Montana is being revised, substantially modifying the earlier map by Ross et al. (1955), and thus a map of the Montana portion of the northern Bighorn Basin has not been included. The map portrays most of the 240 molluscan localities known from the Clark's Fork Basin. The dots on this map may represent more than one locality, as many localities are closely juxtaposed both geographically and stratigraphically. Triangles represent less geographically well-fixed localities. Another 100 or so localities occur throughout the rest of the Bighorn Basin to the south and to the north into Montana. Superimposed on this map is the boundary line of the Clarkforkian and Wasatchian deposits, as interpreted by Gingerich (1989a).

of Gingerich and others can be presented only as intervals of likely occurrence (see figure 16.3, table 16.3).

The Willwood Formation contains most of the section and includes most of the localities discussed in this report. The Willwood Formation, which generally can be interpreted as the "Wasatch" Formation of earlier literature, ranges from late Paleocene to early middle Eocene in age. The Willwood Formation was named by Van Houten (1944) for red-banded strata occurring between the Shoshone River and the northeastern slope of McCullough Peaks, Park County, Wyoming. The Willwood Formation conformably overlies the Fort Union Formation and underlies the Tatman Formation. The underlying formational contact with the Fort Union Formation is not isochronous; the age of the contact becomes older to the north in the Clarks Fork Basin (Gingerich, personal communication 1994).

Nonmarine Molluscan Research in the Bighorn Basin

As with many areas in the Western Interior of the United States, collection of nonmarine mollusks in the Bighorn Basin commenced a little over 100 years ago as part of early coal and land use studies by the U.S. Geological Survey. Although nonmarine mollusks were widely discussed in the establishment of a framework for Western Interior chronostratigraphy, little was published of the extensive data upon which these conclusions were based, and the extent of the fossils collected went unappreciated except for a few USGS paleontologists. This section briefly documents earlier investigations and places their findings into a current geographic and stratigraphic context. For the convenience of the reader, taxon references to the geography and stratigra-

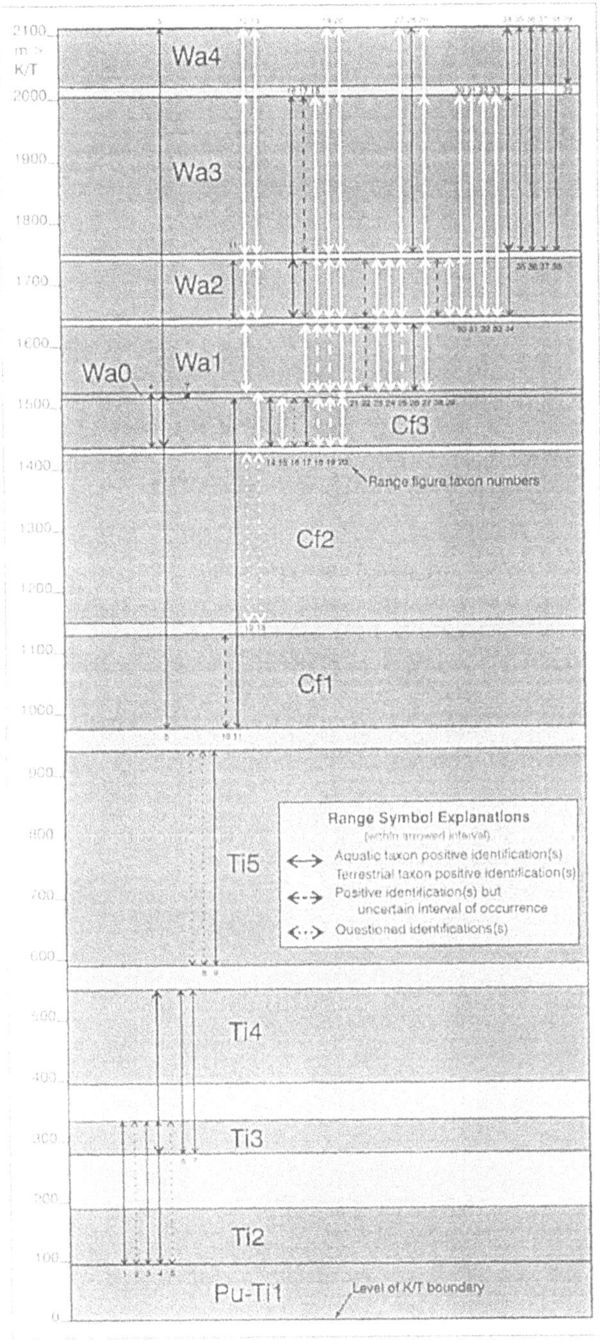

Figure 16.3 Bighorn Basin nonmarine molluscan biostratigraphy as of 1931.

The taxa shown in this figure represent nonmarine molluscan identifications made as of 1931. The bulk of identifications were made by Russell (1931b). Other occurrences were reported by Cockerell and Henderson (1912) and Cockerell (1914a, 1914b; 1915a, 1915b). Each taxon locality is blocked out to its interpreted mammalian age. This range is noted by arrows to the bottoms and tops of a biochron(s). The estimate of the age of the locality was greatly aided by the interpretation of Gingerich (personal communication 1994, 1995) and others (see Acknowledgments). Overlapping estimated locality ranges are denoted by varying sizes of the arrowheads.

phy of localities employ the nomenclature of the original author.

Early USGS Studies

The first mollusks recovered from the Bighorn Basin may have been of Clarkforkian Age. They were collected in 1891 by W. H. Weed from the roof of the number IV coal vein at a Red Lodge Mine in Carbon County, Montana (Weed 1892; L942). C. A. White (in Weed 1892) interpreted the unionids from the faunule as Cretaceous in age. Associated plants were considered by F. H. Knowlton (in Weed 1892) to be of Fort Union types and, thus, distinct from the Laramie flora and comparable, in a broad sense, to the molluscan fauna reported by F. B. Meek and F. V. Hayden from the upper Missouri River of the Williston Basin, North Dakota–Montana (Hartman and Kihm 1992, 1995).

Twentieth-century USGS studies in Carbon County, Montana, resulted in late Paleocene molluscan collections of greater geographic distribution and generally better documentation. Much of the analysis of these collections went unpublished. In 1906, N. H. Darton (1907) collected mollusks from the "lower barren measures" and overlying coal-bearing strata along Wolf Creek (L0532) and near and at Bear Creek (L0533, L0534). Although the assemblages were not considered age-diagnostic at the time, Stanton (in Darton 1907) said the fossils were indicative of the Fort Union Formation, rather than the older Laramie Formation. The latter fossil occurrence (L0534) was the roof of the number 3 coal bed at the Bear Creek mines and will be discussed below. In 1907, the E. G. Woodruff (1907, 1909) crew also discovered mollusks in Paleocene sections of the Fort Union Formation in the Red Lodge coal field along Rock and Bear Creeks. Unpublished collections from the Bear Creek section included Clarkforkian fossils from immediately above the number 3 coal (L4803) and from the Rock Creek section above the number 1 coal (see Woodruff 1909; L5471, p. 96). Stanton and C. A. Fisher, in association with Woodruff, collected mollusks from above the number 1 coal bed at the most western of the Bear Creek mines (L0943). Note that the coal beds were numbered from top (number 1) down. Other unpublished Woodruff Carbon County localities, from south of the Red Lodge coal

Figure 16.3 (continued) The key to the taxon range numbers in this figure is given in table 16.3. Unless otherwise noted, the identifications are as given by Russell (1931b, and unpublished PU-INV labels). A questioned locality number indicates a questioned identification at that locality. A questioned mammalian biochron indicates a suggested age for the locality. See the Locality Register for location data.

Table 16.3
*Bighorn Basin Nonmarine Molluscan Biostratigraphy of 1931

1) *Lioplax nebrascensis:* Ti2–Ti3, L0136
2) *Goniobasis tenucarinata*?: Ti2–Ti3, L0136
3) ?*Goniobasis* sp. undet.: Ti2–Ti3, L0136
4) *Viviparus paludinaeformis:* Ti2–Ti3, L0136; Ti3–Ti4, L0134; Cf3–Wa0, L4110 (Cockerell and Henderson 1912)
5) *Unio wasatchensis:* Ti2–Ti3, L0136?; Cf–Wa, L4282b; Cf–Wa, L4283 (Cockerell 1915b); Cf3–Wa0, L4282a
6) *Campeloma nebrascensis reticulata:* Ti3–Ti4, L0134
7) *Viviparus retusus:* Ti3–Ti4, L0134; Ti5, L0135?; Wa0?, L3952b
8) *Campeloma limnaeiformis*?: Ti5, L0135
9) *Unio* spp. undet. ("suggest *danae*"): Ti5, L0135
10) *Unio priscus*?" Cf1?, L0133
11) *Unio* spp. undet.: Cf, L2309; Wa2, L2313d; Wa2?, L2312a
12) *Oreohelix megarche:* Cf2?, L2310b?; Wa1, L3821; Wa1?, L2312b–c; Wa2, L2313a–b, e.g.; Wa2?, L2312a, d, e; L2314, L2315; Wa2–Wa3, L2316, L3822 (Cockerell 1914b); Wa3–Wa4, L2311, L4284
13) *Oreohelix grangeri:* Cf2?, L2310a?; Cf3–Wa0, L4282a; Wa2, L2313a, d, e, g, h; Wa2?, L2314; Wa3–Wa4, L2311
14) *Unio emargaron:* Cf3, L3952a
15) "*Eucalodium*" *eophilum:* Cf3, L3820
16) *Viviparus leidyi:* Cf3, L3820?; Wa2, L2313a?, g?; Wa2?, L2312e?–f?; Wa2–Wa3 (Cockerell 1915a)
17) *Goniobasis carterii:* Cf3, L3820; Wa2, L2313c, f, h; Wa2?, L2313d; Wa2–Wa3, L3822 (Cockerell 1914a); Wa3?, L2318
18) *Gastrodonta coryphodontis:* Cf3?, L2319; Wa1?, L2312b; Wa2, L2312d; Wa2?, L2312a, L2315; Wa2–Wa3, L3822
19) *Polygyra veternior:* Cf3?, L2319; Wa1?, L2312b; Wa2, L2313a, e; Wa2?, L2312a, L2314, L2315; Wa2–Wa3, L3822; Wa3–Wa4, L2311
20) *Grangerella sinclairi* (includes *G. phenacodorum* fide Russell): Cf3, L3820; Cf3–Wa0, L4110); Wa1, L3952c, Cockerell, 1915a; Wa1?, L2312b; Wa2, L2312d; Wa2?, L2312a, L2314, L2315 (Cockerell 1914a), L2317; Wa2–Wa3, L3822, L4285; Wa3–Wa4, L2311
21) *Polygyra eohippina:* Wa1, L3952c
22) *Physa* sp.: Wa1?, L2312b–e; Wa2?, L2314, L2315
23) *Viviparus wyomingensis:* Wa1, L3821; Wa2, L2313d, f, h, L2315
24) *Grangerella megastoma:* Wa1, L3952c; Wa1?, L2312b; Wa2, L2313d, L2315
25) *Helicina evanstonensis:* Wa1?, L2312b; Wa2, L2313d; Wa2?, L2312d, L2314; Wa3–Wa4, L2311
26) *Physa bridgerensis:* Wa1, L3852c (Cockerell 1915a); Wa3–Wa4, L2311
27) *Gonyodiscus ralstonensis:* Wa1, L3821; Wa1?, L2312b–c; Wa2, L2313d; Wa2?, L2312a, d, L2314; Wa2–Wa3, L3822 (Cockerell 1914b); Wa3–Wa4, L2311
28) *Viviparus* sp. undet.: Wa2?, L2317
29) *Oreohelix* n. sp.?: Wa2, L2313g
30) *Protoboysia complicata:* Wa2, L2313d; Wa2–Wa3, L3822
31) *Oreohelix*? cf. *O. planoconvexus:* Wa2–Wa3, L2316
32) *Microphysula oxyaenae:* Wa2–Wa3, L3822
33) *Microphysula sinoparum:* Wa2–Wa3, L3822
34) *Viviparus jepseni:* Wa2–Wa3, L2316; Wa3–Wa4, L2311
35) *Planorbis storchi:* Wa3–Wa4, L2311
36) *Physa pleromatis:* Wa3–W4, L2311
37) *Unio didymictidus:* Wa3–Wa4, L4283
38) *Unio sinopae:* Wa3–Wa4, L4283; Wa3–Wa4, L2311?
39) *Unio* [*Margaritana*] *clinopisthus:* Wa4, L2320

*Data keyed to taxon ranges presented in figure 16.3.

The identifications presented in this table follow those given by Russell (1931b), but include all of the nonmarine molluscan locality information that was available up through 1931. The interpretation of land-mammal biochrons is derived from Gingerich (1989) and Gingerich (personal communication 1981, 1994, 1995).

field, include L0940 and L1421–L1423. In the same year, other unpublished mollusks were collected by USGS geologists E. E. Powers and E. F. Schramm from near the Montana–Wyoming line (L0944–L0946). These localities have been interpreted as from the Tongue River Member of the Fort Union Formation (Hartman 1984, after Stow 1938), and a reevaluation of the faunules suggests a late Paleocene (late Tiffanian and early Clarkforkian) age.

The only published Woodruff molluscan locality was reported by Stanton (1909; L0483) as also yielding dinosaur bones. Stanton identified the molluscan faunule as a mix of Lancian and Paleocene molluscan taxa, which he regarded as closely related to the molluscan faunas of the Hell Creek and Lance formations of the Williston and Powder River basins of Montana and Wyoming, respectively. The unionid species identified by Stanton are highly distinctive and restricted to the uppermost Cretaceous Series. Unfortunately, a problem with this interpretation is that there are no exposures of the Hell Creek Formation near Red Lodge. Stanton (1909:269, footnote 41) made an invalid reference of this collection to the "Laramie," which was not reported by Woodruff (1909). The fossils, which are not now available for study, may be from Woodruff's Cretaceous collections made in the Lance Formation exposed along the Shoshone River near Cody, in Park County, Wyoming (see Hewett 1914 and Hartman 1984; Localities L1782 and L1785).

In Wyoming, Woodruff collected nonmarine mollusks from Lancian and Paleocene localities in Park County. He recovered fossils from the Fort Union Formation east of Cody (L1784) and from the Willwood Formation in either sec. 12 or 13, T. 57 N., R. 103 W. (L0941), from strata of Clarkforkian Age. Stanton identified specimens from the former locality as representative of a "Fort Union" age (Hewett 1914). The taxa, as recognized by Stanton, were not age-specific. Also in 1907, Stanton and others collected Clarkforkian mollusks from Fort Union strata near the old Corbett Bridge, northeast of Cody, along the Shoshone River (L1425; Hewett 1914).

Other early Wyoming field studies included the collection of Wasatchian mollusks. In 1905, Fisher (1906) collected poorly preserved freshwater mollusks from the gray sandstones in the Wasatch (Tatman of later usage) Formation at Squaw Buttes in southern Park County (L3383). Stanton tentatively interpreted this record as of Eocene age. A Wasatchian to Bridgerian Age was confirmed for this and other mollusks (L4131, L4311) reported by Van Houten (1944) on the basis of identifications made by L. S. Russell. Subsequent study by Bown (1982) showed these mollusks to be of late Wasatchian or earliest Bridgerian Age.

Between 1912 and 1925, USGS geologists collected nonmarine mollusks during the course of a study on the anticlines in the Bighorn Basin (Hewett and Lupton 1917). In 1916, on the west flank of Elk Basin Anticline, C. J. Hares, C. A. Bonine, and W. B. Emery collected shells from the Fort Union Formation in Carbon County, Montana (L947, L1470, L1500) and Big Horn County, Wyoming (L3519, L3982). None of the specimens from these collections have been studied except for the initial identifications by T. W. Stanton.

Contributions by Vertebrate Paleontologists

Beginning in 1910, vertebrate paleontologists began to play an increasingly important role in the collection of nonmarine mollusks in the Bighorn Basin. The invertebrate collections of W. J. Sinclair, Walter Granger, William Stein, G. L. Jepsen, and T. C. von Storch largely follow the history of the acquisition of vertebrate fossils. Thus the history of early Cenozoic vertebrate paleontology in the Bighorn Basin, usefully presented by P. D. Gingerich (1980), is in part a history of invertebrate collection. In earlier studies, molluscan research progressed from the naming of new taxa (e.g., Cockerell's papers) to the drawing of tentative conclusions on the biochronologic utility of mollusks in distinguishing assemblages within and between portions of the Fort Union and Wasatch formations (Russell 1931b). The following sections update and revise, in chronological order, the basic data concerning molluscan localities and faunules presented in early papers to derive our present understanding of the faunal composition of mammalian biochrons in the Bighorn Basin. The faunules are referred to mammalian ages documented by Gingerich in numerous publications. Papers of specific value to this study include Gingerich (1980, 1989a, 1989b), Gingerich et al. (1980), and Gunnell et al. (1993). In addition, Gingerich (personal communication 1981, 1994, 1995) has provided information on the approximate equivalence of earlier reported mammalian localities and UM SC localities and their current biochronologic interpretation. Bighorn Basin viviparid localities previously evaluated by Hartman (1984) are also updated.

Sinclair and Granger

Initial collections by vertebrate paleontologists were reported by Cockerell and Henderson (1912) and Cockerell (1914a, 1914b, 1915a, 1915b) and served, in part, as the foundation for the summary study and new species reported by Russell (1931b, 1934; see figure 16.3 and tables 16.2, 16.3). Cockerell and Henderson (1912) reported on the mollusks collected by Sinclair and Granger (1911, 1912) during their PU-AMNH expeditions of 1910 and 1911. From their collections in 1910, Cockerell and Henderson (1912:230, 231) named *Oreohelix megarche* (plate XXII, figures 4–6) and *Oreohelix grangeri* (plate XXI, figures 5–9). *Oreohelix megarche* was reported from the "Wasatch beds . . . three miles [4.8 km] southeast of Otto (L4284)," which can be reinterpreted to be from the middle Wasatchian NALMA (Zones Wa3–Wa4) of the Willwood Formation in Big Horn County. *Oreohelix grangeri*, along with *Planorbis militaris*, was referred to the "Ralston Beds" without location (L4282a). Russell (1931b:23) stated that the taxon was "said to be from the Ralston (= Clark Fork) beds," but suggested that this record may have been in error on the basis of the rather restricted occurrence of *O. grangeri* to the Gray Bull beds. Present studies show, however, that *Radiocentrum grangeri* is restricted to the middle and upper Clarkforkian Cf2–Cf3 Zones. Without additional information, the location of the type locality is problematic. Also in 1910, "*Goniobasis tenera*" was first reported in the Bighorn Basin by Cockerell and Henderson (1912) from the Wasatch beds south of St. Joe (L4286). This occurrence represents the first record of a sculptured *Elimia* from the Bighorn Basin. Although this material was not reexamined, Russell (1931b) later identified material of similar morphology as *Goniobasis carterii*. The locality can be interpreted as in beds of middle Wasatchian (Biochron Wa4) Age in the Willwood Formation of Big Horn County. Unfortunately, all of the holotypes and some of the hypotypes of the taxa named and mentioned by Cockerell and Henderson (1912) are lost (see table 16.2).

In 1911, Sinclair's collections included specimens identified by Cockerell and Henderson (1912) as *Vivipara paludinaeformis*, *Physa pleromatis*, and *Gastrodonta*(?) *evanstonensis* var. *sinclairi*, v. nov (Cockerell and Henderson 1912:plate XXI, figures 1, 2). These fossils were reported as from the "?Wasatch (Ralston beds)," "about three miles [4.8 km] north of Ralston" (L4110). Cockerell (1915a) tentatively reassigned the latter taxon to the new genus *Grangerella*. Russell (1931b) confirmed this reclassification and introduced the new family Grangerellidae to contain the genera *Grangerella* and *Protoboysia* Cockerell 1914 (see Comments on Taxa). Cockerell and Russell also reported the occurrence of both taxa as from the "Sand Coulee beds," although Russell (1931b) questioned the stratal assignment at the type locality (L4110) of *G. sinclairi*. The Willwood Formation strata about this location are Clarkforkian and earliest Wasatchian in age.

Cockerell and Henderson (1912) also identified "*Vivipara* of the type of *wyomingensis*" and a "*Planorbis* resembling *florissantensis*," along with ganoid fish scales from a Tatman Mountain locality (L4130), which was considered "possibly White River formation." This record is from the Tatman Formation. A similar assemblage was reported from horizons of the Tatman Formation at Tatman Mountain in Park and Big Horn Counties by Rohrer (1964b), Rohrer and Gazin (1965), and Rohrer and Smith (1969). The fossils were identified by D. W. Taylor as Unionidae? gen. and sp. indet., *Sphaerium* sp., *Bellamya* sp. cf. *B. paludinaeformis*, *Bellamya* sp., *Pleurocera* sp. aff. *P. tenera*, *Pleurocera* sp., *Physa*? sp., and *Planorbina pseudoammonius* (L2120–L2125, L2127–L2129), and, as noted above, can be interpreted as being near the transition between the Wasatchian and Bridgerian NALMAs (Bown 1982; see Taylor 1975 for his own revised nomenclature; *Bellamya* to *Viviparus* as per Hartman 1984). Mollusks were also identified by Taylor in the Willwood Formation in the Tatman Mountain area (Rohrer and Gazin 1965; L2117–L2119, L4309). All of these localities are late Wasatchian in age.

In 1912, Granger and Stein discovered important molluscan localities from horizons in the Willwood Formation at various sites in the Clark's Fork Basin of Park County, Wyoming. Because of the significance of their constituent taxa and confusion associated with the interpretation of their location data, these localities will be treated individually. Cockerell (1914a, 1915a:115) first reported three localities relative to the mouth of Pat O'Hara Creek: "three miles (4.8 km) east [L3820], three miles southeast [L3821], and five miles (8 km) southeast [L3822]." Another locality in this general group was reported as from the "head of Big Sand Coulee" (L3952). All four localities can be referred to the Willwood Formation. At "three miles east" (L3820), Cockerell (1915a) named and identified the terrestrial species *Eucalodium eophilum* (Cockerell 1915a:figure 2) and *Grangerella sinclairi* and identified the mesogastropod *Viviparus leidyi* and noted a variety ("A") of the aquatic pulmonate *Physa bridgerensis*. This locality was assigned to the Sand Coulee beds (Cockerell 1915a). Gingerich (1980) reported that the vertebrates from this locality were collected from UM SC-90 (Hartman 1984), which is late Clarkforkian (Biochron Cf3) in age.

At "three miles southeast" (L3821), Cockerell (1914a: plate VIII, figures 1, 2) named *Pyramidula ralstonensis* and identified (Cockerell 1914a, 1915a) *Goniobasis carteri*, *Oreohelix megarche*, *Viviparus leidyi*, a form of *Viviparus wyomingensis*, and a form suggestive of *Helix* or *Macrochlamys nacimientensis* (not found in AMNH collections). This locality was originally assigned to the Ralston and Red Badland beds (Cockerell 1914a) and later to the Sand Coulee beds (Cockerell 1915a). Gingerich (1980) reported that the vertebrates collected from this location are "almost certainly UM locality SC-40" (Hartman 1984:895). The molluscan faunule, which may be derived from various sites in this area, is early Wasatchian (Biochron Wa1) in age.

At "five miles southeast of the mouth of Pat O'Hara Creek" (L3822), Cockerell (1914a, 1914b, 1915a) named a diverse assemblage of terrestrial snails, including *Gastrodonta coryphodontis* (1914a:plate IX, figures 4–6), *Protoboysia complicata* (1914b:figure 1), *Boysia phenacodorum* (1914b:figure 3), *Vitrea sinoparum* (1914b:figure 4), *Thysanophora oxyaenae* (1914b:figure 5), and *Helix veterna veternior* (1915a:figure 3[a–c]), along with identifying *Pyramidula ralstonensis*, *Oreohelix megarche*, and *Grangerella sinclairi*. The faunule included one mesogastropod, a large specimen of *Viviparus leidyi* (Cockerell 1915a). This locality was reported as above the red-banded strata in the Sand Coulee beds. Gingerich (1980) reported that the vertebrates collected from this locality are probably in the vicinity of UM localities SC-54 and SC-87 (Hartman 1984) and are thus early or middle Wasatchian (Biochron Wa2 or Wa3) in age.

The taxa from the "head of Big Sand Coulee" (L3952) were described by Cockerell (1915a, 1915b) as *Unio eomargaron* (1915b:figures 2a, b), *Pleurodonte eohippina* (1915a:figure 4[a–c]), and *Grangerella megastoma* (1915a:figures 1[a–c]). The assemblage, which included *Viviparus retusus*, *Physa bridgerensis*, variety A, and *Grangerella*? *sinclairi* (Cockerell 1915a), was variously referred to as from the Ralston, Clark Fork, or Sand Coulee beds. On the basis of both AMNH label and catalog data and the interpreted usage of bed names in this area, Granger and Stein may have sampled upper Clarkforkian and lower Wasatchian mammalian biozones. A working interpretation of the chronostratigraphic occurrence of the taxa is as follows. The type locality (L3952a) of *Unio eomargaron* Cockerell was in a "coarse friable greenish sandstone" in the "Clark Fork Beds" (Cockerell 1915b:123) and thus may be assignable to a late Clarkforkian (Biochron Cf3) Age. *Viviparus retusus*, which was collected from the Ralston beds (L3952b), may be assignable to the earliest Wasatchian Biochron Wa0 (but was not subsequently identified at this horizon in better documented collections). This taxon, from the same locality, was reported again from the Sand Coulee beds, but this report is possibly in error, being confused with the report of new taxa subsequently collected, possibly from different sites, from the Sand Coulee beds. The remaining taxa were described from collections (L3952c) made in the Sand Coulee beds interpreted as early Wasatchian (Biochron Wa1) in age.

In 1913, two additional molluscan collections were made by Stein. These represent occurrences of certain rarely reported taxa, but, unfortunately, the locality references are not interpretable with much precision. Russell (1931b) reported *Grangerella sinclairi* from 6 miles southeast of Hopkins Ranch (L4285) in Sand Coulee beds, which can be interpreted as probably Biozone Wa2–Wa3. Also, at that time, unpublished lots were collected from the head of Big Sand Coulee (L3952).

The unionids described from the Clarkforkian and Wasatchian deposits of the Bighorn Basin have a rather undistinguished documentary history. *Unio wasatchensis* was named by Cockerell (1914a:105, plate X, figures 4–6) and reported as from "Wasatch; Big Horn Basin, Wyoming. . . . No other details given" (L4282b; see Comments on Taxa). This species is referable to the genus *Plesielliptio* (Taylor in Rohrer and Gazin 1965; Taylor 1975; and this chapter), but its occurrence has not been well documented in the Clarkforkian or lower Wasatchian record. *Unio didymictidis* and *Unio sinopae* were named by Cockerell (1915b:124, 125, figures 3a, 3b, figures 4a–4c, respectively) and reported as from "Wasatch Eocene, Gray Bull beds; north side of Dorsey Creek, nine miles from St. Joe" (L4283), along with numerous specimens of *Unio wasatchensis*. Russell (1931b) again reported that the specimens were unavailable for study. *Unio sinopae* requires the collection of topotypic material to establish its identity. On the basis of mammals collected on the same day as these unionids at a locality of the same description, by George Olsen, the molluscan taxa are middle Wasatchian (Biochron Wa3 or Wa4) in age.

Princeton University and L. S. Russell

Russell (1931b) published the first and only summary of the nonmarine mollusks of the Bighorn Basin (see figure 16.3, table 16.3, and Taylor 1975:table 4). This work was undertaken with the cooperation of Sinclair and G. L. Jepsen while Russell was completing his dissertation studies at Princeton. Earlier publications, which had been on American Museum specimens, were substantially augmented by specimens collected under the direction of Sinclair and Jepsen during the 1927–1929 field seasons, conducted under the auspices of the Princeton University Scott Fund. Their localities were only generally described (e.g., 6 miles east of Powell in Russell 1931b). On the basis of the tremendous number of faunules documented by Gingerich, his students, and colleagues, many of the Sinclair–Jepsen localities can be geographically located with some precision and thus temporally interpreted. The following section is a review of the more salient data derived from the Princeton collections and Russell (1931b) identifications. The localities are presented chronostratigraphically, based on our present interpretation of the mammal ages of the localities, followed by a faunal summary.

Tiffanian localities All of the following localities were reported by Russell (1931b) as from the "middle Fort Union" Formation. The locality reported as "R. 99 W., T. 56 N., S. 11 or 12" (L0136) occurs near UM SC-262 and is early or middle Tiffanian (Biochrons Ti2–Ti3) in age. Russell (1931b) identified *Unio wasatchensis*?, *Viviparus paludinaeformis* (*Campeloma nebrascense nebrascense*, after Hartman 1984), *Lioplax nebrascensis* (including *Goniobasis producta*), and *Goniobasis tenuicarinata*? The locality reported as "south side of Polecat Bench north of Powell" (L0134) is interpreted as middle Tiffanian (Biochron Ti3 or Ti4) in age on the basis of a faunule comparison to the Biochron Ti3–Ti4-age molluscan fauna of the Bullion Creek and Sentinel Butte formations in the Williston Basin (Hartman and Kihm 1995), along with the occurrence of UM SC mammal localities documenting the age in this general area. Russell (1931b:plate 1, figure 9) named the subspecies *Campeloma nebrascensis reticulata* (type locality; *Campeloma nebrascense nebrascense*, after Hartman 1984; see table 16.2) and identified *Viviparus paludinaeformis* (*C. nebrascense* subsp. indet., after Hartman 1984) and *V. retusus*. The locality reported as from a 6.1-m (20-ft) sandstone lens in "south-west of Polecat Bench" (L0135) occurs near UM SC-198 and is interpreted to be late Tiffanian (Biochron Ti5) in age. Russell (1931b) reported unidentified unionid taxa ("*Unio* spp. undt."), *Viviparus retusus*, and "*Campeloma limnaeiformis*"?.

Clarkforkian localities The first attempt to document the molluscan fauna of the Clark Fork beds was by Jepsen (1930) and Russell (1931b). They recognized these beds as the upper part of the Fort Union Formation. The locality reported by Russell (1931b) as "west side of Polecat Bench" (L2309) is too vague to precisely determine its mammalian age based on recent biostratigraphic studies, other than its possible occurrence in strata ranging from late Tiffanian to early Wasatchian in age. Russell (1931b) identified undetermined species of *Unio*. The locality reported as "near mouth of Little Sand Coulée" (L2310) consisted of two collecting sites, referred to as "east of mouth" and "north-east of mouth" (Russell 1931b). These areas are near UM SC-209 or SC-136, which are middle Clarkforkian (Biochron Cf2) in age. These locations are, however, difficult to reinterpret with any precision (Rose, personal communication 1980). Russell (1931b:10, 11) questioningly reported one oreohelicid taxon from each site: "Two species of helicoids are present in the poorly known Clark Fork fauna. One of these (P.U. No. 20016), from east of the mouth of Little Sand Coulée, probably pertains to *Oreohelix grangeri* Cockerell and Henderson, a species said to come from the Ralston (= Clark Fork) beds. . . . The second specimen (P.U. No. 20015), from north-east of the mouth of Little Sand Coulée, is an incomplete, exfoliated shell of a very large snail, possibly *Oreohelix megarche* Cockerell and Henderson. . . . These isolated specimens appear to be fore-runners of the remarkable terrestrial fauna present in the Gray Bull beds."

The locality reported as "six miles east of Powell" (L0133) is "presumably at or near the Ries locality" (Gingerich, personal communication 1981) and is interpreted as early Clarkforkian (Biochron Cf1?) in age. Russell (1931b) identified *Unio priscus*? as from the middle Fort Union Formation. The locality reported from the Sand Coulee beds in "R. 100 W., T. 56 N., sec. 33" (L2319) occurs in the Willwood Formation southwest of UM SC-80 and, thus, may be of late Clarkforkian (Biochron Cf3) Age. Russell (1931b) reported *Gastrodonta corphyodontis*, but PU-INV (Russell 1930, unpublished) records show he also identified *Polygyra veternior* from this locality. If correctly interpreted, this would be the only locality reported by Russell (1931b) as Sand Coulee or Gray Bull Beds that has been reinterpreted as Clarkforkian in age.

Wasatchian localities The following Wasatchian Willwood Formation localities were variously reported as from the Gray Bull and/or Sand Coulee beds (Russell 1931b). The locality reported as "Little Sand Coulee (basin), east of mouth of Pat O'Hara Creek" (L2312) represents a number of separate collecting sites, which were not differentiated by Russell (1931b). Six collections are recognized (L2312a–f). L2312a represents a lot of specimens without any field number designation and is latest Clarkforkian or, more probably, early Wasatchian in age. As identified by Russell (1931b; PU unpublished data), the faunule includes *Unio* sp. indet., *Polygyra veternior*, *Gastrodonta corphyodontis*, *Oreohelix megarche*, *Gonyodiscus ralstonensis*, and *Grangerella sinclairi*. Lots L2312b (not L2312b of Hartman 1984) and L2312c represent "Field nos. 31 and 32 collected July 13, 1929 from near SC-89," which are earliest Wasatchian (Biochron Wa1) or possibly latest Clarkforkian in age (Gingerich, personal communication 1981). The faunule for L2312b–c includes *Helicina evanstonensis*, *Physa* sp., *Oreohelix megarche*, *Gonyodiscus ralstonensis*, *Polygyra veternior*, *Gastrodonta corphyodontis*, *Grangerella megastoma*, and *Grangerella sinclairi* (Russell 1931b; PU unpublished). Lots L2312d, L2312e, and L2312f represent "Field nos. 45, 47, and 56 collected July 15, 1929 from or near SC-12 (west of SC-87)," which are early Wasatchian (Biochron Wa2) in age (Gingerich, personal communication 1981). The faunule of L2312d–f includes *Helicina evanstonensis*, *Viviparus leidyi*? (*V.* sp. cf. *V. meeki* and *V.* sp. indet., after Hartman 1984), *Goniobasis carterii* (not in Russell, an inadvertent omission), *Physa* sp., *Oreohelix megarche*, and *Gonyodiscus ralstonensis* (Russell 1931b; PU unpublished).

The locality reported as "north of divide at head of Little Sand Coulee" (L2313) represents several specific sites or lots that were not differentiated by Russell (1931b). These are treated herein as one general locality near UM SC-2 and are interpreted to be of early Wasatchian (Biochron Wa2) Age. The collection lots, represented by different field numbers (in parentheses below), have varying faunal compositions. The local fauna identified by Russell (1931b) includes *Unio* sp. (64), *Helicina evanstonensis* (64), *Viviparus leidyi*? (*Viviparus* sp. indet., Hartman 1984; 72 and unnumbered), *Viviparus wyomingensis* (*Viviparus* sp. indet., Hartman 1984; 64, 70, and 73), "*Goniobasis carterii*" (59, 70, and 73), *Polygyra veternior* (65 and unnumbered), *Gastrodonta coryphodontis* (64), *Oreohelix megarche* (58, 65, 72, and unnumbered), *Oreohelix grangeri* (72), *Oreohelix* sp. (72), *Gonyodiscus ralstonensis* (64), *Gran-*

gerella megastoma (64), *Grangerella sinclairi* (64), and *Protoboysia complicata* (64).

The locality reported by Russell (1931b) as "R. 102 W., T. 56 N." (L2314) can be more specifically referenced as from near "Camp No. 1" (PU locality catalog), which has been interpreted to be about in the NE sec. 35, T. 56 N., R. 102 W. (D. M. Schankler, personal communication 1979; K. D. Rose, personal communication 1980). Camp No. 1 "was set up on June 27th very near the section corner adjacent to UM locality SC-2" (Gingerich 1980:18; Gingerich et al. 1980). The age of the locality is interpreted to be early Wasatchian (Biochron Wa2). The faunule includes *Helicina evanstonensis*, *Physa* sp., *Polygyra veternior*?, *Oreohelix megarche*, *Oreohelix grangeri*, and *Gonyodiscus ralstonensis* (Russell 1931b). The locality "south of Camp 1 of 1928" (L2317) was omitted from Russell (1931b) but was recorded in the PU catalog and Sinclair's 1929 field notes. Its location is near UM SC-2 and thus probably early Wasatchian (Biochron Wa2) in age. The faunule includes *Grangerella sinclairi* (Russell, unpublished identification) and *Viviparus* sp. indet. (Hartman 1984).

The locality reported as "east of Steins" (L2316) is in the Dorsey Creek area of Big Horn County (not Park; Russell 1931b), and the age of the faunule is early or middle Wasatchian (Biochron Wa2 or Wa3). Two lots are represented in the faunule. The holotype of *Viviparus jepseni* (Russell 1931b:plate 2, figures 9, 10; see table 16.2; reassigned to *Oreoconus* by Taylor 1975 and Hartman 1984, but referred to New Genus P of this report) was collected by Sinclair, while the other lot includes a couple of specimens collected by T. C. von Storch of *Oreohelix megarche* and *Oreohelix* sp. (cf. *Planorbis*? *planoconvexus*; Russell 1931b). *Goniobasis carterii* was reported from "Gray Bull" beds on "Dry Creek" (L2318) by Russell (1931b). The locality is probably "near Emblem" (Schankler, personal communication 1979; Gingerich, personal communication 1981), and may be early to middle Wasatchian in age. Russell (1931b:12) reported *Unio clinopisthus* from "Gray Bull beds" in sec. 34, T. 51 N., R. 95 W. (L2320). Russell added that, on the basis of shell characteristics, the species belonged to the "modern genus *Margaritana*." Taylor (1975) reidentified the specimens as *Plesielliptio priscus*. They were collected by Albert Wardell (of the Wardell Ranch) in 1927 from strata reported to be middle Wasatchian (Biochron Wa4) in age (Gingerich 1989a), but both Gingerich (personal communication 1981) and Rose (personal communication 1980), using the history of mammalian fossil collection in the area as a guide, suggested an early or middle Wasatchian Age. The locality reported as "15 miles west of Basin" (L2311) is difficult to relocate precisely. A middle Wasatchian Age (Biochron Wa3 or Wa4) is suggested for this diverse faunule, which also exemplifies the difference in faunal composition between the Clarkforkian and Wasatchian faunas. Russell (1931b) described as new or identified the following taxa: *Unio sinopae*?, *Viviparus jepseni* (paratype locality), *Helicina evanstonensis*, *Planorbis storchi* (Russell 1931b: plate 2, figures 17–19; type locality; see table 16.3), *Physa bridgerensis*, *Physa pleromatis*, *Polygyra veternior*, *Oreohelix megarche*, *Oreohelix grangeri*, *Gonyodiscus ralstonensis*, and *Grangerella sinclairi*. A collection(s) made in 1929 from the Sand Coulee beds was reported simply as "Park Co., Wyoming" (L2315). The faunule under this designation was incompletely reported by Russell (1931b). Besides *Viviparus wyomingensis*, the following taxa were also identified by Russell (unpublished PU and ANSP records): *Physa* sp. indet., *Polygyra veternior*, *Gastrodonta coryphodontis*, *Oreohelix megarche*, *Grangerella megastoma*, and *Grangerella sinclairi*.

Russell's conclusions Russell (1931b) made the following rather concise statements on the age of the faunas represented by the mammals and mollusks. Note that Jepsen (1930) listed preliminary Russell identifications of the mollusks of the Paleocene and Eocene epochs as part of his review of their mammalian faunas. Figure 16.3 and table 16.3 summarize the biostratigraphy of Paleocene and Eocene faunas known up to and including Russell's taxonomic review. Figure 16.3 places identifications into a modern mammalian biostratigraphic framework and incorporates all of the material that potentially was available for Russell to include in his study. The purpose of Russell's work, it should be noted, was largely to identify and interpret the new specimens added to the molluscan fauna by collections of vertebrate paleontologists in the late 1920s.

Paleocene fossil record: "Numerous fossil shells were collected from the Torrejon [to Tiffanian, this report] portion of the series, which, for the purposes of the present essay, will be designated middle Fort Union." "The Clark Fork mammalian fauna was discovered in the upper portion of this [Fort Union] series and is correlated with the Upper Paleocene (Thanetian)." "A few specimens of mollusks were found in the upper, or Clark Ford [sic] beds" (p. 4). "The fauna known from the middle Paleocene beds is, with one or two exceptions, a typical Paleocene assemblage, both as to the species present and the associations of species in faunules. Such a fauna possesses much closer relationships with the mollusks of the continental Upper Cretaceous formations of the west than it does with known faunas of the Lower Eocene. . . . It also should be noted that only fluviatile and lacustrine genera are present in the middle Fort Union fauna of the present collection. In contrast to the assemblage just discussed, the imperfect shells from the Clark Fork, or upper Fort Union beds, belong to one fluviatile and one terrestrial genus, and strongly suggest species of the Lower Eocene fauna" (Russell 1931b:11).

Eocene fossil record: "The lower Eocene, or so-called Wasatch beds rest upon the Fort Union series. . . . Granger divided the 'Wasatch' into two stages, a lower or Sand Coulée, and an upper or Gray Bull" (Russell 1931b:4). "There is slight difference between the [molluscan] fauna of the 'Sand Coulée' beds and that of the upper Gray Bull. *Goniobasis carterii* and the Viviparidae seem to be more common in the lower horizons, but otherwise the differences can be ascribed plausibly to the chances of collecting. We may consider, therefore, the Mollusca of the Gray Bull stage, using the latter term in the comprehensive sense adopted, as constituting a single, well-marked fauna. The species of this Lower Eocene fauna are, with few exceptions, either confined to the Gray Bull beds, or are known elsewhere only from Eocene strata. As to habitat, almost three-quarters of the genera represented are terrestrial, and if the species and individual specimens are considered, the aquatic element becomes almost insignificant" (Russell 1931b:29).

Russell (1931b:30) speculated that "the cause of this profound change in molluscan fauna is almost certainly to be found in the elevation of the Rocky Mountains. . . . A very important period of orogeny occurred, as shown by unconformities, following the deposition of the Clark Fork stage. This elevation and accompanying change in climate, while probably not profound, were sufficient to drive out or destroy the aquatic Paleocene faunas, and to provide the correct environment for a dominantly terrestrial fauna" (Russell 1931b:30).

After placing Russell's identifications into a current mammalian biochronology for the Clark's Fork Basin, his "aquatic Paleocene fauna" ranges from as old as Biochron Ti2 to possibly as young as Biochron Wa0. Some uncertainty arises from the unspecific nature of location and stratigraphic data. As revised by present studies (see below), this fauna is typical of Biochrons Ti3 and Ti4 molluscan assemblages found in the Bullion Creek and Sentinel Butte formations in the type area of the Fort Union Group (Hartman and Kihm 1995). Subsequent recognition that a few of Russell's "Sand Coulée" localities (1931b) are Clarkforkian in age strengthens his conjecture that the "Clark Fork" beds represent the beginning of the change to a dominantly terrestrial fauna. In addition, a significant proportion of this change had occurred by the early Wasatchian Age, taking place over a short span of time. Additional study, based on new collections, supports the conclusion that the Clarkforkian Age was the time of major faunal change in the terrestrial molluscan record.

Russell and G. G. Simpson localities

Russell (1934) also published new species descriptions from specimens submitted by G. G. Simpson. These were collected from the "Upper Paleocene ('Bear Creek' horizon)" in the Fort Union Formation in Carbon County, Montana, in the northern Bighorn Basin (see USGS Studies above). In 1925 and 1927, J. C. F. Siegfriedt, a physician in Bear Creek, Montana, collected freshwater mollusks from the coal mines near Bear Creek (L1476b). Gingerich (1980:16, 17) recorded the humorous consequences in Siegfriedt's community life that resulted from his discovery of the Bear Creek mammal fauna. In 1932, Simpson and A. C. Silberling, of Harlowton, Montana, collected mollusks from the level of the Bear Creek mammal fauna at the Eagle coal mine (L1478). Russell (1934) named a freshwater bivalve *Elliptio silberlingi* and snail *Goniobasis ursarivulensis*. These assignments were later revised, with the bivalve assigned to *Plesielliptio* (Russell 1934) and the snail synonymized with *Lioplacodes multistriata* (Meek and Hayden 1856; Taylor 1975; Hartman 1984). In addition, Woodring (1926) compared opercula assigned to *Scalez petrolia* to gastropod opercula found at the locality (Hartman 1984). *Scalez*, previously identified as *Viviparus*-like, was included in the Bithyniidae by Taylor (1975).

The Bear Creek horizon has since been interpreted as belonging to the early Clarkforkian (Biochron Cf1) NALMA (Archibald et al. 1987; Gingerich 1989a; K. C. Beard, personal communication 1995). *Plesielliptio silberlingi* is widely known in the Powder River Basin from strata of Clarkforkian Age, on the basis of present studies (see below), but has not been identified elsewhere in the Bighorn Basin. Russell (1931b) described poorly preserved *Unio* specimens as very elongate (L0135) or similar to *Unio danae* (of the Judith River Formation; L2309). These comparisons, especially the latter, which is of Clarkforkian Age, potentially fit the morphology of *P. silberlingi*.

Later USGS Studies

Later USGS investigations continued to focus on resolving stratigraphic problems, using mollusks and mammals as biochronologic aides in correlation. Many of these studies were conducted in the southern part of the Bighorn Basin and in somewhat younger strata. F. B. Van Houten (1944) reported gastropod taxa from two localities from the Tatman Formation in Park and Hot Springs Counties, Wyoming (L4131, L4311), on the basis of identifications by L. S. Russell. In 1946, D. H. Eargle collected mollusks from the Tatman Formation in Hot Springs County, Wyoming (L4113), as part of studies that would result in a preliminary geologic map of the Bighorn Basin (see Andrews et al. 1947). The

Eargle collection was identified and reported by T. C. Yen (1948a) in a brief review of the Eocene freshwater mollusca of Wyoming. Collections were also made by D. W. Taylor and others in 1953, from the Willwood Formation in the Clark's Fork Basin, Park County, Wyoming (L4233, L4234).

In 1961, W. L. Rohrer collected several molluscan localities during the geologic mapping of the Sheep Mountain and Tatman Mountain Quadrangles in Big Horn and Park Counties, Wyoming (Rohrer 1964a, 1964b; L2117, L2119–L2121, L2123–L2127). A few of these localities were also reported by Rohrer (1964b) in a measured section (L2120, L2121), and almost all of them were reported by Rohrer and Gazin (1965) in an examination of the vertebrate–molluscan faunal change between the Willwood and Tatman formations. All of the mollusks reported by Rohrer were identified by D. W. Taylor. In 1962, Rohrer collected another locality (L2155), this time from the Aycross (Pitchfork) Formation, while mapping the Adam Weiss Peak Quadrangle and vicinity, Hot Springs County, Wyoming. The taxa from this locality were identified by D. W. Taylor and reported by Rohrer (1966).

Taylor's contribution

Dwight W. Taylor contributed to our current knowledge of the nonmarine molluscan fauna of the Bighorn Basin by providing the first nomenclatural reassessment of its taxa since the efforts of Russell (1931b). Taylor initiated these studies with the identification of specimens from localities reported by Rohrer (1964b, 1966), and Rohrer and Gazin (1965; see above). Fortunately, as with earlier Bighorn Basin studies, the molluscan identifications continued to be placed into the context of vertebrate studies. These studies were on localities from younger strata than the faunas considered here.

In a USGS Open-File Report, D. W. Taylor (1975) presented a taxonomic review of Russell's (1931b; Jepsen 1930) nonmarine mollusk identifications for the Polecat Bench (Fort Union) and Willwood formations of the Bighorn Basin. Taylor's summary of the taxa of the Willwood Formation was not locality-specific. Table 16.4 presents the nonmarine molluscan fauna of the Bighorn Basin as identified by Russell and revised by Taylor. Taylor (1975) also recorded a few additional localities from the later part of the Wasatchian NALMA in the southern part of the Bighorn Basin.

Taylor (1975) recognized *Plesielliptio priscus*, *Bellamya retusa* (taxon assigned to *Viviparus*), *Campeloma nebrascense*, *C.* sp. cf. *C. vetulum* (see Hartman 1984), and *Cleopatra nebrascensis producta* (taxon assigned to *Lioplacodes*) as a distinctive assemblage in the Fort Union Formation of the northern Great Plains. Using the mammalian temporal control available at the time, he interpreted the age of the assemblage in the Bighorn Basin, as "bracketed between mammals of Torrejonian and Tiffanian age. . . . For the sake of simplicity I have correlated the mollusks as Torrejonian, but they may prove to range into the early Tiffanian also" (Taylor 1975:94). In terms of current mammalian biochronology, this assemblage indicates a middle to late Tiffanian Age.

Taylor (1975:97) recognized the uniqueness of the "molluscan novelties" in the Gray Bull Beds ("earliest Eocene"): "Both the local and the regional distribution of the mollusks can be explained plausibly and simply by inferring that they were associated with two main habitats: savanna, with terrestrial forms that appear for the first time in the geologic record; and forests and streams like those of the preceding Paleocene times, with a Paleocene-like fauna." Russell (1931b:30) had earlier surmised that a change in climate, associated with a change in elevation, resulting from the Laramide Orogeny, "provided the correct environment for a dominantly terrestrial fauna."

Hanley's contribution

John H. Hanley began his contribution to Bighorn Basin nonmarine molluscan paleontology where Taylor had left off, by identifying specimens for stratigraphic studies. For example, in 1978 T. M. Bown's collection from the Aycross Formation in Hot Springs County, Wyoming (L1695), was identified by Hanley and reported by Bown (1982) in his study of the volcaniclastics of the Absaroka Range.

However, Hanley and his field crews contributed most significantly to the current study by rigorously sampling nonmarine mollusks at a large number of localities in the Clark's Fork Basin and in the Montana part of the Bighorn Basin. This work (1976 to 1981) was conducted with the cooperation of Philip D. Gingerich, whose field studies over a number of years had provided Hanley with the best available stratigraphic framework for molluscan biostratigraphy. Hanley's work would remain preliminary, as he commenced other molluscan studies in the Powder River Basin. Because no published or in-house USGS report exists on Hanley's efforts, a brief review of his contribution will be given here. Hanley and his crews collected specimens from about 226 localities and sublocalities. In the Montana part of the Bighorn Basin, they recorded 23 localities from the Fort Union Formation, largely from the greater Red Lodge–Belfry area. Thirteen Fort Union localities were recorded in Wyoming, along with 185 localities from the Willwood Formation. In addition, five collections were made in the Tatman Formation. In Wyoming, all of the Fort Union localities and over 90% of the Willwood localities are on the Cody 15-minute Quadrangle. Hanley prepared one stratigraphic figure that illustrated, primarily, the occurrence of terrestrial

Table 16.4
The nonmarine mollusks of the Fort Union (Polecat Bench) and Willwood Formations after Taylor (1975)

Taylor (1975	Russell (1931b)	Fort Union	Willwood
Freshwater Bivalves			
Plesielliptio priscus	*Unio priscus?; Unio clinopisthus*	X	X
Plesielliptio wasatchensis	*Unio wasatchensis*	?	X
Plesiellipto? sp.	*Unio emargaron*	X	
Freshwater Prosobranchs			
Bellamya retusa	*Viviparus retusus*	X	
Bellamya n. sp. aff. *B. retusa*	*Viviparus retusus*		X*
Bellamya n. sp.	*Viviparus wyomingensis*		X
Bellamya sp.	*Viviparus leidyi?*		X
Campeloma nebascense	*Campeloma nebrascensis reticulata*	X	
Campeloma cf. *C. vetulum*	*Viviparus paludinaeformis*	X	
New Genus A *limneaformis*	*Campeloma limnaeiformis?*	X	
Cleopatra nebrascensis producta	*Lioplax nebrascensis*	X	
Cleopatra tenuicarinata	*Goniobasis tenuicarinata?*	X	
Elimia? n. sp.	*Goniobasis carterii*		X
Freshwater Pulmonates			
Biomphalaria storchi	*Planorbis storchi*		X
Physa bridgerensis	*Physa bridgerensis, P. pleromatis*		X
Terrestrial prosobranchs			
Hendersonia evanstonensis	*Helicina evanstonensis*		X
Dimorphoptychia eohippina	*Polygyra eohippina*		X
Grangerella phenacodorum	*Grangerella sinclairi*		X
Grangerella sinclairi	*Grangerella sinclairi, G. megastoma*		X
Terrestrial Pulmonates			
Oreoconus jepseni	*Viviparus jepseni*		X
Protoboysia complicata	*Protoboysia complicata*		X
"Eucalodium" eophilum	*"Eucalodium" eophilum*		X
Discus ralstonensis	*Gonyodiscus ralstonensis*		X
Gastrodonta coryphodontis	*Gastrodonta coryphodontis*		X
Microphysula oxyaenae	*Microphysula oxyaenae*		X
Microphysula sinoparum	*Microphysula sinoparum*		X
"Polygyra" veternior	*Polygyra veternior*		X
New Genus A *spatiosa*	*Oreohelix megarche?; O. megarche*	X	X
Oreohelix? grangeri	*Oreohelix grangeri?; O. grangeri*	X	X

*= Ralston beds, probably Willwood Formation (Taylor 1975).

taxa by morphotype (figure 16.4; table 16.5). This figure was based on taxa from 70 localities. By modifying Hanley's taxonomic ordering of the taxa to one based on first and last appearances, figure 16.4 represents Hanley's preliminary (circa 1977) interpretation of the biostratigraphy of nonmarine mollusks in the Sand Coulee area of the Clark's Fork Basin. Each vertical line represents a taxon, and 58 are portrayed on this figure in the context of 71 locality (not necessarily stratigraphic) levels. Note that the diagram does not reflect a metrically or temporally controlled organization of taxa. The order is simply Hanley's relative position of localities, some of which were known by Hanley to be at the same stratigraphic horizon. A revised stratigraphic ordering of the localities is presented in figures 16.5–16.7. Revised stratigraphic data for many of Hanley's localities are presented in the Locality Register.

The pattern of ranges reflects well the distinctiveness of the faunas on either side of the Clarkforkian/Wasatchian NALMA boundary, remembering that most of the terrestrial taxa have their lowest occurrence in this interval. Hanley's effort in this work stressed the terrestrial component of the fauna, which has been substantially revised herein. Note that all unnamed mor-

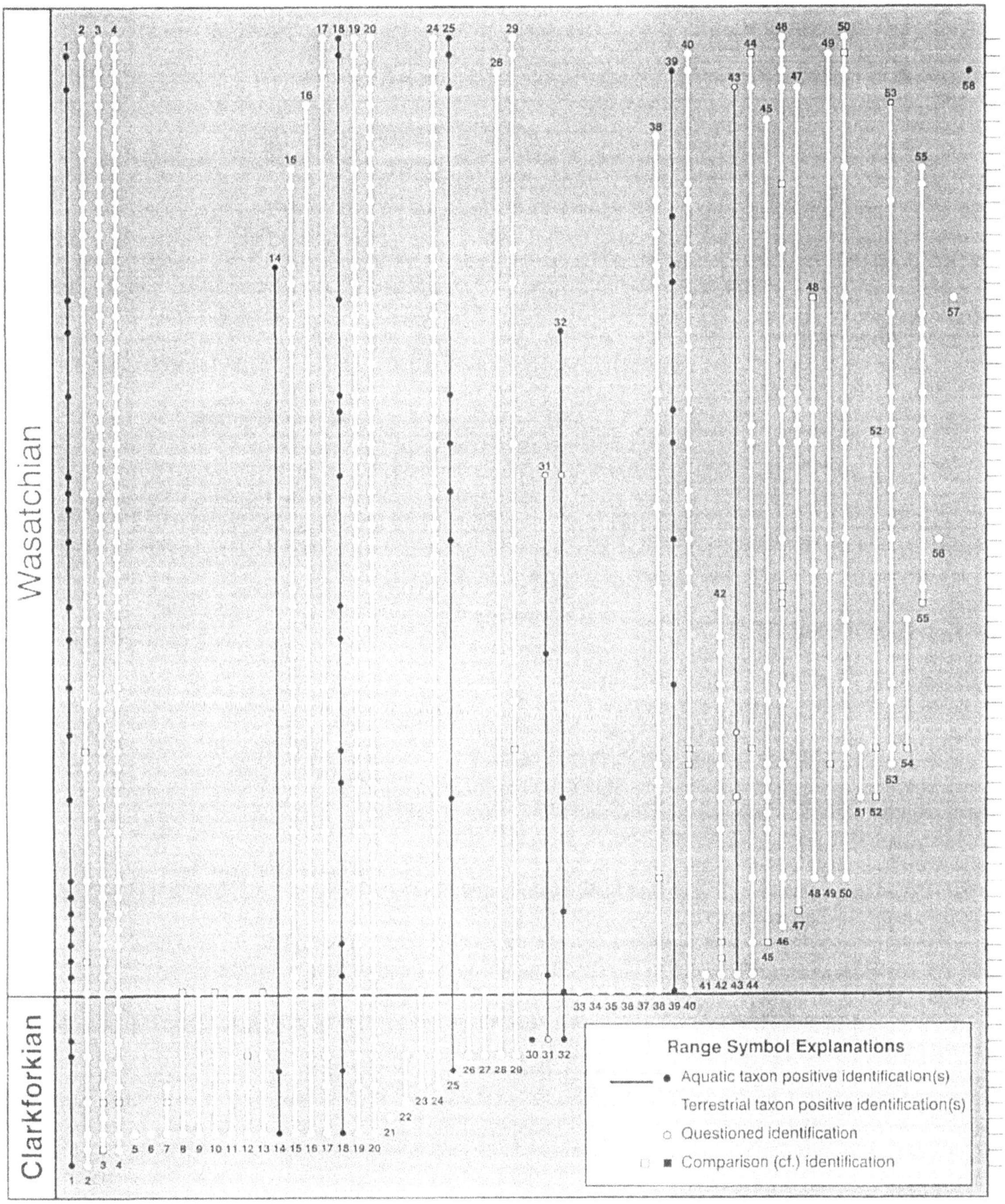

Figure 16.4 Hanley's preliminary stratigraphic ordering of Bighorn Basin nonmarine mollusks.

The taxa shown in this figure represent unpublished nonmarine molluscan identifications by John H. Hanley from collections made in 1976 and 1977. This presentation is derived from a chart made by Hanley titled "Preliminary Biostratigraphy of Terrestrial Gastropoda from the Willwood Formation (Clarkforkian and Wasatchian NALMA equivalents), Sand Coulee Area, Northern Bighorn Basin, Wyoming." His chart also included aquatic gastropods and bivalves, but they were identified generally only to genus. This figure incorporates Hanley's "?" and "cf." identifications. From a personal communication with Hartman, Hanley employed the "?" and "cf." in a way opposite to that used in this chapter. His "?" were intended for potentially distinct taxa, whereas "cf." was used for taxa that might be assigned to the taxon in question except for the quality of preservation. Thus, for the purposes of conformity, figure 16.4 employs "?" and "cf." in the sense of this chapter. Hanley organized his identifications by classification. The taxa are here arranged from first to last occurrence. The stratigraphic ordering (e.g., h01 = horizon 1) of localities is that of Hanley. Note that a number of localities occur at or near the same stratigraphic level. The diagram does not reflect a metrically or temporally controlled organization of taxa. The stratigraphic ordering of some of the localities has been revised and is presented in figure 16.7. The key to the taxon range numbers and localities representing each "horizon" is given in table 16.4. See also the Locality Register for location data.

Table 16.5
*Hanley's Preliminary Identification and Stratigraphic Ordering of Bighorn Basin Nonmarine Mollusks

1) *Viviparus* sp.: (h01) L5808, (h08) L5799, (h09) L5804, (h13) L5757, (h14) L5766, (h15) L5767, (h16) L5765, (h17) L5748a–b, (h18) L5756, (h19) L5749, (h24) L5791, (h28) L5735, (h31) L5786, (h34) L5784, (h36) L5780, (h40) L5727, (h42) L5732, (h43) L5728, (h44) L5714, (h49) L5705, (h53) L5726a, (h68) L5759, (h70) L5762
2) *Grangerella* sp.: (h01) L5808, (h02) L5806, (h03) L5805, (h05) L5810, (h08) L5799, (h12) L5758a–b, (h14) L5766, (h19) L5749, (h26) L5774, (h27) L5775, (h30) L5785, (h35) L5778, (h43) L5728, (h55) L5776, (58) L5781, (h59) L5794, (h60) L5795, (h61) L5796, (h62) L5798, (h64) L5797, (h65) L5724, (h70) L5762, (h71) L5761
3) *Discus ralstonensis:* (h02) L5806, (h03) L5805, (h04) L5812, (h05) L5810, (h06) L5802, (h07) L5809, (h08) L5799, (h12) L5758a–b, (h13) L5757, (h14) L5766, (h15) L5767, (h16) L5765, (h17) L5748a–b, (h18) L5756, (h19) L5749, (h22) L5789, (h23) L5792, (h24) L5791, (h25) L5773, (h26) L5774, (h27) L5775, (h28) L5735, (h29) L5734, (h30) L5785, (h31) L5786, (h35) L5778, (h36) L5780, (h27) L5777, (h38) L5733, (h39) L5732, (h40) L5727, (h41) L5705a, (h42) L5731, (h43) L5728, (h44) L5714, (h45) L5706, (h46) L5705d, (h47) L5705c, (h48) L5705b, (h49) L5705, (h52) L5726, (h53) L5726a, (h54) L5729, (h55) L5776, (h56) L5720, (h57) L5721, (h58) L5781, (h59) L5794, (h60) L5795, (h61) L5796, (h62) L5798, (h63) L5723, (h66) L5695, (h67) L5696, (h68) L5759, (h69) L5760, (h70) L5762, (h71) L5761
4) *Grangerella megastoma:* (h02) L5806, (h03) L5805, (h05) L5810, (h06) L5802, (h08) L5799, (h09) L5804, (h11) L5800, (h12) L5758a–b, (h15) L5767, (h16) L5765, (h17) L5748a–b, (h18) L5756, (h19) L5749, (h22) L5789, (h24) L5791, (h25) L5773, (h26) L5774, (h27) L5775, (h28) L5735, (h30) L5785, (h31) L5786, (h35) L5778, (h36) L5780, (h37) L5777, (h38) L5733, (h39) L5732, (h40) L5727, (h41) L5705a, (h42) L5731, (h43) L5728, (h44) L5714, (h45) L5706, (h46) L5705d, (h47) L5705c, (h48) L5705b, (h49) L5705, (h50) L5707, (h52) L5726, (h53) L5726a, (h54) L5729, (h55) L5776, (h56) L5720, (h57) L5721, (h58) L5781, (h59) L594, (h60) L5795, (h61) L5796, (h62) L5798, (h63) L5723, (h64) L5797, (h65) L5724, (h66) L5695, (h67) L5696, (h68) L5759, (h69) L5760, (h70) L5762, (h71) L5761
5) terrestrial sp. Form X: (h03) L5805
6) terrestrial sp. Form Y: (h03) L5805
7) terrestrial sp. Form T: (h03) L5805, (h05) L5810, (h06) L5802
8) terrestrial sp. Form V: (h03) L5805, (h12) L5758a–b
9) terrestrial sp. Form W: (h03) L5805, (h12) L5758a–b
10) terrestrial sp. Form Z: (h03) L5805, (h12) L5758a–b
11) terrestrial sp. Form R: (h03) L5805, (h06) L5802, (h12) L5758a–b
12) terrestrial sp. Form P: (h03) L5805, (h06) L5802, (h08) L5799, (h12) L5758a–b
13) terrestrial sp. Form S: (h03) L5805, (h06) L5802, (h08) L5799, (h12) L5758a–b
14) Planorbidae gen. and sp. indet.: (h03) L5805, (h07) L5809, (h57) L5721
15) *Helicina evanstonensis:* (h03) L5805, (h12) L5758a–b, (h13) L5757, (h22) L5789, (h24) L5791, (h25) L5773, (h27) L5775, (h30) L5785, (h31) L5786, (h33) L5772, (h36) L5780, (h37) L5777, (h52) L5726, (h53) L5726a, (h54) L5729, (h56) L5720, (h57) L5721, (h62) L5798, (h63) L5723
16) terrestrial sp. Form C: (h03) L5805, (h12) L5758a–b, (h14) L5766, (h15) L5767, (h19) L5749, (h67) L5696
17) Pupillidae gen. and sp. indet.: (h03) L5805, (h06) L5802, (h08) L5799, (h12) L5758a–b, (h22) L5789, (h31) L5786, (h69) L5760, (h71) L5761
18) *Physa* sp.: (h03) L5805, (h07) L5809, (h13) L5757, (h15) L5767, (h25) L5773, (h27) L5775, (h34) L5784, (h36) L5780, (h44) L5714, (h48) L5705b, (h49) L5705, (h55) L5776, (h70) L5762, (h71) L5761
19) *Gastrodonta coryphodontis:* (h03) L5805, (h06) L5802, (h08) L5799, (h12) L5758a–b, (h15) L5767, (h15) L5765, (h19) L5749, (h22) L5789, (h25) L5773, (h27) L5775, (h28) L5735, (h31) L5786, (h33) L5772, (h35) L5778, (h38) L5733, (h40) L5727, (h41) L5705a, (h42) L5732, (h43) L5728, (h45) L5706, (h46) L5705d, (h47) L5705c, (h48) L5705b, (h49) L5705, (h52) L5726, (h53) L5726a, (h54) L5729, (h55) L5776, (h56) L5720, (h57) L5721, (h63) L5723, (h69) L5760, (h70) L5762, (h71) L5761
20) *Oreohelix* spp.: (h03) L5805, (h06) L5802, (h08) L5799, (h09) L5804, (h12) L5758a–b, (h13) L5757, (h14) L5766, (h15) L5767, (h16) L5765, (h18) L5756, (h20) L5754, (h21) L5790, (h22) L5789, (h23) L5792, (h24) L5791, (h25) L5773, (h26) L5774, (h27) L5775, (h28) L5735, (h30) L5785, (h31) L5786, (h33) L5772, (h35) L5778, (h37) L5777, (h38) L5733, (h39) L5732, (h40) L5727, (h41) L5705a, (h42) L5731, (h43) L5728, (h45) L5706, (h46) L5705d, (h47) L5705c, (h48) L5705b, (h49) L5705, (h51) L5718, (h52) L5726, (h53) L5726a, (h54) L5729, (h55) L5776, (h57) L5721, (h58) L5781, (h60) L5795, (h61) L5796, (h62) L5798, (h63) L5723, (h64) L5797, (h65) L5724, (h66) L5695, (h67) L5696, (h68) L5759, (h69) L5760, (h70) L5762, (h71) L5761
21) terrestrial sp. Form O: (h04) L5812
22) terrestrial sp. Form U: (h05) L5810, (h08) L5799
23) terrestrial sp. Form Q: (h06) L5802, (h08) L5799, (h12) L5758a–b
24) *Protoboysia complicata:* (h06) L5802, (h08) L5799, (h12) L5758a–b, (h15) L5767, (h16) L5765, (h17) L5748a–b, (h19) L5749, (h22) L5789, (h27) L5775, (h30) L5785, (h31) L5786, (h38) L5733, (h39) L5732, (h40) L5727, (h41) L5705a, (h43) L5728, (h46)

Table 16.5 *(continued)*

L5705d, (h47) L5705c, (h48) L5705b, (h49) L5705, (h52) L5726, (h53) L5726a, (h54) L5729, (h55) L5776, (h57) L5721, (h58) L5781, (h62) L5798, (h63) L5723, (h65) L5724, (h66) L5695, (h68) L5759, (h69) L5760, (h70) L5762, (h71) L5761

25) *Biomphalaria* sp.: (h07) L5809, (h24) L5791, (h40) L5727, (h43) L5728, (h46) L5705d, (h49) L5705, (h68) L5759, (h70) L5762, (h71) L5761

26) terrestrial sp. Form EE: (h08) L5799, (h12) L5758a–b

26) terrestrial sp. Form GG: (h08) L5799, (h12) L5758a–b

28) terrestrial sp. Form L (h08) L5799, (h12) L5758a–b, (h15) L5767, (h69) L5760

29) *Microphysula* sp. indet.: (h08) L5799, (h27) L5775, (h30) L5785, (h31) L5786, (h40) L5727, (h46) L5705d, (h56) L5720, (h58) L5781, (h70) L5762, (h71) L5761

30) ?*Lioplacodes* sp.: (h09) L5804

31) *Goniobasis tenera:* (h09) L5804, (h13) L5757, (h33) L5772, (h44) L5714

32) Unionidae gen. and sp. indet.: (h09) L5804, (h12) L5758a–b, (h17) L5748a–b, (h24) L5791, (h44) L5714, (h53) L572a

33) terrestrial sp. Form AA: (h12) L5758a–b

34) terrestrial sp. Form BB: (h12) L5758a–b

35) terrestrial sp. Form CC: (h12) L5758a–b

36) terrestrial sp. Form DD: (h12) L5758a–b

37) terrestrial sp. Form FF: (h12) L5758a–b

38) Helicinidae gen. and sp. indet.: (h12) L5758a–b, (h19) L5749, (h25) L5773, (h26) L5774, (h37) L5777, (h42) L5731, (h48) L5705b, (h49) L5705, (h56) L5720, (h57) L5721, (h59) L5794, (h61) L5796, (h65) L5724

39) cf. Planorbidae indet. sp. Form A: (h12) L5758a–b, (h31) L5786, (h40) L5727, (h46) L5705d, (h48) L5705, (h56) L5720, (h57) L5721, (h60) L5795, (h69) L5760

40) *Grangerella sinclairi:* (h12) L5758a–b, (h24) L5791, (h26) L5774, (h27) L5775, (h37) L5777, (h38) L5733, (h40) L5727, (h42) L5731, (h43) L5728, (h46) L5705d, (h48) L5705b, (h49) L5705, (h52) L5726, (h53) L5726a, (h55) L5776, (h58) L5781, (h61) L5796, (h62) L5798, (h69) L5760, (h70) L5762

41) terrestrial sp. Form J: (h13) L5757

42) *Polygyra veternior:* (h13) L5757, (h14) L5766, (h15) L5767, (h22) L5789, (h24) L5791, (h26) L5774, (H27) L5775, (h28) L5735, (h30) L5785, (h31) L5786, (h33) L5772, (h34) L5784, (h36) L5780

43) *Physa bridgerensis:* (h13) L5757, (h24) L5791, (h28) L5735, (h68) L5759

44) terrestrial sp. Form G: (h13) L5757, (h15) L5767, (h19) L5749, (h27) L5775, (h46) L5705d, (h47) L5705c, (h68) L5759, (h70) L5762

45) Polygyridae gen. and sp. indet.: (h15) L5767, (h22) L5789, (h23) L5792, (h24) L5791, (h25) L5773, (h26) L5774, (h30) L5785, (h32) L5787, (h66) L5695

46) terrestrial sp. Form F: (h16) L5765, (h19) L5749, (h26) L5774, (h30) L5785, (h31) L5786, (h36) L5780, (h37) L5777, (h40) L5727, (h46) L5705d, (h48) L5705b, (h49) L5705, (h52) L5726, (h58) L5781, (h62) L5798, (h68) L5759, (h69) L5760, (h70) L5762, (h71) L5761

47) terrestrial sp. Form I: (h17) L5748a–b, (h46) L5705d, (h47) L5705c, (h48) L5705b, (h49) L5705, (h68) L5759

48) terrestrial sp. Form H: (h19) L5749, (h38) L5733, (h46) L5705d, (h48) L5705b, (h55) L5776

49) terrestrial sp. Form D: (h19) L5749, (h26) L5774, (h38) L5733, (h40) L5727, (h70) L5762

50) *Microphysula* sp. Form A: (h19) L5749, (h27) L5775, (h31) L5786, (h35) L5778, (h40) L5727, (h46) L5705d, (h48) L5705b, (h55) L5776, (h63) L5723, (h69) L5760, (h70) L5762, (h71) L5761

51) terrestrial sp. Form M: (h24) L5791, (h27) L5775

52) *Microphysula* sp. Form B: (h24) L5791, (h27) L5775, (h46) L5705d

53) terrestrial sp. Form A: (h26) L5774, (h27) L5775, (h30) L5785, (h31) L5786, (h40) L5727, (h41) L5705a, (h43) L5728, (h46) L5705d, (h47) L5705c,(h48) L5705b, (h49) L5705, (h52) L5726, (h54) L5729, (h55) L5776, (h61) L5796, (h67) L5696

54) terrestrial sp. Form K: (h27) L5775, (h35) L5778

55) terrestrial sp. Form B: (h35) L5778, (h36) L5780, (h43) L5728, (h46) L5705d, (h47) L5705c, (h49) L5705, (h52) L5726, (h53), L5726a, (h54) L5729, (h62) Lt798, (h63) L5723

56) terrestrial sp. Form E: (h40) L5727

57) terrestrial sp. Form N: (h55) L5776

58) *Lymnaea* sp.: (h69) L5760

*Taxa are ordered by taxon number (e.g., 1; see figure 16.4), followed by horizon (e.g., h01) and locality numbers.

Hanley's identifications are given only as a source of reference to his biostratigraphic figure and are otherwise not employed or diagnosed in this chapter.

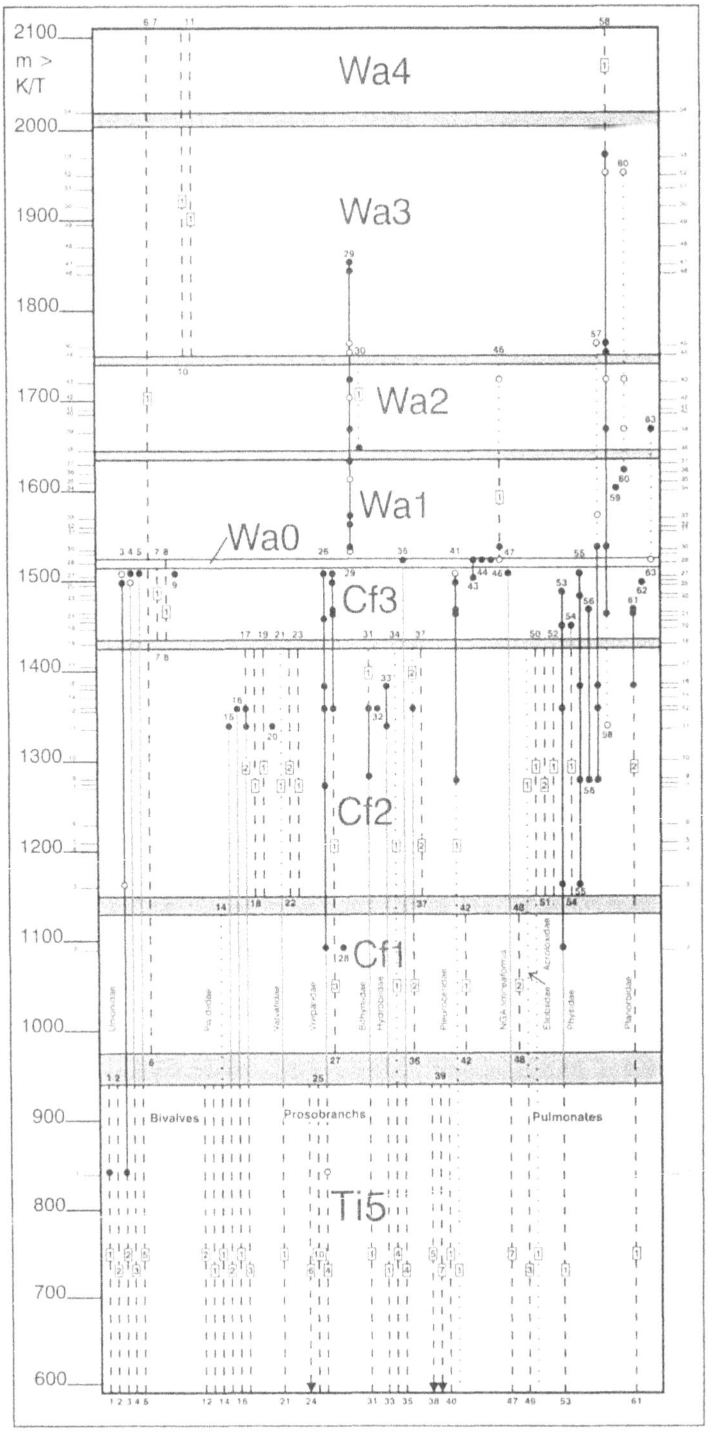

Figure 16.5 Bighorn Basin aquatic molluscan biostratigraphy.

The taxa shown in this figure represent the aquatic molluscan identifications made by Hartman as part of this contribution. The taxa are arranged by family according to their classification (table 16.1). Within each family, taxa are organized from first to last occurrence as they are known from within the Bighorn Basin. The stratigraphic occurrence of each taxon at a locality is shown by a black or open dot (positive and questioned identification, respectively). Dashed ranges indicate that the stratigraphic horizon(s) of a locality(s) is uncertain. The number of localities occurring within the dashed interval is shown in a box on the taxon range. Some species have ranges, represented by a few localities with limited stratigraphic data, in older strata in the Bighorn Basin. The stratigraphic horizon numbers shown on either side of this figure are listed in table 16.10. The taxon range numbers presented in this figure are cross-referenced below with the taxon numbers and names given in table 16.9. Thus Range Number 1 of figure 16.5 is equivalent to Range Number 6 found in figure 16.7 and table 16.9. See the Locality Register for location data.

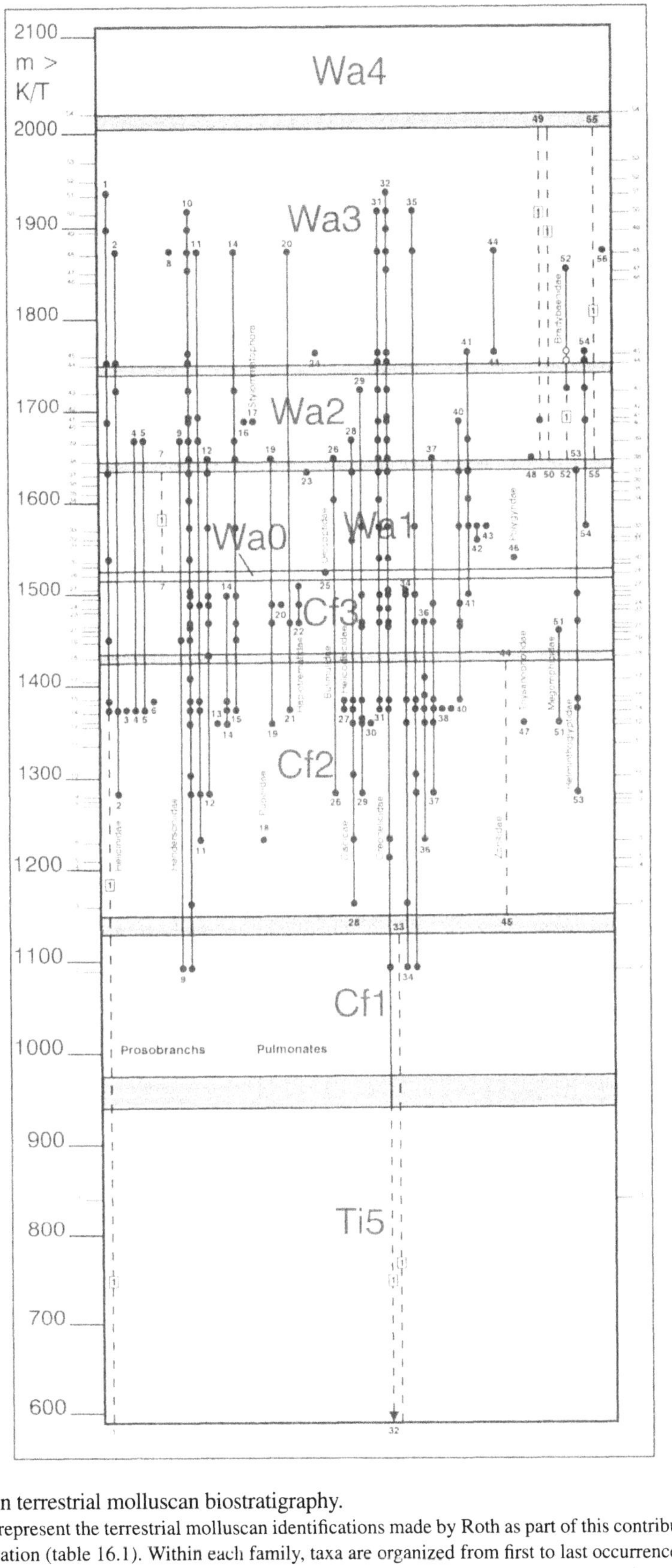

Figure 16.6 Bighorn Basin terrestrial molluscan biostratigraphy.

The taxa shown in this figure represent the terrestrial molluscan identifications made by Roth as part of this contribution. The taxa are arranged by family according to their classification (table 16.1). Within each family, taxa are organized from first to last occurrence as they are known from within the Bighorn Basin. The stratigraphic occurrence of each taxon at a locality is shown by a black or open dot (positive and questioned identification, respectively). Dashed ranges indicate that the stratigraphic horizon(s) of a locality(s) is uncertain. The number of localities occurring within the dashed interval is shown in a box on the taxon range. Some species have ranges, represented by a few localities with limited stratigraphic data, in older strata in the Bighorn Basin. The stratigraphic horizon numbers shown on either side of this figure are listed in table 16.10. The taxon range numbers presented in this figure are cross-referenced below with the taxon numbers given in table 16.9. Thus Range Number 1 of figure 16.6 is equivalent to Range Number 31 found in figure 16.7 and table 16.9. See the Locality Register for location data.

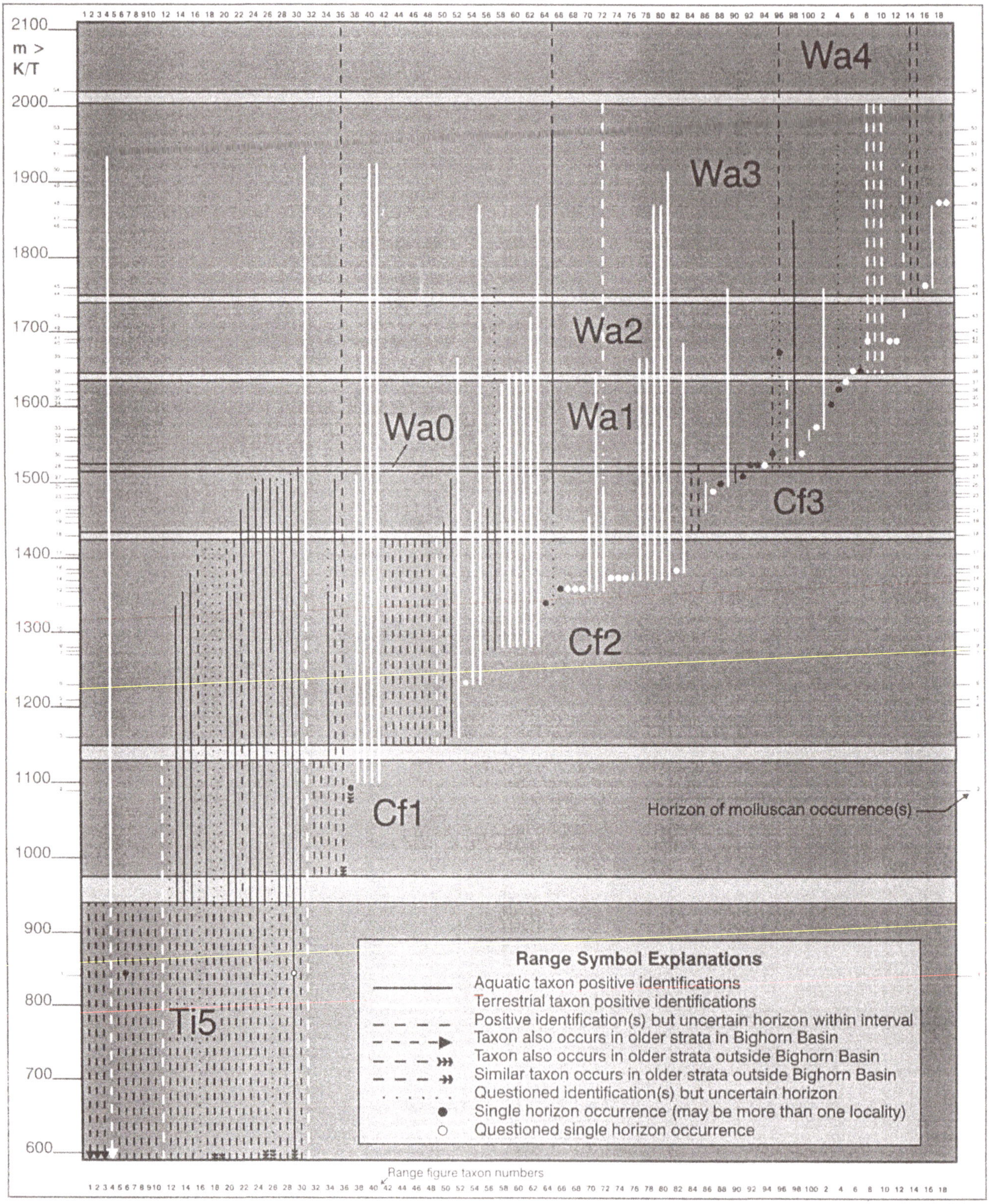

Figure 16.7 Bighorn Basin nonmarine molluscan biostratigraphy.

The taxa shown in this figure represent the nonmarine molluscan identifications made as part of this contribution. The taxa are arranged from first to last occurrence as they are known from the Bighorn Basin. Some species have ranges, represented by a few localities with limited stratigraphic data, in older strata in the Bighorn Basin. In addition, a few taxa occur in older strata elsewhere in the northern Great Plains. As some of these records (e.g., occurrences in the Paskapoo Formation) cannot be tied directly to a mammalian biochronology, they are shown in this figure without extended ranges. The key to the taxon range numbers presented in this figure is given in table 16.9. A questioned locality number indicates a questioned identification at that locality. See table 16.10 for a list of the localities occurring at the stratigraphic horizons shown on either side of the figure. See also the Locality Register for location data.

photypes are of Hanley's usage and remain, for the most part, without diagnosis. Unfortunately, in addition, most of the collections lack identifications, thus limiting our understanding of his taxonomic intentions. Most of the generic identifications of freshwater taxa represent more than one species. Thus Hanley's identifications have not been recognized or utilized in this report.

Hartman (1993b) placed Hanley's identifications into a modern mammalian context. Of the 50 or so taxa that occur in the upper part of the Clarkforkian and lower Wasatchian deposits, about half of these were shown to bear directly on identifying the Clarkforkian/Wasatchian NALMA boundary. The majority of these taxa were pulmonates, denoting a major shift in faunal composition from the Tiffanian NALMA.

Hartman's contribution

Dissertation studies completed by Hartman included a review of the record of Bighorn Basin snails of the mesogastropod family Viviparidae. This work included localities not previously reported, and many identifications have been reevaluated in the present work. Hartman concluded that Tiffanian and early Clarkforkian taxa (regarded as late Paleocene in age at that time) represented a distinctive and relatively diverse viviparid fauna contrasted with the faunas of the late Clarkforkian and early and middle Wasatchian NALMAs (interpreted then as early Eocene in age). On the basis of seven stratigraphically restricted localities, the late Paleocene viviparid fauna included *Campeloma nebrascense*, *Viviparus meeki*, *V. retusus*, *V. haydenianus*, and *V. raynoldsanus*. Although some elements of this fauna have been reported from older strata, the assemblage is distinctive of the late Paleocene Epoch. This fauna includes taxa known from both the Powder River and Williston basins. The viviparid fauna from the Willwood Formation, interpreted to be Eocene in age, was known from 16 localities, but less than half of these were based on positively identified taxa. Two referred taxa, *Viviparus* sp. cf. *V. meeki* and *V.* sp. aff. *V. retusus*, suggest extensions or modifications of late Paleocene forms into the early Eocene Epoch. Poorly preserved specimens of a form similar to *V.* sp. cf. *V. meeki* were also identified from the Bridgerian Aycross Formation. Until additional specimens of these taxa are known from the Bighorn Basin, their biostratigraphic utility is somewhat limited. A form currently considered equivalent to *V.* sp. aff. *V. retusus* is also known from the Powder River Basin from near the stratigraphic horizon of the Roland coal bed (of Baker) and from the Kingsbury Conglomerate (Hartman 1984).

Viviparid specimens from other Fort Union Formation localities of uncertain stratigraphic position were identified as *Viviparus* sp. cf. *V. campaniformis* (L0947) and *Campeloma nebrascense nebrascense* (L0947, L2197). Hanley collected specimens of the former species (interpreted as an undescribed species by Hartman 1984) from a nearby locality (L4373) in the lacustrine member (as recognized by Hickey 1980) of the Fort Union Formation (J. H. Hanley, personal communication 1981). The lacustrine member is considered to be of Tiffanian Age on the basis of paleobotanical evidence (Hickey 1980). *Campeloma n. nebrascense* is best known from Tiffanian strata.

The only named species in the Bighorn Basin considered to represent the Wasatchian NALMA was *Viviparus paludinaeformis*. This taxon is widely known in rocks from the lower and middle Eocene Series (see Hartman 1984). In the Powder River Basin, *V. paludinaeformis* is known primarily from the main body of the "Wasatch" Formation, with most records from near or above the horizon of the Felix coal bed, considered here to be of Wasatchian Age. The only positively identified specimens from the upper Wasatchian Tatman Formation were also identified as *V. paludinaeformis*. Its occurrence in the Tatman was previously published by Rohrer and Gazin (1965).

Hartman (1984) reidentified the following taxa reported from the Willwood Formation. *Viviparus leidyi*? (L2312b), as identified by Russell (1931b), was referred to *V.* sp. cf. *V. meeki* and *V.* sp. indet. *Viviparus wyomingensis*, as reported by Jepsen (1930) and Russell (1931b), was reidentified by Taylor (1975) as *V.* n. sp., was reported by Hartman (1984) as *V.* sp. undet. *Viviparus* sp. aff. *V. raynoldsanus*, as reported by Taylor in Rohrer and Gazin (1965), was reidentified by Taylor (1975) as *V.* n. sp., and reported by Hartman (1984) as *V.* sp. cf. *V. meeki*. Following Taylor (1975), Russell's (1931b) new species, *Viviparus jepseni* (L2316), was tentatively reassigned to the terrestrial snail genus *Oreoconus*. Unevaluated records of *V.* sp. cf. *V. wyomingensis* (Willwood–Tatman undivided) and *V. wyomingensis*? were reported by Cockerell and Henderson (1912) and L. S. Russell in Van Houten (1944), respectively.

Gingerich identifications

Gingerich (1987, 1989b) published identifications and figures of snail specimens from the Clark's Fork Basin. Fossils identified from the middle Clarkforkian NALMA at UM SC-29 (L6280) included *Hendersonia evanstonensis*, *Oreohelix megarche*, cf. *Microphysa* sp., *Discus ralstonensis*, cf. *Pupilla* sp., cf. *Albertanella minuta*, *Grangerella sinclairi*, and cf. *G.? phenacodorum*. Gingerich (1989b) introduced the mammalian Biochron Wa0, representing the earliest Wasatchian NALMA, and reported eight localities bearing the nonmarine mollusk taxa *Elimia* or *Pleurocera* sp., "*Hydrobia*" sp., cf. *Gyraulus* sp., and *Holospira* sp.

Nonmarine Molluscan Biochronology

The composition of nonmarine molluscan faunas through the Paleocene and Eocene epochs (as at other times) is not static. Faunal assemblages changed from one mammalian biochron to another, or over a few biochrons. The following is a brief review of the nonmarine molluscan faunal compositions of the mammalian ages represented in the upper Paleocene and lower Eocene strata of the Bighorn Basin (see table 16.6). The ranges of the taxa mentioned below are shown by family in figure 16.5 (aquatics) and figure 16.6 (terrestrials) and in their totality in figure 16.7. Localities available to interpret these ranges and their stratigraphic horizons are given in tables 16.4 and 16.7–16.10 and in the Locality Register. In addition, faunal comparisons are made to other assemblages during this interval elsewhere in the northern portion of the Western Interior. Brief comments on the taxonomy or diagnosis of selected taxa are given in Comments on Bighorn Basin Taxa.

The Tiffanian NALMA

The molluscan record of the Tiffanian NALMA in the Bighorn Basin spans Biochron Ti2 or Ti3, from which a few taxa are known, to Biochron Ti5, during which a diverse association of aquatic mollusks thrived. In the Bighorn Basin, the Tiffanian fauna is largely known from Carbon County, Montana, which may represent a biofacies more similar to the northern Great Plains than to the Clarks Fork Basin to the south. In general, in the northern Great Plains and intermontane basins, the Puercan and Torrejonian record is dominated by aquatic prosobranchs. Species diversity is generally low, and localities are few. Toward the end of the Torrejonian Age, as represented by the molluscan record in the Crazy Mountains Basin of south-central Montana, both molluscan diversity and general abundance began to increase (Hartman 1989; Hartman and Krause 1993).

By the middle Tiffanian NALMA (Biochron Ti3), molluscan diversity was relatively high, with an increase in aquatic and terrestrial pulmonates (Hartman 1976, 1984, 1988). The Bighorn Basin molluscan record during the early and middle Tiffanian Age can be intervalized only as Biochron Ti2–Ti3. To date, no molluscan localities of an age clearly equivalent to Biochron Ti2 have been reported. The Bighorn Basin fauna of the Ti2–Ti3 biochronal interval can be correlated with similar faunas typically found elsewhere in the northern Great Plains in strata of Biochron Ti3, which include *Viviparus meeki*, *V. retusus*, *Campeloma nebrascense nebrascense*, *Lioplacodes nebrascensis nebrascensis*, *L. n. producta*, *L. tenuicarinata*, New Genus A *limneaformis*, unionids resembling *Plesielliptio priscus* or *P. wasatchensis*, generally small physids, and the oreohelicid New Genus S *spatiosa*.

Biochrons Ti4 or Ti5

Taxa known from localities that may be either Biochron Ti4 or Ti5 equivalent indicate a shift to greater representation of faunal elements inhabiting quieter water, representing, possibly, less bias toward collections made primarily from fluvial deposits. The pleurocerids, sans *L. tenuicarinata*, were present and well documented in Biozone Ti5. Although *C. nebrascense nebrascense* and New Genus A *limneaformis* persisted, with the former in abundance, a new species, derived from the *Viviparus meeki* complex, was present in the form of *Viviparus* sp. cf. *V.* sp. form H (see below). The fauna also includes forms similar to small hydrobiid and valvatid prosobranchs and a new aquatic pulmonate closely resembling *Pleurolimnaea tenuicosta*.

Small snails are uncommon and generally not well preserved in the collections. They include a form similar to *Clenchiella planospiralis* and morphs of *Hydrobia*, referred to as spp. form 1 and form 2, and a form referred to *V.* sp. cf. *V. subumbilicata*. Small physids are also represented. Unionids in the Bighorn Basin are generally poorly preserved. Unionid taxa that may be present in the Bighorn Basin include *Plesielliptio silberlingi*, a similar but apparently distinct morph, and *P.* sp. form 4. Although some of the taxon names introduced by Cockerell (1914a, 1915b) are based on adequate material, types of some species are poorly preserved, and their names are probably worthless as taxon designations (e.g., "*Unio*" *eomargaron*). In addition, the type material of unionids from the Bighorn Basin suffers particularly from poor locality and stratigraphic information. Pisidiid clams have their lowest occurrence in the Tiffanian deposits in the Bighorn Basin, although they are undoubtedly present, if poorly represented, in older Paleocene strata outside the Bighorn Basin. These small to tiny clams are generally only fairly well preserved, and, in all the collections, only a couple of specimens permit examination of hinge dentition, limiting accurate assignment to genus. The form identified in the Ti4–Ti5 zonal interval as *Sphaerium*? sp. form 3 also occurs in strata of Clarkforkian Age.

Biochron Ti5

Biochron Ti5 is well documented by several localities and taxa (figure 16.8, table 16.6). The fauna appears more diverse than in the preceding biochrons, with the addition of more aquatic pulmonates. However, the record of many taxa is not good because many taxa are known only from a few poorly preserved specimens. Introductions include a small representative of *Acroloxus*,

Table 16.6
Nonmarine Molluscan Faunas of Tiffanian–Wasatchian Land–Mammal Bichrons in the Bighorn Basin

Ti2–Ti3 Biochrons
Oreohelicidae
New Genus S *spatiosa*
Pleuroceridae
Liplacodes nebrascensis nebrascensis
Liplacodes nebrascensis producta
Lioplacodes tenuicarinata?
Unionidae
Plesielliptio wasatchensis?
"*Unio*" sp. indet.
Viviparidae
Campeloma nebrascense nebrascense?

Ti3–Ti4 Biochrons
Viviparidae
Viviparus retusus

Ti3–Ti5 Biochrons
incertae sedis
New Genus A *limneaformis*
Physidae
Physa sp. undet.
Pleuroceridae
Lioplacodes tenuicarinata
Viviparidae
Campeloma nebrascense nebrascense
Viviparus meeki
Unionidae
Plesielliptio priscus?

Ti4–Ti5 Biochrons
Ellobiidae
Pleurolimnaea cf. *P. tenuicosta*
Hydrobiidae
Clenchiella cf. *C. planospiralis*?
Hydrobia sp. form 1
Hydrobia sp. form 2
incertae sedis
New Genus A *limneaformis*
Physidae
Physa sp. undet.
Pisidiidae
Sphaerium? sp. form 3
Pleuroceridae
Lioplacodes nebrascensis nebrascensis
Lioplacodes nebrascensis producta
Unionidae
Plesielliptio silberlingi
Plesielliptio cf. *P. silberlingi*
Plesielliptio sp. form 4
Valvatidae
Valvata cf. *V. subumbilicata*
Viviparidae
Campeloma nebrascense nebrascense
Viviparus cf. *V.* sp. form H

Ti5 Biochron
Acroloxidae
Acroloxus minutus?
Acroloxus? sp. indet.
?Bithyniidae
?*Reesidella nana* (operculum)
Ellobiidae
Pleurolimnaea cf. *P. tenuicosta*?
Hydrobiidae
Hydrobia sp. form 1
Hydrobia sp. form 2
incertae sedis
New Genus A *limneaformis*
Oreohelicidae
New Genus S *spatiosa*
Physidae
Physa sp. undet.
Pisidiidae
Eupera? sp. form 4
Sphaerium sp. form 1
Sphaerium sp. form 2
Sphaerium? sp. form 3?
Sphaerium sp. form 5
Planorbidae
planorbid sp. indet.
?planorbid sp. indet.
Pleuroceridae
Lioplacodes nebrascensis nebrascensis
Lioplacodes nebrascensis producta
?*Lioplacodes* sp. form 2
Unionidae
Plesielliptio silberlingi
"*Plesielliptio*" sp. form 4?
"*Unio*" sp. form 1
"*Unio*" sp. form 2
Viviparidae
Campeloma nebrascense nebrascense
Viviparus retusus
Viviparus cf. sp. form H
Viviparus sp. undet.

Ti5–Cf1 Biochrons
Helicinidae
"*Helicina*" *evanstonensis*
Oreohelicidae
cf. New Genus S sp. undet. 1
Pisidiidae
Eupera? sp. indet.
Sphaerium cf. *S. recticardinale*
Pleuroceridae
Lioplacodes tenuicarinata?

Cf–Wa Biochrons
Unionidae
Plesielliptio wasatchensis

Table 16.6 *(continued)*

Cf1 Biochron
Hendersoniidae
Grangerella sinclairi
Crangerella sp. undet.
Hydrobiidae
Hydrobia anthonyi
Hydrobia sp. form 1?
Hydrobia sp. undet.
incertae sedis
New Genus A cf. *limneaformis*
Oreohelicidae
"*Gastrodonta*" *coryphodontis*
New Genus S *spatiosa*
Radiocentrum cf. *R. grangeri*
Physidae
Physa sp. undet.
Pisidiidae
Pisidiidae gen. and sp. indet.
Pleuroceridae
Lioplacodes sp. form 1
Lioplacodes sp. undet.
Viviparidae
Viviparus formosus?
Viviparus retusus
Viviparus sp. form H
Unionidae
Plesielliptio sp. indet.

Cf2 Biochron
Acroloxidae
Acroloxus minutus
Acroloxus radiatulus
Acroloxus cf. *A radiatulus*
Bithyniidae
Reesidella nana (and operculum)
Bulmulidae
cf. *Oreoconus* sp. indet.
Discidae
Discus ralstonensis
Discus cf. *D. ralstonensis*
Discus sp. form I
?Discidae gen. and sp. undet.
Ellobiidae
Pleurolimnaea cf. *P. tenuicosta*
Haplotrematidae
Ancotrema sp. form E
Helicinidae
"*Helicina*" *evanstonensis*
"*Helicina*" cf. "*H.*" *evanstonensis*
"proserpinid" sp. form A
"proserpinid" sp. form B
"proserpinid" sp. form C
cf. *Tozerpina* sp. undet.
Helicodiscidae
Helicodiscus? sp. form H
Helminthoglyptidae
Glypterpes cf. *G. veternior*
Hendersoniidae
Grangerella phenacodorum
Grangerella sinclairi
Grangerella cf. *G. sinclairi*
Grangerella sp. form D
Hendersoniidae sp. undet. (operculum)
Protoboysia complicata
Hydrobiidae
Clenchiella cf. *C. planospiralis*
Clenchiella sp. undet.
Hydrobia anthonyi
Hydrobia sp. form 1?
Hydrobia sp. undet.-indet.
Megomphicidae
Polygyrella cf. *P. parvula*
Oreohelicidae
"*Eucalodium*" *eophilum*
"*Gastrodonta*" *coryphodontis*
New Genus L sp. form K
New Genus S *spatiosa*
New Genus S sp. undet.
Radiocentrum grangeri
Radiocentrum cf. *R. grangeri*
Radiocentrum sp. form J
Physidae
Physa sp. form 1?
Physa sp. form 2
Physa sp. form 5
Physa sp. form 6
Physa sp. form 7
Physa sp. undet.
Pisidiidae
Eupera sp. form 7
Eupera? sp. form 8
pisidiid sp. form 6
Sphaerium sp. form 1
Sphaerium sp. form 2
Sphaerium? sp. form 3
Pisidiidae gen. and sp. undet.
Planorbidae
Planorbinae sp. indet.
planorbid sp. undet.
Pleuroceridae
Lioplacodes tenuicarinata
Lioplacodes sp. undet.
Pupillidae
Albertanella cf. *A.* sp. undet.
Pupillidae gen. and sp. undet.
Thysanophoridae
Microphysula cf. *M. ingersolli*
Unionidae
Plesielliptio sp. indet.
"*Unio*" sp. form 1?

Table 16.6 *(continued)*

Cf2 Biochron *(continued)*
Valvatidae
Valvata cf. *V. bicincta*
Valvata cf. *V. subumbilicata*?
Valvata sp. form 1
Valvata sp. undet.
Viviparidae
Viviparus retusus
Viviparus sp. form H
Viviparus sp. undet.
Zonitidae
?Zonitidae gen. and sp. undet.

Cf2–Cf3 Biochrons
Helicinidae
"*Helicina*" *evanstonensis*
Hendersoniidae
Grangerella sp. undet.
Physidae
Physa sp. form 7

Cf3 Biochron
Discidae
Discus ralstonensis
Discus sp. form I
Haplotrematidae
Ancotrema sp. form E
Haplotrema sp. form G
Helminthoglyptidae
Glypterpes cf. *G. veternior*
Hendersoniidae
Grangerella phenacodorum
Grangerlla sinclairi
Grangerella cf. *G. sinclairi*
Grangerella sp. form D
Protoboysia complicata
incertae sedis
New Genus A *limneaformis*
Megomphicidae
Polygyrella cf. *P. parvula*
Oreohelicidae
"*Eucalodium*" *eophilum*
"*Gastrodonta*" *coryphodontis*
New Genus L sp. form K
New Genus S *spatiosa*
New Genus S cf. *spatiosa*
Radiocentrum grangeri
Radiocentrum cf. *R. grangeri*
Physidae
Physa sp. form 1
Physa cf. P. sp. form 1?
Physa sp. form 5
Physa sp. form 6
Physa sp. undet.
Planorbidae
Planorbella sp. form 1
Planorbinae sp. undet.
Pleuroceridae
Elimia sp. form 1
Lioplacodes sp. undet.-indet.
Lioplacodes tenuicarinata
Pupillidae
Pupillidae gen. and sp. undet.
?Pupillidae gen. and sp. undet.
Unionidae
Plesielliptio sp. form 4
?*Plesielliptio silberlingi*
"*Unio*" *eomargaron*
"*Unio*" sp. form 1
"*Unio*" sp. form 3
"*Unio*" sp. undet.
Viviparidae
Viviparus retusus
Viviparus sp. form H
Viviparus sp. undet.

Cf3–Wa0 Biochrons
Unionidae
"*Unio*" "*wasatchensis*"

Wa0 Biochron
Hydrobiidae
Hydrobia aff. *H.* sp. form 2
Planorbidae
Biomphalaria storchi?
Pleuroceridae
Elimia sp. form 1
Elimia sp. form 2
Elimia sp. form 3?
Elimia? sp. form 4
Urocoptidae
Holospira sp. undet.
Viviparidae
Viviparus sp. indet.

Wa1 Biochron
Bulimulidae
cf. *Oreoconus* sp. indet.
Discidae
Discus ralstonensis
Discus cf. *D ralstonensis*
Discus sp. form I
Haplotrematidae
Haplotrema cf. sp. form G
Helicinidae
Eohipptychia eohippina
"*Helicina*" *evanstonensis*
Helminthoglyptidae
Glypterpes cf. *G. veternior*
Noyo? sp. form N

Table 16.6 *(continued)*

Wa1 Biochron *(continued)*
Hendersoniidae
Grangerella sinclairi
Grangerella cf. *G. sinclairi*
Protoboysia complicata
Oreohelicidae
"*Eucalodium*" *eophilum*
"*Eucalodium*" cf. "*E.*" *eophilum*
"*Gastrodonta*" *coryphodontis*
New Genus S *spatiosa*
New Genus S cf. *spatiosa*
cf. New Genus S sp. undet. 2
Physidae
Physa sp. form 1
Physa sp. form 2
Physa sp. form 3
Physa sp. form 4
Pleuroceridae
Elimia sp. form 3
Elimia sp. indet.
Polygyridae
Neohelix? sp. form M
Unionidae
"*Unio*" sp. indet.
Viviparidae
Viviparus wyomingensis
Viviparus sp. indet.

Wa2 Biochron
?Bradybaenidae
New Genus P *jepseni*
Bulimulidae
cf. *Oreoconus* sp. indet.
Discidae
Discus ralstonensis
Discus cf. *D. ralstonensis*
Discus sp. form I
Helicinidae
"*Helicina*" *evanstonensis*
"*Helicina*" cf. "*H.*" *evanstonensis*
"proserpinid" sp. form B
"proserpinid" sp. form C
Helminthoglyptidae
Noyo? sp. form N
Hendersoniidae
Grangerella sinclairi
Grangerella cf. *G. sinclairi*
Grangerella sp. form D
Grangerella cf. *G.* sp. form D
Grangerella sp. undet.
Protoboysia complicata
Oreohelicidae
"*Eucalodium*" *eophilum*
New Genus L sp. form K
New Genus S *spatiosa*
New Genus S cf. *spatiosa*
Physidae
Physa sp. form 1
Physa sp. form 4?
Planorbidae
Biomphularia storchi
Pleuroceridae
Elimia sp. form 3?
Pupillidae
Pupillidae gen. and sp. undet.
Thysanophoridae
Microphysula oxyaenae
Microphysula cf. *M. oxyaenae*
Stylommatophora
Stylommatophora gen. and sp. undet.
Viviparidae
Viviparus aff. *V. retusus*
Viviparus wyomingensis

Wa2–Wa3 Biochrons
Helminthoglyptidae
Glypterpes veternior
Thysanophoridae
Microphysula sinoparum

Wa3 Biochron
?Bradybaenidae
New Genus P *jepseni*
Discidae
Discus ralstonensis
Haplotrematidae
Ancotrema sp. form E
Haplotrema sp. form F
Helicinidae
"*Helicina*" *evanstonensis*
"*Helicina*" cf. "*H.*" *evanstonensis*
Helicinidae gen. and sp. undet.
Helminthoglyptidae
Noyo? sp. form N
Noyo? cf. *N.*? sp. form N
Hendersoniidae
Grangerella sinclairi
Grangerella sp. form D
Protoboysia complicata
Oreohelicidae
"*Gastrodonta*" *coryphodontis*
"*Gastodonta*" cf. "*G.*" *coryphodontis*
New Genus S *spatiosa*
New Genus S cf. *spatiosa*
Physidae
Physa sp. form 1
Physa sp. form 2?
Physa sp. form 4?
Pleuroceridae
Elimia sp. indet.
Unionidae
"*Unio*" sp. indet.
Viviparidae
Viviparus wyomingensis
Viviparus sp. indet.

Table 16.6 *(continued)*

Wa3–Wa4 Biochrons
?Bradybaenidae
New Genus P *jepseni*
Physidae
Physa sp. form 1
Planorbidae
Biomphalaria storchi
Oreohelicidae
New Genus S *spatiosa*
Unionidae
"Unio" didymictidis
"Unio" sinopae

Wa4 Biochron*
Viviparidae
Viviparus sp. indet.

*Taxa are listed for a particular biochron only if they have been identified at a locality within the biochron. Thus other taxa may also occur through the specified interval.

Families are listed alphabetically. Because of poor stratigraphic and location information, taxa from many pre-Gingerich and Hanley localities cannot be placed precisely into a land-mammal biochron. Thus some taxa are given as spanning more than one biochron. Generally speaking, indeterminate taxa are excluded from the above list except where a taxon of the family is not otherwise represented.

*The fauna listed for biochron Wa4 does not reflect the taxa that probably occur during this time. The scope of this project was intended to include the fauna through the early Wasatchian NALMA and to include Wa3 localities so as to show the transition into the middle Wasatchian.

possibly *A. minutus*, and indeterminate planorbids. Additional new taxa include pisidiids, possibly representing four additional morphologies (see table 16.6), an operculum that may be referable to *Reesidella* (see below), and unionid spp. form 1 and form 2. Taxa that continued into Biochron Ti5 include *Plesielliptio silberlingi*, possibly *P.* sp. form 4 and *Sphaerium*? sp. form 3, *Viviparus retusus*, *V.* sp. cf. *V.* sp. form H, *Campeloma nebrascense nebrascense*, New Genus A *limneaformis*, the taxa of *Lioplacodes nebrascensis*, *Pleurolimnaea* sp. cf. *P. tenuicosta*, undetermined physids, and New Genus S *spatiosa*. Five localities could only be assigned to either Biochron Ti5 or Cf1. These localities apparently add new elements to the fauna, not known, as yet, in either biochron, including cf. New Genus S sp. undet. 1 and *Sphaerium* sp. cf. *S. recticardinale*, but both identifications are based on inadequate material. The identification of *L. tenuicarinata* from this interval is based on poor material, but its presence at this time is likely, as it also occurs in overlying strata of Clarkforkian Age.

Regional Tiffanian correlations

The Powder River Basin contains strata that must be representative of Biochron Ti5 and Clarkforkian NALMA (Hartman 1990, 1992, 1993a, 1993b). Unfortunately, only a few upper Paleocene mammalian localities are known from the Powder River Basin. These include the Olive Locality (Wolberg 1978; Archibald et al. 1987), Newell's Nook (Robinson and Honey 1987), and a locality in Big Horn County, Montana (Rose and Bown 1982). The first two are early and late, but not latest, Tiffanian in age. The third, on the basis of only a few mammal teeth, is probably of Clarkforkian Age. Thus there is minimal mammalian age control through

Table 16.7
Bighorn Basin Aquatic Molluscan Biostratigraphy–Taxon Range Numbers for Figure 5

1 (6)	14 (12)	27 (35)	40 (9)	53 (23)
2 (8)	15 (13)	28 (37)	41 (26)	54 (50)
3 (24)	16 (14)	29 (98)	42 (32)	55 (51)
4 (27)	17 (16)	30 (107)	43 (90)	55 (56)
5 (28)	18 (42)	31 (20)	44 (92)	57 (57)
6 (36)	19 (43)	32 (66)	45 (93)	58 (65)
7 (84)	20 (64)	33 (15)	46 (95	59 (103)
8 (85)	21 (18)	34 (17)	47 (25)	60 (104)
9 (91)	22 (44)	35 (30)	48 (33)	61 (22)
10 (114)	23 (45)	36 (34)	49 (19)	62 (88)
11 (115)	24 (3)	37 (46)	50 (21)	63 (96)
12 (7)	25 (5)	38 (1)	51 (47)	
13 (10)	26 (29)	39 (2)	52 (48)	

Parenthetical numbers refer to taxon range numbers ordered in figure 16.7 and table 16.9.

Table 16.8
Bighorn Basin Terrestrial Molluscan Biostratigraphy–Taxon Range Numbers for Figure 6

1 (31)	12 (59)	23 (105)	34 (38)	45 (49)
2 (63)	13 (67)	24 (116)	35 (40)	46 (99)
3 (73)	14 (72)	25 (94)	36 (54)	47 (69)
4 (77)	15 (79)	26 (60)	37 (61)	48 (106)
5 (78)	16 (111)	27 (76)	38 (74)	49 (108)
6 (82)	17 (112)	28 (52)	39 (75)	50 (109)
7 (97)	18 (53)	29 (62)	40 (83)	51 (70)
8 (118)	19 (71)	30 (68)	41 (89)	52 (113)
9 (39)	20 (87)	31 (81)	42 (100)	53 (58)
10 (41)	21 (80)	32 (4)	43 (101)	54 (102)
11 (55)	22 (97)	33 (11)	44 (117)	55 (110)
				56 (119)

Parenthetical numbers refer to taxon range numbers ordered in figure 16.7 and table 16.9.

Table 16.9
Bighorn Basin Nonmarine Molluscan Biostratigraphy–Taxon Range Numbers and Names for Figures 16.5 to 16.7

1) *Lioplacodes nebrascense nebrascense:* L0136, L4373, L5709, L5711, L5746, L5747
2) *Lioplacodes nebrascense producta:* L0136, L4373, L5709?, L5710, L5711, L5746, L5747, L5891
3) *Campeloma nebrascense nebrascense:* L0134, L0136?, L4373, L5709, L5710, L5711, L5741, L5747
4) New Genus S *spatiosa:* L0136, L4284, L5695, L5696, L5754, L5757, L5758a–b, L5765, L5766, L5767, L5789, L5804, L5831, L5842a, L5846, L5854a, L5854c, L5857, L5860a–b, L5862, L5870, L5876, L5878, L5890, L5904, L6262, L6263, L6264, L6265, L6266, L6268, L6272, L6273, L6281, L6284, L6285, L6286, L6288, L6293, L6311, L6319, L6322, L6330, L6330c, L6464
5) *Viviparus* cf. *V.* sp. form H: L4373, L5710, L5742, L5744?, L5745, L5746, L5747, L5891, L5902, L5903?
6) "*Unio*" sp. form 2: L5868, L5903
7) *Eupera*? sp. form 4: L4373?, L5831
8) *Plesielliptio* cf. *P. silberlingi:* L5711, L5742
9) ?*Lioplacodes* sp. form 2: L6326
10) *Sphaerium*? sp. form 5: L4373
11) New Genus S cf. sp. undet. 1: L5910
12) *Sphaerium* cf. *S. recticardinale:* L5699b
13) *Sphaerium* sp. form 2: L4373?, L5888, L5892
14) *Sphaerium* sp. form 1: L5892, L5906
15) *Clenchiella* cf. *C. planospiralis:* L5742?, L5883, L5888
16) *Sphaerium*? sp. form 3: L4373?, L5831?, L5701, L5702, L5742, L5888, L5906
17) *Hydrobia* sp. form 1: L4373, L5702?, L5711, L5712?, L5742, L5892
18) *Valvata* cf. *V. subumbilicata:* L5701?, L5711
19) *Pleurolimnaea* cf. *P. tenuicosta:* L5701?, L5711, L5892, L5902
20) *Reesidella nana*? (opercula): L5472, L5701, L5895a–b, L5906
21) *Acroloxus minutus:* L5701, L5831?
22) planorbid sp. indet. and undet.: L5701, L5701, L5702, L5805, L5809, L5831, L5859
23) *Physa* sp. undet.: L5742, L5803, L5849, L5850, L5884a, L5885, L5890, L5906
24) "*Unio*" sp. form 1: L0135?, L5868, L5832, L5835?, L5886?, L5903
25) New Genus A *limneaformis:* L0135?, L4373, L5710, L5711, L5741, L5746, L5747, L5835
26) *Lioplacodes tenuicarinata:* L5699b, L5701, L5832, L5835?, L5851, L5859, L5897
27) *Plesielliptio* sp. form 4: L0135?, L5742, L5832?, L5835
28) *Plesielliptio silberlingi:* L4373, L5709, L5711?, L5741, L5746?, L6329
29) *Viviparus retusus:* L0134, L0135?, L3952b, L5808, L5835, L5868?, L5888?, L5890, L5893, L5903, L5906, L5913?, L6300, L6303
30) *Hydrobia* sp. form 2 (and *H.* aff. *H.* sp. form 2-L6248): L4373, L5711, L5892, L5902
31) *Helicina evanstonensis:* L5749, L5757, L5839, L5842a, L5847, L5849, L5878, L5879, L5881, L5910, L6264, L6272, L6286, L6296, L6322
32) *Lioplacodes* sp. form 1: L5712
33) New Genus A cf. "*Lioplacodes*" *limneaformis:* L0855, L0856
34) *Hydrobia anthonyi:* L5694?, L5701?, L5712, L5713?, L5906
35) *Viviparus* sp. form H: L0855, L0856, L5697, L5799, L5832, L5835, L5859, L5906, L6465?
36) *Plesielliptio wasatchensis:* L0136?, L4282b
37) *Viviparus formosus*?: L5890
38) *Radiocentrum* cf. *R. grangeri:* L5758a–b, L5862, L5875a–b, L5883, L5885, L5890
39) *Grangerella* sp. undet.: L5823, L5849, L5890
40) "*Gastrodonta*" *coryphodontis:* L3822, L5758a–b, L5767, L5844a–c, L5846, L5854a, L5878, L5889, L5890, L5895a–b, L6285, L6311
41) *Grangerella sinclairi:* L4110, L5749, L5751, L5803, L5808, L5839, L5842, L5844a–c, L5846, L5849, L5850, L5854a, L5854b, L5860a–b, L5862, L5863, L5875a–b, L5878, L5879, L5881, L5882, L5885, L5889, L5890, L5893, L5899, L5907, L6262, L6263, L6265, L6268, L6269, L6272, L6273, L6274, L6281, L6285, L6286, L6293, L6294, L6311, L6319, L6322, L6330, L6330b, L6464
42) *Eupera* sp. form 7: L5701
43) Pisidiidae gen. undet. sp. form 6: L5702
44) *Valvata* cf. *V. bicincta:* L5701, L5702
45) *Valvata* sp. form 1: L5701
46) *Clenchiella* sp. undet.: L5694
47) *Acroloxus radiatulus:* L5700, L5701

Table 16.9 *(continued)*

48) *Acroloxus* cf. *A. radiatulus:* L5701
49) ?Zonitidae gen. and sp. undet.: L5700
50) *Physa* sp. form 7: L5701, L5849
51) *Physa* sp. form 5: L5835, L5857, L5880?, L5881, L5884a, L5896
52) *Discus* cf. *D. ralstonensis:* L5844a–c, L5875a–b, L5882, L5885, L5889, L5904, L5906, L6262, L6273, L6278
53) cf. *Albertanella* sp. undet.: L5871
54) *Radiocentrum grangeri:* L4282a, L5799, L5847, L5854c, L5904, L5906, L6280, L6294, L6309
55) *Grangerella* sp. form D: L5803, L5823, L5844a–c, L5878, L5895a–b, L5904, L6285, L6293
56) *Physa* sp. form 6: L5851, L5896
57) *Physa* sp. form 2: L5757, L5767?, L5879, L5895a–b?, L5896, L5906, L6330a?
58) *Glypterpes* cf. *G. veternior:* L5749, L5758a–b, L5799, L5802, L5805, L5839, L5842, L5842a, L5854a, L5854c, L5878, L5879, L5893, L5895a–b
59) *Grangerella* cf. *G. sinclairi:* L5696, L5748a–b, L5749, L5758a–b, L5765, L5789, L5799, L5800, L5802, L5803, L5810, L5895a–b
60) cf. *Oreoconus* sp. indet.: L5895a–b, L6268, L6274, L6288
61) New Genus L sp. form K: L5799, L5803, L5839, L5842, L5842a, L5844a–c, L5846, L5875a–b, L5878, L5893, L6268, L6313
62) *Discus* sp. form I: L5806, L5809, L5860a–b, L5895a–b, L5905, L5908, L5908a, L6263, L6265, L6324
63) *Helicina* cf. *H. evanstonensis:* L5895a–b
64) *Eupera*? sp. form 8: L5888, L5906?
65) *Physa* sp. form 1: L2311, L5757, L5762?, L5859?, L5867, L5763?, L5823?, L5888?, L6262, L6263?, L6298, L6317, L6322, L6330a
66) *Reesidella nana:* L5906
67) Hendersoniidae gen. and sp. undet. (operculum): L5907
68) ?Discidae gen. and sp. undet.: L5875a–b
69) *Microphysula* cf. *M. ingersolli:* L5875a–b
70) *Polygyrella* cf. *P. parvula:* L5875a–b, L6303
71) Pupillidae gen. and sp. undet.: L5789, L5799, L5803, L5875a–b
72) *Grangerella phenacodorum:* L3822, L5758a–b, L5844a–c, L5881, L5907
73) "proserpinid" sp. form A: L5844a–c
74) *Radiocentrum* sp. form J: L5842a
75) New Genus S sp. undet.: L6313
76) *Helicodiscus*? sp. form H: L5840, L5878
77) "proserpinid" sp. form B: L5823, L5842a, L5844a–c
78) "proserpinid" sp. form C: L5823, L5847
79) *Protoboysia complicata:* L3822, L5758a–b, L5767, L5789, L5799, L5823, L5841, L5850, L6263, L6265, L6268, L6285
80) *Ancotrema* sp. form E: L5799, L5842a, L5846, L5847, L5854c, L6285
81) *Discus ralstonensis:* L3821, L5757, L5789, L5805, L5823, L5842, L5844a–c, L5847, L5854c, L5857, L5878, L5881, L6263, L6265, L6268, L6269, L6272, L6274, L6281, L6285, L6287, L6311, L6322, L6322, L6330, L6330b
82) cf. *Tozerpina* sp. undet.: L5881
83) "*Eucalodium*" *eophilum:* L3820, L5749, L5765, L5799, L5802, L5803, L5829, L5854c, L5867, L5879, L5881, L6265, L6309
84) "*Unio*" *eomargaron:* L3952a
85) "*Unio wasatchensis*": L4282a
86) *Haplotrema* sp. form G: L5799, L5803, L5835
87) ?Pupillidae gen. and sp. undet.: L5803
88) *Planorbella* sp. form 1: L5758a–b
89) New Genus S cf. *spatiosa:* L5749, L6262, L6265, L6274, L6269, L6330
90) *Elimia* sp. form 1: L6252, L6324
91) "*Unio*" sp. form 3: L5835
92) *Elimia* sp. form 2: L6248?, L6249?, L6250, L6252
93) *Elimia*? sp. form 4: L6249
94) *Holospira* sp. undet.: L6248, L6251, L6253, L6255
95) *Eohipptychia eohippina:* L3952c
96) *Elimia* sp. form 3: L3821, L5757, L5763?, L5899, L6248?, L6316, L6317, L6263?
97) *Biomphalaria storchi:* L2311, L5823, L6249?
98) *Viviparus wyomingensis:* L2312f, L2313?, L2313d, L2313f, L2313g, L2313h?, L2315, L5748a–b, L5749, L5751, L5756, L5757, L5763, L5765?, L5766?, L5767?, L5823, L5899, L6263?, L6265, L6267?, L6283, L6284, L6290, L6292?, L6301?, L6305?, L6316, L6317, L6319, L6322?, L6464

Table 16.9 *(continued)*

99) *Neohelix*? sp. form M: L5757
100) *Eucalodium*" cf. "*E.*" *eophilum:* L6265, L6278
101) New Genus S cf. sp. undet. 2: L5767
102) *Noyo*? sp. form N: L6263, L6265, L6272, L6281, L6322, L6330
103) *Physa* sp. form 3: L6274
104) *Physa* sp. form 4: L5762?, L6262?, L6271?, L6275
105) *Haplotrema* cf. *H.* sp. form G: L5749
106) *Microphysula* cf. *M. oxyaenae:* L6268
107) *Viviparus* aff. *V. retusus:* L5698, L6288
108) *Microphysula oxyaenae:* L3822, L6272
109) *Microphysula sinoparum:* L3822
110) *Glypterpes veternior:* L3822
111) *Grangerella* cf. *G.* sp. form D: L6272
112) Stylommatophora gen. and sp. undet.: L6272
113) New Genus P *jepseni:* L2312e, L2316, L2317, L5736?, L6319, L6263?, L6271, L6271, L6301?
114) "*Unio*" *didymictidis:* L4283
115) "*Unio*" *sinopae:* L4283
116) *Haplotrema* sp. form F: L6330
117) "*Gastrodonta*" cf. "*G.*" *coryphodontis:* L6281, L6285
118) Helcinidae gen. and sp. undet.: L6285
119) *Noyo*? cf. *N.*? sp. form N: L6285

In most cases, indeterminate taxa (e.g., *Viviparus* sp. indet., for which other identifications have been made) were not included in the range figure. Positive and questioned (e.g., L6330, L6285?, respectively) identifications represent the identifications used to construct the range chart (see figure 16.6). See the Locality Register for location and age information for a given locality.

Table 16.10
Stratigraphic Horizons of Bighorn Basin Nonmarine Molluscan Localities*

Hor.	Locality	Formation	Age	m >K/T	Hor.	Locality	Formation	Age	m >K/T
1.00	L5868	Fort Union	Ti5	840.0	14.01	L5839	Willwood	Cf2	1370.0
2.00	L5890	Fort Union	Cf1	1090.0	14.02	L5840	Willwood	Cf2	1370.0
3.01	L5886	Fort Union	Cf2	1160.0	14.03	L5842	Willwood	Cf2	
3.02	L5885	Fort Union	Cf2	1160.0	14.04	L5842a	Willwood	Cf2	
3.03	L5884a	Fort Union	Cf2	1160.0	14.05	L5841	Willwood	Cf2	1370.0
3.04	L5884b	Fort Union	Cf2	1160.0	14.06	L5844a–c	Willwood	Cf2	
4.00	L5874	Willwood	Cf2	1200.0	14.07	L5847	Willwood	Cf2	1370.0
5.00	L5870	Willwood	Cf2	1210.0	14.08	L5846	Willwood	Cf2	1370.0
6.01	L5904	Willwood	Cf2	1230.0	14.09	L6313	Willwood	Cf2	1370.0
6.02	L5871	Willwood	Cf2	1230.0	15.01	L5808	Willwood	Cf2	1380.0
7.00	L6300	Willwood	Cf2	1270.0	15.02	L5805	Willwood	Cf2	1380.0
8.00	L5897	Willwood	Cf2	1275.0	15.03	L5806	Willwood	Cf2	1380.0
8.01	L5896	Willwood	Cf2	1275.0	15.04	L5878	Willwood	Cf2	1380.0
9.01	L5894	Willwood	Cf2	1280.0	15.05	L5876	Willwood	Cf2	
9.02	L5893	Willwood	Cf2	1280.0	15.06	L5883	Willwood	Cf2	1380.0
9.03	L5895a–b	Willwood	Cf2	1280.5	15.07	L5879	Willwood	Cf2	1380.0
10.00	L5889	Willwood	Cf2	1300.0	15.08	L5880	Willwood	Cf2	1380.0
11.00	L5888	Willwood	Cf2	1335.0	15.09	L5882	Willwood	Cf2	1380.0
12.01	L5906	Willwood	Cf2	1355.0	15.10	L5881	Willwood	Cf2	1380.0
12.02	L5907	Willwood	Cf2	1355.0	15.11	L6296	Willwood	Cf2	1380.0
12.03	L5875a–b	Willwood	Cf2	1355.0	16.00	L6280	Willwood	Cf2	1385.0
12.04	L5908	Willwood	Cf2	1355.0	17.00	L6294	Willwood	Cf2	1405.0
12.05	L5908a	Willwood	Cf2	1355.0	18.00	L5810	Willwood	Cf2–Cf3	1430.0
12.06	L5875b	Willwood	Cf2	1355.0	19.01	L5849	Willwood	Cf3	1447.0
13.00	L5905	Willwood	Cf2	1359.9	19.02	L5850	Willwood	Cf3	1447.0

Table 16.10 *(continued)*

Hor.	Locality	Formation	Age	m >K/T	Hor.	Locality	Formation	Age	m >K/T
20.00	L6303	Willwood	Cf3	1455.0	33.03	L5765	Willwood	Wa1	1570.0
21.01	L5859	Willwood	Cf3	1460.0	33.04	L5748a–b	Willwood	Wa1	1570.0
21.02	L5867	Willwood	Cf3	1460.0	33.05	L6265	Willwood	Wa1	1570.0
21.03	L5860a–b	Willwood	Cf3	1460.0	34.00	L6274	Willwood	Wa1	1600.0
22.01	L5799	Willwood	Cf3	1465.0	35.00	L6305	Willwood	Wa1	1610.0
22.02	L5804	Willwood	Cf3	1465.0	36.00	L6275	Willwood	Wa1	1620.0
22.03	L5809	Willwood	Cf3	1465.0	37.01	L6464	Willwood	Wa1	1630.0
22.04	L5851	Willwood	Cf3	1465.0	37.02	L5756	Willwood	Wa1	1630.0
22.05	L5854a	Willwood	Cf3	1465.0	37.03	L5749	Willwood	Wa1	1630.0
22.06	L5854b	Willwood	Cf3	1465.0	37.04	L5751	Willwood	Wa1	1630.0
22.07	L5854c	Willwood	Cf3	1465.0	37.05	L5754	Willwood	Wa1	1630.0
22.08	L6309	Willwood	Cf3	1465.0	37.06	L6273	Willwood	Wa1	1630.0
22.09	L5802	Willwood	Cf3	1465.0	37.07	L6287	Willwood	Wa1	1630.0
23.00	L5857	Willwood	Cf3	1480.0	38.01	L6288	Willwood	Wa2	1645.0
24.01	L5803	Willwood	Cf3	1485.0	38.02	L6268	Willwood	Wa2	1645.0
24.02	L5800	Willwood	Cf3	1485.0	38.03	L5789	Willwood	Wa2	1645.0
25.01	L5758a–b	Willwood	Cf3	1495.0	39.01	L6262	Willwood	Wa2	1665.0
25.02	L5832	Willwood	Cf3	1495.0	39.02	L5823	Willwood	Wa2	1665.0
25.03	L6269	Willwood	Cf3	1495.0	40.01	L5829	Willwood	Wa2	1685.0
25.04	L6270	Willwood	Cf3	1495.0	40.02	L6272	Willwood	Wa2	1685.0
26.01	L5862	Willwood	Cf3	1500.0	41.00	L6293	Willwood	Wa2	1690.0
26.02	L5863	Willwood	Cf3	1500.0	42.00	L6292	Willwood	Wa2	1700.0
26.03	L6324	Willwood	Cf3	1500.0	43.01	L6271	Willwood	Wa2	1720.0
27.01	L5835	Willwood	Cf3	1505.0	43.02	L6263	Willwood	Wa2	1720.0
27.02	L5836	Willwood	Cf3	1505.0	44.01	L5736	Willwood	Wa3	1750.0
27.03	L6277	Willwood	Cf3	1505.0	44.02	L6266	Willwood	Wa3	1750.0
27.04	L6329	Willwood	Cf3	1505.0	44.03	L6322	Willwood	Wa3	1750.0
28.01	L6248	Willwood	Wa0	1520.0	45.01	L6281	Willwood	Wa3	1760.0
28.02	L6249	Willwood	Wa0	1520.0	45.02	L6301	Willwood	Wa3	1760.0
28.03	L6250	Willwood	Wa0	1520.0	45.03	L6330	Willwood	Wa3	1760.0
28.04	L6251	Willwood	Wa0	1520.0	45.04	L6330a	Willwood	Wa3	1760.0
28.05	L6252	Willwood	Wa0	1520.0	45.05	L6330b	Willwood	Wa3	1760.0
28.06	L6253	Willwood	Wa0	1520.0	45.06	L6330c	Willwood	Wa3	1760.0
28.07	L6254	Willwood	Wa0	1520.0	46.00	L6283	Willwood	Wa3	1840.0
28.08	L6255	Willwood	Wa0	1520.0	47.01	L6284	Willwood	Wa3	1850.0
29.00	L6267	Willwood	Wa1	1530.0	47.02	L6319	Willwood	Wa3	1850.0
30.01	L5899	Willwood	Wa1	1535.0	48.00	L6285	Willwood	Wa3	1870.0
30.02	L5757	Willwood	Wa1	1535.0	49.00	L6286	Willwood	Wa3	1895.0
30.03	L6316	Willwood	Wa1	1535.0	50.00	L6311	Willwood	Wa3	1915.0
30.04	L6317	Willwood	Wa1	1535.0	51.00	L6264	Willwood	Wa3	1935.0
31.00	L6278	Willwood	Wa1	1555.0	52.00	L5762	Willwood	Wa3	1950.0
32.00	L6290	Willwood	Wa1	1560.0	53.00	L6298	Willwood	Wa3	1970.0
33.01	L5766	Willwood	Wa1	1570.0	54.00	L6310	Willwood	Wa3	2020.0
33.02	L5767	Willwood	Wa1	1570.0					

*Data keyed to numbered horizons presented in figures 16.5, 16.6, and 16.7.

This table includes only localities to which a relative position of the Cretaceous/Paleogene System boundary is known. The "Hor." code refers to the stratigraphic horizon of the locality. All of the above localities are from the northern portion of the Bighorn Basin, Wyoming. Horizons with more than one locality are denoted by decimals (e.g., 47.02). The stratigraphic level, shown in the last column (m > K/T) as meters above the Cretaceous/Paleogene System boundary, were derived by stated and inferred comparisons to UM SC mammalian localities, from which the metric value was obtained (data from P. D. Gingerich and G. F. Gunnell, personal communication 1994, 1995). Localities with missing metric values have been correlated less precisely to SC localities. Localities not in the above list that are not referrable to a specific stratigraphic horizon were usually intervalized to the biochrons to which they could be referred. This information is retained in the Locality Register.

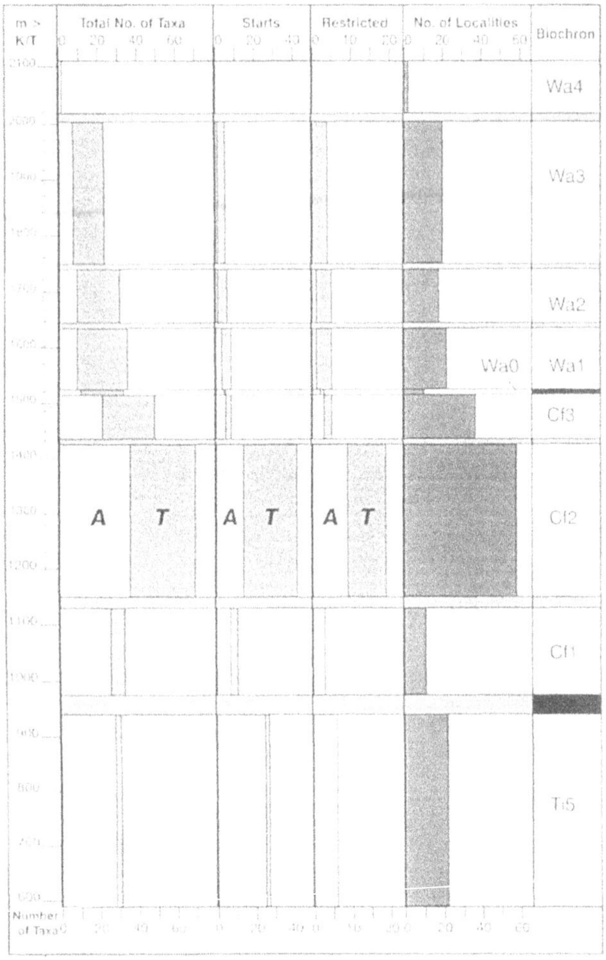

Figure 16.8 Nonmarine molluscan diversity in the Bighorn Basin.

This figure illustrates the changes in diversity of aquatic and terrestrial taxa through time. All histograms use mammalian biochrons as the class interval. The histograms represent (1) the total number of taxa, (2) the number of taxa that began their range in a biochron (Starts), (3) the taxa restricted to a biochron, (4) and the number of localities represented in each biochron. Each biochronal interval for each diversity histogram is divided into two parts, showing the representation of aquatic (A) and terrestrial (T) taxa through time. Note that the scale for the "Restricted" histogram is expanded (doubled) to better display the distribution of taxa. The stratigraphic horizon numbers shown on the left side of this figure are listed in table 16.10.

The histograms clearly show the dominance of aquatic taxa in Biochrons Ti5 and Cf1, their apparent acme during Biochron Cf2, and their reduced diversity from Biochron Cf3 on. The figure also shows that terrestrial taxa diversify during Biochron Cf2, but that they also have a number of taxa restricted to that interval. Terrestrials dominate from Biochron Cf3 on, but the number of originations (new morphotypes) is only slightly greater than that found in aquatic environments. Note also, however, that a larger number of collections have been made in strata of Biochron Cf2. In addition, the good quality of the material of this age has potentially provided a better sampling of the total molluscan biota (see additional comments in text).

the late Paleocene interval on the numerous molluscan taxa known from a large number of localities. However, some molluscan faunal correlations appear possible. Both Powder River Basin mammalian localities occur below the Wall coal bed (after Yen 1948b, but see Hartman 1984). This fauna, which is Biochron Ti3 to Ti4 in age, is characterized by *Plesielliptio priscus*, *Viviparus retusus*, *Campeloma nebrascense nebrascense*, *Lioplacodes nebrascensis nebrascensis*, *L. n. producta*, *L. tenuicarinata*, and New Genus A *limneaformis* and includes early occurrences of *Viviparus* sp. form H and undetermined taxa of *Pisidium* as well. In the overlying Wall–Canyon coal bed interval, *Plesielliptio silberlingi* and *Viviparus raynoldsanus* Meek and Hayden, and species of *Lioplacodes* are found, including *L. multistriata* and *L.* sp. form A, and *L.* sp. undet. A. In this stratigraphic unit, two common taxa are lost, *C. n. nebrascense* and *L. n. producta*. With species assemblages from the Bighorn Basin, this assemblage suggests a mix of Biochrons Ti5 and Cf1 taxa. No new species are found in the thin Canyon–Anderson coal bed interval, but the introduction of taxa similar to those found in the Bighorn Basin suggests that the overlying Anderson–Roland coal bed interval represents a Clarkforkian fauna.

The Clarkforkian NALMA

Biochron Cf1

The early Clarkforkian (Biochron Cf1) molluscan fauna is based on relatively few localities (figure 16.8). However, it includes the first Bighorn Basin occurrence of terrestrial snails (table 16.6), *Grangerella sinclairi*, "*Gastrodonta*" *coryphodontis*, and a form similar to *Radiocentrum grangeri*. The terrestrial fauna also includes New Genus sp. *spatiosa* and an undetermined *Grangerella*. All of these taxa are from a single locality. The occurrence of *G. sinclairi* represents its first occurrence, as well as that of the family Hendersoniidae in the Bighorn Basin. All taxa, except *G.* sp. undet., increased markedly in abundance in Biochron Cf2. During Biochron Cf1, the aquatic taxa were represented by the introduction of *Viviparus* sp. form H, which continued throughout the Clarkforkian Age. Besides this taxon, representing an element in the *Viviparus meeki* complex of like apical whorls, a single specimen very similar to *V. formosus* is noted. *Viviparus retusus*, present at single localities in Tiffanian zones, continues its single locality presence in Zone Cf1. *Hydrobia anthonyi*, another Williston Basin species, appears to occur in the lower Clarkforkian interval on the basis of a few specimens. Other taxa, represented by only a few specimens or occurrences, are listed in table 16.6.

Missing and not found in younger strata in the Bighorn Basin are several aquatic taxa, including *Campeloma nebrascense nebrascense*, *Lioplacodes nebrascensis nebrascensis*, and *L. n. producta*. In the

Powder River Basin, these taxa are found below the Below-Wall coal bed interval.

Biochron Cf2

The molluscan record of the middle Clarkforkian Biochron Cf2 is particularly good in the Bighorn Basin. This biochron, represented by well-preserved specimens from a large number of localities, was a time of a substantial increase in diversity, certainly for terrestrial taxa, but apparently also for aquatic forms (figures 16.6, 16.8). The timing of the introduction of taxa in Biochron Cf2 varied. Nearly all of the taxa known from specific horizons occurred in the later part of the biochron. Many taxa are not known to a specific horizon, leaving doubt as to the exact nature and rate of diversification (figures 16.5, 16.7, and 16.8). Taxa occurring for the first time in the Bighorn Basin include the aquatic prosobranchs *Valvata* sp. cf. *V. bicincta*, *V.* sp. form 1, *Reesidella nana*, *Clenchiella* sp. cf. *C. planospiralis*; the aquatic pulmonate *Acroloxus radiatulus*, *Acroloxus* sp. cf. *A. radiatulus*, *Physa* sp. form 2, *Physa* sp. form 5, *Physa* sp. form 6, *Physa* sp. form 7; the terrestrial prosobranch "*Helicina*" sp. cf. "*H.*" *evanstonensis*, "proserpinid" sp. form A, "proserpinid" sp. form B, "proserpinid" sp. form C, and cf. *Tozerpina* sp. undet., *Grangerella phenacodorum*, *Grangerella* sp. cf. *G. sinclairi*, *Grangerella* sp. form D, *Protoboysia complicata*; and the terrestrial pulmonates *Discus ralstonensis*, *D.* sp. form I, *D.* sp. cf. *D. ralstonensis*, ?Discidae gen. and sp. undet., *Helicodiscus*? sp. form H, *Glypterpes* sp. cf. *G. veternior*, *Ancotrema* sp. form E, *Polygyrella* sp. cf. *P. parvula*, "*Eucalodium*" *eophilum*, New Genus L sp. form K, *Radiocentrum grangeri*, *R.* sp. form J, *Albertanella* sp. cf. *A.* sp. undet., Pupillidae gen. and sp. undet., *Microphysula* sp. cf. *M. ingersolli*, and ?Zonitidae gen. and sp. undet. Along with many of the aquatic taxa above, which indicate quiet water conditions, pisidiid morphologies in Biochron Cf2 include *Eupera* sp. form 7, *Eupera*? sp. form 8, and pisidiid sp. form 6. Their presence, along with the previously reported taxa *Sphaerium* sp. form 1, *Sphaerium* sp. form 2, *Sphaerium*? sp. form 3, and Pisidiidae gen. and sp. undet., and the absence of all but a few poorly preserved unionids suggest a more complex, less fluvially oriented environmental setting.

These introductions of new taxa also represent the first occurrence of terrestrial families, including the Bulimulidae, Discidae, Haplotrematidae, Helicodiscidae, Helminthoglyptidae, Megomphicidae, Pupillidae, and Thysanophoridae and possible first records of the Bradybaenidae and Zonitidae. Biochron Cf2 also included the first good record of the Bithyniidae in the basin. Numerous other taxa have been identified, such as *Valvata* sp. cf. *V. subumbilicata*?, *Physa* sp. form 1?, *Acroloxus minutus*, planorbids, cf. *Oreoconus* sp. indet., and an operculum of a species of Hendersoniidae. Taxa previously reported but now known in some abundance include *Grangerella sinclairi*, "*Gastrodonta*" *coryphodontis*, and *Radiocentrum* sp. cf. *R. grangeri*. Previously noted taxa that are represented in Biozone Cf2 include "*Helicina*" *evanstonensis*, *Viviparus retusus*, *Viviparus* sp. form H, *Lioplacodes tenuicarinata*, *Hydrobia anthonyi*, *Pleurolimnaea* sp. cf. *P. tenuicosta*, and New Genus S *spatiosa*. Taken together or separately, the middle Clarkforkian Age represented the acme of aquatic and terrestrial mollusk species richness.

Biochron Cf3

The late Clarkforkian (Biochron Cf3) NALMA was a time less remarkable for first appearances than for last records of important late Paleocene species (figure 16.5). The following long-ranging prosobranch snail taxa met their demise near the end of the late Clarkforkian Age: *Viviparus retusus*, *Viviparus* sp. form H, *Lioplacodes tenuicarinata*, and New Genus A *limneaformis*. Pulmonates ranging into Biozone Cf2 that are not found in younger strata include the physids spp. form 5, form 6, and form 7, and the terrestrials *Radiocentrum* sp. cf. *R. grangeri*, *Polygyrella* sp. cf. *P. parvula*, and *Glypterpes* sp. cf. *G. veternior*. Also, the best documented record of *Grangerella phenacodorum* ends in Zone Cf3. The exception is the holotype, which has been interpreted, based on available evidence, as belonging to either Biochron Wa2 or Wa3.

Because the taxonomy of the unionid bivalves is less well understood, primarily as a result of poor specimen preservation and limited occurrence, their biostratigraphy is also less certain. *Plesielliptio silberlingi*, *P. wasatchensis*, *P.* sp. form 4, and "*Unio*" sp. form 1 are known best from the Tiffanian deposits. Possible last occurrences of these taxa, however, may be recorded in Zone Cf3. This unit also contains the only record of "*Unio*" sp. form 3 and the type locality of the undiagnosable bivalve "*Unio*" *eomargaron*.

The few first occurrences in Zone Cf3 include New Genus S sp. cf. New Genus S *spatiosa*, *Elimia* sp. form 1, and the first confirmed record of *Physa* sp. form 1. The physid appears to include the identification of *P. bridgerensis* by Russell (1931b). The *Elimia* record is particularly significant even though this taxon may be short-lived, extending only through Biochron Wa0, in that it represents the initial occurrence of a suite of taxa that occurred in the early Wasatchian NALMA. Taxa with short ranges, having their first and last occurrences in Biozone Cf3, include ?Pupillidae gen. and sp. undet., *Haplotrema* sp. form G, and *Planorbella* sp. form 1.

Taxa with older and younger records that encompass a significant portion of the Biochron Cf3 molluscan assemblage include the terrestrial snails "*Helicina*" *evanstonensis*, *Discus ralstonensis*, *D.* sp. form I, *Ancotrema* sp. form E, *Grangerella sinclairi*, *Grangerella* sp. cf. *G. sinclairi*, *Grangerella* sp. form D, *Protoboysia complicata*, "*Eucalodium*" *eophilum*, "*Gastrodonta*" *coryphodontis*, New Genus S *spatiosa*, New Genus L sp. form K, *Radiocentrum grangeri*, and Pupillidae gen. and sp. undet. (see also table 16.6).

Regional Clarkforkian correlations

The middle Clarkforkian Biochron Cf2 should be recognizable elsewhere in the northern Great Plains on the basis of nonmarine mollusks. Where continuous stratigraphic sections are known, such as in the Bighorn and Powder River basins, the interval is recognizable on the basis of aquatic mollusks. The prominence of terrestrial snails in Biozone Cf2 strata in the Bighorn Basin suggests an event that may have regional or broader implications for molluscan biochronology and paleoenvironment (see Summary). The Anderson–Roland coal bed interval in the Powder River Basin is also a level of dramatic species diversification. In order of first appearance, comparable taxa include *Clenchiella planospiralis*, *Reesidella nana*, *Pleurolimnaea tenuicosta*, *Acroloxus radiatulus*, *Valvata bicincta*, *Eupera formosa*, and *Acroloxus minutus*. Other taxa that may prove useful in biochronologic correlation after additional study include pisidiids, valvatids, hydrobiids, planorbids, and physids. The introduction of these taxa occurs stratigraphically near the Anderson coal bed. Faunal resemblance decreases with the introduction of additional taxa towards the Roland coal bed. As a working hypothesis, based on aquatic molluscan taxa, the Cf2/Cf3 biochronal boundary occurs within the Anderson–Roland coal bed interval. This interval, however, clearly lacks the infusion of terrestrial mollusks seen in the middle Clarkforkian deposits of the Bighorn Basin, undoubtedly reflecting paleoenvironmental differences.

Paleocene/Eocene Series boundary correlations

The placement of the Paleocene/Eocene Series boundary in the North American nonmarine section has varied depending on the criteria employed. On the basis of mammals, the boundary horizon has shifted from within the Clarkforkian NALMA at the Cf1/Cf2 zonal boundary (Rose 1979, 1980) to its current placement between the Clarkforkian and Wasatchian NALMAs (Gingerich 1989b). Studies by Wing (1984) have suggested that the Paleocene/Eocene Series boundary may be more accurately correlated with the European Sparnacian/Ypresian Stage boundary if placed in the lower Wasatchian NALMA at the first occurrence of the pollen species *Platycarya platycaryoides* (Roche) Frederiksen & Carpenter. The correlation of the level corresponding to the Paleocene/Eocene Series boundary in North America cannot be directly addressed by the nonmarine molluscan record in the Bighorn Basin. As presented here, the boundary as recognized by mammalian paleontologists is effectively identifiable (excluding Biochron Wa0) on the basis of mollusks in the Bighorn Basin and elsewhere in the Western Interior.

The Wasatchian NALMA

Biochron Wa0

The recently introduced earliest Wasatchian (= early Sandcouleean age) Biochron Wa0 (Gingerich 1989b), even for its apparently short duration, is distinctive. The only occurrence of a urocoptid, *Holospira* sp. undet., in the Bighorn Basin section under study, is in Biozone Wa0. It is, in fact, the only terrestrial snail reported from Biochron Wa0. The genus was already present in the North American Cordillera; earlier species have been reported from the Upper Cretaceous record of Alberta and from the Paleocene deposits of New Mexico (see Comments on Taxa). Gingerich (1989a, 1989b) argued that Biochron Wa0 represented a rarely sampled high floodplain environment. The fact that numerous other land mollusks have otherwise continuous stratigraphic ranges above and below Biozone Wa0 strongly suggests that the unique faunal composition of Zone Wa0 is a facies-controlled phenomenon.

The record of the pleurocerid taxa *Elimia* may also be noteworthy, as several species may occur in the Wa0 zonal interval (figures 16.5, 16.7). At present, the taxonomic study of high-sculptured *Elimia* is a rather nightmarish undertaking. With numerous highly variable taxa (see White 1883) occurring abundantly, but from poorly delimited type localities, a taxonomic revision of the *Elimia* is outside the scope of this chapter. *Goniobasis tenera carteri* and *Goniobasis carterii* (assigned to *Elimia*? n. sp. by Taylor 1975; see table 16.1) have previously been noted in the Bighorn Basin assemblage (Cockerell 1914a; Russell 1931b), ranging from Biozones Cf3 to Wa3 (see table 16.3 for age assessment). For now, morphotypes of *Elimia* have been designated to avoid the problem of referring specimens to vaguely diagnosed species and prompting unjustified biochronologic generalizations. The Biochron Wa0 may contain four *Elimia*. Their distinction as separate species may become unwarranted as additional studies of the morphological variation of *Elimia* are undertaken in biostratigraphically well-controlled sections. These forms differ primarily in the strength and configuration of axial sculpture. The differences are noteworthy, but the extent of the variability is uncertain. *Elimia* sp. form 1 is known from two localities, having its first appearance in Biozone Cf3 and its last in Biozone Wa0. *Elimia* sp. form 2 and *Elimia*? sp. form 4 are known only from Biozone Wa0. *Elimia* sp. form 3, which represents the taxon *Goniobasis carterii* (Russell 1931b), makes a questionable first appearance in Biozone Wa0. A distinctive, very high-spired form, referred to here as *Hydrobia* sp. form 2 (see *Hydrobia* sp., Gingerich 1989b), was recognized in Tiffanian strata. A single occurrence of this taxon is known from Biozone Wa0. It differs slightly in whorl shape and depth of the impression of the suture, but only a single specimen is known. This specimen, although not recognized in figure 16.6 as a separate range, is referred to here as *Hydrobia* sp. aff. *H.* sp. form 2. In addition, a questionable first occurrence of the planorbid *Biomphalaria storchi* is recognized in Biozone Wa0. The only other recognized taxon in this interval is an indeterminate species of *Viviparus*.

Biochron Wa1

The early Wasatchian (= middle Sandcouleean Age) Biochron Wa1 is distinctive in its record of first, exclusive, and last appearances of aquatic and terrestrial taxa (figures 16.7, 16.8). The aquatic prosobranch *Viviparus wyomingensis* made its first appearance in Biochron Wa1 in the Bighorn Basin. This taxon was first described from southwestern Wyoming and appears to have had a rather general distribution during the early Wasatchian Age. The helminthoglyptid *Noyo*? sp. form N also had its first appearance during this time. Taxa with their only, or only unquestioned, record in Zone Wa1 include the aquatic snails *Elimia* sp. form 3 (may occur in older or younger strata), *Physa* sp. form 3, and *Physa* sp. form 4 (may occur in younger strata), and the terrestrial snails *Eohipptychia eohippina*, *Haplotrema* cf. sp. form G, "*Eucalodium*" sp. cf. "*E.*" *eophilum*, cf. New Genus S sp. undet. 2, and *Neohelix*? sp. form M. The latter record is significant in that it probably represents the first occurrence of the family Polygyridae in North America. The record of the helicinid *E. eohippina* represents its type occurrence, as it has not been subsequently identified. During Biochron Wa1, *Glypterpes* sp. cf. *G. veternior* had its last appearance, as possibly *Physa* sp. form 2, which has a questioned younger record. With the end of Biochron Wa1, the apparent spate of morphotypes referable to *Physa* and *Elimia* was over.

A large component of the Biochron Wa1 molluscan assemblage is made up of continuing terrestrial taxa. These include the prosobranchs "*Helicina*" *evanstonensis*, *Grangerella sinclairi*, *G.* sp. cf. *G. sinclairi*, and *Protoboysia complicata* and pulmonates cf. *Oreoconus* sp. indet., *Discus ralstonensis*, *D.* sp. cf. *D. ralstonensis*, *D.* sp. form I, "*Eucalodium*" *eophilum*, "*Gastrodonta*" *coryphodontis*, New Genus S *spatiosa*, and New Genus S sp. cf. New Genus S *spatiosa*. Besides *Physa* sp. form 1, which occurs in younger strata, aquatic bivalves and snails are known only from indeterminate taxa.

Biochron Wa2

In the nature of the molluscan assemblage and rather distinctive record of first and last appearances, Biochron Wa2 (= late Sandcouleean Age) is similar to Biochron Wa1. First records include New Genus P *jepseni*, a possible member of the Bradybaenidae. First appearances of taxa, restricted to the biochron, include the terrestrial *Grangerella* sp. cf. *G.* sp. form D, *Microphysula oxyaenae*, *Microphysula* cf. *M. oxyaenae*, and undetermined stylommatophoran. Taxa not otherwise identified in present collections include the holotypes of *Glypterpes veternior* and *Microphysula sinoparum*. Both species can be delimited only in the interval comprising both Biochrons Wa2 and Wa3. Aquatic snails with a restricted stratigraphic record include *Biomphalaria storchi* (questionably known from Biozone Wa0) and *Viviparus* sp. aff. *V. retusus*. On the basis of available information, the holotype of *B. storchi* may be from Biochron Wa3 or Wa4, but there are no documented records of this age. Last appearances of terrestrial taxa include "proserpinid" sp. form B, "proserpinid" sp. form C, *Grangerella* sp. cf. *G. sinclairi*, cf. *Oreoconus* sp. indet., *Discus* sp. form I, "*Eucalodium*" *eophilum*, New Genus L sp. form K, and Pupillidae gen. and sp. undet. The last appearance of an aquatic taxon is the questioned occurrence of *Elimia* sp. form 3?, which was present in Biochrons Wa0 and Wa1.

Again, the dominant taxa ranging through Biochron Wa2 are terrestrial species (figure 16.6). These include "*Helicina*" *evanstonensis*, "*Helicina*" sp. cf. "*H.*" *evanstonensis*, *Grangerella sinclairi*, *Grangerella* sp. form D, *Protoboysia complicata*, *Discus ralstonensis*, *Discus* sp. cf. *D. ralstonensis*, *Noyo*? sp. form N, New Genus S *spatiosa*, and New Genus S sp. cf. New Genus S *spatiosa*. The continuing aquatic record includes *Viviparus wyomingensis* and *Physa* sp. form 1 and a questioned occurrence of *Physa* sp. form 4. As in Biochron Wa1, unionids and pisidiids were conspicuous in their general absence.

Biochron Wa3

The molluscan record in Biochron Wa3 (early Graybullian age) is accentuated by last occurrences. Most species became at least locally extinct well before the end of Biochron Wa3. Last appearances of important terrestrial taxa include "*Helicina*" *evanstonensis*, "*H.*"

sp. cf. "*H.*" *evanstonensis*, *Grangerella sinclairi*, *G.* sp. form D, *Protoboysia complicata*, *Discus ralstonensis*, "*Gastrodonta*" *coryphodontis*, "*Gastrodonta*" sp. cf. "*G.*" *coryphodontis*, New Genus S *spatiosa*, New Genus S sp. cf. New Genus S *spatiosa*, *Ancotrema* sp. form E, *Noyo*? sp. form N, and New Genus P *jepseni*. Note that the holotype of *Oreohelix megarche* (Biozones Wa3–Wa4), which has been placed in synonymy with New Genus S *spatiosa* (see table 16.2), has an interpreted age of either Biochron Wa3 or Wa4. Thus the range of New Genus S *spatiosa* may possibly have extended into Biochron Wa4. Similarly, the holotype of New Genus P *jepseni* can be delimited only to Biochrons Wa3 and Wa4. Aquatic taxa with apparently last appearances in Biochron Wa3 include *Viviparus wyomingensis* and the physids *Physa* sp. form 1 and questioned records of *Physa* spp. form 2 and form 4. The record of *P.* sp. form 1 may range into Biochron Wa4, based on the inadequate record of *Physa bridgerensis* (Russell 1931b) to which it is presently referred. Taxa that are known only from Biochron Wa3 include the terrestrial species Helicinidae gen. and sp. undet., *Haplotrema* sp. form F, and *Noyo*? sp. cf. *N.*? sp. form N. The occurrence of a few holotypes cannot be delimited beyond suggesting they occur in either Biochron Wa3 or Wa4. Where numerous nontype records show a well-delimited range, these possible range extensions were excluded from figure 16.6. In addition, the holotypes of the species "*Unio*" *didymictidis* and "*Unio*" *sinopae* can only be reported as occurring in either Biochrons Wa3 and Wa4. These taxa have not otherwise been recognized in the collections under study.

Regional Wasatchian correlations

Wasatchian mollusk occurrences are known throughout the Western Interior, but few general studies synthesize the taxonomy and stratigraphy (e.g., La Rocque 1960; Hanley 1974; Taylor 1975). No studies provide the biostratigraphic and temporal control available in the Bighorn Basin through high-quality molluscan collections and detailed mammalian studies. Except for summary statements made below, the Wasatchian record of the Bighorn Basin will be compared only with Wasatchian records of the Powder River Basin found in the "Wasatch" Formation of northeastern Wyoming. As used here, the basal strata of the "Wasatch" Formation occur in the Roland–Arvada coal bed interval and are, most likely, late Clarkforkian in age. This interpretation is confirmed by the Rose and Bown (1982) identification of a single tooth of *Tinimomys graybulliensis* Szalay and teeth of a small plagiomenid. They suggest that the fossil-bearing horizon, stratigraphically about 12 m above the Fort Union/"Wasatch" formational contact, is of Clarkforkian Age (Rose, personal communication 1995).

The transitional beds from Clarkforkian to Wasatchian NALMA in the Powder River Basin include a fairly extensive aquatic assemblage. Several taxa with Tiffanian or Clarkforkian records become extinct in the Roland–Arvada coal bed interval of the basal "Wasatch" Formation. These include (in order from oldest to youngest first appearance) such taxa as *Plesielliptio priscus*, *Lioplacodes tenuicarinata*, New Genus A *limneaformis*, *Viviparus retusus*, *V. raynoldsanus*, *L. multistriata*, *Clenchiella planospiralis*, *C. n.* sp. A (of Taylor 1975), *Reesidella nana*, *Pleurolimnaea tenuicosta*, a physid, *Biomphalaria storchi*, *Hydrobia eulimoides*, pisidiids, and probably *P. silberlingi*. This list comprises solely aquatic taxa, many with numerous records of occurrence. Taxa with records restricted to the Roland–Arvada coal bed interval include (ordered by first appearance) *Plesielliptio* sp. cf. *P. priscus* (of Taylor 1975), *Micropgyrus minutulus*, *Gyraulus militaris*, *Hydrobia warrenana*, and other undescribed aquatic species. Thus the number of taxa that become extinct in the Powder River Basin near the horizon of the Arvada coal bed is significant. On the basis of discussions above and the overlying Wasatchian fauna noted below, these taxa are believed to indicate the end of the Clarkforkian Age.

An early Wasatchian mammalian fauna is known from a few published localities in the Powder River Basin (Delson 1971; Ebaugh 1976; Soister 1968; see Hartman 1984) and work in progress (Peter Robinson, personal communication 1993, 1995). Except for Robinson's work in the southern portion of the basin, these records are of limited biostratigraphic value at this time, other than to indicate that much of the molluscan record is probably of Biochron Wa1 to Wa3 age. Named taxa, all of which are aquatic, with records commencing at or near the base of the main body of the "Wasatch" Formation (above the Arvada coal bed horizon) include *Physa* sp. cf. *P. pleromatis* (after Taylor 1975; 27 m and then 73 m), *Viviparus wyomingensis* (70 m), *Plesielliptio haydeni* (after Taylor 1975; 73 m), *Valvata* sp. aff. *V. lewisi* (after Taylor 1975; 73 m), *Elimia* sp. aff. *E. nodulifera* (after Taylor 1975; 73 m), *Viviparus paludinaeformis* (~94 m), and unnamed morphotypes belonging to the Pisidiidae, Viviparidae, Hydrobiidae, Physidae, and Planorbidae (main body of "Wasatch" Formation). Note again that this fauna is apparently composed exclusively of aquatic taxa. This fauna undoubtedly includes the temporal intervals of Biochrons Wa0 and Wa1 found in the Bighorn Basin, but probable differences in paleoenvironmental settings make extensive comparison difficult. The fauna of the lower portion of the main body of the "Wasatch" Formation in the Powder River Basin is more typical of early Wasatchian assemblages found elsewhere in the Western Interior. Taxa such as *V. paludinaeformis*, *V. wyomingensis*, highly sculptured *Elimia*, and species of inflated *Physa* well demarcate the beginning of Biochron Wa1, if not that of Biochron Wa0. In the Bighorn Basin, these faunal elements would be recognized by the taxa of *Elimia* and *Physa* and the occurrence of *Viviparus wyomingensis*.

A later Powder River Basin Wasatchian fauna encompasses elements of the above assemblage, but also includes new aquatic taxa such as *Valvata* sp. cf. *V. bicincta* (after Taylor 1975) and undescribed species of *Viviparus*, *Acroloxus*, and *Physa*. Of some significance is the first occurrence of terrestrial taxa in this section. Taylor (1975) identified the genus *Oreohelix* in the upper portion of the main body of the "Wasatch" Formation, well above the Felix coal bed. A questioned occurrence of the same genus in the same interval was given in an unpublished USGS report by J. B. Reeside, Jr. (1953). In what are apparently the uppermost strata of the main body of the "Wasatch" Formation, Taylor (1975) also identified unnamed taxa of *Pupoides* and *Discus*. Besides these two genera, terrestrial taxa have also been identified from the Kingsbury Conglomerate of the "Wasatch" Formation. These taxa include *Grangerella*, *Polygrella*, New Genus A *spatiosa* (of Taylor 1975), *Vitrina*, *Carychium*, and a *Helix*? (unpublished report of Reeside 1926). Although the specimens upon which these identifications are based have not been studied as part of the present work, they do suggest a Biochron Wa1 to Wa3 age. The only mammal reported from the Kingsbury Conglomerate was *Diocodexis* (Jepsen 1940), which spans the early Eocene Epoch (Archibald et al. 1987). Additional molluscan study will undoubtedly resolve specific aspects between this emerging terrestrial fauna and that found in the Bighorn Basin.

Comments on Bighorn Basin Taxa

The following taxonomic, stratigraphic, and geographic annotations are the result of the ongoing studies in the Bighorn Basin. These comments are not intended to be exhaustive, as a systematic treatment will follow, but to keynote issues relevant to this chapter.

Unionidae

The unionids from the upper Tiffanian, Clarkforkian, and lower Wasatchian deposits of the northern portion of the Bighorn Basin generally have poorly preserved shell material. This, of course, limits our ability to identify most with confidence. The best shelled material is reported from Locality L4373 in Montana. The identification of fossil unionid genera is based on marginal outline, shell sculpture, and dentition. The use of the genus *Unio* in lower Cenozoic nonmarine deposits of North America traditionally refers to any generally elliptical clam with heterodont dentition and without umbonal sculpture. Specimens with sculpture, either on the umbones or remaining shell, are assigned to other fossil or Recent genera. The modern restricted usage of *Unio* is inappropriate for the Western Interior fossil record, and thus it is placed in quotes (Hartman 1987).

The taxon *Unio eomargaron* is apparently now based only on a holotype consisting of valve exfoliae. Cockerell's (1915b) description and simple outline drawing purported to be of the holotype of this species suggests a comparison to *Plesielliptio* sp. form 4, also from Biozone Cf3 of this report. In contrast to Cockerell's diagnosis, the two specimens assigned to this taxon possess umbonal corrugations and posterolateral ridges. These features are subtle or not preserved on the specimens of the size illustrated by Cockerell.

With *Plesielliptio wasatchensis*, as with other Cockerell unionid taxa from the Bighorn Basin, the stratigraphic and geographic occurrence of type material is only vaguely known. Its horizon was reported as "Wasatch" (Cockerell 1914a). Russell (1931b:13) noted its occurrence in the "Wasatch (i.e., Gray Bull)," but stated that "the only material in the Princeton collection referable to this species is the specimen mentioned from the Fort Union beds. . . . It is possible that the American Museum material actually came from Fort Union beds. Indeed, as pointed out by Cockerell, *U. wasatchensis* is one of a group of species almost, if not entirely, confined to the Upper Cretaceous and Paleocene." The taxon has yet to be documented in the Clarkforkian or lower Wasatchian NALMA. Taylor in Rohrer and Gazin (1965) identified this taxon (L2117) from the Tatman Formation in strata interpreted to be Biochron Wa5 or Wa6 in age (as derived from Gingerich 1989a). As noted by Taylor (1975), a paratype of this species may not be referable to *Plesielliptio* and thus represents a different species (recorded as "*Unio wasatchensis*" in table 16.1). This elongate form was found to be most similar to indeterminate material from Locality L6239 from Biozone Cf3.

Pisidiidae

In the present study, only one pisidiid allows examination of the hinge structure, thus limiting our ability to assign morphotypes to modern genera to which they all certainly belong. Pisidiids are generally represented by only a few specimens at any given locality.

Helicinidae

Bighorn Basin Clarkforkian and lower Wasatchian strata contain the possible last stand of a variety of uncommon neritopsine land snails with apertural barriers ("proserpinid" spp. forms A–C and "cf. *Tozerpina*") that probably represent one or more extinct clades. All are probably undescribed species. Modern-day proserpinine snails, although having apertural barriers, differ from the Bighorn Basin taxa in that the shell is covered with a secondary deposit of enamel, and the peristome remains simple and acute; their similarity to the Bighorn Basin taxa is probably due to convergence.

Helicina evanstonensis has been referred to *Hendersonia* by Pilsbry (1948), Gingerich (1987), and Roth (1986), but it is not a species of *Hendersonia* or a member of Hendersoniidae. Pending further study, we maintain it as "*Helicina*." It ranges through nearly the entire Bighorn Basin section reported here. However, the samples vary in strength of collabral ribbing, so this form taxon may ultimately be split up into two or more taxa. Bishop (1980) proposed the monotypic genus *Eohipptychia* for *Pleurodonte eohippina* Cockerell 1915a and assigned it to the Helicinidae. In spite of a later critique by Solem (1983), we also regard it as a helicinid. None of the specimens we examined was referable to this taxon. Russell's (1931b) assignment of "*P.*" *eohippina* to *Polygyra* was incorrect.

Hendersoniidae

The assignment of grangerellids to Hendersoniidae was proposed by Roth (1992). The Hendersoniidae attained their greatest diversity during the Clarkforkian and Wasatchian ages. The clade appeared earlier and extends to the Recent (in the form of *Hendersonia* in eastern North America and *Waldemaria* in Japan), but in the Bighorn Basin, three, and in some cases four species, occur sympatrically and in great abundance at many localities.

The estimated age of the holotype of *Grangerella phenacodorum* (originally described in the pupilloidean genus *Boysia*) is Biochron Wa2–Wa3. To date, the only positive identifications of this taxon in our material are in Biozones Cf2 and Cf3. *Grangerella* sp. form D is apparently a new species, similar in appearance to *G. sinclairi* and sometimes sympatric with it. It is smaller than *G. sinclairi* but larger than *G. phenacodorum*. *Grangerella* sp. form D appears to have extended through Biochrons Wa2 and Wa3. *Grangerella sinclairi* and *Grangerella megastoma* refer to the same taxon, with *G. sinclairi* having priority. In his paper describing *G. megastoma*, Cockerell (1915a:116) mentioned that *G. sinclairi* was "very abundant." It seems clear that Cockerell regarded the *G. sinclairi* morphotype as the most common *Grangerella* in the Bighorn Basin. Specimens assigned to *G.* sp. cf. *G. sinclairi* represent a form distinguished by Hanley (unpublished) as terrestrial sp. form F (see figure 16.4 and table 16.5). The taxon "Hendersoniidae gen. and sp. undet." is represented by one specimen, a loose operculum closely resembling the opercula of Recent and fossil species of *Waldemaria* (cf. Pierce and Rasmussen 1989:figures 2:2 and 2:5). It probably belongs to a species of *Grangerella*, possibly *G. sinclairi* or *G. phenacodorum*, both of which occur at the same locality.

Valvatidae

Preservation of valvatids is poor, and except for Locality L5701, they are represented by only a few specimens. Additional effort aimed at collecting valvatids from L5701 would aid in providing more meaningful identifications.

Viviparidae

The Viviparidae represent the dominant snail taxon of the Paleocene Epoch and are well represented in the upper Tiffanian and Clarkforkian records of the northern Bighorn Basin, particularly in Montana. Species of *Viviparus* typically have *V. meeki*–type apical structure. Because of the ontogenetic similarity of many species, anything less than good preservation limits one's ability to positively identify many specimens. Therefore, many indeterminate taxa are represented in the collections. Additional studies are necessary to provide a rigorous method for discriminating species of *V.* sp. form H. Although comparably large-sized specimens are present in the Montana portion of the Bighorn Basin, many specimens are smaller than those typically found in the Powder River Basin.

Bithyniidae

The specimens identified as *Reesidella nana* from Locality L5906 are assignable to *Fluminicola protea* Yen (1948b). Taylor (1975) synonymized this taxon with the more widespread *R. nana*. In the Powder River Basin, this taxon ranges from 30 m above the Anderson coal bed to a similar distance above the Roland coal bed. This interval is interpreted herein as middle (Biochron Cf2) and possibly late (Biochron Cf3) Clarkforkian Age. Locality L5906 belongs to Biochron Cf2. Opercula of similar type are known from three localities, one of which includes specimens of *R. nana*, and have tentatively been assigned to that taxon.

Hydrobiidae

Several small Bighorn Basin snails have been assigned to *Hydrobia*. Like "*Unio*," *Hydrobia* in the North American fossil record is a form genus. It may be appropriate, after additional specimens are collected and study undertaken, to remove certain taxa to genera defined on the basis of fossils to limit unwarranted comparison to modern taxa to which they anatomically most likely do not belong. One taxon, *H.* sp. form 2, is distinctive in its multiwhorl, high-spired morphology. This taxon occurs at a locality assigned to Ti4–Ti5 and Ti5 biozonal intervals. A very similar form from Biozone Wa0 was identified by Gingerich (1989b) as "*Hydrobia*" sp. We have referred this specimen to *H.* sp. aff. sp. form 2 inasmuch as its more impressed sutures and less convex whorls are distinctive, but further collections, documenting population variation, may make this discrimination unwarranted.

Pleuroceridae

Elimia taxa in the Bighorn Basin vary in their strength of axial sculpture, showing node or ridge development along specific portions of a whorl. Few specimens are preserved in their entirety, with apical whorls almost always missing, and the external surface layers of the shell are frequently wanting. The form differentiation noted in this chapter is based on variation in sculpture that seems to constitute distinct morphotypes. Variation in these features in modern forms is well known, and such discrimination may later be found to be unnecessary. However, all of the morphotypes except one, *Elimia* sp. form 1, occur in post-Clarkforkian strata, beginning in either Biozone Wa0 or Wa1. Highly sculptured species of *Elimia* are generally a good indication of Wasatchian strata, as similar late Paleocene forms appear to be unknown. The late Clarkforkian (Biochron Cf3) record of *Elimia* sp. form 1 possesses strong and regular axial ridges across the entire whorl.

Family uncertain (incertae sedis)

New Genus A *limneaformis* has been reported from a number of localities that span a considerable amount of time in the Western Interior. Some of the assignments from older strata may be more appropriately referred to a new species in the same family. The specimens from the Bighorn Basin are referable to New Genus A *limneaformis* found in its type area in the Williston Basin. This species has been assigned to a variety of taxa, including both aquatic and terrestrial. Taylor (1975) suggested that this taxon be assigned to the subfamily Lioplacinae within the Viviparidae. This assignment is questioned and under study.

Ellobiidae

The specimens assigned to *Pleurolimnaea* sp. cf. *P. tenuicosta* are generally poor and few in number and may differ from *P. tenuicosta* in sculpture characteristics. The best specimens are from the latest Tiffanian NALMA, but indeterminate specimens of *Pleurolimnaea* from the Clarkforkian NALMA may also be assigned to this species. Additional study may show that Biochron Ti5- to Cf2-aged specimens from the Bighorn, Williston, and Powder River basins should be assigned to the same species.

Acroloxidae

A few specimens of radially sculptured *Acroloxus* occur in Biozones Ti5 and Cf2. Apical features are, however, not always clear. *Acroloxus radiatulus* was identified on the basis of poor material, but apparently not preserving an adapical dimple or impression. Yen (1948b) introduced the species *Palaeancylus radiatus* for specimens with a dimple and possessing radial sculpture. Tozer (1956) employed Yen's taxon for dimpled specimens from the Paskapoo Formation of Alberta and noted that the taxon may be congeneric and conspecific, pending acquisition of material. On the basis of excellent material from Judson localities of Biochron Ti3 age, Hartman (1976) illustrated dimpled and radially sculptured limpets and retained the assignment to *P. radiatus* following Tozer (1956). Taylor (1975:396) stated, "I see no good evidence of more than one species of *Acroloxus* other than *A* n.sp. A." The Bighorn Basin collections appear to contain dimpled specimens and thus are referable to *P. radiatus*. This species record in the Bighorn Basin appears to correlate well with its record in the Powder River Basin. *Palaeancylus* sp. cf. *P. radiatulus* represents a single, incomplete, strongly deformed specimen. Its discrimination is based upon extremely strong radial ridges.

Physidae

Although specimens of *Physa* are common in the Bighorn Basin section, most are not complete. Several morphotypes have been recognized in the present study, in an attempt to determine whether observed morphological differences have temporal or geographic significance. In the absence of a general review of Physidae and owing to the probably large number of incorrect species assignments in the literature, referral of specimens to named taxa was left for future study. Somewhat distinctive physids have been useful for the general recognition of lower Eocene strata. Examples include the specimens illustrated by Russell (1931b) as *Physa bridgerensis* and *P. pleromatis*, which came from the same locality in strata that are either Biochron Ti3 or Ti4 in age. *Physa* sp. form 1 appears referable to one of the illustrated specimens of *P. bridgerensis* (Russell 1931b:figure 14). *Physa* sp. form 1 also appears to be equivalent to an unpublished J. H. Hanley identification of *P. bridgerensis* from Locality L5757. Similar Hanley identifications from younger strata were not examined as part of this study.

Planorbidae

Planorbid specimens from the Bighorn Basin are uncommon, not well preserved, and frequently found in matrix as internal molds. Well-preserved and relatively common specimens were tentatively assigned to *Planorbella* from Locality L5758a. Most specimens are, however, indeterminate. The holotype of *Biomphalaria storchi* is either Biochron Wa3 or Wa4 in age. To date, the only positive identification of this taxon is from Biozone Wa2.

Pupillidae

The collections contain some very minute specimens that later may be assigned to *Albertanella* Russell 1931a and extend the known range of that taxon. The taxon "Pupillidae gen. and sp. undet." includes several different taxa from Biochron Cf2. Identification to genus and species must await further study.

Haplotrematidae

The Clarkforkian (Biochrons Cf2 and Cf3) specimens of *Ancotrema* sp. form E appear to differ slightly from those found in the Wasatchian Biochron Wa3 and, thus, are in need of further study.

Urocoptidae

Holospira is reported from Biozone Wa0 only, and is the only terrestrial mollusk present within it. The genus was already present in the North American Cordillera; earlier species include *Holospira dyeri* Tozer 1956 from the Upper Cretaceous St. Mary River Formation of Alberta and *Holospira grangeri* Cockerell 1914a from the Paleocene Nacimiento Formation of New Mexico (Hartman 1981). The present specimens are internal molds and cannot be identified any more specifically than to genus.

Bulimulidae

The specimens identified as "cf. *Oreoconus*" are poorly preserved but may be assignable to the Bulimulidae. They bear a general resemblance to viviparids, but where the columellar region is preserved, the inner lip rolls over the columella as in the genus *Rabdotus* and the body whorl interrupts the peritreme. The apex tends to be sharp, representing a projecting embryonic shell as in many modern Bulimulidae and unlike, for example, New Genus P *jepseni* (discussed under Bradybaenidae, below).

Helicodiscidae

Helicodiscus? sp. form H is apparently restricted to Biochron Cf2. However, continued study of other samples with minute shells may alter its known stratigraphic range.

Discidae

Discus sp. form I is basically a form taxon for *Discus* samples with a smooth zone on the base where the ribs are obsolete. Specimens are numerous in Biozones Cf2 and Cf3. Other specimens, from Biochrons Wa1 and Wa2, may represent a separate species. The taxon "?Discidae gen. and sp. undet." is represented in the collections by fragments of a probable discid with angulate periphery, domed spire, and threadlike collabral riblets. Its allocation is otherwise uncertain.

Oreohelicidae

Specimens assigned to New Genus S are widespread and numerous in our samples, which represent Biochron Ti5 to Biochron Wa3. Only one species seems to be present, except possibly for specimens in Biozone Ti5. However, because New Genus S *spatiosa* apparently occurs in the Sentinel Butte Formation, which is of Biochron Ti4 and probably Biochron Ti5 age (Hartman and Kihm 1995), the distinction may not be stratigraphically significant.

The estimated age of the holotype of *Oreohelix megarche* Cockerell and Henderson 1912 is Biochron Wa3–Wa4. Following Taylor (1975), we tentatively consider *O. megarche* synonymous with New Genus S *spatiosa*. However, several other taxa that Taylor (1975) synonymized with his "New Genus A *spatiosa*" are distinct (Hartman and Roth, unpublished).

The restriction of *Radiocentrum grangeri* to Biochrons Cf2 and Cf3 is probably real. The distinction between specimens assigned to *Radiocentrum* sp. form J and *R. grangeri* needs further study.

Cockerell's (1914a, 1915a) generic assignments for the taxa "*Gastrodonta*" *coryphodontis* and "*Eucalodium*" *eophilum* are incorrect and are maintained with quotes pending further study. Largely on the strength of embryonic whorl sculpture, we tentatively regard these taxa as oreohelicid, even though their shape (elongate-ovate to ovate-conic and taller than broad) is outside the range of shapes shown by Recent Oreohelicidae.

New Genus L sp. form K is an odd but distinctive taxon and, although common in Biozone Cf2 samples, is apparently unnamed.

Zonitidae

?Zonitidae gen. and sp. undet. is an unnamed taxon represented by two specimens from one locality. It is a semislug (a snail with a reduced shell that cannot contain the whole animal). The greatest present diversity of semislugs occurs in forested, temperate to tropical settings.

Polygyridae

The Bighorn Basin samples contain the probable first appearance of the Polygyridae, as represented by *Neohelix*? sp. form M in Biochron Wa1. Only one specimen of this lot of nine specimens shows the adult lip, but that is very similar to some species of the Recent polygyrid genus *Neohelix*. Emberton (1994) hypothesized the presence of Polygyridae in North America as early as 120–145 Ma, although no pre-Cenozoic fossil record exists.

If correctly identified as *Neohelix*, this occurrence falsifies or at least complicates Emberton's (1994) biogeographic scenario for the evolution of major subgroups of the Polygyridae. The earliest confirmed polygyrid fossils of which he was aware are two unnamed genera from the White River Formation in eastern Wyoming, of Chadronian and Orellan Age, respectively (E. Evanoff and Roth, unpublished data).

Thysanophoridae

The estimated age of the holotype of *Microphysula oxyaenae* is Biochron Wa2–Wa3. The only positive identification of this taxon in our samples is from Biozone Wa2. The known stratigraphic range (Biozone Cf2) of *Microphysula* sp. cf. *M. ingersolli* may expand as more minute specimens are examined. Cockerell's (1914b) taxon *Vitrea sinoparum* is based on poor material, and we have not recognized it in the material studied to date. Russell (1931b), apparently without seeing any new material, reassigned the species to *Microphysula*. Although his reasons for this assignment are not well stated, it is at least as plausible as *Vitrea*.

Megomphicidae

Megomphicidae was proposed 11 months earlier than the family-group name Ammonitellidae and, thus, has priority. Tozer (1956), followed by Taylor (1975) and Roth (1986), referred the Late Cretaceous *Anchistoma parvula* Whiteaves 1885 to *Polygyrella*. The specimens from the Bighorn Basin here called *Polygyrella* sp. cf. *P. parvula* are higher-spired than Cretaceous samples of *P. parvula*. In our samples, the species is restricted to the Cf2–Cf3 zonal interval.

Bradybaenidae

The estimated age of the holotype of New Genus P *jepseni* (originally described in the freshwater prosobranch genus *Viviparus*) is Biochron Wa2–Wa3. Positive identifications of this taxon are from Biozone Wa3. New Genus P *jepseni* is distinctive in its relatively flat to domed embryonic shell with radial sculpture and no clear demarcation between embryonic and postembryonic whorls. Taylor (1975) assigned the species to *Oreoconus*, but it is not a member of Bulimulidae. Bulimulids characteristically have a distinctive embryonic shell with depressed initial one-third whorl (resembling a fist that encloses the thumb) and laterally compressed embryonic whorls. Allocation of New Genus P *jepseni* is uncertain, but it is probably referable to one of the sigmurethran holopod families like Camaenidae or Bradybaenidae, the latter being suggested here. The embryonic whorls of *Oreoconus* have not yet been well described for either the type species, *O. planispira* (Taylor in McKenna et al. 1962), or subsequently referred species (Hanley 1974).

Helminthoglyptidae

The Bighorn Basin collection contains the possible first appearance of the Helminthoglyptidae, with *Noyo*? sp. form N (first seen in Biozone Wa1) and questionably with *Glypterpes* sp. cf. *G. veternior*. The next earliest confirmed occurrence of the Helminthoglyptidae is that of *Greggelix*? *hochbergi* Roth and Megaw 1989 in the Sierra Santa Eulalia, Chihuahua, Mexico (Roth and Megaw 1989). A revised age of between 43.5±1.5 Ma and 39.5±1.5 Ma (Megaw et al. 1994) places that occurrence in the middle Eocene record. "*Helminthoglypta*" *obtusa* Anderson and Hanna 1925 and "*Helminthoglypta*?" *stocki* Hanna 1934, from the Eocene deposits of southern California, are camaenid land snails (Roth 1988). Previous referrals of the genera *Glypterpes* Pilsbry 1893 and *Mesoglypterpes* Yen 1952 to Helminthoglyptidae rest on tenuous readings of nondiagnostic shell characters (Roth 1996). Other fossil species assignable to the genus *Noyo* Roth (1996) include *N. bozemanensis* (Roth 1986) from the Chadronian Dunbar Creek Formation of the Bozeman Group of western Montana and *N. martini* (Hanna 1920) from the Arikareean John Day Formation in Oregon.

The assignment of Cockerell's (1915a) *Helix veterna veternior* to *Glypterpes* follows from *Helix veterna* Meek and Hayden 1862, which is the type species of the genus. However, Cockerell's reasons for calling this taxon a subspecies of *Helix veterna* are not clearly stated. Russell's (1931b) assignment of *Helix veternior* to *Polygyra* is incorrect. Pending examination of the holotype, specimens in our samples are kept separate as *G.* sp. cf. *G. veternior*. The two specimens that extend the range of this taxon into Biochron Wa1 continue to be under study.

A purpose of this chapter is to show the potential and realized utility in the use of nonmarine mollusks for providing a basis for biostratigraphic and biochronologic correlation. In contrast to those who have devoted much effort to the study of mammals and, as a consequence, aided in the development of a model for testing the temporal history of continental strata, few molluscan studies provide the breadth and depth needed to determine their value in biochronologic correlation. The generally simple morphology of molluscan shells and the difficulty of sorting out of ecophenotypic and ontogenetic variation complicate the interpretation of their evolutionary history. However, correlation with other frameworks or biohorizons allows aquatic and terrestrial mollusks to aid in the recognition of North American land-mammal ages. In addition, a mollusk-based stage-scale subdivision of strata in the Bighorn Basin and elsewhere is probably feasible.

Russell (1931b) suggested that a diverse and possibly unique, or at least unusual, molluscan fauna is represented in the Clark's Fork Basin, compared with other probable late to latest Paleocene assemblages known from the Western Interior. Russell's general observation was valid, in light of the fact that few terrestrial species had been described from the Paleocene strata.

Recent studies in North Dakota, Montana, Wyoming, and Saskatchewan have recognized the general structure of the stratigraphic distribution of molluscan taxa through the uppermost Cretaceous and Paleocene strata (Hartman, see references for examples). This interval is dominated by aquatic taxa, consisting largely of freshwater unionids and prosobranchs, often associated with major fluvial systems. Quiet freshwater environments have been sampled less often and thus have contributed in only minor ways, as yet, to the development of a molluscan biostratigraphic framework.

Three geographic areas provide insight into a potentially very interesting story: Locality UM-Sub-Wy 6, Sublette County, Wyoming (Hoback Basin); the Riverdale Locality in McLean County, North Dakota (Williston Basin); and the localities in the Bighorn Basin, Wyoming–Montana detailed in this chapter.

The molluscan faunule from the Hoback Formation Locality UM-Sub-Wy 6 was studied (but never published on) by Moore (1960). Dorr et al. (1977) interpreted this locality as a horizon equivalent to mammalian Locality UM-Sub-WY 21 and 22, which yielded the Battle Mountain local fauna. Archibald et al. (1987) reported it as within Biochron Ti3. Thus the molluscan assemblage at UM-Sub-Wy 6 comprises the first record of a diverse association of terrestrial taxa in the Paleocene record of North America. Moore (1960) reported 17 new species of terrestrial pulmonates. The faunule includes a taxon similar to *Euglandina* (Spiraxidae), pupilliform snails (Pupillidae, Vertiginidae), New Genus T (Hartman et al. 1993; new family?; was erroneously reported as *Oreohelix* by Moore 1960), other undescribed taxa (not referable to *Oreohelix* as per Moore 1960), and a highly diverse assemblage of acicular-shaped species that Moore assigned to *Pseudocolumna*. Although many, if not all, of these latter taxa are assignable to other genera (but of Subulinidae or Ferussaciidae), the diversity recognized by Moore does exist, and most are new species (Roth unpublished data). The aquatic fauna is also diverse. Moore (1960) reported six new species of unionids, one new species of pisidiid, and three new species of hydrobiid. Significantly, Moore identified aquatic taxa representing the molluscan assemblage first identified by Meek and Hayden from the Fort Union Group along the Missouri River in the Williston Basin of western North Dakota and easternmost Montana (see Hartman 1984; Hartman and Kihm 1995): *Viviparus trochiformis* [= *V. meeki*], *Viviparus raynoldsanus*, *Lioplax* (*Lioplacodes*) *tenuicarinata*, *L.* (*L.*)? *limnaeformis*, *L.* (*L.*) sp. cf. *L. nebrascensis producta*, *Micropyrgus minutulus*, *Pleurolimnaea* sp. cf. *P. tenuicosta*, *Aplexa canadensis*, and *Acroloxus minutus*. All of the larger taxa are difficult to positively assign to the identifications of Moore (1960) because of the plastic deformation of the shell form and absence of typical shell surface features (see, in part, Hartman 1984). Clearly, however, the faunule bears a resemblance to the type Fort Union fauna. Present are two or three species of the *Viviparus meeki* complex, *Lioplacodes* species, possibly including *L. tenuicarinata* and *L. nebrascensis*, an elongate hydrobiid, New Genus A *limneaformis*, a form similar in shape to *P. tenuicosta*, but not preserving axial sculpture, physids, and *A. minutus* (representing the distinctive shape, but not preserving shell sculpture). A Biochron Ti3 age for this assemblage is consistent with a Biochron Ti3 to Ti4 age of the fauna in the Williston Basin. Some of the species ranged as young as Biochron Ti5 in the Bighorn Basin.

The Riverdale Locality has yielded the most diverse molluscan faunule in North Dakota, from strata of the Sentinel Butte Formation interpreted to be Biochron Ti4 in age (Hartman et al. 1993; Kihm and Hartman 1991; Hartman and Kihm 1995; Hartman and Roth unpublished). A large number of terrestrial taxa were identified, including several undescribed species. The terrestrial taxa are highly endemic; only New Genus T, which includes the Riverdale taxon *Planorbis planoconvexus* Meek and Hayden 1860, is shared in common with the Hoback Basin. The aquatic aspect of the Riverdale faunule is less distinctive and probably shares the following species with the Sublette County locality: *Lioplacodes nebrascensis*, New Genus A *lim-*

neaformis, *Pleurolimnaea tenuicosta*, and a *Physa* similar to *P. canadensis*. Few terrestrial taxa of the Riverdale Locality are found in the Bighorn Basin. The Riverdale faunule includes species of *Grangerella*, *Discus*, *Radiocentrum*, and forms that are similar to taxa assigned to the Zonitidae and Helminthoglyptidae. The first Bighorn Basin records of *Grangerella* and *Radiocentrum* are in the lower Clarkforkian NALMA (Biozone Cf1), with the subsequent appearance of *Discus* and the Helminthoglyptidae and a questioned occurrence of the Zonitidae in Zone Cf2. The overall picture is one of strong intrabasinal endemism of the terrestrial mollusks. The molluscan record in the Powder River Basin lacks terrestrial representation for the late Tiffanian and Clarkforkian NALMAs. Thus the record preserved in isolated basins suggests varying environmental conditions across the northern Great Plains and associated intermontane basins.

Gunnell et al. (1993) summarized arguments concerning the placement of the Paleocene/Eocene Series boundary in the Bighorn Basin. They suggested that contrary to previous interpretations, the mammalian and palynomorph records effectively place this boundary very near or at the boundary between the Clarkforkian and Wasatchian NALMAs. The authors also noted that across this boundary, significant change was occurring in the structure of terrestrial ecosystems. The record of late Paleocene nonmarine molluscan faunas provides glimpses of distinctive assemblages indicating either shifts in continental landscape or climate or isolated pockets of rarely sampled assemblages. The isolated examples of species-rich terrestrial faunules from Biochrons Ti3 and Ti4 (Hoback and Williston Basin, respectively) may reflect locally differing paleoenvironments. Aquatic taxa are more widespread and provide a basis for temporal comparison. Assemblages of acicular-shaped snails, like those from the Hoback Basin, tend to be found today in moist tropical or subtropical settings. In contrast, the Williston and Clark's Fork basins share a similar faunal composition, which indicates less moist conditions.

Hickey (1980) discussed the environmental changes that he believed took place in the northern Bighorn Basin during the late Paleocene Epoch on the basis of floral interpretations and noted some trends that are worth exploring in future molluscan research. Hickey stated that Tiffanian floras were relatively low in diversity and low in equitability, projecting a mean annual temperature of 10° C. In contrast, the Clarkforkian Age witnessed a trend toward a substantial increase in diversity, equitability, and the percentage of entire-margined leaves. According to Hickey (1980), the floras from the Tiffanian to the Clarkforkian Age shifted from a deciduous broad-leafed forest to forests with a significant broad-leafed evergreen component. The associated interpreted increase in temperature (up to 13.5° C) may indicate a general increase in aridity. The floras of the early Wasatchian Age and their environments of deposition in the central portion of the Bighorn Basin were interpreted by Wing (1980), who stated that the lower 125 m of the Willwood Formation were comparable, in terms of occurrence of carbonaceous layers, to the underlying Fort Union Formation. He suggested that these backswamp deposits strongly resembled their late Paleocene analogues. From the 125 m to the 355 m horizon, tabular carbonaceous shales become less important and eventually cease to occur, representing the end of the backswamp depositional regime. These conditions persisted until near the end of the Wasatchian NALMA. The presence or absence of the depositional settings noted by Hickey (1980) and Wing (1980) through the Tiffanian NALMA and until the end of the Wasatchian NALMA undoubtedly played an important role in the occurrence and distribution of molluscan faunules. However, the changes that occurred in the molluscan faunal composition through this time do not appear to be balanced by the introduction or immigration of taxa reflecting the return of specific aquatic depositional conditions. The loss of certain taxa or their faunal equivalents appears to have been permanent, suggesting important evolutionary changes in the molluscan biota through this time.

The composition of the nonmarine molluscan Paleocene faunas changed through the Paleocene Epoch. A relatively diverse record of taxa, dominated by aquatics but including some rare terrestrial species, can be found in the middle Tiffanian Biozones Ti3–Ti4 of the northern Great Plains. This fauna is largely the one described by Meek and Hayden in the 1850s and 1960s (Hartman and Kihm 1995). Significant compositional changes to this fauna began in the latest Tiffanian NALMA (Biochron Ti5; Bighorn Basin), increased in the early Clarkforkian NALMA (Biochron Cf1), and developed fully, with the presence of a diverse assemblage of terrestrial snails, in the middle Clarkforkian NALMA (Biochron Cf2; see figures 16.7–16.10). The scope of the introduction of new taxa during the middle Clarkforkian NALMA was not repeated in the early Wasatchian NALMA, approximating the Paleocene/Eocene Epoch boundary, as is the case in the Powder River Basin. This pattern of taxon appearance and carryover is shown in figures 16.7 and 16.8. The middle Clarkforkian NALMA (Biochron Cf2) was a time of diversification, apparently for both terrestrial and aquatic species. The loss of aquatic prosobranch taxa continued through the end of the Clarkforkian NALMA. The re-

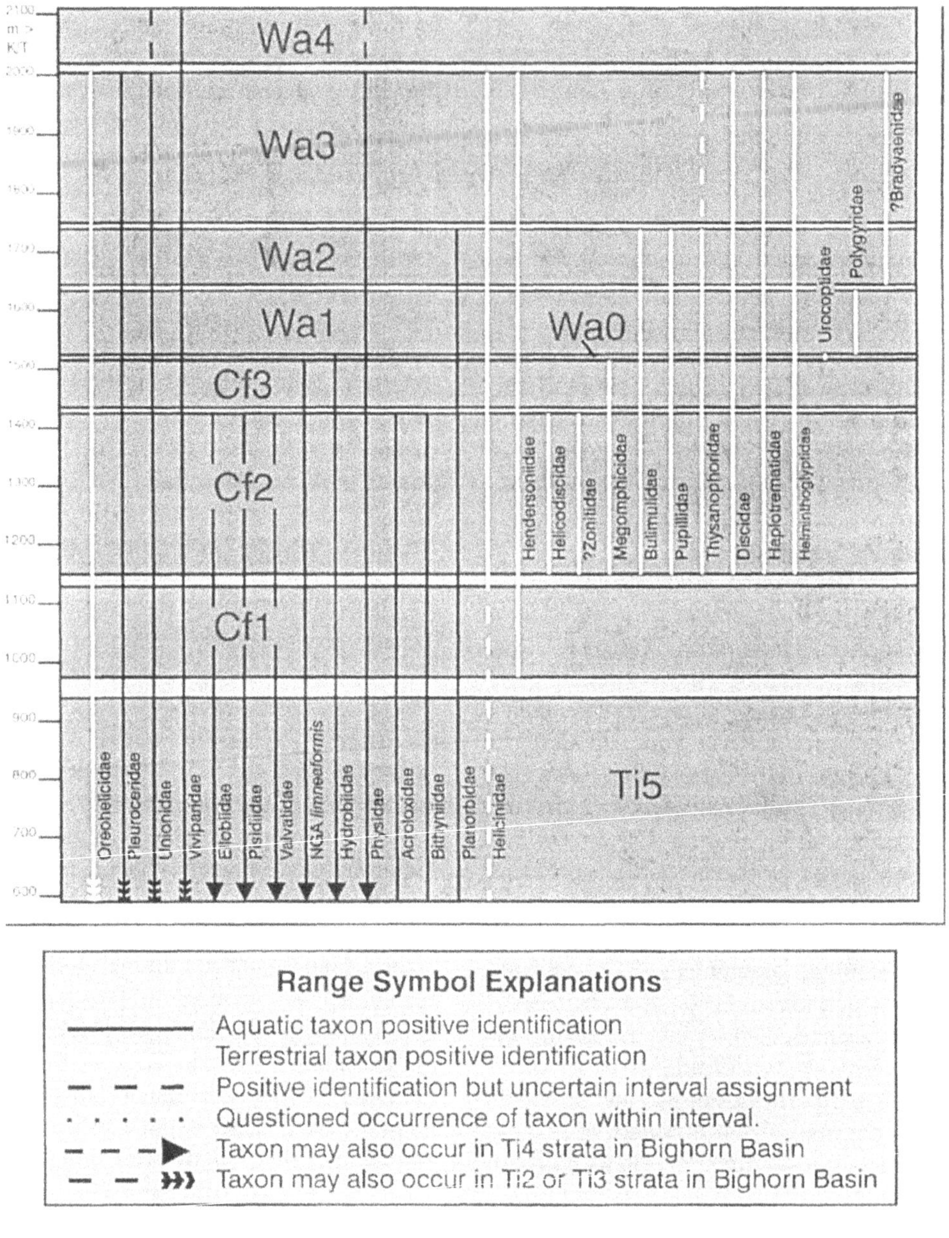

Figure 16.9 Biostratigraphy of nonmarine molluscan families in the Bighorn Basin. This figure intervalizes the occurrence of identified families in the Bighorn Basin.
Range line symbols are slightly different from those in previous figures, indicating, for some taxa, their occurrence in older strata in the Bighorn Basin. Note that most of the terrestrial families have their first appearance in the basin in beds of middle Clarkforkian Age and in younger strata.

duction of aquatic prosobranchs is a distinctive feature of the Wasatchian biota. This pattern continued into the Wasatchian NALMA. However, aquatic species retain their utility in biochronologic correlation. The change in pattern and faunal composition of new taxa is stressed in figures 16.9 and 16.10, which illustrate the timing of the introduction of new families and genera in the Bighorn Basin. The increase in the diversity of terrestrial taxa is clearly the single most important event in molluscan faunal change during this interval, and this change occurred during the middle Clarkforkian Age (Biochron Cf2).

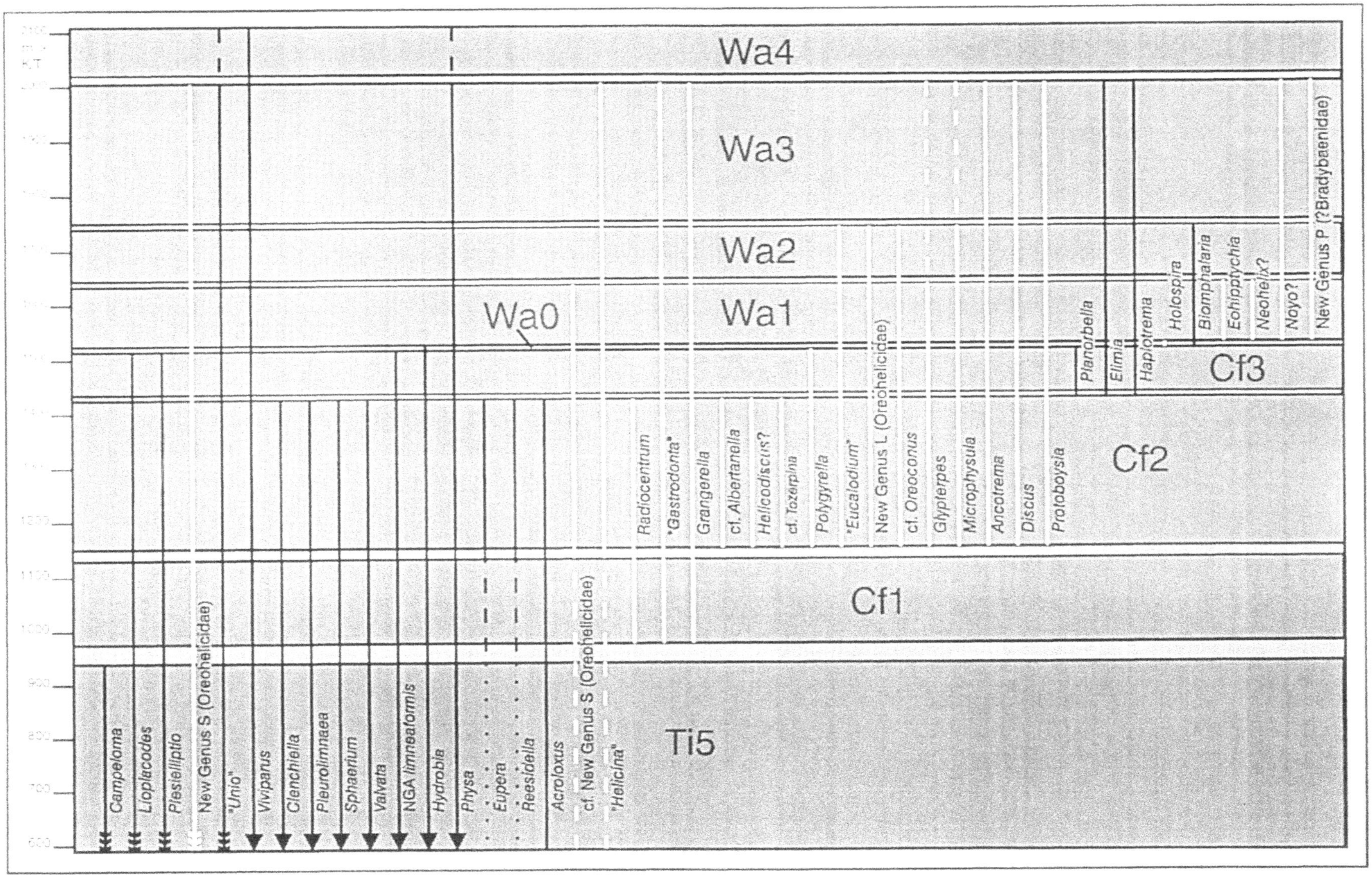

Figure 16.10 Biostratigraphy of nonmarine molluscan genera in the Bighorn Basin. This figure intervalizes the occurrence of identified genera in the Bighorn Basin. Range line symbols (see figure 16.9 for explanations) are slightly different than in previous figures, indicating, for some taxa, their occurrence in older strata in the Bighorn Basin. Note that most of the terrestrial genera have their first appearance in the basin in beds of middle Clarkforkian Age and in younger strata.

ACKNOWLEDGMENTS

We specifically wish to acknowledge the perseverance of Philip D. Gingerich and Gregg F. Gunnell of the University of Michigan Museum of Paleontology for both loans of fossils and information on mammalian localities and fossil collection histories. In studies regarding UMMP, USGS, UCM, PU-INV, and other collections, the cooperation of the following individuals was greatly appreciated: Donald Baird (late of Princeton University), David W. Krause (UMMP and later State University of New York at Stony Brook), Malcolm C. McKenna (AMNH), Peter Robinson (UCM), Kenneth D. Rose (UMMP and later Johns Hopkins University), Loris S. Russell (Royal Ontario Museum), David M. Schankler (Yale University), and Thomas W. Henry, Thomas M. Bown, and Elly Brouwers (USGS-D). We wish also to thank Barry Kues of the University of New Mexico for his conscientious review of the manuscript for this paper. This project was initially supported by grants to Hartman from the Philip M. McKenna Foundation and later by the National Science Foundation, U.S. Bureau of Mines, U.S. Department of Energy, and Energy & Environmental Research Center. This is contribution no. 1657 from the Museum of Paleontology, University of California, Berkeley.

REFERENCES

Andrews, D. A., Pierce, W. G. and Eargle, D. H., 1947. Geologic map of the Bighorn Basin, Wyoming, and Montana, showing terrace deposits and physiographic features. U.S. Geological Survey, Oil and Gas Investigations, Preliminary Map 71, 1 sheet.

Archibald, J. D., Clemens, W. A., Gingerich, P. D., Krause, D. W., Lindsay, E. H. and Rose, K. D., 1987. First North American Land Mammal Ages of the Cenozoic Era. In Woodburne, M. O., (ed.), *Cenozoic Mammals of North America—Geochronology and Biostratigraphy,* University of California Press, Berkeley, Los Angeles, London, p. 24–76.

Bishop, M. J., 1980. Helicinid land snails with apertural barriers. *Journal of Molluscan Studies,* v. 46, p. 241–246.

Bland, T., 1875. Note on the sub-generic character of *Helix jamaicensis* Chemn. and on certain terrestrial mollusks from Haiti; with description of a new species of *Helix* from Colorado. *Annals of the Lyceum of Natural History of New York,* v. 11, p. 146–154.

Bown, T. M., 1982. *Geology, paleontology, and correlation of Eocene volcaniclastic rocks, southeast Absaroka Range, Hot Springs County, Wyoming.* Geological Survey Professional Paper 1201-A, 75 pp.

Clyde, W. C., Stamatakos, J. and Gingerich, P. D., 1994. Chronology of the Wasatchian Land-Mammal Age (early Eocene): Magnetostratigraphic results from the McCullough Peaks Section, northern Bighorn Basin, Wyoming. *Journal of Geology,* v. 102, p. 367–377.

Cockerell, T. D. A., 1914a. Tertiary Mollusca from New Mexico and Wyoming. *Bulletin of the American Museum of Natural History,* v. 33, p. 101–107.

Cockerell, T. D. A., 1914b. Land shells from the Tertiary of Wyoming. *Bulletin of the American Museum of Natural History,* v. 33, p. 323–325.

Cockerell, T. D. A., 1915a. Gastropod Mollusca from the Tertiary strata of the West. *Bulletin of the American Museum of Natural History,* v. 34, p. 115–120.

Cockerell, T. D. A., 1915b. New species of *Unio* from the Tertiary rocks of Wyoming. *Bulletin of the American Museum of Natural History,* v. 34, p. 121–126.

Cockerell, T. D. A. and Henderson, J., 1912. Mollusca from the Tertiary strata of the West. *Bulletin of the American Museum of Natural History,* v. 31, p. 229–234.

Darton, N. H., 1907. Coals of Carbon County, Mont[ana]. *U.S. Geological Survey Bulletin,* v. 316-C, p. 174–193.

Delson, E., 1971. Fossil mammals of the early Wasatchian Powder River local fauna, Eocene of northeast Wyoming. *Bulletin of the American Museum of Natural History,* v. 146, p. 305–364.

Dorr, J. A., Jr., Spearing, D. R. and Steidtmann, J. R., 1977. *Deformation and deposition between a foreland uplift and an impinging thrust belt: Hoback Basin, Wyoming.* Geological Society of America Special Paper 177, 82 pp.

Ebaugh, W. F., 1976. Preliminary surficial and bedrock geologic map of the Big Horn Quadrangle, Sheridan County, Wyoming. U.S. Geological Survey, Miscellaneous Field Studies, Map MF801, 1 sheet.

Emberton, K. C., 1994. Polygyrid land snail phylogeny: External sperm exchange, early North American biogeography, iterative shell evolution. *Biological Journal of the Linnean Society,* v. 52, p. 241–271.

Fisher, C. A., 1906. *Geology and water resources of the Bighorn Basin, Wyoming.* U.S. Geological Survey Professional Paper 53, 72 pp.

Gingerich, P. D., 1980. History of early Cenozoic vertebrate paleontology in the Bighorn Basin. In Gingerich, P. D., (ed.), *Early Cenozoic Paleontology and Stratigraphy of the Bighorn Basin, Wyoming: 1880–1980,* University of Michigan, Papers on Paleontology, v. 24, p. 7–24.

Gingerich, P. D., 1987. Early Eocene bats (Mammalia, Chiroptera) and other vertebrates in freshwater limestones of the Willwood Formation, Clark's Fork Basin, Wyoming. *University of Michigan, Contributions from the Museum of Paleontology,* v. 27, p. 275–320.

Gingerich, P. D., 1989a. Paleocene and Early Eocene of the Bighorn and Clark's Fork Basins, Wyoming. In Flynn, J. J. (ed.), *Mesozoic/Cenozoic Vertebrate Paleontology: Classic localities, contemporary approaches.* Washington: American Geophysical Union, 28th International Geological Congress Field Trip Guidebook T322, p. 47–57.

Gingerich, P. D., 1989b. New earliest Wasatchian mammalian fauna from the Eocene of northwestern Wyoming: Composition and diversity in a rarely sampled high-floodplain assemblage. University of Michigan, Papers on Paleontology, v. 28, 97 pp.

Gingerich, P. D., Rose, K. D. and Krause, D. W., 1980. Early Cenozoic mammalian faunas of the Clark's Fork Basin–Polecat Bench area, northwestern Wyoming. In Gingerich, P. D., (ed.), *Early Cenozoic Paleontology and Stratigraphy of the Bighorn Basin, Wyoming: 1880–1980,* University of Michigan Papers on Paleontology, v. 24, p. 51–64.

GNULEX, 1992. *Stratigraphic nomenclature databases for the United States, its possessions and territories.* U.S. Geological Survey Digital Data Series DDS-6.

Gunnell, G. F., Bartels, W. S. and Gingerich, P. D., 1993. Paleocene–Eocene boundary in continental North America: Biostratigraphy and Geochronology, northern Bighorn Basin, Wyoming. In Lucas, S. G. and Zidek, J., (eds.), *Vertebrate Paleontology in New Mexico,* New Mexico Museum of Natural History and Science, Bulletin 2, p. 137–144.

Hanley, J. H., 1974. *Systematics, paleoecology, and biostratigraphy of nonmarine Mollusca from the Green River and Wasatch Formations (Eocene), southwestern Wyoming and northwestern Colorado.* Ph.D. thesis, Laramie: University of Wyoming, 285 pp.

Hanna, G. D., 1920. Miocene land shells from Oregon. *Kansas University Science Bulletin* 13, p. 1–12.

Hartman, J. H., 1976. *Uppermost Cretaceous and Paleocene nonmarine Mollusca of eastern Montana and southwestern North Dakota.* Master of Science Thesis, Minneapolis: University of Minnesota, 216 pp.

Hartman, J. H., 1981. Early Tertiary nonmarine Mollusca of New Mexico: A review. *Geological Society of America Bulletin,* v. 92, p. 942–950.

Hartman, J. H., 1984. *Systematics, biostratigraphy, and biogeography of latest Cretaceous and early Tertiary Viviparidae (Mollusca, Gastropoda) of southern Saskatchewan, western North Dakota, eastern Montana, and northern Wyoming.* Ph.D. Thesis, Minneapolis: University of Minnesota, 928 pp.

Hartman, J. H., 1987. *The North American Mesozoic nonmarine molluscan record.* Tyrrell Museum of Paleontology (Fourth Symposium on Mesozoic Terrestrial Ecosystems), Occasional Paper, no. 3, p. 106–119.

Hartman, J. H., 1988. Upper Cretaceous and Paleocene nonmarine Mollusca in the Williston Basin. *Proceedings of the North Dakota Academy of Science,* v. 42, p. 29.

Hartman, J. H., 1989. Stratigraphy of uppermost Cretaceous and Paleocene nonmarine Mollusca in the Crazy Mountains Basin, south-central Montana. In French, D. E. and Grabbs, R. F., (eds.), 1989 *Field Conference Guidebook,* Montana Centennial Edition, Montana Geological Society, Geologic Resources of Montana, v. I, p. 163–172.

Hartman, J. H., 1990. Paleocene and lower Eocene nonmarine molluscan biostratigraphy of the Powder River Basin, Wyoming–Montana. *Geological Society of America, Rocky Mountain Section, Abstracts with Programs,* v. 22, p. 14.

Hartman, J. H., 1992. Biochronology of uppermost Cretaceous and lower Tertiary nonmarine Mollusca of the northern Great Plains, U.S.A. and Canada. In Lidgard, S. and Crane, P. R., (eds.), *Fifth North American Paleontological Convention, Abstracts with Programs,* The Paleontological Society, Special Publication No. 6, p. 123.

Hartman, J. H., 1993a. Patterns of diversification of North American Paleogene nonmarine Mollusca in the context of mammalian biochronology. In Catzeflis, F. M., *Paleobiology and Evolution of Early Cenozoic Mammals: A symposium in Honor of Donald E. Russell,* Fourth Congress of the European Society for Evolutionary Biology, Montpellier, France, p. 167.

Hartman, J. H., 1993b. The transition of Nonmarine mollusks across the Paleocene–Eocene boundary in the United States, in a symposium on "The Paleocene–Eocene Boundary." *Journal of Vertebrate Paleontology,* v. 13 (supplement to no. 3), p. 41A.

Hartman, J. H. and Kihm, A. J., 1992. Chronostratigraphy of Paleocene strata in the Williston Basin. In Finkelman, R. B., Tewalt, S. J. and Daly, D. J., (eds.), *Geology and utilization of Fort Union lignites,* Reston: Environmental and Coal Associates, p. 52–75.

Hartman, J. H. and Kihm, A. J., 1995. Age of Meek and Hayden's Fort Union Group (Paleocene), upper Missouri River, North Dakota–Montana. In Hunter, L. D. V. and Schalla, R. A. (eds.), *Seventh International Williston Basin Symposium, 1995 Guidebook,* Billings, Montana, p. 417–428.

Hartman, J. H. and Krause, D. W., 1993. Cretaceous and Paleocene stratigraphy and paleontology of the Shawmut Anticline and the Crazy Mountains Basin, Montana: Road log and overview of recent investigations. In Hunter, L. D. V., (ed.), *Montana Geological Society, Energy and Mineral Resources of Central Montana, 1993 Field Conference Guidebook,* p. 71–84.

Hartman, J. H., Roth, B. and Kihm, A. J., 1993. A diverse assemblage of Paleocene nonmarine mollusks and mammals from the Sentinelle Butte Formation of North Dakota. *Proceedings of the North Dakota Academy of Science,* v. 47, p. 50.

Hewett, D. F., 1914. The Shoshone River section, Wyoming. U.S. Geological Survey Bulletin 541-C, p. 89–113.

Hewett, D. F. and Lupton, C. T., 1917. Anticlines in the southern part of the Big Horn Basin: A preliminary report on the occurrence of oil. *U.S. Geological Survey Bulletin* 656, 192 pp.

Hickey, L. J., 1980. Paleocene stratigraphy and flora of the Clark's Fork Basin. In Gingerich, P. D. (ed.), *Early Cenozoic Paleontology and Stratigraphy of the Bighorn Basin, Wyoming: 1880–1980,* University of Michigan, Papers on Paleontology, v. 24, p. 33–49.

Jepsen, G. L., 1930. Stratigraphy and paleontology of the Paleocene of northeastern Park County, Wyoming. *Proceedings of the American Philosophical Society,* v. 69, p. 463–528.

Jepsen, G. L., 1940. Paleocene faunas of the Polecat Bench Formation, Park County, Wyoming. *Proceedings of the American Philosophical Society,* v. 83, p. 217–341.

Keroher, G. C., 1966. Lexicon of geologic names of the United States for 1936–1960. Part 3, P-Z. *U.S. Geological Survey Bulletin 1200,* p. 2887–4341.

Kihm, A. J. and Hartman, J. H., 1991. Age of the Sentinel Butte Formation, North Dakota. *Journal of Vertebrate Paleontology,* v. 11 (supplement to no. 3), p. 40a.

La Rocque, A., 1960. The molluscan faunas of the Flagstaff Formation of central Utah. *Geological Society of America Memoir* 78, 100 pp.

Love, J. D. and Christiansen, A. C., 1985. Geologic Map of Wyoming (1:500,000). U.S. Geological Survey, 3 sheets.

McKenna, M. C., Robinson, P. and Taylor, D. W., 1962. *Notes on Eocene Mammalia and Mollusca from Tabernacle Butte, Wyoming.* American Museum Novitates, no. 2102, 33 pp.

Megaw, P. K. M., McDowell, F. W. and Roth, B., 1994. Refinement of the age of certain Tertiary land mollusks (Gastropoda: Pulmonata) based on new UPb zircon ages from the Sierra Santa Eulalia, Chihuahua, Mexico. *Malacological Review,* v. 27, p. 109.

Meek, F. B., 1873. Preliminary paleontologic report, consisting of lists and descriptions of fossils, with remarks on the ages of the rocks in which they were found. F. V. Hayden Survey, U.S. Geological [and Geographical] Survey of the Territories, embracing portions of Montana, Idaho, Wyoming, and Utah, 6th Annual Report (for 1872), p. 429–518.

Meek, F. B., 1876. A report on the invertebrate Cretaceous and Tertiary fossils of the Upper Missouri Country. F. V. Hayden Survey, U. S. Geological [and Geographical] Survey of the Territories, Monograph 9, 629 pp.

Meek, F. B. and Hayden, F. V., 1856. Descriptions of new species of Acephala and Gastropoda, from the Tertiary formations of Nebraska Territory, with some general remarks on the geology of the country about the sources of the Missouri River. *Proceedings of the Academy of Natural Sciences of Philadelphia,* v. 8, p. 111–126.

Meek, F. B. and Hayden, F. V., 1857. Descriptions of new species and genera of fossils, collected by Dr. F. V. Hayden in Nebraska Territory, under the direction of Lieut. G. K. Warren, U.S. Topographical Engineer; with some remarks on the Tertiary and Cretaceous formations of the north-west, and the parallelism of the latter with those of other portions of the United States and territories. *Proceedings of the Academy of Natural Sciences of Philadelphia,* v. 9, p. 117–148.

Meek, F. B. and Hayden, F. V., 1860a. Descriptions of new organic remains from the Tertiary, Cretaceous and Jurassic rocks of Nebraska. *Proceedings of the Academy of Natural Sciences of Philadelphia,* v. 12, p. 175–185.

Meek, F. B. and Hayden, F. V., 1860b. Systematic catalogue, with synonyma, etc., of Jurassic, Cretaceous and Tertiary fossils collected in Nebraska [Territory], by the exploring expeditions under the command of Lieut. G. K. Warren, of U.S. Topographical Engineers.

Proceedings of the Academy of Natural Sciences of Philadelphia, v. 12, p. 417–432.

Meek, F. V. and Hayden, F. V., 1862. Descriptions of new Lower Silurian (Primordial), Jurassic, Cretaceous, and Tertiary fossils, collected in Nebraska Territory, by the exploring expedition[s] under the command of Capt. Wm. F. Raynolds, U.S. Top. Engrs.; with some remarks on the rocks from which they were obtained. *Proceedings of the Academy of Natural Science of Philadelphia,* v. 13, p. 415–447.

Moore, R. G., 1960. *A Paleocene fauna from the Hoback Formation, Wyoming.* Ph.D. Thesis, Ann Arbor: University of Michigan, 169 pp.

Pierce, H. G. and Rasmussen, D. L., 1989. New land snails (Archaeogastropoda, Helicinidae) from the Miocene (early Barstovian) Flint Creek Beds of western Montana. *Journal of Paleontology,* v. 63, p. 846–851.

Pilsbry, H. A., 1948. *Land Mollusca of North America (north of Mexico).* Monographs of the Academy of Natural Sciences of Philadelphia, no. 3, p. i–xlvii + 521–1113.

Reeside, J. B., Jr., 1926 (November 4). Reston: U.S. Geological Survey Correspondence (unpublished).

Reeside, J. B., Jr., 1953 (December 10). Reston: U.S. Geological Survey E&R Report (unpublished).

Robinson, L. N. and Honey, J. G., 1987. Geologic setting of a new Paleocene mammal locality in the northern Powder River Basin, Montana. *Palaios,* v. 2, p. 87–90.

Rohrer, W. L., 1964a. Geology of the Sheep Mountain Quadrangle [Big Horn County], Wyoming. U.S. Geological Survey, Geologic Quadrangle Map, GQ-310, 1 sheet.

Rohrer, W. L., 1964b. Geology of the Tatman Mountain Quadrangle [Park and Big Horn Counties], Wyoming. U.S. Geological Survey, Geologic Quadrangle Map, GQ-311, 1 sheet.

Rohrer, W. L., 1966. Geology of the Adam Weiss Peak Quadrangle, Hot Springs and Park Counties, Wyoming. *U.S. Geological Survey Bulletin,* v. 1241-A, p. A1–A39.

Rohrer, W. L. and Gazin, C. L., 1965. *Gray Bull and Lysite faunal zones of the Willwood Formation in the Tatman Mountain area, Bighorn Basin, Wyoming. U.S.* Geological Survey Professional Paper 525-D, p. D133–D138.

Rohrer, W. L. and Smith, J. W., 1969. *Tatman Formation.* Wyoming Geological Association Guidebook, 21st Annual Field Conference, p. 49–54.

Rose, K. D., 1979. *The Clarkforkian Land-Mammal "Age" and mammalian faunal composition across the Paleocene-Eocene boundary.* Ph.D. Thesis, Ann Arbor: University of Michigan, 601 pp.

Rose, K. D. 1980. Clarkforkian Land-Mammal Age: Revised definition, zonation, and tentative intercontinental correlations. *Science,* v. 208, p. 744–746.

Rose, K. D. and Bown, T. M., 1982. New plesiadapiform primates from the Eocene of Wyoming and Montana. *Journal of Vertebrate Paleontology,* v. 2, p. 63–69.

Ross, G. P., Andrews, D. A. and Witkind, I. J., 1955. Geologic map of Montana. U.S. Geological Survey and Montana Bureau of Mines and Geology, 2 sheets.

Roth, B., 1986. Land mollusks (Gastropoda: Pulmonata) from early Tertiary Bozeman Group, Montana. *Proceedings of the California Academy of Sciences,* v. 44, p. 237–267.

Roth, B., 1988. Camaenid land snails (Gastropoda: Pulmonata) from the Eocene of southern California and their bearing on the history of the American Camaenidae. *Transactions of the San Diego Society of Natural History,* v. 21, p. 203–220.

Roth, B., 1992. Identity of the Grangerellidae (Gastropoda: Prosobranchia): A late Mesozoic to Recent family of terrestrial snails. In Giusti, F. and Manganelli, G., (eds.), Abstracts, 11th International Malacological Congress, Sienna 1992, p. 132.

Roth, B., 1996. Homoplastic loss of dart apparatus, phylogeny of the genera, and a phylogenetic taxonomy of the Helminthoglyptidae (Gastropoda: Pulmonata). *Veliger,* v. 39, p. 18–42.

Roth, B. and Megaw, P. K. M., 1989. Early Tertiary land mollusks (Gastropoda: Pulmonata) from Sierra Santa Eulalia, Chihuahua, Mexico, and the origins of the North American arid-land mollusk fauna. *Malacological Review,* v. 22, p. 1–16.

Russell, L. S., 1931a. Mollusca from the Upper Cretaceous and lower Tertiary of Alberta. *Transactions of the Royal Society of Canada,* ser. 3, v. 25, p. 9–19.

Russell, L. S., 1931b. Early Tertiary Mollusca from Wyoming. *Bulletins of American Paleontology,* v. 18, p. 1–38.

Russell, L. S., 1934. New fossil fresh-water Mollusca from the Cretaceous and Paleocene of Montana. *Journal of the Washington Academy of Sciences,* v. 24, p. 128–131.

Simpson, G. G., 1937. The Fort Union of the Crazy Mountain field, Montana, and its mammalian faunas. *Bulletin of the U.S. National Museum,* v. 169, 287 pp.

Sinclair, W. J. and Granger, W., 1911. Eocene and Oligocene of the Wind River and Bighorn Basins. *Bulletin of the American Museum of Natural History,* v. 30, p. 83–117.

Sinclair, W. J. and Granger, W., 1912. Notes on the Tertiary deposits of the Bighorn Basin. *Bulletin of the American Museum of Natural History,* v. 31, p. 57–67.

Soister, P. E., 1968. *Stratigraphy of the Wind River Formation in south-central Wind River Basin, Wyoming.* U.S. Geological Survey Professional Paper 594-A, p. A1–A50.

Solem, A., 1983. Lost or kept internal whorls: Ordinal differences in land snails. *Journal of Molluscan Studies,* Supplement 12A, p. 172–178.

Stanton, T. W., 1909. The age and stratigraphic relations of the "Ceratops Beds" of Wyoming and Montana. *Proceedings of the Washington Academy of Sciences,* v. 11, p. 239–293.

Stow, M. H., 1938. Dating Cretaceous–Eocene tectonic movements in Big Horn Basin by heavy minerals. *Bulletin of the Geological Society of America,* v. 49, p. 731–762.

Stow, M. H., 1946. Dating sedimentation, vulcanism, and orogeny in Beartooth Mountain region, Montana, by heavy minerals. *Bulletin of the Geological Society of America,* v. 57, p. 675–686.

Swanson, R. W., Hubert, M. L., Luttrell, G. W. and Jussen, V. M., 1981. Geologic names of the United States through 1975. *U.S. Geological Survey Bulletin* 1535, 643 pp.

Taylor, D. W., 1975. Early Tertiary mollusks from the Powder River Basin, Wyoming–Montana, and adjacent regions. U.S. Geological Survey, Open-File Report 75-331, 515 pp.

Tozer, E. T., 1956. Uppermost Cretaceous and Paleocene non-marine molluscan faunas of western Alberta. *Geological Survey of Canada Memoir* 280, 125 pp.

Van Houten, F. B., 1944. Stratigraphy of the Willwood and Tatman Formations in northwestern Wyoming. *Bulletin of the Geological Society of America,* v. 55, p. 165–210.

Weed, W. H., 1892. Two Montana coal fields. *Bulletin of the Geological Society of America,* v. 3, p. 301–330.

White, C. A., 1878. Paleontologic papers No. 6: Description of new species of invertebrate fossils from the Laramie group. F. V. Hayden Survey, *U.S. Geological and Geographical Survey of the Territories Bulletin,* v. 4, p. 707–719.

White, C. A., 1882. New molluscan forms from the Laramie and Green River Groups, with discussion of some associated forms heretofore known. *Proceedings of the U.S. National Museum,* v. 5, p. 94–99.

White, C. A., 1883. *A review of the non-marine fossil Mollusca of North America.* U.S. Geological Survey, 3rd Annual Report, p. 403–550.

White, C. A., 1886. On the relation of the Laramie molluscan fauna to that of the succeeding fresh-water Eocene and other groups. *Bulletin of the U.S. Geological Survey,* v. 34, 54 pp.

Whiteaves, J. F., 1885. Report on the Invertebrata of the Laramie and Cretaceous Rocks of the vicinity of the Bow and Belly Rivers and adjacent localities in the North-West Territory. *Geological and Natural History Survey of Canada (Geological Survey of Canada) Contribution to Canadian Palaeontology,* v. 1, p. 1–89.

Wing, S. L., 1980. Fossil floras and plant-bearing beds of the central Bighorn Basin. *University of Michigan, Papers on Paleontology,* v. 24, p. 119–125.

Wing, S. L., 1984. A new basis for recognizing the Paleocene/Eocene boundary in western interior North America. *Science,* v. 226, p. 439–441.

Wolberg, D. L., 1978. *The mammalian paleontology of the late Paleocene (Tiffanian) Circle and Olive localities, McCone and Powder River Counties, Montana.* Ph.D. Thesis, Minneapolis: University of Minnesota, 385 pp.

Woodring, W. P., 1926. Fossil *Viviparus*-like calcareous opercula. Nautilus, v. 40, p. 12–14.

Woodruff, E. G., 1907. Red Lodge coal field, unpublished field notes. U.S. Geological Survey, Field Records Library, acc. 2709.

Woodruff, E. G., 1909. The Red Lodge coal field, Montana. *U.S. Geological Survey Bulletin,* v. 341-A, p. 92–107.

Yen, T.-C., 1946. Paleocene freshwater mollusks from Sheridan County, Wyoming. *American Journal of Science,* v. 244, p. 41–48.

Yen, T.-C., 1948a. Eocene fresh-water Mollusca from Wyoming. *Journal of Paleontology,* v. 22, p. 634–640.

Yen, T.-C., 1948b. *Paleocene fresh-water mollusks from southern Montana.* U.S. Geological Survey Professional Paper 214-C, p. 35–50.

Appendix 16.1
Bighorn Basin Locality Register

Order of locality data: Locality database number (e.g., L0133), county (Co.), state, section (sec.), township (T.), range (R.). NALMA biochron (BIO), Formation (FM), Member (MBR), meters above the Cretaceous/Paleogene System boundary (m), stratigraphic horizon (HOR) (e.g., H21). Reference number(s) (i.e., field number, collection number, institutional number), and reference name, location description, or reference (if source data is not available; see also Notes). Missing data indicate that the data are unavailable or not applicable for the locality.

L0133. Park Co., WY, sec. 20, T. 55 N., R. 98 W. Cf1? BIO, Fort Union FM. Ref name: Six miles east of Powell.

L0134. Park Co., WY, sec., T. 55 N., R. 99 W. Ti3–Ti4 BIO, Fort Union FM. Ref# and name: Sinclair-1929 Field No. 19, South side of Polecat Bench.

L0135. Park Co., WY, sec. 26, T. 57 N., R. 100 W. Ti5 BIO, Fort Union FM. Ref# and name: Sinclair-1929 Field No. 17, Southwest edge of Polecat Bench.

L0136. Park Co., WY, sec. 11, T. 56 N., R. 99 W. Ti2–Ti3 BIO, Fort Union FM. Ref name: Sec. 11 or 12, T. 56 N., R. 99 W.

L0483. Carbon Co., MT, T. 6? S., R. 23? E. La BIO, Hell Creek FM. Ref# Stanton (1909, p. 269–270).

L0532. Carbon Co., MT, sec. 22?, T. 8 S., R. 21 E. Cf1? BIO, Fort Union FM (Tongue River MBR). Ref#: USGS Mz 3897.

L0533. Carbon Co., MT, sec. 4, T. 8 S., R. 21 E. Cf1? BIO, Fort Union FM (Tongue River MBR). Ref#: USGS Mz 3898.

L0534. Carbon Co., MT, sec. 6, T. 8 S., R, 21 E. Cf1 BIO, Fort Union FM (Tongue River MBR). Ref#: USGS Mz 3899.

L0855. Carbon Co., MT, sec. 17, T. 8 S., R. 21 E. Cf1 BIO, Fort Union FM. Ref name: Snake Creek.

L0856. Carbon Co., MT, sec. 21, T. 8 S., R. 21 E. Cf1 BIO, Fort Union FM. Ref name: Grove Creek.

L0940. Carbon Co., MT, sec. 11–12, T. 8 S., R. 21 E. Ti5? BIO, Fort Union FM (Tongue River MBR). Ref#(s). USGS Mz 4963 (Carman/5).

L0941. Park Co., WY, sec. 12–13, T. 57 N., R. 103 W. Cf2–Cf3 BIO, Willwood FM. Ref#(s): USGS Mz 04978 (Woodruff/22).

L0942. Carbon Co., MT, sec. 27, T. 7 S., R. 20 E. Cf1 BIO, Fort Union FM (Tongue River MBR). Ref#: USGS Mz 1542.

L0943. Carbon Co., MT, sec. 36?, T. 7? S., R. 20? E. Cf2? BIO, Fort Union FM. Ref#(s): USGS Mz 5009.

L0944. Carbon Co., MT, sec. 9, T. 9 S., R. 23 E. Ti4? BIO, Fort Union FM (Tongue River MBR). Ref#: USGS Mz 5038 (9-23-02-2).

L0945. Carbon Co., MT, sec. 16, T. 9 S., R. 23 E. Ti4? BIO, Fort Union FM (Tongue River MBR). Ref#(s): USGS Mz 5039 (9-25-07-1).

L0946. Carbon Co., MT, sec. 8. T. 9 S., R. 23 E. Ti4? BIO, Fort Union FM (Tongue River MBR). Ref#(s): USGS Mz 5040 (9-23-07-1).

L0947. Carbon Co., MT, sec. 36, T. 9 S., R. 22 E. Ti4–Ti5 BIO, Fort Union FM (Tongue River MBR). Ref#(s): USGS Mz 9766 (Cz 22569) (Hares/1916-37).

L0948. Carbon Co., MT, sec. 6, T. 7 S., R. 22 E. Ti3–Ti5 BIO, Fort Union FM (Lebo? MBR). Ref#(s): USGS Cz 22779 (Zeller/ZM-1-60).

L1421. Carbon Co., MT, sec. 21, T. 8 S., R. 21 E. Cf1? BIO, Fort Union FM (Tongue River MBR). Ref#(s): USGS Mz 4961 (Carman/3).

L1422. Carbon Co., MT, sec. 35–36, T. 7 S., R. 21 E. Ti5? BIO, Fort Union FM (Tongue River? MBR). Ref#(s): USGS Mz 4962 (Woodruff/4).

L1423. Carbon Co., MT, sec. 14, T. 9 S., R. 21 E. Ti5–Cf1 BIO, Fort Union FM (Tongue River MBR). Ref#(s): USGS Mz 4965 (Carman/7).

Appendix 16.1 *(continued)*

L1425.	Park Co., WY, sec. 12, T. 53 N., R. 101 W. Cf3? BIO, Fort Union FM. Ref#(s): USGS Mz 4994 (Cz 22564; Stanton/1211).
L1470.	Carbon Co., MT, sec. 16, T. 9 S., R. 23 E. Ti5 BIO, Fort Union FM (Tongue River MBR). Ref#(s): USGS Mz 9624 (Cz 22567) (Hares/1916-30).
L1476b.	Carbon Co., MT, sec. 6?, T. 8 S., R. 21 E. Cf1 BIO, Fort Union FM (Tongue River MBR). Ref#(s) and name: USGS Mz 13947 (Cz 22537), Taggart Gulch.
L1478.	Carbon Co., MT, sec. 5, T. 8 S., R. 21 E. Cf1 BIO, Tongue River FM. Ref#(s) and name: USGS Mz 16503 (Simpson/602), Eagle Mine.
L1500.	Carbon Co., MT, T. 7 S., R. 22 E. To–Ti5 BIO, Fort Union FM (Lebo? MBR). Ref#(s): USGS Mz 9773 (Cz 22570) (Hares/1916-36).
L1695.	Hot Springs Co., WY, sec. 22, T. 45 N., R. 100 W. Aycross FM. Ref#(s): USGS Cz D1291NM (TMB-BHB-78-116).
L1782.	Park Co., WY, sec. 22, T. 53 N., R. 101 W. Lance FM. Ref#(s): USGS Mz 4959 (Woodruff/1907-01).
L1784.	Park Co., WY, sec. 23, T. 53 N., R. 101 W. Cf1? BIO, Fort Union FM. Ref#(s): USGS Mz 4973 (Woodruff/1907-17).
L1785.	Park Co., WY, sec. 22, T. 53 N., R. 101 W. Lance FM. Ref#(s): USGS Mz 4976 (Woodruff/1907-20).
L1803.	Carbon Co., MT, sec. 6, T. 8 S., R. 21 E. Ti5–Cf1 BIO, Fort Union FM. Ref# and name: USGS Mz 13459, Eagle Mine.
L2117.	Big Horn Co., WY, sec. 32, T. 51 N., R. 97 W. Wa5–Wa6 BIO, Willwood FM. Ref#(s): USGS Cz 22896 (Rohrer/1961-90).
L2118.	Big Horn Co., WY, sec. 12, T. 50 N., R. 97 W. Wa6 BIO, Willwood FM. Ref#(s): USGS Cz 22897 (Rohrer/1961-13).
L2119.	Big Horn Co., WY, sec. 6, T. 50 N., R. 96 W. Wa6? BIO, Willwood FM. Ref#(s): USGS Cz 22898 (Rohrere/1961-120.
L2120.	Park Co., WY, sec. 34, T. 50 N., R. 98 W. Wa7 BIO, Tatman FM. Ref#(s): USGS Cz 22899 (Rohrer/1961-47).
L2121.	Park Co., WY, sec. 35, T. 50 N., R. 98 W. Wa7 BIO, Tatman FM. Ref#(s): USGS Cz 22900 (Rohrer/1961-48).
L2122.	Park Co., WY, sec. 35, T. 50 N., R. 98 W. Wa7 BIO, Tatman FM. Ref#: USGS Cz 22901.
L2123.	Bighorn Co., WY, sec. 16, T. 50 N., R. 97 W. Wa7 BIO, Tatman FM. Ref#(s): USGS Cz 22902 (Rohrer/1961-76).
L2124.	Bighorn Co., WY, sec. 16, T. 50 N., R. 97 W. Wa7 BIO, Tatman FM. Ref#(s): USGS Cz 22903 (Rohrer/1961-C).
L2125.	Bighorn Co., WY, sec. 14, T. 50 N., R. 97 W. Wa6–Wa7 BIO, Tatman FM. Ref#(s): USGS Cz 22904 (Rohrer/1961-8).
L2126.	Bighorn Co., WY, sec. 14, T. 50 N., R. 97 W. Wa6–Wa7 BIO, Tatman FM. Ref#(s): USGS Cz 22905 (Rohrer/1961-8).
L2127.	Bighorn Co., WY, sec. 20, T. 50 N., R. 97 W. Wa7 BIO, Tatman FM. Ref#(s): USGS Cz 22906 (Rohrer/1961-72).
L2128.	Bighorn Co., WY, sec. 20, T. 50 N., R. 97 W. Wa7 BIO, Tatman FM. Ref#(s): USGS Cz 22907 (Rohrer/1961-72).
L2129.	Bighorn Co., WY, sec 20, T. 50 N., R. 97 W. Wa7 BIO, Tatman FM. Ref#(s): USGS Cz 22908 (Rohrer/1961-72).
L2155.	Hot Springs Co., WY, sec. 30, T. 46 N., R. 100 W. Aycross FM. Ref#(s): USGS Cz 23361 (Rohrer/1962-WR 1).
L2197.	Carbon Co., MT, sec. 27, T. 8 S., R. 22 E. Ti5 BIO, Fort Union FM. Ref: sec. 27, T. 8 S., R. 22 E.
L2309.	Park Co., WY, T. 55? N., R. 100 W. Cf BIO, Fort Union FM. Ref name: Clark Fork beds.
L2310a.	Park Co., WY, T. 56 N., R. 102 W.Cf2? BIO, Fort Union FM. Ref# and name: Sinclair-1929 Field No. 99, east of mouth of Little Sand Coulee.
L2310b.	Park Co., WY, T. 56 N., R. 102 W. Cf2? BIO, Fort Union FM. Ref# and name: Sinclair-1929 Field No. 99, northeast of mouth of Little Sand Coulee.
L2311.	Big Horn Co., WY, T. 51 N., R. 96 W. Wa3–Wa4 BIO, Willwood FM. Ref name: 15 miles west of Basin.
L2312a.	Park Co., WY, sec. 27, T. 56 N., R. 102 W. Wa2? BIO, Willwood FM. Ref name: Little Sand Coulee, east of mouth of Pat O'Hara Creek).
L2312b.	Park Co., WY, sec. 27, T. 56 N., R. 102 W. Wa1? BIO, Willwood FM. Ref# and name: Sinclair-1929 Field No. 31, Little Sand Coulee, east of mouth of Pat O'Hara Creek.
L2312c.	Park Co., WY, sec. 27, T. 56 N., R. 102 W. Wa1? BIO, Willwood FM. Ref# and name: Sinclair-1929 Field No. 32, Little Sand Coulee, east of mouth of Pat O'Hara Creek.
L2312d.	Park Co., WY, sec. 27, T. 56 N., R. 102 W. Wa2? BIO, Willwood FM. Ref# and name: Sinclair-1929 Field No. 45, Little Sand Coulee, east of mouth of Pat O'Hara Creek.
L2312e.	Park Co., WY, sec. 27, T. 56 N., R. 102 W. Wa2? BIO, Willwood FM. Ref# and name: Sinclair-1929 Field No. 47, Little Sand Coulee, east of mouth of Pat O'Hara Creek.
L2312f.	Park Co., WY, see. 27, T. 56 N., R. 102 W. Wa2? BIO, Willwood FM. Ref# and name: Sinclair-1929 Field No. 56, Little Sand Coulee, east of mouth of Pat O'Hara Creek.
L2313.	Park Co., WY, sec. 27, T. 56 N., R. 102 W. Wa2 BIO, Willwood FM, 1720.0 m, HOR 43c. Ref name: North of divide at head of Little Sand Coulee.
L2313a.	Park Co., WY, sec. 27, T. 56 N., R. 102 W. Wa2 BIO, Willwood FM, 1720.0 m. HOR 43c. Ref#(s) and name: Sinclair-1929 Field Nos. 70, 73 [also 72], North of divide at head of Little Sand Coulee.
L2313b.	Park Co., WY, sec. 27, T. 56 N., R. 102 W. Wa2 BIO, Willwood FM, 1720.0 m, HOR 43c. Ref# and name: Sinclair-1929 Field No. 58, North of divide at head of Little Sand Coulee.

Appendix 16.1 *(continued)*

L2313c. Park Co., WY, sec. 27, T. 56 N., R. 102 W. Wa2 BIO, Willwood FM, 1720.0 m, HOR 43c. Ref# and name: Sinclair-1929 Field No. 59, North of divide at head of Little Sand Coulee.

L2313d. Park Co., WY, sec. 27, T. 56 N., R. 102 W. Wa2 BIO, Willwood FM, 1720.0 m, HOR 43c. Ref# and name: Sinclair-1929 Field No. 64, North of divide at head of Little Sand Coulee.

L2313e. Park Co., WY, sec. 27, T. 56 N., R. 102 W. Wa2 BIO, Willwood FM, 1720.0 m, HOR 43c. Ref# and name: Sinclair-1929 Field No. 65, North of divide at head of Little Sand Coulee.

L2313f. Park Co.k, WY, sec. 27, T. 56 N., R. 102 W. Wa2 BIO, Willwood FM, 1720.0 m, HOR 43c. Ref# and name: Sinclair-1929 Field No. 70 L2313, North of divide at head of Little Sand Coulee.

L2313g. Park Co., WY, sec. 27, T. 56 N., R. 102 W. Wa2 BIO, Willwood FM, 1720.0 m, HOR 43c. Ref# and name: Sinclair-1929 Field No. 72, North of divide at head of Little Sand Coulee.

L2313h. Park Co., WY, sec. 27, T. 56 N., R. 102 W. Wa2 BIO, Willwood FM, 1720.0 m, HOR 43c. Ref# and name: Sinclair-1929 Field No. 73, North of divide at head of Little Sand Coulee.

L2314. Park Co., WY, sec. 35, T. 56 N., R. 102 W. Wa2? BIO, Willwood FM. Ref name: T. 56 N., R. 102 W.

L2315. Park Co., WY, T. 56 N., R. 102 W. Wa2?? BIO, Willwood FM. Ref: Park Co., Wyoming.

L2316. Big Horn Co., WY, T. 50 N., R. 95 W. Wa2–Wa3 BIO, Willwood FM. Ref name: east of Steins.

L2317. Park Co., WY, T. 56 N., R. 102 W. Wa2? BIO, Willwood FM. Ref# and name: Sinclair-1929 Field No. 110, South of Camp 1 of 1928.

L2318. Big Horn Co., WY, T. 52 N., R. 96 W. Wa3? BIO, Fort Union FM. Ref name: Dry Creek.

L2319. Park Co., WY, sec. 33, T. 56 N., R. 100 W. Cf3? BIO, Willwood FM. Ref name: Camp No. 2, sec. 33, T. 56 N., R. 100 W.

L2320. Big Horn Co., WY, sec. 34, T. 51 N., R. 95 W. Wa4 BIO, Willwood FM. Ref name: sec. 34, T. 51 N., R. 95 W.

L3383. Park Co., WY, T. 48 N., R. 98 W. Tatman FM. Ref#(s): USGS Mz 3453 (Cz 22560).

L3519. Big Horn Co., WY, T. 54 S., R. 97 E. Cf1–Wa2 BIO. Fort Union? Ref#: USGS Mz 9614 (Hares/1916-14).

L3820. Park Co., WY, sec. 11, T. 56 N., R. 102 W. Cf3 BIO, Willwood FM. Ref name: Three miles east of Pat O'Hara Creek.

L3821. Park Co., WY, sec. 14, T. 56 N., R. 102 W. Wa1 BIO, Willwood FM. Ref name: Three miles southeast of mouth of Pat O'Hara Creek.

L3822. Park Co., WY, sec. 26, T. 56 N., R. 102 W. Wa2–Wa3 BIO, Willwood FM. Ref name: Five miles southeast of mouth of Pat O'Hara Creek.

L3952a. Park Co., WY, sec. 1, T. 55 N., R. 101 W. Cf3 BIO, Willwood FM. Ref name: Head of Big Sand Coulee.

L3952b. Park Co., WY, sec. 1, T. 55 N., R. 101 W. Cf3 BIO, Willwood FM. Ref name: Head of Big Sand Coulee.

L3952c. Park Co., WY, sec. 1, T. 55 N., R. 101 W. Wa1 BIO, Willwood FM. Ref name: Head of Big Sand Coulee.

L3982. Big Horn Co., MT, sec. 2?, T. 54 N., R. 95 W. Ti5–Cf3 BIO, Fort Union? Ref#(s): USGS Mz 9757 (Hares/1916-96).

L4110. Park Co., WY, sec. 2, T. 55 N., R. 100 W. Cf3–Wa0 BIO, Willwood FM. Ref name: Three miles north of Ralston.

L4130. Big Horn Co., WY, sec. T. 50 N., R. 97 W. Wa5–Wa7 BIO, Tatman FM. Ref name: Tatman Mountain.

L4131. Hot Springs Co., WY, sec. 23, T. 48 N., R. 98 W. Tatman FM. Ref: Van Houten (1944, p. 195).

L4233. Park Co., WY, sec. 27, T. 56 N., R. 102 W. Wa2–Wa3 BIO, Willwood FM. Ref#(s): USGS Cz 22653 (Taylor et al., 1953-390).

L4234. Park Co., WY, sec. 36, T. 56 N., R. 103 W. Wa1–Wa3 BIO, Willwood FM. Ref#(s): USGS Cz 22654 (Taylor et al., 1953-391).

L4282a. Park Co., WY. Cf3–Wa0 BIO, Willwood? FM. Ref name: Bighorn Basin, Wyoming.

L4282b. Park Co., WY, Cf–Wa BIO, Willwood? FM. Ref name: Bighorn Basin, Wyoming.

L4283. Big Horn Co., WY, T. 50 N., R. 95 W. Wa3–Wa4 BIO, Willwood FM. Ref name: Nine miles from St. Joe.

L4284. Big Horn Co., WY, sec. 26, T. 51 N., R. 95 W. Wa3–Wa4 BIO, Willwood FM. Ref name: Three miles southeast of Otto.

L4285. Park Co., WY, sec. 23, 26, T. 56 N., R. 102 W. Wa2–Wa3 BIO, Willwood FM. Ref name: Six miles southeast of Hopkins Ranch.

L4286. Big Horn Co., WY, sec. 17, T. 51 N., R. 96 W. Wa4 BIO, Willwood FM. Ref name: One mile south of St. Joe

L4302. Carbon Co., MT, sec. 1, T. 8 S., R. 20 E. Cf1? BIO, Fort Union FM (Tongue River MBR). Ref#(s): USGS Mz 5010 (Stanton/1229).

L4309. Bighorn Co., WY, sec. 20, T. 50 N., R. 96 W. Wa5–Wa6 BIO, Willwood FM. Ref: Rohrer and Gazin (1965), Locality 46.

L4311. Hot Springs Co., WY, sec. 29, T. 47 N., R 97 W. Tatman FM. Ref: Van Houten (1944, p. 195).

L4373. Carbon Co., MT, sec. 27, T. 8 S., R. 22 E. Ti5 BIO, Fort Union FM. Ref#: Hanley/BH29-76.

L4803. Carbon Co., MT, sec. 6, T. 8 S., R. 21 E. Cf1 BIO, Fort Union FM. Ref#(s): USGS Mz 4966 (Woodruff/8).

L5471. Carbon Co., MT, sec. 27, T. 7 S., R. 20 E. Cf2? BIO, Fort Union FM. Ref name: Red Lodge Mine.

L5472. Carbon Co., MT, sec. 27, T. 8 S., RR. 22 E. Ti5 BIO, Fort Union FM. Ref#(s): Hanley/BH22-76 (Hickey 76-054).

L5694. Park Co., WY, sec. 7, T. 57 N., R. 102 W. Cf2 BIO, Willwood FM. Ref#: Hanley/BH07-76.

Appendix 16.1 *(continued)*

L5695.	Park Co., WY, sec. 7, T. 57 N., R. 102 W. Cf2 BIO, Willwood FM. Ref#: Hanley/BH08-76.
L5696.	Park Co., WY, sec. 7, T. 57 N., R. 102 W. Cf2 BIO, Willwood FM. Ref#: Hanley/BH09-76.
L5697.	Park Co., WY, sec. 7, T. 57 N., R. 102 W. Cf2 BIO, Willwood FM. Ref#: Hanley/BH010-76.
L5698.	Park Co., WY, sec. 9, T. 57 N., R. 98 W. Wa2 BIO, Willwood FM. Ref#: Hanley/BH011-76.
L5699a.	Park Co., WY, sec. 21, T. 55 N., R. 98 W. Ti5–Cf1 BIO, Fort Union FM. Ref#(s): Hanley/BH12-76 (USNM 72-141-Hickey).
L5699b.	Park Co., WY, sec. 21, T. 55 N., R. 98 W. Ti5–Cf1 BIO, Fort Union FM. Ref#(s): Hanley/BH12-76 (USNM 72-141-Hickey).
L5700.	Park Co., WY, sec. 16, T. 56 N., R. 101 W. Cf2 BIO, Willwood FM. Ref#: Hanley/BH13-76.
L5701.	Park Co., WY, sec. 16, T. 56 N., R. 101 W. Cf2 BIO, Willwood FM. Ref#: Hanley/BH14-76.
L5702.	Park Co., WY, sec. 27, T. 56 N., R. 101 W. Cf2 BIO, Willwood FM. Ref#: Hanley/BH15-76.
L5703.	Park Co., WY, sec. 34, T. 56 N., R. 101 W. Cf2 Cf3 BIO, Willwood FM. Ref#: Hanley/BH17-76.
L5704.	Park Co., WY, sec. 34, T. 56 N., R. 101 W. Cf2–Cf3 BIO, Willwood FM. Ref#: Hanley/BH18-76.
L5705.	Park Co., WY, sec. 11, T. 55 N., R. 102 W. Wa3 BIO, Willwood FM. Ref#: Hanley/BH19-76.
L5705a.	Park Co., WY, sec. 11, T. 55 N., R. 102 W. Wa3 BIO, Willwood FM. Ref#: Hanley/BH19A-76.
L5705b.	Park Co., WY, sec. 11, T. 55 N., R. 102 W. Wa3 BIO, Willwood FM. Ref#: Hanley/BH19B-76.
L5705c.	Park Co., WY, sec. 11, T. 55 N., R. 102 W. Wa3 BIO, Willwood FM. Ref#: Hanley/BH19C-76.
L5705d.	Park Co., WY, sec. 11, T. 55 N., R. 102 W. Wa3 BIO, Willwood FM. Ref#: Hanley/BH19D-76.
L5706.	Park Co., WY, sec. 11, T. 55 N., R. 102 W. Wa3 BIO, Willwood FM. Ref#: Hanley/BH20-76.
L5707.	Park Co., WY, sec. 11, T. 55 N., R. 102 W. Wa3 BIO, Willwood FM. Ref#: Hanley/BH21-76.
L5708.	Carbon Co., MT, sec. 27, T. 8 S., R. 22 E. Ti5 BIO, Fort Union FM. Ref#: Hanley/BH23-76.
L5709.	Carbon Co., MT, sec. 25, T. 8 S., R. 22 E. Ti4–Ti5 BIO, Fort Union FM. Ref#: Hanley/BH24-76.
L5710.	Carbon Co., MT, sec. 25, T. 8 S., R. 22 E. Ti4–Ti5 BIO, Fort Union FM. Ref#: Hanley/BH25-76.
L5711.	Carbon Co., MT, sec. 25, T. 8 S., R. 22 E. Ti4–Ti5 BIO, Fort Union FM. Ref#: Hanley/BH26-76.
L5712.	Carbon Co., MT, sec. 7, T. 8 S., R. 21 E. Cf1 BIO, Fort Union FM. Ref#(s): Hanley/BH27-76 (Hickey 72-102).
L5713.	Carbon Co., MT, sec. 7, T. 8 S., R. 21 E. Cf1 BIO, Fort Union FM. Ref#(s): Hanley/BH28-76 (Hickey 72-110).
L5714.	Park Co., WY, sec. 11, T. 55 N., R. 102 W. Wa3 BIO, Willwood FM. Ref#: Hanley/BH30-76.
L5718.	Park Co., WY, sec. 11, T. 55 N., R. 102 W. Wa3 BIO, Willwood FM. Ref#: Hanley/BH34-76.
L5720.	Park Co., WY, sec. 11, T. 55 N., R. 102 W. Wa3 BIO, Willwood FM. Ref#: Hanley/BH36-76.
L5721.	Park Co., WY, sec. 11, T. 55 N., R. 102 W. Wa3 BIO, Willwood FM. Ref#: Hanley/BH37-76.
L5723.	Park Co., WY, sec. 11, T. 55 N., R. 102 W. Wa4 BIO, Willwood FM. Ref#: Hanley/BH39-76.
L5724.	Park Co., WY, sec. 11, T. 55 N., R. 102 W. Wa4 BIO, Willwood FM. Ref#: Hanley/BH40-76.
L5726.	Park Co., WY, sec. 10, T. 55 N., R. 102 W. Wa3 BIO, Willwood FM. Ref#: Hanley/BH42-76.
L5726a.	Park Co., WY, sec. 10, T. 55 N., R. 102 W. Wa3 BIO, Willwood FM. Ref#: Hanley/BH42A-76.
L5727.	Park Co., WY, sec. 11, T. 55 N., R. 102 W. Wa3 BIO, Willwood FM. Ref#: Hanley/BH43-76.
L5728.	Park Co., WY, sec. 11, T. 55 N., R. 102 W. Wa3 BIO, Willwood FM. Ref#: Hanley/BH44-76.
L5729.	Park Co., WY, sec. 10, T. 55 N., R. 102 W. Wa3 BIO, Willwood FM. Ref#: Hanley/BH45-76.
L5731.	Park Co., WY, sec. 11, T. 55 N., R. 102 W. Wa3 BIO, Willwood FM. Ref#: Hanley/BH47-76.
L5732.	Park Co., WY, sec. 11, T. 55 N., R. 102 W. Wa3 BIO, Willwood FM. Ref#: Hanley/BH48-76.
L5733.	Park Co., WY, sec. 11, T. 55 N., R. 102 W. Wa3 BIO, Willwood FM. Ref#: Hanley/BH49-76.
L5734.	Park Co., WY, sec. 27, T. 56 N., R. 102 W. Wa3 BIO, Willwood FM. 1750.0 m, HOR 34. Ref#: Hanley/BH50-76.
L5735.	Park Co., WY, sec. 27, T. 56 N., R. 102 W. Wa3 BIO, Willwood FM, 1750.0 m, HOR 33. Ref#: Hanley/BH51-76.
L5736.	Park Co., WY, sec. 27, T. 56 N., R. 102 W. Wa3 BIO, Willwood FM, 1750.0 m, HOR 44a. Ref#: Hanley/BH52-76.
L5738.	Carbon Co., MT, sec. 25, T. 8 S., R. 22 E. Ti4–Ti5 BIO, Fort Union FM. Ref#: Hanley/BH01-77.
L5739.	Carbon Co., MT, sec. 25, T. 8 S., R. 22 E. Ti5? BIO, Fort Union FM. Ref#: Hanley/BH02-77.
L5740.	Carbon Co., MT, sec. 25, T. 8 S., R. 22 E. Ti5? BIO, Fort Union FM. Ref#(s): Hanley/BH03-77.
L5741.	Carbon Co., MT, sec. 25, T. 8 S., R. 22 E. Ti4–Ti5 BIO, Fort Union FM. Ref#: Hanley/BH04-77.
L5742.	Carbon Co., MT, sec. 25, T. 8 S., R. 22 E. Ti4–Ti5 BIO, Fort Union FM. Ref#: Hanley/Bh05-77.
L5743.	Carbon Co., MT, sec. 35, T. 8 S., R. 22 E. Ti5 BIO, Fort Union FM. Ref#(s): Hanley/BH06-77.
L5744.	Carbon Co., MT, sec. 35, T. 8 S., R. 22 E. Ti5 BIO, Fort Union FM. Ref#: Hanley/BH07-77.
L5745.	Carbon Co., MT, sec. 35, T. 8 S., R. 22 E. Ti5 BIO, Fort Union FM. Ref#: Hanley/BH08-77.
L5746.	Carbon Co., MT, sec. 35, T. 8 S., R. 22 E. Ti5 BIO, Fort Union FM. Ref#: Hanley/BH09-77.
L5747.	Carbon Co., MT, sec. 36, T. 8 S., R. 22 E. Ti4–Ti5 BIO, Fort Union FM. Ref#: Hanley/BH10-77.
L5748a–b.	Park Co., WY, sec. 5, T. 55 N., R. 101 W. Wa1 BIO, Willwood FM, 1570.0 m, HOR 33d. Ref#: Hanley/BH11-77.
L5749.	Park Co., WY, sec. 25, T. 56 N., R. 102 W. Wa1 BIO, Willwood FM. 1630.0 m, HOR 37c. Ref#: Hanley/BH12-77.
L5751.	Park Co., WY, sec. 25, T. 56 N., R. 102 W. Wa1 BIO, Willwood FM, 1630.0 m, HOR 37d. Ref#: Hanley/BH14-77.

Appendix 16.1 *(continued)*

L5754.	Park Co., WY, Sec. 25, T. 56 N., R. 102 W. Wa1 BIO, Willwood FM, 1630.0 m, HOR 37e. Ref#: Hanley/BH17-77.
L5756.	Park Co., WY, sec. 25, T. 56 N., R. 102 W. Wa1 BIO, Willwood FM, 1630.0 m, HOR 37b. Ref#: Hanley/BH19-77.
L5757.	Park Co., WY, sec. 24, T. 56 N., R. 102 W. Wa1 BIO, Willwood FM, 1535.0 m, HOR 30b. Ref#: Hanley/BH20-77.
L5758a.	Park Co., WY, sec. 18, T. 56 N., R. 101 W. Cf3 BIO, Willwood FM, 1495.0 m, HOR 25a. Ref#: Hanley/BH21-77.
L5758b.	Park Co., WY, sec. 18. T. 56 N., R. 101 W. Cf3 BIO, Willwood FM, 1495.0 m, HOR 25b. Ref#: Hanley/BH21-77.
L5759.	Big Horn Co., WY, sec. 27, T. 54 N., R. 96 W. Wa4 BIO, Willwood FM. Ref#: Hanley/BH22-77.
L5760.	Park Co., WY, sec. 28, T. 55 N., R 101 W. Wa3 BIO, Willwood FM, 1950.0 m, HOR was 69. Ref# and name: Hanley/ BH23-77, Hackberry Hollow.
L5761.	Park Co., WY, sec. 28, T. 55 N., R. 101 W. Wa3 BIO, Willwood FM, 1950.0 m, HOR was 71. Ref# and name: Hanley/BH24-77, Hackberry Hollow.
L5762.	Park Co., WY, sec. 28, T. 55 N., R. 101 W. Wa3 BIO, Willwood FM, 1950.0 m, HOR 52. Ref# and name: Hanley/BH25-77, Hackberry Hollow.
L5763.	Park Co., WY, sec. 10, T. 54 N., R. 99 W. Wa2 BIO, Willwood FM. Ref#: Hanley/BH26-77.
L5765.	Park Co., WY, sec. 5, T. 55 N., R. 101 W. Wa1 BIO, Willwood FM, 1570.0 m, HOR 33c. Ref#: Hanley/BH28-77.
L5766.	Park Co., WY, sec. 32, T. 56 N., R. 101 W. Wa1 BIO, Willwood FM, 1570.0 m, HOR 33a. Ref#: Hanley/BH29-77.
L5767.	Park Co., WY, sec. 32, T. 56 N., R. 101 W. Wa1 BIO, Willwood FM, 1570.0 m, HOR 33b. Ref#: Hanley/BH30-77.
L5772.	Park Co., WY, sec. 26, T. 56 N., R. 102 W. Wa2 BIO, Willwood FM. Ref#: Hanley/BH35-77.
L5773.	Park Co., WY, sec. 22, T. 56 N., R. 102 W. Wa2 BIO, Willwood FM. Ref#: Hanley/BH36-77.
L5774.	Park Co., WY, sec. 27, T. 56 N., R. 102 W. Wa2 BIO, Willwood FM. Ref#: Hanley/BH37-77.
L5776.	Park Co., WY, sec. 11, T. 55 N., R. 102 W. Wa3 BIO, Willwood FM. Ref#: Hanley/BH 39-77
L5777.	Park Co., WY, sec. 1, T. 55 N., R. 102 W. Wa3 BIO, Willwood FM. Ref#: Hanley/BH 40-77
L5778.	Park Co., WY, sec. 35, T. 56 N., R. 102 W. Wa2 BIO, Willwood FM. Ref#: Hanley/BH 41-77
L5780.	Park Co., WY, sec. 35, T. 56 N., R. 102 W. Wa3 BIO Willwood FM. Ref#: Hanley/BH 43-77
L5781.	park Co., WY, sec. 10, T. 55 N., R. 102 W. Wa3 BIO Willwood FM. Ref#: Hanley/BH 44-77
L5784.	Park Co., WY, sec. 35, T. 56 N., R. 102 W Wa2 BIO Willwood FM. Ref#: Hanley/BH 47-77
L5785.	Park Co., WY sec 26, T. 56 N., R. 102 W. Wa2 BIO Willwood FM. Ref#: Hanley/BH 48-77
L5786.	Park Co., WY sec 26, T. 56 N., R. 102 W. Wa2 BIO Willwood FM. Ref#: Hanley/BH 49-77
L5787.	Park Co., WY, sec. 26, T. 56 N., R. 102 W. Wa2 BIO Willwood FM. Ref#: Hanley/BH 50-77
L5789.	Park Co., WY, sec. 25, T. 56 N., R. 102 W. Wa2 BIO Willwood FM, 1645.0 m, HOR 38c. Ref#: Hanley/BH 52-77
L5799.	Park Co., WY, sec. 29, T. 56 N., R. 101 W. Cf3 BIO Willwood FM, 1465.0 m, HOR 22a. Ref#: Hanley/BH 62-77
L5799.	Park Co., WY, sec. 29, T. 56 N., R. 101 W. Cf3 BIO Willwood FM, 1485.0 m, HOR 24b. Ref#: Hanley/BH 63-77
L5800.	Park Co., WY, sec. 29, T. 56 N., R. 101 W. Cf3 BIO Willwood FM, 1465.0 m, HOR 22a. Ref#: Hanley/BH 62-77
L5802.	Park Co., WY, sec. 28, T. 56 N., R. 101 W. Cf3 BIO Willwood FM, 1465.0 m, HOR 22i. Ref#: Hanley/BH 65-77
L5803.	Park Co., WY, sec. 28, T. 56 N., R. 101 W. Cf3 BIO Willwood FM, 1485.0 m, HOR 24a. Ref#: Hanley/BH 66-77
L5804.	Park Co., WY, sec. 28, T. 56 N., R. 101 W. Cf3 BIO Willwood FM, 1465.0 m, HOR 22b. Ref#: Hanley/BH 67-77
L5805.	Park Co., WY, sec. 7, T. 56 N., R. 101 W. Cf2 BIO Willwood FM, 1380.0 m, HOR 15b. Ref#: Hanley/BH 68-77
L5806.	Park Co., WY, sec. 7, T. 56 N., R. 101 W. Cf2 BIO Willwood FM, 1380.0 m, HOR 15c. Ref#: Hanley/BH 69-77
L5808.	Park Co., WY, sec. 7, T. 56 N., R. 101 W. Cf2 BIO Willwood FM, 1380.0 m, HOR 15a. Ref#: Hanley/BH 71-77
L5809.	Park Co., WY, sec. 34, T. 56 N., R. 101 W. Cf3 BIO Willwood FM, 1465.0 m, HOR 22c. Ref#: Hanley/BH 72-77
L5810.	Park Co., WY, sec. 27, T. 56 N., R. 101 W. Cf2–Cf3 BIO Willwood FM, 1430.0 m, HOR 18. Ref#: Hanley/BH 73-77
L5812.	Park Co., WY, sec. 27, T. 56 N., R. 101 W.Cf2–Cf3 BIO Willwood FM, Ref#: Hanley/BH 75-77
L5823.	Park Co., WY, sec. 3, T. 55 N., R. 101 W. Wa2 BIO Willwood FM, 1665.0 m, HOR 39b. Ref#: Hanley/BH 10-78
L5829.	Park Co., WY, sec. 3, T. 55 N., R. 101 W. Wa2 BIO Willwood FM, 1685.0 m, HOR 40a. Ref#: Hanley/BH 16-78
L5831.	Carbon Co., MT, sec. 27, T. 8 S., R. 22 E. Ti5 BIO, Fort Uniion FM, Ref#(s): Hanley/BH 18-78 (Hickey 76-054).
L5832.	Park Co., WY, sec. 20, T. 56 N., R. 101 W. Cf3 BIO Willwood FM, 1495.0 m, HOR 25c. Ref#: Hanley/BH 19-78
L5835.	Park Co., WY, sec. 19, T. 56 N., R. 101 W. Cf3 BIO Willwood FM, 1505.0 m, HOR 27a. Ref#: Hanley/BH 22-78
L5836.	Park Co., WY, sec. 19, T. 56 N., R. 101 W. Cf3 BIO Willwood FM, 1505.0 m, HOR 27bc. Ref#: Hanley/BH 23-78
L5839.	Park Co., WY, sec. 15, T. 56 N., R. 101 W. Cf2 BIO Willwood FM, 1370.0 m, HOR 14a. Ref#: Hanley/BH 26-78
L5840.	Park Co., WY, sec. 15, T. 56 N., R. 101 W. Cf2 BIO Willwood FM, 1370.0 m, HOR 14b. Ref#: Hanley/BH 27-78
L5841.	Park Co., WY, sec. 15, T. 56 N., R. 101 W. Cf2 BIO Willwood FM, 1370.0 m, HOR 14e. Ref#: Hanley/BH 28-78
L5842.	Park Co., WY, sec. 15, T. 56 N., R. 101 W. Cf2 BIO Fort Union FM, HOR 14c. Ref#: Hanley/BH 29-78
L5842a.	Park Co., WY, sec. 15, T. 56 N., R. 101 W. Cf2 BIO Fort Union FM, HOR 14d. Ref#: Hanley/BH 29A-78
L5844a–c.	Park Co., WY, sec. 15, T. 56 N., R. 101 W. Cf2 BIO Willwood FM, HOR 14f. Ref#: Hanley/BH 31-78[a]
L5846.	Park Co., WY, sec. 15, T. 56 N., R. 101 W. Cf2 BIO Willwood FM, 1370.0 m, HOR 14h. Ref#: Hanley/BH 33-78
L5847.	Park Co., WY, sec. 15, T. 56 N., R. 101 W. Cf2 BIO Willwood FM, 1370.0 m, HOR 14g. Ref#: Hanley/BH 34-78
L5849.	Park Co., WY, sec. 20, T. 56 N., R. 101 W. Cf2–Cf3 BIO Willwood FM, 1447.0 m, HOR 19a. Ref#: Hanley/BH 36-78

Appendix 16.1 *(continued)*

L5850.	Park Co., WY, sec. 20, T. 56 N., R. 101 W. Cf2–Cf3 BIO Willwood FM, 1447.0 m, HOR 19b. Ref#: Hanley/BH 37-78
L5851.	Park Co., WY, sec. 20, T. 56 N., R. 101 W. Cf3 BIO Willwood FM, 1465.0 m, HOR 22d. Ref#: Hanley/BH 38-78
L5854a.	Park Co., WY, sec. 2, T. 55 N., R. 101 W. Cf3 BIO Willwood FM, 1465.0 m, HOR 22e. Ref#: Hanley/BH 41A-78
L5854b.	Park Co., WY, sec. 2, T. 55 N., R. 101 W. Cf3 BIO Willwood FM, 1465.0 m, HOR 22f. Ref#: Hanley/BH 41B-78
L5854c.	Park Co., WY, sec. 2, T. 55 N., R. 101 W. Cf3 BIO Willwood FM, 1465.0 m, HOR 22g. Ref#: Hanley/BH 41C-78
L5857.	Park Co., WY, sec. 10, T. 56 N., R. 102 W. Cf3 BIO Willwood FM, 1480.0 m, HOR 23. Ref#: Hanley/BH 44-78
L5859.	Park Co., WY, sec. 10, T. 56 N., R. 102 W. Cf3 BIO Willwood FM, 1460.0 m, HOR 21a. Ref#: Hanley/BH 46-78
L5860a–b.	Park Co., WY, sec. 11, T. 56 N., R. 102 W. Cf3 BIO Willwood FM, 1460.0 m, HOR 21c. Ref#: Hanley/BH 47-78
L5862.	Park Co., WY, sec. 10, T. 56 N., R. 102 W. Cf3 BIO Willwood FM, 1500.0 m, HOR 26a. Ref#: Hanley/BH 49-78
L5863.	Park Co., WY, sec. 10, T. 56 N., R. 102 W. Cf3 BIO Willwood FM, 1500.0 m, HOR 26b. Ref#: Hanley/BH 50-78
L5867.	Park Co., WY, sec. 11, T. 56 N., R. 102 W. Cf3 BIO Willwood FM, 1460.0 m, HOR 21b. Ref#: Hanley/BH 54-78
L5868.	Park Co., WY, sec. 17, T. 56 N., R. 99 W. Ti5 BIO Fort Union FM, 840.0 m, HOR 01. Ref#: Hanley/BH 55-78
L5870.	Park Co., WY, sec. 6, T. 56 N., R. 101 W. Cf2 BIO Willwood FM, 1210.0 m, HOR 05. Ref#: Hanley/BH 02-79
L5871.	Park Co., WY, sec. 6, T. 56 N., R. 101 W. Cf2 BIO Willwood FM, 1230.0 m, HOR 06b. Ref#: Hanley/BH 03-79
L5874.	Park Co., WY, sec. 1, T. 56 N., R. 102 W. Cf2 BIO Willwood FM, 1200.0 m, HOR 04. Ref#: Hanley/BH 06-79
L5875a.	Park Co., WY, sec. 8, T. 56 N., R. 101 W. Cf2 BIO Willwood FM, 1355.0 m, HOR 12c. Ref#: Hanley/BH 07-79
L5875a–b.	Park Co., WY, sec. 8, T. 56 N., R. 101 W. Cf2 BIO Willwood FM, 1355.0 m, HOR 12c. Ref#: Hanley/BH 07-79
L5875b.	Park Co., WY, sec. 8, T. 56 N., R. 101 W. Cf2 BIO Willwood FM, 1355.0 m, HOR 12f. Ref#: Hanley/BH 07-79
L5876.	Park Co., WY, sec. 7, T. 56 N., R. 101 W. Cf2 BIO Willwood FM, 1355.0 m, HOR 15e. Ref#: Hanley/BH 08-79
L5878.	Park Co., WY, sec. 7, T. 56 N., R. 101 W. Cf2 BIO Willwood FM, 1380.0 m, HOR 15d. Ref#: Hanley/BH 10-79
L5879.	Park Co., WY, sec. 7, T. 56 N., R. 101 W. Cf2 BIO Willwood FM, 1380.0 m, HOR 15g. Ref#: Hanley/BH 11-79
L5880.	Park Co., WY, sec. 7, T. 56 N., R. 101 W. Cf2 BIO Willwood FM, 1380.0 m, HOR 15h. Ref#: Hanley/BH 12-79
L5881.	Park Co., WY, sec. 7, T. 56 N., R. 101 W. Cf2 BIO Willwood FM, 1380.0 m, HOR 15j. Ref#: Hanley/BH 13-79
L5882.	Park Co., WY, sec. 7, T. 56 N., R. 101 W. Cf2 BIO Willwood FM, 1380.0 m, HOR 15i. Ref#: Hanley/BH 14-79
L5883.	Park Co., WY, sec. 7, T. 56 N., R. 101 W. Cf2 BIO Willwood FM, 1380.0 m, HOR 15f. Ref#: Hanley/BH 15-79
L5884a.	Park Co., WY, sec. 32, T. 57 N., R. 101 W. Cf2 BIO Fort Union FM, 1160.0 m, HOR 03c. Ref#: Hanley/BH 16-79
L5884b.	Park Co., WY, sec. 32, T. 57 N., R. 101 W. Cf2 BIO Fort Union FM, 1160.0 m, HOR 03d. Ref#: Hanley/BH 16-79
L5885.	Park Co., WY, sec. 32, T. 57 N., R. 101 W. Cf2 BIO Fort Union FM, 1160.0 m, HOR 03b. Ref#: Hanley/BH 17-79
L5886.	Park Co., WY, sec. 32, T. 57 N., R. 101 W. Cf2 BIO Fort Union FM, 1160.0 m, HOR 03a. Ref#: Hanley/BH 18-79
L5888.	Park Co., WY, sec. 8, T. 56 N., R. 101 W. Cf2 BIO Willwood FM, 1335.0 m, HOR 11. Ref#: Hanley/BH 20-79
L5889.	Park Co., WY, sec. 5, T. 56 N., R. 101 W. Cf2 BIO Willwood FM, 1300.0 m, HOR 10. Ref#: Hanley/BH 21-79
L5890.	Park Co., WY, sec. 26 T. 57 N., R. 101 W. Cf1 BIO Fort Union FM, 1090.0 m, HOR 02. Ref#: Hanley/BH 22-79
L5891.	Carbon Co., MT, sec. 27, T. 8 S., R. 22 E. Ti5 BIO Fort Union FM, HOR 11. Ref#: Hanley/BH 23-79
L5892.	Carbon Co., MT, sec. 27, T. 8 S., R. 22 E. Ti5 BIO Fort Union FM, HOR 11. Ref#: Hanley/BH 24-79
L5893.	Park Co., WY, sec. 6 T. 56 N., R. 101 W. Cf2 BIO Willwood FM, 1280.0 m, HOR 09b. Ref#: Hanley/BH 25-79
L5894.	Park Co., WY, sec. 6 T. 56 N., R. 101 W. Cf2 BIO Willwood FM, 1280.0 m, HOR 09a. Ref#: Hanley/BH 26-79
L5895a–b.	Park Co., WY, sec. 6 T. 56 N., R. 101 W. Cf2 BIO Willwood FM, 1280.5 m, HOR 09c. Ref#: Hanley/BH 27-79
L5896.	Park Co., WY, sec. 35 T. 57 N., R. 100 W. Cf2 BIO Willwood FM, 1275.0 m, HOR 08b. Ref#: Hanley/BH 28-79
L5897.	Park Co., WY, sec. 6 T. 56 N., R. 101 W. Cf2 BIO Willwood FM, 1275.0 m, HOR 08a. Ref#: Hanley/BH 29-79
L5899.	Park Co., WY, sec. 24 T. 56 N., R. 102 W. Wa1 BIO Willwood FM, 1535.0 m, HOR 30a. Ref#: Hanley/BH 01-80
L5902.	Carbon Co., MT, sec. 28 T. 9 S., R. 22 W. Ti5 BIO Fort Union FM, Ref#: Hanley/BH 04-80
L5903.	Park Co., WY, sec. 34 T. 56 N., R. 99 W. Ti5 BIO Fort Union FM, Ref#: Hanley/BH 05-80
L5904.	Park Co., WY, sec. 21 T. 56 N., R. 100 W. Cf2 BIO Willwood FM, 1230.0 m, HOR 06a. Ref#: Hanley/BH 06-80
L5905.	Park Co., WY, sec. 28 T. 56 N., R. 100 W. Cf2 BIO Willwood FM, 1359.9 m, HOR 13. Ref#: Hanley/BH 07-80
L5906.	Park Co., WY, sec. 28 T. 56 N., R. 100 W. Cf2 BIO Willwood FM, 1355.0 m, HOR 12a. Ref#: Hanley/BH 08-80
L5907.	Park Co., WY, sec. 28 T. 56 N., R. 100 W. Cf2 BIO Willwood FM, 1355.0 m, HOR 12b. Ref#: Hanley/BH 09-80
L5908.	Park Co., WY, sec. 28 T. 56 N., R. 100 W. Cf2 BIO Willwood FM, 1355.0 m, HOR 12d. Ref#: Hanley/BH 10-80
L5908a.	Park Co., WY, sec. 28 T. 56 N., R. 100 W. Cf2 BIO Willwood FM, 1355.0 m, HOR 12e. Ref#: Hanley/BH 10A-80
L5910.	Park Co., WY, sec. 33 T. 58 N., R. 101 W.Ti5– Cf1 BIO Fort Union FM, Ref#: Hanley/BH 02-81
L5913.	Park Co., WY, sec. 34 T. 56 N., R. 101 W. Cf2–Cf3 BIO Willwood FM, Ref#: Hanley/BH 16-76
L6248.	Park Co., WY, sec. 10 T. 55 N., R. 100 W. Wa0 BIO Willwood FM, 1520.0 m, HOR 28a. Ref#: UM SC-067
L6249.	Park Co., WY, sec. 10 T. 55 N., R. 100 W. Wa0 BIO Willwood FM, 1520.0 m, HOR 28b. Ref#: UM SC-069
L6250.	Park Co., WY, sec. 10 T. 55 N., R. 100 W. Wa0 BIO Willwood FM, 1520.0 m, HOR 28c. Ref#: UM SC-121
L6251.	Park Co., WY, sec. 14 T. 56 N., R. 102 W. Wa0 BIO Willwood FM, 1520.0 m, HOR 28d. Ref#: UM SC-139
L6252.	Park Co., WY, sec. 10 T. 55 N., R. 100 W. Wa0 BIO Willwood FM, 1520.0 m, HOR 28e. Ref#: UM SC-308
L6253.	Park Co., WY, sec. 13 T. 56 N., R. 102 W. Wa0 BIO Willwood FM, 1520.0 m, HOR 28f. Ref#: UM SC-345
L6254.	Park Co., WY, sec. 13 T. 56 N., R. 102 W. Wa0 BIO Willwood FM, 1520.0 m, HOR 28g. Ref#: UM SC-348

Appendix 16.1 *(continued)*

L6255.	Park Co., WY, sec. 13 T. 56 N., R. 102 W. Wa0 BIO Willwood FM, 1520.0 m, HOR 28h. Ref#: UM SC-349
L6262.	Park Co., WY, sec. 3 T. 55 N., R. 101 W. Wa2 BIO Willwood FM, 1665.0 m, HOR 39a. Ref#: UM SC-001
L6263.	Park Co., WY, sec. 26 T. 56 N., R. 102 W. Wa2 BIO Willwood FM, 1720.0 m, HOR 43b. Ref#: UM SC-002
L6264.	Park Co., WY, sec. 10 T. 55 N., R. 102 W. Wa3 BIO Willwood FM, 1935.0 m, HOR 51. Ref#: UM SC-003.
L6265.	Park Co., WY, sec. 5 T. 55 N., R. 101 W. Wa1 BIO Willwood FM, 1570.0 m, HOR 33e. Ref#: UM SC-004.
L6266.	Park Co., WY, sec. 27 T. 56 N., R. 102 W. Wa3 BIO Willwood FM, 1750.0 m, HOR 44b. Ref#: UM SC-005.
L6267.	Park Co., WY, sec. 32 T. 56 N., R. 101 W. Wa1 BIO Willwood FM, 1530.0 m, HOR 29. Ref#: UM SC-006.
L6268.	Park Co., WY, sec. 25 T. 56 N., R. 102 W. Wa2 BIO Willwood FM, 1645.0 m, HOR 38b. Ref#: UM SC-007.
L6269.	Park Co., WY, sec. 13 T. 56 N., R. 102 W. Cf3 BIO Willwood FM, 1495.0 m, HOR 25b. Ref#: UM SC-010.
L6270.	Park Co., WY, sec. 18 T. 56 N., R. 101 W. Cf3 BIO Willwood FM, 1495.0 m, HOR 25e. Ref#: UM SC-011.
L6271.	Park Co., WY, sec. 27 T. 56 N., R. 102 W. Wa2 BIO Willwood FM, 1720.0 m, HOR 43a. Ref#: UM SC-012.
L6272.	Park Co., WY, sec. 22 T. 56 N., R. 102 W. Wa2 BIO Willwood FM, 1685.0 m, HOR 40b. Ref#: UM SC-013.
L6273.	Park Co., WY, sec. 25 T. 56 N., R. 102 W. Wa1 BIO Willwood FM, 1630.0 m, HOR 37f. Ref#: UM SC-016.
L6274.	Park Co., WY, sec. 24 T. 56 N., R. 102 W. Wa1 BIO Willwood FM, 1600.0 m, HOR 34. Ref#: UM SC-017.
L6275.	Park Co., WY, sec. 23 T. 56 N., R. 102 W. Wa1 BIO Willwood FM, 1620.0 m, HOR 36. Ref#: UM SC-018.
L6277.	Park Co., WY, sec. 20 T. 56 N., R. 101 W. Cf3 BIO Willwood FM, 1505.0 m, HOR 27c. Ref#: UM SC-024.
L6278.	Park Co., WY, sec. 4 T. 55 N., R. 101 W. Wa1 BIO Willwood FM, 1555.0 m, HOR 31. Ref#: UM SC-026.
L6280.	Park Co., WY, sec. 35 T. 56 N., R. 101 W. Cf2 BIO Willwood FM, 1385.0 m, HOR 16. Ref#: UM SC-029.
L6281.	Park Co., WY, sec. 35 T. 56 N., R. 102 W. Wa3 BIO Willwood FM, 1760.0 m, HOR 45a. Ref#: UM SC-031.
L6283.	Park Co., WY, sec. 1 T. 55 N., R. 102 W. Wa3 BIO Willwood FM, 1840.0 m, HOR 46. Ref#: UM SC-033.
L6284.	Park Co., WY, sec. 11 T. 55 N., R. 102 W. Wa3 BIO Willwood FM, 1850.0 m, HOR 47a. Ref#: UM SC-034.
L6285.	Park Co., WY, sec. 11 T. 55 N., R. 102 W. Wa3 BIO Willwood FM, 1870.0 m, HOR 48. Ref#: UM SC-035.
L6286.	Park Co., WY, sec. 11 T. 55 N., R. 102 W. Wa3 BIO Willwood FM, 1895.0 m, HOR 49. Ref#: UM SC-036.
L6287.	Park Co., WY, sec. 23 T. 56 N., R. 102 W. Wa1 BIO Willwood FM, 1630.0 m, HOR 37g. Ref#: UM SC-037.
L6288.	Park Co., WY, sec. 23 T. 56 N., R. 102 W. Wa2 BIO Willwood FM, 1645.0 m, HOR 38a. Ref#: UM SC-038.
L6290.	Park Co., WY, sec. 23 T. 56 N., R. 102 W. Wa1 BIO Willwood FM, 1560.0 m, HOR 32. Ref#: UM SC-041.
L6292.	Park Co., WY, sec. 6 T. 55 N., R. 101 W. Wa2 BIO Willwood FM, 1700.0 m, HOR 42. Ref#: UM SC-046.
L6293.	Park Co., WY, sec. 5 T. 55 N., R. 101 W. Wa2 BIO Willwood FM, 1690.0 m, HOR 41. Ref#: UM SC-047.
L6294.	Park Co., WY, sec. 16 T. 56 N., R. 101 W. Cf2 BIO Willwood FM, 1405.0 m, HOR 17. Ref#: UM SC-053.
L6296.	Park Co., WY, sec. 7 T. 56 N., R. 101 W. Cf2 BIO Willwood FM, 1380.0 m, HOR 15k. Ref#: UM SC-062.
L6298.	Park Co., WY, sec. 10 T. 55 N., R. 102 W. Wa3 BIO Willwood FM, 1970.0 m, HOR 53. Ref#: UM SC-064.
L6300.	Park Co., WY, sec. 36 T. 56 N., R. 100 W. Cf2 BIO Willwood FM, 1270.0 m, HOR 07. Ref#: UM SC-084.
L6301.	Park Co., WY, sec. 26 T. 56 N., R. 102 W. Wa3 BIO Willwood FM, 1760.0 m, HOR 45b. Ref#: UM SC-087.
L6303.	Park Co., WY, sec. 11 T. 56 N., R. 102 W. Cf3 BIO Willwood FM, 1455.0 m, HOR 20. Ref#: UM SC-090.
L6305.	Park Co., WY, sec. 2 T. 55 N., R. 101 W. Wa1 BIO Willwood FM, 1610.0 m, HOR 35. Ref#: UM SC-094.
L6309.	Park Co., WY, sec. 2 T. 55 N., R. 101 W. Cf3 BIO Willwood FM, 1465.0 m, HOR 22h. Ref#: UM SC-102.
L6310.	Park Co., WY, sec. 10 T. 55 N., R. 102 W. Wa4 BIO Willwood FM, 2020.0 m, HOR 54. Ref#: UM SC-112.
L6311.	Park Co., WY, sec. 10 T. 55 N., R. 102 W. Wa3 BIO Willwood FM, 1915.0 m, HOR 50. Ref#: UM SC-114.
L6313.	Park Co., WY, sec. 15 T. 56 N., R. 101 W. Cf2 BIO Willwood FM, 1370.0 m, HOR 14i. Ref#: UM SC-116.
L6316.	Park Co., WY, sec. 19 T. 56 N., R. 101 W. Wa1 BIO Willwood FM, 1535.0 m, HOR 30c. Ref#: UM SC-124.
L6317.	Park Co., WY, sec. 24 T. 55 N., R. 102 W. Wa1 BIO Willwood FM, 1535.0 m, HOR 30d. Ref#: UM SC-125.
L6319.	Park Co., WY, sec. 30 T. 56 N., R. 102 W. Wa3 BIO Willwood FM, 1850.0 m, HOR 47b. Ref#: UM SC-128.
L6322.	Park Co., WY, sec. 1 T. 55 N., R. 102 W. Wa3 BIO Willwood FM, 1750.0 m, HOR 44c. Ref#: UM SC-133.
L6324.	Park Co., WY, sec. 10 T. 56 N., R. 102 W. Cf3 BIO Willwood FM, 1500.0 m, HOR 26c. Ref#: UM SC-138.
L6326.	Park Co., WY, sec. 26 T. 57 N., R. 100 W. Ti5 BIO Fort Union FM, Ref#: UM SC-144.
L6329.	Park Co., WY, sec. 19 T. 56 N., R. 10 W. Cf3 BIO Willwood FM, 1505.0 m, HOR 27d. Ref#: UM SC-153.
L6330.	Park Co., WY, sec. 26 T. 56 N., R. 102 W. Wa3 BIO Willwood FM, 1760.0 m, HOR 45c. Ref#: UM SC-213.
L6330a.	Park Co., WY, sec. 26 T. 56 N., R. 102 W. Wa3 BIO Willwood FM, 1760.0 m, HOR 45d. Ref#: UM SC-213a.
L6330b.	Park Co., WY, sec. 26 T. 56 N., R. 102 W. Wa3 BIO Willwood FM, 1760.0 m, HOR 45e. Ref#: UM SC-213b.
L6330c.	Park Co., WY, sec. 26 T. 56 N., R. 102 W. Wa3 BIO Willwood FM, 1760.0 m, HOR 45f. Ref#: UM SC-213c.
L6464.	Park Co., WY, sec. 25 T. 56 N., R. 102 W. Wa1 BIO Willwood FM, 1630.0 m, HOR 37a. Ref#: UM SC-015.
L6465.	Carbon Co., MT, sec. 5 or 6, T. 8 S., R. 21 E. Cf1 BIO, Fort Union FM. Ref name: AMNH-F1 collections, Eagle Mine.

17

Late Paleocene–Early Eocene Floral and Climatic Change in the Bighorn Basin, Wyoming

SCOTT L. WING

ABSTRACT

Maximum change in floral composition during the Paleocene–Eocene interval in the Bighorn Basin occurred in two steps, the first during the late Clarkforkian (Zone Cf3) and earliest Wasatchian ages (lower part of the lower *Haplomylus–Ectocion* Zone), the second in the mid-Wasatchian Age (*Bunophorus* Interval Zone through lower Lostcabinian interval). The first change in composition resulted largely from an extended period of extinction, the second was the result of a period of accelerated immigration of plants into the Bighorn Basin. The extinction interval spanned the latest Paleocene carbon isotope excursion, but lasted much longer, extending through the middle and later part of Chron C24r (1.0–1.5 my). The extinction began during a period of rising temperature in the later Clarkforkian Age, but continued into the earliest Wasatchian Age, for which time physiognomic and isotopic evidence indicates declining temperatures. The succeeding immigration interval occurred during a period of approximately 0.5 my near the beginning of Chron C24n, and is associated with physiognomic, floristic, and isotopic evidence for climatic warming in the Bighorn Basin. The immigration roughly corresponds to the Cenozoic temperature maximum seen in marine isotopic records.

During the early Eocene Period trends in plant richness correlated roughly with changes in temperature. Richness and temperature declined in the earliest Wasatchian Age (although the minimum of richness may be earlier than the minimum of temperature), and then increased in the early to middle Wasatchian Age. For the latest Paleocene Epoch there is less correlation between plant richness and estimated temperature. Estimated richness is either flat or declining through the Clarkforkian NALMA, once sample size has been controlled for, but there is good evidence for rising temperature. It remains to be explained why richness declined during the late Paleocene warming but increased during the mid-early Eocene warming. Periods of floral and mammalian faunal turnover are not strongly coincident through the late Paleocene–early Eocene interval, but there are three periods that suggest some concordance: an early Clarkforkian increase in immigration, a late Clarkforkian decrease in immigration, and a middle Wasatchian period of high immigration. Conspicuously absent is any floral event equivalent to the earliest Wasatchian immigration in mammals.

Global warming and intercontinental migration have been implicated as major influences on the timing and magnitude of terrestrial biotic change in the late Paleocene and early Eocene Epochs. The paleoclimatic estimates presented here suggest that there was more than one episode of warming near the Paleocene/Eocene Epoch boundary, and that the long, cool interval in between the warmest periods was also an important determinant of composition, turnover rates, and richness in the flora. If the climatic fluctuations recorded in the Bighorn Basin affected the northern mid- to high latitudes generally, then it may be that biotic exchange between the Holarctic continents via high-latitude landbridges took place as a series of pulsed events, rather than in one long, warm period. Resolving these pulses will require continued refinement of the North American record, but also the development of similar fine-scale records on other continents.

The Paleocene/Eocene Epoch boundary has long been recognized as an interval of significant turnover in vertebrate faunas in North America (Wood et al. 1941; Rose 1981a). In the last several decades it also has been identified as a period of climatic warming both on the continents and in the oceans (Wolfe 1978; Miller et al.

1987), and as the time of the major Cenozoic extinction in benthic foraminifera (Pak and Miller 1992). Recent high-resolution stratigraphic studies of stable carbon and oxygen isotopes and clay mineralogy have revealed rapid shifts in temperature, ocean circulation, and climate near the Paleocene/Eocene Epoch boundary that are probably global in scope (Kennett and Stott 1991; Hovan and Rea 1992; Gibson et al. 1993).

Recognition of terrestrial floral change across the Paleocene/Eocene Epoch boundary in North America has developed more slowly. Brown (1962) proposed that the first appearance of the floating fern *Salvinia preauriculata* coincided with the beginning of the Eocene, but did not characterize earliest Eocene floras more fully. Leopold and MacGinitie (1972) argued that middle Eocene floras of the northern Rocky Mountain region were made up of taxa that are today found in areas of seasonally dry climate in the New World, whereas Paleocene floras contained more lineages of mesic, Old World affinities. In the only detailed study of megafloral change across the Paleocene/Eocene Series boundary prior to the 1990s, Hickey (1977) observed 45% extinction of plant taxa across the boundary in the Golden Valley Formation of western North Dakota. The significance of this level of extinction was unclear because there were few floral turnover rates with which to compare it, and because the Golden Valley Formation sections are thin and difficult to correlate with other areas.

The Gulf Coastal Plain is the only region of North America outside the northern Rockies where Paleocene and Eocene megafloras are common. Unfortunately, most of the major megafloral localities occur in stratigraphically isolated clay lenses that are difficult to correlate with one another or with standard marine zonations. A large number of stratigraphically well-constrained palynofloras are known from this region, and these data have been used to demonstrate a substantial but short-lived decline in plant diversity in the terminal Paleocene and earliest Eocene epochs (late Biochron NP9–early Biochron NP10), followed by a sharp recovery in the number of species during the early Eocene Epoch (Frederiksen 1994).

The Paleocene/Eocene Series Boundary in the Bighorn Basin, Wyoming

For many years the Paleocene/Eocene Series boundary in interior western North America was thought to coincide with the transition from Clarkforkian to Wasatchian mammalian faunas (Wood et al. 1941). For a time some workers placed the Paleocene/Eocene Series boundary in the lower Clarkforkian NALMA (Cf1/Cf2 zonal boundary) based on similarities with faunas in the Paris Basin (Gingerich 1980). More recent evidence from palynology and carbon isotope stratigraphy favors placing the Paleocene/Eocene Series boundary slightly above the Clarkforkian/Wasatchian NALMA boundary (Wing 1984a; Koch et al. 1992), as does the co-occurrence in Mississippi of early Wasatchian mammals with dinoflagellates indicative of marine Zone NP9 (Beard et al. 1995).

A distinctive earliest Wasatchian mammalian assemblage, termed the Zone Wa0 fauna, has been recovered in several areas of the Bighorn Basin: the south end of Polecat Bench, Foster Gulch, the north end of the Sand Creek Divide, the Worland East area, the North Butte area, and a locality near Hole in the Ground (figures 17.1–17.4; Gingerich 1989; T. M. Bown and K. D. Rose, personal communication 1994). In all of these areas Zone Wa0 faunas were collected from a <40-m-thick interval containing one or two bright red, deeply weathered paleosols referred to as the "Zone Wa0 red bed." In three separate parts of the Bighorn Basin (Clark's Fork Basin, Worland East area, and North Butte area), mammalian teeth and pedogenic calcium carbonate nodules from the Wa0 zonal interval have carbon isotope ratios much more negative than those immediately higher or lower in the section, which has led Zone Wa0 to be correlated with the terminal Paleocene carbon isotope excursion observed in DSDP and ODP cores (Koch et al. 1992, 1995).

The isotopic correlation proposed by Koch et al. (1992) and the fauna recently reported from eastern Mississippi by Beard et al. (1995) suggest that the NP9/NP10 zonal boundary, sometimes taken as an approximation of the Paleocene/Eocene Series boundary, correlates to a position very low in the Wasatchian NALMA, but not at its base. Based on time-stratigraphic calibrations of marine cores, the carbon isotope excursion is approximately 0.1 to 0.5 my prior to the NP9/NP10 zonal boundary (Kennett and Stott 1991; Berggren et al. 1995). At typical rates of deposition in the Bighorn Basin this amount of time would be represented by 15–80 m of section, indicating that the NP9/NP10 zonal boundary correlates to the lower part of the lower *Haplomylus–Ectocion* Zone of Schankler (1980), near the Wa0/Wa1 zonal boundary of Gingerich (1989), and above the gradational contact of the Fort Union and Willwood Formations in most areas. This correlation of the NP9/NP10 zonal boundary with the Bighorn Basin section is also consistent with the first appearance of pollen of *Platycarya*, which first appears just below the NP9/NP10 zonal boundary on the Atlantic Coastal Plain and at about 90 meters above the Zone Wa0 red bed in the Bighorn Basin (Wing 1984a). (The report of *Platycarya* pollen 30 m below the Zone Wa0 red bed by Wing et al. [1991] was the result of a

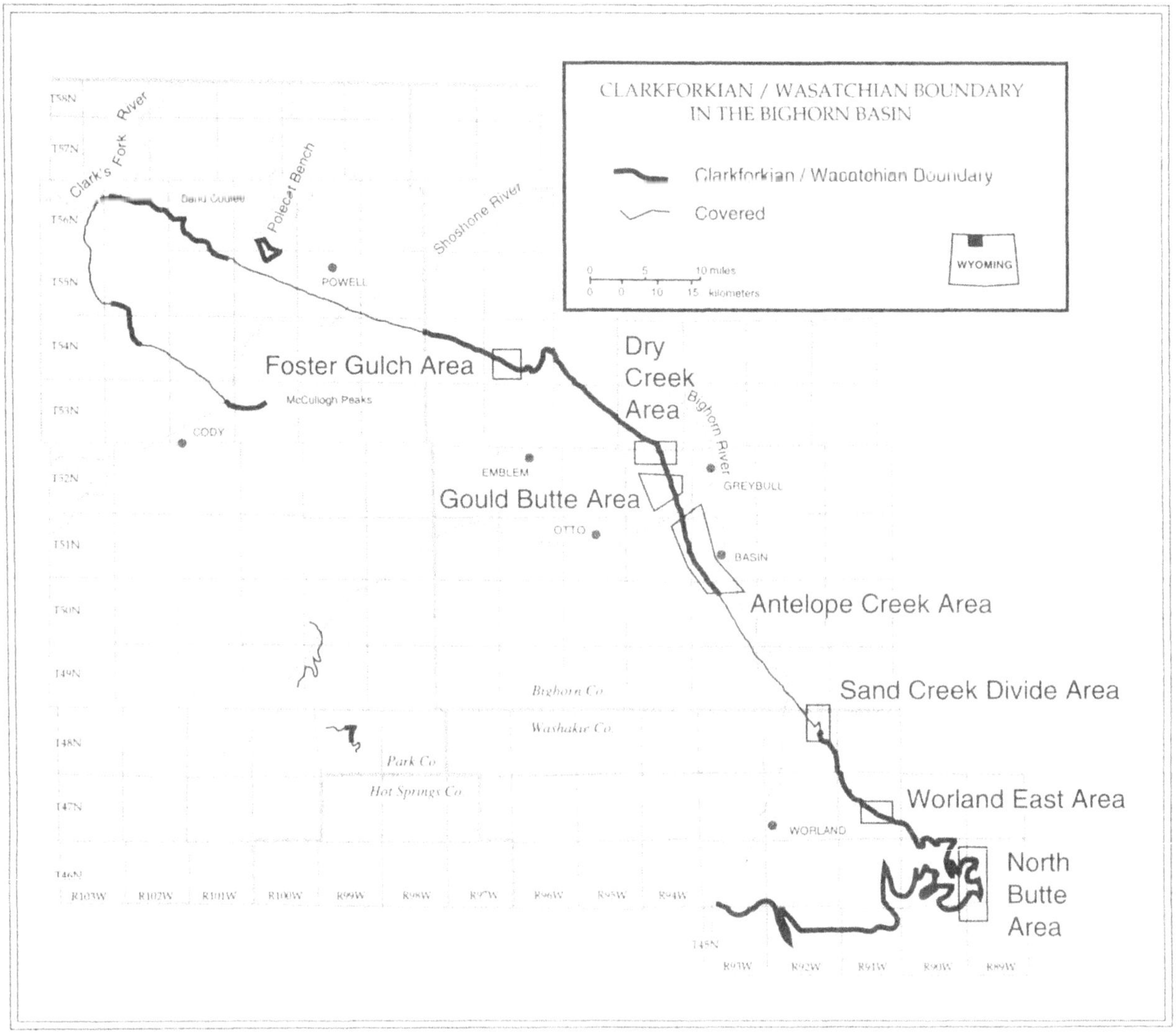

Figure 17.1 Map of the Bighorn Basin showing the trace of the Clarkforkian/Wasatchian NALMA boundary and the seven field areas from which fossil plants were collected for this study.

Map modified from Bown (1975) and Gingerich (1989). Earliest Wasatchian (Zone Wa0) faunas have been found in the North Butte, Worland East, Sand Creek Divide, and Foster Gulch areas, as well as in the Clark's Fork–Polecat Bench area and in the southwestern Bighorn Basin near Hole in the Ground.

stratigraphic miscorrelation of sections in the Elk Creek and Gould Butte areas of the Bighorn Basin.)

If the Paleocene/Eocene Epoch/Series boundary is defined as coeval with the carbon isotope excursion, then it essentially coincides with the mammalian zonal boundary, and in most parts of the central and southern Bighorn Basin, with the lithological transition from the drab Fort Union Formation to the red and variegated mudstones of the Willwood Formation. If the Paleocene/Eocene Series boundary is placed above the P5/P6 zonal boundary (Berggren et al. 1995), then it would correlate to a point more than 100 m above the Zone Wa0 red bed, in the upper part of the lower *Haplomylus–Ectocion* Zone.

In most parts of the Bighorn Basin there is no indication of any significant hiatus in deposition at the Clarkforkian/Wasatchian NALMA boundary, except for the Zone Wa0 paleosols, which may reflect 10–50 thousand years of soil formation, and in the northwestern Bighorn Basin (= Clark's Fork Basin) a thick, multistoried, multilateral sandstone sheet that represents a period of relatively slow sediment accumulation and little fossil preservation (Kraus 1980). In the Gould Butte and Dry Creek areas of the eastern Bighorn Basin, neither the Zone Wa0 red bed nor the sheet sandstone is present, and drab mudstones and carbonaceous shales were deposited through the latest Clarkforkian and into the early Wasatchian ages (figures 17.1, 17.4; Wing and Bown 1985). Continued deposition of paludal sediments in the Gould Butte area has been attributed to local subsidence along the east-west trending Florence Pass lineament (Wing and Bown 1985; Kraus 1992).

Sediments deposited during the late Paleocene and early Eocene epochs in the Bighorn Basin are well exposed over a wide geographic area (figure 17.1). Sediment deposition during this interval was also relatively rapid over much of the Bighorn Basin, resulting in thick sections that offer the possibility of fine-scale temporal resolution of biotic and environmental changes. Typical sedimentation rates in the upper Fort Union and lower Willwood formations were approximately 10–50 cm per thousand years, after compaction, depending on proximity to the axis of the basin and other local factors (Gingerich 1982, 1989; Bown and Kraus 1993). This means that fossil localities, which typically are spaced tens of meters apart, are on average separated by tens of thousands to one hundred thousand years. Bighorn Basin fossils and stratigraphic sections thus offer an opportunity to study the effect of global environmental change on continental biotas with a temporal resolution that is not usually possible prior to the Neogene Period.

Depositional Environments of Plant Fossils

Any attempt to understand long-term temporal change in floras has to confront the effect of depositional environment on floral composition because floral assemblages from different depositional environments tend to have different compositions. Channel and channel margin deposits generally are dominated by durable woody parts and leathery leaves of riparian trees. Levee and crevasse-splay deposits preserve a mixed assemblage of dicotyledonous trees, ferns, and herbaceous monocotyledons. Distal floodplain backswamp deposits characteristically contain a flora of taxodiaceous conifers and a few swamp-living dicotyledonous trees (Hickey 1980; Wing 1984b). Stratigraphic changes in facies types theoretically could affect floral composition, even in the absence of regional or global change in floras. The distribution of fossiliferous facies might also cause the lowest and highest occurrences of taxa to be displaced from their true first and last appearances in the region.

Fortunately, relatively similar environments of deposition prevailed through the late Paleocene and early Eocene epochs in the Bighorn Basin. Although variegated, oxidized paleosol horizons became more common in the early Eocene Epoch, drab intervals 10–30 m thick still occur in the lower 350 m of the Willwood Formation in the central Bighorn Basin. Plant-bearing facies are rare to absent in the Wasatchian part of the Willwood Formation in the northwestern Bighorn Basin (Hickey 1980). Most of the collections used in this study are derived either from levee/crevasse-splay facies or from laterally extensive carbonaceous shale deposits. Plant fossils are common in both of these facies through most of the Fort Union and Willwood formations. Two collections from the upper Clarkforkian deposits of the northwestern Bighorn Basin and all of the assemblages from the *Bunophorus* Interval Zone in the central Bighorn Basin are derived from abandoned channel-fills. Assemblages from abandoned channel-fills typically share taxa with coeval channel and near-channel assemblages, although they also contain floating aquatic and pond margin plants that are rarely preserved in other facies.

Data

A total of 208 fossil plant assemblages were used in this study. Twenty-four Tiffanian and Clarkforkian collections were made by Smithsonian expeditions directed by Leo Hickey and are currently on loan to the Yale Peabody Museum. With the exception of the Princeton Foster Gulch Flora the remainder of the collections were made by Smithsonian Institution or Yale University field parties, and are housed at the sponsoring institutions. All of the Smithsonian and Yale collections were made from small (<3 m × 3 m) quarries in which the fossils were confined to a stratigraphic interval 5–100 cm thick. Many of the laterally extensive carbonaceous shale beds were sampled at multiple sites along strike separated by tens or hundreds of meters. The Princeton Foster Gulch Flora, now also housed at the Smithsonian Institution, was obtained from a stratigraphic interval several meters thick collected along hundreds of meters of outcrop (L. J. Hickey, personal communication 1994). Floral lists for most of the localities are given in Wing et al. (1995).

Most of the plant localities are correlated to others in the same area through bed-tracing and sections measured with a sighting level (figures 17.2–17.4). Stratigraphic positions of about 20 localities have been estimated based on their proximity to localities that have been included in measured sections. Sections in different areas are correlated to one another based on the Zone Wa0 red bed and mammalian biostratigraphy.

Two hundred and seventy of the 297 taxa recognized in this study are based on compression/impression fossils of leaves, and the remainder are based on compressed fruits, flowers, or stems. Because foliar characters are usually sufficient to separate sympatric living species, it is probable that most of the fossil leaf taxa represent species. In some cases a set of closely related species may have produced leaves that cannot be recognized as distinct morphotypes, as is seen in some extant species of *Carya* and *Quercus*. Many of the leaf types

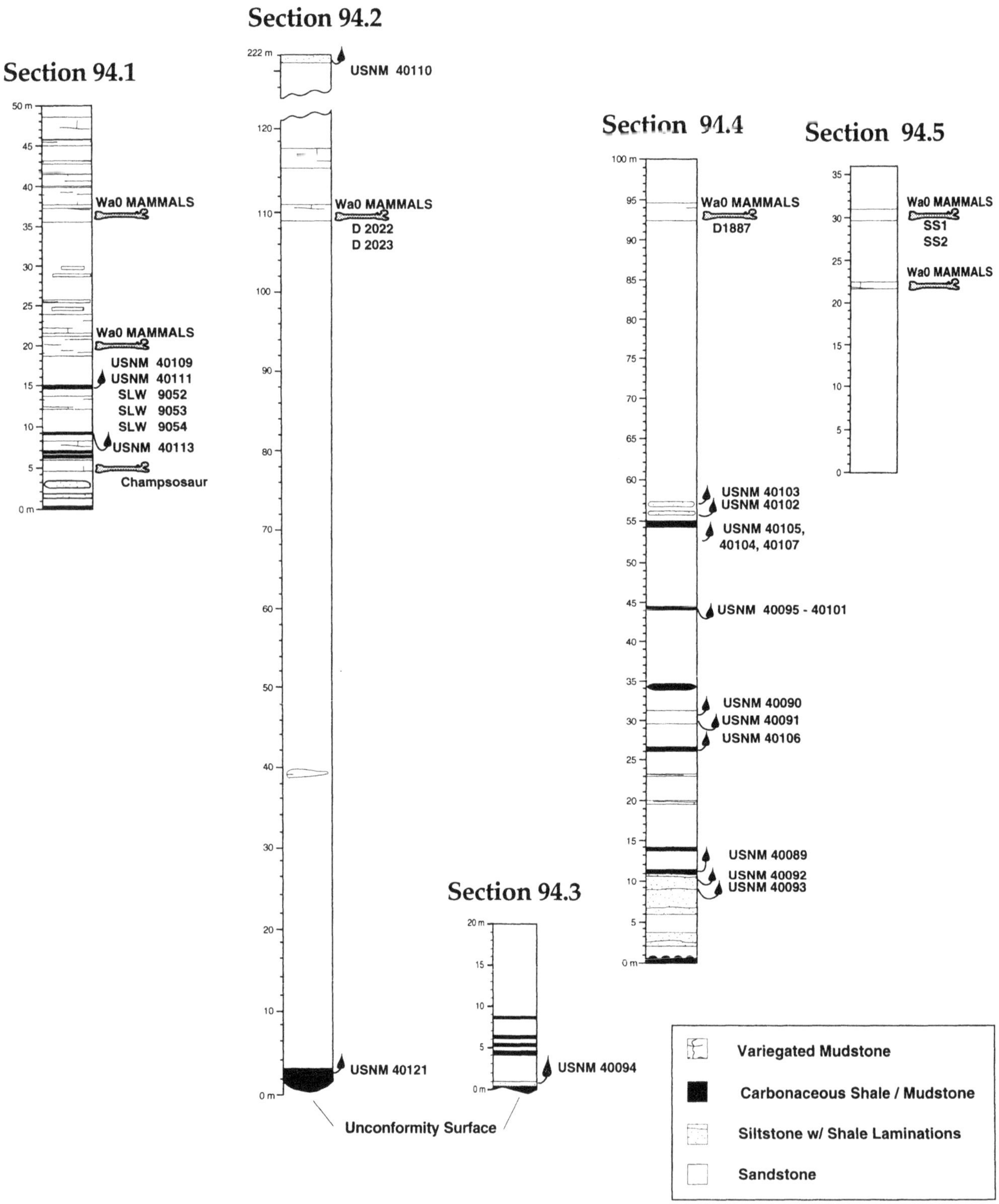

Figure 17.2 Stratigraphic sections of the upper Fort Union Formation and lower Willwood Formation in the vicinity of North Butte, indicating the positions of paleobotanical localities in relation to the Zone Wa0 red bed.

Section WE1

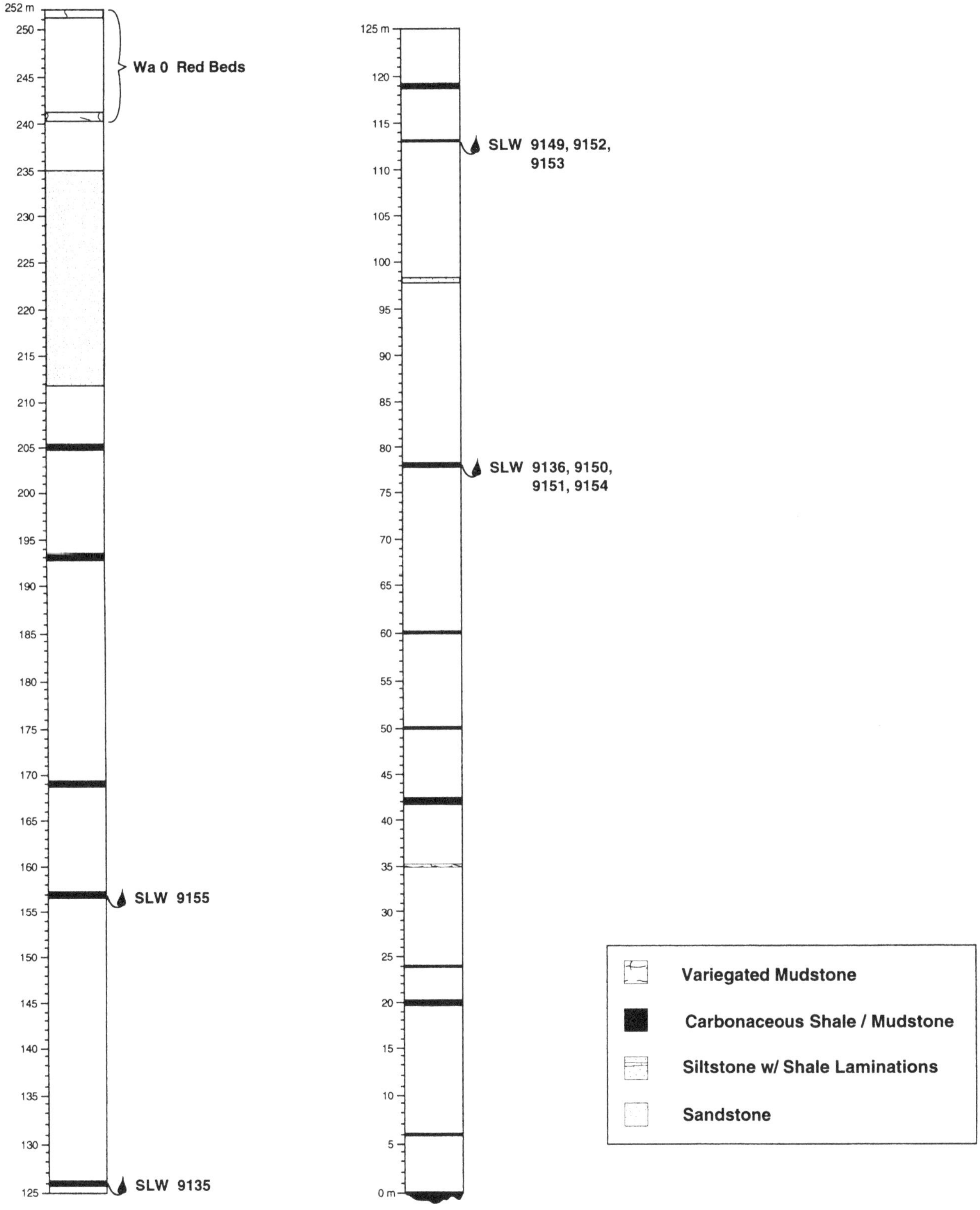

Figure 17.3 Stratigraphic sections of the upper Fort Union and lower Willwood formations in the Worland East area, indicating the positions of paleobotanical localities in relation to the Zone Wa0 red bed.

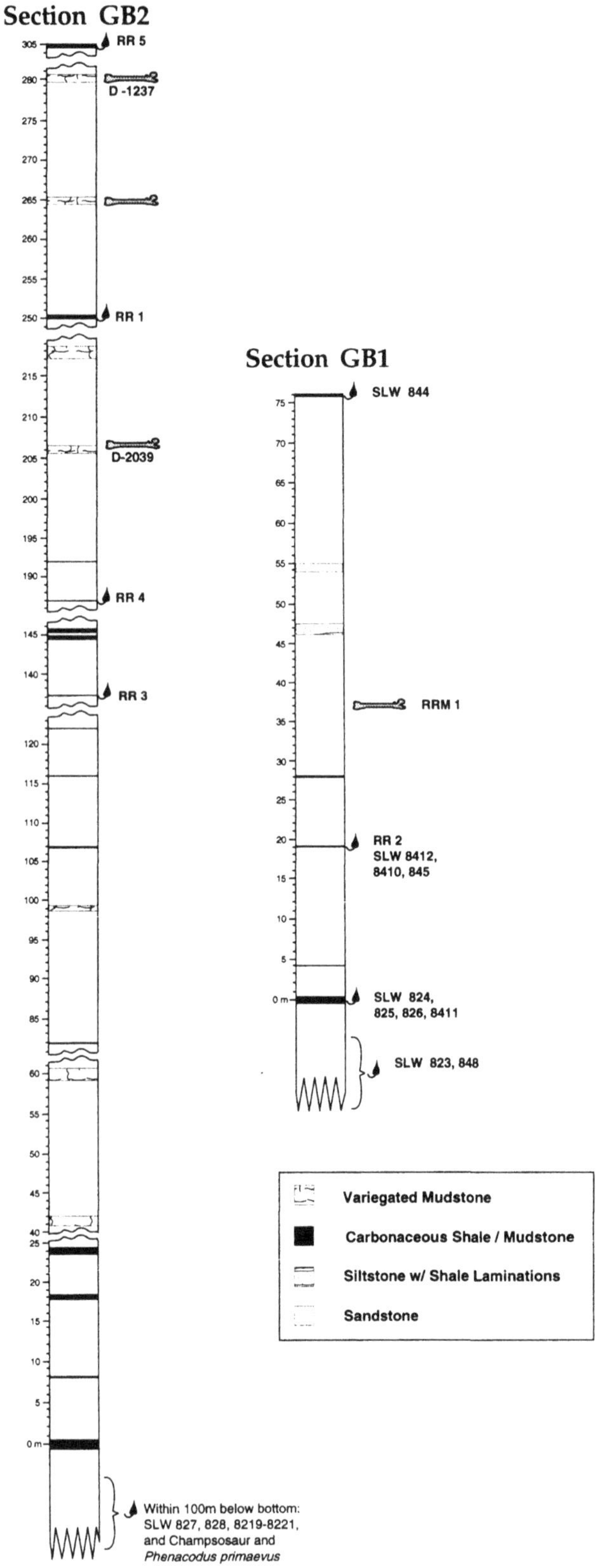

Figure 17.4 Stratigraphic sections of the upper Fort Union and lower Willwood formations in the vicinity of Gould Butte, indicating the positions of paleobotanical localities in relation to Wasatchian mammals, and the absence of the Zone Wa0 red bed in this area. The mammalian faunas in this section are younger than Zone Wa0, except for RRM1, which is a single *Hyracotherium* jaw indicative only of Wasatchian Age.

encountered in this study have not previously been described, and these are referred to by code numbers. Although most of the leaf taxa probably represent single species, higher taxonomic levels are unknown for most. The use of generic names in quotation marks indicates my opinion that the assignment of the fossil to this extant genus is incorrect, but that the name has been validly published. In contrast to dicotyledonous leaves, palm leaves are readily recognized at the family level but difficult to sort to species because they typically occur as small fragments lacking cuticle. I have therefore recognized a single family-level taxon for palm leaves. Fossil taxa based on fruits are presumed to be species-level entities, although no attempt has been made to distinguish subtaxa of legume fruits. The stratigraphic ranges of the 130 taxa that occur at more than one stratigraphic level are shown in figure 17.5. In the discussion that follows, first and last occurrences of taxa refer to their stratigraphic distribution within the Bighorn Basin data set, unless otherwise specified.

Megafloral Biostratigraphy

Previous paleobotanical work in the Bighorn Basin has recognized four floral zones from the Tiffanian through the lower Lostcabinian deposits: the *Magnolia–Glyptostrobus* Zone, corresponding to the lower part of the Tiffanian NALMA, the *Persites–Cornus* Zone, corresponding to the Clarkforkian NALMA, the *Metasequoia–Cnemidaria* Concurrent Range Zone, corresponding to most of the *Haplomylus–Ectocion* Zone, and the *Platycarya* Abundance Zone, corresponding to the lower Lostcabinian NALMA (figure 17.6; Hickey 1980; Wing et al. 1991). Based on the ranges in figure 17.5 it appears that the *Persites–Cornus* Zone can be divided into an upper and lower part, with the upper *Persites–Cornus* Zone lacking the common taxa *"Viburnum" asperum*, *"Viburnum" cupanioides*, *"Viburnum" antiquum*, *Dicotylophyllum anomalum*, and *Dennstaedia americana*, and having *Menispermites parvareolatus*, *Fagopsis greenlandica*, *Populus meigsii*, and *Phoebe* sp. as moderately common components. Betulaceae sp.1, a leaf type associated with fruits called *Paleocarpinus*, also is much more abundant in upper *Persites–Cornus* Zone assemblages.

Although 12 plant taxa have last appearances in the 50 m just below the Zone Wa0 red bed, the survival of both *Persites* and *Cornus* into the basal 20–90 m of the Wasatchian NALMA suggests that the boundary between the upper *Persites–Cornus* Zone and the *Metasequoia–Cnemidaria* Concurrent Range Zone is in the lowermost Wasatchian NALMA (near the top of the Zone Wa0?) rather than at the Clarkforkian/Wasatchian NALMA boundary. There are another 24 last occur-

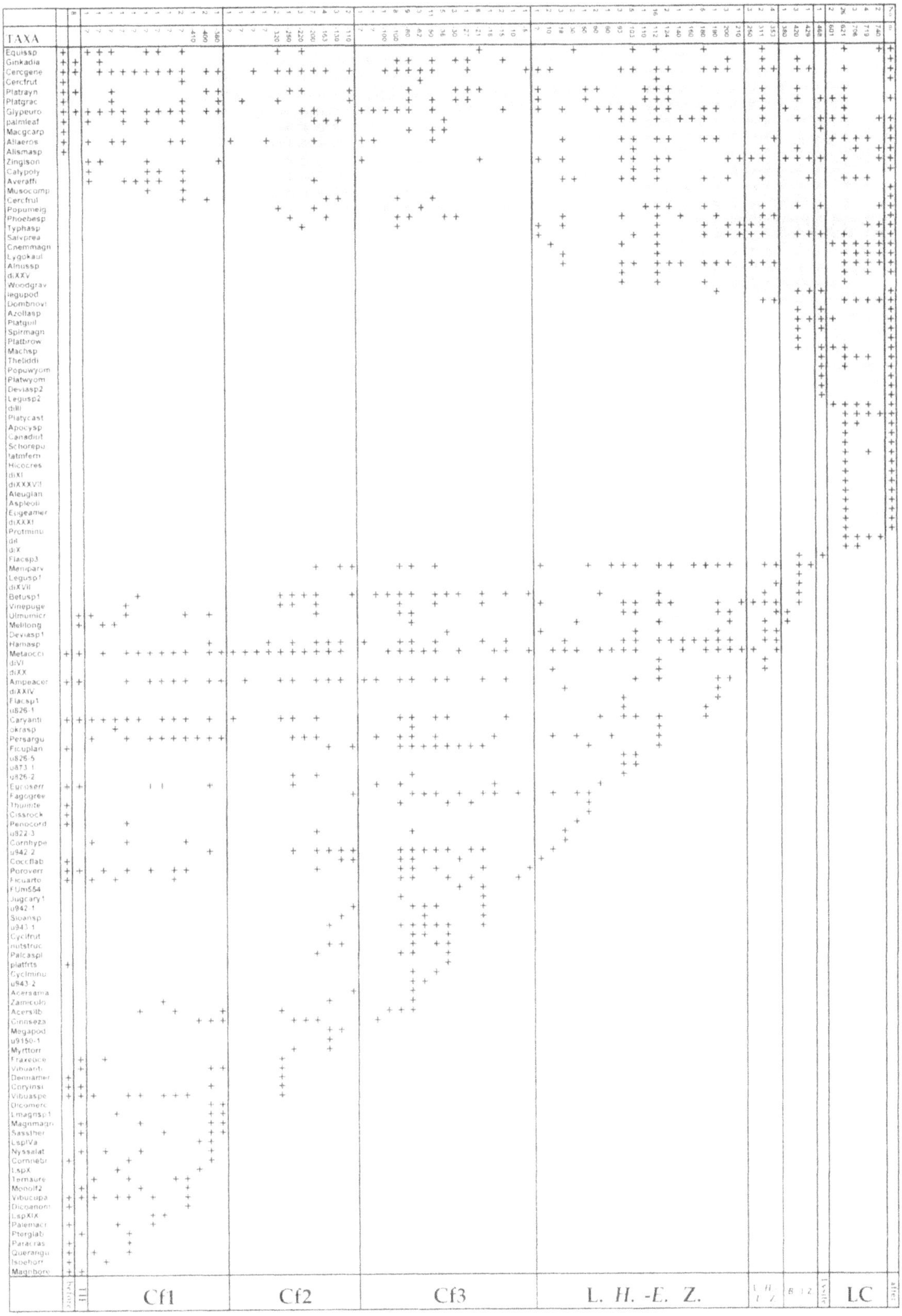

Figure 17.5 Stratigraphic ranges of the 130 plant taxa having occurrences at more than one stratigraphic level. Taxa with last occurrences above or near the top of the section are arranged by first occurrence; taxa with last occurrences below the Lostcabinian interval are arranged by last occurrence. Where shading extends beyond records it indicates known occurrences of the taxon in northwestern Wyoming that extend the stratigraphic range.

No = number of localities at a stratigraphic level, M = number of meters above or below the Zone Wa0 red bed, TIF = Tiffanian NALMA, Cf1 = Zone Cf1 (lower Clarkforkian NALMA), Cf2 = Zone Cf2 (middle Clarkforkian NALMA), Cf3 = Zone Cf3 (upper Clarkforkian NALMA), LHEZ = lower *Haplomylus–Ectocion* Zone, UHEZ = upper *Haplomylus–Ectocion* Zone, BIZ = *Bunophorus* Interval Zone, LC = Lostcabinian interval. Taxa are abbreviated by the first four letters of the genus and species names.

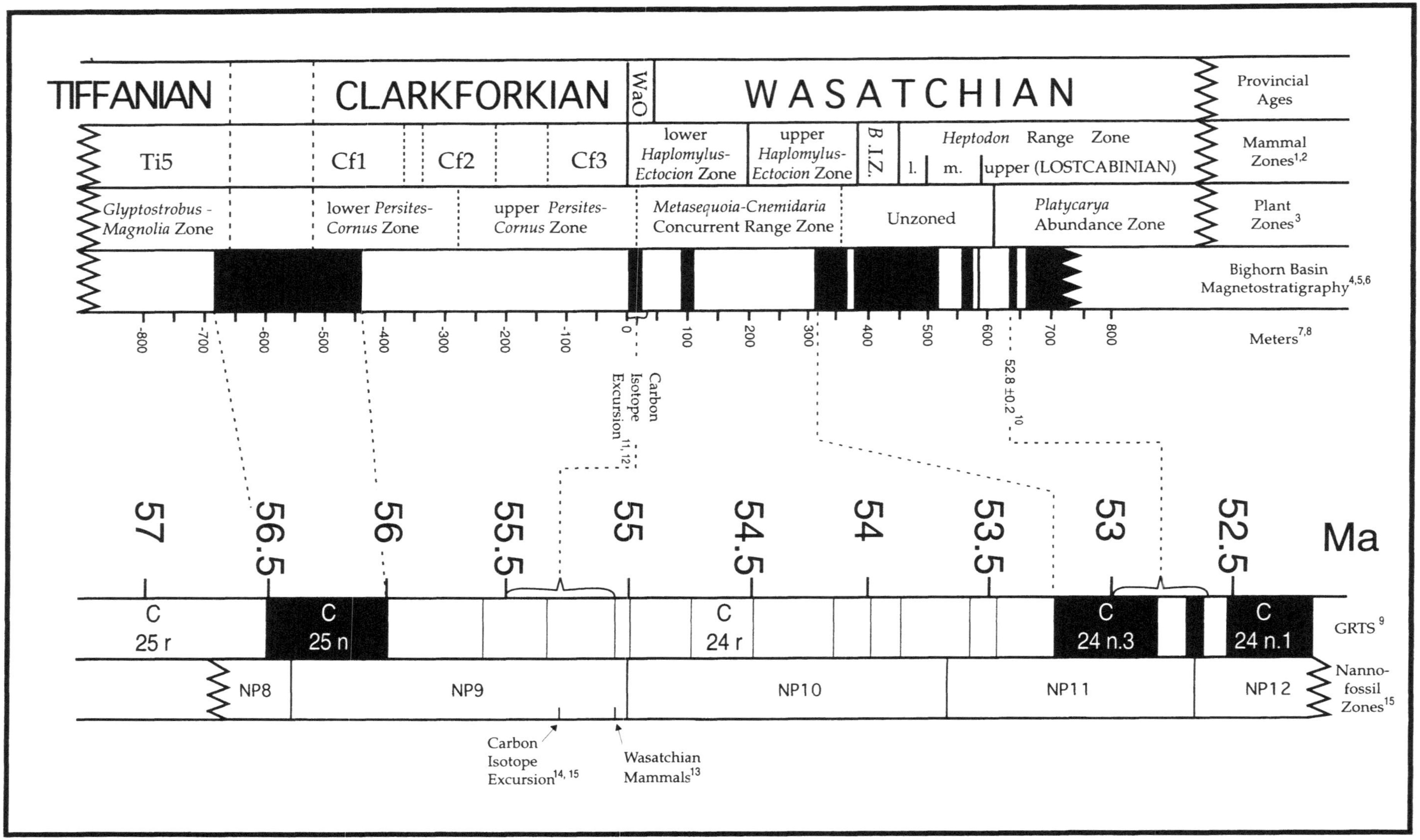

Figure 17.6 Biostratigraphic, magnetostratigraphic, isotopic, and radioisotopic framework for the Paleocene–Eocene sections of the Bighorn Basin.

Numbers indicate sources. 1 = Schankler (1980), 2 = Gingerich et al. (1980), 3 = Hickey (1980) and Wing et al. (1991), 4 = Butler et al. (1981), 5 = Clyde et al. (1994), 6 = Tauxe et al. (1994), 7 = Bown et al. (1994), 8 = Hickey (1980), 9 = Cande and Kent (1992), 10 = Wing et al. (1991), 11 = Koch et al. (1992), 12 = Koch et al. (1995), 13 = Beard et al. (1995), 14 = Kennett and Stott (1991), 15 = Berggren et al. (1995). GRTS indicates the Geomagnetic Reversal Time Scale; thin black lines in the GRTS column are "cryptochrons" of Cande and Kent (1992).

rences of plant taxa in the *Metasequoia–Cnemidaria* Zone, but these appear to be spread out across 300 m of section and offer no obvious division into subzones.

Above the 353 m level of the Elk Creek Section, the floral record becomes more spotty, with only one locality occurring between 353 and 420 m. Even though five localities were collected in the 420–468 m interval, floras from this part of the section have not been used to define a zone. Three of these assemblages are quite diverse (localities SLW882, H, and Pn), but all five are derived from abandoned channel-fill deposits that preserve a somewhat different subset of floodplain vegetation than the preceding backswamp deposits from the 200–350 m interval. Because of the possible strong influence of depositional environment on floral composition, and the absence of fossil plants between 470 and 601 m above the Zone Wa0 red bed (Wing 1984b), it is not desirable to create a floral zone for this interval. Floral assemblages from the 621 m level clearly document the beginning of the *Platycarya* Abundance Zone, but the rare first appearances of common *Platycarya* Abundance Zone taxa such as *"Dombeya" novi-mundi*, *Machaerium* sp., *Thelypteris iddingsii*, and *"Populus" wyomingiana* in the 310–468 m interval suggest that the absence of backswamp deposits in the middle of the Willwood Formation may delay the first appearances of typical late early Eocene to middle Eocene taxa. If tabular carbonaceous backswamp deposits were present throughout the Wasatchian section in the Bighorn Basin, there might be little or no gap between the top of the *Metasequoia–Cnemidaria* Concurrent Range Zone and the bottom of the *Platycarya* Abundance Zone, and the boundary would probably be close to the 500 m level of the Elk Creek section.

Changes in Floral Composition

Floral change across the Paleocene/Eocene Series boundary in the Bighorn Basin began with a pulse of 19 first appearances in Zone Cf1, indicating an important floral immigration event (figure 17.5, table 17.1). The floral immigrants include several taxa that are ubiquitous in late Paleocene floras of the northern Rockies: *Persites argutus*, *"Cinnamomum" sezannense*, *Averrhoites affinis*, *Zingiberopsis isonervosa*, cf. *Vinea pugetensis*, and cf. *Hamamelis* sp. There are 18 last appearances in Zone Cf1, including those of the Cretaceous holdovers *Quereuxia angulata*, *Paranymphaea crassifolia*, and *Dicotylophyllum anomalum*, as well as the common early Paleocene form *"Viburnum" cupanioides*. There are 16 first appearances in Zone Cf2, including the common and distinctive forms *Menispermites parvareolatus*, *Fagopsis greenlandica*, *Populus meigsii*, and *Phoebe* sp. The eight last appearances within Zone Cf2 include the long-ranging and well-defined forms *"Viburnum" antiquum*, *"Viburnum" asperum*, and *Dennstaedia americana*.

The upper Clarkforkian NALMA (Zone Cf3) has 18 last appearances, several of which are common or long-ranging Paleocene forms. Of particular note are the last appearances of *"Ficus" artocarpoides*, *"Cinnamomum" sezannense*, and *Porosia verrucosa*. The seven first appearances within Zone Cf3 are of forms that are undescribed and not known elsewhere, or based on fruits that may not have been consistently noted in older localities. The stratigraphic density of localities in the uppermost Fort Union Formation is insufficient to resolve whether the extinctions were concentrated at specific times within Zone Cf3, but the highest appearances of *Porosia verrucosa* and *"Ficus" artocarpoides* are less than 30 m below the Zone Wa0 red bed.

Twenty additional taxa have their local last appearances within the lower *Haplomylus–Ectocion* Zone, and eight of these are in the basal 50 m, including previously common taxa such as *Eucommia serrata*, *Fagopsis greenlandica*, and *Thuites interruptus*. Other common forms that are restricted to the first 120 m above the Zone Wa0 red bed are *"Ficus" planicostata* and *Persites argutus*. The common late Paleocene form *"Ampelopsis" acerifolia* is last recorded at the 200 m level of the Elk Creek section. Ten first appearances are documented in the lower *Haplomylus–Ectocion* Zone, including three taxa that have long been considered Eocene indicators in the northern Rockies: *Salvinia preauriculata*, *Cnemidaria magna*, and *Lygodium kaulfussi*. All of these index taxa are ferns. Other early Wasatchian first appearances include Leguminosae (based on compressed legume fruits), and an undescribed species of *Alnus*, which is numerically dominant in many lower *Haplomylus–Ectocion* Zone backswamp assemblages.

As sampling density of plants decreases in the upper *Haplomylus–Ectocion* Zone, so do the numbers of first and last occurrences, but two of the five last appearances in this zone are of the formerly abundant backswamp taxa, *Metasequoia occidentalis* and *Hamamelis* sp. One of the two first appearances in the *Haplomylus–Ectocion* Zone is of *"Dombeya" novi-mundi*, which becomes very abundant in lower Lostcabinian backswamp assemblages. In the central Bighorn Basin, where upper *Haplomylus–Ectocion* Zone and younger Eocene sediments are most widely exposed, the dominant environment of preservation for plant fossils begins to shift in this part of the section. Lenticular abandoned channel-fill deposits become more numerous and laterally extensive backswamp carbonaceous shale units are absent (Wing 1984b). Although abandoned channel-fill deposits produce plant fossils through most of the upper

Table 17.1
First and last appearances of common or long-ranging plant taxa

Zone	First Appeaerances	Last Appearances
Cf1	*Zingiberopsis isonervosa*	*"Viburnum" cupanioides*
	Averrhoites affinis	*Quereuxia angulata*
	Persites argutus	*Paranymphaea crassifolia*
	Cornus hyperborea	*Cornus nebrascensis*
	"Cinnamomum" sezannense	*Dicotylophyllum anomalum*
	Acer silberlingii	*"Pterocarya" glabra*
	cf. *Vinea pugetensis*	
	cf. *Hamamelis* sp.	
Cf2	*Menispermites parvareolatus*	*"Viburnum" asperum*
	Phoebe sp.	*"Viburnum" antiquum*
	Populus meigsii	*Dennstaedia americana*
Cf3	*Deviacer* sp. 1	*"Cinnamomum" sezannense*
		Acer silberlingii
		Zamia coloradensis
		Porosia verrucosa
		"Ficus" artocarpoides
LHEZ	*Alnus* sp.	*"Eucommia" serrata*
	Salvinia preauriculata	*Cornus hyperborea*
	Leguminosae	*Persites argutus*
	Cnemidaria magna	*Thuites interruptus*
	Lygodium kaulfussi	*Fagopsis greenlandica*
		"Carya" antiquorum
		"Ficus" planicostata
		"Ampelopsis" acerifolia
UHEZ	*"Dombeya" novi-mundi*	*Metasequoia occidentalis*
		Deviacer sp. 1
BIZ	*Machaerium* sp.	*Ulmus microphylla*
	Thelypteris iddingsii	*"Meliosma" longifolia*
	"Populus" wyomingiana	*Menispermites parvareolatus*
	"Platanus" wyomingensis	cf. *Vinea pugetensis*
LC	*Canavalia diuturna*	*Averrhoites affinis*
	"Eugenia" americana	*Glyptostrobus europaeus*
	Platycarya castaneopsis	
	Aleurites glandulosa	
	"Hicoria" crescentia	
	Apocynaceae sp. 1	

Cf1–3 = Clarkforkian Zones 1–3, LHEZ = lower *Haplomylus–Ectocion* Zone, UHEZ = upper *Haplomylus–Ectocion* Zone, BIZ = *Bunophorus* Interval Zone, LC = Lostcabinian interval.

Paleocene and lower Eocene formations of the Bighorn Basin, in the *Bunophorus* Interval Zone plant remains are preserved exclusively in this environment because backswamp deposits are absent and levee deposits are pedogenically altered.

This facies shift probably influences the precise stratigraphic level for first and last appearances of backswamp taxa in this part of the section, but typical backswamp taxa sometimes are found in channel-fill deposits. For example, *Ulmus microphylla* and *"Meliosma" longifolia*, which are common to dominant in backswamp assemblages from the Clarkforkian and lower Wasatchian deposits, have their last appearances in *Bunophorus* Interval Zone channel-fill lenses, as do *Menispermites parvareolatus*, Betulaceae sp.1, and cf. *Vinea pugetensis*, which are typically found in levee or crevasse-splay deposits lower in the section. These common taxa account for five of the eight last appearances in the *Bunophorus* Interval Zone. There are a total of 12 first appearances in the *Bunophorus* Interval Zone, of which 10 are taxa that range into the Lostcabinian or later Eocene deposits. First appearances also illustrate the potential for typically backswamp taxa to occur in abandoned channel deposits; the first two occurrences of *Machaerium* sp. are in abandoned channel deposits at the 420 and 468 m levels (*Bunophorus* Interval Zone), although in younger, Lostcabinian sediments it is fairly common in laterally extensive backswamp deposits.

The resumption of major backswamp deposition in the lower part of the Lostcabinian section (the 621 m level) is coincident with the first appearances of 21 plant taxa. Seven of these forms are known from multiple localities in the upper part of the Willwood Formation, but are not known from other parts of the northern Rocky Mountain region. The remaining 14 taxa occur in other late Wasatchian to early Bridgerian compression floras such as the Green River Flora (MacGinitie 1969), the Kisinger Lakes Flora (MacGinitie 1974), and the Yellowstone Flora (Knowlton 1899). The most common and widespread of these are *Canavalia diuturna*, *"Eugenia" americana*, *Platycarya castaneopsis*, *Aleurites glandulosa*, and *"Hicoria" crescentia*. Only four last appearances are documented within the Lostcabinian portion of the Willwood Formation, which is less than 200 m thick, and probably represents only the lower one-third to one-half of the total Lostcabinian section. Most of the common Paleocene backswamp taxa are already gone by the base of the Lostcabinian section, but two, *Averrhoites affinis* and *Glyptostrobus europaeus*, may have local to regional last appearances within the lower Lostcabinian section preserved in the central Bighorn Basin.

I have quantified the overall floral similarity of the fossil assemblages in this data set using detrended correspondence analysis (DECORANA) as implemented in the Cornell Ecology Programs (Hill 1979). Because taxa having only a single occurrence do not contribute any information about stratigraphic or temporal change, only the 130 taxa with multiple occurrences were used in the analysis. Although quantitative abundance data are available for some localities, this ordination was performed on presence/absence data in order to be able to include the largest possible number of floral assemblages. The resulting ordination summarizes variation in the composition of the assemblages in a small number of dimensions. The first axis of the ordination places younger assemblages on the left and older ones on the right.

The degree to which the first axis corresponds to change through time in floral composition can be assessed by plotting the axis I scores against stratigraphic level (figure 17.7). There is a range of axis I scores for assemblages even from a single horizon, reflecting lateral variation in floral composition created by spatial heterogeneity of the source vegetation, as well as taphonomic and depositional effects. Clearly superimposed on this variability, however, is the effect of change through time. Taken as a whole there is a fairly linear relationship between stratigraphic level and axis I scores of floral assemblages; the R2 value for a linear regression is 0.70 ($p < .0001$). Careful examination of figure 17.7, however, suggests that the relationship is more a step function than a smooth linear one. Plant assemblages from the upper 100 m of the Fort Union Formation (Zone Cf3) have axis I scores generally ranging from 0 to 380, floras from the 100 to 420 m level of the Willwood Formation (*Haplomylus–Ectocion* Zone) have axis I scores between 390 and 550, whereas floras from the 601 to 740 m (Lostcabinian) interval have scores ranging from 550 to 750. The first major shift in floral composition occurs in the lowest 100 m above the Zone Wa0 red beds (lower half of the lower *Haplomylus–Ectocion* Zone), the second occurs between 468 and 601 m. The first shift does not correspond to a time of reduced sampling density or missing section, but the second shift might appear more gradual if floral samples existed in the 468- to 601-m interval.

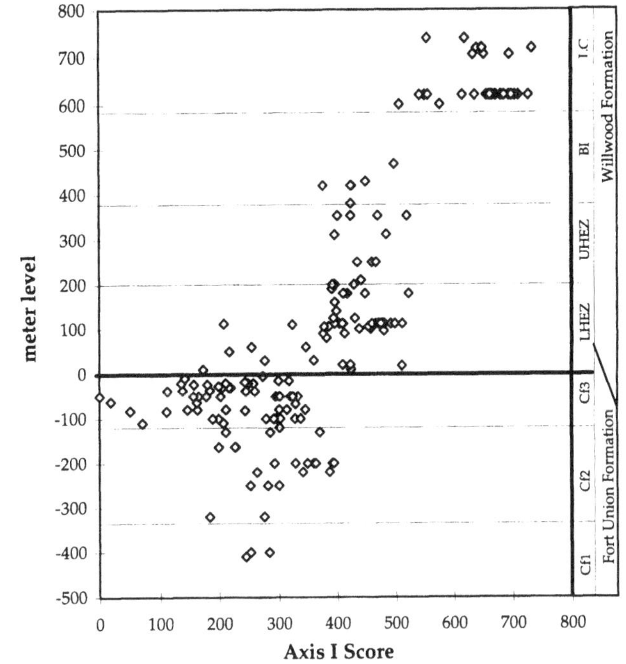

Figure 17.7 DECORANA axis I scores plotted against meter level of each floral assemblage.

Faunal zone abbreviations as in figure 17.5.

Plotting the axis I scores of localities by stratigraphic level also confirms the observation that the major floral shift near the base of the Eocene Series (boundary of the *Persites–Cornus* and *Metasequoia–Cnemidaria* Zones) occurs above rather than at the level of the Zone Wa0 red bed. Assemblages from the lowest 50–100 m above the Zone Wa0 red bed have scores ranging from those typical of plant assemblages in Zone Cf3 to those typical of assemblages from higher in the Wasatchian section. The variable scores for these floras reflect the variable presence in this stratigraphic interval of typical Clarkforkian forms such as *Persites argutus*, *"Ficus" planicostata*, *Fagopsis greenlandica*, and *Cornus hyperborea*, and typical Eocene taxa such as *Lygodium kaulfussi* and *Alnus* sp.

The relatively imprecise correlation of axis I scores of localities with stratigraphic level may result not only from chance variation in the floral composition of any given assemblage, but also from different depositional rates in the multiple local sections. Localities in the basal 100 m above the Zone Wa0 red bed have been collected in the North Butte, Antelope Creek, and Foster Gulch areas, and although the Zone Wa0 red bed appears to be a reliable datum for correlation, there is no way to determine if stratigraphic levels above the baseline in the different areas are precise time equivalents of one another. For localities in the lowest 100 m of section above the Zone Wa0 red bed, axis I scores appear to vary geographically, with the southerly localities (North Butte area) having higher scores than the northerly localities (Foster Gulch area). This pattern raises the possibility of different depositional rates in different areas, or of intrabasinal differences in earliest Wasatchian floral composition. These two potential causes cannot be distinguished with present data.

Although the correlation of floral composition as expressed by axis I scores and stratigraphy is not precise, there are highly significant differences in the mean axis I scores of plant localities from different mammalian zones (figure 17.8). Statistically significant differences exist between the mean axis I scores for plant assemblages from the Tiffanian NALMA and Zone Cf1, between Zones Cf2 and Cf3, between Zone Cf3 and the lower *Haplomylus–Ectocion* Zone, and between the *Bunophorus* Interval Zone and the Lostcabinian interval, but the largest differences are clearly between Zone Cf3 and the lower *Haplomylus–Ectocion* Zone floras, and between the *Bunophorus* Interval Zone and Lostcabinian floras.

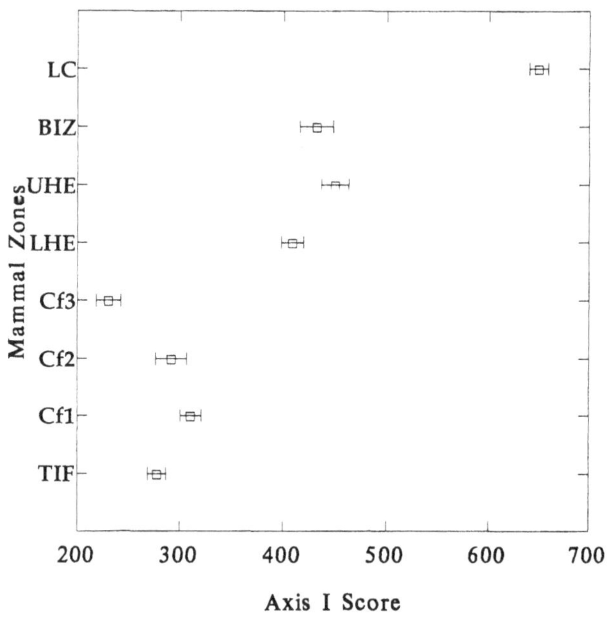

Figure 17.8 Mean DECORANA axis I scores for the plant localities in each mammalian zone. Horizontal bars indicate the standard error of the mean. Note the large offsets in floral composition between Zone Cf3 and the lower *Haplomylus–Ectocion* Zone, and between the *Bunophorus* Interval Zone and the Lostcabinian interval.

Floral Richness and Turnover

Using the ranges of the 130 taxa with multiple appearances I have calculated the number of plant taxa per mammalian zone across the upper Paleocene–lower Eocene interval. Floral taxa were counted using standard range-through assumptions, i.e., taxa were counted as present even if they did not occur in any sample from a zone as long as they were known to be present at both earlier and later times. Both richness and rates of floral turnover can be strongly affected by the completeness of the floral record for a given zone. I have estimated the completeness of the floral record for each zone using a completeness index (*CI*) published by Maas et al. (1995):

$$CI = [Nt/(Nt + Nrt)] \times 100$$

where *Nt* is the total number of taxa documented by the fossils in a zone and *Nrt* is the number of range-through taxa (i.e., those not recovered from the zone but known from higher and lower zones).

The CI shows that floral completeness for the zones used here varies from 69 to 95 out of a possible 100 (table 17.2). The moderately high and uniform indices of floral completeness suggest that differences in sampling intensity and facies do not have a major effect on patterns of floral turnover in this data set, although it is still necessary to consider the effect of sample number on richness. Observed number of range-through taxa ($S_{o\text{-}rt}$ in table 17.2) declines incrementally from 63 in Zone Cf1 to 56 in the lower *Haplomylus–Ectocion* Zone, drops sharply to 38 in the upper *Haplomylus–Ectocion* Zone, then increases to 45 in the *Bunophorus* Interval Zone and 58 in the lower Lostcabinian interval (figure 17.9).

In spite of completeness values between 69 and 95, the observed zonal richness figures should not be taken

Table 17.2
Floral turnover, floral richness, and temperature estimates across the Paleocene-Eocene interval in the Bighorn Basin

	Cf1	Cf2	Cf3	LHEZ	UHEZ	BIZ	LC
Base (Ma)[1]	56.20	55.87	55.53	55.20	53.91	53.30	52.85
Duration (Ma)[2]	0.33	0.33	0.33	1.29	0.47	0.14	0.21
Number plant localities	14	28	54	51	9	6	37
Completeness[3]	87	75	88	95	74	69	76
$S_{o\text{-all}}$	67	58	87	79	30	67	71
$S_{boot5\text{-all}}$	42	22	22	25	22	59	26
$S_{o\text{-rt}}$	63	61	60	56	38	45	58
$S_{o\text{-rt}}$ per my	191	185	182	43	81	321	276
$S_{boot5\text{-rt}}$	38	20	19	22	21	28	22
$S_{boot5\text{-rt}}$ per my	114	61	57	17	44	197	106
First appearance	21	16	7	14	2	12	21
Last appearance	18	8	18	20	5	8	11
Proportion first appearances	0.38	0.26	0.12	0.25	0.05	0.27	0.36
Proportion last appearances	0.33	0.13	0.30	0.36	0.13	0.18	0.19
Mean annual temperature (°C)[4]	12.9	14.1	18.4	18.2/16.4[5]	10.8	15.2	22.2
Number dicotyledonous leaf taxa	39	40	39	42/32	19	50	48

Abbreviations as in table 17.1.
1: Estimated from calibration points in Figure 17.6, and lineaer interpolation between those points based on stratigraphic thickness.
2: Duration of the stratigraphic interval producing plant fossils.
3: Completeness = number sampled range-through taxa/number sampled range-through taxa + number inferred range-through taxa) x 100.
4: Estimated from the equation MAT = 1.41 + 0.306 (percent entire-margined species), standard error = 0.79°C.
5: MAT estimates for the LHE were made separately for the 0 to 100-m interval and the 110 to 200-m interval.

at face value for two reasons: The number of samples per zone varies from 6 to 54 and the estimated durations of the zones range from 0.14 to 1.29 my (table 17.2). Sampling effort can have a strong effect on the richness of fossil plant assemblages (Wing and DiMichele 1995), so some correction must be made for the almost tenfold difference in the number of plant localities in the different intervals. I have made this correction by determining the richness of 250 randomly selected subsets of five localities from each zone. The mean number of plant species recovered from the 250 randomly selected subsets of five sites is shown in table 17.2. The bootstrapping procedure was carried out on the set of taxa with multiple occurrences ($S_{boot5\text{-rt}}$ in table 17.2), and on the whole data set of 294 taxa ($S_{boot5\text{-all}}$). For each of the seven intervals the standard error of the bootstrapped mean is less than one species.

To adjust for the differing durations of the intervals, the observed number of range-through taxa and the bootstrapped means for range-through taxa were divided by the number of years estimated for the zone ($S_{o\text{-rt}}$ per my and $S_{boot5\text{-rt}}$ per my in table 17.2). The duration estimates for the zones are based on the correlations to the timescale shown in figure 17.6 and on linear interpolation between these tie-points based on stratigraphic thickness. The average rock accumulation rate between the Zone Wa0 red bed and the 310 m level of the Elk Creek section of the Willwood Formation is 15.5 cm per thousand years; between 310 and 633 m in the section the rock accumulation rate is 65 cm per thousand years. This large increase in sedimentation rate in the middle of the Willwood Formation is consistent with independent estimates based on paleosol maturity (Bown and Kraus 1993).

The pattern of change in the bootstrapped and temporally adjusted estimates of richness is substantially different from that in the raw data. Estimated richness drops substantially from Zone Cf1 (114) to Zone Cf2 (61), declines slightly to Zone Cf3 (57), then drops sharply again to 17 in the lower *Haplomylus–Ectocion* Zone (figure 17.9). The bootstrapped and time-adjusted estimates rise to 46 in the upper *Haplomylus–Ectocion* Zone, to 321 in the *Bunophorus* Interval Zone, and then drop back to 276 in the Lostcabinian interval.

The differences between the observed and adjusted richness curves are caused by both sample size and interval length. The subdivisions of the Clarkforkian are estimated to have roughly equal durations, but Zone Cf3 is much more heavily sampled than Zones Cf1 and Cf2, which appears to be the primary reason for the stability in richness seen in the raw data (actually a rise when all taxa are included). In the Wasatchian NALMA, however, both sample size and interval duration vary. The low adjusted richness in the lower *Haplomylus–Ectocion* Zone is largely a consequence of the greater length of this interval. The high adjusted rich-

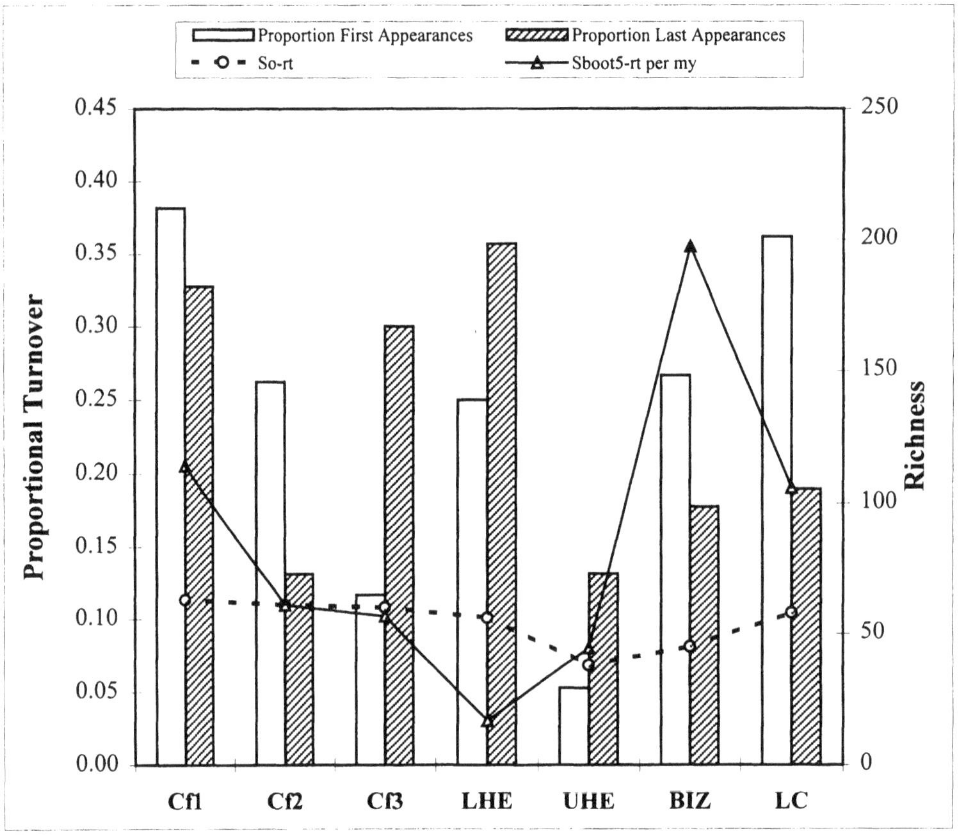

Figure 17.9 Standing richness and per taxon rates of first and last appearance in the flora of the Bighorn Basin across the Paleocene/Eocene Series boundary.

Data are from the taxon ranges in figure 17.5; values are shown in table 17.2. Proportion FA and LA are the proportion of range-through taxa having first and last appearances in the zone. $S_{o\text{-}rt}$ = observed number of range-through taxa, $S_{boot5\text{-}rt}$ per my = mean number of range-through taxa in randomly selected subsets of five localities divided by the duration of the interval. Standard error of the bootstrapped means is <1 species.

ness of the *Bunophorus* Interval Zone is a result of both short duration and the effect of compensating for a small number of samples. Although the exact pattern of change in richness differs between the raw and adjusted figures, both curves show low values in the early part of the Wasatchian NALMA, with higher values earlier and later. The adjusted richness estimate has its minimum in the lower *Haplomylus–Ectocion* Zone, whereas the raw data reach a minimum in the upper *Haplomylus–Ectocion* Zone.

The need to adjust richness statistics to account for differences in duration of intervals can be eliminated by calculating the number of taxa present at each sampling horizon (figure 17.10). Calculating diversity by sampling horizon does not escape problems of differential sampling effort, because levels with more localities are likely to record more species. The effect of differential sampling effort is reduced, however, by using only taxa with multiple occurrences, and by making standard range-through assumptions. The overall pattern of change in richness calculated from sampling horizon data is broadly similar to that seen in the zonal average data: high in Zone Cf3, declining to a low in the upper *Haplomylus–Ectocion* Zone, then increasing back to high levels in the lower Lostcabinian interval. The major differences between the zonal average and meter-level richness patterns are the trough in richness seen between −200 and −300 m (the lower part of Zone Cf2), and the decline in diversity in the uppermost part of the Bighorn Basin Lostcabinian section. The diversity low in Zone Cf2 that shows up in the meter level data reflects the large number of last occurrences in upper Zone Cf1 and lower Zone Cf2, whereas the decline at the top of the section probably results from reduced sampling opportunities and the failure to recognize all of the taxa that range through into the later Lostcabinian interval. Although the Lostcabinian interval is roughly 2.5 my long (Wing et al. 1991), the great majority of the floral localities in the Bighorn Basin Lostcabinian interval are derived from just two widespread fossiliferous horizons, both of which are near the base of the Lostcabinian interval. Perhaps because of the absence of

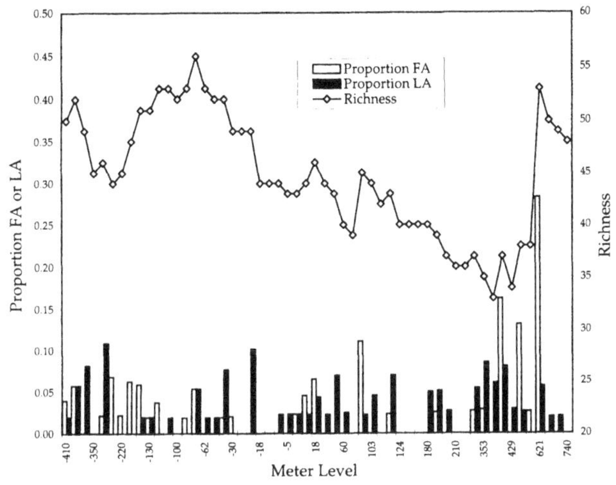

Figure 17.10 Standing richness of range-through taxa and per taxon rates of first and last appearance in the flora of the Bighorn Basin across the Paleocene/Eocene Series boundary calculated by meter level.

Meter levels are above or below the Zone Wa0 red bed. Range data are from figure 17.5. Abbreviations as in figure 17.9.

well-studied later Lostcabinian floras in this area, many taxa known from the lower Lostcabinian interval of the Bighorn Basin are restricted to a single locality, and thus were not included in this analysis of taxa with multiple occurrences. Previous studies of richness not restricted to range-through taxa demonstrated that the Lostcabinian backswamp layer at 621 m had nearly twice the floral diversity of the most diverse single layer in the lower *Haplomylus–Ectocion* Zone (Wing et al. 1995).

The fluctuations in richness across the Paleocene/Eocene Series boundary show the expected relationship with per taxon rates of first and last appearance. In both the zonal averages and in data recorded by meter level, there is a moderately high level of first appearances in Zone Cf1, a predominance of last appearances over first appearances throughout Zone Cf3 and the entire *Haplomylus–Ectocion* Zone, then a strong increase in first appearances in the *Bunophorus* Interval Zone and lowest Lostcabinian interval (figures 17.9 and 17.10).

Standing richness estimates and per taxon rates of first and last appearance can be used together to understand the pattern of floral diversity change through the late Paleocene–early Eocene interval. It appears that a wave of immigrants in Biochron Cf1 increased floral diversity. The decline in floral richness in Zone Cf3 corresponds to the large number of last occurrences and relatively few first appearances. Variations in sampling effort and widely spaced sampling intervals make it difficult to determine the degree to which the last occurrences are concentrated near the Wasatchian/Clarkforkian NALMA boundary, but there is no doubt that the high level of last appearances and the declining richness are maintained into the lower Wasatchian NALMA, in spite of the small pulse of first appearances in the first 100 m above the Zone Wa0 red bed. Sampling in the mid-lower Eocene interval is somewhat problematic because of facies changes and reduced sampling density in the middle and upper Willwood Formation, but it appears that floral diversity begins to recover in the *Bunophorus* Interval Zone as a result of increasing rates of first appearances. Major pulses of first appearances at the 468- and 621-m levels above the Zone Wa0 red bed probably increase floral diversity substantially. Overall the Clarkforkian/Wasatchian age boundary was an interval of high regional extinction and reduced plant diversity lasting 1–2 my, with the lowest levels of diversity occurring in the early Eocene Epoch rather than at the Clarkforkian/Wasatchian age boundary.

Climate Change

Global trends in climate change through the late Paleocene and early Eocene epochs have been established by oxygen isotope analysis of marine microfossils (e.g., Miller et al. 1987). Averaged global data indicate that mid- to high-latitude oceans began to warm in the late Paleocene Epoch (probably Tiffanian Age) and reached a Cenozoic peak in the early Eocene Epoch. Maximum warmth in the marine system appears to have occurred during the part of Chron C24n that corresponds to the 400- to 600-m levels of the Elk Creek Section (*Bunophorus* Interval Zone through Lostcabinian interval; Tauxe et al. 1994). The temperature of ocean deep waters began a slow decline through the later early Eocene Epoch (Miller et al. 1987).

Estimates of mean annual temperature (MAT) for the North American continent based on leaf physiognomy and floristic composition also indicate warming from the late Paleocene into the early Eocene Epoch (Hickey 1977, 1980; Wolfe 1978). The physiognomy of all of the late Paleocene to early Eocene floras in the Bighorn Basin is consistent with mixed deciduous and evergreen broad-leaved forest, but the late Wasatchian floras have a higher proportion of evergreen forms. The taxonomic composition of the younger floras is also consistent with a shift to warmer climates.

Inferred MAT based on leaf margin analysis of Bighorn Basin floras raised the possibility that there was a shorter-term cooling event during the early part of the Wasatchian Age superimposed on the long-term warming from the Paleocene to the Eocene Epoch, but the data were relatively weak (Wing et al. 1991). The MAT estimates presented here (table 17.2, figure 17.11) are also based on leaf margin analysis (Wolfe 1979;

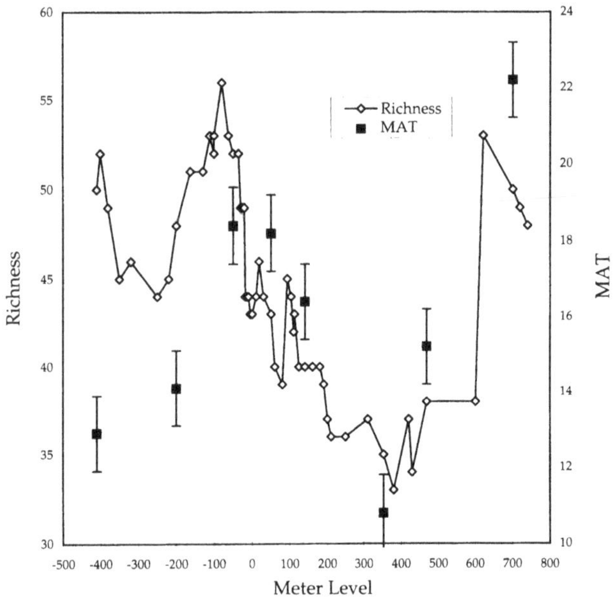

Figure 17.11 Estimated MAT (mean annual temperature) and standing richness of plants across the Paleocene/Eocene Series boundary interval in the Bighorn Basin.

MAT was estimated from the percentage of species with entire-margined leaves using the formula given in table 17.2; vertical bars indicate the standard error of the estimate, which is 0.78° C. Standing richness of range-through taxa is from figure 17.10.

Wing and Greenwood 1993) of all of the dicotyledonous leaf taxa present in each zone, but the sample sizes have been increased over those available to Wing et al. (1991). Because the lower *Haplomylus–Ectocion* Zone represents a much longer interval of time than the other zones, I have analyzed floras between 0 and 100 m separately from those between 110 and 200 m.

The recently collected assemblages from Zone Cf3 and the lower *Haplomylus–Ectocion* Zone strengthen the evidence for a cooling event during the earliest Eocene Epoch (figure 17.11). MAT estimates increase through the Clarkforkian NALMA to a high point of approximately 18° C in Zone Cf3, then decline in the first three estimates for the Wasatchian NALMA, reaching a nadir in the upper *Haplomylus–Ectocion* Zone of approximately 11° C before recovering to about 15° C in the *Bunophorus* Interval Zone and 22° C in the Lostcabinian interval. Using the correlations presented in figure 17.6, the temperature decline corresponds to the period from the latest Paleocene carbon isotope excursion to the early part of Chron C24n, an interval of ~2 my. The warming documented from the upper part of the Elk Creek section was very rapid, with estimated MAT rising 11° C in less than 0.5 my. Many of the new plant taxa appearing in the upper part of the Willwood Formation belong to higher taxa that currently have subtropical to tropical distributions, e.g., Magnoliales, Lauraceae, Myrtaceae, Apocynaceae, and Euphorbiaceae. Locality Pn at the 468 m level is the oldest well-dated locality in North America with a diversity of legumes, including the genus *Machaerium*, a common tree of tropical Latin America. The floristic composition and diversity of Lostcabinian assemblages is consistent with the rapid increase in temperature indicated by physiognomic analyses.

Although there is little marine isotopic evidence for an earliest Eocene cooling event, isotopic data from the Bighorn Basin are consistent with the floral evidence for cooler earliest Eocene climates. Koch et al. (1995) reported oxygen isotopic values in mammalian tooth enamel, freshwater clam shells, and pedogenic carbonates. The most reliable data sources (clams and larger mammals) produced more positive isotopic ratios in the uppermost Clarkforkian NALMA to Zone Wa0 interval, but showed a return to more negative ratios in the lower Wasatchian NALMA above Zone Wa0 but below the *Bunophorus* Interval Zone (Koch et al. 1995). These negative values indicate low oxygen isotope ratios of meteoric water, which are today correlated with cooler terrestrial temperatures (Koch et al. 1995; Koch, personal communication 1995).

Comparison of Floral and Faunal Turnover

Mid- to high-latitude warming has been cited as the key factor permitting interchange of fauna and flora between Asia and North America and Europe and North America across high-latitude landbridges (McKenna 1983; Tiffney 1985; Woodburne and Swisher 1995). The mammalian immigration occurred in two separate phases: the appearance of rodents, tillodonts, *Coryphodon*, and *Haplomylus* at the beginning of the Clarkforkian Age, followed approximately 1 my later by the appearance of artiodactyls, perissodactyls, primates, and hyaenodontid creodonts at the beginning of the Wasatchian Age (Rose 1981a; Gingerich 1989; Maas et al. 1995). These groups do not appear to have mid-Paleocene ancestors in North America, implying that they arrived from a different continent. Several authors have suggested that the late Paleocene–early Eocene mammalian immigrants to North America originated at more tropical latitudes in the Americas, Asia, or Africa (Sloan 1969; Gingerich 1980, 1989; Krause and Maas 1990), and then migrated to middle and high latitudes as global climate warmed in the late Paleocene and early Eocene epochs.

The pattern of turnover in the flora of the Bighorn Basin is not identical to that in the mammalian fauna, but there are points of similarity. Many of the plants that appear in Zone Cf1 belong to higher taxa that are today centered in tropical to subtropical climates and are generally evergreen: the ginger and banana families

(*Zingiberopsis isonervosa, Musophyllum complicatum*), laurel family (*Persites argutus, "Cinnamomum" sezannense*), tea family (*Ternstroemites aureavallis*), and cycads (*Zamia coloradensis*). Maas et al. (1995) have cited the Clarkforkian Age as a time of relatively low first and last appearances in the mammalian fauna, and a period of increasing richness. This appears to be the opposite of the pattern in the flora, but the analysis presented by Maas et al. (1995:figure 4) shows that first appearances decline to very low levels in Zone Cf3 even though it is a relatively well-sampled interval. Last appearances of mammals increased slightly in Zone Cf3 and standing richness decreased slightly. The evenness of relative abundances among species within the fauna also decreased at this time (Rose 1981b). Although the reduction in diversity and increase in last occurrences are much more dramatic in the flora than in the fauna, Zone Cf3 can be characterized for both plants and mammals as a period of few first appearances and declining richness.

The most striking difference between the plant and mammalian patterns of turnover occurs in the lowermost Wasatchian NALMA (Zones Wa0 and Wa1, the latter being equivalent to the lower part of the lower *Haplomylus–Ectocion* Zone). Here mammalian genera have an extraordinarily high rate of first appearances, a low rate of last appearances, and a major increase in richness (Maas et al. 1995; Wing et al. 1995). This provides a strong contrast to the flora, in which last appearances continued to greatly outpace first appearances and richness declined. Although some of the plant genera that first appeared near the beginning of the Wasatchian Age are today typical of subtropical or warmer climates (e.g., *Cnemidaria* in the tree fern family Cyatheaceae, and the floating fern *Salvinia*), others such as *Alnus* and *Lygodium* have wide distributions in temperate and subtropical climates.

Although there are no large turnover peaks in either the flora or fauna in the *Haplomylus–Ectocion* Zone, Schankler (1980) separated the upper and lower parts of the zone on the basis of a number of last appearances of species at about the 200-m level of the Elk Creek section, an event he called Biohorizon A. It has been suggested that Biohorizon A may result from a decrease in sampling density rather than reflecting a real clustering of last appearances (Badgley and Gingerich 1988), but four of the species that disappear at Biohorizon A reappear at the base of the *Bunophorus* Interval Zone, a discontinuity in temporal range that Schankler (1981) attributed to shifting geographic ranges of the taxa in question. Although no similar pattern can be documented in the flora, it may be significant that the gap in the temporal ranges of these mammals corresponds to the nadir in MAT estimated from leaf physiognomy (figure 17.11). Bown et al. (1994) documented rapid body size increases in several lineages of mammals at the level of Biohorizon A.

The next interval of accelerated turnover in the mammalian fauna defines the top of the upper *Haplomylus–Ectocion* Zone. At both the species level (Biohorizon B of Schankler 1980) and the generic level (Wa4/Wa5 zonal boundary of Maas et al. 1995) there is a peak of last appearances. In the flora this "event" cannot be distinguished from the persistent high level of last appearances that prevails from Zone Cf3 through the entire *Haplomylus–Ectocion* Zone.

The top of the *Bunophorus* Interval Zone (Biohorizon C of Schankler 1980) corresponds to a major peak of first appearances of mammalian species in the central Bighorn Basin, but first appearances continue at a fairly high level through much of the upper part of the Elk Creek section (Schankler 1980). Bown et al. (1994) indicated that Biohorizons B and C were nearly coeval events and that many mammalian lineages in the Bighorn Basin experienced increasing morphological variability and species level diversity during the short period of time which separates them. The *Bunophorus* Interval Zone is the first since Zone Cf2 in which floral first appearances are more common than last occurrences, and there is a weak indication that the first appearances in the flora are concentrated in the upper part of the zone.

Between the top of the *Bunophorus* Interval Zone and the base of the Lostcabinian interval there is no plant megafossil record, and in the Lostcabinian interval the density of mammalian fossils decreases dramatically. This makes it difficult to compare the pattern of faunal and floral turnover, but it is important to note that the largest peak of first appearances of plants is just above the base of the Lostcabinian interval. This grouping of first appearances at a single stratigraphic level (621 m) probably reflects the absence of good samples between 468 and 621 m. A more complete record for plants in the "barren" interval might demonstrate a more gradual or stepped pattern of first appearances similar to that seen in the fauna, although in any case this stratigraphic interval represents only about 0.5 my.

The early Eocene Epoch has long been recognized as a period of global warmth, and paleontologists have invoked climatic warming as one factor controlling the intercontinental migration of terrestrial faunas and floras across high-latitude landbridges. However, the pattern and rate of warming during the late Paleocene–early Eocene interval have been defined at only a relatively coarse temporal scale.

Over the last 15 years paleomagnetic, isotopic, biostratigraphic, and radioisotopic studies have increasingly refined the correlation of sections in the Bighorn Basin with global standards. It is now clear that the very thick continental sections in the Bighorn Basin and other intermontane basins of the Rocky Mountains provide a temporally detailed record of climatic and biotic change. Although these continental sections contain many short hiatuses indicated by soil modification of fluvial sediments, typical rates of deposition are 10–50 cm per thousand years, or about two orders of magnitude faster than those seen in deep-sea environments. Where these high rates of deposition are coupled with common preservation of fossils, we should be prepared to interpret closely spaced biotic changes in terms of global (e.g., Milankovitch cycles) and local processes that operate over periods of tens of thousands to hundreds of thousands of years.

The record of floral change presented in this chapter is based on about 50 sampling horizons spread across a period of approximately 3.5 my, yielding an average of 70 kyr between horizons. Physiognomic analyses of paleoclimate require combining many horizons in order to increase the number of species used for the estimate, but the average temporal separation between successive paleoclimatic estimates is still <0.5 my. The previously undetected earliest Eocene cooling revealed by these data may have been missed due to the relatively coarse temporal resolution of previous studies (e.g., Wing et al. 1991). The interpretation of faunal and floral dynamics in terms of climatic and/or sea-level forcing events needs to take into account the possibility of relatively rapid fluctuations in Paleogene global climate, rather than assuming a stately unfolding that may be an artifact of low temporal resolution. Identifying shorter-term climatic changes and their effects on terrestrial organisms will require additional high-resolution studies of thick continental sections; the patterns seen in these studies can be related to global events only by refining correlations between the terrestrial and marine realms.

ACKNOWLEDGMENTS

I would like to acknowledge many years of consultation and advice from fellow workers in the Bighorn Basin early Cenozoic, particularly Thomas Bown, Philip Gingerich, Leo Hickey, and Kenneth Rose. Field work for this project has been supported by the Roland Brown Fund of the Department of Paleobiology, National Museum of Natural History, and by grants from the Scholarly Studies Program of the Smithsonian Institution and the Evolution of Terrestrial Ecosystems Program of the NMNH. I thank Marie-Pierre Aubry, William DiMichele, and Spencer Lucas for reviewing an earlier version of this chapter.

REFERENCES

Badgley, C. and Gingerich, P. D., 1988. Sampling and faunal turnover in early Eocene mammals. *Palaeogeography, Paleoclimatology, Palaeoecology,* v. 63, p. 141–157.

Beard, C., Dawson, M. R. and Tabrum, A. R., 1995. First diverse land mammal fauna from the early Cenozoic of the southeastern United States: the early Wasatchian Red Hot local fauna, Lauderdale County, Mississippi. *Geological Society of America Abstracts with Programs,* p. A-453.

Berggren W. A., Kent, D. V., Swisher, C. C., III and Aubry, M.-P., 1995. A revised Cenozoic geochronology and chronostratigraphy. In Berggren, W. A., Kent, D. V., Aubry, M.-P. and Hardenbol, J. (eds.), *Geochronology Time Scales and Global Stratigraphic Correlation,* Society of Economic Paleontologists and Mineralogists Special Publication No. 54, p. 129–212.

Bown, T. M., 1975. Paleocene and lower Eocene rocks in the Sand Creek–No Water Creek area, Washakie County, Wyoming. Wyoming Geological Association Guidebook of the 27th Annual Field Conference, p. 55–61.

Bown, T. M. and Kraus, M. J., 1993. Time-stratigraphic reconstruction and integration of paleopedologic, sedimentologic, and biotic events (Willwood Formation, lower Eocene, northwest Wyoming, U.S.A.). *Palaios,* v. 8, p. 68–80.

Bown, T. M., Holroyd, P. A. and Rose, K. D., 1994. Mammal extinctions, body size, and paleotemperature. *Proceedings of the National Academy of Sciences of the United States,* v. 91, p. 10,403–10,406.

Bown, T. M., Rose, K. D., Simons, E. L. and Wing, S. L., 1994. *Distribution and stratigraphic correlation of upper Paleocene and lower Eocene fossil mammal and plant localities of the Fort Union, Willwood and Tatman Formations, southern Bighorn Basin, Wyoming.* U.S. Geological Survey Professional Paper 1540, 103 pp.

Brown, R. W., 1962. *Paleocene flora of the Rocky Mountains and Great Plains.* U.S. Geological Survey Professional Paper 375, 119 pp.

Butler, R. F., Gingerich, P. D. and Lindsay, E. H., 1981. Magnetic polarity stratigraphy and biostratigraphy of Paleocene and lower Eocene continental deposits, Clark's Fork Basin, Wyoming. *Journal of Geology,* v. 89, p. 299–316.

Cande, S. C. and Kent, D. V., 1992. A new geomagnetic polarity time scale for the Late Cretaceous and Cenozoic. *Journal of Geophysical Research,* v. 97(B10), p. 13,917–13,951.

Clyde, W. C., Stamatakos, J. and Gingerich, P. D., 1994. Chronology of the Wasatchian Land-Mammal Age (early Eocene): Magnetostratigraphic results from the McCullough Peaks Section, northern Bighorn Basin, Wyoming. *Journal of Geology,* v. 102, p. 367–377.

Frederiksen, N. O., 1994. Paleocene floral diversities and turnover events in eastern North America and their relation to diversity models. *Review of Palaeobotany and Palynology,* v. 82, p. 225–238.

Gibson, T. G., Bybell, L. M. and Owens, J. P., 1993. Latest Paleocene lithologic and biotic events in neritic deposits of southwestern New Jersey. *Paleoceanography,* v. 8, p. 495–514.

Gingerich, P. D., 1980. Evolutionary patterns in early Cenozoic mammals. *Annual Review of Earth and Planetary Sciences,* v. 8, p. 407–424.

Gingerich, P. D., Rose, K. D. and Krause, D. W., 1980. Early Cenozoic mammalian faunas of the Clark's Fork Basin–Polecat Bench area, northwestern Wyoming. In Gingerich, P. D. (ed.), *Early Cenozoic Paleontology and Stratigraphy of the Bighorn Basin, Wyoming.* University of Michigan Papers on Paleontology, v. 24, p. 24–51.

Gingerich, P. D., 1982. Time resolution in mammalian evolution: sampling, lineages, and faunal turnover. In Mamet, B. and Copeland, M. J. (eds.), *Third North American Paleontological Convention, Proceedings,* v. 1, p. 205–210.

Gingerich, P. D., 1989. New earliest Wasatchian mammalian fauna from the Eocene of northwestern Wyoming: Composition and diversity in a rarely sampled high-floodplain assemblage. *University of Michigan Papers on Paleontology,* v. 28, 97 pp.

Hickey, L. J., 1977. Stratigraphy and paleobotany of the Golden Valley Formation (early Tertiary) of western North Dakota. *Geological Society of America Memoir* 150, 181 pp.

Hickey, L. J., 1980. Paleocene stratigraphy and flora of the Clark's Fork Basin. In Gingerich, P. D. (ed.), *Early Cenozoic Paleontology and Stratigraphy of the Bighorn Basin, Wyoming,* University of Michigan Papers on Paleontology, v. 24, p. 33–49.

Hill, M. O., 1979. DECORANA, a FORTRAN program for Detrended Correspondence Analysis and Reciprocal Averaging. Ithaca: Microcomputer Power, 36 pp.

Hovan, S. A. and Rea, D. K., 1992. Paleocene/Eocene boundary changes in atmospheric and oceanic circulation: A southern hemisphere record. *Geology,* v. 20, p. 15–18.

Kennett, J. P. and Stott, L. D., 1991. Abrupt deep-sea warming, palaeoceanographic changes and benthic extinctions at the end of the Palaeocene. *Nature,* v. 353, p. 225–229.

Koch, P. L., Zachos, J. C. and Gingerich, P. D., 1992. Correlation between isotope records in marine and continental carbon reservoirs near the Palaeocene/Eocene boundary. *Nature,* v. 358, p. 319–322.

Koch, P. L., Zachos, J. C. and Dettman, D. L., 1995. Stable isotope stratigraphy and paleoclimatology of the Paleogene Bighorn Basin (Wyoming, USA). *Palaeogeography, Palaeoclimatology, Palaeoecology,* v. 115, p. 61–89.

Knowlton, F. H., 1899. Fossil flora of the Yellowstone National Park. U.S. Geological Survey Monograph 32, p. 651–893.

Kraus, M. J., 1980. Genesis of a fluvial sheet sandstone, Willwood Formation, northwest Wyoming. In Gingerich, P. D. (ed.), *Early Cenozoic Paleontology and Stratigraphy of the Bighorn Basin, Wyoming,* University of Michigan Papers on Paleontology, v. 24, p. 87–94.

Kraus, M. J., 1992. Alluvial response to differential subsidence: sedimentological analysis aided by remote sensing, Willwood Formation (Eocene), Bighorn Basin, Wyoming, USA. *Sedimentology,* v. 39, p. 455–470.

Krause, D. W. and Maas, M. C., 1990. The biogeographic origins of late Paleocene–early Eocene mammalian immigrants to the Western Interior of North America. In Bown, T. M., and Rose, K. D. (eds.), *The dawn of the age of mammals in the northern part of the Rocky Mountain Interior.* Geological Society of America Special Paper 243, p. 71–105.

Leopold, E. B. and MacGinitie, H. D., 1972. Development and affinities of Tertiary floras in the Rocky Mountains. In Graham, A. (ed.), *Floristics and Paleofloristics of Asia and Eastern North America,* Amsterdam: Elsevier, p. 147–200.

Maas, M. C., Anthony, M. R. L., Gingerich, P. D., Gunnell, G. F. and Krause, D. W., 1995. Mammalian generic diversity and turnover in the late Paleocene and early Eocene of the Bighorn and Crazy Mountains Basins, Wyoming and Montana (USA). *Palaeogeography, Palaeoclimatology, Palaeoecology,* v. 115, p. 181–207.

MacGinitie, H. D., 1969. The Eocene Green River flora of northwestern Colorado and northeastern Utah. *University of California Publications in Geological Sciences,* v. 83, p. 1–202.

MacGinitie, H. D., 1974. An early middle Eocene flora from the Yellowstone–Absaroka volcanic province, northwestern Wind River Basin, Wyoming. *University of California Publications in Geological Sciences,* v. 108, 103 pp.

McKenna, M. C., 1983. Cenozoic paleogeography of North Atlantic land bridges. In Bott, M. H. P., Saxov, S., Talwani, M. and Thiede, J. (eds.), *Structure and Development of the Greenland–Scotland Ridge.* New Methods and Concepts, NATO Conference Series, Series IV: Marine Sciences, v. 8, New York, London: Plenum Press, p. 351–399.

Miller, K. G., Fairbanks, R. G. and Mountain, G. S., 1987. Tertiary oxygen isotope synthesis, sea level history, and continental margin erosion. *Paleoceanography,* v. 2, p. 1–19.

Pak, D. K. and Miller, K. G., 1992. Paleocene to Eocene benthic foraminiferal isotopes and assemblages: Implications for deepwater circulation. *Paleoceanography,* v. 7, p. 405–422.

Rose, K. D., 1981a. The Clarkforkian Land-Mammal Age and mammalian faunal composition across the Paleocene/Eocene boundary. *University of Michigan Papers on Paleontology,* v. 26, 197 pp.

Rose, K. D., 1981b. Composition and species diversity in Paleocene and Eocene mammal assemblages: An empirical study. *Journal of Vertebrate Paleontology,* v. 1, p. 367–388.

Schankler, D. M., 1980. Faunal zonation of the Willwood Formation in the central Bighorn Basin, Wyoming. In Gingerich, P. D. (ed.), *Early Cenozoic Paleontology and Stratigraphy of the Bighorn Basin, Wyoming,* University of Michigan Papers on Paleontology, v. 24, p. 99–114.

Schankler, D. M., 1981. Local extinction and ecological re-entry of early Eocene mammals. *Nature,* v. 293, p. 135–138.

Sloan, R. E., 1969. Cretaceous and Paleocene terrestrial communities in western North America. In Yochelson, E. L. (ed.), *Proceedings of the North American Paleontological Convention 1,* Lawrence: Allen Press, p. 427–453.

Tauxe, L., Gee, J., Gallet, Y., Pick, T. and Bown, T., 1994. Magnetostratigraphy of the Willwood Formation, Bighorn Basin, Wyoming: New constraints on the location of the Paleocene/Eocene boundary. *Earth and Planetary Science Letters,* v. 125, p. 159–172.

Tiffney, B. H., 1985. The Eocene North Atlantic land bridge: Its importance in Tertiary and modern phytogeography of the northern hemisphere. *Journal of the Arnold Arboretum,* v. 66, p. 243–273.

Wing, S. L., 1984a. A new basis for recognizing the Paleocene/Eocene boundary in western interior North America. *Science,* v. 226, p. 439–441.

Wing, S. L., 1984b. Relation of paleovegetation to geometry and cyclicity of some fluvial carbonaceous deposits. *Journal of Sedimentary Petrology,* v. 54, p. 52–66.

Wing, S. L. and Bown, T. M., 1985. Fine scale reconstruction of late Paleocene–early Eocene paleogeography in the Bighorn Basin of northern Wyoming. In Flores, R. M. and Kaplan, S. S. (eds.), *Cenozoic Paleogeography of West-Central United States, Denver: Rocky Mountain Section,* Society of Economic Paleontologists and Mineralogists, p. 93–105.

Wing, S. L., Bown, T. M. and Obradovich, J. D., 1991. Early Eocene biotic and climatic change in interior western North America. *Geology,* v. 19, p. 1189–1192.

Wing, S. L. and Greenwood, D. R., 1993. Fossils and fossil climate: The case for equable continental interiors in the Eocene. In Allen, J. R. L., Hoskins, B. J., Sellwood, B. W. and Spicer, R. A. (eds.), *Palaeoclimates and their modeling with special reference to the Mesozoic Era,* Philosophical Transactions of the Royal Society, London B, Biological Sciences, v. 341, p. 243–252.

Wing, S. L. and DiMichele, W. A., 1995. Conflict between local and global changes in plant diversity through geological time. *Palaios,* v. 10, p. 551–564.

Wing, S. L., Alroy, J. and Hickey, L. J, 1995. Plant and mammal diversity in the Paleocene to early Eocene of the Bighorn Basin. *Palaeogeography, Palaeoclimatology, Palaeoecology,* v. 115, p. 117–155.

Wolfe, J. A., 1978. A paleobotanical interpretation of Tertiary climates in the northern hemisphere. *American Scientist,* v. 66, p. 694–703.

Wolfe, J. A., 1979. *Temperature parameters of humid to mesic forests of eastern Asia and relation to forests of other regions of the northern hemisphere and Australasia.* U.S. Geological Survey Professional Paper 1106, p. 1–37.

Wood, H. E., II, Chaney, R. W., Clark, J., Colbert, E. H., Jepsen, G. L., Reeside, J. B., Jr. and Stock, C., 1941. Nomenclature and correlation of the North American continental Tertiary. *Bulletin of the Geological Society of America,* v. 52, p. 1–48.

Woodburne, M. O. and Swisher, C. C., III, 1995. Land mammal high-resolution geochronology, intercontinental overland dispersals, sea level, climate, and vicariance. In Berggren, W. A., Kent, D. V., Aubry, M.-P. and Hardenbol, J. (eds.), *Geochronology Time Scales and Global Stratigraphic Correlation,* Society of Economic Paleontologists and Mineralogists Special Publication No. 54, p. 335–364.

18

Turtles Across the Paleocene/Eocene Epoch Boundary in West-Central North America

J. HOWARD HUTCHISON

ABSTRACT

The approximately 3-my-long interval including the Paleocene/Eocene Epoch boundary incorporates the most rapid rise in diversity of turtles in western North America that is mirrored in western Europe. Warmer polar climates and more equable continental climates allowed immigration into the continental interior from both northern (polar) and southern (tropical) regions. Similar patterns may have promoted passage through continental Asia into western Europe.

Nonmarine turtles of west-central North America exhibited notable fluctuations in diversity (species richness) during the Cenozoic Era (Hay 1908; Hutchison 1980, 1982; Hutchison and Archibald 1986). The strongest increase of diversity in this region was across the Paleocene/Eocene Epoch boundary and during the early Eocene Epoch. This transition is more striking owing to the low level of known diversities in the middle Paleocene Epoch (Hutchison and Archibald 1986) and nearly catastrophic declines at the end of the Eocene (Hutchison 1992). The increase of diversity of reptiles around the Paleocene/Eocene Epoch boundary is generally ascribed to increasing warm and equable climates (Estes and Hutchison 1980; Hutchison 1982). It is across this boundary that the turnover and modernization of the turtle faunas from the Late Cretaceous antecedents reached nearly Recent proportions. Aside from the very earliest Paleocene (early Puercan) turtle fauna, the early Eocene turtle fauna is the most diverse in the Cenozoic Era (Hutchison and Archibald 1986; Hutchison 1992). A combination of events contributed to this rise, including the development of numerous intermontain basins that served as water traps (Lillegraven and Ostresh 1988), increasingly warm and equable climates (Wing et al. 1991; Wing and Greenwood 1993), and favorable physical and climatic conditions in the corridor to the Eurasian faunas.

In North America, the known record of changes in turtles across the Paleocene/Eocene Series boundary (figure 18.1) is concentrated along the Rocky Mountain Province and extends south as far as the Big Bend region of Texas. Within this region the best sampled and most studied area is that of the Big Horn and Wind River basins, followed by that of the Washakie Basin, Wyoming. The former turtle records are particularly well correlated with mammalian and magnetostratigraphic sections. For purposes of the following discussion this north–south region will be referred to as the study area. The exact position of the Paleocene/Eocene Series boundary in midcontinental North America is debatable because of nomenclatural and correlational reasons but lies within the interval encompassed by the Clarkforkian and Wasatchian North American land-mammal "ages" (NALMA; Woodburne 1987). Harland et al. (1990) placed the epoch boundary in early magnetic Chron C24r (early Clarkforkian Age). Because of the correlation debates and the changes in turtle faunas that lag the early Clarkforkian Age, the Clarkforkian and Wasatchian (at least through Zone Wa5) interval will be referred to as the boundary interval.

Cast of Characters

The spectrum of turtles at the family level in the boundary interval is the most diverse in the Cenozoic record of North America, incorporating a combination of archaic and relatively modern families. This diversity is

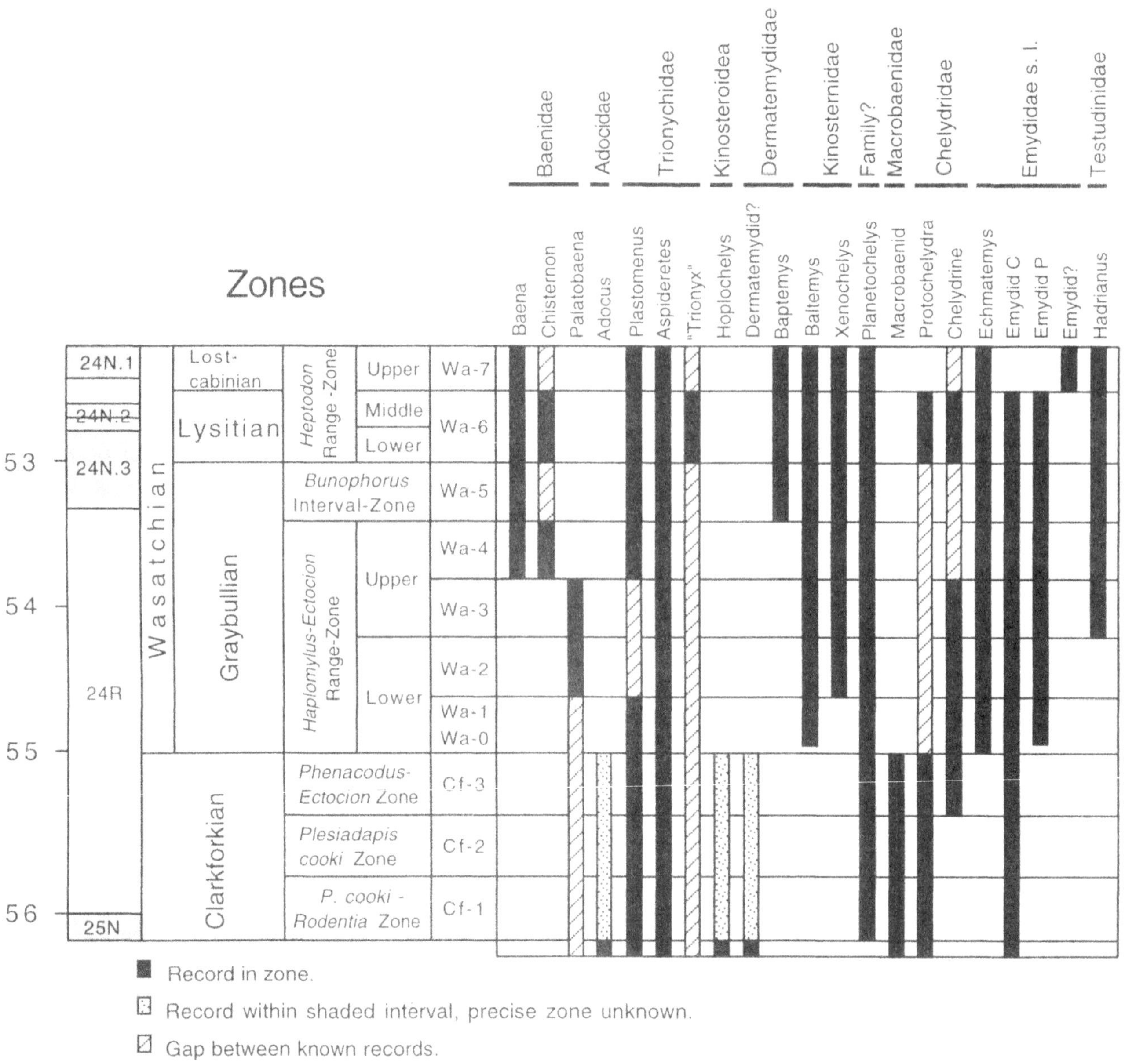

Figure 18.1 Composite range zones of turtle genera in the Clarkforkian and Wasatchian deposits of western North America (Texas to Wyoming).

Zonation after Clyde et al. (1994).

also reflected in the diversity of shell architectural types (figure 18.2) and habitat preferences ranging from fully aquatic (lakes and rivers) to fully terrestrial.

Baenidae (figure 18.2A)

The Baenidae are an archaic group of aquatic riverine turtles that I interpret as bottom-walking and possibly omnivorous to carnivorous. They enjoyed their peak of diversity in the Late Cretaceous Period and earliest Cenozoic Era, declining progressively thereafter, and became extinct by the end of the Eocene Epoch. They appear to be endemic to North America. In the boundary interval they are represented by three monotypic genera. *Palatobaena gaffneyi* Hutchison in Archibald and Hutchison (1979) became extinct in the Wasatchian Age and the genus is represented in pre-Clarkforkian deposits by another species. *Baena arenosa* Leidy 1870 and *Chisternon undatum* (Leidy 1871) appeared in the Wasatchian Age and may be endemic to the study area. Their absence from the early Wasatchian and Clarkforkian Ages may reflect the observation that they are relatively uncommon and difficult to identify without cranial or relatively complete shell material. A sister and possibly ancestral taxon, *Plesiobaena*, is present in pre-Clarkforkian deposits of

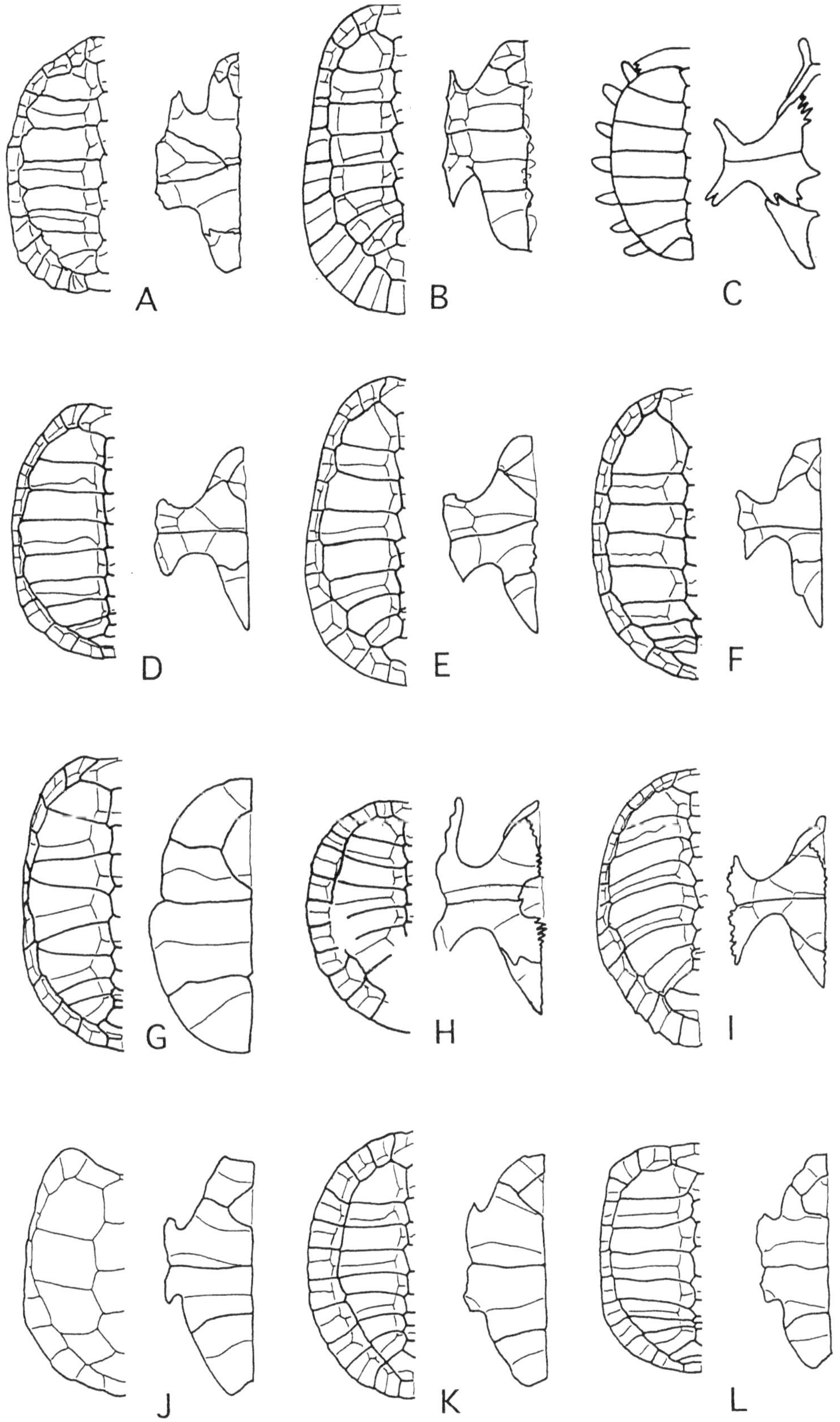

Figure 18.2 Dorsal view of left half of carapace and ventral view right half of plastron of typical North American Paleocene and Eocene families or genera.

A = Baenidae (*Baena* after Hay 1908), B = Adocidae (*Adocus* after Hay 1908). C = Trionychidae (*Aspideretes* after Hay 1908), D = Kinosternoidae (*Hoplochelys* after Gilmore 1920), E = Dermatemydidae (*Baptemys* after Hay 1908), F = Kinosternidae (*Baltemys* after Hutchison 1991), G = Trionychoidae? (*Planetochelys*), H = Macrobaenidae, I = Chelydridae (*Protochelydra* after Erickson 1973, 1984), J = Emydidae (*Emydid* C), K = Emydidae (*Echmatemys* after Hay 1908), L = Testudinidae (*Hadrianus* after Hay 1908). Scales vary.

the area. *Baena* appears to range as far north as Ellesmere Island (personal observation) in Wasatchian deposits.

Adocidae (figure 18.2B)

The Adocidae is a relatively primitive group of the Trionychoidea (Meylan and Gaffney 1989) that enjoyed their greatest diversity and widest distribution in the Cretaceous Period and earlier Paleocene Epoch. These are large herbivorous turtles resembling extant *Dermatemys* in adaptive morphology. In North America they are progressively restricted to the south, with the last records in the Clarkforkian deposits of the Big Bend area.

Trionychidae (figure 18.2C)

This complex group of predaceous and aquatic burrowing turtles apparently originated in Asia in the Early Cretaceous Period and spread to all continents except possibly Antarctica, although they became extinct subsequently in South America and Australia. In North America, the group enjoyed its greatest diversity in the Late Cretaceous Period but was reduced to a single genus by the late Eocene Epoch. In the late Paleocene and Eocene Epochs, the family was represented by at least two subfamilies, Trionychinae and Plastomeninae. The former subfamily contains the form genera *"Aspideretes"* and *"Trionyx." "Trionyx"* as used here includes taxa identified as *Amyda* and *Platypeltis* in older literature. These genera are placed in quotations because they are now interpreted by Meylan (1987) to be restricted to Old World taxa and serve here as shell morphotypes only. The trionychines are characterized by a highly reduced carapace and highly kinetic plastron. The plastominines on the contrary have a fuller carapace and kinesis in adults is limited to the carapacial/plastron contact, the epiplastra, and entoplastron in *Plastomenus*. The Clarkforkian species of *Plastomenus* is distinct from the Wasatchian species.

Kinosternoidae (figure 18.2D)

The Kinosternoidae forms the sister group of the Trionychoidae (Gaffney and Meylan 1988) including the kinosternids and dermatemydids. As here used, it refers to earlier clades or earlier members of these clades of as yet undiagnosed taxa. *Hoplochelys* has been related to the Kinosternidae (Hutchison and Bramble 1981) but for various reasons is not an ancestral taxon. This genus is represented by several nominal taxa that are abundant in New Mexico and Texas in Puercan and Torrejonian deposits, but rare in correlative intervals in the north. The latest record is in Clarkforkian deposits in the Big Bend region of Texas (personal observation). The second group of Kinosternia in the boundary interval is a probable dermatemydid, possibly ancestral to *Baptemys*. It is known only from undescribed material from the Tiffanian and Clarkforkian deposits of the Big Bend region.

Dermatemydidae (figure 18.2E)

Dermatemydids are a small group of moderately large, aquatic, and herbivorous turtles endemic to North and Central America. The dermatemydids apparently originated in this area and remained restricted to it throughout their history. Their origins appear to be southern with Paleocene records in Texas (personal observation) and remaining Cenozoic records in southern North America and Central America. A single species, *Baptemys garmani* (Cope 1872), is represented in the interval under study. The pattern of distribution of *B. garmani* indicates that it dispersed north at least as far as the Big Horn Basin in Wyoming (Hutchison 1980) and southwestern North Dakota (Estes 1988) during the Wasatchian Age and descendent species retreated south again in the early Uintan Age.

Kinosternidea (figure 18.2F)

The Kinosternidea are small but largely carnivorous bottom-walking aquatic turtles. They apparently originated in North or Central America and have been exclusively restricted to the New World throughout their history. The area of continued diversity and probable origin is southern North America and Central America. In the study area, they appeared abruptly during the early Wasatchian Age and were represented by single species in two kinosternine genera, *Baltemys staurogastros* Hutchison 1991 and *Xenochelys lostcabinensis* Hutchison 1991. During the Wasatchian Age, kinosternines extended their range as far north as Ellesmere Island, Canada (Hutchison 1991).

Family incertae sedis (figure 18.2G)

Planetochelys is a highly specialized box turtle that resembles certain emydids in several adaptive respects, but may be a highly specialized trionychoid. The origin of this genus is unknown. The plastron is kinetically bilobed, resembling extant *Terrapene*, and the architecture of the shell and abundance of rodent gnawing marks on the isolated elements suggest that this was a semiterrestrial or terrestrial turtle. *Planetochelys* was originally described from upper Paleocene deposits of Virginia (Weems 1988).

Macrobaenidae (figure 18.2H)

This largely Cretaceous family of chelydrimorph bottom-walking turtles is represented by a single undescribed taxon in the interval under study. The origins of the family are Asiatic, but the family reached North America at least by the Campanian Age, and the same or closely related taxa are represented in Tiffanian deposits of the study area.

Chelydridae (figure 18.2I)

Chelydrids are carnivorous aquatic bottom-walking turtles. This group is generally persistent but poorly represented in the fossil record, probably reflecting the rather solitary behavior known in their modern counterparts. The Chelydridae apparently originated in North America (earliest known in the Turonian Age, personal observation), spreading to Eurasia during the Oligocene Epoch and possibly as early as the Paleocene Epoch (see below). Probable members of the extant chelydrine group began to appear during the Paleocene and Eocene Epochs.

Emydidae s.l. (figure 18.2J–K)

The Emydidae as used here includes the Bataguridae and Emydidae of some authors. This is a diverse group of mostly aquatic turtles which originated in Asia during the Late Cretaceous Period or early Cenozoic Era. Emydids made their first appearance in the form of a single genus, *Emydid* C (figure 18.2J), which occurs in the Torrejonian deposits of North America. This genus of aquatic turtle persisted into the Wasatchian Age. The earliest Wasatchian Age marked the appearance of *Echmatemys* in abundance (figure 18.2K), which persisted as one of the most widely distributed and abundant freshwater turtles in the Eocene Epoch and became extinct in the late Eocene Epoch. *Echmatemys* exhibits no close relationship to *Emydid* C but rather to such Eurasian genera as *Palaeochelys*.

A third and probably a fourth genus appeared later during the Wasatchian Age, neither apparently related to the previous two genera and thus likely being Eurasian immigrants. Still another genus, possibly an early emydine, appeared on Ellesmere Island during the Wasatchian Age (Estes and Hutchison 1980).

Testudinidae (figure 18.2L)

The Testudinidae made a sudden and abundant appearance during the Wasatchian Age in the form of a single species, *Hadrianus majusculus* Hay 1904. Their origins are not certainly known but are probably Asiatic. These large and herbivorous turtles are the most terrestrial of the early Cenozoic turtles.

Discussion

The Paleocene/Eocene Epoch boundary interval marks one of the major transition periods of turtle diversity in North America, presumably because of the rising tropicality and complex tectonic history of the study area. Contributions to this diversity are various and complex, but three or four major factors are apparent: immigrations from the north, immigrations from the south, diversification of endemics, and decline of archaic groups.

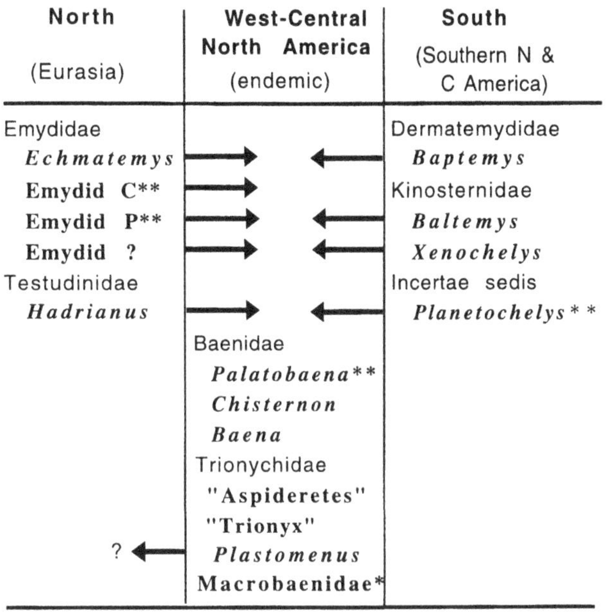

Figure 18.3 Hypotheses of immigration vectors in west-central North America during the Paleocene/Eocene Epoch boundary interval.

* Extinct by end of Clarkforkian Age. ** Extinct by end of Wasatchian Age.

With the warming of the polar regions during the Eocene Epoch (Estes and Hutchison 1980) and land connections to Eurasia (Smith et al. 1994), climatically restricted taxa from southern areas of Eurasia were able to move north and around or over the pole into North America, thence south into North America (figure 18.3). The major groups that appear to have taken this route are the Emydidae and Testudinidae, and both flourished in the Eocene Epoch to become dominant groups in North America during the Neogene Period. *Echmatemys*, *Emydid* C, and a probable emydine are also known from the polar Ellesmere fauna (Estes and Hutchison 1980, and unpublished). A bidirectional dispersal of trionychids may also have occurred, but the systematics of the taxa are not well enough understood to sort this out, with the probable exception of *Plastomenus*, which probably emigrated from North America to Asia during the Wasatchian or Bridgerian Age. One early Bridgerian carettochelyid genus, *Anosteira*, also appears to be an Asian immigrant.

The same warming trends appear to have allowed southern North American turtles to extend their ranges northward. The kinosternids and dermatemydids, which appear to have originated in Central America or southern North America, reached Wyoming during the Wasatchian Age, and at least one kinosternine reached Ellesmere Island. *Planetochelys* is known from upper Paleocene deposits of Virginia (Weems 1988) and Clarkforkian deposits of Colorado and Texas. The relationships of the species are unclear. Whereas the introduction of this taxon into Wyoming may represent an eastern immigrant, a southern origin appears more likely and proximate. It appears that *Planetochelys* had become extinct by the end of the Wasatchian Age.

Some groups are endemic to the study area or are probably derived from endemic ancestors. These include the baenids, the macrobaenids, and probably some if not all of the trionychids.

Despite the general increase in diversification, especially in the Wyoming region, several archaic groups became extinct during the boundary interval. Macrobaenids did not survive beyond the end of the Clarkforkian Age, although the family survived into the Eocene Epoch in Asia (Mlynarksi 1976). *Hoplochelys* and *Adocus* survived latest in the southernmost part of the study area, but were extinct by the end of the Clarkforkian Age. *Emydid* C is widely distributed in the Tiffanian and Clarkforkian deposits of the western North America, ranging from Montana to Texas. Its geographic range in Wasatchian time, however, was restricted to the Big Horn Basin and extended as far north as Ellesmere Island (personal observation). It apparently became extinct by the end of the Lysitian Age on the midcontinent. Although more widely distributed in the Wasatchian deposits of Wyoming, *Emydid* P also appears to have vanished from the record by the end of the Lysitean Age. The general pattern, thus, is a general modernization of the turtle fauna toward the Bridgerian Age.

Comparison with the European Record

The nonmarine (and marine) turtles underwent a similar increase in diversity and modernization of the fauna (figure 18.4) as in North America through the boundary interval (Cernaysian–Grauvian Ages; Broin 1977). Endemics include the Trionychidae and Kallokibotionidae. Some modernization of the fauna includes the extinction of the latter family and *Compsemys*.

The otherwise endemic North American genus *Compsemys* is reported (Broin 1977:16) from the Cernaysian (Thanetian age) and Dormaalian (Ypresian age, in part) deposits, although these records have been questioned by Gaffney and Meylan (1992). If they are valid, *Compsemys* represents a North American immigrant where it is otherwise exclusively known. The

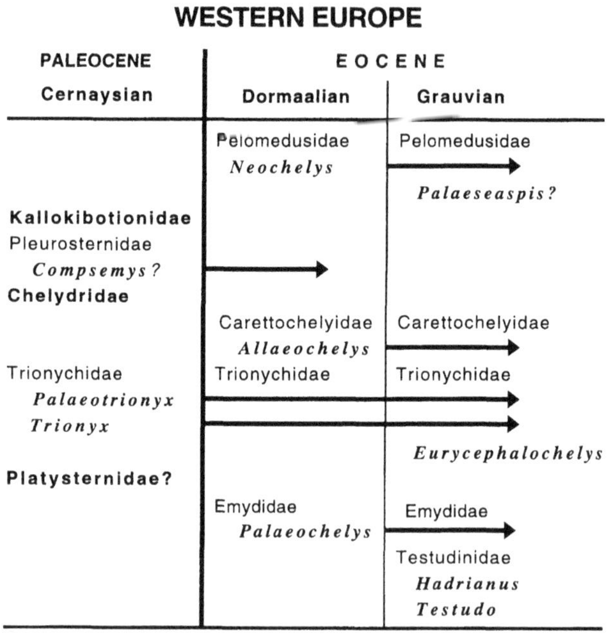

Figure 18.4 Distribution of turtle families and genera in the Paleocene/Eocene Epoch boundary interval in western Europe.

Chelydridae reported by Groessens-van Dyck and Poty (1982) is probably also a North American immigrant. Chelydridae probably became extinct in western Europe after the Cernaysian Age, judging from the rich chelonian Eocene record. Asian or eastern European immigrants (figure 18.5) probably include the emydids and carettochelyids during the Dormaalian Age followed by the testudinids during the Grauvian Age. The relative sequence of appearance and perhaps timing of the emydids and testudinids in Europe is identical to that in North America. The carettochelyids only appeared later

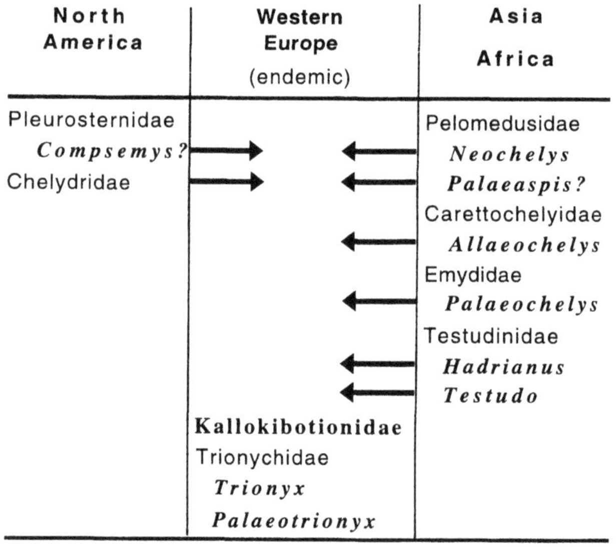

Figure 18.5 Hypotheses of immigration vectors in western Europe during the Paleocene/Eocene Epoch boundary interval.

in North America (early Bridgerian Age; Hutchison 1992) and the genera on the two continents belong to different subfamilies. The genera in these last three families have their earliest records or closest relatives in Asia.

Two possibly African pelomedusids also appeared (figure 18.5), although the family was endemic earlier in western Europe in the form of the marine genus *Taphrosphys* and was also widespread in southern Asia.

Comparison with the Asian Record

The Asian record (figure 18.6) for turtles during the Paleocene/Eocene Epoch boundary interval is poor and poorly constrained biostratigraphically in comparison with North America and Europe. Nevertheless, the most diverse upper Paleocene record of emydids anywhere occurs in the Nongshanian deposits of Asia, and the earliest records of the Testudinidae are reported there (Nessov 1987). The bulk of this record lies in China and Mongolia and much remains to be learned of Central Asian faunas. More importantly, precursor taxa of carettochelyids, emydids, testudinids, and platysternids occur only or earliest in Asia. Trionychids also have their earliest occurrence there, and Asia is currently the center of their greatest diversity. Until the systematics of Paleogene trionychids is more thoroughly analyzed it will not be possible to assess the contribution of Asian trionychids to the boundary issue. With the possible exception of the latter, the expansion of these families into western Europe and North America began in the middle Paleocene Epoch and increased rapidly during the early Eocene Epoch, indicating progressively permissive climatic regimes and geographic corridors.

North America exhibited a marked increase in turtle diversity during the Paleocene/Eocene Epoch boundary interval, apparently owing to warmer polar climates and opportunities for turtles to migrate from the south and for Eurasian turtles to cross over in the north and move south into North America. Western Europe and North America show similar patterns of increase in diversity and in the appearance of emydids and testudinids, probably both immigrants from Asia. Despite the opportunity for direct interchange between western Europe and North America, only one or two endemic taxa (*Compsemys*, Chelydridae) apparently migrated from North America to Europe. The stepped appearance of emydids and testudinids on both these continents may indicate either dispersion from Asia or a track through Europe to North America or vice versa. Several clades with Cretaceous records became extinct in this interval, imparting a more modern appearance to the fauna dominated by emydids, testudinids, kinosternids, and trionychids, as in the modern fauna in southern North America and Central America.

ASIA

PALEOCENE Nongshanian	EOCENE Bumbanian
	Pelomedusidae
Adocidae	
Carettochelyidae	Carettochelyidae
Anosteira	→
	Hemichelys
Trionychidae	Trionychidae
	Aspideretes
	Trionyx
	Amyda
	Macrobaenidae
	Macrobaena
Platysternidae	
Planiplastron	
Lindhomemyidae	
Mongolemys	
Emydidae	Emydidae
Anhuichelys	
Hokouchelys	
Pseudochrysemys	
	"Emys"
	Geoclemys
Testudinidae	

Figure 18.6 Distribution of turtle families and genera in the Paleocene/Eocene Epoch boundary interval in Asia (primarily eastern Asia).

ACKNOWLEDGMENTS

I thank Robert Sullivan, Spencer Lucas, and Marie-Pierre Aubry for constructive reviews of the manuscript. This is contribution No. 1656 of the University of California Museum of Paleontology.

REFERENCES

Archibald, J. D. and Hutchison, J. H., 1979. Revision of the genus *Palatobaena* (Testudines, Baenidae), with the description of a new species. *Postilla,* n. 177, 19 pp.

Broin, F., de., 1977. Contribution à l'étude des chéloniens. Chéloniens continentaux du Crétacé supérieur et du Tertiaire de France. *Mémoires du Muséum national d'Histoire Naturelle,* sér. C, v. 38, 366 pp.

Clyde, W. C., Stamatakos, J. and Gingerich, P. D., 1994. Chronology of the Wasatchian Land-Mammal Age (early Eocene): Magnetostratigraphic results from the McCullough Peaks Section, northern Bighorn Basin, Wyoming. *Journal of Geology,* v. 102, p. 367–377.

Cope, E. D., 1872. On a new genus of *Pleurodira* from the Eocene of Wyoming. *Proceedings of the American Philosophical Society,* v. 12, p. 472–477.

Erickson, B. R., 1973. A new chelydrid turtle *Protochelydra zangerli* from the late Paleocene of North Dakota. *Science Publications, Science Museum of Minnesota,* n. s., v. 2, 16 pp.

Erickson, B. R.,1984. Chelonivorous habits of the Paleocene crocodile *Leidyosuchus formidabilis. Science Publications, Science Museum of Minnesota,* n. s., v. 7, 9 pp.

Estes, R., 1988. Lower vertebrates from the Golden Valley Formation, early Eocene of North Dakota (U.S.A.). *Acta Zoologica Cracoviensia,* v. 31, p. 541–562.

Estes, R. and Hutchison, J. H., 1980. Eocene lower vertebrates from Ellesmere Island, Canadian Arctic Archipelago. *Palaeogeography, Palaeoclimatology, Palaeoecology,* v. 30, p. 325–347.

Gaffney, E. S. and Meylan, P. A., 1988. A phylogeny of turtles. *Systematics Association* spec. v. 35A, p. 157–519.

Gaffney, E. S. and Meylan, P. A., 1992. The Transylvanian turtle, *Kallokibotion,* a primitive cryptodire of Cretaceous age. *American Museum Novitates,* no. 3040, p. 1–37.

Gilmore, C. W., 1920 (1919). *Reptilian faunas of the Torrejon, Puerco, and underlying Upper Cretaceous formation of San Juan County, New Mexico.* U.S. Geological Survey Professional Paper, v. 119, 71 pp.

Groessens-van Dyck, M.-C. and Poty, E., 1982. Le Moliniacien supérieur de Vinalamont. *Bulletin de la Société Liégeoise de Géologie,* v. 91, p.127–134.

Harland, W. B., Armstrong, R. L., Cox, A. V., Craig, L. E., Smith, A. G. and Smith, D. G., 1990. *A Geologic Time Scale 1989.* Cambridge: Cambridge University Press, 263 pp.

Hay, O. P., 1904. On some fossil turtles belonging to the Marsh collection in Yale University Museum. *American Journal of Science,* ser. 4, v. 18, p. 261–276.

Hay, O. P., 1908. Fossil turtles of North America. *Carnegie Institution of Washington,* no. 75, 568 pp.

Hutchison, J. H., 1980. Turtle stratigraphy of the Willwood Formation, Wyoming: Preliminary results. *University of Michigan, Museum of Paleontology, Papers on Paleontology,* v. 24, p. 115–118.

Hutchison, J. H., 1982. Turtle, crocodilian, and champsosaur diversity changes in the Cenozoic of the north-central region of western United States. *Palaeogeography, Palaeoclimatology, Palaeoecology,* v. 37, p. 147–164.

Hutchison, J. H., 1991. Early Kinosterninae (Reptilia: Testudines) and their phylogenetic significance. *Journal of Vertebrate Paleontology,* v. 11, p. 145–167.

Hutchison, J. H., 1992. Western North American reptile and amphibian record across the Eocene/Oligocene boundary and its climatic implications. In Prothero, D. R. and Berggren, W. A. (eds.), *Eocene–Oligocene Climatic and Biotic Evolution,* Princeton University Press, Princeton, p. 451–463.

Hutchison, J. H. and Archibald, J. D., 1986. Diversity of turtles across the Cretaceous/Tertiary boundary in northeastern Montana. *Palaeogeography, Palaeoclimatology, Palaeoecology,* v. 55, p. 1–22.

Hutchison, J. H. and Bramble, D. M., 1981. Homology of the plastral scales of the Kinosternidae and related turtles. *Herpetologica,* v. 37, p. 73–85.

Leidy, J., 1870. (Descriptions of *Emys jaenesi, E. haydeni, Baena arenosa, and Saniwa ensidens*). *Proceedings of the American Philosophical Society* 1870, p.123–124.

Leidy, J., 1871. Remarks on fossil vertebrates from Wyoming. *Proceedings of the American Philosophical Society* 1871, p. 228–229.

Lillegraven, J. A. and Ostesh, L. M., Jr., 1988. Evolution of Wyoming's early Cenozoic topography and drainage patterns. *National Geographic Research,* v. 4, p. 303–327.

Meylan, P. A., 1987. The phylogenetic relationships of soft-shelled turtles (Family Trionychidae). *Bulletin of the American Museum of Natural History,* v. 186, 101 pp.

Meylan, P. A. and Gaffney, E. S., 1989. The skeletal morphology of the Cretaceous cryptodiran turtle, *Adocus,* and the relationships of the Trionychoidea. *American Museum Novitates,* no. 2941, 60 pp.

Mlynarski, M., 1976. Testudines. In Kuhn, O. (ed.), *Encyclopedia of Paleoherpetology,* v. 7, p. i–vi + 1–130.

Nessov, L. A., 1987 (1988). On some Mesozoic turtles of the Soviet Union, Mongolia and China, with comments on systematics. *Studia Geologica Salamaticensia, Studia Palaeocheloniologica,* v. 2, p. 87–102.

Smith, A. G., Smith, D. G. and Funnell, B. M., 1994. *Atlas of Mesozoic and Cenozoic Coastlines.* Cambridge: Cambridge University Press, 99 pp.

Weems, R. E., 1988. Paleocene turtles from the Aquia and Brightseat Formations, with a discussion of their bearing on sea turtle evolution and phylogeny. *Proceedings of the Biological Society of Washington,* v. 10, p. 109–145.

Wing, S. L., Bown, T. M. and Obradovich, J. D., 1991. Early Eocene biotic and climatic change in interior western North America. *Geology,* v. 19, p. 1189–1192.

Wing, S. L. and Greenwood, D. R., 1993. Fossils and fossil climate: The case for equable continental interiors in the Eocene. *Philosophical Transactions of the Royal Society of London,* ser. B, v. 341, p. 243–252.

Woodburne, M. O. (ed.), 1987. *Cenozoic Mammals of North America—Geochronology and Biostratigraphy.* Berkeley, Los Angeles, London: University of California, 336 pp.

19

Mammalian Faunal Composition and the Paleocene/Eocene Epoch/Series Boundary: Evidence from the Northern Bighorn Basin, Wyoming

GREGG F. GUNNELL

ABSTRACT

The northern Bighorn Basin in northwestern Wyoming contains one of the most complete sequences of lower Cenozoic continental sediments known. Large samples of mammalian fossil specimens are available from stratigraphically controlled localities throughout this sequence. A refined local biostratigraphy has allowed assignment of most mammalian localities to specific biostratigraphic zones. Within this biostratigraphic framework, paleobiological range data including standing richness, first appearances, last appearances, and taxonomic compositional changes of mammalian faunas can be studied. Evidence compiled from these data sets indicates that, while there were several episodes of mammalian faunal change, by far the most fundamental change occurred at the boundary between the Clarkforkian and Wasatchian Land Mammal Ages (LMA). Mammalian faunal evidence is only one of a number of lines of biotic evidence useful for the ultimate recognition and placement of the Paleocene/Eocene Epoch/Series boundary. While, in principle, a stratotype section and point for the Paleocene/Eocene Series boundary should be defined lithostratigraphically based on the marine record, in practice such a definition should not ignore terrestrial biotic evidence. The mammalian faunal transition at the Clarkforkian/Wasatchian LMA boundary appears to be nearly temporally coincident with a number of other geologic and biotic changes and suggests that this transition may represent part of a global reorganization of ecosystems. If true, this reorganization could be utilized as partial evidence to delimit the Paleocene/Eocene Series boundary and would, if used in such a fashion within a stratotype framework, make the Paleocene/Eocene Epoch boundary more than a simple point in time.

The northern Bighorn Basin in northwestern Wyoming contains one of the most complete sequences known of lower Cenozoic continental sediments (Gunnell et al. 1993). As such, this area has been the scene of concentrated paleontological and geological fieldwork for over 110 years (e.g., Matthew and Granger 1915, 1918; Van Houten 1944; Gingerich 1980). For the past 20 years, the University of Michigan has been conducting a paleontological fieldwork program concentrating on the middle Tiffanian Land Mammal Age (LMA) through the middle Wasatchian LMA (Gingerich 1976, 1982, 1983a, 1989a, 1989b; Hickey 1980; Kraus 1980; Krause 1980, 1982; Butler et al. 1981, 1987; Rose 1981; Rose and Krause 1982; Winkler 1983; Gunnell 1989; Koch et al. 1992), a time interval almost certainly encompassing the Paleocene/Eocene Epoch boundary (Gunnell et al. 1993). During the course of fieldwork, nearly 40,000 fossil mammal specimens from over 600 discrete localities have been collected.

This chapter summarizes information documenting changes in mammalian faunal composition in the northern Bighorn Basin. Biostratigraphic range data and taxonomic composition of mammalian faunas from the middle Tiffanian through middle Wasatchian sequence are examined. Mammalian faunal evidence supports previous suggestions that, while substantial changes in mammalian taxonomic composition occur throughout the sequence, the single most important documented event for mammalian evolution and faunal turnover occurs at the Clarkforkian/Wasatchian LMA boundary. This LMA boundary occurs near the middle of Magnetochron C24r (Clyde et al. 1994), is coincident with an apparently globally occurring $\delta^{13}C$ isotope spike and possibly a $\delta^{18}O$ isotope spike at approximately 55 Ma

(Koch et al. 1992, 1995), is near or slightly below the nannoplankton NP9/NP10 zonal boundary (Wing et al. 1991), and is slightly older (Wing 1984) or slightly younger (Wing et al. 1991) than the first appearance of *Platycarya* pollen in the continental interior of North America. All of these pieces of evidence suggest that the Clarkforkian–Wasatchian mammalian faunal transi tion may well be part of a pattern of a global change in ecosystems that could signal the end of the Paleocene Epoch or the beginning of the Eocene Epoch.

Northern Bighorn Basin Mammalian Faunas

Fossil mammals were first collected in the Bighorn Basin of northwestern Wyoming in 1880 by Jacob Wortman (Gingerich 1980). The northern Bighorn Basin includes the area north of the Greybull River and is enclosed by the Beartooth and Absaroka ranges to the west and northwest and the Bighorn Range to the east (see Clyde et al. 1994 for a map of the area). Included within this area are the Clark's Fork Basin (with University of Michigan [UM] localities in the Sand Coulee and Polecat Bench stratigraphic sections), the UM Cub Creek section in south-central Montana, the UM Foster Gulch section, and the UM McCullough Peaks section.

The Cenozoic basin-fill sequence present in the northern Bighorn Basin is represented by fluvial mudstones and sandstones of the drab Fort Union Formation overlain by variegated, fluvial mudstones and sandstones of the Willwood Formation (Jepsen 1930; Van Houten 1944; Neasham and Vondra 1972). The Fort Union Formation reaches a maximum thickness of ~3000 m, while the preserved thickness of the Willwood Formation reaches ~800 m. Mammalian faunal units (Wood et al. 1941; Woodburne 1987) represented range from the Puercan through most of the Wasatchian LMA (figure 19.1).

The Puercan and Torrejonian parts of the sequence are represented by single productive localities (Mantua Quarry in the Puercan LMA and Rock Bench Quarry in the Torrejonian LMA) and are compressed into the first 80 m of the local section (Gingerich et al. 1980). The fauna from Mantua Quarry contains 11 mammalian taxa (Jepsen 1930, 1940; Van Valen 1978) representative of Puercan biostratigraphic Zone Pu1 (see Archibald et al. 1987 for delineation of biostratigraphic zones). Rock Bench Quarry contains 58 mammalian species (Rose 1981) assigned to Torrejonian biostratigraphic Zone To3.

The remaining ~1900 m of local section encompasses the Tiffanian, Clarkforkian, and Wasatchian LMAs. Early Tiffanian mammalian samples are scarce in the University of Michigan collections from the northern Bighorn Basin. Three UM localities are known from Tiffanian Zone Ti1 (Cub Creek localities CC-1, 3, 4) and three from Zone Ti2 (Sand Coulee localities SC-263, 264, 332). Mammals are not well represented from any of these localities but the Cub Creek sample contains a Zone Ti1 index fossil, *Plesiadapis praecursor*, while *Plesiadapis anceps*, an index fossil for Zone Ti2, is known from locality SC-263 (Gingerich et al. 1980).

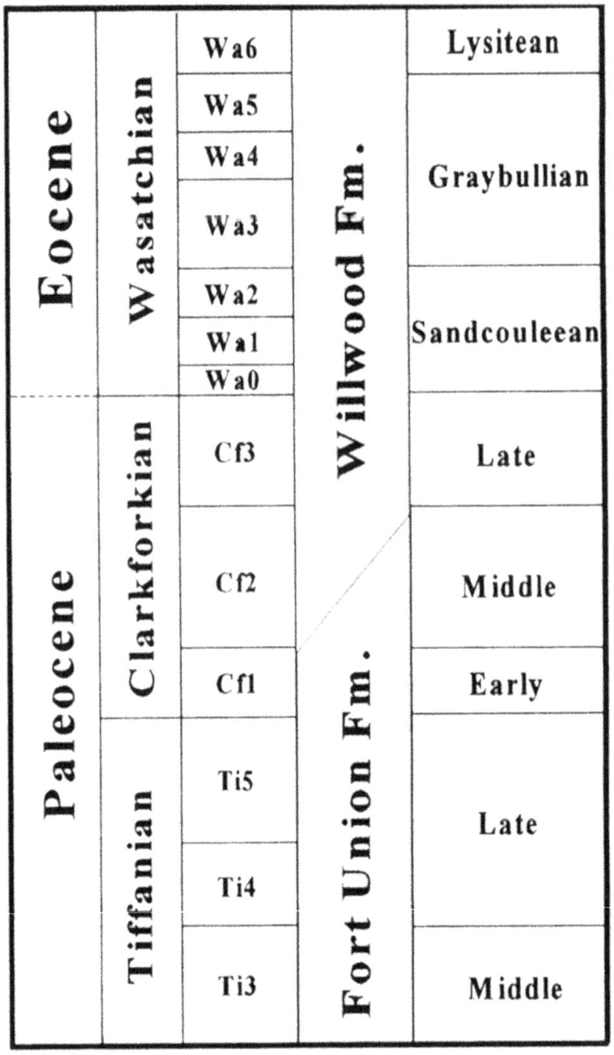

Figure 19.1 Lithostratigraphy and biostratigraphy of the middle Tiffanian through Wasatchian LMA in the northern Bighorn Basin.

Figure summarizes, from left to right, epochs, land mammal ages, biostratigraphic zones, formations, and mammalian subages. Dashed line separating the epochs represents the Paleocene/Eocene Series boundary if recognition is based on biostratigraphic ranges of mammalian taxa.

The middle Tiffanian through middle Wasatchian LMA interval is much better represented in UM collections. Table 19.1 summarizes the numbers of localities and mammalian specimens identifiable to species known from each biostratigraphic zone. The poorly understood and poorly represented Tiffanian Zone Ti6

Table 19.1
Distribution of localities and mammalian specimens in UM collections from the northern Bighorn Basin: biostratigraphic Zones Ti3 through Wa6

Biostratigraphic Zone	Number of Localities	Number of Specimens
Ti3	10	1494
Ti4	15	382
Ti5	52	851
Cf1	42	403
Cf2	82	1477
Cf3	66	499
Wa0	20	289
Wa1	73	1962
Wa2	69	3901
Wa3	31	4017
Wa4	31	1410
Wa5	22	1350
Wa6	49	1321
Totals	567	19,356

(Archibald et al. 1987) is not included in the following analyses. All localities near the Tiffanian/Clarkforkian LMA boundary but below the stratigraphically lowest occurrence of rodents are placed in Zone Ti5, while those stratigraphically above the first appearance of rodents are included in Zone Cf1. Following this interpretation, *Plesiadapis gingerichi* is known to occur both in the latest Tiffanian and the earliest Clarkforkian LMAs.

Northern Bighorn Basin mammalian taxa from the Puercan, Torrejonian, and much of the Tiffanian LMAs are best known from single, productive quarry sites. Such highly productive single localities as Rock Bench, Cedar Point, and Princeton quarries become much less common in Clarkforkian and Wasatchian strata due mainly to changes in depositional environments (Badgley et al. 1995). Most mammalian samples from the Clarkforkian and Wasatchian LMAs are from surface localities that are subject to collecting biases, in addition to the taphonomic sorting biases (the same or different) that affect quarry samples (Winkler 1983; Badgley and Gingerich 1988; Gunnell 1994). Nonetheless, because of the large, well-studied samples available, the Clarkforkian and earliest Wasatchian intervals in the northern Bighorn Basin remain the best sampled record of mammalian succession known spanning this interval.

The analyses that follow are based on University of Michigan collections from Tiffanian biostratigraphic Zone Ti3 through Wasatchian Zone Wa6. This sample includes 567 discrete localities and 19,356 mammalian specimens identifiable to species. Most mammalian groups have been examined at the specific level, providing detailed records unmatched anywhere else. Thus it is not possible to include evidence from other comparable sequences that approach the level of taxonomic resolution available from northwestern Wyoming.

Each specimen included in this analysis is from a discrete, stratigraphically restricted locality. Each locality has been assigned to a specific biostratigraphic zone based on the mammalian faunas contained within that locality. All localities from a single zone have been combined, resulting in an analytically time-averaged mammalian faunal sample representing each zone. Standing richness, first appearances, and last appearances of mammalian orders, families, genera, and species were tabulated based on these time-averaged samples. *Standing richness*, as used in this paper, simply means the number of mammalian taxa present in each zone. A taxon was assumed to have been present in "ghost zones" if it occurred below and above an intervening interval from which it was not represented. First appearances were recorded at the initial appearance of a taxon in the stratigraphic record. Last appearances were recorded at the last interval in which a taxon was represented.

Biostratigraphic zonations were used to subdivide the stratigraphic distribution of localities instead of meter level for several reasons. First, only the Sand Coulee and McCullough Peaks stratigraphic sequences have been measured, so biostratigraphic zones were used to include the Foster Gulch sequence. Second, not all localities in the Sand Coulee and McCullough Peaks sections are included in the measured sections. By using biostratigraphic zones, these localities can be interpolated into the sequence. Use of broader zones instead of more finely controlled metric levels increases sample sizes from each interval and lessens the effects of small sample sizes from some levels, but may mask more detailed faunal changes that appear under finer examination. Comparisons of the results summarized here with those from more detailed sampling intervals (Badgley and Gingerich 1988) or larger samples based on the northern Bighorn Basin and the Crazy Mountain Basin (Maas et al. 1995) reveal minor differences. Differences that do exist are discussed in further detail below.

Standing Richness, Faunal Turnover, and Sampling Biases

Figure 19.2 summarizes the number of mammalian orders, families, genera, and species from the middle Tiffanian through the late Wasatchian LMA. In terms of the number of orders and families present through the sequence (figure 19.2A), there is relative stability with a gradual increase of standing richness into the Wasatchian LMA. Standing richness of genera is more irregular, with peaks in the late Tiffanian (Zone Ti5), the middle Clarkforkian (Zone Cf2), and the early middle

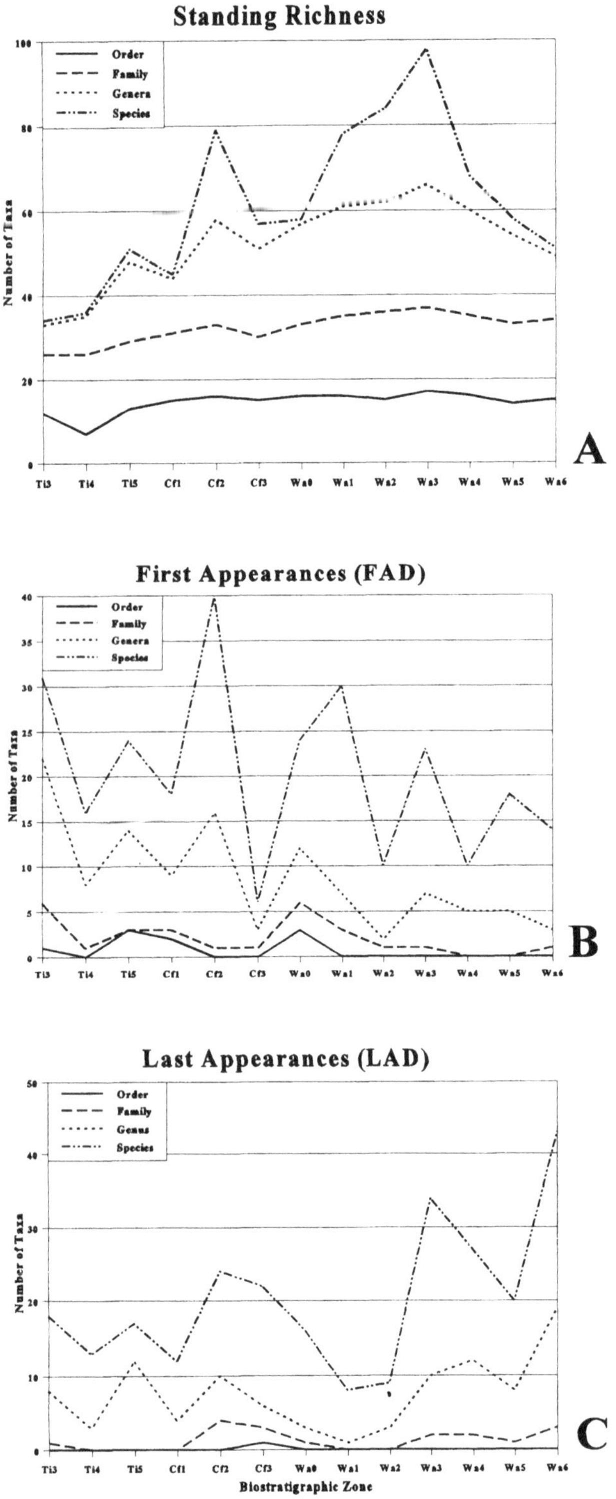

Figure 19.2 Middle Tiffanian through Wasatchian standing richness and faunal turnover based on 19,356 mammalian specimens identifiable to species.

Vertical axis is number of taxa, horizontal axis is time expressed in biostratigraphic zones. (A) standing richness of mammalian orders, families, genera, and species; (B) first appearances (FAD) of mammalian orders, families, genera, and species; (C) last appearances (LAD) of mammalian orders, families, genera, and species.

Wasatchian (Zone Wa3) LMAs. The species pattern mirrors that of the genera but with an increase in species numbers in the middle Clarkforkian LMA and at the base of the Wasatchian LMA, reaching a zenith in Zone Wa3. This pattern of species richness is partly due to taphonomic processes and sampling as the middle Clarkforkian and the Wasatchian LMAs are the best represented and best studied intervals of the sequence.

First appearance (FAD) data by taxonomic rank are presented in figure 19.2B. It should be noted that the oldest (Zone Ti3) and youngest (Zone Wa6) biostratigraphic zones in the local section have been omitted in the following discussions of FADs and last appearances (LADs) as each one is somewhat biased by the lack of well-sampled intervals below Zone Ti3 or above Zone Wa6 (also see Maas et al. 1995). There are two peaks of FADs of new orders, one beginning in Zone Ti5 and lasting through Zone Cf1 and the other at the beginning of Zone Wa0. This pattern is mirrored by FADs among families. FADs of genera and species tend to mirror one another as well, but are obviously more common than FADs among orders or families. The main peaks of FADs among genera occur in Zones Ti5, Cf2, and Wa0-Wa1.

Maas et al. (1995) found that FADs of genera were significantly different from expected background rates of first occurrence in Zones Ti5, Wa0, and Wa1. Based on records from the northern Bighorn Basin, 16 genera have their lowest occurrence in Clarkforkian Zone Cf2 (*Uintacyon*, *Copecion*, *Palaeonictis*, *Plagiomene*, *Worlandia*, *Ceutholestes*, *Leipsanolestes*, *Plagioctenodon*, *Pontifactor*, *Thelysia*, *Palaeanodon*, *Arctodontomys*, *Niptomomys*, *Tinimomys*, *Acritoparamys*, and *Microparamys*). Of these genera, only *Uintacyon*, *Arctodontomys*, *Niptomomys*, and *Palaeanodon* were recognized as first appearing in Zone Cf2 by Maas et al. (1995). Additionally, *Planetetherium* was recognized as first appearing in Zone Cf2 by Maas et al. (1995) when, in fact, it seems to have first appeared in Zone Cf1. Some of these differences are the result of larger samples utilized by Maas et al. (1995), but only *Acritoparamys* and *Leipsanolestes* were reported as occurring *earlier* than in the northern Bighorn Basin. Removing those two genera still leaves 14 new genera first appearing in Zone Cf2 and suggests that this zone is a more important interval of faunal turnover than has been previously thought.

Species FAD peaks occur in the same zones as generic peaks but some differences exist between the records of genera and species. There is a species FAD peak in Zone Wa0, but unlike the generic pattern, species first appearances increase through Zone Wa1. There are also distinct species peaks in first appearances in Zones Wa3 and Wa5 that do not have as prominent counterparts in any other taxonomic levels.

The comparison of FADs of genera and species with those of orders and families shows some noticeable differences as well. Genera and species FADs show a major peak in Zone Cf2, while order and family first appearances are decreasing from Zones Cf1 to Cf2. The only intervals where FADs of all four taxonomic categories show peaks are in Zones Ti5 and Wa0.

Last appearances are shown in figure 19.2C by taxonomic rank. Among orders, there is a single LAD recorded in the uppermost Clarkforkian (Zone Cf3) LMA. Several families last appear in the middle and upper Clarkforkian LMA, but only a single family disappears in Zone Wa0. There is another increase in family LADs occurring from Zone Wa3 through Zone Wa5. An interesting pattern is noticeable among LADs for genera and species. In the Tiffanian and Clarkforkian LMAs, LADs mirror FADs among genera and species. As was the case for FADs, there are LAD peaks for genera and species in Zones Ti5 and Cf2. However, the pattern of FADs and LADs differs for genera and species in the early Wasatchian LMA. Zone Wa0 through Wa1 represent an interval of increase in FADs and a decrease in LADs. LADs again mirror FADs among genera and species from Zone Wa3 through Wa5. This pattern is not substantially different from that documented by Maas et al. (1995).

The overall pattern of mammalian faunal turnover indicates that the late Tiffanian, middle Clarkforkian, and early Wasatchian LMAs were times of increased faunal change. Zones Ti5 and Cf2 reflect times of high faunal turnover in terms of both first and last appearances. Biochrons Wa0 and Wa1 were times of high first appearances and low last appearances. FADs of new mammalian orders, families, genera, and species at the Clarkforkian/Wasatchian LMA boundary reflect a different pattern of mammalian turnover as these first occurrences are not paired with peaks of last occurrences as they are in Biochrons Ti5 and Cf2.

Sampling biases have substantial effects on the patterns of appearance and disappearance documented in the fossil record (Koch 1987; Badgley and Gingerich 1988). In two different analyses based on 20-m intervals (or finer subdivisions) of the Wasatchian sequence in the northern Bighorn Basin (Clark's Fork Basin), Badgley and Gingerich (1988:figures 8 and 9) found a high correlation between sample size and the number of species present at a given interval. Sampling effects are somewhat alleviated in this study by combining collections from biostratigraphic zones (Zones Ti3 through Wa6) into single representative samples from each zone. Figure 19.3A presents a comparison of the number of species (standing richness) versus the number of specimens from each zone. There is a positive correlation between specimen number and species number, but

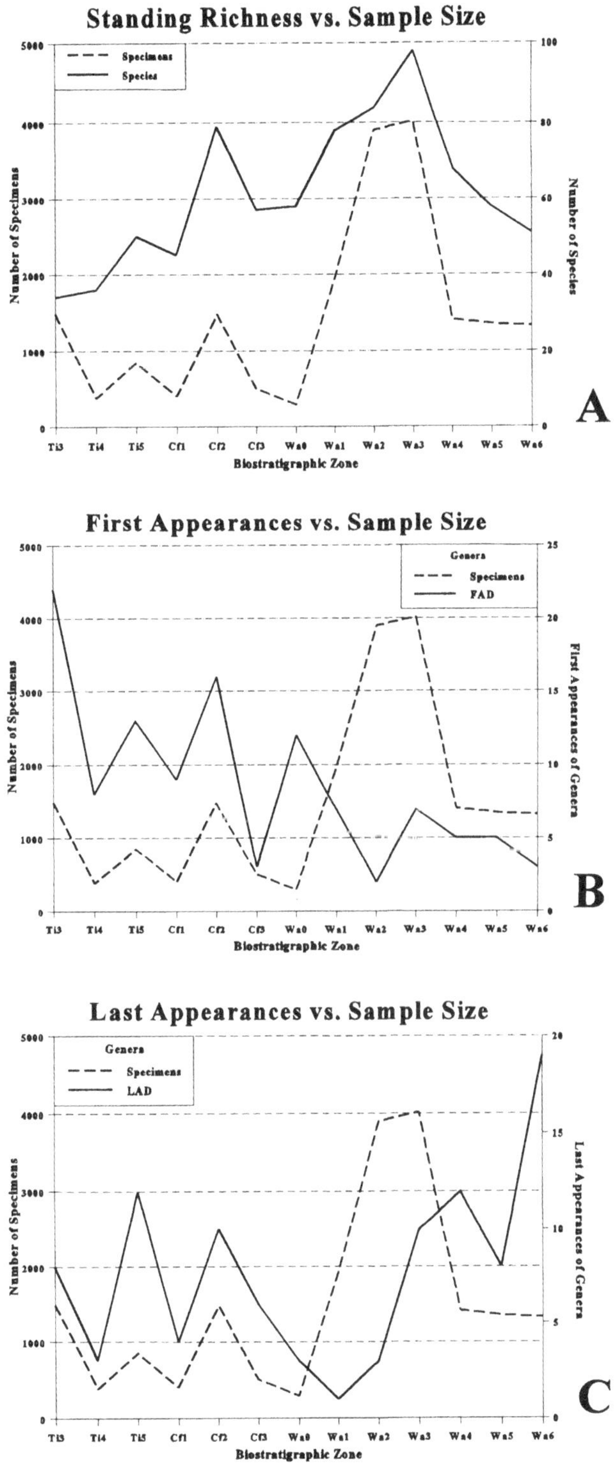

Figure 19.3 Middle Tiffanian through Wasatchian standing richness and faunal turnover compared with sample size based on 19,356 mammalian specimens identifiable to species.

Left vertical axis is sample size, right vertical axis is species number, FADs, and LADs, horizontal axis represents time expressed in biostratigraphic zones. (A) standing richness of mammalian species versus sample size; (B) first appearances of mammalian genera versus sample size; (C) last appearances of mammalian genera versus sample size.

it is weaker than for those samples examined at finer stratigraphic intervals (see also Badgley et al. 1995).

Some discrepancies exist in the correlation of specimen number and standing richness of species. For example, between Zones Ti3 and Ti4 specimen numbers decrease substantially while the number of species increases. The same is true between Zones Cf3 and Wa0. Between Zones Wa2 and Wa3, specimen number increases only slightly while there is a relatively large increase in species number. From Zone Wa4 through Wa6, specimen numbers remain relatively constant while species numbers experience a decrease. Whereas it is clear that sample size does directly affect species number, it is also apparent that some changes in species numbers occur independent of changes in sample size and may reflect "real" faunal events in the biostratigraphic record.

Further analysis of the effects of sample size on faunal turnover suggests that a fundamental change in mammalian faunal composition occurred at the Clarkforkian/Wasatchian LMA boundary. Figure 19.3B plots FADs of mammalian genera versus sample size. As predicted by modeling (Buzas et al. 1982; Koch 1987; Badgley 1990), increases in specimen numbers lead to increases in first appearances throughout the Tiffanian and Clarkforkian LMAs (due to the increased probability of adding rarer taxa as sample sizes increase). In contrast to the predicted pattern, FADs of mammalian genera increase in Zone Wa0 even though sample size decreases to the lowest level (289 specimens) of any interval represented. This may indicate that an influx of new mammalian taxa entered the northern Bighorn Basin at the beginning of the Wasatchian LMA (although sampling from distal floodplain environments in the Zone Wa0 sequence also influences faunal composition; Gingerich 1989b; Gunnell 1994). Interestingly, a substantial increase in specimen number in Zone Wa2 (3901 specimens) does not lead to an increase in first appearances of mammalian genera (the number of species does increase), but a "second wave" of first appearances of genera does occur in Zone Wa3, with specimen numbers holding at high levels (4017 specimens).

Figure 19.3C shows the relationship between LADs of mammalian genera and sample size. In this case the expected pattern is a decrease in LADs with an increase in sample size (Badgley and Gingerich 1988). The opposite pattern seems to hold through the Tiffanian, Clarkforkian, and lower Wasatchian LMAs, with increases in sample sizes coinciding with increases in LADs and decreases in sample size coinciding with decreases in LADs. This is especially true in Zone Wa0 through Wa1, where LADs reach their lowest point in the sequence even though specimen numbers are low, particularly in Zone Wa0. Only above Zone Wa3 (Biohorizon A of Schankler 1980 and Badgley and Gingerich 1988 occurs within Zone Wa3; see Gingerich 1989a) does the expected pattern occur with increased LADs linked to a decrease in sample size. This pattern seems to suggest that Biochrons Ti5, Cf2, and Wa0 do represent periods of "real" mammalian community change.

Figure 19.4 presents scatter plots of the square root of specimen number versus number of species and FADs and LADs of genera. In general, the Tiffanian and late Wasatchian samples tend to have fewer species than other intervals, regardless of the number of specimens (figure 19.4A). Middle Clarkforkian through middle Wasatchian samples tend to have more species regardless of specimen number. In the lowest Wasatchian LMA (Zone Wa0), sample size is the smallest of any interval, yet the number of species represented is relatively high (higher than expected based on a best-fit line projected through the data). In terms of FADs (figure 19.4B) the anomalous position of Zone Wa0 is again reinforced as it has the smallest number of specimens but close to the highest number of FADs of new mammalian genera among all biostratigraphic horizons.

The pattern of LADs (figure 19.4C) seems to be unexplainable simply in terms of specimen number. Intervals represented by the fewest numbers of specimens (Zones Wa0, Ti4, Cf1, and Cf3) also have some of the lowest numbers of LADs. Intervals represented by the highest numbers of specimens (Zones Wa1 through Wa3) have a widespread pattern of LADs. Zones Wa1 and Wa2 have among the lowest numbers of LADs (as expected), while Zone Wa3 has among the highest LADs.

The patterns presented in figures 19.3 and 19.4 suggest that mammalian faunal composition is not simply reflecting sample size differences between zones. Some faunal turnover patterns are probably the result of increasing or decreasing sample size, but some of these patterns clearly reflect actual change in mammalian communities, especially across the Clarkforkian/Wasatchian LMA boundary.

Modern Mammalian Orders and Faunal Turnover

Another clue to the fundamentally different faunal reorganization represented at the Clarkforkian/Wasatchian LMA boundary can be found by examining figure 19.5, which shows the number of extant mammalian orders present through the northern Bighorn Basin sequence. The percentage of specimens representing modern (extant) mammalian orders in the northern Bighorn Basin

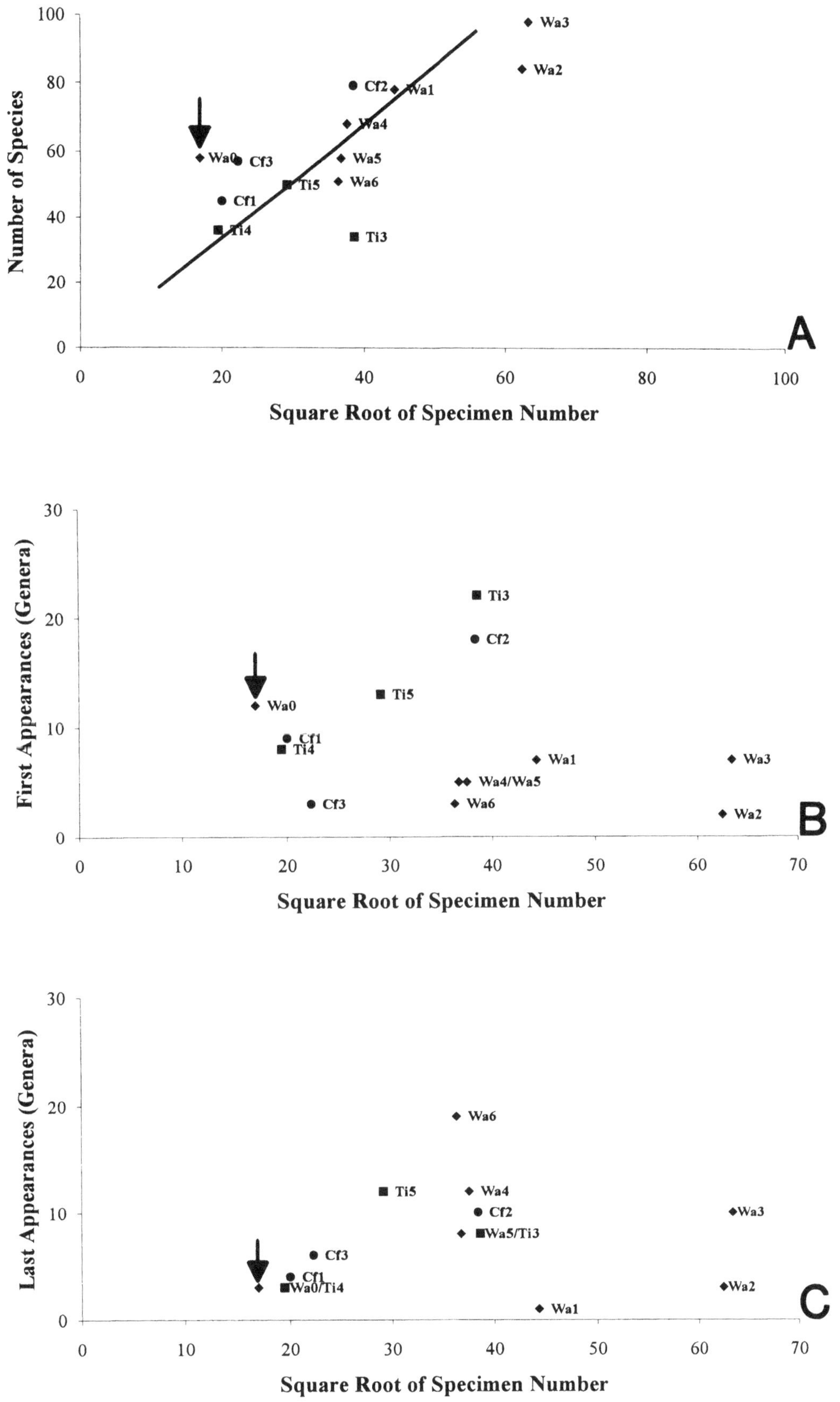

Figure 19.4 Scatter diagrams of square root of number of specimens versus the number of species, first appearances of mammalian genera, and last appearances of mammalian genera.

Arrow indicates Zone Wa0, squares are Tiffanian samples, circles are Clarkforkian samples, and diamonds are Wasatchian samples. Heavy line in A represents best fit line through data (r = 0.78).

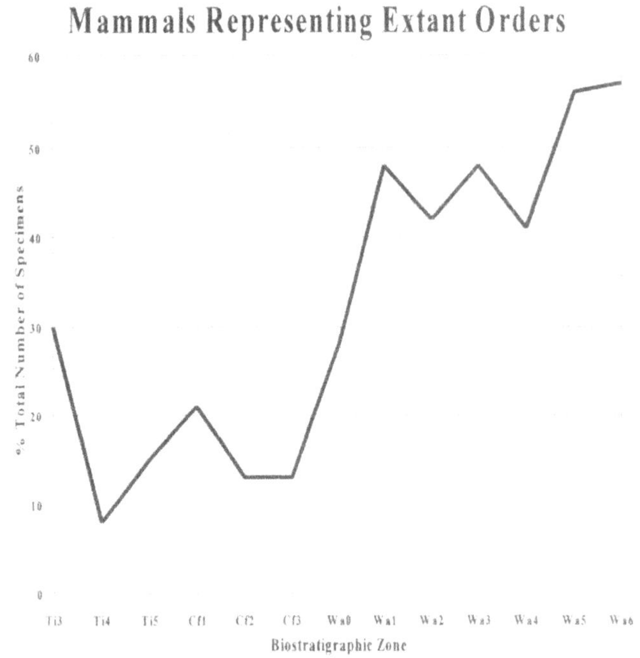

Figure 19.5 Middle Tiffanian through Wasatchian mammalian specimens representing modern orders.

Vertical axis represents percentage of northern Bighorn Basin mammalian specimens belonging to modern (extant) orders. Horizontal axis represents time expressed in biostratigraphic zones.

increases beginning in the Wasatchian LMA. Throughout the Tiffanian and Clarkforkian LMA, the percentage of mammalian specimens representing extant mammalian orders ranges from 8.0% to 30.0% of all specimens, averaging 16.7% through this period. Through the Wasatchian LMA this average rises to 45.7%, ranging from 28% to 57% of all specimens. Not only is there a substantial faunal turnover recorded at the Clarkforkian/Wasatchian LMA boundary, but this turnover differs in character with the influx of a whole new suite of more modern mammals appearing for the first time in the early Wasatchian LMA.

Figures 19.6 and 19.7 present principal components analyses based on the number of mammalian specimens representing selected orders and families. These analyses contrast Tiffanian/Clarkforkian mammalian faunas (figure 19.6) and Clarkforkian/Wasatchian mammalian faunas (figure 19.7). In each case, the most prevalent mammalian taxa are contrasted against less prevalent taxa from each land mammal age. In this manner, each land mammal age can be characterized by a dominant group of mammalian taxa in comparison with another land mammal age.

The Tiffanian LMA was dominated in terms of number of specimens by two orders, Plesiadapiformes (= Proprimates) and Multituberculata (figure 19.6A). A broad, polyphyletic group, referred to as "Insectivora," was also an important component of Tiffanian mammalian faunas. Tiffanian mammalian families (figure 19.6B) were dominated by two plesiadapiforms (Plesiadapidae and Paromomyidae) and one archaic ungulate (= "condylarthrans") family (Arctocyonidae). Less dominant, but still typical, Tiffanian mammalian families included ptilodontid multituberculates, carpolestid plesiadapiforms, palaeoryctid proteutherians, and pantolambdid pantodonts. None of these families are members of any extant mammalian order.

In contrast to the Tiffanian LMA, the Clarkforkian LMA was dominated by the orders "Condylarthra," Palaeanodonta, Mesonychia, Rodentia, and Tillodontia (figure 19.6A). Among Clarkforkian families (figure 19.6B), phenacodontid "condylarthrans," mesonychid mesonychians, paramyid rodents, microsyopid plesiadapiforms, and coryphodontid pantodonts were the most prevalent. Lesser, but still important, Clarkforkian families included prodinoceratid dinoceratans, viverravid and miacid carnivores, oxyaenid creodonts, and arctostylopid notoungulates. Of these families, only paramyids (Rodentia) and viverravids and miacids (Carnivora) are members of extant orders. It is probable that rodents represent immigrants from Asia (Rose 1981; Li and Ting 1985, 1993; Krause and Maas 1990; Li and Chow 1994). The phylogenetic and biogeographic origins of viverravids and miacids remain in doubt as does the precise relationship of these families to the order Carnivora (Van Valen 1966; Flynn and Galiano 1982; Gingerich 1983b; Gingerich and Winkler 1985; Flynn et al. 1988; MacPhee and Novacek 1993; Wyss and Flynn 1993).

Comparison of the Clarkforkian and the Wasatchian LMAs reveals a different pattern of taxonomic contrasts. The Clarkforkian LMA (compared to the Wasatchian LMA) was dominated by the orders Dinocerata, Mesonychia, Notoungulata, and Plesiadapiformes (figure 19.7A). Among families, prodinoceratid dinoceratans, carpolestid and plesiadapid plesiadapiforms, mesonychid mesonychians, and plagiomenid "dermopterans" served to differentiate the Clarkforkian LMA (figure 19.7B) from the Wasatchian LMA. Lesser, but still important, Clarkforkian taxa included arctocyonid and phenacodontid "condylarthrans," pantolambdid pantodonts, and ptilodontid multituberculates. In comparison with the Wasatchian LMA, the Clarkforkian LMA was still dominated by "archaic" characteristically Tiffanian groups such as plesiadapiforms, arctocyonids, and multituberculates, along with characteristic Clarkforkian groups like mesonychians and phenacodontids. None of these taxa are members of extant orders except perhaps plagiomenids but the affinities of this group are in doubt (Gunnell 1989; MacPhee et al. 1989; McKenna 1990).

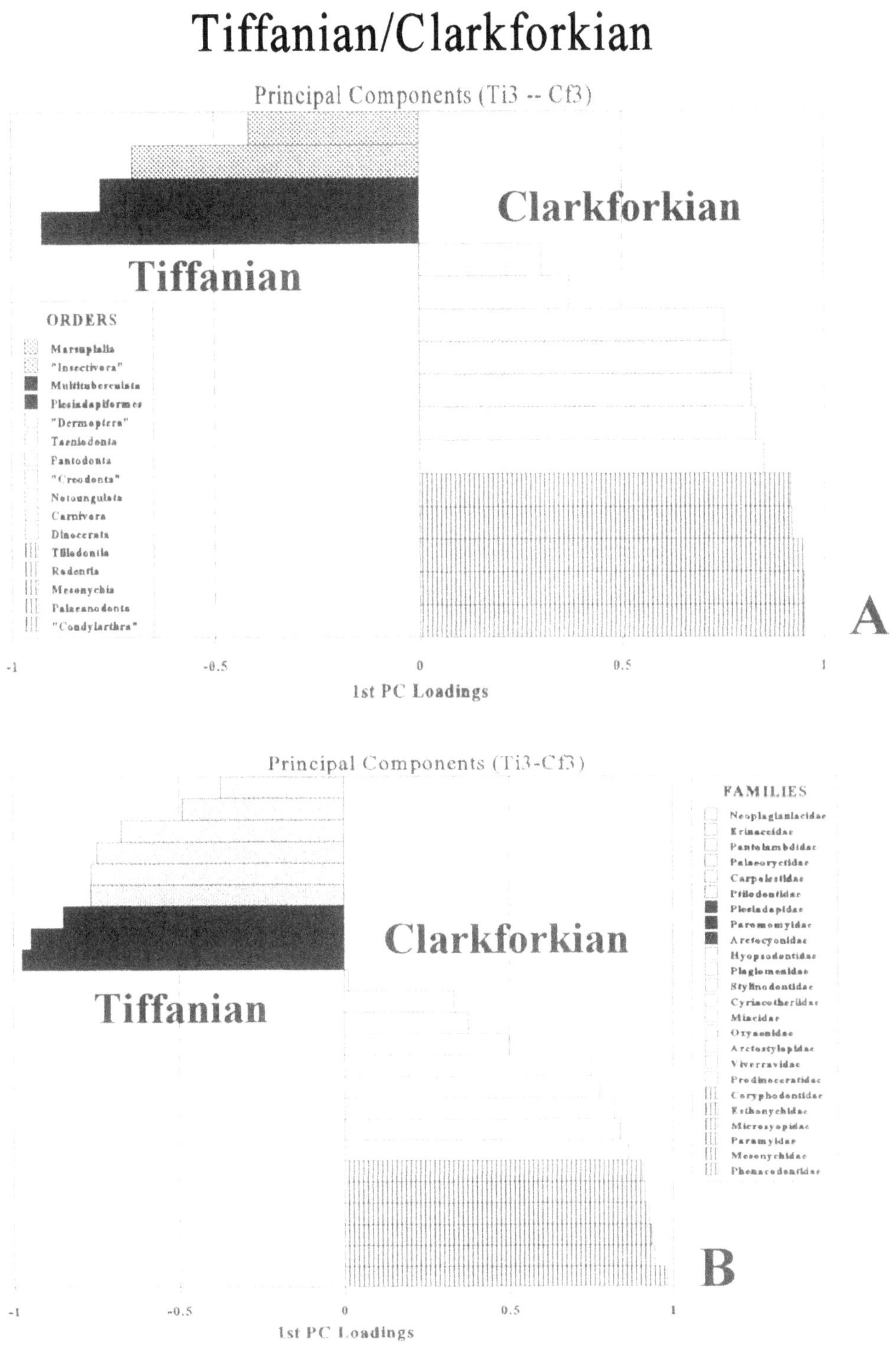

Figure 19.6 Histograms of first principal component loadings (horizontal axis) contrasting Tiffanian and Clarkforkian mammalian faunas.

(A) mammalian orders; (B) mammalian families. Black bars indicate most common Tiffanian orders and families; vertically striped bars indicate most common Clarkforkian orders and families.

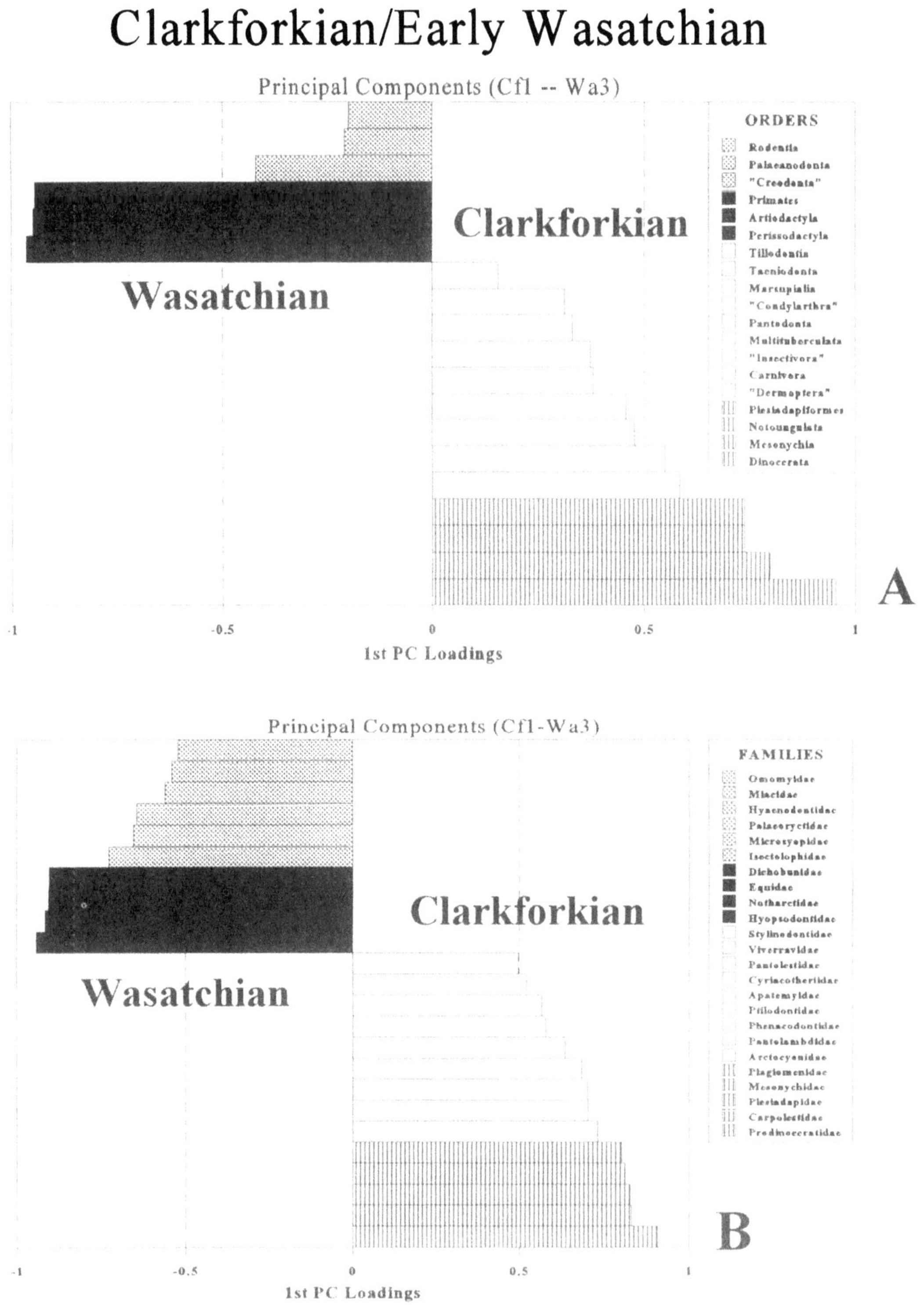

Figure 19.7 Histograms of first principal component loadings (horizontal axis) contrasting Clarkforkian and early Wasatchian mammalian faunas.

(A) mammalian orders; (B) mammalian families. Black bars indicate most common Wasatchian orders and families; vertically striped bars indicate most common Clarkforkian orders and families.

Wasatchian mammalian faunas (compared to the Clarkforkian mammalian faunas) were dominated by the orders Perissodactyla, Artiodactyla, and Primates (figure 19.7A). Among Wasatchian families, hyopsodontids, notharctid primates, equid perissodactyls, and dichobunid artiodactyls were the most distinctive portions of the fauna (figure 19.7B). Lesser, but still important, groups included isectolophid perissodactyls and microsyopid plesiadapiforms. Among these taxa, all but hyopsodontids and microsyopids are members of extant orders.

Mammalian Compositional Changes

Principal component analyses can be used to contrast common and uncommon elements in comparative faunas, but a closer examination of taxonomic composition provides a clearer picture of the changes that occur in faunas through time. Figure 19.8 summarizes changes in taxonomic composition at the ordinal level for five of the most common orders in terms of numbers of specimens through the northern Bighorn Basin sequence. The middle and late Tiffanian LMA were dominated by Plesiadapiformes, which averaged 47.9% of all mammalian specimens through this period. "Condylarths" dominated the Clarkforkian LMA, averaging over 50% of the fauna throughout, and were the most prevalent "order" in the Wasatchian LMA, averaging about 40% of the fauna. Three orders made first appearances in the earliest Wasatchian LMA (Primates, Perissodactyla, and Artiodactyla) and made up a substantial portion of the mammalian fauna through the interval. Totaled together these three orders averaged 37% of Wasatchian mammal faunas.

Figure 19.9 summarizes frequency changes in 18 mammalian families from the middle Tiffanian through late Wasatchian LMA. Plesiadapidae (figure 19.9A) were very common in the Tiffanian LMA, less common in the Clarkforkian LMA, and absent in the Wasatchian LMA. Phenacodontids were relatively common in the

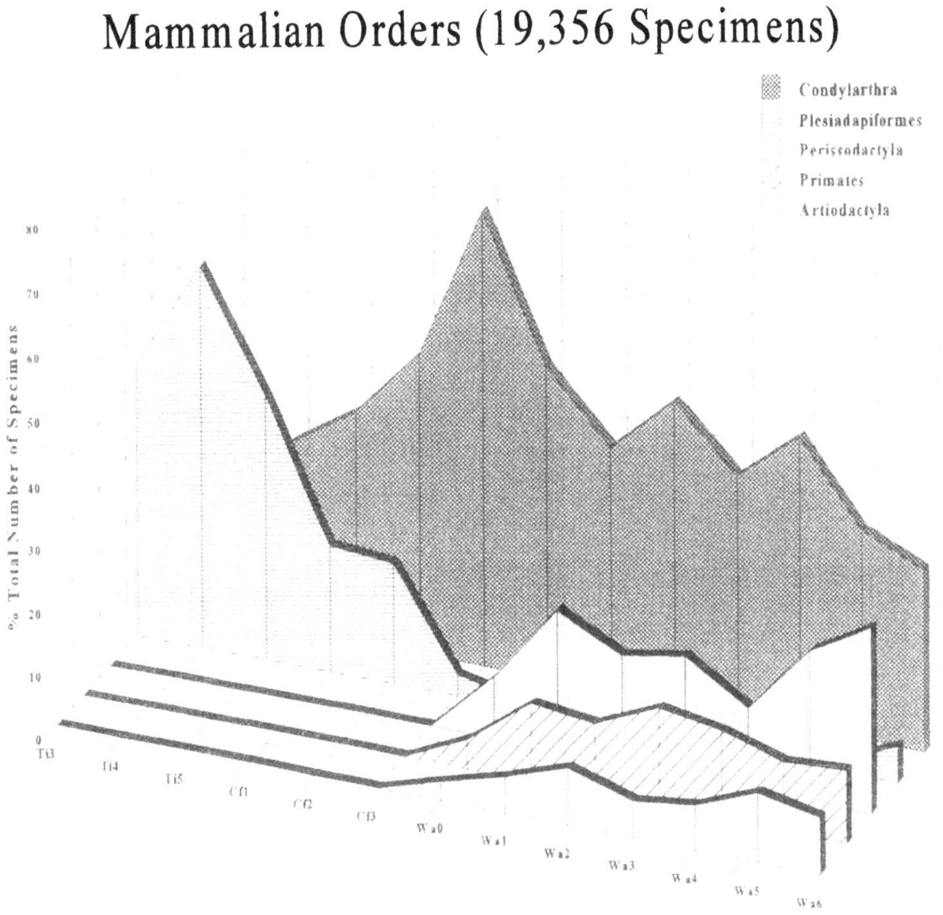

Figure 19.8 Area diagram showing distribution of the most common mammalian orders in the middle Tiffanian through Wasatchian LMA based on 19,356 specimens.

Vertical axis is the percentage of specimens representing each order; horizontal axis is time expressed in biostratigraphic zones.

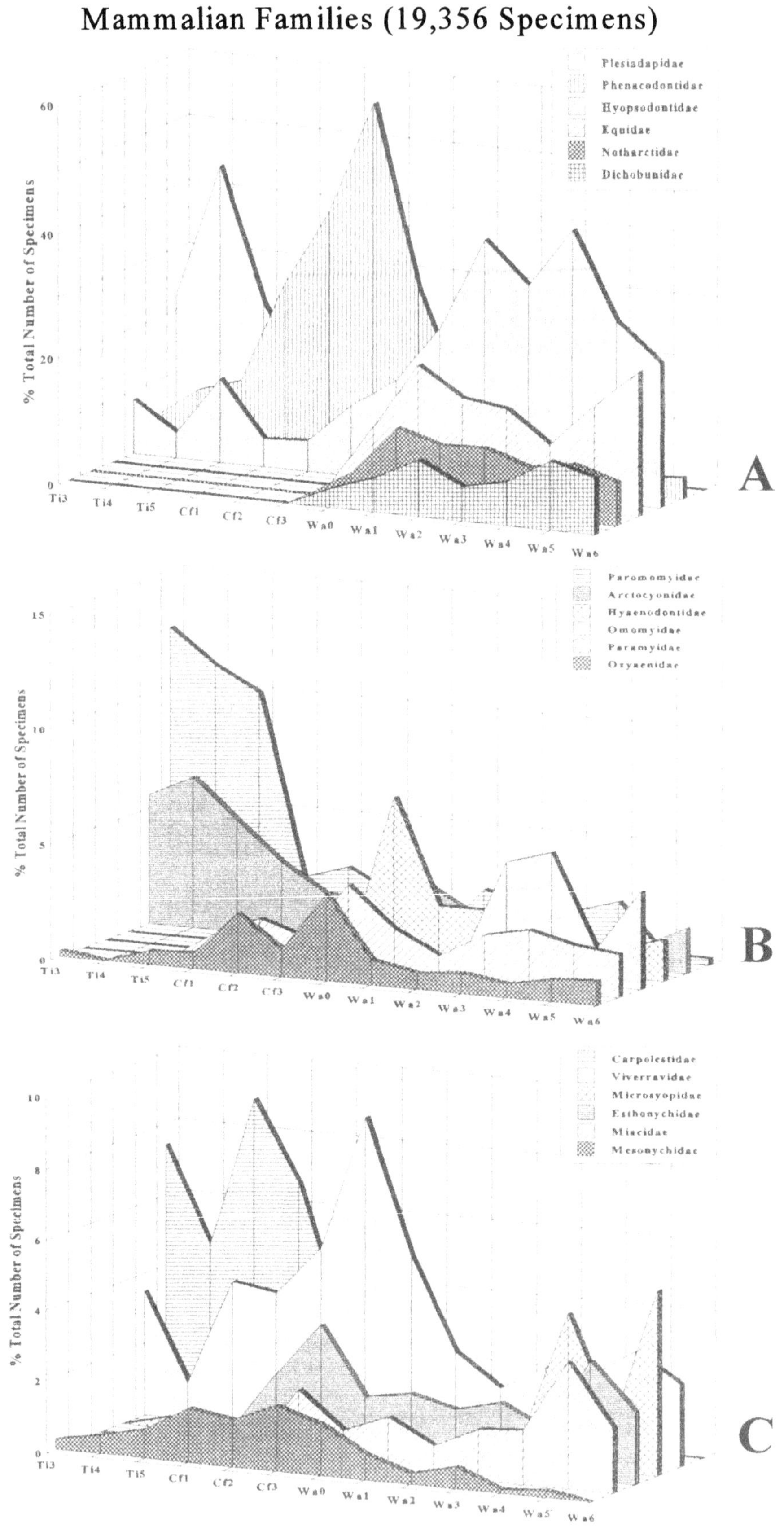

Figure 19.9 Area diagrams of the distribution of biostratigraphically important mammalian families in the middle Tiffanian through Wasatchian LMA.

Arranged from more common families (A) to less common families (C). Vertical axis is the percentage of specimens representing each family; horizontal axis is time expressed in biostratigraphic zones.

Tiffanian LMA and predominant in the Clarkforkian LMA, and became progressively less common through the Wasatchian LMA. Hyopsodontids were relatively uncommon through the Tiffanian and Clarkforkian LMAs but became the dominant mammalian family throughout the Wasatchian LMA. Paramyids made their first appearance in the early Clarkforkian LMA but remained relatively uncommon throughout the Clarkforkian and Wasatchian LMAs. Equids, notharctids, and dichobunids first appeared in the earliest Wasatchian LMA and maintained relatively consistent frequencies throughout the Wasatchian LMA.

Figures 19.9B–C show several less common, but still biostratigraphically important, mammalian families. Of these families, those most predominantly Tiffanian in distribution included paromomyids, carpolestids, and arctocyonids. Paromomyids and arctocyonids were reduced to relatively low frequencies by the beginning of the Clarkforkian LMA, but remained members of the mammalian community throughout the Wasatchian LMA. Carpolestids essentially were gone after the end of the Clarkforkian LMA (Rose 1975, 1981; Bloch and Gingerich 1994).

Viverravids, coryphodontids, and esthonychids were the most common families in the Clarkforkian LMA among these less well-represented taxa. Viverravids were present throughout the sequence but reached their maximum frequency in the middle to late Clarkforkian LMA (Biochrons Cf2–Cf3). Coryphodontids and esthonychids first appeared in the early Clarkforkian LMA and were present throughout the remainder of the sequence at relatively low frequencies. Miacid carnivores appeared in the middle Clarkforkian LMA and maintained relatively low frequency through the Wasatchian LMA. Other relatively common Clarkforkian families included paramyids and oxyaenids.

Several families of less common mammals first appeared in the lower Wasatchian LMA and became quite well represented in the remainder of the sequence. Hyaenodontids and omomyids first appeared in the basal Wasatchian LMA. After an initial radiation, hyaenodontids became stabilized at a lower frequency in the later portion of the Wasatchian LMA. Omomyids remained at relatively low percentages until Biochron Wa3 and then experienced an increase in numbers, followed by a decrease and then a slight increase. Omomyids were relatively well represented in the later part of the Wasatchian LMA.

Two other families, Isectolophidae and Microsyopidae, also became more common towards the end of the Wasatchian LMA. Isectolophids first appeared in Biochron Wa1 and gradually increased in numbers through the Wasatchian LMA. Microsyopids were present in relatively low numbers from the Tiffanian LMA through early Wasatchian LMA but experienced an increase in frequency beginning in Biochron Wa4. Table 19.2 summarizes changes in mammalian faunal composition at important faunal turnover intervals at the ordinal and family level.

Mammalian Trophic Structure

Trophic structure is an important characteristic of mammalian communities. Previous studies (Gunnell et al. 1995) indicated that trophic patterns and the taxonomic makeup of trophic groups changed throughout the northern Bighorn Basin sequence. Figure 19.10 summarizes the taxonomic composition of herbivorous and carnivorous mammals in the northern Bighorn Basin from the middle Tiffanian LMA through the Wasatchian LMA. Among herbivores (figure 19.10A) plesiadapids dominated the Tiffanian LMA and remained common through the middle Clarkforkian LMA. Other relatively common herbivores in the Tiffanian LMA included ptilodontids, hyopsodontids, and phenacodontids. Phenacodontids replaced plesiadapids as the dominant herbivore family in the Clarkforkian LMA. Phenacodontids became progressively less common through the Wasatchian LMA and were replaced by hyopsodontids, equids, and dichobunids as the predominant herbivores of the Wasatchian LMA.

Taxonomic composition and dominance patterns of carnivorous mammals also changed from the Tiffanian through the Wasatchian LMAs (figure 19.10B). Arctocyonid (perhaps more omnivorous than strictly carnivorous) and viverravid carnivores were the dominant carnivorous mammalian groups in the Tiffanian LMA. In the Clarkforkian LMA, viverravids became predominant, along with lesser contributions from oxyaenid creodonts and mesonychid mesonychians. Viverravids remained an important carnivorous group through the Wasatchian LMA with the addition of miacid carnivores (which appeared in the middle Clarkforkian LMA) and hyaenodontid creodonts (which first appeared in the earliest Wasatchian LMA). Oxyaenids remained at relatively low frequency throughout the Wasatchian LMA, as did arctocyonids, while mesonychids decreased in frequency through the Wasatchian LMA. No carnivorous group ever represented more than 9% (viverravids in the late Clarkforkian LMA) of all specimens, as befits their position at the top of the trophic hierarchy.

The major change in mammalian faunas at the Clarkforkian/Wasatchian LMA boundary demonstrated above is also indicated by trophic patterns (Gunnell et al. 1995). In the Tiffanian and Clarkforkian LMAs, both herbivorous and carnivorous mammals were dominated by a single group, with minor contributions from other taxa. Among herbivores, plesiadapids were the domi-

Table 19.2

Change in percentage of total number of specimens for selected mammalian orders and families at selected biostratigraphic zonal boundaries

	Late Tiffanian (Zone Ti5)/Early Clarkforkian (Zone Cf1)			
Order	Family	%Ti5	%Cf1	Change in %
Tillodontia*	Esthonychidae*	0.0	0.5	+0.5
Rodentia*	Paramyidae*	0.0	1.2	+1.2
Pantodonta	Coryphodontidae*	0.0	0.5	+0.5
"Condylarthra"	Phenacodontidae	11.6	28.3	+16.7
" "	Hyopsodontidae	13.3	4.5	−8.8
Plesiadapiformes	Plesiadapidae	22.5	10.4	−12.1
" "	Carpolestidae	8.9	6.7	−2.2
" "	Paromomyidae	10.0	2.2	−7.8
Multituberculata	Ptilodontidae	3.2	0.2	−3.0
	Early (Zone Cf1)/Middle Clarkforkian (Zone Cf2)			
Order	Family	%Cf1	%Cf2	Change in %
Carnivora	Miacidae†	0.0	0.4	+0.4
Rodentia	Paramyidae	1.2	2.0	+0.8
Tillodontia	Esthonychidae	0.5	2.1	+1.6
Pantodonta	Coryphodontidae	0.5	2.2	+1.7
"Condylarthra"	Phenacodontidae	28.3	40.0	+11.7
" "	Hyopsodontidae	4.5	5.0	+0.5
Plesiadapiformes	Carpolestidae	6.7	2.7	−4.0
	Late Clarkforkian (Zone Cf3)/Early Wasatchian (Zone Wa0)			
Order	Family	%Cf3	%Wa0	Change in %
Primates‡	Omomyidae‡	0.0	0.3	+0.3
" "	Notharctidae‡	0.0	4.2	+4.2
Perissodactyla‡	Equidae‡	0.0	10.0	+10.0
Artiodactyla‡	Dichobunidae‡	0.0	3.0	+3.0
Creodonta	Hyaenodontidae‡	0.0	6.9	+6.9
Rodentia	Paramyidae	1.4	3.8	+2.4
"Condylarthra"	Hyopsodontidae	11.2	16.2	+4.8
" "	Phenacodontidae	57.5	29.4	−28.1
Plesiadapiformes	Plesiadapidae	0.8	0.0	−0.8
	Early Wasatchian (Zone Wa0/Wa1)			
Order	Family	%Wa0	%Wa1	Change in %
Perissodactyla	Isectolophidae§	0.0	1.7	+1.7
" "	Equidae	10.0	20.5	+10.5
Primates	Notharctidae	4.2	12.0	+7.8
"Condylarthra"	Hyopsodontidae	16.0	25.0	+9.0
" "	Phenacodontidae	29.4	11.7	−17.7

*First appearance in Zone Cf1.
†First appearance in Zone Cf2.
‡First appearance in Zone Wa0.
§First appearance in Zone Wa1.

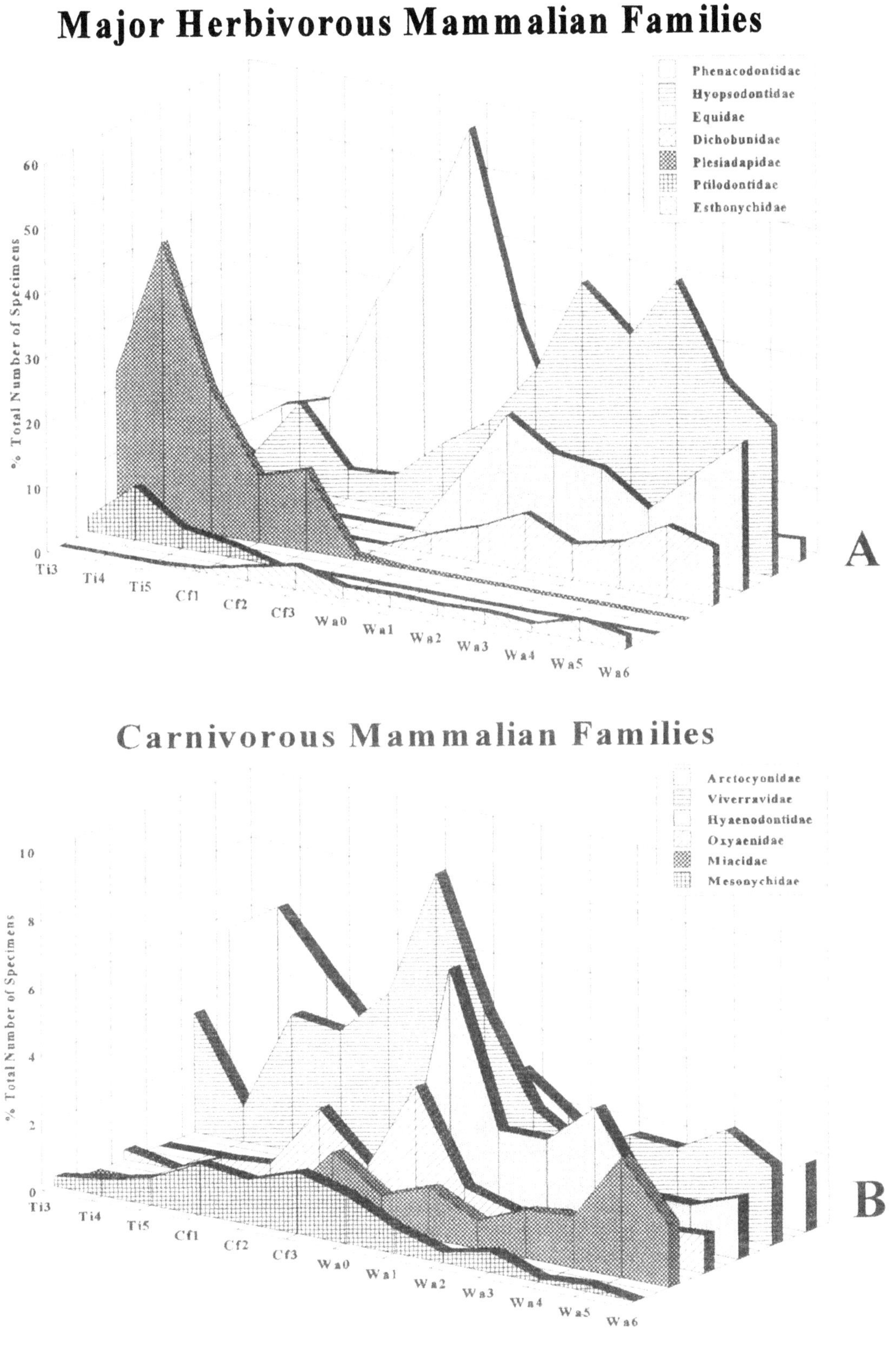

Figure 19.10 Area diagrams showing the distribution of the most common herbivorous mammalian families (A) and carnivorous mammalian families (B) in the middle Tiffanian through Wasatchian LMA.

Vertical axis is percentage of total number of specimens; horizontal axis is time expressed in biostratigraphic zones.

nant Tiffanian group, while phenacodontids became the common herbivorous group in the Clarkforkian LMA. In the Wasatchian LMA, there was less dominance by a single family as several families became important members of herbivorous trophic groups. Hyopsodontids were the most common herbivore in the Wasatchian LMA but equid perissodactyls and dichobunid artiodactyls also made up a substantial portion of the herbivore trophic group.

A similar, but somewhat less striking, pattern is seen among carnivores. Arctocyonids were the dominant carnivorous group of the Tiffanian LMA while viverravid carnivores were the most common carnivorous group in the Clarkforkian LMA. Unlike the pattern seen for herbivores, the Tiffanian LMA also was well represented by viverravids, whereas Clarkforkian carnivores consisted not only of the dominant viverravids, but oxyaenid creodonts, mesonychid mesonychians, and miacid carnivores as well. During the Wasatchian LMA, no carnivore group was dominant over the others although hyaenodontid creodonts underwent a rapid and flourishing radiation beginning in Biochron Wa0. Other important Wasatchian carnivore groups included oxyaenids in the early part of the Wasatchian LMA, viverravids throughout, and miacids throughout but especially later in the Wasatchian LMA.

The Paleocene Series/Epoch was originally defined by Schimper (1874) based on paleobotanical evidence from the Paris Basin. The Eocene Series/Epoch was defined by Lyell (1833) based on the percentage of modern marine invertebrate taxa (mostly bivalves and gastropods) contained in collections from the Paris and London–Hampshire Basins (about 3.5% of 1238 "shells" from these strata were recognized as representing living taxa by Lyell). Since that time, the placement of the boundary between the Paleocene and Eocene Series/Epochs has been contentious. The boundary has been delineated based on a number of different lines of evidence including biostratigraphic (paleobotanical, nannoplanktonic, foraminiferal, and vertebrate), chemostratigraphic (oxygen and carbon isotopes), and other stratigraphic means (sequence stratigraphy, eolian grain size, paleopedomorphology, paleomagnetic stratigraphy, evidence of global or regional tectonism). It has been suggested (Gunnell et al. 1993) that recognition/definition of the Paleocene and Eocene Series/Epochs and placement and correlation of the Paleocene/Eocene Epoch/Series boundary are two separate but related issues. Demarcation and definition of Cenozoic Series/Epochs have traditionally been based on their paleontologic components (Lyell 1833). As such, often there are only minor disagreements concerning what constitutes a "typical Paleocene" (in this context, earlier) or "typical Eocene" (in this context, later) biotic assemblage. Major difficulties occur when attempting to correlate these "typical" assemblages globally or even regionally and from that provide a date (numerical or relative) for placement of the Paleocene/Eocene Epoch/Series boundary. Series/Epoch characterization is most appropriately addressed by paleontological evidence, while placement of boundaries and correlation involve biostratigraphy, magnetostratigraphy, geochemistry, and geochronometry.

Paleontological evidence pertinent to recognition of "typically archaic faunas" (in this context = "typically Paleocene") and "typically modern-aspect faunas" (in this context = "typically Eocene") from continental sediments in the northern Bighorn Basin is predominantly mammalian in nature (but see Bartels 1983; Wing 1984; Wing et al. 1991; Gunnell et al. 1993; Hartman 1993). There is a preponderance of evidence (discussed in detail above) that points to a major change in mammalian faunas at the Clarkforkian/Wasatchian LMA boundary. (1) The earliest Wasatchian LMA (Biochrons Wa0 and Wa1) was a time of high FADs but low LADs, unlike other times of high faunal turnover where high FADs were paired with peaks of LADs (Biochron Ti5 and perhaps Cf2). (2) In contrast to expectations of the effects of sample size, FADs of mammalian genera increased in Biochron Wa0 even though sample size was the smallest of any interval represented in the northern Bighorn Basin sequence. (3) An unexpected (in terms of sampling effects) pattern of LADs seemed to hold through the Tiffanian, Clarkforkian, and early Wasatchian LMAs, with increases in sample sizes coinciding with increases in LADs and decreases in sample size coinciding with decreases in LADs. This is especially true from Biochron Wa0 through Wa1, when LADs reached their lowest point in the interval even though specimen numbers were low, particularly for Biochron Wa0. (4) In the earliest Wasatchian LMA (Biochron Wa0), sample size was the smallest of any interval, yet the standing richness of species represented was relatively high. (5) The percentage of specimens representing modern (extant) mammalian orders in the northern Bighorn Basin increased beginning in the Wasatchian LMA (Biochron Wa0). (6) Wasatchian mammalian faunas were dominated by the orders Perissodactyla, Artiodactyla, and Primates. Among Wasatchian families, hyopsodontids, notharctids, equids, and dichobunids were the most prevalent. Among these taxa, all but hyopsodontids are members of extant orders. (7) Taxonomic makeup and patterns of mammalian trophic groups differed in the Tiffanian–Clarkforkian LMAs versus the Wasatchian LMA. In the Tiffanian and Clarkforkian LMAs, both herbivorous and carnivorous mammals were dominated by a single group, with minor contributions from other taxa. In the Wasatchian LMA, there was less dominance by a single taxon as several families became important members of herbivorous and carnivorous trophic groups.

All of the evidence cited above indicates that a fundamental change in mammalian faunal composition took place at the Clarkforkian/Wasatchian LMA boundary. Archaic mammalian taxa characteristic of the Tiffanian and Clarkforkian LMAs were replaced by modern mammalian groups beginning in the Wasatchian LMA. To be sure, there were other faunal turnovers documented in the mammalian record of the northern Bighorn Basin, but none of these represented as overwhelming a restructuring of mammalian faunas as that at the Clarkforkian/Wasatchian LMA boundary.

Does the Clarkforkian/Wasatchian LMA boundary in the northern Bighorn Basin coincide with the Paleocene/Eocene Series boundary? The presently accepted definition of the Paleocene/Eocene Series boundary suggests that it lies at the base of the London Clay in the London Basin (Berggren et al. 1985a, 1985b; Aubry et al. 1988). The choice of this section to define the boundary is unfortunate for three reasons. First, it does not coincide with the original definition of the Paleocene as proposed by Schimper (1874) based on paleofloras in the Paris Basin, but instead relies on von Koenen's (1885) redefinition of the Paleocene/Eocene Series boundary as lying between the Lignites du Soisonnais and the Sables inférieurs de Cuise in the Paris Basin and between the Woolwich–Reading beds and the London Clay in the London Basin (Clyde et al. 1995). Second, the contact between the Woolwich–Reading beds and the London Clay is unconformable (Knox 1984). Third, it is difficult to correlate this sequence with any terrestrial record using biostratigraphic evidence because of the unconformable nature of the contact between the formations, thus it is difficult to generate support for any particular biotic event being coincident with the Paleocene/Eocene Series boundary as now defined.

Presumably, wherever the Paleocene/Eocene Epoch/Series boundary is eventually placed (IGCP Project 308), it did represent a global perturbation that left tangible geological evidence of its occurrence, even if ultimately the placement of the boundary is determined by historic precedent and consensus of the scientific community. Such a global perturbation, if it was of a sufficient magnitude, should have resulted in changes in global ecosystem that should have been reflected by biotic responses of varying degree and differing temporal onset and extent depending on particular organisms.

It is beyond the scope of this chapter to argue for or against the use of any particular stratigraphic section as the basis for the Paleocene/Eocene Series boundary. However, wherever the boundary is ultimately placed, it will not change the fact that an archaic mammalian fauna was replaced by a more modern mammalian fauna at the Clarkforkian/Wasatchian LMA boundary in the northern Bighorn Basin. A faunal change of the magnitude of the one that occurred at that boundary must have been the result of widespread changes in migrational pathways for terrestrial vertebrates. It is difficult to imagine that such alterations would have occurred without global sea-level and climatic changes contributing to the opening of these migration routes (but see Krause and Maas 1990). It is equally difficult to imagine that, if such global changes in sea level and climate did occur, there would be no evidence of such changes left in the geological record.

From the perspective of terrestrial vertebrate paleontology, a Paleocene/Eocene Series boundary placed at or slightly below the Clarkforkian/Wasatchian LMA boundary in the northern Bighorn Basin seems easiest to accept, given the time-transgressive nature of the sedimentary record and of vertebrate migration. A Paleocene/Eocene Series boundary placed above the Clarkforkian/Wasatchian LMA boundary leaves the more difficult problem of explaining the resultant precocious change of mammalian faunas (clearly a problem to be dealt with by terrestrial biostratigraphers). Ultimate placement of the Paleocene/Eocene Series boundary will rely most heavily on the marine record, but it seems counterproductive to simply relegate terrestrial biotic evidence (despite historic precedent) to a secondary role of correlation after the "golden spike" has been driven.

ACKNOWLEDGMENTS

I wish to thank W. A. Berggren, M.-P. Aubry, and S. G. Lucas for the invitation to participate in this volume and for many stimulating discussions and ideas. I thank C. Badgley, W. S. Bartels, R. J. Burnham, L. S. Collins, P. D. Gingerich, S. G. Lucas, and W. C. Clyde for their insights regarding the Paleocene/Eocene boundary and M.-P. Aubry, D. R. Prothero, and T. E. Williamson for comments on an earlier draft of this manuscript. The evidence presented in this chapter is the result of over 20 years of field work in the Bighorn Basin. I thank all participants (too many to list) who have been involved in this University of Michigan field project over the years. I also wish to extend my gratitude to the people of Powell, Wyoming, especially the Churchill family, for their support during our work.

REFERENCES

Archibald, J. D., Clemens, W. A., Gingerich, P. D., Krause, D. W., Lindsay, E. H. and Rose, K. D., 1987. First North American Land Mammal Ages of the Cenozoic Era. In Woodburne, M. O. (ed.), *Cenozoic Mammals of North America: Geochronology and Biostratigraphy*, University of California Press, Berkeley, Los Angeles, London, p. 24–76.

Aubry, M-P., Berggren, W. A., Kent, D. V., Flynn, J. J., Klitgord, K. D., Obradovich, J. D. and Prothero, D. W., 1988. Paleogene Geochronology: An integrated approach. *Paleoceanography,* v. 3, p. 707–742.

Badgley, C., 1990. A statistical assessment of last appearances in the Eocene record of mammals. In Bown, T. M. and Rose, K. D. (eds.), *Dawn of the Age of Mammals in the Northern Part of the Rocky Mountain Interior, North America,* Geological Society of America Special Paper 243, p. 153–167.

Badgley, C. and Gingerich, P. D., 1988. Sampling and faunal turnover in early Eocene mammals. *Palaeogeography, Palaeoclimatology, Palaeoecology,* v. 63, p. 141–157.

Badgley, C., Bartels, W. S., Morgan, M. E., Behrensmeyer, A. K. and Raza, S. M., 1995. Taphonomy of vertebrate assemblages from the Paleogene of northwestern Wyoming and the Neogene of northern Pakistan. *Palaeogeography, Palaeoclimatology, Palaeoecology,* v. 115, p. 157–180.

Bartels, W. S., 1983. A transitional Paleocene/Eocene reptile fauna from the Bighorn Basin, Wyoming. *Herpetologica,* v. 39, p. 359–374.

Berggren, W. A., Kent, D. V., Flynn, J .J. and Van Couvering, J. A., 1985a. Cenozoic geochronology. *Geological Society of America Bulletin,* v. 96, p. 1407–1418.

Berggren, W. A., Kent, D. V. and Flynn, J. J., 1985b. Paleogene geochronology and chronostratigraphy. In Snelling, N. J. (ed.), *The Chronology of the Geological Record:* Geological Society Memoir No 10, Blackwell, Oxford, p. 141–195.

Bloch, J. I. and Gingerich, P. D., 1994. New species of *Carpolestes* (Mammalia, Proprimates) from Clarkforkian late Paleocene limestones of the Clark's Fork Basin, Wyoming: Teeth, skulls and femur. *Journal of Vertebrate Paleontology,* v. 14 (Supplement to no. 3), p. 17A–17B.

Butler, R. F., Gingerich, P. D. and Lindsay, E. H., 1981. Magnetic polarity stratigraphy and biostratigraphy of Paleocene and lower Eocene continental deposits, Clark's Fork Basin, Wyoming. *Journal of Geology,* v. 89, p. 299–316.

Butler, R. F., Krause, D. W. and Gingerich, P. D., 1987. Magnetic polarity stratigraphy and biostratigraphy of middle–late Paleocene continental deposits of south-central Montana. *Journal of Geology,* v. 95, p. 647–657.

Buzas, M. A., Koch, C. F., Culver, S. J. and Sohl, N. F., 1982. On the distribution of species occurrence. *Paleobiology,* v. 8, p. 143–150.

Clyde, W. C., Stamatakos, J. and Gingerich, P. D., 1994. Chronology of the Wasatchian Land-Mammal Age (early Eocene): Magnetostratigraphic results from the McCullough Peaks Section, northern Bighorn Basin, Wyoming. *Journal of Geology,* v. 102, p. 367–377.

Clyde, W. C., Stamatakos, J. and Gingerich, P. D., 1995. Chronology of the Wasatchian Land-Mammal Age (Early Eocene): Magnetostratigraphic results from the McCullough Peaks Section, northern Bighorn Basin, Wyoming: A reply. *Journal of Geology,* v. 103, p. 464–466.

Flynn, J. J. and Galiano, H., 1982. Phylogeny of Early Tertiary Carnivora, with a description of a new species of *Protictis* from the middle Eocene of northwestern Wyoming. *American Museum Novitates,* no. 2725, p. 1–64.

Flynn, J. J., Neff, N. A. and Tedford, R. H., 1988. Phylogeny of the Carnivora. In Benton, M. J. (ed.), *The Phylogeny and Classification of Tetrapods, v. 2: Mammals,* Systematics Association Special Volume No 35B, Oxford: Clarendon Press, p. 73–116.

Gingerich, P. D., 1976. Cranial anatomy and evolution of early Tertiary Plesiadapidae (Mammalia, Primates). *University of Michigan Papers on Paleontology,* v. 15, 141 pp.

Gingerich, P. D., 1980. History of Early Cenozoic vertebrate paleontology in the Bighorn Basin. In Gingerich, P. D. (ed.), *Early Cenozoic Paleontology and Stratigraphy of the Bighorn Basin, Wyoming: 1880–1980,* University of Michigan Papers on Paleontology, v. 24, p. 7–24.

Gingerich, P. D., 1982. Time resolution in mammalian evolution: sampling, lineages, and faunal turnover. In Mamet, B. and Copeland, M. J. (eds.), *Third North American Paleontological Convention Proceedings,* v. 1, p. 205–210.

Gingerich, P. D., 1983a. Paleocene–Eocene faunal zones and a preliminary analysis of Laramide structural deformation in the Clark's Fork Basin, Wyoming. *Wyoming Geological Association Guidebook* 34, p. 185–195.

Gingerich, P. D., 1983b. Systematics of early Eocene Miacidae (Mammalia, Carnivora) in the Clark's Fork Basin, Wyoming. *Contributions from the Museum of Paleontology, University of Michigan,* v. 26, p. 197–225.

Gingerich, P. D., 1989a. Paleocene and Early Eocene of the Bighorn and Clark's Fork Basins, Wyoming. In Flynn, J. J. (ed.), *Mesozoic/Cenozoic Vertebrate Paleontology: Classic localities, contemporary approaches,* Washington: American Geophysical Union, 28th International Geological Congress Field Trip Guidebook T322, p. 47–57.

Gingerich, P. D., 1989b. *New earliest Wasatchian mammalian fauna from the Eocene of northwestern Wyoming: Composition and diversity in a rarely sampled high-floodplain assemblage.* University of Michigan Papers on Paleontology, v. 28, 97 pp.

Gingerich, P. D., Rose, K. D. and Krause, D. W., 1980. Early Cenozoic mammalian faunas of the Clark's Fork Basin–Polecat Bench area, northwestern Wyoming. In Gingerich, P. D. (ed.), *Early Cenozoic Paleontology and Stratigraphy of the Bighorn Basin, Wyoming: 1880–1980.* University of Michigan Papers on Paleontology, v. 24, p. 51–71.

Gingerich, P. D. and Winkler, D. A., 1985. Systematics of Paleocene Viverravidae (Mammalia, Carnivora) in the Bighorn Basin and Clark's Fork Basin, Wyoming. *Contributions from the Museum of Paleontology, University of Michigan,* v. 27, p. 87–128.

Gunnell, G. F., 1989. *Evolutionary history of Microsyopoidea (Mammalia, ?Primates) and the relationship between Plesiadapiformes and Primates.* University of Michigan Papers on Paleontology, v. 27, 157 pp.

Gunnell, G. F., 1994. Paleocene mammals and faunal analysis of the Chappo Type Locality (Tiffanian), Green River Basin, Wyoming. *Journal of Vertebrate Paleontology,* v. 14, p. 81–104.

Gunnell, G. F., Bartels, W. S. and Gingerich, P. D., 1993. Paleocene–Eocene boundary in continental North America: Biostratigraphy and Geochronology, northern Bighorn Basin, Wyoming. In Lucas, S. G. and Zidek, J. (eds.), *Vertebrate Paleontology in New Mexico,* New Mexico Museum of Natural History and Science Bulletin 2, p. 137–144.

Gunnell, G. F., Morgan, M. E., Maas, M. C. and Gingerich, P. D., 1995. Comparative paleoecology of Paleogene and Neogene mammalian faunas: Trophic structure and composition. *Palaeogeography, Palaeoclimatology, Palaeoecology,* v. 115, p. 265–286.

Hartman, J. H., 1993. The Paleocene–Eocene nonmarine molluscan transition in the United States. *Journal of Vertebrate Paleontology,* v. 13 (Supplement to No 3), p. 41A (Abstracts of papers, 53rd Annual Meeting of Society of Vertebrate Paleontology, Albuquerque, New Mexico).

Hickey, L. J., 1980. Paleocene stratigraphy and flora of the Clark's Fork Basin. In Gingerich, P. D. (ed.), *Early Cenozoic Paleontology and Stratigraphy of the Bighorn Basin, Wyoming: 1880–1980,* University of Michigan, Papers on Paleontology, v. 24, p. 33–49.

Jepsen, G. L., 1930. Stratigraphy and paleontology of the Paleocene of northeastern Park County, Wyoming. *Proceedings of the American Philosophical Society,* v. 69, p. 463–528.

Jepsen, G. L., 1940. Paleocene faunas of the Polecat Bench Formation, Park County, Wyoming. *Proceedings of the American Philosophical Society,* v. 83, p. 217–341.

Knox, R. W. O'B., 1984. Nannoplankton zonation and the Paleocene/Eocene boundary beds of NW Europe: An indirect correlation by means of volcanic ash layers. *Journal of the Geological Society, London,* v. 141, p. 993–999.

Koch, C. F., 1987. Prediction of sample size effects on the measured temporal and geographic distribution patterns of species. *Paleobiology,* v. 13, p. 100–107.

Koch, P. L., Zachos, J. C. and Gingerich, P. D., 1992. Correlation between isotope records in marine and continental carbon reservoirs near the Palaeocene/Eocene boundary. *Nature,* v. 358, p. 319–322.

Koch, P. L., Zachos, J. C. and Dettman, D. L., 1995. Stable isotope stratigraphy and paleoclimatology of the Paleogene Bighorn Basin (Wyoming, USA). *Palaeogeography, Palaeoclimatology, Palaeoecology,* v. 115, p. 61–89.

Kraus, M. J., 1980. Genesis of a fluvial sheet sandstone, Willwood Formation, northwest Wyoming. In Gingerich, P. D. (ed.), *Early Cenozoic Paleontology and Stratigraphy of the Bighorn Basin, Wyoming,* University of Michigan Papers on Paleontology, v. 24, p. 87–94.

Krause, D. W., 1980. Multituberculates from the Clarkforkian Land-Mammal Age, late Paleocene–early Eocene, of western North America. *Journal of Paleontology,* v. 54, p. 1163–1183.

Krause, D. W., 1982. Multituberculates from the Wasatchian Land-Mammal Age, early Eocene, of western North America. *Journal of Paleontology,* v. 56, p. 271–294.

Krause, D. W. and Maas, M. C., 1990. The biogeographic origins of late Paleocene–early Eocene mammalian immigrants to the Western Interior of North America. In Bown, T. M. and Rose, K. D. (eds.), *Dawn of the Age of Mammals in the Northern Part of the Rocky Mountain Interior, North America,* Geological Society of America Special Paper 243, p. 71–105.

Li, C. K. and Chow, M. C., 1994. The origin of rodents. In Tomida, Y., Li, C. K., and Setoguchi, T. (eds.), *Rodent and Lagomorph Families of Asian Origins and Diversification,* Tokyo: National Science Museum Monographs, v. 8, p. 15–18.

Li, C. K and Ting, S. Y., 1985. Possible phylogenetic relationship of Asiatic eurymylids and rodents, with comments on mimotonids. In Luckett, W. P. and Hartenberger, J.-L. (eds.), *Evolutionary Relationships among Rodents,* NATO ASI Series, v. 92, Plenum, New York, p. 35–58.

Li, C. K. and Ting, S. Y., 1993. New cranial and postcranial evidence for the affinities of the eurymylids (Rodentia) and mimotonids (Lagomorpha). In Szalay, F. S., Novacek, M. J. and McKenna, M. C. (eds.), *Mammal Phylogeny: Placentals,* New York: Springer-Verlag, p. 151–158.

Lyell, C., 1833. *Principles of Geology, Being an Attempt to Explain the Former Changes of the Earth's Surface, by Reference to Causes Now in Operation.* London: J. Murray, v. 3, 398 pp. (Facsimile edition, Chicago, University of Chicago [published, 1991]).

Maas, M. C., Anthony, M. R. L., Gingerich, P. D., Gunnell, G. F. and Krause, D. W., 1995. Mammalian generic diversity and turnover in the late Paleocene and early Eocene of the Bighorn and Crazy Mountains Basins, Wyoming and Montana (USA). *Palaeogeography, Palaeoclimatology, Palaeoecology,* v. 115, p. 181–207.

MacPhee, R. D. E., Cartmill, M. and Rose, K. D., 1989. Craniodental morphology and relationships of the supposed Eocene dermopteran *Plagiomene* (Mammalia). *Journal of Vertebrate Paleontology,* v. 9, p. 329–349.

MacPhee, R. D. E. and Novacek, M. J., 1993. Definition and relationships of Lipotyphla. In Szalay, F. S., Novacek, M. J. and McKenna, M. C. (eds.), *Mammal Phylogeny: Placentals,* New York: Springer-Verlag, p. 13–31.

Matthew, W. D. and Granger, W., 1915. A revision of the lower Eocene Wasatch and Wind River faunas. Parts I–IV. *Bulletin of the American Museum of Natural History,* v. 34, p. 1–103, 311–361, 429–483.

Matthew, W. D. and Granger, W., 1918. A revision of the lower Eocene Wasatch and Wind River faunas. Part V. *Bulletin of the American Museum of Natural History,* v. 38, p. 565–657.

McKenna, M. C., 1990. Plagiomenids (Mammalia: ?Dermoptera) from the Oligocene of Oregon, Montana, and South Dakota, and middle Eocene of northwestern Wyoming. In Bown, T. M. and Rose, K. D. (eds.), *Dawn of the Age of Mammals in the Northern Part of the Rocky Mountain Interior, North America,* Geological Society of America Special Paper 243, p. 211–234.

Neasham, J. W. and Vondra, C. F., 1972. Stratigraphy and petrology of the lower Eocene Willwood Formation, Bighorn Basin, Wyoming. *Bulletin of the Geological Society of America,* v. 83, p. 2167–2180.

Rose, K. D., 1975. The Carpolestidae: early Tertiary Primates from North America. *Bulletin of the Museum of Comparative Zoology,* v. 147, p. 1–74.

Rose, K. D., 1981. The Clarkforkian Land-Mammal Age and mammalian faunal composition across the Paleocene/Eocene boundary. *University of Michigan Papers on Paleontology,* v. 26, 197 pp.

Rose, K. D. and Krause, D. W., 1982. Cyriacotheriidae, a new family of Tertiary pantodonts from western North America. *Proceedings of the American Philosophical Society,* v. 126, p. 26–50.

Schankler, D. M., 1980. Faunal zonation of the Willwood Formation in the central Bighorn Basin, Wyoming. In Gingerich, P. D. (ed.), *Early Cenozoic Paleontology and Stratigraphy of the Bighorn Basin, Wyoming: 1880–1980,* University of Michigan Papers on Paleontology, v. 24, p. 99–114.

Schimper, W. P., 1874. Traité de Paléontologie Végétale. Paris: J.B. Baillière, v. 3, 896 pp.

Van Houten, F. B., 1944. Stratigraphy of the Willwood and Tatman Formations in northwestern Wyoming. *Bulletin of the Geological Society of America,* v. 55, p. 165–210.

Van Valen, L., 1966. Deltatheridia, a new order of mammals. *Bulletin of the American Museum of Natural History,* v. 132, p. 3–126.

Van Valen, L., 1978. The beginning of the Age of Mammals. *Evolutionary Theory,* v. 4, p. 45–80.

Von Koenen, A., 1885. Über eine Palaocäne Faune von Kopenhagen. *Abhandlungen Konglige Gesellschaft für Wissenschaft, Gottingen,* v. 32, 128 pp.

Wing, S. L., 1984. A new basis for recognizing the Paleocene/Eocene boundary in western interior North America. *Science,* v. 226, p. 439–441.

Wing, S. L., Bown, T. M. and Obradovich, J. D., 1991. Early Eocene biotic and climatic change in interior western North America. *Geology,* v. 19, p. 1189–1192.

Winkler, D. A., 1983. Paleoecology of an early Eocene mammalian fauna from paleosols in the Clark's Fork Basin, northwestern Wyoming (U.S.A.). *Palaeogeography, Palaeoclimatology, Palaeoecology,* v. 43, p. 261–298.

Wood, H. E., II, Chaney, R. W., Clark, J., Colbert, E. H., Jepsen, G. L., Reeside, J. B., Jr. and Stock, C., 1941. Nomenclature and correlation of the North American continental Tertiary. *Bulletin of the Geological Society of America,* v. 52, p. 1–48.

Woodburne, M. O. (ed.), 1987. *Cenozoic Mammals of North America: Geochronology and Biostratigraphy.* Berkeley, Los Angeles, London: University of California, 336 p.

Wyss, A. R. and Flynn, J. J., 1993. A phylogenetic analysis and definition of the Carnivora. In Szalay, F. S., Novacek, M. J. and McKenna, M. C. (eds.), *Mammal Phylogeny: Placentals,* New York: Springer-Verlag, p. 32–52.

20

Mammalian Faunal Change Across the Paleocene–Eocene Transition in Europe

JEREMY J. HOOKER

ABSTRACT

Mammalian faunal succession in Europe is documented from about 56.5 to 52 Ma, across the interval of the Paleocene–Eocene transition. Changes to the relative dating of some European mammal localities are proposed. A series of northwest European mammal zones is correlated with both the western North American mammalian biostratigraphic scheme and the standard calcareous nannoplankton zonation (NP) within this interval. The major mammalian faunal turnover at the boundary of the Cernaysian and Neustrian European Land Mammal Ages is shown to occur early in Chron C24r. Ecological diversity analysis of key northwest European mammal assemblages is carried out to assess the paleoenvironmental changes through this time interval.

The distinctive nature of the mammalian faunal turnover in Europe around the Paleocene/Eocene Epoch boundary has been known since the last century (e.g., Lemoine 1889), although a key landmark work was that by Teilhard de Chardin (1922). At this time, however, the term *Paleocene* was still not in common use and all the faunas were referred to the Eocene Epoch/Series. Also, the exact source and stratigraphic context of many of the faunas were poorly known. It was not until the 1950s that intensive collecting using screenwashing techniques, especially in the Paris Basin, began to clarify the nature and succession of the faunas (Russell 1968).

It is not easy to correlate land mammal faunas with marine sequences, because the fossil evidence naturally comes from mainly nonmarine strata. However, where facies intercalation or finds in coastal deposits are involved, some direct calibration with dinocyst and calcareous nannoplankton zones is possible and extrapolation to fuller marine sequences of NP (calcareous nannoplankton) and P (planktonic foraminifera) zones has been attempted (e.g., Wing 1984; Hooker 1996b). Where a magnetostratigraphy has been established, as in southern England (Ali et al. 1993, Ellison et al. 1996), indirect correlation suggests that the mammal turnover occurred early in Chron C24r (figure 20.1; Hooker 1996b). This reversed interval spans all the currently favored positions for a revised definition of the Paleocene/Eocene Series/Epoch boundary, of which the mammal event is one of the earliest.

The purpose of this chapter is to investigate the pattern of mammal faunal turnover through this interval. In particular, it attempts to improve mammal correlation within Europe, calibrate European mammal events with both global marine events and with the well-documented mammal events in continental western North America, and interpret faunal change in terms of the broader paleoenvironment.

Some Definition Problems

The main mammalian faunal turnover is recognizable only in the Paris Basin, where it is usually said to occur between the Thanetian and Sparnacian stages. Problems surround the Sparnacian Stage because of its widespread use outside the boundaries of its original concept, including as a land mammal age (e.g., Russell et al. 1982; Savage and Russell 1983). These problems are largely due to a long-term misunderstanding of its age span because of reliance on lithofacies correlation. Laurain et al. (1983) have convincingly demonstrated on dinocyst and charophyte evidence that the type of the Sparnacian Stage, defined on the argiles à lignites facies in the region of Epernay (now termed Argiles à Lignites d'Epernay), is distinctly younger than the argiles à lignites facies in the region of Soissons further to the northwest (termed Argiles à Lignites du Soissonnais). It

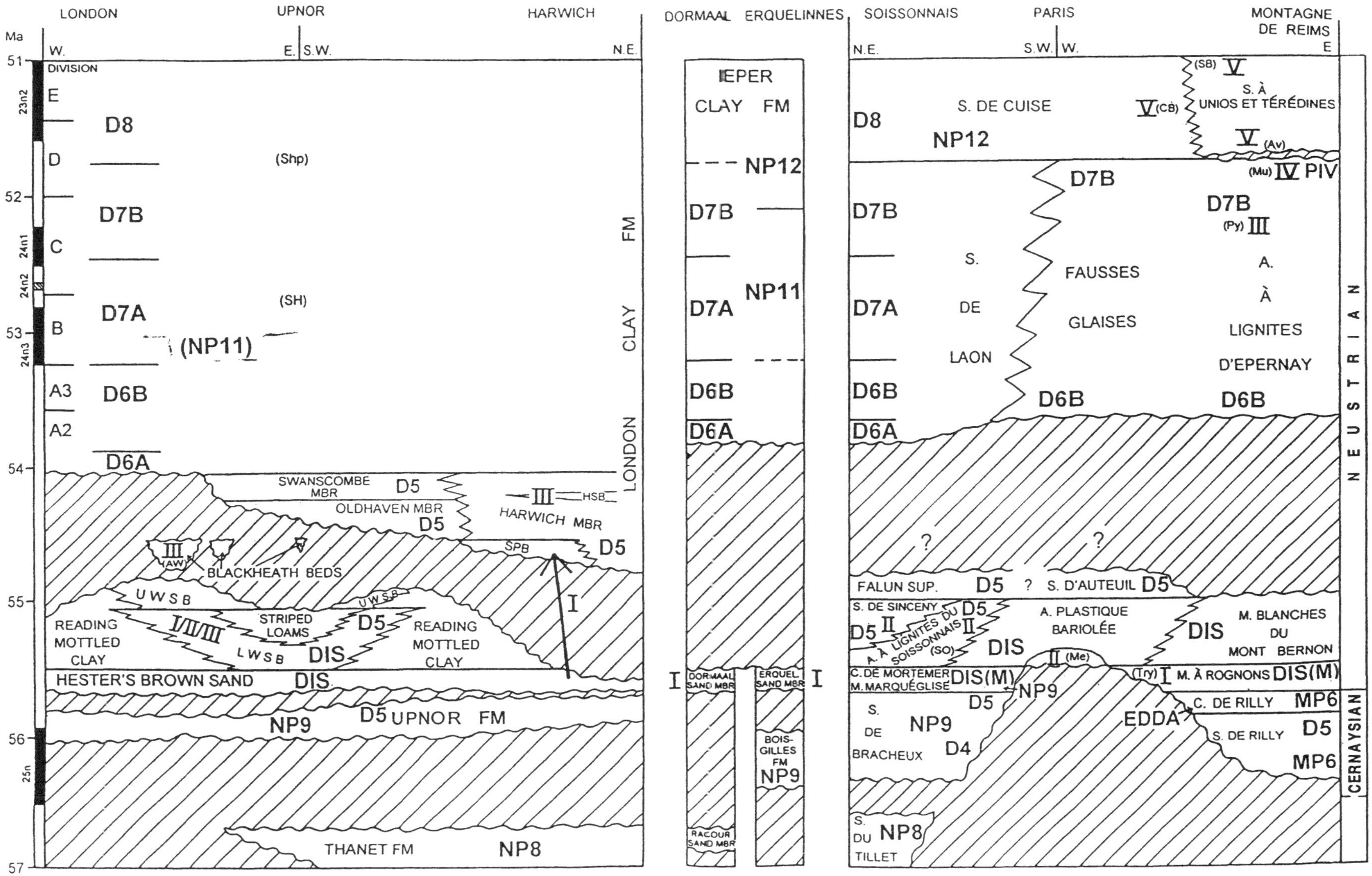

Figure 20.1 Correlation of upper Paleocene and lower Eocene strata in selected parts of the London, Belgian, and Paris Basins.

Calibrated to the geomagnetic time scale of Cande and Kent (1992) and the magnetobiochronologic framework of Berggren and Aubry (1996), modified from Hooker (1996b:figure 3). Magnetostratigraphy has so far been documented only in the London Basin and in the Ieper Clay of the Belgian Basin (Ali et al. 1993; Ellison et al. 1996) and correlation elsewhere is by means of biostratigraphy alone. Chron C24n.2n is shown hatched to indicate that it has not been identified in these basins. The correlation framework relies heavily on the dinocyst zonation (D zones; Costa and Manum 1988; Chateauneuf and Gruas-Cavagnetto 1978; De Coninck 1990; Gruas-Cavagnetto 1968, 1974, 1976; Knox et al. 1983; Laurain et al. 1983; Powell 1992) backed up by calcareous nannoplankton (NP zones; Aubry 1983, is in fact the latter lithostratigraphic unit that is usually used to mean Sparnacian Stage. This unit underlies the Sables de Laon, while its probable age equivalents in the London Basin, the Woolwich and Reading Formations, underlie the London Clay Formation. Both overlying units are marine and dated as Ypresian Stage (global stage). The Argiles à Lignites du Soissonnais have been dated either as of earliest Ypresian Age (e.g., Feugueur 1963) or as of latest Thanetian Age (e.g., Costa et al. 1978), but neither dating has any relevance to the type Sparnacian Stage, which is age equivalent of the reliably Eocene Sables de Laon. Cavelier and Pomerol (1986:257) have proposed emending the Sparnacian to the "pre-Cuisian formations" in the Paris Basin. However, this action has the undesirable effects of excluding the stratotype sequence from the stage's time range and perpetuating a long-standing error of correlation. Moreover, no new stratotype has been proposed. Because of these problems, use of the Sparnacian Stage herein is restricted to its type concept as a local stage and another term is used for the mammal faunas usually so designated.

Fahlbusch (1976) established a series of European Tertiary Land Mammal Ages (ELMA) from the Eocene Series onwards, in parallel with those already in use in North America (NALMA). He proposed Neustrian (Neustrium) for mammal faunas previously termed Sparnacian, in particular to encompass the reference faunas of Dormaal, Mutigny, and Avenay. Neustrian has been little used since its inception and was considered

Figure 20.1 *(continued)*

1986; De Coninck et al. 1981; Siesser et al. 1987; Steurbaut 1990) and charophytes (Riveline 1984). Mammal zones (PE I to PE V; Hooker 1996b) and a relevant reference level (MP6; Brunet et al. 1987) are added to this framework, which they support and, in the lower part of the sequence, help to refine. In particular they support the idea of diachronism of the Argiles à Lignites facies (Cavelier 1987). Although the complete thickness of the London Clay is shown, erosion has reduced its thickness progressively northeastward until only the lower units are present in the Harwich area. Reading Mottled Clay and Hester's (1965) brown sand are informal subdivisions of the Reading Formation; Woolwich Shell Beds and Striped Loams are informal subdivisions of the Woolwich Formation. Upnor Formation (Ellison et al. 1994) replaces "Bottom Bed" of authors. The Harwich, Swanscombe, and Oldhaven Members and the Blackheath Beds have been included in a "Harwich Formation" by Ellison et al. (1994); the term is not used here as it is based on time correlation of the above-named lithofacies, not on strict lithostratigraphy. Major unconformities are indicated by hatching bounded by wavy lines. Abbreviations: A = Argiles, C = Calcaire, HSB = Harwich Stone Band, L = lower, M = Marnes, Mbr = Member, S = Sables, SL = Striped Loams, SPB = Suffolk Pebble Beds, U = upper, WSB = Woolwich Shell Beds. Charophyte zones: EDDA = *Sphaerochara edda*, DIS = *Peckichara disermas*, (M) = range of *P. microcarpa* within DIS, PIV = *P. piveteaui*. Mammal localities: (Av) = Avenay, (AW) = Abbey Wood, (CB) = Condé-en-Brie, (Me) = Meudon (Conglomérat de Meudon), (Mu) = Mutigny, (Py) = Pourcy, (SB) = Sézanne-Broyes, (SH) = Studd Hill, (Shp) = Sheppey, (So) = Soissons.

an unnecessary ELMA by Savage and Russell (1977), who together with many other mammalian biostratigraphers have continued to use Sparnacian for the same span of faunas. However, it is not just paleomammalogists who use Sparnacian, and their concepts are often different. The only way out of the confusion seems to be to abandon the term Sparnacian for mammal faunas, and instead to use Neustrian as an ELMA in its originally defined sense. It is pertinent to point out here that ELMAs (and NALMAs) are not strictly ages in the geochronological sense of Hedberg (1976). Their concept indicates that each is interpreted to define a span of time, but their content and structure (embracing multiple biozones or faunas of mammals) is clearly biostratigraphic (see e.g., Rose 1981a:20–21; Woodburne 1987).

Cernaysian (Lemoine 1889) was established for mammal faunas that immediately predate the Neustrian ELMA and that are represented by the faunas of Cernay, Berru, and the Marnes de Montchenot in the eastern Paris Basin. These faunas are often dated as Thanetian in age. However, it is important to be able to distinguish global time from regional faunal succession. Thus Cernaysian and Neustrian ELMAs are used herein as independent biostratigraphic/biochronologic units.

Brunet et al. (1987) established a series of mammalian reference levels for the European Paleogene System as a more accurate way of dividing up the faunas. These have the prefix MP and are numbered 1–30. Although purporting to be a system of ordering points in time as represented by the total fauna of successive localities, they are usually used like conventional zones, each representing a range of occurrences (see Hooker 1991 and references therein for discussion). The Cernaysian ELMA is represented by Reference Level MP6 and the Neustrian ELMA by Reference Levels MP7 to MP8–9.

Problems over resolution of Reference Level MP8–9 and the definition and time order of some important localities attributed to Reference Level MP7 (Godinot in Brunet et al. 1987; Hooker 1991) have been largely resolved using a combination of parsimony analysis and links to other biostratigraphies (Hooker 1996b). The methodology can be summarized thus: A matrix of taxa and localities has been analyzed phylogenetically using PAUP (Swofford 1990), treating localities as taxa and taxa as apomorphies, to achieve the most parsimonious pattern of relationships between localities on the basis of the taxa shared among them. The result integrates well with the existing dinocyst and charophyte zonations and overcomes the paucity of reliable superpositional data for the Paris Basin mammal localities. In the northern European area, Reference Levels MP7 to MP8–9 have been replaced by five concurrent range biozones. They are in ascending order the *Platychoerops georgei–Cymbalophus cuniculus* Biozone (Zone PE I),

Platychoerops russelli–Teilhardina belgica Biozone (Zone PE II), *Platychoerops daubrei–Cantius eppsi* Biozone (Zone PE III), *Cantius savagei–Arcius fuscus* Biozone (Zone PE IV), and *Donrussellia gallica–Apatemys sigogneaui* Biozone (Zone PE V). They have been numbered for convenience PE I–V, PE standing for Paleocene–Eocene interval, as they span more than one concept of the epoch/series boundary. Thus the major mammalian faunal turnover between the Cernaysian and Neustrian ELMAs can be focused on the transition from Reference Level MP6 to Zone PE I. The major localities attributed to Reference Level MP6 and Zones PE I to PE IV are indicated in appendix 20.1. These cover inclusively the interval from faunas that are definitely Paleocene in age (Cernaysian ELMA) to sufficiently late in the Neustrian ELMA (i.e., the Mutigny locality, high in the Argiles à Lignites d'Epernay, i.e., type Sparnacian) to be considered definitely Eocene in age. Biostratigraphically important taxa in the higher Zone PE V are listed in Hooker (1996b). Detailed arguments regarding the order of these localities and their link with standard calcareous nannoplankton (NP) and with European dinocyst (D) and charophyte zonations can be found in Hooker (1996b). The localities, with their biozonal reference number, can be found on the correlation chart of relevant parts of the London, Belgian, and Paris basins (figure 20.1). Calibration to other biostratigraphies is shown in table 20.1. Essential congruence of the mammalian biozones (and reference levels if treated as zones) with those of both the charophytes and dinocysts is considered to be convincing ev-

Table 20.1

Correlation of northwestern European mammalian, charophyte, dinocyst, and standard calcareous nannoplankton biostratigraphies in the upper Paleocene and lower Eocene Series.

	NP NW	Dinocysts NW	Mammals NW	ELMA	Mammals Locs SW	Charophytes NW
51						
	12	D8	PE V			
					Fournes	
52		D7B	PE IV	N E U S T R I A N	Palette	piveteaui
	11	D7A	PE III			
53	(11)	D6B			Fordones	
54	(10)	D6A			Rians	
		D5				
55						
	(9)		PE II		Silveirinha?	disermas
			PE I			microcarpa
56	9	D5	MP 6	C E R N		edda
		D4				
	8					
57						

Mammalian (Hooker 1996b; Godinot in Brunet et al. 1987), charophyte (Riveline 1984), dinocyst (Costa and Manum 1988), and standard calcareous nannoplankton (NP; Aubry 1983, 1986). The positioning of the southwest European mammal localities follows the discussion in the text. The position of Silveirinha is problematic; it is shown at its potentially oldest, although a level equivalent to Palette is more likely. Dashed boundaries indicate imprecision. NP zones in parentheses indicate extrapolation from outside the area mainly via tephrostratigraphy (Knox 1990). The timescale on the left is in Ma (= mega-annum = millions of years BP) and follows that in figure 20.1. CERN = Cernaysian, ELMA = European land mammal age.

idence that the former represent a sequence of synchronous time intervals.

The Pattern of Change

The Paris, London, and Belgian Basins

Appendix 20.1 lists mammal taxa for key localities in these three basins from Reference Level MP6 to Zone PE IV. Table 20.2 gives figures for the faunal turnovers discussed below. Cernay and Berru (main level, not Lentille) are thought to lie at successive levels within the Sables de Rilly of the Montagne de Reims (Russell 1980). In practice, however, their exact stratigraphic relationships are not so clear because the two localities are separated by a lateral distance of about 1 km and their successions are not identical. Likewise, the four Zone PE I localities probably cover a relatively short time span and are thus likely to differ little in age. For this reason, in order to establish the pattern of change, the faunas respectively of Reference Level MP6 and Zone PE I are combined here. Thirty species-rank taxa have their last appearances in Reference Level MP6, and 36 have their first appearances in Zone PE I (appendix 20.1). Excepting "Nyctitheriidae undet.," only five are seen to survive beyond Reference Level MP6 and some of these with only doubtful species attributions (cf. and aff.). However, this number of 30 last appearances includes authenticated as well as potential pseudoextinctions. Although many taxa from this sequence have been described over the past 30 years, their precise phylogenetic relationships have not usually been established. If one therefore excludes all those Reference Level MP6 taxa that have closer or potentially closer relatives in Zone PE I than do other Reference Level MP6 age equivalent taxa elsewhere in the world, the minimum number of true extinctions drops to 19, or 53% of the fauna.

Of the first appearances in Zone PE I, 25 have no close relatives in Reference Level MP6 and thus are considered to represent dispersals from outside the area. They represent 61% of the total Zone PE I fauna. The remainder are likely to have evolved from Reference Level MP6 taxa in the area. Of the 25 dispersals, 16 have relatives in the Clarkforkian NALMA. The identification of Chron C25n in both the lower Clarkforkian (Butler et al. 1987) and the base of the Upnor Formation (Ellison et al. 1996), and the identity of certain short-ranged taxa in the succeeding Wasatchian Zone Wa0 with ones in Zone PE I, provide strong evidence that the Clarkforkian NALMA predates Zone PE I (Gingerich 1989). The incoming 16 taxa are thus deduced to have dispersed from North America. These figures may change slightly when studies of groups that have yet to be treated systematically in detail are completed (e.g., multituberculates, pantolestids, carnivores, and *Coryphodon*), although they are unlikely to affect the overall significance of the turnover. The remaining nine Zone PE I taxa that represent dispersals are interesting in that they belong to families or orders that also appeared penecontemporaneously in North America and Asia. Their near relatives and their center(s) of origin remain to be established.

The main groups that suffered decimation between Reference Level MP6 and Zone PE I are the order Multituberculata (near total extinction of the European species, but eventual replacement by the incoming North American *Ectypodus*), the paraphyletic order "Condylarthra," especially the family Arctocyonidae, and the "Insectivore" family Adapisoriculidae. The main groups dispersing from North America at this time are the orders Rodentia, Carnivora, Pantodonta, and Apatotheria (the "*Jepsenella*" from Cernay (Szalay 1968) is not an apatothere; personal observation), the families Oxyaenidae (Creodonta), Amphilemuridae (sensu Storch and Richter 1994; Lipotyphla) and Phenacodontidae (Paenungulata), and the genus *Hyopsodus*. The appearances whose centers of origin are obscure relate to the

Table 20.2
Figures for faunal turnover derived from appendix 20.1

Zone or Level	No. of TX	First Appearances: Dispersals: N.Am.	Other	Speciation/Evolution	Last Appearances: Extinctions	Pseud.	Survivals into later Zones
MP6	36				19 (53)	11 (31)	6 (16)
PE I	41	16 (39)	9 (22)	11 (27)	11 (27)	10 (24)	20 (49)
PE II	33	5 (15)	3 (9)	12 (36)	2 (6)	7 (21)	24 (73)
PE III	48	2 (4)	5 (11)	17 (35)	12 (25)	6 (13)	30 (62)
PE IV	44	1 (2)	2 (5)	9 (20)	7 (16)	2 (5)	35 (79)

Figures in parentheses are percentages of the total fauna, the rest are numbers of taxa. All taxa are treated as species, although some are recorded at genus or higher rank. Abbreviations: TX = taxa, N.Am. = from North America, Pseud. = pseudoextinctions and potential pseudoextinctions.

orders Perissodactyla, Artiodactyla, and Primates, and the families Hyaenodontidae (Creodonta) and herpetotheriine "Didelphidae" (Marsupialia; see discussion below).

The scale of turnover in Europe is greater than in the Bighorn Basin, where the minimum true extinction of the Clarkforkian fauna is about 40% (data from Rose 1981a; Gingerich 1983, 1989; Gunnell and Gingerich 1991; Thewissen 1990) and origination in Wasatchian Zone Wa0 of species with no close relatives in the Clarkforkian NALMA is 29% of the total Zone Wa0 fauna (Gingerich 1989, 1993). The origination percentage here may be an underestimate as Zone Wa0 localities are apparently undersampled for very small animals, resulting in a possibly artificial absence of, e.g., marsupials and certain amphilemurid genera that are otherwise known to appear slightly later in the Wasatchian NALMA (Schankler 1980). The lower level of extinction is probably due to the earlier appearances in North America of coryphodontid Pantodonta, Rodentia, *Hyopsodus* (in the Clarkforkian NALMA), Oxyaenidae, Carnivora, Phenacodontidae, and Amphilemuridae (earlier in the Paleocene Epoch), which in Europe combined with the Wasatchian newcomers to intensify the turnover. In the Nemegt Basin, Mongolia, however, the turnover was greater than in Europe, with 67% true extinction of the Nongshanian fauna and 88% of the succeeding Bumbanian fauna having no close relatives in the Nongshanian fauna (data from Dashzeveg 1988).

Further dispersals and extinctions took place in Europe in subsequent Neustrian zones, but they were minor in degree compared to those between Reference Level MP6 and Zone PE I. In Zone PE II, 2 new families appeared, the Peradectidae (Marsupialia) and the Paromomyidae (Plesiadapiformes), both dispersed from North America, while the giant North American mesonychid *Pachyaena gigantea* paid a fleeting visit. Appearing also were the first members of the order Chiroptera (cf. *Icaronycteris*) from an area unknown. In Zone PE III are the first members of the order Tillodontia (family Esthonychidae) to disperse from North America and the first occurrences of ?*Didelphodus* (Cimolestidae), *Hyrachyus* (Hyrachyidae), and *Placentidens* (?Plagiomenidae), whose place of origin is uncertain. The spot occurrence of *Lessnessina* is enigmatic; the genus was described as a periptychid (Hooker 1979), but suggestions of affinities with the Anagalida (Archibald et al. 1983) and Taeniodonta (Estravis and Russell 1992) have also been made. Zone PE III is also marked by the last occurrences in Europe of the Oxyaenidae (*Palaeonictis* and *Oxyaena*), the Pantodonta (*Coryphodon*), the Arctocyonidae (*Arctocyonides* and *Landenodon*) and the Equidae (*Pliolophus*). The ancylopod perissodactyl *Lophiaspis* appeared in Zone PE IV, perhaps dispersing from North America, at the same time as the first significant occurrences of bats (three species).

Southwestern Europe

Unfortunately, only one Thanetian site (Campo, Spain) with only one mammal tooth (?*Paschatherium* sp.) is so far known, although it is well dated, being associated with a Zone NP8 calcareous nannofossil assemblage (Tambareau et al. 1992). In recent years, however, significant Neustrian sites have been discovered or reexcavated in South France and Portugal. The main ones are the Portuguese site of Silveirinha (Estravis 1992) and the French sites of Rians (Godinot 1981), Palette (Godinot et al. 1987), Fordones, and Fournes (Marandat 1991). A minor French site is Monze (Marandat 1991). These faunas are dominated by small forms with bunodont or semibunodont teeth, in particular the genus *Paschatherium*. These taxa are usually referred to either hyopsodontid "Condylarths" (e.g., Godinot 1981) or adapisoricid "Insectivores" (e.g., Van Valen 1967). Of slightly less dominance are rodents, primates, and marsupials. Oxyaenid creodonts, equoid perissodactyls, plesiadapids, and *Coryphodon* are notable absentees. The sites have few species in common except the tapiromorph perissodactyl *Pachynolophus hookeri* (see Hooker 1994b), which occurs at Palette, Monze, Fournes, and Silveirinha (the last recorded as *Hyracotherium* cf. *vulpiceps* by Estravis 1992). They are therefore very difficult to correlate one with another.

They also pose problems for correlation with the northern basins. They have almost universally been attributed to Reference Level MP7 and thought to be older than any faunas yet described from the north. This is based on concepts of primitive evolutionary grade, which in several cases have been disputed, e.g., *Diacodexis antunesi* from Silveirinha (Estravis and Russell 1989; Gingerich 1989) and *Donrussellia* from Palette (Godinot et al. 1987; Gingerich 1989). One of the problems is the similarity of some taxa to those from the Paris Basin locality of Pourcy, which was originally assigned to Reference Level MP7, but which is better placed in Zone PE III (Hooker 1996b). Key southern genera like *Lophiaspis* and *Donrussellia* are first recorded in the Paris Basin in Zones PE IV and PE V, respectively. It could of course be argued alternatively that these genera originated earlier in the south and later dispersed northwards. The most serious shortcoming of dating these localities as early Neustrian in age, however, is the position of the Fordones (Corbières) locality. It is sandwiched between marine strata, which show that it straddles the NP10/NP11 zonal boundary

(Marandat 1991). This position is supported by dinocyst evidence (Gruas-Cavagnetto 1992). This would admittedly place Fordones slightly older than Pourcy, but still well within the span of Zone PE III. Moreover, the paromomyid *Arcius* occurs in all, and the peradectid *Peradectes* in several (absent from Silveirinha and Palette) of the main southern Neustrian localities. The Peradectidae and Paromomyidae have a long Paleocene history exclusive to North America and first occur in the northern European basins in Zone PE II. As the generally accepted route from North America to Europe was via Greenland, the earliest they could have reached southern Europe was also during Biochron PE II. Charophyte occurrences in lower parts of the Corbières sequence yield the successive ranges of *Sphaerochara edda* and *Peckichara disermas*. They correlate respectively with lower and higher parts of Zone NP9 with a small overlap (Massieux et al. 1981). The range of *P. disermas* in the Paris Basin correlates with Zones PE I and II (Riveline 1984; Hooker 1996b), further supporting a higher PE zone for at least Fordones among the southern mammal localities.

If one considers the stratigraphic ranges in the Paris Basin of taxa shared with these southern localities, the position of range overlap in the north should be the best guide to relative age in the south. Thus for Rians, cf. *Didelphodus*, *Ectypodus childei*, *Peradectes louisi*, *Peratherium matronense*, *Cantius eppsi*, and *Paschatherium dolloi* co-occur only in Zone PE III; only *Amphiperatherium goethei* is anomalous, first occurring in Zone PE V. For Palette, *Peratherium constans* and *Lophiaspis maurettei* co-occur only in Zone PE IV. For Fournes, the youngest record for *Arcius fuscus* is in Zone PE IV, while the oldest for *Donrussellia gallica* is in Zone PE V. For Fordones, *Peratherium constans* and *Amphiperatherium brabantense* both range from Zone PE I to Zone PE IV. Silveirinha combines *Lophiaspis*, *Heterohyus*, and *Didelphodus*, whose oldest records in the Paris Basin are in Zone PE IV (for the first two) and PE III (for the last), with *Miacis* cf. *latouri* and *Microparamys* cf. *nanus*, whose typical members are restricted to Zone PE I. Table 20.2 summarizes the correlations proposed here.

Correlation Between Europe and North America

Biostratigraphy

In view of the long-recognized faunal similarities at the genus level (e.g., Savage 1971) between western North American Wasatchian and European Neustrian mammal faunas, implying major interchange, it is strange that few species have been recognized to be in common between the two continents. Until the last decade, this applied only to *Pachyaena gigantea* (see Boule 1903). At the present time (excluding one vagrant bat and one ubiquitous primate), 11 species of land mammal are shared between North America and Europe, thanks to present bridging Arctic sea ice and past intermittent high-latitude land connections during cold phases of the Pleistocene Epoch. A partial explanation for the discrepancy in the early Eocene Epoch could be the relative paucity of intercontinental taxonomic studies, the fragmentary nature of much of the European material, the post-interchange age of the best known faunas and, in the pre–plate-tectonics past, a psychological barrier to acceptance that a fossil species may occur on more than one continent.

Recently, however, the multituberculate *Ectypodus childei*, known from Zones PE III and IV has been recorded from probable Zone Wa2–3 equivalents in the Four Mile area, Colorado (Sloan 1981; Krause 1982 with cf.). Moreover, the hyaenodontid *Arfia junnei* has been identified in the Zone Wa0 sites and the Zone PE I site of Dormaal (Gingerich 1989). It is now recognized at two more Zone PE I sites: Try and Kyson (Hooker 1996b; figure 20.2). The genus is typified by its rela-

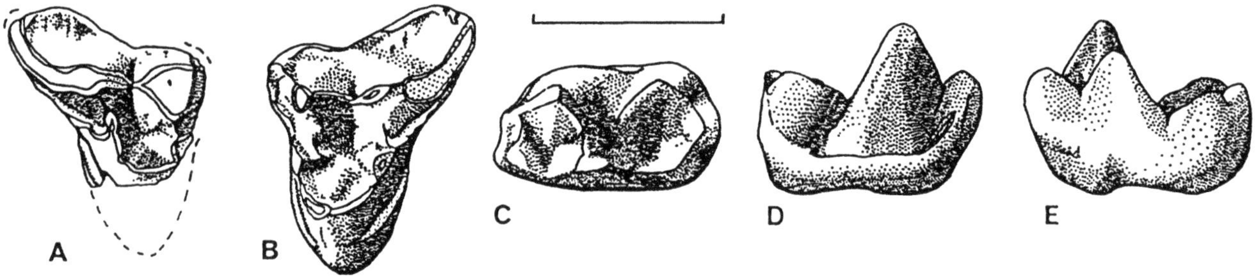

Figure 20.2 Molar teeth of *Arfia junnei* Gingerich.

(A) right M^1 from Try (Marne), France (Muséum National d'Histoire Naturelle, Paris, no. Try-03-L); (B) left M^1 from Zone Wa0, site SC-67, Clark's Fork Basin, Wyoming (University of Michigan Museum of Paleontology, no. 71767); (C–E) right M_1 from the Suffolk Pebble Beds, Kyson, Suffolk, England (Yorkshire Museum, no. YORYM-G-1129). A–C are occlusal views; D is buccal view; E is lingual view. Scale bar = 5 mm.

tively broad talonid and low trigonid on M_1, and widely spaced paracone and metacone and unnotched postmetacrista on M^1. The species is distinguished on size. The similarity, although not identity between Zone Wa0 "*Hyracotherium*" *sandrae* and Zone PE I *Cymbalophus cuniculus* (see Gingerich 1989:figure 39) and between Zone Wa4 *Pliolophus pernix* and Zone PE III *Pliolophus vulpiceps* (see Hooker 1994b) has also been recognized.

Comparison of oxyaenid creodonts from several London and Paris Basin sites with North American species leads me to recognize the presence of two species shared between the two continents. These are *Dipsalidictis transiens*, known from Zone Wa0 to lowest Zone Wa3, which occurs at the French Zone PE II sites of Meudon and Sinceny, and *Oxyaena gulo*, known from Zones Wa2 and 3, which occurs at the English Zone PE III site of Abbey Wood (figure 20.3). The presence of *Dipsalidictis* in Europe was recognized by Gunnell and Gingerich (1991), but the same genus was mistakenly identified at Abbey Wood. The high protoconid, low metaconid, and low, short talonid on the M_2 (figure 20.3J–O) are typical of *Oxyaena* and size matches *O. gulo* closely. Similarly, on M^1, the more distally orientated postmetacrista of *Oxyaena* means a less buccally flared metastylar region and narrower stylar shelf than in *Dipsalidictis* (cf. figure 20.3H–I). The size and morphology of the Meudon and Sinceny *Dipsalidictis* M_1s is so close to those of *D. transiens* as to indicate specific identity. In the Clark's Fork Basin, Wyoming, *Dipsali-*

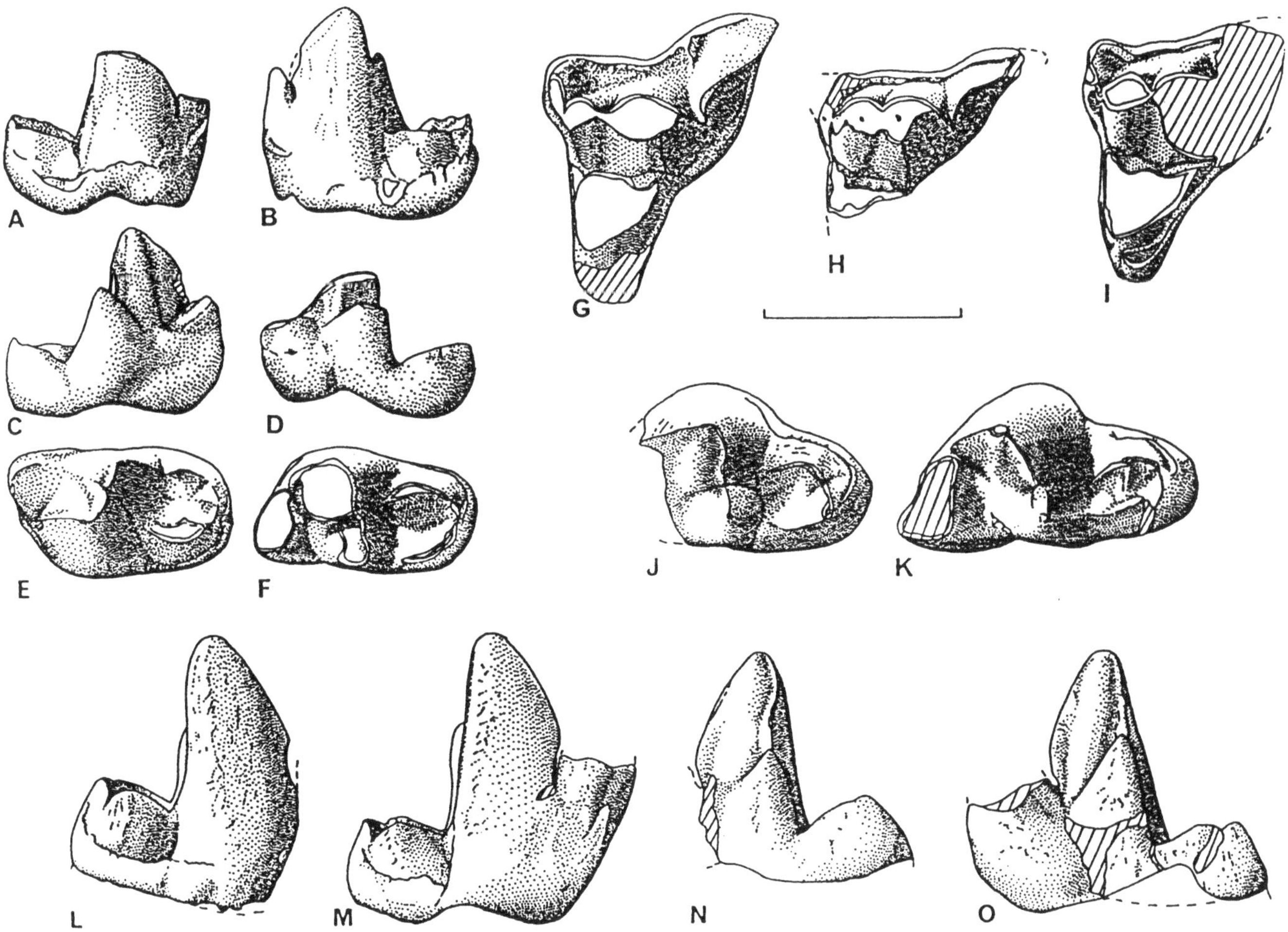

Figure 20.3 Molar teeth of *Dipsalidictis transiens* (Matthew) (A–G) and *Oxyaena gulo* Matthew (H–O).

(A, D, F) right M_1 from between 89–590 foot levels, Willwood Formation, No Water area, Washakie Co., Wyoming (University of Wyoming no. 9759); (B, C, E) left M_1 from Sables de Sinceny, Sinceny (Aisne), France (University of California Museum of Paleontology, no. 83754); (G) left M^1 from the Willwood Formation, Clark's Fork Basin, Wyoming (UMMP.80063); (H) left M^1 from Lessness Shell Bed, Blackheath Beds, Abbey Wood, London (Natural History Museum, London, no. BMNH.M31877); (I) left M^1 from between 89 and 590 foot levels, No Water area, Washakie Co., Wyoming (UW. 9900); (J, L, N) right M_2 from Lessness Shell Bed, Blackheath Beds, Abbey Wood (BMNH.M13778); (K, M, O) right M_2 from the Willwood Formation, Yale–Michigan locality 90c, Clark's Fork Basin, Wyoming (UMMP.63670). E–K are occlusal views; A, B, L, and M are buccal views; C, D, N, and O are lingual views. Scale bar = 10 mm.

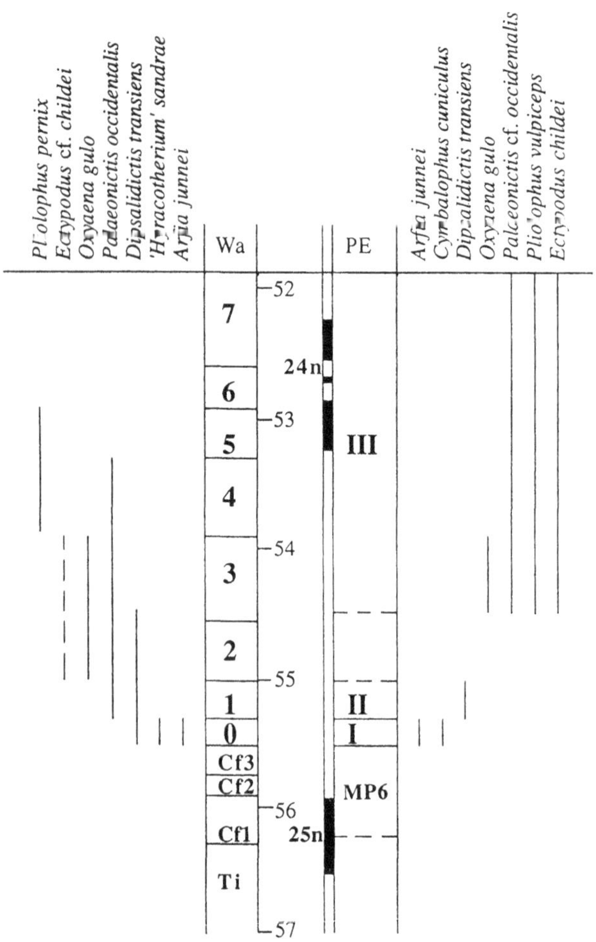

Figure 20.4 Stratigraphic ranges of key Wasatchian/Neustrian mammal species in western North America and Europe, with correlation of North American Wa, Cf, and Ti zones with European PE zones and Reference Level MP6, based on the successive nonoverlapping parts of the ranges of *Arfia junnei*, *Dipsalidictis transiens*, and *Oxyaena gulo*.

The timescale in millions of years BP and the magnetic Chrons C24n and C25n in Europe follow figure 20.1. The record of Chrons C24n and C25n in North America follows Butler et al. (1987) and Clyde et al. (1994)

dictis transiens is succeeded in time (with some range overlap) by *O. gulo* (Gunnell and Gingerich 1991). Whether or not they have an ancestor–descendant relationship (a possibility left open by Gunnell and Gingerich), the same sequential order of the species in Europe as in North America seems to be the strongest evidence yet for detailed correlation between the European Neustrian ELMA and North American Wasatchian NALMA (figure 20.4).

Both Abbey Wood and Pourcy have yielded rare teeth of *Palaeonictis*, distinctly larger than the type material of *P. gigantea*. They compare favorably for size with the low end of the range for North American *Palaeonictis occidentalis*, which has a Wa1 to Wa4 zonal range (Gunnell and Gingerich 1991).

The great similarity of *Pliolophus vulpiceps* from Zone PE III and *P. pernix* from Zones Wa4 to Wa5 suggests a time correlation, but the occurrence of the former at Abbey Wood contradicts the evidence from the oxyaenines. The equoids of Zones Wa1 to Wa3 and Zone PE II are rather dissimilar: "*Hyracotherium*" *grangeri* in Zone Wa1 to lowest Zone Wa3; "*H.*" *aemulor* in higher Zone Wa3 (Gingerich 1991); and *Pliolophus* aff. *vulpiceps*, *Hyracotherium* aff. *leporinum*, and "*Propachynolophus*" sp.1 in Zone PE II (Hooker 1994b). It is therefore possible that intercontinental interchange of equoids may temporarily have ceased after Biochrons Wa0–PE I, until *P. vulpiceps* dispersed to North America during Biochron PE III. *Pliolophus pernix* appears suddenly in the Clark's Fork Basin in Zone Wa4, replacing rather than showing an upward gradation from "*H.*" *aemulor* (Gingerich 1991:figure 12).

No species appear to be in common between higher North American and European zones and it is probably significant that submarine volcanism in east Greenland (Andersen 1988), indicating loss of a land connection with northwestern Europe, began during Chron C24n. The upward extension of two of the ranges in Europe (*Pliolophus vulpiceps* and *Palaeonictis* cf. *occidentalis*) results from occurrences at the site of Pourcy. If part of the Pourcy mammal fauna is indeed reworked from older strata (Cavelier 1987), these range extensions may be artificial. The lack of recognition of dinocyst Zone D7A in either the Fausses Glaises or the Argiles à Lignites d'Epernay (Chateauneuf and Gruas-Cavagnetto 1978; Gruas-Cavagnetto et al. 1980; Laurain et al. 1983) might indicate a hiatus and thus a source of reworking of this age or from the underlying Zone D6B. If so, these two species might have become extinct before Chron C24n.

Similarly, no species are in common between Reference Level MP6 faunas and any fauna in North America. The correlation of Reference Level MP6 to late Tiffanian–early Clarkforkian NALMA on the basis of equivalent evolutionary grade in distinct North American and European plesiadapid lineages (Gingerich 1976; Rose 1981a; Archibald et al. 1987:67) seems unlikely to be reliable in view of the rather different nature of the evolutionary trends in each and their probable distance of relationships (Hooker 1994a:350). The dinocyst and charophyte biostratigraphic evidence, however, does equate Reference Level MP6 with lower Zone NP9 (Hooker 1996b; Massieux et al. 1981), thus covering much of Chron C25n and earliest part of Chron C24r, and approximating the time span of the Clarkforkian NALMA.

The younger dating of Palette proposed here (table

20.1) means that European *Lophiaspis* is no longer the oldest chalicothere, its title being transferred to its close relative, North American *Paleomoropus*, from Zones Wa3, 4 or 5 (see Gingerich 1991:204). The equating of the Zone PE III Pourcy and Zone PE IV Mutigny sites with dinoflagellate cyst Zone D7B (Hooker 1996b) extends upward some of the Paris Basin perissodactyl ranges shown in Hooker (1994b:figure 8), although it does not alter the dispersal models proposed therein.

Calibration to the Carbon Isotope Excursion

A sharp negative carbon isotope excursion has been recognized in the deep sea (Kennett and Stott 1991), located within Zone NP9 and Chron C24r (Berggren and Aubry 1996). It has been calibrated directly through mammal tooth enamel to the mammal succession in the western U.S.A. and located at the base of Zone Wa0 (Koch et al. 1992). Calibrating this carbon isotope excursion to the mammal succession in Europe, however, is not so straightforward. Here, a similar carbon isotope excursion has been recorded at Limay in the Paris Basin in mottled clays with carbonate granules, within the lower part of the Argiles Plastiques Bariolées, about 0.5 m above a sandy calcrete (Sinha and Stott 1994; Stott et al. 1996; Sinha et al. 1996), which is usually termed the Marnes à Rognons. No mammals or other biota occur here. However, only 3 km away at Guitrancourt, the Marnes à Rognons contain the important charophyte marker *Peckichara microcarpa* (Riveline 1984:101–102; Thiry 1981:23). At Try, near Dormans, the Marnes à Rognons are represented by the "conglomérat à Coryphodon" (Laurain and Meyer 1986:110). This is the level from which the Zone PE I mammal fauna (almost entirely represented by surface finds) is presumed to originate, since one element of the fauna is accurately located in situ (Feugueur 1963; see discussion in Hooker 1996b and herein).

Zone PE II mammals occur in the Conglomérat de Meudon at the base of the Argiles Plastiques Bariolées at Meudon. However, at this site, the Marnes à Rognons are missing and the Conglomérat de Meudon rests unconformably on Danian deposits (Russell et al. 1993). Thus the exact level of the Conglomérat de Meudon in a regional context is unclear. It is possible that it may lie some distance above the base of the Argiles Plastiques Bariolées sequence as represented at Limay.

Although indirect and tentative, the available evidence suggests that the carbon isotope excursion recorded at Limay lies between Zones PE I and PE II. This implies that the time correlation of Zone PE I to Zone Wa0 as advocated here is not exact, that the Neustrian ELMA began slightly before the Wasatchian NALMA, that taxa new to both the Neustrian ELMA and the Wasatchian NALMA dispersed westwards from Europe to North America, and that the dispersal event in Europe was not causally related to the warming event evidenced by the isotope excursion. Evidence from the perissodactyls (Hooker 1994b) does suggest a westward dispersal direction for this group. However, the stratigraphic contexts of all the Zone PE I mammal faunas are as yet sufficiently problematic to make accurate calibration to the Limay carbon isotope excursion and thus reliance on any of the above implications very insecure. This is clearly an important area for improving stratigraphic resolution in the future.

Paleoecological Change

Evidence

From their study of modern communities, Andrews et al. (1979) proposed the use of the parameters of body size, foraging locomotor adaptation, and diet to analyze fossil assemblages, thereby reconstructing paleocommunities and habitats. These ecological parameters have been used to demonstrate changing mammalian habitats and in some cases community evolution (Maas and Krause 1994 for the North American Paleocene record; Collinson and Hooker 1987 for the English Eocene and lower Oligocene deposits; Collinson and Hooker 1991 for the Euramerican Cenozoic stratigraphic record; Andrews et al. 1979 for the east African Neogene stratigraphic record). They are applied here to European Cernaysian and Neustrian assemblages, the data being derived from appendix 20.1.

In most cases molar tooth size has been used to calculate body size as previously described (Collinson and Hooker 1987) but because of caveats about using this method directly for carnivorous mammals (Maas and Krause 1994; Van Valkenburgh 1990), these are corrected using estimates provided for individual taxa based mainly on postcranial evidence (e.g., Gunnell and Gingerich 1991; Gingerich and Deutsch 1989).

Locomotor adaptation is poorly known for many Paleocene and early Eocene mammals, although recent major advances have been made in this field thanks to newly collected and/or described material (Thewissen 1991; Rose 1990; Macleod and Rose 1993; Gebo and Rose 1993; Godinot and Smith 1993; Beard 1993; Zhou et al. 1992). Certain family-wide assumptions made earlier for locomotor adaptation (Collinson and Hooker 1987) can thus be modified here and better resolution achieved. There nevertheless remain many Paleocene genera and even families (e.g., Adapisoriculidae, Nyctitheriidae) for which no postcranials have been described. These last are scored here as semiterrestrial on the basis of possible relationships with Lipo-

Table 20.3
Morphological basis for dietary interpretation in appendix 20.1 and figure 20.5.

Insectivore	Molars tribosphenic, tall cusped, narrow basined, premolars relatively unspecialized
Insectivore/Frugivore	Molars small (indicating less than 1 kg body weight), tribosphenic, bunodont and/or broadly basined, posterior premolars may be specialized for puncture crush or shearing crush.
Insectivore–Carnivore/ Frugivore	Molars large (indicating more than 1 kg body weight), tribosphenic, bunodont and/or broadly basined, premolars relatively unspecialized.
Carnivore	Longitudinally oriented, shearing molars or premolars (carnassials) with or without associated unspecialized tribosphenic molars.
Frugivore (fruits, seeds, nuts)	Molars quadricuspate, bunodont.
Herbivore browser	Molars tribosphenic or quadricuspate, with strong cresting, usually of either bilophodont or selenodont pattern.
Frugivore/Herbivore browser	Molars intermediate between the types indicating frugivore and herbivore browser.
Herbivore grazer	As for herbivore browser but hypsodont or hypselodont, with horizontal occlusal surface.

typhla (Gheerbrant and Russell 1989; Butler 1988), although it is recognized that they could have had other adaptations.

The dietary classes essentially follow Collinson and Hooker (1987) although a clearer definition of the morphotypes taken to represent these classes is given. This is similar but not identical to the morphotypes used for this purpose by Maas and Krause (1994) and the intermediate or mixed dietary classes, apparently so abundantly represented in early Paleogene faunas, are characterized too (table 20.3). Each component of a mixed diet is scored in the appropriate class.

Naturally, collecting and taphonomy may impose biases on fossil assemblages and the degree of these should be assessed before drawing conclusions on habitat reconstruction. Most of the assemblages considered here have been collected mainly by screenwashing and thus could be expected to be biased against large animals. However, despite more than a century of surface prospecting of coastal cliff and active quarry exposures of Paleocene and lower Eocene strata in northwestern Europe, relatively few mammals of greater than 1 kg estimated body weight have ever been found. This suggests that the paucity of larger mammals at this time is real. Therefore, a fauna is more likely to be biased by insufficient screenwashing than by insufficient prospecting.

Taphonomic biases here are more difficult to assess, especially as no detailed site studies have been published. A major problem is faunal mixing by either surface collecting or by reworking. The occurrence of Cernaysian species like *Pleuraspidotherium* and *Orthaspidotherium* in the Paris Basin Zone PE I site of Try has been attributed to the former process (Louis et al. 1983). This could alternatively simply be an assumption and it is possible that the genera concerned briefly survived the terminal Cernaysian extinctions (Hooker 1994a, 1996b). Because of uncertainty and the low species count at Try, the site is here omitted from the analyses. Pourcy (Paris Basin) and the Suffolk Pebble Beds localities (London Basin) have probably been affected by reworking. In the former case, this may have caused mixing (Cavelier 1987) although the arguments are not compelling, being based on the co-occurrence of relatively archaic and more modern taxa. In the latter case, the evidence for reworking is stronger (Hooker 1991:80) although there is no apparent evidence of mixing, the fauna closely resembling that of the Belgian Zone PE I site of Dormaal. Pourcy is omitted from the analyses.

The faunas considered here have been recovered from a variety of depositional environments, but few show actual evidence of high-energy water transport in the form of abrasion that might have resulted in size sorting. Those from Dormaal, Erquelinnes, and the Suffolk Pebble Beds localities, however, do; and this is supported by the lithofacies (Casier 1967; de Heinzelin 1979; Hooker 1991). These three faunas have thus been combined here for the ecological analysis in an attempt to redress the possible bias. They are thus labeled PE I in figure 20.5 and in the following few paragraphs for discussive purposes.

Interpretation of the Data

All the faunas have high and/or dominant percentages of the AB size class, the semiterrestrial and scansorial locomotor classes, and the insectivore and frugivore dietary classes. This places them clearly within a closed or forested habitat type (Andrews et al. 1979).

There are some observable time trends in the locomotor and dietary categories (figure 20.5). A dramatic drop in the arboreal percentage from Cernay to Zone PE I is followed by a slight rise in subsequent zones. It reflects the early extinction of most multituberculates and the later appearance of *Peradectes* and reappearance of a few multituberculates. There is an increase in the ter-

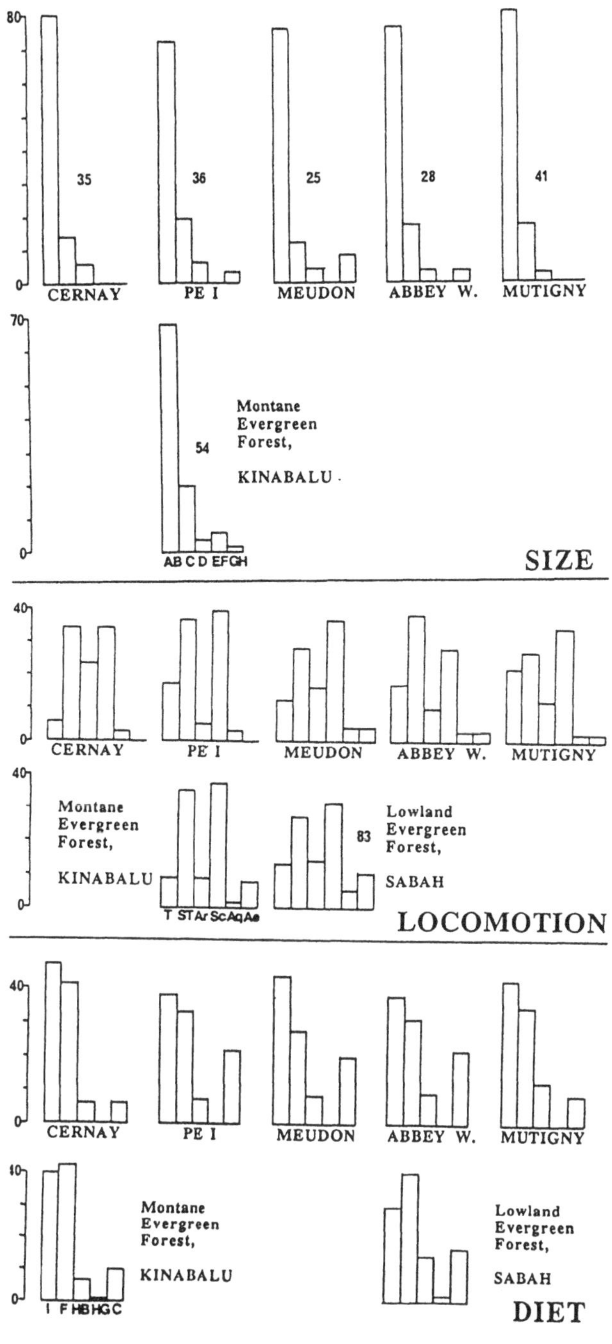

Figure 20.5 Ecological diversity histograms for size, foraging locomotor adaptation, and diet of mammalian faunal assemblages from Cernay, combined Zone PE I sites (Kyson, Ferry Cliff, Erquelinnes, and Dormaal), Meudon (plus Vaugirard), Abbey Wood, and Mutigny. They are compared with equivalent modern histograms for montane evergreen forest of Mt. Kinabalu, Sabah, and for lowland evergreen forest of Sabah as a whole. From Davis (1962), Payne et al. (1985).

The scale is percentage of the total fauna. Figures on right of histograms indicate numbers of species for each locality. Abbreviations: AB = less than 1 kg, C = 1–10 kg, D = 10–45 kg, EF = 45–180 kg, GH = more than 180 kg, T = terrestrial, ST = semiterrestrial, Ar = arboreal, Sc = scansorial, Aq = aquatic, Ae = aerial, I = insectivore, F = frugivore, HB = herbivore browser, HG = herbivore grazer, C = carnivore.

restrial percentage from Cernay to Zone PE I and a maximum is reached at Mutigny. The Zone PE I plot is quite like that of the montane evergreen forest habitat of Mt. Kinabalu, Sabah, except for its larger percentage of terrestrial and absence of aerial classes. The Meudon, Abbey Wood, and Mutigny plots are more like those of the lowland evergreen forest in Sabah, except for their lower percentages of aerial and, at Mutigny, higher percentage of terrestrial species. The Cernay plot seems enigmatic among known modern habitat plots, but may be biased by poor knowledge of locomotor adaptations. The dietary category shows a gradual slight increase in browsing herbivory with time, although the percentage remains relatively low. It reflects the progressive appearance of perissodactyls, *Coryphodon*, and tillodonts and the evolution of a fully herbivorous diet in *Platychoerops*. It also shows a sharp increase in carnivory from Cernay to Zone PE I and a slight reduction between Abbey Wood and Mutigny. This reflects the early sudden appearance of the orders Creodonta and Carnivora and the later extinction of the Oxyaenidae. The similar high percentages of both insectivory and frugivory are close to those of Mt. Kinabalu montane evergreen forest, although the latter differ in having the frugivory rather than the insectivory dominant. The higher browsing percentage at Mutigny is more like that of Sabah lowland evergreen forest, although here the dominance of frugivory over insectivory is even more marked. Andrews et al.'s African evergreen forest plots (1979), whether lowland or montane, typically have a lower percentage of frugivory than insectivory and nearly as high a percentage of browsing herbivory as insectivory. This difference seems to stem from the largely secondary nature of African evergreen forests due to past extensive cultivation (Richards 1952).

Neustrian faunas thus seem to give an ecological signal of equable evergreen forest, with slightly less productivity than a modern southeast Asian one. They also show trends similar to a transition from southeast Asian montane to lowland evergreen forest, which may indicate a climatic warming over this time interval. Such a trend would be consistent with the marine oxygen isotope curve (e.g., Shackleton 1986).

If one combines the information from diet and locomotor adaptation, the differences between the Cernaysian and Neustrian communities appear greater than when these parameters are considered separately. The niche of the abundant Cernaysian semiterrestrial browsing *Pleuraspidotherium* and *Orthaspidotherium* seems not to have been filled by any Neustrian forms. A number of the larger Cernaysian mammals with bunodont cheek teeth must have combined some carnivory with other elements of their diet, but only one, the mesonychid *Dissacus*, seems to have been a specialized flesh-eater, albeit largely a terrestrial scavenger. The incoming of carnassial-bearing groups implying predatory carnivores with varying degrees of tree-climbing capabilities must have meant major new selection pressures on communities that had previously experienced little such predation. This would have been more intense than in North America, where Carnivora and one family of the Creodonta had long existed and the only incoming predatory group was the family Hyaenodontidae (Rose 1981b). Other major Neustrian innovations were terrestrial hooved forms, mainly perissodactyls, artiodactyls, and the pantodont *Coryphodon*, which included much fruit and/or leaves in their diets. The incoming of scansorial insectivorous/frugivorous rodents coincides with the decimation of the multituberculates as in North America, but in the latter continent the turnover was earlier, at the beginning of the Clarkforkian NALMA (Krause 1986). Such a competitive relationship between the two groups in Europe is less certain in view of all the other possible pressures (including tree-climbing carnivores). Moreover, in contrast to North America (Maas et al. 1988), there is less evidence of decimation of Cernaysian plesiadapiforms. The group was much less diverse in the Cernaysian ELMA, although the fact that only one plesiadapid survived beyond Zone PE III and modified its diet from frugivory to herbivory (Hooker 1994a) could be attributable to competition with rodents. The paromomyids, however, increased plesiadapiform diversity and their possible gliding abilities (Beard 1993) may have added a new locomotor strategy (see Runestad and Ruff 1995 for alternative view) in Europe by dispersal from North America during Biochron PE II. Whether this appearance of new niches in the Neustrian ELMA was entirely attributable to the dispersal of new ecological types or reflected also changes in forest structure is unclear.

The southwest European Neustrian localities are considered as a whole to be as yet undersampled compared to those in the North, although Silveirinha has as many species as some northern sites. They have thus not been analyzed, yet a preliminary appraisal of their ecological diversity suggests that they are similar to the northern sites, also indicating forested habitats.

Discussion

Although we have learned a great deal about the succession of late Paleocene and early Eocene mammal faunas in Europe over the past few decades, it is still much less well known than that of western North America, making certain types of interpretation difficult. For instance, the Cernaysian faunas are still almost entirely restricted to one small outlier in the eastern Paris Basin, so that our knowledge of them might be very misleading in terms of Europe as a whole; the overall record is highly punctuated, not continuous; and we still have little idea of the centers of origin and dispersal routes of a number of major mammal groups. These issues are discussed below.

Density of the Record and Nature of the Turnover Between Reference Level MP6 and Zone PE I

To understand the global events involving mammals around the Paleocene/Eocene Epoch boundary, it is necessary to understand first the pattern and timing of events in each area. The sedimentary and mammal record in northwestern Europe is much less complete than in the Bighorn Basin. This is well demonstrated by comparing figure 20.1 with e.g., Rose (1981a:figure 3) and Schankler (1980:figure 2) for an approximately equivalent time span. However, the time interval between Reference Level MP6 and Zone PE I seems relatively short. The record at Montchenot of the dinocyst *Apectodinium parvum* in the Sables de Rilly (Gruas-Cavagnetto 1974), underlying the Marnes de Montchenot (= Calcaire de Rilly) with the highest Reference Level MP6 fauna (Louis in Laurain and Meyer 1986: 117), indicates an age no older than dinocyst Zone D5A (Costa and Manum 1988) or its equivalent Zone Aau (Powell 1992). The base of Zone Aau is recorded in the London Basin high in the Upnor Formation (Powell et al. 1996) within Chron C24r (Ellison et al. 1996), thus no older than 55.9 Ma (Berggren and Aubry 1996). The probable lowest Zone PE I fauna is from the "conglomérat à *Coryphodon*" (= Marnes à Rognons) at Try. The Marnes à Rognons lie a short distance below the record of a carbon isotope excursion at Limay (Sinha and Stott 1994; Stott et al. 1996), which itself has an estimated date of 55.5 Ma in the deep sea (Berggren and Aubry 1996). The maximum time interval between Reference Level MP6 and Zone PE I thus appears to be about 400,000 years.

Given the apparent large scale of the turnover across this short interval (table 20.2), the appearance of new ecological types such as specialized carnivores and the similar number of species in Reference Level MP6 as in Zone PE I, the turnover pattern is more consistent with displacement than replacement. Unfortunately, the degree of change at the critical time is not reliably known, as the youngest Reference Level MP6 and oldest Zone PE I faunas are poor, apparently due to collecting deficiency (8 species for the Marnes de Montchenot and 12 for Try). Nevertheless, the fauna from the Marnes de Montchenot comprises the typical abundant members of Reference Level MP6 with no sign of Neustrian newcomers, although Try may include two notable Reference Level MP6 survivors not known elsewhere in Zone PE I.

Centers of Origin and Dispersal Routes

There has been much speculation on the centers of origin of the main mammal groups that appeared essentially synchronously in Europe, North America, and Asia, namely the orders Artiodactyla, Perissodactyla, Primates, and Chiroptera, and the families Hyaenodontidae (Creodonta) and herpetotheriine "Didelphidae" (Marsupialia). Krause and Maas (1990) have thoroughly investigated the problem of these origins in a temporally broader study of mammalian immigrants into North America. It is normally accepted that low-latitude areas are involved, in view of the absence of Paleocene representatives in any midlatitude faunas in the northern hemisphere, coupled with a short-term warming event associated with the dispersal event at least in North America (Koch et al. 1992). In view of the large amount of published speculation, I offer the following discussion of the often meager evidence for origin of each group to stimulate future collecting efforts (see figure 20.6).

Artiodactyla

Hard evidence is lacking. Late early Eocene *Diacodexis pakistanensis* from India and Pakistan has been said to be the most primitive artiodactyl known, both dentally and postcranially (Thewissen and Hussain 1990). Nevertheless, it is more derived than either *Diacodexis ilicis* (see Gingerich 1989) or *Diacodexis* sp. from Dormaal (Sudre et al. 1983:figure 2) in having a lingually displaced lower first and second molar hypoconulid and a reduced lower molar paraconid. The reduced canines and straight ventral dentary margin also do not represent the most primitive states for the order (Gentry and Hooker 1988). No artiodactyls are known from the Paleocene stratigraphic record of either South America or Africa. They first occurred in Africa during the Oligocene Epoch (Maglio and Cooke 1978) and in South America during the Pliocene Epoch (Webb 1985), which makes these continents unlikely centers of origin. This leaves southern Asia (including India) as the only unsampled area where Paleocene artiodactyls apparently could be found.

Perissodactyla

No perissodactyls are known from the upper Paleocene deposits of Africa or South America. They first occurred in South America during the Pleistocene Epoch (Webb 1985) and in Africa during the Miocene Epoch (Maglio and Cooke 1978), which, as for the artiodactyls, makes these continents unlikely centers of origin. The southern Asian Paleocene stratigraphic record has not been sampled. The phenacodontid *Ectocion* may be the nearest sister taxon to the order (Hooker 1994b). It is known

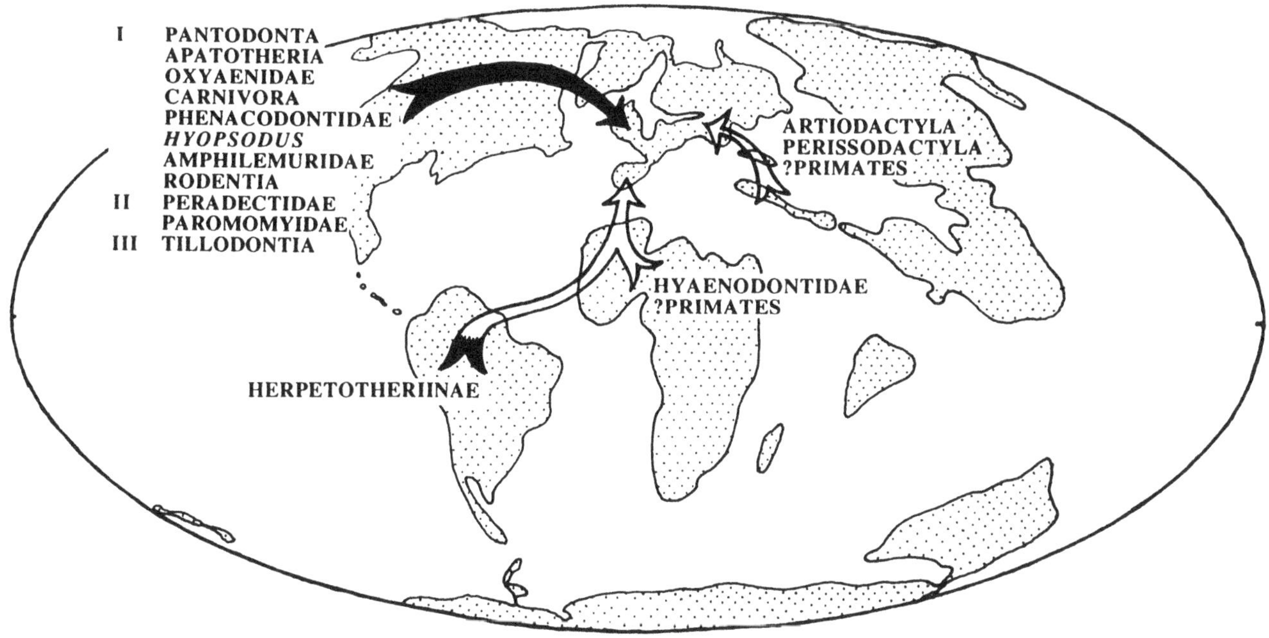

Figure 20.6 Late Paleocene world coastal outline map after Scotese and Golonka (1992).

Land areas stippled. Solid arrow indicates good evidence for dispersal of listed taxa from North America to Europe during the Neustrian ELMA in three phases, corresponding to Zones PE I, II, and III. Hollow arrows indicate possible dispersal trajectories, for which there is little or no direct evidence.

only from North America, where it ranges back to the base of the Tiffanian NALMA (Thewissen 1990). A relationship with other pantomesaxonians, however, is also possible. Perissodactyls appear to have already differentiated into brontotheriids, tapiromorphs, and equoids by the time of their earliest occurrence. Brontotheres are probably overall the most primitive, but are not recorded in strata dated as older than the upper lower Eocene record of Asia and North America. Tapiromorphs are represented in the oldest Wasatchian, oldest Neustrian, and the Bumbanian deposits in all three northern hemisphere continents, but equoids are represented only in the oldest Wasatchian and oldest Neustrian deposits, being absent from Asia until the Miocene Epoch (Hooker 1994b; Ting 1993; Dashzeveg 1988). It is possible that perissodactyls originated in southern Asia, but if so it is difficult to understand how equoids could have been excluded from higher-latitude parts of Asia. Perhaps subordinal differentiation took place in the archipelago of southwestern Asia and dispersal was filtered into both Europe and northern Asia and from there to North America. It is significant that the oldest Neustrian record of northwestern Europe has very primitive members of both Equoidea (*Cymbalophus*) and Tapiromorpha (*Hallensia*), the latter only tentatively differentiated subordinally (Hooker 1994b).

Primates

Only three true primates (i.e., excluding plesiadapiforms) are claimed from the Paleocene: *Decoredon* (Szalay and Li 1986) and *Petrolemur* (Tong 1979) from China and *Altiatlasius* from Morocco (Sigé et al. 1990). Much doubt surrounds the ordinal affinities of *Decoredon*, represented by poorly preserved upper and lower cheek teeth (see Sigé et al. 1990). *Petrolemur* may be an arctostylopid (Cifelli et al. 1989). *Altiatlasius*, a supposed omomyid represented by isolated upper and lower teeth, has upper molars with buccally salient parastyle, distinct stylar shelf, and buccolingually short protocone, character states more primitive than any known primate and comparable in grade to a primitive plesiadapiform like *Purgatorius*. It thus seems too remote phylogenetically to provide evidence of a regional source for the primates that first appeared in northern hemisphere continents. The diversity of North African early Eocene primates including simians (Godinot and Mahboubi 1992) and the possible adapid *Azibius* (Gingerich 1977b) might be construed as support for Africa being the primate center of origin. This would, however, imply a dispersal route to North America and Asia via Europe, which is at odds with the distinctly primitive nature of Bumbanian *Altanius* (Gingerich et al. 1991) unless India was involved, as hypothesized by Krause and Maas (1990). The question will remain unresolved until authenticated Paleocene primates are found.

Chiroptera

The oldest record of European bats is in Zone PE II. The oldest reported bat, *Wyonycteris* from the Clarkforkian NALMA, is more likely to be a nyctithere (Hand et al. 1994; Hooker 1996a). This makes the European record one of the oldest. Identification of Paleocene bats from teeth is likely to be difficult, unless they are associated with postcranials. Until such are discovered, their center of origin will remain a mystery.

Hyaenodontidae

A tentative proviverrine hyaenodontid in the upper Paleocene record of Morocco (Gheerbrant 1995) and a specialized member of the family, *Koholia*, from the lower Eocene of Algeria (Crochet 1988) suggest Africa as a source area for this creodont family.

"Didelphidae" (Herpetotheriinae)

The familial and higher classificatory status of "opossums" is currently unstable. On dental and cranial evidence, Reig et al. (1987) placed *Peratherium* and *Amphiperatherium* (together with North American *Herpetotherium*) in subfamily Herpetotheriinae within the family Didelphidae, but removed *Peradectes* from this family to the Peradectidae. On tarsal evidence, Szalay (1994:418–419) has placed *Peradectes* and *Herpetotherium* (and by implication *Peratherium* and *Amphiperatherium*) in the family Pediomyidae, subfamily Peradectinae, remote from the Didelphidae. Such a close relationship between *Peratherium* and *Amphiperatherium* on one hand and *Peradectes* on the other would imply an origin for the former within North America, on the basis of the long exclusively North American Paleocene history of *Peradectes*. However, dentally, several Paleocene South American marsupials from the Brazilian site of Itaborai (especially *Itaboraidelphis*) are much more similar to *Peratherium* and *Amphiperatherium* than is *Peradectes* (Marshall 1987). It is probably significant that the isolated astragali and calcanea from Itaborai tentatively attributed by Szalay (1994:figures 6.20-ICS5 and 6.22-IAS8) to *Itaboraidelphis* are morphologically close to those from the North American Bridger fauna (Szalay 1994:figure 6.12; although unattributed to a "tooth genus"). Similar isolated undescribed astragali and calcanea occur in the English upper Eocene deposits (personal observation), where the only marsupials represented by teeth are *Amphiperatherium* and *Peratherium*. Whatever the higher classification adopted, the Paleocene teeth and tarsals that are closest to *Amphiperatherium* and *Peratherium* occur in South America, indicating a likely source area for these European genera. The route to Europe at the beginning of the Neustrian could have been either via North America or via Africa (Marshall 1987). The African route has been supported by the Algerian early Eocene supposed marsupials *Garatherium* (Crochet 1984) and *Kasserinotherium* (Crochet 1986). The former has been recently reidentified as an eutherian adapisoriculid (Gheerbrant and Russell 1991), while the latter also has some anomalous features more in keeping with adapisoriculid than marsupial affinities. Nevertheless, the availability of the African route is supported by other vertebrate occurrences (e.g., Storch 1993) in contrast to that via North America.

An improved mammal biostratigraphy in northwestern Europe in the stratigraphic interval spanning the Paleocene/Eocene Series boundary is linked, albeit patchily, with the standard sequence of calcareous nannofossils (NP) and independently can begin to be calibrated with the mammal biostratigraphy of western North America. The Cernaysian–Neustrian turnover occurs within Zone NP9. The major mammalian turnovers at both the Cernaysian/Neustrian ELMA and Clarkforkian/Wasatchian NALMA boundaries, which are closely correlated according to the mammals, appear to coincide with another major biotic event, the benthic foraminiferal extinction, based on cross-correlation through carbon isotope stratigraphy to the Bighorn Basin. Problems still exist, however, in calibrating Neustrian mammals to the carbon isotope excursion in Europe (see above).

The Paleocene/Eocene Series boundary is usually placed at this major turnover by paleomammalogists, whereas it is placed later, at the NP9/NP10 or P5/P6a zonal boundaries, by micropaleontologists (see Berggren and Aubry 1996). The choice is yet to be made, but there is much to recommend the former if it can be conclusively correlated with the benthic foraminiferal extinction, in which case it would be recognizable in both oceanic and terrestrial realms.

Reconstruction of the Zones PE I to PE IV paleocommunities tentatively allows the suggestion of a warming trend consistent with the oxygen isotope record and gives some idea of the ecological changes between the Cernaysian and Neustrian ELMAs. Outstanding gaps in our knowledge, highlighted especially in the discussion above, may best be resolved by the following: more collecting in Zone PE I, to resolve better its stratigraphic relationships; augmenting the youngest Reference Level MP6 faunas; the undertaking of more phylogenetic studies to make a sounder basis for dispersal directions and centers of origin; the carrying out of more morphofunctional studies of teeth, including microwear, to assess ancient diets and of postcranial bones to establish locomotor adaptations; and the publishing of detailed taphonomic studies of individual mammal sites.

ACKNOWLEDGMENTS

I would like to thank the following for access to collections in their care: D. E. Russell (Muséum National d'Histoire Naturelle, Paris), P. D. Gingerich (University of Michigan Museum of Paleontology), M. Godinot (Université des Sciences et Techniques du Languedoc, Montpellier), and P. Ensom (Yorkshire Museum). I am also grateful to Drs. Russell, Gingerich, and Godinot for stimulating discussion and for providing important casts. A. G. Lawson, A. A. Mitchell, D. J. Ward, P. Jeffery, and the Croydon Natural History and Scientific Society kindly presented to the Natural History Museum, London, specimens from Abbey Wood that form herein the basis of new records from this site. D. E. Russell, P. Andrews, and M. E. Collinson critically read the manuscript. N. MacLeod provided practical help with the map. Staff of Bexley Council and English Nature facilitated acces to the Abbey Wood site. This is a contribution from the NHM/UCL-BkB project on Global Change and the Biosphere and Project HO-4 in the NHM Human Origins Programme.

REFERENCES

Ali, J. R., King, C. and Hailwood, E. A., 1993. Magnetostratigraphic calibration of early Eocene depositional sequences in the southern North Sea Basin. In Hailwood, E. A. and Kidd, R. B. (eds.), *High Resolution Stratigraphy,* Geological Society Special Publication, No. 70, p. 99–125.

Andersen, M. S., 1988. Late Cretaceous and early Tertiary extension and volcanism around the Faeroe Islands. In Morton, A. C. and Parson, L. M. (eds.), *Early Tertiary Volcanism and the Opening of the NE Atlantic,* Geological Society Special Publication No. 39, p. 115–122.

Andrews, P., Lord, J. M. and Evans, E. M. N., 1979. Patterns of ecological diversity in fossil and modern mammalian faunas. *Biological Journal of the Linnean Society,* v. 11, p. 177–205.

Archibald, J. D., Rigby, J. K., Jr and Robison, S. F., 1983. Systematic revision of *Oxyacodon* (Condylarthra, Periptychidae) and a description of *O. ferronensis* n. sp. *Journal of Paleontology,* v. 57, p. 53–72.

Archibald, J. D., Clemens, W. A., Gingerich, P. D., Krause, D. W., Lindsay, E. H. and Rose, K. D., 1987. First North American Land Mammal Ages of the Cenozoic Era. In Woodburne, M. O. (ed.), *Cenozoic Mammals of North America: Geochronology and Biostratigraphy,* University of California Press, Berkeley, Los Angeles, London, p. 24–76.

Aubry, M.-P., 1983. Biostratigraphie du Paléogène épicontinental de l'Europe du Nord-Ouest. Etude fondée sur les nannofossiles calcaires. *Documents des Laboratoires de Géologie, Lyon,* v. 89, 317 pp.

Aubry, M.-P., 1986. Paleogene calcareous nannoplankton biostratigraphy of northwestern Europe. *Palaeogeography, Palaeoclimatology, Palaeoecology,* v. 55, p. 267–334.

Baudry, M., 1992. Les tillodontes (Mammalia) de l'Eocène inférieur de France. *Bulletin du Muséum National d'Histoire Naturelle, Paris,* ser. C4, v. 14, p. 205–243.

Beard, K. C., 1993. Origin and evolution of gliding in early Cenozoic

Dermoptera (Mammalia, Primatomorpha). In MacPhee, R. D. E. (ed.), *Primates and Their Relatives in Phylogenetic Perspective,* New York: Plenum Press, p.63–90.

Berggren, W. A. and Aubry, M.-P., 1996. A late Paleocene–early Eocene NW European and North Sea magnetobiochronological correlation network. In Knox, R. W. O'B., Corfield, R. M. and Dunay, R. E. (eds.), *Correlation of the Early Paleogene in Northwest Europe,* Geological Society Special Publication No. 101, p. 309–352.

Boule, M., 1903. Le *Pachyaena* de Vaugirard. *Mémoires de la Société géologique de France,* v. 10, p. 1–16.

Brunet, M., Franzen, J. L., Godinot, M., Hooker, J. J., Legendre, S., Schmidt-Kittler, N. and Vianey-Liaud, M., coordinators, 1987. European reference levels and correlation tables. *Münchner Geowissenschaftlicher Abhandlungen,* ser. A, v. 10, p. 13–31.

Butler, P. M., 1988. Phylogeny of the insectivores. In Benton, M, J. (ed.), *The Phylogeny and Classification of the Tetrapods, v. 2: Mammals,* Systematics Association Special Volume, No. 35B, p. 117–141.

Butler, R. F., Krause, D. W. and Gingerich, P. D., 1987. Magnetic polarity stratigraphy and biostratigraphy of middle–late Paleocene continental deposits of south-central Montana. *Journal of Geology,* v. 95, p. 647–657.

Cande, S. C. and Kent, D. V., 1992. A new geomagnetic polarity time scale for the Late Cretaceous and Cenozoic. *Journal of Geophysical Research,* v. 97(B10), p. 13,917–13,951.

Casier, E., 1967. Le Landénien de Dormaal (Brabant) et sa faune ichthyologique. *Mémoires de l'Institut Royal des Sciences Naturelles de Belgique,* no. 156, p. 1–66.

Cavelier, C., 1987. Diachronisme des faciès et discontinuités dans le Paléogène. Exemple du Thanétien et de l'Yprésien en Ile-de-France. Essai de généralisations et corrélations dans les bassins anglais, belge et parisien. In Cavelier, C. and Lorenz, J. (eds.), *Aspects et Evolution Geologiques du Bassin Parisien,* Bulletin d'Information des Géologues du Bassin de Paris, Mémoire hors série, v. 6, p. 244–270.

Cavelier, C. and Pomerol, C., 1986. Stratigraphy of the Paleogene. *Bulletin de la Société Géologique de France,* ser. 8, v. 2, no. 2, p. 255–265.

Chateauneuf, J. J. and Gruas-Cavagnetto, C., 1978. Les zones de Wetzeliellaceae (Dinophyceae) du Bassin de Paris. *Bulletin du Bureau de Recherches Géologiques et Minières,* série 4, v. 2, p. 59–93.

Cifelli, R. L., Schaff, C. R. and McKenna, M. C., 1989. The relationships of the Arctostylopidae (Mammalia): New data and interpretation. *Bulletin of the Museum of Comparative Zoology, Harvard,* v. 152, p. 1–44.

Clyde, W. C., Stamatakos, J. and Gingerich, P. D., 1994. Chronology of the Wasatchian Land-Mammal Age (early Eocene): Magnetostratigraphic results from the McCullough Peaks Section, northern Bighorn Basin, Wyoming. *Journal of Geology,* v. 102, p. 367–377.

Collinson, M. E. and Hooker, J. J., 1987. Vegetational and mammalian faunal changes in the Early Tertiary of southern England. In Friis, E. M., Chaloner, W. G. and Crane, P. R. (eds.), *The Origins of Angiosperms and their Biological Consequences,* Cambridge: Cambridge University Press, p. 259–304.

Collinson, M. E. and Hooker, J. J., 1991. Fossil evidence of interactions between plants and plant-eating mammals. *Philosophical Transactions of the Royal Society of London,* ser. B, v. 333, p. 197–208.

Costa, L. and Manum, S. B., 1988. The description of the interregional zonation of the Paleogene (D1–D15) and the Miocene (D16–D20). In Vinken, R., compiler, *The Northwest European Tertiary Basin. Results of the International Geological Correlation Programme Project No. 124, Geologisches Jahrbuch,* ser. A, v. 100, p. 321–330.

Costa, L., Denison, C. and Downie, C., 1978. The Palaeocene/Eocene boundary in the Anglo–Paris Basin. *Journal of the Geological Society, London,* v. 135, p. 261–264.

Crochet, J. Y., 1980. *Les Marsupiaux du Tertiaire d'Europe.* Paris: Singer-Polignac, 279 pp.

Crochet, J. Y., 1984. *Garatherium mahboubii* nov. gen., nov. sp., marsupial de l'Eocène inférieur d'El Kohol (Sud-Oranais, Algérie). *Annales de Paléontologie,* ser. Vert.-Invert., v. 70, p. 275–294.

Crochet, J. Y., 1986. *Kasserinotherium tunisiense* nov. gen., nov. sp., troisième marsupial découvert en Afrique (Eocène inférieur de Tunisie). *Comptes-Rendus hebdomadaires des Séances de l'Académie des Sciences de Paris,* ser. 2, v. 302, p. 923–926.

Crochet, J. Y., 1988. Le plus ancien Créodonte africain: *Koholia atlasense* nov. gen., nov. sp. (Eocène inférieur d'El Kohol, Atlas saharien, Algérie). *Comptes-Rendus hebdomadaires des Séances de l'Académie des Sciences de Paris,* ser. 2, v. 307, p. 1795–1798.

Dashzeveg, D., 1988. Holarctic correlation of non-marine Palaeocene–Eocene boundary strata using mammals. *Journal of the Geological Society, London,* v. 145, p. 473–478.

Davis, D. D., 1962. Mammals of the Lowland rain-forest of North Borneo. *Bulletin of the National Museum, State of Singapore,* v. 31, 129 pp.

De Coninck, J., 1990. Ypresian organic-walled phytoplankton in the Belgian Basin and adjacent areas. *Bulletin van de Belgische Vereniging voor Geologie, Brussel,* v. 97, p. 287–319.

De Coninck, J., de Decker, M., de Heinzelin, J. and Willems, W., 1981. L'âge des faunes d'Erquelinnes. *Bulletin de la Société belge de Géologie,* v. 90, p. 121–154.

Ellison, R. A., Ali, J. R., Hine, N. M. and Jolley, D. W., 1996. Recognition of Chron C25n in the upper Paleocene Upnor Formation of the London Basin, UK. In Knox, R. W. O'B., Corfield, R. M. and Dunay, R. E. (eds.), *Correlation of the Early Paleogene in Northwest Europe,* Geological Society Special Publication No 101, p. 185–193.

Ellison, R. A., Knox, R. W. O'B., Jolley, D. W. and King, C., 1994. A revision of the lithostratigraphical classification of the early Palaeogene strata of the London Basin and East Anglia. *Proceedings of the Geologists' Association,* v. 105, p. 187–197.

Estravis, C., 1992. *Estudo dos Mamíferos do Eocénico inferior de Silveirinha (Baixo Mondego).* Dissertacao Grau de Doutor em Geologia, Centro de Estratigrafia e Paleobiologia da Universidade Nova de Lisboa, Lisboa, 247 pp.

Estravis, C. and Russell, D. E., 1989. Découverte d'un nouveau *Diacodexis* (Artiodactyla, Mammalia) dans l'Eocène inférieur de Silveirinha, Portugal. *Palaeovertebrata,* v. 19, p. 29–44.

Estravis, C. and Russell, D. E., 1992. The presence of Taeniodonta (Mammalia) in the early Eocene of Europe. *Ciencias da Terra,* v. 11, p. 191–201.

Fahlbusch, V., 1976. Report on the International Symposium on Mammalian Stratigraphy of the European Tertiary. *Newsletters on Stratigraphy,* v. 5, p. 160–167.

Feugueur, L., 1963. L'Yprésien du Bassin de Paris. Essai de monographie stratigraphique. *Mémoire du Service de la Carte Géologique de France, Paris,* 568 pp.

Gebo, D. L. and Rose, K. D., 1993. Skeletal morphology and locomotor adaptation in *Prolimnocyon atavus,* an early Eocene hyaenodontid creodont. *Journal of Vertebrate Paleontology,* v. 13, p. 125–144.

Gentry, A. W. and Hooker, J. J., 1988. The phylogeny of the Artiodactyla. In Benton, M, J. (ed.), *The Phylogeny and Classification of the Tetrapods, Volume 2: Mammals,* Systematics Association Special Volume No. 35B, Clarendon Press, Oxford, p. 235–272.

Gheerbrant, E., 1995. Les mammifères Paléocènes du Bassin d'Ouarzazate (Maroc) III. Adapisoriculidae et autres mammifères (Carnivora, ?Creodonta, Condylarthra, ?Ungulata et incertae sedis). *Palaeontographica Abt. A,* v. 237, p. 39–132.

Gheerbrant, E. and Russell, D. E., 1989. Présence of the genus *Afrodon* [Mammalia, Lipotyphla(?), Adapisoriculidae] in Europe; new data for the problem of trans-Tethyan relations between Africa and Europe around the K/T boundary. *Palaeogeography, Palaeoclimatology, Palaeoecology,* v. 76, p. 1–15.

Gheerbrant, E. and Russell, D. E., 1991. *Bustylus cernaysi* nov. gen., nov. sp., nouvel adapisoriculidé (Mammalia, Eutheria) Paléocène d'Europe. *Géobios,* v. 24, p. 467–481.

Gingerich, P. D., 1976. Cranial anatomy and evolution of early Tertiary Plesiadapidae (Mammalia, Primates). *University of Michigan Papers on Paleontology,* v. 15, 141 pp.

Gingerich, P. D., 1977a. New species of Eocene primates and the phylogeny of European Adapidae. *Folia Primatologica,* v. 28, p. 60–80.

Gingerich, P. D., 1977b. Radiation of Eocene Adapidae in Europe. *Géobios,* Mémoire spécial 1, p. 165–182.

Gingerich, P. D., 1983. Systematics of early Eocene Miacidae (Mammalia, Carnivora) in the Clark's Fork Basin, Wyoming. *Contributions from the Museum of Paleontology, The University of Michigan,* v. 26, p. 197–225.

Gingerich, P. D., 1989. New earliest Wasatchian mammalian fauna from the Eocene of northwestern Wyoming: Composition and diversity in a rarely sampled high-floodplain assemblage. *University of Michigan Papers on Paleontology,* v. 28, 97 pp.

Gingerich, P. D., 1991. Systematics and evolution of early Eocene Perissodactyla (Mammalia) in the Clark's Fork Basin, Wyoming. *Contributions from the Museum of Paleontology, The University of Michigan,* v. 28, p. 181–213.

Gingerich, P. D., 1993. Early Eocene *Teilhardina brandti:* oldest omomyid primate from North America. *Contributions from the Museum of Paleontology, The University of Michigan,* v. 28, p. 321–326.

Gingerich, P. D. and Deutsch, H. A., 1989. Systematics and evolution of early Eocene Hyaenodontidae (Mammalia, Creodonta) in the Clark's Fork Basin, Wyoming. *Contributions from the Museum of Paleontology, University of Michigan,* v. 27, p. 327–391.

Gingerich, P. D., Dashzeveg, D. and Russell, D. E., 1991. Dentition and systematic relationships of *Altanius orlovi* (Mammalia, Primates) from the early Eocene of Mongolia. *Géobios,* v. 24, p. 637–646.

Godinot, M., 1981. Les mammifères de Rians (Eocène inférieur, Provence). *Palaeovertebrata,* v. 10, p. 43–126.

Godinot, M., 1984. Un nouveau genre de Paromomyidae (Primates) de l'Eocène inférieur d'Europe. *Folia Primatologica,* v. 43, p. 84–96.

Godinot, M. and Mahboubi, M., 1992. Earliest known simian primate found in Algeria. *Nature,* v. 357, p. 324–326.

Godinot, M. and Smith, R., 1993. Affinities and way of life of *Paschatherium* (Condylarthra) from its tarsals. European Society for Evolutionary Biology, 4th Congress, Montpellier, Abstracts, p. 153.

Godinot, M., de Broin, F., Buffetaut, E., Rage, J. C. and Russell, D., 1978. Dormaal: une des plus anciennes faunes éocènes d'Europe. *Comptes-Rendus Hebdomadaires des Séances de l'Académie des Sciences de Paris,* ser. D, v. 287, p. 1273–1276.

Godinot, M., Crochet, J.-Y., Hartenberger, J.-L., Lange-Badré, B., Russell, D. E. and Sigé, B., 1987. Nouvelles données sur les mammifères de Palette (Eocène inférieur, Provence). *Münchner Geowissenschaftliche Abhandlungen,* ser. A, v. 10, p. 273–288.

Gruas-Cavagnetto, C., 1968. Etude palynologique des divers gisements du Sparnacien du Bassin de Paris. *Mémoires de la Société géologique de France,* v. 110, 114 pp.

Gruas-Cavagnetto, C., 1974. Associations sporopolliniques et microplanctoniques de l'Eocène et de l'Oligocène inférieur du Bassin de Paris. *Paléobiologie Continentale,* v. 5, p. 1–20.

Gruas-Cavagnetto, C., 1976. Etude palynologique du sondage de la Défense (Paris). *Revue de Micropaléontologie,* v. 19, p. 27–46.

Gruas-Cavagnetto, C., 1992. Pollens et Dinophycées de l'Ilerdien moyen (Eocène inférieur) de Fordones (Corbières, France). *Cahiers de Micropaléontologie,* v. 6, p. 51–66.

Gruas-Cavagnetto, C., Laurain, M. and Meyer, R., 1980. Paysage végétal et position stratigraphique du sommet des Lignites du Soissonnais dans la Montagne de Reims (Yprésien, Bassin de Paris). *Géobios,* v. 13, p. 947–952.

Gunnell, G. F. and Gingerich, P. D., 1991. Systematics and evolution of late Paleocene and early Eocene Oxyaenidae (Mammalia, Creodonta) in the Clark's Fork Basin, Wyoming. *Contributions from the Museum of Paleontology, University of Michigan,* v. 28, p. 141–180.

Hand, S., Novacek, M., Godthelp, H. and Archer, M., 1994. First Eocene bat from Australia. *Journal of Vertebrate Paleontology,* v. 14, p. 375–381.

Hartenberger, J. L., 1971, Contribution à l'étude des genres *Gliravus* et *Microparamys* (Rodentia) de l'Eocène d'Europe, *Palaeovertebrata,* v. 4, p. 97–135.

Hedberg, H. D. (ed.), 1976. *International stratigraphic guide: A guide to stratigraphic classification terminology, and procedure.* New York: Wiley, 200 pp.

Heinzelin, J. de, 1979. Les "Sables d'Erquelinnes" et leurs ambiguités. In Sigogneau-Russell, D., Les Champsosaures européens: Mise au point sur le champsosaure d'Erquelinnes (Landénien inférieur, Belgique), *Annales de Paléontologie,* ser. Vertébrés, v. 65, p. 94–102.

Hester, S. W., 1965. Stratigraphy and palaeogeography of the Woolwich and Reading Beds. *Bulletin of the Geological Survey of Great Britain,* no. 23, p. 117–137.

Hooker, J. J., 1979. Two new condylarths (Mammalia) from the early Eocene of southern England. *Bulletin of the British Museum of Natural History (Geology),* v. 32, p. 43–56.

Hooker, J. J., 1991. The sequence of mammals in the Thanetian and Ypresian of the London and Belgian Basins. Location of the Palaeocene–Eocene boundary. *Newsletters on Stratigraphy,* v. 25, p. 75–90.

Hooker, J. J., 1994a. A new species of *Platychoerops* (Plesiadapiformes, Mammalia) from the latest Palaeocene of the Paris, London and Belgian Basins. *Géobios,* v. 27, p. 343–352.

Hooker, J. J., 1994b. The beginning of the equoid radiation. *Zoological Journal of the Linnean Society of London,* v. 112, p. 29–63.

Hooker, J. J., 1996a. A primitive emballonurid bat (Chiroptera, Mammalia) from the earliest Eocene of England. *Palaeovertebrata* (in press).

Hooker, J. J., 1996b. Mammalian biostratigraphy across the Palaeocene–Eocene boundary in the Paris, London and Belgian Basins. In Knox, R. W. O'B., Corfield, R. M. and Dunay, R. E. (eds.), *Correlation of the Early Paleogene in Northwest Europe,* Geological Society Special Publication, No. 101, p. 205–218.

Kennett, J. P. and Stott, L. D., 1991. Abrupt deep-sea warming, palaeoceanographic changes and benthic extinctions at the end of the Paleocene. *Nature,* v. 353, p. 225–229.

Knox, R. W. O'B., 1990. Thanetian and early Ypresian chronostratigraphy in south-east England. *Tertiary Research,* v. 11, p. 57–64.

Knox, R. W. O'B., Harland, R. and King, C., 1983. Dinoflagellate cyst analysis of the basal London Clay of southern England. *Newsletters on Stratigraphy,* v. 12, p. 71–74.

Koch, P. L., Zachos, J. C. and Gingerich, P. D., 1992. Correlation between isotope records in marine and continental carbon reservoirs near the Palaeocene/Eocene boundary. *Nature,* v. 358, p. 319–322.

Krause, D. W., 1982. Multituberculates from the Wasatchian Land-Mammal Age, early Eocene, of western North America. *Journal of Paleontology,* v. 56, p. 271–294.

Krause, D. W., 1986. Competitive exclusion and taxonomic displacement in the fossil record: The case of rodents and multituberculates in North America. In Flanagan, K. M. and Lillegraven J. A. (eds.),

Vertebrates, Phylogeny and Philosophy, Contributions to Geology Special Paper 3, Laramie: The University of Wyoming, p. 95–117.

Krause, D. W. and Maas, M. C., 1990. The biogeographic origins of late Paleocene–early Eocene mammalian immigrants to the Western Interior of North America. In Bown, T. M. and Rose, K. D. (eds.), *Dawn of the Age of Mammals in the Northern Part of the Rocky Mountain Interior, North America,* Geological Society of America Special Paper 243, p. 71–105.

Laurain, M. and Meyer, R., 1986. Stratigraphie et paléogéographie de la base du Paléogène champenois. *Géologie de la France,* v. 1986, p. 101–123.

Laurain, M., Barta, L., Bolin, C., Guernet, C., Gruas-Cavagnetto, C., Louis, P., Perreau, M., Riveline, J. and Thiry, M., 1983. Le sondage et la coupe du Mont Bernon à Epernay (Marne). Etude sédimentologique et paléontologique du stratotype du Sparnacien et de la série éocène. *Géologie de la France,* v. 3, p. 235–254.

Lemoine, V., 1889. Considérations générales sur les vertébrés fossiles des environs de Reims et spécialement sur les mammifères de la faune cernaysienne. *Comptes-Rendus des Séances du Congrès International de Zoologie, Paris,* v. 1889, p. 233–279.

Louis, P., Laurain, M., Bolin, C. and Barta, L., 1983. Nouveau gisement de vertébrés dans le Cuisien supérieur de Saint-Agnan (Aisne) et ses relations stratigraphiques avec les autres gisements yprésiens du Bassin parisien. *Bulletin d'Information des Géologues du Bassin de Paris,* v. 20, p. 3–20.

Maas, M. C. and Krause, D. W., 1994. Mammalian turnover and community structure in the Paleocene of North America. *Historical Biology,* v. 8, p. 91–128.

Maas, M. C., Krause, D. W. and Strait, S. G., 1988. The decline and extinction of Plesiadapiformes (Mammalia: ?Primates) in North America: displacement or replacement? *Paleobiology,* v. 14, p. 410–431.

MacLeod, N. and Rose, K. D., 1993. Inferring locomotor behavior in Paleogene mammals via eigenshape analysis. *American Journal of Science,* v. 293-A, p. 300–355.

Maglio, V. J. and Cooke, H. B. S. (eds.), 1978. *Evolution of African Mammals.* Cambridge: Harvard University Press, 641 pp.

Marandat, B., 1991. Mammifères de l'Ilerdien moyen (Eocène inférieur) des Corbières et du Minervois (Bas-Languedoc, France). Systématique, biostratigraphique, corrélations. *Palaeovertebrata,* v. 20, p. 55–144.

Marshall, L. G., 1987. Systematics of Itaboraian (Middle Paleocene) age "opossum-like" marsupials from the limestone quarry at Sao José de Itaboraí, Brazil. In Archer, M. (ed.), *Possums and Opossums:* Studies in Evolution, Chipping Norton, NSW: Surrey Beatty and Sons Pty Ltd, p. 91–160.

Massieux, M., Tambareau, Y. and Villatte, J., 1981. Characées Paléocènes et Eocènes du versant nord des Pyrénées. *Revue de Micropaléontologie,* v. 24, p. 69–82.

Michaux, J., 1968. Les Paramyidae (Rodentia) de l'Eocène inférieur du Bassin de Paris. *Palaeovertebrata,* v. 1, p. 135–193.

Payne, J., Francis, C. M. and Phillipps, K., 1985. *A Field Guide to the Mammals of Borneo.* Kota Kinabalu, the Sabah Society, 332 pp.

Powell, A. J., 1992. Dinoflagellate cysts of the Tertiary System. In Powell, A. J. (ed.), *A Stratigraphic Index of Dinoflagellate Cysts,* London: Chapman and Hall, p. 155–229.

Powell, A. J., Brinkhuis, H. and Bujak, J. P., 1996. Upper Paleocene–lower Eocene dinoflagellate cyst sequence biostratigraphy of southeast England. In Knox, R. W. O'B., Corfield, R. M. and Dunay, R. E. (eds.), *Correlation of the Early Paleogene in Northwest Europe,* Geological Society Special Publication No. 101, p. 145–183.

Reig, O. A., Kirsch, J. A. W. and Marshall, L. G., 1987. Systematic relationships of the living and Neocenozoic American "opossum-like" marsupials (suborder Didelphimorphia), with comments on the classification of these and of the Cretaceous and Paleogene New World and European metatherians. In Archer, M. (ed.), *Possums and Opossums:* Studies in Evolution, Chipping Norton, NSW: Surrey Beatty and Sons Pty Ltd, p. 1–89.

Rich, T. H. V., 1971. Deltatheridia, Carnivora, and Condylarthra (Mammalia) of the Early Eocene, Paris Basin, France. *University of California Publications in Geological Science,* v. 88, 72 pp.

Richards, P. W., 1952. *The Tropical Rain Forest, an ecological study.* Cambridge: Cambridge University Press, 450 pp.

Riveline, J., 1984. Les gisements à charophytes du Cénozoique (Danien à Burdigalien) d'Europe occidentale: Lithostratigraphie, biostratigraphie, chronostratigraphie. *Bulletin d'Information des Géologues du Bassin de Paris, Mémoire hors série,* v. 4, 523 pp.

Rose, K. D., 1981a. The Clarkforkian Land-Mammal Age and mammalian faunal composition across the Paleocene/Eocene boundary. *University of Michigan Papers on Paleontology,* v. 26, 197 pp.

Rose, K. D., 1981b. Composition and species diversity in Paleocene and Eocene mammal assemblages: An empirical study. *Journal of Vertebrate Paleontology,* v. 1, p. 367–388.

Rose, K. D., 1990. *Postcranial skeletal remains and adaptations in early Eocene mammals from the Willwood Formation, Bighorn Basin, Wyoming.* Geological Society of America Special Paper 243, p. 107–133.

Runestad, J. A. and Ruff, C. B., 1995. Structural adaptations for gliding in mammals with implications for locomotor behavior in paromomyids. *American Journal of Physical Anthropology,* v. 98, p. 101–119.

Russell, D. E., 1964. Les mammifères paléocènes d'Europe. *Mémoires du Muséum National d'Histoire Naturelle, Paris,* ser. C, v. 13, p. 1–324.

Russell, D. E., 1968. Succession, en Europe, des faunes mammaliennes au début du Tertiaire. *Mémoires du Bureau de Recherches Géologiques et Minières,* no. 58, p. 291–296.

Russell, D. E., 1980. Sur les condylarthres Cernaysiens *Tricuspiodon* et *Landenodon* (Paléocène supérieur de France). *Palaeovertebrata, Mémoire Jubilaire en hommage à René Lavocat,* p. 127–166.

Russell, D. E., 1981. Un primate nouveau du Paléocène supérieur de France. *Géobios,* v. 14, p. 399–405.

Russell, D. E., Louis, P. and Savage, D. E., 1973. Chiroptera and Dermoptera of the French early Eocene. *University of California Publications in Geological Science,* v. 95, 57 pp.

Russell, D. E., Louis, P. and Savage, D. E., 1975. Les Adapisoricidae de l'Eocène inférieur de France. Réévaluation des formes considérées affines. *Bulletin du Muséum National d'Histoire Naturelle, Paris,* ser. 3, v. 45, p. 129–193.

Russell, D. E., Godinot, M., Louis, P. and Savage, D. E., 1979. Apatotheria (Mammalia) de l'Eocène inférieur de France et de Belgique. *Bulletin du Muséum National d'Histoire Naturelle, Paris,* ser. 4C, v. 1, p. 203–243.

Russell, D. E., Hartenberger, J.-L., Pomerol, C., Sen, S., Schmidt-Kittler, N. and Vianey-Liaud, M., 1982. Mammals and stratigraphy: the Paleogene of Europe. *Palaeovertebrata, Mémoire Extraordinaire,* p. 1–77.

Russell, D. E., Galoyer, A., Louis, P. and Gingerich, P. D., 1988. Nouveaux vertébrés sparnaciens du Conglomérat de Meudon à Meudon, France. *Comptes-Rendus hebdomadaires des Séances de l'Académie des Sciences de Paris,* ser. 2, v. 307, p. 429–433.

Russell, D. E., Bignot, G., Galoyer, A., Guernet, C., Pomerol, C., Riveline, J., Sen, S., Thiry, M. and Tourenq, J., 1993. De la Craie à l'Argile Plastique: Un affleurement remarquable à Meudon près de Paris. *Bulletin d'Information des Géologues du Bassin de Paris,* v. 30, p. 3–9.

Savage, D. E., 1971. The Sparnacian–Wasatchian mammalian fauna, early Eocene, of Europe and North America. *Abhandlungen des hessischen Landesamtes für Bodenforschung,* v. 60, p. 154–158.

Savage, D. E. and Russell, D. E., 1977. Comments on mammalian paleontologic stratigraphy and geochronology; Eocene stages and mammal ages of Europe and North America. *Géobios, Mémoire spécial* 1, p. 47–56.

Savage, D. E. and Russell, D. E., 1983. *Mammalian Paleofaunas of the World.* London: Addison-Wesley Publishing Company, 432 pp.

Savage, D. E., Russell, D. E. and Louis, P., 1966. Ceratomorpha and Ancylopoda (Perissodactyla) from the Lower Eocene Paris Basin, France. *University of California Publications in Geological Science,* v. 66, p. 1–38.

Schankler, D. M., 1980. Faunal zonation of the Willwood Formation in the central Bighorn Basin, Wyoming. *Papers on Paleontology, Museum of Paleontology, University of Michigan,* v. 24, p. 99–114.

Scotese, C. R. and Golonka, J., 1992. PALEOMAP Paleogeographic Atlas, PALEOMAP Progress Report 20. Department of Geology, University of Texas at Arlington.

Shackleton, N. J., 1986. Paleogene stable isotope events. *Palaeogeography, Palaeoclimatology Palaeoecology,* v. 57, p. 91–102.

Siesser, W. G., Ward, D. J. and Lord, A. R., 1987. Calcareous nannoplankton biozonation of the Thanetian Stage (Palaeocene) in the type area. *Journal of Micropalaeontology,* v. 6, p. 85–102.

Sigé, B., Jaeger, J. J., Sudre, J. and Vianey-Liaud, M., 1990. *Altiatlasius koulchii* n. gen. et sp., primate omomyidé du Paléocène supérieur du Maroc, et les origines des Euprimates. *Palaeontographica,* ser. A, v. 214, p. 31–56.

Sinha, A. and Stott, L. D., 1994. New atmospheric pCO_2 estimates from palesols [sic] during the late Paleocene/early Eocene global warming interval. *Global and Planetary Change,* v. 9, p. 297–307.

Sinha, A., Aubry, M.-P., Stott, L. D., Thiry, M., and Berggren, W. A., 1996. Chemostratigraphy of the "lower" Sparnacian deposits (Argiles plastiques bariolées) of the Paris Basin. In Aubry, M.-P. and Benjamini, C. (eds.), *Paleocene/Eocene Boundary Events in Space and Time, Israel Journal of Earth Sciences,* v. 44, p. 223–237.

Sloan, R. E., 1981. Systematics of Paleocene Multituberculates from the San Juan Basin, New Mexico. In Lucas, S. G., Rigby, J. K. and Kues, B. S. (eds.), *Advances in San Juan Basin Paleontology,* Albuquerque: University of New Mexico Press, p. 127–160.

Smith, T. and Smith, R., 1995. Le genre *Dormaalius* QUINET, 1964 de l'Eocène inférieur de Belgique, synonyme du genre *Macrocranion* WEITZEL, 1949 (Mammalia, Lipotyphla). Ministère des Affaires Economiques, Administration des Mines, Service Géologique de Belgique, Professional Paper no. 1994/7, N. 274, p. 1–20.

Steurbaut, E., 1990. Ypresian calcareous nannoplankton biostratigraphy of the Belgian Basin. In Dupuis, C., De Coninck, J. and Steurbaut, E. (eds.), *The Ypresian Stratotype, Bulletin de la Société belge de Géologie,* 97–3/4 (1988), p. 251–285.

Storch, G., 1993. "Grube Messel" and African–South American faunal connections. In George, W. and Lavocat, R. (eds.), *The Africa–South America Connection,* Oxford Monographs on Biogeography, Oxford: Clarendon Press, p. 76–86.

Storch, G. and Richter, G., 1994. Zur Paläobiologie Messeler Igel. *Natur und Museum,* v. 124, p. 81–90.

Stott, L. D., Sinha, A., Thiry, M., Aubry, M.-P. and Berggren, W. A., 1996. Global $\delta^{13}C$ changes across the Paleocene–Eocene boundary: Criteria for terrestrial–marine correlations. In Knox, R. W. O'B., Corfield, R. M. and Dunay, R. E. (eds.), *Correlation of the Early Paleogene in Northwest Europe,* Geological Society Special Publication No. 101, p. 381–399.

Sudre, J., Russell, D. E., Louis, P. and Savage, D. E., 1983. Les Artiodactyles de l'Eocène inférieur d'Europe. *Bulletin du Muséum National d'Histoire Naturelle, Paris,* ser. 4C, v. 5, p. 281–333, 339–365.

Swofford, D. L., 1990. PAUP, Phylogenetic Analysis Using Parsimony, Version 3.0, User's manual. Illinois Natural History Survey, Champaign.

Szalay, F. S., 1968. Origins of the Apatemyidae (Mammalia, Insectivora). *American Museum Novitates,* no. 2352, p. 1–11.

Szalay, F. S., 1994. *Evolutionary History of the Marsupials and an Analysis of Osteological Characters.* Cambridge: Cambridge University Press, 481 pp.

Szalay, F. S. and Li Chuan-Kuei, 1986. Middle Paleocene Euprimate from southern China and the distribution of primates in the Paleogene. *Journal of Human Evolution,* v. 15, p. 387–397.

Tambareau, Y., Russell, D. E., Sigogneau-Russell, D. and Villatte, J., 1992. Découverte de restes de vertébrés dans le Paléocène de Campo (Pyrénées aragonaises). *Bulletin de la Société d'Histoire Naturelle de Toulouse,* v. 128, p. 73–76.

Teilhard de Chardin, P., 1922. Les mammifères de l'Eocène inférieur français et leurs gisements. *Annales de Paléontologie,* v. 11, p. 9–16.

Thewissen, J. G. M., 1990. Evolution of Paleocene and Eocene Phenacodontidae (Mammalia, Condylarthra). *Papers on Paleontology, Museum of Paleontology, University of Michigan,* no. 29, 107 pp.

Thewissen, J. G. M., 1991. Limb osteology and function of the primitive Paleocene ungulate *Pleuraspidotherium* with notes on *Tricuspiodon* and *Dissacus* (Mammalia). *Géobios,* v. 24, p. 483–495.

Thewissen, J. G. M. and Hussain, S. T., 1990. Postcranial osteology of the most primitive artiodactyl *Diacodexis pakistanensis* (Dichobunidae). *Anatomy, Histology, Embryology,* v. 19, p. 37–48.

Thiry, M., 1981. Sédimentation continentale et altérations associées: Calcifications, ferruginisations et silicifications. Les Argiles Plastiques du Sparnacien du Bassin de Paris. *Sciences Géologiques,* v. 64, 173 pp.

Ting Su-yin, 1993. A preliminary report on an early Eocene mammalian fauna from Hengdong, Hunan Province, China. *Kaupia, Darmstädter Beiträge zur Naturgeschichte,* v. 3, p. 201–207.

Tong Yong-shen, 1979. A late Paleocene primate from S. China. *Vertebrata Palasiatica,* v. 17, p. 65–70.

Van Valen, L., 1967. New Paleocene insectivores and insectivore classification. *Bulletin of the American Museum of Natural History,* v. 135, p. 217–284.

Van Valkenburgh, B., 1990. Skeletal and dental predictors of body mass in carnivores. In Damuth, J. and MacFadden, B. J. (eds.), *Body Size in Mammalian Paleobiology: Estimation and Biological Implications,* Cambridge: Cambridge University Press, p. 181–205.

Vianey-Liaud, M., 1986. Les Multituberculés thanétiens de France, et leur rapports avec les Multituberculés Nord-Américains. *Palaeontographica,* ser. A, v. 191, p. 85–171.

Webb, S. D., 1985. Late Cenozoic mammal dispersals between the Americas. In Stehli, F. G. and Webb, S. D. (eds.), *The Great American Biotic Interchange,* New York: Plenum Press, p. 357–386.

Wing, S. L., 1984. A new basis for recognizing the Paleocene/Eocene boundary in western interior North America. *Science,* v. 226, p. 439–441.

Appendix 20.1

Faunal list for major Cernaysian and Neustrian (Zones PE I to PE IV) localities

	MP6		PE I				PE II			PEIII		IV	S	L	D
	C	B	T	E	SP	D	Me	So	Si	AW	P	M			
Cernaysia manueli	*X*												AB	Arb	I/F
Hainina godfriauxi	*X*												AB	Arb	I/F
Adapisoriculus minimus	X												AB	ST	I
Afrodon germanicus	*X*												AB	ST	I
?Palaeictops levei	X												AB	ST	I
Aboletylestes	*X*												AB	ST	I
Pagonomus dionysi	X												AB	Aq	I
"Jepsenella"	*X*												AB	Scan	I
Remiculus deutschi	*X*												AB	ST	I
Adapisorex gaudryi	*X*												AB	Scan	I/F
Arctocyonides trouessarti	*X*												C	ST	I-C/F
Berruvius gingerichi	*X*												AB	Scan	I
Liotomus marshi	X	*X*											AB	Arb	I/F
Neoplagiaulax eocaenus	X	*X*											AB	Arb	I/F
N. copei	X	*X*											AB	Arb	I/F
N. annae	X	*X*											AB	Arb	I/F
N. nicolae	X	*X*											AB	Arb	I/F
N. sylvani	X	X											AB	Arb	I/F
Bustylus cernaysi	X	*X*											AB	ST	I
Plesiadapis tricuspidens	X	X											C	Scan	F
Chiromyoides campanicus	X	X											AB	Scan	F
Arctocyon primaevus	X	*X*											C	ST	I-C/F
Arctocyonides arenae	X	X											C	ST	I-C/F
Landenodon lavocati	X	X											AB	Scan	I-C/F
Tricuspiodon rutimeyeri	X	*X*											D	T	I-C/F
Dipavali petri	X	*X*											AB	Scan	F
Paschatherium (large)	X	X											AB	Scan	I/F
Paschatherium (small)	X	X											AB	Scan	I/F
Louisina mirabilis	X	X											AB	Scan	F
Louisininae undet.	X	*X*											AB	Scan	F
Pleuraspid. aumonieri	X	X	*X*										C	ST	HB
Orthaspidotherium edwardsi	X	X	*X*										AB	ST	HB
Dissacus europaeus & aff.	X					X							D	T	C
Plesiadapis remensis & aff.	X	X	X								*X*		AB	Scan	F
Nyctitheriidae undet.	X				X	X	X	X			X		AB	ST	I
Berruvius lasseroni & cf.		X					X					X	AB	Scan	I
Cymbalophus cuniculus			**X**	X	X								C	T	F/HB
Platychoerops georgei			X	X	X	X							C	Scan	F/HB
Arfia junnei			**X**		X	*X*							C	ST	C/F
Teihardina belgica			**X**	X	X	X	X	X					AB	Arb	I
Arctocyonides sp.			X							*X*			C	ST	I-C/F
Cantius eppsi			**X**		X		X			X	X		AB	Arb	I/F
Coryphodon			**X**	X			X	X	X	X	*X*		GT	T	HB
Multituberculata undet.			X				X	X			X		AB	Arb	I/F
Macrocranion cf. *nitens*			**X**		?	X	X	X		X	X	X	AB	ST	I
Paschatherium dolloi				X	X	X	X				*X*		AB	Scan	I/F
Pontifactor					*X*								AB	ST	I
Chiromyoides sp.					*X*								AB	Scan	F
?Leptictidae undet.					*X*								AB	ST	I
Diacodexis sp. 1					**X**	X							AB	T	I-C/F
Microparamys nanus					**X**	X							AB	Scan	I/F
Paramys sp.					**X**	X							AB	Scan	I/F
Pseudoparamys sp.					**X**	X							AB	Scan	F

Appendix 20.1 (*continued*)

	MP6		PE I				PE II			PEIII		IV	S	L	D
	C	B	T	E	SP	D	Me	So	Si	AW	P	M			
Meldimys sp.					**X**	X		X		?			AB	Scan	F
Microhyus musculus					X	X					*X*		AB	Scan	F
Landenodon woutersi					X	X					*X*		AB	Scan	I-C/F
Hyopsodus wardi					**X**					X	X	*X*	AB	ST	I/F
Miacidae undet.					**X**		X			2		X	AB	Scan	C
Viverravidae undet.					**X**		X					X	AB	ST	C
Pantolestidae undet.					X	X	X			X	X	X	AB	Aq	I
?Prolimnocyon						*X*							C	Scan	C
Miacis latouri						X							AB	Scan	C
Dormaalodon woutersi						*X*							C	ST	C
Macrocranion vandebroeki						*X*							AB	ST	I
Eochiromys landenensis						*X*							AB	Scan	I/F
Apatermys teilhardi						**X**							AB	Scan	I/F
Phenacodus teilhardi						**X**							C	T	F
Adapisoriculus						*X*							AB	ST	I
Palaeonictis gigantea						**X**	X	X					D	ST	C
Hallensia louisi						**X**					X	X	C	T	F/HB
Peratherium constans						**X**						*X*	AB	ST	I/F
Amphiperatherium brabantense						**X**				X		*X*	AB	ST	I/F
*Pachyaena gigantea**							*X*						GH	T	C
Arcius sp.							**X**						AB	Aer	I/F
cf. *Icaronycteris*							**X**						AB	Aer	I
Platychoerops russelli							X	X	?				C	Scan	F/HB
Dipsalidictis transiens							**X**		X				C	Scan	C
Paramys woodi & cf.							X					X	AB	Scan	I/F
Pseudoparamys teilhardi & cf.							X					X	AB	Scan	F
Hyracotherium aff. *leporinum*							X				X	*X*	C	T	F/HB
Apatemys sigogneaui & cf.							X				X	X	AB	Scan	I/F
Paramys ageiensis & cf.							X	X		X	X	L	AB	Scan	I/F
Peratherium matronense							X					L	AB	ST	I/F
Neomatronella							**X**	X		X	X	X	AB	ST	I
Amphiperatherium maximum							**X**			?	X	X	AB	ST	I/F
Peradectes louisi							?	**X**				X	AB	Arb	I/F
Peratherium sp. indet. 1								*X*					AB	ST	I/F
Pliolophus aff. *vulpiceps*								X	X				C	T	F/HB
Apatemys sp.								X		X		*X*	AB	Scan	I/F
Microparamys sp.								X		X		3	AB	Scan	I/F
?Prototomus								**X**		X		X	AB	ST	C
"*Propachynolophus*" sp. 1									X		X	L	C	T	F/HB
Esthonychidae undet.										**X**			C	T	HB
Oxyaena gulo										*X*			C	ST	C
Diacodexis sp. 2										X			AB	T	I-C/F
Lessnessina packmani										*X*			AB	ST	I/F
Omomyidae undet.										X			AB	Arb	I
Eppsinycteris anglica										*X*			AB	Aer	I
Palaeonictis cf. *occidentalis*										X	*X*		D	ST	C
Pliolophus vulpiceps										X	*X*		C	T	F/HB
?Didelphodus										**X**	X	X	AB	ST	I
Arcius fuscus										X	X	X	AB	Aer	I/F
Pantrogna russelli s.s										X	X	L	AB	Scan	I/F
Ectypodus childei										**X**		*X*	AB	Arb	I/F
Phenacodus lemoinei										X	X	X	C	T	F
Paramys pourcyensis											*X*		AB	Scan	I/F
Bunophorus cappettai											X	X	AB	T	I-C/F

Appendix 20.1 (*continued*)

	MP6		PE I				PE II			PEIII		IV	S	L	D
	C	B	T	E	SP	D	Me	So	Si	AW	P	M			
Placentidens lotus											**X**	X	AB	ST	I
Sparnacomys chandoni											X	L	AB	Scan	I/F
Platychoerops daubrei											X	X	C	Scan	HB
Franchaius luciae											X	L	C	T	HB
Apatemys mutiniacus											X	X	AB	Scan	I/F
Franchaius minimus											X	X	C	T	HB
Diacodexis varleti & aff.											X	X	AB	T	I-C/F
Hyrachyus											**X**	L	D	T	HB
Peradectes mutigniensis												X	AB	Arb	I/F
Macrocranion sp.												X	AB	ST	I
Cantius savagei												X	AB	Arb	I/F
?Icaronycteris menui												X	AB	Aer	I
Archaeonycteris brailloni												**X**	AB	Aer	I
Ageina tobieni												**X**	AB	Aer	I
Heterohyus sp.												X	AB	Scan	I/F
Plesiesthonyx munieri												X	C	T	HB
"Propachynolophus" maldani & aff.												X	C	T	F/HB
Lophiaspis maurettei												**X**	D	T	HB
Meldimys louisi												X	AB	Scan	F
Neoplagiaulax aff. *sylvani*												*X*	AB	Arb	I/F

Sources: Baudry (1992), Collinson and Hooker (1987), Crochet (1980), Feugueur (1963), Gheerbrant and Russell (1989, 1991), Gingerich (1976, 1977a, 1989), Godinot (1984), Godinot et al. (1978), Gunnell and Gingerich (1991), Hartenberger (1971), Hooker (1991, 1994a, 1994b, 1996a, 1996b), Krause (1982), Michaux (1968), Rich (1971), Russell (1964, 1980, 1981), Russell et al. (1973, 1975, 1979, 1988), Savage et al. (1966), Sloan (1981), Smith and Smith (1995), Sudre et al. (1983), Thewissen (1990), and Vianey-Liaud (1986).

Records in bold indicate that no close relatives exist at earlier sites on the table (i.e., dispersal into the area). Records italicized indicate that no close relatives exist at later sites in the area (i.e., a true extinction event). Site abbreviations: C = Cernay, B = Berru, T = Try, E = Erquelinnes, SP = Suffolk Pebble Beds localities, D = Dormaal, Me = Meudon, So = Soissons, Si = Sinceny, AW = Abbey Wood, P = Pourcy, M = Mutigny. * = taxon not from Meudon but from the stratigraphically and geographically nearby Argile Plastique bariolées of Vaugirard. N.B.: *Palaeonictis gigantea* and *Coryphodon* recorded from Soissons are from the Argile à Lignites du Soissonnais but not from Soissons itself.

Three columns to right: S = size, L = foraging locomotor adaptation, D = diet. Size classes: AB = less than 1 kg, C = 1–10 kg, D = 10–45 kg, EF = 45–180 kg, GH = more than 180 kg. Locomotor classes: T = terrestrial, ST = semiterrestrial, Arb = arboreal, Scan = scansorial, Aq = aquatic, Aer = aerial. Dietary classes: I = insectivore, F = frugivore, HB = herbivore browser, HG = herbivore grazer, C = carnivore.

21

Fossil Mammals and the Paleocene/Eocene Series Boundary in Europe, North America, and Asia

SPENCER G. LUCAS

ABSTRACT

Mammalian paleontologists have long placed the Paleocene/Eocene (P/E) Series boundary at a perceived significant evolutionary turnover in mammals independent of its position by definition in the standard global chronostratigraphic scale (SGCS), which is the base of the Ypresian Stage. These mammal-based P/E Series boundaries—at the base of the Sparnacian "Stage" in Western Europe, at the base of either the Clarkforkian or the Wasatchian Land-Mammal "Ages" (LMA) in North America and, in Asia, at the base of the Bumbanian (= Lingchan) LMA—are demonstrably older than the base of the Ypresian Stage and thus are of late Paleocene age. In Western Europe (England, Belgium, France), fossil mammals assigned to Reference Level MP7 are of late Paleocene age; the P/E Series boundary is within Reference Level MP8–9 and not associated with a detectable evolutionary turnover in mammals. The two relevant evolutionary turnover events in Western Europe occurred during the late Paleocene (between Reference Levels MP6 and MP7) and during the early Eocene (between Reference Levels MP8–9 and MP10) and delimit the Neustrian LMA. The term *Sparnacian* should not be used for a LMA.

In North America, mammal-based correlation, particularly using coryphodontid pantodonts and plesiadapiforms, indicates that the Clarkforkian LMA correlates to Reference Level MP6 and to the stratigraphic gap between Reference Levels MP6 and MP7, and that the Graybullian SLMA (lower Wasatchian LMA) correlates to Reference Level MP8/9. Therefore, the P/E Series boundary is within the Graybullian interval of the Wasatchian LMA, a correlation supported by magnetostratigraphy, palynostratigraphy, and the identification of the latest Paleocene carbon isotope excursion in lowermost Graybullian strata in Wyoming. The term *Sandcouleean* for earliest Graybullian time should be abandoned as biochronologically unrecognizable.

Less precise mammal-based correlations to Asia suggest the P/E Series boundary is within the Nongshanian LMA. It may coincide with or slightly precede the lowest occurrence of bona fide *Coryphodon* in sections in the Nemegt (Mongolia) and Turpan (Xinjiang Province, China) basins.

Mammalian fossils have long played a significant role in the placement of the Paleocene/Eocene (P/E) Series boundary in nonmarine strata. Indeed, international recognition of the Paleocene was initially based on fossil mammals (Matthew 1937; Simpson 1981). Furthermore, an independent tradition developed long ago among mammalian paleontologists, who identified the P/E Series boundary based wholly on changes in fossil mammals. In this tradition, little regard was often given to whether the mammal-based boundary correlated to the P/E Series boundary as it is defined in the standard global chronostratigraphic scale (SGCS) built on marine biostratigraphy.

In this chapter, I evaluate mammalian biostratigraphy across the P/E Series boundary in Europe, North America, and Asia. The three continents formed a single, Holarctic biogeographic province for mammals during the late Paleocene–early Eocene (e.g., McKenna 1983a, 1983b; figure 21.1). First, I review the mammalian fossil record in these three areas and its internal, mammal-based correlation to provincial biostratigraphies. Then, I correlate all three records to each other and to the P/E Series boundary. In doing so, I attempt to evaluate all mammal-based correlations but place heavy emphasis on correlations based on coryphodontid pantodonts.

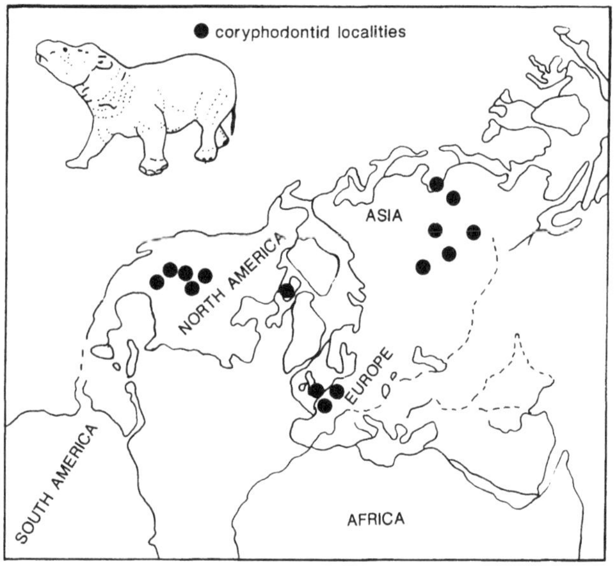

Figure 21.1 Polar projection of the northern hemisphere for late Paleocene–early Eocene time showing principal coryphodontid localities in Europe, North America, and Asia.

These large, hippopotamus-like extinct mammals provide some of the best index fossil mammals for correlation across the P/E Series boundary in Holarctica (Lucas 1984a). For the purposes of this chapter, the P/E Series boundary in the SGCS corresponds to the base of the Ypresian stage, which has an estimated age of approximately 54.6 Ma or 54.7 Ma within Chron C24r on the Integrated Magnetobiochronologic Scale (IMBS: Berggren et al. 1995; Berggren and Aubry 1996; chapter 2).

Some Taxonomic Problems

Disagreements over biostratigraphic correlation often result from taxonomic disagreements. Correlating the stratigraphic (temporal) ranges, the LO (lowest occurrence) or HO (highest occurrence) of identical taxa from separate regions is the basis of sound biostratigraphic correlation. Correlating different taxa because they share the same stage of evolution provides a less sound basis. This is because identifying the same stage of evolution in different taxa relies on correlating separate phylogenies, so it introduces the assumptions behind phylogenetic analysis to the process of correlation. In mammalian paleontology, the most misleading correlations based on stage of evolution rely on the argument that small size is primitive (older) and large size is advanced (younger). Numerous studies attest to the lability of body size in mammalian evolution (many advanced taxa are "small"), so correlation by a size morphocline is suspect. Phylogenetic position has a role to play in biostratigraphic correlation when taxonomic identity is totally unavailable, but it should be a last resort.

A current problem that has undermined the biostratigraphic utility of many mammalian taxa is cladistic taxonomy, which splits lower-level taxa based on their perceived phylogenetic relationships, not on their overall morphological resemblances interpreted within a framework of population variation. For example, Hooker (1984, 1989, 1994) has split the small sample of European specimens he (see Hooker 1980) and previous workers earlier termed *Hyracotherium* into the genera *Cymbalophus*, *Hallensia*, *Hyracotherium*, *Pachynolophus*, *Pliolophus*, and *Systemodon*. The bewildering array of metric and meristic variation present in North American *Hyracotherium* (see, for example, Kitts 1956; Gingerich 1991) encompasses all the characteristics to which Hooker attached cladistic significance and consequently used to split one genus into six. Therefore, I do not employ Hooker's taxonomy of primitive European perissodactyls, all of which I refer to *Hyracotherium*.

Taxonomic identity should be demonstrated by morphological similarity analyzed within the context of population variation. I have termed this approach holomorphologic (Lucas 1989b). Taxa identified by geographic or stratigraphic position are useless to biostratigraphy. Despite the morphological similarity of two samples, some paleontologists have rejected their taxonomic identity because of wide geographic separation (Simons 1963 even argues that to do otherwise is fallacious). In so doing, assumptions about the possible paleogeographic range of a mammalian taxon are made, often with little justification. During the late Paleocene–early Eocene epochs, Europe, North America, and Asia formed a single biogeographic province, so one genus or species could have lived on all three continents. Therefore, no assumptions should be made to limit the possible geographic range of a Holarctic mammal genus or species during this time interval. For example, the morphological cluster that includes the holotype of *Coryphodon eocaenus* is found on all three continents and merits one taxonomic name.

Identifying taxa by their stratigraphic (temporal) position renders them only useful in biostratigraphic correlation by circular reasoning. This is best illustrated by example. Rose (1981:66–67, figure 30), in a study of Clarkforkian and Wasatchian mammals from the Willwood Formation in the Bighorn Basin, Wyoming, found that the range of variation of M_1 size in a sample of specimens of the primate *Phenacolemur* formed a chronocline. Most older (Clarkforkian) specimens have smaller M_1s than do the younger (Wasatchian) specimens, although there is considerable overlap in M_1 size in the Clarkforkian and Wasatchian samples. The specimens are otherwise indistinguishable. Nevertheless, Rose divided the sample into two species using an arbitrary boundary just below the Clarkforkian/Wasatchian

LMA boundary. The specimens above this boundary were assigned to *P. praecox*, and those below the boundary were assigned to *P. pagei*. A holomorphological taxonomic approach, on the other hand, would have assigned all of these specimens to one species. By assigning them to two species, Rose has created an artificial biostratigraphic datum: the LO of *P. pagei* and the HO of *P. praecox* in the Willwood Formation in the Bighorn Basin. Moreover, if a *Phenacolemur* specimen from another basin were identical to specimens in this *Phenacolemur* sample from the Bighorn Basin and had an M_1 size in the middle of the size range of the Bighorn sample, to which species would it belong? And what age is the specimen from the other basin? Obviously, Rose's taxonomy of *Phenacolemur* is impractical, and the resultant biostratigraphic datum is artificial.

Far more problematic is the situation where entire biostratigraphic constructs are created from taxa identified by stratigraphic position (see McKenna et al. 1977 for a critique of this). A case in point is Gingerich's (1989) monograph, in which he resurrected the term *Sandcouleean* to refer to the earliest interval of Wasatchian time represented in Wyoming. Gingerich (1989:table 27) listed 41 mammal species from the Sandcouleean deposits, of which only 10 are restricted to that interval; 24 range into the younger Wasatchian, and 7 are Clarkforkian holdovers. Of the 10 species restricted to the Sandcouleean interval, 6 are named by Gingerich in the monograph based more on stratigraphic position than on morphology.

For example, the Sandcouleean index taxon *Arfia junnei* Gingerich is known from a handful of specimens diagnosed only as being "13% smaller" (Gingerich 1989:33) than Wasatchian *A. zele*. In fact, Gingerich (1989:figure 22) indicates considerable size overlap between teeth of *A. junnei*, *A. zele*, and (younger Wasatchian) *A. shoshoniensis*, which demonstrates that there is no valid size-based diagnosis of *A. junnei*. The sample of *A. junnei* represents specimens that would be called *A. zele* or *A. shoshonensis* had they been found in younger strata.

Three species not named by Gingerich (1989), supposedly restricted to the Sandcouleean interval, are *Meniscotherium priscum* Granger, *Ectocion parvus* Granger, and *Ectoganus bighornensis* Schoch. Williamson and Lucas (1992) synonymized *M. priscum* with *M. chamense*, a taxon that ranges well up into the Wasatchian LMA. *Ectocion parvus* is based on a few specimens, some of which are smaller than *E. osbornianus*, but the distinctiveness of the species is doubtful (West 1976). *Ectoganus bighornensis* was named as a subspecies of *E. copei* by Schoch (1986) based on very few specimens, but I cannot distinguish this taxon on a morphological basis. Finally, the tenth taxon restricted to the Sandcouleean interval is an edentulous jaw fragment that Gingerich (1989) tentatively regarded as a mesonychid or "notoungulate" but which he assigned to cf. *Asiabradypus* sp., a problematic taxon from Asia. I conclude that the Sandcouleean interval is largely an artifact of unworkable taxonomy, not a useful biostratigraphic construct.

Europe

Introduction

The record of fossil mammals in western Europe across the P/E Series boundary is particularly important because the marine stages (Thanetian and Ypresian) used to define the P/E Series boundary in the SGCS have their stratotypes in this region (e.g., Pomerol 1981). It is possible to correlate directly the European upper Paleocene–lower Eocene mammal fossils to these stratotypes and thereby cross-correlate the boundary from marine to nonmarine strata (see below).

These fossil mammals are found in southeastern England, Belgium, and France in what is often referred to as the London–Paris–Belgian Basin (figure 21.2). This is actually part of a larger, northwestern European basin that extended from the Shetland Islands and Atlantic Shelf of Norway into the Carpathians and the Ukraine (Ziegler 1982). This basin connected the North Sea Basin to the northeastern Atlantic between the west-Anglian High and the Amorican Massif. The fossil mammals occur in fluvial, paralic, and, rarely, marine siliciclastics deposited in the narrow southwestern portion of the basin (e.g., Stamp 1921a, 1921b; Davis and Elliot 1958; Feugueur 1963; Gulinck 1965; Hester 1965; Curry 1967; Plaziat 1975; Knox and Harland 1979).

European Mammalian Biostratigraphy

A provincial mammalian biostratigraphy for the Paleogene Series has been established in western Europe using Reference Levels (e.g., Russell et al. 1982; Godinot 1987; figure 21.3). In other words, a characteristic fossil assemblage is identified for each level, and correlation to such assemblages is by presence/absence of taxa or perceived stage of evolution (Savage and Russell 1977). The levels thus are viewed as points in time or datum planes, and they have no boundaries. LMAs exist for the succession of levels (Fahlbusch 1976; Savage and Russell 1983), but they have not been as widely used. Hooker (1996, chapter 20) proposed a range-zone biostratigraphy for the European mammal localities that encompass the P/E Series boundary. However, this biostratigraphy is not based totally on stratigraphic superposition but in part on a consensus

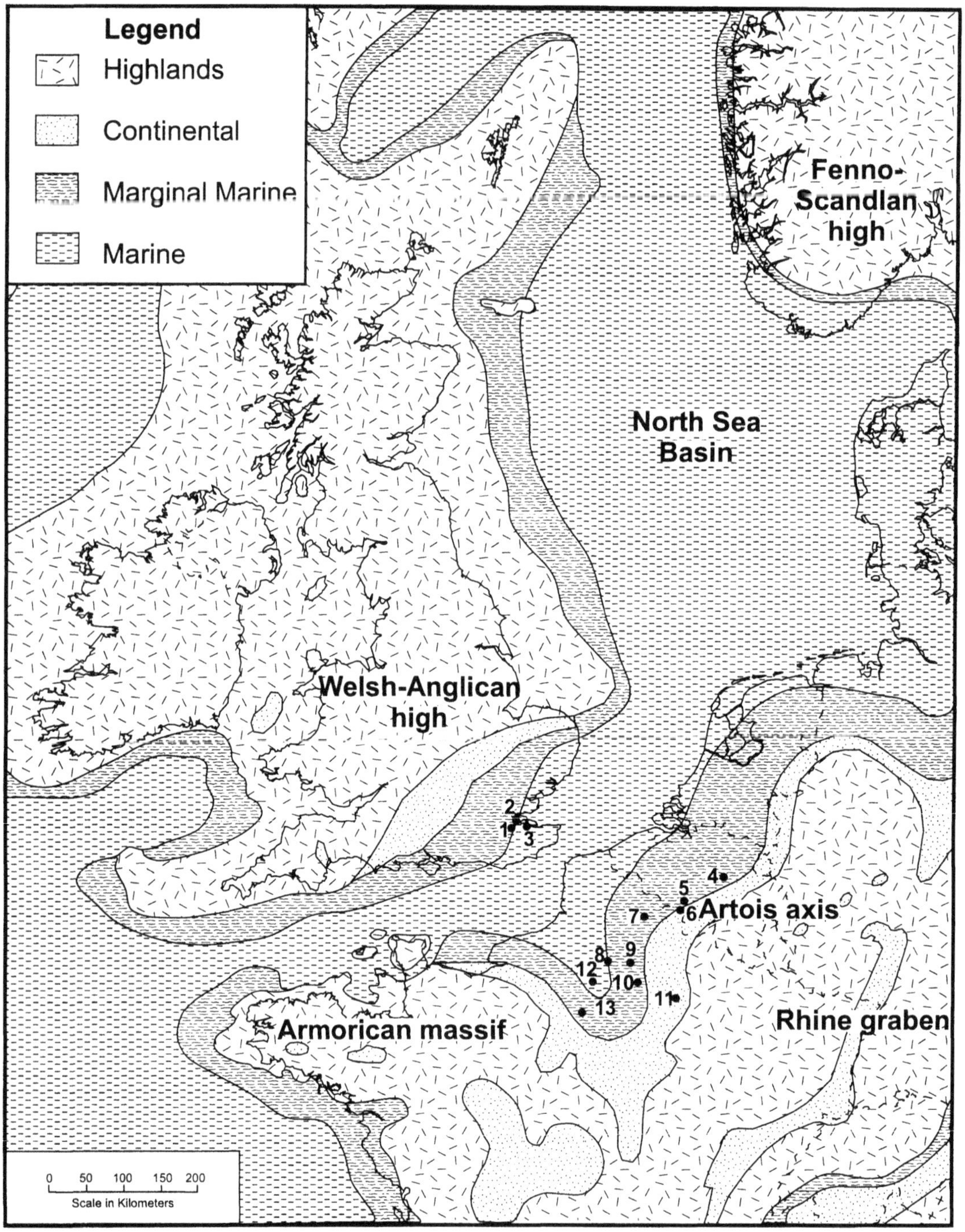

Figure 21.2 Paleogeographic map of western Europe for the late Paleocene–early Eocene epochs (after Ziegler 1982) showing major fossil mammal localities.

1 = Croydon, 2 = London and vicinity, 3 = Abbey Wood, 4 = Orp, 5 = Leval, 6 = Erquelinnes, 7 = Vertain, 8 = Muirancourt, 9 = Sinceny, 10 = Soissons and vicinity, 11 = Epernay and vicinity, 12 = Sarron, and 13 = Meudon and other localities now within the suburbs of Paris.

tree derived from statistical analysis of the mammalian taxa. Many of these taxa are rare and unique to single localities, which undermines the reliability of the statistical analysis and their biostratigraphic significance. Therefore, I do not consider Hooker's new biostratigraphic scheme preferable to the already existing reference levels.

Most of the details of the temporal succession of the standard faunas were placed in a biostratigraphic framework decades ago (e.g., Dollo 1909; Dollo and Teilhard

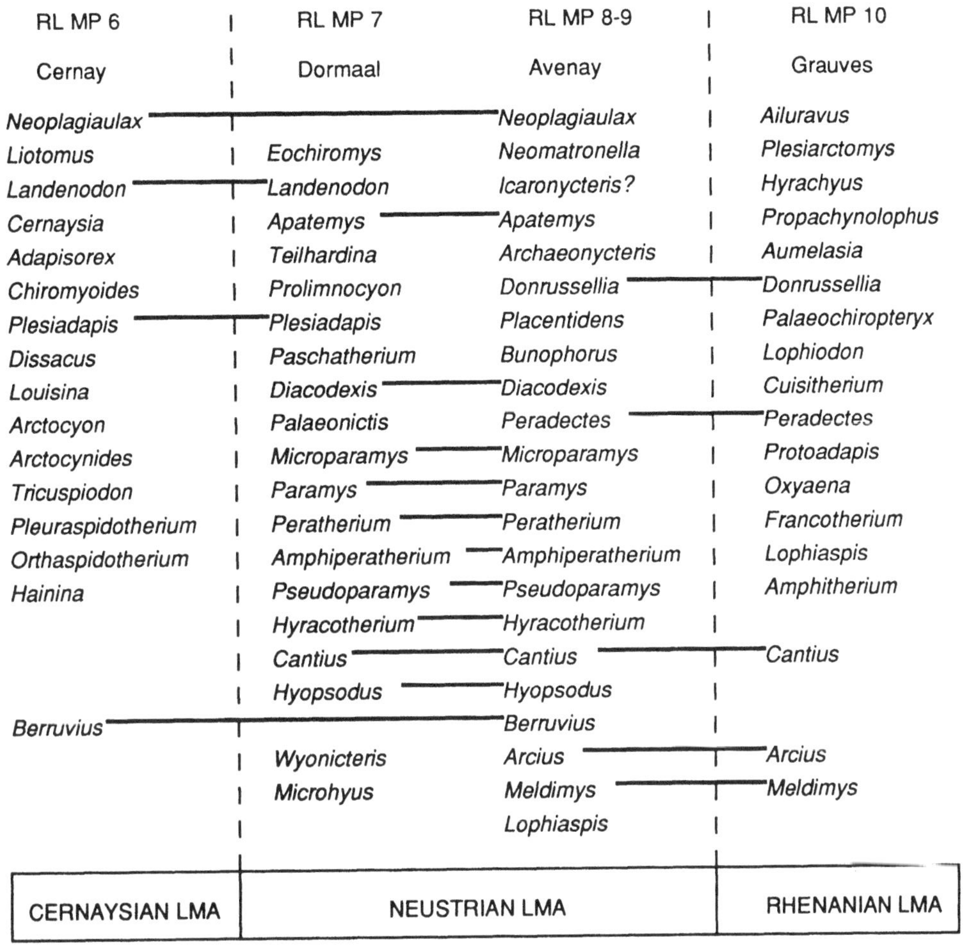

Figure 21.3 Mammalian genera of the type assemblages of Reference Levels MP6 to MP10 in western Europe.

After Godinot (1987); with updates from Hooker (1991), Smith (1995), and Smith et al. (1996). Note: the greatest differences at the generic level are between Reference Levels MP6 and MP7 and between Reference Levels MP8–9 and MP10.

de Chardin 1924; Teilhard de Chardin 1920, 1927; Simpson 1929, 1948a; Rat 1965). The use of Reference Levels is well suited to the European record of fossil mammals where most assemblages are from single beds/localities and thus lack stratigraphic thickness. Furthermore, due to lack of outcrop and rarity of traceable marker beds, most assemblages cannot be superposed or correlated by physical stratigraphy except in a general way.

Four levels are relevant here (figure 21.3)—Reference Level (RL) MP6 (Cernay), RL MP7 (Dormaal), RL MP8–9 (Avenay), and RL MP10 (Grauves). There is consensus on the late Paleocene (Thanetian) age of Cernay, though its exact stratigraphic position and precise age within the Thanetian stage are somewhat uncertain (Berggren and Aubry 1996). The standard fauna and its principal correlative at Mont Berru were collected near the village of Cernay-les-Reims in a 2-m-thick shelly sandstone bed (the "Conglomérat de Cernay") overlain by the Argile Plastique and underlain by lenticular sandstone (Rilly Sands) disconformable on Upper Cretaceous chalk (Lemoine 1880, 1889; Savage and Russell 1983). The Cernay assemblage includes the most primitive European plesiadapiforms (*Chiromyoides*, *Plesiadapis tricuspidens*) and other mammals (e.g., *Dissacus*, *Arctocyon*) of well established Tiffanian age in North America (e.g., Savage and Russell 1983).

At the other end of the spectrum, RL MP10 (Grauves) marks a major change in the mammalian fauna with the appearance of abundant ceratomorph perissodactyls (*Hyrachyus*, *Lophiodon*), the "*Lophiodon* event" of Fahlbusch (1976), which marks the base of the Rhenanian LMA. The Grauves assemblage resembles overlying middle Eocene (Geiseltalian LMA of Franzen and Haubold 1986 and Haubold 1989) mammal assemblages in Europe and it can be well correlated to Bridgerian assemblages in North America (e.g., Savage and Russell 1983). The standard RL MP10 fauna comes from the Marne Valley just southeast of Epernay in the Sables à Unios et Térédines of Cuisian (Ypresian) age (e.g., Aubry 1985). Therefore, the P/E Series boundary is between the levels of Cernay and Grauves, in the interval of Reference Levels MP7 and MP8–9.

The type RL MP7 fauna (Dormaal) is from the upper Landen Formation in Belgium (Godinot et al. 1978; Hooker 1991). It differs from the RL MP8–9 assemblages in the absence of some characteristic RL MP8–9 taxa and retention of *Plesiadapis* (figure 21.3). Reference Level MP8–9 is treated as a single level now, though originally it was believed that the Mutigny (RL MP8) and Avenay (RL MP9) assemblages could be distinguished. The Mutigny and Avenay assemblages are from the Argiles à Lignites north of Epernay. The differences between these two faunas are comparable to the differences between them and RL MP7. Indeed, the differences between the mammal assemblages of RL MP7 and RL MP8–9 are not dramatic (figure 21.3), so both levels should be considered representative of a single LMA, which is best termed the Neustrian LMA (see below).

Sparnacian

Dollfus (1880, 1905) introduced the name *Sparnacian* (based on the Roman name for Epernay), with the Lignites du Soissonais at Mt. Bernon on the southeastern edge of Epernay as its stratotype. The term has generally been used as a stage (e.g., Blondeau et al. 1965; Cavelier and Pomerol 1986), but Savage and Russell (1983) redefined it as an LMA equivalent to the Neustrian LMA of Fahlbusch (1976). Sparnacian time in this sense is equivalent to the temporal range of all the mammalian fossils found in RL MP7 and RL MP8–9 in Western Europe. However, given the many problems the term *Sparnacian* has caused and its varied usage (see Savage 1977; Pomerol 1981; and Savage and Russell 1983 for examples of at least three different usages), it seems unwise to perpetuate the term as the name of an LMA. Fahlbusch's (1976) term *Neustrian* may be better used to refer to the interval that comprises the RL MP7 and RL MP8–9 mammalian assemblages.

England

Hooker and Insole (1980) listed the fossil mammals from southeastern England relevant to the P/E Series boundary, and Collinson and Hooker (1987) and Hooker (1991, 1996, chapter 20) provided recent reviews. Five stratigraphic units produced most of these mammals (ascending): Woolwich Shell Beds, Suffolk Pebble Beds, Oldhaven Beds, Blackheath Beds, and London Clay. The stratigraphic succession of these beds can be

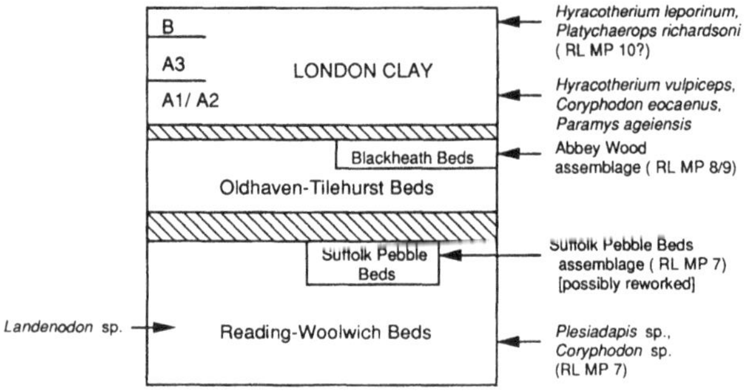

Figure 21.4 Stratigraphic succession and mammal assemblages across the Paleocene/Eocene Series boundary in southeastern England.
Based on data in Hooker 1991.

established directly (figure 21.4), indicating that the marine London Clay is younger than the continental and paralic facies that produce most of the mammal fossils.

The Woolwich Shell Beds have produced only a possible *Plesiadapis* sp. and *Coryphodon* sp. (Bott 1875; Klaasen 1883; Newton 1883). Hooker (1991) reported the possible *Plesiadapis*, but without documentation. The Woolwich Shell Beds at Dulwich produced fragmentary postcrania and an upper canine of *Coryphodon*. The postcrania and the canine are small and may pertain to *C. oweni*. However, the demonstrable sexual dimorphism in canine and body size of *Coryphodon* make it possible that the *Coryphodon* specimens from the Woolwich Shell Beds are female *C. eocaenus* (Lucas 1984a). Indeed, the holotype ulna of *C. "croydonensis"* from the Woolwich Shell Beds near Croydon, Surrey (Park Hill Railway Cutting: Klaasen 1883; Newton 1883) is larger than an ulna from Meudon (Vasseur 1875), which might, on the basis of its association with teeth of *C. oweni*, belong to that species. Thus, the Woolwich ulna may belong to *C. eocaenus*. However, at present only the identification *Coryphodon* sp. can be assigned with certainty to the *Coryphodon* specimens from the Woolwich Shell Beds (Lucas 1984a).

The Blackheath Beds at the famous Abbey Wood Locality have produced the most extensive British fossil mammal assemblage relevant to the P/E Series boundary. Hooker (1991) correlated this to the similar Pourcy assemblage in France (Louis 1963), and thus to RL MP8–9. *Coryphodon eocaenus* and *C. anthracoideus* are known from Abbey Wood (Forster Cooper 1932; Lucas 1984a; figure 21.5).

The Suffolk Pebble Beds have produced an assemblage that may (all or in part) be reworked from older strata. Hooker (1991) correlated it to the Dormaal (RL MP7) assemblage in Belgium. Reworked teeth of *C. eocaenus* and *C. anthracoideus* found in the Pleistocene Red Crag in Suffolk probably were derived from the Suffolk Pebble Beds (figure 21.5). Hooker (1991) men-

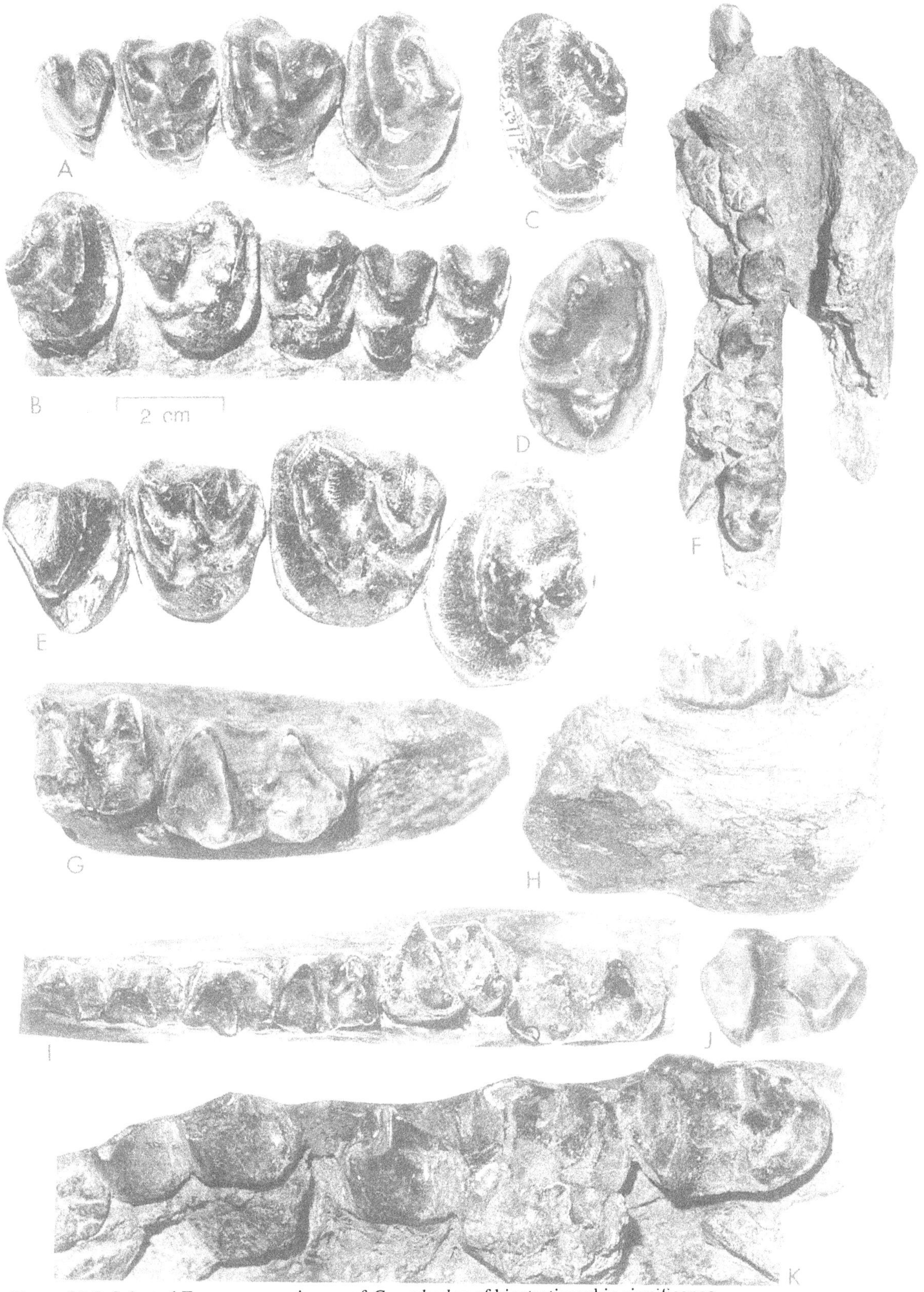

Figure 21.5 Selected European specimens of *Coryphodon* of biostratigraphic significance.

(A) MNHN ARP 10, lectotype of *C. oweni* from Meudon, occlusal view of left P^4–M^3. (B, I) MNHN ARP 8, *C. oweni* from Meudon, occlusal view of right P^3–M^3 (B) and occlusal view of right P_2–M_3 (I). (C) BMNH M 137775, *C. eocaenus* from the Blackheath Beds at Abbey Wood, occlusal view of left M^3. (D) BMNH M, *C. anthracoideus* from the Suffolk Pebble Beds (reworked?), occlusal view of right M^3. (E) MNHN ARP 58, lectotype of *C. anthracoideus* from the Argiles à lignites du Soissonais, occlusal view of left P^4–M^3. (F, K) IRSNM IG 10.018, *C. anthracoideus* from the Landen Formation at Orp-le-Grand, occlusal view of incomplete lower jaw (F) and close-up occlusal view of left P_2–M_3. (G, H) BMNH M 27848, holotype of *C. eocaenus* from Division A? of the London Clay at Harwich off the Essex Coast, right dentary fragment with M_{2-3}, occlusal (G) and labial (H) views. *Institutional abbreviations:* BMNH = The Natural History Museum, London; IRSNM = Institut Royal des Sciences Naturelles, Brussels; MNHN = Museum National d'Histoire Naturelle, Paris.

tioned *Landenodon* from the Tilehurst Beds, but this species ranges from RL MP6 to RL MP8–9 and thus does not provide a precise age.

The lower part of the London Clay (Division A) in the vicinity of Barnet(?), and just off the Essex coast near Harwich, produced teeth of *C. eocaenus*, including the holotype lower jaw fragment (figure 21.5). The *Coryphodon*-producing horizon in the London Clay may be in clays adjacent to the Harwich Stone Band (see Hooker and Ward 1980:4), but this remains uncertain. The Swanscombe Member at the base of the London Clay produced *Paramys ageiensis*, which Hooker correlates to RL MP8–9. The type of *Hyracotherium vulpiceps* (Owen 1858) is from the Harwich Stone Band, still RL MP8–9 according to Hooker (1991). And, at Studd Hill, Herne Bay, Kent, the Division B of the London Clay produced *Hyracotherium leporinum* and *Platychaerops richardsoni*. Hooker (1991) correlated this horizon to RL MP10 based on the possible occurrence of *P. richardsoni* at Grauves, a tenuous correlation.

Belgium

Dormaal is the best known Belgian fossil mammal locality relevant to the P/E Series boundary (Dollo 1909; Dollo and Teilhard de Chardin 1924; Quinet 1969), and as noted above is the type assemblage of RL MP7 (Godinot et al. 1978), but it lacks *Coryphodon*. The fluvio-marine Sables d'Erquelinnes (strata now assigned to the upper Landen Formation) at Erquelinnes (Jeaumont) near the Franco-Belgian border produced an upper premolar and a nearly complete femur that Teilhard de Chardin (1927:28–30) identified as *C. eocaenus*. I have not been able to locate the premolar, but the measurements on this tooth reported by Teilhard de Chardin (1927:28) support assignment to *C. eocaenus*. The Erquelinnes locality also includes *Hyracotherium cuniculus*, so it has been assigned to RL MP7 (Hooker 1991).

At Orp-le-Grand, just south of Dormaal, a partial skull, lower jaw, and skeleton of *C. anthracoideus* (figure 21.5) was collected from lignitic beds in the "Sable fluvio-marin" (= upper Landen Formation) that evidently, and probably unconformably, overlies the Tuffeau de Lincent (Feugueur 1963:figure 96; Lucas 1986). This too appears to be a record assignable to RL MP7.

Similar fluvio-marine strata at Leval in southern Belgium produced two partial femora and a nearly complete astragalus of *Coryphodon*. Although Teilhard de Chardin (1927:29, 30) assigned these bones to *C. eocaenus*, they are indeterminate at the species level.

France

The most extensive fossil mammal assemblages from western Europe relevant to the P/E Series boundary are

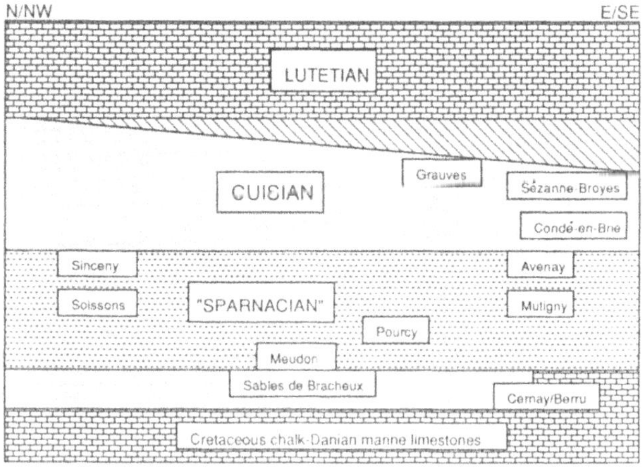

Figure 21.6 French mammal sites of Reference Levels MP6 to MP10 interval and their supposed stratigraphic relationships. After Russell et al. 1982.

from France (figure 21.6). The largest of these, from the Avenay locality, is the Reference Level MP8–9 fauna. This assemblage lacks *Coryphodon*, as does the other large RL MP8–9 assemblage from Mutigny.

Meudon, in suburban Paris, has long been considered by vertebrate paleontologists to be the oldest "Sparnacian" mammal locality because it sits directly on Cretaceous chalk and is overlain by the Argiles Plastique. However, Hooker (1991) argued that its mammals (table 21.1) suggest it is younger than RL MP7 (Dormaal and correlatives) and that Meudon is assignable to RL MP8–9. Thus, Hooker noted that some Meudon taxa (notably *Amphiperatherium maximum*, *Neomatronella*, cf. *Apatemys sigogneaui*, and *Paramys ageiensis*) are typical of RL MP8–9, and that *Platychaerops russelli* from Meudon is more advanced evolutionarily than *Plesiadapis tricuspidens*. (Note that Hooker documented *P. tricuspidens* from the Suffolk Pebble Beds, though like the rest of that assemblage, it may be reworked.) However, several Meudon genera are only found in RL MP7 (*Microhyus*, *Teilhardina*, *Paschatherium*), and the fauna shares many taxa with the Dormaal assemblage (table 21.1). I therefore assign the Meudon assemblage to RL MP7.

Hooker's (1991) reassignment from *Plesiadapis* to *Platychaerops* of a single plesiadapid incisor from Meudon, to which he attached great biostratigraphic significance, also is somewhat equivocal. Clearly the Meudon incisor (figure 21.7) is more derived than *P. tricuspidens* from Mont Berru and the Suffolk Pebble Beds (reworked?) but not as advanced as that of *Platychaerops daubrei* from Mutigny, the characteristic RL MP8–9 plesiadapiform. Therefore, it could be used to assign Meudon either to older RL MP8–9 or to RL MP7.

Abundant *Coryphodon* from Meudon belong to *C. oweni*, a species not found elsewhere in Europe. Specimens of *C. oweni* (figure 21.5) were collected from the

Table 21.1
Comparison of Meudon and Dormaal mammal assemblages

Meudon	Dormal
Peradectes sp.	
Peratherium cf. *P. matronense* Crochet	*Peratherium constans*
Amphiperatherium maximum Crochet	*Amphiperatherium dormaalense*
Paramys cf. *P. woodi* Michaux	*Paramys* cf. *P. woodi* Michaux
P. cf. *P. ageiensis* Michaux	*P.* of *P. pourcyensis*
Paramys sp.	—
Microparamys sp.	*Microparamys nanus*
Pseudoparamys cf. *P. teilhardi* Michaux	*Pseudoparamys lemoinei*
cf. *Apatemys sigogneaui* Russell et al.	*Apatemys* sp.
—	*Eochiromys landenensis*
—	*Entomolestes* cf. *E. nitens*
—	*Dormaalius vandebroeki*
—	*Nycticonodon casieri*
—	*N. caparti*
—	*Wyonicteris richardi*
Palaeonictis sp.	cf. *Palaeonictis* sp.
—	*Oxyaena casieri*
—	*?Palaeosinopa* sp.
—	cf. *Prolimnocyon*
—	*Miacis latouri*
Neomatronella sp.	—
Macrocranion nitens (Matthew)	—
Leptacodon sp.	—
cf. *Icaronycteris* sp.	—
Berruvius cf. *B. lasseroni* Russell	—
Arcius sp.	—
Platychaerops russelli (Gingerich)	*Platychaerops daubrei*
Teilhardina belgica (Teilhard)	*Teilhardina belgica*
Cantius cf. *C. eppsi* (Cooper)	—
—	*Landenodon woutersi*
—	*Landenodon luciani*
—	*Phenacodus teilhardi*
cf. *Paschatherium* sp.	*Paschatherium dolloi*
—	*Microhyus musculus*
—	cf. *Protodichobune* spp.
—	*Diacodexis gigasei*
Hyracotherium sp.	—
Coryphodon oweni Hebert	—

Taxonomic lists from Godinot (1987), Hooker (1991), Smith (1996, and Smith et al. (1996).

Conglomérat de Meudon at three or four localities (Meudon, Passy, Vaugirard, Moulineaux?) now within the suburbs of Paris (Hébert 1856). Some of these teeth, as well as small postcrania that probably pertain to *C. oweni*, may actually have come from argillaceous and lignitic strata less than 1 m above the Meudon conglomerate (see Plante 1870:206, 208), but provenance records are too nondescript to confirm or deny this possibility.

Two localities just north of Epernay (the type locality of the Sparnacian Stage), Pourcy and Cernay (northeast of Reims), purportedly produced specimens of *C. oweni*. Depéret (1906, 1907) originally reported these specimens, but they were not illustrated and cannot be located. The femur and metatarsal from the "Sables et argiles ligniteuses" at Cernay reported by Depéret (1906:443) were assigned by him to *C. oweni* because of their small size. Depéret (1907:11) later mentioned a lower premolar he assigned to *C. oweni* from the "Sables supérieurs" at the "Gite de Pourcy." Depéret (1907) did not describe this tooth or justify its identification, so I identify it only as *Coryphodon* sp. The postcrania from Cernay are also identified as *Coryphodon* sp.

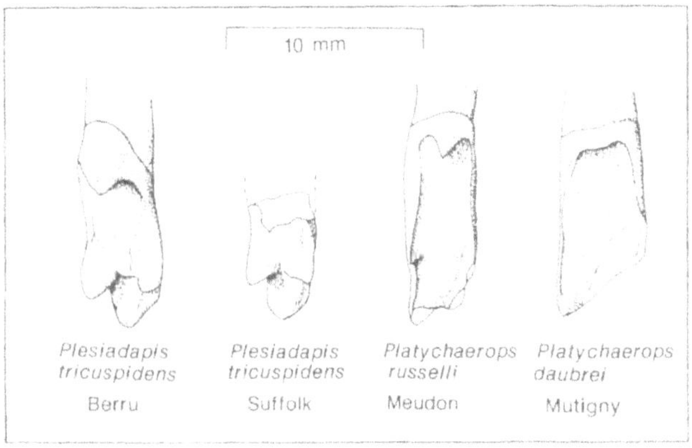

Figure 21.7 The Meudon incisor of *Platychaerops russelli* compared to Reference Level MP7 *Plesiadapis truicuspidens* and Reference Level MP8–9 *Platychaerops daubrei*.
After Hooker 1991.

Other French occurrences of *Coryphodon* (Lucas 1984a) of note are

- Fluvio-marine sands at Vertain in northern France produced the lectotype M^3 and other teeth of *C.* "*gosseleti*" Malaquin, 1899, which Lucas (1984a) considered to be a synonym of *C. anthracoideus*.
- Michaux (1963) followed up an earlier report by Louis and Michaux (1962) by describing a premaxillary fragment from the Sables de Sinceny at Sinceny (west of Laon along the Oise River) that pertains to *C. eocaenus*.
- The Argiles à Lignites in the Soissons (Soissonais) and Laon (Laonnois) districts have produced teeth of *C. eocaenus* and *C. anthracoideus*, including the lectotype (figure 21.5). A single incisor from the Argiles à Lignites(?) at Muirancourt (west of Laon along the Oise River) probably pertains to *C. eocaenus*.
- The Argiles Plastiques in the vicinity of the old railway depot at Sarron along the Oise River have produced teeth of *C. eocaenus* and *C. anthracoideus*.
- Several teeth of *C. eocaenus* and *C. anthracoideus* were collected from the Argiles Plastiques at an unspecified locality (or localities) that may have been in the vicinity of Paris or, more likely, along the Oise Valley.

Correlation Within Europe

My correlation of the European mammalian assemblages to each other (figure 21.8) is very similar to that of Hooker (1991, 1996), with the key difference that I correlate Meudon to RL MP7. Using mammals alone, it is the most precise correlation that can be presented. A more precise correlation of these assemblages is possible using a combination of micropaleontology, sequence stratigraphy, magnetochronology, and radioisotopic dating (Berggren and Aubry 1996, chapter 2). This more precise correlation suggests the following succession (oldest to youngest): Cernay, Dormaal, Meudon = Erquelinnes, Woolwich Shell Beds, Abbey Wood, Pourcy, Division A of the London Clay, Mutigny, Avenay, Division B of the London Clay, and Grauves. Mammalian evidence does not contradict this.

Note that the most significant faunal turnovers in the

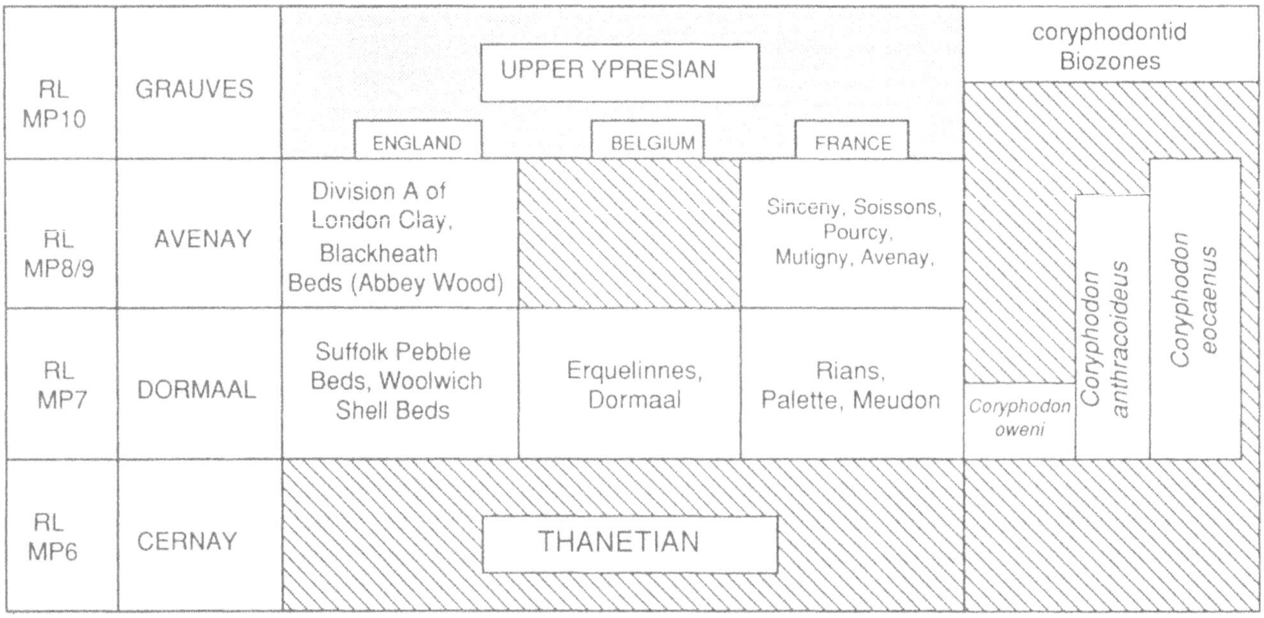

Figure 21.8 Correlation of European mammalian assemblages of the Reference Levels MP7 and MP8–9 interval based on mammals and the resultant *Coryphodon* Biozones.

Western European section occur between RL MP6 and RL MP7 and between RL MP8–9 and RL MP10. The two faunal turnovers readily delimit the Neustrian LMA. At the boundary between RL MP6 and RL MP7, perissodactyls, artiodactyls, chiropterans, euprimates, rodents, and coryphodontid pantodonts appear, and the diversity of multituberculates, condylarths, and plesiadapiforms decreases. Note, though, that this "boundary" is a gap between the Cernay (RL MP6) fauna, which is a single point in Thanetian time, and the RL MP7 assemblages, and the gap certainly accentuates the apparent evolutionary turnover (compare the North American record). At the boundary between RL MP8–9 and RL MP10, pantodonts and multituberculates disappear, and a profusion of lophiodonts dominates the fossil assemblages.

Coryphodontid biozones can be derived from the European record (figure 21.8). *Coryphodon oweni* is restricted to RL MP7, *C. anthracoideus* ranges from RL MP7 through all but the youngest assemblages of RL MP8–9, and *C. eocaenus* occurs essentially throughout RL MP7 and RL MP8–9 assemblages. Indeed, *C. eoacenus* should be considered an index fossil of the Neustrian LMA.

North America

Introduction

Fossil mammals relevant to the P/E Series boundary occur in western North America in synorogenic sediments of the Laramide orogeny (figure 21.9). This orogeny was the result of primarily northeastward movement of the Colorado Plateau so that it impinged upon the basement buttresses of the Rocky Mountain foreland, finally resulting in significant wrench tectonism. Three types of basins formed: (1) Green River–type basins, which are large, equidimensional to elliptical basins with uplifts on three sides and characterized by impounded lacustrine deposition; (2) Denver-type basins, which are elongate, open, asymmetrical synclinal downwarps with an uplift on one side; and (3) Echo Park–type basins, which are narrow, elongate, and of strike-slip origin (Chapin and Cather 1981; see Dickinson et al. 1988 for the same concepts with a different, redundant terminology). The sharpest pulse of Laramide deformation began at about the early/middle Eocene Epoch boundary and formed the majority of the Echo Park–type basins as en-echelon, asymmetric downwarps and tilted and subsided blocks between strands of the wrench-fault system (Chapin and Cather 1981; Lucas 1984b).

Mammalian biostratigraphy relevant to the P/E Series boundary in the western United States is based on the LMAs originally introduced by Wood et al. (1941). These LMAs—Tiffanian, Clarkforkian, Wasatchian, and Bridgerian—were reviewed extensively by Archibald et al. (1987) and Krishtalka et al. (1987). There has never been any doubt that the Tiffanian LMA is of Paleocene age, or of the Eocene age of the Bridgerian LMA, correlations well supported by biostratigraphy, magnetostratigraphy, and/or radioisotopic dating (e.g., Flynn 1986; Archibald et al. 1987; Krishtalka et al. 1987; Lucas and Williamson 1993a; Williamson 1996). Placement of the P/E Series boundary has always been in the Clarkforkian–Wasatchian LMA interval.

Wood (1967) questioned the distinctiveness of the Clarkforkian LMA, but Rose (1980, 1981) has demonstrated its utility. Rose divided the Clarkforkian LMA into three "zones" (also see Archibald et al. 1987). These differentiate lower, middle, and upper Clarkforkian divisions but have been of little use outside of the Clark's Fork Basin because there are few well known Clarkforkian correlatives (see below).

The base of the Wasatchian LMA represents a major faunal turnover in mammalian evolution (e.g., Krause and Maas 1990; Gunnell et al. 1993; Maas et al. 1995). Important first appearances at the base of the Wasatchian include perissodactyls, artiodactyls, euprimates, hyaenodontid creodonts, didymoconids, didelphinine marsupials, several genera of insectivores, the miacid *Miacis*, and the rodent *Microparamys*. This turnover provides the rationale behind placement of the P/E Series boundary at the base of the Wasatchian by many mammalian paleontologists.

Following early studies by Granger (1910, 1914) and Sinclair and Granger (1911), Van Houten (1945) divided the Wasatchian LMA into three faunal zones: Gray Bull (lower Wasatchian), Lysite (middle Wasatchian), and Lost Cabin (upper Wasatchian; figure 21.10). Here, I refer to these concepts by their adjectival forms (Graybullian, Lysitean, and Lostcabinian) as "sub-LMAs" (SLMA) of the Wasatchian LMA. This workable tripartite division of the Wasatchian LMA has been confirmed and refined by detailed stratigraphic collecting and analysis in the "type" areas of the three Wasatchian SLMAs: the Bighorn Basin (Graybullian SLMA; e.g., Bown 1979; Gingerich 1980a, 1980b, 1983, 1989; Schankler 1980; Rose 1981) and the Wind River Basin (Lysitean and Lostcabinian SLMA; Krishtalka and Stucky 1983; Stucky and Krishtalka 1983; Stucky 1984a, 1984b, 1984c; Dawson et al. 1990; Rose et al. 1991; Beard et al. 1992). The composite Graybullian, Lysitean, and Lostcabinian assemblages have been listed (though incompletely) by Savage and Russell (1983).

The Graybullian SLMA is characterized by the co-occurrence of multituberculates, *Homogalax*, *Phenacodus*, *Coryphodon*, *Hyracotherium*, *Hyopsodus*, *Hap-*

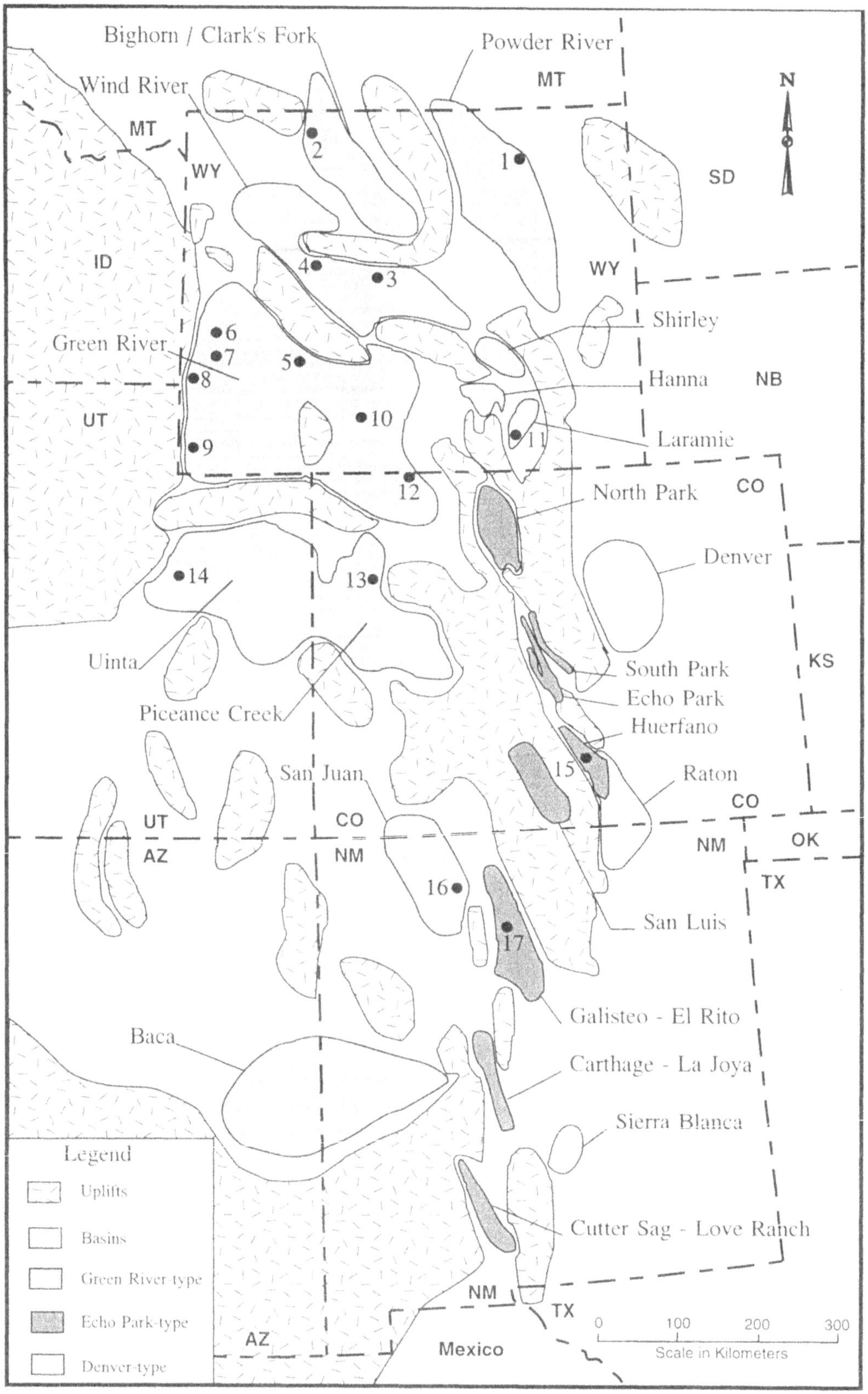

Figure 21.9 Laramide basins of the western United States (after Chapin and Cather 1981) and principal fossil mammal localities of the Paleocene/Eocene Series transition.

1 = Pumpkin Buttes and vicinity, 2 = Bighorn and Clark's Fork Basin localities, 3 = Wind River Basin, 4 = Indian Meadows Formation, 5 = Green River Basin, 6 = Hoback Basin, 7 = LaBarge Area, 8 = Fossil Basin, 9 = Evanston Area, 10 = Washakie Basin, 11 = Laramie Basin, 12 = Four Mile Area, 13 = Piceance Creek Basin, 14 = Uinta Basin, 15 = Huerfano Basin, 16 = San Juan Basin, 17 = Galisteo–El Rito Basin.

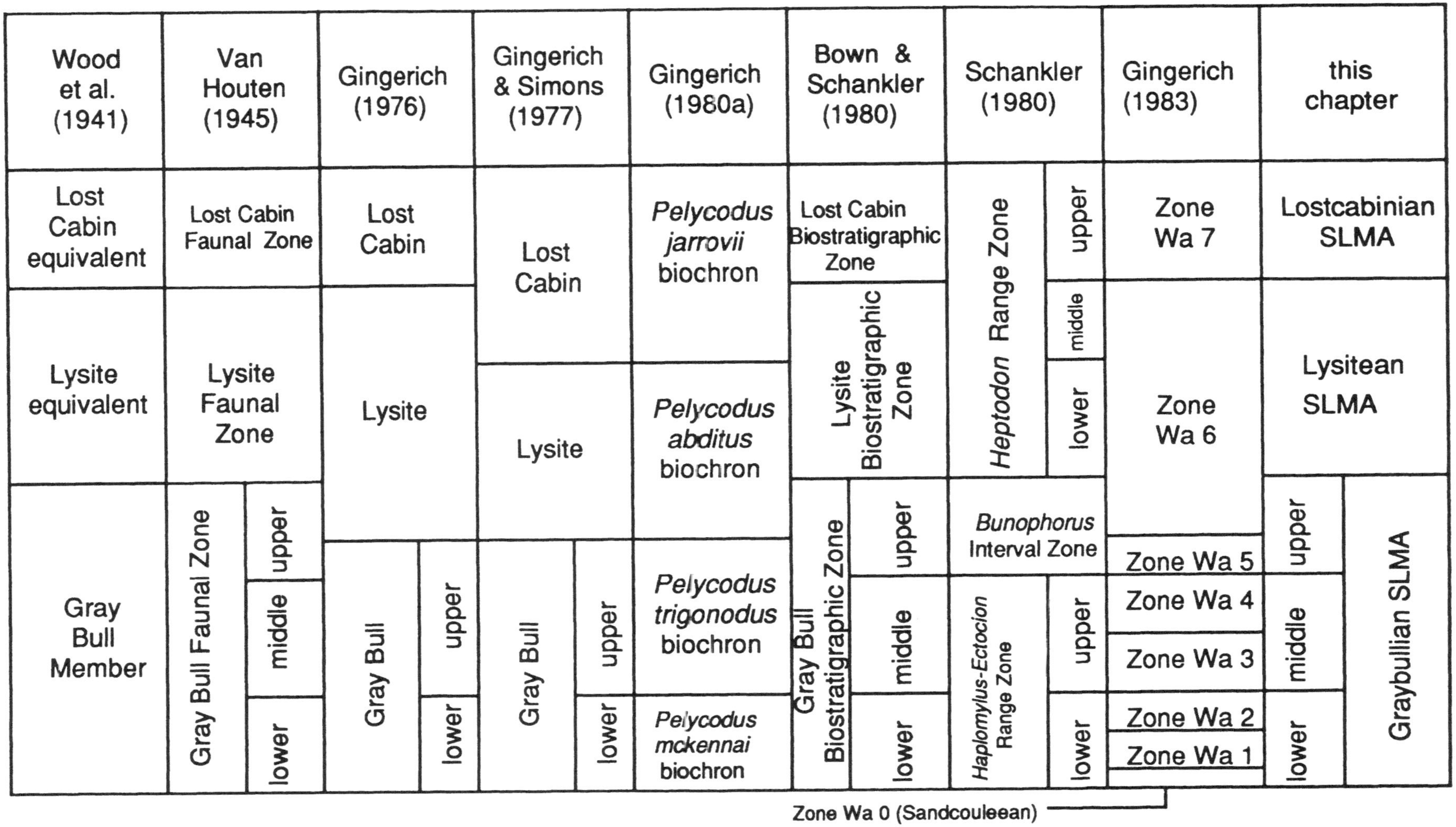

Figure 21.10 Comparison of different subdivisions of the Wasatchian LMA.

lomylus, *Ectocion*, *Pelycodus*, and *Diacodexis*. The assemblage zone of these mammals defines the lower Wasatchian SLMA.

The Lysitean SLMA was defined originally as the *Heptodon* Beds, with both *Homogalax* and *Lambdotherium* absent. LOs of mammals in the Lysitean SLMA are *Heptodon*, *Absarokius*, *Loveina*, and *Hexacodus* (Guthrie 1967; Krishtalka and Stucky 1985). Korth (1984) also listed several rodent species of value in defining the Lysitean SLMA.

The Lostcabinian SLMA was originally based on the fauna of the Lost Cabin Member of the Wind River Formation in the tributaries of Badwater Creek near the village of Lost Cabin, Wind River Basin, Wyoming. The Lostcabinian subage is essentially the time interval represented by the total stratigraphic range of the primitive palaeothere *Lambdotherium* (Guthrie 1971); i.e., it is the *Lambdotherium* biochron. The base of the Lostcabinian SLMA (= end of Lysitean SLMA) is defined by the LO of *Lambdotherium*. Lostcabinian time is also recognized by the LOs of *Shoshonius*, *Antiacodon*, *Orohippus*, *Megalesthonyx*, *Pauromys*, *Sciuravus*, and *Armintodelphys* as well as derived species of *Esthonyx*, *Diacodexis*, *Absarokius*, *Hyopsodus*, *Cantius/Notharctus*, and *Hyracotherium*.

Clark's Fork and Bighorn Basins, Wyoming–Montana

Through the work of Gingerich and his students and collaborators (e.g., Bown 1979, 1980, 1982; Schankler 1980; Rose 1981; Bown and Schankler 1982; Gingerich 1976, 1980a, 1980b, 1983, 1989; Gingerich et al. 1980; Bown et al. 1994), the mammalian biostratigraphy of the Clarkforkian–Wasatchian LMA interval of the Fort Union/Polecat Bench and Willwood Formations in the Clark's Fork and Bighorn Basins, Wyoming and Montana, is the most detailed of this stratigraphic interval. Various biostratigraphic schemes have been used in these rocks (figure 21.11). For the Wasatchian LMA, the scheme used here is to employ the terms *Graybullian* (including *Sandcouleean*), *Lysitean*, and *Lostcabinian* in the Bighorn Basin and other basins essentially as Van Houten (1945) introduced them (Schankler 1980). Their equivalence to other biostratigraphic terms is readily understood from figure 21.11.

The "Big Horn Wasatch" and "Big Horn Wind River" of earlier workers (see, for example, Sinclair and Granger 1911) were formalized lithostratigraphically by Van Houten (1944) as the Willwood Formation. This unit includes about 770 m of rocks deposited during most of the Wasatchian LMA. The Willwood Formation is conformable in most places on rocks variously referred to the Fort Union or Polecat Bench Formations and is divided lithologically into the Elk Creek and Sand Creek facies (Bown 1979).

Schankler (1980), working in the central Bighorn Basin, proposed three biostratigraphic zones for the Willwood Formation. He used a *Haplomylus–Ectocion* Range-Zone for the strata corresponding to the lower and middle subdivision of the Gray Bull Faunal Zone of Van Houten (1945). Schankler's overlying *Bunophorus* Interval-Zone corresponds to Van Houten's upper part of the Gray Bull Faunal Zone. Schankler, however, did not recognize distinct Lysite and Lost Cabin biostratigraphic zones, but instead used a *Heptodon* Range-Zone with lower, middle, and upper divisions for the combined Lysitean–Lostcabinian interval. Indeed, the Lysitean and Lostcabinian SLMAs cannot be as precisely delimited in the stratigraphic succession of the Bighorn Basin as they can in the Wind River and some other basins.

I consider the distribution of *Coryphodon* species in the Bighorn and Clark's Fork Basins (figures 21.11–21.13) to form the principal reference section for the correlation of other *Coryphodon*-bearing deposits in

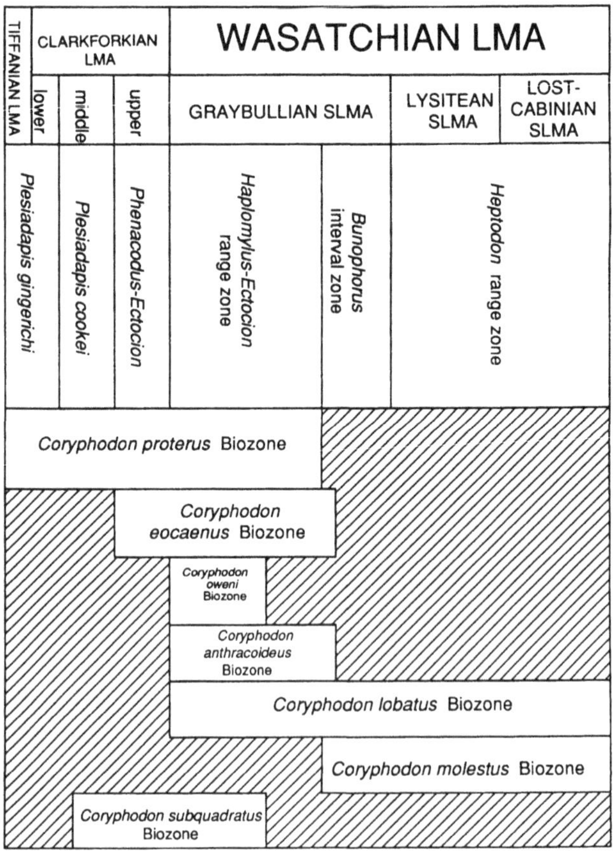

Figure 21.11 *Coryphodon* Biozones derived from the stratigraphic distribution of specimens in the Bighorn and Clark's Fork basins.

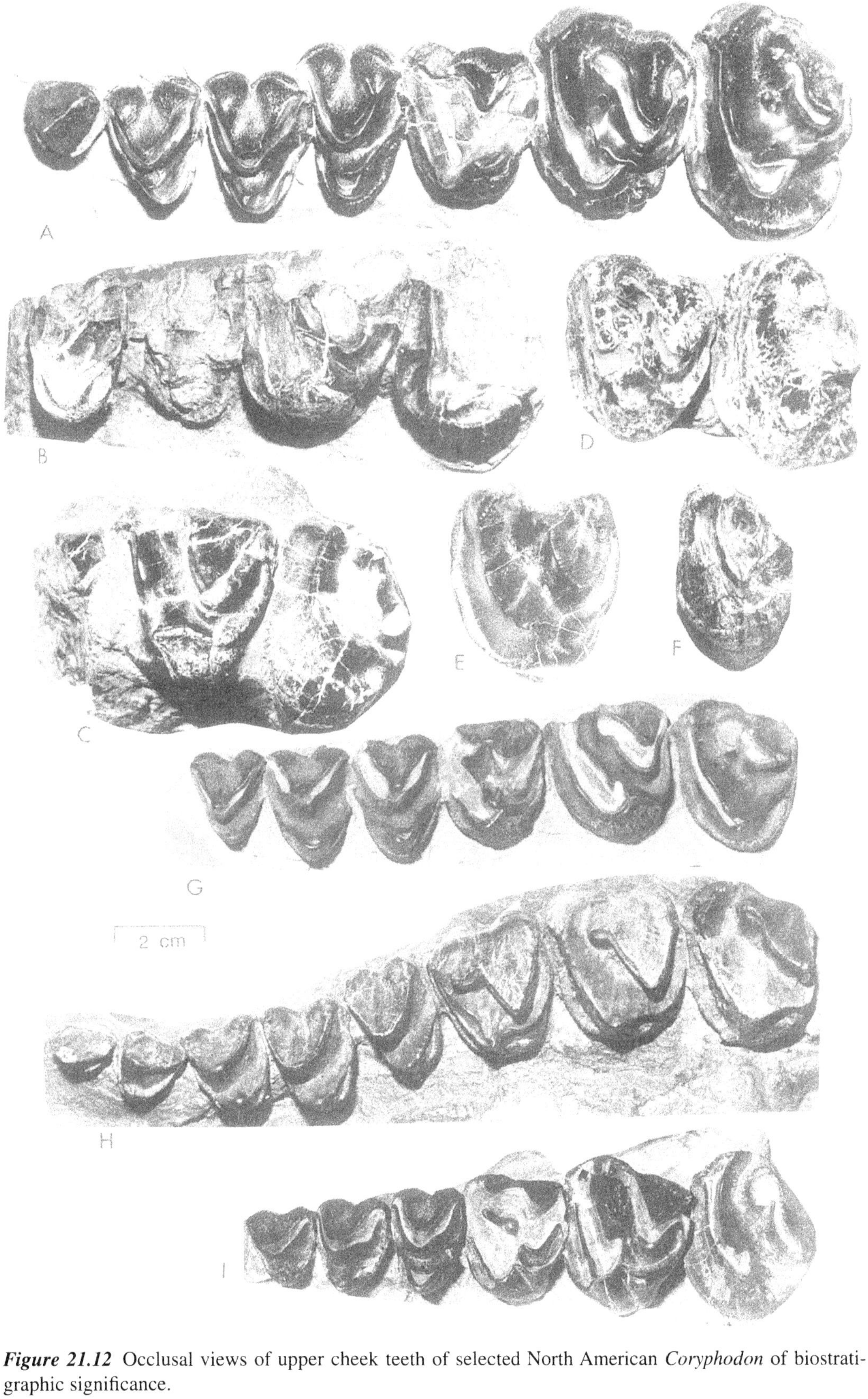

Figure 21.12 Occlusal views of upper cheek teeth of selected North American *Coryphodon* of biostratigraphic significance.

(A) PU 16130, *C. proterus*, left P^1–M^3, Willwood Formation, Bighorn Basin, Graybullian SLMA. (B) YPM 11330, *C. anthracoideus* (holotype of *C. hamatus*), partial left P^4–M^3, Wasatch Formation, Washakie basin, Graybullian SLMA.

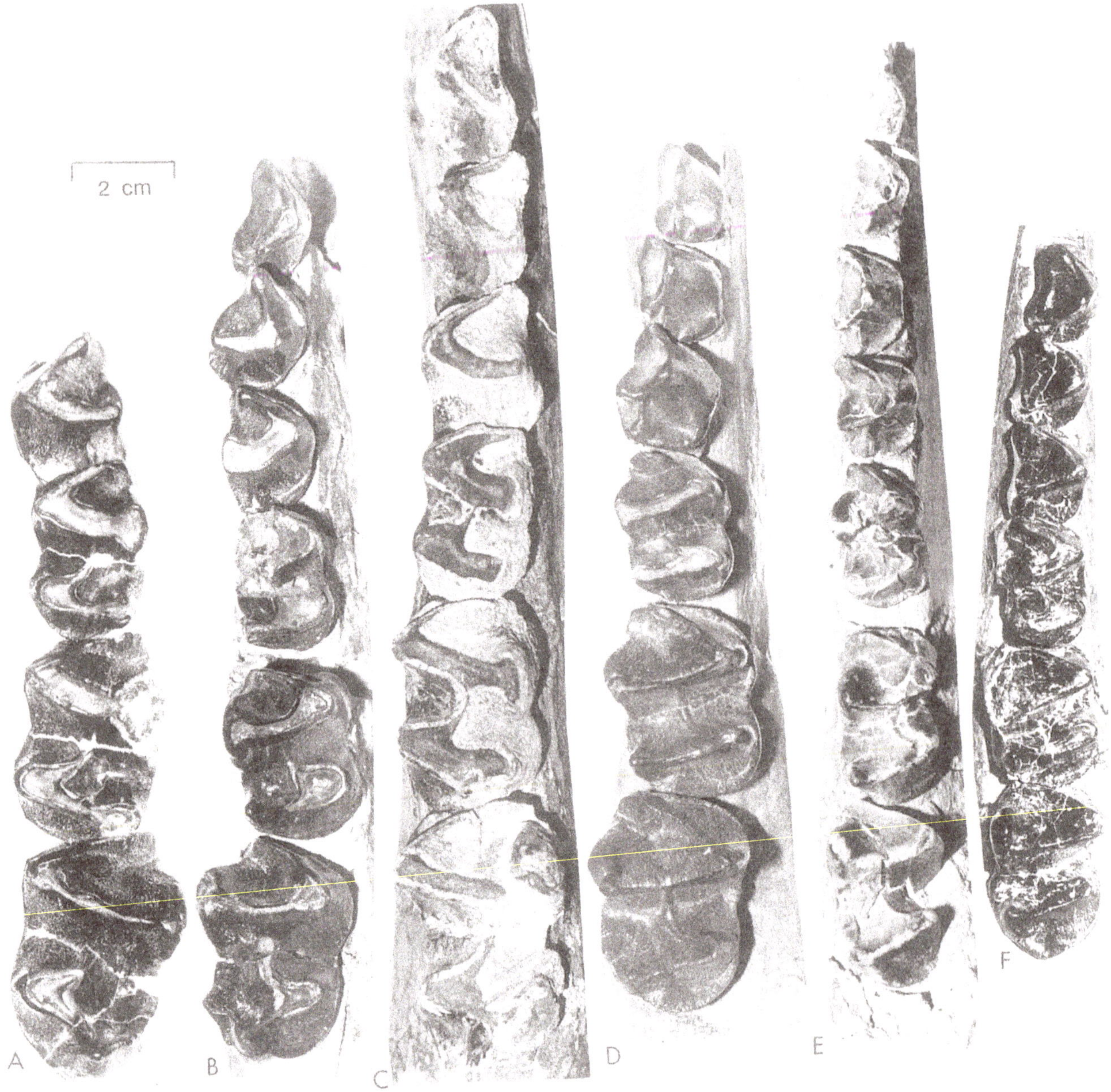

Figure 21.13 Occlusal views of lower cheek teeth of selected North American *Coryphodon* of biostratigraphic significance.

(A) PU 16816, *C. proterus*, left P_4–M_3, Willwood Formation, Bighorn Basin, Clarkforkian LMA. (B) LACM 18139, *C. eocaenus*, right P_2–M_3, Willwood Formation, Bighorn Basin, Graybullian SLMA. (C) FMNH P 15628, *C. lobatus*, left P_2–M_3, DeBeque Formation, Piceance Creek Basin, Graybullian SLMA. (D) AMNH 2976, *C. lobatus* (holotype of *C. ventanus*), right P_2–M_3, Wind River Formation, Wind River Basin, Lysitean SLMA? (E) USNM 1119, *C. molestus* (holotype), right P_1–M_3, Regina Member, San Jose Formation, San Juan Basin, Lysitean SLMA. (F) UALP 7162, *C. simus*, right P_2–M_3, Regina Member, San Jose Formation, San Juan Basin, Lysitean SLMA. Institutional abbreviations: AMNH = American Museum of Natural History, New York; FMNH = Field Museum of Natural History, Chicago; LACM = Natural History Museum of Los Angeles County, Los Angeles; PU = Princeton University, collection now at Yale Peabody Museum, New Haven, Connecticut; UALP = University of Arizona Laboratory of Paleontology, Tucson; USNM = National Museum of Natural History, Smithsonian Institution, Washington, D.C.

Figure 21.12 *(continued)*

(C) AMNH 4300, *C. anthracoideus* (holotype of *C. radians*), left M^{2-3}, Wasatch Formation, Washakie Basin, Graybullian SLMA. (D) PU 13400, *C. proterus* (holotype), left M^{2-3}, Willwood Formation, Bighorn Basin, Clarkforkian LMA. (E) AMNH 4340, *C. subquadratus* (holotype), left M^2, Willwood Formation, Bighorn Basin, Graybullian SLMA. (F) AMNH 4315, *C. eocaenus* (lectotype of *C. armatus*), right M^3 (negative reversed), Wasatch Formation, Washakie Basin, Graybullian SLMA. (G) PU 16830, *C. oweni*, left P^2–M^3, Willwood Formation, Bighorn Basin, Graybullian SLMA. (H) AMNH 16259, *C. molestus*, left P^1–M^3 (note supernumerary premolar), Regina Member, San Jose Formation, San Juan Basin, Lysitean SLMA. (I) AMNH 80757, *C. simus*, left P^2–M^3, Regina Member, San Jose Formation, San Juan Basin, Lysitean SLMA. Institutional abbreviations: AMNH = American Museum of Natural History, New York; PU = Princeton University (collection now at YPM); YPM = Yale Peabody Museum, New Haven, Connecticut.

North America (Lucas 1984a). Uhen and Gingerich (1995) recently reviewed some of the *Coryphodon* specimens from the Clark's Fork basin and presented a taxonomy somewhat different from mine. They used the species names *C. armatus* (= *C. eocaenus*) and *C. radians* (= *C. anthracoideus*) and assigned specimens of *C. oweni* to *C. eocaenus*. When these synonymies are recognized, the stratigraphic ranges of Clark's Fork Basin *Coryphodon* that Uhen and Gingerich (1995) document are the same as mine (figure 21.11).

Magnetic polarity stratigraphy in the Polecat Bench and Willwood Formations of the Clark's Fork and Bighorn Basins (Butler et al. 1981; Clyde et al. 1994) indicates that most of the sampled Willwood Formation (Clarkforkian–Lysitean interval) is of reversed polarity. The uppermost Tiffanian–lower Clarkforkian is a normal polarity interval representing Chron 25n (Butler et al. 1981). Normal polarity intervals in the uppermost Graybullian–Lysitean interval correspond to Chrons 24n.3n and 24n.2n (Clyde et al. 1994).

Wind River and Purdy Basins, Wyoming

The Wind River Basin has produced biostratigraphically significant mammals since Wortman first collected there in 1880. The lower Eocene strata were originally divided into the "Wasatch" and the "Wind River " by Granger (1910). Based on mammalian fossils, he recognized a "?*Coryphodon* Zone" at the bottom in gravelly strata that was referred doubtfully to the "Wasatch" and a "*Lambdotherium* Zone" in the middle and upper part of the section that belonged to his "Wind River." Osborn (1910) recognized an assemblage intermediate between Granger's *Coryphodon* and *Lambdotherium* Zones, characterized by all of the genera common to the two zones of the Big Horn Wasatch Formation, but the absence of *Systemodon* (= *Homogalax*), *Lambdotherium*, and *Eotitanops*. Sinclair and Granger (1911) named the *Lambdotherium* Zone the Lost Cabin Formation and the strata between the *Coryphodon* Zone and the *Lambdotherium* Zone the Lysite Formation.

Tourtelot (1946, 1948, 1953), following Sinclair and Granger (1911), concluded that both the Lysite and Lost Cabin sediments should be considered members of the Wind River Formation. Subsequent workers have recognized Tourtelot's distinctions between the Lysite and Lost Cabin Members (for example, Keefer 1965a; Guthrie 1967, 1971; Love 1978; Stucky and Krishtalka 1982). Lewis and Gazin (in Tourtelot 1953) and Van Houten (1945) confirmed the faunal distinctions established between the two members by Sinclair and Granger (1911). Kelley and Wood (1954) added new taxa to the mammalian fauna of the Lysite Member and discussed the utility of these taxa in biostratigraphic correlation. Keefer (1965a, 1965b, 1970) described the stratigraphy and structural geology of the Wind River Basin, and Guthrie (1967, 1971) reconfirmed the distinct nature of the faunas from the Lysite and Lost Cabin Members.

Recent work on the Wind River mammalian assemblages (Korth 1982; Stucky 1982, 1984a; Stucky and Krishtalka 1982, 1983, 1990; Krishtalka and Stucky 1983, 1984, 1985; Dawson et al. 1990; Rose et al. 1991; Beard et al. 1992) allowed Stucky (1984c) to document the stratigraphic distribution of approximately 120 mammalian species in the upper part of the Wind River Formation in the Red Creek–Deadman Butte area. He recognized two biostratigraphic zones: the *Lambdotherium* Range Zone and the *Palaeosyops borealis* Assemblage Zone. The *Lambdotherium* Range Zone is Lostcabinian in age, and the *Palaeosyops borealis* Assemblage Zone is Gardnerbuttean (the earliest Bridgerian SLMA, introduced by Robinson 1966) in age. Therefore, the Wind River Formation appears to span middle Wasatchian to early Bridgerian time, and three faunal levels—Lysitean, Lostcabinian, and Gardnerbuttean—are now recognized.

Clarkforkian and Wasatchian levels are known in the Wind River Basin if localities at Togwotee Pass, reported but not documented by McKenna (1980b), are included. The oldest localities in the Togwotee Pass area are in strata informally termed "lower variegated sequence" by Love (1947). These Clarkforkian sites yield *Plesiadapis cookei*, *Ectocion*, *Coryphodon proterus*, and *Haplomylus*, among others. Above them, younger, Wasatchian sites are known containing *Hyracotherium*. The overlying "upper variegated sequence" produced an assemblage listed, but not documented, by McKenna (1980b) as ?*Coryphodon* sp., ?*Phenacodus* sp., *Hyracotherium* sp., and ?*Diacodexis* sp. I have examined *Coryphodon* specimens from the "lower variegated sequence" (McKenna 1980b). They are large enough to represent *C. proterus*, but the specimens are too fragmentary to permit species-level identification. If this material is *C. proterus*, it is consistent with the Clarkforkian age assignment advocated for the "lower variegated sequence" by McKenna (1980b) and Rose (1981).

Love (1939) named the Indian Meadows Formation for sediments in the northwestern Wind River Basin that are unconformably overlain by the Wind River Formation. The Indian Meadows Formation (Denison 1937; Love 1939; Keefer 1969) crops out north of Dubois in the northwestern part of the basin. Mammalian fossils include *Esthonyx bisulcatus*, *Oxyaena gulo*, cf. *Pachyaena* sp., *Phenacodus*, *Haplomylus*, *Hyopsodus*, *Coryphodon molestus*, *Hyracotherium*, and *Diacodexis*. This assemblage has been considered to be of early Graybullian age (Denison 1937; Van Houten 1945), but the presence of *C. molestus* suggests a late Graybullian age at the oldest.

The mammalian assemblages (Guthrie 1967, 1971) of the Lysite and Lost Cabin Members of the Wind River Formation typify the Lysitean and Lostcabinian subdivisions of the Wasatchian LMA. *Coryphodon molestus* and *C. lobatus* co-occur in the Lysite and Lost Cabin Members (Lucas 1984a). The occurrence of *C. molestus* and *C. lobatus* in both members is consistent with their stratigraphic distribution in the Bighorn Basin (figure 21.11).

Williston Basin, North Dakota

In southwestern North Dakota (Stark County) at White Butte, the upper part of the Camels Butte Member of the Golden Valley Formation yielded a Wasatchian mammal assemblage listed by Jepsen (1963) and West (1973b). These workers, and Hickey (1977), who studied a florule collected from the lower 10 m of the Camels Butte Member, assigned this member a Graybullian age (also see Archibald et al. 1987; Krishtalka et al. 1987).

Coryphodon molestus is the only species of *Coryphodon* known from the Golden Valley Formation (figure 21.14). Furthermore, David Schankler (personal communication 1982) informed me that representatives of *Cantius*, *Hyopsodus*, *Diacodexis*, *Hyracotherium*, and other mammals of this assemblage that he examined suggest a Lysitean age. This is consistent with the presence of *C. molestus*, so I assign a Lysitean age to the Camels Butte Member of the Golden Valley Formation.

Powder River Basin, Wyoming

Delson (1971) revised the Powder River local fauna from strata of the so-called "Wasatch Formation" near Pumpkin Buttes in Campbell and Johnson Counties, Wyoming. More than 41 species of fossil mammals come from 43 localities spread over a 220-m-thick section. *Homogalax* is present throughout this section, as are *Haplomylus* and *C. eocaenus*. Thus, a Graybullian age seems certain.

Robinson and Ivy (1994) recollected and restudied this interval and stated that it is early Graybullian ("Sandcouleean") in age. Sharp et al. (1964) reported a deciduous fourth premolar of *Lambdotherium* from high on the slopes of Pumpkin Buttes, but Robinson and Ivy (1994) considered it to be another perissodactyl, probably "*Systemodon*" (*Homogalax*).

Green River, Fossil, and Hoback Basins, Wyoming

Fluvial and lacustrine rocks of Eocene age occupy large parts of the Green River Basin (as well as the Washakie and Great Divide Basins) in central Wyoming (figure 21.15). Most of the Eocene rocks belong to the Wasatch and Green River Formations. The Green River Formation is of lacustrine origin, and the Wasatch Formation is of fluviatile origin. Most of the Wasatch Formation underlies the Green River Formation, but the Green River Formation interfingers and intertongues extensively with it, and a complicated lithostratigraphic nomenclature exists (e.g., Krishtalka et al. 1987; Roehler 1991). The Wasatch Formation in central and southwestern Wyoming is divided into two members: the Chappo Member at the base and the unconformably overlying LaBarge Member (Oriel 1961) and three principal tongues, the Niland (Pipiringos 1962), New Fork, and Cathedral Bluffs Tongues (Bradley 1964). The Chappo and LaBarge Members have also been referred to the "Knight Formation" or as the main body of the Wasatch Formation (Veatch 1907; Bradley 1964; Roehler 1991).

In the main body, the West of Elk Mountain local fauna of Gazin (1952, 1962) contains *C. molestus* and *C. anthracoideus*. Gazin (1952, 1962) assigned a Graybullian age to this local fauna, but a close reading of Gazin (1952) reveals that this local fauna is an aggregation of separate stratigraphic levels and localities. Indeed, the presence of *C. anthracoideus* and *C. molestus* suggests that upper Graybullian deposits, or a range of Graybullian and younger horizons, are present here.

In the type area of the Wasatch Formation, near Evanston in southwestern Wyoming, the main body produces a small Lysitean mammalian assemblage that includes *Heptodon*. Clarkforkian and lower Wasatchian mammals have been reported from near LaBarge, on the western side of the basin (Dorr and Gingerich 1980). The La Barge local fauna from the La Barge Member of the Wasatch Formation has long been considered of Lostcabinian age (Gazin 1962). The occurrence of *Coryphodon molestus* and *C. lobatus* in the La Barge fauna is consistent with this age assignment.

In the New Fork–Big Sandy area, the New Fork Tongue is Lostcabinian in age (Gazin 1952; West 1969a, 1969b, 1970, 1973a), and the Cathedral Bluffs Tongue straddles the Wasatchian/Bridgerian LMA boundary (Morris 1954; Gazin 1959; West 1969a, 1969b; Stucky 1984c). The New Fork Tongue assemblages of the Green River Basin were considered by West (1973a) to be of Lostcabinian age, a determination consistent with the presence of *C. molestus* in these assemblages.

Fossil Basin is a separate depositional basin a few kilometers to the west of the western edge of the Green River Basin proper near Kemmerer, Wyoming. It is the site of the world-famous Green River Fish Beds that have produced most of the Green River Lagerstätten. It also includes the type sections of the Knight and Almy Formations of Veatch's (1907) Wasatch Group. Mammalian assemblages of both Graybullian and Lysitean ages have been found at several localities (Gazin 1962;

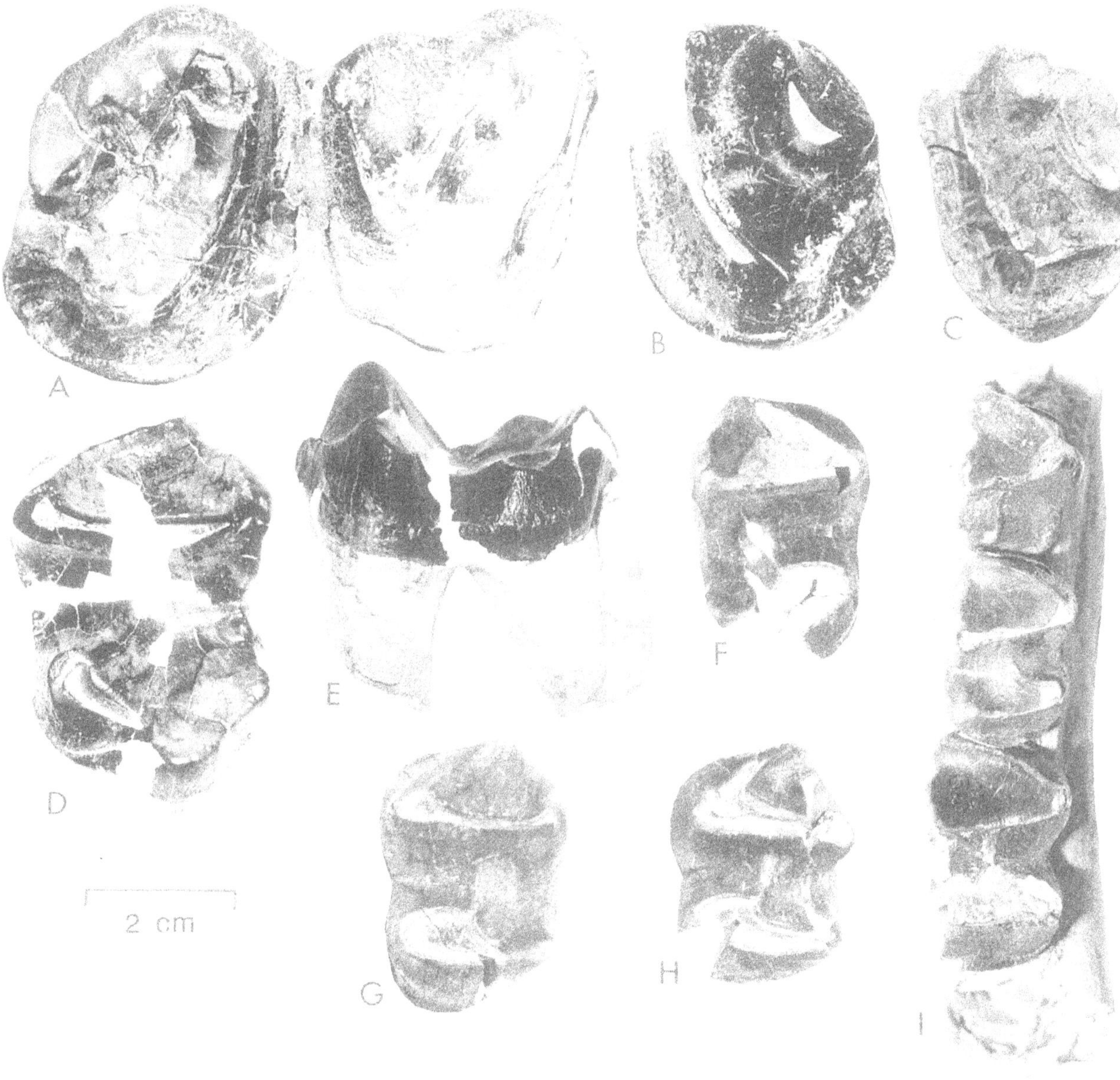

Figure 21.14 Cheek teeth of selected North American *Coryphodon* of biostratigraphic significance.

(A) NMC 30961, *C. lobatus*, right M^{2-3}. (B) NMC 30811, *C. anthracoideus*, left M^3. (C) PU 17981, *C. molestus*, left M^3. (D, E) NMC 30811, *C. anthracoideus*, left M_3. (F–H) PU 16994, *C. molestus*, right M_2 (F), left M_2 (G), and left M_3 (H). (I) PU 17282, *C. molestus*, right P_4–M_2. All views occlusal except E, which is labial. A–B and D–E are from the Margaret Formation, Ellesmere Island (Graybullian SLMA); the remainder are from the Golden Valley Formation, North Dakota (Lysitean SLMA). Institutional abbreviations: NMC = National Museum of Canada, Ottawa; PU = Princeton University, collection now at Yale Peabody Museum, New Haven, Connecticut.

Oriel et al. 1962), and the younger assemblages are physically coincident with the fish-bearing beds. The fish beds are thus Lysitean in age.

The Knight Station local fauna from the main body of the Wasatch Formation in the Fossil Basin was considered by Gazin (1952, 1962) to be of Lysitean age. This age assignment is unlikely because *C. anthracoideus* and *C. subquadratus* are present, suggesting that the Knight Station local fauna is older than Lysitean, possibly of Graybullian age. However, it should be noted that the Knight Station local fauna is also an aggregation of stratigraphic levels and localities (Gazin 1952) that may contain several horizons of different ages.

North of the present Green River Basin is a small physiographic depression, the so-called Hoback Basin. The Clarkforkian–Wasatchian rock units in this basin are the uppermost part of the Hoback Formation and the Pass Peak Formation. A succession of mammalian fossil assemblages that range from Clarkforkian to Lysitean in age has been documented (Dorr 1952, 1958,

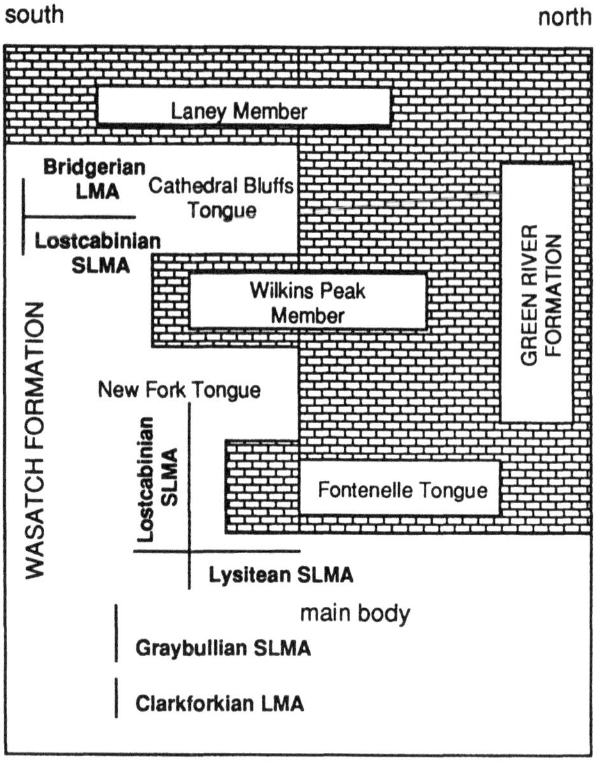

Figure 21.15 Schematic stratigraphic section of the western part of the Green River Basin showing stratigraphic ranges of mammalian assemblages.
In part, after Krishtalka et al. (1987) and Roehler (1991).

1969; Dorr and Steidtmann 1971; Dorr et al. 1977). Dorr (1952, 1956, 1958) recorded four vertebrate localities (University of Michigan localities 2, 7, 10, 20) in the upper part of the Hoback Formation to which he assigned an earliest Wasatchian (Graybullian) age. However, *C. proterus* from Dorr's (1952) locality 7 supports Rose's (1981) assignment of a Clarkforkian age to this locality. Dorr (1969) also concluded that the Pass Peak Formation ranges in age from late Graybullian through Lysitean, and possibly into early Lostcabinian. The *Coryphodon* specimens from localities other than locality 7 in the Hoback Basin are too fragmentary to permit species-level identification.

Washakie Basin, Wyoming–Colorado

The Wasatch Formation, exposed around the flanks of the Washakie Basin and rather extensively in the Great Divide Basin (see Bradley 1964), has yielded numerous mammalian assemblages relevant to the P/E Series boundary. These record all of Clarkforkian and Wasatchian time (Krishtalka et al. 1987), but have not been placed into their precise stratigraphic order.

On the southeastern edge of the basin, in Moffat County, Colorado, are the numerous localities that produce the Four Mile Fauna (McKenna 1960). McKenna (1960) assigned it an early Graybullian age. The occurrence of *C. lobatus* is consistent with this assignment, but does not allow a more precise correlation.

At and near Bitter Creek, Wyoming, the Wasatch Formation has produced fossil mammals from numerous stratigraphic levels (Savage and Russell 1983:figure 4-36). The presence of *C. eocaenus*, *C. anthracoideus*, *C. molestus*, *C. lobatus*, and *C. subquadratus* (e.g., figures 21.12–21.13) supports the idea that the entire Wasatchian LMA, and possibly part of the Clarkforkian LMA, are represented by faunal horizons in this area (Savage and Russell 1983). However, virtually all of the *Coryphodon* specimens are from old localities collected by or for Cope, Marsh, and Gazin prior to World War II, or from University of California localities that have not yet been placed in measured stratigraphic sections. Obviously, more stratigraphic work is needed.

The Dad local fauna (Gazin 1962) from the Niland Tongue of the Wasatch Formation near Baggs, Wyoming, contains *C. molestus*, an occurrence consistent with Gazin's (1962) assignment of a Lysitean age based on analysis of the entire fauna. Other Washakie Basin and Great Divide Basin Wasatchian localities were enumerated by Gazin (1962). Most produce small assemblages with little stratigraphic control that lack age diagnostic specimens of *Coryphodon*.

Laramie Basin, Wyoming

Prichinello (1971) documented a low-diversity assemblage of fossil mammals from the "Wind River Formation" in what she termed the Cooper Lake Basin (= Laramie Basin) in Wyoming. The mammals are from a 30-m-thick interval about 60 m above the base of the "Wind River Formation." Prichinello (1971) considered the assemblage to be Graybullian in age (Wood et al. 1941 had earlier assigned it a middle or late Wasatchian age), which is supported by the presence of *Haplomylus*, *C. eocaenus*, and *C. oweni*.

Uinta Basin, Utah

A specimen of *C. lobatus* from the "Wasatch Formation" near Jensen, Utah, may indicate either a late Graybullian, a Lysitean, or a Lostcabinian age (Lucas 1984a). I know of no other mammals associated with this specimen, so no more precise age can be determined.

Piceance Creek Basin, Colorado

Fossil mammals from the Piceance Creek Basin of northwestern Colorado relevant to the P/E Series boundary were first collected in 1909 and initially published by Patterson (1934, 1935, 1936, 1937, 1939; Patterson and Simons 1958). Kihm (1984) undertook a detailed and extensive study of the Clarkforkian and Wasatchian mammal assemblages from the basin. Unfortunately, this excellent work remains unpublished, so I present a brief summary and evaluation of its biostratigraphic conclusions.

In the Piceance Creek Basin, fossil mammals of Clarkforkian and Wasatchian age occur in rocks referred to as Wasatch, Fort Union, and Ohio Creek Formations by workers of the U.S. Geological Survey (e.g., Cashion 1969; Roehler 1972a, 1972b, 1973a, 1973b; Hail 1973, 1974; Pipiringos and Roselund 1977a, 1977b). However, Wood et al. (1941) used the term *DeBeque Formation*, "to be named and described by Patterson." Patterson did not describe this unit, but Kihm (1984) has, and I use the term as he does.

The DeBeque Formation is at least 2000 m thick and consists of three members (ascending): Atwell Gulch, Molina, and Shire (Donnell 1969). Clarkforkian mammals are known principally from the Atwell Gulch Member, and possible Clarkforkian and definite Wasatchian mammals come from the Shire Member (figure 21.16). Kihm (1984) identified eight distinct biostratigraphic intervals/levels of fossil mammals in the DeBeque Formation. The oldest corresponds to the Plateau Valley local fauna, considered by Kihm to be middle Clarkforkian in age, even though it lacks *Coryphodon*, *Esthonyx*, and rodents. Instead, it contains typical Tiffanian taxa such as *Barylambda faberi*, *Titanoides*, *Lambertocyon ischyrus*, and *Thryptacodon*. *Haplomylus* and *Plesiadapis dubius* are present, as are other Clarkforkian taxa, including *Phenacodus*, *Ectocion*, *Ectoganus*, and *Neoliotomus*. It is possible that the Plateau Valley local fauna is intermediate temporally between typical Tiffanian and typical Clarkforkian assemblages.

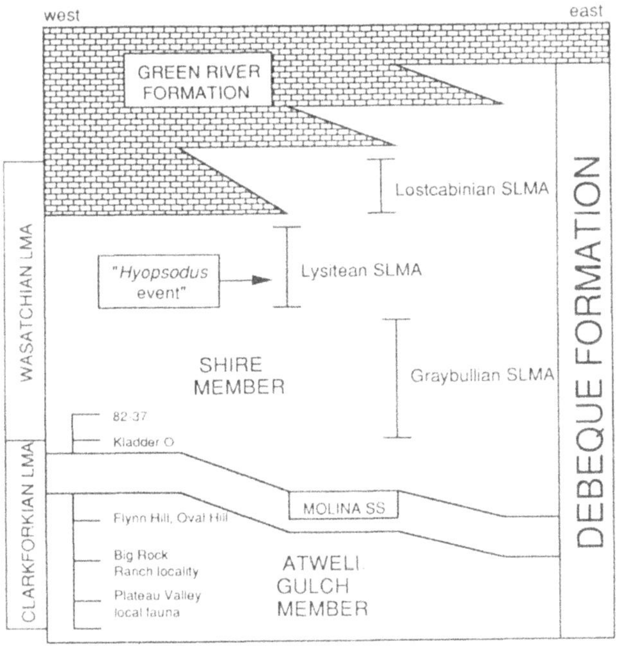

Figure 21.16 Schematic stratigraphic section of the Piceance Creek Basin showing stratigraphic ranges of mammalian assemblages.
After Kihm (1984).

The Big Rock Ranch locality is stratigraphically younger than the level of the Plateau Valley local fauna but only produces *Ptilodus* sp., cf. *Chiromyoides* sp., and ?*Franimys* sp. Kihm (1984) assigned it a middle Clarkforkian age, largely because it is younger than the Plateau Valley local fauna, but it is not age diagnostic.

The Flynn Hill and Oval Hill localities are clearly Clarkforkian in age. *Coryphodon*, *Esthonyx*, and two rodent species appear in this interval, which Kihm assigned a late Clarkforkian age, but I am convinced it is the oldest definitely Clarkforkian assemblage in the basin. *Coryphodon eocaenus* is present at this level. Kihm identified his Kladder O locality as the youngest Clarkforkian site. It produced *Esthonyx* sp., *Palaeonictis* sp., *Phenacodus primaevus*, and *Coryphodon* spp. and could be Graybullian in age.

Graybullian intervals are characterized by *Cantius*, *Haplomylus*, *Phenacodus primaevus*, *Ectocion*, *Homogalax*, and *Esthonyx bisulcatus*. They probably are late Graybullian in age. Kihm tentatively suggested that two Graybullian levels were discernible, but these are not distinct enough taxonomically to be assigned different ages.

The Lysite levels have *Heptodon* (uncommon) and are characterized by the appearance of *Hyopsodus* and *Xenicohippus*. *Lambdotherium* characterizes the Lostcabinian interval. Within the Lysitean SLMA, Kihm identified what he termed the *Hyopsodus* event, which corresponds to the appearance of *H. vortmani* and *H. powellianus*. Kihm regarded these species to be middle Lysitean immigrants into the basin.

Three species of *Coryphodon*—*C. eocaenus*, *C. molestus*, and *C. lobatus*—are present in the DeBeque Formation. The occurrence of *C. eocaenus* in the Atwell Gulch Member is consistent with a Clarkforkian age assignment. The occurrences of *Coryphodon molestus* and *C. lobatus* in the middle and upper parts of the Shire Member are consistent with the Lysitean and Lostcabinian age assignments proposed by Kihm (1984).

Huerfano Basin, Colorado

Robinson (1966) last revised the fossil mammals from the Huerfano Formation of southeastern Colorado. He organized the assemblages into two intervals, one of

Lostcabinian age, the other of early Bridgerian (Gardnerbuttean SLMA) age.

Stucky (1984c) identified four intervals. The oldest encompasses Robinson's localities VIII, IX, and XII, which Stucky (1984c) considered to belong to the Lysitean SLMA, though Robinson (1966) had assigned them to the Lostcabinian SLMA (I cannot resolve this). The next interval (localities IV, VI, and XI = VIIIa) yields abundant *Lambdotherium popoagicum* and is rightly regarded as Lostcabinian in age (Stucky 1984c). Locality VI yields *C. molestus*. The third interval yields *Palaeosyops borealis*, and the youngest interval includes localities I, II, III, and V, which still contain *Coryphodon*.

Although numerous fragmentary specimens of *Coryphodon* have been collected from the Huerfano Formation in the Huerfano Basin (Robinson 1966), identifiable specimens (here assigned to *C. molestus*) are known only from AMNH localities II and VI. Locality VI belongs to the Garcia Canyon local fauna, which Robinson (1966) considered to be of Lostcabinian age, an assignment consistent with the presence of *C. molestus*. Locality II belongs to the Gardner Butte local fauna of Robinson (1966), which he used to define the Gardnerbuttean SLMA, an interval of time he considered transitional between the Lostcabinian SLMA and the Bridgerian LMA. McKenna (1976) and Wallace (1980) underlined the Wasatchian affinities of the Gardner Butte local fauna, and the presence of *C. molestus* further emphasizes this. However, the Gardnerbuttean SLMA is demonstrably younger than the Lostcabinian of the Wind River Basin, and is not just a different faunal facies (see above). The youngest occurrence of *Coryphodon* in North America is in the Gardner Butte local fauna.

San Juan Basin, New Mexico–Colorado

In the San Juan Basin, Wasatchian mammals occur in the San Jose Formation, Tiffanian mammals in the Animas Formation, and Torrejonian mammals in the Nacimiento Formation (figure 21.17). The Nacimiento Formation mammals are the type assemblage of the Torrejonian LMA, and the Animas Formation mammals are typical of the Tiffanian LMA. In the southern (New Mexico) part of the San Juan Basin, Torrejonian mammals are the youngest fossil mammals that occur below deposits with Wasatchian mammals; clearly an unconformity accounts for the absence of a record of Tiffanian and Clarkforkian mammals in this part of the San Juan Basin (Lucas et al. 1981; Williamson and Lucas 1992, 1993; Williamson 1996).

The type Tiffanian fauna is in the Animas Formation in southwestern Colorado. The precise stratigraphic relationship of the Animas with the Nacimiento and San Jose Formations further south is not fully clear. As Smith and Lucas (1991) demonstrated, the Animas Formation is mostly a volcaniclastic piedmont facies laterally equivalent to basinal facies of the Ojo Alamo and Nacimiento Formations further south. However, the unconformity that separates the San Jose and Nacimiento Formations in the southern San Juan Basin becomes a "correlative conformity" northward so that the uppermost Animas may correlate to the lowermost San Jose Formation. Certainly, the type Tiffanian interval of the

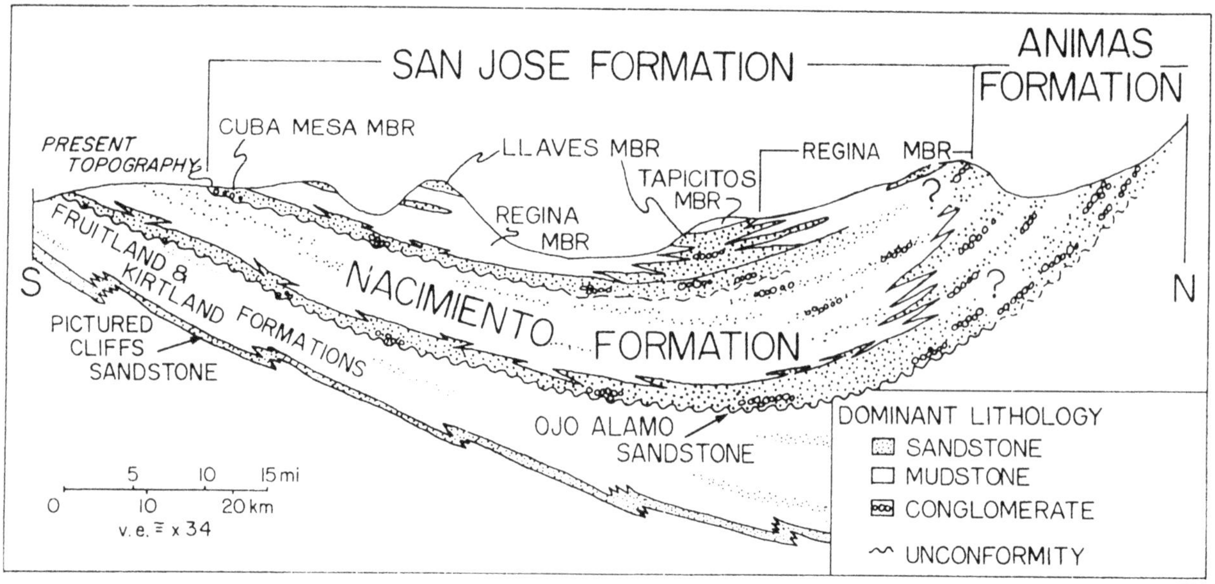

Figure 21.17 Schematic cross-section of the San Juan Basin.
After Smith and Lucas (1991).

Animas Formation must be younger than the Torrejonian mammals of the Nacimiento Formation and older than the oldest Wasatchian mammals from the San Jose Formation.

The San Jose Formation is as much as 630 m thick and consists of five members (Simpson 1948b; Baltz 1967; Lucas 1977; Lucas et al. 1981; Smith 1988, 1992; Smith and Lucas 1991). Interfingered upper tongues of the sandstone-dominated Cuba Mesa Member and the mudstone-dominated Regina Member (150–460 m thick) produce the classic Almagre local fauna, first collected by E. D. Cope in 1874 (Cope 1874, 1877a). The youngest member of the San Jose Formation, the red-bed-mudstone–dominated Tapicitos Member, produces the Largo local fauna, which is noted for its domination by the "condylarth" *Meniscotherium* (Williamson and Lucas 1992). Lucas et al. (1981) listed these assemblages, and Kues (1993) provided a synopsis of subsequent additions.

Age assignments for the Almagre and Largo local faunas have varied. Granger (1914), who first recognized them, considered the Almagre fauna to be Graybullian–Lysitean in age and the Largo fauna to be Lysitean–Lostcabinian in age. Van Houten (1945) and Simpson (1948b) assigned the Almagre fauna to the Graybullian SLMA and the Largo to the Lysitean SLMA. Gingerich and Simons (1977), Gingerich and Gunnell (1979), Lucas (1977), and Froehlich and Reser (1981) assigned both faunas a Lostcabinian age. However, Lucas et al. (1981) and Lucas and Williamson (1992, 1993a, 1993b) assigned both faunas a Lysitean age.

The Largo local fauna is but a depauperate subset of the Almagre fauna. Therefore, the two assemblages cannot be distinguished biochronologically, though superposition demonstrates that the Largo fauna is obviously younger than the Almagre fauna. The Almagre fauna contains several taxa, especially *Heptodon* (misidentified by Froehlich and Reser [1981] as "cf. *Homogalax*"), that indicate it cannot be older than Lysitean in age.

Three species of *Coryphodon*—*C. simus*, *C. molestus*, and *C. lobatus*—are present in the Regina Member of the San Jose Formation in the San Juan Basin (figures 21.12–21.13). These species of the Almagre local fauna are consistent with a Lysitean age. The Largo local fauna, from the Tapicitos Member of the San Jose Formation (above the Regina Member) contains *C. molestus*, and this is consistent with a Lysitean age assignment as well (Lucas et al. 1981). It should be noted that although it is restricted to the Regina Member, *C. simus* is so derived that it is arguably post-Graybullian in age. A more refined biostratigraphy of the San Jose Formation is needed, one that will place the various faunal horizons now lumped into two local faunas into a rigorous stratigraphic framework.

Galisteo–El Rito Basin, New Mexico

In north-central New Mexico, the Galisteo–El Rito Basin is one of the late Laramide Echo Park–type basins. Wasatchian mammals are found in the Diamond Tail and lower Galisteo Formations exposed in this basin, which has 1300 m of Eocene age sediments.

The Diamond Tail Formation has produced a single jaw fragment of *Hyracotherium* sp. (Lucas et al. 1997). Robinson (1957), Lucas (1982b) and Lucas and Kues (1979) reported specimens of *Coryphodon* from the lower part of the Galisteo Formation near Cerrillos (north-central New Mexico). However, these specimens are too fragmentary to allow specific identification. The small Wasatchian Cerrillos local fauna includes *Ectoganus* sp., *Hyracotherium* sp., *Hyopsodus powellianus*, cf. *Homogalax protapirinus*, *Microsyops angustidens*, and *Hyracotherium* sp. It is probably of Lysitean age (Schoch 1981, 1986; Lucas 1982b).

Big Bend National Park, Texas

In the Big Bend National Park, the P/E Series boundary lies in the Black's Peak and Hannold Hill Members of the Tornillo Formation (Schiebout et al. 1987; Schiebout 1995). Fossil mammals from these units are from four distinct stratigraphic intervals (figure 21.18). The lowest produces Tiffanian mammals, including *Titanoides*, *Caenolambda*, *Phenacodus*, *Navajovius*, and *Plesiadapis* (Schiebout 1974, 1995) and may be as young as the upper Tiffanian Ti 5 "Zone" of Archibald et al. (1987).

The upper Hannold Hill Formation produces a Wasatchian assemblage: *Phenacolemur* sp., *Paramys excavatus*, *Coryphodon molestus*, *Hyopsodus miticulus*, *Hyopsodus* cf. *H. wortmani*, *Phenacodus primaevus*, and *Hyracotherium vasacciense* (Wilson 1967; Hartnell 1980a, 1980b; Rapp et al. 1983; Lucas 1989a; Schiebout 1995). *Phenacolemur praecox* is restricted to Clarkforkian–Graybullian taxon deposits in Wyoming (Rose 1981), but it occurs in the Lysitean Regina Member of the San Jose Formation (Almagre local fauna) in the San Juan Basin of northwestern New Mexico (Smith and Lucas 1991). Indeed, all the upper Hannold Hill mammal taxa are known from the Almagre local fauna, and this supports a Lysitean age.

Two intervening levels have produced a handful of specimens. The highest mammal-producing level in the Black Peak's Member yielded the taeniodont *Psittacotherium multifragum*, which suggests a Tiffanian age (Schoch 1986). The other level, low in the Hannold Hill Member, produced *Hyracotherium vassaciense* and *Barylambda jackwilsoni* (originally identified by Schiebout 1974, later misidentified as *Coryphodon* sp.

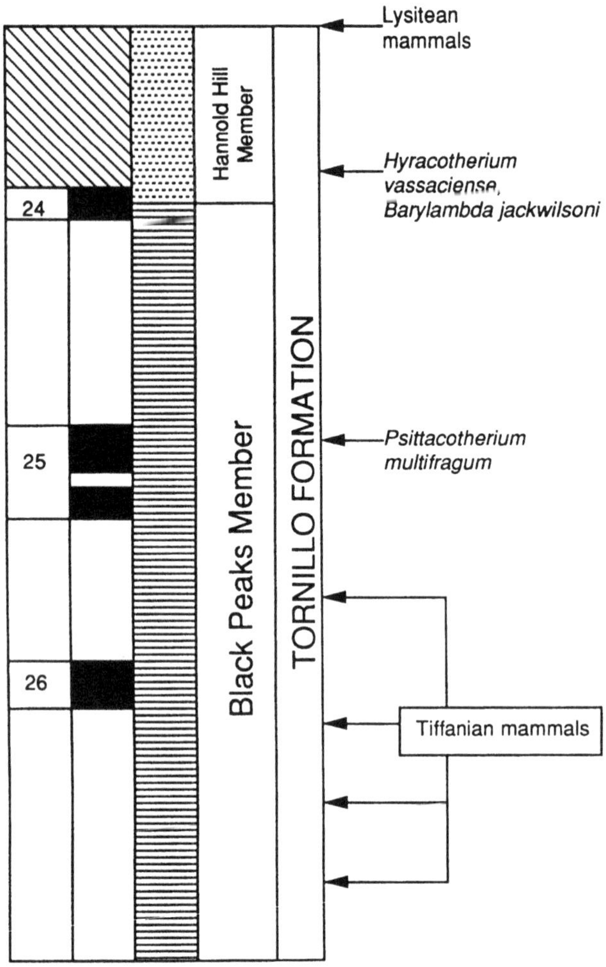

Figure 21.18 Distribution of mammalian assemblages and magnetostratigraphy of the upper Paleocene–lower Eocene succession in Big Bend National Park, Texas.
After Schiebout (1995).

by Schiebout et al. 1987). This interval, based on the occurrence of *Hyracotherium*, must be of Wasatchian age.

Magnetostratigraphy of the Tornillo Formation (Rapp et al. 1983) ties the record of *Psittacotherium multifragum* to Chron C25, which suggests it is of latest Tiffanian or earliest Clarkforkian age and that the underlying Tiffanian mammals are in the Ti3–Ti5 zonal range. Lysitean mammals in the Hannold Hill Member occur above the base of Chron C24r in strata that have not been sampled for magnetostratigraphy.

San Diego County, California

Walsh (1991, 1996) mentioned the undescribed Morena Boulevard local fauna from an unnamed formation east of Pacific Beach in San Diego County, California, and listed *Knightomys*, *Hyopsodus*, cf. *Tubulodon*, and *Meniscotherium*. Williamson and Lucas (1992) identified the *Meniscotherium* as *M. tapiacitum*, a taxon restricted to Graybullian and Lysitean strata.

Baja California del Sur, Mexico

The Las Tetas de Cabra Formation (unnamed formation of Morris 1966, 1968; Novacek et al. 1987) has produced a low-diversity assemblage of Wasatchian mammals that include *Estelestes ensis* Novacek et al., *Prolimnocyon* sp., *Oxyaena* sp., cf. *Hyopsodus* sp., *Ectocion ignotum* Novacek et al., *Phenacodus* sp. cf. *P. vortmani*, *Meniscotherium chamense* (= *M.* cf. *M. priscum* of Novacek et al. 1991: Williamson and Lucas 1992), *Barylambda jackwilsoni* (= Pantolambdidae ?new genus of Novacek et al. 1991), *Esthonyx* sp., *Hyracotherium seekinsi*, *Eotitanops* sp. cf. *E. princeps*, *Diacodexis* sp. cf. *D. gracilis*, Paramyidae?, and *Wyolestes iglesius* Novacek et al. Novacek et al. (1991) did not determine a more precise position of this assemblage within the Wasatchian. The brontothere suggests a late Wasatchian age, but this is contradicted by the presence of *Ectocion* and *Barylambda jackwilsoni*, which may indicate an older age.

The terrestrial mammal-bearing sediments of the Las Tetas de Cabra Formation intertongue with marine strata that produce bivalves and gastropods that form the restricted (middle Ypresian) "Capay Stage" of Givens (1974) and Squires (1984). All fossil mammals occur in an interval of reversed polarity. Flynn et al. (1989) noted that this interval can be identified as either Chron C23r or either of the reversed intervals in Chron C24n.

Sverdrup Basin, Ellesmere Island, Canada

The Margaret Formation of the Eureka Sound Group (Miall 1986; referred to as Member IV of the Eureka Sound Formation in most of the vertebrate paleontological literature) consists of light gray and white sandstone and infrequent thin shale beds with thick (up to 15 m) coal beds. Fossil mammals from the lower part of this unit are found in a small synclinal valley on Ellesmere Island. They include *Hyracotherium* sp., cf. *Ignacius* sp., plagiomenids, *Neoplagiaulax*, Leptictidae, Hyaenodontidae, *Prolimnocyon*, *Viverravus*, cf. *Uintacyon*, a chiromyoid, three species of *Coryphodon*, *Hyrachyus*, and *Lambdotherium* (McKenna 1980a). *Hyrachyus* and *Coryphodon* are the two most abundant mammals. This is a Wasatchian assemblage except for *Hyrachyus*, which normally is taken to mark the base of the Bridgerian. The *Coryphodon* species present suggest a Graybullian age, whereas the occurrence of *Lambdotherium* suggests a Lostcabinian age.

Because of these contradictions, most workers have not attempted a more precise correlation of the Margaret Formation mammals with the Wasatchian SLMAs (West and Dawson 1977; West et al. 1977; McKenna 1980a). Based on Vinson (1981), Hickey et al. (1983) published a magnetostratigraphy of the mammal-bearing strata, which they correlated to Chron C25n and the early part of Chron C24r. Tauxe and Clark (1987) reanalyzed the magnetostratigraphy and could not confirm the magnetostratigraphic results of Hickey et al., which were earlier criticized by Kent et al. (1984) and Flynn et al. (1984). Tauxe and Clark (1987) correlated the mammal-bearing interval to Chron C24r, which is consistent with a mid-Clarkforkian to Graybullian age.

Three species of *Coryphodon*—*C. anthracoideus*, *C. lobatus*, and *C. subquadratus*—are present in the Margaret Formation (West et al. 1977; figure 21.14). This suggests an early to middle Graybullian age, but there are uncertainties. *Lambdotherium* and *Hyrachyus* are also present in the Margaret Formation (West et al. 1977; Hickey et al. 1983), and this suggests a younger age. Obviously, the fossil mammals from the Margaret Formation indicate some diachroneity of mammalian index taxa of the Wasatchian and Bridgerian LMAs, a problem that deserves further attention.

Lauderdale County, Mississippi

Beard and Tabrum (1990) and Beard et al. (1995) reported Wasatchian mammals from near the top of the Tuscahoma Formation and from the overlying Bashi Formation near the Red Hot Truck Stop in Lauderdale County, Mississippi. Only an omomyid primate is known from the Bashi Formation, but the Tuscahoma Formation produced a diverse assemblage that Beard et al. (1995) called the Red Hot local fauna, which includes *Coryphodon*, *Ectocion*, *Haplomylus speirianus*, and *Hyracotherium*. I agree with Beard et al. (1995) that this fauna can be assigned to the Graybullian SLMA.

Planktonic foraminifera from the Bashi Formation belong to Zone P6b (Oliver and Mancini 1980; Mancini 1981). Frederiksen et al. (1982) correlated palynomorphs from the Bashi Formation to Zone NP10. This suggests the Bashi Formation can be assigned an Ypresian age. Beard et al. (1995:18A) stated that the Red Hot local fauna "lies in beds that almost certainly correlate with marine nannoplankton zone NP9." If this correlation is correct, then the Graybullian mammals of the Red Hot local fauna are older than Ypresian and thus of Paleocene age.

Correlation in North America

Correlation of the North American mammal-bearing strata across the P/E Series boundary (figure 21.19) is similar to but more detailed than that of Krishtalka et al. (1987). It relies heavily on the *Coryphodon*-based correlations and the three Wasatchian SLMAs, as outlined above.

Asia

The Asian record of Paleogene mammals is extensive (e.g., Takai 1952; Thenius 1959; Savage and Russell 1983; Li and Ting 1983; Russell and Zhai 1987; Tong et al. 1995). However, only fossil mammals from Kazakhstan, Mongolia, and China are relevant to the P/E Series boundary, although the record from Kazakhstan is spurious. Mammal-bearing deposits occur in a series of foreland basins (figure 21.20) that developed as part of a compressional orogeny resulting from the Paleogene collision of the Indo-Pakistani subcontinent with Asia (Wang 1985; Gu and Chen 1987). There is no evidence of Paleogene mammalian provinciality in Asia (contra Du et al. 1992), so correlations from Kazakhstan to Guangxi Province, China can be firmly established.

Asian Paleogene LMAs

Four sets of Paleogene LMA nomenclature have been introduced for Asia (figure 21.21), by Romer (1966), Li and Ting (1983), Russell and Zhai (1987), and Tong et al. (1995). Romer (1966:table 4) provided a general, global correlation framework between Cenozoic mammal localities and listed Asian LMAs. He made no attempt to define these LMAs, but cited at least one representative locality for each, except for the Ulanbulakian LMA. Romer (1966:332) implied that the LMAs had already been named:

> A series of ages based on the mammals contained in the various North American beds has, however, been established, and approximate correlations can be made between these ages and the European ones. Similar series have been erected for Asiatic and South American deposits. These series of ages are given in Table 1, together with data on the best-known localities, areas, or formations in the various continental areas.

However, a search of earlier literature (e.g., Takai 1952; Thenius 1959) makes it clear that Romer coined these terms.

Li and Ting (1983:appendix 4) introduced a set of what they called "Chinese provincial ages (tentatively)" in a correlation chart where they listed large and small mammals and formations characteristic of each age. They made no explicit attempt at definition, nor did they refer to Romer's (1966) terminology.

Russell and Zhai (1987:appendix) introduced a third set of Asian Paleogene LMAs without reference to the

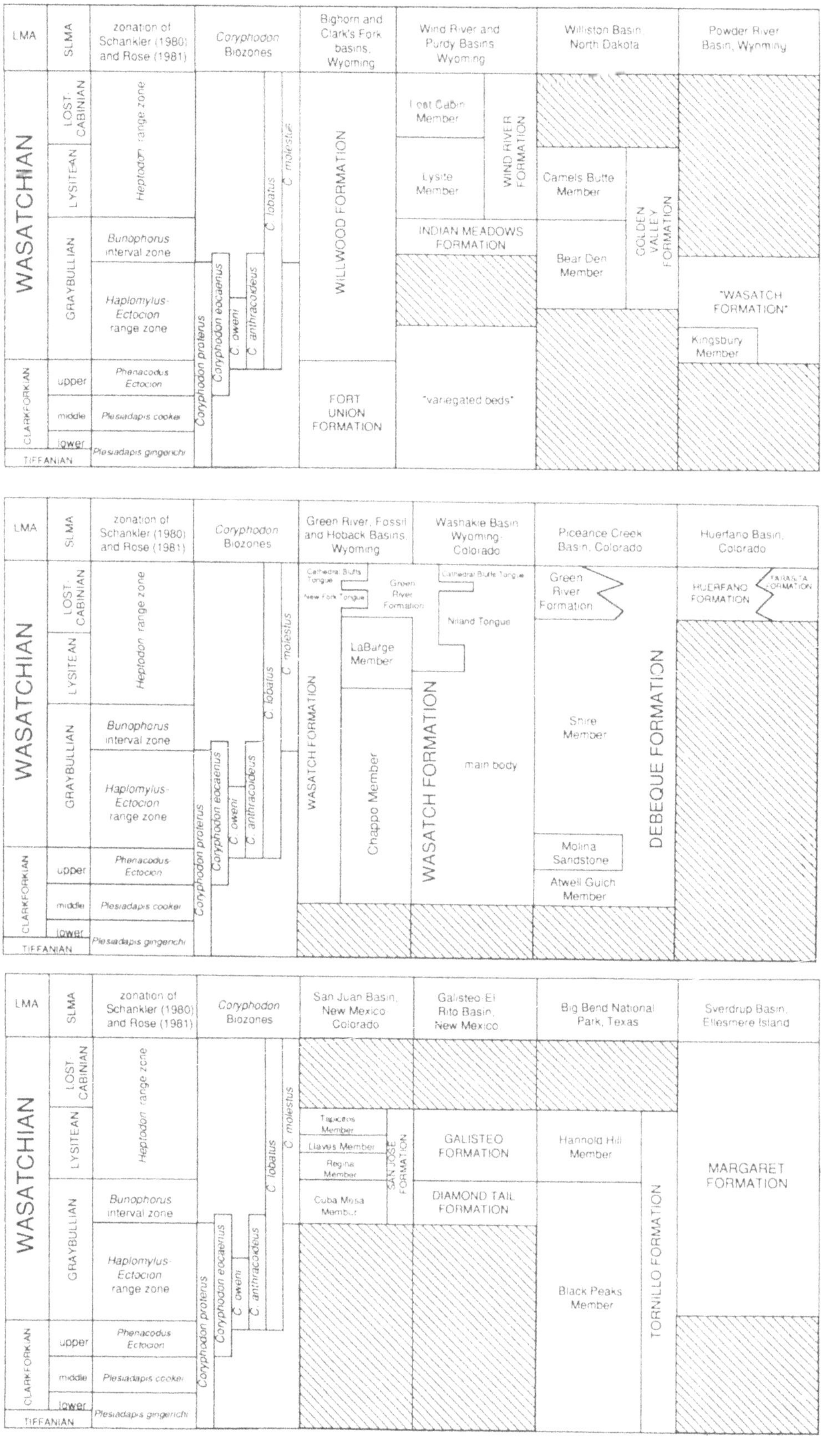

Figure 21.19 Stratigraphic correlation of upper Paleocene–lower Eocene continental strata in North America based on fossil mammals, principally *Coryphodon*.

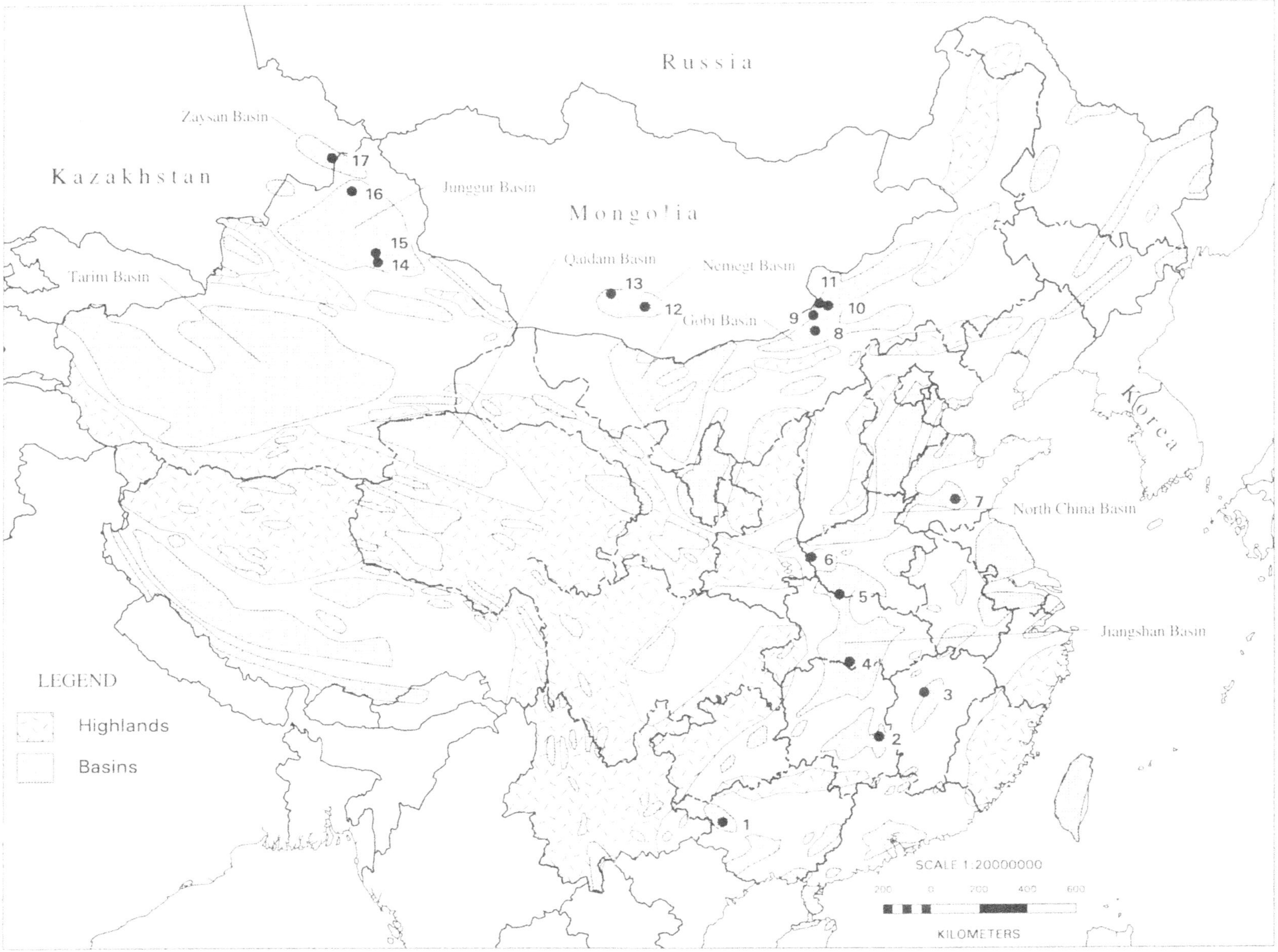

Figure 21.20 Paleogene Basins of eastern Asia and principal fossil mammal localities relevant to the Paleocene/Eocene Series boundary.

1 = Bose Basin, Guangxi Province; 2 = Hengyang Basin, Hunan Province; 3 = Yuanshui Basin, Jiangxi Province; 4 = Yichang District, Hubei Province; 5 = Liguanqiao Basin, Henan Province; 6 = Lushi Basin, Henan Province; 7 = Xintai and Wutu Basins, Shandong Province; 8 = Shara Murun, Nei Mongol Province; 9 = Camp Margetts, Nei Mongol Province; 10 = Arshanto and Irdin Manha, Nei Mongol Province; 11 = Houldjin, Nei Mongol Province; 12 = Kholoboldzhi Nor, Mongolia; 13 = Nemegt Basin, Mongolia; 14 and 15 = Turpan Basin, Xinjiang Province; 16 = Junggur Basin, Xinjiang Province; 17 = Zaysan Basin, Kazakhstan.

Series	Romer (1966)	Li & Ting (1983)	Russell & Zhai (1987)	Tong, Zheng & Qiu (1995)
OLIGOCENE	Hsandagolian	Tabenbulukian	Tabenbulukian	Tabenbulukian
OLIGOCENE	Houljinian	Houldjinian	Shandgolian	Ulantatalian
OLIGOCENE	Ulangochuian	Ulangochuian	Ergilian	Ulangochuan
EOCENE	Sharamurunian, Irdinmanhan	Sharamurunian	Sharamurunian	Naduan Sharamurunian
EOCENE	Arshantan	Arshantan	Irdinmanhan	Irdinmanhan Arshantan
EOCENE	Ulanbulakian	Lingchan	Bumbanian	Lingchan
		Bayanulanian		
PALEOCENE	Gashatan	Nungshanian	Nongshanian	Nongshanian
PALEOCENE		Shanghuan	Shanghuan	Shanghuan

Figure 21.21 The four sets of LMA nomenclature proposed for the Paleogene of Asia.

previous nomenclatures. They listed formations and composite faunas representative of each LMA.

Tong et al. (1995) presented a fourth scheme of Asian Paleogene LMAs (figure 21.21), largely based on a mixture of the nomenclature used by Romer (1966), Li and Ting (1983), and Russell and Zhai (1987). Explicit definitions of each LMA were given, including identification of the "type fauna," correlative faunas, index taxa, and correlation to the SGCS and to the North American LMAs.

Romer's (1966) terminology was little used, and most Chinese workers (e.g., Tong 1989; Wang 1992) did not use the terminology erected by Li and Ting (1983). I advocate use of the terminology of Russell and Zhai (1987), especially because these authors were more explicit in introducing and defining the terms than were Romer (1966) or Li and Ting (1983). Tong et al. (1995) also used much of the terminology of Russell and Zhai (1987), although with minor differences.

I consider the term *Bumbanian* preferable to the term *Lingchan* introduced by Li and Ting (1983) and used by Tong et al. (1995), simply because the mammalian assemblage of the Bumban Member is much more diverse than the mammalian assemblage of the Lingcha Formation (see below). Furthermore, the Bumbanian assemblage includes *Hyracotherium*, *Homogalax*, *Hyopsodus*, *Gomphos*, and rodents, taxa that are useful for long-distance correlations. Also, I use the Irdinmanhan LMA to mark the middle Eocene, and do not now recognize an earlier, Arshantan LMA.

Regardless of the LMA nomenclature adopted, most workers in Asia discriminate three sets of fossil mammal assemblages relevant to the P/E Series boundary (figure 21.22). They are (1) Nongshanian assemblage with the pantodont *Archaeolambda*, the primitive dinoceratan *Prodinoceras*, arctostylopids, *Pseudictops*, and phenacolophids; (2) a rather poorly known Bumbanian assemblage with *Hyracotherium*, *Heptodon*, *Homogalax*, and *Hyopsodus*; and (3) a much better known

Nongshanian LMA	Bumbanian LMA	Irdinmanhan LMA
Lambdopsalis	undescribed multituberculates	no multituberculates
Pseudictops	*Rhombomylus*	———
Prodinoceras	rare *Prodinoceras*	*Gobiatherium*
Archaeolambda	*Asiocoryphodon*	*Eudinoceras*
rare *Coryphodon*	*Heterocoryphodon*	———
Pastoralodon	*Hyopsodus*	———
Mimotona	———	Brontotheriidae
Arctostylops	rare *Arctostylops*	Deperetellidae
Phenacolophidae	*Hyracotherium*	Lophialetidae
———	*Homogalax*	Hyracodontidae
———	*Heptodon*	Amynodontidae

Figure 21.22 Key mammalian taxa of the three LMAs that encompass the Paleocene/Eocene Series boundary in Asia.

Irdinmanhan assemblage with diverse ceratomorphs and the coryphodontid pantodont *Eudinoceras*.

Further subdivisions have thus far proven unconvincing. For example, Tong et al. (1995) and Ting (1998) have suggested that the Lingchan (= Bumbanian) LMA can be divided into three "phases": (1) an older *Cocomys–Orientolophus* phase (Lingcha assemblage); (2) a middle *Changlelestes–Arcitoparamys–"Homogalax"* phase (Wutu assemblage); and (3) a younger *Rhombomylus–Heptodon* phase (Dajian assemblage). However, no stratigraphic superposition documents this, and the evolutionary trajectories that supposedly support it are not documented. Clearly, subdivision of the Nongshanian and Bumbanian LMAs should be possible, but the data are as yet insufficient.

Zaysan Basin, Kazakhstan

Putative late Paleocene mammals were reported by Nessov (1987) and Averianov (1995) from a quarry near Zhylga Railway Station close to Chimkent in southern Kazakhstan. This locality is a single point and therefore of limited biostratigraphic significance.

The oldest well-documented Paleogene mammals known from Kazakhstan—indeed from any part of the FSU—are from the Zaysan Basin in eastern Kazakhstan (figure 21.20). Gabuniya (1961, 1962, 1978, 1984) assigned an early Eocene age to the oldest fossil-mammal–bearing horizons (Chakpaktas and Obayla Formations) in the Zaysan basin, and this age assignment has been retained in the Russian literature (e.g., Shevyryeva 1996). Gabuniya justified this age assignment by identifying cf. *Homogalax* and a primitive miacid. However, both formations produce several land-mammal taxa characteristic of Irdinmanhan time, including *Eudinoceras* (see figure 21.30 below), *Rhodopagus*, *Forstercooperia*, and lophialetid tapiroids (Russell and Zhai 1987). The material Gabuniya assigned to cf. *Homogalax* probably belongs to *Isectolophus bogdulensis* Reshetov 1979, a taxon originally described from the Irdinmanhan at Kholoboldzhi Nor, Mongolia. As Russell and Zhai (1987) indicated, the oldest well-described fossil mammals from the Zaysan Basin are of Irdinmanhan age (also see Lucas et al. 1996).

Red beds below the Chakpaktas Formation produce fossil plants that include *Nordenskioldia*, a genus indicative of a Paleocene age (Ilinskaya 1974; Crane et al. 1991; Lucas et al. 1995). A stratigraphic horizon underlying these plant-bearing red beds contains dinosaur eggshell fragments (Borisov 1963). Middle Eocene mammals (*Eudinoceras*, Lophialetidae) occur in red beds of the Tayzhuzgen Formation at the Zhak Talagay locality in the southwest Zaysan Basin. Thus, Paleocene strata may be present in the Zaysan Basin, but no Paleocene or early Eocene mammals are known. Yakhimovich et al. (1993) published a magnetostratigraphy for the Cenozoic record of the Zaysan basin, but the samples were not thermally demagnetized, so the results are of no significance.

Ulan-Nur Basin, Mongolia

In the Ulan-Nur Basin of Mongolia, the Gashato locality in the Khashat Formation has long represented the oldest Paleocene mammal fauna from Asia (Matthew and Granger 1925; Matthew et al. 1928, 1929; Szalay and McKenna 1971). It is now considered a well-known Nongshanian assemblage that includes *Prodinoceras*, *Pseudictops*, *Phenacolophus*, and *Arctostylops*, among others (see Russell and Zhai 1987:59–60 for a complete faunal list).

From a level stratigraphically above the Gashato fauna in an unnamed formation (corresponding to the "Dalan Dzadgad beds" of Gabuniya et al. 1975; Zhegallo and Shevyryeva 1976), Shevyryeva et al. (1975) reported one specimen of the mimotonid *Gomphos elkema* Shevyryeva. As Russell and Zhai (1987:86–87) indicated, this occurrence suggests a Bumbanian age. Thus, superposition of Nongshanian mammals and the Bumbanian taxon *Gomphos* is demonstrated in the Ulan-Nur Basin.

Nemegt Basin, Mongolia

The Nemegt Basin is the western part of a large intermontane depression in the southern Gobi Altai. Paleogene strata are exposed around the basin axis, where they disconformably overlie the dinosaur-bearing Ne-

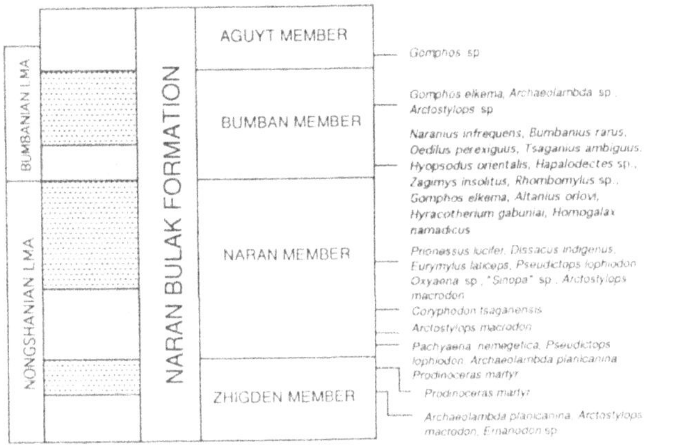

Figure 21.23 Generalized section and distribution of fossil mammals at Tsagan Khushu, Mongolia.
After Dashzeveg (1988).

megt Formation (Jerzykiewicz and Russell 1991). Its middle, Naran Member, produces Nongshanian mammals at several localities (figure 21.23). These mammals include *Prodinoceras*, *Pseudictops*, and *Coryphodon* (a few teeth: figure 21.24; Gradzinski et al. 1969; Reshetov 1976; Badamgarev and Reshetov 1985; Russell and Zhai 1987).

The overlying Bumban Member provided a fossil mammal assemblage upon which Russell and Zhai (1987) based their Bumbanian LMA. Most of this assemblage is from a single, lenticular bed, 0.1–0.3 m thick, which crops out at Tsagan Khushu (Dashzeveg 1988). Taxa are the nyctitheriid insectivores *Bumbanius rarus* Russell and Dashzeveg and *Oedolius perexiguus* Russell and Dashzeveg; the palaeoryctids *Naranius infrequens* Russell and Dashzeveg, *Tsaganius ambiguus* Russell and Dashzeveg; the mixodontid *Zagamys insolutus* Dashzeveg, Flynn, and Russell; the primate *Altanius orlovi* Daszeveg and McKenna; the mimotonid *Gomphos elkema* Shevyryova; the eurymylid *Rhombomylus* cf. *R. turpanensis* Zhai; the rodents *Sharomys parvus* Dashzeveg, *S. singularis* Dashzeveg, *Kharomys gracilis* Dashzeveg, *K. mirandus* Dashzeveg, and *Ulanomys mirificus* Dashzeveg; a hyaenodontid creodont, *Hyopsodus orientalis* Dashzeveg; an unidentified artiodactyl, *Hyracotherium gabuniai* Dashzeveg; *Homogalax namadicus* Dashzeveg; *Arctostylops* sp.; and cf. *Hyracolestes* sp. (Shevyryeva et al. 1975; Dashzeveg 1977, 1979a, 1979b, 1990; Dashzeveg and McKenna 1977; Shevyryeva 1983; Russell and Dashzeveg 1986). Most of these taxa are endemic, but *Gomphos*, *Hyracotherium*, *Hyopsodus*, and *Homogalax* provide a basis for correlation with the Chinese terrestrial stratigraphic record discussed below. Higher levels in the Bumban Member, and in the overlying Aguyt

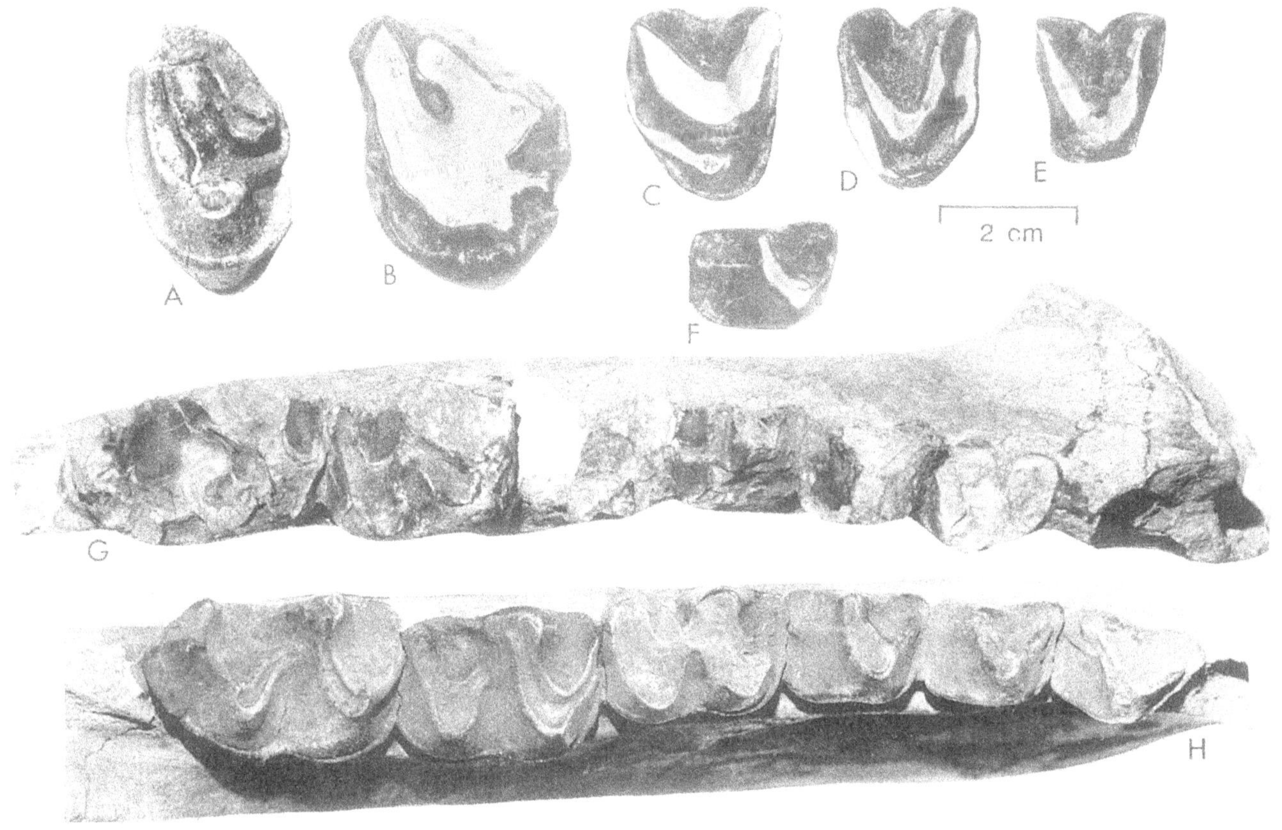

Figure 21.24 Occlusal views of cheek teeth of selected *Coryphodon* from Asia of biostratigraphic significance.
(A) IVPP V 4348, *C. eocaenus*, left M^3, Dabu Formation, Turpan Basin, Xinjiang Province. (B–F) PIN 3102, *C. anthracoideus*, left M^3 (B: holotype of *C. tsaganensis*), right P^4 (C), left P^3 (D), left P^4 (E) and right P_3 (F), Naran Member of Naran Bulak Formation, Nemegt Basin, Mongolia. (G) IVPP uncatalogued, *C. eocaenus*, right P_2–M_3, Yuli Formation, Yuanqu Basin, Shanxi Province. (H) IVPP V 4346, *C. eocaenus* (holotype of *C. dabuensis*), right P_2–M_3, Dabu Formation, Turpan Basin, Xinjiang Province. Institutional abbreviations: IVPP = Institute of Vertebrate Paleontology and Paleoanthropology, Beijing; PIN = Paleontological Institute of the Russian Academy of Sciences, Moscow.

Member, produced *Gomphos*, suggestive of a Bumbanian age.

Turpan Basin, Xinjiang Province, China

The Turpan Basin is a deep topographic depression (up to 154 m below sea level) between the Bogda Shan (north) and the Karlik Shan (south) in western Xinjiang Province, China. Three formations produce mammals relevant to the P/E Series boundary (figure 21.25). The Taizicun Formation in the western part of the basin produced a Nongshanian assemblage that includes *Prodinoceras*, *Archaeolambda*, *Tienshanilophus*, and *Pseudictops* (Tong 1978; Lucas 1982a; Schoch and Lucas 1985). The underlying Subashi Formation produced the dinosaurs *Tarbosaurus* and *Nemegtosaurus* (Dong 1977), and so is of Nemegtian (late Campanian–Maastrichtian) age (Jerzykiewicz and Russell 1991).

The Taizicun Formation overlies the Subashi Formation in the central part of the Turpan Basin near and to the east of Turpan City, from Taizicun into the Huoyan-

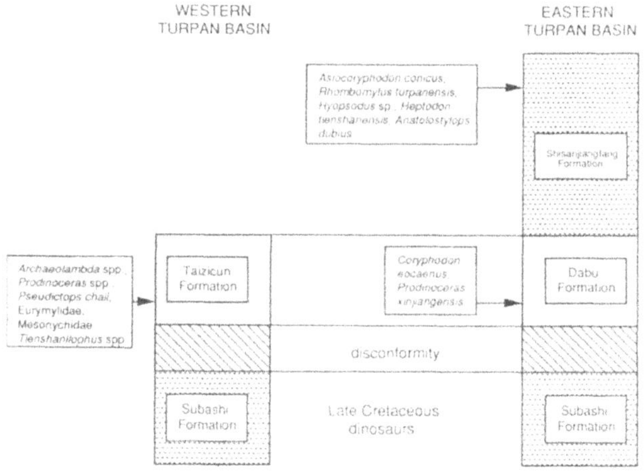

Figure 21.25 Generalized section and distribution of fossil mammals in the Turpan Basin, Xinjiang.
After Russell and Zhai (1987).

shan. But in the northeastern part of the basin, near Dabu and Shisanjianfang, the Dabu Formation overlies the Subashi Formation. Like the Taizicun Formation it is a relatively thin (up to 22 m for Dabu, up to 65 m for Taizicun) unit of gray-white and white sandstones. (I suspect the Taizicun and Dabu are the same lithostratigraphic unit.) The overlying Shisanjiangfang Formation consists of red-bed mudstones up to 272 m thick with fossil mammals found in the interval 18 to 30 m from its top (figure 21.25).

Zhai (1978a, 1978b) described the fossil mammals from the Dabu and Shisanjiangfang Formations. *Coryphodon eocaenus* (= *C. dabuensis* Zhai 1978a; figure 21.24) and *Prodinoceras xinjiangensis* (= *Pyrodon xinjiangensis* Zhai 1978a) are known from the Dabu Formation. *Prodinoceras* is a characteristic Nongshanian taxon, and *Coryphodon* is also known to co-occur with *Prodinoceras* in the Nemegt Basin.

The mammalian assemblage of the Shisanjiangfang Formation consists of *Asiocoryphodon* sp. (figure 21.26), *Rhombomylus turpanensis*, *Hyopsodus* sp., *Heptodon tienshanensis*, and possibly *Anatolostylops dubius* (Zhai 1978b). The Dabu and Shinsanjiang Formations are important because they demonstrate that *Coryphodon* occurs at a lower stratigraphic interval than *Asiocoryphodon*.

Tong (1978) argued that the fossil mammals of the Taizicun Formation are of the same age as the Gashato fauna, given that they share *Prodinoceras* and *Pseudictops*. Furthermore, correlation to the Zhigden Member of the Naran Bulak Formation is well justified by such shared Nongshanian taxa. Zhai (1978a) considered the

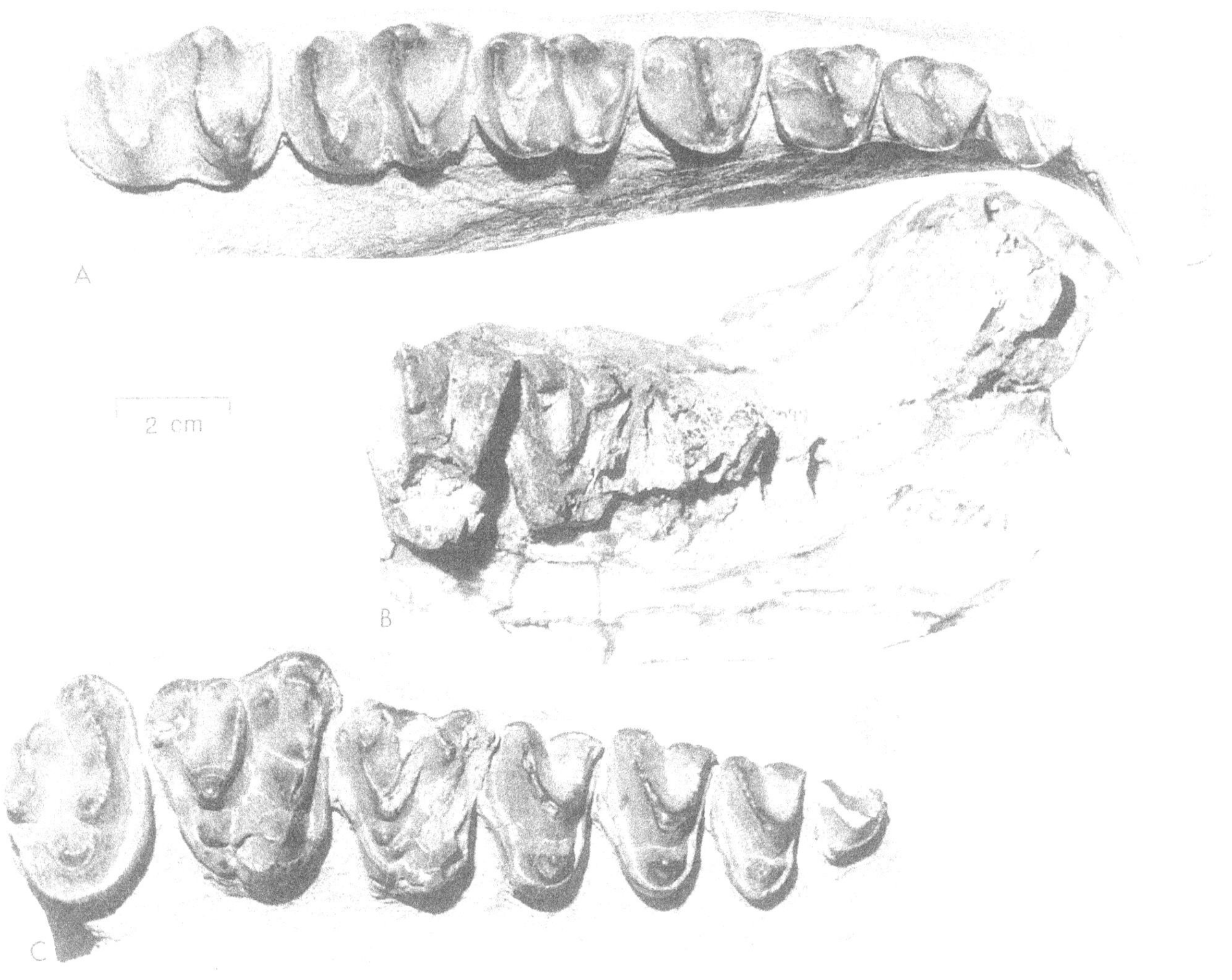

Figure 21.26 Occlusal views of cheek teeth of selected *Asiocoryphodon* from Asia of biostratigraphic significance.

(A, C) IVPP V 5141, *A. conicus* (holotype), right C, P_1–M_3 (A) and right P^1–M^3, Yuhuangding Formation, Liguanqiao Basin, Henan Province. (B) IVPP V 5356, *A.* sp., right maxillary with partial C and P^{2-4}, Shisanjiangfang Formation, Turpan Basin, Xinjiang Province. Institutional abbreviation: IVPP = Institute of Vertebrate Paleontology and Paleoanthropology, Beijing.

Dabu mammals to be younger, but *Coryphodon* and *Prodinoceras* co-occur in the Naran Member of the Naran Bulak Formation. Furthermore, stratigraphic position and similar lithology and thickness suggest that the Taizicun and Dabu Formations are the same lithostratigraphic unit. So, I conclude that the Taizicun and Dabu mammals are of Nongshanian age.

The fossil mammals of the Shisanjiangfang Formation are detectably younger and of Bumbanian age. The co-occurrence of *Rhombomylus turpanensis* and *Hyopsodus* in the Bumban Member (Dashzeveg 1977; Russell and Zhai 1987) supports this. *Asiocoryphodon* is more progressive than *Coryphodon* and never co-occurs with it, so it supports the age assignment (see below).

Nei Mongol (Inner Mongolia) Province, China

The Bayan Ulan locality in the Nomogen Formation produced the pantodont *Pastoralodon*, the uintathere *Prodinoceras*, the anagalid *Pseudictops*, the multituberculate *Lambdopsalis*, and the arctostylopidan *Arctostylops* (Chow and Qi 1978). Russell and Zhai (1987:73) also reported *Lambdotherium*? sp. and *Heptodon*? sp., but Zhai and Meng (1995) stated that the specimen questionably assigned to *Lambdotherium* is an indeterminate perissodactyl, and the questionable *Heptodon* specimen is based on a recording error. If these two taxa are excluded, the Bayan Ulan locality is clearly of Nongshanian Age (Zhai and Meng 1995). Therefore, it cannot be assigned an age intermediate between the Nongshanian and Lingchan LMAs, as did Li and Ting (1983; figure 21.21).

Ligunqiao Basin, Hubei–Henan Provinces, China

The Liguanqiao Basin, located on the border of Hubei and Henan Provinces in China, is a Late Cretaceous–Paleogene intermontane half-graben developed in the eastern Qinling Mountains. Deposits in the basin are dominated by mudstone, sandstone, conglomerate, and marlstone that represent alluvial and lacustrine lithofacies. The Eocene strata have been subdivided into three formations (ascending): the Yuhuangding, Dacangfang, and Hetaoyuan Formations (Cheng and Ma 1990; Ma and Cheng 1991; Du et al. 1992; figure 21.27). The Yuhuangding Formation is well-exposed in an east–west trending outcrop belt, disconformably overlies the Upper Cretaceous Hugang Formation, and conformably underlies the Irdinmanhan Dacangfang Formation. It is subdivided into three (lower, middle, and upper) members (Ma and Cheng 1991).

The lower member, consisting of grayish and pinkish mudstones interbedded with calcareous mudstone, and with grayish yellow and light gray silty mudstone in the upper part, has a thickness of 107 to 252 m. Fossil mammals in this member are *Heterocoryphodon*? sp., *Asiocoryphodon conicus*, and *Heterocoryphodon flerowi* (Lucas and Tong 1987; figure 21.28).

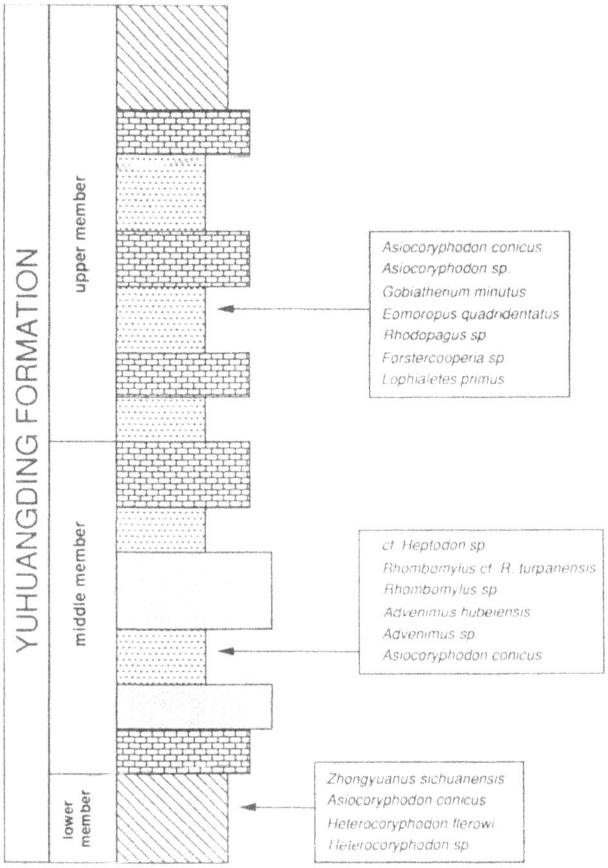

Figure 21.27 Generalized section and distribution of fossil mammals in the Liguanqiao Basin, Henan Province.
After Ma and Cheng (1991).

The middle member consists of grayish white marl intercalated with calcareous silty mudstone and is 224 to 454 m thick. Mammals in this member are cf. *Heptodon* sp., *Rhombomylus* cf. *turpanensis*, *Rhombomylus* sp., *Advenimus hubeiens*, and *Asiocoryphodon conicus*.

The upper member is 56 to 394 m thick and consists of variegated silty mudstone, interbedded with grayish white marl. Mammal fossils include *Asiocoryphodon progressivus*, *Asiocoryphodon* sp., *Gobiatherium minutum*, *Eomoropus*? *zhanggouensis*, *Lophialetes*? *primus*, *Rhodopagus* (= *Yimengia*) sp., and *Forstercooperia*? sp.

Mammal fossils in the Dacangfang Formation include Cocomyidae indet., Carnivora indet., *Mesonyx* sp., Mesonychidae indet., ?*Kuanchuanius* (= *Trogosus*) *danjiangensis*, ?*Eudinoceras* sp., *Eudinoceras* (= *Metacoryphodon*) sp., Coryphodontidae indet., cf. *Palaeosyops* sp., *Lophialetes* sp., cf. *Schlosseria* sp., *Teleolophus* sp., *Rhodopagus* sp. (?), and *Euryodon minimus*. This is clearly an Irdinmanhan assemblage, as determined by Li and Ting (1983), Russell and Zhai (1987), and Ma and Cheng (1991).

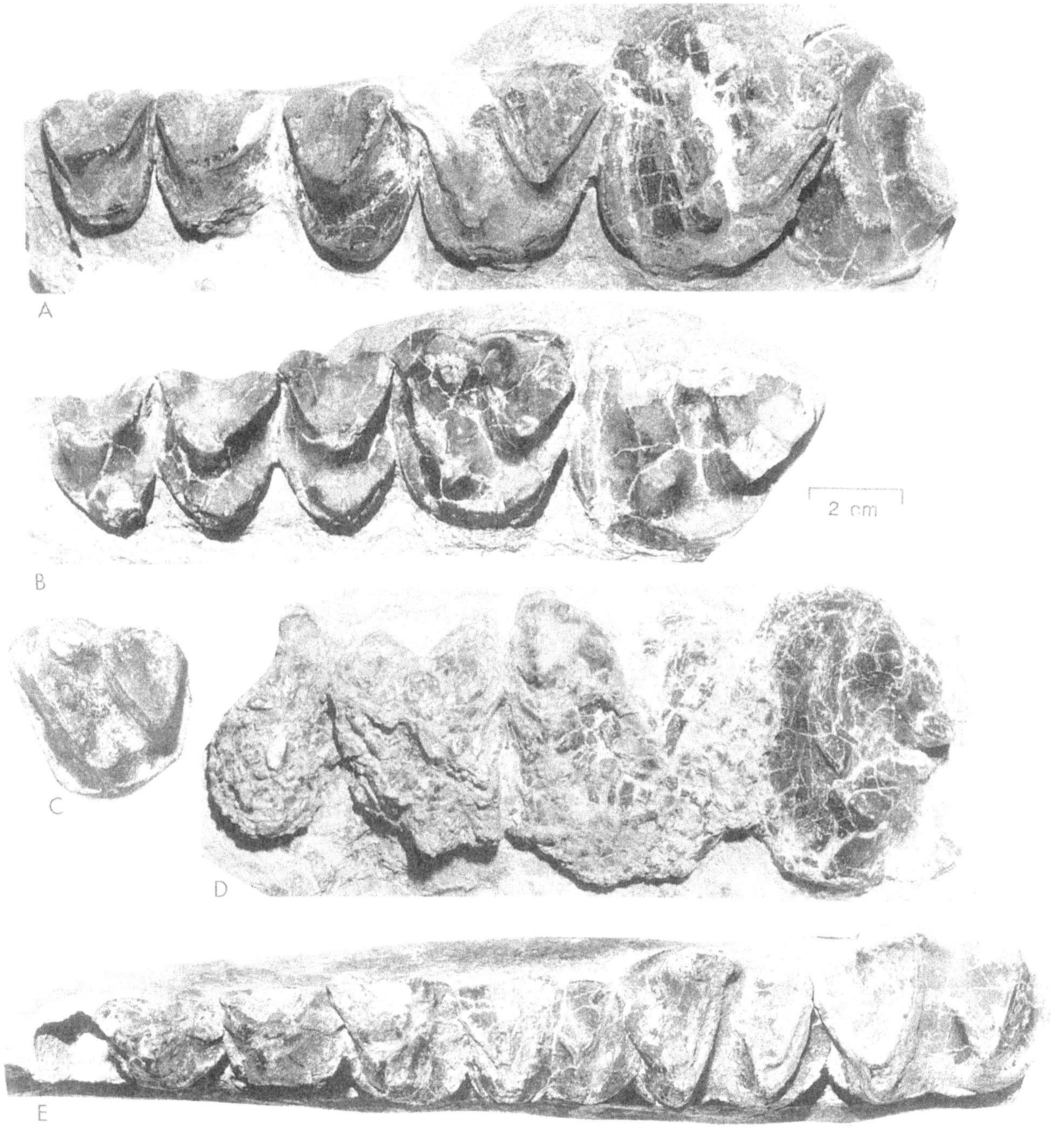

Figure 21.28 Occlusal views of cheek teeth of selected *Heterocoryphodon* from Asia of biostratigraphic significance.

(A) IVPP V 5415, *H. flerowi*, left P^2–M^3, Yuhuangding Formation, Liguanqiao Basin, Henan Province. (B) IVPP V 5147, *H. flerowi*, left P^2–M^2, Yuhuangding Formation, Liguanqiao Basin, Henan Province. (C) IVPP V 927, *H. flerowi* (holotype), left M^1, Wutu Formation, Wutu Basin, Shandong Province. (D) IVPP uncatalogued, *H. flerowi*, left P^4–M^3, Wutu Formation, Wutu Basin, Shandong Province. (E) IVPP V 5146, *H. flerowi*, left P_2–M_3, Yuhuangding Formation, Liguanqiao Basin, Henan Province. Institutional abbreviation: IVPP = Institute of Vertebrate Paleontology and Paleoanthropology, Beijing.

The lower and middle members of the Yuhuangding Formation produce mammals of Bumbanian age. However, the upper member is Irdinmanhan in age. The Yuhuangding and Dacangfang formations in the Liguanqiao Basin are important because they demonstrate the stratigraphic superposition of Bumbanian beds that contain *Asiocoryphodon/Heterocoryphodon* and Irdinmanhan beds that contain *Eudinoceras*.

Wutu Basin, Shandong Province, China

The Wutu Basin in Shandong Province is filled by the Wutu Formation: oil shales, coals, red-bed mudstones, and conglomerates. Chow and Li (1963) described *Homogalax wutuensis* from east of Wutu, and Chow and Li (1965) described *Heptodon niushanensis* from the overlying Niushan Formation. The pantodont *Heterocoryphodon* is also known from the Wutu Formation (figure 21.28).

These Wutu Formation mammals are older than the coal mine locality, which has produced a diverse mammal fauna (31 species), many as yet undescribed. They include neoplagiaulacid multituberculates, epoicotheriid palaeanodonts, carpolestid primates, phenacodontids, paramyid rodents, cf. *Esthonyx*, *Hyopsodus*, *Heptodon*, and *Hyracotherium* (Tong and Wang 1993, 1995; Beard and Wang 1995; Tong and Dawson 1995). As Tong and Wang (1995) noted, this assemblage considerably diversifies the Bumbanian composite assemblage. Described elements are the paramyid rodents *Allagomys oriensis* Tong and Dawson, *Bandaomys zhonghuaensis* Tong and Dawson, *Taishanomys changlensis* Tong and Dawson, *Acritoparamys*? *wutui* Tong and Dawson, and the carpolestids *Chronolestes simul* Beard and Wang and *Carpocristes oriens* Beard and Wang. The Wutu assemblage is clearly Bumbanian in age. It presents potential for organizing a stratigraphic succession of Bumbanian assemblages, though the structural complexities in the mountainous region of Shandong Province may prevent this from being clearly established.

Laian, Anhui Province, China

Rhombomylus laianensis Zhai, Bi, and Yu from the Zhangshanji Formation near Laian, Anhui Province, has been correlated with the Bumbanian interval of the Shisanjiangfang Formation (Zhai et al. 1976). This is probably a Bumbanian record.

Lingcha, Hunan Province, China

In Hunan Province, southern China, near Lingcha and Limuping, the Limuping Formation produced fossil mammals from two stratigraphic intervals (Young 1938; Li et al. 1979; Ting 1993). The lower interval yielded only *Archaeolambda* sp. The upper interval (Institute of Vertebrate Paleontology and Paleoanthropology [IVPP] locality 76003) produced *Cocomys lingchaensis* (Li, Chiu, Yan, and Hsieh), *Matutinia nitidulus* Li, Chiu, Yan, and Hsieh, *Hunanictis inexpectatus* Li, Chiu, Yan, and Hsieh, *Propachynolophus hengyangensis*, *Asiocoryphodon conicus* Xu, and *Hsiangolestes youngi* Zheng and Huang.

At localities near Hengyang, the upper interval produced what Ting (1993) called the Hengdong fauna, which includes *Hsiangolestes youngi*, *Matutinia nitidulus*, *Hapalodectes hetangensis* Ting and Li 1987, *Asiocoryphodon conicus*, and *Hyracotherium hengdongensis*. Ting (1993:206) regarded this assemblage as older than Bumbanian because she estimated *Matutinia* to be more primitive than *Rhombomylus* and the Hengdong *Hyracotherium* (which she named *Orientolophus*) to be more primitive than the *Homogalax* from the Bumban Member and the Wutu Formation. This seems poorly supported, and I consider the Hengdong fauna to be Bumbanian in age, a correlation now agreed to by Ting (personal communication 1996). Note that Lucas (1982a) and Russell and Zhai (1987) included the *Archaeolambda* record with the remainder of the fauna, giving the erroneous impression of a Bumbanian record of *Archaeolambda*.

Ting (1993) erected the new genus *Orientolophus* (type species: *O. hengdongensis*) for a maxillary and mandibular fragment collected from the Lingcha Formation near Hetang. She also suggested that *Hyracotherium gabuniai* from the Bumban Member belongs to this genus. Ting differentiated *Orientolophus* from *Hyracotherium* by its more reduced M^{1-2} protoconule and metaconule and M_{1-3} paralophids and metalophids more bucally (longitudinally) positioned. However, I see little difference between the specimens assigned to *Orientolophus* and *Hyracotherium* specimens illustrated by Kitts (1956), Gingerich (1991), and Hooker (1980; see especially *Hyracotherium cuniculus*, Hooker's figure 1b–c). Thus, I consider *Orientolophus* to be a junior subjective synonym of *Hyracotherium*.

Li et al. (1979) reported fragments of lower cheek teeth including parts of right M_1 and M_2 (IVPP V 5351) from locality 76003 as *Asiocoryphodon*? sp. They definitely belong to *Asiocoryphodon conicus* (Lucas 1984a:357).

Ningjiashan, Jiangxi Province, China

The lower member (Ningjiashan Member) of the Xinyu Formation in Jiangxi Province produced a few fossil mammals, among which are *Heterocoryphodon ninchiashanensis* (Chow and Tung; figure 21.28), *Prodinoceras sinyuensis* (Chow and Tung; see Tong and Lucas 1982 and Schoch and Lucas 1985), *Miacis tenuis* and *Heptodon*? sp. (Chow 1959; Chow and Tung 1962, 1965; Zheng et al. 1975; Gingerich 1983; Schoch and Lucas 1985; Lucas and Tong 1987). The *Prodinoceras* and *Heptodon* specimens are very fragmentary, but the *Heterocoryphodon* is unquestionable and a strong indicator of Bumbanian Age. This is apparently the youngest Asian record of *Prodinoceras*.

Correlation in Asia

Correlation of the assemblages just discussed (figure 21.29) closely resembles that proposed by Russell and

coryphodontid Biozones	KAZAKHSTAN	XINJIANG PROVINCE	MONGOLIA		CHINA		
	Zaysan Basin	Turpan & Junggur Basins	Tsagan Khushu	Kholoboldzhi Nor	JIANGXI PROVINCE	HUBEI PROVINCE	GUANGXI PROVINCE
Eudinoceras	Obayla Formation Chakpaktas Formation	Honglishan Formation		Kholoboldzhi Nor Formation		Dongjun Formation	Donghu Formation
Asiocoryphodon/ Heterocoryphodon		Shisanjiangfang Formation	Naran Bulak Formation: Bumban Member		Xinyu Formation (Ningjiashan Member)		
Coryphodon		Dabu Formation	Naran Bulak Formation: Naran Member				

coryphodontid Biozones	CHINA				
	NEI MONGOL PROVINCE	HENAN PROVINCE	SHANDONG PROVINCE	HUNAN PROVINCE	SHANXI PROVINCE
Eudinoceras	Irdin Manha, Arshanto, Ulan Shireh Formations	Dacangfang Formation	Guanzhuang Formation		
Asiocoryphodon/ Heterocoryphodon		Yuhuangding Formation	Wutu Formation	Lingcha Formation	
Coryphodon					Yuli Formation

Figure 21.29 Coryphodontid-based stratigraphic correlation of the upper Paleocene–lower Eocene continental record of Asia.

Zhai (1987). I base this correlation largely on three Asian coryphodontid Biozones:

- *Coryphodon* Biozone. There are only three occurrences of bona fide *Coryphodon* in Asia: (1) *C. eocaenus* (= *C. dabuensis*) from the Dabu Formation in the Turpan Basin, (2) *C. eocaenus* from the Yuanqu Basin, Shanxi Province (horizon uncertain, but possibly from the Yuli Formation of Lei 1981), and (3) *C. anthracoideus* (= *C. tsaganensis*) from the lower part of the Naran Member of the Naran Bulak Formation at Tsagan Khushu. The occurrence of *Coryphodon* at these localities suggests that they are broadly contemporaneous. The *Coryphodon* species that occur at these localities, *C. eocaenus* and *C. anthracoideus*, co-occur in western Europe and in western North America, so they may be assumed to co-occur in Asia. This supports precise correlation of the Dabu and Yuli Formations with the lower part of the Naran Member of the Naran Bulak Formations. The other fossil mammals known from these units (see above) are consistent with this correlation.
- *Asiocoryphodon/Heterocoryphodon* Biozone. Because fossils of *Asiocoryphodon* and *Heterocoryphodon* co-occur in the Yuhuangding Formation (Henan Province), there is a strong basis for considering localities and horizons that have produced one genus or the other to be broadly contemporaneous. This leads me to correlate the following localities and horizons: (1) Yuhuangding Formation, Liguanqiao Basin, Henan Province (*A. conicus* and *H. ninchiashanensis*), (2) Shisanjiangfang Formation, Turpan Basin, Xinjiang Province (*A. conicus*), (3) Ningjiashan Member, Xinyu Formation, Yuanshi Basin, Jiangxi Province (*H. ninchiashanensis*), and (4) Wutu Formation, Wutu Basin, Shandong Province (*H. flerowi*). Except for the

Wutu Formation, the mammalian fauna from each of these units is small (Li and Ting 1983), but at least some of them (Shisanjiangfang Formation, Yuhuangding Formation) share non-coryphodontid genera (*Rhombomylus*, *Heptodon*) that are consistent with coryphodontid-based correlation.

- *Eudinoceras* Biozone. At present, *Eudinoceras* (figure 21.30) is the most widely distributed coryphodontid in Asia. Therefore, it affords the best opportunity for biostratigraphic correlation in Asia using a coryphodontid. The occurrence of *Eudinoceras* is well documented at the following localities and horizons: (1) Kholoboldzhi Formation at Kholoboldzhi Nor, Mongolia (*E. mongoliensis*), (2) Irdin Manha, Ulan Shireh, and Arshanto Formations at various localities in the Camp Margetts and Shara Murun areas, Nei Mongol Province (*E. mongoliensis*), (3) Honglishan Formation at Xiyiliang, Junggur Basin, Xinjiang Province (*E. mongoliensis*), (4) Chakpaktas and Obayla Formations, Zaysan Basin, Kazakhstan (*E. mongoliensis*), (5) Donghu Formation in the Yichang District, Hubei Province (*E. youngi*), (6) Dongjun Formation, Bose Basin, Guangxi Province (*E. crassum*) and (7) Guanzhuang Formation, Xintai Basin, Shandong Province (specimens of *E. mongoliensis* assigned to "*Metacoryphodon*" by Chow and Qi 1982). Another possible but still undocumented occurrence of *Eudinoceras* is in the Lushi Formation of the Lushi Basin, Henan Province (Tong and Wang 1980). The occurrence of *Eudinoceras* at these localities and horizons suggests that they are broadly contemporaneous. More specifically, the occurrence of *E. mongoliensis* in the Chakpaktas and Obayla Formations, Irdin Manha Formation, Ulan Shireh Formation, and Honglishan Formation supports precise correlation between these units.

The mammalian assemblages from the *Eudinoceras*-bearing units do not contradict the correlations proposed here. However, a complete analysis of these faunas is a complex problem beyond the scope of this study. It is worth noting that Reshetov (1979:table 31), based on his studies of Paleogene tapiroids from Asia, arrived at essentially the same correlation of the *Eudinoceras*-bearing strata that I propose here.

Intercontinental Correlations

Introduction

In strata that encompass the P/E Series boundary, many of the same genera and some of the same species of fossil mammals are found on two or three of the Holarctic continents. This justifies direct mammal-based correlation of the provincial biostratigraphies developed in Western Europe, North America, and Asia. Such correlations have a long tradition, dating back to Cope (1877b), who correlated the North American "Wasatch" to the French "Suessonian" (i.e., "Sparnacian"), a strikingly prescient first approximation. The occurrence of *Coryphodon* and other mammalian taxa in North American Wasatchian assemblages and French "Sparnacian" assemblages was a key element in this correlation, and numerous workers after Cope made reference to this (e.g., Owen 1878; Osborn 1909; Teilhard de Chardin 1921; Simpson 1948a). The correlation of provincial, mammal-based biostratigraphies in Europe, North America, and Asia presented here (figure 21.31) flows logically from the previous sections.

Correlation to the P/E Series Boundary

At present, the P/E Series boundary corresponds to the base of the Ypresian Stage as defined by the Ieper Clay Formation in Belgium (its English equivalent is the London Clay). Berggren and Aubry (1996) have summarized relevant micropaleontologic, magnetochronologic, and radioisotopic dating that establishes the age of the base of the Ieper Clay Formation and have discussed the stratigraphic position of fossil mammal localities assigned to RLs MP6 through MP10. Clearly, all British mammal localities assigned to RL MP7 and RL MP8–9, except the youngest RL MP8–9 occurrences in Division A of the London Clay, are older than the base of the Ieper Clay Formation.

A similar correlation emerges in France, where all RL MP7 localities are older than the base of the Ypresian Stage, as are some RL MP8–9 localities. All relevant Belgian mammal localities belong to RL MP7 and are demonstrably older than the base of the Ieper Clay. The P/E Series boundary thus is within RL MP8–9. More precisely, it is between the levels of Abbey Wood/Pourcy and Division A of the London Clay/Mutigny/Avenay.

This placement of the P/E Series boundary has already been presented (Lucas 1984a, 1989c; Lucas and Williamson 1993) and employed by some workers (Woodburne and Swisher 1995; Berggren and Aubry 1996). However, most mammalian paleontologists locate the P/E Series boundary on the basis of the evolutionary turnover event between Reference Levels MP6 and MP7 (= base of the "Sparnacian"; e.g., Krause and Maas 1990; Gunnell et al. 1993; Tong et al. 1995). This mammal-based boundary is demonstrably older than the P/E Series boundary on the SGCS.

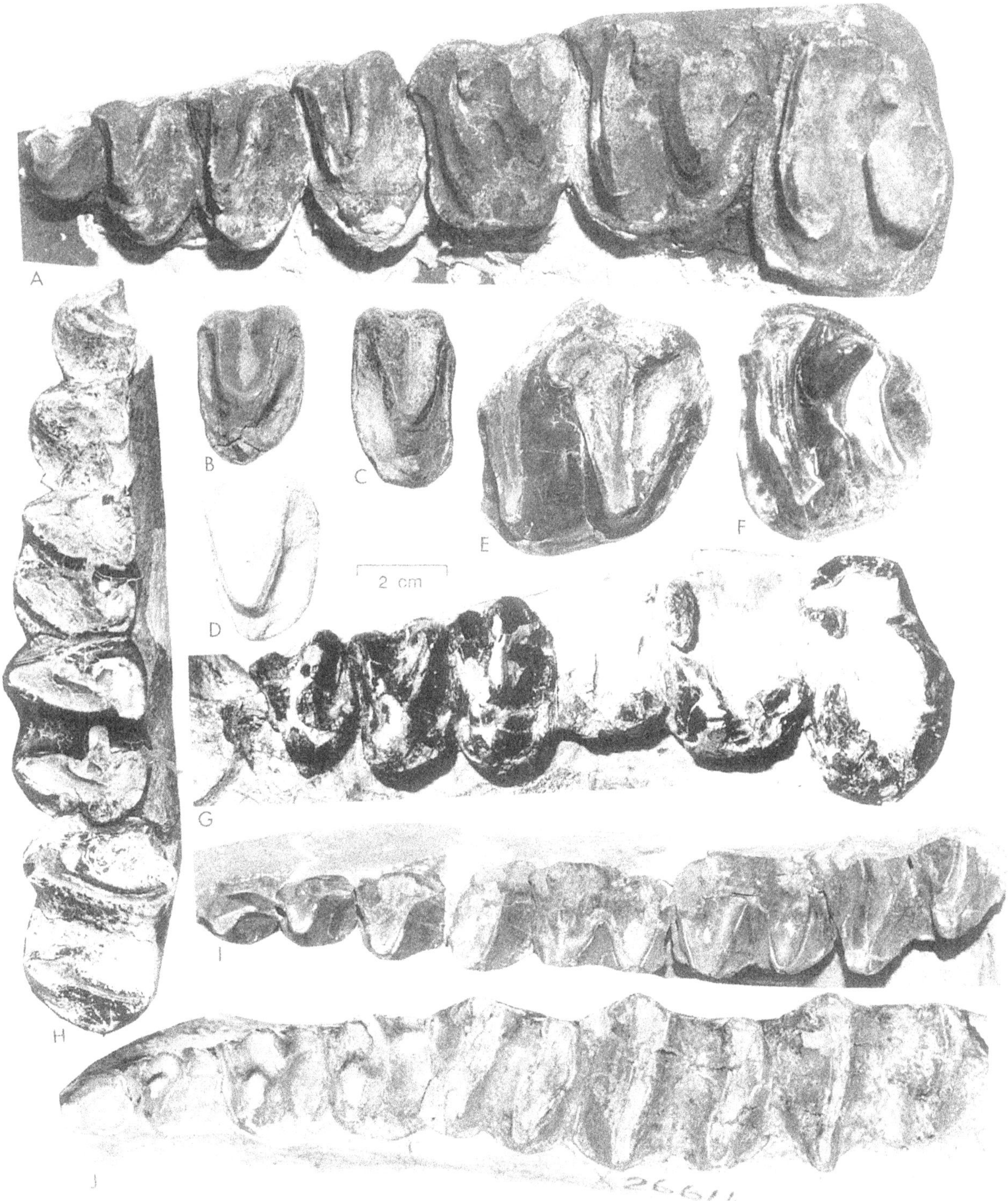

Figure 21.30 Occlusal views of cheek teeth of selected *Eudinoceras* from Asia of biostratigraphic significance.

(A) IVPP V 5697, *E. mongoliensis* (holotype of *Metacoryphodon luminis*), right P^1–M^3 (negative reversed), Arshanto Formation, Nei Mongol Province. (B) AMNH 26638, *E. mongoliensis*, left P^3, Irdin Manha Formation, Nei Mongol Province. (C) AMNH 20102, *E. mongoliensis*, (paratype), right P^4, Irdin Manha Formation, Nei Mongol Province. (D) AMNH 20101, *E. mongoliensis* (holotype), left P^4, Irdin Manha Formation, Nei Mongol Province. (F) IVPP V 4973, *E. crassum*, incomplete left M^2, Dongjun Formation, Bose Basin, Guangxi Province. (F) GIP NZ-1, *E. mongoliensis* (holotype of *E. obailiensis*), left M^3, Obayla Formation, Zaysan Basin, Kazakhstan. (G) AMNH 21744, *E. mongoliensis* (holotype of *E. kholobolchiensis*), left P^2–M^3, Kholoboldzhi Nor, Mongolia. (H) GIP NZ-41, *E. mongoliensis*, left P_2–M_3, Obayla Formation, Zaysan Basin, Kazakhstan. (I) IVPP V 5149, *E. youngi* (holotype), left P_1–M_3, Donghu Formation, Hubei Province. (J) AMNH 26611, *E. mongoliensis*, left P_2–M_3, Irdin Manha Formation, Nei Mongol Province. Institutional abbreviations: AMNH = American Museum of Natural History, New York; GIP = Institute of Paleobiology, Tbilisi, Georgia; IVPP = Institute of Vertebrate Paleontology and Paleoanthropology, Beijing.

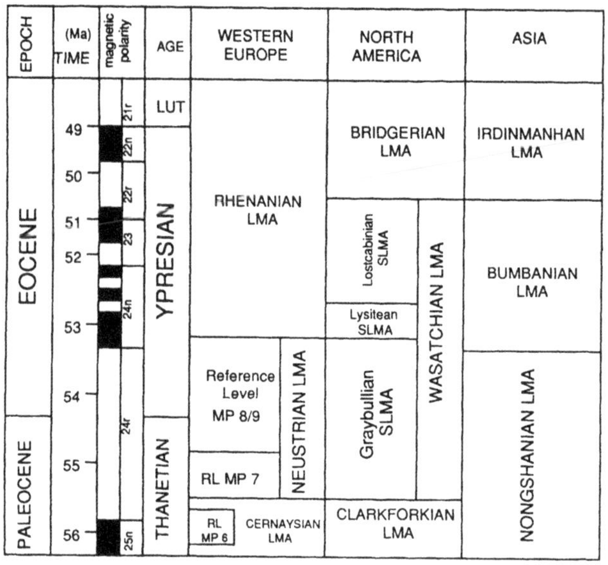

Figure 21.31 Correlation of upper Paleocene–lower Eocene land-mammal "ages" of Europe, North America, and Asia

Intercontinental Correlation: Europe to North America

Simpson (1948a) well summarized the then current status of correlation between the Wasatchian and "Sparnacian":

> It has already been suggested that the various Sparnacian faunules of Europe are not exactly contemporaneous. . . . In general, they appear to be close to the Gray Bull. . . . The conventional Sparnacian may include some Lysite equivalents, but there is no definite evidence of this. (Simpson 1948:10)

Nevertheless, subsequent workers equated the North American Wasatchian LMA and French Sparnacian "Stage" only in a general way (e.g., Russell 1968; Savage 1962, 1971, 1977; McKenna 1972, 1975) and did not accept the more precise correlation of the Graybullian SLMA to the "Sparnacian Stage." In figure 21.31, I use provincial coryphodontid biostratigraphies and mammalian evolutionary turnover events at the base of the Wasatchian and the Neustrian LMAs to correlate the "Sparnacian" and the Graybullian intervals, as did Simpson.

Gingerich (1976; also see Gingerich and Rose 1977) presented a detailed correlation of the "Sparnacian Stage" with the North American Clarkforkian and Wasatchian LMAs. He based his correlation on the taxonomy of plesiadapiforms. According to his work, the basal Sparnacian Conglomérat de Meudon in the Paris Basin is a correlative of the lower part of the Clarkforkian LMA. Rose (1981) revised Gingerich's correlation slightly, but still equated the basal "Sparnacian Stage" with the middle part of the Clarkforkian LMA. Godinot (1982) correlated the Clarkforkian LMA to the lower "Sparnacian Stage" (Reference Level MP7) and the upper "Sparnacian Stage" (Reference Level MP8/9) to the lower Wasatchian LMA.

My correlation (figure 21.31) differs from these plesiadapiform-based correlations because I correlate the Clarkforkian LMA entirely to RL MP6, though it is possible that the upper Clarkforkian LMA is correlative with the older part of MP7 (see below). Because of this difference, I carefully re-examined the plesiadapiform taxonomy of Gingerich (1976) and Rose (1981).

Gingerich (1976) claimed that *Platychaerops* (= *Plesiadapis*) *russelli* from Meudon so closely resembles North American *Plesiadapis cookei* that a correlation of the basal Sparnacian Conglomérat de Meudon with the middle Clarkforkian LMA is justified. Rose (1981) essentially upheld this conclusion. However, I consider *Plesiadapis cookei* to most closely resemble *Plesiadapis tricuspidens* from Cernay and Berru on the basis of incisor morphology. To wit, the upper central incisors of both *Plesiadapis tricuspidens* and *Plesiadapis cookei* have large, medial anterocones, relatively large laterocones and mediocones that are lateral and proximal to the anterocones, laterocones that are slightly larger than the mediocones, large posterocones, and long, narrow crowns. Indeed, *P. cookei* should be considered synonymous with *P. tricuspidens*.

Therefore, the middle Clarkforkian *Plesiadapis cookei* "Zone" of Gingerich (1976) and Rose (1981) is a correlative of their Thanetian *Plesiadapis tricuspidens* "Zone" (i.e., middle Clarkforkian LMA = RL MP6). *Platychaerops russelli* from Meudon has no "evolutionary counterpart" in western North America. These conclusions thus bring the plesiadapid-based correlations into line with the correlations (figure 21.19) presented here. Three further aspects need to be considered:

- The correlation proposed here is also consistent with the palynological correlation established by Wing (1984), who correlated the lowest Wasatchian LMA to encompass the NP9/NP10 nannoplankton zonal boundary, thereby assigning it a late Paleocene age.
- The correlation proposed here indicates that the LO of *C. eocaenus* is older in North America than in Europe. If the two LOs are equated, then the uppermost Clarkforkian LMA correlates with lower RL MP7. This, however, would make the major evolutionary turnover at the base of the Wasatchian and Neustrian LMAs diachronous. It seems more likely that the middle Clarkforkian LMA correlates with RL MP6, and that the late Clarkforkian LMA is not represented in northwestern Europe and corresponds to the gap between Reference Levels MP6 and MP7. Therefore, the LO of *C. eocaenus* is indeed older in North America than in Europe.

- Hooker (1991) documented *Plesiadapis tricuspidens* from a RL MP7 assemblage in the Suffolk Pebble Beds (figure 21.7). Either this *P. tricuspidens* is reworked from a RL MP6 horizon, or the species has a longer temporal range in Europe than previously suspected. I favor the possibility of reworking.

Correlation to Asia

Correlation of the Asian record of mammals to the P/E Series boundary has always been mammal based because of a lack of magnetostratigraphy, other biostratigraphies, and radioisotopic dating. The most conventional correlation places the Nongshanian LMA in the upper Paleocene Series, the Bumbanian LMA in the lower Eocene Series, and the Irdinmanhan LMA in the middle Eocene Series (e.g., Russell and Zhai 1987; Du et al. 1992; Holroyd and Ciochon 1994; Tong et al. 1995). This rests on direct correlation to the North American LMAs, so that Nongshanian LMA = Tiffanian LMA or Tiffanian/Clarkforkian LMAs; Bumbanian LMA = Wasatchian LMA; and Irdinmanhan LMA = Bridgerian LMA.

Dashzeveg (1982b, 1988) provided an excellent example of the rationale behind these correlations. He argued that the Naran and Zhigden Members of the Naran Bulak Formation are Clarkforkian in age, and that the base of the Wasatchian LMA corresponds to the base of the Bumban Member, which he correlates to the Graybullian SLMA. Dashzeveg thus envisioned the P/E Series boundary (= base of Wasatchian LMA) in Asia as corresponding to the appearance of perissodactyls, omomyid primates, rodents and *Hyopsodus*. Ting (1996) supported the same correlation.

In Asia, the classic Gashato Assemblage and its principal correlative, the Taizicun Fauna from the Turpan Basin, lack *Coryphodon* but contain the most primitive uintathere, *Prodinoceras*. This suggests that the Gashato Assemblage is a correlative of the upper Tiffanian to lower Clarkforkian LMAs of western North America, horizons at which *Prodinoceras* (= *Probathyopsis*: Dashzeveg 1982a; Schoch and Lucas 1985) occurs (Rose 1981). The *Coryphodon* occurrences in Asia can be correlated to horizons where *Hyopsodus* and *Hyracotherium* occur (except for the Yuli Formation in the Yuanqu Basin). This is consistent with the correlation proposed here, although a position as old as upper Clarkforkian LMA for the Asian *Coryphodon* horizons would suggest that *Hyopsodus* appeared in Asia earlier than in North America (Rose 1981). The possible occurrence of *Heptodon* in the Yuhuangding Formation (Xu et al. 1979) together with *Asiocoryphodon* and *Heterocoryphodon*, and the fact that the Shisanjiangfang Formation, which contains *Asiocoryphodon*, overlies the Dabu Formation, which contains *C. eocaenus*, supports correlation of the *Asiocoryphodon/Heterocoryphodon*-bearing horizons in Asia with the post-Graybullian Wasatchian interval in western North America. This is essentially the correlation proposed by Li and Ting (1983) based on their analysis of the Paleogene mammals of China. Finally, *Eudinoceras* co-occurs with *Trogosus* (= *Kuanchuanius*: Lucas 1993), *Uintatherium* and *Propalaeotherium parvulum* (= *P. sinense*) in the Guanzhuang Formation in Shandong. This supports a Bridgerian LMA/Lutetian Stage equivalence of the *Eudinoceras*-bearing horizons in Asia, a correlation consistent with Reshetov's (1979) correlation based on his study of the early Cenozoic tapiroids of Asia.

Nevertheless, the correlation proposed here suggests that the evolutionary turnover event apparently took place later in Asia than in Europe and North America. The review of the Asian record above indicates just how patchy mammalian distribution is across the P/E Series boundary. A major hiatus between Nongshanian and Bumbanian assemblages seems highly likely in the sections where they are superposed: the Nemegt and Ulan-Nur basins in Mongolia, the Turpan Basin in Xinjiang Province, and the Hengyang Basin in Hunan Province. Although the hiatus is probably real and corresponds to an unconformity, a lack of collecting cannot be excluded. Furthermore, the turnover event recorded at the base of the Bumbanian LMA is very depauperate compared to the diversity of taxa that first appear in Europe and North America; this underscores the limitations of the Asian record. Magnetostratigraphy would easily resolve this problem.

Comparison to Magnetostratigraphy, Radioisotopic Dating, and Chemostratigraphy

The P/E Series (Thanetian/Ypresian Stage) boundary is placed within Chron C24r (a relatively long interval of reversed polarity) on the global polarity time scale (e.g., Berggren et al. 1995). Magnetostratigraphy of relevant, mammal-bearing strata has been established in Western Europe and North America (Bighorn and Clark's Fork basins, Ellesmere Island, Baja California, Mexico, and Big Bend, Texas). The North American magnetostratigraphies (see above) indicate that the middle Clarkforkian LMA to late Graybullian SLMA spans Chron C24r. An interval with normal polarity in the lower Clarkforkian LMA is correlated to Chron C25n. The upper Graybullian "*Bunophorus* Interval Zone" contains the next normal polarity interval, and that is correlated to Chron C24n.3n. The North American magnetostratigraphy thus constrains the P/E Series boundary to the middle Clarkforkian–middle Graybullian LMAs and is consistent with the correlation presented here.

The radioisotopic chronology of Paleogene ash beds of northwestern Europe (e.g., Knox and Morton 1983,

1988) reviewed by Berggren and Aubry (1996) forms the basis of the numerical chronology used in figure 21.31. Approximate estimated ages (see Berggren and Aubry 1996 for the method) are base of Neustrian LMA = 55.8 Ma, base of Ypresian Stage = 54.6 Ma, and top of Neustrian LMA = 53 Ma.

Krishtalka et al. (1987:table 4.1) reviewed radioisotopic ages relevant to the P/E Series transition in North America. They assigned the Wasatchian LMA a range of about 56.5–57.5 to 51 Ma. Note, however, that no actual North American isotopic dating constrains the older limit, which was determined through magnetostratigraphic correlation to the SGCS. The younger limit of the Wasatchian LMA, however, is reasonably well constrained by numerous lower Bridgerian dated tuffs.

Two K-Ar dates of 54.0 ± 2.4 Ma and 54.3 ± 0.7 Ma from a tuff near the base of the Dawson Arkose in the Denver Basin, Colorado (Soister and Tschudy 1978) lie below so-called early Eocene pollen, but no fossil mammals of Clarkforkian or Wasatchian age are known there. Wing et al. (1991) reported a $^{40}Ar/^{39}Ar$ date of 52.8 ± 0.16 Ma on a tuff in the upper Willwood Formation of early Lostcabinian age. This date does not contradict the correlation advocated here. However, it does indicate that Clyde et al.'s (1994) speculation that the base of the Lostcabinian SLMA lies in Chron C24n.1n is improbable; the base of the Lostcabinian SLMA should correlate with Chron C24n.2n. The Lysitean SLMA thus appears to represent a very short time interval.

The only relevant radioisotopic dates from Asia place little or no constraint on this correlation. Thus, Gabuniya et al. (1975) reported a K/Ar age of 55–56 Ma (= 52.5–54.7 Ma when corrected to new IUGS constants: Dalrymple 1979) for extrusives that underlie the Toruaygyr mammalian Assemblage in Kirgyzstan. The Toruaygyr Assemblage includes *Gobiatherium mirificum*, brontotheres, *Lophialetes expeditus*, *Teleolophus* sp., hyracodontids, and amynodontids (Beliayeva et al. 1974; Reshetov 1979). It is clearly of Irdinmanhan age, and the date thus provides a range of maximum ages for the Irdinmahan LMA and little more.

The date of 50.1 ± 3 Ma (new IUGS) reported by Gabuniya et al. (1975) from the unnamed formation ("Dalan-Dzadgad Beds") above the Khashat Formation in Mongolia is surprisingly young. Clearly, if reliable, it only places a minimum age on the Nongshanian mammal Assemblage at Khashat, and there must be a major temporal gap between the Gashato mammalian fauna and the "Dalan-Dzadgad Beds." Russell and Zhai (1987) considered the date unreliable.

Correlation of mammalian biostratigraphy to the carbon isotope excursion recorded in deep-sea sites (Aubry et al. 1996) is direct because apparently the same carbon isotope excursion has been documented from the lowermost Wasatchian (Zone Wa 0) in the Bighorn Basin (Koch et al. 1992). In the deep-sea sites, the excursion is in the middle part of nannoplankton Zone NP9, so it is of late Paleocene age. This means that the lowermost Graybullian SLMA (Zone Wa0) also is late Paleocene in age.

A Nonmarine Global Stratotype Section and Point (GSSP) for the P/E Series Boundary

Mammalian biostratigraphy provides robust correlations across the P/E Series boundary in the nonmarine strata of Western Europe, North America, and Asia. These correlations are consistent with magnetostratigraphy, radioisotopic dating, and chemostratigraphy. The P/E Series boundary (= base of the Ypresian Stage) is within the Neustrian (= "Sparnacian") LMA in Europe, within the Graybullian interval of the Wasatchian LMA in North America and in the younger part of the Nongshanian LMA in Asia. In Europe and North America, a major turnover event in mammalian evolution is recorded at the base of the Neustrian and the Wasatchian LMAs that predates the P/E Series boundary. In Asia, the same turnover appears to postdate the P/E Series boundary, but probably this is due to the incompleteness of the Asian record.

By definition, the base of the Eocene is the base of the Ypresian Stage on the SGCS. However, a GSSP has not yet been established for the P/E Series boundary. That GSSP may be placed so as to coincide with the base of the Ypresian Stage, or it may correspond to the nannoplankton NP9/NP10 zonal boundary, the planktonic foraminiferal P5/P6 zonal boundary, or the carbon isotope excursion. These are the favored GSSP candidates (Berggren and Aubry 1996), and all of them are younger than the mammalian turnover event. When a GSSP is established for the P/E Series boundary, that boundary will remain within the Neustrian, Wasatchian, and Nongshanian LMAs.

When a GSSP for the P/E Series boundary is established in marine strata, I urge that a "reference" GSSP be established in nonmarine strata. The European and Asian records of nonmarine strata and fossils across the P/E Series boundary are too patchy to provide a suitable GSSP. The most suitable candidate for such a GSSP is in North America in the Bighorn–Clark's Fork Basin of Wyoming–Montana.

ACKNOWLEDGMENTS

NSF Grant DEB-7919681 funded the early phases of this research. Subsequent funding was provided by the National Geographic Society and the New Mexico

Museum of Natural History Foundation. Numerous colleagues aided my studies of coryphodontid and other mammalian fossils on three continents. My understanding of relevant field problems has benefited from the collaboration of many, especially S. M. Cather, R. J. Emry, and T. E. Williamson. M.-P. Aubry, W. A. Berggren, R. J. Emry, J. A., Schiebout, and S. Ting provided helpful reviews of the manuscript. I dedicate this article to the memory of the late Minchen Chow, who opened the door to the world of Chinese Paleogene mammals for me.

REFERENCES

Archibald, J. D., Clemens, W. A., Gingerich, P. D., Krause, D. W., Lindsay, E. H. and Rose, K. D., 1987. First North American Land Mammal Ages of the Cenozoic Era. In Woodburne, M. O. (ed.), *Cenozoic Mammals of North America: Geochronology and Biostratigraphy,* University of California Press, Berkeley, Los Angeles, London, p. 24–76.

Aubry, M.-P., 1985. Northwestern European Paleogene magnetostratigraphy, biostratigraphy, and paleogeography: Calcareous nannofossil evidence. *Geology,* v. 13, p. 198–202.

Aubry, M.-P., Berggren, W. A., Stott, L. and Sinha, A., 1996. The upper Paleocene–lower Eocene stratigraphic record and the Paleocene–Eocene boundary carbon isotope excursion: Implications for geochronology. In Knox, R. W. O'B., Corfield, R. M. and Dunay, R. E. (eds.), *Correlation of the early Paleogene in Northwest Europe,* Geological Society Special Publication No 101, p. 353–380.

Averianov, A. 1995. Nyctitheriid insectivores from the upper Paleocene of southern Kazakhstan (Mammalia: Lipotyphla). *Senckenbergiana Lethaea,* v. 75, p. 215–219.

Badamgarav, D. and Reshetov, V. Y., 1985. Paleontologiya i stratigrafiya Paleogena Zaaltayskoy Gobi [Paleontology and stratigraphy of the Zaaltay Gobi Paleogene]. *Sovmestnaya Sovetskaya–Mongolskaya Paleontologicheskaya Ekspeditsiya Trudy,* v. 25, 104 pp.

Baltz, E. H., Jr., 1967. *Stratigraphy and regional tectonic implications of part of Upper Cretaceous and Tertiary rocks, east-central San Juan Basin, New Mexico.* U.S. Geological Survey Professional Paper 552, 101 pp.

Beard, K. C. and Tabrum, A. R., 1990. The first early Eocene mammal from eastern North America: An omomyid primate from the Bashi Formation, Lauderdale County, Mississippi. *Mississippi Geology,* v. 11, p. 1–6.

Beard, K. C. and Wang, J., 1995. The first Asian plesiadapoids (Mammalia: Primatomorpha). *Annals of the Carnegie Museum,* v. 64, p. 1–33.

Beard, K. C., Dawson, M. R. and Tabrum, A. R., 1995. First diverse land mammal fauna from the early Cenozoic of the southeastern United States: The early Wasatchian Red Hot local fauna, Lauderdale County, Mississippi. *Journal of Vertebrate Paleontology,* v. 15 (supplement to no. 3), p. 18A. (Abstracts of Papers, 55th Annual Meeting of the Society of Vertebrate Paleontology, Pittsburgh, Pennsylvania).

Beard, K. C., Krishtalka, L. and Stucky, R. K., 1992. Revision of the Wind River faunas, early Eocene of central Wyoming. Pt. 12. New species of omomyid primates (Mammalia: Primates: Omomyidae) and omomyid taxonomic composition across the early–middle Eocene boundary. *Annals of the Carnegie Museum,* v. 61, p. 39–62.

Beliayeva, E. I., Trofimov, V. A. and Reshetov, V. J., 1974. Osnovye ehtapy ehvolyutsii mlekopitatushchikh v pozdnem Mesozoe–Paleogene Tsentralnoy Azii [General stages in the evolution of late Mesozoic–Paleogene mammalian faunas in central Asia]. *Sovmestnaya Sovetsko–Mongolskaya Paleontologicheskaya Ekspeditsiya Trudy,* v. 1, p. 19–45.

Berggren, W. A. and Aubry, M.-P., 1996. A late Paleocene–early Eocene NW European and North Sea magnetobiochronological correlation network. In Knox, R. W. O'B., Corfield, R. M. and Dunay, R. E. (eds.), *Correlation of the Early Paleogene in Northwest Europe,* Geological Society Special Publication No. 101, p. 309–352.

Berggren W. A., Kent, D. V., Swisher, C. C., III and Aubry, M.-P., 1995. A revised Cenozoic geochronology and chronostratigraphy. In Berggren, W. A., Kent, D. V., Aubry, M.-P. and Hardenbol, J. (eds.), *Geochronology Time Scales and Global Stratigraphic Correlation,* Society of Economic Paleontologists and Mineralogists Special Publication No. 54, p. 129–212.

Blondeau, A., Cavelier, C., Feugueur, L. and Pomerol, C., 1966. Stratigraphie du Paléogène du Bassin de Paris en relation avec les bassins avoisinants. *Bulletin de la Société Géologique de France,* ser. 7, v. 7, p. 200–221.

Borisov, B. A., 1963. Stratigrafiya verkhnevo myela i paleogyen–nyeogena Zaisanskoi vpadiny [Stratigraphy of the Upper Cretaceous and Paleogene–Neogene of the Zaysan Basin]. *Trudy vsesoyuznovo nauchno-Issledovatelskovo Geologicheskovo Instituta,* v. 94, p. 11–75.

Bott, A., 1875. Tertiary mammalian remains at Dulwich. *The Geologist,* v. 5, p. 302.

Bown, T. M., 1979. Geology and mammalian paleontology of the Sand Creek facies, lower Willwood Formation (lower Eocene), Washakie County, Wyoming. *Wyoming Geological Survey Memoir* 2, 151 pp.

Bown, T. M.,1980. The Willwood Formation (lower Eocene) of the southern Bighorn Basin, Wyoming, and its mammalian fauna. In Gingerich, P. D. (ed.), *Early Cenozoic paleontology and stratigraphy of the Bighorn Basin, Wyoming. 1880–1980,* University of Michigan Papers on Paleontology, v. 24, p. 127–138.

Bown, T. M., 1982. *Geology, paleontology, and correlation of Eocene volcaniclastic rocks, southeast Absaroka Range, Hot Springs County, Wyoming.* Geological Survey Professional Paper 1201-A, 75 pp.

Bown, T. M., Holroyd, P. A. and Rose, K. D., 1994. Mammal extinctions, body size, and paleotemperature. *Proceedings of the National Academy of Sciences of the United States,* v. 91, p. 10,403–10,406.

Bown, T. M., Rose, K. D., Simons, E. L. and Wing, S. L., 1994. *Distribution and stratigraphic correlation of upper Paleocene and lower Eocene fossil mammal and plant localities of the Fort Union, Willwood and Tatman Formations, southern Bighorn Basin, Wyoming.* U.S. Geological Survey Professional Paper 1540, 103 pp.

Bown, T. M. and Schankler, D., 1982. A review of the Proteutheria and Insectivora of the Willwood Formation (lower Eocene), Bighorn Basin, Wyoming. *U.S. Geological Society Bulletin* 1523, p. 1–79.

Bradley, W. H., 1964. *Geology of Green River Formation and associated Eocene rocks in southwestern Wyoming and adjacent parts of Colorado and Utah.* U.S. Geological Survey Professional Paper 496-A, 309 pp.

Butler, R. F., Gingerich, P. D. and Lindsay, E. H., 1981. Magnetic polarity stratigraphy and biostratigraphy of Paleocene and lower Eocene continental deposits, Clark's Fork Basin, Wyoming. *Journal of Geology,* v. 89, p. 299–316.

Cashion, W. B., 1969. *Geologic map of the Black Cabin Gulch quadrangle, Rio Blanco County, Colorado.* U.S. Geological Survey Map GQ-312.

Cavelier, C. and Pomerol, C., 1986. Stratigraphy of the Paleogene. *Bulletin de la Société Géologique de France,* ser. 8, v. 2, no. 2, p. 255–265.

Chapin, C. E. and Cather, S. M., 1981. Eocene tectonics and sedimentation in the Colorado Plateau–Rocky Mountain area. *Arizona Geological Society Digest,* v. 14, p. 173–198.

Cheng, J. and Ma, A., 1990. The new mammalian materials from the Eocene of the Liguanqiao basin in the Qinling Mountains. *Vertebrata PalAsiatica,* v. 28, p. 228–243.

Chow, M., 1959. Eocene vertebrates from Sinyu, Kiangsi Province. *Vertebrata PalAsiatica,* v. 3, p. 106.

Chow, M. and Li, C., 1963. A fossil of *Homogalax* from the Eocene of Shantung. *Scientia Sinica,* v. 12, p. 1411–1412.

Chow, M. and Li, C., 1965. *Homogalax* and *Heptodon* of Shantung. *Vertebrata PalAsiatica,* v. 9, p. 15–21.

Chow, M. and Qi, T., 1978. Paleocene mammalian fossils from Nomogen Formation of Inner Mongolia. *Vertebrata PalAsiatica,* v. 16, p. 77–85.

Chow, M. and Qi, T., 1982. Mammals from the middle Eocene Kuanchuang Formation of Sintai, Shandong. *Vertebrata PalAsiatica,* v. 20, p. 302–313.

Chow, M. and Tung, Y., 1962. Notes on some new uintathere materials of China. *Vertebrata PalAsiatica,* v. 6, p. 368–374.

Chow, M. and Tung, Y., 1965. A new coryphodont from Sinyu, Kiangsi. *Vertebrata PalAsiatica,* v. 9, p. 114–121.

Clyde, W. C., Stamatakos, J. and Gingerich, P. D., 1994. Chronology of the Wasatchian Land-Mammal Age (early Eocene): Magnetostratigraphic results from the McCullough Peaks Section, northern Bighorn Basin, Wyoming. *Journal of Geology,* v. 102, p. 367–377.

Collinson, M. E. and Hooker, J. J., 1987. Vegetational and mammalian faunal changes in the Early Tertiary of southern England. In Friis, E. M., Chaloner, W. G. and Crane, P. R. (eds.), *The Origins of Angiosperms and their Biological Consequences,* Cambridge: Cambridge University Press, p. 259–304.

Cope, E. D., 1874. Notes on the Eocene and Pliocene lacustrine formations of New Mexico, including descriptions of certain new species of vertebrates. *Annual Report of the Geographical Survey West of the 100th Meridian,* p. 115–130.

Cope, E. D., 1877a. Report upon the extinct Vertebrata obtained in New Mexico by parties of the expedition of 1874. *Geographical Survey West of the 100th Meridian,* v. 2, 370 pp.

Cope, E. D., 1877b. The Suessonian fauna in North America. *The American Naturalist,* v. 11, p. 95–99.

Crane, P. R., Manchester, S. R. and Dilcher, D. L., 1991. Reproductive and vegetative structure of *Nordenskioldia* (Trochodendraceae), a vesselless dicotyledon from the early Tertiary of the northern hemisphere. *American Journal of Botany,* v. 78, p. 1311–1334.

Curry, D., 1965. The Palaeogene beds of South East England. *Proceedings of the Geologists' Association,* v. 76, p. 151–173.

Dalrymple, G. B., 1979. Critical tables for conversion of K-Ar ages from old to new constants. *Geology,* v. 7, p. 558–560.

Dashzeveg, D., 1977. O pervoy nakhodkye *Hyopsodus* Leidy 1870 (Mammalia, Condylarthra) v Mongolskoy Narodnoy Respublikye [On the first discovery of *Hyopsodus* Leidy 1870 (Mammalia, Condylarthra) in the Mongolian People's Republic]. *Sovmestnaya Sovyestokovo–Mongolskaya Paleontologicheskaya Ekspeditsiya Trudy,* v. 4, p. 7–13.

Dashzeveg, D., 1979a. Nakhodka girakoteriya v Mongolii [The find of hyracothere in Mongolia]. *Paleontologicheskiy Zhurnal,* v. 3, p. 108–113.

Dashzeveg, D., 1979b. Nakhodka *Homogalax* (Perissodactyla, Tapiroidea) v Mongolii i evo stratigraficheskoye znacheniye [Discovery of *Homogalax* (Perissodactyla, Tapiroidea) in Mongolia and its stratigraphic significance]. *Byulletin Moskovskovo Obshchestvo Ispytately Prirody Otdel Geologicheskiy,* v. 54, p. 105–111.

Dashzeveg, D., 1982a. A revision of the Prodinoceratinae of central Asia and North America. *Paleontological Journal,* v. 16, p. 91–99.

Dashzeveg, D., 1982b. La faune de mammifères du Paléogène inférieur de Naran-Bulak (Asie centrale) et ses corrélations avec l'Europe et l'Amérique du Nord. *Bulletin de la Société Géologique de France,* ser. 7, v. 24, no. 2, p. 275–281.

Dashzeveg, D., 1988. Holarctic correlation of non-marine Palaeocene–Eocene boundary strata using mammals. *Journal of the Geological Society, London,* v. 145, p. 473–478.

Dashzeveg, D., 1990. The earliest rodents (Rodentia, Ctenodactyloidea) of Central Asia. *Acta Zoologica Cracoviana,* v. 33, p. 11–35.

Dashzeveg, D. and McKenna, M. C., 1977. Tarsioid primate from the early Tertiary of the Mongolian People's Republic. *Acta Palaeontologica Polonica,* v. 22, p. 119–137.

Davis, A. G. and Elliott, G. F., 1958. The paleogeography of the London Clay sea. *Proceedings of the Geologists' Association,* v. 68, p. 255–277.

Dawson, M., Krishtalka, L. and Stucky, R. K., 1990. Revision of the Wind River faunas, early Eocene of central Wyoming. Part 9. The oldest hystricomorphous rodent (Mammalia, Rodentia). *Annals of the Carnegie Museum,* v. 59, p. 135–148.

Delson, E., 1971. Fossil mammals of the early Wasatchian Powder River local fauna, Eocene of northeast Wyoming. *Bulletin of the American Museum of Natural History,* v. 146, p. 305–364.

Denison, R. H., 1937. Early lower Eocene mammals from the Wind River Basin, Wyoming. *Proceedings of the New England Zoological Club,* v. 16, p. 11–14.

Depéret, C., 1906. Relations stratigraphiques des faunes de Cernay et de Meudon au Mont de Berru. *Bulletin de la Société Géologique de France,* sér. 4, v. 6, p. 442–443.

Depéret, C., 1907. Sur les progrés récents des connaissances sur les terrains tertiaires inférieurs des environs de Reims. *Bulletin de l'Association Française pour l'Avancement des Sciences,* v. 6, p. 8–13.

Dickinson, W. R., Klute, M. A., Hayes, M. J., Janecke, S. U., Lundin, E. R., McKittrick, M. A. and Olivares, M. D., 1988. Paleogeographic and paleotectonic setting of Laramide sedimentary basins in the central Rocky Mountain region. *Geological Society of America Bulletin,* v. 100, p. 1023–1039.

Dollfus, G. F., 1880. Essai sur l' étendue des terrains tertiaires dans le bassin anglo-parisien. *Bulletin de la Société géologique de Normandie,* v. 6, p. 584–605.

Dollfus, G. F., 1905. Critique de la classification de l'Eocène inférieur (lettre à M. Leriche). *Annales de la Société Géologique de Normandie,* v. 34, p. 373–382.

Dollo, L. 1909. The fossil vertebrates of Belgium. *Annals of the New York Academy of Science,* v. 19, p. 99–119.

Dollo, L. and Teilhard de Chardin, P., 1924. Les gisements de mammifères paléocènes de la Belgique. *Quarterly Journal of the Geological Society of London,* v. 80, p. 12–16.

Dong, Z., 1977. On the dinosaurian remains from Turpan, Xinjiang. *Vertebrata PalAsiatica,* v. 15, p. 59–66.

Donnell, J. R., 1969. Paleocene and lower Eocene units in the southern part of the Piceance Creek Basin, Colorado. *U.S. Geological Survey Bulletin 1274-M,* 18 pp.

Dorr, J. A., Jr., 1952. Early Cenozoic stratigraphy and vertebrate paleontology of the Hoback Basin, Wyoming. *Bulletin of the Geological Society of America,* v. 63, p. 59–94.

Dorr, J. A., Jr., 1956. Post-Cretaceous geologic history in the Hoback Basin area, central-western Wyoming. *Wyoming Geological Association Guidebook* 11, p. 99–108.

Dorr, J. A., Jr., 1958. Early Cenozoic vertebrate paleontology, sedimentation, and orogeny in central western Wyoming. *Bulletin of the Geological Society of America,* v. 69, p. 1217–1244.

Dorr, J. A., Jr., 1969. Mammalian and other fossils, early Eocene Pass Peak Formation, central western Wyoming. *Contributions of the Museum of Paleontology, The University of Michigan,* v. 22, p. 207–219.

Dorr, J. A., Jr. and Gingerich, P. D., 1980. Early Cenozoic mammalian paleontology, geologic structure, and tectonic history in the over-thrust belt near LaBarge, western Wyoming. *University of Wyoming, Contributions to Geology,* v. 18, p. 101–115.

Dorr, J. A., Jr. and Steidtmann, J. R., 1971. Stratigraphic–tectonic implications of a new, earliest Eocene, mammalian faunule from central western Wyoming. *Michigan Academician,* v. 3, p. 25–41.

Dorr, J. A., Jr., Spearing, D. R. and Steidtmann, J. R., 1977. *Deformation and deposition between a foreland uplift and an impinging thrust belt: Hoback Basin, Wyoming.* Geological Society of America, Special Paper 177, 82 pp.

Du, H., Ma, A., Cheng, J. and Wu, W., 1992. Palaeogene biostratigraphic characteristics of red basins in western Henan and its adjacent areas, with an outline of Palaeogene palaeobiogeographic provinces of China. *Acta Geologica Sinica,* v. 5, p. 101–117.

Fahlbusch, V., 1976. Report on the International Symposium on Mammalian Stratigraphy of the European Tertiary. *Newsletters on Stratigraphy,* v. 5, p. 160–167.

Feugueur, L., 1963. L'Yprésien du Bassin de Paris. Essai de monographie stratigraphique. *Mémoire du Service de la Carte Géologique de France, Paris,* 568 pp.

Flynn, J. J., 1986. Correlation and geochronology of middle Eocene strata from the western United States. *Palaeogeography, Palaeoclimatology, Palaeoecology,* v. 55, p. 335–406.

Flynn, J. J., Cipolletti, R. M. and Novacek, M. J., 1989. Chronology of early Eocene marine and terrestrial strata, Baja California, Mexico. *Bulletin of the Geological Society of America,* v. 101, p. 1182–1196.

Flynn, J. J., MacFadden, B. J. and McKenna, M. C., 1984. Land-Mammal Ages, faunal heterochrony, and temporal resolution in Cenozoic terrestrial sequences. *Journal of Geology,* v. 92, p. 687–705.

Forster Cooper, C., 1932. On some mammalian remains from the lower Eocene of the London Clay. *Annals and Magazine of Natural History* (10), v. 9, p. 458–467.

Franzen, J. L. and Haubold, H., 1986. The middle Eocene of European mammalian stratigraphy. Definition of the Geiseltalian. *Modern Geology,* v. 10, p. 159–170.

Frederiksen, N. O., Gibson, T. G. and Bybell, L. M., 1982. Paleocene–Eocene boundary in the eastern Gulf Coast. *Transactions of the Gulf Coast Association of Geological Societies,* v. 32, p. 289–294.

Froehlich, J. W. and Reser, P. K., 1981. First occurrence of *Homogalax* (Mammalia; Perissodactyla; Tapiroidea) in the Regina Member of the San Jose Formation, San Juan Basin, New Mexico. In Lucas, S. G., Rigby, J. K., Jr. and Kues, B. S. (eds.), *Advances in San Juan Basin Paleontology:* Albuquerque: University of New Mexico Press, p. 293–303.

Gabuniya, L. K., 1961. Obailinskaya fauna: dryevnyeishii komplyeks iskopayemykh mlyekopitayushchikh SSSR [Obayla fauna: the oldest complex of fossil mammals from the USSR]. *Soobshcheniya Akademiya Nauk Gruzinskoi SSR,* v. 27, p. 711–714.

Gabuniya, L. K., 1962. Ob ostatkakh mlekopitayushikh iz Obailinskoi svity Zaisanskoi Depressii [On the remains of mammals from the Obailan Svita of the Zaisan Basin]. *Akademia Nauk Gruzinskoi SSSR, Trudy Instituta Paleobiologii,* v. 7, p. 15–28.

Gabuniya, L. K., 1977. Contribution à la connaissance des mammifères Paléogènes du Bassin de Zaissan (Kazakhstan Central). *Géobios, Mémoire Spécial* 1, p. 29–37.

Gabuniya, L. K., 1984. Kratkii obzor faun palyeogyenovykh mlyekopitayushchikh Zaisanskoi vpadiny [Short report on the Paleogene mammals of the Zaysan Basin]. In Gabuniya, L. K. (ed.), *Flora i fauna Zaisanskoi vpadiny* [*Fauna and Flora of the Zaysan Basin*], Tblisi: Akademiya Nauk Gruzinskoi SSR, p. 115–123.

Gabuniya, L. K., Devyatkin, E. V. and Rubensteyn, M. N., 1975. Danniye ob absolyutnum bozraste kainozoiyskikh kontinentalniykh obrazom iiy Azzi i ikh biostratigraficheskoe znachenie [Data on the absolute age of the Cenozoic continental formations of Asia and their biostratigraphic significance]. *Doklady Akademia Nauk SSSR,* v. 225, p. 895–897.

Gazin, C. L., 1952. The lower Eocene Knight Formation of western Wyoming and its mammalian faunas. *Smithsonian Miscellaneous Collections,* v. 117, p. 1–82.

Gazin, C. L., 1959. Paleontological exploration and dating of the early Tertiary deposits in basins adjacent to the Uinta Mountains. *Intermountain Association of Petroleum Geologists Guidebook* 10, p. 133–135.

Gazin, C. L., 1962. A further study of the lower Eocene mammalian faunas of southwestern Wyoming. *Smithsonian Miscellaneous Collections,* v. 144, p. 1–98.

Gingerich, P. D., 1976. Cranial anatomy and evolution of early Tertiary Plesiadapidae (Mammalia, Primates). *University of Michigan Papers on Paleontology,* v. 15, 141 pp.

Gingerich, P. D., 1980a. Evolutionary patterns in early Cenozoic mammals. *Annual Review of Earth and Planetary Sciences,* v. 8, p. 407–424.

Gingerich, P. D. (ed.), 1980b. Early Cenozoic Paleontology and Stratigraphy of the Bighorn Basin, Wyoming. *University of Michigan Papers on Paleontology,* v. 24, 146 pp.

Gingerich, P. D., 1983. Paleocene–Eocene faunal zones and a preliminary analysis of Laramide structural deformation in the Clark's Fork Basin, Wyoming. *Wyoming Geological Association Guidebook* 34, p. 185–195.

Gingerich, P. D., 1989. New earliest Wasatchian mammalian fauna from the Eocene of northwestern Wyoming: Composition and diversity in a rarely sampled high-floodplain assemblage. *University of Michigan Papers on Paleontology,* v. 28, 97 pp.

Gingerich, P. D., 1991. Systematics and evolution of early Eocene Perissodactyla (Mammalia) in the Clark's Fork Basin, Wyoming. *Contributions from the Museum of Paleontology, The University of Michigan,* v. 28, p. 181–213.

Gingerich, P. D. and Gunnell, G. F., 1979. Systematics and evolution of the genus *Esthonyx* (Mammalia, Tillodontia) in the early Eocene of North America. *Contributions from the Museum of Paleontology, The University of Michigan,* v. 25, p. 125–153.

Gingerich, P. D. and Rose, K. D., 1977. Preliminary report on the American Clark's Fork mammal fauna and it correlation with similar faunas in Europe and Asia. *Geobios, Mémoire Special* 1, p. 39–45.

Gingerich, P. D. and Simons, E. L., 1977. Systematics, phylogeny and evolution of early Eocene Adapidae (Mammalia, Primates) in North America. *Contributions from the Museum of Paleontology, The University of Michigan,* v. 24, p. 245–279.

Gingerich, P. D., Rose, K. D. and Krause, D. W., 1980. Early Cenozoic mammalian faunas of the Clark's Fork Basin–Polecat Bench area, northwestern Wyoming. In Gingerich, P. D. (ed.), *Early Cenozoic Paleontology and Stratigraphy of the Bighorn Basin, Wyoming: 1880–1980, University of Michigan Papers on Paleontology,* v. 24, p. 51–68.

Givens, C. R., 1974. Eocene molluscan biostratigraphy of the Pine Mt. area, Ventura County, California. *University of California Publications in Geological Science,* v. 109, p. 1–107.

Godinot, M., 1982. Aspects nouveaux des échanges entre les faunes mammaliennes d'Europe et d'Amérique du Nord à la base de l'Eocène. *Geobios, Mémoire Special* 6, p. 403–412.

Godinot, M., 1987. Mammalian reference levels MP1–10. *Münchner Geowissenschaftliche Abhandlungen A,* v. 10, p. 21–23.

Godinot, M., de Broin, F., Buffetaut, E., Rage, J. C. and Russell, D., 1978. Dormaal: une des plus anciennes faunes éocènes d'Europe. *Comptes-Rendus Hebdomadaires des Séances de l'Académie des Sciences de Paris,* ser. D, v. 287, p. 1273–1276.

Gradzinski, R., Kazmierczak, J. and Lefeld, J., 1969. Geographical and geological data from the Polish–Mongolian paleontological expeditions. *Palaeontologia Polonica,* v. 19, p. 33–82.

Granger, W., 1910. Tertiary faunal horizons in the Wind River Basin, Wyoming, with descriptions of new Eocene mammals. *Bulletin of the American Museum of Natural History,* v. 28, p. 235–251.

Granger, W., 1914. On the names of lower Eocene faunal horizons of Wyoming and New Mexico. *Bulletin of the American Museum of Natural History,* v. 33, p. 201–207.

Gu, Z. and Chen, P., 1987. Paleogene paleogeography and distribution of vertebrate fossils in China. *Acta Palaeontologica Sinica,* v. 26, p. 210–221.

Gulinck, M., 1965. Aperçu général sur les dépôts éocènes de la Belgique. *Bulletin de la Société Géologique de France,* Sér. 7, v. 7, p. 222–227.

Gunnell, G. F., Bartels, W. S. and Gingerich, P. D., 1993. Paleocene-Eocene boundary in continental North America: Biostratigraphy and geochronology, northern Bighorn Basin, Wyoming. In Lucas, S. G. and Zidek, J. (eds.), *Vertebrate Paleontology in New Mexico,* New Mexico Museum of Natural History and Science Bulletin 2, p. 137–144.

Guthrie, D. A., 1967. The mammalian fauna of the Lysite Member, Wind River Formation (early Eocene) of Wyoming. *Southern California Academy of Science Memoirs,* v. 5, 53 pp.

Guthrie, D. A., 1971. The mammalian fauna of the Lost Cabin Member, Wind River Formation (lower Eocene) of Wyoming. *Annals of the Carnegie Museum of Natural History,* v. 43, p. 47–113.

Hail, W. J., Jr., 1973. Geologic map of the Smizer Gulch quadrangle, Rio Blanco and Moffat Counties, Colorado. U.S. Geological Survey Map GQ-1131.

Hail, W. J., Jr., 1974. Geologic map of the Rough Gulch quadrangle, Rio Blanco and Moffat Counties, Colorado. U.S. Geological Survey Map GQ-1195.

Hartnell, J. A., 1980a. *The vertebrate paleontology, depositional environment and sandstone provenance of early Eocene rocks on Tornillo Flat, Big Bend National Park, Brewster County, Texas.* Master of Science thesis, Baton Rouge: Louisiana State University, 174 pp.

Hartnell, J. A., 1980b. Fauna from the Eocene of the Hannold Hill Formation, Big Bend National Park, Brewster County, Texas. *Geological Society of America Abstracts with Programs,* v. 12, p. 179.

Haubold, H., 1989. Die Referentzfauna des Geiseltalium, MP levels 11 bis 13 (Mitteleozän, Lutetium). *Palaeovertebrata,* v. 19, p. 81–93.

Hébert, E., 1856. Recherches sur la faune des premiers sediments tertiaires parisiens (mammifères pachydermes du genre *Coryphodon*). *Annales des Sciences Naturelles,* Sér. 4, v. 6, p. 87–136.

Hester, S. W., 1965. Stratigraphy and palaeogeography of the Woolwich and Reading Beds. *Bulletin of the Geological Survey of Great Britain,* v. 23, p. 117–137.

Hickey, L. J., 1977. Stratigraphy and paleobotany of the Golden Valley Formation (early Tertiary) of western North Dakota. *Geological Society of America Memoir* 150, 181 pp.

Hickey, L. J., West, R. M., Dawson, M. R. and Choi, D. K., 1983. Arctic terrestrial biota: Paleomagnetic evidence of age disparity with mid-northern latitudes during the late Cretaceous and early Tertiary. *Science,* v. 221, p. 1153–1156.

Holroyd, P. A. and Ciochon, R. L., 1994. Relative ages of Eocene primate-bearing deposits of Asia. In Fleagle, J. G. and Kay, R. F. (eds.), *Anthropoid origins,* New York: Plenum Press, p. 123–141.

Hooker, J. J., 1980. The succession of *Hyracotherium* (Perissodactyla, Mammalia) in the English early Eocene. *Bulletin of the British Museum of Natural History (Geology),* v. 33, p. 101–114.

Hooker, J. J., 1984. A primitive ceratomorph (Perissodactyla, Mammalia) from the early Tertiary of Europe. *Zoological Journal of the Linnean Society,* v. 82, p. 229–244.

Hooker, J. J., 1989. Character polarities in early perissodactyls and their significance for *Hyracotherium* and infraordinal relationships. In Prothero, D. R. and Schoch, R. M. (eds.), *The evolution of the Perissodactyla,* Oxford University Press, New York, p. 79–101.

Hooker, J. J., 1991. The sequence of mammals in the Thanetian and Ypresian of the London and Belgian Basins. Location of the Palaeocene-Eocene boundary. *Newsletters on Stratigraphy,* v. 25, p. 75–90.

Hooker, J. J., 1994. The beginning of the equoid radiation. *Zoological Journal of the Linnean Society of London,* v. 112, p. 29–63.

Hooker, J. J., 1996. Mammalian biostratigraphy across the Palaeocene–Eocene boundary in the Paris, London and Belgian Basins. In Knox, R. W. O'B., Corfield, R. M. and Dunay, R. E. (eds.), *Correlation of the Early Paleogene in Northwest Europe,* Geological Society Special Publication No. 101, p. 205–218.

Hooker, J. J. and Insole, A. N., in Hooker, J. J., Insole, A. N., Moody, R. T. J., Walker, C. A. and Ward, D. J., 1980. The distribution of mammals in the English Paleogene. *Tertiary Research,* v. 3, p. 31–45.

Hooker, J. J. and Ward, D. J. in Hooker, J. J., Insole, A. N., Moody, R. T. J., Walker, C. A. and Ward, D. J., 1980. List of localities. *Tertiary Research,* v. 3, p. 3–12.

Ilinskaya, Y., 1974. Trochodendroides Berry, *Nordenskioldia* Heer, *Nyssidium* Heer. In Takhtayan, A. (ed.), *Magnoliophyta Fossilia URSS. I. Magnoliaceae-Eucommiaceae,* Leningrad: Nauka, p. 21–29.

Jepsen, G. L., 1963. Eocene vertebrates, coprolites and plants in the Golden Valley Formation of western North Dakota. *Geological Society of America Bulletin,* v. 74, p. 673–684.

Jerzykiewicz, T. and Russell, D. A., 1991. Late Mesozoic stratigraphy and vertebrates of the Gobi Basin. *Cretaceous Research,* v. 12, p. 345–377.

Keefer, W. R., 1965a. Stratigraphy and geologic history of the uppermost Cretaceous, Paleocene, and lower Eocene rocks in the Wind River Basin, Wyoming. *U.S. Geological Survey Professional Paper* 495-A, 77 pp.

Keefer, W. R., 1965b. Geologic history of Wind River Basin, central Wyoming. *Bulletin of the American Association of Petroleum Geologists,* v. 49, p. 1879–1892.

Keefer, W. R., 1969. General stratigraphy and depositional history of the Fort Union, Indian Meadows, and Wind River formations, Wind River Basin, Wyoming. *Wyoming Geological Association Guidebook* 21, p. 19–23.

Keefer, W. R., 1970. Structural geology of the Wind River Basin, Wyoming. *U.S. Geological Survey Professional Paper,* 495-D, 35 pp.

Kelley, D. R. and Wood, A. E., 1954. The Eocene mammals from the Lysite Member, Wind River Formation of Wyoming. *Journal of Paleontology,* v. 28, p. 337–366.

Kent, D. V., McKenna, M. C., Opdyke, N. D., Flynn, J. J. and MacFadden, B. J., 1984. Arctic biostratigraphic heterochroneity. *Science,* v. 224, p. 173–174.

Kihm, A. J., 1984. *Early Eocene mammalian faunas of the Piceance Creek basin northwestern Colorado.* Ph.D. dissertation, Boulder: University of Colorado, 381 pp.

Kitts, D. B. 1956. American *Hyracotherium* (Perissodactyla, Equidae). *Bulletin of the American Museum of Natural History,* v. 110, p. 1–60.

Klaasen, H. M., 1883. On a section of the lower London Tertiaries at Park Hill, Croydon. *Proceedings of the Geologists' Association,* v. 8, p. 226–249.

Knox, R. W. O'B. and Harland, R., 1979. Stratigraphical relationships of the early Palaeogene ash series of NW Europe. *Journal of the Geological Society of London,* v. 136, p. 463–470.

Knox, R. W. O'B. and Morton, A. C., 1983. Stratigraphical distribution of early Palaeogene pyroclastic deposits in the North Sea Basin. *Proceedings of the Yorkshire Geological Society,* v. 44, p. 355–363.

Knox, R. W. O'B. and Morton, A. C., 1988. The record of early Tertiary N Atlantic volcanism in sediments of the North Sea Basin. In Morton, A. C. and Parson, L. M. (eds.), *Early Tertiary Volcanism and the Opening of the NE Atlantic,* Geological Society Special Publication No 39, p. 407–419.

Koch, P. L., Zachos, J. C. and Gingerich, P. D., 1992. Correlation between isotope records in marine and continental carbon reservoirs near the Palaeocene/Eocene boundary. *Nature,* v. 358, p. 319–322.

Korth, W. W., 1982. Revision of the Wind River faunas, early Eocene of central Wyoming. Pt. 2. Geological setting. *Annals of the Carnegie Museum,* v. 51, p. 57–78.

Korth, W. W., 1984. Earliest Tertiary evolution and radiation of rodents in North America. *Bulletin of the Carnegie Museum of Natural History,* v. 24, p. 1–71.

Krause, D. W. and Maas, M. C., 1990. The biogeographic origins of late Paleocene–early Eocene mammalian immigrants to the Western Interior of North America. In Bown, T. M. and Rose, K. D. (eds.), *Dawn of the Age of Mammals in the Northern Part of the Rocky Mountain Interior, North America,* Geological Society of America Special Paper 243, p. 71–105.

Krishtalka, L. and Stucky, R. K., 1983. Revision of the Wind River faunas, early Eocene of central Wyoming. Part 3. Marsupialia. *Annals of the Carnegie Museum,* v. 52, p. 205–227.

Krishtalka, L. and Stucky, R. K., 1984. Middle Eocene marsupials (Mammalia) from northeastern Utah and the mammalian fauna from Powder wash. *Annals of the Carnegie Museum,* v. 53, p. 31–45.

Krishtalka, L. and Stucky, R. K., 1985. Revision of the Wind River faunas, early Eocene of central Wyoming. Part 7. Revision of *Diacodexis* (Mammalia, Artiodactyla). *Annals of the Carnegie Museum,* v. 54, p. 413–486.

Krishtalka, L., Stucky, R. K., West, R. M., McKenna, M. C., Black, C. C., Bown, T. M., Dawson, M. R., Golz, D. J., Flynn, J. J., Lillegraven, J. A. and Turnbull, W. D., 1987. Eocene (Wasatchian through Duchesnean) biochronology of North America. In Woodburne, M. O. (ed.), *Cenozoic Mammals of North America: Geochronology and Biostratigraphy,* University of California Press, Berkeley, Los Angeles, London, p. 77–117.

Kues, B. S., 1993. Bibliographic catalogue of New Mexico vertebrate fossils. In Lucas, S. G. and Zidek, J. (eds.), *Vertebrate Paleontology in New Mexico,* New Mexico Museum of Natural History and Science, Bulletin 2, p. 199–279.

Lei, Y., 1981. On the subdivision of the early Tertiary and its fauna in the Yuanqu Basin on the Henan–Shansi border. *Bulletin of the Yichang Institute of Geology and Mineral Resources of the Chinese Academy of Geological Sciences,* p. 125–134.

Lemoine, V., 1880. Notice géologique sur les environs de Reims. *Bulletin de la Société Géologique de France,* Sér. 3, v. 9, p. 344–345.

Lemoine, V., 1889. Etude d'ensemble sur les dents des mammifères fossiles des environs de Reims. *Bulletin de la Société Géologique de France,* Sér. 3, v. 19, p. 263–290.

Li, C. and Ting, D. 1983. The Paleogene mammals of China. *Bulletin of the Carnegie Museum of Natural History,* v. 21, p. 1–93.

Li, C., Chiu, C., Yan, D. and Hsieh, S., 1979. Notes on some early Eocene mammalian fossils of Hentung, Hunan. *Vertebrata PalAsiatica,* v. 17, p. 71–82.

Louis, P., 1963. Mammifères de l'Eocène inférieur des environs de Reims et d'Epernay. *80th Congrès de l'Association Francaise pour l'Avancement des Sciences, Reims* 1961, p. 85–90.

Louis, P. and Michaux, J., 1962. Présence de mammifères sparnaciens dans les sablières de Pourcy (Marne). *Comptes-Rendus Sommaires de la Société Géologique de France,* Sér. 7, v. 4, p. 170–172.

Love, J. D., 1939. Geology along the southern margin of the Absaroka Range, Wyoming. *Geological Society of America Special Paper* 20, p. 1–134.

Love, J. D., 1947. Tertiary stratigraphy of the Jackson Hole area, northwestern Wyoming. *U.S. Geological Survey Oil and Gas Investigations Preliminary Chart* 27.

Love, J. D.. 1978. Cenozoic thrust and normal faulting, and tectonic history of the Badwater area, northeastern margin of Wind River Basin, Wyoming. *Wyoming Geological Association Guidebook* 30, p. 235–238.

Lucas, S. G., 1977. Vertebrate paleontology of the San Jose Formation, east-central San Juan Basin, New Mexico. *New Mexico Geological Society Guidebook* 28, p. 221–225.

Lucas, S. G., 1982a. The phylogeny and composition of the order Pantodonta (Mammalia, Eutheria). *Third North American Paleontological Convention Proceedings,* v. 2, p. 337–342.

Lucas, S. G., 1982b. *Vertebrate paleontology, stratigraphy, and biostratigraphy of Eocene Galisteo Formation, north-central New Mexico.* New Mexico Bureau of Mines and Mineral Resources Circular 186, p. 1–34.

Lucas, S. G., 1984a. *Systematics, biostratigraphy and evolution of early Cenozoic Coryphodon (Mammalia, Pantodonta).* PhD. Dissertation, New Haven: Yale University, 648 pp.

Lucas, S. G., 1984b. Correlation of Eocene rocks of the northern Rio Grande rift and adjacent areas: Implications for Laramide tectonics. *New Mexico Geological Society Guidebook* 35, p. 123–128.

Lucas, S. G., 1986. *Coryphodon anthracoideus* (de Blainville, 1846) from the upper Paleocene of Orp-le-Grand, Belgium. *Bulletin de l'Institut Royal des Sciences Naturelles Belges (Science de la Terre),* v. 56, p. 373–381.

Lucas, S. G., 1989a. *Coryphodon* (Mammalia, Pantodonta) from the Hannold Hill Formation, Eocene of Trans-Pecos Texas. *The Pearce-Sellards Series,* no. 46, 16 pp.

Lucas, S. G., 1989b. Stratophenetic and holomorphologic taxonomy of fossil vertebrates. *Journal of Vertebrate Paleontology,* v. 9 (Supplement to no. 3), p. 30A (Abstracts of Papers, 49th Annual Meeting of the Society of Vertebrate Paleontology, Austin, Texas).

Lucas, S. G., 1989c. Fossil mammals and the Paleocene–Eocene boundary in Europe, North America and Asia. *28th International Geological Congress, Washington, D.C., Abstracts,* v. 2, p. 335.

Lucas, S. G., 1993. Pantodonts, tillodonts, uintatheres, and pyrotheres are not ungulates. In Szalay, F. S., Novacek, M. J. and McKenna, M. C. (eds.), *Mammal Phylogeny,* New York: Springer Verlag, p. 182–194.

Lucas, S. G. and Kues, B. S., 1979. Vertebrate biostratigraphy of the Eocene Galisteo Formation, north-central New Mexico. *New Mexico Geological Society Guidebook* 30, p. 225–229.

Lucas, S. G. and Tong, Y., 1987. A new coryphodontid (Mammalia, Pantodonta) from the Eocene of China. *Journal of Vertebrate Paleontology,* v. 7, p. 362–372.

Lucas, S. G. and Williamson, T. E., 1992. Fossil mammals and the early Eocene age of the San Jose Formation, San Juan Basin, New Mexico. *New Mexico Geological Society Guidebook* 43, p. 311–316.

Lucas, S. G. and Williamson, T. E., 1993a. Late Cretaceous–early Eocene vertebrate biostratigraphy and biochronology of the San Juan Basin, New Mexico. In Lucas, S. G. and Zidek, J. (eds.), *Vertebrate Paleontology in New Mexico,* New Mexico Museum of Natural History and Science, Bulletin 2, p. 93–104.

Lucas, S. G. and Williamson, T. E., 1993b. Eocene vertebrates and late Laramide stratigraphy of New Mexico. In Lucas, S. G. and Zidek, J.

(eds.), *Vertebrate Paleontology in New Mexico,* New Mexico Museum of Natural History and Science, Bulletin 2, p. 145–158.

Lucas, S. G., Kordikova, E. G. and Emry, R. J. 1995. Upper Cretaceous–Paleocene stratigraphy and vertebrate paleontology in the Zaysan Basin, Kazakhstan. *Journal of Vertebrate Paleontology,* v. 15 (Supplement to no. 3), p. 41A (Abstracts of papers, 55th Annual Meeting of the Society of Vertebrate Paleontology, Pittsburgh, Pennsylvania).

Lucas, S. G., Schoch, R. M., Manning, E. and Tsentas, C. 1981. The Eocene biostratigraphy of New Mexico. *Geological Society of America Bulletin,* v. 92, p. 951–967.

Lucas, S. G., Emry, R. J., Bayshashov, B. U. and Tyutkova, L. A., 1996. Paleogene mammalian biochronology, Zaysan Basin, Kazakhstan. *30th International Geological Congress, Beijing, Abstracts,* v. 2, p. 84.

Lucas, S. G., Cather, S. M., Abbott, J. C. and Williamson, T. E. 1997. Stratigraphy and tectonics of lower Tertiary strata in the Laramide Galisteo Basin, north-central New Mexico. *New Mexico Geology,* v. 19, p. 89–95.

Ma, A. and Cheng, J., 1991. On biostratigraphic sequences, newly collected mammalian taxa from old localities, additional measuring of four stratigraphic sections and most importantly discovery of several new Eocene localities. *Scientia Geologia Sinica Beijing,* v. 1, p. 21–29.

Maas, M. C., Anthony, M. R. L., Gingerich, P. D., Gunnell, G. F. and Krause, D. W., 1995. Mammalian generic diversity and turnover in the late Paleocene and early Eocene of the Bighorn and Crazy Mountains Basins, Wyoming and Montana (USA). *Palaeogeography, Palaeoclimatology, Palaeoecology,* v. 115, p. 181–207.

Malaquin, A. 1899. Le *Coryphodon gosseleti* (n. sp.) et la faune de l'Eocène inférieur (Landénien–Thanétian) de Vertain. *Annales de la Société Géologique du Nord,* v. 28, p. 257–270.

Mancini, E. A., 1981. Lithostratigraphy and biostratigraphy of Paleocene subsurface strata in southwest Alabama, *Transactions of the Gulf Coast Association of Geological Societies,* v. 31, p. 359–367.

Matthew, W. D., 1937. Paleocene faunas of the San Juan Basin, New Mexico. *Transactions of the American Philosophical Society,* v. 30, 510 pp.

Matthew, W. D. and Granger, W., 1925. Fauna and correlation of the Gashato Formation of Mongolia. *American Museum Novitates,* no. 189, 12 pp.

Matthew, W. D., Granger, W. and Simpson, G. G., 1928. Paleocene multituberculates from Mongolia. *American Museum Novitates,* no. 331, 4 pp.

Matthew, W. D., Granger, W. and Simpson, G. G., 1929. Additions to the fauna of the Gashato Formation of Mongolia. *American Museum Novitates,* no. 376, 12 pp.

McKenna, M. C. 1960. Fossil Mammalia from the early Wasatchian Four Mile fauna, Eocene of northwest Colorado. *University of California Publications in Geological Sciences,* v. 37, 130 pp.

McKenna, M. C., 1972. Was Europe directly connected to North America prior to the middle Eocene? *Evolutionary Biology,* v. 6, p. 179–188.

McKenna, M. C. 1975. Fossil mammals and early Eocene North Atlantic land continuity. *Annals of the Missouri Botanical Garden,* v. 62, p. 335–353.

McKenna, M. C., 1976. *Esthonyx* in the upper faunal assemblage, Huerfano Formation, Eocene of Colorado. *Journal of Paleontology,* v. 50, p. 354–355.

McKenna, M. C. 1980a. Eocene paleolatitude, climate, and mammals of Ellsemere Island. *Palaeogeography, Palaeoclimatology, Palaeoecology,* v. 30, p. 349–362.

McKenna, M. C. 1980b. Late Cretaceous and early Tertiary vertebrate paleontological reconnaissance, Togwotee Pass area, northwestern Wyoming. In Jacobs, L. L. (ed.), *Aspects of Vertebrate History: Essays in Honor of Edwin Harris Colbert,* Flagstaff: Museum of Northern Arizona Press, p. 321–343.

McKenna, M. C., Engelmann, G. F. and Barghoorn, S. F., 1977. Cranial anatomy and evolution of early Tertiary Plesiadapidae (Mammalia, Primates). Philip D. Gingerich, 1976. *University of Michigan Papers on Paleontology,* No. 15, p. 1–141, 12 pls., 42 figs. *Systematic Zoology,* v. 26, p. 233–238.

Miall, A. D., 1986. The Eureka Sound Group (Cretaceous–Oligocene), Canadian Arctic Islands. *Bulletin of Canadian Petroleum Geology,* v. 34, p. 240–270.

Michaux, J. 1963. Presence de *Coryphodon eocaenus* à Sinceny (Aisne) et sa signification stratigraphique. *Comptes-Rendus Sommaires de la Société Géologique de France,* Sér. 7, v. 5, p. 333–334.

Morris, W. J. 1954. An Eocene fauna from the Cathedral Bluffs Tongue of the Washakie Basin, Wyoming. *Journal of Paleontology,* v. 28, p. 195–203.

Morris, W. J., 1966. Fossil mammals from Baja California: New evidence on early Tertiary migrations. *Science,* v. 135, p. 1376–1378.

Morris, W. J., 1968. A new early Tertiary perissodactyl, *Hyracotherium seekinsi,* from Baja California. *Los Angeles County Museum Contributions to Science,* no. 151, p. 1–11.

Nessov, L. A. 1987. Results of search and investigation of Cretaceous and early Paleogene mammals on the territory of the USSR. *Ezhegodnik Vsesoyuznaya Paleontologicheskiy Obshchevstvo,* v. 30, p. 199–218.

Newton, E. T., 1883. Note on *Coryphodon* remains from the Woolwich beds of the Park Hill section, Croydon. *Proceedings of the Geologists' Association,* v. 8, p. 250–254.

Novacek, M. J., Flynn, J. J., Ferrusquía-Villafranca, I. and Cipolletti, R. M., 1987. An early Eocene (Wasatchian) mammal fauna from Baja California. *National Geographic Research,* v. 3, p. 376–388.

Novacek, M. J., Ferrusquía-Villafranca, I., Flynn, J. J., Wyss, A. R. and Norell, M. A., 1991. Wasatchian (early Eocene) mammals and other vertebrates from Baja California, Mexico: The Lomas las Tetas de Cabra fauna. *Bulletin of the American Museum of Natural History,* v. 208, 88 pp.

Oliver, G. E. and Mancini, E. A. 1980. Late Paleocene planktic foraminiferal biostratigraphy of the Tuscahoma Marls in southwest Alabama. *Transactions of the Gulf Coast Association of Geological Societies,* v. 30, p. 467–472.

Oriel, S. S., 1961. *Tongues of the Wasatch and Green River Formations, Fort Hill area, Wyoming.* U.S. Geological Survey Professional Paper 424-B, p. 151–152.

Oriel, S. S., 1962. Main body of Wasatch Formation near La Barge, Wyoming. *Bulletin of the American Association of Petroleum Geologists,* v. 46, p. 2162–2173.

Oriel, S. S., Gazin, C. L. and Tracey, J. I., Jr. 1962. Eocene age of Almy Formation, Wyoming, in its type area. *Bulletin of the American Association of Petroleum Geologists,* v. 46, p. 1936–1937.

Osborn, H. F., 1909. Cenozoic mammal horizons of western North America. *U.S. Geological Survey Bulletin* 361, 138 pp.

Osborn, H. F., 1910. *The Age of Mammals in Europe, Asia and North America.* New York: Macmillan, 635 pp.

Owen, R., 1858. Description of a small lophiodont mammal (*Pliolophus vulpiceps* Owen) from the London Clay near Harwich. *Quarterly Journal of the Geological Society of London,* v. 14, p. 54–71.

Owen, R., 1878. On the occurrence in North America of rare extinct vertebrates found fragmentarily in England. *Annals and Magazine of Natural History,* v. 2, p. 215–223.

Patterson, B., 1934. A contribution to the osteology of *Titanoides* and the relationships of the Amblypoda. *Proceedings of the American Philosophical Society,* v. 73, p. 71–101.

Patterson, B., 1935. Second contribution to the osteology and affinities of the Paleocene amblypod *Titanoides*. *Proceedings of the American Philosophical Society,* v. 75, p. 143–162.

Patterson, B., 1936. Mounted skeleton of *Titanoides* with notes on the associated fauna. *Proceedings of the Geological Society of America* (1935), p. 397–393.

Patterson, B., 1937. A new genus *Barylambda*, for *Titanoides faberi*, Paleocene amblypod. *Field Museum of Natural History Geological Series,* v. 6, p. 229–231.

Patterson, B., 1939. New Pantodonta and Dinocerata from the upper Paleocene of western Colorado. *Field Museum of Natural History Geological,* v. 6, p. 351–384.

Patterson, B. and Simons, E. L., 1958. A new barylambdid pantodont from the late Paleocene. *Breviora,* no. 93, 8 pp.

Pipiringos, G. N., 1962. Uranium-bearing coal in the central part of the Great Divide Basin. *U.S. Geological Survey Bulletin* 1099-A, 103 pp.

Pipiringos, G. N. and Rosenlund, G. C., 1977a. Preliminary geologic map of the Indian Valley quadrangle, Rio Blanco and Moffat Counties, Colorado. U.S. Geological Survey Map MF-836.

Pipiringos, G. N. and Rosenlund, G. C., 1977b. Preliminary geologic map of the White Rock quadrangle, Rio Blanco and Moffat Counties, Colorado. U.S. Geological Survey Map MF-837.

Plante, G., 1870. Sur les lignites inférieurs de l'Argile plastique du Bassin parisien. *Bulletin de la Société Géologique de France,* Sér. 2, v. 27, p. 205–216.

Plaziat, J. 1975. L'Illerdian à l'intérieur du Paléogène languedocian. Ses relations avec le Sparnacien, l'Illerdian sud-pyrénéen, l'Yprésien et le Paléocène. *Bulletin de la Société Géologique de France,* ser. 7, v. 17, p. 168–182.

Pomerol, C., ed., 1981. Stratotypes of Paleogene Stages. *Bulletin d'Information des Géologues du Bassin de Paris,* Mémoire hors série No. 2, 301 pp.

Prichinello, K. A., 1971. Earliest Eocene mammalian fossils from the Laramie Basin of southeast Wyoming. *University of Wyoming Contributions to Geology,* v. 10, p. 73–87.

Quinet, G. E., 1969. Apport de l'étude de la faune mammalienne de Dormaal à la stratigraphie générale du Paléocène supérieur européen et à la théorie synthétique de la molaire mammalienne. *Mémoires de l'Institut Royal des Sciences Naturelles de Belgique,* v. 162, 167 pp.

Rapp, S. D., MacFadden, B. J. and Schiebout, J. A., 1983. Magnetic polarity stratigraphy in the early Tertiary Black Peaks Formation, Big Bend National Park, Texas. *Journal of Geology,* v. 91, p. 555–572.

Rat, V. J., 1965. La succession stratigraphique des mammifères dans l'Eocène du Bassin de Paris. *Bulletin de la Société Géologique de France,* Sér. 7, v. 7, p. 248–256.

Reshetov, V. J., 1976. O pervoi nakhodke roda *Coryphodon* Owen, 1845 (Mammalia, Pantodonta) v Paleogene Mongolskoiy Narodnoi Respublikii [On the first discovery of the genus *Coryphodon* Owen, 1845 (Mammalia, Pantodonta) in the Paleogene of the Mongolian People's Republic]. *Sovmestnaya Sovetsko–Mongolskaya Paleontologicheskaya Ekspeditsaya [Joint Soviet–Mongolian Paleontological Expeditions] Trudy,* v. 3, p. 9–12.

Reshetov, V. J., 1979. Rannyetretichnyye tapirobrazniye Mongolii i SSSR [Early Tertiary tapiroids of Mongolia and the USSR]. *Sovmestnaya Sovetsko–Mongolskaya Paleontologicheskaya Ekspeditsaya [Joint Soviet–Mongolian Paleontological Expeditions] Trudy,* v. 11, p. 1–144.

Robinson, P., 1957. Age of the Galisteo Formation, Santa Fe County, New Mexico. *American Association of Petroleum Geologists Bulletin,* v. 41, p. 757.

Robinson, P., 1966. Fossil Mammalia of the Huerfano Formation, Eocene of Colorado. *Peabody Museum of Natural History, Yale University Bulletin* 21, 95 pp.

Robinson, P. and Ivy, L. D., 1994. Paromomyidae (?Dermoptera) from the Powder River Basin, Wyoming and a discussion of microevolution in closely related species. *Contributions to Geology, University of Wyoming,* v. 30, p. 91–116.

Roehler, H. W., 1972a. Geologic map of the Brushy Point quadrangle, Rio Blanco and Garfield Counties, Colorado. U.S. Geological Survey Map GQ-1018.

Roehler, H. W., 1972b. Geologic map of the Razorback Ridge quadrangle, Rio Blanco and Garfield Counties, Colorado. U.S. Geological Survey Map GQ-1019.

Roehler, H. W., 1973a. Geologic map of the Calf Canyon quadrangle, Garfield County, Colorado. U.S. Geological Survey Map GQ-1086.

Roehler, H. W., 1973b. Geologic map of the Henderson Ridge quadrangle, Garfield County, Colorado. U.S. Geological Survey Map GQ-1113.

Roehler, H. W., 1991. *Revised stratigraphic nomenclature for the Wasatch and Green River formations of Eocene age, Wyoming, Utah and Colorado.* U.S. Geological Survey Professional Paper 1506-B, 38 pp.

Romer, A. S., 1966. *Vertebrate Paleontology.* Chicago: University of Chicago Press (Third edition), 468 pp.

Rose, K. D. 1980. Clarkforkian Land-Mammal Age: Revised definition, zonation, and tentative intercontinental correlations. *Science,* v. 208, p. 744–746.

Rose, K. D., 1981. The Clarkforkian Land-Mammal Age and mammalian faunal composition across the Paleocene/Eocene boundary. *University of Michigan Papers on Paleontology,* v. 26, 197 pp.

Rose, K. D. , Krishtalka, L. and Stucky, R. K., 1991. Revision of the Wind River faunas, early Eocene of central Wyoming. Pt. 11. Palaeanodonta (Mammalia) *Annals of the Carnegie Museum,* v. 60, p. 63–82.

Russell, D. E., 1968. Succession, en Europe, des faunes mammaliennes au début du Tertiaire. *Mémoires du Bureau de Recherches Géologiques et Minières,* no. 58, p. 291–296.

Russell, D. E. and Dashzeveg, D., 1986. Early Eocene insectivores (Mammalia) from the People's Republic of Mongolia. *Palaeontology,* v. 29, p. 269–291.

Russell, D. E. and Zhai, R.-J., 1987. The Paleogene of Asia: Mammals and stratigraphy. *Mémoire du Muséum National d'Histoire Naturelle,* Sér. C (*Sciences de la Terre*), v. 52, 488 pp.

Russell, D. E., Hartenberger, J.-L., Pomerol, C., Sen, S., Schmidt-Kittler, N. and Vianey-Liaud, M., 1982. Mammals and stratigraphy: the Paleogene of Europe. *Palaeovertebrata, Mémoire Extraordinaire,* p. 1–77.

Savage, D. E., 1962. Cenozoic geochronology of the fossil mammals of the Western Hemisphere. *Revista Museo Argentina Nacional Ciencias Zoologicas,* v. 8, p. 51–67.

Savage, D. E., 1971. The Sparnacian–Wasatchian mammalian fauna, early Eocene, of Europe and North America. *Hessischen Landesamt für Bodenforschung,* v. 60, p. 154–158.

Savage, D. E., 1977. Aspects of vertebrate paleontological stratigraphy and geochronology. In Kauffman, E. G. and Hazel, J. E. (eds.), *Concepts and Methods of Biostratigraphy,* Stroudsburg: Dowden, Hutchinson & Ross, Inc., p. 427–442.

Savage, D. E. and Russell, D. E., 1977. Comments on mammalian paleontologic stratigraphy and geochronology; Eocene stages and mammal ages of Europe and North America. *Géobios, Mémoire spécial* 1, p. 47–56.

Savage, D. E. and Russell, D. E., 1983. *Mammalian Paleofaunas of the World.* London: Addison-Wesley , 432 pp.

Schankler, D. M., 1980. Faunal zonation of the Willwood Formation in the central Bighorn Basin, Wyoming. In Gingerich, P. D. (ed.), *Early Cenozoic paleontology and stratigraphy of the Bighorn*

Basin, Wyoming: 1880–1980, University of Michigan Papers on Paleontology, v. 24, p. 99–114.

Schiebout, J. A., 1974. Vertebrate paleontology and paleoecology of Paleocene Black Peaks Formation, Big Bend National Park, Texas. *Bulletin of the Texas Memorial Museum* 24, 88 pp.

Schiebout, J. A., 1995. The Paleocene/Eocene transition on Tornillo Flat in Big Bend National Park, Texas. In Santucci, V. L. and McLelland, L. (eds.), *National Park Service Paleontological Research,* Denver: National Park Service, p. 40–45.

Schiebout, J. A., Rigsby, C. A., Rapp, S. D., Hartnell, J. A. and Standhardt, B. R., 1987. Stratigraphy of the Cretaceous–Tertiary and Paleocene–Eocene transition rocks of Big Bend National Park, Texas. *Journal of Geology,* v. 95, p. 359–371.

Schoch, R. M., 1981, Taxonomy and biostratigraphy of the early Tertiary Taeniodonta (Mammalia, Eutheria): Summary. *Geological Society of America Bulletin,* v. 92, p. 933–941.

Schoch, R. M., 1986. Systematics, functional morphology and macroevolution of the extinct mammalian order Taeniodonta. *Peabody Museum of Natural History, Yale University Bulletin,* v. 42, 307 pp.

Schoch, R. M. and Lucas, S. G., 1985. The phylogeny and classification of the Dinocerata (Mammalia, Eutheria). *Bulletin of the Geological Institutions of the University of Uppsala,* n. s., v. 11, p. 31–58.

Sharp, W. N., McKay, E. J., McKeown, F. A. and White, A. M., 1964. Geology and uranium deposits of the Pumpkin Buttes area of the Powder River Basin, Wyoming. *U.S. Geological Survey Bulletin* 1107-H, p. 541–638.

Shevyryeva, N. S., 1983. Novyye gryzuny rannevo eotsena Zaaltayskoy Gobi (Mongoliya) [New rodents from the early Eocene of the Zaaltay Gobi (Mongolia)]. In Anonymous (ed.), *Materialy IV Vsyesoyuznaya Soveshchaniya,* Leningrad: Nauka, p. 55–58.

Shevyryeva, N. S. 1996. Novyye gryzuny (Rodentia, Mammalia) iz nizhnevo Eotsena Zaysanskoy vpadiny (vostochnyi Kazakhstan) [New rodents (Rodentia, Mammalia) from the lower Eocene of the Zaysan basin (eastern Kazakhstan)]. *Paleontologicheskii Zhurnal,* 1996(1), p. 84–98.

Shevyryeva, N. S., Chkikvadze, V. M. and Zhegallo, V. I., 1975. Novyye dannyye o faunye pozvonochnykh mestonakhozhdeniya Gashato (Mongolskaya Narodnaya Respublika) [New data on the vertebrate faunal locality Gashato (Mongolian Peoples Republic)]. *Soobshcheniya Akademiya Nauk Gruzinskoy SSR,* v. 77, p. 225–228.

Simons, E. L., 1963. Some fallacies in the study of hominid phylogeny. *Science,* v. 141, p. 879–889.

Simpson, G. G., 1929. Paleocene and lower Eocene mammals of Europe. *American Museum Novitates,* no. 354, 17 pp.

Simpson, G. G., 1948a. Evolution, interchange, resemblance of the North American and Eurasian Cenozoic mammalian faunas. *Evolution,* v. 1, p. 218–220.

Simpson, G. G., 1948b. The Eocene of the San Juan Basin, New Mexico. *American Journal of Science,* v. 246, p. 257–282.

Simpson, G. G., 1981. History of vertebrate paleontology in the San Juan Basin. In Lucas, S., Rigby, K., Jr. and Kues, B. (eds.), *Advances in San Juan Basin Paleontology,* Albuquerque: University of New Mexico Press, p. 3–25.

Sinclair, W. J. and Granger, W., 1911. Eocene and Oligocene of the Wind River and Bighorn basins. *Bulletin of the American Museum of Natural History,* v. 30, p. 57–67.

Smith, L. N., 1988. *Basin analysis of the lower Eocene San Jose Formation, San Juan Basin, New Mexico.* Ph. D. Thesis, Albuquerque: University of New Mexico, 166 pp.

Smith, L. N., 1992. *Stratigraphy, sediment dispersal and paleogeography of the lower Eocene San Jose Formation, San Juan Basin, New Mexico and Colorado.* New Mexico Geological Society Guidebook 43, p. 297–309.

Smith, L. N. and Lucas, S. G., 1991. Stratigraphy, sedimentology, and paleontology of the lower Eocene San Jose Formation in the central portion of the San Juan Basin, northwestern New Mexico. *Bulletin of the New Mexico Bureau of Mines and Mineral Resources,* v. 126, 44 pp.

Smith, R., Smith, T. and Sudre, J., 1996. *Diacodexis gigasi* n. sp., le plus ancien Artiodactyle (Mammalia) belge, proche de la limite Paléocène–Eocène. *Bulletin de l'Institut Royal des Sciences Naturelles de Belgique, Sciences de la Terre,* v. 66, p. 177–186.

Smith, T., 1995. Présence du genre *Wyonicteris* (Mammalia, Lipotyphla) à la limite Paléocène–Eocène en Europe. *Comptes-Rendus de l'Academie des Science de Paris,* IIA, v. 321, p. 923–930.

Soister, P. E. and Tschudy, R. M., 1978. Eocene Rocks in Denver Basin. In Pruitt, J. D. and Coffin, P. E. (eds.), *Energy Resources of the Denver Basin,* Denver: Rocky Mountain Association of Geologists, p. 231–235.

Squires, R. L., 1984. Megapaleontology of Eocene Llajas Formation, Simi Valley, California. *Los Angeles County Museum Contributions to Science,* no. 350, p. 1–76.

Stamp, L. D., 1921a. On the beds at the base of the Ypresian in the Anglo–Franco–Belgian Basin. *Geological Magazine,* v. 57, p. 320–346.

Stamp, L. D., 1921b. On cycles of sedimentation in the Eocene strata of the Anglo–Franco–Belgian Basin. *Geological Magazine,* v. 58, p. 409–457.

Stucky, R. K. 1982. *Mammalian faunas and biostratigraphy of the upper part of the Wind River Formation (early to middle Eocene), Natrona County, Wyoming, and the Wasatchian–Bridgerian boundary.* Ph.D. dissertation, Boulder: University of Colorado, 278 pp.

Stucky, R. K. 1984a. Revision of the Wind River faunas, early Eocene of central Wyoming. Part 5. Geology and mammalian biostratigraphy of the upper part of the Wind River Formation. *Annals of the Carnegie Museum,* v. 53, p. 231–294.

Stucky, R. K. 1984b. Revision of the Wind River faunas, early Eocene of central Wyoming. Part 6. Measured sections and localities, Red Creek–Deadman Butte area. *Annals of the Carnegie Museum,* v. 53, p. 295–326.

Stucky, R. K., 1984c. The Wasatchian-Bridgerian land mammal age boundary (early to middle Eocene) in western North America. *Annals of the Carnegie Museum,* v. 53, p. 347–382.

Stucky, R. K. and Krishtalka, L., 1982. Revision of the Wind River faunas, early Eocene of central Wyoming. Pt. 1. Introduction and Multituberculata. *Annals of the Carnegie Museum,* v. 51, p. 39–56.

Stucky, R. K. and Krishtalka, L., 1983, Revision of the Wind River faunas, early Eocene of central Wyoming. Pt. 4. The Tillodontia. *Annals of the Carnegie Museum,* v. 52, p. 375–391.

Stucky, R. K. and L. Krishtalka. 1990. Revision of the Wind River faunas, early Eocene of central Wyoming. Pt. 10. *Bunophorus* (Mammalia, Arctiodactyla). *Annals of the Carnegie Museum,* v. 59, p. 149–172.

Szalay, F. S. and McKenna, M. C., 1971. Beginning of the age of mammals in Asia: The late Paleocene Gashato fauna, Mongolia. *Bulletin of the American Museum of Natural History,* v. 144, p. 273–317.

Takai, F., 1952. A summary of the mammalian faunae of eastern Asia and the interrelationships of continents since the Mesozoic. *Japanese Journal of Geology and Geography,* v. 22, p. 169–205.

Tauxe, L. and Clark, D. R., 1987. New paleomagnetic results from the Eureka Sound Group: Implications for the age of early Tertiary Arctic biota. *Geological Society of America Bulletin,* v. 99, p. 739–747.

Teilhard de Chardin, P., 1920. Sur la succession des faunes mammifères dans l'Eocène inférieur européen. *Comptes-Rendus de l'Académie des Sciences de Paris,* v. 171, p. 1161–1162.

Teilhard de Chardin, P., 1921. Les mammifères de l'Eocène inférieur français et leurs gisements. *Annales de Paléontologie,* v. 10, p. 171–176.

Teilhard de Chardin, P., 1927. Les mammifères de l'Eocène inférieur de la Belgique. *Mémoires du Musée Royal d'Histoire Naturelle de Belgique,* v. 36, p. 1–33.

Thenius, E., 1959. *Tertiär Zweiter Teil.* Stuttgart: Ferdinand Enke Verlag, 328 pp.

Ting, S., 1993. A preliminary report on an early Eocene mammalian fauna from Hengdong, Hunan Province, China. *Kaupia [Darmstädter Beitrège zur Naturgeschichte],* v. 3, p. 201–207.

Ting, S., 1998. Paleocene and early Eocene land mammal ages of Asia. *Bulletin of the Carnegie Museum of Natural History,* v. 34, p. 124–147.

Tong, Y., 1978. Late Paleocene mammals of the Turfan Basin, Sinkiang. *Memoir Institute of Vertebrate Paleontology and Paleoanthropology Academia Sinica* 13, p. 82–101.

Tong, Y., 1989. A review of middle and late Eocene mammalian faunas from China. *Acta Palaeontologica Sinica,* v. 28, p. 663–682.

Tong, Y. and Dawson, M. R., 1995. Early Eocene rodents (Mammalia) from Shandong Province, People's Republic of China. *Annals of the Carnegie Museum,* v. 64, p. 51–63.

Tong, Y. and Lucas, S. G., 1982. A review of Chinese uintatheres and the origin of the Dinocerata (Mammalia, Eutheria). *North American Paleontological Convention Proceedings,* v. 2, p. 551–556.

Tong, Y., and Wang, J., 1980. Subdivision of the Upper Cretaceous and lower Tertiary of the Tantou Basin, the Lushi Basin and the Lingbao Basin of western Henan. *Vertebrata PalAsiatica,* v. 18, p. 21–461.

Tong, Y. and Wang, J., 1993. A new soricomorph (Mammalia, Insectivora) from the early Eocene of Wutu basin, Shandong, China. *Vertebrata PalAsiatica,* v. 31, p. 19–32.

Tong, Y. and Wang, J., 1995. Mammals of the early Eocene Wutu fauna, Shandong Province, China. *Journal of Vertebrate Paleontology,* v. 15 (supplement to no. 3), p. 57A (Abstracts of papers, 55th Annual Meeting of the Society of Vertebrate Paleontology, Pittsburgh, Pennsylvania).

Tong, Y., Zheng, S. and Qiu, Z. 1995. Cenozoic mammal ages of China. *Vertebrata PalAsiatica,* v. 33, p. 290–314.

Tourtelot, H. A., 1946. Tertiary stratigraphy in the northeastern part of Wind River Basin, Wyoming. U.S. Geological Survey Oil and Gas Investigations Chart OC-22.

Tourtelot, H. A., 1948. Tertiary rocks in the northeastern part of the Wind River Basin, Wyoming. In McGrew, P. O. (ed.), *Guidebook for the 3rd Annual Field Conference of the Society of Vertebrate Paleontology in Southeastern Wyoming,* p. 53–67.

Tourtelot, H. A., 1953. Geology of the Badwater area, central Wyoming. U.S. Geological Survey Oil and Gas Investigations Map OM-124.

Uhen, M. D. and Gingerich, P. D., 1995. Evolution of *Coryphodon* (Mammalia, Pantodonta) in the late Paleocene and early Eocene of northwestern Wyoming. *Contributions from the Museum of Paleontology, The University of Michigan,* v. 29, p. 259–289.

Van Houten, F. B., 1944. Stratigraphy of the Willwood and Tatman Formations in northwestern Wyoming. *Bulletin of the Geological Society of America,* v. 55, p. 165–210.

Van Houten, F. B.. 1945. Review of latest Paleocene and early Eocene mammalian faunas. *Journal of Paleontology,* v. 19, p. 421–461.

Vasseur, G., 1875. Sur le cubitus de *Coryphodon oweni. Bulletin de la Société Géologique de France,* Sér. 3, v. 3, p. 181–186.

Veatch, A. C., 1907. Geography and geology of a portion of southwestern Wyoming. *U.S. Geological Survey Professional Paper* 56, p. 87–99.

Vinson, T. E., 1981. *A paleomagnetic study of the Upper Cretaceous to lower Eocene Eureka Sound Formation, Strathcona Fiord, Ellesmere Island, Canada.* Master of Science Thesis, Milwaukee: University of Wisconsin at Milwaukee, 186 pp.

Wallace, S. M. 1980. *A revision of North American early Eocene Brontotheriidae (Mammalia, Perissodactyla).* Master of Science Thesis, Boulder: University of Colorado, 157 pp.

Walsh, S. L., 1991. Eocene mammal faunas of San Diego County. In Abbott, P. L. and May, J. A. (eds.), *Eocene Geologic History of the San Diego Region,* Pacific Section of the Society of Economic Paleontologists and Mineralogists, v. 68, p. 161–178.

Walsh, S. L., 1996. Middle Eocene mammal faunas of San Diego County, California. In Prothero, D. R. and Emry, R. J. (eds.), *The terrestrial Eocene–Oligocene Transition in North America,* Cambridge: Cambridge University Press, p. 75–119.

Wang, B., 1992. The Chinese Oligocene: A preliminary review of mammalian localities and local faunas. In Prothero, D. R. and Berggren, W. A. (eds.), *Eocene–Oligocene Climatic and Biotic Evolution,* Princeton: Princeton University Press, p. 529–547.

Wang, H. (ed.), 1985. *Atlas of the Palaeogeography of China.* Beijing: Cartographic Publishing House, 143 sheets.

West, R. M., 1969a. Geology and vertebrate paleontology of the northeastern Green River Basin, Wyoming. *Wyoming Geological Association Guidebook* 21, p. 77–92.

West, R. M., 1969b. Biostratigraphy of fluvial sediments of the upper Wasatch Formation in the northern Green River Basin, Wyoming. *University of Wyoming Contributions to Geology,* v. 8, p. 184–196.

West, R. M., 1970. Sequence of mammalian faunas of Eocene age in the northern Green River Basin, Wyoming. *Journal of Paleontology,* v. 44, p. 142–147.

West, R. M., 1973a. Geology and mammalian paleontology of the New Fork-Big Sandy area, Sublette County, Wyoming. *Fieldiana Geology,* v. 29, 193 pp.

West, R. M., 1973b. New records of fossil mammals from the early Eocene Golden Valley Formation, North Dakota. *Journal of Mammalogy,* v, 54, p. 749–753.

West, R. M., 1976. The North American Phenacodontidae (Mammalia, Condylarthra). *Milwaukee Public Museum Contributions in Geology and Biology,* no. 6, 78 pp.

West, R. M. and Dawson, M. R., 1977. Mammals from the Palaeogene of the Eureka Sound Formation: Ellesmere Island, Arctic Canada. *Geobios, Mémoire Special* 1, p. 107–124.

West, R. M., Dawson, M. R. and Hutchinson, J. H., 1977. Fossils from the Paleogene Eureka Sound Formation, N.W.T. Canada: Occurrence, climatic and paleogeographic implications. *Milwaukee Public Museum Special Publications in Biology and Geology,* v. 2, p. 77–93.

Williamson, T. E., 1996. The beginning of the age of mammals in the San Juan Basin, New Mexico: Biostratigraphy and evolution of Paleocene mammals of the Nacimiento Formation. *New Mexico Museum of Natural History and Science Bulletin* 8, 141 pp.

Williamson, T. E. and Lucas, S. G., 1992. *Meniscotherium* (Mammalia, "Condylarthra") from the Paleocene/Eocene of western North America. *New Mexico Museum of Natural History and Science Bulletin* 1, 75 pp.

Williamson, T. E. and Lucas, S. G., 1993. Paleocene vertebrate paleontology of the San Juan Basin, New Mexico. In Lucas, S. G. and Zidek, J. (eds.), *Vertebrate Paleontology in New Mexico,* New Mexico Museum of Natural History and Science Bulletin 2, p. 105–135.

Wilson, J. A., 1967. Early Tertiary mammals. *University of Texas, Bureau of Economic Geology Bulletin,* v. 671, p. 157–169.

Wing, S. L., 1984. A new basis for recognizing the Paleocene/Eocene boundary in western interior North America. *Science,* v. 226, p. 439–441.

Wing, S. L., Bown, T. M. and Obradovich, J. D., 1991. Early Eocene biotic and climatic change in interior western North America. *Geology,* v. 19, p. 1189–1192.

Wood, H. E., II, Chaney, R. W., Clark, J., Colbert, E. H., Jepsen, G. L., Reeside, J. B., Jr. and Stock, C., 1941. Nomenclature and correlation of the North American continental Tertiary. *Bulletin of the Geological Society of America,* v. 52, p. 1–48.

Wood, R. C., 1967. A review of the Clark's Fork vertebrate fauna. *Breviora,* no. 257, p. 1–30.

Woodburne, M. O. and Swisher, C. C., III, 1995. Land mammal high-resolution geochronology, intercontinental overland dispersals, sea level, climate, and vicariance. In Berggren, W. A., Kent, D. V., Aubry, M.-P. and Hardenbol, J. (eds.), *Geochronology Time Scales and Global Stratigraphic Correlation,* Society of Economic Paleontologists and Mineralogists Special Publication No. 54, p. 335–364.

Xu, Y., Yan, D., Zhou, S., Han, S. and Zhang, Y., 1979. The subdivision of the red beds of south China. In Anonymous (ed.), *The Mesozoic and Cenozoic Red Beds of South China,* Beijing: Science Press, p. 416–432.

Yakhimovich, V. L., Borisov, B. A., Suleimanova, F. I. and Danukalov, N. F., 1993. Magnitostratigraficheskaya shkala Kainozoya Zaisanskoi kotloviny i ee biostratigraficheskoye obosnovaniye [Magnetostratigraphic scale of the Cenozoic of the Zaysan Basin and its biostratigraphic significance]. Akademiya Nauk Respubliki Bashkortostan Institut Geologii Ufa, 68 pp.

Young, C. C., 1938. An early Tertiary vertebrate fauna from Yuanchu. *Bulletin of the Geological Society of China,* v. 17, p. 413–503.

Young, C. C., 1944. Note on the first Eocene mammal from South China. *American Museum Novitates,* no. 1268, 3 pp.

Zhai, R. 1978a. Two new early Eocene mammals from Sinkiang. *Memoirs of the Institute of Vertebrate Paleontology and Paleonanthropology,* v. 13, p. 102–106.

Zhai, R., 1978b. More fossil evidence favoring an early Eocene connection between Asia and the Nearctic. *Memoirs of the Institute of Vertebrate Paleontology and Paleonanthropology,* v. 13, p. 68–81.

Zhai, R. and Meng, J., 1995. The latest Paleocene Bayan Ulan fauna of Nei Mongol, China. *Journal of Vertebrate Paleontology,* v. 15 (Supplement to no. 3), p. 61A (Abstracts of Papers, 55th Annual Meeting of the Society of Vertebrate Paleontology, Pittsburgh, Pennsylvania).

Zhai, R., Bi, Z., and Yu, Z., 1976. Stratigraphy of Eocene Zhanhshanji Formation with note on a new species of eurymylid mammal. *Vertebrata PalAsiatica,* v. 14, p. 100–103.

Zhegallo, V. I. and Shevyryeva, N. S., 1976. Reviziya geologicheskovo stroyeniuya i novyye dannyye o faune mestonakhozhdeniya Gashat (Paleotsen, MPR) [Revision of geological structure and new data on the Gashato faunal locality (Paleocene, Mongolian Peoples Republic)]. *Sovmestnaya Sovetsko–Mongolskaya Paleontologicheskaya Ekspeditsaya [Joint Soviet–Mongolian Paleontological Expeditions] Trudy* 3, p. 269–279.

Zheng, J., Tung, Y. and Chi, H., 1975. Discovery of Miacidae (Carnivora) in Yuanshui basin, Kiangsi Province. *Vertebrata Palasiatica,* v. 13, p. 96–104.

Ziegler, P. A., 1982. *Geological Atlas of Western and Central Europe.* The Hague: Shell International Petroleum, v. 1, 130 pp., v. 2, 40 sheets.

INDEX

Page numbers followed by *a*, *f*, or *t* refer to information in appendices, figures, or tables.

GPSR Authorized Representative: Easy Access System Europe, Mustamäe tee 50, 10621 Tallinn, Estonia, gpsr.requests@easproject.com

www.ingramcontent.com/pod-product-compliance
Ingram Content Group UK Ltd.
Pitfield, Milton Keynes, MK11 3LW, UK
UKHW050924260926
475591UK00002B/37
9780231102384